Plant and Crop Physiology

Plant and Crop Physiology

Edited by **Jordan Smith**

SYRAWOOD
PUBLISHING HOUSE
New York

Published by Syrawood Publishing House,
750 Third Avenue, 9th Floor,
New York, NY 10017, USA
www.syrawoodpublishinghouse.com

Plant and Crop Physiology
Edited by Jordan Smith

International Standard Book Number: 978-1-68286-344-2 (Hardback)

Printed in China.

Contents

Preface

The main aim of this book is to educate learners and enhance their research focus by presenting diverse topics covering this vast field. This is an advanced book which compiles significant studies by distinguished experts. This book addresses successive solutions to the challenges arising in the area of application, along with it; the book provides scope for future developments.

Plant physiology is an independent discipline dedicated to understanding the functioning of plants. This book aims to bring forth the advanced topics in this field through a compilation of researches and case studies. It explains the prevalent theories and also elucidates the recently discovered plant processes. This interdisciplinary text highlights the significance of crop physiology and its contribution to other field such as plant morphology, phytochemistry, genetics, etc. The book discusses in detail, the techniques for postharvest physiology of crops along with methods for better crop production, such as hydroponic methods. This book will serve as a resource guide for crop physiologists, botanists, horticulturists, researchers, scholars and students.

It was a great honour to edit this book, though there were challenges, as it involved a lot of communication and networking between me and the editorial team. However, the end result was this all-inclusive book covering diverse themes in the field.

Finally, it is important to acknowledge the efforts of the contributors for their excellent chapters, through which a wide variety of issues have been addressed. I would also like to thank my colleagues for their valuable feedback during the making of this book.

Editor

Effect of methyl bromide alternatives on seedling quality, nematodes and pathogenic soil fungi at the Jesup and Glennville Nurseries in Georgia: 2007 to 2008

S. A. Enebak[1]*, T. E. Starkey[2] and M. Quicke[2]

[1]School of Forestry and Wildlife Sciences, 602 Duncan Drive, Auburn University, 36849, USA.
[2]School of Forestry and Wildlife Sciences, Southern Forest Nursery Management Cooperative, Auburn University, AL 36849, USA.

For many years, forest-tree nurseries in the United States have relied on methyl bromide (MBr) soil fumigation to control weeds, pathogenic fungi, insects and nematodes. However, due to the concern over ozone depletion in the stratosphere, finding a soil fumigant alternative for MBr has been a priority for the forest nursery industry since 1991. A large-scale study comparing seven fumigants using operational application techniques and normal nursery management practices over two growing seasons was installed in two forest-tree nurseries in Georgia: the Plum Creek Nursery in Jesup, and the Rayonier Nursery in Glennville. Control of weeds, nematodes, and soilborne fungi was dependent upon the soil fumigant used and cropping season. By the end of the second growing season in 2008, there were no significant differences in seedling densities for any of the soil fumigants tested, as all soil fumigants were similar to MBr. While many seedling producers would prefer to use MBr to produce forest-tree seedlings in perpetuity, each nursery manager will need to identify the best alternative for their nursery conditions. Based on these 2 year trials, the soil fumigants with 100% chloropicrin and Pic + at 336 kg/ha (300 lbs/a) appear to be suitable alternatives for MBr in controlling forest-nursery pests and producing high quality seedlings at the Jesup and Glennville, GA Nurseries.

Key words: Soil fumigation, chloropicrin, soilborne fungi, nematodes, loblolly pine.

INTRODUCTION

Fumigation with methyl bromide (MBr) mixtures has been the most commonly used method for producing high quality, pest-free forest-tree seedlings in the southeastern United States (Jang et al., 1993). Forest nurseries in the United States have relied for many years on MBr soil fumigation to control yellow and purple nutsedge, soil insects, nematodes and soilborne pathogenic fungi. In the southern United States, there are three fungal genera that are of primary concern in the production of pine seedlings. These genera are: *Fusarium*, *Pythium* and *Rhizoctonia,* each of which is associated with seedling root and foliage diseases. Nursery managers have long recognized the importance of MBr to control these soil-borne pathogens in the production of forest-tree seedlings (Henry, 1953; South et al., 1997). Due to the concern over ozone depletion in the stratosphere, the Montreal Protocol under the Clean Air Act has begun a phase-out program for MBr use. Since 1991, finding alternatives for MBr has been a priority for the forest nursery industry. Other materials tested by the Southern Forest Nursery Management Cooperative, without much success, include basamid, hot pepper sauce, steam, hot water and biologicals. Although it may be difficult to find a soil fumigant alternative that is as broad-spectrum as MBr, the nursery industry realizes the importance of testing new compounds, rates and application techniques. Generally, forest-tree nurseries operate on a

*Corresponding author. E-mail: enebasa@auburn.edu.

Table 1. Soil fumigants and rates used in the 2007 to 2008 USDA area wide demonstration plots at the Plum Creek Nursery in Jesup, GA and Rayonier Nursery in Glennville, GA.

Soil fumigant treatment	Rate used	Chemical components
Methyl bromide + chloropicrin	392 kg/ha (350 lb/a)	67% MBr and 33% chloropicrin
Dimethyl disulfide + chloropicrin	113 l/ha (74 gal/a)	79% DMDS and 21% chloropicrin
MBrC 70/30	448 kg/ha (400 lb/a)	70% MBr (98/2) and 30% Solvent A
Pic+	336 kg/ha (300 lb/a)	85% chloropicrin and 15% Solvent A
New Pic+	336 kg/ha (300 lb/a)	85% chloropicrin and 15% Solvent B
Chloropicrin	336 kg/ha (300 lb/a)	100% chloropicrin
Chlor 60	448 kg/ha (400 lb/a)	60% chloropicrin and 40% 1,3-dichloropropene

Solvents A and B are proprietary compounds of Hendrix and Dail, approved by EPA for use in forest-tree nurseries and under full label considerations.

Table 2. Site information for the Jesup and Glennville, GA Nursery fumigation.

Application parameters	Jesup, GA	Glennville, GA
Fumigation	3-Apr-2007	20-Mar-2007
Fumigation type	Broadcast/shank injected	Broadcast/shank injected
Area in trial	2 ha	4 ha
Air temperature range	19 to 31°C	10 to 26°C
Wind speed	5 to 10 km/h	5 to 20 km/h
Soil moisture	8.5%	5.5%
Soil series	Norfolk loamy sand	Tifton loamy sand
Plastic in place	7 days	7 days

3 year cropping system with 2 production years per soil fumigation. Loblolly pine (*Pinus taeda* L.) is the primary species produced in southern forest nurseries (McNabb, 2006) and many soilborne pests can affect its production (Cordell et al., 1989). Despite the phase-out of MBr, forest nursery managers still need soil treatments that are comparable to MBr for control of soil borne pests.

The purpose of these studies is to look at the efficacy of a number of soil fumigants against common soil-borne fungi, weeds, and nematodes and their effect on seedling quality of forest-tree seedlings in the southern United States using operational application techniques under normal nursery management practices over a two year cropping cycle. These studies report a portion of the USDA – ARS Area-wide Pest Management Project for Methyl Bromide Alternatives and are part of a long-term continuing effort by the Southern Forest Nursery Management Cooperative to identify and evaluate soil fumigants. Information from these studies should be used by nursery managers in the southern US to select an MBr alternative that could be useful in the production of forest-tree seedlings in their nurseries.

MATERIALS AND METHODS

Soil fumigation trials were established in two separate forest-seedling nurseries that differed in soil type and production systems in Georgia: the Plum Creek Nursery in Jesup and the Rayonier Nursery in Glennville. The soil fumigates used included MBr and six alternatives that are currently available for large-scale use (Table 1). With the exception of New Pic+, soil fumigants were selected based on results of small plot studies previously conducted by the Southern Forest Nursery Management Cooperative (Starkey et al., 2006). The soil fumigation trial occupied 2 ha out of a total 20 production ha in Jesup and 4 ha out of 32 production ha in Glennville. For each nursery, the various soil fumigants were shank-injected to a depth of 20 to 25 cm using a Raven Flow Meter Control System® (Sioux Falls, SD) and immediately covered with 1 mm High Density Polyethylene Tarp (Cadillac Plastics Inc.) under the weather and soil conditions listed in Table 2. The trial was laid out in nursery sections consisting of nine seedling beds (3.5 m wide) between the irrigation pipelines (20 m) with each seedling bed approximately 182 m long. At the Jesup Nursery, each nursery bed was split into two plots of 90 m each. Each fumigated plot was 3.5 m wide x 90 m long with three seedling beds per treated plot. The experimental design at both nurseries was a randomized complete block that was replicated 5 times at Glennville and 4 times at Jesup. Each nursery sowed a single family of loblolly pine (*P. taeda*) seed in late April 2007 that were lifted in mid November 2007.

The area was fallow during the winter and in March 2008 the beds were prepared for sowing. The second seedling crops' sowing occurred in late April 2008 with seedlings lifted in October/November 2008. With respect to seedling production over the two growing seasons, standard nursery production systems used at each nursery (for example irrigation, fertilization, weed control, insect control, etc.) were allowed to continue within the experimental area.

Seedling quality

The effect of the soil fumigants on seedling densities and growth

Table 3. Loblolly pine seedling density (No./m²), at the end of each cropping season at Jesup and Glennville, GA.

	Jesup, GA Nursery		Glennville, GA Nursery	
Soil treatment	**Nov 2007**	**Nov 2008**	**Oct 2007**	**Oct 2008**
MBr	205 b	215 a	144 a	166 a
Chloropicrin	258 a	226 a	147 a	168 a
Chlor 60	237 ab	237 a	142 a	156 ab
MBrC 70/30	205 b	226 a	135 a	139 b
DMDS + Chlor	205 b	226 a	147 a	139 b
Pic+	205 b	226 a	141 a	153 ab
New Pic+	205 b	237 a	150 a	155 ab
$lsd_{(0.05)}$	43	43	39	23

Within column means followed by the same letter do not differ at P = 0.05 level using Duncan's Multiple Range Test. Target seedling density is 258 seedlings/m² at Jesup, GA and 192 seedling/m² at Glennville, GA.

characteristics were assessed in four subplots (1.2 x 0.3 m) per treatment plot at 7 wks post-sowing, mid-summer (15 weeks post-sowing) and just prior to lifting in the fall (26 weeks post-sowing) in 2007 and 2008, respectively. Twenty-five loblolly pine seedlings per subplot were collected in the fall of each cropping season and returned to the laboratory at Auburn University for analysis. Seedling root collar diameter (RCD), shoot height and seedling dry weight (biomass) was measured for each seedling as well as overall root growth. For root morphology, 10 seedlings per subplot were examined for root length, root surface area, average root diameter and the number of root tips using WinRhizo® software by Regents Instruments Inc., Quebec, Canada.

Soil trichoderma and nematodes

Throughout the two cropping seasons, soil samples were collected from the center seedling bed of each 3-bed plot: at pre-sowing, post-sowing, mid-summer and just prior to seedling lifting in November of each season. Half of each soil sample was plated onto *Trichoderma*-selective media (Elad et al., 1981) and the remaining half was sent to the Soils Laboratory at Auburn University for a quantitative assessment of nematodes.

Soilborne pathogenic fungi

In addition to seedling quality and nematode control, the ability of the soil fumigants to control common soilborne pathogens was examined. Three soilborne pathogens, *Pythium* spp, *Rhizoctonia* spp and *Fusarium* spp, were transferred from the culture collection in the laboratory onto potato dextrose agar (Difco®) and incubated for 2 weeks. Approximately 6 weeks prior to soil fumigation, agar plates of each fungus were used to inoculate bags containing moistened, sterilized oatmeal. The inoculated bags were mixed on a regular basis to encourage fungal growth throughout the oatmeal and left to grow on the laboratory bench at 22°C. Two days prior to soil fumigation, a 300 ml beaker of each oatmeal/fungus mixture was placed into 30 Hubco® soil sample bags (12 x 18 cm), the top folded over, stapled shut, and labeled to each fungus. As the soil fumigation treatments were being applied in the nursery, three Hubco® bags containing the inoculated oatmeal of each soilborne fungus were placed into the soil of each treated plot. To do this, a 45 cm slit was cut into the plastic tarps placed over the fumigated soil and the three bags were buried 15 cm deep. The slit was then sealed with fumigation tape. Six days after soil fumigation, the Hubco® soil bags were removed from each of the soil fumigation plots, placed in a cooler and returned to Auburn University. In the laboratory, each Hubco® soil bag was thoroughly mixed and then opened in a laminar flow hood. From each fungal species, nine oatmeal pieces were placed in groups of three on the following selective media: 1) PARP media for *Pythium* (Kannwischer and Mitchell, 1981); 2) Komada's media for *Fusarium* (Nelson et al., 1993); and 3) Ko and Hora's media (1971) for *Rhizoctonia*. There were three agar plates for each treatment replication.

One week after plating the fumigated oatmeal onto selective agar media, each plate was examined for fungal growth that would indicate the target soilborne pathogens. In addition to the soilborne pathogens, the non-target fungi were also counted. As a control, bags of each soilborne pathogenic fungus not fumigated were also plated out on selective media as described earlier.

RESULTS

Seedling quality

At the Plum Creek Nursery in Jesup, seedling densities at the end of the first growing season in the 100% chloropicrin treatments were significantly greater than the MBr treatments. However, by the end of the second growing season in 2008, there were no significant differences for seedling densities for any of the soil fumigants tested, as all gave similar seedling densities as the standard MBr soil fumigation treatment (Table 3). With respect to the Rayonier Nursery in Glennville, GA, seedling densities for 2007 were far below target levels due to wind damage that occurred shortly after sowing. Taking this seedling reduction into account, there were no significant differences between seedling densities for any of the soil fumigants tested in 2007. While seedling densities for the 2008 seedling crop were generally higher than 2007, they were still below target levels for all soil fumigants tested (Table 4). Unlike the 2007 growing season, where all soil fumigants were similar to MBr, in 2008, soils treated with MBrC 70/30 and DMDS + Chloropicrin had significantly fewer seedlings when

Table 4. Loblolly pine seedling root collar diameter (mm) at the end of each growing season at Jesup and Glennville, GA.

	Jesup, GA Nursery		Glennville, GA Nursery	
Soil treatment	**Nov 2007**	**Nov 2008**	**Oct 2007**	**Oct 2008**
MBr	4.83 ab	4.43 ab	5.34 a	4.93 ab
Chloropicrin	4.65 b	4.70 a	5.33 a	4.62 b
Chlor 60	4.91 ab	4.16 ab	5.41 a	4.58 b
MBrC 70/30	4.99 ab	4.27 ab	5.66 a	4.88 ab
DMDS + Chlor	5.19 a	4.04 ab	5.49 a	5.14 a
Pic+	5.13 a	4.46 ab	5.69 a	4.95 ab
New Pic+	5.13 a	3.97 b	5.55 a	4.91 ab
$lsd_{(0.05)}$	0.37	0.70	0.68	0.44

Means within columns followed by the same letter do not differ at P = 0.05 level using Duncan's Multiple Range Test.

Table 5. Loblolly pine seedling grade by soil fumigation in 2007 and 2008, Jesup, GA.

Soil	2007 Season				2008 Season			
treatment	**Total**	**Grade 1**	**Grade 2**	**Culls**	**Total**	**Grade 1**	**Grade 2**	**Culls**
MBr	205 b	129 ab	65 a	11 a	215 a	64 b	140 a	11 a
Chloropicrin	258 a	107 b	151 b	0 a	226 a	118 a	97 b	11 a
Chlor 60	237 ab	140 a	75 a	11 a	237 a	54 b	161 a	22 a
MBrC 70/30	205 b	129 ab	75 a	0 a	226 a	65 b	140 a	22 a
DMDS + Chlor	205 b	140 a	54 a	11 a	226 a	43 b	151 a	32 a
Pic+	205 b	140 a	43 a	22 a	226 a	75 b	140 a	11 a
New Pic+	205 b	151 a	54 a	0 a	237 a	54 b	161 a	22 a
$lsd_{(0.05)}$	43	22	32	22	43	4	32	22

Means within columns followed by the same letter do not differ at P = 0.05 level using Duncan's Multiple Range Test. Numbers are seedlings per square meter of nursery bed. Nursery target for both growing seasons was 258 seedlings/m^2.

compared to the standard MBr treatment (Table 3). Differences in seedling root collar diameters (RCD) among the soil fumigants tested at Plum Creek were observed (Table 4). Some of the newer chemistries, Pic+, New Pic+ and DMDS had seedlings with significantly larger RCD than seedlings grown in 100% chloropicrin. In 2008, soils fumigated with chloropicrin had significantly larger RCD than New Pic+ (Table 4). When considering seedling grades (Grade 1 = seedlings > 4.69 mm, Grade 2 = seedlings 3.2 to 4.69 mm and Cull = seedlings < 3.2 mm), the proportion of seedlings for each grade was similar across all soil fumigants in 2007 except for chloropicrin. In 2007, chloropicrin was the only soil fumigant tested at this nursery that had more Grade 2 seedlings than Grade 1 seedlings (Table 5).

In 2008 the proportion of Grade 1 seedlings was fewer than in 2007 for all soil fumigants except for chloropicrin (Table 5). Higher seedling densities typically result in a lower mean RCD yielding fewer Grade 1 seedlings per square foot. Chloropicrin-fumigated soils had higher seedling density than other fumigants in 2007, producing a lower percentage of Grade 1 (42%) than Grade 2 (58%) seedlings; yet, with a lower seedling density in 2008, more Grade 1 (50%) than Grade 2 (45%) seedlings were produced. When considering seedling grades at the Rayonier Nursery in Glennville, GA, the proportion of seedlings grown in 2007 for each grade was similar across all soil treatments tested: 77% Grade 1, 22% Grade 2 and 1% Cull. Not surprising, in 2008 the proportion of Grade 1 seedlings declined for all soil treatments with 56% Grade 1, 41% Grade 2 and 3% Cull (Table 6). The lower number of Grade 1 seedlings in 2008 was due to the higher seedling densities. When comparing seedling densities and proportion of Grade 1 seedlings in 2008, two soil fumigants stand out. MBr has a high percentage of Grade 1 seedlings even though it has high seedling density. Soils fumigated with Chlor 60 had a relatively low percentage of Grade 1 seedlings even though those plots had a low seedling density (Table 6). In contrast, soils fumigated with chloropicrin

Table 6. Loblolly pine seedling grade by soil fumigation in 2007 and 2008 Glennville, GA.

Soil treatment	2007 Season				2008 Season			
	Total	Grade 1	Grade 2	Culls	Total	Grade 1	Grade 2	Culls
MBr	144 a	101 a	42 a	1 a	166 a	98 a	65 b	3 bc
Chloropicrin	147 a	110 a	35 a	2 a	168 a	79 ab	82 a	7 a
Chlor 60	142 a	101 a	40 a	1 a	156 ab	70 b	81 a	5 ab
MBrC 70/30	135 a	107 a	27 a	1 a	139 b	79 ab	56 bc	4 ab
DMDS + Chlor	147 a	118 a	28 a	1 a	139 b	93 ab	43 c	3 bc
Pic+	141 a	114 a	25 a	2 a	153 ab	90 ab	60 bc	3 bc
New Pic+	150 a	122 a	27 a	2 a	155 ab	85 ab	65 b	5 bc
lsd $_{(0.05)}$	39	23	17	1	23	16	22	2

Within column means followed by the same letter do not differ at P = 0.05 level using Duncan's Multiple Range Test. Numbers are seedlings per square meter of nursery bed. Target seedling density was 192 seedlings/m^2.

Table 7. Loblolly pine seedling root morphology at the end of each growing season Jesup, GA.

Soil treatment	Root length (cm)		Root surface area (cm^2)		Avg root Dia (mm)		No. of root tips	
	2007	2008	2007	2008	2007	2008	2007	2008
MBr	315 a	243 a	91 a	54 a	0.92 ab	0.72 a	910 a	568 a
Chloropicrin	254 b	249 a	69 b	55 a	0.87 b	0.72 a	730 b	584 a
Chlor 60	274 ab	212 a	78 ab	46 a	0.91 ab	0.72 a	806 ab	532 a
MBrC 70/30	317 a	227 a	91 a	50 a	0.91 ab	0.72 a	943 a	560 a
DMDS + Chlor	299 ab	231 a	87 a	51 a	0.93 a	0.71 a	925 a	579 a
Pic+	277 ab	218 a	84 a	51 a	0.96 a	0.77 a	834 ab	545 a
New Pic+	313 a	232 a	91 a	49 a	0.92 ab	0.70a	941 a	636 a
lsd $_{(0.05)}$	51	54	15	9	0.06	0.10	155	176

Means within columns followed by the same letter do not differ at P = 0.05 level using Duncan's multiple range test.

had a relatively low percentage of Grade 1 seedlings from those plots, but had the highest seedling density. Overall, seedling root architecture and root morphology indicated smaller seedlings in 2008 when compared to seedlings grown in 2007 at Jesup (Table 7). First year soil fumigation results in larger seedlings and these alternative soil fumigants are behaving as expected. In 2007, as far as an MBr alternative, all soil fumigant alternatives performed as well as MBr across all the root morphology measurements at this nursery.

In 2008, there were no significant differences in seedling quality and quantity among any of the soil fumigants that did not contain MBr in its formulation. Overall, the total seedling root length in these trials ranged from 212 to 317 cm, or about 2.1 to 3.0 m of total roots per seedling. Like seedlings at Jesup, seedling root architecture and root morphology indicated smaller seedlings in 2008 when compared to seedlings grown in 2007 in Glennville (Table 8). As far as an MBr alternative, Pic+ performed well across all the root morphology measurements at Glennville with root length and number of root tips significantly better than the other non-MBr fumigants tested.

Soil trichoderma and nematodes

At the end of the first growing season in 2007, nursery soils at Jesup, GA fumigated with MBr had significantly lower levels of *Trichoderma* than Chlor 60 treatments (Table 9). Other soil fumigants, including MBrC 70/30 and chloropicrin, reduced *Trichoderma* levels like that of MBr. By the end of the second growing season (2008), the *Trichoderma* levels within the various soil fumigants tested were all similar to MBr. The effect of soil fumigants on soilborne fungi in 2007 at the Glennville Nursery indicated that soils treated with MBr had significantly lower levels of *Trichoderma* than DMDS + Chloropicrin (Table 9). By the end of the second growing season (2008), the *Trichoderma* levels within the various soil fumigants tested were similar to MBr. Over the course of the 2 year study, each soil fumigant plot was examined five times for both the number and species of nematodes within the soil/seedling interface. Nematode populations within the soil are rarely distributed uniformly across the nursery beds and these studies had a wide range (0 to 687 nematodes /100 cc soil) in numbers and species for all soil fumigants used (Table 10). Because of the

Table 8. Loblolly pine seedling root morphology at the end of each growing season, Glennville, GA.

Soil treatment	Root length (cm)		Root surface area (cm²)		Avg root dia (mm)		No. root tips	
	2007	2008	2007	2008	2007	2008	2007	2008
MBr	292 a	167 a	83 ab	45 ab	0.92 b	0.86 a	715 ab	366 a
Chloropicrin	218 b	152 ab	68 b	40 b	1.02 a	0.85 a	515 c	328 a
Chlor 60	289 a	148 ab	86 a	39 b	0.95 ab	0.87 a	738 ab	303 a
MBrC 70/30	304 a	157 ab	91 a	45 ab	0.98 ab	0.93 a	820 a	325 a
DMDS + Chlor	266 a	170 a	78 ab	48 a	0.95 ab	0.90 a	633 bc	342 a
Pic+	302 a	165 ab	88 a	48 a	0.93 b	0.93 a	735 ab	344 a
New Pic+	264 a	142 b	81 ab	41 ab	0.99 ab	0.94 a	680 b	310 a
lsd $_{(0.05)}$	49	25	17	7	0.08	0.12	137	86

Within column means followed by the same letter do not differ at 0.05 level using Duncan's Multiple Range Test.

Table 9. Post-fumigation recovery of *Trichoderma* spp. from nursery soils in Jesup and Glennville, GA Nurseries in 2007.

	Jesup, GA Nursery			Glennville, GA Nursery		
Treatment	7 weeks	15 weeks	26 weeks	7 weeks	15 weeks	26 weeks
MBr	86 a	117 a	18 bc	137 a	75 bc	90 b
Chloropicrin	42 ab	71 ab	18 bc	56 b	91 b	118 ab
Chlor 60	53 ab	90 ab	64 a	44 b	71 bc	90 b
MBrC 70/30	27 b	42 b	6 c	94 ab	54 c	98 b
DMDS+Chlor	63 ab	84 ab	30 abc	84 ab	146 a	168 a
Pic+	34 ab	61 ab	20 bc	63 b	94 b	108 b
New Pic+	48 ab	93 ab	57 ab	78 ab	68 bc	119 ab
lsd$_{(0.05)}$	55	71	44	67	38	57

Means within columns followed by the same letter do not differ at P = 0.05 level using Duncan's Multiple Range Test. numbers are in colony forming units (CFU) per mg of soil.

variability, there were no differences between treatments within a nursery for any of the soil fumigants tested. Overall, all soil fumigants were effective in reducing nematode populations the first growing season which then increased during the 2 year rotation. One of the more troublesome species on seedling production is the Stunt nematode (*Tylenchorhynchus claytoni*) which appeared during the second cropping season in all soil fumigants tested.

Of the soil fumigants tested, Pic+ had the lowest number of nematodes recovered in soil samples over the course of the study at the Plum Creek nursery. At the Rayonier Nursery, of the fumigants tested, soils treated with Chlor 60 had the fewest nematode numbers.

Soilborne pathogenic fungi

All soil fumigants and rates tested at both nurseries were effective in eliminating target *Pythium, Fusarium* or *Rhizoctonia* that was present on the oatmeal in the Hubo® soil bags. Of the three soilborne pathogens tested, *Fusarium* was recovered more often than the other two fungi (Table 11). When non-fumigated control bags were plated onto their selective media, the individual soilborne fungus was recovered from 100% of all oatmeal groups on each plate. In addition to the soilborne pathogenic fungi, a number of non-target fungi were recovered from the oatmeal. The non-target fungi were primarily saprophytic fungi, *Penicillium, Trichoderma* or *Aspergillus.* DMDS + chloropicrin had the lowest recovery of non-target fungi compared to the other soil fumigants used. The difference in fungal recovery on all the DMDS agar plates for all fungi was apparent even after 7 days, with many agar plates void of any fungal growth. New Pic+ had significantly less non-target *Pythium* and *Rhizoctonia* than chloropicrin.

DISCUSSION

The primary objective of the USDA Areawide MBr Alternative program is to identify possible alternatives to MBr using large-scale, multi-year trials in soils and

Table 10. Soil nematode levels at the end of each growing season in Jesup and Glennville, GA.

Soil treatment	Nematode	Jesup, GA		Glennville, GA	
		2007	2008	2007	2008
	Stunt	8	140	1	39
MBr	Stubby root	39	0	1	0
	Ring	0	0	0	0
	Stunt	20	133	2	25
Chloropicrin	Stubby root	29	1	0	0
	Ring	0	3	0	0
	Stunt	3	123	0	7
Chlor 60	Stubby root	34	0	0	0
	Ring	0	0	0	0
	Stunt	13	145	4	74
MBrC 70/30	Stubby root	19	0	2	0
	Ring	0	1	0	0
	Stunt	33	65	3	18
DMDS+Chlor	Stubby root	34	0	0	0
	Ring	0	1	0	0
	Stunt	46	47	4	16
Pic+	Stubby root	19	0	1	0
	Ring	0	1	1	0
	Stunt	14	47	14	138
New Pic+	Stubby root	36	0	0	0
	Ring	0	0	0	0

Due to the non-uniform distribution of nematodes, there were no differences observed for any treatment x nematode x year, P = 0.05. Numbers are the average of 4 replicates per treatment recovered in 100 cc of soil.

Table 11. Proportion of pathogenic soilborne recovered from inoculated oatmeal exposed to soil fumigants and planted onto selective agar media at Jesup and Glenville, GA, 2007.

	Pythium		Fusarium		Rhizoctonia	
Soil fumigant	Jesup	Glennville	Jesup	Glennville	Jesup	Glennville
MBr	0.00 a	0.00 a	0.00 a	0.00 a	0.00 a	0.02 a
DMDS + Chlor	0.00 a	0.00 a	0.00 a	0.00 a	0.00 a	0.00 a
MBrC 70/30	0.00 a	0.00 a	0.25 a	0.00 a	0.00 a	0.00 a
New Pic+	0.00 a	0.00 a	0.03 a	0.00 a	0.00 a	0.00 a
Pic+	0.00 a	0.00 a	0.04 a	0.00 a	0.00 a	0.00 a
Chloropicrin	0.00 a	0.00 a	0.00 a	0.02 a	0.00 a	0.00 a
Chlor 60	0.00 a	0.00 a	0.01 a	0.06 a	0.00 a	0.00 a
lsd	0.00	0.00	0.31	0.38	0.00	0.03 a

Means within columns followed by the same letter do not differ at P = 0.05 level using Duncan's Multiple Range Test.

conditions throughout the southern U.S. One of the unique aspects of MBr as a soil fumigant is its ability to consistently control weeds, insects, nematodes and fungi across many different growing conditions. Studies conducted within the southern US in forest-tree nurseries have yet to find an MBr alternative that fits those characteristics, and these studies bear that out. For example, seedlings grown in soils fumigated with chloropicrin had the smallest RCD (4.6 mm) for 2007 and the largest RCD (4.7 mm) in 2008 in Glennville. Overall, seedling production and seedling characteristics were dependent upon the soil fumigant used and the cropping season. The true test of an MBr alternative is its performance during the second growing season where treatment differences usually begin to appear as disease, weed, and nematode pressures increase. Based on these trials, when MBr is eventually phased out, those soil fumigants with chloropicrin appear to be the most useful in controlling pests and producing quality seedlings in both Jesup and Glennville forest-tree nurseries. In both instances, seedling densities and root characteristics from plots treated with chloropicrin-based products were similar to MBr. One of the primary reasons for determining the effects of these soil fumigants on root architecture is that a more fibrous root system increases the chance of seedling survival in the field (Hatchell and Muse, 1990; Frampton et al., 2002; Davis and Jacobs, 2005). One soil fumigant, DMDS + chloropicrin, was comparable to MBr in RCD and root morphology characteristics and soilborne *Trichoderma* levels, but had a significant odor problem (garlic or strong propane) that lasted long into the growing season.

The lingering odor with this particular soil fumigant will limit its acceptance by growers as an alternative to MBr. By far the best MBr alternatives tested were Pic+ and New Pic+, with both soil fumigants controlling soilborne pathogens, nematodes, and weeds and producing quality seedlings. The difference between Pic+ and New Pic+ is the solvent used in the application/injection process that reportedly decreases the volatility of chloropicrin and keeps the soil fumigant in place longer. When used as a soil fumigant, both these treatments resulted in similar seedling characteristics (RCD, seedling densities, seedling root morphology, etc) and had similar effects on the levels of the important soilborne fungi, *Trichoderma*. However, it was observed at Glennville, GA that the New Pic+ formulation did not control annual sedge (*Cyperus compressus*). Because of the lack of weed control by New Pic+, this compound has been dropped from further trials in forest-tree nurseries. In addition to the weed problems at Glennville, New Pic+ also had a high incidence of nematodes the second growing season (Table 11). The Rayonier nursery has a long history of nematodes reaching levels that affect seedling production. For these reasons, 1, 3, dichloropropene (Telone®) may need to be used in nurseries with reoccurring nematode pressure. Chlor 60 (containing 40% 1, 3-dichorolpropene) had the lowest nematode levels and may be an option for nurseries that have nematode issues in the second growing season. While the seedling densities and root characteristics with chloropicrin were encouraging, one of the potential pitfalls with using 100% chloropicrin at 336 kg/ha (300 lbs/a) is the buffer zone restrictions under current nursery fumigation practices. If these restrictions limit the use of 100% chloropicrin, then Pic+ with 85% chloropicrin would be the best alternative at both nurseries. The final decision when selecting an MBr alternative needs to take into consideration the ability of the soil fumigant to work under individual nursery soil conditions and the impact the new soil fumigation rules that come into place in 2012 will have on each individual nursery.

Soil fumigation is an effective way to reduce pathogenic soilborne fungi in the nursery that affect seedling production and survival after outplanting. All soil fumigants tested at both nurseries were found to be effective in controlling *Pythium*, *Rhizoctonia* and *Fusarium* when inoculated onto oatmeal. The wide-spread use of MBr has minimized extensive seedling losses due to soilborne pathogenic fungi. Of the fungi tested, *Pythium* still can cause damping-off problems in the early spring and is often limited to areas of poor drainage and standing water. *Rhizoctonia* can appear in nurseries both as root decay and as foliage blight, especially in the second-year crops post fumigation as the fungus increases over the first growing season (Carey and McQuage, 2004). At best, the soil fumigants reduced levels of these pathogens for the two cropping seasons and longer rotations would need to be tested to see if a nursery could get three seedling crops from one fumigation. One of the unique aspects of soil fumigants currently being tested in southern forest nurseries is that they do not completely eliminate beneficial fungi which are needed for seedling growth. In these two nurseries, the soil fumigants tested did not completely eliminate all soilborne fungi. This is important since previous research has shown that *Trichoderma* is an important soilborne fungus necessary for proper pine seedling growth (Bailey and Lumsden, 1998; Dong et al., 1987; Papavizas, 1985; Mousseaux et al., 1998; Samuels, 1996). In these trials, the population levels of non-target soilborne fungi rebounded quickly with all soil fumigants used as previous Nursery Cooperative research has shown that *Trichoderma* is not as sensitive to MBr as other soil fumigants (Carey et al., 2005; Starkey and Enebak, 2008). In contrast, dazomet, a soil fumigant tested by the Nursery Cooperative for several years, significantly reduced the levels of beneficial fungi which remained after two growing seasons (Starkey et al., 2006). While many nursery managers would prefer to use MBr in perpetuity to grow forest-tree seedlings, MBr will eventually be unavailable and each nursery manager will need to identify the best alternative for their nursery conditions.

These trials and others indicate that, while not the perfect replacement, seedling production is still possible without MBr if compounds such as chloropicrin are used and managers pay close attention to weed and nematode pests that are less susceptible to chloropicrin than MBr.

REFERENCES

Bailey BA, Lumsden RD (1998). Direct effects of *Trichoderma* and *Gliocladium* on plant growth and resistance to pathogens. *In Trichoderma* and *Gliocladium*, Edited by G.E. Harman and C.P. Kubicek. Taylor and Francis, Inc., Bristol, 2: 185–204.

Carey WA, McCraw D, Enebak SA (2005). Seedling production by seed treatment and fumigation treatment at the Glennville Regeneration Center in 2004. Auburn Univ. South. For. Nursery Manage. Coop., Res. Rep., 05-04. Auburn, AL, p. 5.

Cary WA, McQuage K (2004). Control of Rhizoctonia foliage blight by fungicides and fumigation: Lower application rates and fumigation effects. Auburn Univ. South. For. Nursery Manage. Coop., Res. Rep. 04-03. Auburn, AL, p. 4.

Cordell CE, Anders R, Hoffard WH, Landis TD, Smith RS, Toko HV (1989). Forest nursery pests. Wash (DC): Agric. Handb., 680: 184.

Davis AS, Jacobs DF (2005). Quantifying root system quality of nursery seedlings and relationship to outplanting performance. New For., 30: 295-311.

Dong LF, Mang XY, Mang HG (1987). Breaking seed dormancy of *Pinus bungeana* Zucc. with Trichoderma-4030 inoculations. New For., 1: 245-249.

Elad Y, Chet I, Henis, Y (1981). A selective medium for improving quantitative isolation of *Trichoderma* spp. from soil. Phytoparasitica, 9: 59-67.

Frampton J, Isik F, Goldfarb B (2002). Effects of nursery characteristics on field survival and growth of loblolly pine rooted cuttings. South. J. Appl. For., 26: 207-213.

Hatchell GE, Muse, HD (1990). Nursery cultural practices and morphological attributes of longleaf pine bare-root stock as indicators of early field performance. USDA For. Serv. Res. Pap. SE., p. 277.

Henry BW (1953). A root rot of southern pine seedlings and its control by soil fumigation. Phytopathology, 43: 81-88.

Jang E, Wood WS, Dorschner K, Schaub J, Smith D, Fraedrich S, Hsu H (1993). Methyl bromide phase out in the US: Impact and alternatives. In: USDA Workshop on alternatives for methyl bromide. Crystal City, VA.

Kannwischer ME, Mitchell DJ (1981). Relationship of numbers of spores of *Phytophthora parasitica* var. *nicotianae* to infections and mortality of tobacco. Phytopathology, 71: 69-73.

Ko WH, Hora FK (1971). A selective medium for the quantitative determination of *Rhizoctonia solani* in soil. Phytopathology, 61: 707.

McNabb KL (2006). Forest tree seedling production in the South for the 2005-2006 planting season. Auburn Univ. South. For. Nursery Manage. Coop., Tech. Note, 06-02: 10.

Mousseaux MR, Dumroese RK, James RL, Wenny DL, Knudsen GR (1998). Efficacy of *Trichoderma harzianum* as a biological control of *Fusarium oxysporum* in container-grown Douglas-fir. New For., 15: 11–21.

Nelson PE, Toussoun TA, Marasas WOF (1993). Fusarium species: An illustrated manual for identification. The Pa. State Univ. Press. University Park.

Papavizas GC (1985). *Trichoderma* and *Gliocladium*: Biology, ecology, and potential for biocontrol. Annu. Rev. Phytopathol., 23: 23–54.

Samuels GJ (1996). *Trichoderma*: A review of biology and systematics of the genus. Mycol. Res., 100: 923–936.

South DB, Carey WA, Enebak SA (1997). Chloropicrin as a soil fumigant in forest nurseries. For. Chron., 73: 489-494.

Starkey TE, Enebak SA (2008). Indian Mound Nursery, Texas: Methyl bromide alternative study 2005-2007. Auburn Univ. South. For. Nursery Manage. Coop., Res. Rep. 08-07. Auburn, AL, p. 11.

Starkey TE, Enebak SA, McCraw D (2006). Seedling quality and weed control with methyl bromide and methyl iodide at the Glennville Regeneration Center 2005-2006. Auburn Univ. South. For. Nursery Coop., Res. Rep. 06-05. Auburn, AL, p. 5.

2

Households' exploitation of non-wood forest products (NWFPs) in Okitipupa Local Government Area of Ondo State, Nigeria

A. S. Oyekale* and D. O. Ajesi

Department of Agricultural Economics, University of Ibadan, Ibadan, Nigeria.

Exploitation of NWFPs poses serious threat to sustainable forest management in Nigeria. This study analyzed the driving forces for exploiting forest products in Okitipupa Local Government area of Ondo State, Nigeria. Data were obtained through structured questionnaires that were randomly administered to farmers. The data were subjected to descriptive analysis and Tobit regression. Results show that monetary worth of monthly collections of NWFPs is ₦12,167.50 and fuel wood has the highest average monthly collection worth ₦ 3,576.50. The Tobit regression shows that community variable sex, age of house heads number of dependant farming as primary occupation, having alternative means of meeting the needs for NWFPs and dependants income show statistical significance ($p<0.05$). It was recommended that efforts to reduce degradation of NWFPs must consider development of research into provision of alternatives to some of the resources, reduction in demographic pressure and keen consideration of some cultural gender factors.

Key words: Degradation, forest products, exploitation, Ondo state.

INTRODUCTION

Nigerian forest is renewable natural resource that provides timber for domestic and industrial uses, food for man and wild animals, protective covers for environmental resources such as soil and recreational facilities for tourist attraction (Popoola, 1992; Nathaniel and Adebobola, 2001). Cognizance has been taken of the importance of forest resources to the socio-economic and political development of the Nigerian people (NEST, 1991). Beside timbers, forest contains some other useful products like chewing sticks, wrapping leaves, medicinal plant, camwood, sheabutter, sheanut, gum arabic, Niger Gutta, Bees wax, snail, mushroom, etc. These products are referred to as minor forest products, because of inadequate documentation of their trade in international markets. However, with appropriate market development, foreign exchange can be earned from these minor forest products (Osemeobo and Ujor, 1999). Conceptually, the definition of NWFP follows the classification given by FAO (1999) where these are referred to as all biological materials (other than wood) which are extracted from natural forests for human use as well as services derived from forests and allied land uses. Osemeobo and Ujor (1999) also submitted that NWFPs s are defined as forest materials derived from soil mineral, water, fauna and flora resources other than round wood (sawn wood). The NWFPs are classified into non-wood and non-timber products. The non-wood products are derived from wild animals, herbs, leaves, latex, gum, resins, ropes, fruits, seeds, fungi, fodder, forage, gravel, clay, limestone and natural salt. The woody but non-timber products include poles, fuel-wood, charcoal, rattan canes, sponge, chew sticks, bamboo's and others. These NWFPs make vital contributions to the well-being of the rural poor by providing food materials which are essential dietary supplements during off farm seasons and periods of low agricultural production (NEST, 1991; Arnold and Perez, 2001; Lan et al., 2002).

Availability of NWFPs in any Nigerian ecological zone is a function of the richness of the vegetation, in terms of

*Corresponding author. E-mail: asoyekale@yahoo.com.

the structure of species occurring in it and the use of biotic species by the rural population. However, it had been noted that data collection on NWFPs is constrained by many factors. These include information loss, poor torage facilities, inadequate inventory, high rate of poaching, low level forest management and mode of resource harvesting. Despite these set-backs, improvement in data collection can be attained through intensive research, involvement of local communities in

the conservation of NWFPs and partnership in the development of forest resources (Osemeobo and Ujor, 1999). It should be noted that Nigeria natural resources, whether renewable or non-renewable, have suffered significant exploitation by man, particularly since the 20th century due to increase in agricultural land frontier and pressure on vegetation for provision of food and income (Agboola, 1979; FORMECU, 1995). The consumption pattern of NWFPs is highly influenced by socio-cultural and economic factors. Some of the key factors include taboos, social acceptance, level of utilization, demand and tenure issues. In most cases, NWFPs are fully recognized at local level and are harnessed to meet household food security and as dependable sources of income. Also, trade in NWFPs is organized both at the rural and urban areas. However, NWFP respond to market failures because they are not reflected in GDP calculations. Prices of NWFP fluctuate but they are higher during the off season than in the producing season. The middle men fix prices for the goods and sustain the trade both in rural and urban areas (Osemeobo and Ujor, 1999). The sustainable conservation and management of NWFP have been difficult to streamline in Nigerian forestry because over 90% of the NWFP respond to market failures (Osemeobo and Ujor, 1999). The worth of NWFP in Nigeria is difficult to estimate as the prices placed on them are often localized and vary from one area to another. Moreover, the value placed on NWFP is often lower than the cost of replacement (Osemeobo, 1992). Hence, the value of NWFPs is at best social costs; because the markets have failed to reflect the full value of NWFP to under-pricing market prices do not reflect their full marginal opportunity cost to society (Salau, 1991).

By implication because most NWFP are treated as free goods, they are overused, degraded and hence they respond to market failures. The response of agricultural policies to economic reforms and trade policies towards market oriented regimes has increased pressure on the utilization of the Nigerian NWFP (World Bank, 1996). For example while other sub-sector within agriculture are gaining more lands from natural forest. NWFPs are loosing the land base and their recognition. Most of the NWFP are treated as intangibles with free access and little restriction (Osemeobo and Ujor, 1999). This paper has the objective of identifying the driving forces for exploiting NWFPs in Okitipupa Local Government area of Ondo State. The first working hypothesis states that with increasing population (proxied by the number of dependants) does not lead to increasing demands for NWFPs. The second hypothesis states that income from NWFPs does not contribute significantly to household income.

MATERIALS AND METHODS

The study area and sampling procedures

Okitipupa Local Government is one of the local councils in Ondo State. It is bounded in the North by Odigbo Local Government, east by Irele Local Government, south by Ilaje Ese Odo Local Government and west by Ogun State. The forest areas across the Local Government areas are distinctively marked with high density of oil palm trees and timbers. Agriculture is the main occupation of people in the villages. The instrument for data collection is structured questionnaire administered to respondents on a random selection basis. We depended on memory recall since it was impossible for having panel data set. Data were obtained from randomly selected towns and villages like Okitipupa, Ode- Aye, Igbotako, Ilu Tuntun, Idepe, Igodan and Igbodogo. A total of 120 questionnaires were administered with proportionality to the estimated number of farming households. However, 20 were rejected for inconsistent information.

Estimated model

The Tobit model was used to estimate the parameters due to zero values that some of the respondents have on the amount of forest products collected (Gujarati, 1995). The model is as stated:

$$F_i = \alpha_0 + \alpha_i \sum_{i=1}^{13} X_i + v_i$$

Where F_i is the monetary worth of all the NWFPs collected (₦), α's are the estimated parameters; X_1 is the community dummy (Okitipupa = 1, 0 otherwise); X_2 is the sex of house head (male = 1, 0 otherwise); X_3 is the age of house head; X_4 is the marital status of respondent (single = 1, 0 otherwise); X_5 is the family type (monogamous = 1, 0 otherwise); X_6 is the number of children; X_7 is the number of other dependent relatives; X_8 is the educational status (no formal education = 1, 0 otherwise); X_9 is the primary occupation (farming = 1, 0 otherwise); X_{10} is the have alternatives to forest products dummy (yes = 1, 0 otherwise); X_{11} is the monthly income of respondent (₦); X_{12} is the total monthly income of other working members in the household (₦); X_{13} is the number of land areas cultivated (hectares) and v_i is the stochastic error term.

RESULTS AND DISCUSSION

A descriptive analysis of the socio-economic characteristics of the respondents reveals that 87% of them were males. This shows that most of the households in Okitipupa Local Government Area are still headed by males. Household female headship as revealed in the study resulted from divorce (4%), separation (4%) and state of being single (5%). Average age is 52.79 years with coefficient of deviation of 25.10%. This reveals that majority of the people were already getting old. The oldest was 88 years, while the youngest was 25 years. About 30% of the respondents did not attain any formal education and 73% indicated that farming was their primary occupation. Average number of children of the

Table 1. Average monthly NWFPs collected availability and accessibility in Okitipupa Local Government Area.

NWFPs	Monthly collection (₦)	Coefficient of variation	Availability	Accessibility	Non exploiters (%)
Fuel wood	3576.50	173.29	89.00	85.00	18.00
Bush meat	2505.00	260.76	81.00	71.00	30.00
Chewing stick	724.40	252.67	81.00	76.00	35.00
Fruits and nuts	1101.80	240.92	78.00	72.00	41.00
Gums and resins	313.33	510.33	65.00	62.00	75.00
Herbs and medicinal plants	818.50	174.02	85.00	75.00	32.00
Mushrooms	372.40	211.50	75.00	70.00	54.00
Honey	421.80	234.05	74.00	68.00	64.00
Snail	1259.40	195.30	86.00	74.00	39.00
Leaves and vegetables	738.50	204.10	84.00	75.00	34.00
Poles	339.00	479.40	59.00	38.00	88.00
Total collection	12167.50	154.15	77.91	69.64	17.00

Table 2. A Tobit analysis of the socio-economic determinants of NWFP exploitation in Okitipupa.

Variables	Parameter	T-statistics
Constant	33,431.42***	3.08
Community	-10,029.41**	-2.16
Sex	-11,666.53**	-2.19
Age	-441.56**	-2.13
Marital status	-10,786.65	-1.59
Family type	2,458.03	0.52
Household size	245.68	0.35
Number of dependant	1,670.41**	2.41
No formal education	1,464.16	0.30
Primary occupation	11,949.57**	2.12
Forest alternatives	-9,636.54**	-2.12
Income	-0.12	-0.08
Dependants' average income	0.51**	2.02
Land areas cultivated	299.30	0.56

*** statistically significant at 1% and ** statistically significant at 5%.

respondent is 6.77, with coefficient of variation of 57.11%, while average number of dependent relatives was 2.72 with coefficient of variation of 121.64%. Table 1 shows that for all the NWFPs, the average monetary worth of monthly collections is ₦12, 167.50 with 17% not collecting any. Out of the individual NWFPs, fuel wood has the highest average monthly collection worth ₦ 3,576.50, while gums and resins has the lowest (₦ 313.33). There is wide variation in the amount of NWFPs collected by the people as revealed by the coefficient of variation. This is due to the fact that not all the people really have access to these products. The highest variability is recorded for poles (479.40%), while the lowest is recorded for total NWFPs (154.15%). The NWFP with highest availability as reported by the respondents is fuel wood (89%) while the lowest is poles (59%). Accessibility is also highest for fuel wood (85%), while pole also has the lowest accessibility of 38%. Out of the individual NWFPs, fuel wood has the lowest percentage non exploiters (18%) and pole has the highest of 88%. Table 2 shows the results of the Tobit regression. It shows that the likelihood ratio chi-square is statistically significant ($p<0.01$). The community variable is statistically significant ($p<0.05$). This implies that respondents from Okitipupa gather NWFPs, whose monetary values are lower by ₦10,029.41. This is expected because forests are not readily availability in growing peri-urban center. The sex variable is also statistically significant ($p<0.05$). This shows that households headed by males gather NWFPs that are lower in monetary values by ₦11, 666.53. This is also expected because fuel wood gathering, which dominates the NWFPs is culturally reserved for women.

When the households are headed by women due to migration or death of the man, the tendency is that dependency on forest products increases. The parameter for the age of house heads shows statistical significance ($p<0.05$). This implies that when the age of house heads increases by 1 year, the monetary worth of NWFPs that will be gathered will decline by ₦441.56. This is expected because aged house heads may not have enough strength for gathering forest products that are daily depleted, making them available in farther places. The dependant variable shows statistical significance ($p<0.05$). This shows that increasing the number of dependants increases the monetary worth of the NWFPs gathered by ₦1, 670.41. The first hypothesis is hereby rejected. Also, the parameter of farming as primary occupation is statistically significant ($p<0.05$). This shows that involvement in farming as primary occupation increases the monetary worth of NWFPs gathered by ₦11, 949.57. The parameter possession of alternatives to NWFPs shows statistical significance ($p<0.05$). This implies that those who have alternative means of meeting their needs for NWFPs gather NWFPs that are lower in monetary worth by ₦9, 636.54. While the incomes of house heads do not significantly influence NWFPs exploitation, the dependants' income is positively correlated with the monetary worth of the NWFPs that the households gather. This is expected because in many rural settings, it is the youths that take gathering of forest products as an hobby for personal income. The second hypothesis is also rejected.

CONCLUSIONS AND RECOMMENDATIONS

NWFPs in some Nigerian rural areas suffer significant exploitation, which has resulted into their non-availability as some already go into extinction. This study attempts to evaluate whether households have access to these products and analyze some factors that can promote their utilization. The policy issues derived from the findings of the study are hereby discussed: first, female headship increases exploitation of NWFPs. Policies to reduce degradation of the Nigerian forest resources must take cognizance of the peculiar gender factors associated with dependence on the forest. Occurrences in the nation that increases female household headship will have some detrimental effects on the forest resources in the rural areas. Efforts to address the problem must therefore seek to provide enabling environment where men are willing and able to take responsibility for living with and providing for their homes. Secondly, policies to reduce fertility in the rural areas must be formulated. The study reveals that presence of many dependants necessitates exploitation of NWFPs. This is even corroborated by the positive relationship that exists between the income of the dependants and the worth of the NWFPs gathered. The government needs to provide appropriate sensitization in the rural areas about the need for having few number of children. However, development of low-cost technology for quicker performance of some farm activities is necessary because some farmers have many children in order to have enough hands to be involved in farming operations. Finally, provision of alternative means of meeting the need for gathering NWFPs will reduce exploitation of forest products. This is necessary because availability of kerosene at affordable price may reduce dependence of people on fuel wood. The government should ensure promotion of research into the alternatives that exist for NWFPs. For instance, provision of rural health center can reduce dependence on herbs and availability of toothpaste can reduce dependence on chewing sticks.

In order to safeguard the future of future generation by avoiding complete exploitation of NWFPs, rural farmers have to be educated on the ways of domesticating some wildlife like grass cutter, rabbits and snails. This is an income-generating venture that has the potential of reducing exploitation pressure on the forest.

REFERENCES

Agboola SA (1979). Agricultural atlas of Nigeria, Oxford University Press, Ibadan, 148 p.

Arnold JEM, Perez MR (2001). Can non-timber forest products match tropical forest conservation and development objectives? Ecol. Econ., 39: 437-447.

Food and Agriculture Organization (FAO) (1999). Non-wood forest products for rural income and sustainable forestry. FAO Publication Division, Rome.

FORMECU (1995). An assessment of land use and vegetation changes in Nigeria 1978 - 1995. Geomatics, Ontario, Canada.

Gujarati DN (1995). Basic Econometrics McGraw-Hill International Ed. 838p.

Lan LV, Ziegler S, Grever T (2002). Utilization of forest products and environmental services in Bach Ma National Park, Vietnam Internet file retrieved from http://www.mekong-protected-areas.org/vietnam/docs/bach_ma_forest_products.pdf.

NEST (Nigeria Environmental Study Action /Team) (1991). Nigeria's Threatened Environment: A National Profile. Nigeria: NEST.

Nathaniel IT, Adebobola N (2001). "The Effects of Poverty in Conservation of Biodiversity: The Nigeria Experience" Science in Africa Magazine. Internet file retrieved from www.scienceinafrica.co.za/2001/nov/nigeria.htm.

Osemeobo GJ (1992). Land use issues on wild plant conservation in Nigeria, J. Environ. Manage., 36: 17-26

Osemeobo GJ, Ujor G (1999). Non-Wood Forest Products in Nigeria. EC/FAO ACP Data Collection project Technical Report AFDCA/TN/06 FAO: Rome. Internet file retrieved from http://www.fao.org/documents/show_cdr.asp?url_file=/DOCREP/003/X6695E/X6695E06.htm.

Popoola LA (1992). An invited paper to be presented at the International enabling conference. "The green check initiative" Environmental forum 23-26 April, 2000.

Salau A (1991). Ecology and Economics: Economic Strategies for Natural Resources Conservation in Nigeria, 8p.

World Bank (1996). Poverty in the Midst of Plenty: The challenge of growth with inclusion in Nigeria. A World Bank poverty assessment report, World Bank, Washington, D.C.

3

Effect of root zone salinity on mineral nutrition and growth of beri (*Zizyphus mauritiana* lam) and jaman (*Eugenia jambolana* lamk)

Asrar Sarwar[1] and Zahoor Aslam[2]

[1]Research Officer, Rural Business Development Center, Lok Sanjh Foundation, Faisalabad, Pakistan.
[2]Project Director NABSAB, Ministry of Agriculture, Islamabad, Pakistan.

Excess of salts in the root zone of plants adversely affects their growth. It may result in loss of stands, reduced rates of plant growth, reduced yields and in severe cases, total crop failure. Salinity limits water uptake by plants by reducing the osmotic potential and thus the total potential of the soil water, additionally certain salts may be specifically toxic to plants or may upset nutritional balance, if they are present in excessive amount or proportion. The problem of increasing salinity implicates the selection and development of salt tolerant plants, in addition to other preventive and management measures. The present study intends to illustrate the effect of root zone salinity on the survival, growth, ionic activity and nutritive value of Beri (*Zizyphus mauritiana* lam) and Jaman (*Eugenia jambolana* Lamk). Experiments were conducted in pots filled with gravel applied with treatment solutions containing nutrients and salts to produce salinity levels of ECi-5, 10, 15, 20 and 25 dSm^{-1}. The relative dry matter production separately for browseable material, shoot and root against salinity was calculated as mean of four replications determine after 30 days of transplantation. Three parts of the plant showed almost similar growth reduction; however, the reduction of root growth being slightly more severe for both the plants.

Key words: Soil salinity, root zone salinity, Indus basin, growth, nutritive value, Beri, Jaman.

INTRODUCTION

Salinization of soil and ground water is a serious land degradation problem in semi arid and arid regions of Pakistan. It is estimated that nearly 6 million hectares (35 to 40% of total irrigated area) are affected with salinity, which has reduced 25% production potential of the Indus basin and it was estimated that two million ha are abandoned due to severe salinity (Wolters and Bhutta, 1997). The extent keeps on changing due to the dynamic nature of the problem about 75% of the population and about half of the Gross National Product (GNP) of Pakistan are directly or indirectly related to the agricultural sector and due to the presence of this shallow and saline groundwater, about 40,000 ha are annually abandoned within the Indus basin due to secondary salinization (WAPDA, 1989). The main causes of increasing threat are poor irrigation and drainage practices, which is depriving the farmers from productive resources and thus, posing negative impacts to their livelihood (Lamkbers, 2003).
Pakistan has given due significance to this issue after the country came into being. A complete and durable solution of the problem of salinity in Pakistan lies in leaching and drainage of excessive salts from the surface and root zone of soil. However, this approach is expensive, energy intensive and time consuming. It is possible saves costs of energy resources. The alternative method to restores saline land is saline agriculture technology. This approach is cost effective. This is basically a pro-poor approach to enhance income of poor farmers, who normally leave the land barren otherwise.

This approach emphasized on using highly saline water and lands on a sustained basis through the profitable and integrated use of genetic resources including plants, animals, fish and insects, and improved agricultural

*Corresponding author. E-mail: asrarsarwar@gmail.com.

practices. Management practices of these soils include chemical amendments, organic matter, mineral fertilizers and judicious selection of the salt tolerant forages and grasses. This approach attempts to promote bio-reclamation techniques using salt tolerant plants, bushes, trees and fodder grasses and production utilization of this biomass. Plants particularly trees are commonly referred as biological pumps and play an important role in the overall hydrological cycle in a given area. Studies in Pakistan showed that, highly saline waters could be used to grow salt tolerant fodder grasses to improve the quality and quantity of livestock (Lennard and Qureshi, 1998; Hussain et al., 1990).

Ziziphus mauritiana (locally knows as Beri) and *Eugenia jambolana* lam (locally known as Jaman) family Rhamnaceae and Myrtaceae respectively, are multi purpose plants of semi arid regions. Besides fruit, these plants also yield timber of marginal value, brushwood, fuel wood and leaf fodder (Pareek, 2001). Leaves and other browseable material of these plants are the primary source of mineral for small browsing animals like goats, sheep and lambs. Under saline condition, mineral ion interaction in the external may effect in the internal requirement of elements essential for plant growth and development (Gratten and Grieve, 1999). The availability, uptake and partitioning of mineral ions within the plants are controlled by numerous environmental factors, including concentration and composition of solutes in the growing medium. These imbalances influence the plants color, texture and mineral nutrition value. The present study intends to illustrate the effect of root zone salinity on growth, survival and chemical composition of different parts of Beri and Jaman plants.

MATERIALS AND METHODS

The experiments were conducted at Nuclear Institute for Agriculture and Biology (NIAB), Faisalabad, Pakistan from July till October 2005, in net house salinity environmental division. Seeds of the Beri and Jaman were collected from the local market. Seeds were sown in plastic packets filled with soils, silt and farm yard manure (FYM) at the ratio of 1:1:1 at NIAB, Faisalabad. Beri (*Z. mauritiana*) was sown on 20th July and Jaman (*E. jambolana* lamk) on 22 August, 2006. The soil of plants was dipped in the water for few minutes and then gently removed from the plants with hands. The plants were transplanted at the age of four weeks into pots. The setup consists of 24 sets of plastic pots for each experiment with four replications for each treatment, which were filled with thoroughly acid washed quartz gravel 5 to 25 mm dia (Qureshi et al., 1997).

One week after transplanting, the plants were exposed to salinity. The salinity of the medium was gradually increased by ECi-5 dSm^{-1} increment at one-day interval, until it reached the highest level. The pots were irrigated with water of six different qualities. The control treatment with half strength Hoagland nutritive solution (Hoagland and Arnon, 1950) with electric conductivity (ECi) of 1 dSm^{-1} and five salt treatments, containing ECi- 5,10,15,20 and 25 dSm^{-1} respectively, obtained by adding Na_2SO_4, $CaCl_2$, $MgCl_2$ and NaCl at the ratio of 10:5:1:4. To obtain a subject level of salinity in the pots, NaCl with concentration of 200 g/L of water was added as needed. The pH of the solution was uncontrolled and ranged from 6.5 to 7.5. In the absence of the drainage, salinity was measured with the help of the portable EC meter. To assure that target ion concentrations were maintained, irrigation water was changed twice in a week. Plants were aerated three times in a day by pulling the plants containing pot out of the irrigated water.

At the end of the experiment plants were harvested from the each treatment and divided into three parts, brows able material, shoot and root. The plant parts were washed with distilled water and dried. The shoot, root length and fresh weight of browseable material, shoot and root were determined on all treatments after 30 days of transplanting. Plants dry matter was measured after drying at 70°C for 48 h. The dried parts of the plants were grind separately and were wet digested in HNO_3 and $HClO_4$. Mn^{2+}, Fe^{2+} and Cu^{2+} were determined with the help of Atomic Absorption Spectrophotometer (AASPM), Na^+, Ca^{2+} and K^+ by flame photometer (Jenway PFP7) and P by spectrophotometer. Water content (%) was determined from fresh and dry weight measurements of plants. Mineral nutrient content of plants was calculated from the ion concentration (meq per 100 g dry weight basis). Data were analyzed using MSTATC software and means were tested, using Tukey's standardized range test (SAS institute Inc., 1997) at 5% level of significance ($p<0.05$).

RESULTS AND DISCUSSION

Chemical analyses of the collected data were analyzed using MSTATC software. The results revealed that Na^+ and K^+ concentration in Beri were significantly increased with increase in salinity. Both plants showed burning and darkening of leaves at higher salinity levels (20 to 25 dSm^{-1}). The cause of this disorder was not identified. However, high concentration of Na^+ in the plants tissues may be toxic for the plant survival. For most plants, the presence of higher concentration of Na^+ limits growth and can severely depress acquisition of K^+ and Ca^{2+}. In this study, minimum Na^+ concentration (11.32 meq/100 g dry weight) was recorded in control while maximum was found in (158.4 meq/ 100 g dry weight) ECi-20 dSm^{-1} (Table 1). Browseable material Na^+ in Beri increased significantly as the salinity (ECi) increased from 1.0-10 dSm^{-1} (Table 1). While ion concentration in the shoots and roots of Beri were 50% less then browseable material at higher salinity levels as shown in Table 1. The Jaman was a weaker accumulator of Na^+ than the Beri. The relatively high concentration of Na^+ in the moderately saline irrigation water (ECi-10 dSm^{-1}) resulted in substantially higher Na^+ contents for these plants as compared to the control values.

Under saline condition, maintenance of adequate level of K^+ is essential for plants survival. Higher sodium contents in irrigation water not only interfere with K^+ uptake by the roots but also may disrupt the integrity of root membrane and alter the selectivity of root system K^+ over Na^+ (Grattean and Grieve, 1999). External K^+ concentration was constant in the study. As a result, K^+ concentration in the plants significantly decreased with increase in the salinity, as shown in Tables 1 and 2.

Calcium is an important component of plant nutrition and it plays a vital physiological role in the metabolism of plants. The imbalances of minerals in the substrate also

Table 1. Effect of root zone salinity on chemical composition of Beri.

Element	Plant parts	Control	ECi-5	ECi-10
	BM	501.35^{ns}	492.45^{ns}	458.82^{ns}
Tissue water	Shoot	292.86^{ns}	282.19^{ns}	276.09^{ns}
	Root	478.80^{ns}	567.56^{ns}	448.68^{ns}
	BM	11.32^{c}	41.10^{b}	72.05^{a}
Na^+	Shoot	17.50^{b}	37.92^{a}	41.44^{a}
	Root	23.05^{b}	48.56^{a}	50.56^{a}
	BM	68.31^{ns}	62.20^{ns}	52.56^{ns}
K^+	Shoot	34.08^{a}	23.35^{b}	21.85^{b}
	Root	39.15^{ns}	34.58^{ns}	32.70^{ns}
	BM	15.44^{ns}	12.73^{ns}	8.73^{ns}
Ca^{2+}	Shoot	5.81^{ns}	8.18^{ns}	8.73^{ns}
	Root	19.96^{a}	15.59^{ab}	11.70^{b}
	BM	21.2^{5a}	7.46^{b}	4.31^{b}
P	Shoot	1.58^{ns}	5.07^{ns}	7.76^{ns}
	Root	17.64^{ns}	17.92^{ns}	6.97^{ns}
	BM	2.25^{ns}	1.82^{ns}	1.28^{ns}
Fe^{2+}	Shoot	0.70^{ns}	5.30^{ns}	2.88^{ns}
	Root	6.93^{ns}	3.34^{ns}	2.06^{ns}
	BM	0.30^{ns}	0.22^{ns}	0.09^{ns}
Mn^{2+}	Shoot	0.05^{b}	0.398^{a}	0.23^{ab}
	Root	0.33^{ns}	0.25^{ns}	0.16^{ns}
	BM	0.12^{ns}	0.10^{ns}	0.11^{ns}
Cu^{2+}	Shoot	0.12^{ns}	0.12^{ns}	0.11^{ns}
	Root	0.13^{ns}	0.11^{ns}	0.11^{ns}

Within columns, means followed by different letter are significantly different at the 0.05 probability level according to Tukey's standardized range test.

adversely affect Ca^{2+} availability to plants (Lauchi, 1990; Gratten and Grieve, 1999). The Ca^{2+} concentration, which were adequate for plants growth under non-saline condition may be nutritionally inadequate and growth limiting factor under saline conditions. The Ca^{2+} status of the plant was strongly influenced by ions composition of the external medium, the presence of the other salinizing ions in the substrate may reduce the Ca^{2+} activity and availability to the plants (Saurez and Grieve, 1998; Willumsen et al., 1996). Excessive amount of Na^+ may displace the Ca^{2+} from its extracellular binding sites with plant organ to further disrupt Ca^{2+} acquisitions, uptake and transport. The Ca^{2+} concentrations in the root of Beri and shoot of Jaman were significantly decreased as salinity level increased from 1.0-15 dSm^{-1}.

The Ca^{2+} concentrations in other parts of the plants were found statistically non-significant. The influence of salinity on P accumulation in crop plants is variable and depends upon the plant and experimental conditions (Champagnol, 1979). In many cases, salinity decreased the P concentration in plant tissue (Sharpley et al., 1992); in others, salinity increased P or had no effect. It is not surprising that these differences among studies occur, since P concentrations vary widely in different experiments and other nutrient interactions could be occurring simultaneously. Analysis of data showed that, salinity also significantly decreased P in browseable material of Beri. The browseable material P concentrations in the Beri varied from (21.25 to 4.31 meq/100 g dry weight) basis as salinity increased from 1.0 to 15 dSm^{-1} (Table 1). Fe^{2+} acts as catalyst in the production of the chlorophyll. It is imperative in respiration, affecting the iron-porphyrin protein complex, which is a carrier of oxygen and activator of oxygen in the plants. Reports on the influence

Table 2. Effect of root zone salinity on chemical composition of Jaman.

Element	Plant parts	Control	ECi-5	ECi-10
Tissue water	BM	472.74^{ns}	548.77^{ns}	405.27^{ns}
	Shoot	198.20^{ns}	297.96^{ns}	275.77^{ns}
	Root	188.42^{ns}	211.78^{ns}	241.11^{ns}
Na^+	BM	29.55^{b}	45.42^{b}	79.32^{a}
	Shoot	11.85^{b}	29.25^{b}	55.50^{a}
	Root	8.60^{ns}	12.83^{ns}	23.40^{ns}
K^+	BM	13.13^{a}	7.54^{b}	11.10^{ab}
	Shoot	15.38^{a}	9.80^{b}	12.88^{ab}
	Root	10.63^{ns}	13.93^{ns}	14.95^{ns}
Ca^{2+}	BM	17.47^{ns}	15.50^{ns}	23.35^{ns}
	Shoot	23.45^{a}	4.45^{ab}	0.79^{b}
	Root	16.07^{ns}	15.35^{ns}	12.63^{ns}
P	BM	3.28^{ns}	7.98^{ns}	1.08^{ns}
	Shoot	23.46^{ns}	4.45^{ns}	0.80^{ns}
	Root	59.61^{ns}	4.62^{ns}	0.75^{ns}
Fe^{2+}	BM	2.83^{ns}	1.90^{ns}	1.90^{ns}
	Shoot	0.77^{ns}	3.47^{ns}	3.47^{ns}
	Root	3.55^{ns}	4.04^{ns}	4.04^{ns}
Mn^{2+}	BM	0.07^{ns}	0.04^{ns}	0.07^{ns}
	Shoot	0.02^{ns}	0.01^{ns}	0.03^{ns}
	Root	0.05^{ns}	0.05^{ns}	0.08^{ns}
Cu^{2+}	BM	0.12^{b}	0.12^{b}	0.14^{a}
	Shoot	0.13^{ab}	0.12^{b}	0.14^{a}
	Root	0.13^{ns}	0.13^{ns}	0.14^{ns}

Within columns, means followed by different letter are significantly different at the 0.05 probability level according to Tukey's standardized range test.

of salinity on the iron (Fe^{2+}) concentration in plants are as inconsistent as those that concern Zn^{2+} and Mn^{2+} concentration. Statistical analysis showed that the effect of salinity on the Fe^{2+} contents was non-significant (Table 1) in both the plants.

Although differences were found in the literature regarding the effect of salinity on shoot Mn^{2+} concentration in barley, the differences may be explained in part, by the composition of the salinizing salts (Suhayda, 1992). The irrigation with saline water significantly decreased the Mn^{2+} contents in the shoot of Beri while effect of salinity on the concentrations of Mn^{2+} in Jaman was found statistically non-significant (Table 2). The result of the study further indicated that, Cu^{2+} was found statistically non-significant in Beri. However, the most of Cu^{2+} were accumulated in the browseable material and shoot of the Jaman. The results obtained showed that, shoot growth of both plants progressively decreased with increase in the salinity levels of the root zone. However, this decrease was not of the same extent for both Beri and Jaman. Relative salt tolerance is often quantified as the salt level results in a 50% reduction in shoot growth (yield), on alternatively the threshold salinity level where plants show decline, followed by the rate of or slope yield reduction (Mass and Hoffman, 1977; Carrow and Duncan, 1998).

However, the different species behave differently with respect to their salt tolerance. The 50% reduction in dry weight of Beri and Jaman (as compared to control) were found at ECi level 11 and 10 dSm^{-1} respectively. Overall, plant growth and dry matter production was higher in Beri than in Jaman. The total dry matter production of the

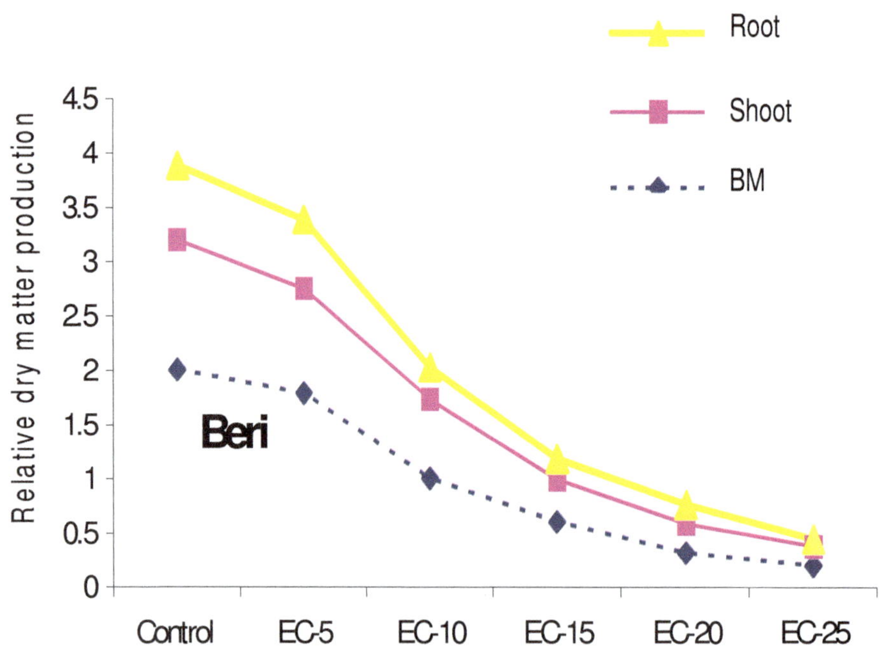

Figure 1. Relative dry matter production of browseable material, shoot and root as affected by salinity.

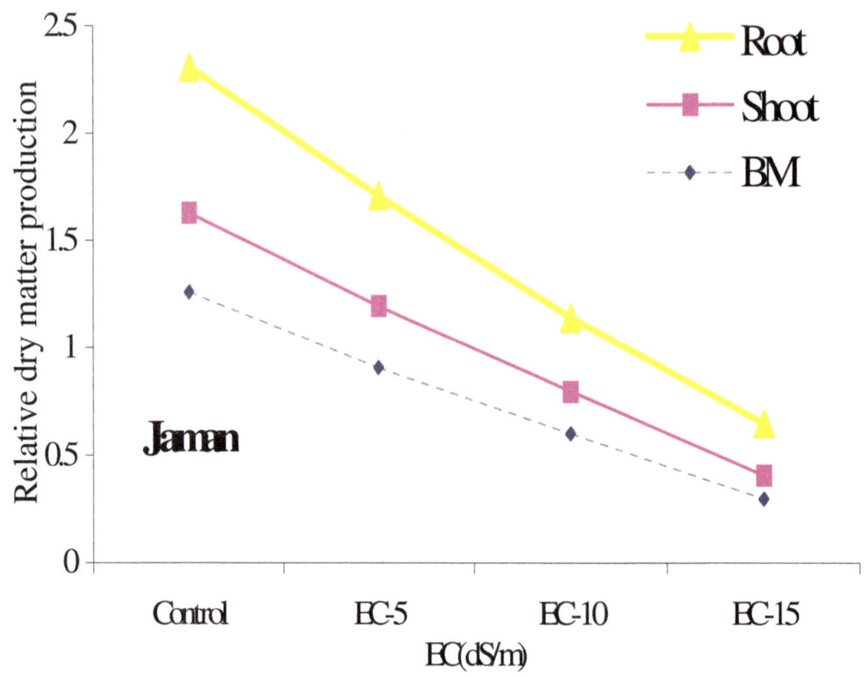

Figure 2. Relative dry matter production of browseable material, shoot and root as affected by salinity.

Jaman was approximately 72% of that of Beri. Another difference between Beri and Jaman was regarding the distribution of dry matter. The ratio between the roots and the total dry matter production was found approximately 0.2 for Beri and 0.3 for Jaman. Increases in the salinity affected the growth and dry matter production in both the plants. Figures 1 and 2 presents the relative dry matter production separately for browseable material, shoot and root against salinity, calculated as mean of four replications determined after 30 days of transplantation.

Table 3. Mineral ion concentration per 100 g dry weight of different salt tolerant plants grown on moderately saline soils. Published values are given for comparison.

Crop	Kallar grass	Lana	Kochea	Mesquite leaves	*Atriplex lentiformis*	*Atriplex amnicola*
Na^+	8.09	24.88	20.53	8.78	23.44	21.77
K^+	39.86	48.46	58.95	55.01	54.86	63.12
Ca^{2+}	15.11	22.20	21.25	41.31	32.13	29.24
P	6.81	8.20	7.42	8.52	8.52	6.32
Fe^{2+}	1.43	1.63	1.50	1.55	1.64	1.42
Mn^{2+}	0.57	0.33	0.029	0.014	0.021	0.010
Cu^{2+}	0.021	0.022	0.020	0.028	0.022	0.021

Source: Data from twenty years of NIAB.

Three parts of the plant showed almost similar growth reduction, however, the reduction of root growth being slightly more severe for both the plants.

Conclusion

Comparison with other salt tolerant species showed that both plants could be used for the browsing of the small animals (Table 3). The use of moderately saline irrigation waters for the production of leaf fodder appears to be a viable option for growers, in areas where the supplies of high quality water are not available. The crop yield can undoubtedly maximize under field conditions, by refinement in the culture and management practices. Such as use of gypsum and farmyard manure and by improving fertilization regime and timing.

REFERENCES

Sharpley AN, Meisinger JJ, Power JF, Suarez D.L (1992). Root extraction of nutrients associated with long-term soil management, Advances in Soil Science (B. Stewart, ed.). Springer-Verlag, pp. 151-217.

Carrow RN, Duncan RR (1998). Salt affected turf sites-assessment and management. Chelsea, Michigan: Ann Arbor Press, pp 83-99.

Champagnol F (1979). Relationships between phosphate nutrition of plants and salt toxicity, Phosphorus Agri. C. 76: 35-43

Gratten, S. R. and C.M. Grieve. 1999. Salinity-mineral nutrient relation in horticulture crops. Sci. Hortic. 78: 127-157.

Hussain T, Timmer V, Akram H, Yaqub A, Aslam M, Gilani G, Abbas MA (1990). Brackish water management strategies for crop production. Proceedings of the Indo-Pak workshop on soil salinity and water management. Pakistan Agricultural Research Council, Islamabad, Pakistan, 2: 369-381

Hoagland DR, Arnon DI (1950). The water culture method for growing plants without soil. California Agriculture experiment circular Pp347.

Lamkbers H. (2003). A key environmental issue in southern Australia. Dry land salinity. Plants Soil, p. 257.

Lauchi A. (1990). Calcium, salinity and the plasma membrane. The Am. Soc. Plant Physiol. Symp. Series, 4: 26-35.

Mass EV, Hoffman GJ (1977). Rop salt tolerance- current assessment. J. Irrigation Drainage Division. ASCE, 103: 115-132.

Pareek OP (2001). Ber. International Center for under Utilized Crops, Southampton University, Southampton, UK.

Qureshi RH, Barrett-Lennard EG (1998). Saline Agriculture for Irrigated Land in Pakistan: A Handbook. Australian Center for International Agricultural Research (ACIAR). Canberra, Australia.

Qureshi RH, Salim M, Aslam Z, Sandhu G (1997). An improved gravel culture technique for salt tolerance studies on plant. Pak. J. Agric. Sci., 14: 11-8.

SAS Institute, Inc, 1997. SAS/STAT Software: Changes and enhancement through Release Cary, NC. 6.12.

Suhayda CG. Wang X, Redmann RE (1992). Trace metal composition of Hordeum species is altered by salinity, Salinity and Sustainable Agriculture, Prairie Salinity Publication No. 1, Swift Current Research Station, Saskatchewn

Suarez DL, Grieve CM (1998). Predicting cations ratios in corn from saline solution composition. J. Exp. Bot. 39: 605-612.

Twenty Years of NIAB (1992). Fourth five year report on research and other activities. Internal Report, 77/II.

Water and Power Development Authority (WAPDA) (1989). PC-II Performa for monitoring salinity control and reclamation Projects, SCARPs for the period 1989-93. SMO Publication No. 100. Lahore, Pakistan, 60p.

Willumsen JK. Petersen K, Kaack K (1996). Yield and blossom-end rot of tomato as affected by salinity and cation activity ratios in the root zone. J. Hortic. Sci. 71: 81-98.

Wolters W, Bhutta MN (1997). Need for integrated irrigation and drainage management, example of Pakistan. Proceedings of the ILRI symposium, Towards Integrated Irrigation and Drainage Management, Wageningen, The Netherlands.

Production, processing and marketing of ginger in Southern Ethiopia

Endrias Geta[1]* and Asfaw Kifle[2]

[1]Haramaya University, Ethiopia.
[2]Areka Agricultural Research Center, SNNPRS, Ethiopia.

The cultivation of ginger was started in Ethiopia during 13th century when Arabs introduced it from India to east Africa (Janson, 1981). In Ethiopia it is limited mostly in the wetter regions of Southern Nations, Nationalities and Peoples Regional State (SNNPRS) and some parts of western Oromia. Most of the commercial production is practiced in SNNPRS by farmers within Kambata-Tambaro, Wolaita and Hadiya zones. Despite its uses and contribution to the livelihoods of smallholders and to economy of the country, there has been limited attempt to promote its production and marketing through improvement of pre-and post harvest handling of the crop. This paper highlights the results of a survey and a review of secondary sources of information in sample woredas from ginger belt of country with the objective of identifying ginger value chain problems and potentials in southern Ethiopia. The study identified that there are high potential ginger growing areas in SNNPRS located in Wolaita, Kambata-Tambaro and Hadiya zones that tend to specialize in production and marketing of the crop. However, production and marketing of ginger in these areas is constrained by low quality varieties, poor pre- and post-harvest handling practices, low product market prices and lack of access to markets.

Key words: Ethiopia, ginger, marketing, processing.

INTRODUCTION

Ginger is a commercially produced horticultural crop in SNNPS, Ethiopia. The producing areas in SNNPRS are said to be ginger belts in Ethiopia where much of the country's production and marketing activities are located. Even though these areas were known as major suppliers of ginger in the country, limited attention has been given to expand production for betterment of smallholder farmers engaged in production and marketing activities. Efforts to generate improved technologies were limited to agronomic practices with no concerted effort to improve product quality, which is highly influenced through its value chain from pre-harvest to postharvest management practices including processing of final products. Moreover, no effort has been made to assess farmers' pre and post-harvest management practices that could be used as a benchmark for improvement works targeting product quality and sustainable supply. Factors that determine performance of ginger marketing and reduce benefits to farmers engaged in its production were not adequately understood by policy makers and agricultural extension workers. This survey was carried out in major ginger producing areas of the region with the objectives of identifying pre and post-harvest management practices, processing techniques and marketing actors and practices in ginger value chain. It was found out that farmers practice their long-existed management practices in selecting varieties to be planted, undertaking field management of the crop, processing and marketing. Low market prices due to poor product qualities, cyclical variations in product price, lack of improved processing and storage facilities were found to be major ginger production and marketing constraints.

RESEARCH METHODS

This paper was based on the information obtained through reconnaissance survey undertaken in ginger growing areas of SNNPRS, Ethiopia in February, 2010. Two administrative zones, that is, Wolaita and Kambata-Tambaro, where farmers specialize in

*Corresponding author. E-mail: geta.endrias@yahoo.com.

Table 1. Ginger area, production and productivity in southern Ethiopia.

Administrative zone	Woreda	Area (ha)	Production (qt)	Yield (qt/ha)
Hadiya	West Badawacho	423	67680	160
Wolaita	Boloso Bombe	7000	1120000	160
Kambata-Tambaro	Hadaro-Tunto, Qacha Bira	8223	1308652	159
Dawro	Gena Bosa	1500	225000	150
Kafa	Gimbo	55	8800	160
Shaka		379	60640	160
Bench Maji	Sheko	187	29920	160
Konta		473	75680	160
Total		18240	2896372	

Source: BOARD.

Table 2. Extent of ginger cultivating weredas of SNNPRS.

Woreda	Total area production (ha)	Percentage of ginger producing farmers	Total production (qt)
Hadaro Tunto	3986	80	637760
Boloso Bombe	5000	85	600000

Source: BOARD, 2008.

ginger production were purposely selected for the study. Consecutively, two ginger specializing districts from these zones were selected based on the discussions made with the zonal departments of agriculture. From each district selected, four ginger producing villages were covered by the survey. A qualitative survey method, using semi-structured checklists and informal survey tools such group discussions, personal interviews and filed observations, was administered to collect the required information. Field survey data were supplemented with information obtained through review of published sources. The data collected were summarized and discussed using a descriptive approach which is a qualitative research format.

RESULTS AND DISCUSSION

Overview of ginger production in SNNPRS

In Ethiopia ginger is cultivated under sub-optimal conditions with rain fall often less than 1500 mm per year and at lower temperatures (Jansen, 1981). However, reasonable yields have been recorded under sub-optimal conditions of the country. As indicated in Table 1 in 2006/2007, about 2896372 quintals of fresh ginger was produced from an area of 18,240 ha indicating regional average rhizome yield of 160 qt/ha (BOARD, 2008).

Table 2 shows that about 32% of the total arable lands are allotted to ginger production indicating that ginger was most extensively produced crop of Hadaro Tunto wereda. Same is true with Boloso Bombe woreda with only slight variation in the proportion of land used for ginger production. An estimated total yield of 1237760 quintals of fresh ginger rhizomes in the year 2008 was obtained from the total area of 8986 ha of land from the two major ginger cultivating weredas.

According to farmers of the areas, there had been a higher and more extensive production of ginger in ginger belt of the region before the last 3 to 4 years. However, in the subsequent years, a sudden fall of the price shortly occurred followed by a high production, which resulted in low market demand of the produce. Consequently, farmers could not get benefit from ginger production for the last 3 to 4 years. As a result, a total production and the size of land allotted by farmers to ginger production are declining at alarming rate. Currently, almost all resource poor farmers in Wolaita area are shifting to the production of a high yielding newly released taro variety and maize by abandoning ginger production. Farmers argue that growing ginger commercially is full of uncertainty if there is no market guarantee based on a certain agreement. Once it lacks market demand, the only alternative would be dumping, because no part of it goes to household consumption, unlike food crops.

Pre-harvest management practices

Land preparation

Most ginger growing farmers of the region start plowing the land allocated for ginger production at the end of the rainy season (from late September to early October) after harvesting previous crop. Plowing the same unit of land will be repeated for about 3 to 5 times. By doing so, seeds of weeds get exposed to the sun periodically and most of them die until dry-planting of ginger takes place 2 to 3 months after land preparation (February/January). The land that has been pulverized to a fine tilth at the end of the rainy season when the soil is at moist condition will

Table 3. Characteristics of different ginger vernaculars cultivated in SNNPRS.

Cultivar group	Vernacular	Unique characteristics
Local	'Feminine Wolaita'	Highly palmated rhizomes Large sized rhizomes More productive
	Masculine Wolaita	High fiber content Highly pungent Large number of prominent roots
Introduced	Bilbo	One prominent root
	Bolbo	2 to 3 prominent roots

conserve its moisture for 2 to 3 months. Depending on the soil type and the moisture content of the soils as well as the capacity of the farmer to use oxen power, the frequency of tillage ranges between 3 and 10. Plowing the land just at the end of the rainy season allows dry season planting of ginger. It enables early planting of the crop to take advantage of using the total amount of the rainfall distributed in the growing season and the crop will complete its normal growth cycle, which in turn makes it more productive. Conversely, late planting, as the result of late onset of the rain, results in significant yield reduction as indicated by farmers.

Variety selection

Farmers identify two groups of ginger varieties in SNNPRS, that is, local and introduced, each group comprising two varieties with distinct morphological characteristics. In Wolaita, farmers classify the local cultivars as 'Masculine Wolaita' and 'Feminine Wolaita'. However, in Kambata, farmers identify one local cultivar known as 'Hargema', that shows great resemblance with the cultivar 'Masculine Wolaita'. In both administrative zones, the local cultivars have been under production since time immemorial. However, cultivars categorized in the introduced group were introduced to the area in the late 1990s. According to farmers, these two introduced cultivars have been disseminated through informal seed system. Most farmers are substituting the local cultivars with the new introductions. Each quarter of ginger production of the region tends to produce distinctive types of cultivars. The different characteristics/ requirements of the major ginger vernaculars currently cultivated in SNNPRS as pointed out by the farmers are presented in (Table 3).

As can be observed from (Table 3), farmers also identify quite a lot of variations among the different cultivars for pre-harvest requirements and post-harvest characteristics as well as differences in commercial values.

Method of propagation and management of planting materials

Ginger is propagated by planting pieces or portion of rhizomes which have at least one good bud. In SNNPRS, farmers developed their own traditional management method of ginger planting material. It involves two methods that is, underground storage method and indoor storage method. The former method is not a common practice but the latter is the popular seed management method that most farmers are practicing in the region. With the first method, ginger rhizomes of the previous crop are kept in the ground without being harvested until the time of planting. During planting time, the crop is harvested, rhizomes are cut into pieces of the required sizes and are immediately planted. In this case, it may take 2 to 3 months for the emergence of the new plants. The second method of propagation involves the following steps. First, rhizomes of the previous crop are harvested in November. Secondly, rhizomes are cut into pieces of the required sizes (usually to 2.5 to 5 cm length) and are placed mostly in plastic sacks, and finally, sacks containing the setts of ginger are stored in residential houses. After 15 days to one month of cutting the rhizomes into pieces, the cut pieces tend to sprout in the sacks where they are stored. Once sprouting is initiated, dry or wet planting will take place. Farmers indicate that the rate of germination of rhizomes using indoor seed storage method is very fast.

Farmers in the region plant larger pieces of ginger rhizomes. The reason is that larger pieces will be re-harvested as 'Kania' together with the newly developed rhizomes and are economically valuable when they are sold in mixture with the new harvest. However, no evidence has been reported on the quality of 'Kania' to be used as input for preparing any sort of ginger final products (that is, ginger powder, oleoresin, and volatile oil). A study by Hailemichael and Tesfaye (2008) in Tepi, Ethiopia indicated that seed rhizome size significantly affected dry rhizome yield. However, there have been no recommendations on the effect of the management of planting material on yield and quality of ginger in the study area. Similarly, the effect of size of the cut pieces of rhizome on the yield of ginger also remains to be unaddressed.

Spacing, planting depth and time of planting

Size of cut pieces is one of the major factors that affect seed rate in ginger. Ginger growing farmers of the region also use approximately similar seed rate with that of research recommendation which falls in the range of 17 to 28 qt/ha. Because of its biennial nature, ginger needs to be planted as early as possible in the growing season to exploit the limited moisture in its growing areas in the region. It is customary to plant ginger in the dry months on plots of land which have been prepared during wet season with the objective of conserving water for dry planting. To make use of the conserved water, deep planting is essential. Long ago, Janson (1981) reported a planting depth of 5 to 10 cm used for ginger in Ethiopia. Planting depth is one of the most limiting agronomic factors that affect productivity of crops. Ginger producing farmers of SNNPRS prefer dry planting in December and January on the unit of lands that have been pulverized to a fine tilth during wet season for the purpose of moisture conservation. Farmers declare that this joint effect of planting time associated with the indicated tillage practice increases productivity. In general, seed rate, planting depth and time of planting need to be explored in order to exploit the genetic potential of the ginger varieties.

Fertilizer use

Farmyard is a major fertilizer used for ginger production in SNNPRS. However, it cannot support large scale production since the availability of farmyard manure is limited. On the other hand, the rate of its application for optimum yield per unit area has also not been determined for ginger. In addition, farmers complain that working with farmyard manure is laborious. For large scale production to be undertaken outside the homestead, the cost of transportation of farmyard manure is also very high.

Some farmers who plant ginger at large scale in the region also use compost as a source of nutrient for producing ginger making from all kinds of organic materials such as crop residues, kitchen wastes, garden cuttings, and manure. Compost supplies nutrients at the right time in required quantities. Compost is especially useful for improving the soil structure and fertility.

Coffee pulp and husk have been the other sources of nutrients used by the farmers that produce ginger at commercial level in the region. They purchase the coffee pulp and husk from the wet and dry coffee processing plants operating in their vicinity and even from the neighboring woredas and transport to their farm when they speculate high market demand for their ginger product.

Although the price of chemical fertilizers is unaffordable, and chemical fertilizers do not improve the soil structure, some farmers producing ginger at commercial level use DAP when they expect that they will get good prices. Almost all farmers do not use UREA in ginger production as they perceive that "UREA dries the soils". Time of application is variable with season of planting. During dry planting, farmers do not apply chemical fertilizers. They apply the fertilizer by broadcasting approximately 3 to 4.5 months of planting during the cultivation at the rate of approximately 100 kg/ha when the rhizomes emerge. However, farmers do not use chemical fertilizers in small scale cultivation, and where prices are fluctuating. It has been a usual practice in ginger production to give a combination of organic and chemical fertilizers. However, the correct proportion of both nutrient sources for optimum yield/quality need to be defined. Similarly, research needs to be conducted to determine the correct type, rate and time of application of the chemical fertilizers for the optimum yield/quality as well as to analyze the economic aspect of fertilizer application. In India farmers apply manure 25 to 30 tons/ha or apply compost at planting, and N 36, P, 16, and K 66KG/ha (Janson, 1981). The recommended rate of chemical fertilizers at Tepi Sub Research Center for an optimum yield of ginger was 375 kg/ha urea and 175kg/ha DAP.

Weeding and pest control

In Ethiopia, farmers follow non-chemical weed management practices that include deep plowing of the seedbed at the end of the rainy season and frequent cultivation. Before the season of planting comes, farmers begin to prepare seedbed at the end of the rainy season (between late September and early October) plowing the same unit of land as frequently as 3 to 5 times. By doing so, seeds of weeds get exposed to the sun periodically and most of them die until dry-planting will take place 2 to 3 months after land preparation (January to February). Pests of ginger are root-knot nematodes, and shoot borers (Purseglove, 1972). So far, in Ethiopia, no disease and pest records of ginger have been reported.

Harvesting

In Ethiopia, it is not habitual to harvest ginger based on the end use. In SNNRS, for dried and fresh ginger, mature rhizome with a full aroma, flavor and pungency are harvested after planting when the leaves begin to turn yellow. Harvesting is usually manual taking no care whether the outer epidermis or the rhizomes are damaged or not. Two forms of ginger harvesting are identified in SNNPR, that is, one-season harvesting and perennated harvesting. One-season harvesting involves harvesting of matured rhizomes within one growing season. In SNNPR, usually harvesting of ginger begins 8 to 9 months starting from late September when the leaves start turning yellow and the stems stop growing, which, may extend to March or until the onset of the 'belg' rain. Within the same growing season, farmers sometimes deliberately extend the time of harvesting speculating high market price. Sometimes ginger is grown as a perennial crop in SNNRS. In this case farmers deliberately leave the matured ginger rhizomes in the field for two or more successive seasons without harvesting.

Postharvest management practices

Postharvest management practices of ginger involve several activities. These include preparation of dry ginger, storage and processing. Farmers' practice in the study areas were identified and presented as follows.

Preparation of dry ginger

In Ethiopia, quality-reducing traditional drying methods have been used for the preparation of dried ginger. Actors of these methods include farmers, retailers and some wholesalers. The major actors of the drying process are individual farmers followed by the retailers. The contribution of wholesalers, who sometimes use cemented floor, to the total dry ginger preparation is insignificant.

Farmers spread the harvested ginger rhizomes on dusty and dirty floors without cleaning and trimming roots from the rhizomes to prepare a dried ginger or sometimes leave it on farm after harvesting until it loses some percentage of moisture and spread on floor for full drying. Sun drying of ginger is a laborious process which could take 30 to 45 days depending on the type of variety and weather condition. Brilliant and stable sunshine condition at Areka, for example, in most cases is not attainable due to some irregular showers of rain, which leads to mold development mostly in the introduced cultivars. For this reason, it is habitual to break the rhizomes into tiny pieces to enhance fast drying. Breaking the rhizomes into smaller pieces has its own negative impact on the quality of the dried product by decreasing the volatile oil content of the rhizome on one hand, and enhancing the development of the Aflatoxins on the other. Similarly, peeling off and damaging of the rhizome skin during the process of harvesting and drying should be minimized. While drying, the rhizomes need to be frequently overturned to maintain uniformity of the rhizomes until the product attains the moisture content of 7 to 12%, which is required for dry ginger.

Washing the rhizomes before drying, that is practiced by a single wholesaler company the areas, is unthinkable by farmers in the region. Surprisingly, no significant price variation per unit volume was made between products that have subjected to drying on the cemented and non-cemented floor. As far as the volume of ginger that a farmer owns and water availability for washing the rhizomes is concerned, washing rhizomes at individual farmer's level would be impractical. Therefore, it will be essential to design the strategy, which possibly involves cooperatives and/or interested investors. The cooperatives or the investors may collect fresh ginger rhizomes from individual farmers on a given price agreement and they may add value to the product by washing and drying or they can sellout the rhizomes after washing to any enterprise/company for further processing.

In general, drying process employed in the region so far for the preparation of dried ginger is too primitive and, hence could not maintain the product's quality. Consequently, its contribution to the livelihood of ginger growers and to the economy of the country has been insignificant. Thus, it calls for the joint effort of the actors including farmers, private sector, GOs and NGOs in order that quality dried ginger product might be supplied and become competitive in the global market.

Storage

Storage provides protection against weather, moisture, insects, micro organisms, rodents and any type of infestation and contamination. In SNNPRS, farmers often sellout their fresh ginger immediately after harvesting especially, when there is a high market demand. As a matter of fact, the ginger market has always been so much unstable and farmers are enforced to keep their produce in the storage for some period after drying until they get remunerative price. Producers store dried ginger in their houses using various types of traditional structures. Therefore, they use it as a means of value addition by changing time of supply of the produce. Moreover, storage is practiced to make use of the advantage of storage as 'saving cash in bank'. So far, no improved structure has been developed for ginger storage in the production area.

It is customary to store dried ginger either in bulk or in woven plastic sacks in non cemented dusty rooms, sometimes, for up to 10 years, depending on the type of the variety and the market demand. However, such storage condition not at all fulfills the basic requirements for safe and scientific storage. It is evident that the quality of ginger decreases with long storage period leading to decline in essential oil content. Nevertheless, some farmers continue storing their ginger for up to 10 years. Therefore, Post-harvest management of ginger like, preparation of dried ginger and storage are need immediate attention.

Processing

In Ethiopia, few companies are engaged under two categories of ginger processing activities. The first category is preparation of dried ginger and, the second is extraction of ginger oleoresin. The former has been mainly performed by individual ginger producing farmers with a negligible portion going to one private enterprise situated near Bombe, Wolaita, that is, drying ginger on cemented floor after washing. The latter has been carried out by two factories known as Ethiopian Spices Extraction Factory (formerly owned by the government and has recently been privatized), and KASK (privately owned and now on the hands of a Bank due to liability). The two companies (Ethiopian Spices Factory and KASK), which are engaged in oleoresin extraction could not operate to their full capacity due to the following reasons:

1. The dried ginger has been full of impurity as it is prepared traditionally by local farmers.
2. There is low percentage of oleoresin content per unit volume of ginger as a result of perenniated/extended harvest, long storage period, over drying and/or re-drying due to dampness.
3. The factories which were designed to extract pepper oleoresin formerly require high cost for modification of parts.

Currently, both factories are not functional due to transition. Even if they operate to their full capacity, it would remain insignificant as they absorb only the small fraction of the volume of the ginger yield produced per year in the region.

Apart from these, the Ethiopian Essential Oils Research Center sometimes extracts essential oils from different essential oil yielding plant spices for research purposes. Otherwise, no company in the region or at the country level was engaged in practical ginger processing activity.

Marketing of ginger products

The issue of marketing is of great importance for produces of ginger in the areas. This is due to the fact that almost all the total amount of ginger produced by small-scale farmers is supplied to market with negligible amount locally consumed. However, ginger growing farmers have been severely affected by fluctuating market prices over years. Hence, production of ginger in terms of area has been growing in response to good market prices and declining in response to depressing prices. Thus, it can be realized that market signals are very determinant factors dictating ginger production in the areas.

Marketable products of ginger

Three types of ginger products are supplied to market in these areas. The first one is fresh ginger which is usually supplied to market in rainy season for immediate cash demand. It is also supplied in dry season as assemblers collect it for re-sell to large traders after sun-drying. Dry season fresh ginger supply also targets local farmers who demand it as a planting material. Long ago, Ethiopia was exporting some fresh ginger to Egypt, Saudi Arabia and Yemen. The fresh Ethiopian ginger was reported to be a good product with a nice color, and quality which can be stored quite long (Janson, 1981). Currently Ethiopia is not exporting fresh ginger. Exporting of fresh ginger demands high quality ginger varieties and early time of harvesting as well as high financial capacity and transportation technology.

The second type of marketable ginger product is dried ginger. In Ethiopia, dried ginger is favorite ginger product exchanged in large volumes by all market participants at different stages of marketing from local assembling to export market. Dried ginger for this purpose is harvested 8 to 9 months after planting when there is mature rhizome with a full aroma, flavor and pungency. Further delay not only increases fiber but also decreases oil and oleoresin content.

The third type of marketable ginger products are extracted ginger products. These include ginger powder, essential oils, oleoresin and others. The extent to which rhizomes are treated prior to drying directly affects fiber and volatile oil content. Removal of the cork skin reduces fiber content but enhances oil loss through rupture of the surface oil-cells; thus, cleanly peeled ginger generally has a lower oil and fiber content (Pruthi, 1998; Weiss, 2002). The benefits from extracting and marketing these products are going unreached these times by the country.

Ginger production and marketing constraints

Several ginger production and marketing constraints were identified by farmers during the survey. Farmers recall that one quintal of fresh ginger was equivalent to that of about 11 quintals of maize before three years when this study was conducted. On the contrary, at present, the value of 20 quintals of fresh ginger is less than a value of one quintal of maize. This indicates that the market price of ginger is declining and that of food crops is rising. At present, a quintal of fresh ginger is sold for birr 20 to 30, which hardly surpasses the transportation cost to have a quintal arrived at the market. Moreover, as ginger crop is high-input user such as: high frequency of weeding and high input requirements including seed and fertilizers, the high cost of production could result in loss if market prices are below the expected level as there are irregularities. In addition, due to lack of facilities for ginger drying, long time is required to get dry ginger by using traditional practices that lead to development of fungi during wet weather condition contributing to low market price.

Generally, the market price of ginger is highly unstable.

In some years it rises up as high as Birr 2250 per quintal of dried ginger as opposed to less than Birr 150 per quintal when the price falls down. In the past, according to farmers, it was familiar to resume high price after some years of price fall. They argue that recently, the price change is uncertain and difficult to predict that would hamper further ginger production and storage in the region.

Marketing malpractice in ginger marketing has been a common practice, especially when the price of ginger becomes higher. It is practiced by both the producers and traders. Two forms of malpractices by producers in some areas of ginger production have been recognized. These are mixing a fresh ginger with a damp soil, and a dried ginger with special stones having exactly the same physical appearance with that of the dried ginger rhizomes. Marketing malpractices by regional traders is reflected mostly in altering weighing balance adjustments to their advantages. For example, their weighing balances for a quintal are commonly adjusted to read 125 to 130 kg for fresh ginger and 110 to 115 kg for dried ginger. Moreover, the unit of mass measurement known as 'Feresula', commercially an equivalent of 17 kg, is accepted as 20 kg by the regional traders when they buy the ginger from the farmers while when they sellout the product to the subsequent levels of traders, they restore it to 17 kg. Farmers have no role in deciding on the price of their produce. It is the traders through illegitimate brokers who decide the price of ginger. As a result, farmers have no choice than accepting the price set by a few traders. Some commercial farmers who attempted to transport their produce directly to the wholesalers to the central market had been put to tragedy by the action of brokers. If brokers get information that a farmer has transported the produce to the central market, they tend to shortly communicate each other and immediately block the farmer not to sellout the produce to the wholesaler. If the produce remains unsold, the farmer will be obliged to pay higher cost for hoarding the product. The existence of middlemen along the ginger marketing chain not only reduced the producers' share, it also exposed farmers to additional expenses by blocking the flow of information.

CONCLUSIONS AND RECOMMENDATIONS

Ginger has been produced as an important commercial horticultural crop in southern Ethiopia. Farmers produce different varieties and apply traditional management practices. External support in promoting improved varieties and management practices is almost non-existent. Ginger product markets are very volatile and price fluctuations create a disincentive for farmers in the production of the crop. Thus, researchers have to assist farmers in indentifying improved varieties with desirable market traits, appropriate agronomic and post-harvest management practices including drying methods, sorting and grading techniques, as well as processing technologies. Extension workers and other development practitioners have join hands with the farmers in addressing marketing problems such as: easing barriers to entry into markets by organizing ginger producers into producers and marketing cooperatives.

REFERENCES

BOARD (SNNPRS Bureau of Agriculture and Rural Development) (2008). Unpublished data.

Hailemichael G, Tesfaye K (2008). The effects of seed rhizome size on the growth, yield and economic return of ginger (*Zingiber officinale* Rosc.). Asian J. Plant Sci., 7: 213-217.

Jansen PCM (1981). Spices, Condiments and Medicinal Plants in Ethiopia, their Taxonomy and Agricultural Significance. Wageningen PUDOC., pp. 1-132.

Pruthi JS (1998). Spices and Condiments. Nation Boo Trust, A-5 Green Park, New Delhi, pp. 147-154.

Purseglove JW (1972). Tropical crops: Monocotyledons, 1st edition, Longman group Limited, UK, London, pp. 52-54.

Weiss EA (2002). Spice Crop, 1st ed., CABI Publishing, UK, pp. 316-338.

Fruit yield of virus-resistant transgenic summer squash in simulated commercial plantings under conditions of high disease pressure

Ferdinand E. Klas[1,2]*, Marc Fuchs[1] and Dennis Gonsalves[1,3]

[1]Department of Plant Pathology and Plant-Microbe Biology, Cornell University, New York State Agricultural Experiment Station, Geneva, NY, 14456, USA.
[3]Anton de Kom University van Suriname Faculteit der Technologische Wetenschappen, Universiteitscomplex Leysweg, Building XVII, P. O. Box 12292, Paramaribo Suriname, USA.
[3]USDA-ARS, Pacific Basin Agricultural Research Center, 64 Nowelo Street, Hilo, HA 96720, USA.

Fruit yield of transgenic crookneck summer squash ZW-20 resistant to Zucchini yellow mosaic virus (ZYMV) and Watermelon mosaic virus (WMV) and of a susceptible nontransgenic lineage of the same genotype was compared over two consecutive years. Field trials relied on small-scale plantings that reflected commercial settings under conditions of severe disease pressure by ZYMV and WMV with infection achieved via aphid-mediated inoculation from virus source border plants. Across all trials, all transgenic plants were highly resistant to ZYMV and WMV, and the majority (79%, 331 of 421) produced 3 to 9 fruits per plant. In contrast, all control plants had severe systemic symptoms and the majority (80%, 336 of 421) produced 0 to 4 fruits per plant. In addition, all fruits of transgenic squash ZW-20H and ZW-20B were of marketable quality whereas most fruits of nontransgenic controls (96%, 947 of 989) were unmarketable. Differences in fruit number (P = 0.0001) and fruit weight (P = 0.0001) between transgenic and conventional squash plants were significant but not between ZW-20H and ZW-20B plants (P = 0.933 and P = 0.964, respectively). This is the first report on a comparative analysis of fruit yield of transgenic versus conventional summer squash under conditions approaching commercial plantings in which high infection rates of ZYMV and WMV were achieved via indigenous aphid populations.

Key words: Fruit yield, transgenic, high disease pressure, summer squash, virus-resistant

INTRODUCTION

Virus-resistant transgenic summer squash (*Cucurbita pepo* spp. *ovifera* var. *ovifera*) have been successfully developed (Tricoli et al., 1995) and commercialized in United States (Gaba et al., 2004; Johnson et al., 2007). Transgenic squash line ZW-20 was deregulated in 1994 (Medley, 1994). This was the first disease-resistant transgenic crop to receive exemption status in the United States. Plants of transgenic squash ZW-20 express the coat protein (CP) gene of Zucchini yellow mosaic virus (ZYMV) and Watermelon mosaic virus (WMV) and are highly resistant to single and mixed infection by ZYMV and/or WMV, as shown by multiple field trials in different locations (Arce-Ochoa et al., 1995; Clough and Hamm, 1995; Fuchs and Gonsalves, 1995; Fuchs et al., 1998; Klas et al., 2006; Schultheis and Walters, 1998; Tricoli et al., 1995). Transgenic plants restrict ZYMV and WMV to chlorotic dots or blotches primarily in lower leaves when subjected to infection by these two viruses (Fuchs and Gonsalves, 1995; Fuchs et al., 1998; Klas et al., 2006). In contrast, nontransgenic squash develop severe systemic foliar symptoms consisting of mosaic, deformation, shoestringing and stunted growth following infection by ZYMV and WMV.

Two types of plants were identified during the development of transgenic line ZW-20 (Tricoli et al., 1995).

*Corresponding author. E-mail: fklas@yahoo.com.

Some ZW-20 plants remained asymptomatic or developed chlorotic dots upon infection by ZYMV or WMV and contained the H insert of the WMV CP gene. These plants were labeled ZW-20H (Tricoli et al., 1995). Other ZW-20 plants remained asymptomatic or developed chlorotic dots or blotches upon infection by ZYMV or WMV and lacked the H insert of the WMV CP gene. Theseplants were labeled ZW-20B (Tricoli et al., 1995). A spatiotemporal analysis indicated that, despite localized mild symptoms, transgenic ZW-20H and, to a lesser extent,ZW-20B do not readily serve as virus source for secondary spread of ZYMV and WMV (Klas et al., 2006). Plants of transgenic squash ZW-20, whether ZW-20H or ZW-20B, produce similar yield of marketable fruits as conventional summer squash under conditions of low, if any, disease pressure (Clough and Hamm, 1995). In contrast, under conditions of severe disease pressure by ZYMV and WMV, transgenic squash ZW-20 yield significantly more marketable fruits than conventional controls (Arce-Ochoa et al., 1995; Clough and Hamm, 1995; Fuchs and Gonsalves, 1995; Schultheis and Walters, 1998; Tricoli et al., 1995).

Most of field experiments designed to compare the fruit yield of virus-resistant transgenic and virus-susceptible conventional summer squash were carried out on complete randomized block designs with mixtures of different genotypes (Arce-Ochoa et al., 1995; Fuchs and Gonsalves, 1995; Fuchs et al., 1998; Schultheis and Walters, 1998; Tricoli et al., 1995). This approach provides valuable insight into the performance of test plants but does not reflect conditions of commercial settings for which plantings of a given genotype are uniform. Furthermore, comparative yield analyses have often been carried out under conditions of high disease pressure achieved by mechanically inoculating a few test plants in the greenhouse prior to transplanting and establishing them uniformly among test plants. These virus-infected plants served as primary inoculum source from which secondary infection takes place through aphid-mediated spread. This approach achieved 44 to 100% infection in nontransgenic plants (Clough and Hamm, 1995, Fuchs and Gonsalves, 1995; Fuchs et al., 1998) but did not reflect natural conditions of virus infection. Indeed, aphids essentially transmit viruses to plants in commercial fields from outside sources and subsequently spread them within fields. Some resistance screening experiments relied exclusively also on indigenous aphid populations to spread viruses, with no mechanically inoculated plants as primary virus source. These conditions achieved a 53% infection rate in conventional plants (Arce-Ochoa et al., 1995). Other field experiments reproduced commercial conditions with no virus source plants and insecticides applied to control aphid populations; infection rates reached 65% in nontransgenic controls (Schulteis and Walters, 1998).

To date, no information is available on the comparative fruit yield analysis of transgenic versus conventional squash under field conditions approaching commercial settings for which a 100% virus infection rate is achieved in susceptible plants via aphid inoculation from virus source plants located outside of experimental plantings. In this paper, we present a comparative analysis of fruit production of transgenic summer squash ZW-20 and conventional summer squash in small-scale field settings simulating commercial fields with high virus infection rates achieved through aphid-mediated inoculation

MATERIALS AND METHODS

Plant materials

Transgenic crookneck summer squash line ZW-20 was developed by *Agrobacterium tumefaciens*-mediated transformation. It contains the CP genes of ZYMV and WMV (Tricoli et al., 1995). Transgenic hybrids ZW-20H and ZW-20B, resulting from crosses of transgenic plants, which were homozygous for the two CP transgenes, and an untransformed parent, were used in this study. The commercial conventional cultivar Pavo, which is susceptible to ZYMV and WMV, and has the same genetic background as the two transgenic hybrids, was used as control.

Field layout

Six squash field plots were established over two consecutive years at the Crittenden farm, Cornell University, New York State Agricultural Experiment Station in Geneva, NY (Klas et al., 2006). In 1994, two field plots were planted under permits issued by APHIS-USDA (U.S. Department of Agriculture's - Animal and Plant Health Inspection Service), one with transgenic ZW-20B plants and one with nontransgenic control plants. The two field plots were established 20 m apart with 8 rows, 2 m apart, and 27 plants per row at a 1 m within-row distance. In 1995, four field plots were planted, two with transgenic plants (one with ZW-20H and one with ZW-20B) and two with nontransgenic control plants (Control 1 and Control 2).

The four field plots were established 20 m apart and consisted of 10 rows with 20 plants each. The between- and within-row distances were the same as in 1994. Permits were not required in 1995 because transgenic line ZW-20 was deregulated in December of 1994. The six field plots were surrounded by a single row of nontransgenic controls, which were mechanically inoculated with ZYMV or WMV prior to transplanting (Fuchs and Gonsalves, 1995). Mechanically inoculated plants were transplanted in alternating groups of four ZYMV-infected and four WMV-infected plants. Infected plants were used as primary virus inoculum for aphid-vectored infections. No insecticide was used throughout the two growing seasons for efficient virus spread by indigenous aphid populations.

Disease progress

Virus incidence was monitored by visual observation of symptoms every 3 to 4 days and by double antibody sandwich (DAS) enzyme-linked immunosorbent assay (ELISA) using immunoglobulins specific to ZYMV and WMV that were produced in our laboratory. Tissue from leaves in positions 3 to 5 at the apical end of each test plant was sampled three times during the growing seasons and tested in DAS-ELISA (Fuchs and Gonsalves, 1995). Samples were considered positive if their optical density reading at 405 nm (OD_{405nm}) were at least twice the values of the healthy nontransgenic controls (Klas et al., 2006).

Figure 1. Resistance of transgenic crookneck summer squash ZW-20 to infection by ZYMV and WMV, and fruit production. (A) Squash ZW-20H or ZW-20B (second to fourth row from the left) surrounded by mechanically-inoculated conventional squash (left row and first plant of the second row), (B) Marketable fruit production of summer squash ZW-20H or ZW-20B, (C) Close-up of a marketable fruit of a transgenic (left) and of a five unmarketable fruits of conventional summer squash (right), and (D) Unmarketable fruit production of conventional summer squash.

Fruit yield and data analyses

In 1994, fruit yield was assessed three times, irrespective of fruit size. In 1995, fruit yield was determined nine times at two to three-day intervals from August 7 to August 25. In both years, fruits of each test plant were visually rated for marketability, counted, and weighed. Fruit marketability was scored based on shape, size (12-18 cm) and absence of virus symptoms. Cull fruits were distorted, too small in size at maturity, and showed partial or complete discoloration, and green streaks or dots.

The fruit production frequency distribution was composed for each genotype in both growing seasons. Also, the cumulative number of fruits was plotted against time for each genotype. Data on 1995 fruit number and weight were summarized and subjected to paired *t*-test using the SPSS 8.0 software package for assessing significant differences between pairs of transgenic (ZW-20B and ZW-20H) and conventional (Control 1 and Control 2) squash plant genotypes.

RESULTS

Resistance to ZYMV and WMV in transgenic summer squash

The reaction of transgenic and conventional summer squash plants to infection by ZYMV and WMV in 1994 and 1995 was reported previously (Klas et al., 2006). Briefly, transgenic squash ZW-20H and ZW-20B were highly resistant in both growing seasons (Figure 1A). Plants of both transgenic lines were either asymptomatic or developed mild symptoms that consisted of chlorotic dots or blotches mainly on old leaves. ZYMV and WMV could be detected by DAS-ELISA in symptomatic but not in asymptomatic leaf tissue. At the end of the two growing seasons, some transgenic plants reacted positively to ZYMV (9-36%), WMV (2-17%), or ZYMV and WMV (4-12%) (Klas et al., 2006). In contrast, conventional squash were highly susceptible to infection by ZYMV and WMV, and showed severe systemic symptoms. The two target viruses were readily detected in leaves of nontransgenic plants by DAS-ELISA. At the end of the two growing seasons, 95-100% of the control plants reacted positively to ZYMV and WMV, and the rate of mixed infection by ZYMV and WMV was high (57-86%) (Klas et al., 2006).

Total fruit production and frequency distribution:

In 1994, fruits were harvested at 38, 65, and 82 days post-planting (dpp) because the major emphasis of the work was on spatio-temporal virus spread rather than fruit production. Therefore, most fruits were oversized (25 to 30 cm), in particular for the two last harvest dates.

Table 1. Cumulative fruit yield of transgenic ZW-20H and ZW-20B, and nontransgenic squash over two consecutive years.

		Total fruits[a]			Marketable fruits[b]		
Year	Genotype[c]	N[d]	Number	Weight (kg)	Number	Total weight (kg)	Weight/fruit (kg)[e]
1994	ZW-20B	143	567	313.2	567	313.2	na
	Control	140	389	121.8	42	26.4	na
1995	ZW-20H	139	790	150.3	790	150.3	0.19
	ZW-20B	139	798	149.5	798	149.5	0.19
	Control 1	139	443	62.3	0	0	0
	Control 2	142	217	30.4	0	0	0

[a]Fruits were harvested three times in 1994 and nine times at 2-3 days intervals in 1995. Fruits of each test plant were counted and weighed. [b]Marketable fruits were asymptomatic and 12-18 cm in size. [c]Transgenic ZW-20B and ZW-20H express the CP gene of ZYMV and WMV. Crookneck summer quash cv. Pavo was used as control. [d]Number of plants tested per genotype. [e]na: not application because fruits were not harvested at 2-3 days interval according to agricultural practice

Nevertheless, transgenic ZW-20B produced 1.5 times more fruits than nontransgenic controls (P = 0.006) (Table 1). In 1995, fruits were harvested at regular time intervals, e.g. 31, 33, 35, 38, 40, 42, 45, 47, and 49 dpp, and transgenic ZW-20 plants produced 1.8-3.7 times more fruits than controls (Table 1). Differences between transgenic and conventional squash were significant (P = 0.0001) but not between transgenic ZW-20H and ZW-20B (P = 0.933) (Table 1).

Analysis of the frequency distribution of fruit production in 1994 indicated that the majority of transgenic ZW-20B plants (85%, 121 of 143) produced 3 to 6 fruits while most nontransgenic controls produced only 1 to 4 fruits (91%, 127 of 140) (Figure 2A). In 1995, most transgenic ZW-20H (80%, 111 of 139) and ZW-20B (71%, 99 of 139) plants produced 3 to 9 and 4 to 9 fruits, respectively (Figure 2B). A few plants of both transgenic genotypes even had 11 to 13 fruits. In contrast, most nontransgenic plants produced 1 to 4 fruits in Control 1 (68%, 95 of 139) and 0 to 2 fruits in Control 2 (80%, 114 of 142) fields (Figure 2B).

Fruit weight, size and marketability

In 1994, transgenic ZW-20B plants yielded 2.6 times more than conventional squash plants in terms of fruit weight. In 1995, transgenic ZW-20H and ZW-20B yielded 2.4-5 times more than controls (Table 1). Differences between transgenic ZW-20H and nontransgenic plants in Control 1 (P = 0.05) and Control 2 (P = 0.0016) were significant. Similarly, differences between transgenic ZW-20B and nontransgenic plants in Control 1 (P = 0.001) and Control 2 (P = 0.0001) were significant. Differences between ransgenic ZW-20H and transgenic ZW-20B were not significant (P = 0.964), confirming an equivalent performance of these two genotypes.

In terms of fruit size, some fruits of transgenic ZW-20B (33%, 62 of 190) had a marketable size (12-18 cm) at the first harvest (38 dpp) but not (0%, 0 of 268) at the second harvest (65 dpp) dates in 1994. Fruits of the latter harvest date were oversized. In contrast, only 23% (42 of 185) of the fruits of conventional plants reached a marketable size at the first harvest but none (0%, 0 of 186) at the second harvest. In 1995, when harvest was done at 2-3 days intervals according to commercial practice, fruits of transgenic ZW-20H and ZW-20B plants were of marketable size (Figure 1B) with an average weight of 0.19 kg (Table 1). In addition, all transgenic fruits were asymptomatic and marketable (Figure 3). In contrast, only a few fruits of conventional plants (21%, 93 of 443 in Control 1 and 16%, 35 of 217 in Control 2) reached marketable size at maturity with an average weight of 0.14 kg (Table 1). However, none of them was marketable (Figure 3) because they were symptomatic (Figure 1C and 1D). Therefore, in contrast to the 1994 season, no marketable fruit was harvested from nontransgenic plants in 1995. This result reflected a difference in infection pressure between both years with 99% of the control plants being severely symptomatic already at 25 dpp in 1995 and only 40% of them at 30 dpp in 1994 (Klas et al. 2006).

DISCUSSION

The cumulative production of marketable fruits of transgenic summer squash ZW-20 was significantly higher as compared to conventional summer squash of the same genotype in small-scale field settings simulating commercial plantings that were established under conditions of high disease pressure by ZYMV and WMV achieved via aphid-mediated inoculation. These findings achieved, for the first time, under simulated commercial conditions confirm previous reports on the performance of transgenic squash ZW-20 (Arce-Ochoa et al., 1995; Clough and Hamm, 1995; Fuchs and Gonsalves, 1995; Fuchs et al., 1998; Schultheis and Walters, 1998; Tricoli et al.,

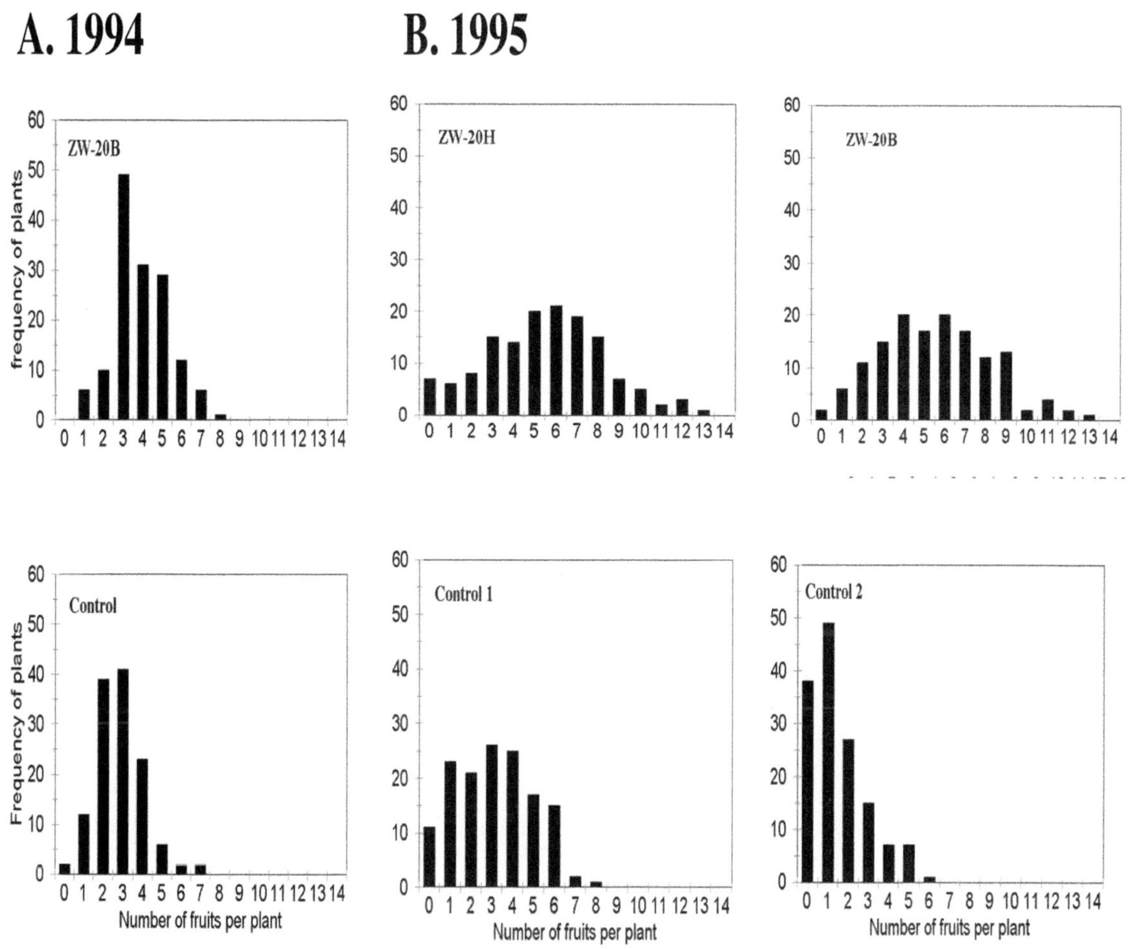

Figure 2. Frequency distribution of the number of fruits per plant for (A) transgenic ZW-20B and nontransgenic controls in 1994, and (B) transgenic ZW-20H, transgenic ZW-20B, and nontransgenic controls in 1995.

1995).

Earlier studies on the performance of virus-resistant transgenic summer squash relied on small-scale field experiments in which different genotypes were mixed (Tricoli et al., 1995; Fuchs and Gonsalves, 1995; Fuchs et al., 1998). Therefore, they did not reflect commercial settings. Also, some of the earlier studies were based on conditions of low, if any, disease pressure (Arce-Ochoa et al., 1995; Schultheis and Walters, 1998). Eventually high disease pressure was achieved by mechanical or aphid-mediated inoculation (Clough and Hamm, 1995; Fuchs and Gonsalves, 1995; Fuchs et al., 1998; Tricoli et al., 1995) but none of the earlier studies relied on conditions of severe disease incidence exclusively achieved through virus infection via indigenous aphid populations from primary virus source plants located outside of the experimental fields. Therefore, to the best of our knowledge, our study is the first to report on a comparative yield assessment of virus-resistant transgenic and virus-susceptible conventional summer squash in settings that simulate commercial fields and natural conditions of virus infection with no insecticides to control indigenous aphid populations. These conditions are important for a good appreciation of the practical horticultural performance of virus-resistant transgenic squash plants.

In our study, a 95-100% infection rate was achieved in control plants. In spite of this very high disease pressure of ZYMV and WMV, fruits of transgenic squash ZW-20 exhibited no symptoms and maintained a marketable value, confirming previous findings (Clough and Hamm, 1995; Fuchs and Gonsalves, 1995; Tricoli et al., 1995). In addition, transgenic ZW-20B plants had a 2.4 to 8-fold increase in weight of marketable fruits compared to controls. Transgenic ZW-20H was also superior to conventional plants in terms of marketable fruit production.

In 2008, squash cultivars were grown on approximately 18,130 hectares in the United States. The seven major squash producing states were Florida (19% of U.S. output), Michigan (15%), California (13%), Georgia (12%), New York (9%), New Jersey (8%) and North Carolina (8%), producing 83% of the nation's squash crop. Yield losses of summer squash to aphid-transmitted viruses are often ranging from 20 to 80% and estimated

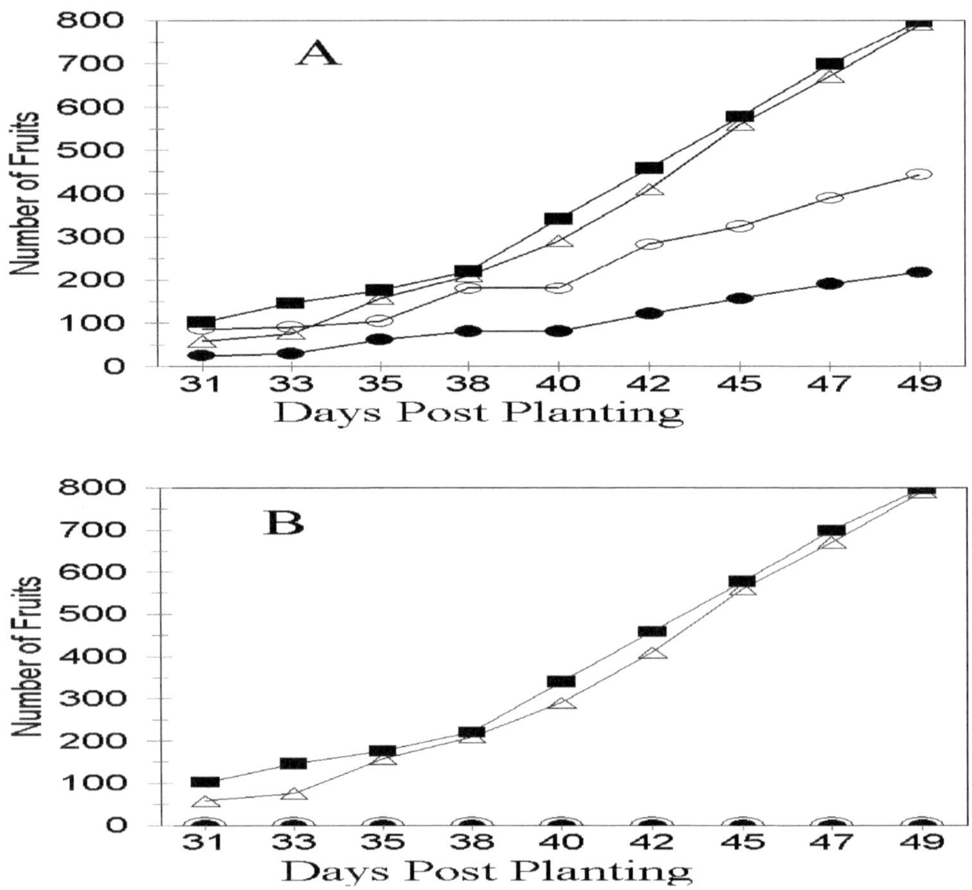

Figure 3. Cumulative number of (A) total and (B) marketable fruits of transgenic squash ZW-20H (-△-) and ZW-20B (-■-) plants, and nontransgenic squash (Control 1 -O- and Control 2 -●-) plants in 1995.

to $2.6 million in Georgia in 1997 (Gianessi et al., 2002). Traditionally, control of aphid-borne viruses in squash is attempted by cultural practices, including delayed transplanting relative to aphid vector flights, use of film mulch to repel aphid vectors, and application of stylet oil in combination with insecticides to reduce aphid vector populations (Blancard et al., 2004; Gaba et al., 2004; Perring et al., 1999; Zitter et al., 1996). In Georgia, it is estimated that 10 applications of stylet oil and insect-cides are made routinely per acre to control aphids and, hence, limit virus transmission (Gianessi et al., 2002). The selection and use of squash cultivars is also a critical component for virus disease management. A number of squash cultivars with virus resistance have been developed successfully through conventional breeding (Munger, 1983; Provvidenti, 1993). However, cultivars with a high degree of resistance to multiple viruses have not been successfully developed yet through conven-tional breeding approaches (Gaba et al., 2004; Tricoli et al., 1995). This is mainly due to genetic incompatibility between donors and recipients of resistance genes, and linkage of virus resistance genes to undesirable traits. The adoption of virus-resistant transgenic summer squash cultivars is steadily increasing since 1996. In 2006, the adoption rate was estimated to 24% (3,025 hectares) in five of the seven major producing states with an average rate of 70% in New Jersey, 20% in Florida and Georgia, 7% in North Carolina and 5% in Michigan (Johnson et al., 2007). The first virus-resistant transgenic squash line, denoted ZW-20, and cultivars derived thereof were deregulated in the United States in 1994 (Medley, 1994). Transgenic line ZW-20H was used as one of the parents to develop commercial F1 summer squash cultivars, including crookneck cvs. Freedom II (Tricoli et al., 1995) and Prelude II, straightneck cv. Patriot II, and zucchini cvs. Independence II and Declaration II. In addition to transgenic squash ZW-20, another transgenic line, denoted CZW-3, was developed (Tricoli et al., 1995). Transgenic line CZW-3 expresses the CP genes of CMV, ZYMV, and WMV, and is highly resistant to single or mixed infections by these three aphid-borne viruses (Tricoli et al., 1995; Fuchs et al., 1998; Schultheis and Walters, 1998). Transgenic line CZW-3 and cultivars derived thereof were deregulated in 1996 (Acord, 1996) and commercialized thereafter as semi-crookneck cv. Destiny III, green-stem straightneck cv. Conqueror III, straightneck cv. Liberator III, and zucchini cvs. Judgement III and Justice III.

As shown in our study, the use of virus-resistant transgenic summer squash not only prevents crop failure due to severe virus incidence but also provides sustained production even under conditions of severe virus infection. Furthermore, lessons from the cultivation of virus-resistant transgenic summer squash over more than a decade in the United States and extensive safety assessment studies have provided compelling evidence that the technology is an efficient, safe and sustainable approach to control devastating virus diseases (Fuchs and Gonsalves, 2007).

ACKNOWLEDGEMENTS

We are grateful to D. M. Tricoli from Seminis Vegetable Seeds Inc. for providing us with seeds of transgenic and nontransgenic squash. We thank Sheri Ecker-Day and the late George Mittak for their excellent help. This work was partially supported by a competitive grant from USDA's Biotechnology Risk Assessment Grant Program (no 95-33120-1878).

REFERENCES

Acord BD (1996). Availability of determination of nonregulated status for a squash line genetically engineered for virus resistance. Federal Register, 61: 33484-33485.

Arce-Ochoa JP, Dainello F, Pike LM, Drews, D (1995). Field performance comparison of two transgenic summer squash hybrids to their parental hybrid line. Hort. Sci., 30: 492-493.

Blancard D, Lecoq H, Pitrat M (2004). Cucurbit diseases: Observation, identification and control. Wiley and Sons, New York, NY. 304 pp

Clough GH, Hamm PB (1995). Coat protein transgenic resistance to watermelon mosaic virus and zucchini yellow mosaic virus in squash and cantaloupe. Plant Dis., 79: 1107-1109.

Fuchs M, Gonsalves D (1995). Resistance of transgenic hybrid squash ZW-20 expressing the coat protein genes of zucchini yellow mosaic virus and watermelon mosaic virus to mixed infections of both potyviruses. Bio/Technol., 13: 1466-1473.

Fuchs M, Gonsalves D (2007). Safety of virus-resistant transgenic plants two decades after their introduction: Lessons from realistic field risk assessment studies. Annu. Rev. Phytopathol., 45: 173-202.

Fuchs M, Tricoli DM, Carney KJ, Schlesser M, McFerson JR, Gonsalves D (1998). Comparative virus resistance and fruit yield of transgenic squash with single and multiple coat protein genes. Plant Dis., 82: 1350-1356.

Gaba V, Zelcer A, Gal-On A (2004). Cucurbit biotechnology - The importance of virus resistance. *In vitro* Cell Dev. Biol. Plant., 40: 346-358.

Gianessi LP, Silvers CS, Sankula S, Carpenter JE (2002). Virus resistant squash. In: Plant Biotechnology: Current and potential impact for improving pest management in U.S. agriculture.. An analysis of 40 case studies. National Center for Food and agricultural policy, Washington, DC. .75 pp.

Johnson SR, Strom S, Grillo K (2007). Quantification of the impacts on US agriculture of biotechnology-derived crops planted in 2006. National Center for Food and Agriculture policy. http://www.ncfap.org.

Klas FE, Fuchs M, Gonsalves D (2006). Comparative spatial spread overtime of Zucchini yellow mosaic virus (ZYMV) and Watermelon mosaic virus (WMV) in transgenic squash expressing the coat protein genes of ZYMV and WMV, and in nontransgenic squash. Transgenic Res., 15: 527-541.

Medley TL (1994). Availability of determination of nonregulated status for virus resistant squash. Federal Register, 59: 64187-64189.

Munger HM (1983). Breeding for viral resistance in cucurbits. p. 8-43.In M.Kyle (ed) : Resistance to viral diseases of vegetables: Genetics and breeding, Timber Press, Portland, OR.278 pp .

Perring TM, Gruenhagen NM and Farrar CA (1999). Management of plant viral disease through chemical control of insect vectors. Annu. Rev. Entomol., 44: 457-481.

Provvidenti R (1993). Resistance to viral diseases of cucurbits, p. 8-43. In M. Kyle (ed): Resistance to viral diseases of vegetables. Timber Press, Portland,OR.ISBN 10-0-88192-256-0

Schulteis JR, Walters SA (1998). Yield and virus resistance of summer squash cultivars and breeding lines in North Carolina. HortTechnol., 8: 31-39.

Tricoli D, Carney KJ, Russell PF, McMaster JR, Groff DW, Hadden KC, Himmel PT, Hubbard JP, Boeshore ML, Quemada HD (1995). Field evaluation of transgenic squash containing single or multiple virus coat protein gene constructs for resistance to cucumber mosaic virus, watermelon mosaic virus 2 and zucchini yellow mosaic virus. Bio/Technol., 13: 1458-1465.

Zitter TA, Hopkins DL, Thomas CE (1996). Compendium of cucurbit diseases. APS Press.

Aqueous ozone in the root zone: Friend or foe?

Thomas Graham*, Ping Zhang and Michael Dixon

School of Environmental Sciences, University of Guelph, Ontario, N1G 2W1, Canada.

Aqueous ozone ($O_{3(aq)}$) solutions were applied to the rockwool substrate of hydroponically cultured tomato and cucumber plants. Single applications of high concentration solutions (0, 5, 10, 15, 20 mg/L), as well as repeated application of lower concentration solutions (0, 2, 4, 6 mg/L), had no impact on leaf area and shoot dry weight accumulation. Repeated $O_{3(aq)}$ applications were also applied to cucumber plants inoculated with *Pythium aphanidermatum*. Pathogen levels were significantly reduced in all treatments containing $O_{3(aq)}$. The reduction in pathogen numbers did not necessarily affect plant productivity.

Key words: *Pythium aphanidermatum*, irrigation water reuse, oxygenation, tomato, cucumber, rockwool, hydroponics, phytotoxicity.

INTRODUCTION

In nearly all of the world's major greenhouse and nursery production regions, water is now the limiting resource. Managers face water supply challenges in the form of restrictions, competing uses, deteriorating quality (e.g. salinity, chemical contamination etc.), and rising costs associated with accessing reliable supplies (Bouwer, 2000). These challenges have fostered a shift towards the collection and reapplication of irrigation waters (Bouwer, 2000; Richard et al., 2006). Although this makes good use of a limited resource, it contributes to a second major production challenge in greenhouse and nursery systems, namely disease proliferation.

In absence of a system to treat the recovered water, growers risk disease proliferation via the reapplication of contaminated solutions. Many options are available for treating the recovered solutions, including filtration, heat, surfactants, ultraviolet radiation, and chemical disinfection (Cayanan et al., 2008; Ehret et al., 2001). Aqueous ozone ($O_{3(aq)}$) is also an option in some greenhouse and nursery settings (Ehret et al., 2001; Graham et al., 2009), as it is a proven water disinfection technology with over 100 years of application experience from which to draw.

Although a proven technology, widespread adoption of $O_{3(aq)}$ as an irrigation water remediation tool has been slow due to actual and perceived limitations. The first limitation is the cost and complexity of the systems, which currently limits the use of ozone to larger operations. This being said, continual advances in ozone generation and dissolution technologies may soon address these barriers. A second major limitation is the fact that ozone is a known phytotoxic gas. This phytotoxicity has been clearly demonstrated by many studies over the past 50 years that have examined plant responses to troposphere ozone enrichment (Bell and Treshow, 2002).

Although gaseous ozone can be phytotoxic at low concentrations (Bergmann et al., 1999), in aqueous solution, the mass transfer physics and chemical stability are much different than in the free gas state (Gottschalk et al., 2000). This difference is often overlooked when developing treatment applications for irrigation systems, thus hindering the development of alternative disease management protocols that do not suffer from the afflictions of standard commercial pest control strategies. Unlike commercial pesticides, $O_{3(aq)}$ does not leave a residual nor is the development of pathogen resistance likely as ozone reacts with diverse cellular constituents (Guzel-Seydim, 2004).

Ozone is unstable in solution; any ozone that has not reacted with chemical or biological contaminants reverts to diatomic oxygen (Beltrán, 2004), which in itself has potential for improving crop performance (Zheng, 2007; Drew, 1997). Growers incorporating ozone into their irrigation management strategy typically allow the ozone to dissipate or actively remove it prior to distribution to the

*Corresponding author. E-mail: tgraham@uoguelph.ca.

crop. This removal is carried out as a prudent action to avoid any potential crop damage resulting from ozone off-gas. This prudence is particularly justified in overhead irrigation systems where significant off-gassing can occur, which if not properly managed can cause foliar damage (Graham et al., 2009). When applied directly to the growth substrate (e.g. drip) this risk is greatly reduced, as the solutions are not exposed to the bulk atmosphere. The little information that is available regarding the direct application of $O_{3(aq)}$ to growth substrate, suggests that the phytotoxic potential may be overestimated and the use of $O_{3(aq)}$ may hold promise for diversifying irrigation management options (Ohashi-Kaneko et al., 2009; Sloan and Engelke, 2005).

EXPERIMENTATION

Experiments were conducted to develop an initial understanding of the potential for using $O_{3(aq)}$ solutions as a component of a greenhouse or nursery irrigation management plan. Tomato (*Solanum lycopersicum* L. cv Trust F1) and cucumber (*Cucumis sativus* L. cv. Serenade F1) plants were grown in rockwool hydroponic culture and subjected to $O_{3(aq)}$ irrigation regimes in isolation or in combination with a pathogen (*Pythium aphanidermatum*) challenge. The objectives were: (1) to determine if $O_{3(aq)}$ applied directly to a rockwool hydro-ponic substrate suppressed productivity (as measured by leaf area and dry matter accumulation); and (2) determine if $O_{3(aq)}$ applied directly to a rockwool growth substrate can reduce the incidence of *P. aphanidermatum* .

RESULTS AND DISCUSSION

Although these studies are limited in scope, there was clear indication that aqueous ozone could be applied directly to the surface of rockwool growth substrate in both tomato and cucumber hydroponic culture without adversely influencing growth (Figure 1). It was also evident that some level of pathogen suppression was achieved through the application of $O_{3(aq)}$, although the connection between reduced pathogen presence and the maintenance of plant performance was not definitive (Figure 2).

In the first two studies, 2 L aliquots of solutions containing high $O_{3(aq)}$ concentrations (5, 10, 15, 20 mg/L) were applied in a single dose to the root zones of tomato and cucumber plants. The results (Figures 1A-D) clearly showed that there were no discernible effects on growth as determined by the leaf area and dry matter accumulation. These results were somewhat unexpected as the concentrations employed were excessive in comparison to typical water treatment applications. These same concentrations, when applied as a foliar drench, elicit varying degrees of phytotoxicity (data not shown) (Graham et al., 2009). During treatment application, the drainage was collected and the ozone residual was measured. In all cases, very low (<0.03 mg/L) or no ozone remained in the solution after passage through the rockwool block (as measured by standard indigo methods, (Hach Co. Loveland CO, USA)). This was an indication that the majority of ozone reacted with some component of the root – substrate complex (including micro-organisms, and chemical constituents). Visual examination of the root mass revealed no symptomatic evidence of root browning or other damage indicative of oxidative stress.

Transient applications, even at the very high concentrations employed in the first two studies, may not have been sufficient to cause any visible phytotoxicity symptoms. It was theorized that a mature plant with an established root zone ecosystem would have a significant buffer (against ozone damage) in the form of accumulated organic compounds and micro-organisms. Repeated applications may overwhelm this buffer capacity. In a third study, the concentrations applied were reduced to 0, 2, 4, 6 mg/L $O_{3(aq)}$ but were applied twice daily for six days. Once again, the results (Figures 1E-F) indicated that there was no loss in production even with these frequent ozone applications. Recent work by Ohashi-Kaneko et al. (2009), in which ozone was applied to tomato plants once per week for three weeks at 1.5 mg/L, supports these findings.

Given that no phytotoxic responses were observed during these studies and combined with the potential for improved productivity shown by others (Ohashi-Kaneko et al., 2009; Sloan and Engelke, 2005), the question is then one of disease control benefits. If $O_{3(aq)}$ does not cause any significant negative growth responses then it may have potential to be used in the control of disease vectors in the root zone (McDonald, 2007). In a third series of short experiments, cucumber plants were inoculated with *P. aphanidermatum* and subjected to a series of ozone treatments. The results are summarized in Figure 2.

Although the application of $O_{3(aq)}$ clearly reduced the presence of *P. aphanidermatum* in both studies (Figures 2A and D), it did not necessarily result in a maintenance of productivity under intense disease pressure (Figures 2E-F). Given the limited scope and the failure for the studies to corroborate one another in terms of producti-vity maintenance (Figures 2B-C, E-F), only limited inferences can be made on the efficacy of $O_{3(aq)}$ to prevent production loss due to disease. The pathogen load was clearly reduced in both studies (Figures 2A and D), which suggests that ozone is reaching the roots; however, the treated plants in the higher dose study did not maintain the production levels observed in the control plants. In this case, $O_{3(aq)}$ was either ineffective against the establishment of the pathogen (as inferred from SDW and LA data), or the combination of biotic and abiotic stressors (disease and high $O_{3(aq)}$) acted to suppress productivity. This is not to say that disease control and maximum productivity could not be achieved under a more rigorous treatment protocol, as evidenced in Figures 2A-C, but rather that the complexities of the systems should not be underestimated.

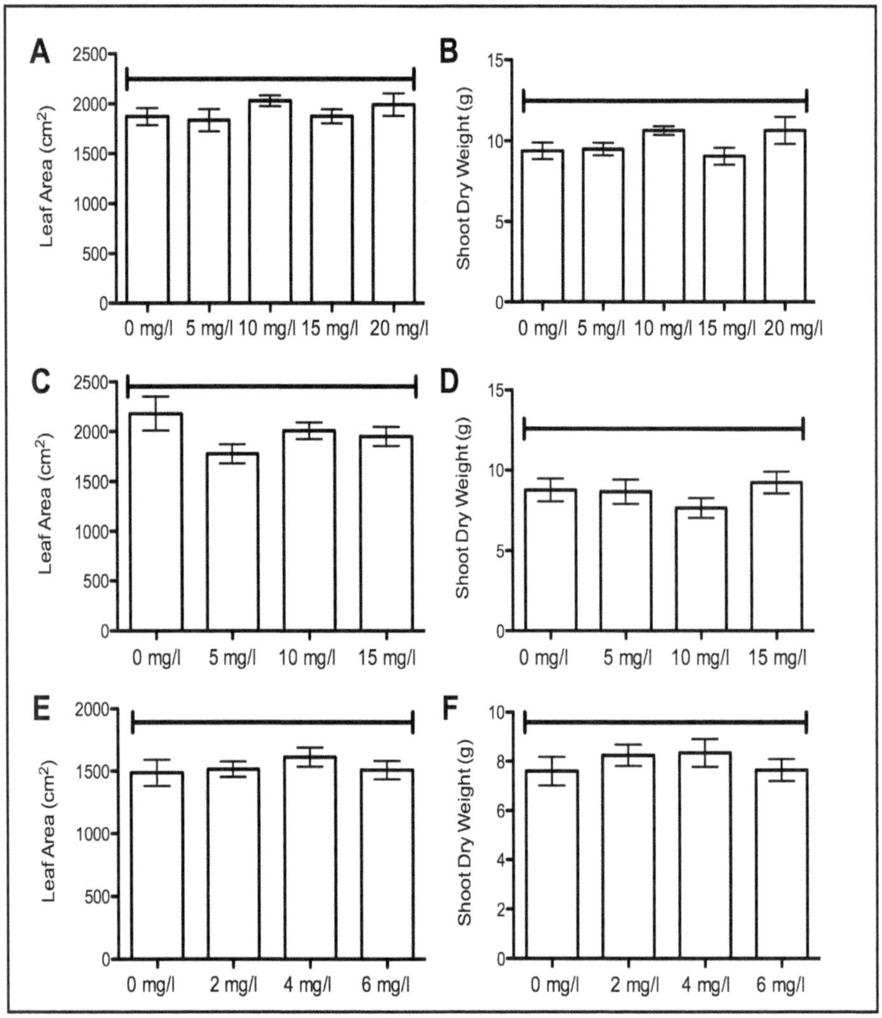

Figure 1. Leaf area (LA) and shoot dry weight (SDW) response of tomato and cucumber to direct applications of aqueous ozone to the rockwool (Grodan Delta-10 Gro-blocks 10x10x 6.5cm) growth substrate: A-B) Response of six week old tomato plants to a one time root zone $O_{3(aq)}$ application. Each plant received a 2 litre aliquot from one of five $O_{3(aq)}$ solutions (0, 5, 10 15, 20 mg/l). The solutions were poured over the rockwool cube at an average rate of 1 l/min. Plants were grown for an additional 12 days before being destructively analysed (n=6); C-D) Response of six week old cucumber plants to a one time root zone $O_{3(aq)}$ application. Each plant received a 2 litre aliquot from one of four $O_{3(aq)}$ solutions (0, 5, 10 15 mg/l). The solutions were poured over the rockwool cube at an average rate of 1 l/min. Plants were grown for an additional 10 days before being destructively analysed (n=5); E-F) Response of six week old tomato plants to twice daily (10:00 and 17:00) root zone applications of a 1 litre $O_{3(aq)}$ solution (0, 2, 4, 6 mg/l). Treatments commenced when the plants were six weeks old and continued for 6 days, after which the plants were grown for an additional 7 days before being destructively analysed: A-F) Columns falling under the same horizontal line are not statistically different at $p<0.05$; error bars are +/- standard error of the mean (one-way ANOVA with Tukey's post test, GraphPad Prism ver. 5.0c for Mac, GraphPad Software, San Diego, Calif. USA). Ozone solutions were prepared using an oxygen-fed (90-95% O_2) corona discharge ozone generator (CD1500P, Clearwater Tech., San Luis Obispo, CA., USA) and a Shaw Mixer™ ozone mass transfer system (Purification Research Technologies Incorporated, Guelph, Ontario, Canada). Ozone concentrations were measured with a dissolved ozone sensor (Q45H, ATI, Collegeville, PA, USA) calibrated against the indigo method (Bader and Hoigne, 1981). LA was determined using a leaf area meter (LI-3100C, Li-Cor, Lincon, NE). SDW was determined after drying all samples to a constant mass. All plants were grown in a research greenhouse at the University of Guelph.

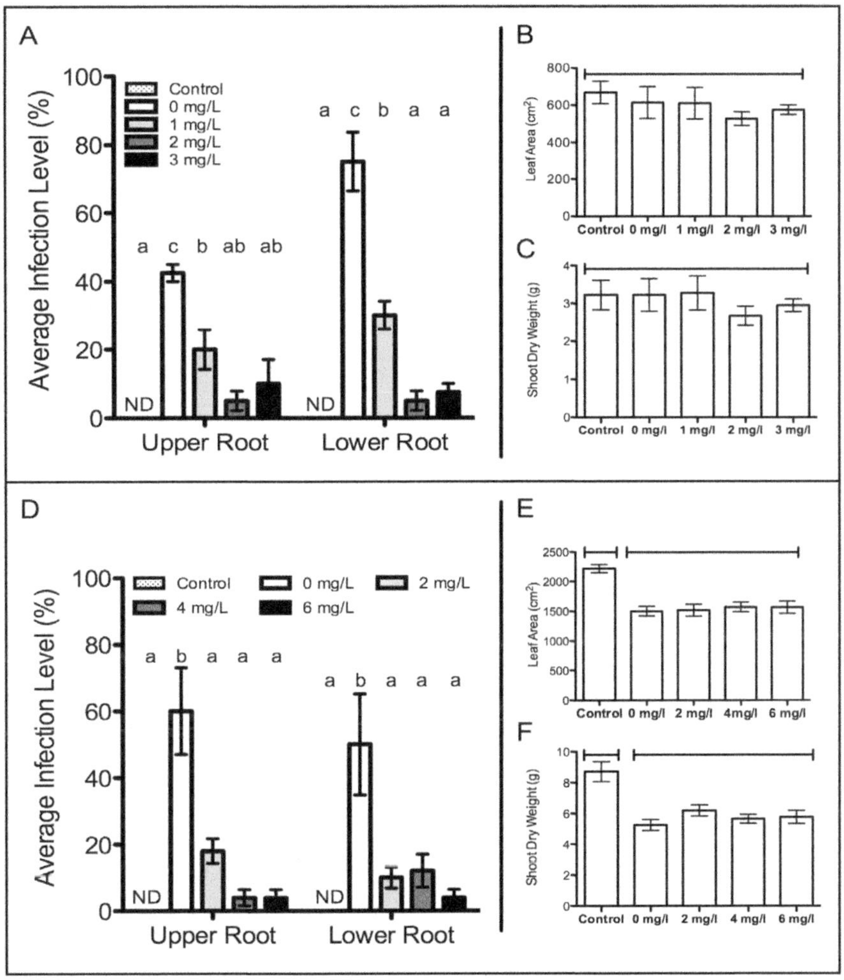

Figure 2. Response of cucumber plants grown in rockwool growth media (Grodan Delta-10 Gro-blocks 10x10x 6.5cm), previously inoculated with *P. aphanidermatum*, to application of $O_{3(aq)}$: A) Average infection level (percent of root segments sampled) in the upper and lower half of the root mass for treatments consisting of a Control (not inoculated, no ozone), 0, 1, 2, 3 mg/l $O_{3(aq)}$. The plants were inoculated and allowed to stand for several hours before the first treatment application. One litre aliquots (per plant) were applied to the rockwool substrate twice daily (1 l/min; 10:00 and 17:00) for 14 days (n=4); B-C) Leaf area and shoot dry weight response to the treatments described in A.; D) Average infection level (percent of root segments sampled) in the upper and lower half of the root mass for treatments consisting of a Control (not inoculated, no ozone), 0, 2, 4, 6 mg/l $O_{3(aq)}$. The plants were inoculated and allowed to stand for several hours before the first treatment application. One litre aliquots (per plant) were applied to the rockwool substrate twice daily (1 l/min; 10:00 and 17:00) for 14 days (n=5); E-F) Leaf area and shoot dry weight response to the treatments described in D; A-F) Columns falling under the same horizontal line or those having the same letter appearing above it as other columns in the group are not significantly different at $p<0.05$; error bars are +/- standard error of the mean (one-way ANOVA with Tukey's post test, GraphPad Prism ver. 5.0c for Mac, GraphPad Software, San Diego, Calif. USA). Ozone solutions were prepared using an oxygen-fed (90-95% O_2) corona discharge ozone generator (CD1500P, Clearwater Tech., San Luis Obispo, CA., USA) and a Shaw Mixer^{TM} ozone mass transfer system (Purification Research Technologies Incorporated, Guelph, Ontario, Canada). Ozone concentrations were measured with a dissolved ozone sensor (Q45H, ATI, Collegeville, PA, USA) calibrated against the indigo method (Bader and Hoigne, 1981). LA was determined using a leaf area meter (LI-3100C, Li-Cor, Lincon, NE). SDW was determined after drying all samples to a constant mass. All plants were grown in a research greenhouse at the University of Guelph. *Pythium aphanidermatum* inoculum was prepared by selection on P5 media followed by a propagation phase in V8 media, which was then applied as a root drench. ND - Pathogen Not Detected.

Given these results, it is reasonable to assume that routine application of $O_{3(aq)}$ could prevent the establishment of pathogens in rockwool hydroponic culture. In the presented studies, the plants were specifically inoculated with a rich source of *P. aphanidermatum* to elicit a disease response. In an actual production system such a deliberate and strong disease exposure would likely be rare. Frequent applications of $O_{3(aq)}$ could therefore have some potential for disease prevention in commercial settings; a potential that warrants further investigation.

The results presented justify a more thorough evaluation of the direct application of $O_{3(aq)}$ to the growth substrate of common greenhouse and nursery crops. Focus should be given to the determination of concentration and application frequency thresholds as well as investigations into the influence of substrate type and application timing (to take advantage of diurnal stomatal states). Clearly the phytotoxicity dynamics of ozone in aqueous solution is different than that of the gas phase. This knowledge opens the door to a range of horticultural $O_{3(aq)}$ applications related (primarily) to irrigation system maintenance and pest management.

ACKNOWLEDGEMENTS

The authors would like to thank Canadian Hydrogardens Ltd. for graciously supplying the seeds for these studies, the Ontario Centres of Excellence for the funding to support this research, and Purification Research Technologies Incorporated (PRTI) for supplying the ozone systems.

REFERENCES

Bader H, Hoigne J (1981). Determination of ozone in water by the indigo method. Water Res., 15(4): 449-456.

Bell JNB, Treshow M eds. (2002) Air Pollution and Plant Life – 2nd ed. John Wiley & Sons Ltd., West Sussex, England, pp. 69-118.

Beltrán FJ (2004) Ozone Reaction Kinetics for Water and Wastewater Systems. Lewis Publishers, New York, USA. pp. 358.

Bergmann E, Bender J, Weigel HJ (1999). Ozone threshold doses and exposure-response relationships for the development of ozone injury symptoms in wild plant species. New Phytol. 144: 423-435.

Bouwer H (2000). Integrated water management: emerging issues and challenges. Agric. Water Manage., 45: 217-228.

Cayanan DF, Zheng Y, Zhang P, Graham T, Dixon M, Chong C, Llewellyn J (2008). Sensitivity of Five Container-grown Species to Chlorine in Overhead Irrigation Water. Hort. Sci. 43(6): 1882-1887.

Drew MC (1997). Oxygen deficiency and root metabolism: Injury and acclimation under hypoxia and anoxia. Annu. Rev. Plant Phys., 48: 223-250.

Ehret DL, Alsanius B, Wohanka W, Menzies JG, Utkhede R (2001). Disinfestation of recirculating nutrient solutions in greenhouse horticulture. Agronomie, 21: 323-339.

Gottschalk C, Libra JA, Saupe A (2000). Ozonation of Water and Waste Water – A Practical Guide to Understanding Ozone and its Applications. Wiley-VCH, Weinheim, Germany. pp. 189.

Graham T, Zhang P, Zheng Y, Dixon, MA (2009). Phytotoxicity of aqueous ozone on five container-grown nursery species. HortScience. 44(3): 774-780.

Guzel-Seydim ZB, Greene AK, Seydim AC (2004). Use of ozone in the food industry. LWT Food Sci. Technol., 37(4): 453-460.

McDonald GV (2007) Ozone (O3) efficacy on reduction of *Phytophthora capsici* in recirculated horticultural irrigation water. Texas A&M University. Thesis. pp.121.

Ohashi-Kaneko K, Yoshii M, Isobe T, Park J-S, Kurata K, Fujiwara, K (2009). Nutrient solution prepared with ozonated water does not damage early growth of hydroponically grown tomatoes. Ozone-Sci. Eng., 31(1): 21-27.

Richard S, Zheng Y, Dixon M (2006). To recycle or not to recycle? Greenhouse Canada, December: 20-25.

Sloan JJ, Engelke MC (2005). Effect of ozonated water on creeping bentgrass growth in a sand medium. HortTechnology. 15(1): 148-152.

Zheng Y, Wang L, Dixon M (2007). An upper limit for elevated root zone dissolved oxygen concentration for tomato. Sci. Hort., 113: 162-165.

7

Pollen Isolation Method Affects Interspecific Hybridization in Eucalyptus

Girijashankar V

ITC R&D Centre, Shameerpet, Hyderabad, Andhra Pradesh, India.
Corresponding author. E-mail: vgirija_shank@yahoo.com

The ultimate test for utility of a pollen sample is the seed yield following control pollination (CP). In the present study, two different methods of eucalyptus pollen isolation (PI) namely: wet-lyophylisation (WL) and conventionally dry-sieving (DS) were followed. The pollen thus isolated was used to carry interspecific hybridization in order to evaluate and compared the efficiency the above two methods of PI in eucalyptus. Apart from testing the efficiency of the above two methods of PI, CP was carried to generate new combinations of alleles towards developing hybrid eucalyptus for commercial use. Wide-interspecific hybridization was performed between important eucalyptus species following a partial diallel mating design. Pollen from three paternal species namely E. globulus (5 trees), E. pellita (1 tree) and E. urophylla (1 tree) were used to CP three elite commercial clones of E. camaldulensis. Except in few occasions, the percent capsule set and number of seeds obtained per 10 capsule (≥30 seeds) with the PI through WL method was comparatively high and significantly different from conventional DS method (≤30 seeds) at $P<0.01$. The results of this experiment prove that the method of PI affects the outcome of wide-interspecific hybridisation in eucalyptus.

Key words: Wet-lyophylisation, pollen isolation, control pollination, interspecific hybridization, eucalyptus.

INTRODUCTION

Eucalyptus is the second most widely planted (20 million hectares) forestry tree species in the world (GIT Forestry, 2009). Among the available 700 species and hybrids of eucalyptus, *Eucalyptus camaldulensis* Dehn forms one of the major sources for paper pulp, fuel wood and timber (Eldridge et al., 1993). This species is used in eucalyptus breeding programmes because of its wide adaptability to grow under drought and rain-fed conditions in tropical and semi-arid regions of the world (Butcher et al., 2009).

The wood properties of eucalyptus are known to vary between species, which could be exploited through introgression of useful alleles from other species into the desired ones towards enhancing wood yield and quality. Interspecific hybridization of eucalyptus can address this requirement through CP of elite trees having desirable traits of interest. Hodgson (1975) and Van Wyk (1977) reported the use of fresh eucalyptus flowers in carrying CP. Conventionally; pollen is isolated by dry sieving (DS) method which involves brushing of clipped or unclipped stamens of mature flowers on to nylon sieve. This method is largely followed because of its simplicity. Recently, it has been reported that the use of sieved pollen in breeding program of *Acacia mangium* can result in outcross contamination up to 19.1%. Further, this process substantially decreases the working labour productivity when compared to the use of the entire inflorescences from the male parent (8.7%) as the pollen applicator (Griffin et al., 2010).

Earlier, Griffin et al. (1982) proved the efficiency of VP method of PI in CP studies of *E. regnans* and reported improvement in mean seed yield per capsule. However, this method was associated with problems such as clogging during filteration as well as sticking of pollen to filter papers. To overcome these shortcomings associated with Griffins VP method and conventional DS methods, we developed and reported Wet-lyophilisation (WL) method of PI. The ultimate test of pollen utility is the capsule set and seed yield following control pollination.

Figure 1. Wide-interspecific hybridization of eucalyptus. a) Eucalyptus pollen collected through conventional DS method and viewed at random under 10X objective of compound microscope. The occurrence of dead cell mats and sparsely distributed pollen can be seen. Scale = 120 µm. b) Emasculated flowers of E. camaldulensis clone 27, 71 and 3 (left to right) used as maternal parents in interspecific hybridization studies. Variation in style length, shapes and flower size can be observed. c) Control pollinated flowers, bagged on E. camaldulensis clone 71. d) Difference in hybrid seed size starting with smaller seed of clone 3 (left) followed by larger seed belonging to clone 27 (center) and medium sized seed of clone 71 (right).) e) Variation in capsule size among the three E. camaldulensis clones namely 27, 71 and 3 (left to right) which resulted in variation in seed set and percent capsule set. f) E. camaldulensis x E. globulus hybrid seedlings showing variations in number of cotyledons and shapes - intermediary to their parents. g) Difference in the cotyledon shape of hybrid seedlings. Lower: hybrid seedling showing E. globulus character - bilobed, chordate, heart shape. Upper: hybrid seedlings cotyledon showing characteristic E. camaldulensis shape – kidney or bean shape.

In the present study, in order to compare the efficiency between WL and DS methods of PI, wide-interspecific hybridization was carried between 4 important species of eucalyptus. The widely adaptable drought tolerant tropical species *E. camaldulensis,* was used as maternal parent while *Eucalyptus globulus* (a temperate species), *Eucalyptus pellita* and *Eucalyptus urophylla* were used as the pollen donors. The efficiency of PI methods was evaluated in terms of percent capsule set and seed set /10 capsules.

MATERIALS AND METHODS

Pollen isolation methods

The WL method involves tender grinding of stamens twice in water followed by filtration, centrifugation and lyophilization as described in Girijashankar (2010). This method is also depicted as a flow chart in Figure 2. DS method of PI on the other hand, involves clipping of stamens from mature, unopened flowers followed by shade drying and tender rubbing of the same stamens against a 70 µm pore size nylon sieve which results in filtration of fine pollen powder. This pollen filtrate may often result in dead cell mats and debris (Figure 1a).

Plant materials

Eucalyptus sp. belonging to three different sections of the major subgenus, *Symphyomyrtus,* was used in the present study. Pollen was collected separately following both methods (DS and WL) from mature, unopened flowers of *E. globulus* (section: *Maidenaria*), *E. pellita* (section: *Transversaria*) and *E. urophylla* (section: *Transversaria*). The pollen thus collected is used to cross pollinate three elite commercial clones of *E. camaldulensis* (section: *Exsertaria*). The trees and species used were spread across two South Indian states namely Andhra Pradesh and Tamilnadu. Interspecific hybridization was carried at Andhra Pradesh State Forestry Research Center, Mulugu, Medak (district), Andhra Pradesh, India.

Control pollination and interspecific hybridization

For comparing the efficiency of WL and DS methods of PI, interspecific hybridization was carried out with pollen collected separately from 5 trees of *E. globulus* and one tree each from *E. pellita* (EP-HP-26) and *E. urophylla* (single tree pollen: EU-S6). CP was carried out with mature flowers of *E. camaldulensis* (belonging to ITC, Badhrachalam) elite clones namely 3, 27 and 71 (15 years old) (Figure 1b). For each clone, the best two out of the crossing trees were selected for hybridization experiments. This is because within the same clone, each individual tree in the orchard showed

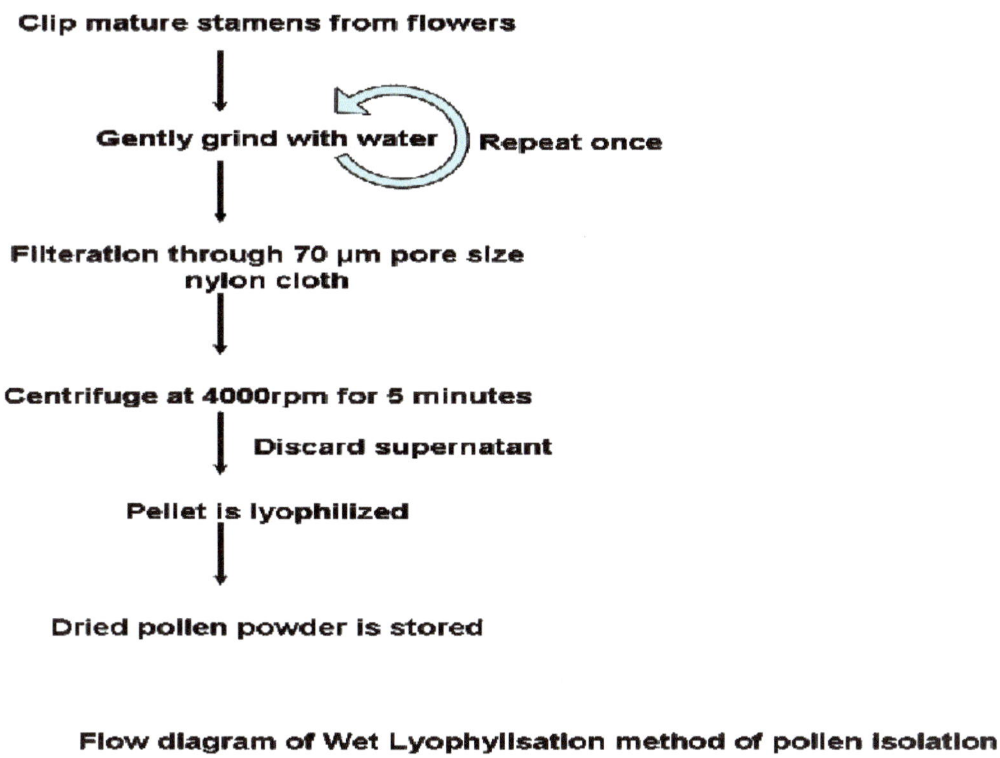

Figure 2. Flow diagram of Wet-Lyophylisation method of pollen isolation.

variation in the crossing rates which was revealed during our earlier experiments. Thus, three maternal clones were crossed with seven pollen donors, eventually giving rise to a total of 21 different combinations of wide-interspecific hybridizations (Table 1).

For CP, the matured unopened flowers are carefully emasculated. Later, they are washed with a sharp jet of water using a squeeze bottle and immediately bagged in a cotton cloth bag (30 X 10 cm) (Figure 1c). On the fourth day, after emasculation, the cotton bags were opened and the receptive stigmas which turned pinkish with sugary secretion were dipped in the respective pollen powders and bagged immediately. The capsules were harvested ≥100 days after CP. The number of capsules and seeds were manually collected and counted (Figures 1d and e).

Statistical analysis

Paired T test was performed in order to ascertain the efficiency of WL and DS methods of PI. In this study of wide-interspecific hybridizations, a total of 42 combinations of CP were evaluated (21 each for WL and DS methods, respectively). This lead to the development of 21 different families of hybrid seed as mentioned in Table 1. Each value in Table 1 is an average of 5 bags, that is, 5 replications per each combination of CP. Further, a minimum of 20 receptive flowers were control pollinated and maintained per bagged inflorescence. Thus, a minimum of 100 receptive flowers (5 bags X 20 flowers/bag) were control pollinated with pollen from each paternal tree. The seed set was expressed as number of seeds obtained per 10 capsules because not all the receptive flowers that were control pollinated could set seed (few capsules contain only chaff). As reported by Meddings et al. (2003), success of interspecific hybridization (*E. camaldulensis* x *E. globulus*) was confirmed by observing cotyledon shapes of hybrid seedlings which were intermediate to that of the respective parent species (Figures 1f and g).

RESULTS

Interspecific hybridization

WL method of PI resulted in enhanced capsule set (significantly different at $P<0.01$) over the conventional DS method. With few exceptions, the number of seeds obtained per 10 capsules was also found to be significantly different between the two methods of PI at $P<0.01$ (Table 1). The average number of seeds obtained per 10 capsules was less in the DS method when compared to the WL method of PI. The number of seeds obtained per capsule following DS method is in the range of 20-30 seeds per 10 capsules. On the other hand, using the same crossing combinations, WL method resulted in 30 to 40 seeds per 10 capsules. Among the 21 combinations of crosses, it was observed that few capsules did not set seed, while few capsules dried or withered off, irrespective of the pollen source or method of collection. As such, the withered capsules were not taken into account for analysis. The percent capsule set and number of seeds/10 capsules was high for clone 27 which has comparatively larger flowers and developed bigger capsules.

Contrary to this, lowest values for the above traits was observed in case of clone 3 which produced smaller

Table 1. Comparison of pollen isolation methods and their effect on interspecific hybridization.

Paternal species and tree number	Maternal parent	Wet lyophilisation		Dry sieving	
	E. camaldulensis	% capsule set	Hybrid seeds /10 capsules	% capsule set	Hybrid seeds /10 capsules
E. globulus 1	Clone No. 3	42 ± 9	32	35 ± 8	23
2		52 ± 7.1	32	26 ± 8	26
3		38 ± 4.3	30	24 ± 6.5	25
4		42 ± 5.4	24	33 ± 7.5	20
5		45 ± 6	30	22 ± 4.5	20
1	Clone No. 27	95 ± 1.2	36	70 ± 8.5	33
2		94 ± 1.7	39	65 ± 7.6	31
3		92 ± 2.3	39	59 ± 7.5	36
4		84 ± 2.8	31	61± 7	33
5		100 ± 0.0	40	76± 6.4	31
1	Clone No. 71	63 ± 4.1	33	60± 4.5	20
2		70 ± 9	38	62± 8.6	21
3		54 ± 6.3	31	50± 3.4	19
4		52± 4.2	30	46 ± 5	31
5		55 ± 9	34	46± 3.1	30
E. pellita EP-HP-26	Clone No 3	75 ± 8.5	23	15± 8	22
EP-HP-26	Clone No 27	93 ± 2.5	31	37± 5	31
EP-HP-26	Clone No 71	76 ± 8.1	35	35± 4.3	24
E. urophylla EU-S6	Clone No 3	75 ± 9.1	26	43± 8.0	22
EU-S6	Clone No 27	96 ± 3.3	34	65± 4.2	27
EU-S6	Clone No 71	95 ± 2.4	31	70± 5.4	23
	T cal	6.68	5.85		
	DF	20	20		
	P<0.01	S	S		

Paired T test: T cal-values between WL and DS methods of pollen isolation affecting the % capsule set and number of hybrid seeds obtained/10 capsules. Hybrid seeds were manually counted from 10 randomly selected capsules.

flowers and capsules (Figures 1b and e and Table 1). The size of the flower and capsule from the three female clones used in the present study was in the order of 27>71>3 (Figures 1b and e). The percent capsule set and the size of hybrid seeds also (in general) followed a similar pattern (Figure 1d). In this experiment, it was observed that the size of the capsule as well as the hybrid seeds was dependent on the maternal parent that is *E. camaldulensis*.

Generally, *E. camaldulensis* blooms twice a year, with the first flowering occurs between September and January (winter flowering) while the second bloom occurs during February to May (summer flowering). After carrying interspecific hybridization for two consecutive years, it has been observed that the percent capsule set and the number of seed/10 capsules was high on winter flowers, that is, September to January. In contrast, summer flowers result in higher flower drop, capsule withering and empty capsule formation that eventually lead to failure of our earlier hybridization attempts.

A maximum of 11 hybrid seeds per individual capsule were developed in *E. camaldulensis* x *E. globulus* cross combination. Open pollination in these three *E. camaldulensis* clones resulted in 10 to 24 seeds per capsule, while the capsule set was in the range of 75 - 100%. These observations have been recorded during winter flowering. However, CP and wide-interspecific hybridization which involves manual handling resulted in a reduced capsule set (35 to 70% in most cases) as well as decrease seed set per capsule (2 to 4 seeds per capsule). Furthermore, the success of interspecific hybridization between *E. camaldulensis* x *E. globulus* in the present study was confirmed by the cotyledon shapes which were observed to be inter-mediate between the parent species (Figure 1f). In few occasions, the hybrids seedlings completely developed either the *E. globulus* cotyledon shape (bilobed, heart and chordate shape) or the maternal parent *E. camaldulensis* shape (bean, kidney) (Figure 1g).

DISCUSSION

Hybridization efforts toward eucalyptus germplasm enhancement are hindered due to asynchronous flowering and incompatibility barriers (Savva et al., 1988). In order to overcome these barriers, different PI methods have been developed to collect pollen from desirable trees. Earlier, we have reported that WL method is efficient when compared to conventional DS method of PI. WL method resulted in a reduction of debris content in pollen powder and en-hanced pollen deposition per stigma along with minimal affect on pollen viability and quantity being isolated (Girijashankar, 2010).

As in the present study, partial diallels have been used earlier in eucalyptus breeding programmes towards improving cellulose and wood quality traits (Bison et al., 2007). It is known that eucalyptus interspecific hybridisation via CP is a labour-intensive, costly exercise and produces lower seed yields (Horsley et al., 2010; Meddings et al., 2003). Similar observations have been made between natural open pollination and CP in our interspecific hybridization experiments.

In this study, the pollen collected by WL method resulted in enhancing the numbers of seeds obtained per capsule as well as the percent capsule set when compared to the DS method (P<0.01). Earlier, Griffin et al. (1982) carried CP studies using *E. regnans* and reported that the pollen collected through the VP method which also involves immersion of pollen in water (as in the case of WL method) did not affect pollen viability. Further, they also reported that PI, through VP method, resulted in an enhancement of pollen deposition per stigma along with a mean increase of 0.5 seed/capsule. Similarly, WL method of PI also improved the eucalyptus breeding efficiency when compared to the DS method of PI by enhancing the number of seeds / 10 capsules and percent capsule set. Thus, Griffins VP method, as well as WL method of PI stands proof for the increased eucalyptus breeding efficiency.

Resource competition has been shown to exist between ovules within an individual capsule and the presence of larger flowers can increase the chance of the capsule set in eucalyptus (Griffin et al., 1987; Suitor et al., 2007). Similar observations were recorded in our experimentation especially with clone 27 which produced larger flowers and capsules that resulted in higher seed set (3.8) and percent capsule formation (90%) when compared to clone 3 with smaller flowers (2 to 3 seeds/ capsule and 40 to 60% capsule set), irrespective of the pollen viability and paternity.

In eucalyptus, increase in the number of pollen tubes reaching the ovary can enhance viable seed formation (Trindade et al., 2001). It is not possible for all the pollen that land on the stigmatic surface to successfully fertilize the ovules or all the ovules that get fertilized can form seeds. In eucalyptus, only few pollen tubes travel successfully till the end of the stylar canal and eventually fertilize the ovules (Suitor et al., 2007). Earlier works in eucalyptus have proven that genotypes allowing higher proportion of pollen tubes to grow till the base of the style have higher fertilization rates and increased reproductive success (Suitor et al., 2007).

Proteins and lipids on the surface of the pollen grain are involved in the pollen–stigma interaction (Stephenson et al., 1997; Hulskamp et al., 1995). Treating pollen with organic solvents (toluene, carbon tetra chloride, etc.) proved to reduce cross compatibility barriers among woody trees like willow (Stott, 1984) but at the cost of reduced pollen viability (Kopp et al., 2002). In the present study of PI i.e. WL method, mixing the pollen with water or in other terms use of water as solvent for extracting pollen resulted in the development of a yellow colour to water that was finally discarded. It is hypothesize that the water used in the WL method of PI removes some of the chemical molecules present on the pollen grains which may play an important role in eucalyptus interspecific hybridization. Stanley and Search (1971) reported that immersion of tree pollen in water can cause elution of proteins from pollen walls. Unlike the above mentioned reports, Girijashankar (2010). These observations on pollen viability between WL and DS methods of PI. These observations on pollen viability are in agreement with earlier findings of Griffin et al. (1982) and Heslop and Heslop (1985) who reported that eucalyptus pollen are resistant to immersion in water.

Reproductive success of *E. camaldulensis* and *E. globulus* in terms of number of seeds/capsule and % capsule set are known to get influenced by physical and physiological parameters which in turn are dependent on genotype, season and moisture availability (Suitor et al., 2007; Varghese et al., 2009; Butcher et al., 2009). However, success of interspecific hybridization between *E. camaldulensis* x *E. globulus* was rarely reported (Meddings et al., 2003). In the present study, considering the cotyledon shape of hybrid seedlings (intermediate between the two parents) as an excellent way of checking hybrid status (Meddings et al., 2003), we report that our CP experiments successfully resulted in a generation of new hybrids. In such a rare crossing combination, we could develop 15 families of hybrid eucalyptus and also show that the method of PI does affect percent capsule set as well as the number of hybrid seeds developed per capsule.

For successful interspecific hybridization, especially with *E. camaldulensis* (if used as a maternal parent) in tropical and semi-arid climates, it is beneficial to follow the WL method of pollen collection and winter flowering season for improving hybrid seed yields. Based on this study, we conclude that wet-lyophylisation method of PI is beneficial when compared to the conventional DS method followed in eucalyptus breeding programme. WL method of PI needs to be further evaluated using other species and genera in order to confirm its efficiency across different regions.

ACKNOWLEDGMENTS

The guidance and technical advices rendered by Samresh Dwivedi are highly appreciated. The authors are thankful for the technical advices and constant support rendered by Hareesh TS, Sunita Lakshmi, Raj Kumar and Kamal Kannan towards eucalyptus pollen isolation and interspecific hybridization studies.

REFERENCES

Bison O, Magno APR, Gabriel D (2007). Combining ability of elite clones of *Eucalyptus grandis* and *Eucalyptus urophylla* with *Eucalyptus globulus*. Gene. Mol. Biol., 30(2): 417-422.

Butcher PA, McDonald MW, Bell JC (2009). Congruence between environmental parameters, morphology and genetic structure in Australia's most widely distributed eucalypt, *Eucalyptus camaldulensis*. Tree Gene. Genom., 5(1): 189-210.

Eldridge K, Davidson J, Harwood C, Wyk G (1993). Eucalyptus domestication and breeding. Clarendon, Oxford, pp. 60–71.

Girijashankar V (2010). Effect of eucalyptus pollen isolation methods on pollen viability, debris content, quantity isolated and pollen density per stigma. J. Pl. Breed. Crop Sci. 2(9): 273-279.

GIT Forestry (2009). http://git-forestry.com/download_git_eucalyptus_map.htm

Griffin AR, Ching KK, Johnson KW, Hand FC, Burgess IP (1982). Processing eucalyptus pollen for use in controlled pollination. Silvae Genetica, 31(5-6): 198-203.

Griffin AR, Morgan GF, Fripp YJ (1987). Preferential outcrossing in *Eucalyptus regnans* F. Muell. Austr. J. Bot., 35: 465-475.

Griffin AR, Tran Duc Vuong, Harbard JL, Wong CY, Brooker C, Vaillancourt RE (2010). Improving controlled pollination methodology for breeding *Acacia mangium* Willd. New For., 40: 131–142.

Heslop HJ, Heslop HY (1985). Germination of stress tolerant *Eucalyptus* pollen. J. Cell Sci., 73: 135-157.

Hodgson LM (1975). Some aspects of reproductive behaviour in *E. grandis*. D. Sc. Thesis, Dept General Botany, University of Pretoria, South Africa.

Horsley TN, Johnson SD, Myburg AA (2010). Comparison of different control-pollination techniques for small-flowered *Eucalyptus*. New For., 39(1): 75-88.

Hulskamp M, Kopczak SD, Horejsi TF, Kihl BK, Pruitt RE (1995). Identification of genes required for pollen-stigma recognition in *Arabidopsis thaliana*. Plant J., 8(5): 703–714.

Kopp FR, Charles AM, Patricia RDN, Lawrence BS, Lawrence PA (2002). Collection and storage of pollen from Salix (Salicaceae). Am. J. Bot., 89(2): 248–252.

Meddings AR, McComb AJ, Michael CC, Sandra RT, Richard AM (2003). *Eucalyptus camaldulensis* × *E. globulus* hybrids. Australian J. Bot., 51(3): 319–331.

Savva M, Potts BM, Reid JB (1988). The breeding system and gene flow in *Eucalyptus urnigera*. In: Pollination '88, University of Melbourne, Parkville (eds RB Knox, MB Singh L Troini), pp. 176–182.

Stanley RG, Search RW (1971). Pollen protein diffusates. In. Pollen: Development and Physiology. Ed. J Heslop-Harrison. Butterworths, London.

Stephenson AG, Doughty J, Dixon S, Elleman C, Hiscok S, Dickinson HG (1997). The male determinant of self-incompatibility in *Brassica oleracea* is located in the pollen coating. Plant J., 2(6): 1351–1359.

Stott KG (1984). Improving the biomass potential of willow by selection and breeding. *In* K. Perttu [ed.], Ecol. Manage. For. Biomass Prod. Syst., 15: 233–260.

Suitor S, Potts BM, Brown PH, Gracie AJ, Gore PL (2007). Factors limiting capsule set, seed set and reproductive success in *Eucalyptus globulus* seed orchards. In: IUFRO conference on *Eucalyptus* and diversity: balancing productivity and sustainability. Durban, South Africa, 22-26, October.

Trindade H, Boavida C, Borralho N, Feij JA (2001). Successful fertilization and seed set from pollination on immature non-dehisced flowers of *Eucalyptus globulus*. Annal. Bot., 87: 469-475.

Van Wyk G (1977). Pollen handling, controlled pollination and grafting of *Eucalyptus grandis*. South Afr. J. For., 101: 47-53.

Varghese M, Ravi N, Kamalakannan R, Chris HE (2009). Effect of silvicultural treatments on growth, fertility and capsule traits in seedling seed orchards of *Eucalyptus camaldulensis* and *E. tereticornis*. New For., 37: 99–107.

8

Development of a low-cost tensiometer driven irrigation control unit and evaluation of its suitability for irrigation of lychee trees in the uplands of Northern Thailand in a participatory approach

Satid Pinmanee[1], Wolfram Spreer[1,2]*, Klaus Spohrer[2], Somchai Ongprasert[1,3] and Joachim Müller[1,2]

[1]The Uplands Program, Faculty of Agriculture, Chiang Mai University, Chiang Mai, Thailand.
[2]Institute of Agricultural Engineering, University of Hohenheim, Stuttgart, Germany.
[3]Department of Soil Resources and Environment, Mae Jo University, Chiang Mai, Thailand.

In upland areas of Northern Thailand, lychee (*Litchi chinensis*, Sonn.) is one of the predominant fruit crops. Fruit development takes place during the dry season (January until June). Therefore, high lychee yields can only be obtained under irrigation. As water is an increasingly scarce resource in the hillsides, strategies for more efficient water use are fundamental for a sustainable increase in agricultural production. An Automatic Tensio Control (ATC) unit was developed to use soil matric potential to open and close the inlet valve of a micro-irrigation lateral. In order to test the functioning of the ATC in the field, four different irrigation regimes have been tested in cooperation with local farmers on 10 year old lychee trees on a commercial orchard in Mae Sa Mai, close to Chiang Mai, Thailand. Testing the ATC on station and under field conditions, it has proven its reliability in maintaining a favorable moisture regime in the soil. The irrigated lychee trees responded to ATC irrigation with a higher yield and higher fruit weight compared to the other treatments. ATC works without electric components, so that it can be operated in areas without electricity. Furthermore, it is assembled from cheap and simple materials. Thus, it can be easily copied and adapted to different agro-ecological and socio-economic environments.

Key words: *Litchi chinensis*, automatic irrigation, appropriate technology, soil moisture.

INTRODUCTION

This study was carried out in the Mae Sa Valley, in the upland area North of Chiang Mai. Lychee (*Litchi chinensis* Sonn.), the predominant fruit crop of the area is mainly cultivated by farmers belonging to the Hmong ethnic minority. Fruit development takes place during the hot dry season (January until June). Therefore high lychee yields can only be obtained under irrigation. As water is an increasingly scarce resource in the hillsides, strategies for more efficient water use are fundamental for a sustainable increase in agricultural production. In the present, irrigation scheduling is often based on experience and observation of the soil moisture in the upper horizon. Furthermore, fields are scattered on steep slopes and are not accessed on a regular base. Thus, irrigation intervals are often sub-optimal to guarantee good supply of the plants without excessive water consumption. Irrigation control devices are still not commonly used in upland agriculture, especially in remote places without public power supply.

Nowadays, the use of soil-moisture sensors is the state or the art of optimizing irrigation scheduling and reduces water losses (Pathan et al., 2007). Automatic irrigation is widely used all over the world in intensive agricultural

*Corresponding author. E-mail: spreerwo@uni-hohenheim.de.

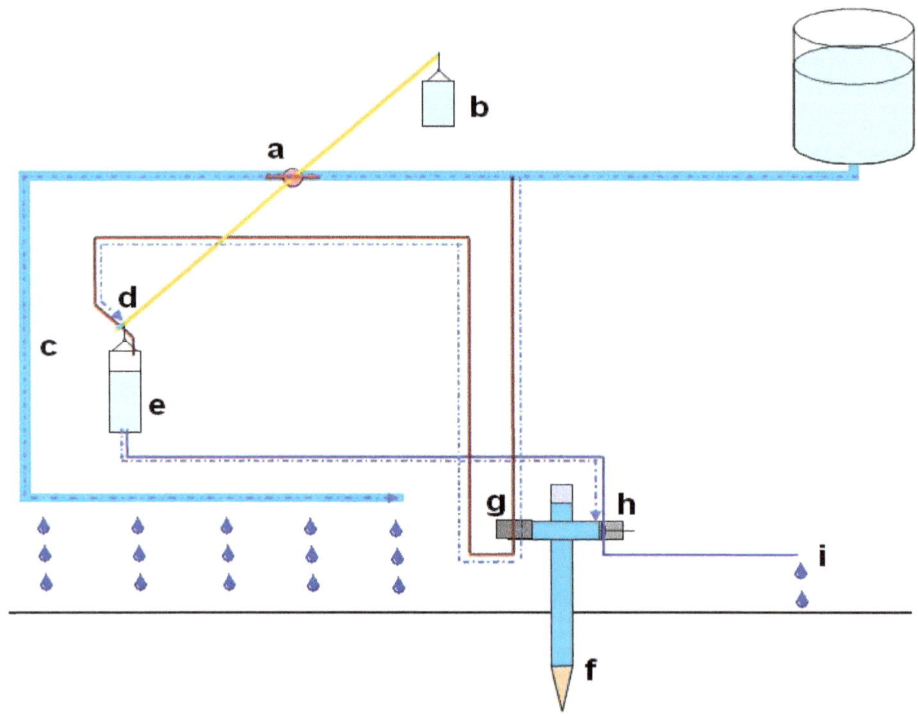

Figure 1. Schematic of the ATC functioning: (a)Operation valve, (b)counterweight, (c) main line and drip line, (d) overflow protection valve, (e) control cylinder, (f) tensiosensor, (g) inlet valve, (h) outlet valve and (i) outflow.

production. Automatic irrigation control units are based on measuring soil, climatic and plant parameters. In many cases models are employed to calculate the actual amount of irrigation water needed (Agulia, 2003). Tensiometers are frequently used as sensors that directly operate irrigation control units, because the plant water requirement can be directly attributed to the soil matric potential (Al-Amoud and Mohammed, 1995; Smajstrla and Locascio, 1996). Especially in drought stress sensitive annual plants like tomato tensiometer driven units guarantee an adequate supply of irrigation water (Munoz-Carpena et al., 2005). However, due to expensive sensors or data processing units, mostly powered by electricity, they are restricted to areas with access to the public electricity grid, or need a separate power supply which makes them more expensive and more vulnerable to malfunctions under field conditions (Chesmore, 2001),

Mechanical devices to control irrigation based on soil matric potential have been developed (Peterson et al., 1993; Meron et al., 2001). The pressure range determines the requirements to sturdiness and processing of materials. Fittings and pre-fabricated pieces are expensive or not readily available in developing countries. Possible wearing parts are difficult to purchase as spare parts.

The aim of this study was to develop a full-automatic tensiometer-based irrigation control unit, which is completely built from local materials and can be operated without electrical power supply. The device was tested to monitor its functioning under controlled conditions and on a farmer's orchard in the upland areas close to Chiang Mai, Thailand. After discussion with the farmers the device was compared to common place irrigation strategies, practised in the region. Yield and fruit quality analysis after one irrigation season were evaluated.

'Participation' was understood as the involvement of all individuals and groups who are directly and indirectly affected by our research activities (Neef et al., 2006). The research process was designed in a transparent way with the aim to increase the sustainability of research activities and improve their results. Farmers were consulted and invited to work out the problem statement together with the scientists. The testing of the device in the practise was done in close cooperation with the farmers and results were discussed. Thus, during the whole process of development, testing and evaluation a high degree of participation was achieved.

MATERIALS AND METHODS

The functioning of the Automatic Tensio Control (ATC) is displayed in Figure 1. The central functioning part of the ATC is the tensiosensor, which is introduced into the rootzone. This unit is connected to a lever which opens and closes the operation valve, mounted on the main line. A counterweight on one end of the lever keeps the operation valve closed. If the soil dries out and the soil matric potential in the rootzone surpasses 450 mbar, the inlet valve of the tensio sensor opens and water from the main line flows into the control cylinder. When the control cylinder is full with water, its weight presses the lever down. The operation valve opens and at the same time the overflow protection valve is closed, to prevent water from flowing into the control cylinder. In this position with the main valve open, water runs into the irrigation system. When the soil matric potential drops below 300 mbar, the outlet valve of the tension-sensor opens and water flows from the bottle into the outflow. When the control cylinder is empty, the counterweight closes the operation valve.

For on-station testing one unit of the ATC was installed in a concrete container of 1 m^3 volume, where one lychee tree was planted in substrate taken from a nearby orchard. The ATC was connected to a 16 mm PE pipe on which four drippers (Netafim JR8) were mounted in a complete circle 20 cm around the tree trunk. The set up was sheltered from rain by use of a transparent plastic roof. A pressure sensor was connected to the tension-unit of the ATC in order to monitor the soil matric potential. The main valve was connected to a 30 V power source by an electric contact, which closed a circuit, when the valve was open. Both soil-water-tension and time of opening of the valve were recorded by a datalogger in a 10 min interval.

For the field testing, four different irrigation regimes have been tested with 10 year old lychee trees on a smallholder farmer's orchard in Mae Sa Mai, close to Chiang Mai, Thailand, at 18.55°N, 98.53°E and an altitude of ca. 800 m a.s.l.. The heavy, loamy soil (Acrisol) and the steep slope of the experimental field well represent the prevailing agro-ecological conditions in the zone for which the automatic tensio control (ATC) was developed. Irrigation started after flowering of the trees in February and was continued until harvest in June. From March on several rainfall events made irrigation temporarily obsolete. All trees on the experimental plot were supplied by four pressure compensated drippers (Netafim JR8) mounted on 16 mm PE pipe. Four lines of seven trees were watered by one 100 L barrel each, located 10 m above the trees, filled from a water basin nearby through a gravity pipeline with a flow meter and a float switch at the barrel inlet. With a ball valve at the outlet of each barrel, all lines could be irrigated independently.

Local lychee farmers in the area predominantly belong to the Hmong ethnic minority. They mainly rely on their experience in scheduling irrigation and often underlie severe water shortage during the dry season. To ensure that the results would be meaningful to them, the treatments were designed considering the wishes expressed by the owner of the orchard. Thus, the following treatments have been applied

Tensiometer control

A tensiometer was installed at a depth of 30 cm within the rootzone of the first tree in the row. The farmer was advised to irrigate at a soil matric potential above 400 mbar and stop at below 200 mbar. Therefore, a manometer was mounted as reading unit and the respective values were identified by different coloured stickers. This setup was chosen as a locally available soil moisture monitoring devise as it is produced in Thailand.

Visual control

Irrigation was scheduled by the owner of the orchard following his traditional criteria, such as weather observation, manual and visual .appraisal of top soil moisture and leaf water status. Irrigation was restricted by overall water availability, as prevailed for the rest of the orchard. Due to the farmer's limited time, some irrigation was carried out by the author of this study in agreement with him.

ATC

One unit was installed at the first tree in the row, with the sensor cup at 30 cm in the rootzone. The operation valve was connected to the inlet of the lateral line to automatically irrigate within the above described range in soil matric potential.

Excess irrigation

Based on the owner's observation that a lychee tree close to his basin produced better yield at higher quality than other trees in his orchard, he requested to include an excessively irrigated treatment into the experiment. In this treatment the valve to the drip line was always kept opened, so that the barrel only served as small buffer and pressure stabilizer. However, water was always flowing, when available on the orchard.

Each monitoring tree was divided into eight equal sectors for monitoring yield and fruit quality. One sector was harvested per treatment and complete yield was determined by a mechanical scale. The water use efficiency (WUE) was calculated as the quotient of yield (kg) and total water consumption (m^3). Due to the rainfall WUE could not be attributed to irrigation water only, as proposed by FAO (Doorenbos and Kassam, 1979). In each sector 15 fruits were randomly collected to determine fruit weight (electronic scale), size (vernier caliper) and water content (gravimetric method). Total soluble solids (TSS) content in the fruit flesh was measured using a portable digital refractometer (Atago® PR-32, Saitama, Japan).

RESULTS AND DISCUSSION

Figure 2 shows the functioning of the ATC under controlled conditions. Whenever the valve opened a 30 V current was measured. Depending on the evapotranspiration the irrigation intervals changed over the monitoring period. Whenever the soil matric potential amounted to more than 450 mbar, the ATC activated the main valve and irrigation took place. The graph shows the precise reaction of the control valve to the soil matric potential, keeping it within the predetermined conditions.

Also under field conditions the ATC worked without disturbances. The functioning of the unit was checked each time when the other treatments had to be irrigated, but no change in the setting had to be undertaken. Farmers positively responded to this observation. However, they considered further trials necessary, before they would depend on an automatic control device.

The amount of irrigation that was applied by the ATC was higher than in the two manually irrigated treatments. However, with 41.56 kg/ tree the calculated yield surpassed those treatments to the extent that WUE did not decrease. Interestingly, the overall supply to the tree considering rainfalls did not vary much between visual control and ATC with 2.79 and 3.39 m^3/tree, respectively.

It can thus be assumed, that the higher yield response to ATC was based on the temporal distribution of water availability.

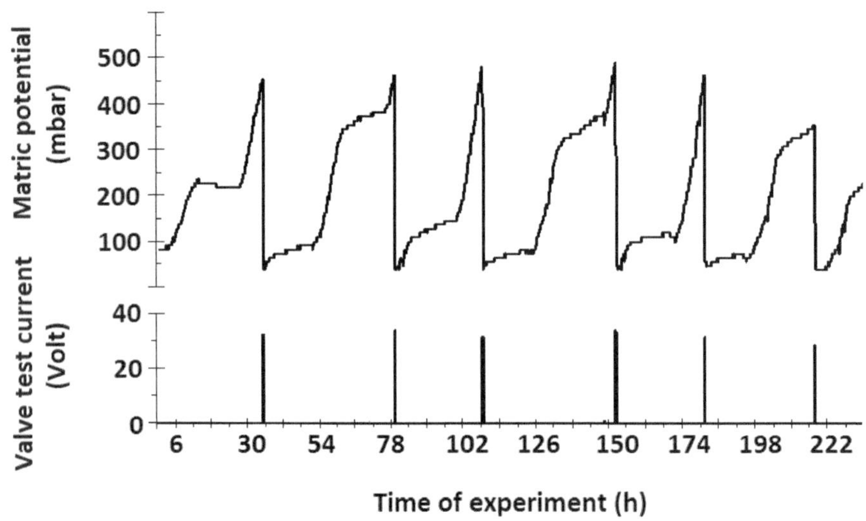

Figure 2. Monitoring of the ATC over one week. The soil matric potential is given in mbar. At the opening of the main valve an electric circuit delivered a test voltage of ca. 30 V.

Table 1. Water consumption, yield and water use efficiency of ATC as compared to farmer practise, tensiometer controlled irrigation and overirrigation in lychee trees. Same letters indicate no statistical difference at P < 0.05.

Treatment	Water availability (m^3 $tree^{-1}$)			Yield/tree (kg)	WUE
	Irrigation	Rain	Total		
Tensiometer	0.31	2.57	2.88	29.19 *bc*	10.18 *a*
Visual control	0.22	2.57	2.79	36.72 *ab*	12.94 *a*
ATC	0.82	2.57	3.39	41.56 *a*	12.27 *a*
Excess irrigation	2.19	2.57	4.76	25.66 *c*	5.34 *b*
$LSD_{0.05}$				0.54	3.37
SE				0.89	0.60

Table 2. Fruit quality parameters of lychees from ATC irrigated trees as compared to farmer practise, tensiometer controlled irrigation and overirrigation. Same letters indicate no statistical difference at P < 0.05.

Treatment	Avg. fruit weight (g)	Edible parts (% of fruit)	TSS in edible parts (°Brix)	Fruit water content (%)
Tensiometer	16.45 *b*	62.49 *b*	15.62 *b*	80.28 *b*
Visual control	16.35 *b*	62.82 *b*	16.60 *a*	83.36 *a*
ATC	17.80 *a*	64.23 *a*	15.25 *b*	82.76 *a*
Excess irrigation	16.84 *ab*	62.45 *b*	13.86 *c*	82.43 *ab*
$LSD_{0.05}$	0.98	1.61	0.75	2.21
SE	0.18	0.29	0.13	0.40

The excess irrigation treatment showed significantly lower yields lead, leading to an extremely low WUE (Table 1). The average fruit size and the share of edible parts were highest under ATC irrigation. However, these fruits had a rather low Brix value. With respect to those quality parameters, the conventionally irrigated (visual control) fruits had the most favorable properties: high water content with a high concentration of TSS (Table 2).

A possibly positive impact of light to moderate Drought stress on fruit quality has to be further investigated for lychee

CONCLUSION

Testing the ATC on station and under field conditions, it has proven its reliability in maintaining a favourable moisture regime in the soil. The irrigated lychee trees responded to ATC irrigation with a high yield. Even though the small sample size and the restricted possibilities for yield measurement under field conditions could put a question mark on the statistical relevance of the findings, the results illustrate the clear advantages of the ATC:

Automatic functioning: Without additional maintenance in the field, the only labour requirement is that for installation of the device. In further studies the actual labour saving effect has to be quantified and expressed as opportunity costs. Furthermore, it needs to be considered that when farmers have to survey irrigation, they tend to extend irrigation times in order to decrease the changeover times. This is an additional source of sub-optimal allocation of water.

Optimal irrigation water allocation: As compared to the traditional irrigation which is based on rather superficial observation of soil, plant and climate parameters in combination with the farmer's experience, the ATC relies purely on maintaining a constant soil matric potential in soil. It is therefore more precise and reacts more immediate to changes in the soil moisture. Even when soil moisture is controlled by measuring devices, e.g. tensiometers, fluctuations are generated due to the observation interval.

Among the mentioned positive effects, the full automatic functioning of the device was considered most important by the farmers. The lower labor requirement enables them to dedicate more time to other farm activities. However, even after one season of operation, farmers expressed concern about the reliability of the device and further tests have been requested for. The higher yield compared to the visual control, as traditionally practiced by the farmer, was not considered substantial.

As a conclusion, the ATC an automatic device has proofed its functionality for fruit tree irrigation. It works without electric components, so that it can be operated in areas without electricity. Furthermore, it is assembled from cheap and simple materials. Thus, it can be easily copied and adapted to different conditions. Further tests need to be carried out in order to evaluate its long-term reliability and its adaptability to different agro-ecological, as well a socio-economic environments.

ACKNOWLEDGEMENT

This work was financed by Deutsche Forschungsgemeinschaft (DFG) in the frame of the collaborative research program "Sustainable rural development in mountainous regions of Southeast Asia" (SFB 564).

REFERENCES

Agulia (2003). Development of an automatic irrigation control system for outdoor vegetable growing. Verlag Ulrich E. Grauer, Beuren, Germany. 3-86186-434-7 (in Germany).

Al-Amoud AI, Mohammed FS (1995). Automatic irrigation scheduling by remote controlled tensiometers under arid climatic conditions. Int. Agric. Eng. J., 4 (3), 131-145.

Chesmore ED (2001). Advances in electronic technology for sensing and monitoring in agriculture and horticulture. Acta Hortic., 562: 141-147.

Doorenbos J, Kassam AH (1979). Yield response to water. Food and Agricultural Organization of the United Nations, Rome, Italy.

Meron M, Hallel R, Bravdo R, Wallach R (2001). Tensiometer actuated automatic micro irrigation of apples. Acta Hortic., 562: 63-69.

Munoz-Carpena R, Dukes MD, Li YC, Klassen W (2005). Field comparison of tensiometer and granular matrix sensor automatic drip irrigation on tomato. Hort Technol., 15: 584-590.

Neef A, Heidhues F, Stahr K, Sruamsiri P (2006). Participatory and Integrated Research in Mountainous Regions of Thailand and Vietnam: Approaches and Lessons Learned. J. Mountain Sci., 3(4): 305-324.

Pathan SM, Barton L, Colmer TD (2007). Evaluation of a soil moisture sensor to reduce water and nutrient leaching in turfgrass (*Cynodon dactylon* cv. Wintergreen). Aust. J. Exp. Agric., 47: 215-222.

Peterson DL, Glenn DM, Wolford SD (1993). Tensiometer-irrigation control valve. Appl. Eng. Agric., 9(3): 293-297.

Smajstrla AG, Locascio SJ (1996). Tensiometer-controlled, drip-irrigation scheduling of tomato. Appl. Eng. Agric., 12(3): 315-319.

Efficiency of seed extraction methods in the control of loss and low seed quality in egusi melon (*Colocynthis citrullus* L.)

Ogbonna, P. E.* and Odo, L. C.

Department of Crop Science, University of Nigeria, Nsukka, 410001, Nigeria.

This study determines the effect of methods of seed extraction on seed loss and viability on Egusi melon (*Colocynthis citrullus* L). Three methods of fruit breaking; vertical cut with knife, horizontal cut with knife and hitting with wooden club were tested to determine their effect on seed damage. Three covering materials; grass straw, black polyethylene sheet and no cover were also tested to determine their effect on the number of days to fruit fermentation and seed germination. The result showed that percentage seed loss resulting from vertical cutting with knife was 16% and was significantly higher than 3.4% seed loss arising from horizontal cut with knife and 0% from hitting with wooden club. It was also discovered that fruits covered with grass straw took 8.67 days to complete fruit fermentation while under black polyethylene cover it occurred at 12 days. Fruits that were left uncovered fermented at 9.33 days. Seed germination was also significantly affected by covering materials. Seed germination of 85.3, 80.0 and 30.0 were recorded on seeds obtained from fruits that were fermented with no cover, grass straw and black polyethylene cover respectively.

Key words: Seed extraction, germination, seed loss.

INTRODUCTION

The Egusi melon *Colocynthis citrullus* L. is a member of the cucurbitaceae plant family (Jeffrey, 1980). It is a tropical crop grown for its seeds. It is a native of Africa and is very popular in most African country especially Nigeria. The seed is rich in oil and protein, containing high levels of most essential amino acids (Nwokolo and Sim, 1987; Ogbonna, 2007). The egusi seed after peeling is ground and used in preparing soup and stew. The seed is also roasted and eaten as snack. Fermented egusi seed also served as protein rich condiment for preparing soup. The oil when extracted is edible and is used in frying. Girgis and Said (1968), reported that the oil contains high percentage of unsaturated fatty acid and is recommended for diet intended to reduce high blood cholesterol.

It has been reported that Egusi melon requires less effort and input to produce and yields as high as 900 kg/ha has be recorded (Ogbonna, 2007). The crop is therefore a cheap source of protein and oil in human diet.

However processing of Egusi fruit to extract the seeds is very tedious and requires plenty water. The fruit is first, made to ferment to make the extraction of the seed easy. To quicken fermentation the fruits are split open and later covered, some times, with grass straw in the field. Techniques adopted in splitting the fruits vary. Seed damages are bound to occur during the breaking process. However the degree of damage differs with the methods. Seed loss during processing is a phenomenal issue in crop production. Losses in both quantitative and qualitative attributes of seeds have been reported (Desai, 2004; Sinh et al., 2002,). Against this background the objectives of this study are to determine differences in seed loss associated with different methods of breaking Egusi melon fruit and the effect of different covering materials on rate of fruit fermentation and seed germination in Egusi melon.

MATERIALS AND METHODS

This study was conducted at the Department of Crop Science Research farm, University of Nigeria, Nsukka in 2009. The area is

*Corresponding author. E-mail: ogbonnaptr@yahoo.com.

located on latitude 06°52' North and longitude 07°24' East and an altitude of 447 m above sea level.

Seeds of Nsukka Local cultivar of egusi melon obtained from the Department of Crop Science seed collection were used.

An area of land measuring 26.5 by 14 m was ploughed and harrowed and later prepared into seed beds. The seeds were sown in April 14, 2009 at the spacing of one by one metre. NPK 20:10:10 fertilizer at the rate of 200 kg/ha was applied to the crops at 4 weeks after planting. The plot was also clean weeded. At maturity, fruits were harvested from fifty plants selected at random. These were counted, weighed and recorded. The number of fruits/plant and weight fruits/plant were estimated by dividing by the number of plants (50).

Determination of proportion of seed losses due to methods of fruit breaking

The experiment was laid out in a complete randomized design (CRD) with six replications. The treatments include three fruit breaking methods; vertical cut with knife (VCK), horizontal cut with knife (HCK) and breaking by hitting with wooden club (HC). Eighteen hips of equal sized fruits were arranged according to the treatment design in a cleared area in the field. Each hip was made up of twenty fruits. Each treatment was replicated six times, meaning that each treatment was applied to six of the eighteen hips. This was done at random. Number of damaged seeds/fruit was obtained by counting the number of damaged seeds after fruit breaking in each hip. This was divided by twenty (20) which is the number of fruits/hip. Number of undamaged seeds/hip was also obtained by counting the seeds after processing. Number of undamaged seeds/ fruit was estimated by dividing number of undamaged seeds/hip by twenty. Total number of seeds/fruit was estimated by adding number of damaged seeds/fruit and number of undamaged seeds/fruit. Percentage seed loss was estimated from the above values.

Determination of effect covering materials on days to fruit fermentation and seed germination

The experiment was laid out in a completely randomized design (CRD) with six replications. The treatments comprised of three covering materials; grass straw (GC), black polyethylene sheet (PC) and no cover (NC). Eighteen hips of equal sized fruits were arranged according to the experimental design in a cleared area of the field. The fruits were cut horizontally into two halves with knife. Six hips were separately covered with grass straws. Another six hips were covered separately with black polyethylene sheets while the remaining six hips were not covered. The hips were inspected on daily basis. Number of days taken to complete fermentation was recorded for each hip.

After fermentation the seeds were extracted per hip and washed separately with copious quantity of water. The seeds were dried under the sun and later stored in black polyethylene bags. Ten seeds were taken from each of the eighteen seed samples for germination test. Eighteen plastic Petri-dishes were washed thoroughly and two layers of filter papers were placed in each of them. About 10ml of distilled water was poured on the filter paper. Each ten seeds sample was placed in a Petri-dish. Thereafter the Petri-dishes were placed on the laboratory bench. Percentage seed germination was obtained by daily count of the number of germinated seeds/Petri-dish up to ten days after plating the seeds. This was divided by ten (total number of seeds in a Petri-dish) and multiplied by 100.

Data analysis

Data collected were subjected to analysis of variance according to the procedure described by Steel and Torrie (1980). Zero values were transformed using the square root transformation method. Detection of differences between means was done using the F-LSD method by Obi (2007).

RESULTS

Records obtained from the fifty sample plants showed mean number of fruits/plant of 2.5. After processing, mean number of seeds/fruit of 168 and weight of seeds/fruit of 21.08 g were also recorded. Estimated seed yield per ha was 568 kg.

The result of the effect of fruit breaking methods on number of damaged seeds/fruit revealed significant difference ($p < 0.05$) between the methods. Highest number of damaged seeds/fruit was observed in vertical cutting with knife. This was significantly higher than number of damaged seed resulting from horizontal cutting with knife and hitting with wooden club. No seed was damaged when breaking was done with wooden club (Table 1). Weight of damaged seeds/fruit maintained the same trend (Table 2). Number and weight of damaged seeds/plant and percentage seed loss followed the same pattern (Tables 3, 4 and 5). The observation made on the effect of covering materials on days to fermentation revealed that covering the hips with grass straws after fruit breaking resulted to early fermentation. This was followed by leaving the hips bare (no cover). Covering with black polyethylene sheet significantly ($p < 0.05$) delayed fermentation when compared with other covering materials (Table 6). Covering material also significantly ($p < 0.05$) affected seed germination (Table 7). Highest percentage seed germination was obtained from seed processed from fruits fermented without covering. No covering, however, had statistically the same effect on seed germination with covering with grass straws. Seeds from black polyethylene covered hips gave the least seed germination after ten days.

DISCUSSION

The essence of breaking the Egusi melon fruits is to create easy entry for micro organisms that will initiate fermentation of the endocarp tissue for easy extraction of seeds. Makannjuola (1972) had noted that before deciding on the best method of pod breaking that it will be necessary to understand the physical and mechanical properties of the fruit as well as the geometrical arrangement of the seeds in the endocarp. Breaking the fruit by cutting vertically and horizontally led to seed losses. However seed loss recorded from vertical cut with knife was higher. This was due to the positioning of the ellipsoidal flat seeds at almost right angle to the vertical axis of the fruit. In this position, however only few seeds were injured by cutting horizontally.

Breaking by hitting with wooden club resulted to no

Table 1. Effect of methods of Fruit breaking on number of damaged seeds/fruit.

	Methods of breaking fruit		
Replications	**Vertical cut with knife**	**Horizontal cut with knife**	**Hitting with wooden club**
1	28.90	5.63	0.00
2	29.68	6.02	0.00
3	26.30	4.88	0.00
4	30.05	6.51	0.00
5	29.60	5.98	0.00
6	26.65	5.36	0.00
Mean	28.53	5.73	0.00

F-LSD ($p < 0.05$) for comparing treatment means = 10.50.

Table 2. Effect of methods of Fruit breaking on weight of damaged seeds/fruit (g).

	Methods of breaking fruit		
Replications	**Vertical cut with knife**	**Horizontal cut with knife**	**Hitting with wooden club**
1	3.61	0.71	0.00
2	4.01	0.86	0.00
3	3.12	0.56	0.00
4	4.21	0.93	0.00
5	3.78	0.81	0.00
6	3.53	0.63	0.00
Mean	3.71	0.75	0.00

F-LSD ($p < 0.05$) for comparing treatment means = 1.70.

Table 3. Effect of methods of Fruit breaking on number of damaged seeds/plant.

	Methods of Breaking Fruit		
Replications	**Vertical cut with knife**	**Horizontal cut with knife**	**Hitting with wooden club**
1	72.25	14.08	0.00
2	74.20	15.05	0.00
3	65.75	12.20	0.00
4	75.13	16.28	0.00
5	74.00	14.95	0.00
6	66.63	13.40	0.00
Mean	71.33	14.33	0.00

F-LSD ($p < 0.05$) for comparing treatment means = 35.60.

seed damage; rather the hitting is expected to further enhanced fermentation by softening the endocarp tissue. It will, however, be pointed out that cutting with knife appears to be less laborious and takes less time to achieve when compared with hitting with club. To avoid seed losses, breaking by hitting with wooden club is recommended. Okolo et al. (2002) had reported that bruising seeds during processing results to pathogenic deterioration of the seeds and suggested the design of an efficient melon deppoding machine. Covering of the broken fruits is made to hasten fermentation by creating an environment that will be conducive to the micro organisms to cause fermentation. Covering with grass straws was found to take the least number of days to achieve fermentation while covering with black polyethylene sheet, took the highest number of days to bring about fermentation of the fruits. The result appears to suggest that grass straws cover created a favourable environment for the micro organisms responsible for fermentation of fruits. Black polyethylene sheet on the other hand delayed fermentation because the environment was not very suitable for the micro

Table 4. Effect of methods of Fruit breaking on weight of damaged seeds/plant (g).

	Methods of breaking fruit		
Replications	**Vertical cut with knife**	**Horizontal cut with knife**	**Hitting with wooden club**
1	9.03	1.78	0.00
2	10.03	2.15	0.00
3	7.80	1.40	0.00
4	10.30	2.32	0.00
5	9.45	2.02	0.00
6	8.83	1.58	0.00
Mean	9.24	1.88	0.00

F-LSD ($p < 0.05$) for comparing treatment means = 5.33.

Table 5. Effect of methods of Fruit breaking on percentage seed loss (%).

	Methods of breaking fruit		
Replications	**Vertical cut with knife**	**Horizontal cut with knife**	**Hitting with wooden club**
1	16.36	3.68	0.00
2	16.78	3.82	0.00
3	15.92	2.88	0.00
4	17.80	3.88	0.00
5	16.54	3.78	0.00
6	16.20	3.26	0.00
Mean	16.60	3.40	0.00

F-LSD ($p < 0.05$) for comparing treatment means = 12.86.

Table 6. Effect of covering material on number of days to fruit fermentation.

	Covering materials		
Replications	**Polyethylene cover**	**Grass straw cover**	**No cover**
1	12.00	8.00	9.00
2	13.00	9.00	9.00
3	12.00	9.00	10.00
4	12.00	8.00	9.00
5	11.00	9.00	10.00
6	12.00	9.00	9.00
Mean	12.00	8.67	9.33

F-LSD ($p < 0.05$) for comparing treatment means = 2.64.

organisms. Since black colour absorbs heat, it is therefore likely that the black polyethylene cover absorbed and disseminated heat to the covered materials to a level that impeded the activity of the micro organisms. It may also be attributed to poor aeration of the covered fruits. If for instance aerobic micro-organisms were involved, their activities will also be reduced. The condition of the polyethylene covered fruits was so severe that hips that were left bare fermented earlier. It will also be pointed out that covering apart from enhancing fermentation is also aimed at preventing birds from feeding on the seeds. Maynard (2001) had reported that inherent seed qualities as well as the storage conditions are factors that determine the viability of seeds. Post harvest handling, processing and storage have been indicted in loss in seed viability (Desai, 2004; Sinha et al., 2002). Germination test measured in percentage seed germination has been used for quality assessment in seeds. This study has shown that fruit fermentation using grass straws cover and no cover

Table 7. Effect of covering material on percentage seed germination (%).

	Methods of breaking fruit		
Replications	Polyethylene cover	Grass straw cover	No cover
1	20.00	80.00	80.00
2	10.00	90.00	90.00
3	30.00	70.00	90.00
4	30.00	80.00	85.00
5	10.00	80.00	80.00
6	20.00	80.00	90.00
Mean	20.00	80.00	85.83

F-LSD ($p < 0.05$) for comparing treatment means = 42.20.

produced seeds with higher percentage germination than seeds from black polyethylene cover. This implies that covering with black polyethylene sheet must have reduced seed viability. This appears to confirm the earlier assumption that black polyethylene cover might have absorbed and disseminated heat to the covered fruits. Excessive heat probably reduced seed viability.

Seed losses and poor seed quality are problems associated with inappropriate and inefficient processing and storage methods. It is therefore recommended that melon fruit breaking by hitting with wooden club and covering with grass straws will reduce seed loss, enhance fruit fermentation and also ensure high seed viability.

REFERENCES

Desai BB (2004). Seed Handbook: Biology, Production, Processing and Storage. 2nd ed., p. 787.

Girgis P, Said F (1968). Lesser known Nigeria edible oils and fats. I. Characteristics of melon seed oils. J. Sci. Food Agric,. 19: 615-616.

Jeffrey C (1980). A Review of the *Cucurbitaceae* J. Linn. Soc., 81: 233-247.

Makanjuola GA (1972). Study on some of the physical properties of melon seeds. J. Agric. Eng. Res., 17(1): 128-137.

Maynard DN (2001). Water melon, Characteristics, Production and Marketing. Alexanderia VA ASHS press, p. 234.

Nwokolo E, JS Sim (1987). Nutritional assessment of defatted oil meal of melon (*Colocynthis cutrullus* L.) and fluted pumpkin (*Telfairia occidentalis* Hook) by chick assay. J. Sci. Food Agric., 38: 237-246.

Obi IU (1986). Statistcal Methods of Detecting Differences Between Treatment Means. SNAAP Press, Nigeria Ltd., p. 45.

Ogbonna PE, Obi IU (2007). Performance of Genotype, in Crosses of Egusi melon (*Colocynths Citrullus* L.). Agro-Sci., 6(1): 17-24.

Ogbonna PE, Obi IU (2006). Effect of poultry manure and planting date on the growth and yield of Egusi melon (*Colocynthis Citrullus* L.) in the Nsukka plains of South Eastern Nigeria. Samaru J. Agric. Res., 16: 74-76.

Okolo SA, Agun BJ, Jimoh AA (2002). Design and fabrication of melon washing machine. World J. Biotechnol., 3(2): 481-486.

Sa-Nguansak T, Suchada V (1999). Minimizing quality losses during maize seed processing in postharvest management. In sustainable technology development in crop production Dentscher Tropentag 1999 in Berlin.

Sinha JP, Vishwakarma MK, Atwal SS, Sinha SN (2002). Quality improvement in rice (*Oryza sativa L.)* seed through mechanical seed processing. India J. Agric Sci., 72(11): 643-647.

Steel RGD, Torrie JH (1980). Principles and Procedures of Statistics 2nd ed. McGraw-Hill Book Co.Inc. NY., p. 633.

10

Deterioration of Acacia in western Butana plain, Sudan

Samir Mohamed Ali Hassan Alredaisy* and Mohamed Ahmed Haj Ali Zubair

Department of Geography, Faculty of Education, University of Khartoum, 406 Omdurman, Sudan.

Acacia of the western Butana plain of the Sudan has undergone severe deterioration during last decades as demonstrated by a survey carried out in 2005. The majority of Acacia is non-succulent which are differing spatially, by soil type and in cover; frequency; abundance and density. There was a climax for Acacia in the study area prior to 1850, while its deterioration took place through four successive stages. The ecosystem carrying capacity of the study area was adequate to support Acacia up to the year 1900 when human exploitation of Acacia was balancing with its growth up to that year. Ecological disequilibrium started directly after 1970 and manifested by declining ecosystem carrying capacity and thus deterioration of Acacia. This deterioration of Acacia appears to be linked with the reduction in average annual rainfall. This paper demonstrates that, though important, this factor is not the sole, and may not even be the main factor involved. Government investment policy and associated population activities including, for example agricultural expansion which contributed by 40% in Acacia deterioration and fuel wood by 30%, are shown to have great significance, and cast doubt to any belief that a return to higher rainfall levels would reverse the current deterioration processes.

Key words: Acacia, deterioration, Butana, rainfall fluctuation, Sahel zone, rainfed agriculture human factors, community education.

INTRODUCTION

Acacia is a Greek word means "thorns", which is a Genus for trees and shrubs of Mimosaceane family. They are drought resisting trees where there are more than 300 types of Acacia in Australia, 1000 types in America while there are more than 400 types in Africa and Asia with restricted types in arid and semi arid parts. Many types of Acacia exist together such as Simaroubaceae, Capparideceae and Tamaricaceae and Rhamnaceae. They are identical and sensitive to climate and soil factors and therefore, they occur in forms of dense or sparse forests, except those ones subjected to human interference. There are many types of succulent; non-succulent and evergreen perennials Acacia. Acacias generally grow in the range of 200 to 600 mm rainfall with different adaptations. Succulent Acacia store water in their stems for times of water deficiency while the non succulent Acacia extend their roots deep into the soil searching for water, and the evergreens have adapted to resist drought (Walton, 1979; Hills, 1966; Clarke, 1954).

Acacia species generally took the umbrella shape. They used to grow in whatever topographic characteristic were, but due to change of climatic conditions they mostly confined to depressions where water is expected.

Acacia species are mostly found in arid and semi arid parts of the world, but the most suitable places for their distribution lie between 10° north and south of the Equator up to the Cancer and Capricorn. Recent studies (Mensching, 1988; Kamur, 1995) show that the home of Acacia is Sudan, Nigeria, Mali, Pakistan and India. In Africa it extends from the Senegal to Red sea across the Sahelian and Soudao zones. Generally, adaptation methods of Acacia depend on rainfall amount, temperature, topography, soil type and plant abundance.

Many studies confirmed the ecological benefits of Acacia in the improvement of soil fertility, soil protection and retention of water as well as reducing erosion effects and heat. *Acacia albida* raises soil fertility and increase Bulbrush millet (Dukhun) and sorghum vulgarize (Dura) production in the Gezira plain of the Sudan (Obeid and Sif Eldin, 1970). Also it increases nitrogen content of soil and cation exchange by 50% (Dancette and Poulain, 1969).

*Corresponding author. E-mail: samiralredaisy@yahoo.com.

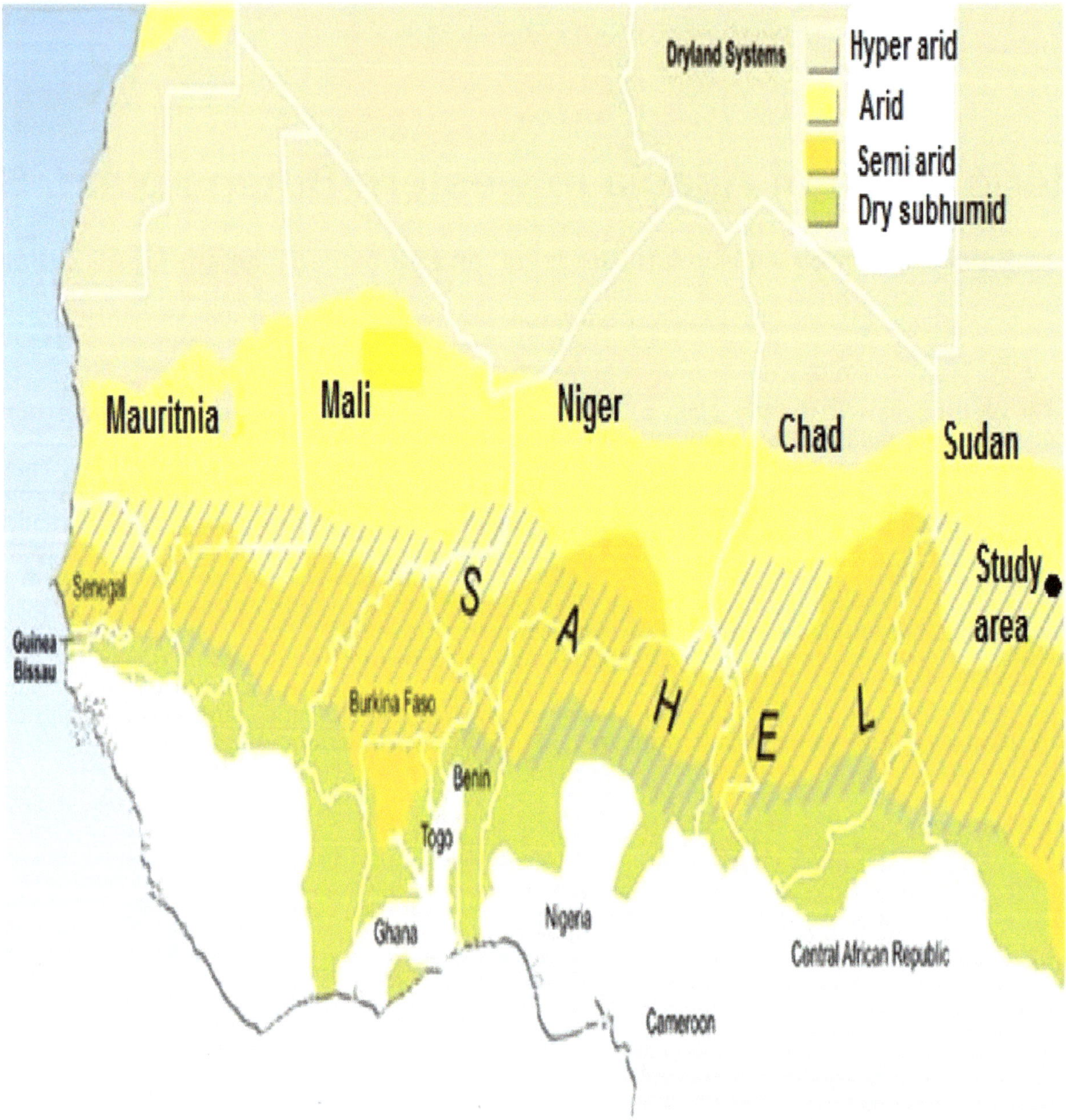

Figure 1. Dry land system and location of the study area in the Sahel.

The present investigation assesses spatial variability of Acacia deterioration and the contribution of natural and human factors that resulted in that deterioration in western Butana area. This deterioration suggests a rundown of the of natural resources in the region. Furthermore, this area was classified by the UN Conference of 1977 as being under "very high risk" of desertification (Figure 1). This paper examines deterioration of Acacia that has occurred within this region by looking at the results of a survey carried out in 2005 and the notification of studies covering the period from 1928 to 1999.

DATAS AND METHODS

The study area of western Butana of Sudan is remarkably lying between the Blue Nile in the west and Sahara desert in the north, at 14 to 16°N and 33 to 35°E. The area is chosen because it is one of the most hit areas by Acacia deterioration in Sudan. Physically, the study area is part of the Butana region which is a plain surface intermitted by dispersed hills covered with alluvium. Topography of Butana includes three major units. Firstly, highlands and isolated mountains in the southeast. Secondly, plain area dominating the area and characterized by clayey soil (45 to 80% clay particles) either flat or slow sloped. Thirdly, Wadis (valley) area including depositional areas around seasonal rivers like Atbara and Rahad. Most of the area is underlain by Basement Complex of Tertiary.

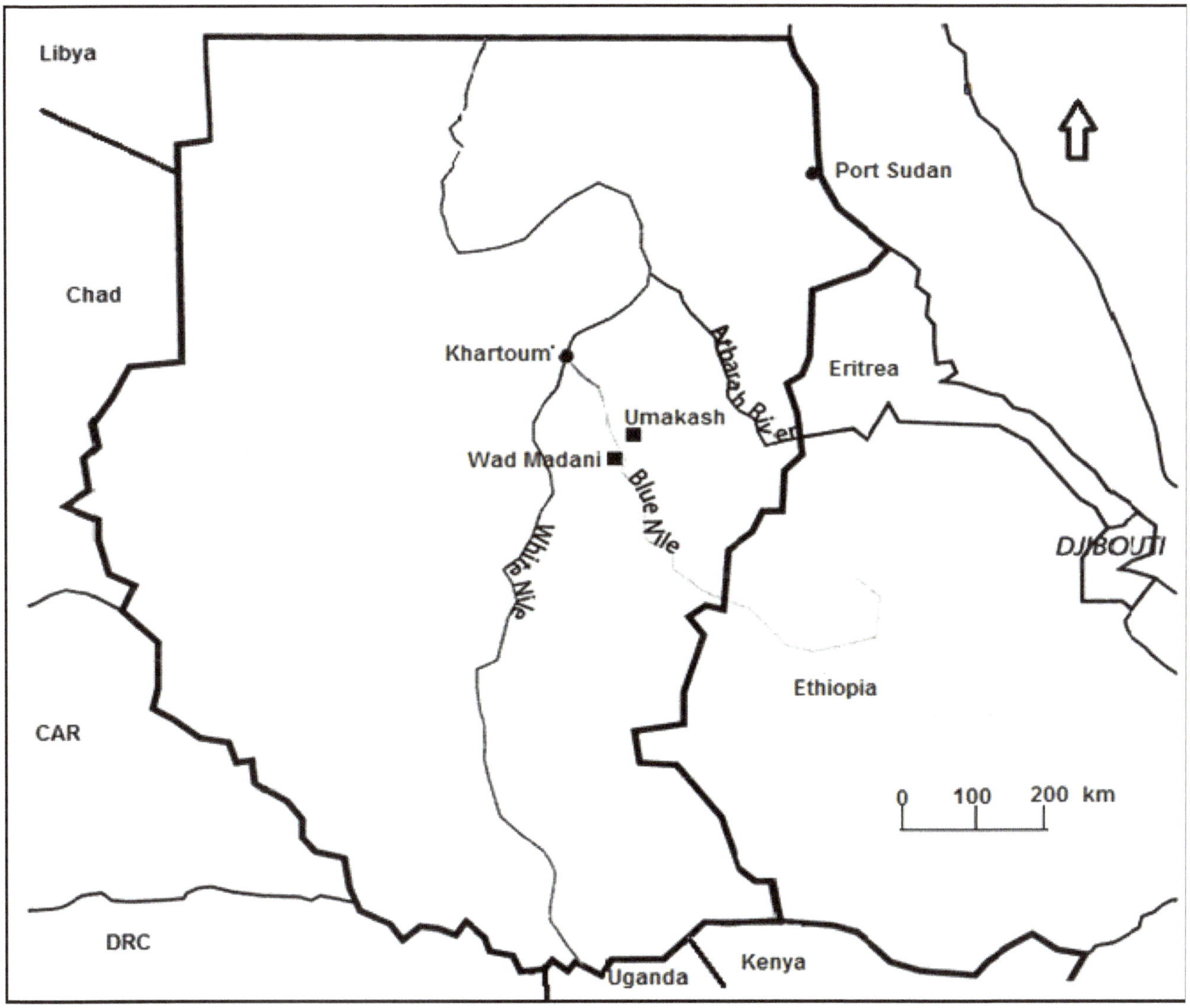

Figure 2. Location of the study area as indicated to by Umakash. Source: After Department of Survey, Sudan 2008.

Basalts both of which provide little water except in the detrital material around the occasional hills and small supplies to be found along joints in the rock (Davies, 1964). Two distinctive climatic belts are found in Butana area. The first one is semi arid climate found in the north and northwest and characterized by summer seasonal rains during July to October. The second one is a wet climate found in the eastern and southern parts of the state with average rainfall of 500 to 900 mm/year and maximum mean temperature of 47°C.

This is an experimental study designed to trace Acacia deterioration in western Butana plain of Sudan. The type of the experiment is integrated including Acacia and human data. Data for Acacia is collected by dividing the study area into three main zones indicated to by A, B and C letters according to density and diversity of Acacia and location with respect to the Blue Nile (Figure 2). Each zone is divided into squares. 5 to 7 squares are chosen for each direction line, with distance interval of 2 to 3 km. The midpoint at which the eight major and minor directions meet together is specified at the village of Um Akash (Figure 2), from which data are collected towards that direction. The area of each square is 10,000 m^2 (100 × 100 m). Then vegetative data is collected following laws for arid lands fieldwork on vegetation survey which requires that, for each plant type in every square number of species for every plant type; % of vegetative cover for each plant type; height of a plant; % of frequency; type abundance; type density and number of plants should be satisfied. Following Walton (1979), Hills (1966) and Clarke (1954), Acacia cover in the study is classified into succulent perennials, non-succulent perennials and evergreens.

Data for soil samples are done by collecting 21 samples from 7 sites (3 for each site). Composite samples are done by collecting sufficient number of sub-samples supposed to be reprehensive to the original population. These sub samples are mixed together to produce one composite homogeneous sample. The stage of judgment sampling is done from Ahamda forest from areas ranging into Acacia density from dense through sparse to bare of Acacia, with soil depths in ranges of 0 to 15 cm and 15 to 30 cm to collect 30 samples from 15 sites. Five soil properties were investigated including pH; E.C; O.C; Na and Ca and the physical properties of clay – silt- sand.

Data for human factors is collected through a sample size of 200 individuals, distributed into 12 villages representing 13% of total number of villages in the study area and also representing 10 to 15% of total population in each village. In addition, interviewing

Table 1. Occurrence of Acacia by zone and soil type; cover; frequency; abundance, density and life cycle in western Butana plain, Sudan.

Local name	Latin name	Zone			Soil type			Cover, frequency, density and abundance by zone												By total area			
								Cover			Frequency			Abundance			Density						
		A	B	C	Clayey	Sandy	Loamy	A	B	C	A	B	C	A	B	C	A	B	C	Cover	Frequency	Abundance	Density
Sunut	*A. nilotica* ^	+	-	-	-	-	+	2.8	-	-	18.1	19.3	-	3.5	-	-	0.6	-	-	1.5	4.2	3.5	0.04
Sial	*A. tortillis vor raddiana*^	+	+	+	+	+	+	24. 1	5.9	23.5	90.9	0.8	41.8	5.9	1.4	4.6	5.4	0.2	1.9	19.3	24.6	4.5	1.9
Talieh	*A. seyal vor fistula*^	+	-	-	+	-	+	1.6	-	-	18.1	0.7	-	2.0	-	-	0.4	-	-	0.8	4.2	2.0	0.04
Kitir	*A. mellifera*^	-	-	+	-	-	+	-	-	3.0	-	-	25.0	-	-	1.0	-	-	3.0	0.7	6.4	1.0	0.06
Hashab	*A. seyel vor sayal*^	+	-	+	-	+	-	-	-	-	-	0.4	-	-	-	-	-	-	-	-	-	-	-
Haraz	*A. albida**	+	-	-	-	-	+	0.8	-	-	18.1	8.0	-	1.0	-	-	0.2	-	-	0.4	4.2	1.0	0.04
Laaot	*A. nobica**	+	+	-	+	+	+	9.8	4.2	8.2	63.6	-	16.6	3.4	0.6	4.0	2.2	0.2	0.7	8.0	25.5	3.0	0.8
Samur	*A. tortillis vor tortillis*^	-	+	+	-	+	-	-	-	-	-	4.8	-	-	-	-	-	-	-	-	-	-	-
Higlig	*Balanites aegyptiaca**	+	+	-	+	+	+	4.0	10.2	-	63.6	-	-	1.4	1.2	-	0.9	0.5	-	4.8	36.2	1.3	0.5
Sarah	*Maerua crassifolia*^	-	+	-	-	+	-	-	-	-	-	2.3	-	-	-	-	-	-	-	-	-	-	-
Tundub	*Capparis deciduas*^	-	+	-	+	-	-	1.6	2.5	2.0	18.1	0.6	16.6	2.0	1.5	2.0	0.4	0.1	0.3	2.3	8.1	1.8	0.2
Sidir	*Z. spina Christi"*	+	-	-	-	-	+	-	1.7	1.0	-	3.0	8.3	-	1.0	1.0	-	0.08	0.08	0.6	6.4	1.0	0.06

Life cycle: * Succulent, ^ non Succulent , " evergreen.

with eldest farmers and herders to collect data concerned with past vegetation in the study area, as well as some government officials are carried out.

Measurement methods included quantitative methods to calculate Acacia coverage, frequency, density and abundance.

Frequency (%) = no. of squares for one type (A) × 100/total number of squares under study = number %.

Cover (%) = no. of one type × 100 / total number of Acacias = number %.

Density (type/ a square) = number of individual of one type (A) / total number of the studied squares = number of Acacia/square.

Abundance (type/square) = number of individuals of one type (A) /number of squares where that type (A) is found.

In addition, some descriptive statistical methods are applied on human data. Latin names equivalent with local names for *Acacia* species were obtained from Faculty of Forestry, University of Khartoum.

RESULTS

Spatial patterns

The majority of Acacia is non- succulent which are differing spatially, by soil type and in cover; frequency; abundance and density (Table 1). Zone A has the highest occurrence of *Acacia* species, followed by Zones B and C respectively. *Acacia tortillis vor raddiana* and *Acacia nilotica* dominate Zone A which lies close to the Blue Nile, forming closed forests such as Rufa'a, Dalawat, Hibaika and Ahamda forests. Zone B lies in the middle of the study area which exposed to heavily and extensive traditional rainfed agriculture where by now there is one Acacia tree species per five feddans (1feddan = 1.038 acres). They are mostly of Balanites aegyptiaca. Zone C has little Acacia occurrence, mostly of *Acacia mellifera.* Differences on Acacia occurrence are depicted by soil types (Table 1). Acacia is diverse on clayey soil and although of that, they are less occurring On sandy soil, Acacia is more diverse and. dispersive but, less denser. On the loamy soil, Acacia nilotica are abundant with few *Acacia seyel vor fistula*.

Table 1 also depicts cover; frequency; abundance and density of Acacia in the study area. Although these statistics are low in Zone A, excepting forests close to the Blue Nile, they are the highest compared to the other two zones. In Zone A, *Acacia tortillis vor raddiana* ranked high by 24.1% cover, 90.9% frequency, 5.9 abundance and 5.4 density. Fieldwork indicated that (Figure 3),

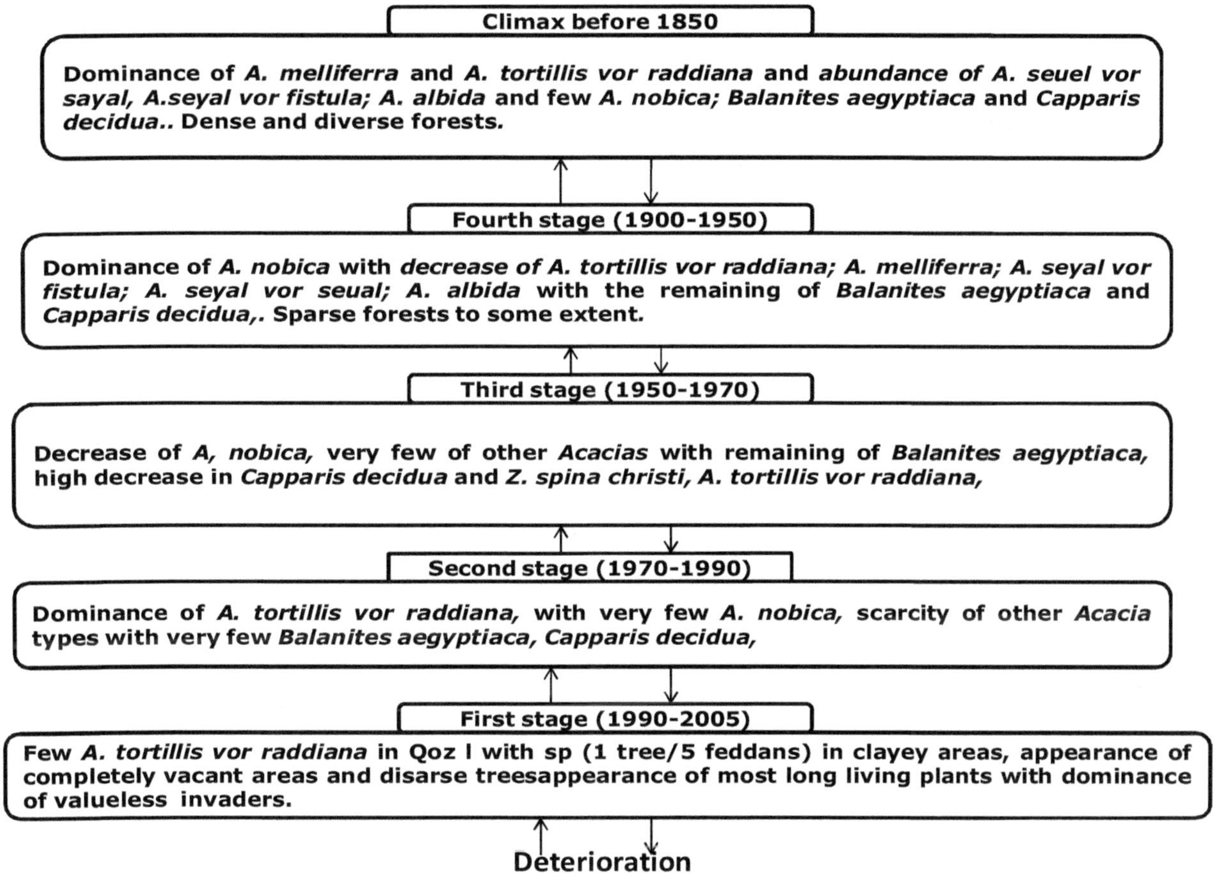

Figure 3. stages of Acacia deterioration in western Butana plain, Sudan.

Acacia trees which were dominating in the past in this zone were not only A. *tortillis vor raddiana,* but they were *Acacia nobica, A. albida, Balanites aegyptiaca* and all of *Acacia seuel vor seyal, Maerua crassifolia*, *A. tortillis vor tortillis*; *A. seyel vor fistula*, all of which are very few at the present time (Table 1). Zone B is dominated by B. aegyptiaca and *A. tortillis vor raddiana*, while the remaining Acacia trees did represent more than 5% covering. *A. tortillis vor raddiana* and *B. aegyptiaca*, *Ziziphus spina christi, A. seyel vor sayal, A. seyel vor sayal*, *Maerua crassifolia*, *A. nobica* and *Capparis deciduas* are very rare while they were dominant in the past (Figure 3). *Acacia* species dominating clayey and sandy soils is *A. tortillis vor raddiana* by 19.3% cover and 24.6% frequency. In zone c, *A. tortillis vor raddiana* dominates sandy soil by 23% cover, 41.3% frequency, 4.6 abundance and density by 1.9. It is followed by *A. nobica* with 8.2% cover, 16.6 % frequency, 4 abundance and 0.7 density. This zone had experienced *A. mellifera* in the past which formed forests while by now representing only 3% of the total Acacia cover. *A. seyal vor seyel, A. albida* and *A. seyel vor fistula* have disappeared. Also *Blepharis edulis, Arianthema pentandra, Cymbopogen nervatus* and *Pennisetum pdystachyum* have deeply deteriorated.

Deterioration of Acacia

Based on studies carried out from 1928 to 1999 and the fieldwork (2005), there was a climax for Acacia in the study area prior to 1850, while its deterioration took place through four successive stages (Figure 3). During the climax, abundant and diverse Acacia trees were dominating in forms of dense forests. During the fourth stage, from 1900 through 1950, *A. nobica* was dominating, while many other Acacia trees such as *A. tortillis vor raddiana* decreased although they were dominating during the climax and there seems to be a general decrease in the number of Acacia trees. During this stage, according to Richard (1928) as indicated by Harrison (1955), Butana area was very dense with varieties of forests and there were abundant *A. mellifera*, *A. nobica*, *A. seyel vor fistula*, *A. seyel vor seyel* and *A. albidia.* The dominant Acacia was *B. edulis* which is highly nutritive to animals. Such *Acacia* species were covering vast areas of Sufiaak; Raida; Jebel Mundara and Abaytor. Smith (1966) indicated that *A. melifera,*

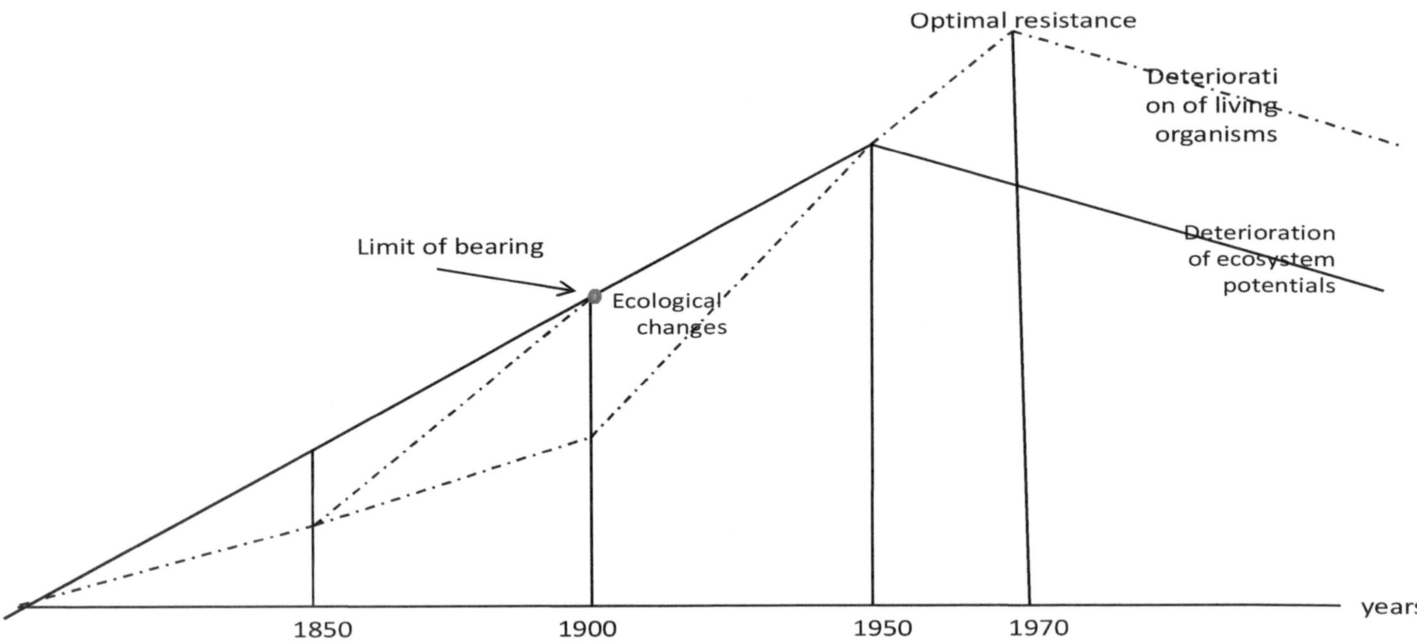

Figure 4. Ecosystem carrying capacity and Acacia resistance and deterioration.

A. nobica, A. tortillis vor raddiana, A, sayel vor seyal and C. deciduas as dominating and they were main sources for animal feeding.

During the third stage, from 1950 through 1970, still there was continuous decrease of *A. nobica* and other types of Acacia which became very few. Harrison (1955) mentioned that the dominant Acacia at that time was *B. edulis,* besides abundant *A. melifera*, *M. crassifolia, A. seyel vor fistula, C. deciduas.* During the 1960s, Lebon (1965) indicated the state of vegetation cover as very dense and diverse around Khartoum and Abu Delaig " close to the study area "where *A. tortillis vor tortillis* was dominating.

During the second stage from 1970 through 1990, there was one Acacia type dominating, while there was scarcity of major other types. According to fieldwork results, Acacia sharply deteriorated soon following 1970 where huge bare areas with wide sparsely trees formed the landscape. Bashar (1985) comparatively studied the statistics of Harrison (1955) and his own statistics in 1985, and concluded that Acacia had deteriorated in terms of density, cover and frequency following 1960. Butana became almost bare of trees where the remaining ones had confined to valleys and depressions, while *A. mellifera* occurred in plains and was replaced by *A. nobica* after being subjected to fire for agricultural purposes. Bashar (1985) further added that, trees which were replacing the old ones were unpalatable, where *Calotropis procra* had replaced *A. mellifera. A. seyel vor fistula* and *A.seyel vor seyel,* for example. Generally there is decline in Acacia by cover, frequency, abundance and density in clayey areas and dominance of low nutritive value *Acacia* species (Pflaumbaum, 1994).

These successive stages of Acacia deterioration (Figure 3) are expressed as ecosystem carrying capacity to support Acacia in western Butana plain (Figure 4).

The ecosystem carrying capacity of the study area was adequate to support Acacia up to the year 1900. Following that year, environmental changes spanning over 50 years had affected that ecosystem carrying capacity, up to 1950, by when that ecosystem carrying capacity collapsed, while Acacia continued to resist these ecological changes up to 1970. Since 1970 Acacia severely deteriorated.

Environmental and human factors

Ecological disequilibrium in the study area started directly after 1970 (Figure 4), and manifested by declining ecosystem carrying capacity and thus deterioration of Acacia. Environmental factors responsible for Acacia deterioration in the study area might include successive droughts, declining rainfall, retreat of isohyets and temperature rising. Climatic variability has an impact on the Acacia dynamics (Joaquín and Gabriel, 2000). High annual rain variability (ca.35%) and the very high evaporation rate, ca. 3 mm per day in the dry season were real in the study area (Farouk and Abu Sin, 1982). Rainfall ranges between 100 to 600 mm. However, average annual rainfall for the period 1922 through to1994 was 408.9 mm. During 1994, rainfall records in

Sahel rainfall from 1900 to 2007 averaged over June, July, August, September, and October

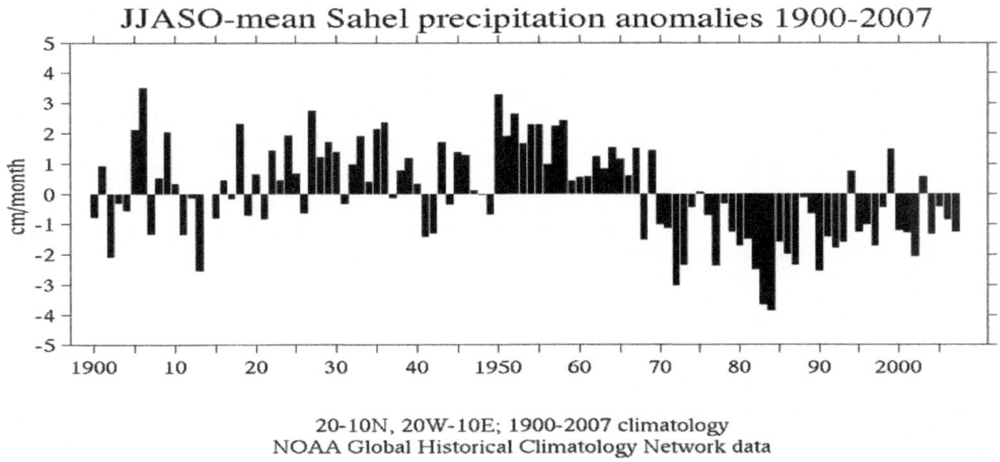

Figure 5. Precipitation anomalies in the Sahel, 1900 to 2007. Source: Stewart (2009).

some major stations in Butana area, for example, were 777 mm. in Gedarief, 669.5mm in Wd el Houri and 616 mm in Gedambalia while it was 600 mm in Hawata (Meteorology Office-Gedarief, 1994). Elagib (2010) studied the rainfall seasonality index (SI), precipitation concentration index (PCI) and modified Fournier index (MFI) for rainfall erosivity for the hyper-arid region of Sudan consisting of monthly rainfall measurements spanning over 60 years, 1945 to 2007, for three index meteorological stations, two on the Nile corridor and one on the Red Sea coast. The region is characterized by high year-to-year variability in rainfall leading to extreme seasonality/irregular distribution of rainfall over the year. Although prevalent diminishing rainfall amounts have been witnessed, there are marked tendencies for some months to become wetter, indicating changing intra-annual rainfall variability and thus monthly rainfall erosivity. No statistically significant trends were observed in rainfall seasonality and concentration during the common data period of 1945 to 2007. Cases of high and very high erosion powers were detected. A significant decreasing trend in erosivity is shown for one inland station (Elagib, 2010).

According to Hulme (1990) rainfall depletion has been most severe in semi-arid central Sudan between 1921 to 1950 and 1956 to 1985 annual rainfall has declined by 15%, the length of the wet season has contracted by three weeks, and rainfall zones have migrated southwards by between 50 and 100 km. This depletion has been due more to a reduction in the frequency of rain events rather than to a reduced rainfall yield per rain event. Ayoub (1999) compared long-term rainfall in four sub regions in Sudan and showed that rainfall decline had been in the magnitude of 30-40%. The western parts of the Sudan (Kordofan and Darfur) experienced extreme rainfall anomalies than the eastern and central parts (Gedarief "Butana" and Damazin), and had suffered greater periods of desiccation than the eastern and central parts. The decadal rainfall means showed below average rainfall for the last three decades in all these sub regions. Hulme (1990) assessed the role of an upper troposphere synoptic feature of importance in modulating surface rainfall over Sudan in the eastern Sahel. The Tropical easterly Jet (TEJ) provides an example of an inter-regional circulation feature linking the Sahelian and Southeast Asian monsoons and ultimately, perhaps, forced by ENSO-related anomalies in the Sahelian zone. Recent work by meteorologists and oceanographers (Stewart, 2009) has shown that much of the recent year-to-year changes in Sahel rainfall are forced by changes in sea-surface temperature in the Gulf of Guinea (on the equator near the prime meridian) and by El Niño in the Pacific. When the gulf is warm, the Intertropopical Convergence Zone shifts south away from the Sahel reducing the African monsoon that draws moist air into the Sahel. Longer term changes in rainfall from decade to decade are forced by changes in sea-surface temperature in the western Indian and tropical Atlantic oceans. When these areas are cool, Sahel rainfall increases (Figure 5). At the same time a period of severe drought led to large-scale environmental degradation.

Fieldwork results depicted changes in soil chemical and physical properties; appearance of compact salty soils, ultra alkalinity; poor organic matter content. There is increase in Na rate; Ph range was 8-8.5 tending towards more alkalinity; E.C at depth of 0-45 cm was 0 to 1.1, while at depth of 45-90 it was 0 to 1 millimoze. Organic Carbon was very low and there is an increase in Calcium. Topography of the study area is also changing, manifested by its distraction; increase of dust storms, damping of valleys and depressions; villages and water reservoirs and destruction of wild life.

As far as human factors responsible for Acacia deterioration are concerned, our fieldwork results revealed that in 1993, population of Butana were 1076430, density was 50.8/km^2. In 2005, average family size was 6.6 while population annual growth rate was 2.23%, birth rate was 37.7% and migration constituted 21% among the studied population. Age structure depicted children less than 15 yrs old as constituting 50.2%; adults as 44%, elders as 5.5% of the total population. Females were 46.3% and males 53.7%. This population consumes fuel wood which contributed by 40% for deterioration of Acacia in the study area according to our fieldwork survey. A family of 7 persons living close to the Blue Nile consumes on average 8 pounds/day of fuel wood when the weight of a tree is estimated at 75 pounds. This means that such a family consumes 3.3 trees/month or 40 trees /year.

On the other hand, a family of the same size living remote from the Blue Nile, when the weight of a tree is estimated at 50 pounds, consumes half of a tree per day, 15 trees /month or 180 tree/year. People particularly prefer *A. mellifera, A. nilotica* and *A. seyal vor fistula* for fueling. Stebbing (1972) mentioned that Rufa'a area supplemented Khartoum by 1000 ton/year of fuel wood in 1936 and increased to 5000 ton/year in 1956 of *A. mellifera*, *A. nilotica* and *A. seyal vor fistula.*

Concerning agricultural expansion, fieldwork survey revealed that it had contributed by 30% for Acacia deterioration. Agriculture started since 1940s as it was just estimated after the Second World War that of the 100000 tons of Dura marketed annually in the Sudan, 30000 tons came from the Gedaref "Butana" area (Jefferson, 1949). This diverted people from grazing to agriculture when rainfalls were abundant and high production created competition among people to open new agricultural lands far exceeded their affordability to cultivate, where more than 20 feddans/person was individual holding (Fieldwork, 2005). All these activities had cleared up Acacia from Butana area (Abu Sin, 1970). Similarly Grazing highly contributed by 20% for Acacia deterioration. For building of animal fences, which are renewed annually, people prefer thorny *A. mellifera* and *Z. spina christi* in particular and they also used to fill forage gab during summer by grazing and logging of Acacia branches, fruits, leaves and grazing on small shrubs and offshoots. Building purposes contributed by 7% for Acacia deterioration particularly Tamarix tenx "Athal" while use of Acacia for Folk crafts and for making of agricultural tools contributed by 3% for Acacia deterioration.

DISCUSSION

Several studies depicted the ecological deterioration of vegetation cover in Sudan where dense tropical forests were used to cover most areas of the country (Stebbing, 1972). The Sahara desert was flourished with animal and plant life where fossil trees, buried valleys are evident. Many clues are evident that tropical forests were near to Cairo up to mid Pleistocene. Walton (1979), Nichelson (1978), studied paleoclimate of the Sahelian zone and explained that there were warm, rainy and humid periods and there were also dry periods. Before 18,000 years ago, these areas witnessed a drought period which led to formation of sand dunes and drying up of lakes. In the last 12,000 years climatic conditions changed to heavy rains, while the last 3,000 years rainfall decreased which exposed lands to erosion and deposition processes when recent sand dunes are formed.

The three focal sub areas of western Butana plain are distinctly different in Acacia distribution by occurrence; soil type; cover; frequency; abundance and density, so the identification of the factors that characterize those places provides insight into environmental and human factors and help to target control and amelioration strategies. Based on the analysis, the pattern of Acacia distribution by zones in 2005 may indicate several different underlying processes. Firstly, zones may be near a common place for human exploitation, where high affected parts are in places with conditions that are especially well suited for that such as areas close to the Blue Nile where high population concentration is expected. Secondly, the Acacia deterioration may be related to past and concurrent socio-economic conditions that are reflected in agricultural expansion and fuel resources use. Thirdly, Acacia deterioration may occur where official government control was not sufficient.

In the first instance, the most affected areas by Acacia deterioration are located where people were close enough to an area abundant with Acacia. The type of environment in which people live is partly defined by the landscape and housing characteristics. Housing characteristics are a result of available domestic building materials produced locally and mostly of forests produce. Landscape of the Butana is that suitable for both irrigated and rainfed agriculture. Irrigated agriculture included all the schemes close to the Blue Nile and Rahad river such as Gunied Sugar production scheme which dates back to 1964 and the Rahad agricultural schemes which dates back to 1975. This type of agriculture had cleared up huge lands rich enough with Acacia for agriculture use. Mechanized rainfed agriculture which started during World War 2 was done similarly. In this rainfed sector, land ownership politicization dates back to Sudan's division by colonial administrators in 1923 into tribal homelands. The strong relationship between a tribe and its homeland has allowed the major tribes to use and monopolize the natural resources within their homeland. Competition over land for agricultural and exploitation purposes were further politicized by the 1970 Unregistered Lands Act (Ayoub, 2006). The legislation entitling the government to use force in safeguarding "its" land and encouraging the accumulation of land by a

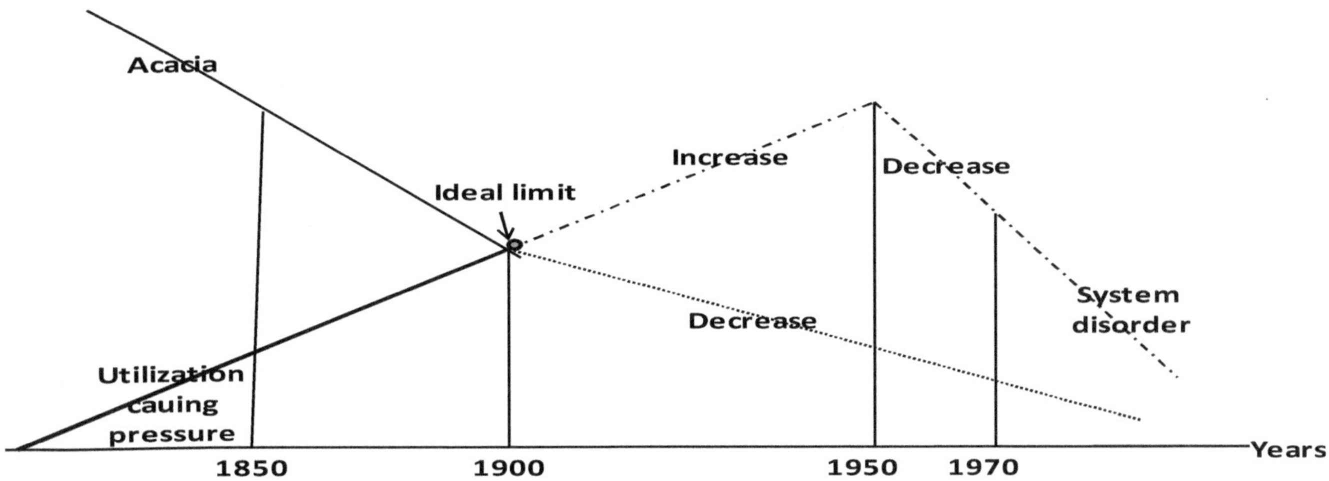

Figure 6. Human exploitation of Acacia versus its deterioration.

minority of rich local and foreign investors. The 1970 Act also enabled the government to implement a development policy based on the expansion of the agricultural sector, especially mechanized farming, and by 2005 the total area under mechanized farming had increased fifteenfold (Ayoub, 2006).

In addition, vast tracts of land in Butana area have been allotted to private capital investments since the 1990. Investment Act, substantially cutting rural communities' rights to land. Mechanized farming remains a major source of sealing off nomadic routes, water points and pastures, fostering a culture of land-grabbing. The processes behind this agriculture exploitation have led to extensive deterioration of Acacia in the region which is noticed by (Figure 6), which is further supported by Figure 4.

Human exploitation of Acacia was balancing with its growth up to the year 1900 (ideal limit). Following that year, human exploitation continuously increased till the year 1950, while Acacia started declining. Human exploitation decreased respectively after the year 1950, due to the general ecosystem disturbance where Acacia is no longer adequately available. As Acacia communities change, so will the associated micro-organism, fungus and animal species. Joaquin et al. (2000) explored the validity of three responses of Acacia to increased soil erosion, reduction of Acacia cover, number of species and reduced substitution of species in the intensely eroded Eocene marls of the Prepyrenees in north east Spain. Acacia degradation explained 48% of the species number variance. In the later stages of degradation a significant substitution of species was not observed, only a lower frequency of occurrence of several species that appeared in the whole set. Through the process of degradation, 47% of species displayed significantly reduced frequencies as degradation increased, none showed a significant increase in frequency. It is concluded that there are no characteristic species in these Acacia communities that survive in the severely eroded marls. Among the few species that had increased in frequency, most only colonized favorable micro-environments (Joaquin et al., 2000). The solutions for replacement of the cut off Acacia in Butana area were not always adequate to keep on with Acacia strategies for conservation. Closed Acacia and replanted Acacia forests were created suffered from a general inattention. These forests likely proved an excellent place for logging.

Social factors are very complex and are related to environmental factors. People without adequate assurance of income or who are unaware of the possibility that Acacia cutting and logging will lose their most precious wealth might be under poverty line or may be less likely to have adequate income. This may be reflected in the locations of most affected areas by Acacia deterioration and may play a role in the identifiable patterns and factors. The most disappeared *Acacia* species may be biased toward more areas of higher socio-economic populations. It is also possible that the social factors of education and occupation are only associated with the Acacia exploitation indirectly due to their correlation with environmental factors, or there is a real increased risk from behaviors that are linked to these two factors.

Social and environmental factors are interrelated making it more difficult to create a clear cut and concise limit in the study area. Population density is related to areas close to water sources, and higher income populations are more likely to live in urban places demanding for forests and animal products. The human demographic variables are related to general behaviors towards Acacia exploitation and deterioration. The specific importance of the background of Arab Nomads in the study area needs further exploration. By way of comparison, tree logging has also brought out a potential link to Arab Nomads.

Conclusions

The example of deterioration of Acacia in western Butana plain of the Sudan can be a showcase for whole the country. Deterioration of Acacia in Sudan has cut part of the GNP of the country and severely affected rural economies living there. It caused food gab, tribal conflicts, outside migration and changing ethnic and labor force structure of urban and agricultural areas of Sudan.

The introduction of irrigated and mechanized agriculture into the Butana area since World War 2 is a situation that should not be replicated in the future, unless good plans that consider conservation of the environment were put. The Government and local population have now adjusted measures to better control *of Acacia* species and coordination of efforts and the recognition of the importance of this coordination are positive outcomes of the awareness of Acacia deterioration. While the ongoing deterioration of Acacia is a risk to human economic benefits, the natural dangers by climatic change to human population most likely has increased and the susceptibility of ecological degradation has been increased either through drastic reduction in numbers of Acacia cut down annually, so the number in the near future will be replicated. Improved surveillance of *Acacia* species across the region as well as vigilance in reduction in Acacia bare grounds through re-plantation will have a significant impact on reducing the risk for deterioration. Places where agricultural development occurred during the rapid growth of the 1950s may benefit in particular from greater attention to reducing deteriorating habitats.

This analysis of the spatial patterns of Acacia and factors for its deterioration resulted in many further questions and firm conclusions. The environmental and human factors are the keys to pinpointing hotspots and will be the focus of future work. Because the systematic data collection needed to detect the timing, intensity and location of environmental and human activity did not occur across Sudan, this aspect of the ecology is limited. The assessment of Acacia deterioration cannot be made outside of the context of the rural environment in which it is present. It is clear that the propensity for the deterioration to be more present in some places is greater than for others, but the effect of control also is important. Further research in the response of Acacia to the environmental risk and a qualitative assessment of the environmental and social factors related to exploitation and control of Acacia should also be a matter of concern. Acacia logging control combined with appropriately targeted community education outreach are key factors to reduce Acacia deterioration. However, recognition of the continued development of coupled climate-Acacia models worldwide will facilitate the exploration of a broad range of global change issues, including the potential role of Acacia feedbacks within the climate system, and the impact of climate variability and transient climate change on the terrestrial biosphere where *Acacia* specie are significant for Butana and for Sudan which is famous with Gum Arabic, a precious produce of A. *seyel vor sayal* "Hashab".

REFERENCES

Abu SME (1970). The regional geography of Butana north of the railway. Unpublished M.A. of Geography, University of Khartoum. Sudan.

Ayoub M (2006). Land and conflict in Sudan.cr@c-r.org.

Ayoub TA (1999). Land degradation, rainfall variability and food production in the Sahelian zone of the Sudan. United Nations Environment Program. Nairobi. Kenya.

Bashar M (1985). some suggestions for curing the problem of desertification. J. Tasahour, 3: 12-24 Sudan.

Clarke L G (1954). Elements of ecology. John Wiley and sons. London.

Dancette C, Poulain JF (1969). Influence of *Acacia albida* on pedioclimatic factors and crops yield. J. Afr. Soils, 14: 140-187.

Davies HRJ (1964). An agricultural revolution in the African tropics, the development of mechanized agriculture on the clay plains of the Republic of Sudan. J. En Soc. Geografie, pp. 101-108.

Elagib NA (2010). Changing rainfall, seasonality and erosivity in the hyper-arid zone of Sudan. J. Appl. Climat., pp. l39-144.

Farouk DA, Abu SMA (1982). Water supply problems in the Butana region-central Sudan with special emphasis on Jebel Qeili area. GeoJournal, 6(1): 15-16.

Harrison MN (1955). Report on grazing survey of the Sudan. Library of the school of Oriental and Arican studies. Durham University, UK.

Hills ES (1966). Arid land geographical approach. Methuen and Co. Ltd. London.

Hulme M (1990). The Changing Rainfall Resources of Sudan. Transactions of the Institute of British Geographers, New Ser., 15(1): 21-34

Jefferson JHK (1949). Sudan's grain supply. Sudan notes and records, 30: 77-98.

Joaquín GC, Gabriel MM (2000). Effects of soil erosion on the floristic composition of *Acacia* communities on marl in northeast Spain. J. Acacia Sci., 11(3): 329-336.

Kamur GE (1995). Modern concept of ecology. Improved edition, India.

Lebon JHG (1965). Land use in the Sudan. Metropolis Press. London.

Mensching HG (1988). Land degradation and desertification in the Sahelian zone. In Whitehead E E,

Nichelson SE (1978). Climate variation in the Sahel and other African regions during the past five centuries. J. Arid Environment.1:3-34.

Obeid M, Sif Eldin A (1970). Ecological study of the vegetation of the Sudan, the effect of simulated rainfall distribution at isohyets on the regeneration of *Acacia Senegal (L)* wild on clay and sandy soil. J. Appl. Ecol. 8:203-209.

Pflaumbaum H (1994). Rangeland carrying capacity in the Butana. Animal research and development. 39. Institute of Scientific Cooperation. Germany.

Smith RL (1966). Ecology and field Biology. Harper and Row. New York.

Stebbing EP (1972). The creeping desert in the Sudan and elsewhere in Africa. Forest Department. Khartoum. Sudan.

Stewart R (2009). Oceanography in the 21st Century – An Online Textbook. URL:http// oceanworld.tamu.edu/ocean.401.

Walton K (1979). Arid Lands. London Press.

Biocontrol approach to management of greenpeach aphid *Myzus persicae* in garden peas for a sustainable ecosystem

Ochieng, S. O.[1] and Nderitu, P. W.[2]

[1]Jomo Kenyatta University of Agriculture and Technology, P. O. Box 62000-00100 Nairobi, Kenya.
[2]Department of Seed, Crop and Horticultural Sciences, Moi University, P. O. Box 1125-30100 Eldoret, Kenya.

Green peach aphid has been a significant pest of garden peas in Kenya for many years, because of its capacity to transmit viruses. Despite advances in integrated pest management, and frequent use of insecticides, the industry is still plagued by the insect. Adding to the problem is the fact that the peach aphid has proved to be resistant to various insecticides; there is a need to shift emphasis on biological control agents and softer chemicals. The aim of this study was to evaluate hard and soft chemical products in their ability to control *Myzus persicae* and their effects on aphid parasitoids. The treatments evaluated were dish washing soap with insecticidal properties, Teepol®, neem based insecticide, Achook® and a pyrethroid, Karate®. Efficacy was evaluated by taking aphid and parasitized aphid counts on 2 cm portion of the shoot tip/leaves of selected shoots before treatments were applied, once a week for four weeks. The treatments were applied at the concentrations recommended by the manufacturers. The three insecticides controlled the aphids. There was no significant difference in aphid numbers between the three insecticides. The population of aphids on the control was high. Except for the Karate treatment, the other two insecticides tested did not have an adverse effect on aphid parasitoid. Teepol and Achook promise to be useful agents for controlling green peach aphids in garden peas and at the same time are friendly to the aphid parasitoid.

Key words: Integrated pest management, *Pisium sativum*, parasitoid, *Myzus persicae*, biocontrol.

INTRODUCTION

Garden pea, *Pisum sativum* is a cool season annual crop produced worldwide for human consumption and animal feed. It originated in the near East and Mediterranean Regions, and has been grown since early Neolithic times (Food, Agriculture and Environment, 2003). In recent years, garden pea has become a major horticultural export in Kenya and represents 40 to 50% of vegetable production (Mrskos and Muehlbauer, 2000). It is consumed for its high protein content, vitamins and minerals such as phosphorus. In Kenya, garden peas are grown in Central, Eastern, Rift Valley and Western provinces. Major production constraints for peas include; adverse environmental conditions such as frost, drought and excessive heat, diseases such as powdery mildew, downy mildew, fusarium wilt and rot, crop pests such as thrips, leaf miner, and white flies. The green peach aphids are a major pest for peas (Kraft and Pfleger, 2001). Aphids can inflict various kinds of damages on a crop, nymphs and adults extract food materials from a plant leading to retarded growth. They extract large quantities of sap which reduces leaf photosynthesis, affecting production. Pathogenic viruses are also introduced through aphid feeding (Van Schett, 2003). The green peach aphid, *Myzus persicae* is probably the most notorious pest of garden peas, because of its wide host range, worldwide distribution and number of viral diseases that it vectors for example pea venation mosaic virus, pea leaf roll virus, pea streak virus and red clover vein mosaic virus (Kraft and Pfleger, 2001). Various methods have been attempted in the control of aphids. These include using a strong jet of water from a hose which dislodges aphids, but it has not been practical for the soft stemmed annuals

*Corresponding author. E-mail: silasochieng200@yahoo.com.

Plate 1. Arrangement of experimental plants.

(Pauline et al., 1992), use of ash mixed with water in vegetable kitchen gardens, intercropping of garden peas with lemon grass and onions but this has not been practical in large scale production and the use of both hard and soft chemicals. Soft chemicals include those with short-lived residues, have a natural origin either from plants, animals or even water, they are the least toxic chemicals. Hard chemicals are synthetic and have persistent residues. Insecticidal soap is also classified as a least toxic chemical since it has non-persistent residues. Soaps only control aphids present on the day they are sprayed, so applying non-persistent insecticides like soap may provide more effective long term control (Cranshaw, 1996). Biological control using parasitoids has also been attempted for example. In Washington; aphid- attacking parasitic wasps have been successfully used in the control of green peach aphid in potatoes (Pike et al., 1999, 2000). Up till now, parasitoids have been subjected to substantial toxic sprays in potatoes, which have limited their occurrence and activity in the crop (Pike et al., 2000). Various synthetic chemicals from different classes have been used in control of green peach aphid in potatoes. These include; pyrethroids: deltamethrin, permethrin, resmethrin; organophosphates: malathion, guthion; and carbamates: lindane and kelthane. Research shows that all these insecticides induced aphid outbreaks once multiple applications of insecticides from the same class were used. This was attributed to their selective killing of the predators and parasites that naturally keep green peach aphid population under control (Robert et al., 1998).

In the past twenty years, the concern has been to increase yield of vegetables by all means to meet the high demand regardless of the impact on the environment and sustainability of the agro ecosystem. Therefore, farmers rely on use of chemical pesticides to control aphid pest (Ushirika, 2007). In attempt to curb the pests' population with synthetic pesticides evidence of irrational use appeared and seems to be responsible for many environmental hazards and contamination risks of vegetable products (Abdelrahman et al., 1994). Repeated use of pesticides has led to pest resistance to pesticides, pest resurgence, pesticide substitution and lethal effects on non-target organisms including human as well as environmental pollution (Norris et al., 2003). Due to these limitations, there was need to find alternative control measures with different modes of action that would be effective, user and environment friendly.

METHODOLOGY

Experimental site

The research was conducted at Chepkoilel Campus in Uasin Gishu District, Rift Valley Province. Annual rainfall is between 900 to 1000 mm, an altitude of 2180 asl, longitude 35° 15'E and latitude 0° 30'N, annual mean temperature is 17°C. The experiment was conducted under a shade under temperatures varying from 22 to 32°C daily.

Experimental design and treatments

The experimental design was a completely randomized design (CRD) whereby there were four treatments with three replications (Plate 1). Treatments consisted of: dish washing soap, Teepol® with insecticidal properties, at a rate of 3 ml of Teepol in 1 L of water; botanical insecticide which is a neem product, *Azadiracta indica* Achook® 0.15% EC at a rate of 1 ml of Achook in 1 L of water and Chemical control using Karate® 1.75 EC at a rate of 1 ml of lamdacyhalothrin in 4 L of water and a control where no insecticide was

Plate 2. Clean Karate treated plants with no aphids nor parasitized aphids.

applied.

Planting

Planting was done in perforated polythene pots, each pot was filled with soil at a ratio of 2:1:1:1 of normal soil, murram, manure and sand respectively. Five seeds of garden peas were planted per pot (Plate 1).

Crop maintenance

Weeding was done twice; 2nd and 5th week after planting. Prevalent weeds were couch grass and Macdonald eye. Diseases were not prevalent, only damping off affected the plants at the early seedling stage; this was controlled by rouging of infected plants. Irrigation was done when required using a watering can. Top dressing using Easygro® foliar feed was done on the 2nd, 4th and 6th week after planting. This was done by dissolving 25 g of fertilizer in 20 L of water and spraying on the leaflets of the plant by use of a hand sprayer. Inoculation of aphids was done on the 4th week after planting. Aphids were collected from a garden pea farm (Can ken farm, Eldoret) by picking the leaflets that were highly infested and put them in a jam jar whose lid had a freeze net to let in air. Inoculation was done on the same day; approximately 20 aphids were introduced per pot by brushing them off from the leaflets to the various pots using a brush with soft bristles. The aphids were allowed 4 days for multiplication before application of the insecticides. Treatment application was done on the 5th week from planting, and there after on weekly intervals until the 7th week after planting.

Data collection

Data was collected on two plants per pot that had been tagged. The number of live aphids (small green insects) and parasitized aphids (cocoon inside the aphid cuticle that is hard, leathery brown to golden yellow casing known as a mummy or parasitized aphid) were recorded on 2 cm of the young growing shoot since aphids were densely populated at that point. This was done on all the pots in various treatments on a weekly basis starting from the 5th week till the 8th week after planting.

Data analyses

Data were subjected to 'analysis of variance' using the statistical package SPSS for windows 10.0. Means that were significantly different were separated using Duncan multiple range test. The significance differences were identified at 95% level of confidence.

RESULTS

Plants treated with Karate caused total mortality of aphids and they had no parasitization (Plate 2). There was high infestation of aphids on all the garden pea plants on the various treatments by the beginning of the 6th week prior to insecticide application, leading to high parasitization (Plate 4). Plants not treated with any insecticide were heavily infested by aphids on the seventh week after planting; the plants turned yellow to light brown, and then dried up by the 8th week (Plate 3). Another observation noted was that *Aphidius ervi* a parasitoid of the green peach aphid was seen flying in between garden pea plants showing compatibility of some insecticides with biocontrols (Plate 5). In general there was a significant difference in aphid numbers between the treated plants and the control (Table 1). There was no significant difference in the number of aphids on plants treated with Karate, Teepol and Achook. In the 6th week after planting, the various insecticides were sprayed. Teepol, Achook and Karate did not show significance in the reduction of aphid numbers but they differed significantly

Plate 3. High infestation of aphids on a plant in control.

from the control ($p<0.05$). However, Karate caused total mortality of aphids after the first spray of the insecticide (Plate 2). In the 7th week after planting, plants treated with Teepol, Karate and Achook had no significance difference in the number of aphids. In the 8th week all the treatments were not significantly different in their reduction of aphid numbers (Table 1). On the 5th week no insecticides had been sprayed, the number of parasitized aphids was not significantly different across the treatments. After insecticide application, the number of parasitized aphids was significantly different between control and the other treatments ($p < 0.05$), from week 6 to 8. Plants treated with Karate had no parasitized aphids (Table 2) (Plate 2).

DISCUSSION

In Kenya garden peas, *P. sativum* is an important vegetable crop both for local consumption and export produce. The most important pest for garden peas is green peach aphid (Kraft and Pfleger, 2001). These pest's population exploded as a result of the wide use of synthetic insecticides in the late 1950s (Dittrich and Ernst, 1990). In this investigation, soft and hard chemicals were evaluated for possible use in management of green peach aphid in garden peas. The evaluation of performance of a botanical insecticide Achook®, dish washing soap Teepol® and a pyrethroid Karate® was based on aphid and parasitized aphid counts in the treated plots as well as the control. In the 5th week aphid and parasitized aphid numbers were high and there was no significant difference between the treatments since no insecticides had been applied. On the 6th week aphid numbers reduced in plants treated with Teepol and Achook while those treated with karate caused 100% mortality of aphids (Plate 2). Due to the reduction in aphid numbers, parasitized aphids also reduced in numbers in plants treated with Teepol and Achook while plants treated with Karate had no parasitism. In the control aphid population increased, at these stage aphids were stabilizing and clinging on the plant since no treatment had been applied (Plate 4). Parasitized aphids on the other hand increased on the control, this finding was also reported by Norris et al. (2003) that parasitic wasp, *A. ervi* parasitizes aphids best when they are under high population (Plate 4). On the 7th week aphid numbers reduced in plants treated with Teepol and Achook while Karate treated plants had neither aphids nor parasitization. Aphid numbers on the control reduced due to the high numbers of aphids on the previous week that had caused extensive sucking of plant sap, plants turned yellow to light brown to dark brown then drying up by the 8th week (Plate 3). These findings are similar to those reported by Petit and Smilowitz (1982) aphid sucking causes water stress, wilting and retarded growth. Aphids started developing wings since plant sap had reduced and therefore there was a decline in the number of aphids on the 7th week. These findings are similar to those reported by van Schelt (2003) where aphids develop wings and migrated once there is reduced plant quality. Parasitization of aphids was at its peak on the 7th week since the previous week the aphids were at their peak hence the parasitoid is very active when the aphid population is high (Tables 1 and 2). On the 8th week aphid numbers were not significantly different across the

Plate 4. High aphid numbers with high parasitization.

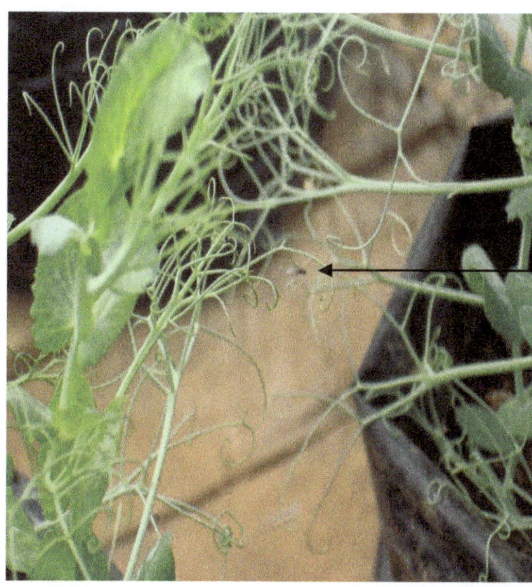

Plate 5. A parasitic wasp (*Aphidius ervi*) flying in-between the treatments.

Table 1. Effect of Achook; a botanical insecticide, Teepol; dish washing soap, Karate; a pyrethroid on green peach aphid (*Myzus persicae*) numbers over time.

Treatment	Weeks after planting			
	5	6	7	8
Control	76.67[a]	100.33[b]	34.67[b]	11[a]
Karate	86.33[a]	0[a]	0[a]	0[a]
Teepol	74.67[a]	18.33[a]	13.33[ab]	2[a]
Achook	106.33[a]	9.33[a]	2.33[a]	0[a]

Figures labeled with the same lower case letter indicate no significance while those with different letters indicate significant difference ($p<0.05$, Duncan multiple range test).

Table 2. Effect of Achook; a botanical insecticide, Teepol; dish washing soap and Karate; a pyrethroid on aphid parasitization over time.

Treatment	Weeks after planting			
	5	6	7	8
Control	11.67^{a}	30^{b}	42.67^{b}	35^{b}
Karate	7.33^{a}	0^{a}	0^{a}	0^{a}
Teepol	6.33^{a}	2.67^{a}	5.33^{a}	3.33^{a}
Achook	12.67^{a}	3^{a}	8^{a}	2.33^{a}

Figures labeled with the same lower case letter indicate no significance while those with different letters indicate significant difference ($p<0.05$, Duncan multiple range test).

treatments. Karate treated plants had no aphids while plants treated with Teepol and Achook had very low numbers of aphids however control plants were not significantly different from the insecticide treated plants. This was due to the reduction in plant quality and therefore aphids had developed wings and flown away and only a few remained. On the other hand Karate treated plants had no parasitism while Teepol and Achook had low parasitization, since aphid numbers were also low.

On the control, parasitized aphid numbers were significant in comparison with the treated pots, this was because the parasitic wasp does not depend on the plant quality directly however, parasitized aphid numbers had decreased by the 8th week due to low aphid numbers. Highest mortality of aphids was recorded on pots treated with Karate®, with no parasitized aphids (Plate 2). This shows that Karate® is highly effective in control of aphids but research shows that under repeated application of the chemical, it will lead to green peach aphid resistance. It also wipes out the natural enemies that provide long term control of the aphid (Flint, 1999). Recent research done on the active ingredient of karate, Lamda cyhalothrin states that the cotton aphid has already built resistance in cotton (Jackson, 2001). Dish washing soap with Insecticidal properties, Teepol® has also been found to be effective in control of aphids. It was less toxic to parasitoids; these findings are similar to those obtained by Cranshaw (1996). Insecticidal soaps are considered selective insecticides because of their minimal adverse effects on other organisms. Lady beetles, green lace-wings, pollinating bees and most other beneficial insects are not very susceptible to soap sprays. Research done to find environment friendly insecticides and harmless to non-target organisms to control *M. persicae* in peach orchards, stated that insecticidal soap reduced aphid population by 19.1% and it was not toxic to coccinellids; beneficial organisms of peach. A commercial based insecticide from neem product *A. indica*, Achook® was found to be effective in the control of aphids but there was no significance in the aphid number reduction com-pared to those in plants treated with Teepol and karate. Therefore neem is effective in the control of aphids, this finding are similar to those obtained by Nisbet et al. (1993). Use of a commercial formulation of Neem (RD-Repelin) successfully deterred aphids attempting to land, probe or oviposit on brinjals. Schmutterer (1990) reported reduced fecundity and longevity in aphids treated with Neem seed extract. On the other hand Achook® was least toxic to parasitoids; this was also reported by Copping (2001). Neem extracts are usually safe for beneficial organisms, such as bees, predators and parasitoids, mammals, and for the environ-ment. Azadiractin is reported to be relatively harmless to bees, spiders, lady beetles, parasitoid wasps and adult butterflies (Stark and Walter, 2001).

Conclusion

According to this investigation Neem based insecticide (Achook®) and Dish washing soap with insecticidal properties (Teepol®) can be good insecticidal options for Integrated Pest Management Program in the control of green peach aphid (*M. persicae*) in garden peas. They are selective, present a less negative impact on the ecosystem and works in association with biological control organisms.

REFERENCES

Abdurrahman AA (1994). Prospects of vegetables IPM implementation in Sudan. In: Dabrowski, Z.T (Ed.) Integrated vegetable crop management in Sudan ICIPE, Science press, Nairobi, Kenya 33.

Armstrong CM (1983). Working to Breed a Better Predator. New Alchemy Q., 14:14-15.

Booth DC, Conner GM, Smiley ET, Fraedrich BR (1990). Control of the crapemyrtle aphid Insect. Accarac.Tests, 15:331.

Boughton JA, White AJ, Capinera JL (2006). Effects of elicitor-induced plant resistance against arthropod pests in tomatoes.Econ. Entom. J., 65: 881-882.

Boydston R, Seymour M (2000). Volunteer potato outlook for 2000.Wash. Potato Commission Spud topics, 45 (15).

Copping LG (2001). The Bio Pesticides Manual. Second Edition. British crop Protection counsel.

Cranshaw WS (1996). Evaluation of peach aphid (*Myzus persicae*) infestations and insecticidal soap treatments on peach trees in the Great smoky mountains national park, North Carolina. USDA Forest Service, Forest Pest Management Report 85-3-1.

Dawson GW, Griffiths DC, Merritt LA, Mudd A, Pickett JA, Wadhams LJ, Woodcock CM (1990). Aphid semiochemicals—a review, and recent advances on the sex pheromone. Chem. Ecol. J., 16: 3019-3030.

Dittrich UK, Ernst GH (1990b). Chemical control and insecticide resistance in aphids. In: Gerling, D (ed), Aphids, their bionomics, pest status and management. Intercept Hants England. pp. 263-285

Flint ML (1999). Pests the Garden and Small Farm: A Grower's Guide to Using Less Pesticide, 2nd ed. Oakland: University of California Agricultural and Natural Resources Publication No. 3332.

Food Agriculture & Environment (2003). Food Agricultural Environment Panel of experts on Integrated Pest control on vegetables 5th Oct. 15-25 2003. Rome. Italy. FAO-UN meeting rep.

Georghiou RM (1990).Control of green peach aphid and potato leaf roll virus. University of Minnesota Extension Service and the State Seed Potato Programs of the States of Minnesota and North Dakota. Available at http://ipmworld.umn.edu/alert98.htm.

Gibson RW, Pickett JA, Dawson GW, Rice AD, Strible MF (1984). Effects of aphid alarm pheromone derivatives and related compounds on non- and semi- persistent plant virus transmission by *Myzus persicae*. Ann. Appl. Biol., 104: 203-209.

Jackson TM (2001). Detection and monitoring for resistance in arthropods based on biochemical characteristics, In: G. Georghiou and T. Saito [eds], Pest Resistance to Pesticides. Plenum Press, New York. pp. 99-116

Jacobson M (1990). Review of neem research in the United States. In: Locke, JC and Lawson, RH (Eds). Neem's potential in pest management programs. Proceedings of the USDA Neem workshop, Beltsville, M.D USDA, ARS 86, pp.4-14.

Kraft JM, Pfleger FL (2001). Compendium of pea diseases and pests, second Edition. American Phytopathological Society, St.Paul, Minnesota. p 67.

Lowery DT, Sears MK (1986). Effect of exposure to the insecticide azinphosmethyl on reproduction of green peach aphid (Homoptera: Aphididae). Econ. Entom. J., 79: 1534-1538.

Marco S (1993). Incidence of non persistently transmitted viruses in pepper sprayed with whitewash, oil, and insecticide, alone or combined. Plant Dis., 77: 1119-1122.

Marrkula M, Tittanen K (1997). Use of the predatory midge aphidoletes aphidimyza (*Diptera, Cecidomyiidae*) against aphids in glass house cultures. Proc. Symp. XV. int. cingr. Ent. Wash. D. C.US.D.A., Agri.Res.Serv.ARS-NE85: 43-44.

McLeod PJ, Steinkraus DC, Correll JC, Morelock TE (1998). Prevalence of *Erynia neoaphidis* (Entomophthorales: Entomophthoraceae) infections of green peach aphid (Homoptera: Aphididae) on spinach in the Arkansas River Valley. Environ. Entom. J. 27: 796-800.

Mrskos M, Muehlbauer FJ (2000). Survey of pea and faba bean cultivars. State Institute for Agriculture Supervision and Testing, Brno, Czech Republic. (In Czech).

Newberry GD, Thornton RE (2000). Response of natural enemies to the green peach aphid in different plant cultures. Environ. Entom. J., 10:375-378.

Nielsen DG (1990). Evaluation of biorational pesticides for use in arboriculture: J. Arboric., 16: 82-83.

Nisbet AJ, Woodford JT, Strang RC, Connoly JD (1993). Sytemic anti-feedant effects of azadirachtin on the peach-potato aphid *Myzus persicae* .Entomol. Exp. Appl., 68:87-98.

Norris RF, Caswel-chen EP, Kogan JD (2003). Concepts in Integrated Pest Management. Prentice-Hall of India, New Delphi. pp 586.

Palumbo JC, Reyes HJ, Mullis HT, Amaya CH, Ledesma AL, Carey L (2001).Neonictinoids and Azadiractin in lettuce: comparison of application methods for control of lettuce aphids. University of Arizona 2001 Vegetable Report.

Pauline RW, E1-Agamy FM, Haynes KF (1992). Susceptibility of the pea aphid to the insecticide; Imidacloprid and a predator in the presence of synthetic aphid alarm pheromone. Econ. Entom.J. 85:794-798.

Pike KS, Stary PT, Miller D, Allison L, Boydston, Graft G (1999). Host range and habitats of the aphid-parasitoid *Diaeratiella rapae* (M'intosh) (Hym, Aphididiiae) in Washington State. Environ. Entoml. J. 28:61-71.

Pike KS, Stary P, Miller T, Graft G, Allison D, Boydston L (2000). Aphid parasitoids (Hymenoptera, Braconidae, Aphidinae) of Northwest USA. Proceedings of the Entomology Society Wash. 102: (In press).

Pinese B, Lisle AT, Ramsey MD, Halfpapp KH, de Faveri S (1994). Control of aphid-borne papaya ringspot potyvirus in zucchini marrow (*Cucurbita pepo*) with reflective mulches and mineral oil-insecticide sprays. J.Pest Mgnt., 40: 81-87.

Petitt FL, Smilowitz Z (1982). Green peach aphid feeding damage to potato in various plant growth stages. Econ. Entom. J., 75: 431-435.

Powell DM, Mondor WT (1976). Area control of the green peach aphid on peach and the reduction of potato leaf roll virus. Am. Pot. J., 53: 123-139.

Robert S, David R, Radcliffe T (1998). Department of Entomology University of Minnesota, St.Paul.10 August, 1998 www: <http://ipmworld.edu/alert98.htm>.

Schmutterer H (1990). Properties and Potential of natural pesticides form the neem tree, Azadiracta indica. Ann. Rev. Entomol., 35:271-297.

Shean B, Cranshaw WS (1991). Differential susceptibilities of green peach aphid (Homoptera: Aphididae) and two endoparasitoids (Hymenoptera: Encyrtidae and Braconidae) to pesticides. Econ. Entom. J., 84: 844-850.

Stark JD, Walter JF (2001). Stage and age influence on susceptibility of *Coccinella septempunctata* after direct exposure to Neemix, a neem insecticide. Econ. Entom. J., 90(5):1102-1105.

van Schett HF (2003). Entomology Course; Morphology and Biology of Insects and Mites, Koppert Biological Systems, Netherlands.

Weinzierl R, Henn T (1991) Alternatives in Insect management: Biological and Biorational Approaches. North Central Regional Extension. Publication 401.

Wyman JA, Toscano NC, Kido K, Johnson H, Mayberry KS (1979). Effects of mulching on the spread of aphid-transmitted watermelon mosaic virus to summer squash. J. Econ. Entomol., 72: 139-143.

12

The effects of different concentrations of indole-3-butyric acid (IBA) on leafy stem cuttings of four tropical timber species

Egbe E. A.*, Chuyong G. B., Fonge A. B., Tata B. L. and Tabot P. T.[1]

Department of Plant and Animal Sciences, Faculty of Science, University of Buea, P. O. Box 63, Buea, Cameroon.

The effects of different concentrations of indole-3-butyric acid (IBA) on the rooting of leafy stem cuttings of *Albizia zygia*, *Blighia welwitschii*, *Lophira alata* and *Pterocarpus soyauxii* were investigated using non-mist propagators. Four hormone concentrations were used; 16000, 8000, 4000 and 0 μg ml^{-1} (control). Results indicated that *Albizia*, *Blighia* and *Lophira* cuttings rooted best when treated with 16000 μg ml^{-1} IBA (80, 100 and 40% respectively) and their control cuttings did not root. *Pterocarpus* stem cuttings treated with 8000 and 4000 μg ml^{-1} had 100% rooting while those treated with 16000 μg ml^{-1} and control was 80%. The highest root number was noted in cuttings of the various species treated with 16000 μg ml^{-1} IBA and there were significantly differences only from treatments of *Blighia* and *Pterocarpus* species. Stem cuttings treated with IBA at 16000 μg ml^{-1} also gave the fastest rooting time which ranged from 18 to 78 days. Stem cuttings of *Blighia*, *Albizia* and *Lophira* species treated with16000 μg ml^{-1} IBA also had the longest roots (2.48, 1.23 and 1.45 cm respectively). The longest root length of *Pterocarpus* was observed in cuttings treated with 8000 μg ml^{-1} of IBA (3.72 cm). Vegetative production of seedlings of these species is essential for reforestation programmes.

Key words: Timber species, indole -3- butyric acid, leafy stem cuttings, rooting time, percentage rooting.

INTRODUCTION

Tropical forests contain much terrestrial biodiversity, provide food, shelter, health care, protect water and soil resources store carbon in biomass and maintain the delicate composition of the atmosphere (Neuwinger, 2000). One very important material provided by forests is wood which is used in construction, as fuel, in making furniture and other implements. Tropical timbers are preferred as a source of wood because of their natural durability and good working properties (Miller and Wiedenhoeft, 2002). Despite these important uses, tropical forests are threatened by unsustainable land and resource use (De Capua, 2005). The greatest constraint to forest regeneration projects is the lack of good planting material (materials that are viable and free from pest and diseases). For example, the viability of *Terminalia* spp. is about 0.1%, *Balakata baccata* 25%, *Michelia baillonii* 31% and *Erythrina subumbrans* 39% (Mostacedo and Fredericksen, 1999). In a study on seed viability, seed of some tree species were found to loss viability rapidly; *Swietenia macrophylla* seeds were viable for only 6 weeks. Viability is also affected by storage conditions, thus, seeds of *Araucaria hunsteinii* will die if allowed to dry out (Blakesleya et al., 2002). Seed availability can be a problem in tropical trees. For example, *Triplochiton scleroxylon* trees only fruits once in 4 to 7 years (Howland and Bowen, 1977) and the seeds are only viable for 2 to 3 weeks while *Lophira alata* would fruit once every 5 or even more years (Bonner;1992). Many other tree species have a long period of juvenility before they become sexually mature and capable of flowering and fruiting, and this can have a range of 5 to 50 years. The tree species *Baillonella toxisperma* would start flowering at the age of 50 to 70 years but regular fructification would

*Corresponding author. E-mail: egbe1@yahoo.com.

start later when the bole is about 70 cm in diameter (Louppe, 2005). Furthermore, tree seeds typically suffer from pre-germination mortality caused by herbivores, diseases or other pathogens.

Another constraint to tree regeneration is pest attack. An example of this is the seedling growth of *Milicia excelsa* and *Milicia regia* that is seriously affected by gall-forming *Phytolyma* spp. (iroko gall flies) which kill the seedling apex (Ofori et al., 1996). Teak seeds are attacked by a number of fungi especially *Aspergillus* spp. and *Fusarium* spp. These fungi also caused both pre-emergence and post-emergence damping-off disease on seeds of *Albizia lebbeck* and *Rhizoctonia solani* (Mostacedo and Fredericksen, 1999). Thus, there are many reasons why it is important to screen some of these important species for their amenability to vegetative regenerate using juveniles stem cuttings. Vegetative propagation provides a unique opportunity of avoiding the problem of poor seed viability and recalcitrant seeds which are predominant in tropical tree species. It is also an assembly of germplasm for base populations to maintain certain desirable characteristics (Amri, 2010; Leakey and Simons, 2000). The four species used for this study were: *Albizia zygia* (D. C.) Macbr (small leaf), *Blighia welwitschii* (Hiern) Radlk, *Lophira alata* Banks ex C. F. Gaertn (Azobe or iron wood) and *Pterocarpus soyauxii* Taub (African Padouk or cam wood). They were selected based on commercial, conservational and cultural importance. *A. zygia* produces a moderately durable timber, is a nitrogen fixer, thus important in agro-forestry. Its wood is used for indoor construction, light flooring, furniture, canoes, implements, carving, veneer and plywood. It is commonly used as firewood and for charcoal production (Burkill, 1995). The wood pulp is suitable for paper production (Kpikpi, 1992). Its bark produces gum that is useful in the food, cosmetic and pharmaceutical industries (Odeku, 2005). In traditional medicine, bark sap is instilled in the eyes to treat oph-thalmia (Neuwinger, 2000). *B. welwitschii* (Sapindaceae) produces class four timbers and can be planted as an ornamental; thus it is a goods candidate for the greening of towns. Its seeds are used as fish poison and are not toxic to other organisms. Its bark is used in the treatment of measles (Neuwinger, 2000).

L. alata (Ochnaceae) produces a very durable timber that is highly commercialized (Laird, 1999). It is popular for heavy construction work, harbour works and railway sleepers (UNEP-WCMC, 1999). The species is used locally in traditional medicine, for treating backache, toothache, respiratory and stomach problems and yellow fever (Neuwinger, 2000). The leaves are used in termite control, and oil from the seeds is used as food and to make ointments and soaps (Laird, 1999). Its timber, commonly called "Azobe" is rich in silica, resistant to marine borer, and it has been classified as vulnerable on the IUCN Red List (IUCN, 2009). *P. soyauxii* (Fabaceae) produces a very durable timber and is classified as vulnerable on the IUCN Red List (IUCN, 2009). It is highly commercialized (Laird, 1999). The wood contains alkaloids and is also rich in tannins (Richter and Dallwitz, 2000). It is a preferred wood for furniture, carving, canoes, stools, musical instruments and agricultural implements (Laird, 1999). The bark has antifungal activity against some pathogenic fungi. The ground stem is of cultural and medicinal importance associated with childbirth and marriages (Laird et al., 1997). The low cost macropropagation is a solution to forestry and agroforestry projects and for the conservation of biodiversity especially threatened tree species.

MATERIALS AND METHODS

Study site

This study was carried out in the shade house (87.6% full sunlight, measured with digital Luxmetre (MS 6610 Mastech®) at the University of Buea, Cameroon. Buea is situated in the Southwest region of Cameroon between latitude 3° 57' and 4° 27' N and longitude 8° 58' and 9° 25' E (Tchouto et al., 1997). It has a humid tropical climate with a mean annual rainfall of about 2500 mm, most of which is between June and September (unpublished data). The mean annual temperature is 28°C, relative humidity average ranges from 80 to 86% and annual sunshine between 900 to 12000 h per annum (Ndam, 1998). The soil type is basically volcanic, rich in mineral content but due to continuous cropping, the soil fertility has reduced drastically.

Preparation of leafy stem cuttings and planting medium

Seeds of *B. welwitschii*, *A. zygia* and *P. soyauxii* were sown in seed boxes filled with top soil. The germinated seeds were pre-transplanted into black polybags filled with top forest soil as the potting medium. The seedlings of these species were raised in the nursery and the main stems were used for stem cuttings when they were 7 months old with mean height range of 45.5 to 65.8 cm depending on the species. Wildlings of *L. alata* that were about 1 year old were collected from the Limbe Botanic Garden. The mean temperature in the shade was about 22°C and the seedlings were watered three times a week. Leafy stem cuttings were made from firm but young stems in slants using a secateur. 64 stem cuttings of about 6 cm were collected from each of the four species. The number of stem cuttings per shoot of the seedling ranged from 4 to 6 and these were one node leafy cuttings. The uppermost part of the stem was cut and discarded. A total of 256 one node leafy stem cuttings were used and there were four replicates per treatment. The leaves on the stems were reduced with a scissor to less than half their sizes with an area of about 25 cm^2. This was to reduce the transpiration rate of the stem cuttings. Coarse sand was boiled at 100°C for 8 h and allowed to cool for 24 h in the laboratory. It was then used to fill small perforated trays. The rooting hormone Indole-3-butyric acid (IBA) was used and different concentrations were prepared as described by Leakey et al. (1990). The IBA formulation was prepared using 95% ethanol diluted to 50% with distilled water. The various concentrations were: 16000, 8000, 4000 and 0 µg ml^{-1} (control) 50% diluted alcohol only. The cut ends of one node leafy cuttings were dipped into the solutions according to the different concentrations and fan to evaporate the alcohol and thus increasing the concentration of IBA at the cutting base. A hole was made in

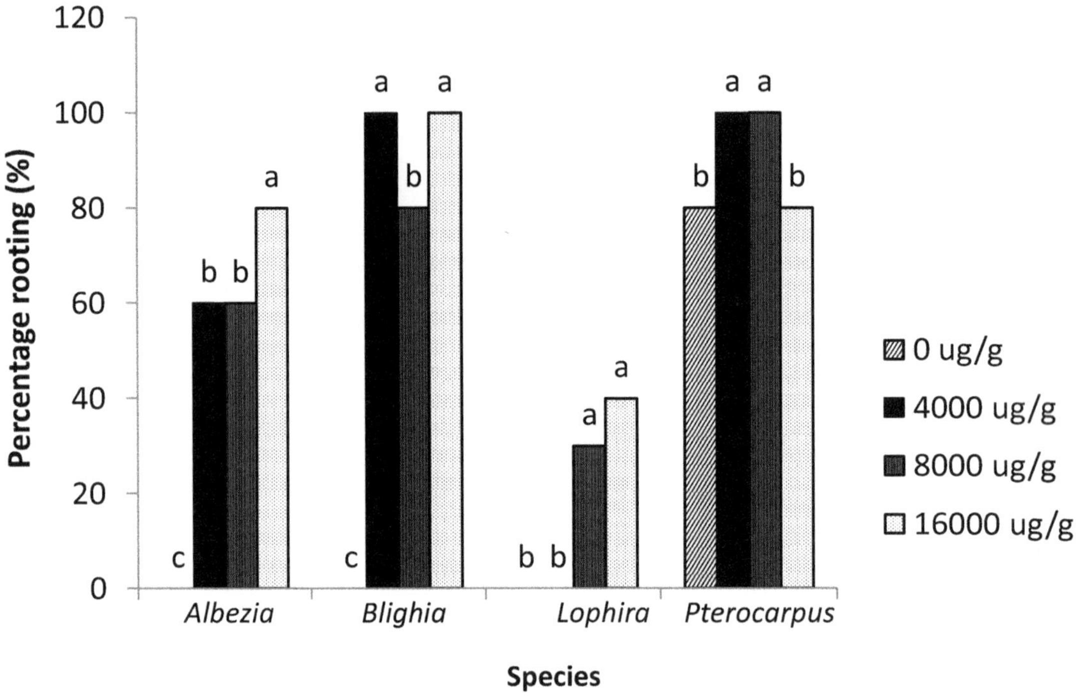

Figure 1. The effect of various hormonal treatments on rooting percentage of leafy stem cuttings of the various timber species. Values grouped by the same letter for each species are not significantly different at P< 0.05.

the rooting medium (sterilized sand) with a pencil and the leafy stem cutting inserted into it. Stem cuttings of each treatment were inserted in a separate chamber of the trays.

A non–mist propagator (Leaky et al., 1990) of dimensions length 100 × 60 × 40 cm was used for this experiment. The leafy stem cuttings were sprayed with fine drops of water after every 48 h using a 2 L capacity hand sprayer. There were 16 cuttings per treatment and a total of 64 per species. The cuttings were evaluated for rooting every 7 days for 90 days. The non-mist propagator is placed in a shade with about 60% of the full sun light and mean temperature of 20°C. The percentage of rooted leafy stem cutting for each treatment was calculated as a percentage of the total leafy stem cuttings used. The length of the longest root was measured with a ruler to the nearest 1.0 mm from the point of emergence to where it ends (when all treatments had either rooted or decayed). Diameter of roots was measured with an electronic digital caliper (Shenzhen®, to the nearest 0.1 mm) and root number was counted. The means of these were all calculated. Rooting time is the number of days it took the leafy stem cuttings to root (presence of root primodium on the leafy stem cuttings) with root diameter and length of at least 0.2 and 5 mm respectively. Analysis of variance was carried out on the resulting data using the MINITAB Version 15 statistical package to assess the best hormonal concentration. All analyses were carried out at 95% interval.

RESULTS

The effects of hormonal concentrations on the percentage rooting of leafy stem cuttings

Rooting was affected by IBA concentration in all species (Figure 1). Leafy stem cuttings of *Albizia* treated with 16000 µg ml^{-1} of IBA rooted the most (80%) while control did not root and this significantly different at P≤ 0.05. *Blighia* leafy stem cuttings had the highest rooting percentages of 100% with IBA concentration of 16000 and 4000 µg ml^{-1} while the control had 0% and these were significantly different. *Lophira* leafy stem cuttings had the highest rooting percentage in cuttings treated with 8000 and 16000 µg ml^{-1} IBA (40%) while those treated with 4000 and 0 µg ml^{-1} did not root. Leafy stem cuttings of *P. soyauxii* treated with 8000 and 4000 µg ml^{-1} IBA each gave the highest percentage rooting (100%) and the lowest was in cuttings treated with 16000 µg ml^{-1} of IBA and control (80%) and these were significantly different (P≤0.01) (Figure 1).

The effects of IBA concentrations on the number of roots on leafy stem cuttings

The mean number of roots of leafy stem cuttings per treatment was affected by the different concentrations of IBA (Figure 2). *Albizia* cuttings treated with 16000 µg ml^{-1} of IBA had the highest mean number of roots (5) compared to those treated with 8000 and 4000 µg ml^{-1} IBA which had the least number of roots (4 roots),

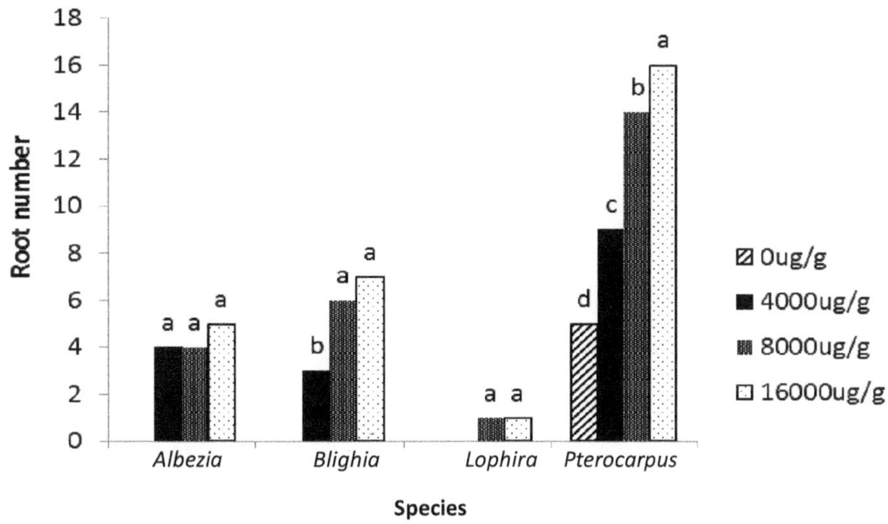

Figure 2. The effect of various hormonal treatments on root number of leafy stem cuttings of timber species. Values grouped by the same letter for each species are not significantly different at P< 0.05

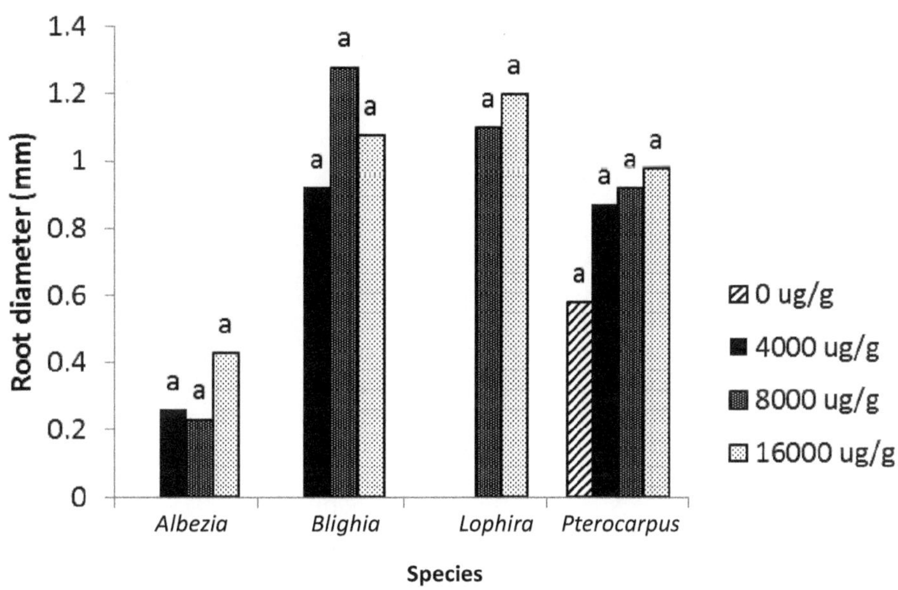

Figure 3. The effect of various hormonal treatments on root diameter of leafy stem cuttings. Values grouped by the same letter for each species are not significantly different at P< 0.05.

although the numbers did not vary significantly at P≤0.05. Leafy stem cuttings of *Blighia* and *Pterocarpus* treated with 16000 µg ml^{-1} of IBA had the highest mean root number of 7 and 16 respectively. Untreated cuttings of *Pterocarpus* had the least root number (5) and it was significantly different between the different treatments (P≤ 0.05) while *Blighia* cuttings treated with 4000 µg ml^{-1} had the least root number (Figure 2).

The effects of hormone concentrations on root diameter of rooted cuttings

Root diameter was unaffected by IBA treatments (Figure 3). The root diameter had a range of 0.23 to 1.28 mm and the maximum and the least diameter were observed in *Blighia* and *Albizia* leafy stem cuttings treated with 8000 µg ml^{-1} IBA respectively.

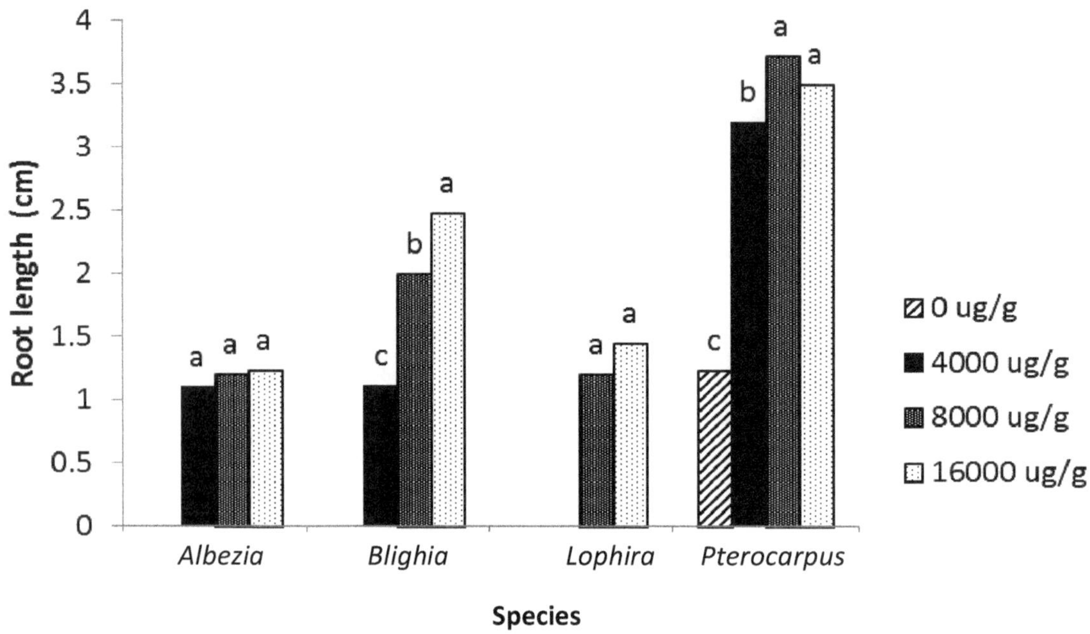

Figure 4. The effect of various hormonal treatments on root length of leafy stem cuttings of timber species. Values grouped by the same letter for each species are not significantly different at P< 0.05.

The effects of hormone concentrations on root length of rooted cuttings

Root length was affected by the different concentrations of IBA (Figure 4). Leafy stem cuttings of *Albizia*, *Blighia* and *Lophira* treated with 16000 µg ml^{-1} IBA had the longest mean roots length (1.23, 2.48 and 1.45 cm respectively) at the end of 90 days. The least root lengths for *Albizia* and *Blighia* species was recorded in cuttings treated with 4000 of IBA with values of 1.10 and 1.11 cm respectively while the shortest root length for *Lophira* species was noted in cuttings treated with 8000 µg ml^{-1} IBA. These lengths did vary significantly between treatments at P≤0.05 only for *Blighia* species. The mean longest roots on *Pterocarpus* leafy stem cuttings were those treated with 8000 µg ml^{-1} IBA (3.72 cm) and the shortest were noted in the control (1.23 cm) and this was significantly different between the treatments at P≤0.05.

The effect of hormone concentrations on the rooting time

The speed of rooting varied between species and of leafy stem cuttings of the different species. Leafy stem cuttings of *Albizia*, *Blighia*, *Lophira* and *Pterocarpus* treated with 16000 µg ml^{-1} IBA had the least rooting time (25, 38, 72 and 18 days respectively) and there was significant difference between the treatments at P≤0.05 except for *Lophira* that was not significantly different for the IBA concentrations (8000 and 1600 µg ml^{-1} IBA). The highest rooting time for *Lophira* cuttings was 78 days while that of *Pterocarpus* was recorded in control (38 days) (Figure 5).

DISCUSSION

Auxins are known to promote adventitious root development of stem cuttings, through their ability to promote the initiation of lateral root primodia and enhance transport of carbohydrates to cuttings' bases (Hartmann et al., 1990). IBA induced rooting in all the species. This is in conformity with Abdullah et al. (2005) who observed between 65% (in 0.4% IBA) and 15% (in the control) rooting in (*Baccaurea sapida*) cuttings treated with different IBA concentrations. Cuttings of *Pterocarpus* not treated with IBA rooted successfully. This suggests that *Pterocarpus* leafy stem cutting had high levels endogenous auxins. A similar observation was reported by Magingo and Dick (2001) who recorded rooting percentages of 86 and 59% in cuttings of *Pterocarpus angolensis* and *Brachystegia spiciformis* respectively, without auxins. Successful rooting without hormones has also been reported in *Shorea macrophylla* (Lo, 1985) and *Nauclea diderrichii* (Leakey, 1990). Leafy stem cuttings of *B. welwitschii*, *A. zygia* and *L. alata* did not root in the absence of IBA. Probably, the amount of endogenous auxins within these species is not enough to initiate cell division. This is in conformity with reports of Mesén et al. (1997) who reported that the cuttings *of Cordia alliodora* failed to root in the absence of auxin treatment. The root number, root diameter and root length for the different

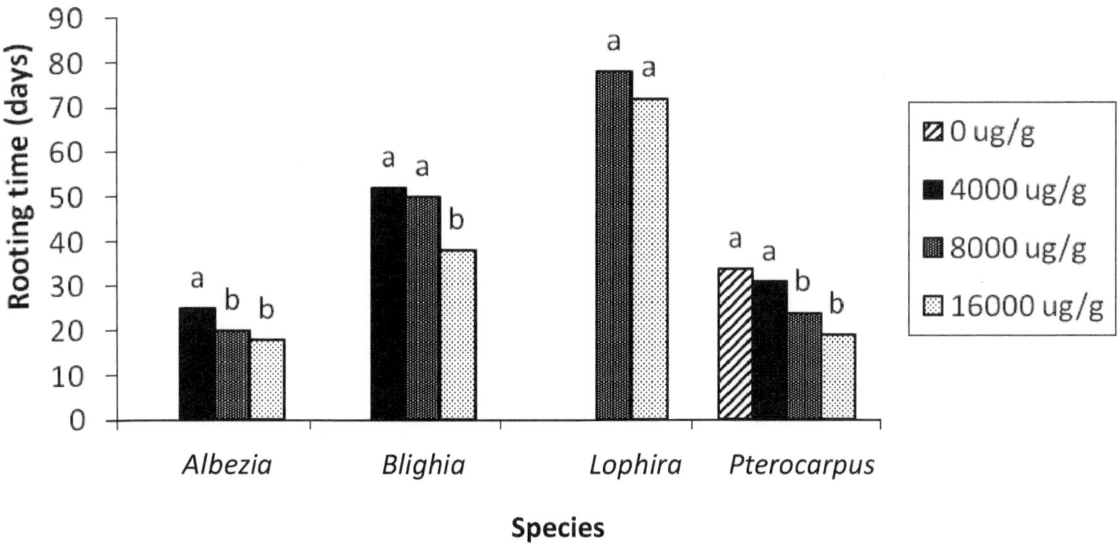

Figure 5. The effect of various hormonal treatments on rooting speed leafy stem cuttings of timber species. Values grouped by the same letter for each species are not significantly different at P< 0.05.

treatments did not vary significantly for the hormone treated stem cuttings of *B. welwitschii* and *A. Zygia.* The various IBA treatments did not affect the rate of rooting of *B. welwitschii*. These observations are in contrast to the findings of Abdullah et al. (2005) who registered significant differences in the number of root developed and root length with different hormone concentrations.

However, results obtained with seedlings of *P. soyauxii* show that the various IBA concentrations had significant differences on root length, root number and rate of rooting and it is in conformity with the findings of Abdullah et al. (2005). This was probably due to difference in the amount of polysaccharides hydrolyzed.

Conclusion

The hormone indole-3-butyric acid induced rooting in the different species studied. *Pterocarpus* cuttings can root successfully without the use of hormones. The IBA concentrations of 16000 and 8000 µg ml^{-1} were good for the rooting of *A. zygia, B. welwitschii* while 16000 µg ml^{-1} IBA was best for the rooting of *L. alata*. The production of vigorous and healthy seedlings by vegetative means would enhance forest management through enrichment planting especially for species that are threatened or those that reproduces once in 5 years or more and those that have poor recruitment rates from seeds.

REFERENCES

Abdullah ATM, Hossain MA, Bhuiyan MK (2005). Propagation of Latkan (*Baccaurea sapida* Muell. Arg.) by Mature Stem Cutting Research. J. Agric. Biol. Sci., 1: 129-134.

Amri E (2010). Viable options and factors in consideration for low cost vegetative propagation of tropical trees. Int. J. Bot., **6**: 187-193.

Blakesleya D, Elliott S, Kuarak C, Navakitbumrung P, Zangkum S, Anusarnsunthorn V (2002). Propagation framework of tree species to restore seasonally dry tropical forest: Implications of seasonal seed dispersal and dormancy. For. Ecol. Manag., 164:31-38

Bonner FT (1992). Seed Technology: A Challenge for Tropical Forestry. Tree Planters, 43:142-145

Burkill HM (1995). The Useful Plants of West Tropical Africa. 2nd Edition. Volume 3, Families J–L. Royal Botanic Gardens, Kew, Richmond, United Kingdom. p 857.

De Capua J (2005). The Congo Basin Forest Summit held February 4th and 5th in Brazzaville. Report on Congo Basin Washington, US. p. 11.

Howland P, Bowen MR (1977). *Triplochiton scleroxylon* K. Schum. and other West African Tropical hardwoods. West African Hardwoods Improvement Project Research Report 1971-1977. For. Res. Institute Nig., pp. 154.

Kpikpi WM (1992). Wood structure and paper-making potentials of *Ricinodendron heudelotii* and *Albizia zygia* in relation to *Gmelina arborea*. Nig. J. Bot., 5: 41–50.

Laird SA (1999). The management of forests for timber and non-wood forest products in Central Africa. In: Sunderland, T.C.H., Clark, L.E. and Vantomme, P. Eds. NonWood Forest Products of Central Africa: Current Research Issues and Prospects for Conservation and Development. Food and Agriculture Organisation, Rome. p. 413.

Laird SA, Betafor M, Enanga M, Fominyam C, Itoe M, Litonga E, Mafani J, Menyoli J, Meseke J, Mukete W, Motia M, Ndumbe P, Nkefor J, Nning J, Ndam N, Sunderland T, Tchouto P, Wana M (1997). Medicinal Plants of the Limbe Botanic Garden. Limbe Botanic Garden, Cameroon. Oxford University Press. p. 378

Leakey RRB (1990). *Nauclea diderrichii*; rooting of stem cuttings, clonal variation in shoot dominance and branch plagiotropism. *Trees*, 4:164-169

Leakey RRB, Mesén JF, Tchoundjeu Z, Longman KA, Dick JM, Newton A, Matin A, Grace J, Munro RC, Muthoka PN (1990). Low-technology techniques for the vegetative propagation of tropical trees Commonwealth For. Rev., 69:247-257.

Leakey RRB, Simons AJ (2000). When does Vegetative Propagation Provide a Viable Alternative to Propagation by Seed in Forestry and Agroforestry in the Tropics and Sub-Tropics. In: The Procurement of

Forestry Seeds in Tropical and Subtropical Countries-The Example of Kenya Contributions to Forest Sciences, Wolf, H. and J. Albrecht (Eds.). Verlag, India, pp: 67-81.
Lo YN (1985). Rooting of cuttings in relation to the propagation medium. Combined Proceedings of the International Plant Propagators' Society, 35: 472-485.
Louppe D (2005). *Baillonella toxisperma* Pierre In: Louppe D., Oteng-Amoako AA., Brink M. (Editors). Prota 7(1) Timbers/Bois d'œuvre 1. [CD-Rom] PROTA, Wageningen, Netherlands.
Magingo FSS, Dick JMP (2001). Propagation of two Miombo woodland trees by leafy stem cuttings obtained from seedlings. Agrofor, Syst., 51: 49-55.
Mesén F, Newton AC, Leakey RRB (1997). Effect of indole-3-butyric acid on the rooting of leafy cuttings of *Cordia alliodora.* For. Ecol. Manag., 92:45-54.
Miller R, Wiedenhoeft A (2002). CITES Identification Guide – Tropical Woods: Guide to the Identification of Tropical Woods Controlled under the Convention on International Trade in Endangered Species of Wild Fauna and Flora. Handbook. USDA Forest Service and Animal and Plant Health Inspection Service, Canada, p. 23.
Mostacedo CB, Fredericksen TS (1999). Regeneration status of important tropical forest tree species in Bolivia: assessment and recommendations. For. Ecol. Manag., 124:263-273.
Ndam N (1998). Trees regeneration vegetation dynamics and the maintenance of biodiversity on Mount Cameroon: The relative impact of natural and human disturbance. PhD thesis, University of Wales. pp. 635
Neuwinger HD (2000). African Traditional Medicine: A Dictionary of Plant Use and Applications. Medpharm Scientific, Stuttgart, Germany. p. 589.
Ofori DA, Newton AC, Leakey RRB, Grace J (1996). Vegetative propagation of *Millicia excelsa* by leafy stem cuttings: effects of auxin concentration, leaf area and rooting medium. For. Ecol. Manag., 84:39-44.
Richter HG, Dallwitz MJ (2000). Commercial timbers: descriptions, illustrations, identification, and information retrieval.' In English, French, German, and Spanish, pp. 10-14.
Tchouto P, Edwards I, Cheek M, Ndam N, Acworth J (1997). Mount Cameroon cloud forest. Paper presented at XVth AEFAT Congress. p. 27.
UNEP-WCMC (1999) Contribution to an Evaluation of Tree species using the new CITES listing criteria. UNEP-World Conservation Monitoring Centre, Cambridge. p. 143.

Growth habit, plant density and weed control on weed and root yield of sweet potato (*Ipomoea batatas* L.) Areka, Southern Ethiopia

Tenaw Workayehu*, Waga Mazengia and Legesse Hidoto

Awassa Agricultural Research Center, P. O. Box 366, Awassa, Ethiopia.

Sweet potato is mostly left unweeded or weeded lately. Effect of variety, plant density and weeding frequency on weed and tuber yield was studied from 1992 to 1994. Three varieties (Tis 1499: erect and early; Tis 2498: long vine, spreading and early, and Koka 6 intermediate), two weeding frequencies (30 to 40 (W1) and 70 days after sprout (DAS) (W2), and four plant densities (5, 7, 10, and 12.5 plants m^{-2}) were laid out in factorial arrangement in randomized complete block design with three replications. The major weed was broad leaf covering 95% of the total weeds, *Capsella rubella*, *Guizotia scabra* and *Plantago lanceolata* being the dominant weeds. In 1992 and 1993, mean weed density in Tis 2498 was 17 and 18% less compared with Tis 1499 and Koka 6, respectively. Weeds at higher plant densities were 11.8% less compared with the lowest density. All the cultivars produced more yield in 1994, but Tis 2498 and Koka 6 yielded 35.4 and 8.3% less than Tis 1499, respectively. Weeding twice in the first two years significantly increased the mean yield by 46.2%. Weeding of Tis 1499 and Koka 6 twice significantly increased the yield by 23.2% each but Tis 2498 did not respond to weeding. The cultivars Tis 1499 and Koka 6 yielded 22.7 and 18.8 t ha^{-1}, more at 7 and 10 plants m^{-2}, respectively, but Tis 2498 did not respond. Weeding once and twice within the range of 7 to 12.5 plants m^{-2} had no significant yield variation indicating weeding once would suffice at higher plant density. Varieties with high population having erect and intermediate growth habits can be weeded twice. Growth with spreading canopy structure and plant density could be used as means to reduce weed infestation. This will save farmers time and labor and thus, breeders should focus in developing cultivars with spreading canopy structure with high yield.

Key words: Areka, Ethiopia, canopy structure, plant density, sweet potato, weed density, weed infestation.

INTRODUCTION

Sweet potato (*Ipomoea batatas L.*) is one of the major horticultural crops widely grown in the Eastern and Southern regions of Ethiopia. At the national level, the total area coverage of sweet potato was 53,311.65 ha (CACC, 2006). It grows in the lowland and mid altitude areas of the Southern region and occupies more than 29,523.93 ha of land (CACC, 2003). It is grown both in *belg* and *meher* seasons covering large hectares in *meher* (320,823 and 766,495 ha, respectively). The report shows that sweet potato shares 34.1% of the total area and 36.8% of the total regional volume of root crop production. It is indicated that 11.5% of the regional root crop area was grown with sweet potato resulting in 5.12% of the regional root crop production in *belg*.

It is used as a source of food and income to the poor and needy farmers in sweet potato growing areas. In addition, the different parts are used for various purposes: the root for home consumption and sale, the aboveground part for planting material, sale and feed for livestock, and as a soil conservation mechanism. Thus, about 81.5% of the total production of sweet potato is used for home consumption, 9.6% for sale, 6.1% for animal feed, 0.64% for planting material, 0.42% for wages in kind and 1.83% for other uses (CACC, 2003).

The potential yield of sweet potato in research goes up to 50 t ha^{-1}; but on station and on-farm research with

*Corresponding author. E-mail: ttenaw@yahoo.com.

improved management practices, it indicated a yield ranging from 17.5 to 35.0 t ha^{-1}. However, because of poor management practices, the yield under farmers' condition is low, on average, 7.6 to 7.9 t ha^{-1} (CACC, 2003), while others reported the root yield as low as 5 t ha^{-1} (Degu and Workayehu, 1990; Degu et al., 1991). The report of Bureau of Agriculture (BoA) of Sidama zone indicated that the yield under farmers' management ranges between 6.5 and 9.0 t ha^{-1} which is far below the potential yield. CACC (2007) report also shows that the yield at the national level varies between 7.3 and 8.1 t ha^{-1}. A plant density of 8 plants m^{-2} is under practice while the optimum plant density recommended was between 3.3 and 5.6 plant m^{-2}. The low productivity is attributed to low yielding cultivar, weed competition, low or high plant density, little fertilizer application, and insect pests (CACC, 2003). According to CACC (2003), weeds account for 11.64% of the total damages of sweet potato production. A yield loss of 87 to 98.9% was recorded if sweet potato is left unweeded; even early or late weeding reduced the yield (Awassa progress report, 1991).

Tillage, hand weeding, and multiple cropping systems are of the many alternatives available for weed control. Hand weeding and chemicals are effective to control weeds (ICARDA, 1984). Availability, costliness, efficacy, and its effect on human health are problems to consider when using herbicides. In sweet potato producing areas of the region, land preparation is carried out using traditional ox-drawn implement with few plowings due to shortage of oxen (>50% of the farmers in the region have no ox, CSA, 2003). Weed control measures such as increased plant density and cultivar selection can be exploited. Weeds could be suppressed either by increasing plant density or using cultivars with better competitive ability and provide an optimum yield per unit area. Cultivars differ in their ability to suppress weeds and some studies indicate varieties differing in canopy structure showed no significant effect on weed infestation (Tenaw et al., 1997). On the contrary, Taye and Tanner (1997) reported some varieties to better compete and suppress weeds due to their canopy structure. Olofsdotter et al. (1999) reported varietal differences in suppressing the infestation of *Echinochloa crus-galli* weed and pointed a 15 to 50% of the variation in weed suppression among cultivars due to plant height. Weed interference in a given cultivar is determined by the allelopathic effect combined with competitive ability of the cultivar. Olofsdotter et al. (1999) also noted that shoot and root parameters could account for the differences among cultivars in reducing weed infestation. Other researchers (ICARDA, 1984; Salonen, 1992; Seavers and Wright, 1999) indicated leaf area, plant height, and number of tillers, canopy structure and development as factors that contribute to better crop competitiveness with weed.

Increased seed rate reduces weed growth and weed infestation tends to increase with decreased seed rate (ICARDA, 1984). Various research reports show increased grain yield because of reduced weed competition that resulted from increased plant density (ICARDA, 1984; Elmore, 1998). Higher plant density can reduce tillering that increases the ability of cultivar competitiveness (Kim et al., 1999) while it has a variable effect on leaf area (Kim et al., 1999; Lisson et al., 2000). High plant density can slow down crop growth rate and reduce leaf area index that has a relation in enhancing the competitiveness of the cultivar (Lisson et al., 2000). So far, there is no research information on the relative competitiveness of sweet potato cultivar and plant density on weed population/infestation. This study was initiated to evaluate the effect of cultivar with different canopy structure, plant density and seasonal variation on weed infestation and root yield of sweet potato.

MATERIALS AND METHODS

Site description

The study was conducted at Areka Agricultural Research Center for three consecutive seasons, 1992 to 1994. The soil is reddish brown with clay loam texture, which is classified as Nitosol. The soil has a pH of 5.45 (acidic), TN 0.204 g kg^{-1} (low), P_2O_5 3 mg kg^{-1} (low) and CEC 24.61 cmol kg^{-1} soil (medium) (Landon, 1984). The location has two crop growing seasons, mainly *belg and mehir* that begin from February to May and June to September, respectively.

Treatments

Three cultivars of sweet potato, Tis 1499 erect/vertical growth and, short vine with low vegetative growth and early maturing; Tis 2498 spreading/, long vine with high vegetative growth, and early maturing; and Koka 6 an intermediate with medium vegetative growth and medium maturing, four plant densities (5, 7, 10, and 12.5 plants m^{-2}), and two weeding frequencies [W1-one time weeding 30 to 40 days after sprout (DAS), and W2-weeding twice 30 to 40 and 70 DAS] were used. Plant density was arranged by varying row spacing (1.0, 0.7, 0.5, 0.4 m) and maintaining plant spacing constant (0.20 m). Factorial arrangement in randomized complete block design replicated three times was laid out and the plot size was 4 m wide and 4 m long.

Crop management

Planting dates were June 13 and 4 of 1992 and 1993, respectively, and May 26, 1994. The land was plowed with tractor. Fertilizer was applied at the rate of 18 N and 46 kg P_2O_5 ha^{-1}, respectively, DAP was used as source of the nutrients.

Data collection

Weed types and density were identified using four quadrat counts per plot using iron square of 25 × 25 cm. Data on weed type, weed m^{-2}, leaf area (1994 only), branching, plant population at harvest, and root yield were collected. Leaf area, using destructive sampling, was measured by taking 10 fully expanded leaves per plot.

Table 1. Amount of precipitation in the growing season and long-term rainfall (1988 to 1991) in Areka, Ethiopia.

Month	Season			
	1992	**1993**	**1994**	**Long-term**
May	191.6	310.3	279.8	141.2
June	68.2	94.1	173.0	136.4
July	174.3	269.7	221.7	236.4
August	331.3	185.7	273.9	189.9
September	224.5	117.6	117.9	199.8
October	188.7	94.2	6.7	119.2
Total	1178.6	1071.6	1073.0	1022.9

Statistical analysis

Economic analysis of sweet potato was conducted using yield and the current mean market price of sweet potato (US$ 0.53 kg^{-1}). Variable costs were determined for each treatment with: labor required for W1 and W2 of 20 and 25 person-day ha^{-1}, respectively, at US$ 1.16 $perosn^{-1}$ day^{-1}; labor for cultivation of 20, 25, 30, and 35 person-day for a density of 50×10^3, 70×10^3, 100×0^3 and 125×10^3 plants ha^{-1}, respectively, at US$ 1.16 $person^{-1}$ day^{-1}; and labor for harvesting of 20, 25, 30 and 35 person/day for the aforementioned plant densities, respectively, at US$ 1.13 $person^{-1}$ day^{-1}. Other costs were considered constant across treatments. Net income (NI) was determined as the difference of gross income and variable costs (Babatunde, 2003). Because of the interaction effect of year with variety and plant density, combined analysis over years was conducted after test of homogeneity (Gomez and Gomez, 1994). Log transformation was made for weed infestation and tuber yield of sweet potato to stabilize the variances. Analysis of variance was conducted using SAS statistical package (SAS, 2000) and Tukey test of significance to differentiate treatment means. Correlation and stepwise multiple regression analysis were carried.

RESULTS

Rainfall

Total rainfall in 1992 was above the long-term average (15% more); however, amount of rain in some decades of June, July and September was low with poor distribution. In addition, the heavy rainfall in August and September caused water logging. The rainfall pattern was thus below normal in June (50%) and July (26.3%) and above normal in August (174.5%) and September (112.4%). The precipitation in June, July, August, and September of 1993 was low, which negatively affected the sprout (germination) and physiological maturity. The distribution was poor in June and July and was below normal in June (31%) and September (41%), and normal in July and August. In the year 1994, amount of precipitation was low in June, July, and September but the distribution was good (Table 1). Overall, weather variations in the years affected the effect of cultivar, plant density and weed control on weed infestation, crop growth and yield.

Weed types

The dominant weed species were broadleaf covering 90, 100 and 95% of the total weed populations in 1992, 1993 and 1994 seasons, respectively. More weed population was observed in 1993 followed by 1994. The major weed species during the growing seasons were *Plantago lanceolata, Capsella rubella,* and *Guizotia scabra,* which accounted for 16 to 27% of all weeds present in all years (Table 2).

Weeds m^{-2}

Effect of variety on weed was dependent on seasonal variation (Table 3). Weed population in Tis 1499 was 53.7 and 48.1% less in 1992 and 1994, respectively, compared with that in 1993 (Table 4). In 1992, weed infestation in Tis 2498 was 16.4 and 23.5% less than those of Tis 1499 and Koka 6, respectively, while weeds in Tis 2498 during 1993 were 17.7 and 15.8% less than Tis 1499 and Koka 6, respectively; however, all varieties showed no significant variation in weed population in 1992 and 1994. Weed infestation in 100×10^3 plants ha^{-1} was 11.3 and 12.3% less compared with the density of 50 and 70×10^3 plants ha^{-1}, respectively, though not significantly different from the highest plant density indicating significantly less weed as plant density increased (Table 5). Plant density was negatively associated with weed infestation indicating a negatively linear relationship (r = -0.328**, n=72).

Branch number

Plant density, cultivar and weeding practice had a significant effect on branches per plant. The first three plant densities had more branches and yet no significant variation while significantly fewer branches were observed at the highest plant density (12.5 plants m^{-2}) (Table 6). On the other hand, Tis 1499 and weeding once had significantly more branches. Branching was positively associated with root yield (r = 0.332, n=72).

Leaf area

Leaf area was significantly affected by variety and its

Table 2. Weed types and density m^{-2} (number) observed during the experimental seasons, 1992 to 1994, Areka.

Weed type		
Broad leaf	**Other weeds**	**Grass spp.**
Capsella rubella	*Galinsoga parviflora*	*Cyperus* spp.
Guizotia scabra	*Leuscas* spp	*Setaria* spp.
Plantago lanceolata	*Commelina benghalensis*	
Tagates minuta	*Anagalis* spp	
Ageratum conizoides	*Crinum* spp.	
	Isolepsis spp.	
	Bidens pilosa	
	Portulacea spp.	

Table 3. ANOVA showing the effect of source of variation on weed and measured parameters of sweet potato.

Source of variation	Weed density (no.)	Branch no.	Plant density (no.ha^{-1})	Leaf area (cm^2)	Root yield (kg ha^{-1})
Year (Y)	**	-	*	-	**
Weeding frequency (W)	ns	ns	ns	ns	**
Y*W	ns	ns	**	ns	**
Variety (V)	ns	*	ns	**	**
Plant density (Pd)	*	**	**	ns	**
Y*Pd	ns	-	*	-	**
W*Pd	ns	ns	ns	ns	ns
Y*V	**	-	ns	-	**
V*Pd	ns	ns	ns	**	**
V*W	ns	ns	ns	ns	**

ns, *, and ** shows not significant, and significant at 5 and 1% probability level, respectively.

Table 4. Interaction effect of year and variety on total weed population, weeds m^{-2}.

Season	Variety		
	Tis 1499	Koka 6	Tis 2498
1992	183^{cd}	200^{cd}	153^{d}
1993	395^{a}	386^{a}	325^{b}
1994	205^{cd}	188^{cd}	232^{c}

Same letter in a column and row shows no significant difference at 5% probability level.

Table 5. Effect of plant density on weed population m^{-2}.

Plant density/plants ha^{-1}	Weed density m^{-2}
50×10^3	265^{ab}
70×10^3	268^{a}
100×10^3	235^{b}
125×10^3	238^{ab}

Same letter in a column shows no significant difference at 5% probability level.

interaction with plant density. Varieties responded differently to the different plant densities. Koka 6 and Tis 2498 had high leaf area at 7 and 5 plants m^{-2}, respectively, but no significant variable among plant densities

Table 6. Effect of plant density, variety and weeding frequency on number of branches plant^{-1}, Areka (1993 to 1994).

Plant density no. (ha^{-1})	Branching (no.)	Variety	Branching (no.)	Weeding frequency	Branching (no.)
50 × 10^3	5.8^a	Tis 1499	6.2^a	W1	5.8^a
70 × 10^3	5.5^a	Tis 2498	5.1^b	W2	5.2^b
100 × 10^3	6.1^a	Koka 6	5.3^b		
125 × 10^3	4.7^b				

W1-one weeding at 30-40 DAS, W2-two weedings at 30 to 40 and 70 DAS. Same letter in each column shows no significant difference at 5% probability level.

Table 7. Interaction effect of variety and plant density on leaf area (cm^2) 1994, Areka.

Plant density (plant ha-1) (×103)	Variety		
	Tis 1499	Koka 6	Tis 2498
50	48.6^{de}	51.8$^{c\text{-}e}$	81.40^a
70	49.3^{de}	66.4$^{a\text{-}c}$	60.9$^{b\text{-}d}$
100	42.5^e	63.8$^{b\text{-}d}$	71.2^{ab}
125	49.2^{de}	61.3$^{b\text{-}d}$	59.9$^{b\text{-}d}$

Same letter across a column and row shows no significant difference at 5% probability level.

in Tis 1499 (Table 7). On average, leaf area of Tis 2498 was significantly 44.3 and 12.5% higher than Tis 1499 and Koka 6, respectively. Leaf area had a negative relationship with total weeds m^{-2} showing the higher the leaf area the lower was the weed density.

Root yield

Effects of weeding and variety on tuber yield were dependent on seasonal variation (Table 3). In 1992 and 1993, weeding twice produced 44 and 48% more yield than weeding once, respectively, while in 1994 there was no significant yield variation between weeding frequencies (Table 8). About 6.2% of the total variation in tuber yield was attributed to weeding practice as indicated by the regression equation: tuber yield (q ha^{-1}) = 41.43+26.70W (R^2 = 0.062**), where 'W' is weeding frequency. All varieties produced high tuber yield in 1994 in particular Tis 1499 and Koka 6 while in 1992 significant yield variation was not observed among cultivars. On the other hand, the yield of Koka 6 did not significantly vary in the first two years. Tuber yield of Tis 1499 was 46.1 and 29.5% more than Koka 6 in 1992 and 1993, respectively, but the yield of Tis 2498 was 26.4, 19.7, and 46% less compared with Tis 1499 in 1992, 1993, and 1994, respectively.

Varieties responded differently to the different weeding practices and plant densities. There was an increase in yield as density increased from 5 to 10 plants m^{-2} although significant variation between 7 and 10, and 7 to 12.5 plants m^{-2} was not observed in Tis 1499 and Koka 6, respectively (Table 9). The yield of Tis 2498 did not show significant variation among plant densities despite the high yield from 10 plants m^{-2}. The result showed that a population of 10 plants m^{-2} produced significantly high tuber yield although it was not significantly different from 7 plant m^{-2}. Effect of weeding on tuber yield was dictated by variation in variety and plant density (Table 10). Weeding of Tis 1499 and Koka 6 twice increased the yield by 22.5 and 23.9%, respectively, while weeding practice did not have a significant effect on tuber yield of Tis 2498. Although weeding once and twice produced more as plant density increased (up to 10 plants m^{-2}), further increase reduced the yield. Although weeding twice increased tuber yield by 23.6, 18.2, 15.4, and 5.1% for the 5, 7, 10 and 12.5 plants m^{-2}, respectively, compared with weeding once, the variation within the range of 7 to 12.5 plants m^{-2} was not statistically significant in both weeding practices.

Economic benefit

There was an increase in net benefit as plant density increased from 50 × 10^3 to 100 × 10^3 plants ha^{-1} for Tis1499 and Koka 6 while the optimum economic return for Tis 2498 was between 70 ×10^3 and 100 ×10^3 plants ha^{-1} (Table 11). There was 5.5 and 4.8% more net benefit from 100 × 10^3 plants ha^{-1} than the 70×10^3 plants ha^{-1} for Tis 1499 and Koka 6, respectively. Weeding Tis 1499 and Koka 6 twice, significantly increased the economic benefit by 22.5 and 23.9% more, compared with one weeding, respectively, but the income from Tis 2498 was not significant between weeding practices.

DISCUSSION

Weed infestation was dictated by variation in year,

Table 8. Interaction effect of season with weeding frequency and variety on tuber yield of sweet potato (Mg ha^{-1}), Areka.

Variable	Year		
	1992	1993	1994
Weeding frequency			
W1	7.00 ^{d}	10.20^{c}	27.74^{a}
W2	10.08 ^{c}	15.13^{b}	26.53^{a}
Variety			
Tis 1499	10.58^{de}	14.76^{bc}	31.27^{a}
Koka 6	7.24^{ef}	11.40^{de}	33.25^{a}
Tis 2498	7.79^{ef}	11.85^{cd}	16.89^{b}

W1-weeding once at 30-40 DAS, W2-weeding twice at 30 to 40 and 70 DAS. Same letter in each column and row of each factor shows no significant difference at 5% probability level.

Table 9. Interaction of variety and plant density on tuber yield of sweet potato (Mg ha^{-1}), Areka.

Plant density (no.ha^{-1})	Variety		
	Tis 1499	Koka 6	Tis 2498
50 × 10^3	14.89^{b-d}	12.65^{d}	11.37^{d}
70 × 10^3	22.08^{a}	18.36 $^{a-c}$	12.56^{d}
100 × 10^3	23.31^{a}	19.24^{ab}	12.90^{cd}
125 × 10^3	15.19^{b-d}	18.95^{ab}	11.87^{d}

Same letter across a column and row shows no significant difference at 5% probability level.

Table 10. Interaction effect of weeding frequency with variety, and plant density on tuber yield of sweet potato (Mg ha^{-1}) Areka.

Variable	Weeding frequency	
	One time:30-40 DAS	Two times: 30-40 and 70DAS
Variety		
Tis 1499	16.96^{bc}	20.78^{a}
Koka 6	15.45^{cd}	19.15^{ab}
Tis 2498	12.53^{de}	11.82^{e}
Plant density (no.ha^{-1})		
50 × 10^3	11.60^{c}	14.34^{bc}
70 × 10^3	16.19^{ab}	19.14^{a}
100 × 10^3	17.16^{ab}	19.81^{a}
125 × 10^3	14.96^{bc}	15.72^{ab}

W1-weeding once at 30-40 DAS (days after sprout), W2-weeding twice at 30 to 40 and 70 DAS. Same letter across a column and row for each factor shows no significant difference at 5% probability level.

variety, and plant density which positively or negatively affected the performance of crops and weeds. The high weed infestation in 1993 was due to unfavorable weather in some months during the growing season that reduced the competitive ability of the crop. There was better light interception for weed growth to compete more for soil nutrients and moisture because of poor crop growth. More weeds were observed on varieties having erect (Tis 1499) and intermediate (Koka 6) growth habits, which had 10 and 8.9% more weeds than spreading type (Tis 2498), respectively. The low weed infestation in Tis 2498 could be attributed to its canopy structure and leaf area, which might contribute to better competition for light, moisture, and nutrient (Olofsdotter et al., 1999). The finding showed that varieties with erect and intermediate types need weeding twice while the spreading type can

Table 11. Effect of plant density, variety and weeding frequency on net economic benefit of sweet potato, Eb ha^{-1} (Ethiopian birr).

Variable	Variety		
	Tis 1499	Koka 6	Tis 2498
Plant density			
50 × 10^3	89340	75900	68220
70 × 10^3	132480	110160	75360
100 × 10^3	139860	115440	77400
125 × 10^3	91140	113700	71220
Weeding frequency			
Weeding once: 30-40 DAS	101500	92440	74920
Weeding twice: 30-40 and 70 DAS	124355	1145750	70595

Cost: Weeding: 20 and 25 man-days ha^{-1} and 13 Eb (Ethiopian birr) man-day^{-1}: Harvesting: 20, 25, 30, and 35 man-days ha^{-1} and 15 Eb man-day^{-1}: Cultivation: 20, 25, 30, and 35 man-days and 13 Eb man-day^{-1}. Current sweet potato market price: 6 birr kg^{-1}: Planting material cost: 0.15 Eb $cutting^{-1}$. 1USD=11.25 Eb (Ethiopian birr).

be weeded once. Thus, suppression of early weed growth resulted in less weed infestation (Johri et al., 1992; Mt Pleasant et al., 1992). Faster growth and better ground cover was achieved by Tis 2498 with leaf area ranging from 60 to 81 cm^{-2} compared with Tis 1499 (42-49 cm^{-2}) and Koka 6 (52-66 cm^{-2}).

The root and aboveground biomass could account for differences among cultivars in suppressing weeds, probably reducing the competitiveness of weeds for moisture, nutrients and light (Olofsdotter et al., 1999). The low weed population (11.8% less) at higher plant densities (10 and 12.5 plants m^{-2}) indicated weed suppression as plant density increased probably due to less light interception due to less space for weed growth, and this is in agreement with the finding of (Mt Pleasant et al. 1992 and ICARDA 1984).

Where there was high plant density, there prevailed intra and inter row competition for light, moisture and nutrients, and this resulted in less number of branches, which is in line with the findings of Kim et al. (1999). There was a positive association of branching with tuber yield (r = 0.332**) indicating more branches associated with more tuber yield.

A decrease in leaf area observed as plant density increased was in line with the findings of Kim et al. (1999) and Lisson et al. (2000). The non-significance of weeding frequency on leaf area could be due to at earlier crop growth stage weeding reduced weed competition, which gave an advantage to better growth and use the resources available. It may be concluded that the spreading type may produce more leaf area in the lowest density that may favor growth.

Weather condition affected plant population at harvest. Where there was better amount and distribution, there was better plant sprout (emergence), which finally resulted in higher plant population at harvest. Weeding twice coupled with better season increased harvestable plant population.

Weed infestation in varieties with erect and intermediate growth habits was more but when weeded, produced significantly higher tuber yield. On the other hand, the spreading cultivar, which suppressed weed infestation, had fewer yields than Tis 1499. The smaller yield loss observed in Tis 2498 compared with other varieties might be attributed to its competitiveness with weed.

The high yield was also attributed to increased plant density and weeding frequency coupled with varietal differences in canopy structure. The high yield in Tis 1499 and Koka 6 was their response to weeding frequency, plant density, and season. At each weeding practice, there was an increase in yield as the season progressed due to better amount and distribution of rainfall. The finding showed that weeding the spreading cultivar once at earlier growth stage would suffice due to the crop becoming more competitive and coupled with good distribution of rainfall. Favorable weather provided better crop growth and development, and less weed infestation thus leading to better crop yield.

Overall, a cultivar with spreading canopy structure better competes with weed and showed no significant difference between weeding frequencies. Increased plant density also reduced weed infestation. A population density of 100 × 10^3 plants ha^{-1} for the three varieties and weeding of Tis 1499 and Koka 6 twice was economical. Cultivars with spreading structure and increased plant density could be exploited as means for weed suppression, which saves time and labor of the farmer. Thus, breeders should focus in developing cultivars with spreading growth habit and high yielding, and this tempts breeders to look into this opportunity to help farmers use their time and labor effectively.

REFERENCE

Awassa Progress Report (1991). Research Progress reports

1988-1990.
Babatunde FE (2003). Intercrop productivity of Roselle in Nigeria. Afr. Crop Sci. J., 11(1): 43-47.
Central Agricultural Census Commission (CACC) (2006). Ethiopian Agricultural Sample Enumeration, 2001/02 (1994 EC): Report on the preliminary results of area, production and yield of temporary crops (Meher season, private peasant holdings).
Central Agricultural Census Commission (CACC) (2003). Ethiopian Agricultural Sample Enumeration, 2001/02 (1994 EC): Report on the preliminary results of area, production and yield of temporary crops. (Meher season, private peasant holdings). Part I.
Central Agricultural Census Commission (CACC) (2003). Ethiopian Agricultural Sample Enumeration, 2001/02 (1994 EC): Results for Southern Nations, Nationalities and Peoples' region (SNNPR), Statistical Report on Farm Management Practices, Part III B. Addis Ababa, Ethiopia.
Central Agricultural Census Commission (CACC) (2003). Ethiopian Agricultural Sample Enumeration, 2001/02 (1994 EC): Results for Southern Nations, Nationalities and Peoples' region (SNNPR), Statistical Report on area and production of crops. Part II B.
Elmore RW (1998). Soybean cultivar response to row spacing and seeding rates in rainfed and irrigated environments. J.Prod. Agric., 11(3): 326-331.
Getahun D, Mwangi W, Tenaw W, Grisley B (1991). Areka area mixed farming zone, North Omo region: Diagnostic survey. Research report no. 15. Institute of Agricultural Research, Addis Ababa, Ethiopia.
Getahun D, Tenaw W (1990). Initial results of informal survey. Areka mixed farming zone, Wolayita district, Sidamo region. Working paper no. 11. Institute of Agricultural Research, Addis Ababa, Ethiopia.
Gomez K, Gomez AA (1976). Statistical Procedures for Agricultural Research Second Edition. IRRI book. John Wiley Sons. New York, Brisbane.
ICARDA (1984). Annual report 1983.Allepo, Syria, pp. 5-12.
Johri AK, Singh G, Sharma D (1992). Nutrient uptake by wheat and associated weeds as influenced by management practices. Tropical Agriculture (Trinidad), 69 (4): 391-393.
Kim BOK, Kim HHo, Ko JK, Shin HT (1999). Effect of plant density and nitrogen level on growth and yield in heavy panicle weight type of Japonica rice. Korean J.Crop Sci., 44(2): 106-111.
Landon JR (1984). Booker Tropical Soil Manual: handbook for soil survey and agricultural land evaluation in the tropics and subtropics. Booker Agriculture International Limited, USA.
Lisson SN, Mendham NJ, Carberry PS (2000). Development of a Hemp simulation model 3.The effect of plant density on leaf appearance, expansion and senescence. Australian J. Exp. Agric., 40(3): 419-423.
Mt Pleasant J, McCollum RE, Coble HD (1992). Weed management in a low input cropping system in the Peruvian Amazon region. Trop. Agric. (Trinidad), 69 (3): 250-259.
Olofsdotter M, Navarez D, Rebulanan M, Streibig JC (1999). Weed-suppressing rice cultivars–does allellopathy play a role? Weed Res., 39(6): 441-454.
SAS Institute (2000). SAS User's guide, Statistics version 8.2 ed. SAS Inst., Cary, NC, USA.
Salonen J (1992). Efficiency of reduced herbicide doses in spring cereals of different competitive ability. Weed Res., 32(6): 483-491.
Seavers GP, Wright KJ (1999). Crop canopy development and structure influence on weed suppression. Weed Res., 39(4): 319-328.
Taye T, Tanner DG (1997). Grass weed competition and calculated economic threshold densities in bread wheat in Ethiopia. Afr. Crop Sci. J., 5(4): 371-384.
Tenaw W, Beyenesh Z, Waga M (1997). Effect of variety, seed rate and weeding frequency on weed infestation and grain yield of haricot bean. In: Fassil Reda and Tanner, D.G. (eds.). Proceedings of the second and third annual Conference of the Ethiopian Weed Science Society. Addis Ababa, Ethiopia, pp. 61-69.

In vitro tuberization of glory lily (*Gloriosa superba* L.)

Anandhi Selvarasu* and Rajamani Kandhasamy

Department of Medicinal and Aromatic Crops, Horticultural College and Research Institute, Tamil Nadu Agricultural University, Coimbatore-641 003, Tamil Nadu, India.

***Gloriosa superba* L., a climber belonging to the family, Colchicaceae is a high value medicinal plant cultivated in Tamil Nadu. Colchicine is one major tropalone alkaloids of medicinal importance which cures gout, cancer, rheumatism. Vegetative propagation through V or L shaped tubers is a commonly followed practice but is considered slow with poor multiplication ratio of 1: 1 every year. The cost involved towards planting material (800 kg of tubers/acre) alone accounts to 2.0 lakhs at Rs. 250 per kg of tuber prevailing for the last three years. In this context, a suitable protocol for *in vitro* tuberisation using non-dormant tubers of *Gloriosa superba* L. on Murashige and Skoog medium supplemented with various concentrations growth regulators have been reported in the present study. Sprouted tuber node explants were sterilized with 70% ethanol for 30 s followed by 60 s in $HgCl_2$ which reduced the contamination percentage (8.00%). MS medium supplemented with 4.0 mgL^{-1}BAP and 1.0 mgL^{-1} NAA recorded the highest response for primary tuber (100%) and secondary tuber (100%) formation. This also recorded the maximum number of tuber (1.77) from single explants. GA_3 (1.0 mgL^{-1}) was observed to be vital for the elongation of shoot whereas, IAA (1.0 mgL^{-1}) in combination with IBA (0.5 mgL^{-1}) was effective for induction of roots on MS medium.**

Key words: Glory lily, ms medium, tuber node, shooting, rooting, primary tubers.

INTRODUCTION

Gloriosa superba L., commonly known as Kalahari (Hindi), glory lily, superb lily and tiger lily is an export oriented medicinal plant and a member of the Colchicaceae family. *G. superba* is propagated by 'V' or 'L' shaped tubers which sprouts during the rainy season. Vegetative propagation through tubers is a commonly followed practice but is considered slow with poor multiplication ratio of 1: 1 every year (Krause, 1986). Through micro propagation technique, elite plants are produced with high multiplication rates (1: 8) which also produce disease-free planting materials (Seemanti et al., 2007). Suprio (2011) reported that *in vitro* regeneration by multiple auxiliary shoot induction from sprouted tuber and seedlings derived from shoot tip explants was successful, particularly with the use of TDZ at 0.1 mg/L concentration which induced up to 14.1 auxiliary shoots per explants. Rapid multiplication through micro propagation is an attempt towards aiming for higher multiplication ratio and for development of quality planting materials in the present study.

*Corresponding author. E-mail: sanandhijasmin@gmail.com.

MATERIALS AND METHODS

Mature tubers of *G. superba* were collected from healthy plants from farmer's field, Mulanoor of Tirupur district, Tamil Nadu. The sprouted tuber node was the explants, that is, 1 cm long portion of tuber containing 3 to 5 mm long whitish apical buds. The explants were excised and washed thoroughly with detergent solution (Teepol 0.1%) followed by 3 to 4 washings with sterile distilled water. After that, explants were rinsed with 70% ethanol for 30 s, and was then washed with sterile distilled water for about 3 times followed by washing with 0.1% $HgCl_2$ solution at different duration (35, 40, 45, 50, 55 and 60 s) and washed 3 to 5 times in sterile water. The explants excised in appropriate size were implanted in

Table 1. Effect of duration of exposure of sterilants on tuber nodal explants of *Gloriosa superb.*

Treatment		Tuber node		
$HgCl_2$ concentration (%)	Duration of exposure (s)	Contamination (%)	Survival (%)	Mortality (%)
0.1	35	98.00(83.08)	2.00(8.13)	0.00(0.74)
0.1	40	90.00(71.58)	10.00(18.44)	0.00(0.74)
0.1	45	86.00(68.08)	10.00(18.44)	4.00(11.54)
0.1	50	35.00(36.27)	62.00(51.95)	3.00(9.97)
0.1	55	9.00(17.46)	91.00(72.67)	0.00(0.74)
0.1	60	8.00(16.43)	80.00(63.46)	12.00(20.27)
	Mean	54.33(48.82)	42.00(38.85)	3.17(7.33)
	SE (d)	1.953	1.2502	0.1472
	CD (0.05)	4.2567	2.7240	0.3207

*Values in parenthesis are arcsine transformed.

the MS medium. The different growth regulators like BAP (3.0, 4.0, 5.0 mgL^{-1}), GA_3 (0.5 and 1.0 mgL^{-1}) and NAA (0.5, 1.0 mgL^{-1}) were used in the medium both separately and in combinations.

RESULTS AND DISCUSSION

In the present study on *G. superba*, the sterilization procedure reduced the degree of contamination of explant (tuber node) to 8.00% when it was surface sterilized with 70% ethanol for 30 s followed by 60 s in $HgCl_2$. When the herbaceous part of explants was treated with lower concentration of sterilants with lower period of exposure, the soft tissues were not affected (Table. 1). This also supports the results of Custers and Bergervoet (1994), Sivakumar and Krishnamoorthy (2000) and Sivakumar et al. (2003) in *G. superba,* that surface sterilization using 0.1% mercuric chloride for 5 and 10 min respectively produced contamination free cultures. Similar reports were also given by Hassan and Roy (2005).

Micro tuber formation

Finnie and Vanstaden (1989) reported that multiple productions of tubers were important rather than shoot *in vitro.* The highest response (100%) to *in vitro* tuber formation in glory lily in the present study was observed on MS medium supplemented with 5.0 mgL^{-1} BAP +1.0 mgL^{-1} NAA. Similar report was published by Sivakumar et al. (2003) who stated that auxin and cytokinin were the effective inducers of micro tuberization in different yam species cultured *in vitro*, although, certain types of these regulators are more effective than others for increasing micro tuber induction frequencies. For example, naphthalene acetic acid (NAA) and kinetin (KN) have been shown to promote micro tuber induction in *G. superba* (Suprio, 2011) and *Dioscorea bulbifera* (Ammirato, 1982) and *Dioscorea alata* (Mantell and Hugo, 1989).

In the present study, response to micro tuber formation was higher only in the medium supplemented with NAA at 1.0 mgL^{-1}. Similar results were given by Alizadeh et al. (1998), in which NAA at 1.0 and 2.0 mgL^{-1} respectively, showed promotive effect but the effect was short termed and less than one tuber was obtained from single explants in the course of 12 weeks on MS medium supplemented with 2.0 mgl^{-1} NAA.

In the present study, medium supplemented with BAP and NAA resulted in maximum number (1.77) of micro tuber formation. The medium supplemented with BAP and NAA gave maximum (100%) response to primary and secondary tuber formation (Table 2). This result is in accordance with Islam et al. (2004) who stated that BAP and NAA yielded a better response of *in vitro* micro rhizome induction in *Curcuma longa*. Nayak (2000) and Sharma and Singh (1995) also reported that 5.0 mgL^{-1} BAP in *Curcuma aromatica* and 8.0 mgL^{-1} BAP in ginger, respectively enhanced micro rhizome production. Iglesias et al. (1999) reported, that the combination of BA and NAA 0.5 µM gave better response of bulblet formation *in vitro* in *Hyacinthoides paivae*.

Micro rooting

In the present study, maximum value for response to rooting (49.41%), number of roots (3.06), root length (1.90 cm) was observed in the MS medium supplemented with 1.0 mgL^{-1} IAA + 0.5 mgL^{-1} IBA, while the medium with either IAA or IBA alone exhibited a poor response for the aforementioned characters (Table 3). This results are in accordance with Sayeed and Shyamal (2005) who stated that rooting in regenerated shoots of *G. superba* was achieved at 90% when the excised shoots were cultured individually on root induction medium, consisting of half strength MS salts with 1.0 mgL^{-1} IAA + 0.5 mgL^{-1} IBA. Similarly Suprio (2011) also reported that NAA (0.5 mg/L) and IBA(0.1 mg/L) was the best *in vitro* rooting medium for *G. superba*.

Table 2. Effect of growth regulators on response to primary and secondary tuber formation, days for tuber formation and number of tubers formed per explants.

Treatment no.	Growth regulators (mgL^{-1})			Response to tuber formation (%)				Days for tuber formation (days)		Number of tubers per explants
	GA_3	BAP	NAA	Primary tuber		Secondary tuber		Primary tuber	Secondary tuber	
T_1	-	-	-	61.66	(51.80)	0	(0.55)	33.11	0.00	0.77
T_2	0.5	-	-	35.00	(36.27)	0	(0.55)	31.11	0.00	0.44
T_3	1.0	-	-	56.66	(48.86)	0	(0.55)	31.55	0.00	0.66
T_4	-	5.0	-	43.33	(41.15)	0	(0.55)	21.77	0.00	0.55
T_5	1.0	5.0	-	68.33	(55.97)	78.33	(43.73)	21.00	54.55	1.22
T_6	-	10.0	-	41.33	(41.16)	0	(0.55)	21.55	0.00	0.55
T_7	1.0	10.0	-	75.00	(60.31)	60	(50.95)	22.55	55.66	1.33
T_8	-	4.0	1.0	95.00	(79.36)	86.66	(69.24)	21.55	50.33	1.44
T_9	-	5.0	1.0	100.00	(89.45)	100	(89.45)	19.88	45.11	1.77
Mean				64.25	(56.04)	33.51	(28.46)	24.89	22.86	0.97
SE (d)				4.1070		10.5942		1.4399	8.5751	0.2100
CD (0.05)				8.6286		22.2578		3.0253	18.0159	0.4412

*Values in parenthesis are arcsine transformed.

Table 3. Effect of growth regulators on response to rooting (%), days for rooting (days), no. of roots and length of root (cm).

Treatment no.	Growth regulators (mgl^{-1})		Response to rooting (%)		Days for rooting (days)	Number of roots per plantlet	Root length (cm)
	IAA	IBA					
T_1	0.5	-	41.33	(40.00)	27.94	1.46	0.76
T_2	0.5	0.5	43.18	(41.07)	22.65	1.80	1.20
T_3	0.5	1.0	46.00	(42.70)	18.99	1.96	1.76
T_4	1.0	-	46.26	(42.85)	15.83	1.86	1.26
T_5	1.0	0.5	49.41	(44.65)	13.60	3.06	1.90
T_6	1.0	1.0	46.78	(43.15)	17.14	2.63	1.83
Mean			45.49	(42.40)	19.36	2.133	1.45
SE(d)			0.7887		2.9799	0.3226	0.3388
CD(0.05)			1.7185		6.4928	0.7029	0.7383

*Values in parenthesis are arcsine transformed.

REFERENCES

Alizadeh S, Mantell, Viana AM (1998). *In vitro* shoot culture and microtuber induction in steroid yam *Dioscorea composita* Hemsl. Plant Cell Tissue Org. Cult., 53: 107-112.

Ammirato PV (1982). Growth and morphogenesis in cultures of the monocot yam, *Dioscorea*. In: A. Fujiwara, Editor, Plant Tissue Culture, Maruzen, Tokyo, pp. 169–170.

Custers JBM, Bergervoet JHW (1994). Micropropagation of Gloriosa: Towards a practical protocol. Sci. Hortic., 57: 323-334.

Finnie IF, Vanstaden J (1989). *In vitro* propagation of Sandersonia and Gloriosa. Plant Cell Tissue Org. Cult., 19: 151-158.

Hassan AKMS, Roy SK (2005). Micropropagation of *Gloriosa superba* L. through high frequency shoot proliferation. Plant Tissue Cult., 15(1): 67-74.

Iglesias I, Feijoo MC, Feknandez SF (1999). Temperature and

photoperiod influences on *in vitro* bulb formation of *Hyacinthodes paivae.* Acta Hortic., 486: 339-342.
Islam MA, Kloppstech K, Jacobsen HJ (2004). Efficient procedure for *in vitro* micro rhizome induction in *Curcuma longa* L.(Zingiberaceae)-A medicinal plant of tropical asia. Plant tissue Culture, 14(2): 123-134.
Krause J (1986). Production of gloriosa tubers from seeds. Acta Hortic., 177: 353-360.
Mantell SH, Hugo SA (1989). Effects of photoperiod, mineral strength, inorganic ammonium, sucrose and cytokinin on root shoot and microtuber development in shoot cultures of *Dioscorea alata* L. and *D. bulbifera* L. yams, Plant Cell Tissue Org. Cult., 16: 23–37.
Nayak S (2000). *In vitro* multiplication and microrhizome induction in *Curcuma aromatica* Salisb. Plant Growth Regular. 32: 41-47.
Sayeed HAKM, Shyamal KR (2005). Micropropagation of *Gloriosa superba* L. Through High Frequency Shoot Proliferation. Plant Tissue Cult., 15(1): 67-74.
Seemanti G, Biswajit G, Sumita J (2007). *In vitro* tuberization of *Gloriosa superba* L. on basal medium. Sci. Hortic., 114(3): 220-223.
Sharma TR, Singh BM (1995). *In vitro* microrhizome production in *Zingiber officinale* Rosc. Plant Cell Rep., 15: 274-277.
Sivakumar G, Krishnamurthy KV (2000). Micropropagation of *Gloriosa superba* L—an endangered species of Asia and Africa, Curr. Sci., 78 (1): 31–32.
Sivakumar G, Krishnamurthy KV, Rajendran TD (2003). *In vitro* corm production in *Gloriosa superba* L., an ayurvedic medicinal plant, J. Hortic. Sci. Biotechnol., 78 (4): 450–453.
Suprio S (2011).Studies on micropropagation and *in vitro* production of medicinally important bio-molecules and their enhancement in *Gloriosa superba* Linn. Ph.D. Thesis submitted to Tamil Nadu Agricultural University, Coimbatore

Allometric equations for estimating biomass in naturally established *Prosopis* stands in Kenya

Gabriel M. Muturi[1,2]*, Jason G. Kariuki[1], Lourens Poorter[2] and Godefridus M. J. Mohren[2]

[1]Kenya Forestry Research Institute, P. O. Box 20412, 00200-Nairobi, Kenya.
[2]Forest Ecology and Forest Management Group, Wageningen University, P. O. Box 47, AA Wageningen, The Netherlands.

Forty five *Prosopis* stems of 2.5-18.0 cm diameter at breast height (DBH) were sampled at Nadapal along the Turkwel riverine forest for development of biomass and volume prediction equations for naturally established stands. Basal diameter (D_{30}), DBH and heights were measured, felled trees and their volumes, fresh and dry weights determined. Linear and power models were evaluated for volume and biomass prediction through regression analysis of measured tree parameters. Power models yielded better results than linear models in volume and biomass prediction, with D_{30} and DBH being more reliable than height. Validation of models at two sites in Marigat and Bura, revealed strong significant correlations between predicted and measured tree biomass and volumes, suggesting effectiveness of the models in biomass prediction across sites. Subsequently, model development and model validation data were pooled to develop national models. Basal diameter was found to be the best variable in the development of power models for biomass and volume prediction across the country. When logarithmically transformed, biomass and volume per tree had strong significant linear relationship with basal diameter, and are accordingly recommended for quick biomass and volume estimation in the field.

Key words: Biomass, Bura, Marigat, Nadapal, Turkwel, power models, linear models.

INTRODUCTION

Wood biomass is the main source of energy in Kenya, with notable localized deficits (Kirubi et al., 2000; Okello et al., 2001; Ng'etich et al., 2009). Provision of wood fuel was one of the main objectives of *Prosopis* species introduction in the country (Mwangi and Swallow, 2008). This objective is not yet fully realized, despite the successful establishment of the species in the initial trials (Maghembe et al., 1983; Olukoye et al., 2003) and their subsequent spread into other unintended areas, as a result of natural seed dispersal by livestock, wild animals and water (Mwangi and Swallow, 2008; Mworia et al., 2011). Recently, the government gave a policy direction for utilization of *Prosopis* species to control their further spread in areas where they are invasive (GOK, 2007). Exploitation of the species for wood fuel, as initially intended, provides excellent opportunities for invasion control and reducing wood biomass energy deficits.

However, management guidelines, particularly for biomass quantification are yet to be developed. The species composition of natural stands is also poorly understood because introduced species such *Prosopis chilensis* Stunz, *Prosopis juliflora* (Sw.) and *Prosopis pallida* Kunth are difficult to differentiate because of similar morphology and formation of hybrids of intermediate phenotypes (Saidman et al., 1996; Pasieznik et al., 2001; Landeras et al., 2006). These species could be coexisting in any of the natural stands.

Allometric equations are used to predict tree and stand biomass, based on easily measured tree variables such height, diameters and crown. Normally, developed equations are specific to species, sites, tree age and management (Lott et al., 2000; Claesson et al., 2001; Kairo et al., 2009), thus limiting their generalized transferability.

*Corresponding author. E-mail gmuturi@kefri.org, gabriel.muturi@wur.nl.

Although biomass prediction models were developed in the early stages of *Prosopis* species introduction, the models were for localized use in *P. juliflora* plantation trial (Maghembe et al., 1983) and comparison of species growth within species screening trials (Rosenschein et al., 1999). Therefore, their application in naturally established *Prosopis* species stands is limited by species composition, sites variations and the lack of any prescribed management in the naturally established stands.

Basal or stump diameter, height, crown area and depth are the commonly used biomass predictors in young trees and multistemmed trees and shrubs (Maghembe et al., 1983; Eshete and Stahl, 1998; Kariuki et al., 2007; Kaonga and Bayliss-Smith, 2010; Zeng et al., 2010). For *Prosopis* species, basal diameter is one of the most commonly used variable for biomass prediction (Maghembe et al., 1983; Elfadl et al., 1989; Duff et al., 1994) and height inclusion found to improve biomass prediction models (Padron and Navarro, 2004). Although basal diameter is not always specified in literature (Maghembe et al., 1983; Duff et al., 1994), diameter measured at 30 cm above the ground (D_{30}) is more frequently used as basal diameter (Eshete and Stahl, 1998, Padron and Navarro, 2004; Kariuki et al., 2007) and was thus adopted for this study. Furthermore, accurate tree height and crown measurement may not be feasible in naturally established dense *Prosopis* stands, because of interlocking tree canopies. The objective of this study was therefore to evaluate the suitability of diameters (D_{30} and DBH) as variables for biomass estimation in naturally established *Prosopis* stands.

MATERIALS AND METHODS

Sites descriptions

Sites for this study were identified at Katilu and Nadapal during a related study on prediction of *Prosopis* species invasion in Kenya (Muturi et al., 2010). The structure of DBH for *Prosopis* stands in the two sites was evaluated using sixteen intensive sample plots (Barnett and Stohlgren, 2003). Subsequently, Nadapal site was selected for allometric models development. Two more sites: - Marigat and Bura were selected for validation of the developed models. All the sites are within arid areas of Kenya that are characterized by low erratic rainfall, high temperatures and high evapotranspiration potentials, leading to high soil moisture deficits (Sombroek et al., 1982). However, Nadapal and Bura are riverine sites where microclimate is modified by forests, periodic flooding and river flow. Rainfall in all the sites is bimodal with peaks around April and November. Mean annual rainfall along the Turkwel riverine forest ranges from 500 mm upstream to less than 200 mm downstream, with high inter-annual variations (Reid and Ellis, 1995; Stave et al., 2005). The mean rainfall at Katilu is higher (≈ 350 mm) than at Nadapal (≈ 200 mm). Temperatures in the vast arid areas adjacent to Turkwel riverine forest ranges from 28-40°C. At Marigat, the mean annual rainfall is estimated at ≈500 mm with mean temperatures of 24-34°C (Ekaya et al., 2001; Kipkorir et al., 2002). Bura has a mean annual rainfall of 370 mm and mean daily temperature of 28°C (Maingi and Marsh, 2006). The geographical locations of the sites are shown in Figure 1.

Tree sampling

Thirty trees from the site selected for allometric equations development were randomly sampled and their height, diameters (D_{30} and DBH), fresh weight and volumes measured in the field. The trees were sampled to represent the known range of diameters for the species in the area, at specified intervals of 2 - 3 cm classes. Tree height was measured with a measuring rod, while diameters were measured using a diameter tape. The choice of basal diameter measurement point (30 cm from ground surface) was influenced by the multistemming of stumps, as most *Prosopis* trees were found to branch at beyond 30 cm from the ground. As the number of merchantable stems (DBH ≥2.5 cm) per stump were occasionally greater than 1, the number of stems used in model development was 45. Each measured tree or merchantable stem was felled and branches for volume and biomass measurement cut off at diameters of ≥1.5 cm (the lower limit for woody material considered suitable for use as fuel wood). Several tree discs or segments totaling about 1kg were sampled at 2.5 m intervals from the base of each tree upwards for dry weight determination. For multistemmed trees, only one stem was sampled and uniformity of tree characteristics assumed for the other stems. The actual weight of all the sampled tree components was then determined with a spring balance and sample volumes determined immediately using water displacement method. In this method, a bucket was placed in a large container, water filled to the brim and tree samples submerged in water. Displaced water was collected in the container where the bucket was placed and the volume of water displaced by each sample measured with graduated containers.

For determination of whole tree weight and volume, trees were cut into small sizes immediately after felling, depending on the depth of the bucket used in water displacement. Tree segments of weights that could be easily lifted were fastened together with a sisal twine and weighed with a spring balance until the entire tree materials were exhausted; and their volumes determined using the aforementioned water displacement method. Volumes and weights were then recorded separately for each tree or merchantable stem.

Samples collected for moisture content determination were kept under field conditions while in field, and at room temperature after each day, until the field sampling was complete. Samples were then taken to KEFRI laboratory, weighed and oven dried at 105°C for 72 hours and re-weighed daily until constants weight was achieved for each sample as previously described (Maghembe et al., 1983; Padron and Navarro, 2004). Percentage dry weight of the samples was determined after the drying process and the values obtained for each tree sample used to convert fresh weight into dry weight for the respective tree. The procedure was repeated for trees sampled at Marigat and Bura for model validation. The characteristics of trees in the three sites are summarized in Table 1.

Modeling

Square and log transformed (Duff et al., 1994; Padron and Navarro, 2004; Alvarez et al., 2011) and untransformed (Maghembe et al., 1983; Padron and Navarro, 2004) basal diameters have been used for *Prosopis* biomass prediction depending on the species and nature of the stand studied. Therefore, transformed and untransformed data were evaluated for model development in the current study, using linear and power regression between tree height, D_{30} and DBH as the independent variables and tree volume or biomass as the dependent variables. Goodness of fit and model comparisons were evaluated using coefficient of determination (R^2), the standard error (SE) and F values (Sokal and Rohlf, 1981). Models with high R^2, low se and high F value were selected and used to predict volume and tree biomass for trees sampled at Marigat and Bura. Subsequently, linear regression was used to test the effectiveness of selected models in predicting tree volume and

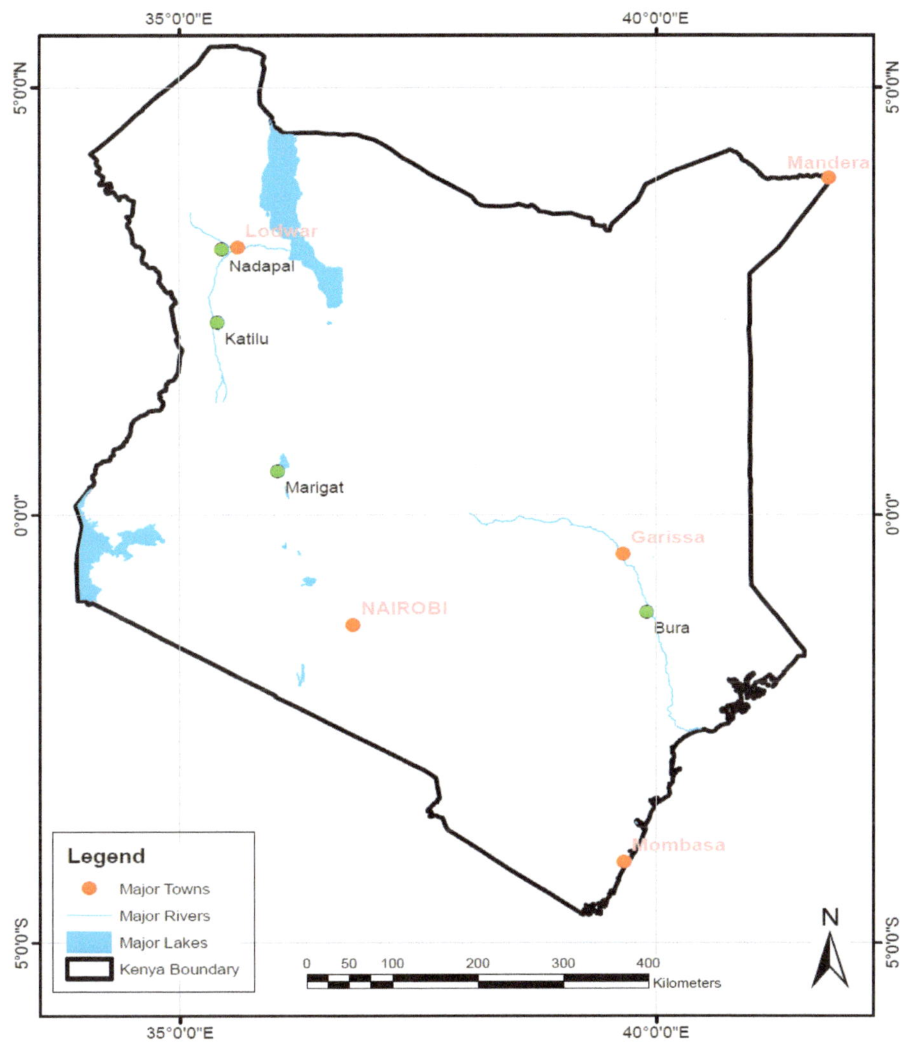

Figure 1. Approximate geographical locations of data collection sites (circular symbols) in Kenya. Nadapal, Katilu and Bura are riverine sites, whereas Marigat is a flat low lying site that receives seasonal runoff.

Table 1. Summary of tree variables measured at the three sites.

Site	Number of trees	D_{30} (cm)	DBH (cm)	Height (m)	Dry weight (kg/tree)
Nadapal	30(45)	2.8 - 18.5	2.5 - 18.0	3.6 - 10.4	0.50 - 180.1
Marigat	11	3.6 - 18.0	2.5 - 17.5	3.9 - 12.0	1.10 - 148.8
Bura	10	2.6 - 18.0	2.5 - 16.8	6.3 - 11.5	0.76 - 134.0

Range of basal diameter (D_{30}), diameter at breast height (DBH) and dry weight of *Prosopis* trees sampled at Nadapal, Marigat and Bura. At Nadapal, 30 trees were sampled but individual stems were 45 because some trees multistemmed.

biomass using measured volumes and weights as the independent variables and those predicted as the dependent variables. The best models based on R^2 and standard error were selected, the data from the three sites pooled and variables in the selected models used to develop overall models for predicting tree biomass and volumes in naturally established stands. The selected models were tested in predicting tree biomass along the Turkwel using data collected from the DBH structure evaluation plots. Statistical variation of selected models was evaluated with paired t-test, using predicted tree weights or volume for diameter models (D_{30} and DBH) and diameters and multiplicative (H*D) models. All statistical analyses were carried out using SPSS.

RESULTS

The DBH distribution structures for stands at Katilu revealed

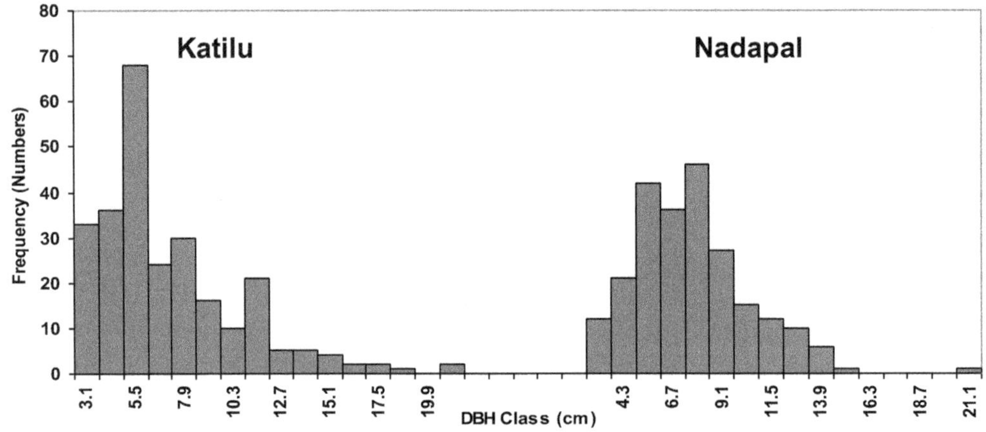

Figure 2. Structure of *Prosopis* diameter at breast height (DBH) at Katilu and Nadapal.The dbh structure at Nadapal was a near normal distribution compared to the dbh at Katilu, hence a stand at Nadapal was selected for biomass and volume prediction models development.

a mixed trend while a near normal distribution trend was found at Nadapal (Figure 2). The Nadapal site was therefore considered more suitable for model development. Preliminary evaluation of model development using transformed and untransformed data could not justify data transformation as reliable models were obtained with untransformed data. Untransformed data was therefore used for allometric equations development. Power models ($y = aX^b \pm e$) were stronger than linear models ($y = aX + b \pm e$) in relating tree volumes, fresh weight and dry weights to tree diameters and height (Table 2). This fact is exemplified by comparison of height and D_{30} multiplicative power ($Y = aHD_{30}{}^b$) and linear ($Y = aHD_{30} + b$) models for fresh weight, volume and dry weight (Table 2). From this comparison, both R^2 and F values were higher in power than in linear models, while a contrary trend was observed for standard error. Compared to diameter models, height models were weak on all comparison attributes (Table 2). Statistical comparisons of predicted tree biomass and volumes revealed insignificant differences between diameters models and between comparable diameter and multiplicative models (t-test data not shown).

The diameter and multiplicative models selected for validation at Marigat and Bura were all effective in correlating predicted and measured tree weights and volumes, with DBH model yielding the best results (Table 3). Multiplicative models revealed mixed results (Table 3).

Prediction was slightly better at Bura than Marigat, when models parameters were compared between the two sites (Table 3). However, this was statistically insignificant as exemplified by the overlapping linear models fitted between predicted and measured dry weight and volume, using D_{30} models (Figure 3).

Results for models developed using pooled data from all sites is shown in Table 4. From the table, D_{30} models were superior to DBH models according to model evaluation parameters (R^2, F value and SE), and compared favourably with the corresponding multiplicative models. Linear D_{30} models were therefore developed using log transformations for biomass and volumes estimation in the field (Figure 4), while the power dry weight model was used to estimate dry biomass in plots used for DBH structure evaluation. The predicted biomass ranged from 769 to 18,953 Kg per ha, depending on variation of mean plot basal diameter.

DISCUSSION

Prosopis stands in Kenya can be described as very dynamic in terms of diameter distribution structures (Figure 2). This observation may be associated with invasion trends, a spontaneous process that can vary with site conditions. For example, Katilu is an irrigated agriculture area, where invasion is primarily driven by continuous farm abandonment unlike Nadapal where abandonment was incidental, according to local informants and a previous study (Muturi et al., 2010). From the current study, it was apparent that basal diameter was consistently the best variable for predicting both tree volume, and tree biomass, a finding that was consistent with previous studies on *Prosopis* allometric models (Maghembe et al., 1983, Elfadl et al., 1989, Duff *et al.,* 1994). Although multiplicative models were slightly better than basal diameter models (Table 2), multiplicative models lead to a higher error in models validation (Table 3), a factor that can be attributed to the challenges of accurate height measurement in closed canopy *Prosopis* stands. Since DBH models were also highly effective in biomass and volume prediction, the study confirms the suitability of diameters in volume and biomass prediction in naturally established *Prosopis* stands where canopy and height cannot be accurately measured.

Table 2. Models for tree volumes and weight prediction for tree samples collected at Nadapal.

Equation	R^2	SE	F	P-|value
Fresh weight equation				
$Fw = 0.0004H^{5.9471}$	0.76	0.64	138.6	<0.01
$Fw = 0.1283D_{30}^{2.5194}$	0.97	0.21	1641.3	<0.01
$Fw = 0.2772DBH^{2.3624}$	0.94	0.33	646.1	<0.01
$Fw = 0.0134H*D_{30}^{1.9096}$	0.98	0.17	2589.5	<0.01
$Fw = 0.0276H*DBH^{1.812}$	0.95	0.29	847.3	<0.01
$Fw = 1.3554H*D_{30} - 37.071$	**0.92**	**16.3**	**494.0**	**<0.01**
Volume equation				
$V = 0.0002H^{6.3337}$	0.74	0.72	122.7	<0.01
$V = 0.0664D_{30}^{2.7308}$	0.98	0.20	2047.0	<0.01
$V = 0.1578DBH^{2.5446}$	0.93	0.37	576.8	<0.01
$V = 0.0059H*D_{30}^{2.0608}$	**0.98**	**0.20**	**2116.4**	**<0.01**
$V = 0.0134H*DBH^{1.9464}$	0.93	0.35	666.9	<0.01
$V = 1.1298 H*D_{30} - 29.898$	**0.94**	**11.7**	**666.4**	**<0.01**
Dry weight equation				
$Dw =0.0001H^{6.3237}$	0.75	0.70	131.3	<0.01
$Dw = 0.0483D_{30}^{2.6906}$	0.97	0.24	1398.7	<0.01
$Dw = 0.1114DBH^{2.5164}$	0.93	0.38	559.6	<0.01
$Dw = 0.0044H*D_{30}^{2.0372}$	**0.98**	**0.21**	**1844.2**	**<0.01**
$Dw = 0.0096H*DBH^{1.9293}$	0.94	0.34	695.9	<0.01
$Dw = 0.7833H*D_{30} - 21.619$	**0.89**	**10.7**	**381.9**	**<0.01**

Power models were better than linear models based coefficient of determination (R^2), standard error (SE) and F value (F), as exemplified by HD_{30} models (in bold) for fresh weight, volume and dry weight prediction.

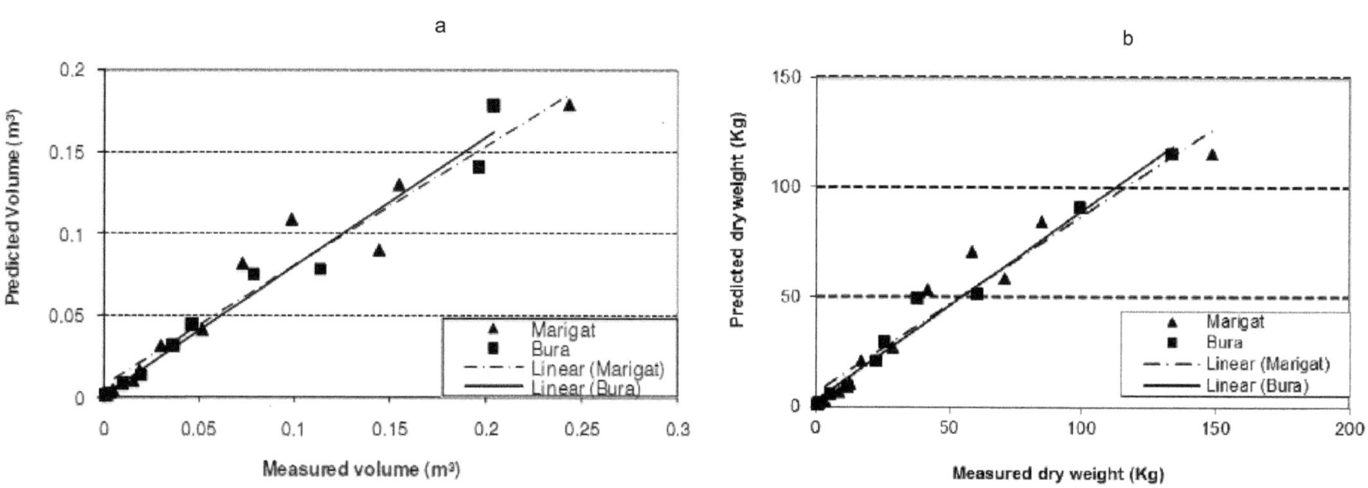

Figure 3. Scatter plots between measured and predicted tree volume (a) and tree dry weight (b). Scatter plots between measured and predicted tree volumes (a) and dry weight (b) for Marigat and Bura. Prediction was based on basal diameter (D_{30}) models. The fitted linear regression lines overlapped in both cases.

Trees sampled from Nadapal were representative of trees found in the other two *Prosopis* stands in Kenya, as evident from good model biomass and volumes prediction at Marigat and Bura (Table 3), despite the distance between the three sites (Figure 1). The slight difference in model validation outputs could be attributed to site variation, since models are sensitive to site conditions (Kairo et al., 2009), among other factors.

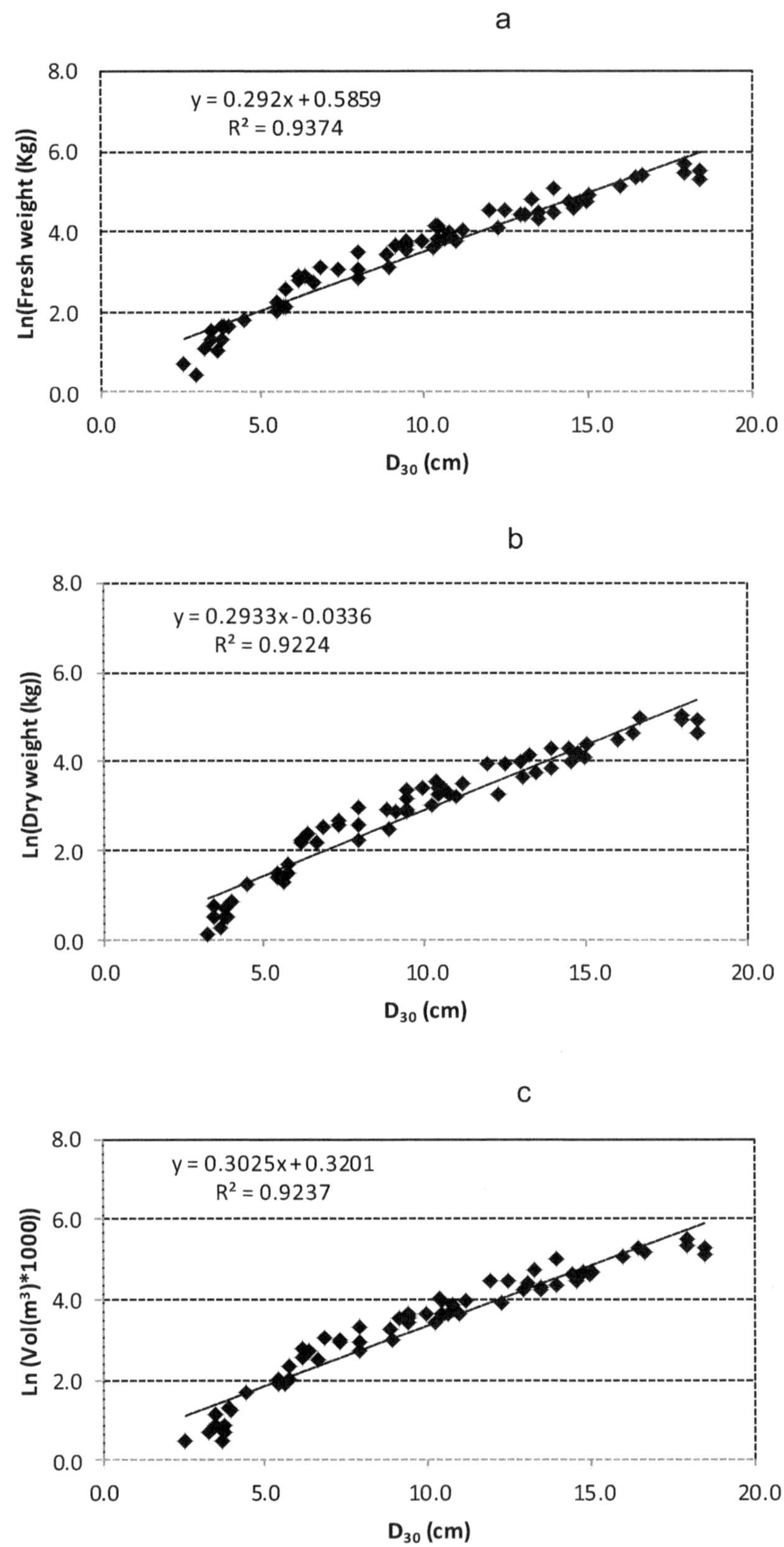

Figure 4. Relationship between basal diameter (D_{30}) and logarithmically transformed (Ln) fresh weight (a), dry weight (b) and volume (c), and basal diameter in *Prosopis* trees. For each case, a linear model (y=ax ±b) and coefficient of determination (R^2), and linear regression model lines are included in the scatterplot.

Table 3. Results of models validation at Marigat and Bura.

Parameter used	Marigat			Bura		
	R^2	SE	F	R^2	SE	F
Fresh weight (Kg)						
D_{30}	0.92	17.9	95.4	0.98	9.7	381.7
H^*D_{30}	0.93	30.9	106.6	0.97	19.2	299.1
DBH	0.97	13.7	273.5	0.99	7.1	1010.0
H*DBH	0.92	36.3	105.6	0.98	17.0	463.3
Volume (m^3)						
D_{30}	0.93	16.1	109.0	0.97	10.4	304.9
H^*D_{30}	0.89	38.6	65.3	0.96	24.6	172.7
DBH	0.98	10.9	403.6	0.99	8.1	707.9
H*DBH	0.89	44.7	67.5	0.97	23.0	240.1
Dry weight (Kg)						
D_{30}	0.94	10.1	116.2	0.98	5.9	402.8
H^*D_{30}	0.94	18.4	130.2	0.99	8.8	602.9
DBH	0.98	8.0	318.8	0.99	5.3	700.9
H*DBH	0.94	22.4	120.5	0.99	7.2	1062.8

Coefficient of determination (R^2), standard error (SE) and F values (F) for linear regression between predicted and measured tree volumes and weights for samples collected at Marigat and Bura. The measured and predicted weight or volume were based on either D_{30}, H^*D_{30}, DBH or H*DBH as the model variable. For all cases $P < 0.01$.

Compared to Marigat, Bura is a riverine site and sites conditions at Bura may compare favourably with those of Nadapal than site conditions between Marigat and Nadapal. However, the sensitivity of developed models to sites appeared negligible because of overlaps on fitted linear prediction models for both sites (Figure 3). Also site effects were mitigated by pooling all the data in development of final models (Table 4) and (Figure 3).

Biomass and volumes of sampled trees were lower than what was recorded for trees of comparable basal diameters under plantation conditions at the early stages of *Prosopis* introduction to Kenya (Maghembe et al., 1983), despite the use of a lower utilizable diameter branch limit in the current study. The possible explanation for this deviation could be variations in management, sites conditions, the species studied and lack of clarity in point of basal diameter measurement in the previous study. While the species studied under plantation condition was positively identified as *P. juliflora* and their initial management well prescribed (Maghembe et al., 1983), the trees that were sampled in the current study could be a mixture of *P. chilensis, P. juliflora* and *P. pallida*, following multiple introductions of *Prosopis* species, subsequent seeds exchange and seed dispersal by livestock and wildlife (Barrow, 1980; Herlocker et al., 1980; Paetkau, 1980; Mworia et al., 2011). The naturally established stands are also not subjected to any management.

Compared to three sites where trees were sampled in this study, Mombasa has an annual rainfall ≈1200mm (Maghembe et al., 1983), which is higher than in the sites of the current study. It is likely therefore, that soil moisture in the study sites was more limiting to tree growth than in Mombasa.

Stands densities in plots sampled for DBH structure evaluation was within the range found in areas of *Prosopis* invasions (van Klinken et al., 2006). Although comparative studies of biomass in invaded riverine ecosystems is generally lacking, the potential contribution of *Prosopis* to biomass energy was evident from the observed dry weight range. The predicted biomass provides a basis for planned resource exploitation, by selecting trees and or stands that have best fuel wood based on stem diameters.

Conclusions

The trees selected for model development were representative enough for trees found in other *Prosopis* stands in Kenya, as evident from the capacity of developed models to predict tree weights and volume at Bura and Marigat, the distance between sites notwithstanding. By pooling the data for model development and validation, the potential for site sensitivity was mitigated. Therefore, the developed models have a wide application in predicting *Prosopis* biomass and volume in natural stands, particularly in Kenya. Since diameter models were better than height models and height inclusion in multiplicative models did not result to significant model

Table 4. Power models for biomass and volume prediction developed with pooled data from Nadapal, Marigat and Bura.

Predicted parameter	Allometric equation	R^2	SE	F	P Value
D_{30} (cm)					
Fresh weight (Kg)	$Y=0.132X^{2.5301}$	0.98	0.21	2457.6	<0.01
Volume (m^3)	$Y=0.00008X^{2.7058}$	0.98	0.23	2506.1	<0.01
Dry weight (Kg)	$Y=0.0507X^{2.6759}$	0.97	0.23	2418	<0.01
DBH (cm)					
Fresh weight (Kg)	$Y=0.2539X^{2.3909}$	0.96	0.28	1385.0	<0.01
Volume (m^3)	$Y=0002X^{2.5497}$	0.95	0.32	1236.3	<0.01
Dry weight (Kg)	$Y=0.1067X^{2.5142}$	0.95	0.33	1087.7	<0.01
HD_{30}					
Fresh weight (Kg)	$Y=0.01395X^{1.8855}$	0.98	0.19	2948.9	<0.01
Volume (m^3)	$Y= 0.000007X^{2.0147}$	0.98	0.21	2785.0	<0.01
Dry weight (Kg)	$Y= 0.005X^{1.9797}$	0.96	0.27	1714.5	<0.01

Coefficient of determination (R^2), standard error (Se) and F value (F) of allometric equations developed from pooled data that was collected at Nadapal, Marigat and Bura. The equations were derived with basal diameter (D_{30}), diameter at breast height (DBH) and both Height (H) and D_{30} (HD_{30}).

improvement, the use of diameter for biomass and volume prediction is recommended. A choice can be made between use of D_{30} and DBH in biomass and volume estimation, depending on stand characteristics, because the difference between models prediction outputs were insignificant between the two diameters. However, use of D_{30} is highly recommended based on model evaluation parameters. The logarithmic linear D_{30} models recommended for biomass and volume estimation in the field are:

(a) Ln(Fresh weight (Kg)) = $0.292D_{30} + 0.59$ ($R^2 = 0.94$),
(b) Ln(Dry weight (Kg)) = $0.2933D_{30} - 0.03$ ($R^2 = 0.92$) and
(c) Ln(Volume (m^3)*1000) = $0.3025D_{30} + 0.32$ ($R^2 = 0.92$).

ACKNOWLEDGEMENTS

This study was financially supported by NUFFIC PhD grant No CF3671/2007 and KEFRI research grant to Gabriel Muturi. In the field Margaret Kuria, Simon Waweru, Daniel Ngasike, Cheptumo Relimoi, Koech, Stephen Kariuki, Peter Mwai and John Kimondo supported data collection and field logistics. Local communities in Turkana, Marigat and Bura were very cooperative and supportive during field data collection. Laban Maiyo assisted in production of sites location map, Mr. Bernard Kamondo provided valuable editorial inputs to an earlier version of the manuscript and anonymous reviewers helped to shape the paper to its current state. We are grateful for wide range of support provided by the above mentioned institutions and individuals.

REFERENCES

Alvarez JA, Villagra PE, Villalba R, Cony MA, Alberto M (2011). Wood productivity of *Prosopis flexuosa* DC woodlands in the central Monte: Influence of population structure and tree-growth habit. J. Arid Environ., 75: 7-13.

Barnett DT, Stohlgren TJ (2003). A nested-intensity design for surveying plant diversity. Biodivers. Conserv., 12: 255-278.

Barrow EGC (1980). " Multipurpose trees, current experience and future possibilities, in Buck, L. (editor) Proceedings on Kenya National Seminar on Agroforestry. 12-22 October 1980, Nairobi, pp. 499-505.

Claesson S, Sahlen K, Lundmark T (2001). Functions for biomass estimation of young *Pinus sylvestris, Picea abies* and *Betula* spp. from stands in northern Sweden with high stand densities. Scand. J. For. Res., 16: 138-146.

Duff AB, Meyer JM, Pollock C, Felker P (1994). Biomass Production and Diameter Growth of 9 Half-Sib Families of Mesquite (*Prosopis-Glandulosa* Var *Glandulosa*) and a Fast-Growing Prosopis-Alba Half-Sib Family Grown in Texas. For. Ecol. Manage., 67: 257-266.

Ekaya WN, Kinyamario JI, Karue CN (2001). Abiotic and herbaceous vegetation characteristics of an arid rangeland in Kenya. Afr. J. Range Forage Sci., 18:125-129.

Elfadl M, Gronski S, Asah H, Tipton A, Fulbright TE, Felker P (1989). Regression Equations to Predict Fresh Weight and 3 Grades of Lumber from Large Mesquite (*Prosopis-Glandulosa* Var *Glandulosa*) in Texas. For. Ecol. Manage., 26:275-284.

Eshete G, Stahl G (1998). Functions for multi-phase assessment of biomass in *acacia* woodlands of the Rift Valley of Ethiopia. For. Ecol. Manag., 105: 79-90.

Kairo JG, Bosire J, Langat J, Kirui B, Koedam N (2009). Allometry and biomass distribution in replanted mangrove plantations at Gazi Bay, Kenya. Aquat. Conserv.: Mar. Freshwat. Ecosyst., 19: S63-S69.

Kaonga ML, Bayliss-Smith TP (2010). Allometric models for estimation of aboveground carbon stocks in improved fallows in eastern Zambia. Agroforest, Syst., 78: 217-232.

Kariuki JG, Machua J, Luvanda AM, Kigomo JN, Muindi FK, Macharia EW (2007). Baseline survey of woodland utilization and degradation around Kakuma Refugee Camp. KEFRI/JOFCA Report, p. 40.

Kipkorir EC, Raes D, Massawe B (2002). Seasonal water production functions and yield response factors for maize and onion in Perkerra, Kenya. Agric. Water Manage., 56: 229-240.

Kirubi C, Wamicha WN, Laichena JK (2000). The effects of woodfuel consumption in the ASAL areas of Kenya: the case of Marsabit forest. Afr. J. Ecol., 38: 47-52.

Landeras G, Alfonso M, Pasiecznik NM, Harris PJC, Ramirez L (2006). Identification of *Prosopis juliflora* and *Prosopis pallida* accessions using molecular markers. Biodivers. Conserv., 15: 1829-1844.

Lott JE, Howard SB, Black CR, Ong CK (2000). Allometric estimation of above-ground biomass and leaf area in managed *Grevillea robusta* agroforestry systems. Agroforest. Syst., 49: 1-15.

Maghembe JA, Kariuki EM, Haller RD (1983). Biomass and nutrient accumulation in young *Prosopis juliflora* at Mombasa, Kenya. Agrofor. Syst., 1:313-321.

Maingi JK, Marsh SE (2006). Composition, structure, and regeneration patterns in a gallery forest along the Tana River near Bura, Kenya. For. Ecol. Manage., 236: 211-228.

Muturi GM, Mohren GMJ, Kimani JN (2010). Prediction of *Prosopis* species invasion in Kenya using geographical information system techniques. Afr. J. Ecol., 48: 628-636.

Mwangi E, Swallow B (2008). *Prosopis juliflora* invasion and rural livelihoods in Lake Baringo area of Kenya. Conserv. Soc., 6:130-140.

Mworia JK, Kinyamario JI, Omari JK, Wambua JK (2011). Patterns of seed dispersal and establishment of the invader *Prosopis juliflora* in the upper floodplain of Tana River, Kenya. Afr. J. Range Forage Sci., 28: 35-41.

Ng'etich KA, Birech RJ, Kyalo D, Bett KE, Freyer B (2009). Caught between Energy Demands and Food Needs: Dilemmas of Smallholder Farmers in Njoro, Kenya. J. Agr. Rural Dev. Trop., 110: 23-28.

Okello BD, O'Connor TG, Young TP (2001). Growth, biomass estimates, and charcoal production of *Acacia drepanolobium* in Laikipia, Kenya. For. Ecol. Manag., 142: 143-153.

Olukoye GA, Wamicha WN, Kinyamario JI (2003). Assessment of the performance of exotic and indigenous tree and shrub species for rehabilitating saline soils of Northern Kenya. Afr. J. Ecol., 41: 164-170.

Padron E, Navarro RM (2004). Estimation of above-ground biomass in naturally occurring populations of *Prosopis pallida* (H.& B. ex. Willd.) HBK in the north of Peru. J. Arid Environ., 56: 283-292.

Paetkau P (1980). Tree seed distribution report – August 1979 –July 1980. pp. 603-609 in Buck, L. (editor) Proceedings on Kenya National Seminar on Agroforestry. 12-22 October 1980, Nairobi.

Pasiecznik NM, Felker P, Harris PJC, Harsh LN, Cruz G, Tewarri JC, Cadoret K, Madonado LJ (2001). The *Prosopis juliflora-Prosopis pallida* complex: A monograph. HDRA Coventry, UK, p. 162.

Reid RS, Ellis JE (1995). Impacts of Pastoralists on Woodlands in South Turkana, Kenya - Livestock-Mediated Tree Recruitment. Eco. Appl., 5: 978-992.

Rosenschein A, Tietema T, Hal DO (1999). Biomass measurement and monitoring of trees and shrubs in a semi-arid region of central Kenya. J. Arid Environ., 42: 97-116.

Saidman BO, Vilardi JC, Pocovi MI, Acreche N (1996). Genetic divergence among species of the section Strombocarpa, genus *Prosopis* (Leguminosae). J. Genetics, 75: 139-149.

Sokal RRFJ, Rohlf FJ (1981). Biometry: the principles and practice statistics in biological research. San Francisco: W.H. Freeman, p. 859.

Sombroek WG, Braun HMH, van Der Pouw (1980). Exploratory soil map and agro-climatic map of Kenya. Scale 1:1,000,000. Kenya Soil Survey Nairobi, p. 56.

Stave J, Oba G, Stenseth NC, Nordal I (2005). Environmental gradients in the Turkwel riverine forest, Kenya: Hypotheses on dam-induced vegetation change. For. Ecol. Manage., 212: 184-198.

van Klinken RD, Graham J, Flack LK (2006). Population ecology of hybrid mesquite (*Prosopis* species) in Western Australia: how does it differ from native range invasions and what are the implications for impacts and management? Biol. Inv., 8: 727-741.

Zeng HQ, Liu QJ, Feng ZW, Ma ZQ (2010). Biomass equations for four shrub species in subtropical China. J. For. Res., 15: 83-90.

Tests of direct sowing of the durum wheat under pivot in the Saharan zone: Influence on yield components

M. O. Kheyar*, B. Mouhouche and M. Mahdi

Département du Génie Rural, Ecole Nationale Supérieure Agronomique El Harrach Alger (Algérie).

Our experiment is mainly based on the comparison of two methods of tillage leading of cereal cultivation under pivot in the arid zone, a case of El Meniâa (wilaya of Ghardaia). To achieve this, we have tested the technique of direct sowing in the Saharan zone by applying a random block experimental device with two treatments, T1 and T2, representing, respectively, a direct sowing and a sowing preceded by soil tillage by means of a double passage of the tine cultivator. The results show that the obtained yields reveal no significant difference between the two treatments, which means that the leading of cereal cultivation under pivot can do without a prior tillage.

Key words: Arid zones, cereal cultivation, pivot, tillage, direct sowing, yield.

INTRODUCTION

The intensive cereal cultivation leading in the highlighted perimeters in the arid zones (Adrar, El Oued, Ghardaia and Ouargla) constitutes an interesting palliative to the cereal production deficiency in the humid and sub-humid zones of Algeria. In these zones, climatic and edaphic conditions are extreme; therefore, the leading of winter cereal cultivation irrigated by pivot is totally different from that practised for the pluvial cereal cultivation in the North of Algeria.

Indeed, the climate is excessively arid and highly evaporating; thus, the prevailing winds have a negative impact on shallow, very fragile soils (Daoud and Halitim, 1994). Also, land exploitation in these regions is not without troubles, particularly in the section of soil handling mechanization (Kheyar et al., 2006; Amara et al., 2007).

Accordingly, the preparation of the seedbed can properly be envisaged only with instruments that do not turn the soil as the chisel and the cultivator. This practice which requires tine tools is similar to the simplified cultivation technique or minimum work. Nevertheless, the use of the tine cultivator comes along with an on-surface stone rise prejudicial to the successful achievement of the sowing and emphasizes wind erosion of the finest particles, thus entailing a decline of the soil fertility.

*Corresponding author. E-mail: m.kheyar@ensa.dz, kheyar2000@yahoo.fr.

Numerous comparative tests have shown that, in general, cereal yields are not so much influenced by the cultivation technique, but rather by the quality of the sowing (depth and regularity). This is confirmed by the work of Anken et al. (1999) which showed that cereal crops can bear a much reduced tillage, even none. For Gardner et al. (1999), quoting many authors, they indicate that the cereal yields are little affected by the tillage techniques.

Under pivot, when the production conditions are gathered (normally leached salts, rational irrigation, optimization of the fertilization and weed control), the direct sowing, regarding the numerous advantages that it offers, isn't it the alternative solution to the current method of leading?

It is thus to this question that we intend to bring an answer through field tests.

MATERIALS AND METHODS

Our experiment took place in El Meniâa (El Golea), an oasis in the South of Algeria situated at 270 km in the Southwest of Ghardaïa. Its geographical details are 30° 34' of latitude North and 0° 52' of longitude East, and 397 m of altitude (Figure 1).

The region of El-Meniâa belongs to the hyper-arid bioclimatic zone with low and erratic rainfall, of about 79 mm per year. The temperatures are very high and in strong daily amplitudes, the insolation and the luminosity are very intense, therefore, the evapotranspiration is significant. The frequent winds of the North can reach velocities from 11 to 15 m/sec. In this austere environment, the agricultural activity is inconceivable without irrigation

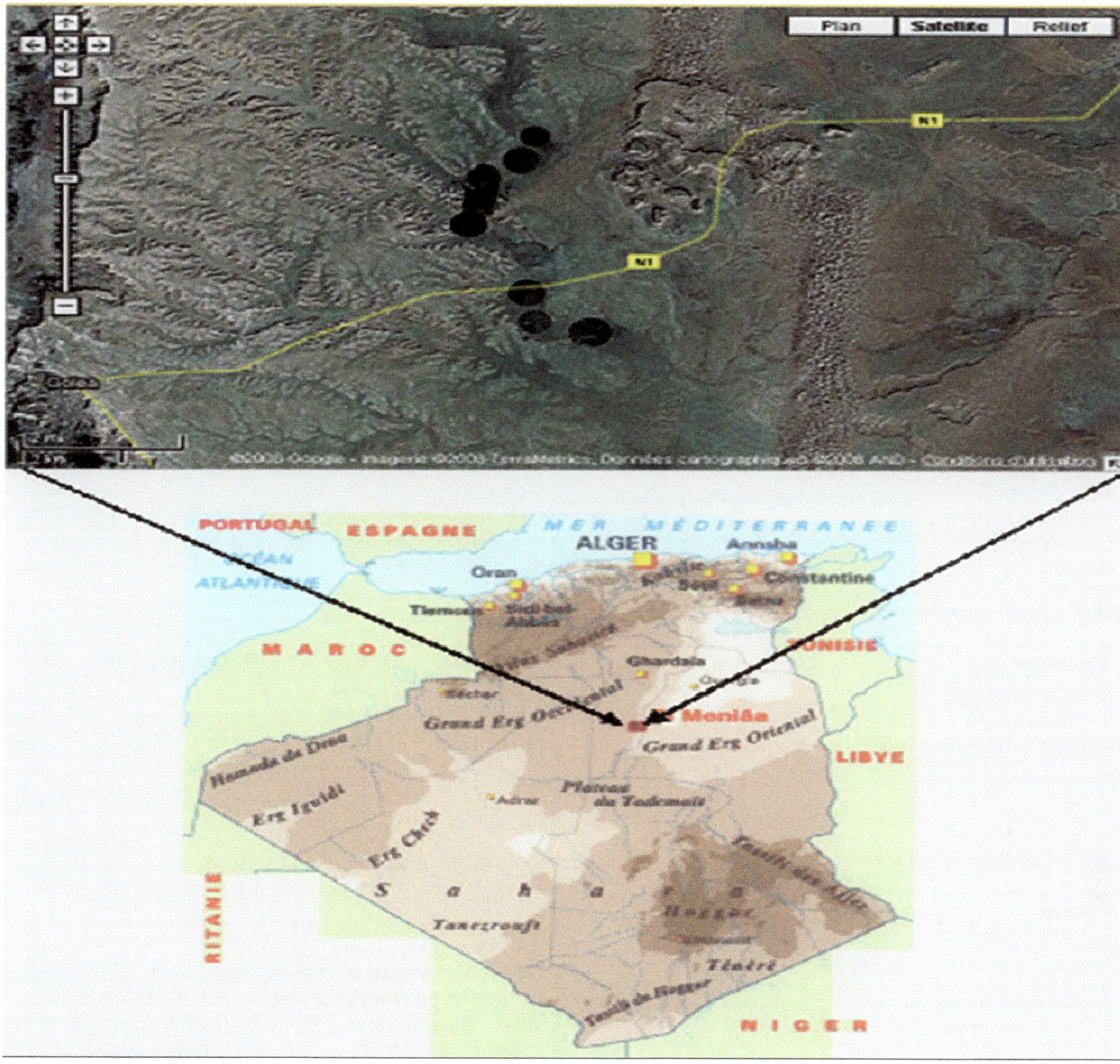

Figure 1. Location of test site.

and without an effective protection of the soil against wind erosion.

Field tests were conducted on pivot 8 where the soil is characterized by a sandy texture (99% of sand, 1% of silt, and 0% of clay), an average to low rate of stone loading in some places and a weak cover by the rests of the previous cultivation (barley).

In addition to adjusting the sowing depth, both seeders were tested in a fixed position to verify the uniformity of the distribution and to adapt the rotation speed of the distributing elements to the recommended seedling dose. The experimental device is a device with two treatments (a conventional sowing noted as CS and a direct sowing noted as DS) and three repetitions. Each elementary plot has a length of 100 m and a width of 18 m. A border of 0.5 m separates two adjacent plots. All relative data to the test are given in Table 1.

For each elementary plot and on small areas of 1 m^2, we have estimated to the three-leaf stage, the population density (raised stems/m^2) and to the maturity stage, the average height of stems, average root length, number of ears/m^2, the number of grains/ear and the thousand grain weight (TGW). The crop yield (q/ha) is the cumulative weight of grains in a small area per hectare.

RESULTS AND DISCUSSION

The regular follow-up of the plots from sowing to harvest has enabled the collecting of results in relation to crop growth and to yield components.

Table 1. Data relating to the test.

Operation	Conventional sowing	Direct sowing
False sowing	Pre-irrigation to help the removal of weeds	
Weeding	Pre-sowing herbicide : Round up	
Soil preparation	Two crossed passages with a tine cultivator allowance	None
Fertilization	Centrifugal fertilizer spreader	
Dose	Two proportions of 92 units of P_2O_5 (TSP 46%) One proportion of 100 units of K_2O (potassium sulphate 50%). Proportion of 280 units of nitrogen as UAN (urea, ammonia, nitrate).	
Seeder	On line seeder 'AGRIC' Ploughshare burying elements	Direct seeder 'VADERSTAD' Milled disc burying elements
Characteristics	Density: 350 (seeds/m²); TGW : 55 g ; Germinative faculty : 93% Dose : 214 kg/ha Sown population : 349158 plants/ha Sowing speed : 6 km/h	

Table 2. Sowing qualitative indications.

Parameter	Conventional sowing	Direct sowing	$p<0.05$	Statistical significance
Sowing depth (cm)	3. 35	2.25		
Covering rate %	97	85	0.195065	NS
Number of raised stems /m²	208.93	197.76	0.407759	NS

NS : not significant.

Table 3. Effect of the type of sowing on the growth.

Parameter	Conventional sowing	Direct sowing
Stem height (cm)	54.58	56.07
Root length (cm)	12.83	12.21
Standard deviation	2.04	2.17

Effect on the number of raised stems

Table 2 synthesizes the results of the qualitative indications of sowing (depth, seed covering rate and rising rate). For a sowing density of 350 seeds per m^2, the rising percentage in the conventional sowing is 59% (209 feet) compared with that of the direct sowing which is only 52% (198 feet). In direct sowing, the shallow depth of location (2.25 cm) and the insufficient seed covering (bad closure of the sowing line) by the burying elements of the direct seeder have contributed in reducing the rising percentage. Over 15% of seeds per m^2 have not been covered and remain on the surface exposed to predators (in particular crows which dig up the superficial or badly covered seeds). The speed of movement of the seeder, even if it is not very important (6 km/h), seems to be the main cause of this situation.

Effects on plant growth

The comparison of the results (Table 3) of plots conducted in conventional sowing and in directs sowing, the stem heights and the root length, do not show a significant difference. The analysis of variance, likewise, shows that the sowing method had no significant effect on both parameters measured.

As regards the development of the root system and in the presence of soils which are little fertile and which are very poor in organic matter, the lengths of the roots obtained are approximately identical and the maximum value does not exceed 13 cm. Field observations have shown that it is the spatial distribution of roots only that differs between the two sowing methods.

In direct sowing (undisturbed soil), the roots tend to develop horizontally. These results confirm those obtained by Dwyer et al. (1996) and Carof (2006) who report a significant concentration of roots on the surface in the no-till technique. This concentration is attributable to an increase in resistance of the soil to root penetration (Breune, 1997). This phenomenon is particularly marked

Table 4. Effect of the sowing method on the yield and its components.

Studied variable	Sowing type		p<0.05	Statistical significance
	Conventional	Direct		
Ears/m²	403.1	407.093	0.804066	NS
Grains/ear	29.03	30.34	0.573323	NS
Grains/m²	11702	12348	0.407759	NS
TGW (g)	48.83	52.94	0.090570	NS
Yield (q/ha)	56.35	65.38	0.076692	NS

NS: not significant.

Table 5. Working time to carry out the operations.

Technique	Tillage (h/ha)	Sowing (h/ha)	Total time (h/ha)
Conventional sowing	0.8	0.8	1.6
Direct sowing	00	0.8	0.8

in the presence of sandy soils susceptible to settling (Lesturgez, 2005). On the other hand, in the conventional sowing, initial soil disturbance by the cultivator tines has created a structural state favourable to the growth of roots that arrive up to the base of the searched area by the ploughing tool to reach the water and the nutrients (Nicou, 1974). Therefore, the roots follow the easiest path to assure their growth.

Effect on the yield and its components

The results presented in Table 4 indicate that the sowing method had no influence on the yield components; the TGW shows a difference of 4 g between the two types with a value of 53 g for the direct sowing against 49 g for the conventional sowing.

Concerning the number of ears in the m^2, and despite the weak density of plot population carried out in the direct sowing, the number of ears obtained is slightly higher (407 ears / m^2) than the conventional sowing (403 ears / m^2). The same observation was made by Harrad (2003) on similar soils. It seems that the weak density of population at the rise helps improve the number of ears per m^2. It is the same for the voluntary settling of the sowing line. The 'settling' pneumatic wheels equipping the direct seeder have enabled to assure a better contact of the seed with the soil, a contact which would also be responsible for the number of ears obtained.

Even if the statistical analysis confirms that the sowing type had no significant effect on the yield, it appears for the reasons previously mentioned that the direct sowing has enabled obtaining relatively better grain yield, thanks to an ear population, a number of grains per ear and a TGW higher than the conventional sowing; at least as regards our test conditions.

Evaluation of the execution time of the work

Table 5 gives the real estimation of the working time devoted to each cultivation operation according to the sowing type. For an average speed of movement of 6 km/h and considering only the available equipment at the farm (a tractor with 4 wheel drive, 3 m seeders, and a 4 m tine cultivator), the results which are limited to the execution time of the work only (labor not included) indicate that the time taken to complete the sowing in the conventional technique is double that of the direct sowing. Notwithstanding the operations common to both techniques (weeding and fertilization), we can deduce that limiting the interventions to the sole seeder helps to decrease the number of passages and to reduce soil settling (Faure, 1981). From an economic standpoint, even if the price of the direct seeder is high (USD 25 000), investment is offset by the time of executing the work much less than the conventional seeding. The time savings will translate into reduced fuel consumption, reduced average costs of mechanization charges and production of wheat (Mrabet, 2001).

Conclusion

Our preliminary results have shown that in a Saharan environment (light textured soils, very poor in organic matter), the disturbance of the soil by the tine cultivator does not provide any additional benefits with regard to the no-tillage represented by the direct sowing, particularly in terms of yield. On the other hand, the direct sowing has enabled the obtaining of a relatively better yield (65.38 q / ha against 56.35 q / ha) thanks to an ear population (407) and a higher number of grains per ear (30.34).

From the techno-economic point of view, the farm size (over 100 ha) combined with topography (flat land), to the nature of the soil (full with stones in general), is a motivation for the adoption of the direct sowing which enables reduce the mechanization costs and preserve the soils from physical degradation related to their work.

REFERENCES

Amara M, Kheyar MO, Harrad F (2007). Effet de la succession des machines agricoles pour la mise en place d'un blé sous pivot sur le développement des racines et conséquences sur le rendement. Sciences, Technol. Dév., 1: 22- 30.

Anken T, Irala E, Ammann H, Heusser J (1999). Travail du sol et mise en place des cultures Le travail du sol minimal convient parfaitement au blé d'automne Rapport FAT No. 534 Tänikon Suisse, 8p.

Breune I (1997). Le pénétromètre: un outil de détermination de la qualité structurale des sols Mémoire pour l'obtention du grade de maître ès sciences Université Laval, 88p.

Carof F (2006). Fonctionnement de peuplements en semis direct associant du blé tendre d'hiver (Triticum aestivum L.) à différentes plantes de couverture en climat tempéré. Thèse de doctorat Institut National Agronomique Paris .Grignon, 115p.

Daoud Y, Halitim A (1994). Irrigation et salinisation au Sahara algérien. Sécheresse, 5: 151-60.

Dwyer LM, Ma BL, Stewart D, Hayhoe WHN, Balchin D, Culley JLB, McGovern M (1996). Root mass distribution under conventional and conservation tillage. Can. J. Soil Sci., 76: 23-28.

Faure A (1981). Etude des tassements du sol. Tracteur et machines agricoles, 731: 31- 85.

Gardner C M.K, Laryea K.B, Unger P.W (1999). Soil physical constraints to plant growth and crop production land and water development division FAO Rome, 98p.

Harrad F (2003). Contribution à l'établissement d'un itinéraire technique pour la mise en place du blé dans les zones sahariennes (Adrar): Effet de la succession des outils aratoires sur le développement de la plante. Mémoire de magister INA Alger, 107p.

Kheyar MO, Amara M, Harrad F (2007) Mécanisation de la céréaliculture algérienne constat et perspectives de développement. Annales de l'INA, 28(1) : 2.

Lesturgez G (2005). Densification des sols sableux sous culture mécanisée Cas du Nord-est Thaïlandais thèse de doctorat Université Henri Poincaré Nancy (France), 164p.

Mrabet R (2001). Le semis direct : Potentiel et limites pour une agriculture durable en Afrique du Nord. Centre de développement sous-régional pour l'Afrique du Nord, Nations Unies, Commission Économique pour l'Afrique (CEA/TNG/CDSR/AGR), 32p.

Nicou R (1974). Contribution à l'étude et à l'amélioration de la porosité des sols sableux et sablo- argileux de la zone tropicale sèche conséquences agronomiques l'agronomie tropicale, 29: 15.

17

Assessment of regeneration potential of *Hibiscus sabdariffa* L. under established ecosystems in Cameroon

Anjah G. M.[1]*, Ogunsanwo O. Y.[2], Jimoh S. O.[2], Forjoh J. N.[1] and Tsombou F. M.[1]

[1]Department of Plant Biology, University of Dschang, Bp 67, Dschang, Cameroon.
[2]Department of Forest Resources Management, University of Ibadan, Nigeria.

***Hibiscus sabdariffa* is a non-timber-forest-plant (NTFP) that has the potential of producing fibers good in paper-making. Germination and growth potentials of *H. sabdariffa* under farmland, fallowland, plantation and forest ecosystems were studied. Seeds of *H. sabdariffa* were sown to assess germination percentage, germination rate, germination time and velocity coefficient on established plots of each ecosystem. Seedlings were randomly selected to measure growth and yield parameters of height, diameter, leaf number, fresh and dry weights. Data were subjected to ANOVA and means separated by LSD at P = 0.05. Values of germination percentage for fallowland (17.5) were significantly different from those of farmland (19.4) and plantation (19.1) ecosystems, while natural forest (18.9) had an interactive significance with farmland, plantation and fallowland. There was no significant difference between germination times among the ecosystems, but the coefficient velocity and rate of germination among the ecosystems were significantly different. There was a significant variation in growth and yield values among the ecosystems. However, only seedlings on farmland and fallowland survived with a survival rate of 96 to 86%, respectively. *H. sabdariffa* can thus, be sustainably regenerated on farmland and fallowland ecosystems for the extraction of stem fibres for paper-making.**

Key words: Non-timber-forest-plants, fibres, *Hibiscus sabdariffa,* ecosystems, growth, germination, yeild, regeneration, sustainability.

INTRODUCTION

All biotic and abiotic factors together within a certain area form an ecosystem (Edmonds, 1991). The structure and seasonality of the associated vegetation, complement with ecological data such as elevation, humidity, drainage, salinity of water and characteristics of water bodies are each determining modifiers that separate partially distinct sets of species (Vreugdenhil et al., 2003). In all ecosystems, there is a cycling of events from germination, growth, development and decomposition (Acquaah, 2004). Plant germination demarcates the transition from the seed being dependent on food sources from the mother plant, to an independent plant capable of taking up nutrients and growing independently (Schmidt, 2000). Hence, germination is the last link in the chain of seed handling processes. Germination proceeds when growth emerges from a period of dormancy leading to the development of a new plant (Raven et al., 2005). Environmental factors affecting germination and seedling establishment in different ecosystems may be positively correlated. However, spatial and temporal fluctuations in local variation in predation, competition and edaphic conditions can cause considerable discordance between seed bank longevity, germination and subsequent

*Corresponding author. E-mail: ngracemendi@yahoo.com.

seedling survival (Mukhtar, 2008). This in turn can lead to weak correlations between seed germination parameters and adult plant distribution.

Elfry et al. (1980) observed that photoperiodic effect was quite significant on reactions of roselle varieties. Kumar and Echekwa (1985) observed that germination for five roselle (*Hibiscus sabdariffa* L.) varieties was quite successful and seedling emergence started on the 4th day after sowing. Growth on the other hand, is often thought of simply as an increase in size (Acquaah, 2004) and is also affected by both biotic and abiotic factors. These changes in form and structure could be measured at various levels of growth using a variety of growth parameters of which are stem height, diameter and leaf number (Elfry et al., 1980). This pattern follows both for timber and NTFP. The ecosystem of West and Central Africa are rich in natural resources and have tremendous biodiversity particularly wood and non wood species providing food, fuel, fibre, timber medicines, construction and building materials (Ayuk et al., 1999). *H. sabdariffa* is an NTFP of the tropical region that have contributed to rural poverty alleviation. It is cultivated by the local population for bast fibre obtained from the stem (Ogunsanwo and Scotannd, 1989). The fibre which stands at about 1.5 m long is used for cordages and as substitute for jute in the manufacture of burlap (Duke, 2003). Concern for its development for pulp and paper is not unconnected with this use. Attempts have been made on the potential of these and related species to supply long time pulp (Oluwadare and Scotannd, 2007). These efforts have produced encouraging results. For instance research has shown that *H. sabdariffa* has potential to supply long fibres with most fibre having length between 2.1 to 5.0 mm which is categorised under long fibre (Oluwadare and Scotannd, 2007). However, research on the evaluation of germination and growth potential under established ecosystems are not known nor documented.

The ecosystems at which this species can best be established for sustainably regenerated for continuous supply of fibres are not known. Therefore the objective of this study is to assess the sustainable regeneration of *H. sabdariffa* by evaluating its germination, growth and yield potential under farmland, plantation, fallowland and forest ecosystems, respectively.

MATERIALS AND METHODS

Study site

This work was carried out in Tombel locality situated in the tropical coastal rainforest South West Region of Cameroun at an altitude of 2064 m and lying between 4° 45 N and 9° 40 E (www.encater, 2008). Tombel is highly volcanic with an extinct plutonic mountain characterised by a wet period of nine months of abundant rainfall and 3 months of dryness. It receives an annual rainfall of 2,500 mm and has an average annual temperature of about 21.20°C (http://www.pgdrn – gtz-Cameroun.org).

Site preparation

This work was carried out between February and October, 2009. The different ecosystems were identified and experimental sites leased out by owners. Field guides were present to ease accessibility into respective ecosystem plots. Plots of 10 × 10 m were demarcated using a metric tape on each established ecosystem which included a natural forest, a cocoa plantation, a farmland from which crops had just been harvested and abandoned or fallowland which had not been cultivated for close to five years. These ecosystems were cleared using a machete and ploughing of the soil was done with a hoe.

Germination

Seeds of *H. sabdariffa* were bought from the markets. One hundred seeds were sown per plot of each ecosystem at a planting distance of 50 cm. Holes of 2.5 cm were drilled using drillers on the ridges and one seed hand-dropped into one hole (Figure 1). These holes were then slightly covered with soil. The number of seeds that germinated per ecosystem was counted cumulatively and the germination percentage, germination rate, germination time and coefficient velocity were calculated. Calculation of velocity coefficient was by Labouriaul (1983). Coefficient velocity of germination:

$$CVG = \left(\sum_{i=1}^{k} fi \Big/ \sum_{i=1}^{k} fixi \right) 100$$

Where fi = number of newly germinated seedling on day i, Xi = number of day from sowing and K = last germination time.

Growth

Measurement of growth parameters of *H. sabdariffa* commenced one month after seeds were sown. 10 uniform seedlings were randomly selected from germination beds and tagged for growth measurements. Weeding was done manually during this period once every month. The plants were consecutively disinfected at the 2nd, 8th, and 22nd weeks of growth by spraying with Beauchamp 72% WP fungicide (50 g/16l of H_2O) and Kunfu 5% WP insecticide (10 g/16l of H_2O) using Berthoud sprayer to control fungi and insect attacks, respectively. Watering was done on a daily basis during the first two months (since it was just the beginning of the rains). Measurement of the stem length and diameter were done using a regulated tape and the number of leafs counted cumulatively.

Yield

Eighteen weeks after planting, each plot was divided into four subplots in order to vary the harvesting intensity. The subplots were denoted subplots 1, 2, 3, and 4, respectively. Subplot 1 was harvested firstly. This constituted the 25% harvesting intensity. Twenty-one weeks after planting, subplot 2 was harvested, which constituted harvesting at 50% intensity. Subplots 3 and 4 were harvested 25 and 28 weeks respectively after planting, constituting harvesting intensities of 75 and 100%, respectively. The harvesting

Figure 1. Seeds of *H. sabdariffaba* ready for planting.

Table 1. Percentage seed germination of *H. sabdariffa* in different ecosystems.

Ecosystem	Germination (%)		
	Days after planting		
	4	5	6
Forest	67.5	20.5	6.5
Farmland	78.0	10.5	7.5
Abandoned	61.5	21.0	5.0
Plantation	74.5	15.5	5.5

was done using sharp cutters specially designed. Entire crop for each subplot of each ecosystem were harvested and carried to the laboratory where the stem, leaf and fruit components were separated and weighed using an analytical top loading electronic balance (Ohams) to determine the maximum fresh weight yields. Harvesting at different intensities of 25, 50, 75 and 100% was carried out to determine the optimum level of yield and sustainability of plant in the different ecosystems.

Statistical analysis

Data collected were subjected to analysis of variance (ANOVA) as described by Cochran and Cox (1957) using statistical Microsoft Office Excel (2007). The treatment means were separated using least significant difference (LSD) for statistical significance at 5% confidence interval ($P<0.05$). Germination values were obtained as described by Oni and Gbadamasi (1998). Graphs were plotted to show the evolution of the growth parameters over time for each plant using Microsoft Office Excel (2007).

RESULTS

Germination

The seeds of *H. sabdariffa* started emerging from the soil 4 days after they were sown in all four ecosystems. The highest germination percentage was observed on day 3 of 78% for the farmland ecosystem (Table 1). No significant difference was observed between the farmland (19.4) and plantation (19.1) ecosystems; but these were significantly different from the fallowland (17.5) ecosystem. The fallowland ecosystem was not significantly different with the forest ecosystem (18.9) as seen on Table 2. Table 3 shows the values of germination time, germination rate, velocity coefficient of all the four ecosystems. The velocity coefficient was significantly different from the means of the germination

Table 2. Mean values for seed germination percentage variations of *H. sabdariffa* in ecosystems.

Ecosystems	Seed germination
Farm	19.4[a]
Plantation	19.1[a]
Natural forest	18.9[ab]
Fallowland	17.5[b]

Means values with the same letter from the different ecosystem are not significantly different at $P<0.05$.

Table 3. Mean values for germination time, rate and velocity coefficient.

Ecosystem	Germination parameters		
	Time (days)	Rate	Velocity coefficient (%)
Forest	3.4	0.19	29.8
Farmland	3.3	0.19	30.6
Abandoned	3.4	0.75	29.8
Plantation	3.3	0.19	30.6

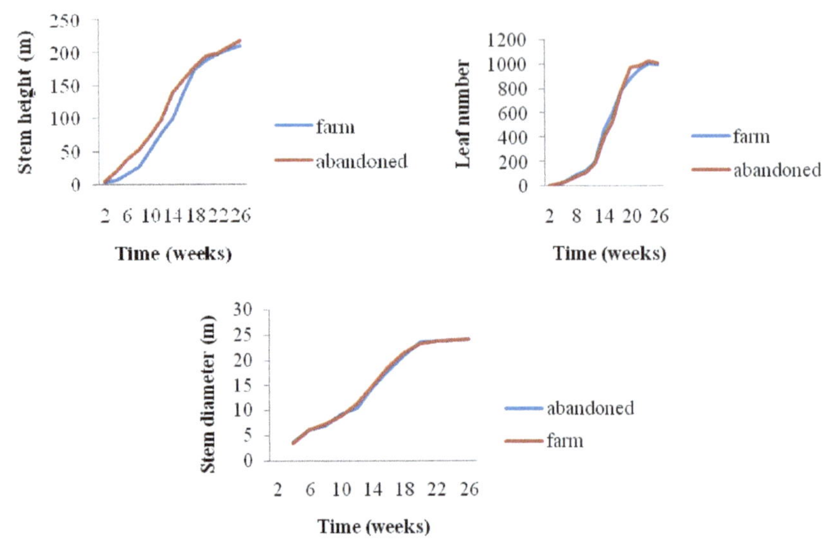

Figure 2. Growth evolution of stem height, leaf number and stem diameter of *H. sabdariffa* in farmland and fallowland ecosystems.

time and germination rate while the values of the means of germination rate were significantly different among the ecosystems.

Growth

Growth parameters of *H. sabdariffa* were observed from April to September as observed in Figure 2. These values varied among all four ecosystems. Only the plants in the farmland and fallowland ecosystems survived or were able to complete their life cycle, that is, germination, growth and the production of seeds. None of the plants in the forest and plantation ecosystems survived after the 4th week of germination. The plants in the farmland and fallowland ecosystems completed their life cycles 5 months after germination. Growth was slow for the first 6 weeks after germination but this rapidly became

Table 4. Mean values for growth parameters of *H. sabdariff* in ecosystems.

Ecosystem	Stem height (cm)	Leaf number	Stem diameter (mm)
Fallowland	121.66[a]	474.60[a]	14.47[a]
Farmland	106.91[a]	1167.47[a]	14. 23[a]
Plantation	2.24[b]	1.04[b]	0.16[b]
Forest	0.75[b]	0.23[b]	0.00

Means with the same letter are not significantly different at P<0.05 using LSD.

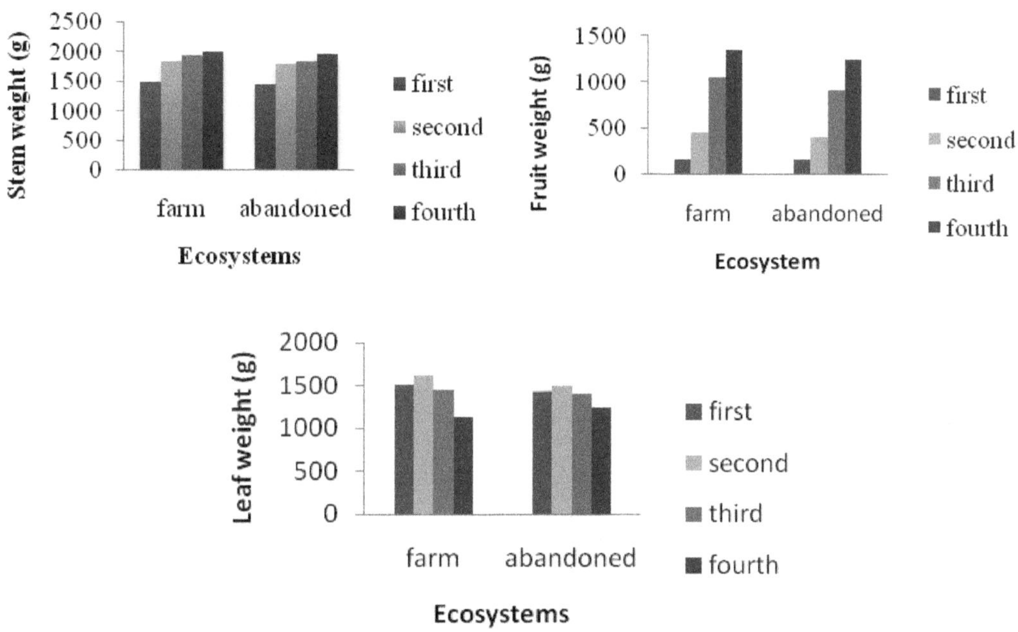

Figure 3. Stem, fruit and leaf yields of *Hibiscus sabdariffa* in farmland and plantation ecosystems.

exponential as from the 10th to about the 28th week of growth. At P < 0.05, no significant differences was observed between the values of stem height, leaf number, and stem diameter for fallowland and farmland ecosystems but these values were both significantly different from the plantation and forest ecosystem for stem height, leaf number and stem diameter at the earlier phases of observation (Table 4).

Yield harvest

Figure 3 shows the histograms for stem, fruit and leaf yields of *H. sabdariffa* in farmland and plantation ecosystems. The stems harvested at 25% intensity at subplots 1 had the least weight in both the farmland and fallowland ecosystems. These weights increased for subplots 2, 3, and 4 at harvesting intensities of 50, 75, and 100%, respectively. The weight of the fruits harvested also increased progressively from 25, 50, 75 and 100% harvesting intensities. The farmland and fallowland ecosystems recorded mean stem and fruit weights which were significantly different at P <0.05 though stem and fruit yields were higher for farmlands (2003 g for stem yield and 1350 g for fruit yield), than for fallowland (1956 g for stem yield and 1280 g for fruit yield). The farmland and fallowland ecosystems recorded leaf weights which were not significantly different at P <0.05 (Table 5).

DISCUSSION

The results of germination of *H. sabdariffa* seeds in this study are in agreement with those observed by Kumar and Echekwa (1985) for this species. He confirmed the observation that seedlings of *H. sabdariffa* emerged from the soil 3 or 4 days after sowing. Elfry et al. (1980) concluded that fewer or more days may be required for seedlings to emerge under ideal soil moisture and

Table 5. Yield weight variation of *H. sabdariffa.*

Ecosystem	Stem weight (g)	Fruit weight (g)	Leaf weight (g)
Farmland	1826.25[a]	751.50[a]	1421.75[a]
Fallowland	1765.75[b]	674.75[b]	1388.25[a]

Mean values with the same letter superscript are not significantly different for the ecosystems at P<0.05.

temperature as was the case in all four ecosystems where 61.5 to 78% of seedlings emerged from the soil three days after sowing in all four ecosystems with an average germination time of 3.31 days. More than half of the total number of seeds germinated on the third day after sowing indicating that the environmental factors, temperature, water, oxygen and light were favorable (Raven et al., 2005). The germination percentages of the seeds of *H. sabdariffa* were not significantly different at P< 0.05. The different soil compositions in the different ecosystems had little or no effect on the germination percentage because all four ecosystems recorded very high germination percentages with very slight differences which varied from 87% in fallowland ecosystem to 96% in the farmland ecosystem. This was in accordance with Edmonds (1991) who postulated that *H. sabdariffa* requires rich loamy soils although they may grow on a variety of soils including new and old alluvium and lateritic loam as was the case of our experimental site. The values for stem height of *H. sabdariffa* were significantly different at P< 0.05 in all four ecosystems. Only the plants in the farmland and fallowland ecosystems survived or were able to complete their life cycle having an average height of 210 cm which is slightly different from 204 cm obtained by Morton (1987).

The plants in the forest and plantation ecosystems respectively did not survive after 4 weeks of growth probably due to the fact that *H. sabdariffa* is a heliophyte, thus needing sufficient light for growth (Elfry et al., 1980). Thus, owing to shade effect created by the trees in the forest and the plantation ecosystems, these seedlings could not survive. The leaf number and stem diameter, just like the stem height evolution, also varied significantly in farm and fallowland ecosystems. This can be explained by the fact that *H. sabdariffa* plants are not adapted to grow under shade due to the physiological nature of their leaf. Only 2% of the light incident on the forest canopy reaches the floor (Edmonds, 1991). The leaves of heliophytes are not adapted to conduct photosynthesis at this low light intensity (Raven et al., 2005), which is necessary for growth hence obviously led to the wilting of the *H. sabdariffa* seedlings at this stage in the plantation and natural forest ecosystems. Duke (2003) observed that *H. sabdariffa* started producing flowers as from 120 to 130 days after sowing which is in accordance with our observation. This is in accordance with Elfry et al. (1980) that *H. sabdariffa* exhibits marked photoperiodism. This is true because the plant in the farmland and fallowland ecosystems started producing flowers in mid- September with short day-length of 11 h. Kumar (1985) stated that *H. sabdariffa* has two distinct phases of growth like other crops. These are the vegetative and reproductive phases. During the vegetative phase the energy of the plant is concentrated on the development and maturation of vegetative parts like stem, leaves and roots. At this early phase, the plant absorbs water at a rapid rate and often takes in most of the nitrogen, potassium and phosphorus that is utilized for growth and development. At the reproductive phase, there is the formation of flowers and the development of fruits and seeds of the plant leading to a drop in the number of leaves and slowing down in stem height.

Plant yield harvested after the 18th week (25% intensity) of growth had the lowest stem, leaf and fruit yields in both farmland and abandoned farmland ecosystems. This varied through the 50, 75 and 100% harvesting intensities for all the growth features. The leaf yield harvest increased from the 25% harvest intensity to the 50% harvesting intensity. As from the 50% harvesting intensity, this leaf yield dropped probably due to the fact that during this period the number of leaves that fell off the plant was higher than the number of leaves that developed from the plant. At P<0.05, the plant yield significantly varied in both ecosystems. This result were reported by Duke (2003) who concluded that the early harvest of *H. sabdariffa* gives low yield while late harvesting yields a substantial quantity and quality of stem fibers and fruits.

However; higher yield parameters for the farmland compared to fallowland ecosystems is probably due to the fact that continuous tillage, in addition to mixing and stirring of soil breaks up aggregates and exposes organo+mineral surfaces otherwise inaccessible to decomposers (Post and Kwon, 2000). Also, Alvarez and Alvarez (2000) observed that organic matter enriches the soil providing a good percentage of nitrogen for plant rapid growth. Meanwhile, on abandoned farmlands; the soil texture is not sufficiently aerated and mineral nutrients are not readily available to plants; thus in conclusion, *H. sabdariffa* exhibited very significant values in growth parameters in the farmland and fallowland ecosystems and since there were significant variations at different harvesting intensities, this species can be sustainably regenerated for long periods in the farmland and fallowland ecosystems.

As for germination, *Hibiscus* can germinate in all the ecosystems, that is, farmland, fallowland, plantation and forest. However, being a heliophyte, the possibility of transplanting the germinated seedlings and the period of transplating to appropriate ecosystem can be investigated.

REFERENCES

Acquaah G (2004). Sexual Propagation. Horticulture Principles and Practices. 2nd Eds. Prentice-Hall of India Private Limited, India. pp. 81-138.

Alvarez R, Alvarez CR (2000). Soil organic matterpools and association with carbon mineralization kinetics. Soil Sci. Soc. Am. J., 64(1): 184-189.

Ayuk ET, Duguma B, Franzel S, Kengue J, Mollet M, Tiki-Manga T, Zenkeng P (1999). Uses, management and economic potentials of *Garcinia kola* and *Ricinodendron heudelotii* in the humid lowlands of Cameroon. J. Trop. For. Sci., 11(4): 740 -761.

Cochran WG, Cox CM (1957). Experimental Designs. John Wiley and Sons, New York.

Duke JA (2003). Handbook of phytochemical constituents of GRAS Herbs and other economic plants. CRB Press, Baton Rouge, USA.

Edmonds JM (1991). The distribution of *Hibiscus sabdariffa* L. Section Furcaria in Tropical East Africa. Systematic and Ecogeographic studies on crop gene poles, 6, IBPGR, Rome Italy.

Elfry MM, Khafaga ER, Prinz D (1980). Studies on the photoperiodic reaction of roselle (H. sabdariffa L. var. sabdariffa).Angewandte Botanik. 54(5/6): 279 – 85.

http:// WWW.encata, 2008.

http:// WWW.pgdrn –gtz –Cameroun.org. 2006.

Kumar VN, Echekwa CA (1985). Yeild components of five roselle (*H. sabdariffa* L.) varieties in northern Nigeria. Easr Afr. Agric. For. J., 51(2): 105 – 112.

Labouriaul LG (1983). Some effects of deuterium oxide on the isothermal germination of tomato seeds. Boletin de la Sociedad Venezolana de Ciencies Naturales, 38:153-166.

Morton J (1987). Roselle. Fruits of warm climates. Miami, FL. pp. 281–286.

Mukhtar (2008). Effect of some plant growth regulators on the growth and nutritional value of H. sabdariffa L. (Red roselle). Int. Jor. P. App. Scs., 2(3): 70 – 75.

Ogunsanwo OY, Oluwadare AO (1999). The application of palmeric equation in predicting the strength properties of hand sheets from mixed pulps. J. Trop. For. Res., 13(1): 81 – 86.

Oluwadare AO, Scotannd OA (2007). The relationship between fibers characteristics and pulp sheet properties of *Leucaena leucocaphala* (lam) De wit. Middle-East J. Sci., Press, 2(2): 63-68.

Oni O, Gbadamasi AE (1998). Progeny variations in seedlings of Dacryodes edulis. G. Don. J. Trop. For. Res., 14: 38-47.

Post MW, Kwon CK (2000). Soil carbon sequestration and Land-Use change: Processes and Potential; Global Change Biol., 6: 317-328.

Raven PH, Ray FE, Susan EE (2005). Biology of Plants, 7th Edition. New York: W.H. Freeman and Company Publishers, pp. 504-508.

Schmidt L (2000). Danida Forest Seed Centre. In: 'Guide to Handling of Tropical and Subtropical Forest Seed' DFSC, p. 511.

Vreugdenhil D, Terborgh J, Cleef AM, Sinitsyn M, Boere GC, Archaga VL, Prins HHT (2003). Comprehensive Protected Areas System Composition and Monitoring, IUCN, Gland, Switzerland, p. 106.

Botanical formulation and extracts based on plant leaves and flower, a substitute for toxic chemical and waxes for shelf life extension and quality retention of apple cv Starking Delicious in India

Sandeep K. Chauhan[2*], Thakur K. S.[1], Jawa N. K.[2] and Thakur K. P.[2]

[1]Department of Postharvest Technology, Dr Y.S. Parmar University of Horticulture and Forestry, Nauni-173 230, India.
[2]FHEL, CONCOR, Government of India.

Pre-harvest treatments consisted of nimbecidine, neem azal, neem gold, with Bavistin, as control. The plant leaves/flowers used as coatings were neem leaf extract, drake leaf extract, spearmint leaf extract, marigold flower extract, and semperfresh. Neem, melia, mentha, walnut, banna, basooti, and camphor were used as cushioning materials in packages. Freshly harvested fruits were subjected to the above treatments and were kept under refrigerated storage (1±1°C) for analysis at a monthly interval up to 6 months. Nimbecidine (1.5%) was found better in reducing physiological loss in weight (PLW) and retaining fruit firmness, whereas, neem azal (2.0%) was found effective in retaining maximum total soluble solids (TSS) content, starch iodine rating and pectin content at the end of 6 months storage period. Among the treatments of extracts of plant leaves/flowers, fruits treated with 20% drake leaf extract proved to be most effective in reducing weight loss, whereas, maximum retention of firmness was recorded in fruits treated with 20% neem leaf extracts. Drake and neem leaf extracts also retained maximum TSS content. Minimum decrease in starch content maximum pectin content in the fruits and were recorded with 20% neem leaf extract. Leaf extract was highly effective in reducing spoilage as no spoilage was recorded under this treatment. Fruits cushioned with camphor leaves were superior over non-cushioned fruits in retaining most of quality characteristics.

Key words: Neem (*Azadirachta indica*), plant extract, drake (*Melia azedarach*), spearmint, marigold, banna (*Vitex negundu*), basooti (*Adhatoda vasica*), camphor, apple.

INTRODUCTION

In India apple is the most important fruit of Himalayan region. Despite spectacular progress made in the area and production, it is estimated that 20 to 30% of the total production is lost during the post-harvest handling period due to lack of proper handling and storage facilities. With the ever-increasing demand for good quality fruits, which are free from fungicides and pesticides residues, the growers are forced to produce quality fruit, especially after the removal of restrictions on international trade. In the absence of highly capital intensive handling infrastructure like precooling, refrigerated transport and controlled atmospheric storage, growers will have to depend upon on alternative, simple and low cost technique till such facilities are created.

From the last few fears the use of various chemicals and waxing material at the pre and post harvest stages is becoming popular among growers in order to enhance the shelf life of fruits. However, the use of these substances has their own limitations, as some of them are believed to be ecologically unsafe and economically not viable beside leaving their residue on the fruit surface,

*Corresponding author. E-mail: sandeep.chauhan@rediff-mail.com

which may have the direct effect on human health. Additionally some of them may be associated with the changes of aroma of the fruit. In order to overcome these short comings there is a urgent need of substances which are of biological origin with growth regulating, fungicidal, insecticidal properties (Dhaliwal and Arora 1996). Grainge et al. (1984) have documented and classified a number of plants belonging to various families having growth regulating, fungicidal properties and plants like neem, melia, mentha ,lantana are under active investigation for use as a plant protection agents. Already neem based formulations are available in the market. Owing to its various effects, azadirachtin is considered as the most active principal substance in neem, which has growth regulating, fungicidal and insecticidal properties (Schmutters, 1990).

The present study was undertaken with the objective to study the effect of pre- harvest treatment of apple cv starking delicious, with three neem based formulations, fresh leaves and their extracts on the storage quality of apples.

MATERIALS AND METHODS

Neem based formulations were tried in 2003/2004 and 2004/2005 and the experiment was laid out in well maintained commercial orchard situated 2100 m above msl in villages of Khaneti (Kotgarh) district, Shimla, Himachal Pradesh, India. Thirty well grown, uniform and 15 - 20 years old " starking delicious " apple trees raised on seedling rootstocks were selected. The trees were maintained under a uniform schedule of cultural operations and subjected to pre-harvest treatments of 3 commercial neem based formulations viz., nimbecidine, neem azal and neem Gold, 20 - 25 days harvest. Each treatment was replicated 3 times with each replication being applied individually to separate trees. The treatments were nimbecidine (0.5, 1.0, 1.5%), neem azal (1.0, 1.5, 2.0%), neem gold (0.5, 1.0, 1.5%), and bavistin (0.05%) was used as control. The entire fruits from individual trees were harvested manually and only sound, medium sized fruits were selected. The fruits were directly packed in corrugated fibre boxes (CFB) carton with paper moulded trays and were immediately transported to the laboratory for observing changes in fruit quality stored under refrigerated (1±1 °C) conditions. While in second experiment coating materials were prepared from extracts of leaves and flowers which have been used traditionally for preventing spoilage in different crops. Aqueous extracts of plant materials were prepared as per the method described by Gakhukar (1996) and Sharma et al. (1997). The method consisted of collection of fresh leaves/flowers, shade drying and grinding to powder form in an electric blender. The powdered material was stored in HDPE bottles till use. Aqueous solutions were prepared by soaking a known weight of the powdered material in an equal quantity of water and keeping it overnight. The extract was separated with the help of muslin cloth and it was considered to be of 100% strength, which was diluted by adding appropriate quantity of distilled water to make up the desired concentration. Guar gum was also added to all the coating solutions at the rate of 2.0%. Semperfresh, a waxing material generally used commercially for waxing of apples, was used to compare the effectiveness of leaf and flower extracts. The aqueous extracts prepared and used were neem leaf (*Azadirachta indica*) extract 10, 20% drake (*Melia azedarach*) leaf extract 10, 20%, spearmint leaf (*Mentha spicata*) extract 10, 20%, marigold flowers (*Tagetes erecta*) extract 10, 20% and semperfresh (control, 1.5%). For the application of postharvest coating treatments uniform, unblemished medium sized fruits were selected and washed in clean tap water. After air-drying the fruits were coated with the above extracts by dipping them for 5 min. The coated fruits were placed on newspaper sheet for drying in shade for half an hour at room temperature and also to remove excess coating materials. Immediately after drying the fruits were kept with respect to their treatments under refrigerated (1±1 °C) storage.

In third experiment fresh leaves of plants known to posses antimicrobial properties were collected from different parts of Himachal Pradesh during the fruit season 2003/2004 and 2004/2005. 25 g of fresh leaves were placed in CFB cartons (32 x 18 × 16 cm) of 3.5 kg capacities uniformly distributed from top to bottom of the trays along with the freshly harvested fruit and kept immediately after treatments under refrigerated storage (1±1 °C). The leaves used were neem/camphor (*Cinnamomum camphora*), Banna/China/caste tree (*Vitex negundu*), basooti (*Adhatoda vasica*) leaves and control (uncushioned). Observations regarding PLW, firmness, TSS, starch -iodine rating, pectin and spoilage of fruits were recorded at monthly interval during the storage period of 6 months. PLW were weighed on a physical balance, fruit firmness were measured with an Effigi Penetrometer, TSS with the help of Erma hand refrectrometer, pectin was measured by Carra and Haynis methods as described by Ranganna (1986), starch iodine rating was measured by as described by Phillips and Poapst (1959), whereas fruit spoilage was calculated on percentage basis.

RESULTS

Experiment 1: Effect of neem based formulations

Increase in PLW during storage duration was observed under all neem based treatments though it was relatively less than that observed in control fruits (Table 1). Nimbecidine (1.5%) was most effective in reducing PLW in comparison to other treatments.

Loss in flesh firmness (Table 1) was lowest in 1.5% nimbecidine followed by 1.0% nimbecidine and 2.0% neem azal. Better firmness observed in nimbecidine treated fruit could be due to the direct effect of azadirachtin present in neem formulation on pectin molecules.

TSS in general increased as the storage period advanced upto 120 days, registering a gradual decline thereafter (Table 2). Although, definite treatment effects were not discernible, yet the treatments 2.0% neem azal and 1.5% nimbecidine demonstrated superiority by recording higher TSS at the end of 6 months storage.

Starch iodine rating (Table 2) indicated decline trend in the starch content of fruits with an increase in storage duration under all the treatments. However, 2.0% neem azal and 1.5% nimbecidine resulted in minimum loss of starch content in fruit, such an affect may be attributed to the effect of active substances especially azadirachtin present in neem formulation slowing down the changes in constituents of fruit as a result of slower ripening changes.

It was observed that pectin content (Table 2) showed a gradual decline with an advancement of storage duration under all treatments. Among various formulation 2.0% neem azal retained maximum pectin content.

Spoilage (Table 3) is one of the most important criteria

Table 1. Effect of pre harvest treatments of commercial neem based formulations on the physiological loss in weight (PLW) and fruit firmness (N) of Starking Delicious apples during storage at 1±1 °C.

Treatment (T)	Storage interval in days													
	Physiological loss in weight (%)							Fruit Firmness (N)						
	30	60	90	120	150	180	Mean	30	60	90	120	150	180	Mean
T_1: Nimbecidine (0.5%)	1.56 (1.12)	2.60 (1.61)	3.73 (1.93)	4.83 (2.19)	6.06 (2.46)	6.62 (2.57)	4.23 (2.05)	74.99	71.52	67.61	65.30	58.62	48.31	64.41
T_2 : Nimbecidine (1.0%)	1.50 (1.22)	2.52 (1.58)	3.69 (1.92)	4.78 (2.18)	5.88 (2.42)	6.46 (2.54)	4.13 (2.03)	75.26	71.88	68.28	65.79	58.94	48.79	64.81
T_3: Nimbecidine (1.5%)	1.4 (1.20)	2.43 (1.55)	3.59 (1.89)	4.66 (2.15)	5.79 (2.40)	6.24 (2.49)	4.03 (2.01)	75.62	72.41	68.50	66.19	59.43	49.24	65.21
T_4: Neem Azal (1.0%)	1.5 (1.24)	2.82 (1.67)	3.74 (1.93)	4.89 (2.21)	6.13 (2.47)	6.72 (2.59)	4.31 (2.07)	74.73	71.43	67.30	65.03	58.09	48.53	64.18
T_5: Neem Azal (1.5%)	1.50 (1.22)	2.71 (1.64)	3.68 (1.91)	4.78 (2.18)	6.08 (2.46)	6.61 (2.57)	4.22 (2.05)	74.90	71.75	67.74	65.39	58.22	48.71	64.45
T_6: Neem Azal (2.0%)	1.44 (1.20)	2.61 (1.61)	3.59 (1.89)	4.68 (2.16)	6.04 (2.45)	6.48 (2.54)	4.15 (2.03)	75.35	72.10	68.01	65.65	58.76	49.11	64.81
T_7: Neem Gold (0.5%)	1.75 (1.32)	2.79 (1.67)	3.89 (1.97)	4.91 (2.21)	6.12 (2.47)	6.71 (2.59)	4.36 (2.08)	74.55	71.48	67.12	63.65	58.05	47.86	63.78
T_8: Neem Gold (1.0%)	1.67 (1.29)	2.75 (1.65)	3.86 (1.96)	4.83 (2.19)	6.09 (2.46)	6.67 (2.58)	4.31 (2.07)	74.90	71.70	67.43	64.01	58.49	48.71	64.18
T_9: Neem Gold (1.5%)	1.60 (1.26)	2.66 (1.63)	3.77 (1.94)	4.60 (2.14)	5.94 (2.43)	6.61 (2.57)	4.20 (2.04)	75.17	71.88	67.79	64.18	60.45	49.06	64.72
T_{10}: Bavistin (0.05%)	1.69 (1.30)	2.87 (1.69)	3.85 (1.96)	4.83 (2.19)	6.35 (2.51)	7.03 (2.65)	4.44 (2.10)	75.17	72.15	69.26	64.41	58.31	47.95	64.54
Mean	1.57 (1.25)	2.68 (1.63)	3.74 (1.93)	4.77 (2.18)	6.05 (2.45)	6.62 (2.57)		75.08	71.84	67.92	64.94	58.71	48.71	
Initial value								**76.86**						
$CD_{0.05}$														
T				0.02							0.04			
I				0.01							0.03			
TXI				0.05							0.09			

Figures in parenthesis are square root transformed values.

in determining the effect of neem formulations in retaining the storage quality of apple and spoilage was reduced significantly under all treatments. There was no spoilage up to 3 months of storage in treated fapples. Apple spoilage was (0.4%) in Neem gold and 3.6% in Bavistin treated fruit in 4th month. Among treatments 1.5% Nimbecidine and 2.0% Neem Azal decreased the spoilage considerably.

Experiment 2: Effect of leaf extracts

Various plant leaf/flower extract treatments caused significant reductions in PLW (Table 4) and the most effective treatment in this regard was coating with 20%. Drake leaf extract (T_4) where minimum physiological loss in weight (3.50%) was recorded, although it was statistically at par with T_6 (3.52%) and T_3 (3.56%), respect-tively. However, the maximum mean physiological loss in weight (3.89%) was recorded in fruits treated with 10% Marigold flower extract (T_7), which was closely followed by the treatment of 1.5% Semperfresh (T_9) and 20% Marigold flower extract (T_8).

Among various plant leaf/flower extracts, 20% Neem leaf extract (T_2) was the most effective in (Table 4) retaining higher mean fruit firmness (67.16 N) and it was followed by T_1 (66.76 N) and T_8 (66.50 N), respectively, whereas, minimum and significantly lower fruit firmness (65.07 N) was recorded under 1.5% Semperfresh (T_9) which also served as a control treatment for the experiment.

From the data, TSS (Table 5) contents, in general increased as the storage period progressed up to 120 days and thereafter, it followed a declining pattern during subsequent storage. Among various coating treatments with plant extracts maximum mean TSS (12.65°B) was recorded when 20% Drake leaf extract (T_4), was applied, although it was statistically at par with the treatments T_2 (12.62°B) and T_6 (12.60°B), respectively. The lowest mean value (12.32°B) for

Table 2. Effect of pre harvest treatments of commercial neem based formulations on the total soluble solids* (°Brix), starch iodine rating and pectin content of Starking Delicious apples during storage at 1±1 °C.

Treatment (T)	Storage interval in days																				
	Total soluble solids (oBrix)							Starch iodine rating*							Pectin**						
	30	60	90	120	150	180	Mean	30	60	90	120	150	180	Mean	30	60	90	120	150	180	Mean
T1: Nimbecidine (0.5%)	10.73	12.64	13.13	14.26	13.09	12.4	12.7	4.13	4.44	5.07	5.25	6.37	0.745	5.45	1.7	1.61	1.51	1.36	0.89	0.72	1.29
T2 : Nimbecidine (1.0%)	10.66	12.57	13.09	14.16	13.41	12.85	12.79	4.08	4.34	4.94	5.13	6.32	7.42	5.37	1.72	1.64	1.54	1.4	1.02	0.8	1.35
T3: Nimbecidine (1.5%)	10.61	12.46	13.02	14.13	13.77	12.99	12.83	4.04	4.29	4.9	5.1	6.29	7.39	5.33	1.74	1.67	1.57	1.42	1.05	0.85	1.38
T4: Neem Azal (1.0%)	10.76	12.7	13.18	14.29	13.1	12.35	12.73	4.1	4.41	5.06	5.22	6.32	7.4	5.42	1.74	1.7	1.61	1.45	1	0.85	1.39
T5: Neem Azal (1.5%)	10.72	12.67	13.12	14.28	13.38	12.41	12.76	4.05	4.35	5.01	5.12	6.27	7.37	5.36	1.76	1.71	1.65	1.48	1.02	0.87	1.41
T6: Neem Azal (2.0%)	10.64	12.54	13.09	14.23	13.84	12.84	12.86	4.02	4.27	4.96	5.09	6.22	7.27	5.31	1.78	1.73	1.68	1.55	1.17	0.96	1.47
T7: Neem Gold (0.5%)	10.75	12.74	13.11	14.3	13.1	11.96	12.66	4.24	4.43	5.19	5.35	6.42	7.59	5.53	1.69	1.63	1.6	1.36	0.99	0.87	1.35
T8: Neem Gold (1.0%)	10.68	12.62	13.09	14.2	13.69	12.06	12.72	4.21	4.4	5.14	5.29	6.36	7.5	5.48	1.71	1.66	1.63	1.38	1	0.91	1.38
T9: Neem Gold (1.5%)	10.65	12.54	13.03	14.16	13.84	12.37	12.76	4.18	4.36	5.11	5.23	6.29	7.4	5.42	1.73	1.69	1.65	1.45	1.03	0.93	1.41
T10: Bavistin (0.05%)	10.73	12.65	13.21	14.29	13.1	10.92	12.48	4.35	4.44	5.26	5.45	6.58	7.66	5.62	1.74	1.65	1.57	1.41	0.96	0.57	1.31
Mean	10.69	12.61	13.1	14.23	13.43	12.31		4.14	4.37	5.06	5.22	6.34	7.44		1.73	1.66	1.6	1.42	1.01	0.83	
***Initial value**	10.22							4.08							1.86						
CD0.05																					
T				0.05							0.01							0.02			
I				0.04							0.01							0.01			
TxI				0.13							0.03							0.05			

*, Rating on the basis of starch test guide; **, As % calcium pectate.

Table 3. Effect of pre harvest treatments of commercial neem based formulations on the fruit spoilage (%) of Starking Delicious apples during storage at 1±1 °C.

Treatment (T)	Storage interval in days						
	Fruit spoilage (%)						
	30	60	90	120	150	180	Mean
T_1 Nimbecidine (5%)	0.00 (0.00)	0.00 (0.00)	0.00 (0.00)	0.00 (0.00)	0.00 (0.00)	2.00 (1.41)	0.33 (0.57)
T_2 Nimbecidine (1.0%)	0.00 (0.00)	0.00 (0.00)	0.00 (0.00)	0.00 (0.00)	0.00 (0.00)	0.80 (0.89)	0.13 (0.36)
T_3 Nimbecidine (1.5%)	0.00 (0.00)	0.00 (0.00)	0.00 (0.00)	0.00 (0.00)	0.00 (0.00)	0.00 (0.00)	0.00 (0.00)
T_4 Neem Azal (1.0%)	0.00 (0.00)	0.00 (0.00)	0.00 (0.00)	0.00 (0.00)	2.40 (1.59)	4.10 (2.02)	1.08 (1.03)
T_5 Neem Azal (1.5%)	0.00 (0.00)	0.00 (0.00)	0.00 (0.00)	0.00 (0.00)	0.00 (0.00)	0.00 (0.00)	0.00 (0.00)
T_6 Neem Azal (2.0%)	0.00 (0.00)	0.00 (0.00)	0.00 (0.00)	0.00 (0.00)	0.00 (0.00)	0.00 (0.00)	0.00 (0.00)
T_7 Neem Gold (0.5%)	0.00 (0.00)	0.00 (0.00)	0.00 (0.00)	0.40 (0.60)	2.41 (1.55)	4.94 (2.22)	1.29 (1.13)
T_8 Neem Gold (1.0%)	0.00 (0.00)	0.00 (0.00)	0.00 (0.00)	0.00 (0.00)	1.20 (1.09)	2.83 (1.68)	0.67 (0.81)
T_9 Neem Gold (1.5%)	0.00 (0.00)	0.00 (0.00)	0.00 (0.00)	0.00 (0.00)	0.00 (0.00)	0.00 (0.00)	0.06 (0.03)
T_{10} Bavistin (0.05%)	0.00 (0.00)	0.00 (0.00)	0.00 (0.00)	3.63 (1.90)	5.85 (2.42)	9.26 (3.04)	3.12 (1.76)
Mean	0.00 (0.00)	0.00 (0.00)	0.00 (0.00)	0.41 (0.63)	1.18 (1.08)	2.43 (1.55)	

$CD_{0.05}$: T = 0.26; I = 0.20; TxI = 0.66; Figures in parenthesis are square root transformed values.

Table 4. Effect of post harvest treatments of various plant extracts on the physiological loss in weight (PLW) and fruit firmness (N) of Starking Delicious apples during storage at 1±1 °C.

Treatment (T)	Storage interval in days (2003-04)													
	Physiological loss in weight (%)							Fruit firmness (N)						
	30	60	90	120	150	180	Mean	30	60	90	120	150	180	Mean
T_1 Neem leaf extracts (10%)	1.73 (1.31)	2.39 (1.54)	3.08 (1.75)	4.55 (2.13)	5.15 (2.26)	5.23 (2.28)	3.68 (1.91)	76.02	74.55	70.10	66.54	60.89	52.44	66.76
T_2 Neem leaf extracts (20%)	1.63 (1.27)	2.32 (1.52)	3.03 (1.70)	4.36 (2.10)	5.03 (2.24)	5.12 (2.25)	3.59 (1.89)	77.31	74.68	70.55	66.72	60.98	52.75	67.16
T_3 Drake leaf extracts (10%)	1.62 (1.26)	2.27 (1.53)	3.03 (1.74)	4.31 (2.07)	5.02 (2.24)	5.15 (2.26)	3.56 (1.88)	75.88	74.01	69.48	65.34	60.45	51.33	66.10
T_4 Drake leaf extracts (20%)	1.58 (1.25)	2.17 (1.50)	2.99 (1.72)	4.24 (2.05)	4.99 (2.23)	5.04 (2.24)	3.50 (1.87)	76.15	74.10	69.57	65.61	60.67	51.60	66.28
T_5 Spearmint leaf extracts (10%)	1.60 (1.26)	2.35 (1.53)	3.01 (1.73)	4.33 (2.08)	4.95 (2.24)	5.31 (2.30)	3.59 (1.89)	75.84	73.70	68.10	65.25	60.49	50.93	65.70
T_6 Spearmint leaf extracts (20%)	1.63 (1.26)	2.31 (1.51)	2.95 (1.71)	4.26 (2.06)	4.85 (2.20)	5.16 (2.27)	3.52 (1.87)	76.02	74.10	69.66	65.47	60.63	51.29	66.19
T_7 Marigold flower extract (10%)	1.77 (1.33)	2.40 (1.54)	3.14 (1.77)	4.63 (2.15)	5.26 (2.29)	6.16 (2.48)	3.89 (1.97)	76.06	74.15	69.97	66.10	60.67	51.06	66.32
T_8 Marigold flower extract (20%)	1.72 (1.30)	2.37 (1.53)	3.07 (1.75)	4.53 (2.12)	5.17 (2.29)	6.07 (2.48)	3.82 (1.96)	76.24	74.73	70.14	66.23	60.45	51.24	66.50
T_9 Semper fresh (1.5%)	1.68 (1.29)	2.35 (1.53)	3.05 (1.74)	4.49 (2.11)	5.18 (2.27)	6.42 (2.53)	3.86 (1.96)	76.06	73.88	70.01	65.16	60.45	45.06	65.07
Mean	1.66 (1.28)	2.32 (1.52)	3.04 (1.74)	4.42 (2.10)	5.08 (2.25)	5.53 (2.35)		76.19	74.19	69.70	65.83	60.63	50.84	
Initial value								78.84						
$CD_{0.05}$														
T				0.06							0.04			
I				0.04							0.03			
TxI				0.13							0.09			

Figures in parenthesis are transformed values

TSS content was recorded in 1.5% Semperfresh treated fruits and it was significantly lower in comparison to all other treatments.

Data reveal that there was a gradual and continuous decline in starch content (Table 5) of fruits with a progressive increase in storage duration. Amongst treatments maximum reduction in starch content of fruit was recorded as a result of the application of 1.5% Semperfresh (T_9), although it was statistically at par with the results of the treatment with 10% Marigold flower extract (T_7). At the same time minimum reduction in starch content was recorded when 20% Neem leaf extract (T_2) was applied and it was followed by the treatments T_6 and T_4, respectively.

In retaining pectin(Table 5) the most effective treatment in this regard was coating with 20% Neem leaf extract (T_2) as it retained maximum mean pectin content (1.45%) in fruits and it was closely followed by T_4 (1.42%) and T_6 (1.41%), with the differences among them being non-significant. The lowest mean pectin content (1.34%) was recorded in 1.5% Semperfresh (T_9)

Spoilage (Table 6) due to rotting did not occur under any of the treatments during the first four storage intervals. Thereafter, some spoilage was recorded under most of the treatments and a significant increase was observed as the storage period increased to 180 days. Among treatments, 20% Spearmint leaf extract (T_6) proved to be highly effective in reducing fruit spoilage as no spoilage was recorded under this treatment. It was followed by the treatment consisting of coating with 20% Neem leaf extract (T_2) On the other hand, maximum mean fruit spoilage (1.42%) was recorded with coating 1.5% Semperfresh (T_9).

Experiment 3: Effect of cushioning by leaves in packages

From the data (Table 7) it is evident that non-cushioned control fruits (T_8) exhibited maximum mean PLW (4.39%). which was significantly higher than that recorded in all other treatments. It was reduced to varying extents by the use of different cushioning treatments. The most effective treatment in minimizing moisture loss was camphor leaves (T_5) in which only 3.97% PLW was recorded and it was followed jointly by the treatments T_2 and T_6 (4.01%), although there

Table 5. Effect of post harvest treatments of various plant extracts on the total soluble solids* (^{o}Brix), starch iodine rating and pectin content of Starking Delicious apples during storage at 1±1°C.

Treatment (T)	Storage interval in days (2004-05)																				
	Total soluble solids (oBrix)							Starch iodine rating*							Pectin**						
	30	60	90	120	150	180	Mean	30	60	90	120	150	180	Mean	30	60	90	120	150	180	Mean
T1 Neem leaf extract (10%)	10.46	12.1	13.09	14.13	13.48	12.07	12.55	4.26	4.91	5.22	6.19	6.3	7.39	5.71	1.71	1.62	1.53	1.42	1.08	0.97	1.39
T2 Neem leaf extract (20%)	10.4	12.04	13.06	14.11	13.71	12.39	12.62	4.23	4.82	5.16	6.18	6.26	7.3	5.65	1.78	1.65	1.58	1.5	1.12	1.07	1.45
T3 Drake leaf extract (10%)	10.46	12.1	13.09	14.12	13.47	12.06	12.55	4.31	4.93	5.26	6.26	6.36	7.43	5.75	1.73	1.67	1.56	1.44	1.06	0.94	1.39
T4 Drake leaf extract (20%)	10.38	12.04	13.05	14.14	13.74	12.56	12.65	4.29	4.88	5.24	6.23	6.28	7.31	5.7	1.76	1.71	1.59	1.47	1.08	0.98	1.42
T5 Spearmint leaf extract (10%)	10.43	12.08	13.08	14.14	13.37	11.83	12.49	4.33	4.93	5.18	6.22	6.31	7.4	5.73	1.76	1.64	1.55	1.42	1.06	0.87	1.38
T6 Spearmint leaf extract (20%)	10.4	12.05	13.04	14.1	13.59	12.38	12.6	4.26	4.89	5.14	6.18	6.28	7.35	5.68	1.78	1.68	1.57	1.46	1.08	0.9	1.41
T7 Marigold flower extract (10%)	10.52	12.13	13.21	14.19	13.38	11.89	12.56	4.36	4.96	5.32	6.37	6.45	7.53	5.83	1.7	1.63	1.51	1.39	1.03	0.83	1.35
T8 Marigold flower extract (20%)	10.45	12.08	13.15	14.13	13.79	11.93	12.59	4.31	4.91	5.26	6.24	6.38	7.42	5.75	1.74	1.66	1.55	1.42	1.09	0.87	1.39
T9 Semper fresh (1.5%)	10.16	12.12	13.08	14.16	13.23	10.91	12.32	4.29	4.95	5.33	6.32	6.5	7.68	5.84	1.73	1.67	1.54	1.43	1.06	0.59	1.34
Mean	10.44	12.08	13.1	14.13	13.53	11.99		4.29	4.91	5.23	6.24	6.34	7.42		1.74	1.66	1.55	1.44	1.07	0.89	
***Initial value**	10.11							4.07							1.82						
CD0.05																					
T				0.05							0.04							0.04			
I				0.04							0.03							0.03			
TxI				0.11							0.1							0.09			

*, Rating on the basis of starch test guide; **. As % calcim pectate.

Table 6. Effect of post harvest treatments of various plant extracts on the fruit spoilage (%) of Starking Delicious apples during storage at 1±1°C

Treatment (T)	Fruit spoilage (%)						
	30	60	90	120	150	180	Mean
T_1 Neem leaves extract (10%)	0.00 (0.00)	0.00 (0.00)	0.00 (0.00)	0.00 (0.00)	0.40 (0.63)	1.20 (1.09)	0.26 (0.50)
T_2 Neem leaves extract (20%)	0.00 (0.00)	0.00 (0.00)	0.00 (0.00)	0.00 (0.00)	0.00 (0.00)	0.40 (0.63)	0.06 (0.03)
T_3 Drake leaves extract (10%)	0.00 (0.00)	0.00 (0.00)	0.00 (0.00)	0.00 (0.00)	0.80 (0.89)	2.00 (1.41)	0.46 (0.06)
T_4 Drake leaves extract (20%)	0.00 (0.00)	0.00 (0.00)	0.00 (0.00)	0.00 (0.00)	0.00 (0.00)	1.60 (1.26)	0.26 (0.05)
T_5 Spearmint leaves extract (10%)	0.00 (0.00)	0.00 (0.00)	0.00 (0.00)	0.00 (0.00)	0.40 (0.63)	2.00 (1.41)	0.39 (0.06)
T_6 Spearmint leaves extract (20%)	0.00 (0.00)	0.00 (0.00)	0.00 (0.00)	0.00 (0.00)	0.00 (0.00)	0.00 (0.00)	0.00 (0.00)
T_7 Marigold flower extract (10%)	0.00 (0.00)	0.00 (0.00)	0.00 (0.00)	0.00 (0.00)	2.00 (1.41)	3.27 (1.80)	0.87 (0.90)
T_8 Marigold flower extract (20%)	0.00 (0.00)	0.00 (0.00)	0.00 (0.00)	0.00 (0.00)	1.60 (1.26)	3.25 (1.60)	0.80 (0.89)
T_9 Semper fresh (1.5%)	0.00 (0.00)	0.00 (0.00)	0.00 (0.00)	0.00 (0.00)	3.20 (1.78)	5.37 (2.31)	1.42 (1.13)
Mean	0.00 (0.00)	0.00 (0.00)	0.00 (0.00)	0.00 (0.00)	0.84 (0.91)	2.12 (1.45)	

$CD_{0.05;}$ T=0.01; I=0.01; TxI= 0.2; figures in parenthesis are square root transformed values.

Table 7. Effect of fresh leaves as cushioning material in packages on the physiological loss in weight (PLW) and fruit firmness* (N) of Starking Delicious apples during storage at 1$\pm$1°C

Treatment (T)	Storage Intrervals in days													
	Physiological loss weight (PLW)							Fruit firmness (N)						
	30	60	90	120	150	180	Mean	30	60	90	120	150	180	Mean
T_1 Neem leaves(*Azadirachta indica)*	1.59 (1.26)	2.64 (1.62)	3.66 (1.91)	4.22 (2.05)	5.92 (2.43)	6.22 (2.49)	4.04 (2.01)	74.01	70.59	67.12	62.98	58.00	48.97	63.61
T_2 Spearmint leaves(*Mentha spicata)*	1.57 (1.25)	2.62 (1.61)	3.62 (1.90)	4.19 (2.04)	5.87 (2.42)	6.18 (2.48)	4.00 (2.00)	73.44	70.68	67.08	63.21	57.74	48.84	63.47
T_3 Drake leaves(*Melia azedarach)*	1.58 (1.26)	2.64 (1.62)	3.65 (1.91)	4.23 (2.05)	5.94 (2.43)	6.14 (2.47)	4.03 (2.01)	73.39	70.41	66.99	63.12	58.05	49.37	63.56
T_4 Walnut leaves *(Juglans regia)*	1.56 (1.25)	2.63 (1.62)	3.64 (1.90)	4.22 (2.05)	5.92 (2.43)	6.28 (2.50)	4.04 (2.01)	73.57	70.68	67.12	63.16	58.00	49.11	63.61
T_5 Camphor leaves*(cinnamomum camphora)*	1.64 (1.28)	2.66 (1.63)	3.67 (1.91)	4.28 (2.06)	5.76 (2.40)	5.90 (2.42)	3.98 (1.99)	73.21	70.55	67.79	63.52	58.49	50.57	64.05
T_6 Banna leaves *(Vitex negundu)*	1.62 (1.25)	2.66 (1.63)	3.58 (1.89)	4.21 (2.05)	5.93 (2.43)	6.22 (2.49)	4.04 (2.01)	72.95	69.79	67.08	63.87	58.18	49.55	63.56
T_7 Vasaka leaves*(Adhatoda vasika)*	1.67 (1.29)	2.62 (1.61)	3.63 (1.90)	4.23 (2.05)	6.11 (2.47)	6.19 (2.48)	4.07 (2.02)	73.13	69.97	68.01	63.21	58.58	49.68	63.74
T_8 Control	1.71 (1.30)	2.74 (1.65)	3.85 (1.96)	4.84 (2.20)	6.12 (2.50)	6.28 (2.59)	4.35 (2.08)	73.57	69.88	67.21	62.85	47.50	40.52	60.23
Mean	1.61 (1.25)	2.65 (1.63)	3.65 (1.91)	4.30 (2.07)	6.01 (5.97)	6.17 (6.23)		73.39	70.32	67.30	63.25	56.85	48.35	
***Initial value**								76.33						
$CD_{0.05}$														
T	0.01										0.06			
I	0.01										0.05			
TxI	0.03										0.15			

Figures in parenthesis are square root transformed values.

were no significant difference among any of the treatments in which cushioning material was used.It is clear that among various treatments use of camphor leaves (T_5) exhibited the lowest decline in fruit firmness (Table 7) thereby, resulting in significantly higher mean fruit firmness (64.05 N) and it was followed by T_7 (63.74) and then jointly by T_1 and T_4 (63.61 N). However, the lowest mean value of (60.23 N) was recorded in fruit that were kept un-cushioned (T_8).

Among fresh leaves significantly higher mean TSS (Table 8) contents (13.09^{o}B) were recorded in response to cushioning with Camphor leaves (T_5) which was followed jointly by T_1 and T_2 (12.95^{o}B) and then by T_6 (12.91^{o}B). After the 90 days storage period there was a faster decline in TSS contents of non- cushioned control fruits (T_9) as a results of which they exhibited the lowest value after 180 days of storage (9.42^{o}B) and also the lowest mean value (12.20^{o}B) over this period.

There was a gradual and continuous decline in the starch content (Table 8) of fruits under all the treatments as the starch content is inversely proportional to the numerical value for starch-iodine rating. However, the decline was slower in fruits packed with various fresh leaves as cushioning material. The decrease was relatively faster in untreated control fruits (T_8) and as such these fruits exhibited maximum reduction in starch content by the last sampling date and also the maximum mean reduction in starch content. At the same time, minimum reduction in starch content was recorded in fruits cushioned with Drake leaves (T_3) and it was followed by T_7 and T_1 and jointly by T_2 and T_6, respectively.

It was observed that there was a gradual decline in pectin content (Table 8) of fruit during the entire storage duration, the decline being more pronounced in control fruits (T_8) as compared to that in fruits cushioned with leaves, where it was relatively slow and more gradual. It was observed that cushioning with fresh leaves decreased the rate of pectin degradation and therefore, enabled the fruits to retain higher pectin content. Among fresh leaves, maximum pectin (1.29%) was recorded in fruits cushioned with Spearmint leaves (T_2) and it was followed by T_7 (1.28%) and T_3 (1.27%) and then jointly by T_1 and T_5 (1.26%) in that order. The lowest mean pectin content (1.06%) was recorded in non-cushioned control fruits (T_8). The control fruits also exhibited the lowest pectin content (0.50%) on the last sampling date.

A perusal of the pooled data reveals that during

Table 8. Effect of fresh leaves as cushioning material in packages on the total soluble solids* (^{o}Brix) of Starking Delicious apples during storage at $1\pm1^{o}C$

Treatment (T)	Storage Intervals in days																				
	Total soluble solids (TSS)							Starch iodine rating*							Pectin content**						
	30	60	90	120	150	180	Mean	30	60	90	120	150	180	Mean	30	60	90	120	150	180	Mean
T1 Neem leaves(Azadirachta indica)	11.23	12.66	13.74	14.83	13.12	12.12	12.95	4.11	4.23	4.75	5.62	6.37	7.22	5.38	1.64	1.57	1.41	1.18	0.95	0.84	1.26
T2 Spearmint leaves(Mentha spicata)	11.27	12.78	13.89	14.87	12.89	12.02	12.95	4.18	4.24	4.62	5.59	6.45	7.3	5.39	1.67	1.61	1.38	1.3	0.91	0.78	1.29
T3 Drake leaves(Melia azedarach)	11.18	12.54	13.69	14.69	12.63	11.9	12.77	4.11	4.32	4.69	5.76	5.9	7.02	5.3	1.66	1.63	1.37	1.29	0.87	0.81	1.27
T4 Walnut leaves (Juglans regia)	11.29	12.54	13.68	14.57	12.65	12.07	12.8	4.08	4.26	4.6	5.71	6.32	7.84	5.46	1.63	1.54	1.37	1.32	0.85	0.76	1.25
T5 Camphor leaves(cinnamomum camphora)	11.34	12.62	13.93	14.91	13.45	12.25	13.09	4.12	4.29	4.66	5.68	6.42	7.54	5.45	1.65	1.56	1.33	1.31	0.87	0.79	1.26
T6 Banna leaves (Vitex negundu)	11.29	12.74	13.94	14.97	12.66	11.88	12.91	4.12	4.24	4.59	5.63	6.32	7.46	5.39	1.64	1.51	1.33	1.23	0.84	0.71	1.21
T7 Vasaka leaves(Adhatoda vasika)	11.21	12.77	13.92	14.93	12.73	11.85	12.9	4.07	4.2	4.49	5.73	6.38	7.05	5.32	1.62	1.52	1.4	1.3	1.03	0.83	1.28
T8 Control	11.41	12.98	14.91	13.81	10.7	9.42	12.2	4.53	4.88	6.11	6.28	6.52	8.77	6.18	1.54	1.39	1.22	1.13	0.61	0.5	1.06
Mean	11.28	12.7	13.96	14.7	12.61	11.69		4.16	4.33	4.81	5.75	6.33	7.52		1.64	1.54	1.36	1.25	0.86	0.75	

*,Rating on the basis of starch test guide; **, As % calcim pectate.

Table 9. Effect of fresh leaves as cushioning material in packages on the fruit spoilage (%) of Starking delicious apples during storage at 1 ± 1 °C.

Treatment (T)	Storage interval in days (2003-04)						
	Fruit spoilage (%)						
	30	60	90	120	150	180	Mean
T_1 Neem leaves*(Azadirachta indica)*	0.00 (0.00)	0.00 (0.00)	1.25 (1.11)	3.14 (1.77)	5.22 (2.28)	6.89 (2.62)	2.75 (1.65)
T_2 Spearmint leaves*(Mentha spicata)*	0.00 (0.00)	0.00 (0.00)	1.87 (1.36)	3.18 (1.78)	5.27 (2.29)	6.98 (2.64)	2.88 (1.09)
T_3 Drake leaves*(Melia azedarach)*	0.00 (0.00)	0.00 (0.00)	1.25 (1.11)	3.16 (1.77)	5.22 (2.28)	7.58 (2.75)	2.86 (1.69)
T_4 Walnut leaves *(Juglans regia)*	0.00 (0.00)	0.00 (0.00)	2.50 (1.58)	4.48 (2.12)	6.04 (2.45)	8.57 (2.92)	3.59 (1.89)
T_5 Camphor leaves*(Cinnamomum camphora)*	0.00 (0.00)	0.00 (0.00)	0.00 (0.00)	1.25 (1.11)	2.51 (1.58)	4.54 (2.13)	1.38 (1.17)
T_6 Banna leaves *(Vitex negundu)*	0.00 (0.00)	0.00 (0.00)	1.25 (1.11)	3.16 (1.77)	5.22 (2.28)	8.31 (2.88)	2.99 (1.72)
T_7 Vasaka leaves*(Adhatoda vasika)*	0.00 (0.00)	0.00 (0.00)	2.50 (1.58)	3.84 (1.95)	5.99 (2.44)	8.51 (2.91)	3.47 (1.86)
T_8 Control	0.00 (0.00)	1.25 (1.11)	3.79 (1.94)	5.26 (2.29)	7.64 (2.76)	13.57 (3.68)	5.25 (2.29)
Mean	0.00 (0.00)	0.15 (0.39)	1.80 (1.34)	3.43 (1.85)	5.38 (2.31)	8.11 (2.84)	

CD0.05; T=0.09; I=0.07; TxI=0.19; figures in parenthesis are square root transformed values.

the first 30 days there was no spoilage (Table 9) of fruits under any treatment. However, thereafter a slight increase in spoilage was recorded under some treatments and after 120 days spoilage was observed under all the treatments. Among cushioning treatments with fresh leaves, Camphor leaves (T_5) delayed the appearance of spoilage symptoms till 120 days and exhibited the lowest mean fruit spoilage (1.38%) and therefore was the most effective treatment in controlling spoilage during the entire storage duration of 180 days. It was followed by the treatments T_1 (2.75%) and T_3 (2.86%), respectively. On the other hand maximum fruit spoilage (5.25%) was recorded in non-cushioned control (T_8) fruit and it was significantly higher in comparison to all other treatments.

DISCUSSION

Experiment 1: Effect of neem based formulations

During the present investigations increase in physiological loss in weight (PLW) with an increase in storage duration was observed under all treatments consisting of neem based formulations though it was relatively less than that observed in control fruits. It is a well known fact that with an increase in storage duration the respiratory and transpiratory losses keeps on increasing which results in loss of metabolites and moisture ultimately resulting in lower fruit weight (Wilkinson, 1965; Wills et al., 1980; Singh and Rana, 1992).

The degree of flesh firmness has been reported to be a fairly good index of storability and stage of ripening in apple as in many other fruits. It is one of the most crucial factors in determining the post-harvest quality and physiology of apple (Burg, 1962). With a decrease in fruit firmness the tissue rigidity decreases as a result of hydrolysis of intercellular pectin and also due to a decrease in cell turgor pressure because of an increase in permeability of cell membrane to water. The decrease in both the components of fruit firmness appears to contribute to tissue softening (Pollard, 1975). Retention of better firmness in Nimbecidine treated fruit can be attributed to the direct effect of azadirchtin, a principle active compound presented in neem formulations, on pectin molecules which are believed to regulate the calcium and pectin integrity, thereby lowering chances of its breakdown during storage (Kleeberg, 1996). Softening of fruits is caused either by breakdown of insoluble protopectin into soluble pectin or by hydrolysis of starch (Mattoo et al., 1975). It may also be due to cellular disintegration and microbial attack leading to increased membrane permeability (Wills et al., 1980).

During the present study it was observed that in response to pre-harvest treatments the amount of total soluble solids (TSS), contents in general increased as the storage period advanced up to 120 days, registering a gradual decline thereafter. The increase in fruit soluble solids during storage may result from an increase in the concentration of organic solutes as a consequence of water loss (Ryall and Pentzer, 1982). It may possibly be also due to the numerous anabolic and catabolic processes taking place in the fruit, preparing it for senescence. Hydrolysis of starch yielding mono-disaccharides could be one of the reasons for the increase in TSS.

Starch-iodine rating indicated a declining trend in the starch content of fruits with an increase in storage duration under all the treatments. The loss of starch in apple fruit during storage may be due to its hydrolysis into sugars (Wills et al., 1980; Priest and Lougheed, 1981) and many workers have reported similar observations earlier (Kleeberg, 1996 and Singh et al., 2000). However, 2.0% Neem Azal and 1.5% Nimbecidine resulted in minimum loss of starch content in fruit. Such an affect may be attributed to the effect of active substances present in neem formulations slowing down the changes in constituents of fruit as a results of slower ripening changes.

It was observed that pectin content showed a gradual decline with an advancement of storage duration under all treatments. The loss in pectin content may be due to the breakdown of pectin during storage as claimed by Sandhu et al (1990). The gradual decrease in pectin content with the advancement of storage period might be due to the action of pectin degrading enzymes present in fruit (Nara et al., 2001). Azadirchtin a principle compound in neem formulations is reported to retard the destrification of pectin thereby slowing down its breakdown, resulting in higher pectin content in such fruits (Gakhukar, 1996; Kleeberg, 1996; Ozdemir et al., 1996; Singh et al., 2000).

Spoilage of fruits due to rotting was decreased considerably with the use of various neem formulations especially 1.5% Nimbecidine and 2.0% Neem Azal. Similar observations have also been reported earlier by Chai et al. (1990) and Singh et al. (2000) in apple and mango fruits, respectively. Reduction in spoilage due to rotting with use of neem formulations may be attributed to the presence of the principle compound azadirchtin which has the ability to check the growth of microbes that are responsible for causing rotting and also to its ability to reduce the rates of respiration and transpiration in fruits (Gakhukar, 1996). Similar results were also revealed by Singh et al. (1993) who found effective control of microbe infestations in banana fruits by the use of formulation of neem.

Effect of leaf extracts

During the present investigation, an increase in physiological loss in weight (PLW) was observed in all treatments with an increase in storage duration. Singh and Rana (1992) reported that respiration and transpiration are the two important processes, which are mainly responsible for water loss from fruits and vegetables. The loss of pectin substances from the middle lamella of the cell wall is perhaps the key step in the ripening process that leads to the loss of cell wall integrity of fruits (Gross and Sams, 1984) and consequently leads to softening. During thepresent studies treatments with different leaf/flower extracts was observed to retain higher pectin content in fruits which might explain retention of better firmness in such fruits. The present findings are in conformity with the findings of Ozdemir et al. (1996); Hwang et al. (1998) and Singh et al. (2000) who have also reported similar retention in firmness of fruits as a result of coating with waxes and plant leaf extracts at low temperature storage

The increase in TSS and sugar contents during the earlier part of storage may be due to the hydrolysis of insoluble polysaccharides into simple sugar. Such changes are expected to be slower and more gradual when the metabolism, of the commodity is slowed down by the application of various treatments. With the decrease in metabolism the rate of utilization of stored metabolites is also slowed down thereby resulting in retention of higher levels of these constituents. Increase in TSS content in apple fruit during storage has previously been reported to be due to hydrolysis of polysaccharides and dehydration of fruit (Fidler et al., 1973; Gross and Sams' dehydration of fruit (Gross and Sams, 1984; Suni et al., 2000)

There was an increase in starch-iodine rating, indicating disappearance or loss of starch in fruits under all the treatments during storage. Minimum reduction in starch content of fruit was recorded with 20% Neem leaf extract whereas other coating materials were effective to a lesser extent in this regard. Such an affect of coating treatments may be attributed to the slower ripening changes as the metabolism of fruits can be expected to be slower when the fruits are treated and stored in conditions which are not conducive to enhance ripening. Hardisty (1975) reported that starch-iodine test was best to indicate the quantity of starch is apple. Panovo (1975) considers the decrease in starch content to be a good indicator of the degree of maturation.

In general, the pectin content in fruit exhibited a continuous decline with an increase in storage duration under all coating treatments with 20% Neem leaf extract being the most effective in retaining higher pectin content. The subsequent loss in pectin content may be due to breakdown of pectin during storage as claimed by Sandhu et al. (1990). The gradual decrease in pectin content with the advancement of storage period might be the result of pectic enzyme activity on natural pectin in the fruit. These findings are also supported by the observations of Nara et al. (2001) who observed similar changes in the pectin content of stored apple fruit.

Spoilage is one of the most important criteria in determining the effect of extracts in retaining the storage quality of apple and it was observed that under all the treatment spoilage was reduced significantly. Among treatments 20% Spearmint leaf extract followed by Drake and Neem were found to be most effective in reducing spoilage. Similar findings on reduction of spoilage with botanical extracts have also been reported by various workers (Chai et al., 1990; Singh et al., 2000; Bhardwaj and Sen, 2003). Gakhukar (1996) reported that botanical extracts have the capability to act as antifeedent and antirepallent thereby inhibiting the pathogenicity of various microorganisms.

Effect of cushioning by leaves in packages

During the present study lower physiological loss in weight (PLW) of fruits was observed in response to cushioning with camphor and spearmint leaves. Reduction in weight loss due to these treatments might be therefore due to their effect on slowing down of physiological processes responsible for weight loss (Sharples and Johnson, 1977). Favorable effects of fresh leaves on lowering down the water vapour losses in fruits have also reported by Samanta and Prasad (1996).

During the present study, decrease in fruit firmness during storage was observed to be lower in fruits that were cushioned with camphor and vasaka leaves. Tzoutzoukou and Bourances (1997) reported a direct correlation between fruit firmness and calcium content in apple fruit as calcium is an integral part of the cell wall and combines with pectic acid to form calcium pectate and thus helps is maintaining the cellular integrity of fruit. It was observed that with use of fresh leaves as cushioning material in packages, the total soluble solids (TSS), total sugars and reducing sugar contents increased during the initial storage period up to 120 days whereas, in non-cushioned fruit this increase was observed only up to 90 days. After these periods a decline was observed in the constituents during the remaining storage period. The initial increase might be due to the fact that during storage starch get hydrolyzed into mono and disaccharides which leads to an increase in TSS and sugars contents (Aly et al., 1981) and after complete hydrolysis of starch no further increase in these constituents occur but subsequently a decline due to utilization as primary substrates for respiration (Wills et al., 1980). Among the treatments cushioning of camphor leaves (T_5) in general retained higher amount of these constituents during storage as compared to control fruits.

Decrease in starch-iodine rating was observed with an increase in storage duration. Drake leaves retained maximum starch content in apple fruits during 180 days of storage. Such an effect may be attributed to ability of menthol to act as an antisenscent agent (Grainge et al., 1984). The loss of starch in apple fruit during storage may also be predicted due to its hydrolysis to sugars (Wills et al., 1980; Priest and Lougheed, 1981) which acts an intermediatory substrate for catabolic and anabolic pathway.

Pectin content in general showed a decline with an increase in storage duration under all the treatments. However, fruits treated with Spearmint leaves retained higher pectin contents throughout storage. The decrease in pectin levels during storage may be due to its breakdown during storage as claimed by Sandhu et al. (1990). The slower and gradual decrease in pectin content of fruits cushioned with spearmint leaves during storage may be attributed to lower respiration rates, proteolysis and tissue breakdown in such fruits, as menthol is believed to deactivate the removal of galacturonic acid residues from the pectin molecule (Jazzar and Hammad, 2003) thereby retaining higher pectin content in fruits.

Apple fruits show considerable spoilage, even during storage at low temperature. Cushioning with camphor leaves was found to be the most effective treatment in reducing spoilage during storage. Volatiles generated by these botanicals are reported to destroy incipient infection on fruit, which may cause rotting during storage (Saxena et al., 1981). Similar observations have also been reported by Gakhukar (1996), Singh et al. (2000); Bhardwaj and Sen (2003).

REFERENCES

Aly MM, ZA, Gamy El, RH Biggs (1981). Ethylene production and firmness of peach and nectarine fruits as related to storage. Proc Florida State Horti Soc. 94:291-294.

Bhardwaj RL, Sen NZ (2003). Zero energy cool chamber storage of Mandarin (Citrus reticulata cv. Nagpur Santra). J. Food Sci. Technol. 40:669-672.

Burg SP (1962). The physiology of ethylene formation. Annu. Rev. Plant Physiol. 13:265-302.

Chai YL, Dana BO, Deery NC (1990). Shelf- life extension of Michigan apples using sucrose polyester. J. Food Proc. Preserv. 15:197-214.

Dhaliwal GS, Arora R (1996). Principle of insect-pest management. National Agriculture Technology Centre, Ludhiana, India. p. 374.

Fidler JC, Wilkinson BG, Ednex KL, Sharples RO (1973). Respiration rate in climacteric fruit. Research Review. Commun. Agric. Bull. England 19:30-45.

Gakhukar RT (1996). Commercial and industrial aspects of neem based pesticide. Pestology 22(10):15-32.

Grainge MS, Ahmed WC, Mitchel, Hylen LW (1984). Plant species reportedly possessing pest control properties, an EWC/UN data base. Resource System Institute, EWC Honoly College of Tropical Agriculture and Human Resouces, UWV of Hawali.

Gross KC, Sams CE (1984). Changes in the cell wall neutral sugar composition during fruit ripening. A special survey. Phytochemistry 23:2457-2461.

Hardisty SE (1975). How to pick apples at the right time. Himachal Hortic 18(1-4):52-55.

Hwang YS, Kim YA, Lee JC (1998). Effect of postharvest application of chitosan and wax and ethylene scrubbing on quality changes in stored 'Tsugaru' apples. J. Korean Soc. Hort. Sci. 38(5):579-582.

Jazzar C, Hammad EAF (2003). The efficiency of enhanced aqueous extract of *Melia azadirach* leaves and fruits integrated with the *Camptoyles reuteri* releases against sweet potato white fly nymphs. Bull. Insectol. 56(2):269-275.

Kleeberg H (1996). The Neem Azal conception: Future possibilities of the use of neem in biological and integrated pest management. In: Neem and Enviornment, Singh RP, Chavi MS, Raheja RK (eds). Oxford and IBH Publ .Co. Pvt. Ltd., New Delhi. p. 875-882.

Mattoo AK, Murata T, Pantastico ErB, Ogata K, Chachin K, Phan CT.(1975). Chemical changes during ripening and senescence. In: Postharvest Physiology, Handling and Utilization of Tropical and Subtropical Fruits and Vegetables (ed. Er B Pantastico). AVI, New York. pp. 103-127.

Nara K, Kato Y, Motomura Y (2001). Involvement of terminal-arabinose and galactose pectic compounds in mealiness of apple fruit during storage. Postharvest Biol. Technol. 22(2):141-150.

Ozdemir AE, Kaskan N, Ayar IT, Dumdar O (1996). Effect of Semperfresh treatments on the post-harvest physiology of cold stored apples cv.Golden Delicious. Turkey. J. Agric For. 19:11-15.

Panovo R (1975). Changes in the flesh consistency and starch content in the fruit of some apple cultivars during ripening. Gradinarska i lozerska Nauka 12:32-40.

Phillips WR, Poapst PA (1959). Storage of apple. Canadian Department of Agriculture Bulletin No. 776.

Pollard IE (1975). Pectinolytic enzyme activity and changes in water potential components associated with internal breakdown in McIntosh apple. J. Am. Soc. Hort. Sci. 100:647-649.

Priest KL, Lougheed EC (1981). Evaluation of apple maturity using the starch-iodine test. Factsheet No. 81.625, Ministry of Agriculture and Food, Ontario, Canada.

Ranganna S (1986). Handbook of analysis and quality control for fruit and vegetable products. 2nd edn Tata McGraw Hill Pub. Co, New Delhi.

Ryall AL, Pentzer WT (1982). Handling, Transportation and Storage of Fruit and Vegetables. AVI Publication Co., Inc

Samanta RK, Prasad MV (1996). Indegenous post harvest technology, National Research Centre on Spices at Culcutta, India.

Sandhu KS, Bhatia BS, Shukla FC (1990). Effect of lye treatment on the quality of Kinnow juice. Indian J. Hortic. 47:55-59

Saxena R, Liquido NJ, Justo HB (1981). Natural pesticides from the neem tree (*Azadirachta indica*). Proceeding of International Neem Conference Rattach Engem, Germany.

Schmutters H (1990). Properties and potential of natural pesticides from neem tree *Azadirachta indica*. Ann. Rev. of Entomol. 35: 271-279

Sharma RN, Tare V, Vartak PH (1997). Bio-activity of some natural oils from Himalayan region on different insect. Indian J. Sci. Res. 1(1):16-18

Sharples R, Johnson DS (1977). The influence of calcium on sensecence changes in apples. Ann. Appl. Biol. 85:450-453

Singh JN, Acharya P, Singh BB (1993). Effect of GA and plant extracts on storage behaviour of mango *(Manifera indica)* cv. Langra. Haryana J. Hortic. Sci. 29(3-4):140-143.

Singh JP, Rana GS (1992). A note on the effect of post-harvest treatments on the physico-chemical changes during storage of sweet orange cv. 'Blood Red' fruits in ZECC. Haryana J. Hort. Sci. 21(3):221-223.

Suni M, Nyman M, Eriksson NA, Bjork L, Bjork I (2000). Carbohydrate composition and content of organic acids in fresh and stored apples. J. Sci. Food Agric. 80(10):1538-44.

Tzoutzoukou CG, Bourances DL (1997). Effect of preharvest application of calcium on the postharvest physiology of apricot fruit. J. Plant Nutr. 20(2-3):295-309.

Wilkinson BG (1965). Some effects of storage under different conditions of humidity on the physical properties of apple. J. Hort. Sci. 40:58-65.

Wills RBH, Bembridge PA, Scott KJ. 1980. Use of flesh firmness and other objective tests to determine consumer acceptability of Delicious apple. Aust J. Exp. Agric. Anim. Husb. 20:252-256.

Comparative survey on the essential oil composition from the seeds and flowers of *Foeniculum vulgare* Mill. from Kerman province

M. Moghtader

Department of Biodiversity, Institute of Science and High Technology and Environmental Sciences, Graduate University of Advanced Technology, Kerman, Iran.

This study was designed to examine the chemical composition of the essential oil of *Foeniculum vulgare* Mill. (apiaceae). The chemical composition of the essential oil from seeds and flowers of *F. vulgare* from Kerman province, Iran was obtained by the hydrodistillation method, analyzed by gas chromatography and gas chromatography mass spectrometry. Twenty four compounds, accounting for 99.75% of the total oil with 2.75% oil yield were identified in the essential oil of the seeds. The main constituents of the essential oil were *trans*-anethole (49.64%), Fenchyl acetate (14.21%), estragole (8.67%), fenchone (6.37%) and limonene (4.23%). Twenty six compounds, accounting for 98.28% of the total oil with 1.96% oil yield were identified in the essential oil of the flowers. The main constituents of the essential oil were *trans*-anethole (55.68%), Fenchyl acetate (11.56%), estragole (6.87%), fenchone (4.43%) and limonene (3.18%).

Key words: *Foeniculum vulgare* Mill., *trans*-anethole, essential oil composition, gas-chromatography/mass-spectrometry (GC/MS).

INTRODUCTION

Fennel is one of the precious medicinal plant that are widely used in pharmacy, perfume, cosmetic, hygienic and food industries (Zargari, 1995). Fennel has a long history of herbal use and is a commonly used as household remedy, being useful in the treatment of a variety of complaints, especially those of the digestive system. The seeds, leaves and roots can be used, but the seeds are most active medicinally and are the part normally used. An infusion is used in the treatment of indigestion, abdominal distension, stomach pains etc. An essential oil obtained from the seed is used in aromatherapy. Fennel contains anethole, which can explain some of its medical effects. The essential oil is bactericidal, carminative and stimulant (Huxley, 1992).

Fennel (*Foeniculum vulgare*), is a plant species in the genus *Foeniculum* (treated as the sole species in the genus by most botanists). It is a member of the family Apiaceae (formerly the Umbelliferae). It is a hardy, perennial, umbelliferous herb, with yellow flowers and feathery leaves. It is generally considered indigenous to the shores of the Mediterranean, but has become widely naturalized elsewhere (particularly, it seems, areas colonized by the Romans) and may now be found growing wild in many parts of the world, especially on dry soils near the sea-coast and on river-banks. Florence fennel or finocchio is a selection with a swollen, bulb-like stem base that is used as a vegetable. Fennel is used as a food plant by the larvae of some Lepidoptera species including the mouse moth and the anise swallowtail.

Fennel is an evergreen perennial herb. It is erect, glaucous green, and grows to heights of up to 2.5 m, with hollow stems. It is hardy to zone 5 and is not frost tender. The flowers are hermaphrodite (have both male and female organs) and are pollinated by Insects. The plant is self-fertile. The leaves grow up to 40 cm long; they are finely dissected, with the ultimate segments filiform (threadlike), about 0.5 mm wide. (Its leaves are similar to those of dill, but thinner.) The flowers are produced in terminal compound umbels 5 to 15 cm wide, each umbel section having 20 to 50 tiny yellow flowers on short pedicels. The fruit is a dry seed from 4 to 10 mm long, half as wide or less, and grooved (Blamey and Grey-Wilson, 1989). On account of its carminative properties, fennel is chiefly used medicinally with purgatives to allay their side effects, and for this purpose forms one of the ingredients of the well-known compound liquorice powder.

The content of essential varies strongly (0.6 to 6%); fruits in the center of an umbel are generally greater, greener and stronger in fragrance. Time of harvest and climate are also important. The essential oil of *F. vulgare* plant has been studied in Iran and other countries but the chemical composition of the essential oil of *F. vulgare* grown in Kerman province is yet to be determined. In the present work we have analyzed the chemical composition of the seeds and the flowers of *F. vulgare* Mill. that grows in Kerman province in Iran and then the results were compared with various origins in other countries.

MATERIALS AND METHODS

Plant material collection and isolation of their essential oil

The seeds and flowers of *F. vulgare* plant were obtained from plants grown in a village in Kerman province, Iran at full flowering stage in August 2012. The samples were cleaned in shade condition to prevent volatility of the plant material constituents and to keep the natural color of the sample fixed. Then they were air-dried and were powdered using a milling machine and kept in a cool dry place until ready for extraction of the essential oil. Afterwards, essential oil was taken from 150 g of the powdered sample in hydrodistillation method with the help of Clevenger set for three hours. Following the sample oils were dried with anhydrous sodium sulfate and kept in sterile sample tubes in refrigerator. The oil yields from seeds and flowers were calculated.

Analysis of essential oil

Gas chromatography (GC)

GC analysis was performed using a model HP-439 gas chromatograph equipped with column CP Sil. 5CB in 25 m length, internal diameter of 0.25 mm and film thickness 0.39 µm. Oven temperatures was from 60 to 220°C at a rate of 7°C slope per minute. Injector temperature was 280 centigrade and detector (FID) temperature was 270°C and carrier gas was helium.

Gas chromatography/mass spectrometry

In order to analyze and identify the combinations forming the essential oil, the Chromatograph gas set attached to a mass spectrometry, Model Hewlett Packard-5973 was used. The conditions of analysis and specifications of the GC/MC set were as follows: Capillary column HP 5MS in 60 m length, internal diameter of 0.25 mm and layer thickness of 0.25 mm, thermal program of oven (3 min) in 60°C, then 60 to 220°C with a 6°C slope per minute, then 3 min in 220°C, the temperature of place of injection 280°C, gas conveying helium, the speed of gas move 1.0 ml per minute, the ratio of fission 1 to 43, the rate of injection 0.1 ml, temperature of the reservoir of ionization 230°C, ionization mode EI, Ionization energy 70eV. The series of normal Alkans C8-C17 were also injected to the set under the same condition with that of essential oil injection to calculate restrictive index (RI) of components of essential oil. The restrictive index of components of the sample was calculated by using a computerized program. Finally, the components of essential oil was identified by comparing the mass spectrums obtained with the existing standard mass spectrums at electronic library of Wiley 2000 (DVD-ROM) existing in absolution software of gas-chromatography/mass-spectrometry (GC/MS) set and calculation of standard restrictive index in accordance with C_8-C_{17} Alkans and comparing them with the existing standard figurers in references (Adams, 2001).

RESULTS AND DISCUSSION

The identified combinations in essential oil, restrictive index (RI), and quantitative percentage of the compounds from seeds and flowers are listed in Table 1. The study of the analysis of *F. vulgare* Mill. essential oil under investigation showed that twenty four compounds, accounting for 99.75% of the total oil with 2.75% (v/w) oil yield were identified in the essential oil of the seeds. The main constituents of the essential oil were *trans*-anethole (49.64%), Fenchyl acetate (14.21%), estragole (8.67%), fenchone (6.37%) and limonene (4.23%) with 83.12% constituting the highest percentage of essential oil. Also from twenty six compounds, accounting for 98.28% of the total oil with 1.96% (v/w) oil yield were identified in the essential oil of the flowers. The main constituents of the essential oil were *trans*-anethole (55.68%), Fenchyl acetate (11.56%), estragole (6.87%), fenchone (4.43%) and limonene (3.18%) with 81.72% constituting the highest percentage of essential oil.

The quality and quantity of the materials forming *F. vulgare* Mill. essential oil had some differences and similarities with the cases reported in other regions. The studies of the ingredients of the essential oil of botanical populations with ecological and genetic differences can be of great importance in identifying the variety of essential oil inside the population of specie. It seems that the geographical origin of *F. vulgare* Mill. greatly influences the oil quality. The essential oil of *F. vulgare* Mill. plant has been widely studied in Iran and other countries but the chemical composition of the essential oil of *F. vulgare* grown in Kerman province is yet to be determined. In present study, results showed the major oil constituents of the seeds and the flowers of *F. vulgare*

Table 1. Combinations identified in the essential oil of *F. vulgare* Mill.

Compound name	(RI)	Seed (%)	Flower (%)
α-Thujene	926	1.21	0.68
α-Pinene	935	2.36	1.35
Camphene	957	0.28	-
Sabinene	978	0.78	1.49
β-Pinene	987	1.45	1.67
Myrcene	992	1.08	-
α-Phellandrene	998	0.24	0.45
P-Cymene	1025	-	1.65
Limonene	1035	4.23	3.18
1,8-Cineole	1038	1.37	0.47
(E)-β-Ocimene	1045	-	0.28
γ-Terpinene	1067	-	0.19
Fenchone	1079	6.37	4.43
Terpinolene	1085	0.46	-
Camphore	1138	0.96	-
α-Terpineol	1203	0.14	1.25
Estragole	1218	8.67	6.87
Dihydro carveol	1229	-	0.85
Trans-caveole	1235	-	0.24
Fenchyl acetate	1247	14.21	11.56
Cis-anethol	1252	3.67	2.24
Trans-cinnamaldehyde	1258	0.21	-
Trans-anethole	1269	49.64	55.68
Methyl acetate	1278	-	0.32
α-Terpinyl acetate	1302	0.32	-
Piperitenone	1329	0.37	1.32
β-Elemene	1352	0.33	0.12
β-Caryophyllene	1425	-	0.47
α-Humulene	1466	0.67	-
Germacrene D	1482	0.34	0.46
Elemol	1521	0.39	0.67
Spathulenol	1548	-	0.16
α-Cadinole	1567	-	0.23
Total		99.75	98.28

The indexes of restrictive have been calculated by injecting the mixture of normal hydrocarbons (C8-C17) to HP-5MS column.

Mill. from Kerman province, Iran were *trans*-anethole and Fenchyl acetate. Extensive research has been conducted on the chemical composition of volatile oils of fennel major constituents of sweet fennel (var. *dulce*) and bitter fennel (var. *vulgare*) to include anethole, estragole, and fenchone plus an additional 18 compounds extracted in the monoterpene fraction of the fruit. The minor constituents accounted for 1 to 5% of the total oil of the volatile oil and included: α-pinene, camphene, β-pinene, α-phellandrene, myrcene, limonene, β-phellandrene, γ-terpinene, *cis*-ocimene, terpinolene, and *P*-cymene.

Significant differences in oil composition between var. *dulce* and *vulgare* were noted (Karlsen et al., 1969). In a comparison study of 13 fennel (*F. vulgare* Mill. var. *vulgare*) populations of different relatively stable and variable characteristics were distinguished from evaluating the correlation matrix of the morphological and chemical properties from two successive years and the medium strength correlation has been found only between the level of essential oil accumulation and the size of seeds (Bernáth et al., 1996). In a research seed oil composition of *F. vulgare* was reported to be influenced by growing region (Arslan et al., 1989).

In a study the chemical composition of stems, leaves and seeds of *F. vulgare* Miller from two Spanish locations (Santander and Aranjuez) were investigated. Two different chemotypes have been found in the seeds: in one chemotypes methyl chavicol (54.9%), fenchone (24.6%) in the seeds collected in Santander; and in other chemotypes (E)-anethole (54.9 to 38.1%), fenchone (22.2 to 34.4%) in the seeds gathered in Aranjuez. The oil from the leaves collected in Aranjuez (April, June) was found to contain methyl chavicol (12.3 to 12.1%), α-phellandrene (9.4 to 27.2%), limonene (25.3 to 18%) and fenchone (19.4 to 18.3%) as major constituents. The oil from the stems gathered in Aranjuez (September) was characterized by a high content of (E)-anethole (17.4 to 3 to 3%), α-pinene (9.7 to 14.4%), α-phellandrene (24.3 to 31.4%), p-cymene (11.5 to 5 to 2%), limonene (11.6 to 15%) and fenchone (9 to 1 to 17 to 5%) (García-Jiménez et al., 2000).

In a research the main identified components of the flower and unripe and ripe fruit oils of *F. vulgare* ssp. *piperitum* were estragole (53.08, 56.11, and 61.08%), fenchone (13.53, 19.18, and 23.46%), and *α*-phellandrene (5.77, 3.30, and 0.72%) (Özcan et al., 2007). It is reported that the major oil constituent of sweet fennel fruit was *trans*-anethole, estragole, fenchone, *P*-anisaldehyde and limonene (Katsiotis, 1988). Volatile oil extracts of fennel seeds were *trans*-Anethole (anise, licorice), estragole (anise, licorice, sweet), fenchone (mint, camphor, warm), and 1-octen-3-ol (mushroom) were the most intense odor compounds detected in fennel extracts (Daz-Maroto et al., 2005). The major compounds of Ground fennel (*F. vulgare* Mill.) seeds in Montenegro were *trans*-anethole (68.6 to 75.0%) and (62.0%), methylchavicol (5.09 to 9.10%) and (4.90%), fenchone (8.40 to 14.7%) and (20.3%) (Damjanović et al., 2005).

In a study, main compound of the essential oil of fennel leaves, flowers, fruits and stems were (E)-anethole varied from 50.1% in fruits to 73.6% in stems. Limonene (1.5 to 20.7%), fenchone (3.6 to 15.9%), α-pinene (1.9 to 19.5%) and α-phellandrene (2.4 to 8.8%) were the other quantitatively important constituents (Venskutonis et al., 1996). It is reported that *trans*-Anethole was the main volatile compound of fennel (*F. vulgare* Mill.) (Díaz-Maroto et al., 2006). In other study main composition of annual sweet fennel (*F. vulgare* var. *dulce* Mill.), were *Trans*-anethole, fenchone, estragole and limonene represented 79.2% of the herb and 93.1% of the fruit (Embong et al., 1977).

ACKNOWLEDGEMENTS

This study was conducted under the financial support of the Department of Biodiversity, Institute of Science and High Technology and Environmental Sciences, Graduate University of Advanced Technology under number contract (1.1044) in Kerman. Hereby, the executive and colleagues of the research project expresses their thanks for the support extended to them.

REFERENCES

Adams RP (2001). Identification of essential oil components by gas chromatography mass spectroscopy. Illinois Allured Publication Corporation.

Arslan N, Bayrak A, Akgul A (1989). The yield and components of essential oil in fennels of different origin (*Foeniculum vulgare* Mill.) grown in Ankara conditions. Herba Hungarica 28(3):27-31.

Bernáth J, Németh É, Kattaa A, Héthelyi É (1996). Morphological and Chemical Evaluation of Fennel (*Foeniculum vulgar* Mill.) Populations of Different Origin. J. Essent. Oil Res. 8(3):247-253.

Blamey M, Grey-Wilson C (1989). Flora of Britain and Northern Europe. Lubrecht & Cramer Ltd; 1St Edition edition. P. 544.

Damjanović B, Lepojević Z, Živković V, Tolić A (2005). Extraction of fennel (*Foeniculum vulgar* Mill.) seeds with supercritical CO_2: Comparison with hydro distillation. Food Chem. 92(1):143–149.

Daz-Maroto MC, Hidalgo IJ, Eva Snchez-Palomo E, Prez-Coello MS (2005). Volatile Components and Key Odorants of Fennel (*Foeniculum vulgar* Mill.) and Thyme (*Thymus vulgaris* L.) Oil Extracts Obtained by Simultaneous Distillation–Extraction and Supercritical Fluid Extraction. J. Agric. Food Chem. 53(13):5385–5389.

Díaz-Maroto MC, Pérez-Coello MS, Esteban J, Sanz J (2006). Comparison of the Volatile Composition of Wild Fennel Samples (*Foeniculum vulgar* Mill.) from Central Spain. J. Agric. Food Chem. 54(18):6814–6818.

Embong MB, Hadziyer D, Molnar S (1977). Essential oils from spices grown in Alberta. Fennel oil (*Foeniculum vulgare* var. *dulce*). Can. J. Plant Sci. 57(3):829-837.

García-Jiménez N, Péerez-Alonso MJ, Velasco-Negueruela A (2000). Chemical Composition of Fennel Oil, *Foeniculum vulgar* Miller, from Spain. J. Essent. Oil Res. 12(2):159-162.

Huxley A (1992). The New RHS Dictionary of Gardening. Grove's Dictionaries; Paper and slipcase edition. p.3000.

Karlsen J, Baerheim Svendsen A, Chingova B, Zolotovitch G (1969). Studies on the fruits of *Foeniculum* species and their essential oil. Planta Med. 17:281-293.

Katsiotis ST (1988). Study of different parameters influencing the composition of hydrodistilled sweet fennel oil. Flav. Fragr. J. 4:221-224.

Özcan MM, Chalchat JC, Arslan D, Ateş A, Ahmet Ünver A (2007). Comparative Essential Oil Composition and Antifungal Effect of Bitter Fennel (*Foeniculum vulgar* ssp. *piperitum*) Fruit Oils Obtained during Different Vegetation. J. Med. Food 9(4):552-561.

Venskutonis PR, Airidas Dapkevicius A, van Beek TA (1996). Essential Oils of Fennel *(Foeniculum vulgar* Mill.) from Lithuania. J. Essent. Oil Res. 8(2):211-213.

Zargari A (1995). Medical plants, 5th Edition, Tehran University Press. 2:553.

Analysis of woodfuel supply and demand balance in Kiambu, Thika and Maragwa districts in central Kenya

Githiomi J. K.[1]*, Kung'u J. B.[2] and Mugendi D. N.[2]

[1]Kenya Research Institute-Forest Products Research Centre, P.O. Box 64636-00620 Nairobi, Kenya.
[2]Department of Environmental Sciences, Kenyatta University, P.O. Box 43844-00100 Nairobi, Kenya.

This study was undertaken in Kiambu, Thika and Maragwa districts of Central Kenya and it aimed at developing micro-level wood energy plan that would act as a case study for future decentralized wood energy plans in Kenya. Primary data was collected from households, service/production institutions comprising of factories/industries, schools, restaurants, hospitals and prisons using structured questionnaires. The study established a deficit between supply and demand balance of -41.7, -45.6 and -50.1% of woodfuel in 2006 for Kiambu, Thika and Maragwa districts respectively. The strategies suggested in micro-level wood energy plans to curb these deficits include allocation of gazetted plantation forest to fuel wood production, increase of improved stove technology, increase of on-farm tree land area and use of alternative energy sources. The combination of all these strategies gave a surplus of 22,903, 46,947 and 32,409 tons of woodfuel in Kiambu, Thika and Maragwa districts, respectively by 2018. The study recommended implementation of the identified strategies aimed at reducing the huge deficit between supply and demand, enhancing inter-institutional collaboration in all sectors related to wood energy development, developing clear policies to guide charcoal and firewood production and marketing and having regular wood energy surveys.

Key words: Central Kenya, wood energy, planning, woodfuel deficit.

INTRODUCTION

It is estimated that about 90% of Kenyan rural households use woodfuel either as firewood or charcoal (MoE, 2002). Woodfuel meets over 93% of rural household energy needs whilst charcoal is the dominant fuel in urban households (Theuri, 2002; Kituyi, 2008). Besides being the standard cooking fuel for the majority of Kenyan households, fuel wood is also an important energy source for small-scale rural industries such as tobacco curing, tea drying, brick making, fish smoking and bakeries among others (Githiomi, 2010). Despite the importance of wood energy in the country's economic development, woodfuel data on supply and demand are scarce and characterized by a high degree of uncertainty that makes it difficult to undertake relevant wood energy planning and policy formulation. This has led to lack of integration of wood energy into national level planning exercises which are important in formulating national policies that are used in allocation of government investment and resources and prioritization of development objectives and targets. The scarcity and uncertainty of wood energy data is due to the fact that it is mainly handled in the informal sector and does not pass through monetized economy like in the case of liquefied petroleum gas (LPG), kerosene and electricity which are alternatives to wood energy. The national energy planning by ministry of energy concentrates more on these conventional fuels while forestry planning by Ministry of Wildlife and Forestry focus more on supply of commercial wood and conservation of protected areas (Republic of Kenya, 2002a). These two key sectors deny biomass energy the comprehensive consideration it deserves due to the role it plays in supply of energy to majority of Kenyans.

Lack of sustainable wood energy production planning has lead to scarcity and over-exploitation of natural resources and environmental degradation. This is

*Corresponding author. E-mail: josephgithiomi@ymail.com.

Table 1. Sampled schools with feeding programme in Kiambu, Thika and Maragwa districts in 2006.

District	Division	Number of schools selected in each school category					
		PD	PB	PBD	SD	SB	SBD
Kiambu	Lari	16	-	-	22	06	-
	Kiambu municipality	11	03	-	04	03	02
	Kiambaa	16	02	-	10	04	02
	Limuru	13	-	-	11	05	01
	Kikuyu	30	-	02	16	07	01
	Githunguri	15	03	01	19	08	01
Thika	Municipality	-	1	-	-	09	-
	Ruiru	19	02	04	12	10	03
	Gatundu	06	-	02	19	11	04
	Kamwangi	-	-	-	18	08	-
	Kakuzi	-	-	-	22	08	-
	Gatanga				22	08	-
Maragwa	Maragwa	03	01	06	10	04	01
	Kandara	04	01	05	23	08	11
	Kigumo	-	-	02	15	02	04
	Makuyu	04	-	03	09	03	02

supported by past studies by Akinga (1980) and Ministry of Energy (2002) which despite being two decades apart showed widening gap between supply and demand as biomass resource was being depleted faster than the rate of replenishment. A decentralized area based approach to planning is essential to properly understand wood energy situations as dictated by varied socio-economic factors between the districts that lead to formulation of proper site-specific wood energy plans, policies and their implementation strategies to ensure sustainable supply. This study therefore aimed at generating data that will contribute to developing micro-level wood energy plan and its implementation strategies which will act as a case study for future decentralized wood energy planning in Kenya.

METHODOLOGY

This study was carried out in Kiambu, Thika and Maragwa districts of Central Kenya. The districts were purposively sampled on the basis of various factors which include diverse wood production systems (farmlands, plantations, natural forests), diverse ecological conditions and population densities among others. The selected districts had rainfall range from 800 to 1400 mm, 500 to 2500 mm and 900 to 2700 mm for Kiambu, Thika and Maragwa districts respectively (Republic of Kenya, 1997a, 1997b, 2002b). The study was designed to generate and analyze the wood energy parameters that were essential in developing wood energy plans for the selected area. The study involved assessing the wood energy resource availability and supply through consideration of wood production systems (plantation forests, natural forests and trees on farms), woodfuel consumption by different sectors and utilization technologies. The target groups in this study were households, institutions, industries, service providers (for example, hotels and kiosks). The study on wood energy production, consumption and supply involved sampling techniques where the sample group was selected following appropriate sampling techniques to obtain information on variables of interest to a larger population. This permitted inferences as to the behavior of the studied variables within the population.

The data obtained from the sample was used to make future projections to forecast energy demand and supply which were used in obtaining demand-supply balances that were used in developing a sustainable wood energy plan for the study area. Sampling of the households was done using multistage stratified random sampling technique beginning with stratification sampling procedures as outlined by Lee-Ann and Martin (1997). Each of the three districts under study was stratified according to the weights in socio-economic and climatic activities as indicators. Following this procedure, at least 40% of the divisions with relatively homogeneous characteristics in each district were sampled. This ensured heterogeneity was well captured and represented. Similar procedure was followed for selection of administrative locations and sub-locations. In total, 200 households were sampled in each district. Primary and secondary schools with school feeding programme in each district of study were listed and clustered into primary day (PD), primary day and boarding (PDB), primary boarding (PB), secondary day (SD), secondary day and boarding (SDB) and secondary boarding (SB). Forty percent (40%) of the samples in each cluster were selected randomly from random tables. This facilitated selecting random sample units from each cluster while all the sampled schools had the same opportunity of being selected. Based on this sampling criterion, various categories of schools were selected from each of the three districts as shown in Table 1.

Sampling of local food processing services (food kiosks, butcheries and hotels) was done within each sampled division. A total of 80 questionnaires were distributed across sampled units per district based on preliminary reconnaissance survey. A total sampling procedure was employed for tea factories and service institutions like prisons and hospitals within each district as they were found to be few. The data was collected using different structured and semi-structured questionnaires for each sector to provide the required information. Supply/demand analysis of wood energy consumption was undertaken and the results used in forecasting several scenarios over time range of 10 years.

Formulas used in projections

Population projection

The following formula shown was obtained from district planning offices which were provided by Central Bureau of Statistics (Republic of Kenya, 2001) to project population data together with the relevant population percentage growth rate:

$P_{n+1} = P_n e^{rt}$

Where: P_n = base population; P_{n+1} = projected population; r = the population growth; t = the time period of projection.

In this formula the population growth rate for the three districts (Kiambu 1.8%, Thika 0.8% and Maragwa 0.8%) were again used to project growth as the new growth rate is expected after the next census in 2009.

Projected numbers of households (PNH)

This was computed by dividing the projected population and the household size of 1999 census that is:

PNH = PPG/HZ

Where: PNH is the projected number of households, PPG is the projected population growth and HZ is the household size.

Projected woodfuel consumption

This was computed using the following formula:

$$PWC = r * PNH$$

Where: PWC is the projected woodfuel consumption, *r* is the mean rate of consumption in tonnes of woodfuel per household and PNH is the projected number of households.

Sustainable supply

Sustainable supply was computed using the formula:

$$SS = AAP * H$$

Where: *SS* is the sustainable supply of woodfuel from protected forests and farm lands, *AAP* is the average annual productivity of the forest stand which varied for example, closed forest- (1.3 m^3/year); woodlands (0.64 m^3/year), bushlands (0.44 m^3/year), wooded grasslands (0.25 m^3/year), grasslands (0.08 m^3/year), farmlands (1.44 m^3/year), forest plantations (19.9 m^3/year) (MoE, 2002). *H* is the land area in hectares under trees.

Expected area under trees

This was computed using the following formula:

$$EAT = \mu * p * NH$$

Where: *EAT* is the expected area under trees, μ is the mean area in ha under trees per given household, *p* is the proportion percentage expected to be under trees on the basis of individual household land size, *NH* is the projected number of households.

Amount of wood equivalent to carbonized charcoal

The formula used to calculate this was as follows:

$$Qtywd = \frac{AMTC}{r}$$

Where: *Qtywd* is the quantity of wood used for making charcoal, *AMTC* is the amount of charcoal carbonized, *r* is the rate of recovery (in this case the r for earth kiln is 20%).

Expected amount of wood to be saved under improved cooking technologies

This was computed using the formula:

$$ASWd = AUr - AUIr$$

Where: ASW_d is the expected amount of wood to be saved under improved use of cooking stoves, *AUr* is the amount of wood used under low efficiency (r = 10%) like in three stone stove, *AUIr* is the expected amount to be used under improved stoves with high efficiency (r = 30%).

Development of wood energy scenarios

In developing wood energy scenarios, the current situation at the time the survey was conducted in 2006 was first assessed and used as a base year. The factors likely to change in the future like the population were used to develop projections. The forecasting of wood energy demand and supply in the future was done as one of the processes of wood energy planning. Through these forecasts, a plan was developed with wood energy demand-supply balancing which proposes and determines a series of related developments and intervention strategies related to wood energy in order to sustainably bridge the gap caused by increasing wood energy demand. The approaches or strategies that were used to address the wood energy deficit were either aimed at reducing demand or increasing supply. These included increasing the land under trees for woodfuel production in both gazetted forests and farm lands, improving wood energy utilization technologies and encouraging fuel switch to conventional fuels. The wood energy demand and wood supply were built in this case study using projection formula shown earlier to develop various scenarios from the data that was taken in 2006 from Kiambu, Thika and Maragwa districts. First, an overview of the current situation in 2006 was created by specifying the data at the starting year and a basic scenario developed assuming a continuation of the same current trend.

A comparative consideration of different scenario combinations which gave the best options for wood energy demand and supply scenarios were considered and the interventions were evaluated using the developed scenarios.

RESULTS AND DISCUSSION

Woodfuel consumption among the districts was simulated

Table 2. Projections of population, supply of fuel wood and its consumption by different sectors in Kiambu district.

Year	2006	2008	2010	2012	2014	2016	2018
Projected population	843,917	874,852	906,920	940,164	974,627	1,010,352	1,047,388
Number of individuals expected to use charcoal	59,074	61,239	63,484	65,811	68,223	70,724	73,317
Expected area (ha) under trees on-farm	81,176	83,451	85,809	88253	90,787	93,414	96,138
Expected wood supply in tons from on-farm	116,894	120,169	123,565	127,085	130,734	134,517	138,438
Firewood consumption by households in tons	230,670	239,126	247,891	256,978	266,398	276,163	286,286
Equivalence of wood for charcoal consumed by households in tons	2,120	2,198	2,278	2,362	2,449	2,538	2,631
Firewood consumption by tea factories in tons	9,441	14,400	15,600	17,500	19,385	21,339	23,361
Firewood consumption by schools	23,102	24,882	26,689	29,502	32,315	35,128	37,941
Woodfuel consumption by prisons in tons	120	125	137	156	169	177	181
Woodfuel consumption by hospitals in tons	120	129	135	149	157	169	177
Woodfuel consumption by restaurants in tons	31	39	46	55	63	73	81
Farmland supply of woodfuel on basis of consumption	109,120	112,110	115,210	118,424	121,756	125,209	128,790
Total woodfuel expected to be used in ton	265,606	280,901	292,779	306,704	320,938	335,590	350,661
Sustainable supply of woodfuel	154,904	157,894	160,995	164,208	167,540	170,994	174,574
% deficit	-41.7	-43.8	-45.0	-46.5	-47.8	-49.0	-50.2

Table 3. Projections of population, supply of fuel wood and consumption by different sectors in Thika district.

Year	2006	2008	2010	2012	2014	2016	2018
Projected population	682,905	693,919	705,111	716,483	728,039	739,782	751,713
Number of individuals expected to use charcoal	58,046	58,983	59,934	60,901	61,883	62,881	63,895
Expected area (ha) under trees on-farm	275,643	276,599	277,570	278,557	279,560	280,580	281,615
Expected wood supply in tons from on-farm	191,418	192,082	192,757	193,442	194,139	194,847	195,566
Firewood consumption by households in tons	330,741	336,075	341,496	347,004	352,601	358,288	364,066
Equivalence of wood for charcoal consumption in tons	5,583	5,673	5,764	5,857	5,952	6,048	6,145
Firewood consumption by tea factories in tons	7,709	10,002	13,901	13,490	14,500	16,484	18,569.0
Firewood consumption by schools	36,828	39,012	41,866	43,768	45,671	47,573	49,476
Woodfuel consumption by prisons in tons	547	555	564	578	590	600	612
Woodfuel consumption by hospitals in tons	84	108	128	142	165	205	230
Woodfuel consumption by restaurants in tons	318	376	434	473	523	580	626
Farmland supply of woodfuel on basis of consumption	177,062	177,676	178,300	178,934	179,579	180,233	180,899
Total woodfuel expected to be used in ton	381,812	391,803	404,155	411,314	420,003	429,779	439,726
Sustainable supply of woodfuel	207,627	208,291	208,965	209,651	210,347	211,055	211,774
% deficit	-45.6	-46.8	-48.3	-49.0	-49.9	-50.9	-51.8

from base year 2006 to 2018 using the principles of long range energy alternative program (LEAP). The simulation used percentage population growth rate of 1.8, 0.8, 0.8% and household hold size of 3.9, 3.8 and 4.3 for Kiambu, Thika and Maragwa districts, respectively (Republic of Kenya, 2001). The projection results are shown in Tables 2, 3 and 4. The population and wood energy consumption increased from base year 2006 to projected year 2018. Similar projections were done for schools, factories, prisons, hospitals and restaurants where all these sectors expected woodfuel shortage in the future. The trend on woodfuel consumption, sustainable supply and resulting deficits without any interventions (business as usual scenario) for both supply and demand show an overall increase in consumption of woodfuel in all the districts. By considering energy balances which involves matching energy demand and supply, several options for intervention were developed both at demand and supply sides. The supply and demand intervention strategies developed in section 3.1, 3.2, 3.3, and 3.4 are aimed at bridging the wide gap between the sustainable supply and consumption, and are based on the base case

Table 4. Projections of population, supply of fuel wood and consumption by different sectors in Maragwa district.

Year	2006	2008	2010	2012	2014	2016	2018
Projected population	410,315	416,933	423,657	430,490	437,433	444,489	451,658
Number of individuals expected to use charcoal	16,412	16,677	16,946	17,219	17,497	17,779	18,066
Expected area (ha) under trees on-farm	93,477	93,974	94,478	94,992	95,513	96,043	96,581
Expected wood supply in tons from on-farm	134,607	135,322	136,049	136,788	137,539	138,301	139,077
Firewood consumption by households in tons	201,951	205,208	208,518.	211,881	215,298	218,771	222,299
Equivalence of wood for charcoal consumption in tons	803	816	830	843	857	870	884
Firewood consumption by tea factories in tons	7,616	9,019	10,212	12,334	14,548	16,564	18,230
Firewood consumption by schools	30456	32,897	35932	39738	43545	47352	51158
Woodfuel consumption by prisons in tons	360	409	473	534	589	645	699
Woodfuel consumption by hospitals in tons	56	73	82	93	101	109	120
Woodfuel consumption by restaurants in tons	87	98	111	122	130	138	147
Farmland supply of woodfuel on basis of consumption	129,222	129,909	130,607	131,316	132,037	132,769	133,514
Total woodfuel expected to be used in ton	241,330	248,523	256,160	265,547	275,070.	284,451	293,541
Sustainable supply of woodfuel	155,204	155,562	155,925	156,295	156,670	157,052	157,439
% deficit of woodfuel	-50.1	-51.4	-52.7	-54.2	-55.7	-57.0	-58.2

scenario. The base case scenario assumed all existing parameters influencing wood energy consumption remained unaltered. The analysis used 2006 as the base year for projection purpose as the data used was collected in that year. Energy demand dynamics such as response to scarcity and prices were not considered in making projections. The supply/demand of wood energy across all sub-sectors has been marked by an acute imbalance.

The simulations as provided in Tables 2, 3 and 4 showed for instance in 2008 total sustainable supply of woodfuel was 157,894, 208,291 and 155,562 tons for Kiambu, Thika and Maragwa, respectively. Similarly, on the same year, the expected consumption was 280,901, 387,826 and 248,523 tons resulting to a deficit of -43.8, -45.6 and -51.4% for Kiambu, Thika and Maragwa districts, respectively. The household sector was found to be consuming the highest amount of fuel wood followed by schools. The consumption is projected to increase further to 350,661, 439,726 and 293,541 tons of fuel wood by 2018 as compared to 174,574, 211,774 and 157,439 tons of the expected sustainable supply in the same year leading to a deficit of -50.2, -51.8 and -58.2% for Kiambu, Thika and Maragwa districts, respectively by 2018. The stated projections provide a clear indication of high deficit of fuel wood if all energy sectors continue using woodfuel without any interventions. If the "business as usual" remains as indicated in projected years, the implication is that various sectors would obtain their woodfuel from other sources outside the district or will over-harvest the available stock which will likely accelerate deforestation in the district leading to high insufficiency of fuel wood supply in future. This would not only impact on their economic activities but also environmental degradation and low land productivity as the trees are very valuable in stabilizing soil, reducing water runoff, providing habitats and carbon dioxide fixation. The woodfuel deficit may also lead to households using farm wastes which would further lead to soil degradation.

Results of demand and supply of wood energy provided an opportunity to identify possible gaps in wood energy which can be addressed through development of appropriate interventions to sustain the ever increasing demand of woodfuel. In development of efficient wood energy plan, analysis of demand and supply balance of woodfuel was used to create possible scenarios from base value (business as usual). The scenario strategies were aimed at mitigating the difference between sustainable supply and consumption. Therefore, in this study, four scenarios were identified and simulated to provide possible alternatives of developing an effective wood energy plan for each of the three districts. These were: allocation of 16, 19 and 21% of gazetted plantation forests in Kiambu, Thika and Maragwa districts, respectively for fuelwood production. The second strategy was establishment of woodlots on household farms depending on their farm sizes. Third strategy was to increase adoption of efficient technology devices used in households as well as other sectors of wood energy. Finally, the fourth strategy was to increase use of other alternative sources of energy like LPG, kerosene and electricity. It was envisaged that a combination of these strategies/scenarios involving supply and demand strategies would provide correct interventions in meeting the woodfuel gaps and to some extent provide a surplus if the plan was to be implemented.

The descriptions of these strategies/scenarios and their projected contributions in reducing the woodfuel deficits in each of the three districts are presented in the subsequent sections.

Intervention strategy based on increasing plantation for woodfuel production in Kiambu, Thika and Maragwa districts

The first strategy suggested allocation of 16, 19 and 21% of the gazetted plantation forests in Kiambu, Thika and Maragwa districts, respectively. The strategy envisioned sustainable supply of firewood through increase of plantations for woodfuel development at an equal proportionate rate which is equivalent to 299 ha for the next 10 years from 2009 (two years needed for necessary planning), will inject about 5,790, 2,985 and 5,510 tons of fuel wood annually in each of the three districts, respectively. This will imply that in 2018 about 47,689, 23,880 and 44,080 tons of fuel wood could have been produced for use by various sectors leaving a deficit of -36.6% (128,397.9 tons), -46.4% (204,072 tons) and -43.2% (126,669 tons) for Kiambu, Thika and Maragwa districts, respectively. This may be attained if fast growing and well managed plantation trees species like *Eucalyptus species*, *Acacia mearnsii*, *Acacia melanoxylon*, *Grevillea robusta* among others are planted. This strategy can be achieved through leasing out forest land to private investors like Kenya Tea Development Agency (KTDA) who have been expressing interest of leasing land for firewood supply to their factories.

In other parts of the world, planting of plantations for supply of energy has been done in some countries like India, Indonesia and South Korea in Asia and Rwanda, Sudan and Ethiopia in Africa (FAO, 2001). This is supply-side intervention which is aimed at enhancing supply of woodfuel from plantation forests.

Intervention strategy based on increasing the farmland area under tree production in Kiambu, Thika and Maragwa districts

Strategy two aims at establishing on-farm woodlots which depends on household land sizes. The strategy envisages that for those households with less than two acres of land, at least 0.1 acres will be under trees throughout the years providing an average sustainable supply of wood biomass of about 6,571, 20,586 and 17,064 tons per year for Kiambu, Thika and Maragwa districts, respectively. On the other hand, for those farmers who have 2 to 3 acres of land, it was anticipated that 0.2 to 0.3 ha will be under trees resulting to sustainable supply of 27,024, 21,235 and 17,640 tons annually for each of the three districts in that order. Equally for farmers that will be having at least 4 acres of land size, it was envisioned that at least 0.6 ha will be under trees resulting to sustainable supply of wood biomass at an average of 25,223, 59,765 and 21,517 tons per year for Kiambu, Thika and Maragwa, respectively. Overall, this strategy may be achieved if annual planting programme is keenly introduced to the mentioned categories of land sizes among the various households. This may drastically reduce woodfuel deficit from 128,397 (-36.6%) to 31,746 tons (-9.1%) in Kiambu district, 204,072 (-46.4%) to 81,730 tons (-18.6%) in Thika district and 131,051 (-52%) to 35,125 tons (-12%) in Maragwa district by 2018. Thus, the summation of all sustainable wood biomass less industrial wood biomass (20%) shows that this scenario of woodlots development/increase of tree planting on farm will reduce a lot of pressure on wood energy. The government intends to increase the forest cover to 10% by 2030 (Republic of Kenya, 2007) and there are efforts to enact law requiring land owners to plant trees on at least 10% of their land holdings. If enacted, the increase of land area under trees will be easily implemented. This intervention strategy is supply oriented as it aims to increase woodfuel supply from farmlands.

While promoting the tree planting on farms, the preference of tree species and obstacles to tree planting which were identified in this study should be considered. The introduced farm forestry technologies should reflect the needs of diverse farm units. This is due to the fact that different households have different needs and therefore introducing different strategies will allow farmers to choose an appropriate option instead of having to decide whether or not to adopt a single technique (Raintree, 1981). The farm forestry technique to be promoted for the household production should incorporate local knowledge on tree growing to maximize benefits. The tree planting in the farms would diversify the productivity of the farm making the ecosystem more resilient and sustainable therefore improving the livelihoods of the households. One of the obstacles to tree planting identified in this study is lack of seedlings. It is important that nurseries are responsive to the local requirements through producing seedlings that are in demand. To increase the on farm tree planting to meet the demand from various sectors, a strategy of outgrower tree schemes can be developed by tea factories and other service institutions. The schemes will involve contractual partnership between the partners (land owners and the tea factories/service institutions). The scheme can be arranged such that the growers are provided with technical advice on forestry practices which are essential to the success of outgrower schemes (Arnold, 1997) and the planting material.

Following a contractual agreement, a clear management plan is essential to ensure effective implementation aimed at achieving the target for long-term viability. This will ensure guaranteed market for the farmers and stable source of woodfuel by the tea factories and service institutions.

Intervention strategy based on increasing adoption of efficient utilization technologies in Kiambu, Thika and Maragwa districts

The third strategy (adoption of efficient technology devices) will contribute to reduction of woodfuel deficits by assuming that the households that were using three stone fire with an efficiency of 10% will slowly shift to more efficient technologies like improved firewood and charcoal stoves (like kuni mbili and Kenya Ceramic Jikos). These improved technologies are reported to have an efficiency of over 30% (Kammen, 1994), which means if adopted; there will be a substantial reduction in the demand of woodfuel. This scenario envisaged that by 2018, 80% of the households will have adopted the mentioned improved technologies at an increasing rate of 10% each year as from 2011. This implies that a significant amount of woodfuel will be saved hence reducing the consumption projected. This is a demand side intervention strategy as it aims at reducing consumption of woodfuel. Implementation of this strategy in combination with the earlier two strategies (strategy 3.1 and 3.2) will reduce consumption of woodfuel by at least 53,558, 70,427 and 51,974 tons leading to 21,811 (surplus of 7.3%), 70,427 (deficit of -2.6%) and 16,848 (surplus of 5.7%) in Kiambu, Thika and Maragwa, respectively, by 2018. This projected reduction of firewood consumption as a result of adoption of efficient technologies forms a realistic approach.

In addition to savings in woodfuel consumption, the improved stove has an added advantage of reducing smoke and health hazards like acute respiratory infection (Kammen, 1994). This strategy needs creation of awareness on indoor air pollution, particularly for women and children which can be achieved through organizing capacity building programmes like trainings aimed at developing stove production skills. A practical example of a successful case of building capacity in women groups in western Kenya has been reported where 11,000 improved Upesi stoves are made and sold per year (www.practicalaction.org/household_energy).

Intervention strategy based on promotion of increased use of alternative sources of energy in Kiambu, Thika and Maragwa districts

Finally, the demand oriented option which aims at reducing the amount of woodfuel consumed through use of alternative sources of energy was considered. If the rural and urban households slowly combine the use of LPG, kerosene stoves and electrical appliances, the use of woodfuel would be reduced thus minimizing the pressure on wood biomass for domestic use. This will reduce the consumption of wood biomass by 8, 16 and 11% in Kiambu, Thika and Maragwa districts, respectively, by 2018. If this scenario is applied in combination with the other stated three intervention strategies, a surplus of 22,903, 46,947 and 32,409 tons in the three districts respectively will be realized. This strategy is supported by the energy policy of 2004, which promotes the use of cleaner fuels like LPG through subsidies (MoE, 2004). As an example the government removed 'value added tax' on LPG in the Kenya Government budget of June 2004 to encourage the use of LPG. The government has also enforced harmonization of different types of LPG regulators which previously reduced competition in the market by restricting a customer to using one brand of LPG. While the use of alternative fuels sounds fairly realistic, it is likely to be faced with difficulties in price fluctuations of LPG, kerosene and electricity bills which have been on the increase in the recent times. Other barriers that hider their wider use include supply distribution, high initial cost of appliances among others.

The switch to conventional fuels like kerosene and LPG will occur as a wider change on social and economic conditions. This decentralized wood energy planning case study intends to provide bases for future wood energy planning in other Kenyan districts which should be integrated to district development committees (DDCs). District planning committees should be involved in monitoring wood energy plan implementation. This study was aimed at getting solutions to wood energy deficit through matching demand and supply in the three districts. The strategies suggested to meet the deficits were either demand-oriented (like the improving utilization technologies to reduce demand) or supply-oriented (through increasing the area under trees or substituting supply with alternative fuels). The combinations of both supply and demand oriented solutions were adopted to give a sustainable supply of the woodfuel. The production, distribution and consumption of wood energy are site specific as the demand and supply vary per region. The spatial variation among the districts should be taken into account for decentralized wood energy planning. With decentralized planning, the specific and unique characteristics of the demand and supply of energy per district are considered. For effective planning, there is need for wood energy database development which should have wood energy resource data, and supply/demand data that are useful in developing wood energy plans. There is also need for technical and financial support for development of capacity in wood energy data collection.

The capacity building can be done to the existing government structures in Kenya Forest Service (KFS), Ministry of Energy and Ministry of Agriculture through training in wood energy data collection.

ACKNOWLEDGEMENT

Authors would like to thank the District Forest Officers of Kiambu, Thika and Maragwa districts for assisting in data collection and KEFRI management for supporting the study.

REFERENCES

Akinga AW (1980). Woodfuel Survey in Kenya. Forestry Department, Ministry of Natural Resources report, Nairobi, Kenya. p. 42.

Arnold M (1997). Trees as outgrower crops for forest industries. Experience from Philippines and South Africa. Rural Development Network, Overseas Development Institute. London, United Kingdom. p. 19.

Food and Agriculture Organization (2001). Plantation and wood energy, Forest plantation thematic papers. Rome, Italy. Accessed on 11/2/09 from: http://www.fao.org//DOCREP/004/AC125E/ac125e00..html

Githiomi J (2010). Micro-level wood energy planning for Kiambu, Thika and Maragwa districts; a case study for decentralized wood energy plan in Kenya. PhD Thesis, Department of Environmental Studies, Kenyatta University, p. 176.

Kammen DM (1994). From Energy Efficiency to Social Utility: Lessons from Cookstove Design, Dissemination and Use. Accessed January 20th 2006 from: http://www.undp.org/seed/energy/policy/ch-5.htm.

Kituyi E (2008). Policy Proposal for Sustainable Consumption and Production of Energy in Kenya: A Memorandum to the National Environment Policy Committee. Renewable Energy Technology Assistance Programme, Policy Brief Series No. 2, 2008. Nairobi, Kenya. p. 8.

Lee-Ann CH, Martin AB (1997). Surveying Natural Population. Columbia University Press, New York. pp. 113-152.

Ministry of Energy (2002). Study on Kenya's Energy Demand, Supply and Policy Strategy for Households, Small scale Industries and Service Establishments. Kamfor Consultants, Nairobi, Kenya. p. 133.

Ministry of Energy (MOE) (2004). Sessional Paper No. 4 on Energy. Government printers, Nairobi, Kenya. p. 60.

Raintree JB (1981). Bio-economic consideration in the design of agroforestry intercropping systems. ICRAF, Nairobi.

Republic of Kenya (1997a). Kiambu District Development Plan. Government Printers, Nairobi, Kenya. p. 156.

Republic of Kenya (1997b). Thika District Development plan 2002-2008. Government Printers, Nairobi, Kenya. p 165.

Republic of Kenya (2002b). Maragwa District Development Plan 2002-2008: Effective Management for Sustainable Economic Growth and Poverty Reduction. Government Printers, Nairobi, Kenya. p. 59.

Republic of Kenya (2007). Kenya Vision 2030. Government printers, Nairobi, Kenya. p. 12.

Republic of Kenya (2001). 1999 Population and Housing Census: Counting our people for development. Government Printers, Nairobi, Kenya. p. 159.

Republic of Kenya (2002a). National Development Plan 2002-2008: Effective Management for Sustainable Economic Growth and Poverty Reduction. Government Printers, Nairobi, Kenya. p. 79.

Theuri K (2002). Woodfuel Policy and Legislation in Kenya; A paper presented during the Regional Workshop on Woodfuel Policy and Legislation held in ICRAF, Nairobi, Kenya. p. 6.

21

Evaluation of methyl bromide alternatives on loblolly pine production and seedling quality over three growing seasons at the Pine Hill Nursery in Camden, Alabama

Scott A. Enebak[1], D. Paul Jackson[2], Tom E. Starkey[1] and Marie Quicke[1]

[1]Southern Forest Nursery Management Cooperative, Forest Health Dynamics Laboratory, School of Forestry and Wildlife Sciences, 602 Duncan Drive, Auburn University, Alabama, 36849, USA.
[2]Department of Agricultural Sciences, Louisiana Tech University, P. O. Box 10198, Ruston, LA 71272, USA.

Forest-tree nurseries in the United States have relied on methyl bromide (MBr) soil fumigation to control weeds, fungi, and nematodes. Due to the world-wide phase-out of MBr use, finding a soil fumigant alternative for MBr has been a priority for the forest nursery industry. A large-scale study comparing seven fumigants using operational application techniques and normal nursery management practices over three growing seasons was installed to determine the effects of MBr alternatives on seedling quality and quantity. Seedling densities at the end of the first growing season were similar across all soil fumigants tested. The newer chemistries, Pic+ and dimethyl-disulfide (DMDS), had seedlings with similar root collar diameters (RCDs) as MBr during the first and second growing seasons. Seedling RCDs declined for all treatments over the 3-year rotation due to the buildup of soilborne pests and weeds. Seedling root architecture and root morphology were similar for all soil fumigants. At the end of each growing season, nursery soils had similar levels of *Trichoderma* and nematode populations across all soil fumigants tested. These trials indicate that, while not the perfect replacement in all nursery soils, seedling production is still possible without MBr if compounds such as chloropicrin are used and managers pay close attention to weed and nematode pests that are less susceptible to chloropicrin than MBr.

Key words: Soil fumigation, chloropicrin, soilborne fungi, nematodes, *Pinus taeda*.

INTRODUCTION

Fumigation with methyl bromide (MBr) mixtures has been the most commonly used method for producing high quality, pest-free forest-tree seedlings in the southeastern United States (Jang et al., 1993). Forest nurseries in the United States have relied for many years on MBr soil fumigation to control yellow (*Cypress esculentus*) and purple (*Cypress rotundus*) nutsedge, soil insects, nematodes and soilborne pathogenic fungi (Cordell et al.,

1989). Forest-seedling nursery managers have long recognized the importance of MBr to control these soilborne pathogens in the production of forest-tree seedlings (South et al., 1997). Due to the concern over ozone depletion in the stratosphere, the Montreal Protocol under the Clean Air Act began a phase-out program for MBr use in 2005 (Jang et al., 1993). Since 1991, finding alternatives for MBr has been a priority for the forest nursery industry. Other materials tested by the Southern Forest Nursery Management Cooperative, without much success, include basamid, hot pepper sauce, steam, hot water, and biologicals that included bacteria and fungi (Mousseaux et al., 1988; Enebak et al., 1999). Although it may be difficult to find a soil fumigant alternative that is as broad-spectrum as MBr, the nursery industry realizes the importance of testing new compounds, rates and application techniques.

The southeastern region represents the main softwood producing area of the United States and accounts for approximately 64% of the total timber harvested in the nation (Smith et al., 2001). Timberland and land ownership patterns have shifted to non-industrial private land, while forest industry timberland holdings have fallen in correspondence with the increase in non-industrial private land gains. Despite the recent trend, approximately 55% of the total forest industry timberlands are in the southeastern region (Smith et al., 2001). Loblolly (*Pinus taeda* L.) and slash (*Pinus elliottii* Engelm.) pine are the primary timber species grown in the southeastern United States, representing the second largest softwood group by volume (Smith et al., 2001). Loblolly pine has a number of favorable characteristics that have attributed to its widespread planting in the southeast. The species is ideal for site restoration as it easily regenerates and is the hardiest of the southern pines in terms of its reproductive ability and rapid growth rate on diverse sites. This adaptability has led to it being planted extensively across the southeast, both on and off sites preferred by the species. These characteristics made loblolly pine ideal for the heavily eroded soils of the early 20th century, and the diverse sites that the forest industry manages today. With genetically improved loblolly pine available, the opportunity for expansive plantings on a full range of site conditions, with genetically similar trees has become the new norm in the southern United States (Byram et al., 2005).

Because of the species' wide-spread use and value to landowners, loblolly pine is the primary species produced in southern forest nurseries (McNabb and VanderSchaaf, 2003) and many soilborne pests can negatively affect their production (Cordell et al., 1989). Despite the phase-out of MBr, forest nursery managers still need soil treatments that are comparable to MBr for control of soilborne pests. As part of the USDA-ARS Areawide Pest Management Project for MBr Alternatives, these studies are part of a long-term continuing effort to identify and evaluate soil fumigants against common nursery pests. Information from these studies should be used by nursery managers in the southern United States to select an MBr alternative that assists with the production of forest-tree seedlings in their particular nursery. The identification of a soil fumigant using operational application techniques under normal nursery management practices will allow continued reforestation programs important to the economy of the southern United States.

MATERIALS AND METHODS

Soil fumigants

Included in this trial at the Pine Hill Nursery, Wilcox County, Alabama, (32° 03' 53" N; 87° 20'57" W) were eight soil fumigants: six were selected from previous Southern Forest Nursery Management Cooperative fumigation studies (Enebak et al., 2012) and two were rates of iodomethane, which had never been tested in the production of forest-tree seedlings (Table 1). The fumigants were shank-injected on 23 March, 2009. Iodomethane treatments were covered with virtually impermeable film (VIF), while all other treatments were covered with a 1 mm high density polyethylene (HDPE) Tarp (AEP Industries Inc.) as a broadcast/flat tarp (Table 2). The experimental area occupied approximately 2.5 ha and the trial was laid out in six nursery sections that consisted of nine seedling beds between the irrigation pipelines with each bed being approximately 170 × 1.2 m. The experiment was a randomized complete block design with each treatment being 85 m long and replicated four times. Each 9-bed nursery section included 3 soil fumigation treatments (3 nursery beds per treatment). Because of the different plastic requirements, the iodomethane treatments could not be placed with the other soil fumigants and thus, were analyzed as a separate experiment. After fumigation and in each of the three growing seasons (2009 - 2011), a single family of loblolly pine seed was sown to the area at 236 seed/m^2. Seedlings in the trial area were maintained using standard nursery cultural practices (fertilization, irrigation, pest management, etc.) until lifting took place each winter (November to December). After lifting the seedlings at the end of each 9-mo growing season (November to December), the treatment area was left fallow and prepared for sowing the following spring (April).

Seedling quality and quantity

The effect of the soil fumigant treatments on seedling densities and growth characteristics were assessed in four subplots (1.2 × 0.3 m) per treatment plot at 7 weeks post-sowing, mid-summer (15 weeks post-sowing) and just prior to lifting in the fall (26 weeks post-sowing) in 2009, 2010 and 2011, respectively. Twenty-five loblolly pine seedlings per subplot were collected in the fall of each cropping season and returned to the laboratory at Auburn University for analysis. Seedling root collar diameter (RCD), shoot height and seedling dry weight (biomass) and root weight ratio (RWR) were measured for each seedling as well as overall root morphology. For root morphology, 10 seedlings per subplot were examined for total root length, total root surface area, average root diameter and the number of root tips using a flat-bed scanner and analysis of the root image using WinRhizo® software developed by Regents Instruments Inc., Quebec, Canada.

Soilborne *Trichoderma* and nematodes

Throughout the three cropping seasons, 20 soil samples (15 cm soil probe) were systematically collected from the center seedling bed of each 3-bed plot: at pre-sowing, post-sowing, mid-summer and

Table 1. MBr alternative soil fumigants and rates used in the 2009 - 2011 demonstration trial at the Pine Hill forest-tree nursery in Camden, AL.

Treatment	Application rate (kg/ha)	Plastic type[y]	Fumigant components
Pic+	336	HDPE	85% chloropicrin and 15% solvent
Chlor 100	336	HDPE	100% chloropicrin
MBr	392	HDPE	67% MBr and 33% chloropicrin
MBC 70/30	448	HDPE	70% MBr (98/2) and 30% solvent
Chlor 60	448	HDPE	60% chloropicrin and 40% 1,3-dichloropropene (Telone)
DMDS + Chlor	113 (l/ha)	HDPE	79% dimethyl disulfide and 21% chloropicrin
[z]Midas 98/2	112	VIF	98% iodomethane and 2% chloropicrin
Midas 50/50	180	VIF	50% iodomethane and 50% chloropicrin

[y]HDPE, High density polyethylene; VIF, virtually impermeable film. [z]The Midas treatments were considered a separate trial from the other soil treatments due to the plastic requirement.

Table 2. Site information for the 2009 - 2011 demonstration trial at the Pine Hill forest-tree nursery in Camden, AL.

Trial parameter	Camden, AL
Fumigation date	23 March, 2009
Fumigation type	Shank injected; broadcast/flat tarp
Experimental area	2.5 ha
Air temperature	16 - 25°C
Wind speed	5 - 10 km/h
Soil moisture	7.6%
Soil series	Lenoir silt loam
Plastic in place	14 days

Table 3. Loblolly pine seedling density at lifting over three growing seasons (2009 - 2011) for the demonstration trial at the Pine Hill forest-tree nursery in Camden, AL.

Fumigant	Seedling density (m^2)		
	2009	2010	2011
MBr	224[ax]	269[a]	226[a]
Chlor 100	226[a]	253[a]	222[a]
MBC 70/30	227[a]	277[a]	280[a]
Chlor 60	221[a]	234[a]	252[a]
Pic+	228[a]	250[a]	236[a]
DMDS + Chlor	210[a]	252[a]	237[a]
[y]$LSD_{(0.05)}$	(31)	(78)	(65)
[z]Midas 50/50	203[a]	253[a]	209[a]
Midas 98/2	196[b]	253[a]	200[a]
$LSD_{(0.05)}$	(5)	(75)	(33)

[x]Means (within a column) followed by the same letter are not significantly different based on Duncan's multiple range test ($p \leq 0.05$). [y]Least significant differences are in parenthesis. [z]The Midas treatments were considered a separate trial from the other soil treatments due to the plastic requirement.

just prior to seedling lifting at the end of each growing season. Half of each soil sample was plated onto Trichoderma-selective media (Elad et al., 1981) to determine soilborne fungi levels and the remaining half was sent to the Soils Laboratory at Auburn University for a quantitative assessment of nematode populations using a rapid centrifugal flotation technique (Jenkins, 1964).

RESULTS

Seedling quality and quantity

At the Pine Hill Nursery in Alabama, seedling densities at the end of the first growing season in 2009 were similar across all soil fumigants tested, ranging from 210 to 228 seedling/m^2 (Table 3). Similar results were observed for the 2010 and 2011 growing seasons for the soil fumigants used under the HDPE plastics. In contrast, soils treated with Midas had significantly fewer seedlings than the MBr treatments in 2009 (Table 3). However, by the end of the second and third growing seasons in 2010 and 2011, there were no significant differences in seedling densities for any of the soil fumigants tested, as all treatments gave similar seedling densities as the standard MBr soil fumigation treatment.

Differences in seedling RCDs among the soil fumigants tested at the Pine Hill Nursery were observed during the final growing season with MBr-treated soils producing significantly larger seedling diameters over all other soil fumigants tested (Table 4). Some of the newer chemistries, Pic+ and dimethyl-disulfide (DMDS), had seedlings with similar RCDs as MBr during the three growing seasons. Seedling RCDs declined for all treatments over the 3-year rotation (Table 4).

Seedling root architecture and root morphology as measured by root length, surface area, root diameter and root tips were similar for all soil fumigants over the three cropping seasons (Table 5). Generally, first year soil fumigation results in larger seedlings. However, as far as an MBr alternative, all soil fumigant alternatives performed as well as MBr across all the root morphology measurements. Like that of RCD, and time since soil fumigation, root characteristics tended to decrease over the three growing seasons (Table 5).

The RWR is defined as the weight of the roots divided

Table 4. Loblolly pine seedling RCD at lifting over three growing seasons (2009 - 2011) for the trial at the Pine Hill forest-tree nursery in Camden, AL.

Fumigant	Root collar diameter (mm)		
	2009	2010	2011
MBr	4.21^{ax}	4.01^{a}	3.89^{a}
Chlor 100	4.33^{a}	4.12^{a}	3.46^{b}
MBC 70/30	3.97^{a}	3.73^{a}	3.14^{b}
Chlor 60	4.46^{a}	4.13^{a}	3.34^{b}
Pic+	4.04^{a}	4.02^{a}	3.43^{b}
DMDS + Chlor	4.26^{a}	4.18^{a}	3.45^{b}
$^{y}LSD_{(0.05)}$	(0.60)	(0.54)	(0.42)
^{z}Midas 50/50	4.83^{a}	3.92^{a}	3.83^{a}
Midas 98/2	4.37^{a}	3.60^{a}	3.77^{a}
$LSD_{(0.05)}$	(1.87)	(1.92)	(0.93)

^{x}Means (within a column) followed by the same letter are not significantly different based on Duncan's multiple range test ($p \leq 0.05$). ^{y}Least significant differences are in parenthesis. ^{z}The Midas treatments were considered a separate trial from the other soil treatments due to the plastic requirement.

Table 5. Loblolly pine seedling root morphology at lifting over three growing seasons (2009 - 2011) for the demonstration trial at the Pine Hill forest-tree nursery in Camden, AL.

Fumigant	Root morphology											
	Root length (cm)			Root surface area (cm^2)			Root diameter (mm)			Root tips (#)		
	2009	2010	2011	2009	2010	2011	2009	2010	2011	2009	2010	2011
MBr	183^{ax}	167^{a}	138^{a}	51^{a}	53^{a}	44^{a}	0.89^{a}	1.02^{a}	1.02^{a}	398^{a}	457^{a}	299^{a}
Chlor 100	168^{a}	164^{a}	133^{a}	49^{a}	55^{a}	43^{a}	0.95^{a}	1.09^{a}	1.05^{a}	350^{a}	377^{a}	274^{a}
MBC 70/30	158^{a}	156^{a}	112^{a}	45^{a}	47^{a}	36^{a}	0.93^{a}	0.97^{a}	1.03^{a}	374^{a}	405^{a}	242^{a}
Chlor 60	169^{a}	163^{a}	116^{a}	49^{a}	52^{a}	37^{a}	0.93^{a}	1.03^{a}	1.04^{a}	328^{a}	381^{a}	244^{a}
Pic+	154^{a}	161^{a}	126^{a}	44^{a}	49^{a}	39^{a}	0.95^{a}	0.99^{a}	1.01^{a}	319^{a}	398^{a}	262^{a}
DMDS + Chlor	162^{a}	185^{a}	112^{a}	47^{a}	57^{a}	36^{a}	0.94^{a}	1.00^{a}	1.03^{a}	328^{a}	440^{a}	235^{a}
$^{y}LSD_{(0.05)}$	(27)	(38)	(32)	(9)	(16)	(9)	(0.09)	(0.11)	(0.08)	(74)	(123)	(69)
^{z}Midas 50/50	217^{a}	181^{a}	149^{a}	62^{a}	57^{a}	47^{a}	0.94^{a}	1.00^{a}	1.02^{a}	463^{a}	354^{a}	298^{a}
Midas 98/2	202^{a}	159^{a}	145^{a}	57^{a}	52^{a}	46^{a}	0.92^{a}	1.04^{a}	1.04^{a}	435^{a}	300^{a}	313^{a}
$LSD_{(0.05)}$	(359)	(127)	(142)	(54)	(10)	(30)	(0.78)	(0.67)	(0.52)	(531)	(440)	(278)

^{x}Means (within a column) followed by the same letter are not significantly different based on Duncan's Multiple Range Test ($p \leq 0.05$). ^{y}Least significant differences are in parenthesis. ^{z}The Midas treatments were considered a separate trial from the other soil treatments due to the plastic requirement.

by the total seedling weight with an optimum seedling having a RWR of >27% with higher RWR having better survival after outplanting. The RWR of seedlings grown in the different soil treatments resulted in differences in seedling root ratios only during the first growing season (Table 6). Of all the soil fumigants tested, Chlor 60 had the lowest RWR of all the treatments in 2009. However, treatments had similar RWRs in 2010 and 2011 (Table 6). While none of the treatments resulted in the optimum RWR, a number of factors affect RWR, including the time of lifting, growing density, the time of root pruning, irrigation and fertilization. For these trials, none of the MBr alternatives were detrimental to root growth that could affect seedling survival after out planting.

Soilborne *Trichoderma* and nematodes

At the end of each growing season, nursery soils at the Pine Hill Nursery in Alabama had similar levels of *Trichoderma* across all soil fumigants tested (Table 7).

Nematode populations within the soil are rarely

Table 6. Loblolly pine seedling RWRs at lifting over three growing seasons (2009-2011) for the demonstration trial at the Pine Hill forest-tree nursery in Camden, AL.

Fumigant	Root weight ratio (%)		
	2009	2010	2011
MBr	12.6[abx]	15.7[a]	14.7[a]
Chlor 100	12.4[ab]	15.4[a]	15.9[a]
MBC 70/30	12.4[ab]	13.6[a]	14.7[a]
Chlor 60	11.2[b]	15.3[a]	14.5[a]
Pic+	12.0[ab]	14.9[a]	15.0[a]
DMDS + Chlor	14.0[a]	15.1[a]	15.1[a]
[y]$LSD_{(0.05)}$	(2.3)	(1.9)	(2.4)
[z]Midas 50/50	13.3[a]	16.3[a]	14.5[a]
Midas 98/2	17.1[a]	17.1[a]	16.3[a]
$LSD_{(0.05)}$	(41.3)	(1.6)	(13.0)

[x]Means (within a column) followed by the same letter are not significantly different based on Duncan's Multiple Range Test ($p \leq 0.05$). [y]Least significant differences are in parenthesis. [z]The Midas treatments were considered a separate trial from the other soil treatments due to the plastic requirement.

Table 7. Number of *Trichoderma* colony forming units (CFUs) from soils collected over three growing seasons (2009-2011) for the demonstration trial at the Pine Hill forest-tree nursery in Camden, AL.

Fumigant	*Trichoderma* (CFUs/mg soil)					
	2009		2010		2011	
	June	December	June	November	June	November
MBr	82.5[ax]	101.0[ab]	99.0[a]	59.0[ab]	79.0[a]	76.0[a]
Chlor 100	59.8[a]	179.0[a]	111.0[a]	38.3[b]	95.2[a]	76.5[a]
MBC 70/30	36.3[a]	135.0[ab]	112.5[a]	82.0[a]	80.7[a]	76.2[a]
Chlor 60	26.3[a]	111.8[ab]	188.8[a]	46.5[b]	82.7[a]	69.0[a]
Pic+	22.5[a]	53.5[b]	124.3[a]	32.5[b]	55.0[b]	76.8[a]
DMDS + Chlor	35.5[a]	77.3[ab]	92.3[a]	32.5[b]	92.6[a]	97.2[a]
[y]$LSD_{(0.05)}$	(67.4)	(109.9)	(107.3)	(32.2)	(17.7)	(30.9)
[z]Midas 50/50	123.0[a]	56.3[a]	187.0[a]	74.7[a]	94.7[a]	85.4[a]
Midas 98/2	28.3[b]	34.8[a]	101.7[b]	116.7[a]	75.4[a]	94.0[a]
[y]$LSD_{(0.05)}$	(51.9)	(44.4)	(36.3)	(36.8)	(21.2)	(16.1)

[x]Means (within a column) followed by the same letter are not significantly different based on Duncan's multiple range test ($p \leq 0.05$). [y]Least significant differences are in parenthesis. [z]The Midas treatments were considered a separate trial from the other soil treatments due to the plastic requirement.

distributed uniformly across the nursery beds and these studies had a wide range (0 to 67 nematodes /100 cc soil) in numbers and species for all soil fumigants used (Table 8). Because of the variability, there were no differences between treatments for any of the soil fumigants tested.

Overall, all soil fumigants were effective in eliminating nematode populations the first growing season, which were maintained during the 3-year rotation. There was an increase during the second growing season, but populations did not increase into the third season (Table 8). One of the more troublesome species on seedling production is the stunt nematode (*Tylenchorhynchus claytoni*) which appeared only during the second cropping season. Of the soil fumigants tested, Chlor 60 was the only soil fumigant not to have any nematodes recovered in soil samples over the course of the study at the Pine Hill Nursery.

DISCUSSION

The primary objective of the USDA Areawide MBr

Table 8. Average number of nematodes per 100 cubic centimeters (cc) of soil over three growing seasons (2009 - 2011) for the demonstration trial at the Pine Hill forest-tree nursery in Camden, AL.

Fumigant	Nematodes (# per 100 cc of soil)					
	2009		2010		2011	
	June	December[x]	June	November[y]	June[z]	Nov
MBr	0	29	0	0	0	0
Chlor 100	0	0	0	47	67	0
MBC 70/30	0	0	0	65	0	0
Chlor 60	0	0	0	0	0	0
Pic+	0	0	0	12	0	0
DMDS + Chlor	0	26	0	0	0	0
Midas 50/50	0	17	0	0	0	0
Midas 98/2	0	28	0	43	0	0

[x]Nematodes recovered in December, 2009 were stunt nematodes. [y]Nematodes recovered in November, 2010 were stubby root nematodes. [z]Nematodes recovered in June, 2011 were spiral nematodes.

alternative program is to identify possible alternatives to MBr using large-scale, multi-year trials in various soil types and environmental conditions throughout the southern United States. One of the unique aspects of MBr as a soil fumigant is its ability to consistently control weeds, insects, nematodes and fungi across many different growing conditions and nursery soil types. Studies conducted within the southern United States in forest-tree nurseries have yet to find an MBr alternative that fits all of these characteristics (Enebak et al., 2012). The true test of an MBr alternative is its performance during the second and third growing season where treatment differences usually begin to appear as disease, weed, and nematode pressures increase. Based on these trials, those soil fumigants with chloropicrin appear to be the most useful in controlling pests and producing quality seedlings as seedling densities and root characteristics were similar to MBr.

One of the primary reasons for determining the effects of these soil fumigants on root architecture is that a more fibrous root system increases the chance of seedling survival in the field (Hatchell and Muse 1990; Frampton et al., 2002; Davis and Jacobs, 2005). One soil fumigant, DMDS + chloropicrin was comparable to MBr in RCD, root morphology characteristics and soilborne *Trichoderma* levels, but had a significant odor problem (that is, garlic or strong propane) that lasted long into the growing season. The lingering odor with this particular soil fumigant may limit its acceptance by growers as an alternative to MBr. The best MBr alternatives tested were Pic+ and Chlor 60 with both soil fumigants controlling nematodes and producing quality seedlings similar to MBr (Tables 3, 4, 5 and 6).

The application of Midas® (iodomethane and chloropicrin) in these trials was the first large-scale use of this compound on the production of forest-tree seedlings in the United States. Considered a drop-in replacement by the US Environmental Protection Agency, the US Department of Agriculture and the European Union, the compound was applied under VIF at less than half the rates of the other soil fumigants. The reduced active ingredient was done for two reasons: 1) the iodomethane was expensive (twice the cost of chloropicrin per ha) and 2) the VIF film increased the effectiveness of the compound by limiting the off-gassing. While the compound showed promise as an MBr alternative with respect to seedling quality and soilborne pest control, Arysta Life Science removed Midas® from the North American market in 2012 due to increasing costs and environmental pressures to re-examine the compound's registration in the US.

In addition to the weed pressures, some nursery soils have a history of nematodes reaching levels that affect seedling production (Cram and Fraedrich, 2005). For these reasons, 1, 3-dichloropropene (Telone®) may need to be used in nurseries with reoccurring nematode pressure. The soil type at the Pine Hill Nursery (Lenoir silt loam) has a higher soil density than that of most forest-tree nurseries and generally has not resulted in production issues due to nematode populations. Chlor 60 (containing 40% Telone®) had the lowest nematode levels and may be an option for nurseries that have nematode issues in the second growing season. While the seedling densities and root characteristics with chloropicrin were encouraging, one of the potential pitfalls with using 100% chloropicrin at 336 kg/ha is the buffer zone restrictions under current soil fumigation practices in the US. If these restrictions limit the use of 100% chloropicrin, then Pic+ with 85% chloropicrin would be the best alternative to MBr. The final decision when selecting an MBr alternative at individual nurseries should take into consideration the soil fumigant's ability to work under individual nursery soil conditions and the impact from new soil fumigation rules that came into place in 2012.

Soil fumigation is effective for reducing pathogenic soilborne fungi in the nursery that can positively affect seedling production and survival after outplanting. In other trials, the soil fumigants tested were found to be effective in controlling Pythium, Rhizoctonia, and Fusarium when inoculated onto oatmeal (Enebak et al., 2012).The wide-spread use of MBr has minimized extensive seedling losses due to soilborne pathogenic fungi. Pythium still can cause damping-off problems in the early spring and is often limited to areas of poor drainage and standing water. The soil-type at the Pine Hill Nursery is conducive to damping-off caused by Pythium early in the growing season. Rhizoctonia can appear in nurseries both as root decay and as a foliar blight, especially in second-year crops post-fumigation as the fungus increases over the first growing season (Carey and McQuage, 2004). Soilborne pathogens did not appear to affect seedling production in this 3-year trial, however, nursery managers remain uncertain for what the future holds once MBr is unavailable.

One of the unique aspects of soil fumigants is that they do not completely eliminate the beneficial soilborne fungi which are needed for seedling growth, especially *Trichoderma* which is necessary for pine seedling growth (Bailey and Lumsden, 1998; Mousseaux et al., 1998). In these trials, the population levels of non-target soilborne fungi rebounded quickly. Likewise, all soil fumigants used in previous Nursery Cooperative research has shown that *Trichoderma* is not as sensitive to MBr as other soil fumigants (Carey et al., 2005). In contrast, dazomet, a soil fumigant tested by the Nursery Cooperative for several years,significantly reduced the levels of beneficial fungi which remained after two growing seasons (Starkey et al., 2006).

While many nursery managers would prefer to use MBr in perpetuity to grow forest-tree seedlings, MBr will eventually be unavailable and each nursery manager will need to identify the best alternative for their nursery conditions. These trials at the Pine Hill Nursery indicate that seedling production is still possible if compounds such as chloropicrin are used and managers pay close attention to weed and nematode pests that are less susceptible to chloropicrin than MBr.

REFERENCES

Bailey BA, Lumsden RD (1998). Direct effects of *Trichoderma* and *Gliocladium* on plant growth and resistance to pathogens. In *Trichoderma* and *Gliocladium*, vol. 2.Edited by G.E. Harman and C.P. Kubicek.Taylor and Francis, Inc., Bristol, Pa. pp.185–204.

Byram TD, Mullin TJ, White TL, van Buijtenen JP (2005). The Future of Tree Improvement in the Southeastern United States: Alternative Visions for the Next Decade. Southern J. Appl. For. 29:88-95.

Carey WA, McCraw D, Enebak SA (2005).Seedling production by seed treatment and fumigation treatment at the Glennville Regeneration Center in 2004. Auburn University Southern Forest Nursery Management Cooperative Research Report 05-04.Auburn, AL. P. 5.

Carey WA McQuage K (2004). Control of Rhizoctonia foliage blight by fungicides and fumigation: Lower application rates and fumigation effects. Auburn University Southern Forest Nursery Management Cooperative, Research Report 04-03. Auburn, AL. P. 4.

Cordell CE, Anders R, Hoffard WH, Landis TD, Smith RS, Toko HV (1989). Forest nursery pests. Wash (DC): Agricultural Handbook. 680:184.

Cram M, Fraedrich SW (2005). Management options for control of a stunt and needle nematode in southern forest nurseries. In: Dumroese RK, Riley LE, Landis TD tech. coords. 2005. National proceedings: Forest and Conservation Nursery Associations; 2004 July 12–15; Charleston, NC; Proc. RMRS-P-35. Fort Collins, CO: U.S. Department of Agriculture, Forest Service, Rocky Mountain Research Station.

Davis AS, Jacobs DF (2005). Quantifying root system quality of nursery seedlings and relationship to out planting performance. New Forests. 30:295-311.

Elad Y, Chet I, Henis Y (1981). A selective medium for improving quantitative isolation of *Trichoderma* spp. from soil. Phytoparasitica 9:59-67.

Enebak SA, Wei G, Kloepper JW (1998). Effects of plant growth-promoting rhizobacteria on loblolly and slash pine seedlings. For. Sci. 44:139-144.

Enebak SA, Starkey TE, Quicke M (2012).Effect of methyl bromide alternatives on seedling quality, nematodes and pathogenic soil fungi at the Jesup and Glennville Nurseries in Georgia: 2007-2008. J.Hort.For. 4:1-7.

Frampton J, Isik F, Goldfarb B (2002). Effects of nursery characteristics on field survival and growth of loblolly pine rooted cuttings. South. J.Appl. For. 26:207-213.

Hatchell GE, Muse HD (1990). Nursery cultural practices and morphological attributes of longleaf pine bare-root stock as indicators of early field performance. USDA For. Serv. Res. Pap. SE. P. 277.

Jang E, Wood WS, Dorschner K, Schaub J, Smith D, Fraedrich S, Hsu H (1993). Methyl bromide phase out in the US: Impact and alternatives, In: USDA workshop on alternatives for methyl bromide. Crystal City, VA.

Jenkins H (1964). A rapid centrifugal flotation technique for separating nematodes from soil. Plant Disease Reporter 48:692.

McNabb K, VanderSchaaf C (2003). A survey of forest seedling production in the south for the 2002-2003 planting season. Technical Note 2003-01. Southern Forest Nursery Management Cooperative, Auburn University P. 16.

Mousseaux MR, Dumroese RK, James RL, Wenny DL, and Knudsen GR (1998). Efficacy of *Trichoderma harzianum* as a biological control of *Fusarium oxysporum* in container-grown Douglas-fir.New Forests. 15:11-21.

South DB, Carey WA, Enebak SA (1997). Chloropicrin as a soil fumigant in forest nurseries. For. Chron. 73:489-494.

Smith WB, Vissage JS, Darr DR, Sheffield RM (2001). Forest Resources of the United States, 1997. U.S. Dep. Agric., Forest Service, Gen. Tech. Rep. NC. P. 219.

Starkey TE, Enebak SA, McCraw D (2006). Seedling quality and weed control with methyl bromide and methyl iodide at the Glennville Regeneration Center 2005-2006. Auburn University Southern Forest Nursery Management Cooperative. Research Report 06-05. Auburn, AL. P. 5.

Chemical composition of the essential oil from *Ziziphora clinopodioids* Lam. from Iran by means of gas chromatography-mass spectrometry (GC-MS)

A. R. Sardashti[1]*, J. Valizadeh[2] and Y. Adhami[1]

[1]Department of Chemistry, Faculty of Sciences, University of Sistan and Baluchestan, P. O. Box 98167 Zahedan, Iran.
[2]Department of Biology, Faculty of Sciences, University of Sistan and Baluchestan, P. O. Box 98167, Zahedan, Iran.

The essential oil from aerial parts of *Ziziphora clinopodioids* plant was extracted by hydro distillation with efficiency of 1.86% (w/w). The components were characterized by means of gas chromatography–mass spectrometry (GC/MS) using kovat's retention indices. Thirty-two components of the essential oil which composed 98.18% of the total oil were characterized. The major components were pulegone (61.67%), cis-caran-trans-2-ol (12.66%), 1, 8-cineole (10.23%) and 2-β-pinene (2.16%).

Key words: *Ziziphora clinopodioides*, essential oil, pulegone, (-) cis-caran-trans 2-ol, 1, 8-cineole, 2-β-pinene.

INTRODUCTION

The genus Ziziphora belongs to the family Labiatae consisting of four species that is widespread all over Iran. *Ziziphora clinopodioides* Lam. with the common Persian name "kakuti-e kugi" is an endemic species that grows wild in Iran, Afghanistan, Iraq and Azerbaijan (Talish). *Z. clinopodioids* Lam. is an edible medicinal plant with leaves, flowers and stems that are frequently used as wild vegetable or additive in foods to offer aroma and flavor (Zargari, 1993). The plant is used as stomachic, anti-fever, anti-inflammatory, sedative and flavoring agent in Iranian folk medicine. The chemical composition of the essential oil from *Z. clinopodiodes* growing wild in Bakhteyari province was examined by Sajjadi et al. (2005). Twenty-two components were identified in which pulegone was 53.2%, p-metha-1-en-1-ol was 21.4%, 1, 8-cineole was 10.3% and β-pinene was 1.6% (Sajjadi et al., 2005). Meral et al. (2002) in their research about Ziziphora taurica subsp. clinopodioides in Turkey showed the components which have the higher percentage including pulegone (81.86%), limonene (4.48%) and piperitenone (2.30%). In another research undertaken by Verdian-rizi (2008), *Z. clinopodioides* Lam. was collected from North of Iran and the essential oil from this plant with outcome of 0.9% w/w was extracted and 26 components were recognized including 97.62% of total essential oil. The most important components were pulegone (36.45%), piperitenone (19.12%), and menth-2-en-1-ol (5.31%).

*Corresponding author. E-mail: asardashti65@gmail.com.

MATERIALS AND METHODS

Plant materials

The aerial parts of *Z. clinopoclioides* were collected at the flowering stage from the Taftan area in Baluchestan region Southeastern of Iran in June, 2005. Its scientific name was adopted from the Karaj Herbarium, Institute of Iran (Mozaffarian, 1996).

Isolation of essential oil

The aerial parts (20 g) were dried at 25°C in the shade and subjected to hydro distillation using a Clevenger-type apparatus for 4 h obtained an essential oil which was collected in Hexane-solvent and dried with anhydrous sodium sulphate, weighed and stored at 4°C in the dark until use. The efficiency of extraction was 1.861±0.002 w/w and Relative Standard Deviation (RSD) for four extractions were 3.49 ppt.

Gas chromatography–mass spectrometry (GC/MS) analysis

The analysis of the essential oils was performed using a Hewlett-

Table 1. Chemical composition of essential oil from *Ziziphora clinopodioides* Lam.

Compounds	Retention indices	Percentage (%)
Phellandrene	927	0.09
α-Pinene	934	1.47
Camphene	948	0.46
B-D-xylopyranose	968	0.05
Sabinene	974	1.36
2-β-Pinene	976	2.16
β-myrcene	990	0.74
Ethyl-amyl carbinol	997	0.24
Cyclohexene,1-methyl-4-(1-methylidene)	1015	0.08
1,8 -cineole	1034	10.23
Trans-ocimene	1044	0.3
Γ-Terpinene	1058	0.15
(IR)-(+)-Trans-is limonene	1070	1.03
α-Terpinolene	1084	0.33
Cyclopropane, pentyl	1130	0.14
(-)Cis-Caran-trans-2-ol	1155	12.66
Cis-Cyclohexanone-5-methyl-2-(1-methyl ethyl)	1164	0.34
Isopulegone	1175	1.78
3-Cyclohexene-1-methanol / /α/ α4 / - trimethyl	1206	0.21
Pulegone	1253	61.67
Cyclohexene -1-one/3-methyl-6(1-methylethyl) 2-	1262	0.07
Z-Citral	1274	0.05
6-Octen-1-ol/3/7-dimethyl-formate	1279	0.05
Bicyclo[2/2/1 hetan-2-ol/1/7/7-trimethyl-acetate(15-endo)	1286	0.11
1/1-Dimethyl-3-methylidene-2-vinylcyclo hexane	1303	0.18
phenol/4-ethenyl-2-methoxy	1316	0.4
2-Cyclohexan-1-one/3-methyl-6-(1-methylethylidene	1341	1.09
Cyclohexanone/2/5-dimethyl-2-(1-methyletheny)	1349	0.18
Trans-Caryophyllene	1413	0.18
Germacrene D	1475	0.06
(-)Caryophyllene oxide	1576	0.3
(3R*/4S*)-3-(2-Nitro-4-methoxyphenyl)-4-(4-hydroxphenyl)hexane	2144	0.02

Compounds presented in order of elution from the HP-5Ms capillary column; retention index relative to n-alkanes on the HP-5Ms capillary column; percentage based on FiD peak area normalization.

Packard 6890 network GC system, equipped with a 60 m* 0.25 mm id, 0.25 µm HP-5 Ms capillary column, and a HP 5973 mass selective detector. Helium was the carrier gas at 1 ml/min. Injector and MS transfer line temperatures were 250 and 260°C respectively. Column temperature was set at 40°C for 1 min, and then programmed from 40 to 250°C at a rate of 3°C/min, and finally held isothermally for 20 min for GC/MS detection. An electron ionization technique was used with ionization energy of 70 eV; retention indices were calculated by using the retention times of C8 to C26 n-alkenes which were injected into the oil at the same chromatographic conditions based on Vanden's method (Van den Dool and Kartz, 1963).

Identification of compounds

The linear retention indices for all the compounds were determined by co-injection of the sample with a solution containing the homologous series of C8 to C26 n- alkanes. The individual constituents were identified by their identical retention indices, referring to known compounds from the literature (Adams, 1995) and also by comparing their mass spectra with either the known compounds or with the Wiley 7 mass spectral database.

RESULTS AND DISCUSSION

The identified components and their percentages are listed according to their elution order (Table 1). Identification of the components was based on GC retention indices (Vanden and Kartz, 1963). The fragmentation pattern of the mass spectra was computer-matched with Wiley GC/MS library and compared with those reported in other literature s (Adams, 1995). As

Table 2. Chemical composition of essential oil of Z. *clinopodioides* Lam by chemical class.

Chemical composition by chemical class	Percentage (%)	Number of compounds
Monoterpene hydrocarbons	8.17	11
Oxygenated monoterpenes	88.1	9
Sesquiterpene hydrocarbons	0.24	2
Oxygenated sesquiterpenes	0.30	1
Other hydrocarbons compounds	0.32	2
Other oxygenated compounds	1.05	7
Total monoterpenoids	96.2 7	20
Total sesquiterpenoids	0.54	3
Total	98.18	32

shown in Table 2, 32 components which composed 98.18% of the total oil were identified. The major components were pulegone (61.67%), cis-caran trans-2-ol (12.66%), 1, 8-cineole (10.23%) and 2- β -pinene (2.16%). A qualitative comparison of the oil constituents of *Z. clinopodioides* with those of other species reported showed varying compositions. The results of this investigation indicated that the essential oil of aerial parts of *Z. clinopodioides* Lam. was characterized by high contents of oxygenated monoterpenes (88.10%) according to Meral et al. (2002).

Conclusion

An analysis of the essential oil of *Z. clinopodioides* Lam. by GC/MS showed that the components that have the highest percentages are: pulegone, (-) cis-caran-trans-2-ol and 1, 8-cineole. These compounds have anti-bacterial properties and can be used in making anaesthetic and antiseptic drugs. Other important compounds are α-pinene which is found only in the essential oil as a result of strong light influence.

ACKNOWLEDGEMENT

The authors are grateful to Professor Issa Yavari for his cooperation in this work.

REFERENCES

Adams R P (1995). Identification of essential oil components By gas chromatography /mass spectroscopy, Allured Publishing. Corporation: Carol stream, IL.

Meral GE, Koyalioglu S, Ozturk B (2002). Essential oil composition and antioxidant activity of endemic Z .taurica subsp. Clinopodioides, Fitoterapia 73(7-8):716-8.

Mozaffarian V (1996).Dictionary of Iranian plant Names farhang moser, Tehran Press. p. 591.

Sajjadi SE, Ghassemi N, Baluchi M (2005). Chemical composition of essential oil from aerial parts of *Ziziphora clinopodioides* Lam. Pajoohesh VA sazandeghi. (58):1-9.

Van den Dool H, Kartz PD (1963). A generalization of the retention index system including linear temperature programmed gas liquid partition chromatography. Chromatography 11:463-471.

Verdian-rizi M (2008).Essential oil composition and Biological Activity of *Z. Clinopodioides* American-Eurasian J. Sustain. Agric. 2(1):69-71.

Zargari A (1993). Iranian Medicinal plant, volume 4, Tehran University press. p. 103-104.

Effect of organic seed pelleting on seed storability and quality seedling production in biofuel tree species

P. Srimathi[1], N. Mariappan[2*], L. Sundaramoorthy[2] and M. Paramathma[3]

[1]Department of Seed Science and Technology, Seed Centre, TNAU, Coimbatore-641003, Tamil Nadu, India.
[2]Vanavarayar Institute of Agriculture, Pollachi, Coimbatore-642103, Tamil Nadu, India.
[3]Centre of Excellence for Biofuel, Forest College and Research Institute, TNAU, Mettupalayam-641103, Tamil Nadu, India.

***Jatropha curcas* and *Pongamia pinnata* are the promising biofuel species worldwide that contained 35 to 45% seed oil, respectively. Due to the reason of high oil content of these seeds are liable to getting easily deteriorated by fungal and insects during storage and that could hold the limitation of spreading through afforestation. To resolve this problem, the present study was carried out with *J. curcas* and *P. pinnata* seeds pelleted with leaf powders of *Azadirachta indica*, *P. pinnata* and *Adhatoda vasica* in order to study the effects of botanical pelleting storability and production of quality seedlings at nursery. This experiment was conducted at Department of Seed Science and Technology, Tamil Nadu Agricultural University, Coimbatore, Tamil Nadu, India during 2009 to 2010. The results revealed that *J. curcas* and *P. pinnata* seeds pelleted with pungam and neem leaf powder showed the maximum germination percentage and growth characters at nursery, respectively. From the study, it was also observed that *Jatropha* seeds pelleted with pungam leaf powder recorded 169% higher germination and *P. pinnata* seeds pelleted with *A. indica* leaf powder registered 92% higher germination than control after 9 months of storage. We conclude from this study that, botanical leaf powder seed pelleting not only improve the longevity of seeds through protecting the seeds from fungal and insect attack but also improve the seed and soil relationships through enriching the rhizosphere region of seed to produce better growth and development.**

Key words: Jatropha, pungam, leaf powder pelleting, seed storage, growth, development.

INTRODUCTION

Jatropha curcas and *Pongamia pinnata* seeds are having a strong trend to deterioration because of their high oil content, demanding special attention and care during storage for the maintenance of their viability and vigour. Such aspect can be also considered valid for seeds of physic nut (*J. curcas*), which belongs to the Family Euphorbiaceae and pungam (*p. pinnata*) it belongs to the family Leguminaceae has stood nationwide for its potential potential production of biodiesel, due to the high oil content in the seeds about 40 and 35 to 45%, respectively (Beniwal and Chauhan, 2011).To address the continuing need for fuel in India, research on *J. curcas* and *P. pinnata* seeds are being done to explore its potential for producing bio-diesel.

Seed pelleting is applied to seeds to improve the outer structure into uniformly round seeds, which enable precision planting. Seed pelleting with plant parts can augment the seed quality, hence improving the seedling value and reduce cost of transplanting (Govinden and Levantard, 2008). At present we need to approach ecofriendly ideas and alternatives that recent time, the use of different plant parts and their byproducts has given the impression to bean effective alternative to noxious chemical insecticides and fungicides for controlling various insect pests and diseases in storage. As many as 2400 plant species in the world has been identified that have properties and biological activity against awide range of pests (Grainge and Ahmed, 1988).

Seed management techniques are employed to invigourate the seed by modifying the physiological stamina of the seed. Seed pelleting is one such physical treatment given to seed to invigorate the seedling vigour. Some of the common benefits of pelleting are uniformity in seed size, precision planting, uniform stands with reduced seed rate, more insect and disease resistance, better performance under stress conditions and additional nourishment to the seedlings (Peterhalmer, 2003). In addition it is also employed to add needy substance to the individual seed so that the seeds get invigourative effect on absorption of such materials, at initial watering by enriching the rhizosphere region of each and every seed as nutritive without physiological modification of seed but by simple physical alterations of the seed (Suma, 2005).

During the storage seeds or grains that might be infected by fungi which cause a deteriorate its viability, discoloration, various biochemical changes, heating and mustiness, loss in weight, and production of toxins when it is consumed by human and animals may injurious to their health (Arya and Perello, 2010). In many cases, fungi infecting seeds are seed-borne pathogens, they play an important role in the transmission of numerous pathogenic species to seedlings as well as to the soil (Dharmaputra et. al, 2009).

The oil seeds such as soybean, sunflower, and peanut are contributing much of the world's biodiesel production, thus oilseeds species currently beingwidely studied. Though, little importance has been given to the processes of seed storage, which constitutes a major problem for agriculture (Tekrony, 2006). There are several factors that are known to influence the progress of deterioration during seed storage but high temperatures and humidity during storage increase the deterioration speed of seeds (Pukacka et al., 2009).

Uses of different chemicals are costly and may cause natural hazard, whereas botanicals are less costly, easily available to the farmers, safe to handle and they can prepare easily. Comparative study of botanicals helps to choose the suitable one for storing the seeds. The study on seed pelleting in agriculture especially in field crops were well exploited but in case of tree species it is very scanty. Hence, there is an urgent need to develop appropriate seed pelleting technology for enhancing the growth and development of Jatropha and Pungam since these two tree species are very important biodiesel yielding trees. So keeping all the above facts in view, an attempt was made with Jatropha and pungam seeds to study the efficacy of botanical seed pelleting on storage and quality seedling production.

MATERIALS AND METHODS

Seed sources

The bulk seeds of *J. curcas* from Mettupalayam (11.3° N 76.95° E) and *p. pinnata* from Pillur (11.24251° N 76.96009° E) seed source of Karamadai, Coimbatore district of Tamil Nadu, India.

Experimental materials

The collected seeds were pelleted with leaf powders *viz.,* pungam (*P. pinnata*), neem (*A. indica*), and adathoda (*Adhatoda vasica*) leaf powders at 300 g $^{kg-1}$ of seed, as filler material using 10% maida as common adhesive (250 to 300 ml/kg^{-1}) and were dried in shade to reduce the moisture content to 9%.

Seedling quality parameters

The treated seeds were evaluated for their seedling quality *viz.,* germination (%), root length (cm), shoot length (cm), fresh weight of 10 seedlings^{-1} (g), dry matter production of 10 seedlings^{-1} (g) under ambient room conditions (RH 95 ± 2 % and 25 ± 2°C). Vigour index values were also computed by using the formula; Vigour index 1 = Germination (%) x Total seedling length (cm) as per Abdul–Baki and Anderson (1973).

Storage of seeds

The seed were stored in cloth bags under the ambient conditions of Coimbatore (11°1'6 N 76° 58'21 E and 320 MSL) for a period of nine months and were observed for seed quality characters as mentioned above room conditions.

Statistical analysis

The data gathered were statistically scrutinized as per Panse and Sukhathme (1999) under F test of significance for understanding the level of significance among the seed treatments, seed and seedling quality characters.

RESULTS AND DISCUSSION

In the present investigation, significant mean performance for seedling growth were recorded by seeds pelleted with leaf powders of pungam (*P. pinnata*), neem (*A. indica*) and adathoda (*A. vasica)* as filler material using 10% maida as common adhesive (250 to 300 ml/kg) in jatropha and pungam at all periods of storage (Tables 1 and 2). There was a significant effect in terms of seedling quality characters on *J. curcas* germination using different botanicals (Table 1). The germination

Table 1. Influence of botanical seed pelleting on storability of Jatropha seeds.

Botanicals (B)	Periods of storage in Months(M)				
	Initial	After 3months	After 6months	After 9 months	Mean
			Germination (%)		
Neem	77 (61.13)	61(51.57)	48(44.03)	32(34.13)	55 (47.71)
Pungam	70(56.83)	69 (56.40)	48(43.67)	35(36.17)	56 (48.27)
Adathoda	78(62.27)	62 (51.83)	47(43.07)	25(29.93)	53 (46.78)
Control	67(54.83)	42 (40.57)	25(29.93)	13(34.13)	37 (39.87)
Mean	73(58.76)	59(50.09)	42(40.17)	26 (33.59)	
CD (P=0.05)	M 3.0*		B 3.0*		MB NS
Botanicals (B)			**Shoot length (cm)**		
Neem	28.5	26.3	22.9	23.1	25.2
Pungam	25.6	28.2	23.2	26.1	25.8
Adathoda	28.1	28.1	24.9	23.9	26.3
Control	27.7	21.7	22.3	22.3	23.5
Mean	27.5	26.1	23.3	23.8	
CD (P=0.05)	M NS		B NS		MB 3.0
Botanicals (B)			**Root length (cm)**		
Neem	7.8	6.2	7.3	6.7	7.0
Pungam	7.5	8.2	6.3	7.7	7.4
Adathoda	7.9	6.6	6.3	5.9	6.7
Control	6.9	6.4	5.5	5.0	6.0
Mean	7.5	6.8	6.4	6.3	
CD (P=0.05)	M NS		B NS		MB 1.4
Botanicals (B)			**Fresh weight 10 seedlings^{-1}(g)**		
Neem	49.7	47.7	47.2	38.5	45.8
Pungam	47.8	47.3	44.0	39.0	44.5
Adathoda	48.3	43.9	44.1	38.4	43.7
Control	47.4	41.5	34.8	34.8	39.6
Mean	48.3	45.1	42.5	37.7	
CD (P=0.05)	M 2.3*		B 2.3*		MB NS
Botanicals (B)			**Dry weight 10 seedlings^{-1}(g)**		
Neem	4.6	4.4	3.9	3.6	4.1
Pungam	3.7	4.6	3.3	3.9	3.9
Adathoda	4.4	4.4	3.2	3.7	3.9
Control	3.3	3.6	3.1	3.1	3.3
Mean	4.0	4.3	3.4	3.6	
CD (P=0.05)	M 0.3*		B 0.3*		MB NS
Botanicals (B)			**Vigour index**		
Neem	2786	1995	1460	938	1795
Pungam	2314	2524	1405	1182	1856
Adathoda	2820	2138	1446	735	1785
Control	2306	1183	693	368	1138
Mean	2556	1960	1251	806	
CD (P=0.05)	M 152.2*		B 152.2*		MB 304.5*

*, Significant at P=0.05 level; NS, non-significant; M-period of storage in months; B, botanicals; MB, interaction between months of storage and botanicals.

Table 2. Influence of botanical seed pelleting on storability of pungam seeds.

Botanicals (B)	Periods of storage in Months(M)				
	Initial	3	6	9	Mean
			Germination (%)		
Neem	92(73.1)	70(56.9)	55(47.9)	23(28.3)	60(51.5)
Pungam	89(70.7)	68(55.6)	56(48.2)	19(25.8)	58(50.1)
Adathoda	95(76.5)	69(56.0)	50(45.0)	15(22.8)	57(50.1)
Control	84(66.4)	53(46.7)	44(41.3)	12(19.6)	48(43.5)
Mean	90(71.7)	65(53.8)	51(45.6)	17(24.1)	
CD (P=0.05)	B 3.9*		M 3.9*		BM NS
Botanicals (B)			**Shoot length (cm)**		
Neem	36.7	29.6	27.7	24.5	29.6
Pungam	32.1	26.2	27.0	23.8	27.3
Adathoda	33.8	27.7	27.1	27.8	29.1
Control	31.1	30.8	25.5	21.6	27.2
Mean	33.4	28.6	26.8	24.4	
CD (P=0.05)	B NS		M 2.0*		BM NS
Botanicals (B)			**Root length (cm)**		
Neem	17.3	13.8	13.4	11.5	14.0
Pungam	17.8	13.2	12.8	12.6	14.1
Adathoda	16.0	13.3	11.6	17.7	14.6
Control	16.8	13.7	12.3	10.8	13.4
Mean	17.0	13.5	12.5	13.1	
CD (P=0.05)	B NS		M 1.5*		BM 3.0*
Botanicals (B)			**Fresh weight 10 seedlings (g)**		
Neem	58.3	46.6	49.2	45.9	58.3
Pungam	59.4	51.2	49.4	41.6	59.4
Adathoda	54.6	38.7	49.6	37.8	54.6
Control	55.1	44.3	47.7	43.5	55.1
Mean	56.8	45.2	49.0		56.8
CD (P=0.05)	B 3.2*		M 3.2*		BM NS
Botanicals (B)			**Dry weight 10 seedlings (g)**		
Neem	14.0	10.9	12.3	11.5	14.0
Pungam	13.6	11.8	12.1	10.8	13.6
Adathoda	13.1	10.0	11.9	10.7	13.1
Control	13.0	11.3	11.5	10.7	13.0
Mean	13.4	11.0	11.9		13.4
CD (P=0.05)	B NS		M 0.7*		BM NS
Botanicals (B)			**Vigour index**		
Neem	4932	3035	2082	880	4932
Pungam	4441	2660	2206	692	4441
Adathoda	4701	2805	1933	681	4701
Control	4024	2359	1640	373	4024
Mean	4524	2714	1965		4524
CD (P=0.05)	B 286.9*		M 286.9*		BM NS

*, Significant at P=0.05 level; NS, non-significant; M-period of storage in months; B, botanicals; MB, interaction between months of storage and botanicals.

percentage was the highest in the seeds pelleted and stored with pungam leaf powder that recorded 35% followed by neem and adathoda were recorded 32 and 25%, respectively after nine months of storage. Untreated seeds showed only 13% germination thus revealed that 13, 11, and 7 time lower performance than pungam, neem and adathoda leaf powder pelleting. Similarly root length (cm), shoot length (cm), fresh weight (%), dry weight (%) and vigour index expelled significantly higher value in pungam pelleted seeds followed by neem and adathoda than control. However all pelleted seeds performed better than that of the unpelleted seeds due to their influence in enriching the efficiency of rhizosphere soil for production of quality seedlings. The better performance of pungam leaf powder might be because it acts as a wick in absorbingregulating and correcting the soil moisture availability and thus enhanced the better seed soil relationship (Lu et al. 1983). The leaf powders also contains gibberellins like substances, the saponins and micronutrients the zinc, which have synergistically activated to form the indole acetic acid (IAA).

In addition the chlorophyll molecules of leaf powders, amino acids and humic acid present in the soil rhizosphere might have acted as a chelating agent and activated the growth and development of pungam leaf powder pelleted seeds into robust seedling at nursery. The energy of the seed recorded through the seedling vigour parameters also highlighted that pungam leaf powder followed by neem leaf powder pelleting served as the better seed invigourative treatment. However, Bharathi (1999) expressed that seed pelleting with *Albizia amara* enhanced the seedling quality characters and helps in improving the germination of seeds under stress condition such as soil salinity. Gurunathan et al. (2006) also revealed that seed pelleting with pungam leaf powder at 300g kg^{-1} of seed followed by Azospirillum at 100g kg^{-1} of seed maximized quality seedling production at nursery. The storage studies with the pelleted seed revealed that the seed pelleted with pungam leaf powder could be stored up to six months with 48% germination.

It was observed that, *P. pinnata* seeds pelleted with neem leaf powder leaf powder at 300 g kg^{-1} of seed maximized the seedling germination percent to 25% which was 92% higher than the controlat the end of 9 months of storage period. The order of performance of other pelleting treatments such as pungam and adathoda compared to control the germination percent was improved 58 and 25%, respectively after nine months storage (Table 2). The better performance of neem leaf powder might be because it contains quercetin (flavonoid) and nimbosterol (β-sitosterol) as well as number of liminoids (nimbin and its derivatives), these properties are having antibacterial and antifungal properties that may lengthen seed longevity in storage. Neem products may also repel insects, stop their feeding, inhibit reproduction and cause other interruptions (Schumuttere, 1990).

Pal and Basu (1995) reported that wheat seeds treated with neem leaf powder at 2 g kg^{-1} of seed recorded maximum germination and seedling vigour after 7 months of storage underambient condition. Sharma (1995) recommended treatment of maize seeds with neem leaf powder (100 g/kg^{-1}) and ash (10 g/kg^{-1}) for effective reduction in percent seed damage during storage. Ogunwolu and Odunlani (1996) reported that cowpea seeds treated with neem leaf powder (3 g/kg^{-1}) seeds showed reduced bruchid infestation after five months of storage. Anil et al. (1998) observed that soybean seeds treated with neem products and untreated seeds did not show any significant differences in germination. But seed treatment controlled association of mycoflora up to 120 days of storage period.

Hossain et al. (1999) observed that soybean seeds coated with neem leaf powder showed excellent control of seed borne disease of *colletotrichum dematum* and maintained better seed health during the storage period. Arati (2000) reported that Bengal gram seeds treated with neem leaf powder recorded higher germination (65.91%) and vigour index (1282) compared to control at the end of 10 months of storage period. Maraddi (2002) observed that cowpea seeds treated with neem leaf powder (5 g/kg^{-1}) recorded higher germination (39.5%) and vigour index (1072) compared to control (34.2% and 864, respectively) at the end of 10 months of storage period. Khatun et al. (2011) reported that the lentil seeds were treated with neem leaf powder and stored till next planting time; it has improved the seed quality, such as moisture content, germination capacity, root length, shoot length of the seedlings and vigour index.

The pelleted seeds stored up to nine months under ambient conditions also exposed the better performance of leaf powder pelleting treatment in storage highlighting consistent efficacy of pelleting treatment on production on quality seedlings. The increment in germination *J. curcas* with pelleting treatments ranged between 169 to 92% compared to control and *Pongamia pinnata* seeds pelleted with neem leaf powder increased the range of germination per cent from 92 to 25% compared to un-pelleted seeds. The vigour parameters of stored seeds also found to be influence due to pelleted seeds compared to un-pelleted control seeds. Thusthe study revealed that seed pelleting with pungam and neem leaf powder at 300g kg^{-1} of seed maximized the seedling quality characters of *J. curcas* and *P. pinnata* initially and on storage.

Conclusion

We conclude from this study that, botanical leaf powder seed pelleting are not only improve the longevity of seeds through protect the seeds from fungal and insect attack but also improve the seed and soil relationships through enriching the rhizosphere region of seed to produce better growth and development.

ACKNOWLEDGEMENTS

The authors are grateful to National Oilseeds and Vegetable Oils Development (NOVOD) Board for funding this scheme. They also express their thanks to the staff members of Seed Science and Technology, TNAU for completion of this research work.

REFERENCES

Abdul-Baki AA, Anderson JD (1973). Vigour determination in soybean seed by multiple criteria. Crop Sci. 13:630-633.

Anil KN, Gaur A, Sunk SSK, Devkumar C (1998). Performance of neem products on storability of soybean. Seed Res. 26(2):138-146.

Arati P (2000). Influence of containers and seed treatment on storability of chickpea. M.Sc. (Agric.) Thesis. University of Agricultural Sciences, Dharwad.

Arya A, Perello AE (2010). Management of fungal plant pathogens. CAB International. P. 388.

Beniwal RS, Chauhan R (2011). *Pongamia pinnata* as an alternative source of renewable energy. APA News (Asia Pacific Agroforestry Newsletter) 38:19.

Bharathi A (1999). Studies on handling, management and storage of neem seed (*Azadirachta indica* A. Juss). Ph.D. Thesis. Tamil Nadu Agricultural University, Coimbatore.

Dharmaputra OS, Worang RL, Syarief RM (2009). The Quality of Physic Nut (Jatropha curcas) Seeds Affected by Water Activity and Duration of Storage. Microbiol. Indonesia 3(3):139–145

Govinden SJ, Levantard M (2008). Comparative studies of seed priming and pelleting onpercentage and mean time to germination of seeds of tomato (*Lycopersicon esculentum* Mill.). Afr. J. Agric. Res. 3(10):725-731.

Grainge M, Ahmed S (1988). Handbook of Plants with Pest Control Properties. John Wiley and Sons. New York, P. 470.

Gurunathan N, Srimathi P, Paramathma M, Kumaran K, Parthiban KT (2006). Seed pelleting for production of quality seeding in *Jatropha curcas*. Abstr. XII National Seed Seminar, 24-26, Feburary 2006, ANGRAU, Hyderbad.

Hossain I, Suratuzzaman M, Khalil MI (1999). Seed heath of soybean and control of seed born fungi with botanicals. Bangladesh J. Train. Develop. 12(2):99-105.

Khatun A, Kabir G, Bhuiyan MAH, Khanam D (2011). Effect of preserved seeds using different botanicals on seed quality of lentil. Bangladesh J. Agric. Res. 36(3):381-387.

Lu S, Ming D, Jiang T (1983). A preliminary report on the effect of $CaCO_3$ pelleted seeds of Chinese milk vetch on its yield. Shanghai Agric. Sci. Technol. 6:9170-9173.

Maraddi B (2002). Influence of growth regulators on seed yield and quality and seed treatments on storabilityof cowpea cv. C-252, M.Sc. (Agri.) Thesis. University of Agricultural Sciences. Dharwad.

Ogunwolu EO, Odunlani AT (1996). Suppression of seed bruchid (*Callosobruchus maculates*) development and damage on cowpea (*Vigna ungiculata*) with *Zantoxylemz antoxyloider* L. (Rutaceae) root bark powder when compared to neem seed powder and primiphos methyl. Crop Protect. 15:603-607.

Pal P, Basu RN (1995). Effect of powdered chilli, turmeric and neem leaf on pre and post storage germinability of wheat seed. Indian Agric. 57(4):267-271.

Peterhalmer (2003). Enhancing seed performance for better yield and quality. Asian Seed Planting Mater. 10(2):4-6.

Pukacka S, Ratajczak E, Kalemba E (2009). Non-reducing sugar levels in beech (*Fagus sylvatica*) seeds as related to withstanding desiccation and storage. J. Plant. Physiol. 166:1381–1390.

Schumuttere H (1990). Properties and potential of natural pesticides from the neem tree Azadirachta indica. Annu. Rev. Entomol. 35:271-297.

Sharma RK (1995). Neem leaf powder and cobash against *Rhyzopertha dominica* (F), in stored maize. Indian J. Entomol. 57:15-17.

Suma N (2005). Studies of seed quality enhancement techniques in sesamum (*Sesamum indicum.* L) cv. Co 1 in sesamum. M.Sc. Thesis. Tamil Nadu Agricultural University, Coimbatore.

Tekrony DM (2006). Seeds: the delivery system for crop science. Crop Sci. 46:2263–2269.

Consumer attitudes for new plant species using the example of *Trachycarpus wagnerianus*

P. Lampert*, K. Menrad and Schoeps J.

Department of Horticulture and Food Engineering, University of Applied Sciences Weihenstephan-Triesdorf, Germany.

This article reviews the purchase behaviour of consumers for plants and the attitude towards horticultural novelties. The palm, *Trachycarpus wagnerianus* especially, is the main focus of this study. Based on a survey of 408 customers in Bavaria during autumn 2009, the results gave an impression of how consumers inform themselves about novelties in horticulture as well as, their attitude to *T. wagnerianus*. Within the standardized questionnaire, a conjoint measurement was applied in order to investigate the impact of the different characteristics of the palm on consumers' choice. Drawing all factors of the conjoint measurement into consideration, the *T. wagnerianus* has slight advantages when compared to the common *Trachycarpus fortunei*.

Key words: Conjoint analysis, consumer research, palms, purchasing behaviour.

INTRODUCTION

The horticultural market in Germany represents the biggest market in Europe with revenues of around 9.5 billion € for flowers and plants in 2009 (Lerch, 2009). Especially the sales of indoor and outdoor plants increased between the years 2004 to 2009. Indoor plants had a market share of 52.3% in 2008 and outdoor plants represent a share of 28.9% (Lerch, 2009). One of the reason for growing sales of horticultural plants is the fact that gardening has become trendier in Germany during the previous years (Lerch, 2009) Many people enjoy their terrace and garden as a comfortable enlargement of their house which is regarded as a kind of living room in the fresh air (Lerch, 2009). The garden turns out to be a place for recreation as well as, a social meeting point. The number of gardens, terraces and balconies is still increasing in Germany, which is an important factor for potential future growth especially, for sales of outdoor plants (Lerch, 2009). Within the outdoor segment, palms for garden use are one of the horticultural plants that capture the Mediterranean trend. *Trachycarpus* species especially, have been imported in increasing numbers mostly from China in recent years (Schrage, 2008). They are not just used as pot plants, but a number of palms have been planted in the open ground even in Germany. Not just in botanical gardens but also in increasing numbers in private gardens, although, they must be protected against low temperatures in most regions of Germany during winter (Doerken and Steinecke, 2009). Palms mostly planted in the open ground in Netherlands are *Trachycarpus fortunei* and *Chamaerops humilis* (Bennink, 2009). Up till now, this market trend towards palms is mostly captured by online shops which offer a large variety of so-called cold hardy palms. For example, Google offers a number of 29,500 hits of German internet sites to this topic. Meanwhile, a few specialized wholesalers for palms and other exotic plants entered the market as well (Fuchs, 2009). In addition to the rather widely spread *T. Fortunei*, another form of *Trachycarpus* is on its way into the German market: *Trachycarpus wagnerianus*. In plant systematic, it was announced by Govaerts and Dransfield (2005) that *T. wagnerianus* is not an owned species but a synonym of *T. fortunei*. Nevertheless, the name *T. wagnerianus* is still used to show the difference to the common *T. fortunei*. *T. wagnerianus* is supposed to be more suitable for Central European climate conditions. Its advantages are a smaller, stronger leaf with a short leafshaft which leads to a higher wind resistance and a more compact habitus (Bennink, 2009). But still, *T. wagnerianus* is rather

*Corresponding author. E-mail: p.lampert@wz-straubing.de

unknown among the consumers in Germany. Within a students' project at the University of Applied Sciences Weihenstephan-Triesdorf, the purchase behaviour with respect to horticultural novelties and especially, *T. wagnerianus* was investigated in a consumer survey in 2009. To date, only a few published studies are available in this field. For example, Berghage and Wolnick (2000) made a survey on consumers' preferences for flower and leaf colour of new *Guinea Impatiens* during a garden festival in Pennsylvania where the participants of this survey showed a preference for bright solid colours (in particular, red or red-violet) and bicolour flowers. In addition, the age of the respondent was correlated to the probability of purchasing new *G. Impatiens*, with customers aged above 40 years being more likely to purchase such plants (Berghage and Wolnick, 2000). Posadas et al. (2006) analyzed the levels of liking and willingness to pay for selected garden Chrysanthemum in Mississippi in 2003. For that purpose, nine Chrysanthemum were presented to 579 survey participants in three different pot sizes. The level of liking of the nine cultivars was estimated by using the ordered Logit method, while the willingness to pay model for the nine cultivars was analyzed by using Tobit method (Aldrich and Nelson, 1984). Respondents preferred 5 specific cultivars with female respondents being willing to pay more for the cultivars. In addition, respondents who previously bought chrysanthemum reported higher likings for the cultivars but they were not willing to pay more for them. With respect to socio-demographic attributes of the respondents, age and household size, the pot size of the cultivars had a negative effect on the respondents' willingness to pay for the cultivars (Posadas et al., 2006).

Hudson and Griffin (2004) examined consumer willingness to pay for cut flower product attributes using a choice-based experiment through mail survey. This study showed that consumers preferred mixed colour arrangements (compared to single colour arrangements) and that they were willing to pay a price premium for cut flower products that were guaranteed as produced in the local area (that is, in this case, the state of Mississippi). In their study in Taiwan, Huang and Yeh (2009) concluded that "curiosity fulfilment" is a major factor for the purchase of flowers. The term "curiosity fulfilment" includes a strong consumers' preference for novelties in flowers (Huang and Yeh, 2009). Altogether, there is a lack of studies in scientific literature which capture consumer behaviour and willingness to pay for new outdoor plants. Thus, this study was initiated in order to analyze these aspects with respect to *T. wagnerianus.*

METHODOLOGY

A standardized questionnaire was developed for the consumer survey which consists of four major parts such as:

1. Purchase behaviour for horticultural products in general
2. Attitude towards novelties in horticultural products
3. Attitude towards palms and the new *T. wagnerianus*
4. Socio-demographic data

In scientific literature, conjoint measurement is regarded as a suitable tool for measuring the different attributes and their influence on the consumers' preferences. It is also used to implement market segmentation and product positioning (Green and Krieger, 1991). One requirement for a conjoint analysis is to have a product which consists of several different characteristics (Albers at al., 2009). The global aim of conjoint measurement is to investigate part-worth functions for the attributes of a good in doing evaluation on the consumers' preferences (Green and Srinivasan, 1978). The part-worth functions are measured on the individual level in a traditional conjoint analysis Green and Krieger, 1991). As a result, preference heterogeneity among consumers, if present, can be found when using this method (Green and Krieger, 1991). But not only are the part-worth functions of the individual consumers also important. Even more, the total consumer benefit is the objective to the researcher (Backhaus et al., 2006). The consumer preferences analyzed in a conjoint measurement are a substantial information base for the pricing and product policy when marketing a new product (Albers et al., 2009). A traditional conjoint analysis was done in our study as one option of the de-compositional methods. In case there are just two parameter values of the goods' characteristics, the so-called Trade-off method is used within a conjoint experiment. If there are more parameter values of interest in the product to be analyzed, the Full-Profile method is the first choice for collecting the required information in the conjoint experiment. Thus, we used the latter method for our study. In horticulture, there are very few studies that dealt with conjoint analysis in order to gain knowledge about consumer preferences. For example, a conjoint measurement was used to investigate consumer preferences for geraniums (Behe et al., 1999), residential landscapes or Tabletop Christmas trees (Behe et al., 2005a, b). Also for roses (Prince et al., 1980) and for rhododendrons (Gineo, 1990), this method was applied.

Backhaus et al. (2006) suggested five steps for a conjoint measurement. In the first step, certain characteristics of the good must be chosen and the data collection design must be developed in the next step. After that, data collection is realized within a survey in which consumer preferences of the fictive products are measured. Then, the individual part-worth functions are statistically estimated from this data pool. In the last step, the part-worth functions of the individual respondents are aggregated if necessary. Backhaus et al. (2006) further suggested using a low number of product characteristics in the conjoint experiment. In our study, four characteristics of *T. wagnerianus* had to be measured which had different parameter values. "Cold hardiness" of this palm was divided into "possible", "possible with protection" and "impossible". "Leave withstand against wind" had the values "low" and "high" and the "price levels" were set to 129, 159, and € 189 respectively. The characteristic "plant habitus" had the parameter values "compact" and "wide". With the help of these characteristics, a differentiation to the *T. fortunei* could be made. Therefore, these parameters were chosen for the conjoint measurement. After collecting the required data, conjoint analysis defines the overall consumer preference for a particular product, in this case, the palm *T. wagnerianus,* as the sum of the part-worth for each parameter value. Backhaus et al. (2006) suggested that there should be less than 20 fictive products (stimuli) in a full study. As we would have had 2 × 2 × 3 × 3 = 36 stimuli in our study, it was reduced to 11 fictive products with the help of the software SPSS version 17 (Statistical Package for the Social Science). For the fictive products, 11 coloured photographs were used showing either a *T. wagnerianus* or *T. fortunei*. With the help of these pictures that were integrated in the questionnaire, the surveyed people were asked to build a rank order from 1 to 11 whereas, number one would be their first choice of buying. Card numbers 1 to 9 are those cards that are being tested. Card numbers

10 and 11 are so-called hold-out cards. They are in the experiment designed to check the validity of the statistical results of the experiment and therefore, are not used to estimate the part worth functions.

The questionnaire was spread out by students of our university. Berekhoven et al. (2006) postulates to have a strong focus on the universe as the asked people can have a major impact on the results. In contrary to other branches, it is quite hard to identify the universe of people who are interested in plants and gardening and in particular, to those people who are potentially interested in new varieties of palms. So, the students manually handed out the questionnaires in different parts all over Bavaria and distributed them mostly to homes with garden whose inhabitants were supposed to have a bigger affinity to gardening than others. Altogether, a total number of 2,000 questionnaires were spread out which represents 0.37% of the 538,000 garden owners in the federal state of Bavaria, Germany (Beiersdorf, 2009) at the end of October 2009. Most of the questionnaires were spread out in the area around Freising and Munich and a few were distributed in the Northern part of Bavaria with a focus on the city of Nuremberg.

RESULTS AND DISCUSSION

General characteristics of the sample

408 usable filled-in questionnaires were sent back which equals to a response rate of 20.4%. A majority of 69.9% of the respondents were females and between forty-one and sixty years of age (53.4%). The majority of female buyers of horticultural products which was found in our survey is supported by a series of studies in different countries (Yue and Behe, 2008; Schmahl, 2008; Posadas et al., 2006; Menrad and Hardung, 2006; Menrad and Fink, 2005; Batt and Pool, 2004; Mortelmans and Damen, 2001; Behe and Wolnick, 1991a, b). The same relates to the majority of buyers of horticultural products aged 50 years or more in Germany (Menrad and Fink, 2005; Fink, 2004). More than three-quarter (75.5 %) of the surveyed people live in their own homes and 86.7% have a garden. This means that people who live in rented homes have access to a garden. 40.2% of the participants had a monthly net income of more than € 3,000 per household and 57% stated at least a university entrance allowance. Compared to the entire Bavarian population where the average net income per household is € 3.172 (Albers et al., 2009), our sample is representative in this point. However, the level of education is above average. In Germany, only 24% had university entrance allowance (Aldrich and Nelson, 1984). The respondents either live mostly in small villages below 5,000 inhabitants (35.9%) or in cities from 100,000 up to 1 million inhabitants (43.8%). 36 people (9.1%) live in a town from 5,000 up to 10,000 inhabitants and 11.1% or 44 people live in large cities with a population of more than 1 million people.

Purchase behaviour for horticultural products

29.4% of our respondents buy horticultural products at least monthly in a nursery which makes this outlet the most popular distribution channel for horticultural products. Garden centres are frequented at least monthly by 14.1%, grocery stores by 17.5%, DIY-stores by 14% of the people. In contrary, tree nurseries are rarely or never frequented by 79.7% of the people as well as, Online Shops in which 93% of the surveyed people rarely or never buy horticultural products. According to Lerch (2009), the most important outlets for plants are still traditional nurseries and florist shops with a market share of 29.3% which is congruent to our results. 33.5% of the respondents of our survey spend between € 100 and € 199 per year for plants and flowers and just 10.7% spend below € 50 per year. 34.5% of the surveyed people spend above € 200 per year. The average spending on flowers and plants in Germany was € 116 per capita in 2008 (Lerch, 2009).

As shown in Figure 1, the price-benefit-ratio, high quality plants and well-to-understand plant care instructions are the most important purchasing criteria for our respondents. But also, expert advice is at least an important factor for 72.1% of the surveyed people. Other studies also indicated the high relevance of good plant quality and the personal advice as important purchase criteria for horticultural products in Germany (Schmahl, 2008; Elgner, 2008; Kittemann, 2007; Menrad and Fink, 2005; Rothenburger et al., 1996). Concerning garden design, 74.4% of the surveyed people like to set their own course. Just 6.2% of the respondents prefer exotic and extraordinary plants which are not often used so far. A garden that is close to nature is preferred by 64.1% of the sample's respondents.

Information seeking behaviour for horticultural novelties

17.6% of the respondents inform themselves regularly about horticultural novelties but the big majority of 62% only does this from time to time. 20.1% have no interest in new plants. Basic information about plants is gathered personally while shopping (56.1%) followed by specific gardening magazines (26.2%), TV (24.5%) and specific books (17.6%). Postage mailings (10.3%) and E-mail Newsletters (2.2%) have limited relevance for gathering information. 41% of the surveyed people read a gardening magazine sometimes and 13.3% read such a magazine regularly. From the regular readers of gardening magazines, 47.2% inform themselves regularly about horticultural novelties while 50.9% of the regular readers are interested in new plant products only partially. However, the regular readers of gardening magazines spend significantly more money on plants. 47.1% of the people who read gardening magazines regularly spend annually € 200 or more on plants. From the people who never read a magazine (45.6% of all respondents) just 26.3% spend 200 € or more per year

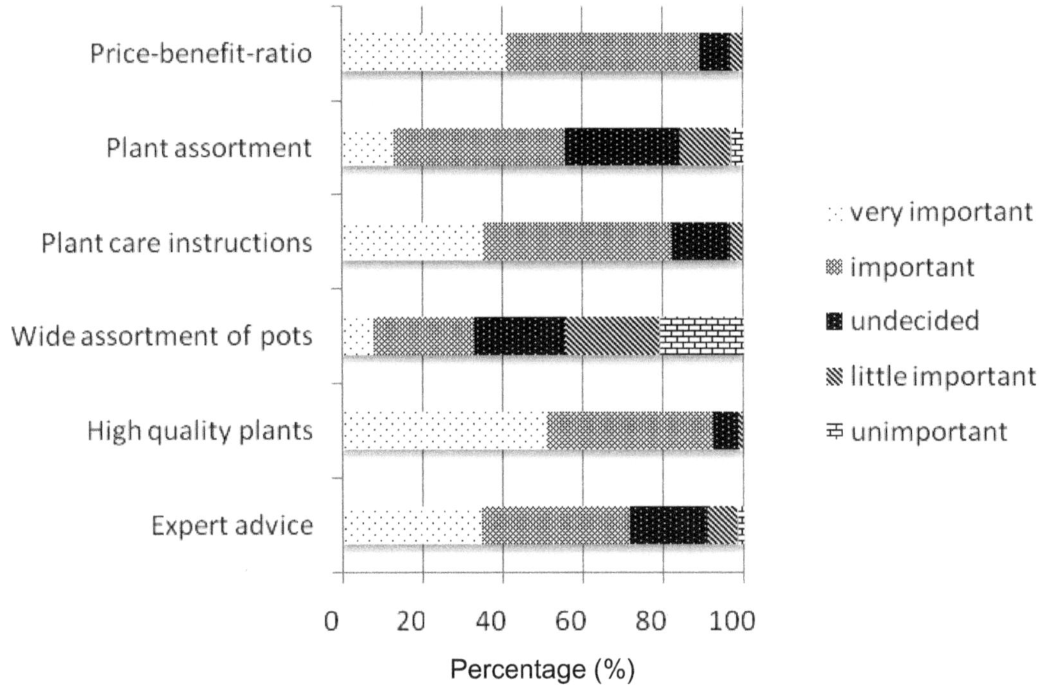

Figure 1. Criteria for plant purchase (n = 408).

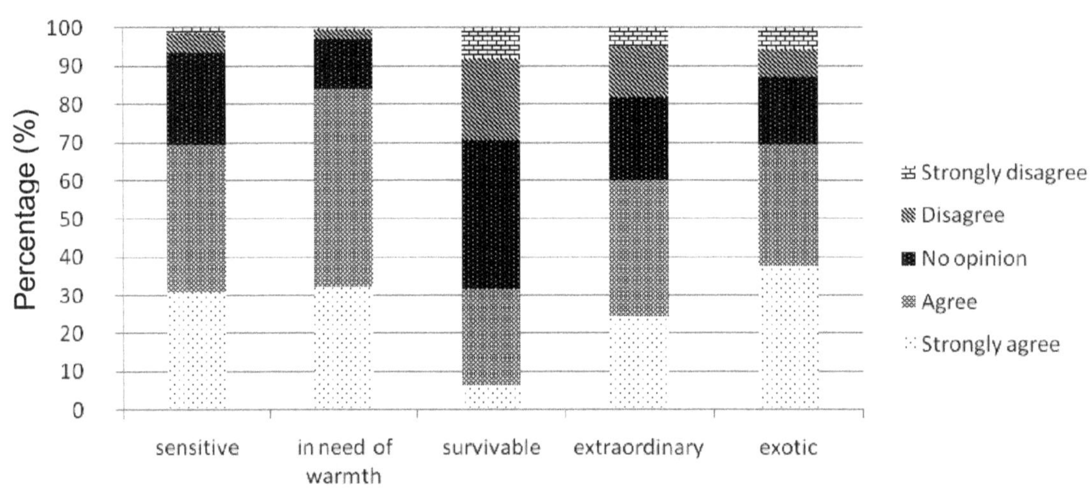

Figure 2. Consumers' opinion about palms in gardens (n = 408).

on plants. 52.9% of all respondents of our survey sometimes watch a TV show about gardening, 10.4% regularly and 36.7% never watch such a show.

Attitudes and buying behaviour towards garden palms and the palm *Trachycarpus wagnerianus*

For most of the surveyed people, palms are a symbol for holidays in Southern countries (61.1%) and only 11% of the respondents feel that palms are expensive. In recent years, 63.7% of the surveyed people have already bought a palm. Buying palms is not popular with older people. 71.9% of the people older than 70 years have never bought a palm whereas in the age group between 41 to 50 years, a percentage of 71.2% has bought one. In general, a high household income increases the willingness to buy palms significantly while 75.3% of the group with a monthly net income of € 4,000 per household have already bought a palm and only 50% of the respondents with a monthly net income below € 1,000 did so in previous years. As shown in Figure 2, 84.2% of

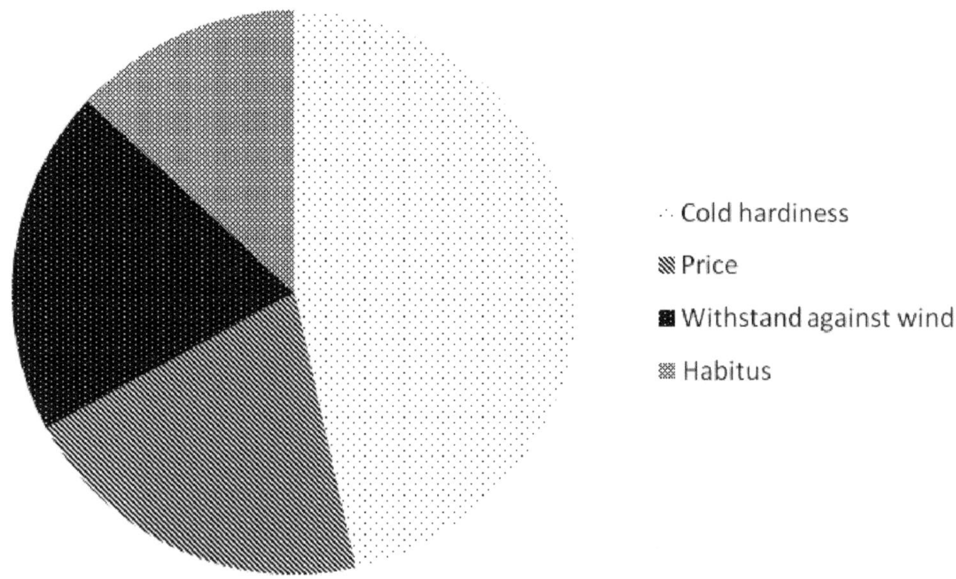

Figure 3. Relative importance of the plant characteristics (n = 293).

Table 1. Part-worth functions of the palm's characteristics (n = 293).

Overall benefit			
Characteristics of the palm	**Parameter value**	**Part-worth functions**	**Standard error**
Price	Price 1: € 129.-	0.533	0.070
	Price 2: € 159.-	0.167	0.070
	Price 3: € 189.-	-0.699	0.070
Habitus	Compact	-0.187	0.053
	Wide	0.187	0.053
Withstand against wind	High	0.802	0.053
	Low	-0.802	0.053
Cold hardiness	Impossible	-2.187	0.070
	Possible with protection	0.357	0.070
	Possible	1.830	0.070
Constant		4.795	0.055

the respondents believed that palms in gardens need warmth and 69.5% think that they are sensitive. But also 31.6% of the respondents stated that palms in gardens are survivable and 76.9% of this group already bought a palm.

To get an insight to how the different attributes of the new palm *T. wagnerianus* influence consumers' choice, a conjoint measurement as described previously was included in the survey which was carried out by 293 respondents. When analysing the relative importance of the plant's characteristics for purchasing a palm, the cold hardiness was the most important factor with a relative importance of 46.5%. But with both palm varieties not being completely winter-hardy, this factor cannot be used in the argumentation pro *T. wagnerianus.* As seen in Figure 3, the factor "price" is not of that high importance than one could have expected. The relative importance of the price is at 20.6%. That is, close to the high withstand against wind with an importance of 19.8%. The habitus of the palm is with 13.1% of low importance.

Regarding the part-worth functions of the analyzed palm characteristics (Table1), the strongest impact on the consumers' choice has the characteristic "cold hardiness" with the parameter value "possible" with a positive part worth of 1.830. This shows that the surveyed consumers have a rather positive attitude and a certain preference

Table 2. Comparison of consumer estimations of different palm varieties.

Variable	Preferred parameter value	Part-worth function	Given parameter value	Part-worth function	Given parameter value	Part-worth function
Cold hardiness	Possible	1.830	Hardy with protection	0.357	Hardy with protection	0.357
Withstand against wind	High	0.802	High	0.802	Low	-0.802
Price	€ 129.-	0.533	€ 189.-	-0.699	€ 159.-	0.167
Habitus	wide	0.187	Compact	-0.187	Wide	0.187
Constant factor		4.795		4.795		4.795
Aggregated part-worth functions		8.147		5.068		4.704

for a winter-hardy palm which – ideally – can survive the winter in open ground in their garden. When analyzing the "withstand against wind" the surveyed consumers prefer "high" with a part-worth value of 0.802. Concerning the price of the palm, the highest part-worth with a value of 0.533 was calculated for the lowest price of € 129 while the highest price of € 189 achieves a negative part-worth value. This result is in line with the expected negative linearity of higher prices of products as suggested by consumer theory. A wider habitus was judged as more positive than a compact habitus (Table 1). According to the results of our survey, the optimal palm would get aggregated part-worth functions of 8.127 (Table 2). But this palm was not available on the German market so far, since it represents a simulation of the ideal plant. When the given characteristics of the palms which are available on the German market were compared, it was seen that *T. wagnerianus* and its characteristics are more preferred by the surveyed people than those of *T. fortunei.* Altogether, *T. wagnerianus* gets an aggregated part-worth function of 5.068 compared to 4.704 of *T. fortunei* (Table 2). The reason therefore, is the higher withstand against wind of *T. wagnerianus* which is preferred by the surveyed consumers. Therefore, this characteristic can be used in marketing activities for *T. wagnerianus* as this characteristic is one of the big differences to *T. fortunei* and positively evaluated by the respondents. It can also be easily shown and explained at the point of sale to the customers.

Conclusions

The results of our survey gave an insight in consumer behaviour and preferences with respect to new varieties of palms on the market of horticultural plants in Germany. Despite limitations in the representativeness of the sample regarding for example, area covered, education level and income status of the respondents, the study contributes in identifying major factors which influence consumer behaviour and preferences in this field. The identified consumer preferences as well as, the collected data concerning the used information channels for new plant varieties support the development of target-groups oriented marketing activities for these plants. Furthermore, they help to develop appropriate information and communication activities in this field. Further research activities could complement the collected information in particular concerning consumers' willingness to pay for specific palm varieties and the required information when promoting this type of plants as well as, the interest of plant retailers in new plant varieties. Considering all results together, it can be expected that *T. wagnerianus* has a good chance to win market shares in the market of horticultural plants in Germany. In traditional nurseries, it could be a suitable product as consumers appreciate expert advice which can be ideally provided in this shop type with respect to the palm and its special characteristics. As shown in other consumer studies, customers of horticultural products in Germany regard personal advice provided in traditional nurseries as one major advantage of this distribution channel compared to garden centres or DIY stores (Menrad and Fink, 2005; Rothenburger et al., 1996). To increase the publicity and availability of *T. wagnerianus*, a suitable activity could be to spread the palms amongst plant-retailers as consumers mainly inform themselves about horticultural novelties while shopping in a store. Thus, *T. wagnerianus* needs to achieve a certain distribution level so that potential consumers can have a chance to get familiar with this rather unknown plant. Palms in common are regarded as a strong symbol for holidays. Considering the "homing"-trend, palms for gardens can offer holiday-feeling to peoples' homes and catch up to the trend to "Mediterranean gardens" in Germany as well. Though *T. wagnerianus* is not colder hardy than *T. fortunei,* it has the advantage of the high withstand to wind which makes it the first choice for customers even with a higher price as shown in the results of the conjoint experiment in our study.

REFERENCES

Albers S, Klapper D, Konrad U, Walter A, Wolf J (2009). Methodik der empirischen Forschung. 3rd Edition. Gabler Verlag, Wiesbaden. pp. 413-432.

Aldrich J, Nelson FD (1984). Linear models, probability, logit and probit models. Sage Publications, Newbury Park. pp. 82-84.

Backhaus K, Erichson B, Plinke W, Weiber R (2006). Multivariate nalysis - An application-oriented introduction.11th edition. pringer Verlag, Berlin: pp. 543-601

Batt PJ, Pool J (2004). Consumer preferences for cut flowers in Western Australia. Acta Hort. (ISHS) 655: 81-88.

Behe BK, Hardy J, Barton S, Brooker J, Fernandez T, Hall C, Hicks R, Hinson R, Knight P, McNiel R, Page T, Rowe B, Safley CD, Schutzki R (2005a). Landscape plant material, size and design sophistication increase perceived home value. J. Env. Hort. 23(3): 127-133.

Behe BK, Nelson R, Barton S, Hall C, Safley CD, Turner S (1999): Behe BK, Nelson R, Barton S, Hall C, Safley CD, Turner S (1999). Consumer preferences for geranium flower colour, leaf variegation and price. HortScience 34: 740-742.

Behe BK, Walden RM, Duck MW, Cregg BM, Kelley KM (2005b). Consumer preferences for tabletop Christmas trees. HortScience 40(2): 409-412.

Behe BK, Wolnick DJ (1991a). Type of floral product purchased and demographic characteristics and floral knowledge of consumers. HortScience 26(4): 414-416.

Behe BK, Wolnick DJ (1991b). Market segmentation of Pennsylvania floral consumers by purchase volume and primary retail outlet. HortScience 26(10): 1328-1331.

Beiersdorf H (2009). Neue Gartenbau-Statistik Bayern. Bavarian Ministry of Agriculture, Munich.

Bennink P (2009). Palms in the Netherlands: Moisture more dangerous than frost. Tuin & Landschap 16-2009: 30-31.

Berekhoven L, Eckert W, Ellenrieder P (2006). Market Research - Methodological principles and practical applications. 11th edition.

Doerken V, Steinecke H (2009). Frost damages after the winter 2008/09. Gartenpraxis pp. 36-39.

Elgner N (2008). Retail nurseries - finding the right way. Deutscher Gartenbau. 62(13), 16-17.

Fink C (2004). Expectations of consumers when buying horticultural products: what the customer expects and what retail offers Diploma thesis FH Weihenstephan.

Fuchs U (2009). Hardy palms from young professionals. Deutscher Gartenbau Produktion & Handel. pp. 16-19.

Gabler, Wiesbaden. Berghage RD, Wolnick DJ (2000). Consumer colour preference in New Guinea Impatiens. Hortic. Technol., 10(1): 206-208.

Ginea WM (1990). A conjoint/logit analysis of nursery stock purchases. N.E.J. Agr. Res. Econ., 19: 49-58.

Govaerts R, Dransfield J (2005). World checklist of palms: 1-223. The Board of Trustees of the Royal Botanic Gardens, Kew.

Green P, Krieger A (1991). Segmenting markets with conjoint analysis. October 1991. J. Mark., 55: 20-31.

Green P, Srinivasan V (1978). Conjoint analysis in consumer research: Issues and outlook. September 1978. J. Consum. Res. 5: 103-123.

Huang LC, Yeh TF (2009). An analysis of floral consumption values and their difference for genders and geographic regions. Hortic. Technol., 19(1): 101-107.

Hudson D, Griffin E (2004). Market potential for „Mississippi Grown“ cut flowers. Mississippi Agricultural and Forestry Experiment Station, Mississippi State University. Bulletin. p. 1140.

Kittemann F (2007). Customer satisfaction measurement of various retail outlets for potted plants, nursery products and perennials. Diploma thesis FH Weihenstephan.

Lerch C (2009). Living Green Industry Focus. BBE Retail Experts, Cologne. Menrad K, Fink C (2005) Retail - customers in DIY markets and horticultural retailers.. Deutscher Gartenbau. 59 (1): 34-35.

Menrad K, Hardung P (2006). Who buys how? Deutscher Gartenbau. 60(24:, 40-41.

Mortelmans D, Damen S (2001). Attitudes on commercialisation and anti-commercial reactions on gift-giving in Belgium. J. Consum. Behav., 1(2): 156-173.

Posadas BC, Coker CH, Fain G, Knight PR (2006). Consumer survey of selected garden chrysanthemum cultivars in Mississippi. Hortic. Technol., 16(3): 539-543.

Prince TL, Robertson JL, Chatfield LH (1980). Factors affecting the marketability of roses. J. Amer. Soc. Hortic. Sci., 105: 388-393.

Rothenburger W, Reibstirn K, Orth UR (1996). Cost and benefit of customer service in horticultural retail stores. In: Acta Hort. (ISHS). 429: 111-118.

Schmahl S (2008). Why are men so reluctant to buy at the garden center? In: Grüner Markt., 4: 14-15.

Schrage R (2008). Plant Protection Service Annual Report 2008 Landwirtschafts-kammer Nordrhein-Westfalen , Muenster.

Yue C, Behe BK (2008). Estimating U.S. Consumers' choice for floral retail outlets. Hortic. Sci., 43: 764-769.

Pythium species and cold storage affect the root growth potential and survival of loblolly (*Pinus taeda* L.) and slash pine (*Pinus elliottii* Engelm.) seedlings

D. Paul Jackson*, Scott A. Enebak and David B. South

School of Forestry and Wildlife Sciences, Auburn University, Alabama 36849, USA.

When bareroot Southern pine (*Pinus*) seedlings are lifted in the fall and placed in cold storage for more than one week, survival is generally lower than when seedlings are lifted and stored during the winter. The combination of root wounding at the time of lifting with the presence of *Pythium* in the soil, the cool, moist conditions in cold storage encourage *Pythium* growth that results in seedling mortality after outplanting. Loblolly pine (*Pinus taeda* L.) and slash pine (*Pinus elliottii* Engelm.) seedlings were inoculated with either *Pythium dimorphum* or *Pythium irregulare*, cold stored for 3 weeks, and placed in a hydroponic system (aerated aquariums). Seedling Root growth potential (RGP was measured as the number, length, volume, surface area, and diameter of new roots and survival was recorded 4 months after outplanting. Both *Pythium* species reduced the number of new roots and root length, root surface area, and root volume in slash pine. *P. irregulare* reduced the number of new roots, root length, and root surface area and *P. dimorphum* reduced the number of new roots on loblolly pine. Although, *P. irregulare* lowered RGP, it did not reduce loblolly pine survival. In contrast, *P. irregulare* inoculation did reduce the survival of slash pine seedlings after outplanting.

Key words: *Pythium dimorphum*, *Pythium irregular*, root collar diameter; bareroot.

INTRODUCTION

Cold storage of pine (*Pinus*) seedlings is a common practice in forest tree nurseries of the Southern US when seedling demand is low or when weather conditions are poor for outplanting. Typically, seedlings are lifted from nursery beds and placed in cold storage (1 to 5°C) for 2 to 3 days if outplanted immediately ("hot planting") or for periods of one to several weeks (long-term storage). In some cases, lifting bareroot seedlings in the fall (before the winter solstice) has resulted in poor outplanting survival after long-term storage (Kahler and Gilmore, 1961; Dierauf, 1976; Hebb, 1982; Venator, 1984). However, seedling survival tends to improve when Seedlings are lifted and stored for the same duration during the winter months (Kahler and Gilmore, 1961; Hebb, 1982). The lifting season generally spans from October to the end of February, but can extend into March. The ability to cold store bareroot seedlings in the fall could allow seedlings to be outplanted earlier when field conditions are more favorable for root growth compared to the warmer, drier months of spring (Garber and Mexal, 1980).

Two theories exist to explain why bareroot pine seedling survival is poor following lifting and cold storage during the fall season. The first theory involves the lifting of seedlings that have non-dormant buds or that have not been exposed to a certain number of chilling hours. Chilling hours is defined as the exposure of seedlings to above-freezing temperatures (< 8°C) for a specific period of time, and after this time period, seedling buds may be classified as dormant (May, 1984). Carlson (1991) claimed that successful cold storage was dependant on meeting a chilling hour requirement and seedlings having

*Corresponding author. E-mail: dpj0001@auburn.edu.

Abbreviations: RGP, Root growth potential; **RCD,** root collar diameter.

dormant buds, which is why seedlings are generally "hot planted" after fall lifting and not stored. However, container-grown loblolly pine (*Pinus taeda* L.) (Boyer and South, 1985) and longleaf pine (*Pinus palustris* Mill.) (Pickens, 1998) have been successfully stored and outplanted in the fall without exposure to a certain number of chilling hours. These findings suggest that poor outplanting survival after storage may not be related to either chilling hours or bud dormancy but to another factor.

The second theory for poor seedling survival after storage may occur during the lifting process, where seedling roots can be torn and wounded as they are lifted from nursery beds, and as a result, provide infection sites for soil-borne pathogens such as *Pythium*. After seedlings are lifted, they are sprayed or dipped into a combination of water and superabsorbent gel to prevent desiccation in seedling storage bags (May, 1984). Excess water can pool in seedling bags and along with the cool storage temperatures (1 to 5°C), may provide an ideal environment for certain species of *Pythium* to multiply.

Jones et al. (1992) recovered *Pythium* on bareroot longleaf pine seedlings held in storage for 6 weeks. In their study, treating longleaf pine with combinations of metalaxyl and benomyl resulted in greater than 90% survival, while those that did not receive fungicides had less than 20% survival. Another study by Sun (1996) revealed that inoculating longleaf pine seedlings with increasing levels of *Pythium dimorphum* Hendrix and Campbell caused increases in seedling mortality after storage. This may have been the first study to show that a *Pythium* species could kill cold stored pine seedlings after outplanting.

Pythium is an oomycete or "water mold" commonly found in nursery soils that causes "fine feeder root disease" on young pine seedlings, which are important for nutrient and water absorption (Kelley and Oak, 1989). Seedling survival after outplanting depends on the production of new fine feeder roots. Root growth potential (RGP), is a measure of seedling quality based on new root production and can be defined as the ability of a seedling to initiate and grow new roots within a prescribed time period in an environment that is optimum for root growth (Simpson and Ritchie, 1996). Thus, root growth potential is often used to determine seedling performance potential before outplanting. Methods to determine RGP involve culturing seedling roots in either soil, hydroponic, or aeroponic systems (Rietveld, 1989), and the number and/or size of new white roots produced in these systems are then used to quantify RGP. A study has shown that RGP of loblolly pine steadily declined after 9 weeks of storage for fall-lifted seedlings, but improved after lifting and 9 weeks of storage in the winter (DeWald and Feret, 1988). Another study revealed that shortleaf pine (*Pinus echinata* Mill.) seedlings lifted in the winter had the greatest RGP after 4 weeks of storage compared to seedlings lifted and stored prior to the winter solstice (Hallgren and Tauer, 1989). In addition, the term "December dip" was coined by South (1999), which described unexplained seedling mortality observed in loblolly pine outplanted during the month of December. The "dips" in loblolly pine survival were more prevalent after cold storage and a link to decreases in seedling RGP was speculated as the reason (South, 1999).

The objective of this study was to quantify the effects that *P. dimorphum* and *Pythium irregulare* Buisman had on root growth potential, diameter growth, and survival of loblolly and slash pine (*Pinus elliottii* Engelm.) seedlings after long-term cold storage. *P. dimorphum* was chosen based on the results from inoculations on stored longleaf pine seedlings (Sun, 1996). *P. dimorphum* was first isolated from diseased loblolly pine roots in Louisiana (Hendrix and Campbell, 1971) and has since been isolated by Ho (1986) and Asiegbu et al. (1996). Root inoculations with *P. dimorphum* have reduced shoot height of Norway spruce (*Picea abies*) seedlings, and the pathogen has also been recovered from benomyl-treated seedlings (Borja, 1995). *P. irregulare* is not known as a storage pathogen but was chosen based on being one of the most common damping-off pathogens in nurseries (Hendrix and Campbell, 1973) and also to serve as a baseline for comparison to *P. dimorphum*. It was first discovered in the Netherlands on pea roots and cucumber seeds (Van der Plaats-Niterink, 1981) and has been associated with outplanting mortality of Mexican weeping pine (*Pinus patula*) seedlings in South Africa (Linde et al., 1994). This study was designed to test the null hypothesis that the root growth potential and survival of loblolly and slash pine seedlings are not affected by inoculations with *Pythium* species and cold storage. To date, the null hypothesis (that is, no effect on survival of stored seedlings) has not been rejected for any pathogen other than *Pythium* species.

MATERIALS AND METHODS

Pythium inoculum

P. dimorphum (ATCC 22843) and *P. irregulare* (ATCC 10951) were obtained from American Type Culture Collection (ATCC®, Manassas, VA). Both *Pythium* species were aseptically transferred from the ATCC vials to oatmeal agar (Kim et al., 2005). From the advancing margin of the fungal mycelium, three 0.5 cm disks of each *Pythium* spp. were again transferred to additional oatmeal agar plates to eventually use as seedling inoculum. Prior to inoculation, 1,190 g of oatmeal and 400 ml of distilled water were combined in two autoclavable bags, mixed thoroughly, autoclaved, and allowed to cool for 24 h. *Pythium* inoculum and agar were cut into pieces with one autoclaved bag receiving three agar plates of *P. dimorphum* while the other bag received three agar plates of *P. irregulare*. The oatmeal/*Pythium* inoculum was mixed every 12 h and stored at room temperature for 10 days prior to seedling root inoculations.

Seedling inoculations

Bareroot half-sib family loblolly and slash pine seedlings were obtained from Smurfit-Stone Corporation's Rock Creek Nursery near Brewton, AL on December 12, 2008. The soil texture at the nursery was loamy sand and the seedlings were grown in first year fumigated ground (2008). Prior to inoculations, loblolly and slash pine remained in cold storage (4 to 5°C) for 8 weeks at Auburn University.

On February 10, 2009, seedlings were subjected to four inoculation treatments: 50 and 200 g of *P. dimorphum* oatmeal inoculum and 50 and 200 g of *P. irregulare* oatmeal inoculum. Non-inoculated seedlings (controls) were dipped into a bucket of water without inoculum as the fifth treatment in the study. Inoculations began by weighing out the desired treatment amount of oatmeal inoculum on a scale and mixing in 11 L of water. Seedlings remained immersed in the bucket for approximately 5 s. Buckets were emptied, rinsed, and filled with a fresh inoculum mixture after each seedling bundle was inoculated. Inoculated seedlings were immediately placed in separate plastic bags (49 L) and put in cold storage (4 to 5°C) for 3 weeks. Three replications of each treatment were inoculated and the experimental unit consisted of 15 loblolly or slash pine seedlings (a replication or bag in storage). In all, 225 each of loblolly and slash pine seedlings were used in the study.

Root growth potential

After 3 weeks in storage, loblolly and slash pine seedlings were placed in an aerated hydroponic system as described by Palmer and Holen (1986) for 28 days. Fifteen aquariums (38 L) (5 treatments × 3 replications), which allowed seedling roots to be suspended in water contained an experimental unit of loblolly and slash pine seedlings each (15 of each) for each of the five treatments.

The aquariums were placed in a Randomized Complete Block Design (RCBD) on three greenhouse tables. Seedling root collar diameters (RCD) were measured on day 1 and 28 with calipers on the main stem. At day 28, seedling survival was recorded and RGP quantified by measuring the number, length, volume, surface area, and diameter of new white root tips using a WinRhizo™ root scanner and computer software (Regent Instruments, Inc., Quebec City, Quebec, Canada).

Seedling survival

After scanning, seedlings were returned to their appropriate location in the aquariums. Seedling roots were not allowed to dry out during the time they were removed from the aquariums for scanning and returned to the aquariums. On day 34 (April 9, 2009), the seedlings were removed from the aquariums and hand planted in a randomized complete block design at 0.3 × 0.3 m spacing on an outplanting site composed of sand at Auburn University. Seedling survival was monitored for 4 months.

Statistical analyses

Analyses of means were conducted using a General Linear Model (GLM) in SAS statistical software (9th ed., SAS Institute, Cary, NC). Means of each experimental unit for each dependant variable were analyzed using Analysis of Covariance (ANCOVA), where initial root collar diameter (before placing seedlings in the hydroponic system) was included into the analysis to factor out any differences in seedling size (South et al., 1989). Orthogonal contrasts were performed using combined levels of *P. dimorphum* versus controls and *P. irregulare* versus controls. Data for each pine species were analyzed separately.

RESULTS

Loblolly pine

Inoculation with *P. dimorphum* reduced the number of new root tips on loblolly pine seedlings (Table 1). Non-inoculated seedlings had an average of 80 new roots, while seedlings inoculated with *P. dimorphum* had an average of 51 new roots. Aside from new root production, *P. dimorphum* did not affect any other RGP measurement (root length, surface area, volume, or diameter). *P. dimorphum* did not affect loblolly pine root collar diameter (Table 1).

Loblolly pine inoculated with *P. irregulare* had an average of 38 new white roots, which were 42 fewer roots than non-inoculated seedlings (Table 1). *P. irregulare* reduced root length and surface area by 15 cm and 2.84 cm^2, respectively, but did not affect root volume or diameter. Root collar diameter was also not affected by *P. irregulare*.

Seedling survival was not affected by root inoculation with *P. dimorphum* or *P. irregulare* after 28 days in the aquariums and 4 months after outplanting (Table 1). Despite reductions in new root growth by both *Pythium* species, only three of the 225 loblolly pine seedlings died during the RGP trial (data not shown).

Slash pine

P. dimorphum reduced the number of new white roots and root length, root surface area, and root volume of slash pine seedlings (Table 2). Non-inoculated seedlings had 174 new roots, whereas *P. dimorphum* inoculated seedlings averaged 81 new roots. *P. dimorphum* inoculated seedlings had an average reduction of 45 cm of root length, 11.64 cm^2 of root surface area, and 0.25 cm^3 of root volume when compared to non-inoculated seedlings (Table 2). Slash pine root collar diameter growth was not affected by *P. dimorphum*.

P. irregulare inoculated slash pine produced an average of 25 new roots, which were 149 fewer roots than non-inoculated seedlings (174 new roots) (Table 2). Compared to non-inoculated seedlings, *P. irregulare* inoculations reduced average root length, root surface area, and root volume by 70 cm, 17.38 and 0.39 cm^3, respectively. The effect from *P. irregulare* on all of the root growth potential variables was more severe than those for *P. dimorphum* with the exception of root diameter, which was unaffected by both oomycetes. *P. irregulare* inoculations reduced slash pine root collar diameter (Table 2). There was a two-fold reduction in RCD (- 0.17 mm) for *P. irregulare* inoculated seedlings when compared to non-inoculated seedlings (+ 0.15 mm).

During the 28-day RGP trial, *P. irregulare* inoculated slash pine survival was 85% as compared to 100% for *P. dimorphum* inoculated seedlings (Table 2). Four months following outplanting, seedling survival for *P. irregulare*

Table 1. Loblolly pine root growth potential (RGP), root collar diameter (RCD), and seedling survival as affected by *Pythium* treatments, analysis of covariance for RGP and RCD, and analysis of variance for survival.

Treatment	Root growth potential					Root collar diameter			Seedling survival	
	New roots (#)	Root length (cm)	Root surface area (cm^2)	Root volume (cm^3)	Root diameter (mm)	RCD before[z] (mm)	RCD after[y] (mm)	RCD growth (mm)	Survival 1[x] (%)	Survival 2[w] (%)
Control	80	24.67	5.06	0.09	0.71	4.38	4.55	0.16	97	43
P. dimorphum 50 g	52	11.16	2.31	0.04	0.77	4.33	4.34	0.00	97	46
P. dimorphum 200 g	51	18.81	4.88	0.11	0.84	4.64	4.75	0.11	100	15
P. irregulare 50 g	43	11.09	2.33	0.04	0.75	4.42	4.47	0.05	97	41
P. irregulare 200 g	33	8.33	2.12	0.05	0.80	4.53	4.52	-0.01	100	8
LSD[v]	(31)	(14.67)	(3.44)	(0.07)	(0.16)	(0)	(0.21)	(0.21)	(4)	(33)
Factor	P > F									
Covariate	0.8748	0.4191	0.2038	0.1078	0.2201	——[u]	0.0004	0.4879	——	——
Replication	0.1271	0.2717	0.3518	0.3788	0.0364	——	0.3964	0.3964	0.0256	0.0949
Treatment	0.0493	0.1381	0.2167	0.3407	0.7782	——	0.3108	0.3108	0.4609	0.3438
Control vs. *P. dimorphum*	0.0241	0.0766	0.1716	0.3759	0.2421	——	0.1366	0.1366	0.4774	0.3406
Control vs. *P. irregulare*	0.0053	0.0202	0.0397	0.0873	0.4209	——	0.0726	0.0726	0.4774	0.1774
P. dimorphum vs. *P. irregulare*	0.2159	0.2957	0.2527	0.2311	0.5895	——	0.6216	0.6216	1.0000	0.5841

[z] = RCD before the 28-day RGP trial, [y] = RCD after the 28-day RGP trial, [x] = survival after the 28-day RGP trial, [w] = survival 4 months after outplanting, [v] = least significant difference ($\alpha = 0.05$), [u] = no P-value due to covariate factor.

Table 2. Slash pine root growth potential (RGP), root collar diameter (RCD), and seedling survival as affected by *Pythium* treatments, analysis of covariance for RGP and RCD, and analysis of variance for survival.

Treatment	Root growth potential					Root collar diameter			Seedling survival	
	New roots (#)	Root length (cm)	Root surface area (cm^2)	Root volume (cm^3)	Root diameter (mm)	RCD before[z] (mm)	RCD after[y] (mm)	RCD growth (mm)	Survival 1[x] (%)	Survival 2[w] (%)
Control	174	75.02	19.93	0.44	0.85	4.78	4.93	0.15	100	68
P. dimorphum 50 g	67	22.34	6.66	0.17	0.98	4.64	4.54	-0.10	100	26
P. dimorphum 200 g	95	36.79	9.92	0.22	0.85	4.75	4.80	0.05	100	64
P. irregulare 50 g	16	2.42	0.59	0.01	0.85	4.80	4.56	-0.24	82	0
P. irregulare 200 g	35	7.53	1.96	0.04	0.84	4.72	4.63	-0.09	88	13
LSD[v]	57	26	6.75	0.16	0.19	0	0.26	0.26	14	41
Factor	P > F									
Covariate	0.7595	0.6918	0.7592	0.8215	0.1072	——[u]	0.0068	0.4012	——	——
Replication	0.1209	0.0923	0.0550	0.0463	0.6994	——	0.5162	0.5162	0.1507	0.3759

Table 2. Contd.

Treatment	0.0022	0.0018	0.0017	0.0028	0.2262		—	0.0619	0.0619		0.0659	0.0168
Control vs. *P. dimorphum*	0.0039	0.0024	0.0031	0.0073	0.2100		—	0.1021	0.1021		1.0000	0.1716
Control vs. *P. irregulare*	0.0002	0.0002	0.0001	0.0002	0.9962		—	0.0126	0.0126		0.0298	0.004
P. dimorphum vs. *P. irregulare*	0.0124	0.0157	0.0095	0.0084	0.1346		—	0.1383	0.1383		0.0121	0.0158

^{z} = RCD before the 28-day RGP trial, ^{y} = RCD after the 28-day RGP trial, ^{x} = survival after the 28-dayRGP trial, ^{w} = survival 4 months after outplanting, ^{v} = least significant difference (α = 0.05), ^{u} = no P-value due to covariate factor.

inoculated seedlings decreased to 7% as compared to 45% for *P. dimorphum* inoculated slash pine. The lack of root production on seedlings inoculated with 50 and 200 g was consistent with low seedling survival at each inoculum level (0 and 13%, respectively) after outplanting. However, even though RGP of *P. dimorphum*inoculated slash pine was less than non-inoculated seedlings, seedling survival was not affected.

DISCUSSION

Root growth potential is a measurement used to determine early seedling performance (survival) potential prior to outplanting. A seedling's ability to grow an adequate root system is important for good establishment in the field. Due to the need for soil moisture immediately after outplanting (Bronnum, 2005), the number of new roots produced by a seedling for the uptake of water may be the most important measure of RGP. Our results suggest that if loblolly or slash pine roots are infected with *P. dimorphum* or *P. irregulare* in the nursery, the production of new roots could be reduced after cold storage. The effects from *P. dimorphum* and *P. irregulare* inoculations on RGP were independent and differed depending on pine species. *P. irregulare* was consistently more virulent to loblolly (in terms of RGP) and slash pine (in terms of RGP and survival) than *P. dimorphum*. Slash pine RGP and survival was affected more by inoculations with both *Pythium* species than loblolly pine. Variations in genotype between loblolly and slash pine might account for these differences. Another study has shown independent effects from bacterial inoculations by demonstrating a decrease in root growth for loblolly and slash pine seedlings that received *Bacillus subtilis* compared to *Bacillus pumilus* and *Bacillus sphaericus* (Enebak et al., 1998). Their study also showed other response differences between the two pine species by loblolly pine experiencing increases in root length and biomass while slash pine did not.

Despite reductions in RGP of loblolly and slash pine seedlings from both *Pythium* species, only slash pine inoculated with *P. irregulare* experienced reductions in outplanting survival when compared to non-inoculated seedlings. Genotypic differences in pathogen susceptibility might also explain the ability of loblolly pine to recover from reduced RGP.

Seedling root collar diameter (RCD) is another indicator of seedling quality that is relatively quick to assess based on seedling morphology. A larger RCD usually indicates a healthier pine seedling that is correlated with higher survival and growth compared to smaller seedlings (South et al., 1985). Loblolly pine RCD was unaffected by either *Pythium* species, but *P. irregulare*-inoculated slash pine RCD was reduced during the 28-day RGP test by an average of 0.32 mm as compared to non-inoculated slash pine. This is the first report of a reduction in slash pine RCD after inoculations with *P. irregulare*, cold storage, and 28 days in a hydroponic system. The inoculation treatment may have reduced the ability of seedlings to uptake water, and this may have caused the RCD to shrink. A similar reduction in RCD due to inoculation with *Pythium* has been reported for longleaf pine (Jackson et al., 2008). In that study, the RCD of longleaf pine inoculated with *P. dimorphum* and *P. irregulare* decreased by 0.42 and 0.32 mm respectively, compared to non-inoculated seedlings. DeWald and Feret (1988) reported loblolly pine survival to be greater than 80% when outplanted with only 4 new roots. It could be that *P. irregulare* inoculated slash pine survival was affected more by reductions in RCD than the number of new roots (16 to 35) after outplanting.

Conclusions

We rejected the null hypothesis: root growth potential and survival of loblolly and slash pine seedlings is not affected by *Pythium* inoculation prior to cold storage. Both *P. dimorphum and P. irregulare* reduced loblolly and slash pine root growth potential, and after outplanting, only *P.*

irregulare inoculated slash pine experienced decreases in survival. In addition, the root collar diameter of *P. irregulare* inoculated slash pine was reduced during the RGP trial. This may be the first report of such effects from *Pythium* species on loblolly and slash pine seedlings after cold storage.

ACKNOWLEDGEMENTS

The authors would like to thank the scientist that took time to review the initial manuscript. We wish to thank Doug Shelburne and Smurfit-Stone Corporation for providing the seedlings used in the experiment and Barry Brooks for his assistance and the Southern Forest Nursery Management Cooperative at Auburn University for funding this research.

REFERENCES

Asiegbu FO, Lonneborg A, Johansson M (1996). Chitin and glucans detected in the cell walls of *Pythium dimorphum*-an oomycetous fungus. European J. For. Pathol., 26: 315-321.

Bronnum P (2005). Preplanting indicators of survival and growth of dessicated *Abies procera* bareroot planting stock. Scand. J. For. Res., 20: 36-46.

Borja I (1995). Root dieback and defense responses in Norway spruce seedlings infected with the fungal pathogen, *Pythium dimorphum*. Agricultural University of Norway, PhD Diss., p. 126.

Boyer JN, South DB (1985). Dormancy, chilling requirements, and storability of container-grown loblolly pine seedlings. Auburn University, Southern Forest Nursery Management Cooperative Research Note, 23: 12.

Carlson WC (1991). Lifting, storing, and transporting southern pine seedlings. Chapter 16, Forest Regeneration Manual, M.L. Duryea and P.M. Dougherty, eds., Kluwer Academic Publishers, Norwell, MA. p. 433.

DeWald LE, Feret PP (1988). Changes in loblolly pine seedling root growth potential, dry weight, and dormancy during cold storage. For. Sci., 34(1): 41-54.

Dierauf TA (1976). Fall planting and storage using non-dormant seedlings. Proc. Southeastern Area Nurseryman's Conference, pp. 31-36.

Enebak SA, Wei G, Kloepper JW (1998). Effects of plant growth-promoting Rhizobacteria on loblolly and slash pine seedlings. For. Sci., 44(1): 139-144.

Feret PP, Kreh RE, DeWald LE (1985). Root growth potential of stored loblolly pine seedlings. Proc. Third Biennial Southern Silvicultural Research Conference, Gen. Tech. Rep., SO-54: 18-24.

Garber MP, Mexal JG (1980). Lift and storage practices: their impact on successful establishment of southern pine plantations. New Z. J. For. Sci., 10(1): 72-82.

Hallgren SW, Tauer CG (1989). Root growth potential, first-year survival, and growth of shortleaf pine seedlings show effects of lift date, storage, and family. South J. Appl. For., 13: 163-169.

Hebb EA (1982). Cold storage of Choctawhatchee sand pine seedlings: Effect of lifting date and length of storage. South J. Appl. For., 6(4): 229-231.

Hendrix FF, Campbell WA (1971). A new species of Pythium with spiny oogonia and large chlamydospores. Mycologia, 63: 978-982.

Hendrix FF, Campbell WA (1973). *Pythiums* as plant pathogens. Annual Rev. Phytopathol., 11: 77-98.

Ho HH (1986). *Pythium dimorphum* from rhododendron. Mycopathologia, 93: 141-145.

Jackson DP, Gilliam C, South D, Enebak S, Eakes DJ (2008). The effect of *Pythium* and cold storage on the root growth potential and root collar diameter of longleaf pine (*Pinus palustris*) seedlings. Proc. Southern Nursery Association Conference, 53: 374-378.

Jones JP, Pantone DJ, Barnett JP, Brissette JC (1992). The relationship between fungal population development and root damage of cold-stored longleaf pine seedlings. Louisiana State University Agricultural Center, Bulletin. 832: 23.

Kahler LH, Gilmore AR (1961). Field survival of cold stored loblolly pine seedlings. Tree Planters' Notes, 45: 15-16.

Kelley WD, Oak SW (1989). Damping-off, p. 118-119. In: C.E. Cordell, R.L. Anderson, W.H. Hoffard, T.D. Landis, R.S. Smith, Jr., and T.V. Harvey (eds.). Forest Nursery Pests. USDA Forest Service, Agricultural Handbook No. 680: 184.

Kim YK, Xiao CL, Rogers JD (2005). Influence of culture media and environmental factors on mycelia growth and pycnidial production of *Sphaeropsis pyriputrescens*. Mycologia, 97(1): 25-32.

Linde C, Kemp GHJ, Wingfield MJ (1994). *Pythium irregulare* associated with *Pinus* seedling death on previously cultivated lands. Plant Dis., 78: 1002-1005.

May JT (1984). Southern Pine Nursery Handbook. U.S. Department of Agriculture, Forest Service, Southern Region.

Palmer L, Holen I (1986). The aquarium tester-a fast, inexpensive device for evaluating seedling quality. Tree Planters' Notes, 37(3): 13-16.

Pickens B (1998). Long-term storage of container longleaf pine. Tech Update, North Carolina Division of Forest Resources. 5(1): 1.

Rietveld WJ (1989). Evaluation of three root growth potential techniques with tree seedlings. New For., 3: 181-189.

Simpson DG, Ritchie GA (1996). Does RGP predict field performance? A debate. New For., 13: 249-273.

South DB, Boyer JN, Bosch L (1985). Survival and growth of loblolly pine as influenced by seedling grade: 13-year results. South J. Appl. For., 9(2): 76-81.

South DB, Larsen HS, Williams HM, Boyer JN (1989). Use of seedling size as a covariate for root growth potential studies. Auburn University Southern Forest Management Cooperative, Technical Note Number, 34: 8.

South DB (1999). The December dip of loblolly pine. Proc. Forest and Conservation Nursery Associations. Gen. Tech. Rep. SRS., 25: 14-17.

Sun X (1996). Biological control of Pythiaceous fungi causing death of longleaf pine seedlings during cold storage. Louisiana State University, Baton Rouge, PhD Diss., p.135.

Van der Plaats-Niterink AJ (1981). Monograph of the Genus Pythium. Stud. Mycol., 21: 242.

Venator CR (1984). Evidence of the presence of a lifting window for loblolly pine nursery seedlings. Proc: Southern Nursery Conference-Eastern Session, Asheville, NC, pp. 184-190.

26

Banana domestication on the Arabian Peninsula: A review of their domestication history

Khair Tuwair Said Al-Busaidi

Directorate General of Agriculture and Livestock Research, Ministry of Agriculture and Fisheries, Crop Production Research Centre, Fruit Research Department, P. O. Box 343 Postal Code 615, Samail, Sultanate of Oman.

Bananas are one of domesticated crops prehistoric in Arabian region. However, domestication history of this crop in this region did not receive attention by researchers. The strategic location of the Arabian Peninsula at the intersection of several main Old World trade routes has made it an incubator for bananas coming from different continents of the world. Tracing the history of bananas in this region and understanding the drivers of their diversity under inhospitable conditions may help maintain and further develop banana cultivation in similar areas of the world, especially where this issue has not been extensively studied. Domesticated banana cultivars presently found in the Arabian Peninsula are not indigenous to the region. The Islands of Southeast Asia and Southern China are the strongest candidate to be the primary sites of banana domestication, the subsequent phases of cultivation and translocation to other parts of the world during the prehistoric era. The Gulf ports, in particular Omani coastal ports and Indian Ocean trade effectively contributed to the exchange of plant genetic resources including bananas between the Arabian Peninsula areas and the Indian subcontinent, Africa and China. There is a relationship between ancient coastal ports and the presence of a variety of banana subgroups in some areas of the Arabian Peninsula such as Oman.

Key words: Maritime routes, ancient ports, linguistic, archaeological, translocation, banana history.

INTRODUCTION

Bananas have a long history of domestication and the role of mankind in their diffusion throughout the tropical and subtropical regions of the world has made them one of the most important fruit crops. Over the centuries, bananas have contributed to the stability of rural communities where they have provided work for farmers and reduced farmer migration to cities in search of alternate livelihoods (De Langhe et al., 2009). Eighty-seven percent of bananas produced globally are consumed locally (Biodiversity International, 2012). Even in many countries of the Arabian Peninsula, bananas play an important role as a food crop, although this arid region has an ecologically unfavourable climate for banana cultivation. Banana production in Arabian countries accounts for 2% of the total world banana production and 1.5% of total area harvested (FAOSTAT, 2010).

Over the past millennia, the banana domestication process has undergone different stages involving exploitation, hybridizations, somatic mutations and cultivation inside the natural habitat (De Langhe et al., 2009).

Subsequent steps involved the dispersal of domesticated varieties outside their natural habitats, to different geographical regions of the world. This led to often unpredicted genetic changes in banana varieties (Buerkert et al., 2009; De Langhe et al., 2009). Given the serious challenges in particular biotic and abiotic stresses facing sustainable banana production, in addition to local

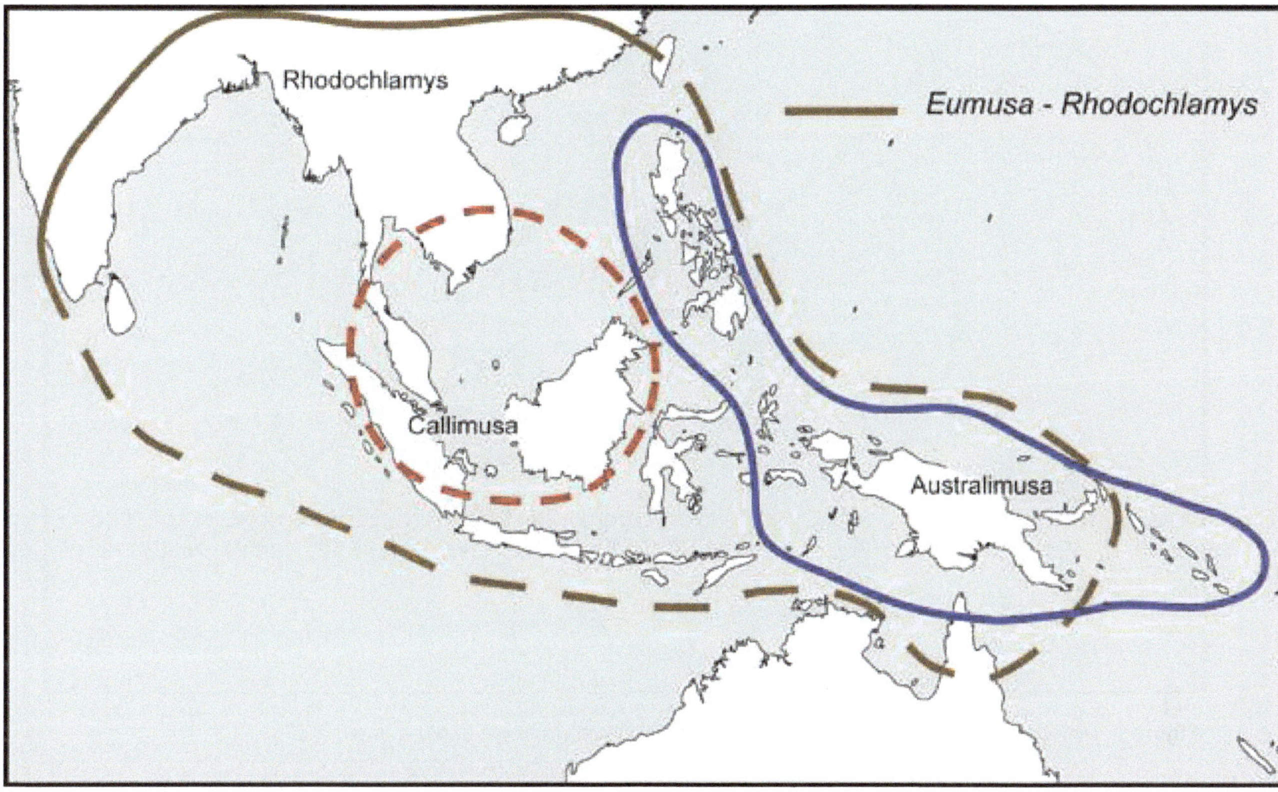

Figure 1. Distribution of *Musa* genus in East Asia (Simmonds, 1962, altered from De Langhe, 2009).

problems such as oasis modernization (Buerkert et al., 2009; Al-Saady et al., 2010), understanding and tracing banana diversity in the Arabian Peninsula is important. The diverse cultivars found in this region of the world include specially adapted varieties, not found anywhere else in the world. This study highlights the history of banana domestication in the Arabian Peninsula areas, particularly introduction and cultivation by reviewing papers that discuss linguistic, genetic and archaeological evidence as well as maritime routes that were perhaps used to introduce bananas to the Arabian Peninsula.

BANANA DOMESTICATION IN THE WORLD

Classification and distribution of banana species

The exact date of banana domestication is still subject to speculation, but recent multidisciplinary evidence indicates that the first domestication stage took place about 4500 years before present (BP) (De Langhe and de Maret, 1999; De Langhe et al., 2009). However, based on the archaeological evidence from New Guinea (Perrier et al., 2011), cultivation of domesticated banana varieties started about 6500 years BP. Human contact and migration played a crucial role in the domestication of banana varieties outside their natural habitat (Mindzie et al., 2001; Vrydaghs et al., 2009; Perrier et al., 2011).

Banana belongs to the family Musaceae which includes three genera; Asian and African *Ensete*, Asian *Musella* and East Asian *Musa* (De Langhe et al., 2009; Perrier et al., 2011). All edible bananas belonged to the genus *Musa* (Perrier et al., 2011). Simmonds (1962) divided the genus of *Musa* into 4 sections (Figure 1): Eumusa which covers all of East Asia, except Eastern Melanesia, Rhodochlamys which is spread along the monsoonal mainland of Southeast Asia, Australimusa which is distributed from south-eastern Indonesia and the southern Philippines to Melanesia, and Callimusa which is presented in Southern Vietnam, Malaysia, Borneo and Sumatra. According to Simmonds and Shepherd (1955), most of the edible diploid and triploid bananas are formed by inter- and intra-specific crosses of *Musa acuminate* (A) and *Musa balbisianana* (B) and are classified into the groups representing both their ploidy and species composition, that is AA, AAA, AAB and ABB. Fuller and Boivin (2009) mentioned that most plantains belonged to the AAB group and most desert bananas to AAA.

Secondary and tertiary distribution followed (Figure 2) and produced seven recognizable geographical areas each with a high density of specific cultivars (De Langhe, 2009):

(1) AA and AAA cultivars in the triangle between Indonesia, the Philippines and Melanesia (red line), with an exceptional density of AA cultivars in and around New

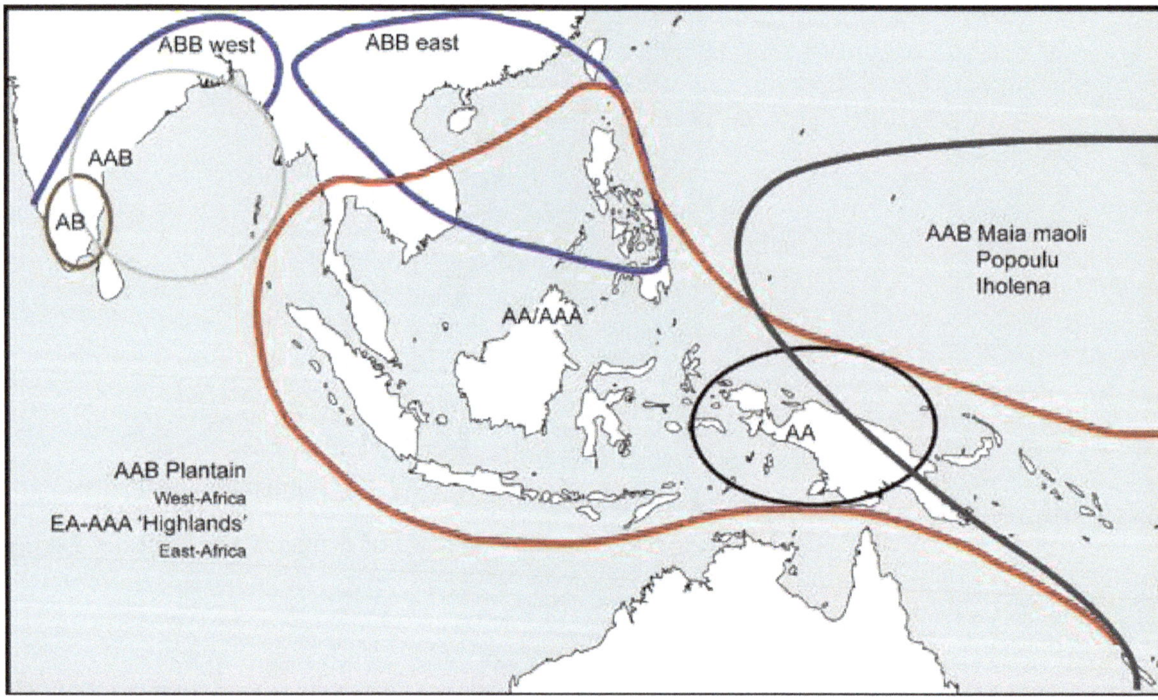

Figure 2. Distribution of the main banana cultivar groups (taken from De Langhe et al., 2009).

Guinea (black oval);
(2) Highland AAA bananas in the Great Lakes region of East Africa (East African Highland Bananas: EA-AAA 'Highlands'; area not shown on map in Figure 2);
(3) AAB plantains in the rainforest zone of Africa (area not shown on map in Figure 2);
(4) AAB Maia Maoli-Popoulu-Iholena cultivars in Oceania (dark grey line);
(5) AB (brown oval) and other AAB (pale grey circles) cultivars in South India;
(6) Eastern ABB cultivars in the Philippines and Vietnam (blue line - closed);
(7) Western ABB cultivars in Northeast and South India (blue line - open).

Linguistic and cultural evidence indicates that West Africa was the earliest centre for growing AAB plantains, while specimen of the AAA group were the first grown in East Africa (De Langhe and Maret, 1999; De Langhe et.al., 2009; Blench, 2003). The great diversity of cultivars in these areas and Iron Age phytoliths of banana found in Cameroon support this evidence (Mbida et al., 2000).

Origin and migration routes of banana species from natural habitat to other continents

Recent DNA and fingerprinting analyses of more than 400 wild and cultivated accessions, in addition to samples taken from Cameroon and Nigeria to better represent the diversity found in Africa, revealed that the islands of Southeast Asia and Western Melanesia likely are the main centre for the hybridization between different *M. acuminata* subspecies which generated edible diploids cultivar subgroups (AA cultivars; Perrier et al., 2011). Banana domestication passed through two stages: first the translocation from wild to edible diploids by hybridization of *M. acuminata* during the Holocene in New Guinea (Simmonds, 1962; Perrier et al., 2011) and subsequently the development of edible triploids (AAA) from edible diploids (AA) via spontaneous triploidizations (Perrier et al., 2011). Most domesticated bananas are triploid, including the widely distributed commercial Cavendish group (Perrier et al., 2011). Lastly, Perrier et al. (2011) used genetic, linguistic and archaeological data to determine the locations of banana groups. They suggested three contact areas of *M. acuminata* subspecies where the development of domesticated diploids took place: the Northern contact area with *malaccensis, microcarpa* and *errans* in South East Asia, Borneo and the Philippines; the eastern contact area with *errans* and *banksii* between the Philippines and New Guinea; and the Southern contact area with *banksii, zebrina* and *microcarpa* located between New Guinea and Java. It seems clear that the islands of Southeast Asia are the origin for all bananas and subsequently the different groups migrated to different areas of the world (Perrier et al., 2011). Based on this, Perrier et al. (2011) suggested two independent introduction events for triploid subgroups to Africa:

(i) AAA Mutika Lujugira and associated AAcv moved from the region between New Guinea and Java (Southern contact area) to East Africa;
(ii) AAB African Plantains moved from the Philippines and New Guinea (Eastern contact area) to Africa.

These hypotheses were also supported by Blench (2009) who mentioned that based on the botanical and linguistic evidence West African plantains (AAB) arrived from Southeast Asia earlier than 3000 BP. Similarly, Fuller and Madella (2009) hypothesized that the major diffusion of banana cultivars occurred in the later Iron Age, 2000 years BC. This is based on textual sources and historical linguistics from South Asia and China.

Geography of the Arabian Peninsula and early agriculture

The Arabian Peninsula is natural point of contact between the continent of Africa from the West and the Levant and Europe from the north and the continent of Asia from the east (Boivin and Fuller, 2009). In most of the Arabian Peninsula, the desert climate is considered the most important agro-ecological factor to determine landuse. The Arabian Peninsula is located between two main rainfall patterns: the winter rains of the Mediterranean region and summer monsoon rains (Boivin and Fuller, 2009). Digging wells and extracting water from aquifers supported agriculture in lowland oases and coastal regions which have insufficient rainfall (Edens, 1993; Blau, 1999). The later development of the *aflaj* system (Magee, 2005) leads to increased use of oases for agriculture and allowed the cultivation of different crops. Agricultural and accompanying maritime activities in the Arabian Peninsula have been triggered to a large extent by changes in climatic conditions over the Holocene period.

The early and middle Holocene (9000 to 2500 before Christian, BC) periods were characterized by relatively high rainfall intercepted by pronounced dry periods, in particular between 4000 to 4500 years BC (Boivin and Fuller, 2009). During this period, settlements were established in the desert areas such as the Eastern Sahara or Arabian Desert from northern Egypt to Eritrea in the south and parts of Sudan and Ethiopia, which were in habitated by Egyptian/Sudan groups (Hassan, 1997; Fuller, 1998).The Thar desert or Great Indian Desert (arid region in the north-western part of the Indian subcontinent) was settled by Mesolithic groups from India / Pakistan (Fuller, 2006). Similarly, the An Nafud or Al-Nefud settlement (located in the Northern part of the Arabian Peninsula) was established and with vegetables and fruits were grown (Boivin and Fuller, 2009).

In the mid-Holocene (6000 to 5900 BC), the conditions were wetter than at present and this was also the case for An-Nafud in the north of the Arabian Peninsula (Lézine et al., 1998). Potts (2008) mentioned that in the seventh millennium BC, Arabia was more attractive to the people than the Levant and Mesopotamia and this might have contributed to their migration to Arabia and the associated transfer of cultivated plant species from the Levant to Arabia. Evidence from northern Oman, the United Arab Emirates and the An-Nafud region indicates a period of aridity around 3800 BC called the 'Dark Millennium' during which even well established settlements collapsed (Uerpmann, 2003). Additionally, evidence from the Awafi palaeolake in the United Arab Emirates indicates that during the period from 3900 BC and 3200 BC, two arid periods led to a general decline of vegetation and its disappearance in Eastern Arabia (Parker et al., 2004, 2006).

Introduction of crops to the Arabian Peninsula in the Bronze age

The fertile periods in the early and middle Holocene experienced by the Arabian Peninsula and the establishment of agricultural settlements such as An-Nafud encouraged the entry of various agricultural crops including banana. Hammer et al. (2009) reported that about 21.3% of plant species in the Arabian Peninsula came from South and Southeast Asia, 20.6 % from the Near East and East Mediterranean and 15.4% from South America. This data supports the view of Boivin and Fuller (2009) who considered that most of the plants and animals in the Arabian Peninsula region are not native to this region. Despite the limited archaeological evidence of agriculture in the Arabian Peninsula the presence of agricultural equipment (Potts, 1994), of ancient (3200 BC) irrigation systems in the highlands of Yemen (Harrower, 2008) and the great areas of wheat and barley cultivation in some areas of the Arabian and Persian Gulf region indicate that agriculture has been practiced in this region for a long time. Taking into accounts the geographical distribution of banana genotypes (Figure 2), it seems that the Arab/Swahili and Malagasy-Malay (after 600 AD) trade played a vital role in the early dispersal of banana across the Indian Ocean and Red Sea to the Arab world. Similar banana varieties are found in Oman and Egypt (Castillo and Fuller, 2012). This trade route dispersal theory is supported by the discovery of remains of banana peels from the Arab trading period in Quesir al-Qadim located on the coast of the Red Sea (Van Der Veen, 2011).

In general, data indicate that the early third millennium saw the introduction of different crops including banana from Africa, South and Southeast Asia to the Arabian region (Boivin and Fuller, 2009). This has been confirmed by intensive presence of bread wheat in the Persian Gulf and the presence of the same variety in Asia particularly in the Indus region (Fuller, 2003, 2006). This indicates the starting of crop migration from South Asia to the

Arabian Peninsula. Also, in Yemen, Egypt and Nubia emmer wheat was dominant until the first millennium BC (Fuller, 2004), suggesting genetic exchange between the Arabian Peninsula and North Africa.

According to Robinson (1996), banana is mentioned in the Holy Koran as the 'tree of paradise' and the name of the genus (*Musa*) is derived from the Arabic word Mouz. Ancient Egyptian drawings already show bananas and it is believed also that the Assyrian civilization, which extended its authority to the Nile, introduced bananas to the Middle East (Attif and Muhammad, 2000). In 327 BC, the first accurate description of bananas appeared in Greek books after the invasion of India by Alexander the Great; however, it is believed that Arabs introduced bananas from India to the Middle East and North Africa (Attif and Muhammad, 2000). Kinder and Hilgemann (1974) mention that the arrival of Muslims in India, where a great variety of bananas exists, may have contributed to the introducing of bananas to the Arabian Peninsula.

Despite the long and ancient history of banana cultivation in the Arabian Peninsula, the archaeological and historical evidence is still limited. The exact origin, entry date and routes of banana introduction to this region remain a source of speculation. Since edible bananas reproduce vegetatively and not by seeds, their spread is particularly difficult to track (Vrydaghs et al., 2009). Potts (1994) mentioned that banana arrived in south-eastern Arabia by the 9th Century. Historical and cultivation evidence suggests that the origin of the banana planted in Dhofar and Yemen described in Medieval times (Varisco, 1994) was New Guinea/Indonesia (Kennedy, 2008) and the Valley of the Indus (Fuller and Madella 2001). Archaeobotanical evidence from Oman (Muweilah, Mleiha, Hili Bat, Ras al-Hamra, Ras al-Jinz), United Arab Emirates (Dalma, Umm an-Nar, Tell, Abraq, Muweilah, Mleiha, and Rumeilah and Yemen (Sabir, Hajar Bin Humeid, Hajar al-Tamrah, Haja al-Rayhani, Baraqish, Raybun and Khawlan sites: al-Raqlah, Jubabat al-Juruf, Wadi Yanaiim, Dhamar sites: Hayt al-Suad, al-Massanah) indicates that plants were domesticated in Arabian Peninsula between 1500-5000 BC, (Boivin and Fuller, 2009) and bananas could therefore have been introduced even earlier than in the 9th century BC.

Banana genetic diversity on the Arabian Peninsula

Despite of drought periods experienced by the Arabian Peninsula in successive millennia, bananas have not vanished. This was previously noted in the report of a survey by De Langhe (2002) in some Arabian countries. He pointed out the existence of large banana genetic diversity in this region. The bananas found in the Middle East (Jordan, Egypt and Oman), belong to the subgroups AA, AAA, AAB, AB and ABB (De Langhe, 2002). The crop may have undergone some modification over time, which made it more adapted to the arid regional conditions. Genetic mutations and human practices such as cultivation of banana under the shade of date palm in the Interior Governorate of Oman to provide humid microclimate could be one of the factors which have contributed to the survival of this crop in this region. As mentioned previously, Southeast Asia is the main source of banana; however the bananas present on the Arabian Peninsula today did not necessarily come directly from Southeast Asia, but may have travelled through Africa or India, before reaching the habitats on the Arabian Peninsula. However, this dispersal theory remains controversial. Buerkert et al. (2009) hypothesize that the AAA cultivar (cv. 'Umq Bir') recently discovered in the Upper Tiwi Valley of Oman reached there via East Africa, most likely Zanzibar, Madagascar or the Comoros islands where many AA and AAA cultivars are available.

De Langhe and Maret (1999) and Kennedy (2008) believe that *Musa sapientum,* which was planted in Dhofar and Yemen in antiquity, was introduced through Guinea/Indonesia, while others believe it came through the Indus Valley (Fuller and Madella, 2001). More than 31 accessions of banana were planted in 1997/1998 at the Salalah Agriculture Research Station of Dhofar Governorate, Oman. The origins of these accessions are the Comoros Islands, Zanzibar and India (De Langhe, 2002). Recently, nine hybrid cultivars (FHIA) were introduced by INIBAP and evaluated under southern (Dhofar Governorate) and northern Omani (North Al-Batinah Governorate, Sohar) conditions with respect to yield and tolerance to biotic and abiotic stresses, especially Sigatoka and Panama disease as well as salt and heat stress. When banana varieties are transferred from one area to another, their names sometimes remain the same and in this case it is easy to trace them (De Langhe, 2002). However, sometimes the names are changed immediately after arrival to a new place or after different generations which makes it more difficult to trace and identify them linguistically (Perrier et al., 2011). For example, in Yemen banana are called 'Al-Mawaz Al-Hindi' which means Indian Banana. In Egypt banana is also called 'Hindi'. In Oman, the 'Somali' banana variety may have been introduced from Somalia by individuals while the 'Malindi' variety might have entered Oman from the town of Malindi located northeast of Mombasa, on the Indian Ocean. Similarly, the 'Zanzibar' variety likely is from the Island of Zanzibar (Tanzania). The 'Fardh' variety belongs to the Mysore group possibly having been introduced from near the town of Mysore in India, the plantain 'Kenya' from Kenya and 'Abubaker Pilipino' from the Philippines.

The role of the Arabian Peninsula in inter-regional and inter-continental exchange of crops

There has been a lot of discussion about the role of the Arabian Peninsula's maritime ports in the transmission of plant materials, including banana, to different regions of the world. This role is supported through ancients tombs discovered in Bahrain and Kuwait, dating back to the second millennium BC and through Geniza records, pointing to an early contact of Arabs with South Asia (Nizami, 1994). The long presence of Arabs in India is reflected by the use of the word *Hindi* as the suffix to Arabic terms such as Mauz Hindi, Ud Hindi and Tamar Hindi (Nizami, 1994). Watson (1983) suggests that medieval Arab trade played a vital role in introducing banana to East Africa and Madagascar. This was confirmed by Sauer (1952) and Cleuziou and Tosi (1989) who believed that the Arabian Peninsula served as an intermediate region for the transmission of plant and animal materials from Asia and Africa throughout antiquity. Oman was the Gulf country to produce frankincense and copper and both trading goods likely played a major role in the inter-regional plant species exchange within in Arabian Peninsula as well outside Arabia (Hammer et al., 2009).

The ancient trade between Gujarat and Arabia was particularly important during the second millennium BC and is considered to have triggered crop exchanges between Africa and South Asia (Boivin and Fuller, 2009). The availability of African crops in Gujarat and Baluchistan during second millennium BC provides evidence that maritime contact between Gujarat and Oman and Dilmon extended to Yemen and Africa (Boivin and Fuller, 2009). The discovery of African crop species at 33 archaeological sites in South Asia, dating back to the Middle Bronze Age (2000 BC) and Iron Age (300 BC) indicates that the exchange of crops between the two continents went both ways (Fuller, 2003; Cooke et al., 2005). However, Boivin et al. (2009) believed that African crops did not transfer to the Arabian Peninsula in the Bronze Age due to the absence of agricultural communities during that time, unlike at the coastal region of Gujarat. However, Vansina (1990) believed that the introduction of AAB banana to the Upper Nile Great Lakes region of Africa occurred not from the coast but through North West Africa (Atiff and Mohamed, 2000).

THE ROLE OF MARITIME ROUTES IN TRANSFERRING PLANTS TO ARABIA

The Indian Ocean

The transfer and exchange of plants outside their natural habitats was not limited only to the land roads, but maritime routes played an essential role in the transmission process. However, the role of maritime routes in the transport of crop plants to new areas is still subject to much speculation. Blench (2003) suggested three maritime routes to transfer plants species between Africa and India. The first route was between Northwest India and Egypt, across Iran, the second was a shipping route (The 'Sabaean Lane') that linked Oman, Egypt, India and Africa, across the Sea Red and the Indian Ocean, and the third route run between the West Indian coast and East Africa across the open sea. The Arabian Peninsula has played an important role in maritime trade since the Bronze Age and therefore also in the exchange and dispersal of plant genetic resources across the Indian Ocean (Fuller et al., 2011). Furthermore, archaeological discoveries in the Arabian region support this hypothesis: Chinese coins found at Al-Qualify in eastern Saudi Arabia give evidence of the role of Arabian Gulf ports in ancient trade with the Far East (Cribb and Potts, 1996). Moreover, Fuller and Boivin (2009) viewed that the north-western Indian Ocean was one of the earliest long-distance maritime routes in the world. It allowed the significant exchange of livestock (e.g. cattle) and crops like bananas and taro (*Colocasia esculenta*) as early as 2000 BC. Also, the latter authors stated that the first biological exchange was in the Bronze Age / Chalcolithic (3000–1200 BC) along the circumference of the Arabia Peninsula and northern Indian Ocean where domesticated species were transferred between the Savannahs of India and northeast Africa and Yemen. Fuller and Boivin (2009) supported their view by claiming that at the end of the 3rd and 2nd millennium BC, there was a crop transfer between eastern Africa and South Asia through the Indian Ocean and along the southern coast of Arabia. This shows that the coastal ports contributed greatly to the exchange of plant resources between continents. The maritime ports between the Indian subcontinent (Figure 3) and Oman established in the middle of the 3rd millennium BC fostered the cultural and commercial relations and also contributed to the exchange of plant materials (Al Jarow, 2011; Al-Wagad, 2011). This is supported by archaeological discoveries in the port of Samahram in the Dhofar Governorate were Indian statues, pieces of pottery and coins dating from the 1st millennium BC (Albright, 1982) were unearthed as were Indian potteries in the port of Sohar (Al Jarrow, 2011) (Figure 3).

An intensive maritime trade between Gujarat and Arabia in the 2nd millennium BC is considered to be the starting point of plant species exchange between Asia and Africa. The evidence for this is the availability of African crops in Baluchistan during that period. This indicates that maritime trade between Gujarat and Oman was strong and extended to the west of the Arabian Peninsula towards Yemen and Africa (Fuller and Boivin, 2009). Blench (2003) mentioned that in the early medieval period, plants were introduced into West Africa by Arabs. Boivin and Fuller (2009) mentioned that the Indian Ocean was controlled by Arabic traders and hence these might be responsible for introducing and exchanging crops between eastern Africa Southeast Asian and India across Indian Ocean and along the southern coast of Arabia in the 2nd millennium BC. Also, Sauer (1952) and McMaster (1962) mentioned that the eastern coasts of the Arabian Peninsula contributed to the transfer of Southeast Asian bananas to Africa.

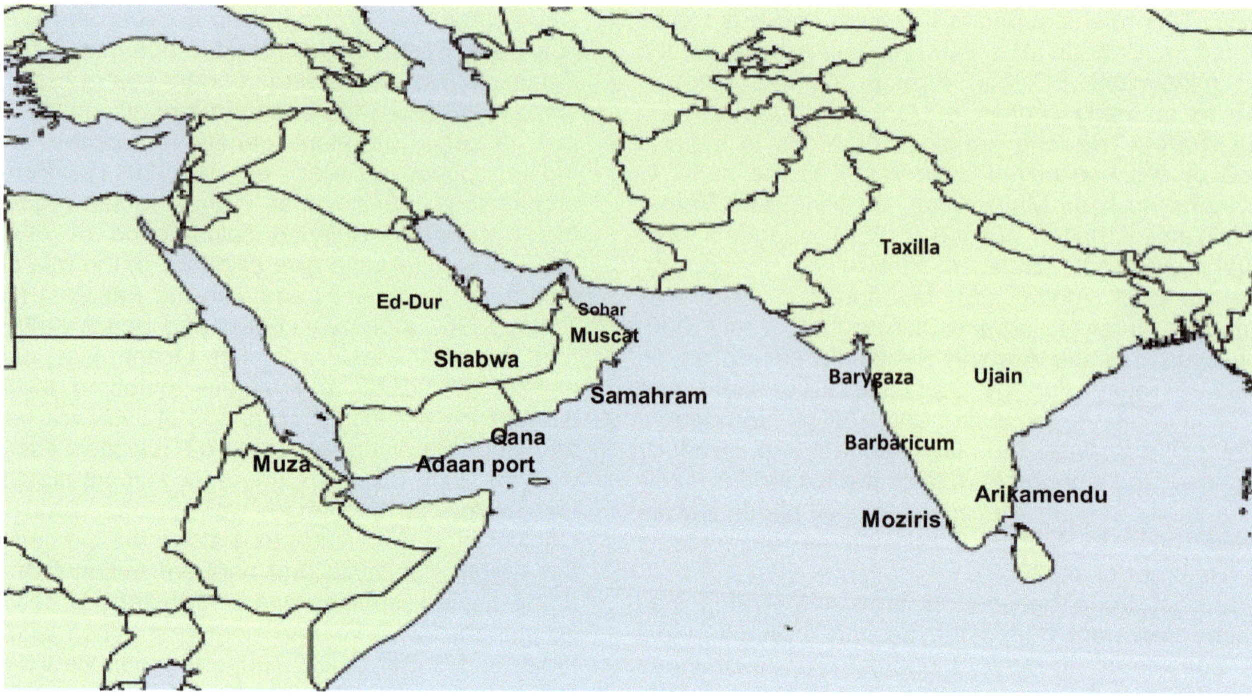

Figure 3. Ancient Indian ports with trade links to ports in the southern Arabian Peninsula (altered form Al Jarow, 2011).

Over centuries Omani coastal were a route of transit trade between the countries of the Far East (India, China and the East Indies) and the Arabian Peninsula countries as well as Iraq, the Mediterranean Sea countries and East Africa. Muftah (2011) also suggested that commercial ships in their trip between the coastal Omani ports and India to and on to China might have used three different maritime routes (Figure 3). The first route starts in Basra in Iraq, heads towards the Eastern coast of the Gulf, stops in Ciraf (Bu Shar) in Southern Iran, then reaches Sohar and Muscat to cross the Indian Ocean to Coolum Meli south of Almalbar on the Indian coast. The second route stretches from Muscat to Polien on the Indian coast, then continues to Serindep (Sri Lanka) and Kelah port and finally reaches Khanfo in China. The third route begins in the Omani ports of Dhofar and Merbat, goes to Kalikot or Coolum Meli and then directly to China (Figure 4).

Archaeological discoveries at Ras al-Jinz and Ras al-Had in the Sharqia North Governorate of Oman, dating back to the third millennium BC, yielded remains of boats and Indian rings from the Bronze Age to the 5th millennium BC (Jūtīlir et al., 1983). This provides evidence for ancient relations between Oman and the civilizations of Mesopotamia, Sindh and Africa. Michael (1994) mentioned commercial maritime routes to India from the Strait of Harmuz, through Ras al-Had in northern Oman directly to the Eastern Indian coast. This sea route linking Oman and India may have greatly contributed to the direct existence of banana diversity in Dhofar and coastal cities such as Sohar and Tiwi (Figure 5). This hypothesis is supported by Muftah (2011) who mentioned that India, China and Southeast Asian Islands were the main sources of commodities to the coastal Omani ports and then to other regions of Arabian Peninsula. He also added that bananas and coconuts (*Cocos nucifera*) were the most important fruit crops imported from India to Oman in Medieval times.

The Red Sea trade: Incense routes

Most scholars agree that the 'Land of Punt', a mining region in southern Egypt played an essential role in the maritime trade in the Red Sea during the 3rd millennium. Electrum, slaves, and particularly frankincense (*Boswellia sacra*) and myrrh (*Commiphora myrrha*) were traded from Punt via the Red Sea (Boivin and Fuller, 2009). The Arabian frankincense species is native to Dhofar in Oman and Hadhramout in Yemen, while other species of frankincense are native to northern and western Ethiopia and Eritrea (*B. papyrifera*), some areas of Sudan and West Africa (Boivin and Fuller, 2009), Somalia (Hepper, 1969) and the Island of Socotra (Boivin and Fuller, 2009). The genus of Myrrh tree (*Commiphora*) is native to southern and eastern Ethiopia, Somalia, Yemen, southwest Saudi Arabia and the coastal plains of Oman (Boivin and Fuller, 2009). All of these countries also cultivate banana and it is therefore likely that there was an intensive exchange of banana germplasm between these countries. Archaeological evidence from Barbar, Umm-an-Nar, Tall Abraq, Hili, Wadi Suq, Ras al-Hamra,

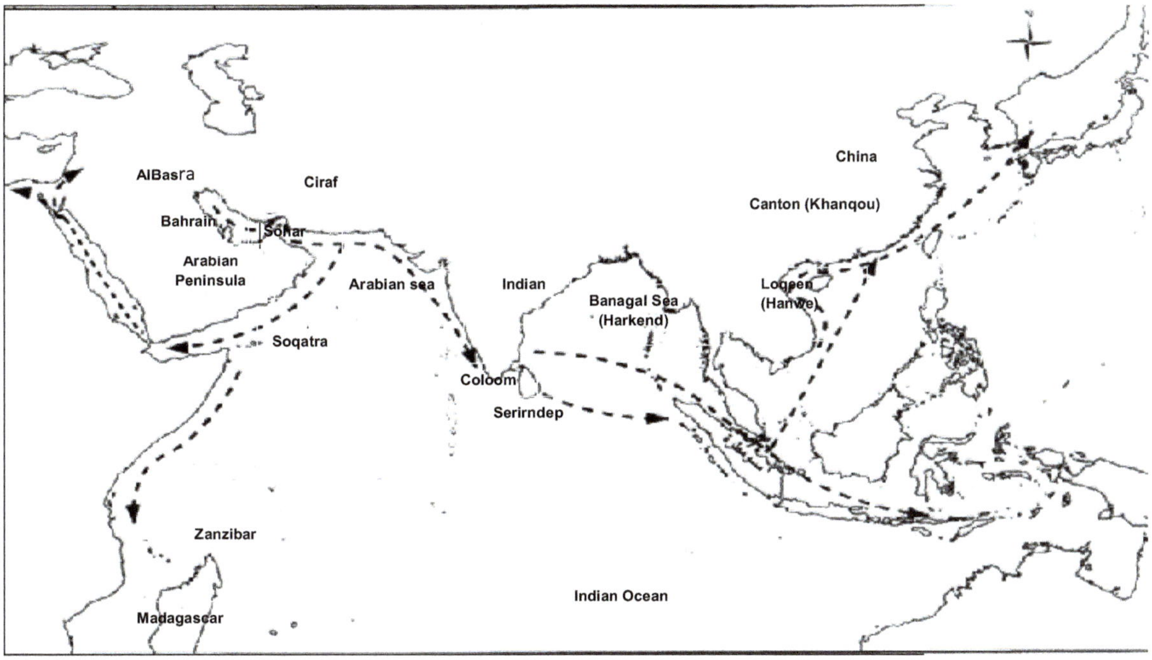

Figure 4. Trade routes and sea ports between the Arabian Peninsula, India and China (altered from Al-Wagad, 2011; Muftah, 2011).

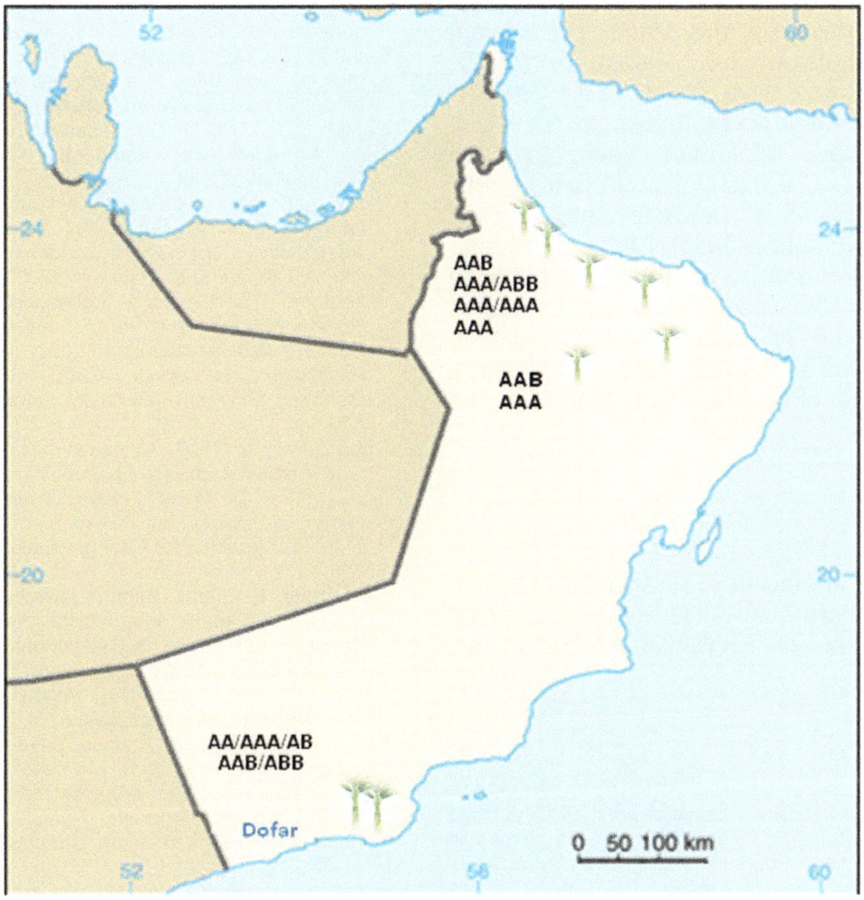

Figure 5. Regional distribution of banana cultivars groups in Oman (De Langhe, 2002; Al-Saady, 2010).

Ras al-Hadd, Ras al-Jinz, and as-Suwayh confirm the prosperity of the Red Sea trade during the 3rd millennium (Boivin and Fuller, 2009). It can therefore be hypothesized that trade of frankincense and myrrh via the Red Sea contributed to the exchange of plant genetic sources such as bananas between Africa, India and the Arabian Peninsula.

CONCLUSIONS

While the debate on the origins of banana domestication is on-going, evidence presented in this chapter suggests that the islands of Southeast Asia and Southern China are the primary sites of banana domestication followed by subsequent phases of pre-historic cultivation and translocation to other parts of the world. Owing to the desert climate conditions in many areas of the Arabian Peninsula and based on linguistic, genetic, archaeological and maritime route data, it can be concluded that domesticated banana cultivars presently found in this area are not indigenous to the region. Humankind maintained these varieties over the ages, despite unfavourable climatic conditions. Whether domesticated banana varieties were introduced directly from their natural habitat or arrived via Africa or India cannot be clearly determined at this stage. The ancient commercial relations between the Arabian Peninsula, East Africa and India may have played a key role in introducing banana to the Arabian Peninsula. The Gulf ports, in particular those of Oman, seem to have effectively contributed to the exchange of plant genetic resources including bananas between the Arabian Peninsula and the Indian subcontinent, Africa and China. Oman's strategic location and its wealth of frankincense and copper were factors that made it an important transit centre for the exchange of plant genetic resources, including different banana subgroups. Further studies are needed to verify the truth of banana domestication on the Arabian Peninsula.

ACKNOWLEDGEMENT

The authors invaluable appreciation also goes to Pro. Dr. Buerkert and Dr. Alexandra Zum Felde, Kassel University for their scientific advice and English correction of this investigation.

REFERENCES

Al Jarow AS (2011).Trade relationships between ports of Oman and ports of the Indian subcontinent (In light of Greek and Romanian resources and archaeological evidence). International Symposium: Oman and India. Prospects and civilization. The Omani Studies Centre and the Office for External Cooperation at Sultan Qaboos University, Muscat, Oman.

Al-Saady NA, Al-Lawati AH, Al-Subhi AM, Khan AI (2010). Evaluation of Genetic diversity in Omani banana cultivars (*Musa* cvs.) using AFLP markers. J. Plant Sci. 5(4):402-413.

Albright FP (1982). The American Archaeological Expedition in Dhofar, Oman. American Foundation for the Study of Man (AFSM). Foundation of the Study of Man, Vol. 6. Washington D.C., USA.

Al-Wagad MMAH (2011). Shipping routes between Oman and India in light of Arabic geographical resources. International Symposium: Oman and India. Prospects and Civilization. The Omani Studies Centre and the Office for External Cooperation at Sultan Qaboos University, Muscat, Oman.

Attif MI, Mohammed NHKH (2000). Banana. Its plantation, management and production. 1st edition, Alexandria, Egypt. 460 p.

Biodiversity International (2012). Sustainable production. http://bananas.biodiversityinternational.org/.

Blau S (1999). Of water and oil: Exploitation of natural resources and social change in eastern Arabia. In: Gosden, C. and Hather, J. (Eds.). The prehistory of food: Appetites for change 83-98. Routledge, London, England.

Blench R (2003). The movement of cultivated plants between Africa and India in prehistory. In: Neumann, K., Butler, A. and *Kahlheber, S.* (Eds): Food, Fuel and Fields. Progress in African Archaebotany. Africa Praehistorica 15. Köln (Heinrich–Barth–Institute). pp. 273-292.

Blench R (2009). Banana and plantations in Africa: Re-interpreting the linguistic evidence. Ethnobot. Res. Appl. 7:363-380.

Boivin N, Fuller DQ (2009). Shell middens, ships and seeds: Exploring coastal subsistence, maritime trade and the dispersal of domesticates in and around the ancient Arabian Peninsula. J. World Prehistory 22:113-180.

Boivin N, Blench R, Fuller DQ (2009). Archaeological, linguistic and historical sources on ancient seafaring: A multidisplinary approach to the study of early maritime contact and exchange in the Arabian Peninsula, In: Petraglia, M and Rose, J, (Eds.). The evolution of human populations in Arabia. Paleo-environments, prehistory and genetics. 18:251-278.

Buerkert A, De Langhe E, Al Khanjari S (2009). Ecology and morphological traits of an ancient *Musa* accuminata cultivar from a mountain oasis of Oman. Genetic Resour. Crop Evol. 56:609-614.

Castillo C, Fuller D (2012). Bananas: the spread of a tropical forest fruit as an agricultural staple. In: The Oxford Handbook of the Archaeology of Diet (in press).

Cleuziou S, Tosi M (1989). The southeastern frontier of the ancient Near East. In K. Frifelt and P. Serenson (Eds.), South Asian Archaeology Curzon Press: London. pp. 15-48.

Cooke M, Fuller DQ, Rajan K (2005). Early historic agriculture in southern Tamil Nadu: Archaeobotanical research at Mangudi, Kodumanal and Perur. In: U. Franke-Vogt and J. Weisshaar (Eds.), South Asian Archaeology 2003: Proceedings of the European Association for South Asian Archaeology Conference, Bonn, Germany, 7th–11th July 2003. Linden Soft, Aachen, Germany. pp. 329-334.

Cribb J, Potts D (1996). Chinese coin finds from Arabia and the Arabian Gulf. Arabian Archaeol. Epigraphy 7:108-118.

De Langhe E, De Maret P (1999). Tracking the banana: its significance in early agriculture. In C. Gosden and J. Hather (Eds). The Prehistory of Food .Appetites for Change. Rutledge, London, England. pp. 377-398.

De Langhe E (2002). Banana Diversity in the Middle East (Jordan, Egypt and Oman). International Network for the Improvement of Banana and Plantain (INIBAP), Montpellier, France, 56 p.

De Langhe E, Vrydaghs L, De Maret P, Perrier X, Denham T (2009). Why bananas matter: An introduction to the history of banana domestication. Ethnobot. Res. Appl. 7:165-177.

Edens C (1993). Indus-Arabian interaction during the Bronze Age: A review of evidence. In G. L. Possehl (Ed.), Harappan civilization: A recent perspective (2nd ed). New Delhi, Oxford and IBH.

FAOSTAT (2010). Food and Agriculture Organization of the United Nations. Statistics Division, http://faostat.fao.org (data accessed on 17.02.2012).

Fuller DQ (1998). Palaeoecology of the Wadi Muqaddam: A preliminary report on the significance of the plant and animal remains. Sudan Nubia 2:52-60.

Fuller DQ (2003). African crops in prehistoric South Asia: A critical

review. In: Neumann, K., Butler, A. and Kahlheber, S. (Eds.). Food, Fuel and Fields. Progress in African Archaeobotany. Heinrich-Barth Institut, Cologne, Germany. pp. 239-271.
Fuller DQ (2004). Early Kushite agriculture: Archaeobotanical evidence from Kawa. Sudan Nubia 8:70-74.
Fuller D Q (2006). Agricultural origins and frontiers in South Asia: A working synthesis. J. World Prehistory 20:1-86.
Fuller DQ, Madella M (2009). Banana cultivation in South Asia and East Asia: A review of the evidence from archaeology and Linguistics. Ethnobot. Res. Appl. 7:333-351.
Fuller DQ, Madella M (2001). Issues in Harappan archaeobotany: Retrospect and prospect. In: Settar, S. and Korisettar, R. (Eds.). Indian archaeology in Retrospect, Vol. II. Protohistory. Manohar, New Delhi, India. pp. 317-390.
Fuller DQ, Boivin N (2009). Crops, cattle and commensals across the Indian Ocean: Current and potential archaeobiological evidence. Etudes Ocean Indian 43:13-46.
Fuller DQ, Boivin N, Hoogervorst T, Allaby R (2011). Across the Indian Ocean: the prehistoric movement of plants and animals. Antiquity 85(328):544-558.
Hammer K, Gebauer J, Al Khanjari S, Buerkert A (2009). Oman at the cross-roads of inter-regional exchange of cultivated plants. Genet. Resour. Crop Evol. 56(4):547-560.
Harrower M (2008). Hydrology, ideology, and the origins of irrigation in ancient southwest Arabia. Current Anthropology 49(3): 497-510.
Hassan FA (1997). Holocene palaeoclimates of Africa. Afr. Archaeological Rev. 14(4): 213-230.
Hepper FN (1969). Arabian and African frankincense. J. Egypt. Archaeol. 55:66-72.
Jūtīlir JR, Firth N, Huston SS (1983). Preliminary research in ancient mining in the Sultanate of Oman. Our heritage. Ministry of Heritage and Culture, Oman 41 p.
Kennedy J (2008). Pacific bananas: Complex origins, multiple dispersals? Asian Perspect. 47(1):75-94.
Kinder H, Hilgemann W (1974). Atlas of World History Vol. 1. From the Beginning to the Eve of the French Revolution. Penguin, Harmondsworth, England. 299 p.
Lézine AM, Saliége JF, Robert C, Wertz F, Inizian ML (1998). Holocene lakes from Ramlat as-Sab'atayn (Yemen) illustrate the impact of monsoon activity in southern Arabia. Quat. Res. 50:290-299.
Magee P (2005). The chronology and environmental background of Iron Age settlement in Southeastern Iran and the question of the origin of the Qanat irrigation system. Iranica Antiqua 40:217-231.
Mbida CM, Van Neer W, Doutrelepont H, Vrydaghs L (2000). Evidence for banana cultivation and animal husbandry during the first millennium BC in the forest of southern Cameroon. J. Archaeol. Sci. 27:151-162.
McMaster DN (1962). Speculations on the coming of the banana to Uganda, J. Trop. Geogr. 16:57-69.
Michael R (1994). The Archaeology of the Arabian Gulf (500-323 BC). 1st ed. Microwave Technology Series (Gebundene Ausgabe), Routledge, ISBN 0415032687, New York, NY, USA. 388 p.
Mindzie MC, Doutrelepont H, Vrydaghs L, Swennen RL, Swennen RJ, Beeckman H, De langhe E, Maret P (2001). First archaeological evidence of banana cultivations in central Africa during the third millennium before present. Veget. Hist. Archaeobot. 10:1-6.
Muftah AMAF (2011). Ports of the coast of Oman and their role in the prosperity of trade between Oman and India in the Islamic era (200-800 AH). International Symposium: Oman and India. Prospects and Civilization. The Omani Studies Center and the Office for External Cooperation at Sultan Qaboos University, Muscat, Oman.
Nizami KA (1994). Early Arab contact with South Asia. J. Islamic Stud. 5(1):52-69.
Parker AG, Eckersly L, Smith, MM, Goudie AS, Stokes S, Ward S, White K, Hodson MJ (2004). Holocene vegetation dynamics in the northeastern Rub' al-Khali desert, Arabian Peninsula: A phytolith, pollen and carbon isotope study. J. Quat. Sci. 19(7):665-676.
Parker AG, Goudie AS, Stokes S, White K, Hodson MJ, Manning M, Kennet D (2006). A record of Holocene climate change from lake geochemical analyses in south-eastern Arabia. Quat. Res. 66(3):465-476.
Perrier X, De Langhe E, Donohue M, Lentfer C, Vrydaghs L, Bakry F, Careel F, Hippolyte I, Horry JP, Jenny C, Lebot, V, Risterucci AM, Tomekpe K, Doutrelepont H, Ball T, Manwaring J, Maret P, Denham T (2011). Multidisplinary perspectives on banana (*Musa spp*) domestication. PNAS 108(28):11311-11318.
Potts DT (2008). Arabian Peninsula. In: D. Pearsall (Ed.). Encyclopedia of archaeology. pp 827-834. Elsevier, New York, NY, USA.
Potts DT (1994). Contributions to the agrarian history of Eastern Arabia II: The cultivars. Arabian Archaeol. Epigraphy 5:236-275.
Robinson JC (1996). Bananas and Plantain. Cambridge University Press, Cambridge, UK, 238 p.
Sauer JD (1952). Agricultural Origins and Dispersals. American Geographical Society, New York, NY, USA.
Simmonds NW (1962). The Evolution of the Bananas. Tropical Science Series. Longmans, London, England.
Simmonds NW, Shepherd K (1955). The taxonomy and origins of the cultivated bananas. J. Linnaean Soc. (Botany) 55:302-312.
Uerpmann M (2003). The dark millennium: Remarks on the final atone age in the Emirates and Oman. In: Potts, D.T., Naboodah, H. and Hellyer, P. (Eds.), Archaeology of the United Arab Emirates. Proceedings of the First International Conference on the Archaeology of the UAE. Trident, London, England. pp. 74-81.
Vrydaghs L, Ball T, Volkaert H, Houwe IVD, Manwaring J, De Langhe E (2009). Differentiating the Volcaniform Phytoliths of Bananas: *Musa acuminata*. Ethnobot. Res. Appl. 7:165-177.
Van der Veen M (2011). Consumption, Trade and Innovation: Exploring the Botanical Remains from the Roman and Islamic Ports at Quseir Al-Qadim, Egypt. Africa Magna Verlag, Frankfurt, Germany 303 p.
Vansina J (1990). Paths in the Rainforest: Towards a History of Political Tradition in Equatorial Africa. University of Wisconsin Press, Madison, WI, 438 p.
Varisco DM (1994). Medieval agriculture and islamic science. The Almanac of a Yemeni Sultan. University of Washington Press. USA. 349 p.
Watson AM (1983). Agricultural innovation in the Early Islamic world. Cambridge Cambridge University Press, Cambridge, England, pp 272.

Response to γ radiation and storage temperature on quarantine pests of alphonso mango

M. K. Yadav and B. R. Parmar

Department of Horticulture, N. M. College of Agriculture, Navsari Agricultural University, Dandi Road, Navsari-396450, India.

Application of food irradiation for sanitary and phytosanitary purposes to ensure food safety and quality and facilitate international trade, including quarantine measures related to fresh products. The irradiation technique is now emerging as a demanding tool for controlling insect-pests. This technique has been a successful tool against a number of pest species. In this regard, the study was conducted to evaluate four (0.00, 0.20, 0.40 and 0.60 kGy) gamma irradiation doses and similar storage temperature (Ambient, 9 and 12°C and CA storage at 12°C) against the quarantine pests of alphonso mango fruit. The fruits were exposed to gamma radiation for different doses from the source of ^{60}Co. The two year studies revealed that there were no stone weevil and fruit fly incidence in any treatment (gamma irradiation or storage temperature) during study, except unirradiated fruits stored at ambient temperature; recorded fruit fly infestation.

Key words: Fruit fly, gamma irradiation, quarantine pests, stone weevil, storage temperature.

INTRODUCTION

Mangoes (*Mangifera indica* L.; family Anacardiacae) are in over 90 countries worldwide. Asia accounts for 77% of global mango production and the Americas and Africa account for 13 and 19%, respectively (Pereira et al., 2010). It has been a long journey, both for food irradiation research and development as well as the emperor of fruits, the Indian mango, to make it to one of the world's most wanted markets, the USA. India is the global leader in mango production and it is also one of the most important tropical fruit of Asia with a high demand in the world market (Tharanathan et al., 2006). International trade, particularly of mango, is subjected to quarantine regulations in most of the importing countries because of the presence of fruit flies (*Bactrocera dorsalis*, Hendel) and stone weevil (*Strenochetus mangiferae*) in India. United States, Japan, Australia and New Zealand have previously prohibited import of mango from India. Current official quarantine procedures used for treating fruits and plant materials for export markets are vapour heat treatment, chemical fumigation with ethylene dibromide or methyl bromide or combinations of fumigation and cold treatment (Thomas, 1993). However, none of these treatments are effective against the mango stone weevil. The use of ethylene dibromide for quarantine fumigation of fruits is banned in USA because of the evidences of its carcinogenic and mutagenic effects in test animals. Methyl bromide has phytoxic effect on many fruits.

The usefulness of gamma irradiation as an alternative to chemical fumigation has been demonstrated against both fruit flies and mango stone weevil. Since the weevil develops and completes its life cycle inside the mango seed, conventional, chemical or physical treatments are not effective against this insect. Irradiations was also found to inhibit the development stages of mango fruit fly and the melon fly (*Dacus cucurbitae*, coquillot) commonly found infesting mango. Irradiation disinfestations of mangoes were found to be superior to more conventional methods of disinfestations, such as ethylene dibromide fumigation or vapour heat treatment (Yadav et al., 2013). Irradiations are a physical process for the treatment of foods akin to conventional process like heating or freezing. It prevents food poisoning, reduces wastage to contamination and at the same time preserves quality (Mahindru, 2009). The issues related to quarantine and phytosanitary are the major stumbling blocks to trade, both national and international (Yadav et al., 2010). Therefore, the study conducted for find out the balance between required effective dose and tolerance of fruit to irradiation under various storage temperatures.

MATERIALS AND METHODS

Material and preparation

The export grade mangoes of cv. Alphonso were harvested from orchard of Navsari Agricultural University, Navsari, Gujarat, India. The selected mangoes from Class I as per the quality parameters specified and described in "post harvest manual for mangoes" published by Agricultural Production and Export Development Authority (Anonymous, 2007). These were sorted for uniformity in size, maturity and freedom from defects. The fruits were kept in plastic crates with cushioned material and transported to cold storage, Post Harvest Technology Unit of ASPEE College of Horticulture and Forestry, Navsari Agricultural University, Navsari (Gujarat) India. Than after, fruits were again sorted to remove those with spotty and having bad appearance. The individual fruit weight was from 250 to 350 g. The selected fruits wash with chlorine water and after drying fruits packed in corrugated fibre board (CFB) boxes cushioned with tissue paper. The dimensions of CFB box were 370 × 275 × 90 mm and gross weight of box with fruits was 3.0 kg. One CFB box having nine fruits for each treatment, each treatment replicated thrice. The packed boxes kept in cold storage at 12°C for 8 h for pre-cooling treatment. The time gap between harvesting and pre-cooling was not more than 6 h.

Radiation treatment

After pre-cooling, fruits transport to irradiation treatment in air conditioned vehicle. It was carried out at ISOMED plant, Board of Radiation and Isotope Technology, Sir Bhabha Atomic Research Centre, Mumbai, India. The fruits were exposed to gamma radiation for different doses from the source radio isotope ^{60}Co with energy 1.33 MeV. The source rack with its ^{60}Co pencils was raised from the storage pool by automated control system. The product packages are then moved around the radiation source in such a way that they are exposed to radiation for predefined time. There were four irradiation doses that is, I_1 -0.00 kGy (Unirradiated), I_2 -0.20kGy, I_3 -0.40kGy and I_4 -0.60kGy. The time gap between pre-cooling and irradiation was not more than 9 h. After irradiation, fruits immediately transported to cold storage of university in air conditioned vehicle.

Storage conditions

The CFB boxes were kept in cold storage at different temperature as per storage temperature treatments. The treatments are S_1- Ambient (27±2°C, 60-70% RH, S_2-9°C, 90% RH, S_3-12°C, 90% RH and S_4- Control atmospheric storage (12°C, O_2 2%, CO_2 3% and RH 90%). The infestation of fruit fly and stone weevil were studied.

Measurement protocols for incidence of fruit fly and stone weevil

The fruits cut with the help of sharp knife and visually observed the larvae of the fruit fly at the end of storage. The number of fruit fly attacked fruits were counted and expressed as percentage. For stone weevil, stone of fruits cut with the help of sharp knife and visually observed the stone weevil at the end of storage. The number of stone weevil attacked fruits were counted and expressed as percentage.

Data analysis

Two years data obtained from experiment was statistically analyzed using ANOVA for completely randomizes design with factorial concept. Significance differences among treatments were compared using the Fisher's analysis of variance at the 5% probability level, technique as given by Panse and Sukhatme (1967). The data were subjected to appropriate transformation (arcsine) when necessary to meet the assumptions of normality.

RESULTS AND DISCUSSION

The fruit fly incidence was significantly influenced by the irradiation and storage temperature individually as well as in interaction (Table 1). It can be seen from the data there was no fruit fly incidence in fruits exposed to gamma rays. Fruits stored at 9 and 12°C and in CA (12°C) storage also showed no fruit fly incidence. Only unirradiated ambient stored fruits showed 9.28% fruit fly incidence. The irradiated fruits stored at ambient temperature as well as the unirradiated fruits stored at 9 and 12°C and in CA (12°C) storage showed fruit fly free fruits might be due to sensitivity of the insects to irradiation, the sensitivity being directly proportional to reproductive activity and inversely proportional to the degree of differentiation. Thus, irradiation had lethal effect on dividing cells during embryonic development in eggs. Therefore irradiated fruits had less infestation compared to unirradiated fruits (Udipi and Ghugre, 2010). Similar results also observed by Manoto et al. (1987) and Srivastava (1997) in mango; Singh and Pal (2007) in guava and Rivera and Hallman (2007) in papaya. In conclusion view of point the fruit fly incidence was not

Table 1. Response to irradiation and storage temperature against mango fruit fly (*Bactrocera dorsalis*, Hendel) and stone weevil (*Strenochetus mangiferae*) percent.

Fruit fly (*Bactrocera dorsalis*, Hendel)															
	I year					**II year**					**Mean**				
Source	I_1	I_2	I_3	I_4	Mean	I_1	I_2	I_3	I_4	Mean	I_1	I_2	I_3	I_4	Mean
S_1	0.00	0.00	0.00	0.00	0.00	10.12 (18.55)	0.00 (1.65)	0.00 (1.65)	0.00 (1.65)	2.5 (6.64)	9.28 (5.60)	0.00 (0.82)	0.00 (0.82)	0.00 (0.82)	2.94 (2.02)
S_2	0.00	0.00	0.00	0.00	0.00	0.00 (1.65)	0.00 (1.65)	0.00 (1.65)	0.00 (1.65)	0.00 (1.65)	0.00 (0.82)	0.00 (0.82)	0.00 (0.82)	0.00 (0.82)	0.00 (0.82)
S_3	0.00	0.00	0.00	0.00	0.00	0.00 (1.65)	0.00 (1.65)	0.00 (1.65)	0.00 (1.65)	0.00 (1.65)	0.00 (0.82)	0.00 (0.82)	0.00 (0.82)	0.00 (0.82)	0.00 (0.82)
S_4	0.00	0.00	0.00	0.00	0.00	0.00 (1.65)	0.00 (1.65)	0.00 (1.65)	0.00 (1.65)	0.00 (1.65)	0.00 (0.82)	0.00 (0.82)	0.00 (0.82)	0.00 (0.82)	0.00 (0.82)
Mean	0.00	0.00	0.00	0.00		2.53 (6.64)	0.00 (1.65)	0.00 (1.65)	0.00 (1.65)		2.94 (2.02)	0.00 (0.82)	0.00 (0.82)	0.00 (0.82)	
Source		I	S	I X S		I		S		I X S	I		S		I X S
S. Em ±		-	-	-		0.001		0.001		0.002	0.001		0.001		0.012
CD (P≤0.05)		-	-	-		0.003		0.003		0.006	0.003		0.003		0.036

N.B.: There was no incidence of stone weevil (*Strenochetus mangiferae*) in both years. Figure in parenthesis indicates ARC SINE transformed value Where I = irradiation, S = Storage temperature, CA = Control atmospheric storage.

observed in the fruits subjected to gamma irradiation whereas; 9.28% was observed in fruits which was kept without irradiation. No fruits were found with stone weevil infestation.

Conflict of Interests

The author(s) have not declared any conflict of interests.

ACKNOWLEDGEMENTS

Thanks are due to Dr. R. Chander, Dr. L. N. Bandi, Dr. A. Shrivastva and Shri Jyotis, Board of Radiation and Isotope Technology, BARC, Mumbai and Dr. N. L. Patel, Dean, ACHF, NAU, Navsari (India).

REFERENCES

Anonymous (2007). Post harvest manual for mangoes. Agricultural Production and Export Development Authority, Ministry of Agriculture. Government of India.

Mahindru SN (2009). Food preservation and irradiation. APH publishing corporation, New Delhi.

Manoto EC, Rosario MS, Del-Rosario, MS (1987). Mango disinfestation strategies in relation to quarantine requirements in the Philippines. In: Plant quarantine and phytosanitary Barriers to trade in the ASEAN, 9-10 Dec, pp. 193-207.

Panse VG, Sukhatme PV (1967). Statistical Methods for Agricultural workers, ICAR, New Delhi.

Pereira T, Tijskens, LMM, Vanoli M, Rizzolo A, Eccherzerbini P, Torricelli A, Spinelli L, Filgueiras H (2010). Assessing the harvest maturity of Brazilian mangoes. EW Hevett et al.(Ed..) ,Proc IS on Post Harvest Pacifica 2009. Acta Hort. 880.

Rivera ZT, Hallman GJ (2007). Low dose irradiation phytosanitary treatment against. *Mediterranean* fruit fly (Diptera: Tephritidae). Florida Entomol. 90(2):343-346. http://dx.doi.org/10.1653/0015-4040(2007)90[343:LIPTAM]2.0.CO;2

Singh SP, Pal PK (2007). Post harvest fruit fly disinfestation strategies in rainy season guava crop. Acta Hort. 375:591-596.

Srivastava RP (1997). Mango quarantine pests export and rules. International Book Distributing Co, Lucknow, India.

Tharanathan RN, Yashoda HM, Prabha TN (2006). Mango (*Mangifera indica* L), 'the king of fruits' – An overview. Food Rev. Int. 22:95-123. http://dx.doi.org/10.1080/87559120600574493

Thomas P (1993). Irradiation of foods. In: Chadha, KL and Pareek OP (Eds.). Advances in Horticulture, Malhotra Publishing House, New Delhi. pp. 1810-1835.

Udipi SA, Ghurge PS (2010). Applications of food irradiation. In: Udipi, SA and Ghugre PS (Eds.), Food irradiation, Agrotech Publishing academy, Udaipur. pp. 40-71.

Yadav MK, Patel NL, Hazarika A, Chaudhary PM (2010). Physiological changes in Kesar mango as influenced by treatment and storage. Hort. J. 23(1):16-17.

Yadav MK, Patel NL, Parmar BR, Nayak D (2013). Evaluation of physiological and sensory changes of kesar mango (*Mangifera indica* L.) influenced by ionizing radiation and storage temperature. SAARC Agric. J. 1(2):69-80.

Comparative analysis of on-farm timber conversion systems in Kenya

George Muthike[1], Christopher Kanali[2] and Douglas Shitanda[3]

[1]Kenya Forestry Research Institute (KEFRI) Forest Products Research Centre, P. O. Box .64636-00620, Kenya.
[2]Department of Biomechanical and Environmental Engineering, Jomo Kenyatta University of Agriculture and Technology, Nairobi, Kenya.
[3]MultiMedia University of Kenya, Kenya.

Exploitation of trees on private farms in Kenya is mainly by chain, bench and pit sawing systems, whose efficiency has not been well established. The sawing systems were evaluated for their performance using three species: *Eucalyptus saligna*, *Grevillea robusta* and *Prosopis juliflora*. Data was subjected to analysis of variance and mean comparisons. Results showed significant differences (*$P<0.05$*) among the sawing systems. Chainsaw system recovered more logs from the trees with a log recovery rate of 76%, which was significantly higher than those of 65 and 63% for bench and pit saws respectively. Timber recovery rates of 39.8, 35.9 and 30.0% for pit, bench and chain saws respectively were observed. The chainsaw system produced the highest rate of timber output (0.23 m^3/man-hour), differing significantly from benchsaw (0.17) and pit saw (0.08). Chain and bench saws recorded significantly different fuel consumption rates of 8.41 and 14.86 $m^3 l^{-1}$ respectively. Timber sawn with the chain saw had the highest dimensional deviation (+/-5.53 mm), which was significantly higher than deviations for benchsaw (+/-3.55 mm) and pitsaw (+/-2.44 mm). The study shows that, although chainsaw recorded the highest timber production rate per man hour, it had the lowest timber recovery. Timber sawn with this system had the highest dimensional deviations. Bench saw was more economical for sawing large diameter logs, with chainsaw performing better for small diameter logs. The three systems recorded very low recovery of timber and are thus not suitable for processing timber, in this error of diminishing tree cover unless they are improved.

Key words: Efficiency, sawing systems, timber recovery, production rate, size, deviation.

INTRODUCTION

Forest cover in Kenya has been decreasing in the past 20 years and is estimated to be about 6% (FAO, 2010). Some of the major factors that contributed to this situation are illegal harvesting and conversion of forest land to other uses. Inefficient logging and processing to a lesser extent has also contributed to the same. This decrease led to partial ban on logging in state forests by Kenyan government in 1999 (FAO, 2005). The ban led to the closure of saw mills and a serious shortage of sawn timber in the country, contributing to high timber prices. Due to the timber shortage, farms have became a major source of timber, sawn using small-scale sawing systems that include chain, bench and pit saws (Holding et al., 2001; Holding-Anyonge and Roshetko, 2003).

Bench saws are used mainly in areas where saw logs are available in relatively large quantities and trees are easily accessible by a tractor. The initial cost of this system is high, thus limiting its affordability. Pit sawing is an old and cheap technology used in areas where trees grow in steep terrain. Pit saws are considered slow and uneconomical to operate (Muthike et al., 2010).

Chain saw, which initially was used for felling and cross cutting wood, has been adopted for sawing dimensional timber because it is portable, easy to operate relatively cheaper to acquire and less limited by terrain (Pasiecznik and Brewer, 2006; Popoola, 2010). While in some countries, chainsaw milling is a legal and important sub-sector of the forest industry, supporting rural livelihoods, in many others, it is associated with illegal forest activities. In such countries, different scenarios exist. Some of them are; the system is illegal, permitted under some regulations, restricted to domestic use only while in others it is only permitted for small-scale commercial production. Even when it is authorized, its activities are often difficult to monitor due to the large number of people involved and the small size and mobility of its operations (Wit et al., 2010).

Despite various studies (Holding et al., 2001; Holding-Anyonge and Roshetko, 2003; Marfo, 2010) showing that chainsaw milling system is inefficient, its use has continued to rise unabated, gradually replacing other timber milling systems (Pasiecznik, 2010). There have only been a handful of studies in Kenya reporting on the efficiency of timber milling systems and these have only reported timber recovery while ignoring efficiency and other critical aspects of timber sawing.

Conversion efficiencies of sawing systems are very important for the management of sustainable production of sawn timber resources in a contemporary world (Pasiecznik, 2010).

In order to efficiently utilize the limited timber resources in Kenya, there is need to establish the efficiencies for timber sawing systems used. Factors to investigate are log and sawn timber recovery, timber yield and fuel consumption rates, which are key elements in production of timber (Owusu et al., 2011). Similarly, there is a need to study the factors affecting recovery and introduce appropriate measures to improve them. This study was therefore conducted in order to address the information gaps on the relative efficiencies of different timber sawing systems.

MATERIALS AND METHODS

Selection of timber species, sawing systems, equipment and operators

Timber species used for this study included two species commonly grown and converted into sawn timber on the farms (*Eucalyptus saligna* and *Grevillea robusta*). A third species, *Prosopis juliflora*, was added to the list to represent some of the difficult to saw hard wood species currently being promoted for planting in the dry areas in Kenya. Mature trees of *Eucalyptus* and *Grevillea* species were sampled from farms, while Prosopis trees were obtained from a research plot in a dryland district. Due to scarcity of mature trees on farms as a result of high demand, the oldest available Grevillea and Eucalyptus trees with relatively good stem form and minimum defects were sampled and felled using a chainsaw.

A well maintained chainsaw, fitted with a new chain and sharpened appropriately was used in chain sawing. Similarly, for bench sawing, a locally fabricated saw bench, driven by a tractor was used (Table 1). A trained and experienced sawyer operated the chainsaw and also acted as the lead sawyer for the bench saw. Two operators with over 5 years experience in pit sawing were sourced from the field to operate the pit saw system during the study.

Timber sawing and data collection

After each tree was felled, total tree height was measured from the bottom to the tip of the crown using a measuring tape. The total merchantable bole length from the butt to the first or second forking point depending on the stem characteristics was taken to determine the length of the useful portion of the tree. The trees were then randomly sampled and three trees marked for each sawing system.

Each tree in the respective group was crosscut into logs of lengths that are acceptable in the timber market while minimizing on defects. Logs with the smallest diameter and length that a particular sawing system can economically convert were chosen and the rejected ones set aside. The volume of the logs obtained in each case and volume of the rejected ones were determined to evaluate the log recovery for each sawing system.

Diameter measurements were taken at the butt, top and at every 0.5 m along the length of the log using a diameter tape and used to compute the mean diameter for each log. Log lengths were measured to the nearest 0.1 of a meter using a measuring tape. Log volume was then computed using Smalian Formula (ISO, 1983); (Equation 1) where V_L is the merchantable log volume, L is the merchantable log length, D is the log mean diameter and π is a constant.

$$V_L = \pi D^2L/4 \quad (1)$$

All timber sawing was done on the felling site in order to minimize cost. Monkey jacks were used in turning the logs to facilitate processing. A stop watch was used to time the operations. All timber sawing involved the through and through method and produced timber of nominal market dimensions. The volume of the resultant sawn timber was computed as shown in Equation 2, where b is timber breadth, d is timber depth, l is timber length (ISO, 1983). System timber recovery rate ($R\%$) was computed by considering the relationship between log volume (V_L) and volume of the resultant sawn timber (V_T) as shown in Equations (3).

$$V_T = \Sigma\ (bdl) \quad (2)$$

$$R\% = 100V_T/V_L \quad (3)$$

For both chainsaw and bench saw systems, the fuel tank capacities were recorded. At the beginning of each sawing operation, the machine fuel tank was filled to capacity. The machine was operated until all the fuel was used then refuelled and the records updated accordingly. After sawing, the tank was emptied and the balance re-measured to determine the volume of fuel used in litres to the first decimal place. Pit sawing did not involve measurement of fuel since the system is operated manually.

Three pieces of sawn timber from every log sawn using each system were randomly sampled. On each of these pieces, timber thickness was measured at the ends and at every 0.5 m along the length. This data was used to compute the mean size deviation from the pre-set dimension for the timber. Three other pieces were

Table 1. Specification for the sawing systems used.

Sawing system	Make/ model	Engine power (HP)	Fuel type	Saw type and Kerf (mm)
Chain saw	Husqvarna 365	4.6	Petrol/oil mix	9.5 (chain)
Bench saw	MF 135 Tractor (3 years old)	20	Diesel	7.5 (circular saw blade)
Pit saw	N/A	Manual	Manual	3 (flat blade)

randomly sampled from each of the species and tested for wood density at 12 to 15% moisture content.

Economic analysis

A simple economic analysis of the sawing systems was performed using the generated data, prices of production inputs like fuel and lubricants, labour and trees. Data generated based on different log diameters and shapes for the different sawing systems and timber species. Further still, analysis was focused on the specific effects of timber species and the size of sawn timber, owing to their effects on timber production, recovery rates and therefore to the cost of production.

Data analysis

The data were analyzed using Microsoft Excel and SPSS software. The effects of the experimental factors; wood species and sawing systems, both in isolation and in combination; on log recovery, sawn timber recoveries, fuel consumption rate, sawn timber production rate and timber size deviation were determined. Mean comparisons were performed for individual factors in isolation and in combination at 95% confidence level.

RESULTS AND DISCUSSION

Log conversion characteristics

Individual species log conversion rates were found to significantly vary with the sawing system used and had the chainsaw system recovering more logs than the other two systems (Table 3). The higher log conversion rate of the chainsaw system can be attributed to the ability of the chainsaw to process smaller log diameters (15 cm) where other systems could not.

High log conversion rate by chainsaw system could be attributed to the ability of the system to convert very small diameter and poorly (crooked) shaped logs than the other two systems (Samuel et al, 2007). In a study using different tropical hardwood species and various sawing systems in Ghana, Owusu et al. (2011), reported that chain sawing (as used in this study) recorded higher (mean 75%) log conversion rate than wood mizer system (68%). Similar results are reported in the same country by Marfo (2010). However, both studies reported that sawyers tended to leave out smaller diameter logs unprocessed due to abundance of wood in tropical forests unlike Kenya where trees are in short supply and sawyers aim at maximizing on the available materials.

The mean log conversion rate for all species sawn with pit saw was significantly lower than when sawn with the other two systems. This is attributed to the mode of operation for pit sawing. The log being sawn must be at least 2.4 m long to accommodate one of the operators to stand on one end while sawing the other end.

Low log conversion rate observed for *P. juliflora* and partly *G. robusta* was attributed to the species stem characteristics. The *P. juliflora* trees used for this study were sourced from a plantation that was planted in the 1980s for drought mitigation in the dry lands. These trees lacked management and grew into bushes with a few dominant trees towering above the rest. The stems were therefore small diameter, crooked and branchy, thus lowering the merchantable bole length and diameter. *G. robusta* trees from farms similarly lacked management attention due to farmers' limited knowledge on tree growing. This resulted into short merchantable bole length due to poor pruning and thin top logs compared to Eucalyptus, which is self pruning and develops straight and more or less cylindrical stems.

Sawn timber recovery

Timber recovery differed from system to system. Chainsaw system consistently recorded the lowest mean timber recovery for all the three species (Table 4).

Among the tree species, *E. saligna* recorded the highest mean sawn timber recovery for the respective sawing systems, while *P. juliflora* had the lowest timber recoveries. Similar trends were observed when the other sawing systems were used for the respective species. This could be attributed to the log characteristics (form and diameter size) of the respective species.

These results however differ from on-farm timber recovery results reported by Frimpong-Mensah (2004), which reported timber recovery rates ranging from 22 to 28% for a variety of hard wood species sawn using chainsaws. A related study (Gyamfi and Adu, 2009) reported on-farm recovery rates ranging from 28 to 45% for the same species sawn using small-scale circular saw mills, which is within the range reported for bench saw in this study. Such variations may be due to differences in sawyer skills and experience. In a study in Kenya, Samuel et al. (2007) pointed out that operators' level of skill and experience significantly contributed to both recovery and surface quality of sawn timber, although it is not clear whether the logs used in that study had similar diameter and form as those used in the current study.

Table 2. Characteristics of the tree species used in the study (Density is at 12-15% moisture content).

Scientific name	Mean tree Age (years)	Mean tree height (m)	Mean merchantable bore length (m)	Mean diameter (dbh) (m)	Mean density (kg/m^3)
Eucalyptus saligna	25	16.72	12.54	0.36	639.67
Grevillea robusta	20	11.70	7.35	0.29	529.99
Prosopis juliflora	27	6.67	2.66	0.21	864.67

Table 3. Log conversion rate for the sawing systems ($p<0.05$).

Sawing system/species	Percentage Log Conversion Rate			
	Eucalyptus saligna	*Grevillea robusta*	*Prosopis juliflora*	Mean
Chain saw	84.16	76.72	66.41	75.76
Bench saw	73.95	63.43	57.95	65.11
Pit saw	72.35	62.10	53.34	62.60

Table 4. Timber recovery for the three sawing systems ($P<0.05$).

Sawing system	Timber Recovery (%)			Mean
	E. saligna	*G. robusta*	*P. juliflora*	
Chainsaw	32.43	29.85	27.84	30.04
Bench saw	38.15	36.45	33.09	35.90
Pit saw	42.38	38.85	38.03	39.75

Timber production rate

The volume of sawn timber produced per unit time of system operation varied from system to system. Chain saw system consistently recorded the highest mean sawn timber production rates per operator for all the three species (Table 5). The bench saw system emerged the fastest among the three sawing systems and thus had the highest timber production rate. However, it should be noted that the system used at least six laborers unlike chain saw and pit saw systems which required two operators each. On conversion of the production rates per man-hour, chain saw system performed better except for *E. saligna* timber. Pit saw recorded the lowest sawn timber production rate for all the three species sawn. This was attributed to the manual operation of the system. High timber production rate achieved with *E. saligna* for all the sawing systems could be attributed to large diameter and good stem form as well as moderate density (640 kg/m^3), which makes it relatively easy to saw. Timber production rate for *G. robusta* was lower than *E. saligna* despite its lower density (527 kg/m^3). Small diameter logs with poorer stem form associated with *P. juliflora* wood as well as its high density (865 kg/m^3) may have been responsible for the low sawn timber production rate. An earlier study in Kenya reported similar trends for *E. saligna* and *G. robusta* wood sawn using chainsaw (Samuel et al., 2007). In Nigeria, Popoola (2010) reported that timber production rate and recovery increased with increase in log diameter and stem quality for a variety of species.

Fuel consumption

Fuel consumption by chain and bench saw systems differed for different wood species with bench saw recording higher mean fuel consumption (Table 6). Pit saw system was manually operated, therefore did not use fuel.

These differences in fuel consumption rate follow the same pattern shown by wood density for the respective species (Table 2). More fuel was consumed in sawing high density wood than when sawing lower density species. The rate of fuel consumption for sawing *P. juliflora* and *E. saligna* with chainsaw system however did not differ significantly although the species densities were significantly different. These results are similar to those reported by Owusu et al. (2011). High density wood poses higher resistance to sawing tools and therefore requires more fuel to drive the cutting tools through it (De Lasaux et al., 2004; Damnyag and Darko, 2009).

Size deviation

Timber size differed among wood species and among sawing systems (Table 7). On the overall, chainsaw recorded the highest mean timber size deviation.

Table 5. Timber Production rates for the three sawing systems as used to process different tree species ($P<0.05$).

Sawing system	Timber production rate (m^3/h)			
	E. saligna	*G. robusta*	*P. juliflora*	Mean
Chainsaw	0.21	0.24	0.16	0.20
Bench saw	0.26	0.18	0.12	0.19
Pit saw	0.09	0.08	0.07	0.08

Table 6. Fuel Consumption rates for the three sawing systems ($P<0.05$).

Sawing system	Timber recovery (%)			Mean
	E. saligna	*G. robusta*	*P. juliflora*	
Chainsaw	32.43	29.85	27.84	30.04
Bench saw	38.15	36.45	33.09	35.90
Pit saw	42.38	38.85	38.03	39.75

Table 7. Size deviation in timber sawn using different sawing systems ($P<0.05$).

Sawing system	Timber size deviation (mm)			Mean
	E. saligna	*G. robusta*	*P. juliflora*	
Chainsaw	5.26	5.33	5.99	5.53
Bench saw	3.54	3.20	3.90	3.55
Pit saw	2.42	2.20	2.70	2.44

This could be attributed to the mode of operation (freehand), machine vibration due to its engine characteristics (2-stroke), removed depth gauges and machine weight (8 kg). Timber size deviations for bench saw could be attributed to inconsistencies due to manual in-feeding of the logs during sawing. Low mean size deviation for pit saw system could be attributed to the mode of operation of the system.

Pit saw is operated by two operators, pushing and pulling the saw blade up and down, following lines drawn on the surface of the log being sawn. The operator standing on top of the log directs the saw blade to avoid wavering. Since the operation is manual, the cutting speed is slow and saw vibration is minimal. These reduce possibilities of the saw deviating from the pre-marked cutting line. This improves timber recovery and lowers size deviation.

Among the wood species, *P. juliflora* recorded the highest timber size deviation, while *G. robusta*. Although the mean size deviations did not differ significantly, there was a clear trend which showed that timber size deviation increased with wood density for all sawing systems. This could have been a factor of harder wood causing saws to deviate from the sawing line as a result of friction of the cutters as they bite into the wood. Other inherent differences in timber properties could also have contributed to variations in timber dimensions from the pre-set sizes. Timber defects like knots for example have been shown to interfere with smooth cutting of wood due to the change in density and grain direction around the knot area (Fehr and Pasiecznik, 2006).

Economic analysis

The cost of production per cubic meter of sawn timber differed among the sawing systems. Chainsaw system recorded the lowest mean production and the highest mean margin (Table 8). The profit margins obtained for each sawing system differed among timber species. Bench saw system performed well with large diameter and good form logs of *Eucalyptus saligna* wood. This provided more wood for bench saw to process within a shorter time and therefore more economical. The system however performed poorly with *P. juliflora*, a factor of both small diameter and high density. Chain saw system recorded higher profit margins when used to saw *G. robusta* and *P. juliflora* than bench saw. This was as a result of the system ability to process both low quality and high density logs better than bench saw. Pit sawing recorded the lowest profit margins. This was due to the long time engagement of labour to process logs manually.

With such returns, bench saw system can be more

Table 8. Mean Production cost, income and margins for *Grevillea robusta* timber sawn with the different sawing systems.

System	Production cost (m^3 *US$*)			Income (m^3)			Profit margin (m^3)		
	G. robusta	*E. saligna*	*P. juliflora*	*G. robusta*	*E. saligna*	*P. juliflora*	*G. robusta*	*E. saligna*	*P. juliflora*
Chain saw	65.23	74.02	116.12	103.20	120.35	153.18	37.97	46.33	37.06
Bench saw	76.55	67.59	132.67	103.20	120.35	153.18	26.65	52.76	20.51
Pit saw	81.24	86.15	141.24	103.20	120.35	153.18	21.96	34.2	11.94

*1US$ = Ksh 81.00.

profitable with large diameter logs of good form, which provide more wood for processing within a shorter time. Chainsaw system would be more preferred for smaller diameter logs of poorer quality. The system can also be good for the hard-to-saw tree species, particularly those dry land species with high densities.

Conclusions

The higher timber production rate for large diameter logs and low size deviation obtained with the bench saw suggests that this system would be appropriate for timber producers on farms. However, due to the high fuel consumption and labour requirement, the bench saw system becomes less economical than chainsaw, and particularly for small diameter logs and in situations where trees are few and scattered as in the case on farms today.

Chain sawing system shows higher profit margins when used to process smaller diameter logs. This makes chainsaw system a better choice for timber processors on farms. However, the low timber recovery and non-uniform timber sizes, associated with this sawing system make it a poor performer, especially when timber is desired for specialized uses. Unless a better sawing system is developed to replace chainsaws, there is need to improve timber recovery and timber size uniformity to make chainsaw more acceptable, since the system is the most commonly used for processing timber on farms despite its shortfalls.

Despite a relatively higher timber recovery rate compared to other systems, pit sawing was the most uneconomical amongst the three timber sawing systems. The pit sawing system is also not suitable for processing short and small diameter logs because of the potential risk of injury to the saw operators. It is therefore no a surprise that pit saw system is being replaced by chainsaw system in recent years, possibly due to the combined effects of poor economic performance and decreasing size of logs.

ACKNOWLEDGEMENTS

The research results reported in this paper were generated with funding from Kenya Forestry Research Institute under Farm Forestry-GoK research grants. The authors are grateful for the support, which has been maintained over the last 5 years. They specifically expresses their thanks to David Munene, the KEFRI machine operator for his dedication to work during the entire process of data collection. Additionally, pit saw operators and all other persons who participated in the study are acknowledged for their input.

REFERENCES

Damnyag L, Darko Obiri B (2009). Investment in Chainsaw milling compared to improved chainsaw milling in eight forest districts of Ghana. EU-Chainsaw project. Tropenbos International.

De Lasaux MJ, Spinelli R, Shelley JR, Hartsough RB (2004). Using a small-log portable sawmill to reduce fuel reduction treatment cost on small parcels. USDA Forest Service Economic Action Program, 01-DG-11051150-021.

FAO (2010). Global Forest Resources Assessment. The Forest Resources Assessment Programme. Country report, Kenya. FRA 2010/107. Rome, 2010.

FAO (2005). Realizing the economic benefits of agroforestry: experiences, lessons and challenge. In *State of the World's Forests 2005*, Rome. pp. 88–95.

Fehr C, Pasiecznik NM (2006). The potential of chainsaw milling for improving rural livelihoods in Uganda and DR Congo. Policy brief. HDRA, Coventry, UK. P. 2.

Frimpong-Mennsah K (2004). Chainsaw lumber production is wasteful, fact or myth? National workshop on Chainsaw lumber production in Ghana, A necessary evil? Tropenbos International. P. 4.

Gyamfi R, Adu GA (2009). A pilot study on sawn wood conversion efficiency in selected sawmills in Ghana. National workshop on Chainsaw lumber production in Ghana, A necessary evil? Tropenbos International. P. 6.

Holding-Anyonge C, Roshetko JM (2003). Farm-level timber production: orienting farmers towards the market. Unaslyva 212 9540:48-56.

Holding C, Njuguna P, Gatundu C. (2001). Farm sourced timber: the restructuring of the timber industry in Kenya – opportunities and challenges. Forest Extension, IUFRO, Vienna, Austria.

ISO (1983). Measurement of log sizes and determination of volume. International Standards Organization.

Popoola L (2010). Chainsaw milling in Nigeria. (in), Wit, Marieke and Jinke van Dam (eds.). (2010). *Chainsaw milling: supplier to local markets.* European Tropical Forest Research Net work. Tropenbos International, Wageningen, the Netherlands. pp. 22-226.

Marfo E (2010). Chainsaw milling in Ghana: Context, drivers and impacts. (in), Wit, Marieke and Jinke van Dam (eds.). (2010). *Chainsaw milling: supplier to local markets.* European Tropical Forest Research Net work. Tropenbos International, Wageningen, the Netherlands, pp. 22-226.

Muthike GM, Shitanda D, Kanali C L, Muisu FN (2010). Chainsaw milling in Kenya. (in), Wit, Marieke and Jinke van Dam (eds.). (2010).

Chainsaw milling: supplier to local markets. European Tropical Forest Research Net work. Tropenbos International, Wageningen, the Netherlands, pp. 22-226.

Pasiecznik N (2010), Sawmilling with chainsaws: a technical overview. (in), Wit, Marieke and Jinke van Dam (eds.). (2010). *Chainsaw milling: supplier to local markets.* European Tropical Forest Research Net work. Tropenbos International, Wageningen, the Netherlands. xxii + 226pp.

Owusu FW, Appiah JK, Damnyag L, Blay D (2011). Comparative analysis of recovery efficiencies of some milling techniques in Ghana. Ghana J. Forest. 27:87-100.

Pasiecznik NM, Brewer MCM (2006). Chainsaw milling, an appropriate technology to increase benefits from on-farm trees. Demonstration, poster and presentation. In: Benefits of Farm Woodlands and Agroforestry in UK Lowlands. Third Annual Meeting of the Farm Woodland Forum, 27-28 June 2006. Northmoor Trust, Oxford, UK.

Samuel JH, Pasiecznik NM, Fehr C (2007). The Potential of Chainsaw Milling Outside Forests. Summary report with economic and policy case studies from East Africa. HDRA, Coventry, UK. P. 66.

Wit M, Jinke van Dam (eds.). (2010). *Chainsaw milling: supplier to local markets.* Tropenbos International, Wageningen, the Netherlands. Xxii + 226 pp.

29

Innovative approach for assessing sustainability of the medicinal plant-*Gardenia gummifera* Linn. F.

Dolly Parmar Kotwal

Government Post Graduate College, Barwani, M. P. India.

Less known medicinal plants were studied in Bhopal and Sehore districts, M.P. India to suggest appropriate strategies for their sustainable management. The study area was rich in medicinal plants but due to their extensive uses and habitat degradation, these were less available. An innovative approach has been developed to assess the status of certain less known medicinal plants among which *Gardenia gummifera* was one. The approach comprise of six categories and 25 parameters namely: habitat (5 parameters), ecology (3 parameters), biology (6 parameters), use/trade (3 parameters), collection (4 parameters), and legal and institutional (4 parameters). These were arrived after study of relevant literature, discussion with the knowledgeable persons and field situation in the study area. All the parameters have been ranked into three scoring levels namely: low, medium and high with scoring of 1, 2 and 4 marks respectively. The species was studied and ranked as vulnerable based on these parameters. The primary source of data was field observations on the selected species and interviewing the stakeholders' that is, herbal practitioners, primary collectors, dealers and manufacturers. Attempts were made to know the degradation of natural habitats, biotic disturbances and exploitation. *G. gummifera* scored minimum marks (46%) which indicate its rareness in the area and over exploitation. Attempts have been made to suggest strategy for the conservation and sustainable utilization including training to build capacity of primary collectors.

Key words: Assessment, categories, parameters, threatened, medicinal plants, value addition.

INTRODUCTION

The people in India have tremendous passion for medicinal plants. In the oral tradition, the local communities in every ecosystem right from the Trans Himalayas down to the coastal plains have discovered the medicinal uses of thousands of plants occurring in their surroundings. They use them for a wide range of health related applications from common cold to memory improvement, treatment of poisonous snakebites as cure for muscular dystrophy and enhancement for general immunity of the body.

There are several traditional rituals (ethno-botany) highlighting the importance of certain plants. On festivals the flowers or fruits are offered to the Gods or Deities. Some of these are- Paradise flower, Plants of Bible, worship of Basil (Tulsi) or Lakshmi, the Goddess of prosperity on Lotus. Yet ethno-botany is generally taken as science of relationship of only primitive or aboriginal people with plants (Jain and Mitra, 1997). All traditional systems of medicine had their roots in ethno-botany (Jain and Mitra, 1997). The word ethno botany was applied to such knowledge by Harshberger (1895). India probably has one of the richest plant medicine cultures in the world. It is a culture that is of tremendous contemporary relevance because, it can on one hand ensure health security to millions of people and on the other it can provide new and safe herbal medicines to the entire world.

There are around 25000 plant based formulations used in folk medicine known to rural communities over India. Around 10000 formulations are available in indigenous medical texts, many of which are yet to be sufficiently tapped (Basu and Kirtikar, 1918). The science of traditional medicine flourished in India for quite a long time yet for the while it was subdued under the impact of modern medicine. The coming of chemical revolution in world and boom of synthetic products including the synthetic medicines, the faith in and the popularity of traditional herbal medicines gradually declined but now it is upcoming.

ASSESSMENT OF THREATS ON MEDICINAL PLANTS

Detailed studies on the distribution and utilization pattern of selected medicinal plant species was done in Sehore and Bhopal Districts of Madhya Pradesh, India. Efforts have been made to prioritize medicinal plants of conservation concern beginning with south India, and subsequently for northern and central India also, using conservation assessment and management planning (CAMP) methodology for assessment of threat status in accordance with the IUCN red list categories. The result of these CAMP workshops for South India enlists about 112 medicinal plant taxa under various degrees of threats ranging from "threatened" to "extinct in the wild". Similar threat assessment exercises were carried out for Northern India's medicinal plant species also. The CAMP workshops for high altitude Himalayan medicinal plants of Jammu and Kashmir and Himanchal Pradesh have assigned Red list categories to 42 taxa of temperate region.

The 1997 IUCN red list of threatened plants published by the IUCN Species Survival Commission presents a shocking picture of nearly 34000 species, or 12.5% of the world's flora, facing extinction. Using the same proportion for recorded medicinal plants, India's nearly 10000 native medicinal plant species are expected to fall the into threatened category. This indicates that special conservation measures are required for these plants failing which these are likely to become extinct and the world will be deprived of their usefulness.

The causes of decline

The cause of decline in the distributional area and the population of these plants may be natural such as extreme climatic conditions like drought, flood, cold, hot, diseases, problems in regeneration etc. More useful species are over exploited and become threatened.

Nearly 90% of medicinal plants used by the Indian Industry are collected from the wild. More then 70% of the collections involve destructive harvesting from the wild use of the parts like flowers, fruits, seeds, roots, bark, wood, stem and the entire plant (herbs). These pose definite threats to the sustainable survival of medicinal plants (Kotwal and Bhattacharya, 1999).

Due to popularity, drug companies (under the license) manufacture many of the Ayurvedic medicines. As a result there is increase in global demands for natural products over the last two decades or so. The entrepreneurs were under pressure to meet the demands for which they resorted to unsustainable practices. These include premature harvesting, destructive harvesting, over harvesting, lack of quality control, compromise on quality of raw material etc. All these contribute in making the Ayurveda medicinal practice un-popular.

The trade in medicinal plants

The demand for the raw material is much more than the availability. Due to either unavailability or less availability of some of the medicinal plants, which either have become extinct or have become rare, their cost has become quite high. The manufacturers find it difficult to use such species in formulation as the price of the medicine manufactured from rare/costly plant would be high. This compels the manufacturers to add a substitute, which is not prescribed in the texts. The substituting material may not be beneficial and at times may be harmful. Adulteration makes the medicines either less effective or ineffective (Ram, 2000).

Though Ayurveda specifies usage of 1200 to 1500 medicinal plants, contemporary Ayurvedic Industry consume around 550 to 600 medicinal plants, out of these around 90% are procured from wild areas mostly the forest land. Another category of medicinal plants consists of species which grow in wild as well as in agriculture lands. Some species are grown commercially on large scale to meet the demand of Ayurvedic industry. Medicinal plants procured from cultivated fields amount for less then 10% of total medicinal plants in trade.

The Government of India banned export of some 29 species of medicinal plants. Various studies being done over a period has indicated risk of extinction of valuable species due to high industrial demand. The results of a series of conservation assessment and management planning workshops (CAMP) held in various parts of the country to assess the threat status of important medicinal plants are alarming. Several species were found in endangered category. This signifies that the distribution range and population of most of the useful plants have critically reduced and if conservation efforts are not strictly implemented then these are likely to become extinct. The turnover of the Indian herbal drug industry was Rs. 2300 crore during 1997 and reached around Rs. 4000 crore by year 2000. The total estimated quantity of exports was 36200 tones in the year 2000. It is estimated that this can be raised to Rs. 3000 crore by 2005 and to Rs. 10000 crore by 2010 as per estimates of Planning Commission (2000). At present the share of India in world Herbal Market is quite insignificant due to the lack of organizations both at cultivator / collector as well as industry level (Bisen, 2001).

Sustainable harvesting

Sustainability is very important aspect particularly in production and harvesting of medicinal plants. Most of the medicinal plants growing in nature are over exploited. Without doubt some of the high value herbs are threatened with extinction in the wild. There are several medicinal herbs, which are considered as economically profitable in harvesting from the wilderness. Some are relatively abundant, whereas many others are moderately available and few important species are concentrated only in certain pockets, which are mostly threatened. Several species are in high demand by the traders. Competition has resulted in the early and over harvest of few species from the more accessible sites. The bulk is picked when plants are only 2/3 or even half of their potential age and size. The optimal time of harvest for most species is immediately after the seeds reach maturity. Harvesting of mature plants in appropriate extent and by proper method would result in more yield of better quality. Despite these benefits, collection frequently takes place much earlier. Competition between collectors is one of the main causes of pre mature harvesting on open access land that is, forest. Large scale harvesting before maturity reduces the size of population (density) of exploited species.

The study site

The study was conducted in Bhopal and Sehore districts of Madhya Pradesh, India. Bhopal is capital of the state and situated on 23° 16' north latitude and 77° 25' east longitude. The total area of the district is 2778.0 sq. km and the reported human population is 1836784. Sehore is the parent district of Bhopal, which was formed much later, then its formation as the state capital. It is about 35 km West of Bhopal. The area of Sehore district lies between latitudes 23° 33' to 23° 38' 52' N and longitudes 76° 36' to 76° 59'. The total area of the district is 6563.68 sq. km. Both districts have tropical dry deciduous forests (Gazetteer, 1999).

Objectives of the study

1. To study the traditional uses of less known medicinal plants.
2. To suggest appropriate strategies for research development and sustainable management.

METHODOLOGY

At the initial stage it was very difficult to precisely decide the names of less known plants of the study area from which detailed studies were made.

A list of important medicinal plants of the study area was prepared. This was based on frequent field visits, study of literature (Hains, 1916) and discussion with herbal practitioners. Thereafter the species for detailed studies were selected. The criteria for selection were based on occurrence of the species in different habitats, their importance and relatively less known for traditional uses in the area. The web site of Google (Internet web site) was searched on the internet for all the important medicinal plants of the study area. The number of searches was categorized in three categories namely:

1) Less known = up to 1000
2) More known = 1001 to 3000
3) Most known ≥ 3000

The "rapid vulnerability assessment" (RVA) method (Jennifer et al., 2001) collects information to identify species, resources and sites that may be at risk of over exploitation. Although it was very difficult to elicit the required information from the local herbal practitioners, still attempts were made to get some information on various aspects of relatively less known medicinal plants of the study area. A questionnaire and data format was specially developed for the study covering following aspects: Local, vernacular and botanical names, description, geographical distribution, habitat suitability and habitat specificity, biotic disturbances, distribution and abundance of selected medicinal plant, growth, flowering, fruiting, regeneration, pests and diseases, use and trade, collection related aspects, legal and institutional aspects. Suggestions are given for research development and sustainable management. For assessing sustainability of medicinal plant, the following six categories were recognized (Table 1).

A. Habitat - 5 parameters (Habitat suitability, habitat specificity, climate (rainfall, temperature), biotic disturbances (Grazing, fire), protection level).
B. Ecology - 3 parameters (Occurrence, distribution and abundance).
C. Biology - 6 parameters (Growth, flowering, fruiting, seeding, regeneration, pests and diseases.
D. Use/trade - 3 parameters (Plant parts used, other uses (wood, tannin, fiber etc.), Use/trade).
E. Collection - 4 parameters (Collection stage, collection extent, collection method, collecting people).
F. Legal and institutional - 4 parameters (Legal protection, institutions (Societies), awareness, capacity building).

Thus a total of 25 parameters were identified covering all aspects to assess the sustainability. These were arrived after study of relevant literature, discussion with the knowledgeable persons and field situation in the study area. All the parameters have been ranked into three levels of scoring namely: low, medium and high with

Table 1. Parameters for assessing sustainability of *Gardenia gummifera*. Scoring: Low = 1, Medium = 2 and High = 4.

S/N	Parameters	Low (1 mark)	Medium (2 marks)	High (4 marks)
A. Habitat				
1	Habitat suitability	Less suitable	----	----
2	Habitat specificity	Specific	----	----
3	Climate (Rainfall, temperature)	----	----	Normal
4	Biotic disturbances (grazing, fire)	---	Frequent	----
5	Protection level	----	Good	----
Sub-Total		2	4	4
Habitat 10 marks out of maximum 20 (50.0%)				
B. Plant ecology				
6	Occurrence	Wild	----	----
7	Distribution	Fragmented	----	----
8	Abundance	Scanty	----	----
	Sub-Total	3	----	----
	Scoring 3 marks out of 12 (25.0%)			
C. Plant biology				
9	Growth	Poor	----	----
10	Flowering	----	Good	----
11	Fruiting	----	Good	-----
12	Seeding	Poor	----	-----
13	Regeneration	Poor	----	----
14	Pests and diseases	----	----	Few
Sub- Total		3	4	4
Scoring 11 marks out of 24 (45.8%)				
D. Use/Trade				
15	Plant parts used	----	Gum –Bark , terminal buds	-----
16	Other uses (Wood, tannin, fiber etc.)	----	----	1use
17	Use/trade	----	Localised	---
Sub-Total		----	4	4
Scoring 8 marks out of 12 (66.7%)				
E. Collection				
18	Collection stage	----	Sub- Mature	----
19	Collection extent		----	Least
20	Collection method	----	Partly destructive	----
21	Collecting people	----	Un-Organised	---
Sub-Total		----	6	4
Scoring 10 marks out of 16 (62.5%)				
F. Legal, institutional				
22	Legal protection	Poor	----	----
23	Institutions (Societies)	Few	----	----
24	Awareness	Scanty	----	----
25	Capacity building	Least	----	----
Sub-Total		4	----	----
Scoring 4 marks out of 16 (25.0%)				

scoring of 1, 2 and 4 marks respectively. The species were studied and ranked according to these parameters. Based on these considerations 10 species were selected for detailed study among which *Gardenia gummifera* was one.

***Gardennia gummifera Linn. F.* (Family Rubiaceae)**

Other names for *G. gummifera Linn. F.* are *Hing patri*, *Dikamali*, *Gurudu*, *Cambi gum tree*, *Kamarri*, *Manchi bikki*, *Tella – manga*, *Kambili ppicin*, and *Cittubikke* for Sanskrit, Hindi, Oriya, English, Gujarati, Telgu, Tamil and Kanada, respectively. Distribution of *G. gummifera Linn. F.* occurs throughout India in Tropical Natural Forests. In Madhya Pradesh state it is one of the constituent species of natural forests and found in almost all the districts. In the study area it was observed in Samardha, Berasia, Budni and Bhopal forest ranges.

Habitat

The plant is a small tree or shrub and grows in rocky areas in slightly undulating terrain along with other species. It occurs in specific habitats having well drained shallow soil with exposed rocks at places usually on plateaus, which are not too dry. It has specific habitat requirements as mentioned previously. Normal range of maximum and minimum temperature and rainfall in the area is quite suitable for the species. The deficit rains has resulted in drying of some plants during peak of summer. Some plants were partly dried and re-sprouted during rainy season. The plant grows only in natural forests, which are mainly managed by Government. The biotic disturbance in the form of accidental forest fires, livestock grazing and illicit cutting of trees is of common occurrence in the study area. Since the species grows only in natural forests and therefore it gets some protection better then those growing in wastelands.

Plant ecology

Under this heading are three parameters of occurrence, distribution and abundance of the species are considered. The plant grows only in the natural forests and not cultivated. However it has a very good potential for growing in gardens because of its beautiful white fragrant flowers beside medicinal value for its gum after which it derives its scientific name. The distribution of the species is fragmented at least within the study area. It can be seen in specific areas only and not found for a considerable distance indicating its fragmented distribution. The extent of occurrence is scanty because it can be seen only at few places.

Plant biology

A large handsome shrub with white bark, sub-sessile shining simple leaves which are apposite and decussately arranged. Leaf size is 3 to 8 cm long, and 12 to 16 pairs of secondary nerves. Flowers are large (2 to 3 cm) with white fragrant. Fruit is ovoid (2 to 4 cm) with fleshy mesocarp, and is edible. Flowering occurs during June to July. The plant flowers every year during onset of monsoon rains. However sporadic flowering takes place during other months of the year. Fruiting is during August to October. The fruiting medium has 20 to 40 fruits per bush.

Seeding is medium in extent. Some fruits dry pre-maturely owing to very hot and dry weather conditions. The regeneration is usually by seeds, which are spread by birds that feed on the fruits. Some times mature fruits fall on the ground releasing seeds, which germinate under favorable conditions. Growth is slow. In the study area most of the plants were in bushy form.

Use/ trade

Only gum resin is medicinally useful which comes on the tip of the branches. Other parts of the plant do not have any specific use except the flowers. However, the local people along with other similar material collect the dried plants as fuelwood.

Collection

The gum resin at the tip of branches is collected by breaking the terminal bud usually by hand. In the study area the collection was negligible. In the event of collecting the gum resin, the terminal buds are affected. The gum resin from main stem is also collected.

The method of collection is partly destructive affecting the terminal buds and sometimes the main stem in case of collection on large scale. Only some local people sponsored by local herbal practitioners collect gum resin. Un-organised local people make collections. In the study area commercial collection does not take place probably because of less availability of plants.

Legal, institutional

The species has no specific legal protection as in case of other species listed in Wildlife (Protection) Act, CITES, negative list of export etc. However, the habitat has some level of protection because it grows in Government forest. Most of the people are not much aware about the importance and use of this species. There are no societies specifically for collection of this or any other medicinal plant. At present there are no efforts towards any capacity building measures for sustainable collection and utilization of medicinal plants.

Therapeutic properties and uses (Parts used- gum)

Local use

It is useful in excretion of intestinal parasitic worms. When given orally with local cow urine for hook worms, 1 g of gum + 6 g cumin + 20 g and cow urine excretes the worms.

Other uses

The gum locally known as Dinkamali (*G. gummifera*) is used to treat stomach ailments in humans. In the market it costs 50 rupees for 100 g. Dinkamali is also used as spray on soybean crops to combat *Heliothis armigera* and on vegetable crops such as cauliflower, tomato and cabbage to combat other small insects. The process of preparation is the same as for asafetida, except that the mixture is boiled for 25 to 30 min (CSIR, 1986) (Figure 1).

Gardenia flowers are not raised by the Indian floriculturists, although some species are found in gardens. The young shoots of *G. gummifera* are the source of Dikamali gum, used to rub on the gums of infants when teething. The gum is antiseptic, carminative, expectorant, sore, spasm, stimulant, vermifuge, repellant (Figure 2).

RESULTS AND DISCUSSION

Detailed observations on 25 parameters of medicinal plant are given in Table 2.

The species has scored 46 marks out of 100. It has scored fewer marks under habitat, plant ecology, collection, use and legal, institutional parameters based on the scoring it is ranked as vulnerable.

Habitat

Habitat is the home of living organisms. The medicinal

Figure 1. Gardenia gummifera – A plant.

Figure 2. Gardenia gummifera – A flower.

plants are affected by the habitat. The *G. gummifera* has scored 50% marks, which indicates the causes of it being so rare. Because of its medicinal values, it is in demand, due to which it is over exploited. It is difficult to meet the demand only from the wilderness. Cultivation of the species is the alternative but standard methods of commercial cultivation not readily available.

Plant ecology

A minimum of 41.7 marks were scored by *G. gummifera*. The species occurs only in wilderness areas and not cultivated in the study area. It has fragmented distribution and scanty occurrence in specific habitats.

Plant biology

The intrinsic factors within the plant body play important role in growth. These are flowering, fruiting, seeding, regeneration and resistance to pests and diseases. The species under reference was studied in respect of these parameters. *G. gummifera* scored 45.8% marks. The

Table 2. Scoring of marks for sustainability of *Gardenia gummifera* under various parameters.

Category	Parameters	Total marks	Marks scored	%
A	Habitat	20	10	50.0
B	Plant ecology	12	5	41.7
C	Plant biology	24	11	45.8
D	Use/Trade	12	8	66.7
E	Collection	16	10	62.5
F	Legal, institutional	16	4	25.0
Total	---	100	46	---

Ranking; Critical = 25, Vulnerable = 25 to 50 and Not vulnerable = > 50.

species has poor regeneration due to different reasons. The species occurs only in wilderness areas and not cultivated. Only few mature fruits fall on ground and the natural regeneration is poor.

Use/trade

Various parts of plants have different uses. In some species only one part is useful while in others more parts are useful. The species has scored a 66.7% marks due to use of only one part on local scale only.

Collection

Useful plants and plant parts are collected. There are three main aspects under this. These are stage of collection, extent of collection and method of collection. The score of the species is 62.5%, which is slightly more, then limits of vulnerability. It is because of less availability of the required material and collection at local level only.

Legal institutional

This is also a very important parameter that affects various aspects of the medicinal plants. Certain species that come under threatened categories of IUCN, in appendices of CITES, listed in negative list of export and in Wildlife (Protection) Act have legal protection with provisions of punishment. *G. gummifera* has no such legal status. The species has scored 25% marks due to no effective legal protection.

RESEARCH DEVELOPMENTS

The natural regeneration of the species is poor. It has slightly specific type of habitat requirements. Very little can be practically done in its natural habitat on these aspects except reduction of biotic disturbances:

1) The species has slow growth because of intrinsic factors. It produces few mature seeds, which are the only means of regeneration. The fruits are eaten by birds and thus an attraction for them. It does not propagate in nature by vegetative means. Looking to the medicinal importance and also aesthetic value the plant has the potential for growing in gardens and in that eventuality other means of propagation can be developed.
2) Some awareness can be created among the people about its use and importance. It will be worthwhile to organize the local people in the form of societies for sustainable utilization of medicinal plants.

Conservation and sustainable utilization

It is necessary to develop appropriate programme for conservation and sustainable utilization of medicinal plants particularly for the health care of rural poor and also to help in their livelihoods. For the state of Madhya Pradesh 35 species of medicinal plants and 16 species of aromatic plants has been prepared as priority list and a business plan has been developed for these species (Government of Madhya Pradesh, 2002).

Attempts have been made to suggest strategy for the conservation and sustainable utilization. Suggestions for training to build capacity of primary collectors in terms of time, methods and quantity to be collected. It has been observed that poor returns to dependant population also accelerate over exploitation of the useful species. Ethical approach may lead to sustainable utilization.

The medicinal plants are the principle health care resources for majority of the people living in rural areas. Their demand is increasing day-by-day. Most of the material is harvested from wilderness areas, mainly the forests. Only few species are cultivated. The unregulated collection and trade of medicinal plants has serious implications on the survival of many species, which are threatened. Beside, it is also important to develop appropriate harvesting standards based on scientific studies (FRLHT, 1997, 1999).

Techniques for commercial cultivation of prioritized species that are in high demand but short supply from wilderness also need to be developed and the technology should be transferred from lab to the land for wider use. There is need to emphasize *in-situ* conservation of medicinal plants and their sustainable harvesting involving local people. Some of the Joint Forest Management Committees may be motivated for this.

It is very important to impart necessary training to the concerned local people for technical issues pertaining to conservation, cultivation, value addition and marketing of medicinal plants. Value addition is very important aspect for fetching better returns to primary collectors. The values of processed products may increase by 8 to 10 times than the raw products.

Marketing a medicinal plant is ticklish due to poor storage, poor transport and lack of definite consumers/purchasers. Due to uncertainty of marketing, some medicinal plants are not collected. Some of the manufacturers do not prefer to purchase processed material due to apprehension of adulteration. The buyers dictate marketing.

The deterioration in quality leads to poor marketing. The manufacturers and other stakeholders in the sector need to ensure "quality standards" in collection, value addition, transport, storage and processing of end products to ensure the efficacy of Ayurvedic medicines. Attempts should be made to develop standards for certification of medicinal plants as "Organic or Natural products". Similarly, cultivation of medicinal plants should also be done without use of pesticides or chemical fertilizer so as to ensure their medicinal properties and standard quality.

Conclusion

The utilization of medicinal plants in the remote past was sustainable. The traditional ethics were followed which imposed several restrictions on time of collection, method of collection, extent of collection etc. The practitioners followed the traditional ethics, and as such, the natural habitats were maintained and the natural regeneration was ensured with continued availability of medicinal plants.

But in the recent times the traditional ethics are not followed possibly due to high demand and the greed. This has reversed the situation from plenty to scanty. This is the situation with respect to most of the natural resources. Therefore appropriate strategies covering aforementioned parameters need to be developed and implemented. The innovative approach of assessment of all parameters for each species is quite holistic covering all the aspects. This has the potential to assess weaknesses and gaps in the present position and practices with definite indications for sustainable management of medicinal plants.

Conflict of Interests

The author(s) have not declared any conflict of interests.

REFERENCES

Basu BD, Kirtikar KR (1918). Indian Medicinal Plants (Revised), Reprint 1984.

Bisen SS (2001). Marketing of forest based crude drugs and herbs. Mimeograph. Tropical Forest Research Institute (TFRI), Jabalpur.

CSIR (1986) Useful plants of India. Council of Scientific and Industrial Research, New Delhi.

FRLHT (1997). Medicinal plants of India. Guidelines for National Policy and conservation Programmes. FRLHT, Bangalore.

FRLHT (1999). The key role of forestry sector in conserving India's medicinal plants – concepts and operational futures. P. 26.

Gazetteer (1999). Bhopal and Sehore districts. Govt. of Madhya Pradesh.

Government of Madhya Pradesh (2002). Business plan for Medicinal and Aromatic plants of Madhya Pradesh. Dept. of Biodiversity and Biotechnology and M.P. State Forest Produce Federation. P. 48.

Hains HH (1916). Descriptive list of important trees, shrubs and economic herbs of Southern circle, Central provinces. The Pioneer Press, Allahabad.

Harshberger JW (1895). Some new ideas. Philadelphia Evening Telegraph.

Jain and Mitra (1997). Contribution to Indian Ethnobotany 3rd revised edition. Edtd. By S. K. Jain. Scientific Publishers India. Jodhpur. P. 322.

Jennifer LG, Wong K Thornber N, Baker N (2001). Resource assessment of Non-wood forest products- Experience and Biometric principles. F A O Rome.

Kotwal D (2002). Ethno-botanical research development of less known medicinal plants of Bhopal and Sehore Districts. Ph. D. thesis, Barkatullah University, Bhopal. India.

Kotwal PC, Bhattacharya P (1999). Management strategies for conservation, cultivation and utilization of medicinal plants. Reading material for Management Development programme (MDP). Indian Institute of Forest Management (IIFM), Bhopal, Madhya Pradesh.

Ram P (2000). Unsustainable extraction of Bio- diversity raises question on future of Indian system of medicines. IIFM New latter Vol. III (3), September 2000.

Comparative evaluation of the essential oil composition from the leaves and flowers of *Hyssopus officinalis* L.

M. Moghtader

Department of Biodiversity, Institute of Science and High Technology and Environmental Sciences, Graduate University of Advanced Technology, Kerman, Iran.

The chemical constituents of essential oil were obtained from the fresh leaves and flowers of *Hyssopus officinalis* L. (Lamiaceae) by hydro distillation method analyzed by gas chromatography and gas chromatography mass spectrometry. Hydro distillation method was used to extract the essential oil. Thirty-five compounds were identified, accounting for 92.13% of the total oil with 0.75% (v/w) oil yield in the essential oil of the leaves. The main constituents of the essential oil were Iso pinocamphone (38.47%), Pinocomphone (13.32%), n-decane (8.67%) and Pinocarvone (5.34%). Thirty-six compounds were identified, accounting for 98.68% of the total oil with 1.38% (v/w) oil yield in the essential oil of the flowers. The main constituents of the essential oil were Iso pinocamphone (40.25%), Pinocomphone (14.92%), n-decane (8.63%) and Pinocarvone (6.76%).

Key words: *Hyssopus officinalis* L., iso pinocamphone, volatile oil, gas chromatography mass spectrometry (GC/MS).

INTRODUCTION

Hyssop (*Hyssopus officinalis*, synonym *Hyssopus decumbens*) is a herbaceous plant of the genus Hyssopus native to Southern Europe, the Middle East, and the region surrounding the Caspian Sea. *H. officinalis* L. is an evergreen Shrub growing to 0.6 m (2ft) by 0.6 m (2ft in). This plant can be grown for ground cover when spaced about 45 cm apart each way. It is hardy to zone 7. It is in leaf 12-January. It is the in flower of July to September, and its seeds ripen from August to October. The flowers are hermaphrodite (have both male and female organs) and are pollinated by Bees. Hyssop is a brightly colored shrub or subshrub that ranges from 30 to 60 cm in height. The stem is woody at the base, a number of straight branches are grown from it. Its leaves are lanceolate, dark green in color, and from 2 to 2.5 cm long. During the summer, the plant produces bunches of pink, blue, or, more rarely, white fragrant flowers. These give rise to small oblong achenes. The species as a whole is resistant to drought, and tolerant of chalky, sandy soils. It thrives in full sun and warm climates. The plant is commonly used by beekeepers to produce a rich and aromatic honey. Herb hyssop leaves are used as an aromatic condiment. The leaves have a lightly bitter taste due to its tannins, and an intense minty aroma. It is used moderately in cooking due to its intensity. The herb is also used to flavor liqueur, and is part of the official formulation of Chartreuse. The plant also includes the chemicals thujone and phenol, which give it antiseptic properties (Van Wyk and Michael, 2004). A strongly aromatic flavor, somewhat like a cross between sage and mint, it has fallen out of flavor in the recent years. It has a positive effect when used to treat bronchitis and respiratory infections, especially where there is excessive mucous production (Chevallier, 2001). Hyssop can irritate the mucous membranes, so it is the best given after an infection has peaked, when the herb's tonic action encourages a general recovery (Chiej, 1984). The leaves and flowering tops are antiseptic, antitussive, astringent, carminative, diaphoretic, emmenagogue, expectorant, pectoral, sedative, stimulant, stomachic, tonic and vasodilator. It is commonly used as an aromatic herb and medicinal plant (Mills and Bone, 2005). The plant can be harvested when in full flower and dried for later use. A tea made from the leaves is used in the treatment of flatulence, stomach-aches, upper respiratory tract infections, coughs in children etc. A poultice made from the fresh herb is used to heal wounds. The essential oil is used in aromatherapy. This oil should not be used on the people who are highly strung as it can cause epileptic symptoms. The essential oil should not be used internally except under professional supervision.

Hyssop can be grown as a dwarf hedge, it responds well to trimming in the spring. The growing plant attracts cabbage white butterflies away from brassicas. Another report says that hyssop attracts cabbage white butterflies and should not be grown near cabbages. An essential oil from the leaves is antiseptic and also used in perfumery and as a food flavoring (Huxley, 1992). It has a particularly fine odor and is much valued by perfumers. Average yield of the oil is about 0.6%. Yields from the blue-flowered variety are circa 1 to 1.5% of essential oil, the red-flowered variety yields about 0.8%, whilst the white-flowered form yields 0.5% of essential oil. The plant was formerly used as a strewing herb (Huxley, 1992) and is also used in pot-pourri. A tea made from the leaves is useful for controlling bacterial plant diseases. The essential oil of hyssop is widely used for food, pharmaceutical and cosmetic industries throughout the world. Therefore, it is very important to know the chemical characteristics of the oil for economic use and enhanced performance of the end products. This study was carried out to determine the essential oil of *H. officinalis* (L.) (Lamiaceae) collected from wild in the Khalkhal, Southeast Ardabil Province in Iran. The essential oil of *H. officinalis* plant has been studied in Iran and other countries but the chemical composition of the essential oil of *H. officinalis* has not been determined in Khalkhal, Iran. In the present work we have analyzed the chemical composition of the leaves and the flowers of *H. officinalis* L. that grow in Khalkhal and then the results were compared with various origins in other countries.

MATERIALS AND METHODS

Plant material collection and isolation of their essential oil

The leaves and flowers of *H. officinalis* were obtained from Khalkhal in Ardabil Province from 1843 m height, at full flowering stage in June 2012. The samples were cleaned in the shade condition to prevent volatility of the plant material constituents and to keep the natural color of the sample fixed. Then they were air-dried and powdered using a milling machine and kept in a cool dry place until ready for extraction of the essential oil. Afterwards, essential oil was extracted from 150 g of the powdered sample using hydro-distillation method with the help of Clevenger set for three hours. Following, the oil samples were dried using anhydrous sodium sulfate and kept in sterile sample tubes in refrigerator. The oil yields of leaves and flowers were calculated.

Analysis of essential oil

Gas chromatography

GC analysis was performed using a model of HP-439 gas chromatograph equipped with column CP Sil. 5CB in 25 meters length, internal diameter of 0.25 mm and film thickness 0.39 µm. Oven temperature was between 60 and 220°C at a rate of 7°C slope per minute. The injector temperature was 280°C and detector (FID) temperature was 270°C while the carrier gas was helium.

Gas chromatography/mass mass spectrometry

The chromatograph gas set was attached to a Mass Spectrometry to analyze and identify the combinations forming the essential oil, Model Hewlett Packard-5973 was used for this purpose. The conditions of analysis and specifications of the GC/MC set were as follows: Capillary column HP 5MS in 60 meters length, internal diameter of 0.25 mm and layer thickness of 0.25 µm, thermal program of oven (3 min) in 60°C, then 60 to 220°C with a 6°C slope per minute, then 3 min in 220°C, the temperature of place of injection 280°C, gas conveying helium, the speed of gas move 1.0 milliliter per minute, the ratio of fission 1 to 43, the rate of injection 0.1 µl, temperature of the reservoir of ionization 230°C, ionization mode EI, with an Ionization energy of 70eV. The series of normal Alkans C8-C17 were also injected to the set under the same condition with that of essential oil injection to calculate Restrictive Index (RI) of components of essential oil. The RI of components of the sample was calculated by using a computerized program. Finally, the components of essential oil were identified by comparing the mass spectrums obtained using the existing standard mass spectrums at electronic library of Wiley 2000 existing in Absolution software of GC/Ms set and calculation of standard RI in accordance with C_8-C_{17} Alkanes and by comparing them with the existing standard figures in references (Adams, 2001).

RESULTS AND DISCUSSION

The identified combinations of essential oil, RI, and quantitative percentage of the compounds from seeds and flowers are listed in Table 1. The study of the analysis of *H. officinalis* L. essential oil under investigation showed that thirty-five compounds, accounting for 92.13% of the total oil with 0.75% (v/w) oil yield were identified in the essential oil of the leaves. The main constituents of the essential oil were *Iso* pinocamphone (38.47%), Pinocomphone (13.32%), n-decane (8.67%) and Pinocarvone (5.34%) with 65.8% constituting the highest percentage of essential oil. Also from thirty-six compounds, accounting for 98.68% of the total oil was identified in the essential oil of the flowers with 1.38% (v/w) oil yields. The main constituents of the essential oil were *Iso* pinocamphone (40.25%), Pinocomphone (14.92%), n-decane (8.63%) and Pinocarvone (6.76%) with 70.56% constituting the highest

Table 1. Combinations identified in the essential oil of *Hyssopus officinalis* L.

Compound name	(RI)	Leave (%)	Flower (%)
α-thujene	928	0.32	0.59
α-pinene	935	1.14	1.27
Sabinene	978	0.97	0.28
β-pinene	987	1.78	1.64
Myrcene	992	-	0.61
n-decane	997	8.67	8.63
δ-3-carene	1012	0.85	1.23
P-cymene	1029	0.45	0.68
Limonene	1035	0.47	0.33
1,8-cineole	1038	0.56	1.69
(Z)-β-ocimene	1045	1.27	0.95
(E)-β-ocimene	1054	0.75	0.48
Linalool	1107	0.53	0.67
Camphor	1136	0.65	0.42
Pinocomphone	1158	13.32	14.92
Pinocarvone	1165	5.34	6.76
Iso pinocamphone	1178	38.47	40.25
Terpinen-4-ol	1192	1.78	1.89
α-terpineol	1203	0.25	0.34
Myrtenol	1212	1.75	1.65
Carvon	1236	0.27	0.44
Cumin aldehyde	1245	0.68	0.45
Piperitone	1263	1.65	1.34
Thymol	1289	0.45	0.89
Carvacrol	1306	2.54	2.97
Eugenol	1348	0.89	0.54
β-bourbonene	1375	1.63	1.27
Methyl eugenol	1396	0.21	0.46
α-gurjunene	1417	0.39	0.34
β-caryophyllene	1425	0.32	0.41
α-humulene	1466	0.18	-
Allo aromadendrene	1467	0.67	0.24
Germacrene D	1482	0.57	0.68
Elemol	1537	0.53	0.78
Spathulenol	1568	-	0.16
Caryophyllene oxide	1582	0.48	0.67
α-bisabolol	1675	1.35	1.76
Total		92.13	98.68

The indexes of restrictive have been calculated by injecting the mixture of normal hydrocarbons (C8-C17) to HP-5MS column.

percentage of essential oil. The quality and quantity of the materials forming *H. officinalis* L. essential oil had some differences and similarities with the cases reported in other regions. The studies of essential oils ingredients for botanical populations with ecological and genetic differences can be of great importance in identifying the variety of essential oils inside the population of species. It seems that the geographical origin of *H. officinalis* L. greatly influences the oil quality. The essential oil of *H. officinalis* L. plant has been widely studied in Iran and other countries but the chemical composition of the essential oil of *H. officinalis* grown in Ardabil province is yet to be determined. The present study, results showed the major oil constituents of the leaves and the flowers of *H. officinalis* L. from Ardabil province, Iran were *Iso* pinocamphone and Pinocomphone. In a study they were the chemical compositions of essential oil of aerial part of *H. officinalis* obtained from the hydrodistillation. The chemical composition of essential oil was determined using gas chromatography/mass spectrometry (GC/ MS).

Thirty nine components were identified in *H. officinalis* oil that include thymol (18.95%), fl-bisabolol (10.62%), carvacrol (7.73%), n–Dodecan (5.23%), caryophyllene (4.96%), ortho–acetanisol (4.72%), camphor (3.47%), cumin aldehyde (3.22%) and spathulenol (3.02%) as major components in essential oil (Dehghanzadeh et al., 2012). In a research the plants of *H. officinalis* ssp. *officinalis* genotype raised through seeds sown in early December 1997 flowered in May 1998. The essential oil yields obtained upon hydrodistillation of above ground parts, harvested in May, were 0.25% on fresh herbage weight basis and 1.18% on dry herbage weight basis.

The GC and GC–MS analysis of the essential oil led to the identification of 21 compounds representing 95.6% of the oil, having seven monoterpene hydrocarbons (32.3%), five oxygenated monoterpenes (60.5%) one phenol (0.2%) and six sesquiterpene hydrocarbons (0.35%). The major constituents of the camphorous odoured oil were pinocamphone (49.1%) >β-pinene (18.4%) >isopinocamphone (9.7%) (Garg et al., 1999). The oil obtained by hydrodistillation from the aerial parts of *H. officinalis* L. subsp. *angustifolius* (Bieb.) Arcangeli from Turkey was analyzed by GC–MS. Thirty-four components were characterized, representing 91.0% of the total components detected. The major constituents were identified as pinocarvone (36.3%), pinocamphone (19.6%), *β*-pinene (10.6%), 1,8-cineole (7.2%) and isopinocamphone (5.3%) (Ozer et al., 2005). The essential oil from *H. officinalis* grown in Spain was examined by GC and GC/MS. The oil was characterized by a high content of 1,8-cineole (52.89%) and β-pinene (16.82%) as the main components (Vallejo et al., 1995).

In a comparison study of Hyssop (*H. officinalis* L.), native to the Caucasus, North Western Iran, Turkish North Eastern Black Sea region, and Southern Anatolia, it is a highly valued medicinal plant. The experiment was conducted to find the effect of harvesting at different blooming stages of the plant on fresh and dry herbage yield, dry leaf yield, essential oil content, and essential oil components.

In total, twenty-nine components were identified in hyssop essential oil by GC/MS. Isopinocamphone was the dominating component (47.9 to 51.4%) in all analyzed oil samples. The results clearly demonstrated that oil contents are seriously affected by the environmental conditions and stage of blooming, with the highest oil yield and oil contents at the post-blooming stage (Kizil et al., 2008). In a study the essential oil was obtained (0.66%) from the aerial plant parts of wild *H. officinalis* L., collected around Petnjica (Montenegro). It was examined by a combination of GC and GC/MS. Fifty-seven constituents were found, out of which the major ones were methyl eugenol (38.3%), limonene (37.4%) and β-pinene (9.6%) (Gorunovic et al., 1995). In a research the essential oils were obtained from wild growing *H. officinalis* L. ssp. *aristatus* (Godr.) Briq. At two stages of development, they are found to be very similar in composition with 1,8-cineole (48.2 and 39.6%), isopinocamphone (16.3 and 29.2%) and β-pinene (11.4 and 39.6%) as major constituents. The essential oil obtained commercially from cultivated *H. officinalis* contains larger amounts of isopinocamphone (40.2%), pinocamphone (10.3%), and β-pinene (14.2%), but no traces of 1,8-cineole (Tsankova et al., 1993). The essential oils from different parts of hyssop (*H. officinalis* L.) were investigated by means of GC and GC-MS at three developmental stages of plant. 15 other terpenes were detected besides the main components pinocamphone, camphor and β-pinene, that included isopinocamphone, α and β-phellandrene, germacrene D, and some derivatives of myrtenol. The sesquiterpene alcohol hedycaryol was found to be converted to elemol during GC and MS analysis. As compared with the essential oil content (0.03 to 0.16% of the fresh plant material), the glycosidic bound volatiles were present in lower concentrations (0.01 to 0.06%).

The glycosidic fraction was hydrolysed by means of Pectinol C and β-glucosidase yielding among others octan-3-ol, linalol, cis-nerolidol, benzyl alcohol, phenylethanol, eugenol and o-vanillin. The bicyclic terpenes myrtenol and verbenol could only be detected in small amounts as glycosides of the leaves. This fact gives reason for doubt about a direct connection between the glycosidic bound volatiles and the biogenesis of the essential oil components in hyssop (Schulz and Stahl-Biskup, 1991). Three forms of hyssop *H. officinalis* L., f. *cyaneus* Alef. f. *ruber* Mill, and f. *albus* Alef. exist in the wild of Yugoslavia that were multiplied and cultivated. The *cyaneus* form, characterized by its blue flowers, yielded between 4.9 and 5.8 tonnes of fresh plant material per hectare, and essential oil in yields ranging from 0.65 to 0.75%. The pink-flowered *ruber* form and the white flowered *albus*, respectively, yielded 3.9 to 5.1 tonnes/ha and 4.5 to 6.5 tonnes/ha of fresh plant material in yields of 0.7 to 1.1% and 0.6 to 1.0%. The chemical analysis of different batches of oils produced shows that they are mainly composed of *cis* and *trans*-pinocamphone, and pinocarvone, together with lesser amounts of germacrene D, bicyclogermacrene, elemol and spathulenol (Chalchat et al., 2001). Three Italian strains of hyssop wildly grown in different natural habitats of the Abruzzi region (Central Italy) and classified as *H. officinalis* L. subsp. *aristatus*, were characterized on the basis of their essential oil composition. The oils were obtained by steam distillation of the fresh aerial parts of the plants and analyzed by GC and GC/MS. Thirty-three compounds were identified in this connection. Relevant differences in the quantitative composition were observed among the oils so that the existence of three different chemotypes could be realized.

In particular, one of the strains was characterized by high contents of myrtenol (32.6%) and β-pinene (193%), another contained β-pinene (24.7%) and 1,8-cineole (231%) as the main components, while the third was very rich of methyl eugenol (439%) and limonene (15.9%).

Whereas the latter two strains were similar in composition of oils from Spain and Montenegro, respectively, the former possessed a very unusual oil composition (Piccaglia et al., 1999).

In a study the leaves gathered near Khandiza (Uzbek SSR) in 1986 yielded 0.34% essential oil, that is, only half of the quantity expected by literature. Its composition was investigated by GC and chromato-mass spectrometry, and is tabulated in comparison with that of Zotov (1974). The main constituents were identified as pinocamphone (71%), β-pinene (8.6%) and 1,8-cineole (6.4%). Sabinene content was only 1.3%, with no limonene or isopinocamphone (respectively 16.3, 13.8 and 12.1% according to Zotov) (Dzhumaev et al., 1990). The chemical composition of *H. officinalis* (Lamiaceae) essential oil grown in southeastern Spain was analyzed by GC-MS.

The study is focused on chemical heterogeneity of different oil batches and their extraction yield due to the high relevance of this species in the world market, cultivated under irrigation and non-irrigation conditions and with different harvesting dates.

All essential oil samples have two main terpene compounds which are pinocamphone and *iso*-pinocamphone, accounting for approximately 35 to 40% of the total oil contents. Other relevant compounds were identified, with *β*-pinene, which accounted for 10 to 17% contribution to the total composition, standing out. Significant differences between their volatile compositions have been observed between treatments, being limonene, (E)-*β*-ocimene, pinocarveol, *α*-pinene and *β*-phellandrene the compounds that contributed most to discrimination.

It was also observed that the irrigation period is the most favorable for the cultivation of hyssop in this region, especially for batch 7 which gives the highest extraction yield and the best EO quality (Moro et al., 2011). A study was carried out to determine the essential oil of *H. officinalis* (L.) (*Lamiaceae*) collected from the wild of Southeast Anatolian, Turkey.

Chemical compositions of hydrodistilled essential oils obtained from hyssop leaves were analyzed by GC-MS. It was determined that hyssop essential oil contained *iso*pinocamphone (57.27%), (-)-β-pinene (7.23%), (-)-terpinen-4-ol (7.13%), pinocarvone (6.49%), carvacrol (3.02%), *p*-cymene (2.81%) and myrtenal (2.32%) as its major components (Kizil et al., 2010). The composition of *H. officinalis* L. essential oil was analyzed by GC and GC/MS. The main components of the oil were cis-pinocamphone (42.9%), trans-pinocamphone (14.1%), germacrene-D-11-ol (5.7%) and elemol (5.6%) (Mitic and Dordevic, 2000).

ACKNOWLEDGEMENTS

This study was conducted under the financial support from the Department of Biodiversity, Institute of Science and High Technology and Environmental Sciences, Graduate University of Advanced Technology, Kerman, Iran. Hereby, the executive and colleagues of the research project expresses their thanks for the support extended to them.

REFERENCES

Adams RP (2001). Identification of essential oil components by gas chromatography mass spectroscopy. Illinois Allured Publication Corporation.

Chalchat JC, Adamovic D, Gorunovic MS (2001). Composition of Oils of Three Cultivated Forms of *Hyssopus officinalis* Endemic in Yugoslavia: f. albus Alef. f. cyaneus Alef. And f. ruber Mill. J. Essent. Oil Res. 13(6):419-421.

Chevallier A (2001). Encyclopedia of Medicinal Plants. Dorling Kindersley Publishing Australia. 336 p.

Chiej R (1984). The Macdonald Encyclopedia of Medicinal Plants. London, Macdonald & Co., P. 274.

Dehghanzadeh N, Ketabchi S, Alizadeh A (2012). Essential Oil Composition and Antibacterial Activity of *Hyssopus officinalis* L. Grown in Iran. Asian J. Exp. Biol. Sci. 3(4):767-771.

Dzhumaev KhK, Zenkevich IG, Tkachenko KG, Tsibul'skaya IA (1990). Essential oil of the leaves of *Hyssopus seravschanicus* from South Uzbekistan. Chem. Nat. Comp. 26(1):101-102.

Garg SN, Naqvi AA, Singh A, Ram G, Kumar S (1999). Composition of essential oil from an annual crop of *Hyssopus officinalis* grown in Indian plains. Flav. Fragr. J. 14(3):170-172.

Gorunovic MS, Bogavac PM, Chalchat JC, Chabard JL (1995). Essential Oil of *Hyssopus officinalis* L., Lamiaceae of Montenegro Origin. J. Essent. Oil Res. 7(1):39-43.

Huxley A (1992). The New RHS Dictionary of Gardening.

Kizil S, Hasimi N, Tolan V, Kilinc E, Hakan Karatas H (2010). Chemical Composition, Antimicrobial and Antioxidant Activities of Hyssop (*Hyssopus officinalis* L.) Essential Oil. Not. Bot. Hort. Agrobot. Cluj. 38(3):99-103.

Kizil S, Toncer O, Ipek A, Arslan N, Saglam S, Saglam KhM (2008). Blooming stages of Turkish hyssop (*Hyssopus officinalis* L.) affect essential oil composition. Acta Agric. Scand. 58(3):273-279.

Mills S, Bone K (2005). The Essential Guide to Herbal Safety. Philadelphia, Elsevier Churchill Livingstone. P. 376.

Mitic V, Dordevic S (2000). Essential oil composition of *Hyssopus officinalis* L. cultivated in Serbia. Facta universitatis. Phys. Chem. Technol. 2(2):105-108.

Moro A, Zalacain A, Mendoza JH, Carmona M (2011). Effects of Agronomic Practices on Volatile Composition of *Hyssopus officinalis* L. Essential Oils. Mole. 16(5):4131-4139.

Ozer H, Sahin F, Kilic H, Gulluce M (2005). Essential oil composition of Hyssopus officinalis L. subsp. angustifolius (Bieb.) Arcangeli from Turkey . Flav. Fragr. J. 20(1):42-44.

Piccaglia R, Pace L, Tammaro F (1999). Characterization of essential oils from three Italian ecotypes of Hyssop [*Hyssopus officinalis* L. subsp. aristatus (Gordron Briq.]. J. Essent. Oil Res. 11(6):693-699.

Schulz G, Stahl-Biskup E (1991). Essential oils and glycosidic bound volatiles from leaves, stems, flowers and roots of *Hyssopus officinalis* L. (lamiaceae). Flav. Fragr. J. 6(1):69-73.

Tsankova ET, Konaktchiev AN, Genova EM (1993). Chemical Composition of the Essential Oils of Two *Hyssopus officinalis* Taxa. J. Essent. Oil Res. 5(6):609-611.

Vallejo MCG, Herraiz JG, Pérez-Alonso MJ, Velasco-Negueruela A (1995). Volatile Oil of *Hyssopus officinalis* L. from Spain. J. Essent. Oil Res. 7(5):567-568.

Van Wyk V, Michael W (2004). Medicinal Plants of the World. P. 177.

Effect of *Aloe vera* gel coating on postharvest quality and shelf life of mango (*Mangifera indica* L.) fruits Var. 'Ngowe'

Ochiki Sophia, Gesimba Morwani Robert and Wolukau Joseph Ngwela

Department of Crops, Horticulture and Soils Sciences, Egerton University, P. O. Box 536-20115, Egerton-Kenya.

Mango is a highly perishable fruit and high post-harvest losses occur in Africa. In order to address this problem, 4 concentrations of *Aloe vera* gel (AG) (0, 25, 50 and 75%) and chitosan (1%) were tested at two temperature levels (room temperature 15-22°C and 13°C) to determine their effect on the postharvest life of mango (var.'Ngowe'). The experimental design was a 5 by 2 factorial experiment embedded in a complete randomized design with three replications. Data were recorded on titratable acidity, fruit colour, ascorbic acid and anthracnose disease incidence. The results showed that at both temperatures 50 and 75% Aloe concentrations significantly increased the shelf life evidenced by reduced decrease in titratable acidity. Fruit colour and ascorbic acid were also maintained for longer periods in these treatments. Findings of this study demonstrate the potential of using *Aloe vera* gel at 50% as a coating for improved postharvest shelf life and maintaining quality of mango fruits hence reduced postharvest losses.

Key words: *Aloe vera* gel, postharvest shelf life, mango.

INTRODUCTION

Mango (*Mangifera indica* L.) is the most economically important fruit of the Anacardiaceae family (Tharanathan et al., 2006). World trade in mangoes has been increasing over the years, and both exports from Kenya and local consumption is growing. The world market continues to become more price-competitive in spite of postharvest challenges e.g. losses caused by diseases (HCDA, 2011). Mango is one of the most popular fruits all over the world as it has an attractive color, delicious taste and excellent nutritional properties. However, mango fruits are climacteric and ripen rapidly after harvest, this limits their storage, handling and transport potential (Lalel et al., 2003).

Mango is an easy access to post-harvest disease infection and production and consumption imbalances after harvesting lead to considerable losses (Zeng et al., 2006). Therefore, scientists are working towards prolonging the shelf life of the fruit by slowing down the ripening process while maintaining quality and flavor. Fruit coating after harvesting is becoming popular in this respect (Gill et al., 2005). However, possible health risks associated with the residue of the coating materials like fungicides are reducing the scope of coatings. Edible coatings have no residue associated risks and are possible alternative options (Ergun and Satici, 2012).

The use of *Aloe vera* gel has drawn interest in the food industry (Arowora et al., 2013). *A. vera* based edible coatings have been shown to prevent loss of moisture and firmness, control respiration rate and development and maturation, delay oxidative browning, and reduce microorganism proliferation in fruits such as sweet cherry, table grapes and nectarines (Valverde et al., 2005; Matinez-Romero et al., 2005; Ahmed et al., 2009). In addition to the traditional role of edible coatings as a barrier to water loss and delaying fruit senescence, the new generation coatings are being designed for incorporation and/or for controlled release of antioxidants, nutraceuticals, chemical additives and natural antimicrobial agents. It has also been reported that *A. vera* extracts possess antimicrobial activity against gram positive and gram negative bacterial pathogens (Adetunji, 2008).

The use of *A. vera* gel as an edible surface coating has been reported to prolong the shelf life and to delay changes in parameters related to deterioration of quality in sweet cherry and table grapes (Martinez- Romero et al., 2006; Serrano et al., 2006), yet no studies have demonstrated the use of *A. vera* natural plant extract based on its antifungal properties on enhancement of shelf life and quality of mango fruits. Therefore, this study was conducted with the objective of evaluating the effects of the different *A. vera* gel on postharvest life of mango fruits.

MATERIALS AND METHODS

Research site

The postharvest study was carried out in a laboratory at Egerton University, Njoro, Kenya. The laboratory lies at a latitude of 0° 23' South, longitude 35° 35' East, altitude of approximately 2,238 m a.s.l in the Lower Highland 3 (LH3) agroecological zone (Jaetzold and Schmidt, 1983). The laboratory records average maximum and minimum temperatures of 19 to 22°C and 5 to 8°C, respectively (Egerton Metrological Station, 2009).

Materials

Mango

The variety 'Ngowe' was used. 'Ngowe' is a popular mango variety in Kenya, which has little fibre and has excellent eating quality but it is susceptible to anthracnose. All the fruits that were used in this study were acquired from a grower in Masii in Machakos County, Kenya. The fruits were harvested at the mature green stage having no visible blemish. The fruits were transported to the laboratory on the same day.

Aloe vera

Leaves of *A. vera* were harvested from Lare in Nakuru County, Kenya. Only the fully extended mature leaves were harvested. The leaves were then stored in plastic papers and transported to the laboratory within same day.

Chitosan

Crushed chitosan powder food grade was purchased from Kobian Chemicals Company, Nairobi.

Preparation of coating solutions

Aloe gel was obtained from fresh aloe leaves, the matrix was separated from the outer cortex of the leaves and the colourless hydroparenchyma homogenized in a blender. The resulting mixture was filtered using Whatman filter paper number 100 to remove the fibres. The liquid constituted fresh *A. vera* gel. The gel matrix was pasteurized at 70°C for 45 min. For stabilizing, the gel was cooled immediately to an ambient temperature and 4.5 g of ascorbic acid was added; 4.5 g of citric acid was then added to adjust the pH to 4. To prepare chitosan coating, 1% Chitosan (Kobian Chemical Co.) was dissolved in a 0.5% glacial acetic acid and distilled water. The pH value of the chitosan solution was then adjusted to 5.6 using 0.1 M NaOH.

Application of treatments and experimental design

The coating solutions were: aloe gel (0%) as a negative control, aloe gel (25%), aloe gel (50%), aloe gel (75%), and chitosan (1%) as a positive control. Fresh fruits were dipped completely into the coatings solutions at room temperature for 25 min. The fruits were allowed to drain and then dried at room temperature to allow a thin film layer to be formed on the fruits. The fruits were then stored at room temperature and at 13°C. Mature, green fruits, without any visible blemish, were selected and the pedicels were removed. The fruits were then randomly divided into eight lots of twenty fruits each. The first lot constituted the positive control and was coated with chitosan. The second, third, fourth and fifth lots were coated by dipping completely in *A. vera* gel at concentrations of 0, 25, 50 and 75% respectively and stored at room temperature and at 13°C. The experiment was laid out as a 5 by 2 factorial experiment embedded in a completely randomized design with three replications. Various parameters were evaluated at 4 day intervals until the overall acceptability became unsatisfactory for each lot of samples (the fruit was considered as waste when it is infected by disease.

Titrable acidity (TA)

TA was determined by titrating 100 ml of juice against sodium hydroxide having concentration of 0.1 N (AOAC 2000) and expressed as the percentage of citric acid per 100 g fresh mass. Individual mango fruit from each treatment was ground in a blender to obtain freshly prepared juice which was then filtered using filter papers.

Colour

Peel color was measured at the equator on opposite cheeks of the fruit. Flesh color was measured in the center of one cut cheek. Both peel and flesh colours were measured using portable whiteness colourimeter (WSD-3 TYPE). Measurements were recorded using standard Hunter L a b chromatic system and were expressed as lightness (L), greenness (-a), redness (+a) yellowness (+b), blueness (-b) colour space coordinates. The instrument was calibrated with a standard white ceramic tile and black tile and set up for D65 as illuminate and a 10° observer angle.

Vitamin C content

This was determined by titrating 10 g of mixed pulp sample against

Table 1. Effect of *Aloe vera* gel on Anthracnose severity and anthracnose disease incidents.

Treatment	Severity index		Anthracnose incidence (%)
	% skin area	scores	
Chitosan	12.5[a]	2	45
0% Aloe	63.6[a]	4	93
25% Aloe	57.1[a]	4	80
50% Aloe	50.0[a]	3	73
75% Aloe	45.6[a]	3	60

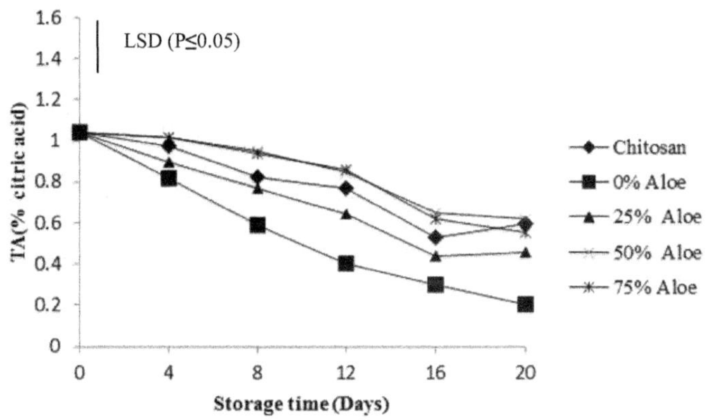

Figure 1. Titratable acidity of mango fruits var. 'Ngowe' as affected by *A. vera* gel coatings

the standard 2, 6 dichlorophenol dyes following the procedure outlined in AOAC (2000).

Degree and rate of anthracnose incidents

Anthracnose severity was assessed by measuring the diameter of anthracnose lesions on mango fruits and ranked by use of scale 1–5 where 1=0% of fruit surface rotten, 2=1–25%, 3=26–50%, 4=51–75% and 5=76–100%)

Data analysis

The data collected was subjected to Analysis of Variance (ANOVA) at $P \leq 0.05$, using PROC GLM code of SAS (version 9, 2005) and means for significant treatments separated using the Tukey's Honestly Significant Different Test at $P \leq 0.05$.

RESULTS AND DISCUSSION

Titratable acidity

Mango fruits with coating presented a statistically higher titratable acidity (TA) during storage in spite of the slight decrease observed (Figure 1). TA increased during storage in all treatments but the rate of increase in treated fruits was comparatively slower compared to the control. At day zero there was no significant difference ($P \leq 0.05$) between the treatments (Table 1). The initial TA was 1.04% citric acid. In day four, fruits coated with 50 and 75% *A. vera* gel concentration had a significantly lower TA value compared with those coated with 0% *A. vera* gel.

On day-8, fruits coated with 50% *A. vera* gel had a significantly lower TA values compared with those coated with other fruit coating treatments. On day-12, the highest TA was observed in fruits coated with 75% *A. vera* gel and the lowest readings were recorded for fruits coated with 0% *A. vera* gel. At day sixteen of the storage period, the highest TA was observed on fruits coated with 50% *A. vera* gel and the lowest readings were recorded for fruits coated with 0% *A. vera* gel. At day at the end of storage period (twenty days), fruits coated with 50% *A. vera* gel had the highest TA while the control had the lowest TA value.

Generally TA was maintained for those fruits coated with 50 and 75% *A. vera* gel. TA decreased gradually in all treatments but the rate was slower in treated fruits compared to negative control. There was a lower decrease in the titratable acidity for *A. vera* and chitosan coated mangoes. *A. vera* gel and chitosan coating must have modified internal atmosphere thus reducing ripening and maintenance of the TA (Nabigol and Asghari, 2013). Reduction in TA for uncoated fruits is due to conversion of acids into sugars and their further utilization in the metabolic processes of the fruit. Doreyappa and Huddar (2001) reported the similar pattern in different varieties of mango fruits stored at 18 to 34°C. They observed a series of physico-chemical changes during ripening and the major changes were decrease in acidity. The acidity of the fruit is an important character to determine its quality and acceptability. Very high or very low values of the acidity are not recommended for good fruit.

Fruit colour

The coatings were effective on peel colour change of the mangoes stored under room temperature conditions and 13°C (colour change during ripening for this variety of mango is from green to yellow). Color changes in peel are presented as L*, a*, b* and were expressed as lightness (L*), greenness (-a*), yellowness (+b*), colour space coordinates. Fruits from each treatment for both

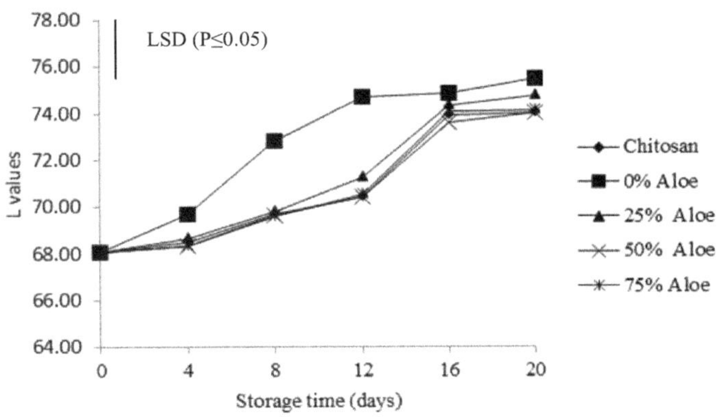

Figure 2. Effect of different *A. vera* gel concentrations on L* value of the peel colour of mango fruits var. 'Ngowe'.

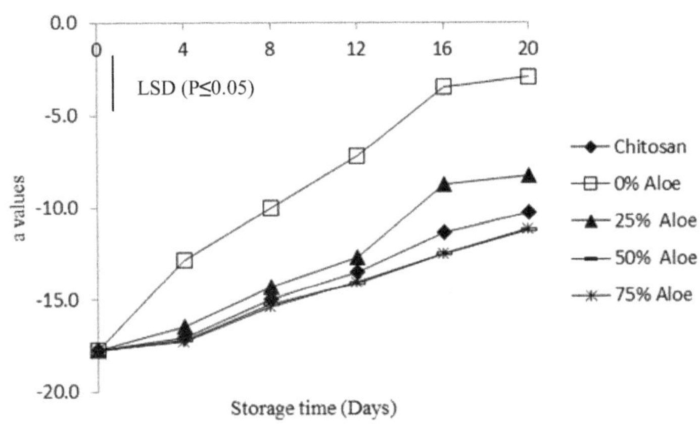

Figure 3. Effect of different *Aloe vera* gel concentrations on Chromatic a* of the peel color of mango fruits var. 'Ngowe'.

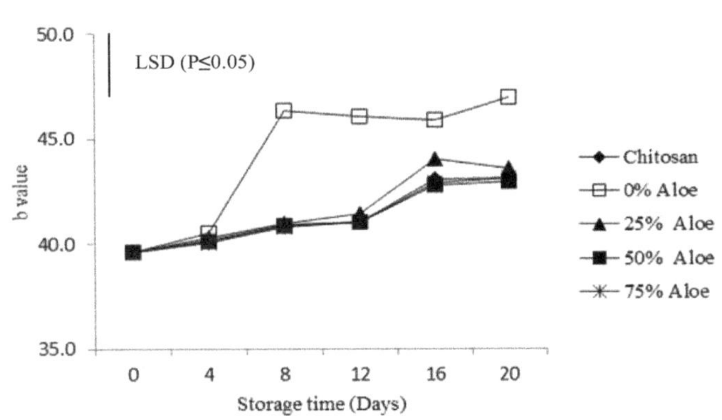

Figure 4. Effect of different *Aloe vera* gel concentrations on Chromatic b* of the peel color of mango fruits var. 'Ngowe'.

trials registered some changes in chromatic L*, a* and b* colour values during the storage period (Figures 2, 3 and 4 respectively).

Lightness (L*) increased significantly (P≤0.05) during storage, changes in L* value for control (L* values increased from 68.1 to 75.5) which was higher than what was recorded in the coated fruits (Figure 2). L* values increased over time irrespective of the coatings from 68.1 to 75.5 for negative control, from 68.1 to 74.8 for fruits coated with 25% *A. vera* gel, from 68.1 to 74.0 for 50% *A. vera* gel coated fruits, from 68.1 to 74.1 for 75% *A. vera* gel coated fruits and from 68.1 to 74.0 for chitosan coated fruits while no significant difference (P≤0.05) were recorded for chitosan and 50% *A. vera* gel coated treatments at the end of the storage period.

Chromatic a* value from mango fruits also increased over time irrespective of the treatments. There was a gradual significant increase (P≤0.05) in peel a* value beginning on day eight (Figure 3). Before the storage, the a* value was -17.7, and after the storage the value reached to -2.9 for negative control fruits, to -8.3 for 25% *A. vera* gel coated fruits, to -11.2 for 50% *A. vera* gel coated fruits,-11.2 for 75% *A. vera* gel coated fruits and to -10.3 for chitosan coated fruits. The increase in a* value was however, slower for fruit coated with 50 and 75% *A. vera* compared to control or 25% *A. vera* gel treatments.

Chromatic b* value similar to L* and a* value for fruits slightly increased over time regardless of the coatings, and it was significantly higher (P≤0.05) on day eight for the negative control (Figure 4). Initial b* value was 39.6, afterwards the value gradually increased, reaching to 46.9 for negative control fruit, to 43.5 for 25% *A. vera* gel coated fruits, to 42.9 for 50% *A. vera* gel coated fruits, 43.1 for 75% *A. vera* gel coated fruits and to 43.1 for chitosan coated fruits. The increase in a* value was however, slower in fruits coated with 50 and 75% *A. vera* gel and chitosan compared to negative control.

Generally the peel color of the mango fruits coated with 50 and75% *A. vera* gel was significantly less developed than those coated with other treatments. The coatings were effective on flesh colour change of the mangoes under room temperature conditions and 13°C (colour change during ripening for this variety 'Ngowe' is from green to yellow). Color changes in peel are presented as L*, a*, b* and were expressed as lightness (L), greenness (-a), blueness (+a), yellowness (+b), colour space coordinates. Fruits from each treatment for both trials registered some changes in chromatic L*, a* and b* colour values during the storage period (Figures 5, 6 and 7 respectively).

The L* (lightness) decreased during storage, L* values for control decreased from 88.1 to 72.5 which was lower than the coated fruits (Figure 5). L* values decreased over time irrespective of the treatments from 88.1 to 72.5 for negative control, from 88.1 to 75.8 for 25% *A. vera* gel coated fruit, 76.0 for 50% *A. vera* gel coated fruits, 76.6 for 75% *A. vera* gel coated fruits and 76.0 for chitosan coated fruits. There was no significant difference (P≤0.05) between fruits coated with chitosan and 50% *A.*

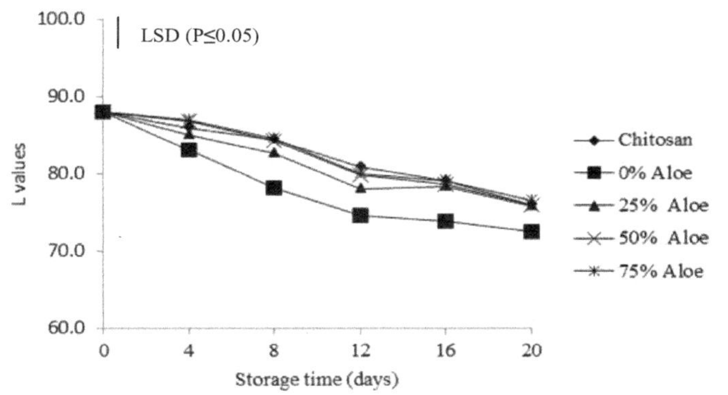

Figure 5. Effect of different *Aloe vera* gel concentrations on Chromatic L* value of the flesh color of mango fruits var. 'Ngowe'.

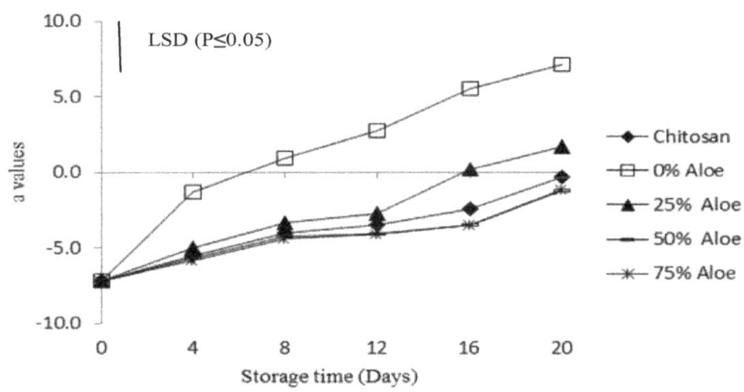

Figure 6. Effect of different *Aloe vera* gel concentrations on Chromatic a* of the flesh color of mango fruits.

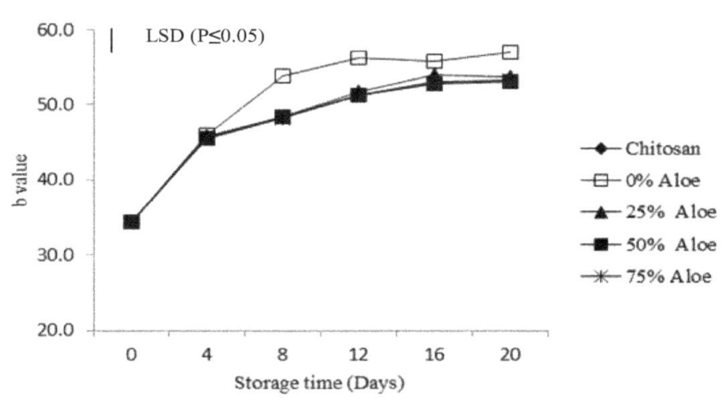

Figure 7. Effect of different *Aloe vera* gel concentrations on Chromatic b* of the flesh color of mango fruits var. 'Ngowe'.

vera gel coated treatments at the end of the storage period.

Chromatic a* value from mango fruits also increased over time irrespective of the treatments. There was a gradual increase in flesh a* value beginning on day eight (Figure 6). Before storage, a* value was -7.2, and after the storage the value increased to 7.1 for negative control fruits. It also increased to 1.7 for fruits coated with 25% *A. vera* gel, -1.2 for fruits coated with 50% *A. vera* gel, -1.2 for 75% *A. vera* gel coated fruits and to -0.3 for chitosan coated fruit. The increase in a* value was however, slower for fruit coated with 50 and 75% *A. vera* compared to negative control or 25% *A. vera* gel treatments. There was a significant difference ($P \leq 0.05$) among the treatments at the end of the storage period.

Chromatic b* value gradually increased over time regardless of the treatments, and it was significantly higher ($P \leq 0.05$) on day eight for fruits coated with 0% *A. vera* gel (Figure 8). Initial b* value was 34.4 and it gradually increased, reaching to 57.0 for fruits coated with 0% *A. vera* gel, to 53.6 for 25%, to 53.0 for 50% , 53.2 for 75% *A. vera* gel and to 53.2 for chitosan coated fruits. The increase in b* value was however, slower for fruit coated with 50 and 75% *A. vera* and chitosan compared to 0% *A. vera* gel. There was a significant difference ($P \leq 0.05$) among the treatments at the end of the storage period.

Generally the flesh color of the mango fruits coated with 50 and75% *A. vera* treatments was significantly less developed than flesh colours in the other treatments. Color is related to the presence of different pigments. Changes in colour are mainly due to chlorophyll transformation into other pigments and to the synthesis of carotenoids and anthocyanins. Chlorophyll retention was higher in the fruits coated with *A. vera* gel and chitosan coatings and least of it was seen in the uncoated fruit. *A. vera* gel treatment and chitosan delayed the green colour loss on the fruit skin. Coatings applied to fruit act as a barrier, altering permeability to gases. This results in increased internal CO_2 contents, slowing down the external and internal colour change of the fruit in return delaying chlorophyll degradation and carotenoid synthesis (Ergun and Satici, 2012). Similar results of colour retention in coated fruit had been reported in carambola fruits (Neeta et al., 2013).

Ascorbic acid

The ascorbic acid content in the mango fruits decreased significantly during the ripening storage period (Fig.8). A significant decrease ($P \leq 0.05$) in ascorbic acid contents was observed in all the treatments. Initially, the ascorbic acid was 26.6 mg/100 g. At day four, there was significant difference ($P \leq 0.05$) between the negative control (0% *A. vera* gel) and the other treatments but there was no significant difference among 50 and 75% *A. vera* gel concentrations and those coated with 1% chitosan (the positive control). At day eight, 50% *A. vera gel* was the most effective in reducing decrease in ascorbic acid followed by chitosan while the 0% aloe had the least ascorbic acid. At day twelve, there was

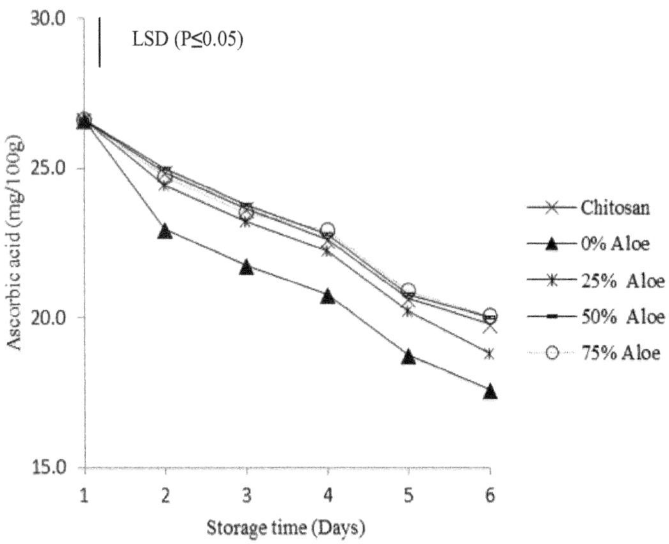

Figure 8. Ascorbic acid of mango fruits var. 'Ngowe' as affected by *Aloe vera* gel coatings.

significant difference between the control and the rest of the treatments, 75% *A. vera* gel had the highest ascorbic acid.

At day sixteen, there were significant effects between the control and the other treatments 25, 50 and 75% *A. vera* gel and chitosan treatments. The negative control had the lowest ascorbic acid while75% *A. vera* gel had the highest ascorbic acid value. At day twenty, 0% *A. vera* gel had the lowest ascorbic acid while 75% *A. vera* gel had the highest ascorbic acid among the other treatments. Generally, all treatments had a decrease in ascorbic acid from the initial value. There was a gradual decrease in all days although the decrease was reduced by *A. vera* treatments. Ascorbic acid is one of the most abundant antioxidants present in fruits. Results suggest that *A. vera* gel and chitosan coating caused lower losses of antioxidant capacity by the end of storage, when coated fruits were compared to the negative control. Application of *A. vera* gel and chitosan modified internal atmosphere, more concentration of CO_2 resulting to lower concentration of O_2 hence the oxidation process was retarded which caused reduction in conversion of ascorbic acid to dehydro ascorbic acid. Lal et al. (2003) reported similar results on mango using various chemicals under different storage period.

Anthracnose incidence

None of the *A. vera* gel concentration tested inhibited the growth of *Colletotrichum gloeosporioides* as compared to the control, and grew almost similarly with all treatments through the 7-day incubation period. Mango fruits treated with 1% chitosan had the lowest disease severity index. The highest fungicidal effect was observed in those mangoes coated with 1% chitosan.

Conclusion AND Recommendation

Findings of this study demonstrate the potential of using *A. vera* gel as a coating for improved postharvest shelf life and maintaining quality of mango fruits hence reduced postharvest losses. The results showed that at both temperatures, 50 and 75% aloe concentrations significantly increased the shelf life evidenced by reduced decrease in titratable acidity. Fruit colour and ascorbic acid were also maintained for twenty days in these treatments. Since *A. vera* is an edible plant, does not pose any environmental hazard and is easily available in Kenya and other tropical regions, *A. vera* at 50% concentration can be used as an alternative fruit coating for mangoes.

Conflict of Interest

The authors have not declared any conflict of interest.

AKNOWLEDGEMENTS

Author wish to express his sincere thanks for the financial support received from National Council of Science Technology and Innovation (NCST) without which this work would not have been a success. Am also grateful to the Department of Crops, Horticulture and Soils of Egerton University for their support.

REFERENCES

Ahmed MJ, Singh Z, Khan AS (2009). Postharvest Aloe vera jel-coating modulates fruit ripening and quality of 'Arctic Snow' nectarine kept in ambient and cold storage. Int. J. Food. Sci. Tech. 44: 1024-1033.

Arowora KA, Williams JO, Adetunji CO, Fawole O B, Afolayan SS, Olaleye OO, Adetunji J Ogundele BA (2013). Effects of Aloe Vera Coatings on Quality Characteristics of Oranges Stored Under Cold Storage. Greener J. Agric. Sci. 3:39–47.

Ergun M, Satici F (2012). Use of Aloe Vera Gel as Biopreservative for " Granny Smith " and " Red Chief " Apples. J. Ani. Plant Sci. 22(2):363-368.

HCDA (2011). Horticultural Crops Development Authority. 6:1–8.

Lalel HJD, Singh Z, Chye S (2003). Aroma volatiles production during fruit ripening of " Kensington Pride ". Mango, P. 27.

Martinez-Romero D, Alburquerque N, Valverde JM, Guillen F, Castillo S, Valero D (2006). Postharvest sweet cherry quality and safety maintenance by Aloe vera treatment: a new edible coating. Postharvest Biol. Technol. 39:93–100. http://dx.doi.org/10.1016/j.postharvbio.2005.09.006

Nabigol A, Asghari A (2013). Antifungal activity of Aloe vera gel on quality of minimally processed pomegranate arils. Int. J. Agro. Plant Prod. 4:833-838.

Neeta BG, Manu LC, Rao TVR (2013). Effect of edible coatings on quality and shelf life of carambola (*Averrhoa carambola* L.) fruit during storage. J. Food Sci. Technol. DOI 10.1007/s13197-013-0988-9.

Serrano M, Castillo S, Valverde J, Martianez-Romero D, Guillean F, Valero D (2006). Use of Aloe vera gel coating preserves the functional properties of table grapes. Agric. Food Chem. 54:3882–3886.http://dx.doi.org/10.1021/jf060168p PMid:16719510

Tharanathan RN, Yashoda HM, Prabha TN (2006). Mango (*Mangifera indica* L .), " The King of Fruits "— An Overview Mango (*Mangifera indica* L .). Food Rev. Int. pp. 37-41.

Valverde JM, Valero D, Martinez-Romero D, Guillen F, Castillo S, Serrano M (2005). Novel coating based on Aloe vera gel to maintain table grape quality and safety. Agric. Food Chem. 53:7807–7813http://dx.doi.org/10.1021/jf050962v PMid:16190634

32

Physiological responses of six tomato (*Lycopersicon esculentum* Mill.) cultivars to water stress

Zuzana Jureková[1], Kristína Németh-Molnár[1], Viera Paganová[2]*

[1]Faculty of European Studies and Regional Development, Slovak University of Agriculture in Nitra, Marianská 10, 949 01 Nitra, Slovakia.

[2]Faculty of Horticulture and Landscape Engineering, Slovak University of Agriculture in Nitra, Tulipánová 7, 949 76 Nitra, Slovakia.

The aim of the present study was to investigate the physiological responses of six tomato (*Lycopersicon esculentum* Mill.) cultivars to water stress. To this end, plants were exposed to slow dehydration at the third unfolded leaf stage for 23 days. The relative water content (RWC), leaf area and leaf L-proline were determined at 10, 17 and 23 days after treatment application. Our results showed that during slow dehydration, the leaf RWC declined in all studied genotypes, whereas L–proline accumulated. A statistically significant effect of the sampling date (water stress duration) on RWC values was also observed. In addition, the differences in proline content were significantly influenced by tomato genotype, sampling date and the level of substrate saturation. Putting all these together, the results of this study indicate that the adaptive potential of the studied genotypes was expressed in a different relationship between the relative water content and growth of the leaf area. However, three of the tomato genotypes exhibited reduced growth in leaf area in response to the decreased RWC, whereas other tomato genotypes retained a balanced RWC accompanied by further growth of the leaf area.

Key words: Tomato, genotypes, growth, water stress, L-proline, relative water content.

INTRODUCTION

Higher plants, which are continuously exposed to intense or moderate water deficit, use morphological, cellular and molecular adaptive mechanisms to survive (Bohnert and Jensen, 1996; Zhu et al., 1997; Hasegawa et al., 2000; Pedrol et al., 2000). Plant responses to drought include changes of growth intensity in shoots and roots, and an acceleration of the plant development (Lannucci et al., 2000, Ma et al., 2006). In addition, drought has been demonstrated to be extremely negative at the reproductive stage of plant development (Ludlow and Muchow, 1990; Ma et al., 2006). Therefore, plants tend to avoid dangerous periods by changing the intensity of the physiological processes and the position of organs, or by leaf rolling (Fernandez and Castrillo, 1999; Kadioglu and Terzi, 2007; Saglam et al., 2008).

It has been demonstrated that for many plant species, a decrease in leaf water potential (-1.0 MPa) induces an osmoregulatory mechanism through the accumulation of some primary and secondary metabolites, such as: amino acids, amides, carbohydrates and salt cations (Serraj and Sinclair, 2002; Zhu, 2002; Ashraf and Foolad, 2006). Among these osmoprotective substances, the amino acid L-proline has been shown to be accumulated under stress conditions (Taylor, 1996; Ben-Rouina et al., 2006), particularly under drought conditions, and many other plants have also displayed significant increases in L-proline levels (Delauney and Verma, 1993; Hasegawa et al., 2000; Adejare et al., 2006; Umebese, 2008). In experiments with transgenic plants, it has been shown that plants overexpressing the Δ^1-pyrolin-*5*-carboxylate synthetase, a gene involved in proline biosynthesis (Kavi-

*Corresponding author. E-mail: viera.paganova@uniag.sk.

Table 1. Characteristics of the studied tomato genotypes.

Tomato variety	Origin	Description
Premium	Collection of genetic sources in Slovak Republic	Determinate, bush tomato, main season variety,
Denár	Collection of genetic sources in Czech Republic	Determinate, bush tomato, semi-fast ripening variety
Hana	Collection of genetic sources in Czech Republic	Semi-determinate tomato, semi-fast ripening variety
Robura	Collection of genetic sources in Slovak Republic	Determinate, bush tomato, late variety
Moldeny	Collection of genetic sources in Slovak Republic	Bush tomato, semi-fast ripening variety
UC-82		Determinate, bush tomato, fast ripening variety

Kishor et al., 1995; Nanjo et al., 1999; Hong et al., 2000), have a good capacity to tolerate drought. In our opinion, proline indicates plant stress and it shows a considerable adaptive plant response against stress. The present work is focused on the study of selected physiological responses of tomato genotypes and the determination of the adaptation model to drought during the process of slow dehydration.

MATERIALS AND METHODS

In this study, six tomato (*Lycopersicon esculentum* Mill.) genotypes, namely: Hana, UC-82, Premium, Denar, Moldeny and Robura, were used. The genotypes were selected according to the heterogeneity of their growth type and in terms of tomato fruit ripening (Table 1); although tomato seeds were pre-germinated on Petri-dishes. In the second decade of March, the root tips were detected and seeds were sown to seedbeds and placed in greenhouse. As such, the seedlings germinated within one week.

At the second unfolded leaf stage, 30 days old plants were individually transplanted in containers (5 L contents) with sowing substrate (pH = 5.0 – 6.5, content of salts < 1.5 g/l, N = 50 - 300 mg /l, P_2O_5 = 80 - 300 mg/l, K_2O = 80 – 400 mg/l), after which the content of dry mass (38%) and water (62%) in the substrate was analysed at the beginning of the experiment. During the experiment, the average daily air temperature was 21.9°C and the average air humidity was 77.8%. Before treatment application, the substrate was maintained at 70% of its full water capacity. At the vegetative growth stage, 30 tomato plants of each genotype were divided into two groups (15 individuals in each group). For the first one, plants were watered at only 40% of their full water capacity (stress) for 23 days; while for the second group, plants were retained at 70% substrate saturation (control). However, the water regime was maintained by daily watering of the individual plants to constant weight. From each tomato genotype, 10 plants (5 stressed and 5 controls) were analysed and their samples were taken from the control and stressed plants at the 10th, 17th and 23rd day after treatment application. After exposure period of the treatment, the flowers appeared on lateral shoots. The following parameters were analysed: content of water in the whole plant and its vegetative organs, leaf area (A) in square meters (m^2) determined from leaf scanning using the Corel SCAN 8 software and Corel OCR-TRACE 8, content of L-proline in leaves determined colorimetrically in the toluene extract according to Bates et al. (1973) after tissue homogenisation and filtration. Excitation (E) was determined at 519 nm using a Merck spectrophotometer (NOVA 400). The relative water content (RWC%), which represents the relationship between the content of water in the plant organs and the content of water during their full turgescence was calculated according to the method of Gonzáles et al. (2001), while the relative growth rate of the leaf area ($RGR_A = \ln A_2 - \ln A_1 / t_2 - t_1$) was calculated according to the method of Květ (1971). However, Hunt (1982) and Kolb and Steiner (1990) used the same basic calculation for the mentioned parameter.

The data obtained in this study were analyzed using multifactor ANOVA, and the mean values were compared using the Tukey test at a confidence interval of 95% and a significance level of $p \leq 0.05$. The regression analysis on 95% significance level was applied for assessment of the associations between RWC content and leaf area (A). However, all assessments were carried out using the statistical software package, Statgraphics Centurion XV.

RESULTS

The study of the physiological responses of six tomato genotypes growing in conditions with good and low water supply revealed the extreme heterogeneity of the analysed genotypes in their responses to water stress. Cultivar efficiency retains or only slightly reduces the intensity of physiological processes under conditions of water stress and hence can be a significant indicator of their drought tolerance. According to Umebese et al. (2009), water stress reduces the shoot elongation rate of tomato and amaranth in vegetative growth stages and decreases shoot biomass production of both species in the reproductive growth stage.

The influence of tomato cultivar, water regime and stress duration (sampling date) was analysed on values of the relative water content (RWC) in tomato leaves. The differences in the RWC values between cultivars and treatment (stress and control) were not statistically significant (Table 2). The RWC was significantly influenced by the sampling date (water stress duration). From the beginning of the experiment (0-day), the leaf RWC decreased significantly with the lowest value (77.08%) obtained at the 17th day of the treatment, while on the next sampling date, a moderate increment of the average RWC to 82.33% was recorded (Table 3).

One plausible cause of these changes is the accumulation of L-proline in plant leaves (Jureková et al., 2003; Gubiš et al., 2006). The differences in proline content are significantly influenced by genotype, sampling date and the level of the substrate saturation (Table 4). The results of multifactorial analysis of variance confirmed the

Table 2. Multifactor analysis of variance for parameter RWC of different tomato cultivars.

Source of variation	Sum of squares	Df	Mean square	F-value	P-value
Main factor					
A: Cultivar	100.589	5	20.1178	1.74	0.1872 n.s.
B: Sampling date	1470.28	3	490.093	42.28	0.0000 ***
C: Water regime	45.2408	1	45.2408	3.90	0.0669 n.s.
Interaction					
AB	179.763	15	11.9842	1.03	0.4747 n.s.
AC	109.569	5	21.9138	1.89	0.1558 n.s.
BC	39.4408	3	131.469	1.13	0.3670 n.s.
Residual	173.879	15	11.5919		
Total	2118.76	47			

Table 3. Average values of RWC (%) and Tukey-test at a significance level of p ≤0.05. Data are given for particular tomato cultivars, sampling dates and different levels of substrate saturation.

Cultivar	N	Average	Sampling date	N	Average
Premium	8	82.92 a	0th day	12	91.27 c
Moldeny	8	83.81 a	10th day	12	78.45 ab
Hana	8	83.78 a	17th day	12	77.08 a
Denar	8	80.19 a	23rd day	12	82.33 b
Robura	8	80.50 a	Water regime	*N*	*Average*
UC-82	8	82.49 a	70%	24	83.25 a

Table 4. Multifactor ANOVA of L-proline content in the leaves of tomato cultivars.

Source of variation	Sum of squares	Df	Mean square	F-value	P-value
Main factor					
A: Cultivar	6.81352	5	1.3627	3.99	0.0167 *
B: Sampling date	5.50345	3	1.83448	5.37	0.0103 *
C: Water regime	4.26021	1	4.26021	12.48	0.0030 **
Interaction					
AB	6.6102	15	0.44068	1.29	0.3136 n.s.
AC	2.21904	5	0.443808	1.30	0.3156 n.s.
BC	4.94688	3	1.64896	4.83	0.0151 *
Residual	5.12038	15	0.341358		
Total	35.4737	47			

statistically significant interactive effect of sampling date and water regime on proline content (Table 4).

The significant differences in the average values of proline content were confirmed for cultivars Moldeny and Denar (Table 5). In comparison to other studied cultivars, Denar showed quite a high average value of proline, which indicates the highest expressive adaptation of this tomato genotype to stress conditions. A similar adaptation mechanism was confirmed for all studied genotypes in the variant with lower substrate saturation (40%), which accumulated an average proline content of 0.99 μmol. g^{-1} of fresh mass (Table 5).

As part of the study of proline accumulation in relation to the duration of water deficit, a significant increase of proline content was confirmed on the 17th day of the experiment. This date was duplicated in the assay of the changes in water content of the leaves. At the 23rd day of treatment, the content of proline decreased to an average

Table 5. Average values of L-proline (μmol.g^{-1}) and Tukey-test at significance level of p ≤0.05. Data are given for particular tomato cultivars, sampling dates and different levels of substrate saturation.

Cultivar	N	Average	Sampling date	N	Average
Premium	8	0.30 ^{ab}	0thday	12	0.52 ^{a}
Moldeny	8	0.22 ^{a}	10thday	12	0.50 ^{a}
Hana	8	0.80 ^{ab}	17thday	12	1.28 ^{b}
Denar	8	1.25 ^{b}	23rdday	12	0.48 ^{a}
Robura	8	1.05 ^{ab}	Water regime	*N*	Average
UC-82	8	0.54 ^{ab}	40	24	0.99 ^{b}
			70	24	0.40 ^{a}

Table 6. Correlation (R-squared values) between RWC and the leaf area (A) of three tomato genotypes with balanced growth of the leaf area.

Cultivar	Parameter	RWC (%)
Premium	A	98.40***
Robura	A	84.01**
UC-82	A	77.89*

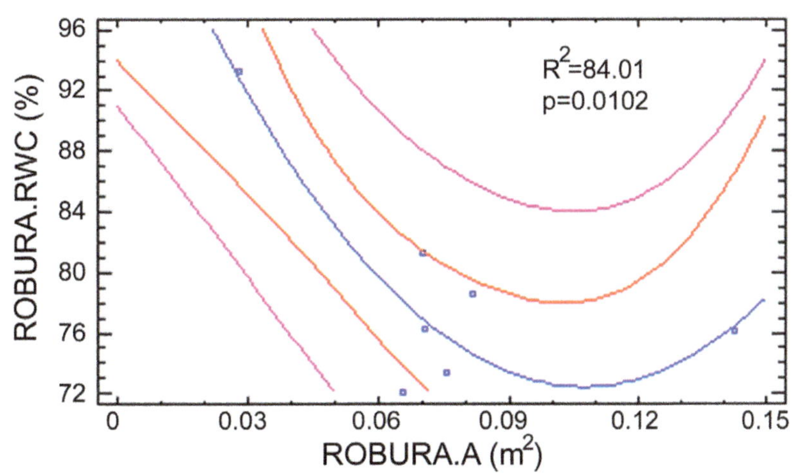

Figure 1. Polynomial correlation between relative water content RWC) and leaf area (A) of tomato cultivar ROBURA growing in conditions of differentiated water regime. Blue line: fitted regression model; red lines: 95% confidence intervals for mean values of RWC; pink lines: 95% prediction limits for new observations.

value of 0.48 μmol.g^{-1}, which was similar to that at the beginning of the experiment.

Plants grown at a low level of substrate saturation (40%) had significantly higher proline content as compared to the control plants (Table 5). During the 23-day period of water deficit, differences in leaf area growth were recorded for particular genotypes. On the 10th day of water stress, the most sensitive reactions were recorded on cultivars Hana, Denar and Moldeny, which reduced the leaf area growth. However, the final leaf area depletions were 45.3, 43.5, and 31.7% for Hana, Denar and Moldeny cultivars, respectively.

Premium, Robura and UC-82 cultivars did not reduce the growth of their leaf area (A). For these genotypes, quite a significant correlation between the leaf area (A) and RWC was observed (Table 6). The relationship can be described with a second order polynomial (Figures 1, 2 and 3). For the mentioned cultivars, there is an evident decrease in the RWC with increasing leaf size. Later, RWC values became stable and did not change with later

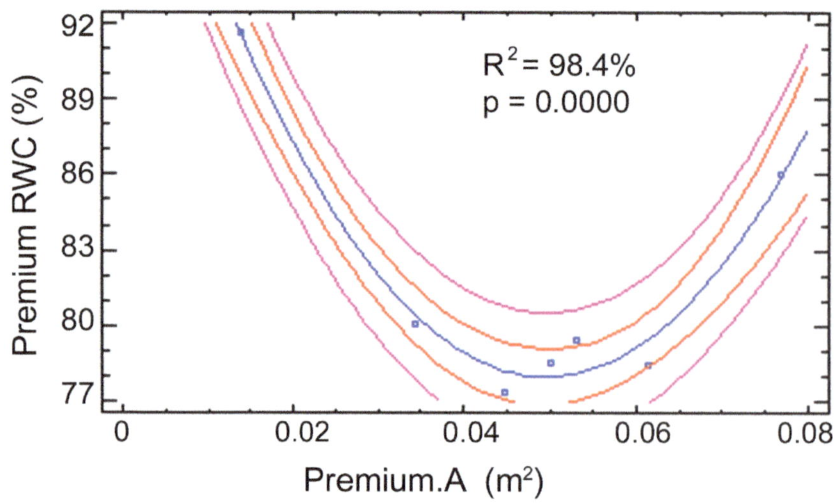

Figure 2. Relationship between relative water content (RWC) and leaf area (A) of tomato cultivar Premium growing in conditions of differentiated water regime. Blue line: fitted regression model; red lines: 95% confidence intervals for mean values of RWC; pink lines: 95% prediction limits for new observations.

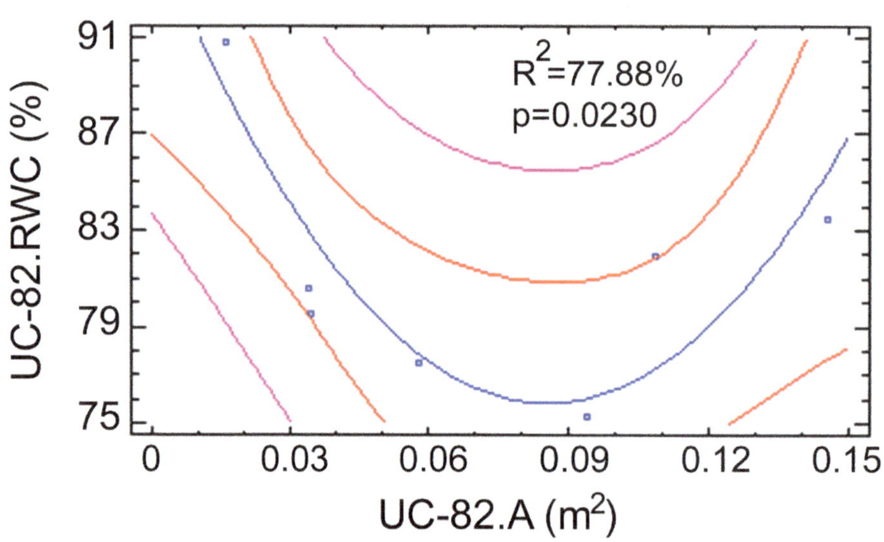

Figure 3. Relationship between relative water content (RWC) and leaf area (A) of tomato cultivar UC-82 growing in conditions of differentiated water regime. Blue line: fitted regression model; red lines: 95% confidence intervals for mean values of RWC; pink lines: 95% prediction limits for new observations.

growth of the leaf area (Figures 1, 2 and 3). Besides the plant responses already mentioned, low values of proline in leaves accompanied by an increase of the relative water content (RWC) and relative growth rate of the leaf area (RGR_A) were recorded after 17 days of water deficit. As such, the highest values of RGR_A were obtained in the interval between the 17th and 23rd day of drought for the six tomato genotypes.

DISCUSSION

Some genotypes of cultural plants have a specific growth and productive potential. Its realisation depends on environmental conditions. There is an increasing demand for identification of such genotypes tolerant to unfavourable factors including drought. However, the exact identification of genetic conditioning in plant reactions to drought has not yet been successful. Many papers have studied the osmoprotectant accumulation, in which the parameter has been used in selection for tolerance to environmental stresses (Ozturk and Demir, 2002; Hsu et al., 2003; Kavi-Kishore et al., 2005; Ashraf and Foolad, 2007). The results of the experiments with six tomato cultivars in conditions of increasing drought, however, revealed some physiological reactions of tomato cultivars, which may be regarded as significant traits of adaptation to water deficiency.

Reduction of the relative water content (RWC) in leaves of all the studied genotypes was detected during slow plant dehydration. Differences in this parameter among the studied genotypes were not significant. Depending on the duration of the plant hydration, the lowest values of relative water content were measured on the 17th day of the experiment. However, these values were maintained on 85 to 78%, which correlated (Gonzáles and Gonzáles-Vilar 2001) with changes in the composition of the tissues and some alterations in the relative rates of photosynthesis and respiration.

There is much information in the literature about osmotic adjustment as a result of osmoprotectant accumulation in plants (Hsu et al., 2003; Nayyar and Walia, 2003; Kavi-Kishore et al., 2005). For instance, proline accumulation is considered to be a part of osmotic adjustment, which allows turgor maintenance, decrease in water potential and an increase of water uptake from the soil (Ali et al., 1999). In our experiments, the accumulation of L-proline was observed in leaves of tomato plants with reduced RWC values. Proline content was influenced by tomato genotype, sampling date and water deficiency. However, statistically significant differences in proline content among genotypes were recorded (Tables 3 and 4). The question is whether the differences mentioned are symptoms of stress, or a result of genotype adaptation to stress. Garcia et al. (1997) and Lutts et al. (1999) consider proline accumulation to be rather a symptom of damage, than an indicator of plant resistance. According to our findings, the significant influence of the time factor (stress duration) and dehydration on proline content confirms its role in plant adaptation to water stress as postulated by Taylor (1996) and Hasegawa et al. (2000).

According to some authors, the adaptive potential of some particular plant species includes curbing of water losses, achieved by a reduction of both the leaf area growth rate (Hsiao, 2000) and the permeability of the cuticle towards water (Riederer, 2001), a closing of stomata and a reduction in the transpiration rate (Tardieu et al., 1996).

Our results refer to genotype conditioning in the mentioned plant responses. Under conditions of water stress, the three genotypes (Hana, Denar and Moldeny) reduced relative water content (RWC) and growth rate of the leaf area. Under the same conditions, Premium, Robura and UC-82 cultivars maintained quite a number of elevated values of leaf area, while the relative water content (RWC) in their leaves was significantly correlated with the leaf area. After reduction of RWC at the beginning of the experiment, these genotypes maintained a stable RWC which did not change following growth of the leaf area. Thus, this study confirms the significant differences in physiological responses among tomato genotypes and their adaptation to water deficiency. However, relative water content correlated with leaf area growth and proline accumulation, and these correlations can be regarded as indicators of genotype drought tolerance.

Conclusions

Due to the current climate changes exemplified by longer drought periods and higher temperatures during the growing season, a description of the principles and processes of plant adaptation to unfavorable conditions is essential. According to results obtained in our experiments with tomato cultivars, all genotypes studied responded to water deficiency with physiological responses on the level of plant water regime, growth response and osmoregulation. However, some tomato cultivars like Premium-determinate (main season variety), Robura-determinate (late variety) and UC-82 determinate (fast ripening variety) had significantly correlated RWC values and growth of leaf area. For the three tomato genotypes studied (Hana, Denar and Moldeny, which are all determinate semi-fast ripening varieties), the growth of the leaf area was limited by reduction of relative water content (RWC) in the leaves; whereas the other genotypes, after initial reduction, maintained stable RWC. Nonetheless, the values of this parameter did not change with growth of the leaf area.

Fundamentally, statistically significant differences in the average L-proline content were noted between the tomatoes' genotypes studied. Also, a significant influence of stress duration (sampling date) and substrate saturation (water regime) was confirmed. The highest average proline content was found in the cultivar Denar, which displays the most evident adaptation of a tomato genotype (vs. other cultivars) to stress conditions through osmotic adjustment. A similar mechanism of plant adaptation was also confirmed for other tomato cultivars growing in conditions with lower substrate saturation.

ACKNOWLEDGEMENTS

This study was supported by the Grant Agency of the Ministry of Education, Slovak Republic. The authors wish to acknowledge the Slovak Academy of Sciences (VEGA grant No. 1/0426/09). The adaptability and vitality of plants as a criterion of their utilization in urban spaces and landscape).

REFERENCES

Ali M, Jensen CR, Mogensen VO, Andersen MN, Henson IE (1999). Root signalling and osmotic adjustment during intermittent soil drying sustain grain yield of field grown wheat. Field Crop Res., 62: 35-52.

Adejare FB, Umebese CE (2008). Water stress induces cultivar dependent changes in stomatal complex, yield and osmotic adjustments in *Glycine max* L. Int. J. Agric. Res., 3: 287-295

Ashraf M, Foolad MR (2007). Roles of glycine betaine and proline in improving plant abiotic stress resistance. Environ. Exp. Bot., 59(2): 206-216

Bates LS, Waldrew RR, Teare ID (1973), Rapid determination of free proline for water-stress studies. Plant Soil, 39:113-127.

Ben–Rouina B, Ben-Ahmed CH, Athar H (2006). Water relations, proline accumulation and photosynthetic activity in olive tree (*Olea europea L.* cv"Chemlali") in response to salt stress. Pak. J. Bot., 38(5): 1397-1406.

Bohnert HJ, Jensen RG (1996). Strategies for engineering water-stress tolerance in plants. Trends Biotechnol., 14: 89-97.

Delauney AJ, Verma DPS (1993). Proline biosynthesis and osmoregulation in plants In: Plant J., 4: 215-223.

Fernandez D, Castrillo M (1999). Maize leaf rolling initiation. Photosynthetica, 37:493-497.

Garcia AB, de Almeida EJ, Iyer S, Gerats T, Van Montagu M, Caplan AB (1997). Effects of osmoprotectans upon NaCl stress in rice. Plant Physiol., 115: 159-169.

Gonzáles I, Gonzáles-Vilar M (2001), Determination of relative water content. In : Reigosa, M.J., Handbook of Plant Ecophysiology TechniquesKluwer Academic Publishers, Netherlands. pp. 207-212.

Gubiś J, Jureková Z, Dragúňová M, Hudcovićová M, Lichtnerová H, Dokupil T (2006). Transformed tobacco plants with increased proline content, Biotechnology, Scientific Pedagogical Publishing.

Hasegawa PM, Bressan RA, Zhu JK, Bohnert HJ (2000), Plant cellular and molecular responses to high salinity. Ann. Rev. Plant Physiol., 51: 463-499.

Hong Z, Lakkineni K, Zhang Z, Verma DPS (2000). Removal of feedback inhibition of Δ^1- pyrroline-5-carboxylate synthetase results in increased proline accumulation and protection of plants from osmotic stress. Plant Physiol., 122: 1129-1136.

Hsiao TC (2000). Leaf and Root Growth in Relation to Water Status. HortScience, 35(6): 1051-1058.

Hsu SY, Hsu YT, Kao CH (2003). The effect of polyethylene glycole on proline accumulation in rice leaves. Biol. Plant. 46: 73-78.

Hunt R (1982). Plant growth curves. Edvard Arnold, London, pp. 2-247.

Lannucci A, Rascio A, Russo M, Di Fonso N, Martiniello P (2000). Physiological responses to water stress following a conditioning period in berseem clover. Plant Soil, 223: 217-227.

Jureková Z, Molnárová K, Lichtnerová H, Dragúňová M (2003). Production of free proline in genotypes of tomato (*Lycopersicon esculentum* Mill.). In:Nové poznatky z genetiky a šľachtenia poľnohospodárskych rastlín. Zborník referátov. Piešťany, pp. 63-66.

Kadioglu A, Terzi R (2007). A dehydration Avoidance Mechanism : Leaf rolling. Bot. Rev., 73 (4): 290-302.

Kavi-Kishor PB, Hong Z, Miao GH, Hu CAA, Verma DPS (1995). Overexpression of Δ^1- pyrroline-5 carboxylate synthetase increases proline production and confers osmotolerance in transgenic plants. Plant Physiol., 108: 1387-1394.

Kavi-Kishor PB, Sangham S, Amrutha RN, Laxmi PS, Naidu KR, Rao KRSS, Rao S, Reddy KJ, Theriappan P, Sreenivasulu N (2005). Regulation of proline biosynthesis, degradation, uptake and transport in higher plants : its implications in plant growth and abiotic stress tolerance. Curr. Sci., 88: 424-438.

Kolb TE, Steiner KC (1990). Growth and biomass partitioning of Northen Red Oak and Yellow-Poplar Seedlings: Effect of Shading and Grass root Competition. For. Sci., 36(1):34-44.

Květ J, Nečas J, Ondok JP (197). Metody růstové analýzy. Studijní informace ÚVTIZ, Praha, pp. 112 s.

Ludlow MM, Muchow RC (1990). A critical evaluation of traits for improving crop yields in water limited environments. Adv. Agron. 43: 107-153.

Lutts S, Majerus V, Kinet,JM (1999). NaCl effects on proline metabolism in rice (*Oryza sativa*) seedlings. In: Physiol. Plantarum, 105: 450-458.

Ma Q, Nickman SR, Turner DW (2006). Response of osmotic adjustment and seed yield of *Brassica napus* and *B.juncea* to soil water deficit at different growth stages. Aust. J. Agric. Res., 57: 221-226.

Nanjo T, Kobayashi M, Yoshiba Y (1999), Biological function of proline in morphogenesis and osmotolerance revealed in antisense transgenic *Arabidopsis thaliana*. Plant J., 18: 185-193.

Nayyar H, Walia DP (2003). Water stress induced proline accumulation in contrasting wheat genotypes as affected by calcium and abscisic acid. Biol. Plant. 46: 275-279.

Ozturk L, Demir Y (2002). *In vivo* and *in vitro* protective role of proline. Plant Growth Regul., 38: 259-264.

Pedrol N, Ramos P, Reigosa MJ (2000). Phenotypic plasticity and acclimation to water deficits in velvet-grass: a long-term greenhouse experiment. Changes in leaf morphology, photosynthesis and stress-induced metabolites. J. Plant Physiol., 157: 383:393.

Riederer M (2001). Physiology and ecology of plants under stress. In: J. Exp. Bot., 52:4.

Saglam AA, Kadioglu R, Terzi R, Saruhan N (2008). Leaf rolling and biochemical changes in them in post-stress emerging *Ctenanthe setosa* plant under drought conditions. Russ. J. Pl. Physiol., 55: 48-53.

Serraj R, Sinclair TR (2002). Osmolyte accumulation: can it really help increase crop yield under drought condition?. Plant Cell Environ., 25: 331-341.

Tardieu F, Davies WJ (1996). Root-shoot communication and whole plant regulation of water flux. In: Water deficit in Plants, Bios. Sci. Publ., pp.147-162.

Taylor CB (1996). Proline and water deficit. Ups, downs, ins and outs. Plant Cell, 8: 1221-1224.

Umebese CE, Olatimilehin TO, Ogunsusi TA (2009). Salicylic acid Protects Nitrate Reductase Activity, Growth and Proline in Amaranth and Tomato Plants during Water deficit. Am. J. Agric. Biol. Sci., 4(3): 224-229

Zhu JK (2002). Salt and Drought Stress Signal Transduction in Plants. Ann. Rev. Plant Biol., 53: 247-273.

33

Propagation of Cabralea canjerana by mini-cutting

Eliseo S. Gimenes[1], Paula Kielse[1], Kelen L. Haygert[1], Frederico D. Fleig[2], Daniel E. Keathley[3] and Dilson A. Bisognin[1]

[1]Department of Crop Science, Federal University of Santa Maria, RS, Brazil.
[2]Department of Forestry Science, Federal University of Santa Maria, RS, Brazil.
[3]Department of Horticulture, Michigan State University, East Lansing, MI, USA.

Canjerana (*Cabralea canjerana* (Vell.) Martius) is a tree species indigenous to Brazil that when grown and managed in plantation systems are of great ecological and economic importance. Due to the difficulty of producing seminal seedlings, we examine the possibility of vegetative propagation by evaluating the rooting potential of canjerana mini-cuttings with different concentrations of indolbutyric acid (IBA) and substrate combinations. Mini-cuttings were treated with 2000 mg/L of IBA and planted in commercial substrate; coarse sand; carbonized rice husks; and a combination of the two. Apical and nodal mini-cuttings were treated with 0, 1000, 2000 and 3000 mg/L of IBA and planted in a combination of commercial substrate, coarse sand and carbonized rice husks. A mini-clonal hedge was formed with three clones of canjerana to evaluate mini-stump productivity and mini-cutting rooting. The combination of commercial substrate, coarse sand and carbonized rice husks maximized mini-cuttings rooting. Nodal mini-cuttings had higher rooting capability than apical ones. The application of 3000 mg/L of IBA improved rooting differentiation and growth of canjerana mini-cuttings. Canjerana clones differ in rooting capability and survival rates in vegetative propagation systems, but the use of a mini-cutting propagation system is a feasible production technique for this important species.

Key words: Vegetative propagation, miniclonal hedge, mini-cutting rooting, indolbutyric acid.

INTRODUCTION

Cabralea canjerana (Vell.) Martius, known as canjerana in southern Brazil, is a native tree species that belongs to the Meliaceae family. Canjerana is a common species in some Atlantic Forest fragments, and it is very important for forest regeneration programs in degraded areas (Nobrega et al., 2008). The wood is considered to be one of the most valuable, with excellent quality characteristics and high resistance to the attack of xylophage insects. The stem bark can be used for the extraction of a commercially important red dye. Both the stem bark and the roots have medicinal properties and can be used as a purgative, febrifuge, abortifacient, antidyspeptic, astringent and emetic (Carvalho, 2006). Extracts from leaves and seeds can affect the development of *Ascia monuste orseis* larvae, reducing leaf consumption in cabbage (Mata and Lomonaco, 2013).

The seeds of canjerana present recalcitrant storage behavior, with a drastic reduction in germination after 15 days of beneficiation (Grunennvaldt et al., 2014), which is an obstacle for the production of seminal seedlings. Vegetative propagation enables the production of plantlets in any season and the establishment of specific and favorable combinations of important characters, as well as the possibility of fixing superior genetic interactions and high levels of heterozygosity (Zobel, 1993; Gaspar et al., 2005). As canjerana provides wood and other products, the development of efficient clonal propagation techniques for superior genotypes will increase management and productivity of commercial plantations, thereby contributing to local economies.

The use of mini-cuttings is one of the most recent techniques for vegetative propagation. The mini-cutting technique can enhance propagation efficiency and was developed to overcome the rooting problems of some adult plants of the genus *Eucalyptus* (Wendling et al., 2003). Mini-cuttings from rejuvenated propagules are used for regeneration and mass production of woody species plantlets (Hartmann et al., 2011). In some clones of *Eucalyptus* the mini-cuttings from rejuvenated plant material develop a better root system than is found in mini-cuttings from non-rejuvenated tissue (Wendling and Xavier, 2005). Research results with other tree species indicate the viability of using rooted mini-cuttings for commercial propagation (Dumroese et al., 2003; Xavier et al., 2003; Wendling et al., 2007; Lima et al., 2009; Silva et al., 2010; Dias et al., 2012).

In the plantlet production from mini-cuttings, the substrate plays a fundamental role in root initiation and growth. It keeps the mini-cuttings in an upright position, provides a matrix for needed water and nutrients, and improves root differentiation and growth. Because of these specific but multifaceted requirements, the best substrate is usually composed of a variety of different materials that fulfill these physical functions and facilitate the hormonal stimulation of the desired growth responses, as well as providing availability of needed assimilates for rooting and plant health (Fonteno et al., 1981; Smart et al., 2003). Frequently in forest species, the inclusion of exogenous indolbutyric acid (IBA) has been shown to improve root differentiation (Wise et al., 1985; Rosier et al., 2004; San José et al., 2012). However, mini-cuttings of some woody species have been shown to root even without the application of IBA (Silva et al., 2010; Ferreira et al., 2010).

The choice of the substrate for the production of plantlets should take into consideration physical factors, such as texture and density, which interfere with aeration, water retention capacity and aggregation of the substrate (Wendling and Gatto, 2002). A good substrate must provide sufficient porosity to allow good aeration and have a high water retention capacity (Hartmann et al., 2011). Recommendations in the literature vary from an addition of 20 to 40% or a reduction of 60 to 80% of porous material (carbonized rice husks, coconut fiber, pine needles, compost, etc.) for the substrate composition for different species and propagation conditions (Wendling and Gatto, 2002). Substrates with air-filled pore space of 30%, with total porosity of 85% and water availability between 24 and 40% were considered as ideal conditions for seedling production (Schimitz et al., 2002), but no such information was found for mini-cuttings rooting. The three individual substrates, carbonized rice husks, commercial substrate, and coarse sand, and their combinations exhibit different physical properties from those considered ideal by Schimitz et al. (2002) for seedlings production. Carbonized rice husks increase the total porosity and air-filled pore space in the combination of substrates, but not to the optimum level suggested by Schimitz et al. (2002). This probable explains why the best substrate is usually obtained by using a mixture of components to ensure nutritional balance and diverse microbial flora (Wendling and Gatto, 2002). Given the ecological and economic importance of canjerana and the difficulty of producing seminal seedlings, this study was intended to examine the possibility of vegetative propagation by mini-cuttings. The objective was to evaluate the rooting potential of canjerana mini-cuttings with different concentrations of IBA and substrate combinations.

MATERIALS AND METHODS

The study was carried out in an acclimatized greenhouse of polycarbonate 10 mm panels with maximum temperature set to 32°C. For the experiments, one-year-old seedlings of canjerana were placed in plastic pots containing 300 cm^3 of a commercial substrate (organic pine bark base) and submitted to drastic pruning (coppice) to produce the sprouts that were used for mini-cutting preparation after 60 days. The coppice seedlings were manually irrigated twice a week with a nutrient fertilizer solution with 50% of the salt concentrations as described by Wendling et al. (2007). The mini-cuttings were grown in a vertical position in polyethylene trays (55 × 34 × 15 cm) containing the substrate and maintained in a greenhouse with a system of intermittent misting, mean air temperature of 25°C, and relative humidity of approximately 80%. The high relative humidity was maintained by nebulizers, triggered by a timer. After evaluating rooting and survival in a mist chamber, the rooted mini-cuttings were acclimatized in a greenhouse for 30 days and used for the formation of the mini-clonal hedge.

The first experiment evaluated the effect of the substrate on mini-cutting rooting and survival. The bases of single bud mini-cuttings, 1.5 to 2.0 cm long containing a half leaflet, were immersed for 10 s in a 50% water/ethanol solution containing 2000 mg/L of IBA and placed in one of the four substrates: commercial (organic pine bark base); coarse sand; carbonized rice husks; and the combination of equal proportions by volume of commercial substrate, coarse sand and carbonized rice husks. Samples of these substrates were collected, dried at 65°C for 48 h and submitted for physical analysis (Fermino, 2003). Carbonized rice husks and the mixture of commercial substrate, coarse sand and carbonized rice husks were the substrates with air-filled pore space that came closer to the optimal value of 30% (Table 1). After 60 days of cultivation, mini-cuttings were evaluated for rooting and survival and the number and total length of roots (cm). The experiment was a complete random design with five replicates of four mini-cuttings.

The second experiment evaluated the type of mini-cuttings (apical and nodal) of 1.5 to 2.0 cm in length containing a half leaflet. The bases of mini-cuttings were immersed for 10 s in a 50% water/ethanol solution containing either 0, 1000, 2000 or 3000 mg/L of IBA. The control treatment consisted of a 50% water/ethanol solution. A mixture of equal proportions by volume of commercial substrate, coarse sand and carbonized rice husks was used as substrate. The rooting and survival of the shoots and the number

Table 1. Physical properties of the substrates used for the evaluation of the rooting capacity of canjerana.

Substrates[1]	Dry density (kg/m^3)	Total porosity (%)	Aeration space in saturated substrate (%)	Water availability (%)
Comercial	335	50	15.4	5.4
Coarse Sand	1505	27	12.6	11.5
CRH	167	41	31.4	2.8
Comercial+CS+CRH	56	43	23.9	4.4

[1]Commercial: organic pine bark base; CS: coarse sand; CRH: carbonized rice husks.

Table 2. Percentage survival and rooting and number and length of roots of mini-cuttings of canjerana treated with 2,000 mg/L of indolbutyric acid and grown in different substrates after 60 days of cultivation.

Substrates[1]	Survival (%)	Rooting (%)	Number of roots	Length of roots (cm)
Comercial	70.0[b2]	15.0[b]	1.8[a]	3.3[a]
Coarse sand	70.0[b]	45.0[b]	1.8[a]	2.5[a]
CRH	85.0[b]	30.0[b]	1.8[a]	2.7[a]
Comercial+CS+CRH	100.0[a]	75.0[a]	1.7[a]	3.0[a]
Mean	81.2	41.2	1.8	2.9
CV (%)	14.1	39.4	14.4	8.6

[1] Commercial: organic pine bark base; CS: coarse sand; CRH: carbonized rice husks. [2] Means values followed by the same letter in a column are not significantly different by the Tukey's test at the probability of 5%.

and total length of roots (cm) per mini-cutting were evaluated after 60 days of cultivation. The experiment was a 2 x 4 factorial (apical and nodal mini-cuttings by IBA concentrations) in a complete random design with five replicates of four mini-cuttings.

The third experiment evaluated the productivity of individual mini-stumps and rooting of mini-cuttings of three clones of canjerana. The mini-clonal hedge was established in a soilless system subirrigated with nutrient solution (Bandinelli et al., 2013). Twelve rooted mini-cuttings (complete plantlets) per single stock plant of canjerana were planted at 10 x 10 cm spacing in one tray to form the mini-clonal hedge and given a 15 min daily irrigation with a nutrient solution at 50% concentration of salts as described by Wendling et al. (2007). The pH of the nutrient solution was maintained between 5.5 and 6.0 and an electrical conductivity in 1.5 dS m^{-1}. The sprouts of three consecutive harvesting times of the three clones were used to prepare mini-cuttings of 1.5 to 2.0 cm in length containing a half leaflet. The mini-cuttings were treated with 3000 mg/L of IBA solution for 10 s and planted in a combination of equal proportions by volume of commercial substrate, coarse sand and carbonized rice husks. The number of mini-cuttings per mini-stump was recorded and the percentage of survival and rooting were evaluated at 60 days after planting. The experiment was a complete random design with eight replicates of six mini-cuttings.

Data were submitted to analysis of variance and for those variables with significant differences ($p \leq 0.05$), treatment means were compared by Tukey test or polynomial regression, as appropriate. For purposes of analysis, percentage data were transformed to arcsin$\sqrt{x/100}$ and counting data to $\sqrt{x+0.5}$ to attend the statistical presuppositions. All analysis well done with the ESTAT (UNESP - Jaboticabal) program.

RESULTS AND DISCUSSION

The different substrates affected the rooting and survival percentages of canjerana mini-cuttings ($p \leq 0.05$). The combination of commercial substrate, coarse sand and carbonized rice husks resulted in the highest percentage of mini-cutting rooting (Table 2). The number and length of roots per mini-cutting did not differ significantly across substrates. On average, rooted mini-cuttings produced 1.8 roots with an average total length of 2.9 cm. In the experiment using individual substrates (soil, sand, carbonized rice husks, peat, decomposed residue of black wattle bark) and their combinations, only the sand showed a value of available water within the ideal range for its use as a substrate for seedlings production, which implies that this material can ensure high water availability (Schimitz et al., 2002). The increased survival in combined substrate (Table 2) may reflect the effect of the coarse sand resulting in increased water availability to the mini-cuttings of canjerana. For *Maytenus ilicifolia* (Mart. ex Reissek) a higher rooting percentage (94.3%) was obtained with mini-cutting cultivation in pure commercial substrate (Lima et al., 2009). Rooting of *Calophyllum brasiliense* (Camb.) mini-cuttings was lower in commercial substrates (bark composted pine) and carbonized rice husks when compared with vermiculite (Silva et al., 2010). Clearly then, the optimal substrate varies according to species and conditions of rooting, and substrate evaluation is essential in optimizing rooting protocols given their determinant role in defining the range of root induction. Mini-cuttings from nodal segments showed a higher percentage of rooting and an increased number and length of roots in comparison to

Table 3. Percentage of survival and rooting and number and length of roots of apical and nodal canjerana mini-cuttings after 60 days of cultivation.

Mini-cuttings	Survival (%)	Rooting (%)	Number of roots	Total length of roots (cm)
Nodal	80.0^{a1}	50.0^{a}	2.2^{a}	6.3^{a}
Apical	74.0^{a}	17.0^{b}	1.4^{b}	3.0^{b}
Mean	77.0	34.0	1.8	4.6
F value^{2}	0.2^{ns}	5.3**	4.5**	36.6**
CV (%)	22.7	57.1	21.0	13.6

[1] Means values followed by different letters in a column are significantly different by the F test at the indicated probability. [2] (^{ns}) no significant, (*) significant at 5% of probability, and (**) significant at 1% of probability.

Table 4. Percentage survival and rooting and number and length of roots of mini-cuttings of canjerana treated or untreated with indolbutyric acid (IBA) after 60 days of cultivation.

Mini-cuttings	Surviv^{a}l (%)	Rooting (%)	Number of roots	Tot^{a}l length of roots (cm)
Treated with IBA	79.3^{a1}	43.3^{a}	2.0^{a}	5.2^{a}
Untreated	70.0^{a}	5.0^{b}	1.1^{b}	2.6^{b}
Mean	74.6	24.1	1.5	3.9
F value^{2}	1.3^{ns}	12.9**	14.8**	70.0**
CV (%)	17.8	52.1	13.6	10.5

[1] Means values followed by different letters in a column are significantly different by the F test at the indicated probability.
[2] (^{ns}) no significant, (*) significant at 5% of probability, and (**) significant at 1% of probability.

the apical ones (Table 3). The increment in rooting percentage of nodal segments was 2.9 folds. No significant difference in survival was observed between apical and nodal mini-cuttings. Interestingly, there were no significant differences ($p \geq 0.05$) in the responses of nodal versus apical explants to the IBA treatments, so data from the apical and nodal mini-cuttings were combined in Table 4. Given the lack of IBA treatment effects in this experiment, the differences found between nodal and apical mini-cuttings may be associated with the levels of carbohydrates, amino acids and other substances that promote adventitious rooting of certain tissues of the mini-cuttings, as noted by Hartmann et al. (2011). Thus, the mini-cuttings from nodal segments may have more appropriate levels of reserves than the apical ones, which would improve rooting capability. In addition, cells that have hormones or other compounds that confer an endogenous potential for root formation, such as auxin, quickly react to specific stimuli, like light and temperature (Sorin et al., 2005). It is possible that mini-cuttings from nodal segments have a better competence for rooting, but it does not affect mini-cutting survival.

The mini-stumps of canjerana had adequate nutritional level and the mini-cuttings for comparisons between clones were taken from the same sprouting period, with differences thus being an indication of existing genetic variation between canjerana clones, as expected. These are the first results of canjerana vegetative propagation by mini-cuttings and clearly indicate the effect of nodal and apical mini-cuttings on root initiation and growth. These differences are very important in developing a system for mass production of plantlets, because of economic and labor usage issues. It is possible that these differences are minimized with the advancing age of the shoots and the size of the mini-cuttings (greater than 2.0 cm for example), which should be the aim of future studies.

In other species, such as *Calophyllum brasiliense* (Camb.), apical and nodal mini-cuttings did not differ for rooting (Silva et al., 2010). These results were attributed to the good nutritional status of stock plants of the mini-clonal hedge, as well as to the physiological age of the two types of mini-cuttings, which makes them similar in the degree of juvenility and tissue lignification. In a study with mini-cuttings of *Anadenanthera macrocarpa* (Benth) Brenan, apical mini-cuttings were more responsive to rooting than nodal ones (Dias et al., 2012), being attributed to the greater degree of juvenility and less lignification of apical tissues (Xavier et al., 2003). Furthermore, apical mini-cuttings have a higher concentration of endogenous rooting promoters by virtue of their proximity to both the sites of auxin synthesis and to less differentiated tissues, potentially resulting in increased dedifferentiation of cells to their meristematic condition, which is essential for root initiation (Gehlot et al., 2014).

The application of IBA increased ($p \leq 0.05$) the percentage of rooting and the length and number of roots per canjerana mini-cutting (Table 4). The application of IBA did not significantly affect the survival of the mini-

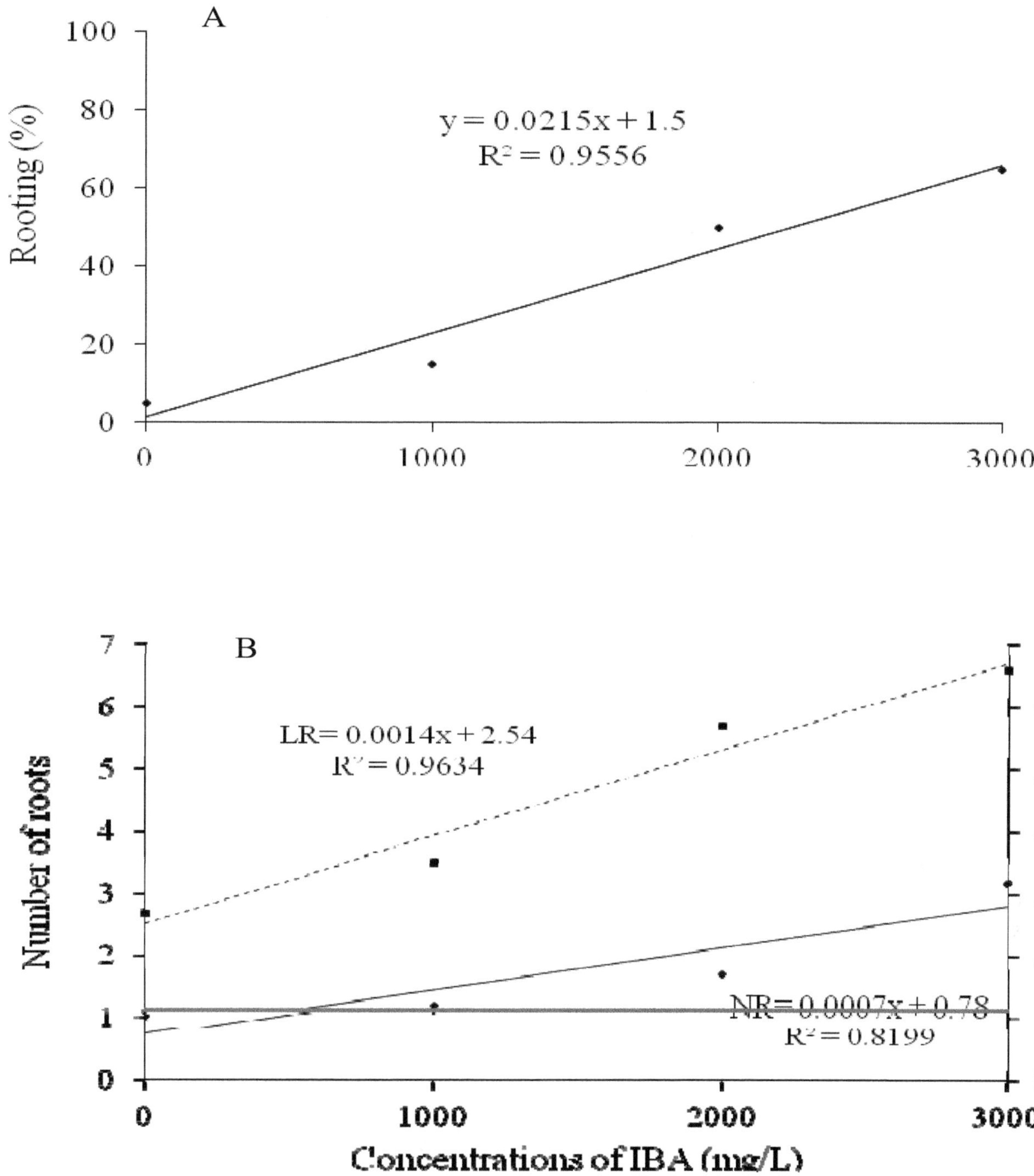

Figure 1. Rooting percent (A) and number (NR) and length of roots (LR) (cm) (B) of mini-cuttings of canjerana treated with different concentrations of indolbutyric acid (IBA) after 60 days of cultivation.

cuttings. Increasing the concentration of the IBA treatment improved the percentage of rooting and increased both the number and length of roots (Figure 1). At 60 days, canjerana mini-cuttings treated with IBA showed a higher percentage of survival and rooting, as well as increased root growth compared with mini-cuttings with no IBA treatment (Table 4). Auxin application enhances root formation in species with low rooting response, as governed by either genotype or physiological stage (Xavier et al., 2003). It is possible that low rooting response is caused by low content of endogenous auxin, which often requires the exogenous application of growth regulators to establish the competence and determination of target cells (Taiz and Zeiger, 2008). Therefore, canjerana fits in this group of species, because the application of IBA increased both

Table 5. The average number of mini-cuttings (NM) and percentages of survival (S) and rooting (R) of mini-cuttings of canjerana harvested at three different harvested dates from mini-stumps formed from mini-cuttings that were rooted in a soilless system subirrigated with nutrient solution.

Clones	Harvest I			Harvest II			Harvest III		
	NM	S (%)	R (%)	NM	S (%)	R (%)	NM	S (%)	R (%)
SM1	2.5[a1]	84[c]	53[b]	2.7[a]	73[c]	69[a]	2.8[a]	72[a]	72[b]
SM3	2.4[a]	98[a]	50[b]	2.5[a]	79[b]	51[b]	2.6[a]	55[b]	52[b]
SM13	1.8[a]	93[b]	67[a]	2.9[a]	90[a]	68[a]	3.0[a]	81[a]	80[a]
Mean	2.2	92	57	2.7	81	63	2.8	69	68
CV (%)	12.04	3.69	3.84	11.05	3.47	4.37	10.52	3.23	4.99

[1] Means values followed by the same letter in a column are not significantly different by the Tukey's test at the probability of 5%.

rooting percentage and the number and length of the new roots (Table 4 and Figure 1), which explains the higher rooting of canjerana mini-cuttings treated with IBA.

The observed differences in average root length may be associated with the increase in rooting competence promoted by IBA application that resulted in a rapid differentiation of the tissues for the formation of adventitious roots in canjerana mini-cuttings. In this study, only the nodal mini-cuttings treated with 2000 or 3000 mg/L of IBA had already developed adventitious roots at 30 days after cultivation (data not shown). The use of IBA resulted in an increased percentage of rooting and both the number and length of roots in the canjerana mini-cuttings (Figure 1). Also, the use of IBA favored rooting and did not affect the survival of mini-cuttings (Table 4), indicating that there was no phytotoxic effect in any of the tested concentrations (Figure 1). Mini-cuttings responded to an increase in IBA concentration up to 3000 mg/L, which can be considered a high concentration for mini-cuttings. At that concentration, 65% of mini-cuttings rooted, with an average of 3.2 roots per mini-cutting which averaged 6.6 cm in length at 60 days following cultivation. This contrasts with some previous mini-cutting studies in which the use of IBA concentrations above 2000 mg/L had negative effects on rooting due to the high degree of juvenility of the new shoot (Titon et al., 2003). It is thought that this is due to the fact that the juvenile mini-cuttings already have tissues with an endogenous hormonal balance that is favorable for rooting, which then leads to no (Silva et al., 2010; Ferreira et al., 2010; Wendling et al., 2010) or even a negative response to exogenous application of IBA (Xavier et al., 2003).

The production of mini-cuttings per mini-stump did not differ among the clones of canjerana that were evaluated in this study. The lowest production of mini-cuttings per mini-stump was 1.8 in the first harvest period and the highest was 3.0 in the third harvest (Table 5). The survival of mini-stumps was 95% and did not differ among harvest dates or clones (data not shown). In mini-cuttings from the mini-clonal hedge grown in a soilless system using subirrigation with a nutrient solution, the average production of mini-cuttings per harvesting date was 2.2 for the first harvest period, 2.7 for the second harvest period, and 2.8 for the third harvest period. These values were higher than has been reported for *Cedrella fissilis* (Vell.) (Xavier et al., 2003) and lower than those observed in *Ilex paraguariensis* (St. Hil.) (Wendling et al., 2007). There were significant differences in rooting ($p \leq 0.05$) among clones and harvest dates. The rooting percentage was 50% or greater in all treatments and showed an increase during the successive harvest periods. Interestingly, survival percentage dropped as rooting percentage increased (Table 5). In a study with *Liquidambar styraciflua* (L.), the survival percentage of rooted minicuttings was influenced by relative humidity and sunlight during the acclimatization period in the greenhouse and outdoor conditions (Wendling et al., 2010). Therefore, increasing the harvesting period of canjerana mini-cuttings can enhance the productivity of mini-stumps, and the maintenance of mini-cuttings in the greenhouse conditions may increase survival and rooting, as mentioned by Wendling et al. (2010). The decline in survival was particularly evident in the clone SM3. This interplay between survival, rooting percentage, and clone indicates the need for tailoring production systems specifically for each clone that will be propagated.

The results of this study clearly show that vegetative propagation by mini-cuttings is a feasible alternative for the mass production of canjerana plantlets for commercial plantation establishment (Figure 2). Apical and nodal mini-cuttings with 1.5 to 2.0 cm in length containing a half leaflet may be planted in media consisting of the same proportions of commercial substrate, coarse sand and carbonized rice husks. Clearly, though, nodal mini-cuttings have a higher competence for rooting. Based on this work, mini-cuttings should be treated with 3000 mg/L of IBA for the greatest rooting efficiency. Given clonal differences in these studies, it is possible that mini-cuttings with different amounts of reserves and/or from other clones might have a greater competence for rooting or higher survival rates during the propagation process, which would facilitate the mass production of plantlets, possibly with lower

Figure 2. Shoots in mini-stump of canjerana (A), mini-cutting (nodal segment) rooted with different concentrations of indolbutyric acid (IBA) (B), seedling with four months of age (C), acclimatized seedling in tube (D), rooted mini-cuttings being acclimatized in pots (E), mini-clonal hedge in a soilless system (coarse sand as substrate) subirrigated with nutrient solution (F), and mini-stumps with shoots (G).

concentrations or even without the application of IBA. Once the rooted mini-cuttings are acclimatized in the greenhouse they may be transferred to either the mini-clonal hedge or to a full sun area for further acclimatization (Figure 2). The new plantlets can then be effectively planted in the field for growth and wood quality evaluations. This work shows that the use of mini-cuttings is a feasible technique for mass production of canjerana plantlets for plantation establishment from selected clones in order to enhance management and growth of this important species.

Conflict of Interest

The authors have not declared any conflict of interest.

REFERENCES

Bandinelli MG, Bisognin DA, Gnocato FS, Mambrin RB, Sausen D, Nicoloso FT (2013). MS salt and sucrose concentrations in the *in vitro* multiplication and acclimatization of potato. Hort. Brasil. 31:242-247. Available from http://www.scielo.br/ pdf/hb/v31n2/11.pdf

Carvalho PER (2006). Espécies arbóreas brasileiras. Embrapa informação tecnológica; Colombo: Embrapa Floresta 2:628.

Dias PC, Xavier A, Oliveira LS, Paiva HN, Correia CG (2012). Vegetative propagation of angico-vermelho (*Anadenanthera macrocarpa* (Benth.) Brenan) through cuttings and mini-cuttings techniques. Rev. Árvore. 36(3):389-399. Available from http://dx.doi.org/10.1590/S0100-67622012000300001

Dumroese RK, Wenny DL, Morrison SJ (2003). Propagation protocol for container willows and poplars using mini-cuttings. N. P. J. 4(2):137-139. Available from http://npj.uwpress.org/content/4/2/137.full.pdf

Fermino MH (2003). Métodos de análise para caracterização física de substratos para plantas. Tese (Doutorado), Departamento de Horticultura e Silvicultura. Universidade Federal do Rio Grande do Sul. P. 89. Available from http://www.lume.ufrgs.br/bitstream/handle/10183/2269/000367173.pdf?sequence=1

Ferreira BGA, Zuffellato-Ribas KC, Wendling I, Koehler HS, Nogueira AC (2010). Minicutting of *Sapium glandulatum* (Vell.) Pax with the use of indolebutyric acid and naphtalene acetic acid Ciênc. Florest. 20(1):19-31. Available from http://www.gepe.ufpr.br/pdfs/miniestaquia_de_sapium_glandulatum.pdf

Fonteno WC, Cassel DK, Larson RA (1981). Physical properties of three container media and their effect of Poinsettia growth. J. Am. Soc. Hort. Sci. 106(6):731-746.

Gaspar MJ, Borralho N, Gomes AL (2005). Comparison between field performance of cuttings and seedlings of Eucalyptus globulus. Ann. For. Sci. 62:837–841. Available from http://www.afs-journal.org/articles/forest/pdf/2005/08/F5096.pdf

Gehlot A, Gupta RK, Tripathi A, Arya ID, Arya S (2014). Vegetative propagation of *Azadirachta indica*: effect of auxin and rooting media on adventitious root induction in mini-cuttings. Adv. For. Sci. 1(1):1-9. Available from http://200.129.241.78/ojs/index.php/afor/article/view/1357/1194

Grunennvaldt RL, Cantarelli EB, Salamoni AT (2014). Armazenamento e viabilidade de sementes de *Cabralea canjerana* (Vell.) Mart. Com. Sci. 5(1):98-105. Available from http://comunicata.ufpi.br/index.php/comunicata/article/view/321/227

Hartmann HT, Kester DE, Davies FT, Geneve RL (2011). Plant propagation: principles and practices. New Jersey: Prentice Hall. P. 928.

Lima DMD, Tanno GN, Purcino M, Biasi LA, Zuffellato-Ribas K, Zanette F (2009) Rooting of espinheira-santa (*Maytenus ilicifolia* Mart. ex Reissek) minicuttings on different substrates. Ciênc. agrotec. 33(2):617-623. Available from http://dx.doi.org/10.1590/S1413-70542009000200040

Mata RFF, Lomonaco C (2013) Toxicity, deterrence and repelence of aqueous extracts of *Cabralea canjerana* ssp. polytricha (Meliaceae) on Ascia monuste orseis (Lepidoptera), the cabbage catterpillar. Rev. Árvore 37(2):361-368. Available from http://www.scielo.br/pdf/rarv/v37n2/18.pdf

Nobrega AMF, Valeri SV, Paula RC, Silva SA (2008). Natural regeneration in forest remainders and reforested areas of Mogi-Guaçu river floodplain in Luiz Antônio county, São Paulo, Brazil. Rev. Árvore. (32):909-920. Available from http://www.scielo.br/pdf/rarv/v32n5/16.pdf

Rosier CL, Frampton J, Goldfarb B, Blazich FA, Wise FC (2004). Growth stage, auxin type, and concentration influence rooting of stem cuttings of fraser fir. Hortsci. 39(6):1397-1402. Available from http://hortsci.ashspublications.org/content/39/6/1397.full.pdf+html

San José MC, Romero L, Janeiro LV (2012). Effect of indole-3-butyric acid on root formation in *Alnus glutinosa* microcuttings. Silva Fenn. 46(5):643–654. Available from http://m.metla.eu/silvafennica/full/sf46/sf465643.pdf

Schimitz JAK, Souza PVDD, Kampf AN (2002). Physical and chemical properties of substrates with mineral and organic origin for growth of potted plants. Cienc. Rural. 32(6):937-944. Available from http://dx.doi.org/10.1590/S0103-84782002000600005

Silva RL, Oliveira MLD, Monte MA, Xavier A (2010). Guanandi (*Calophyllum brasiliense*) clonal propagation through minicuttings. Agron. Costarricense, 34(1):99-104. Available from http://www.redalyc.org/articulo.oa?id=43617800010

Smart DR, Kocsis L, Walker MA, Stockert C (2003). Dormant buds and adventitious root formation by *Vitis* and other woody plants. J. Plant Growth Regul. 21:296-314.

Sorin C, Bussell JD, Camus I, Ljung K, Kowalczyk M, Geiss G, McKhann H, Garcion C, Vaucheret H, Sandberg G, Bellini C (2005). Auxin and light control of adventitious rooting in arabidopsis require ARGONAUTE1. Plant Cell 17:1343-1359. Available from http://www.plantcell.org/content/17/5/1343.full.pdf+html

Taiz L, Zeiger E (2008). Fisiologia vegetal. 4 ed. Porto Alegre: Artmed. P. 820.

Titon M, Xavier A, Otoni WC, Gonçalves GRD (2003). Minicuttings and microcuttings rooting of *Eucalyptus* grandis W. Hill ex Maiden. clones as affected by IBA. Rev. Árvore 27(1):1-7. Available from http://dx.doi.org/10.1590/S0100-67622003000100001

Wendling I, Gatto A (2002). Substrato, adubação e irrigação na produção de mudas. Viçosa: Aprenda Fácil. P. 145.

Wendling I, Xavier A (2003). Serial minicuttings technique as a rejuvenation method for *Eucalyptus* clones. Pesq. Agropec. Bras. 38(4):475-480. Available from http://www.scielo.br/pdf/pab/v38n4/a05v38n4.pdf

Wendling I, Ferrari M, Dutra LF (2005). Produção de mudas de corticeira-do-mato (*Erythrina falcata* Bentham) por miniestaquia a partir de propágulos juvenis. Comunicado Técnico. Colombo: Embrapa Florestas, P. 130. Available from http://www.bibliotecaflorestal.ufv.br/bitstream/handle/123456789/6285/Comunicado_Tecnico_130.pdf?sequence=1

Wendling I, Xavier A (2005). Influence of the serial minicutting technique on rooting vigor of *Eucalyptus grandis* clones. Rev. Árvore. 29(5):681-689. Available from http://www.scielo.br/pdf/rarv/v29n5/a03v29n5.pdf

Wendling I, Dutra LF, Grossi F (2007). Production and survival of *Ilex paraguariensis* minicuttings and ministumps cultivated in semi-hydroponic system. Pesq. Agropec. Bras. 42(2):289-292. Available from http://www.scielo.br/pdf/pab/v42n2/19.pdf

Wendling I, Brondani GE, Dutra LF, Hansel FA (2010). Mini-cuttings technique: a new ex vitro method for clonal propagation of sweetgum. New Forest. 39:343-353 Available from http://www.researchgate.net/publication/232062592_Mini-cuttings_technique_a_new_ex_vitro_method_for_clonal_propagation_of_sweetgum

Wise FC, Blazich FA, Hinesley LE (1985). Propagation of *Abies fraseri* by softwood stem cuttings. Can. J. For. Res. 15:1172-1176. Available from http://www.nrcresearchpress.com/doi/pdf/10.1139/x85-190

Xavier A, Santos GAD, Oliveira MLD (2003). Rooting of stem and leaf minicuttings in the vegetative propagation of Cedro-Rosa (*Cedrela fissilis* Vell.). Rev. Árvore. 27(2):139-143. doi: 10.1590/S0100-67622003000200003

Zobel B (1993). Vegetative propagation in production forestry. J. For. 90:29-34.

Distribution and regeneration status of *Vitex payos* (Lour.) Merr. in Kenyan drylands

James Munga Kimondo[1] , Jacob Godfrey Agea[2], Clement Akais Okia[3], Andrew Woiso Dino[4], Refaat Atalla Ahmed Abohassan[5], Jackson Mulatya[1] and Zewge Teklehaimanot[6]

[1]Kenya Forestry Research Institute, P. O. Box 20412-00200, Nairobi, Kenya.
[2]College of Agricultural and Environmental Sciences, Makerere University, P. O. Box 7062 Kampala, Uganda.
[3]ICRAF Uganda Country Office P. O. Box 26416 Kampala, Uganda.
[4]Department of Biological Science, Sokoine University of Agriculture, P. O. Box 3038, Morogoro, Tanzania.
[5]Faculty of Meteorology, Environment and Arid Land Agriculture, King Abdulaziz University, P. O. Box 80208 Jeddah, 21589 Saudi Arabia.
[6]School of Environment, Natural Resources and Geography, Bangor University, Bangor-Gwynedd LL57, UK.

We investigated the population structure and regeneration status of *Vitex payos* (Lour.) Merr. in Kenyan drylands. The study quantified the spatial distribution pattern of *V. payos* tree populations in their natural range; and assessed their regeneration status to determine the stability of the populations Woodlands and farm inventories were conducted in Mbeere, Mwingi and Kitui districts of the Eastern Province of Kenya. The nearest-neighbour sampling method was used to determine tree density and distribution of *V. payos* in the study sites. The number of seedlings and saplings were counted. The diameters, crown diameter, and tree heights of sampled trees were measured. These morphological parameters were summarized on per hectare basis. The patterns of distribution of *V. payos* trees showed an aggregation of trees on farms and bushes. Tree densities ranged from 1.6 on farmlands to 20.3 trees per ha in the woodlands. The expected mean distances between nearest neighbouring trees were higher than the observed values on all sites, confirming that *V. payos* trees were more aggregated than randomly dispersed. The mean tree heights varied from < 5 m to > 9 m cross study sites. The sampled populations were dominated by trees (55%) within the range of 10 to 20 cm dbh. Highhest numbers of seedlings from all origins were recorded in the bushes in Kitui and Mbeere districts 101 and 78 seedlings per hectare, respectively. Sapling population densities were generally low.

Key words: Edible wild plants, dryland resources, indigenous fruit trees, Vitex payos, wild fruit trees, Kenya.

INTRODUCTION

Past agricultural activities are some of the major factors that have influenced the present day species composition among the various natural vegetation types in Kenya(Turner et al., 1998). These activities include cultivation of food crops, planting of exotic tree species and overstocking of livestock that interfere with the regeneration of natural vegetation through continuous trampling and/or browsing of young plants as well as the reproductive parts of adult plants - flowers and fruits (Turner et al., 1998). In particular, the planting of exotic species for different end-use products including fruits currently poses a major threat to indigenous species such as *Vitex payos*, yet many of these indigenous species are considered slow growers and therefore, often not preferred by farmers for on-farm planting (Dunn, 1991). The increasing human population exploiting a given environment, land fragmentation and climate change also contribute to the overall decline in indigenous vegetation (Wanyonyi, 2012; Oroda, 2011; Morris, 2010).

V. payos like other indigenous fruit trees are commonly found scattered on farms in Kenya (Jama et al., 2008; Ekesa et al., 2009). Small scale farmers retain naturally regenerated *V. payos* trees on arable fields, grazing lands, homesteads and along farm boundaries for various purposes such as fruits, apiculture, timber, shade, soil amelioration, firewood among others (Jama et al., 2008). The species regenerates naturally by seed, root suckers and coppicing (Maundu et al., 1999; Mbora et al., 2008). Root suckers often develop from injured roots especially during land preparation for agricultural purposes (Mbora et al., 2008) or construction of soil conservation structures such as terraces. Due to heavy browsing of the regenerates, especially during the dry season, majority of them remain stunted for long periods. Regenerates established from seeds often suffer heavy mortality because of combined effects of browsing and drought before their root systems are fully developed (Luoga et al., 2004). On the contrary, regenerates that grow from root suckers and stumps benefit from already established root systems and therefore tolerate the extreme environmental vagaries with less threat of mortality (Kauter et al., 2003). Trees in woodlands are often small and suffer from the dry weather due to excess competition. This is unlike on the farms where trees are properly protected through weeding thereby reducing competition for both nutrients and water resulting in bigger trees. However, due to the principal purpose of the land is agricultural, only a few trees are retained. These trees also maintain their leaves for a longer period even after the on-set of the dry season.

The objectives of this study were therefore to: quantify the spatial distribution pattern of *V. payos* tree populations in their natural range; analyze the size distribution pattern of *V. payos* trees under different land uses; and assess the regeneration status of the species in order to determine the stability of the populations. It was hypothesized that the tree spatial distribution in the natural range of the species is not different among locations and land uses. Secondly, that the density of seedling, sapling and mature trees are each randomly distributed on farms and bushlands with their populations decreasing with increasing age, a J-inverse distribution, indicating stable regenerating populations.

MATERIALS AND METHODS

Study area

The study was conducted in Mbeere, Mwingi and Kitui Districts which are located in the arid and semi-arid (ASAL) parts of Kenya. Mbeere is situated at between latitude 00°20' and 00° 50' S and longitude 37°16' and 37° 56' E with a mean annual rainfall of between 640 and 1100 mm, while Mwingi is situated at between 00°03' and 01° 12' S and 37°47' to 38°57' E and receives 600 to 1100 mm mean annual rainfall. Kitui District lies between 00°04' and 03°00' S and 37°45' to 39°00' E and receives a mean annual rainfall of between 500 and 1000 mm. The rains have bimodal distribution from March to May (long rains) and October to December (short rains) with peaks in April and November, respectively. Minimum and maximum temperatures in the study districts are fairly constant with values of 21.6 and 34.8°C in Mbeere, 22.8 and 37.2°C in Mwingi, 21.9 and 36.6°C in Kitui, respectively. These areas are predominantly under subsistence crop production practiced by low resourced farmers, who also rear large herds of livestock under free range grazing system. These areas are also characterized by low soil fertility and experience frequent droughts and famine.

Data collection

Nearest-neighbour sampling method, in which the distance and direction of the 100 sampled trees were measured from a randomly selected individual tree to its nearest neighbour, was used to determine tree density and distribution of *V. payos* in the three study sites (Sparks et al., 2002). However, several trees were found to be each other's nearest neighbour and use of these distances twice could decrease sample variance (Donnelly, 1978). Furthermore, the trees at the edge of the imaginary plot could not have trees outside the fixed number as their nearest neighbours. To overcome the bias as a result of these two conditions, Donnelly's correction factor was used (Donnelly, 1978). Tree densities in circular plots could be estimated as $\rho = n/\pi r_d^2$, where ρ is tree density (trees m^{-2}), n is the sample size and r_d is the distance (m) of the uttermost tree from plot centre, assuming the area occupied by the n trees was circular. In each of the three study areas, two major types of land use, woodlands and on farm, were randomly selected for the survey of tree density, size class distribution and regeneration status of *V. payos*.

In each land use type, one mature *V. payos* tree (≥ 10 cm dbh) was randomly selected and used as the starting tree in all the three sites. Moving in a circular manner around the first tree and while increasing the radius, other 99 neighbour trees were identified, located and marked, keeping the area compact. All the tree positions were recorded using a GPS (GPS 12XL Garmin Olathe KS, USA). Bearings and horizontal distance to the nearest neighbour were also recorded between all trees in a plot using compass and meter tape. These recorded distances and bearings from the starting tree and from one tree to the others were used to estimate the spatial distribution and tree population density in each study site.

The diameters at breast height (DBH), tree heights and crown diameters of all the 100 sampled trees were measured. DBH was measured with a diameter tape, height with a telescopic pole and crown diameter was derived as the average of two perpendicular crown diameters through the crown vertical projection.The area occupied by the 100 sampled trees was mapped graphically on a grid scaled at 50 × 50 m for the study of species regeneration. All the 0.25 hectare grids were allocated a number to allow random selection of plots to sample for seedlings and saplings. Ten percent of these 0.25 ha plots were randomly selected. Within these plots, the seedlings (≤ 1 m height) and saplings (> 1 m height but dbh ≤ 10 cm) were counted. Seedlings and saplings were classified as originating from seed, root suckers or stump coppices. They were distinguished as follows: those from seed generally had one stem at the ground level, those from root suckers were characterized by multiple stems at the base while those from stump coppice had visible stumps. When there were doubts regarding the origin of an individual seedling, root system was partially excavated to reveal the stump or the old root. The number of regenerates of each origin was used to calculate the density (stems ha^{-1}).

Data analysis

The area occupied by the sample trees was more of a polygon with many sides as prescribed by outer trees. Consequently, the area of the plot, A in m^2, was obtained by tracing the coordinates of the outer edge trees and calculating the area of the triangles and rectangles. The lengths of the sides of the polygon prescribed by the outer trees were summed to give the perimeter *L* of the plot. The tree density was therefore computed as ρ= n/A. The mean distance between nearest neighbours was computed as $r_A = (\sum r_i)/n$, where r_i is distance of the ith tree from its nearest neighbour. The mean distances that would be expected, r_E, for each plot if the trees were randomly distributed and with the same tree density as the computed value and the variance, σ_r^2 were calculated using the Clark and Evans (1954) formula, $r_E = 1/2\sqrt(\rho)$ and $\sigma_r^2 = 0.0683/\rho$, where r_E is the expected mean distance to the nearest neighbour in an infinitely large population with a random distribution of density ρ - (where the unit of measurement used in the calculation of ρ must be the same as that used in measuring *r* – in this case m and m^2).

However, with Donnelly (1978) correction factor, the corrected mean nearest neighbour distance (r_c) and the variance (σ_c^2) are,

$$r_c = r_E + (0.0514 + 0.041/\sqrt{n})*(L/n) \text{ and } \sigma_c^2 = \sigma_r^2 + 0.037\sqrt{(1/\rho)}*(L/n),$$

Where L is the sample plot boundary length (as a polygon) and other variables as defined earlier.

Assuming spatial randomness of the trees, the standard error of the corrected expected mean nearest neighbour distance (s_r),

$$s_r = \sqrt{(\sigma_r^2/n)}$$

such that the *p* value of deviation from the expected mean distance is obtained as (Campbell, 1996):

$$c = (r_A - r_c)/s_r$$

The ratio *R* of r_A/r_c was used as a measure of the extent to which the sampled populations deviated from a randomly distributed population of the same density. Where the trees are heavily aggregated, the *R* ratio tends to 0 as the distances between most neighbouring trees are close to or equal to 0 (Clark and Evans, 1954). At the other extreme, where trees are regularly and uniformly distributed, the *R* ratio is greater than 1 (Clark and Evans, 1954). Besides comparing the spatial distribution of trees between plots, descriptive statistics, non-parametric analyses and t-tests were calculated using SPSS for windows Version 16 to make comparisons between study sites (districts) and land uses (bush and farm). To compare individual districts where significant differences were found, Tukey's post hoc test was applied (Kleinbaum and Kupper, 1978; Dancey and Reidy, 2007).

The tree morphological parameters of height, diameter at breast height and crown diameter were summarized on per hectare basis and checked for normality using Levene test (Dancey and Leidy, 2007). Trees were assigned into dbh classes (10 ≤ 19.9 cm, 20.0 ≤ 29.9 cm, 30.0 ≤ 39.9 cm, 40.0 cm and above); height classes (≤ 5.00 m, 5.01 ≤ 7.00 m, 7.01 ≤ 9.00 m, and 9.01 m and above), and crown diameter classes (≤ 3 m, 3 - 5 m, 5 - 7 m, 7 - 9 m, and 9 m and above). Populations were compared using analysis of variance (ANOVA) at 95% confidence interval. Tukey's post hoc test allowed pair wise comparison of the mean heights, dbh and crown diameters of individual populations. To determine those populations that were actually different, a Mann Whitney U test was used to compare the two land uses while a Kruskal Wallis test was used to compare the districts. Seedlings and saplings were summarized on a per hectare basis from the 0.25 ha plots and ANOVA done to compare the different seedling origins while the land uses were compared using Student t test. Histograms were presented to show the regeneration trend among seedlings, saplings and trees for each population.

RESULTS

V. payos tree population structure

Spatial distribution patterns of mature *V. payos* trees showed that they were not uniformly distributed across their natural range (Figure 1). Trees were aggregated both on cultivated farms as well as in bushlands. Tree densities in the study areas ranged from 1.6 trees per hectare on cultivated land in Mwingi District to 20.3 trees in bush/fallow land in Kitui District (Table 1). Tree densities were higher in bushlands than on cultivated land in all districts. The expected mean distances between nearest neighbouring trees were higher than the observed values on all sites under the two land uses, an indication that trees in these areas were more aggregated than randomly dispersed. This is further supported by the *R* ratio values of the three villages which are less than 1 and the negative standard variate of the normal curve (c) (Table 1). Distribution of trees was positively skewed as there were few isolated trees sampled that were far from the others (Figure 1 and Table 1).

Comparison of the six populations of *V. payos*, gave an *F* value of 5.97 indicating that the six populations were significantly different at $p = 0.05$. Seperation of the means revealed significant Chi-square values of 152.605 for districts and 156.799 for study plots both with an associated probability value of $p < 0.001$. Trees in Mbeere were relatively more dispersed while trees in Kitui were more aggregated (28.28 ± 1.60 m for Mbeere; 17.26 ± 1.74 m for Mwingi; and 6.29 ± 0.57 m for Kitui trees). In the study plots, the mean nearest neighbour distance ranged from 5.10 ± 0.7 m in on farm land in Kitui to 33.56 ± 2.47 m in farm lands in Mbeere District. There were no differences between combined data for each land use from the three districts with *Z* value of -0.505 and *p* value of 0.613.

At 95% confidence interval, the population structure

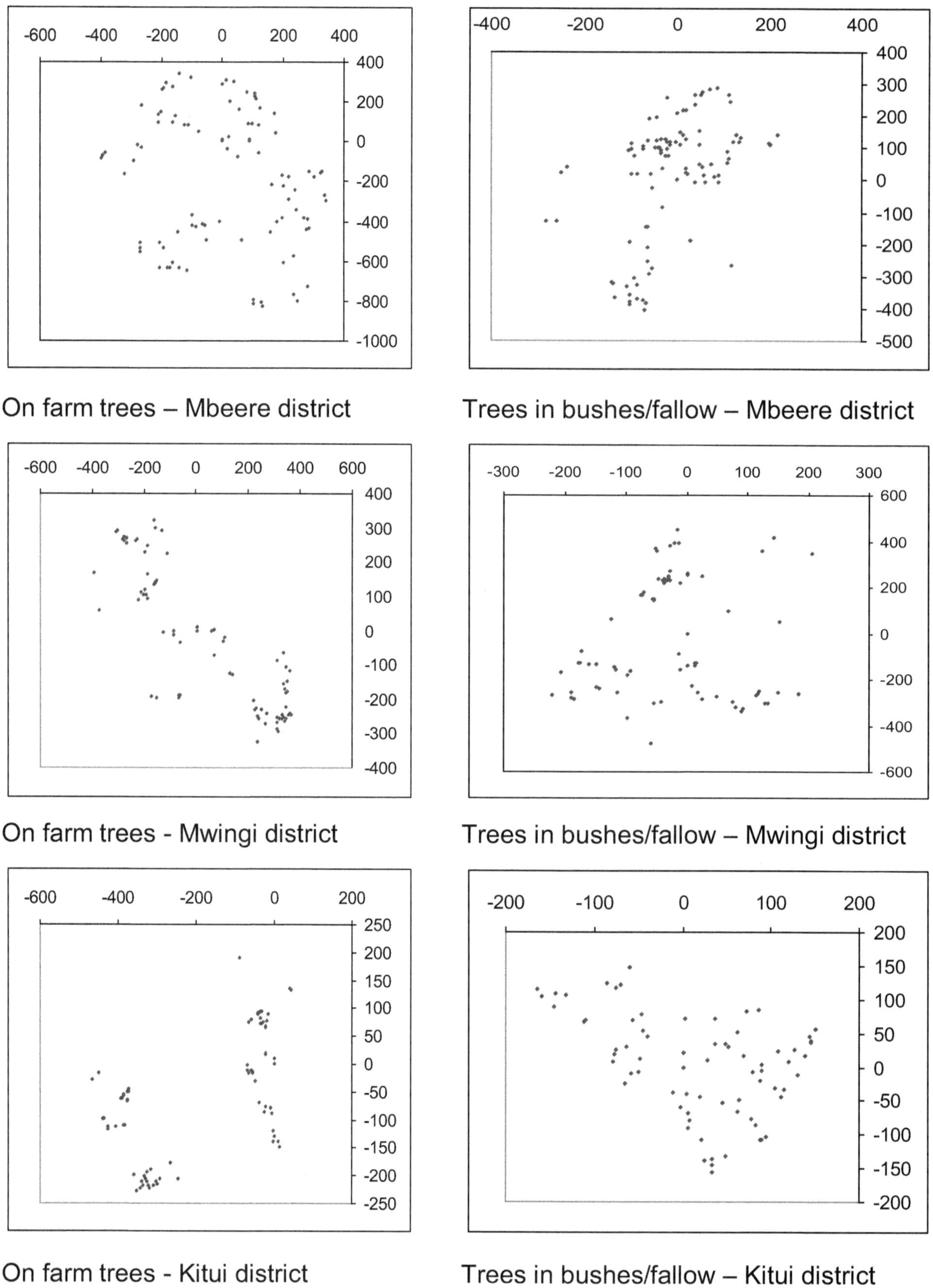

Figure 1. Spatial distribution pattern of mature *V. payos* trees on arable land and bushland in villages in Mbeere, Mwingi and Kitui Districts, Kenya (distances are in metres, 0,0 represents the center tree).

Table 1. Comparison of *V. payos* tree population densities and spatial distribution in arable and bushland in Mbeere, Mwingi and Kitui districts of Eastern province Kenya

Statistics	Mwingi		Kitui		Mbeere	
	On farm	Bushes/fallow	On farm	Bushes/fallow	On farm	Bushes/fallow
Area occupied by sample trees as a polygon (m2)	283119.5	267361.5	137157	49275.5	610279	210422.5
Number of trees sampled (n)	100	100	100	100	100	100
Sample population density (ρ)	0.000353	0.000374	0.000729	0.00203	0.000164	0.000475
Sample tree density per hectare (m)	3.53	3.74	7.29	20.29	1.64	4.75
Total distance between nearest neighbour (Σri)	1386.56	2045.43	509.86	728.11	3356.32	2279.46
Mean distance between nearest neighbours (rA) = (Σri)/n	13.86	20.45	5.10	7.28	33.56	22.79
Skewness of distribution of mean nearest neighbor	1.74	1.94	1.82	1.01	0.72	2.23
Expected mean distance between nearest neighbour in a large randomly distributed population of ρ density (rE =1/2√ρ)	26.61	25.85	18.52	11.10	39.04	22.94
Length of the sample plot boundary L	2132	2204	1470	925	2892	1799
Corrected expected mean nearest neighbour distance rc rc= rE+(0.0514+0.041/√n)*(L/n)	27.79	25.99	19.34	11.61	40.65	23.94
R ratio (measures the degree to which observed distribution deviates from a randomly distributed population of same density), [R = (ra/rc)]	0.50	0.79	0.26	0.63	0.86	0.99
σrE= 0.26136/√ (n. ρ): standard error of mean distance to nearest neighbour in a randomly distributed population of density ρ.	1.39	1.35	0.97	0.58	2.04	1.20
Variance of normally distributed distances in a random populations, $\sigma r^2 = 0.0683/\rho$	193.48	170.59	93.69	33.65	416.46	143.79
Variance of corrected mean nearest neighbour distances (σc2) [$\sigma c2=\sigma r^2+0.037\sqrt{(1/\rho)}*(L/n)$]	235.47	42.17	113.83	41.25	500.02	174.33
Standard error of the expected mean nearest neighbour distance sr = $\sqrt{(\sigma r^2/n)}$	1.39	1.31	0.97	0.58	2.04	1.20
Standard variate of the normal curve (c) = (rA – rE)/(σrE)	-10.02	-4.10	-14.68	-7.47	-3.48	-0.96

of the *V. payos* in Kitui District, irrespective of land use, were not significantly different.

The same was observed for sample plots in Mwingi District. However, trees on farms in Mbeere District were significantly more dispersed than those in bushes. Pair wise comparisons of the mean nearest neighbour distances between individual study plots using the Mann Whitney test showed all were significantly different (*Z* values -2.649 to -9.213, *p* values 0.01 to < 0.001) except between those in Mwingi; those in Kitui and the ones in bushes in Mwingi and Mbeere Districts.

Tree morphological characteristics

Tree height

Tree heights were not significantly different between the two land use practices across districts. Over 40% of trees from Kitui District were in the ≤ 5 m and 5 – 7 m height classes each, but 40% of those in Mbeere District were in the 5 – 7 m and 30% in the 7 – 9 m classes (Figure 2). Trees from Mwingi District were mainly in the intermediate height classes between these extremes. Mean heights differed significantly between individual study plots (Table 2). The trees in Kitui District were significantly shorter while trees in Mbeere were the tallest. In all districts, the trees on farms were taller than those in bushes (Table 2). Analysis of variance revealed that the tree heights in the three districts were significantly different from each other at 95% confidence interval (Table 2).

Diameter at breast height (dbh)

V. payos populations sampled were dominated by

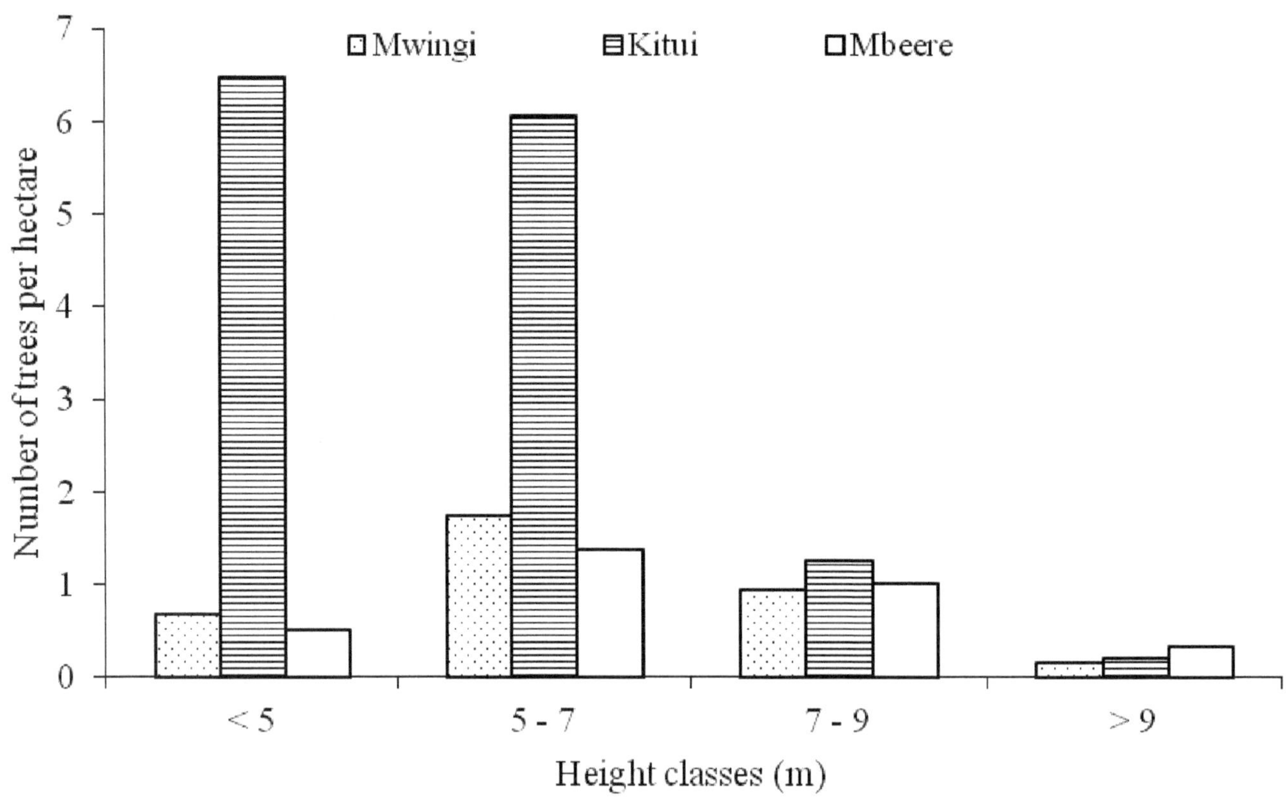

Figure 2. Height class distribution of *V. payos* trees in Mbeere, Mwingi and Kitui Districts in Eastern province, Kenya.

Table 2. Mean *Vitex payos* tree heights, dbh and crown diameter in three districts in Eastern province, Kenya.

District	Height mean (m)	Mean dbh (cm)	Mean crown diameter (m)
Kitui	5.25 ± 0.10^{a}	18.0 ± 0.4^{a}	4.86 ± 0.10^{a}
Mwingi	6.16 ± 0.11^{b}	18.1 ± 0.5^{a}	5.50 ± 0.12^{b}
Mbeere	6.80 ± 0.13^{c}	24.4 ± 0.6^{b}	5.81 ± 0.13^{b}

trees within the range of 10 to 20 cm diameter with over 55% of the trees (Figure 3). In Kitui and Mwingi Districts, the 10 - 20 cm diameter class contributed 60 and 76% of their population on farms and bushes, respectively. Trees with diameters above 30 cm made up to 8 and 7% of the trees in the two districts respectively. There was no clear difference in terms of tree diameters on farms and in bushes in the two districts. Trees in Mbeere in the 10 - 20 cm diameter class accounted for 25% of the population while trees with diameters higher than 30 cm were more on farms (36%) as compared to bushland (20%). Diameters differed significantly among plots in Mbeere with trees from this district differing significantly from those of the other two districts at p = 0.05 (Table 2). Mean diameters were higher in *V. payos* trees found in bushlands in Mwingi and Kitui Districts while the trees found on the farms in Mbeere had higher diameters than in those that were in the bushlands.

Tree crown diameter

Over 77% of trees in all populations were in 3 – 5 m and 5 – 7 m crown diameter classes. Among trees in bushland of Kitui District, 67% of the tree population had crown diameters of up to 5 m while those in cultivated sites had 51% of their population with crown diameters in the same range. The two study plots had 8% of the trees with crown diameters of more than 7 m. In Mwingi District, 38% of the trees had crowns with diameter of up to 5 m while 17% of the tree population had above 7 m crown diameters. There were more trees in the bushes with high crown diameters at 20% than on farms (15%). In Mbeere District, there were more trees with crown diameters less than 5 m (41%) and above 7 m (26%) on the farms than in the bushes (23%). Analysis of variance revealed significant differences between tree crown diameters in districts (Table 2). There were no significant

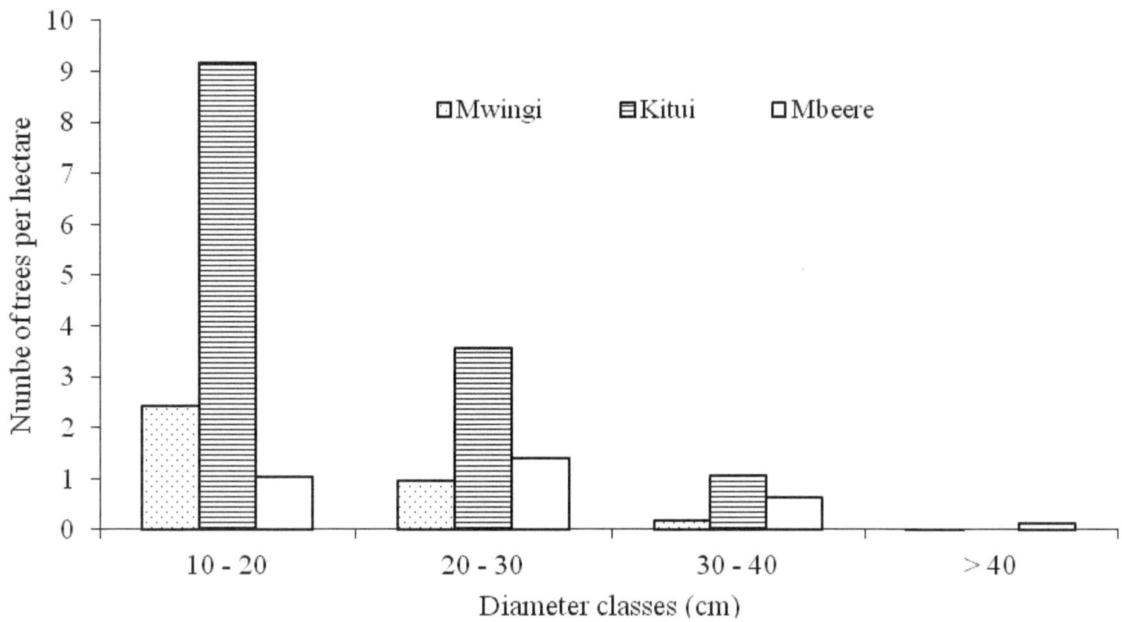

Figure 3. Diameter class distribution of *V. payos* trees in Mbeere, Mwingi and Kitui Districts, Kenya

Table 3. Mean number of *V. payos* seedlings per hectare from different land uses three districts in Eastern province, Kenya.

District	Land use	Origin of seedlings			
		Seed	Root	Stump/coppice	Combined
Mbeere	On farm	0.04[a]	0.04[a]	5.02[a]	6.00[a]
Mwingi	On farm	1.45[a]	6.55[bc]	2.91[a]	10.91[ab]
Mwingi	Bushland	4.08[ab]	1.06[ab]	1.06[a]	8.00[ab]
Kitui	Bushland	14.67[bc]	38.67[d]	47.33[c]	100.7[c]
Kitui	On farm	17.33[bc]	4.44[bc]	7.56[ab]	29.03[b]
Mbeere	Bushland	47.56[c]	16.44[cd]	13.78[b]	77.78[c]

Mean number of seedlings in a column followed by same letter are not significantly different at $p = 0.05$.

differences between tree crown diameters of trees on farms and in the bushlands. However, the trees on farms had slightly bigger crown diameters in Mwingi and Kitui Districts. Mbeere trees in bushes had bigger crowns than those on farms though the difference was minimal.

Regeneration of *V. payos*

Seedling populations

The highest numbers of seedlings from all origins were recorded in bushlands in Kitui and Mbeere Districts with combined 101 and 78 seedlings per hectare, respectively. The farm site in Kitui had 29 seedlings per hectare. Other study sites had fewer seedlings, with less than 10 individuals from all three sources, that is, seed, stumps and root sprouts. Sixty one percent of seedlings in bushlands in Mbeere were mainly from seed while 46.7% of those in bushlands in Kitui were from stumps and 38.5% from root suckers. The farm site in Kitui had a moderate population of seedlings (17 seedlings per hectare) of seed origin, which were more than those in bushlands (15 seedlings per hectare). On the farms in Mwingi, there were seedlings from root suckers that were protected by farmers. There were seven seedlings per hectare from stumps on farms compared to two seedlings from seed origin per hectare. In bushlands in Mwingi, regeneration was very low as was the case on farm in Mbeere District. In these study plots, regeneration was very low irrespective of the seedling origin with six and eight seedlings per hectare respectively (Table 3).

Analyses of variance revealed that there were significant differences between seedling numbers from individual origins and combined origins between the sample plots (Table 3). In the Mwingi bushlands, there

Table 4. Mean number of *V. payos* seedlings per hectare from Mwingi, Mbeere and Kitui districts in Eastern province, Kenya.

District / land use	Origin of seedlings			
	Seed	**Root**	**Stump/coppice**	**Combined**
Mwingi	3.05 ± 1.68^{a}	4.19 ± 2.15^{a}	2.29 ± 0.76^{a}	9.52 ± 2.81^{a}
Mbeere	22.74 ± 8.37^{b}	8.00 ± 3.28^{ab}	9.26 ± 2.27^{b}	40.00 ± 11.22^{b}
Kitui	16.27 ± 5.07^{b}	18.13 ± 6.33^{b}	23.47 ± 9.13^{b}	57.87± 14.63^{b}

Mean ± (sem) number of seedlings followed by same letter are not significantly different at p = 0.05.

Table 5. Mean number of saplings per hectare from different origins in three districts in Eastern province, Kenya (± sem).

District	Land use	Origin of saplings			
		Seed	**Root**	**Stump/coppice**	**Combined**
Kitui-	On farm	0	3.56 ± 2.05	7.56 ± 5.23	11.11 ± 6.79
Mbeere	On farm	0	0	1.2 ± 0.85	1.2 ± 0.85
Mwingi	On farm	0.36 ± 0.36	1.09 ± 1.09	2.18 ± 1.46	3.64 ± 2.50
Mbeere	Bushland	0.44 ± 0.44	1.78 ± 1.35	4.44 ± 2.62	6.67 ± 3.64
Mwingi	Bushland	2.8 ± 1.34	1.2 ± 0.85	2 ± 1.37	6 ± 2.40
Kitui	Bushland	3.33 ± 2.17	19.33 ± 12.33	12.67 ± 4.43	35.33 ± 18.17
p value		0.055	0.145	0.163	0.216

were high regeneration from seed while in Kitui bushlands, there were more seedlings regenerated from the root sprouts and coppices (Table 3). There were significant differences in seedling population from all origins among the three districts. Among seedlings of seed origin, Mwingi had the lowest population (6 seedlings per hectare) while Mbeere had the highest (48 seedlings per hectare) (Table 4). However, among seedlings from root sprouts and coppices, Mwingi had the lowest population (8 and 4 seedlings per hectare) while Kitui District had the highest number (43 and 55 seedlings per hectare respectively). Overall Kitui had highest seedling population though not significantly different from those in Mbeere. Combining all the districts, seedlings from all origins had significantly higher densities in bushes (55.36±11.35) than on farms (14.8±3.92).

Sapling populations

Sapling population densities were low, ranging from one sapling per hectare on farms in Mbeere District to 35 saplings per hectare in bushlands in Kitui. The saplings were generally few on farms with an overall mean of five saplings against a mean of 16 saplings in bushlands from the three study districts. In all locations, saplings of coppice origin dominated except in bushlands in Kitui where those from root suckers were dominant and in bushlands in Mbeere where saplings of seed origin were dominant (Table 5). On farms in Kitui and Mwingi Districts, there were no saplings of seed origin encountered, while on farm in Mbeere, only saplings of coppice were encountered. Analyses revealed no significant differences between saplings from all origins between the study sites (Table 5). Between the land uses, mean number of saplings of seed origin differed significantly (p = 0.009) while those from root suckers, coppices and all sources combined were not statistically different. Although there were no significant difference of sapling from root sprouts and stump coppices between on farm and bushlands, the number was higher in the bushlands in all cases. Comparison of the combined population of seedlings in the two land uses, saplings and trees, revealed a normal population trend though weak in bushlands (Figure 4). On cultivated sites, a reverse trend was observed: on farms in Mwingi District, there were more saplings than seedlings; while in Kitui and Mbeere Districts study sites, there were more trees than saplings (Figure 4).

DISCUSSION

Population structure

V. payos trees on farms were normally clustered in aggregates based on the farmers' acceptance of the species, the past land use practices and the current level of utilization of the trees and their products. Some farmers have left some trees in clusters on the farm for commercial harvesting of the fruits while others spared

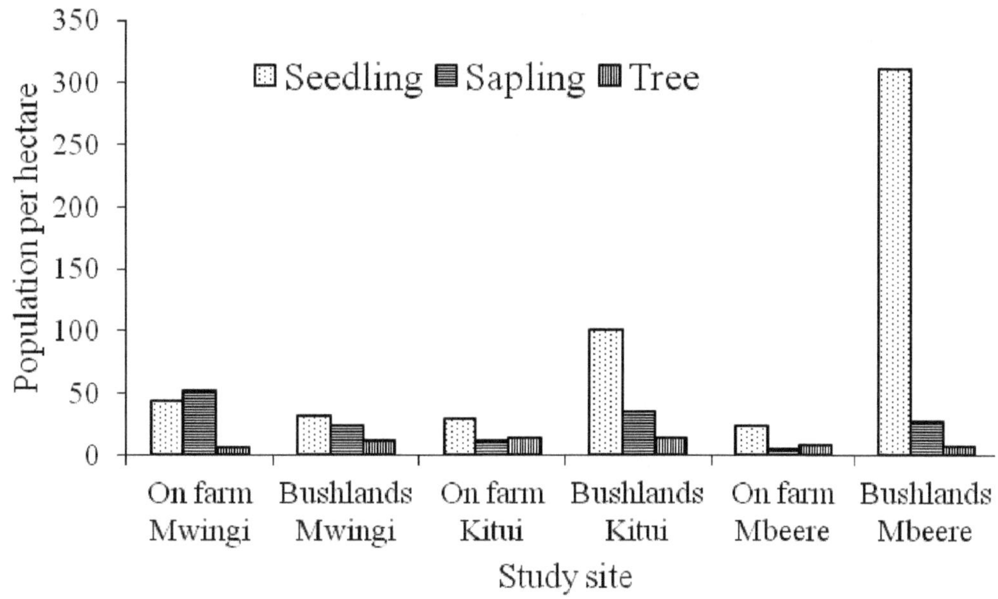

Figure 4. *V. payos* seedlings,saplings and trees in three districts in Eastern province, Kenya.

individual trees for provision of subsistence fruits to their households. Other trees were retained to provide sites for beehive keeping on the farms or shade for animals during dry season. Tree population densities expressed on a per hectare basis showed that there were fewer trees on farms than in bushes in all districts. In Mbeere, where the fruits are widely consumed with no sales, there were correspondingly fewer trees per hectare on farms than in bushlands. While in Mwingi District, where some farmers were involved in sale of fruits, the density of *V. payos* on the farm and in bushes were more or less equal. Farmers in this district were divided on the actual position of these fruits as commodity of trade. While some farmers left their fruits to waste on the ground, a few collected and took to markets. As a result, the maintenance and appreciation of the species was moderate, thus the low density of trees. In Kitui, the density of *V. payos* trees was higher than the other places; and higher in bushlands than on farms. Kitui study area was situated between two local markets, Ikanga and Mutomo, where these fruits are sold when in season. Thus traders frequented the area and many farmers were involved in on-farm sale of *V. payos* fruits.

The spatial distribution of the species did show aggregated clumps of trees scattered on farms. This indicated selective removal of trees and/or poor regeneration of the species. Among the six study populations, trees aggregation was high in Kitui and on farm sites in Mwingi with mean distances between nearest neighbouring trees being less than 15 m. The cause of this situation was retention of clumps of trees by individual farmers, rather than all sampled trees being close together on farms in the two districts. *V. payos* trees in Kitui District bushlands were scattered closely together as a stand in woodland whose farming activities were minimal and the owners were of advanced age. This clumped spatial distribution may be related to the past cutting for firewood and charcoal and the extensive coppicing and suckering ability of species.

In the villages with high mean distances between trees, several conserved trees were isolated on farms away from their nearest neighbouring trees (e.g. Mbeere District bushlands). While He et al. (1997) pointed out that the spatial distribution of a population shifts from highly aggregated to more random or uniform with succession, the tree population in the study areas was very low. Further, both abiotic (e.g. drought) and biotic factors (e.g. grazing, cutting, poor germination) may have greatly contributed to the current distribution pattern. According to Jama et al. (2008), the density of indigenous fruit trees often decreases with increasing human population pressure. Tree distribution in Mbeere was sparser than in the other two districts. The use level of the *V. payos* trees in Mbeere was low with only a few trees retained by individual farmers for provision of subsistence fruits and setting of beehives. Their crowns were said to be incompatible with both tobacco and 'khat' (*Catha edulis*) crops on farms, two cash crops extensively cultivated in Mbeere study area.

The height structure of *V. payos* showed that there is a consistent size class distribution trend in Mbeere and Mwingi Districts, which differed from that of Kitui. Stands in Kitui were predominantly in an area that experienced less precipitation than the other areas. The area formed the extreme edge of natural range of species to the south. It was therefore possible that the trees were not in their optimal conditions in terms of moisture, and thus the stunted growth. The diameter growth of the trees also followed the same trend, with tree sizes increasing as the environmental conditions (rainfall) improves from Kitui through Mwingi to Mbeere. Seedlings and saplings are by virtue of their size, greatly influenced by intensity, magnitude and frequency of disturbance they are subjected to (Khan et al., 1987; Duchok et al., 2005). In the present study, destruction of seedlings occurred mainly in the cultivated farmlands partly due to little appreciation of these trees by the farmers. During land preparation and weeding of crops, the seedlings and saplings of *V. payos* were normally uprooted together with other weeds. In the field, seedlings originating from seed could fail to grow due to soil compaction or destruction of seed before germination. Preliminary observations of the fruits of *V. payos* revealed that most of them contained few viable seed, either due to poor pollination or to fungal infection of embryo immediately after pollination (Kimondo, 2010).

Duchok et al. (2005) showed that germinated seedlings of *Illicium griffithii* were mainly found below fruiting trees. *V. payos* seedlings germinated below canopies of fruiting trees in bushes and cultivated areas. Nevertheless, in the bushes, these sites were normally the resting places for animals in the field especially during the dry hot season. Animals trampled upon the young seedlings causing heavy mortality. Kindeya (2003) quoted in Ogbazghi et al. (2006) reported that the seedlings in grazing fields often suffer as high as 90% mortality. Young seedlings were also prone to death due to effects of drought. In the present study, saplings on farms were frequently cut back to reduce the area they occupied thereby curtailing their development into mature trees. During the dry season, green foliage of *V. payos* saplings provided material for covering traditional earth kilns for charcoal production (Kimondo, 2010). In the fields, livestock also heavily browsed on crown of saplings while others were broken. Transition of saplings into trees was therefore very low. This seems to imply that there was generally poor natural regeneration status of *V. payos* in all the study areas, though the situation was more critical on the farms.

Variation of trees in size class distribution in natural stand should ideally follow a normal distribution. Many factors could cause some deviations in different locations. Ogbazghi et al. (2006) indicated that the deviation could be caused by the pattern of regeneration and anthropogenic factors assuming local environmental conditions are the same. In Mwingi, it was noted that *V. payos* trees compete for space on the farms with several introduced fruits such as citrus, mangoes, avocados and guava among others. The exotic fruits have ready market locally, and they fetch higher prices than the indigenous fruits. As such, indigenous fruit trees were cut to create space for agricultural crops and exotic fruit trees. The occurrence of burnt stumps of indigenous trees, including *V. payos* on farms near traditional earth charcoal kilns indicated the aim was to obtain charcoal and suppress the tree coppicing. There was marginal utilization of *V. payos* fruits in Mbeere mainly by children. The high income realized from both 'khat' and tobacco necessitated cutting of most *V. payos* and other indigenous trees to minimize competition of resources such as light, water and nutrients. High germination rate of *V. payos* seed in bushland in Mbeere occurred on land whose owners were absent (Kimondo, 2010). Fallen ripe fruits generally remained on the ground with minimal collection. After rains, these germinated directly below the tree. However, during the dry season, animals browsed and trampled the young seedlings resulting in minimal translation into saplings.

In a stable tree population, the number of seedlings should be higher than that of saplings and saplings numbers higher than that of trees (Duchok et al., 2005). The situation in Mbeere and Mwingi may without any efforts to plant or conserve the regenerates of *V. payos*, lead to the species becoming rare. The fact that most seedlings and saplings originated from root suckers and coppices is not healthy for the survival of *V. payos* in the study areas. It was therefore concluded that in such a situation, future planting stocks could be clones of existing trees as few seedlings from the seeds do mature into productive trees (Neke et al., 2006). The benefits associated with new genetic combinations shall thus be limited where regeneration through seed remains suppressed. This could also eventually result in regenerating trees close to parent trees, thus hindering exploitation of potentially new niches.

Crown diameter of *V. payos* ranged from 3 to 9 m though trees with greater diameters were observed especially on cultivated or fallow sites. This implied that with appropriate tending of trees and reduction of competition, it is possible to increase the crown volume and therefore fruit production. The large sized tree crowns in Mbeere could have been influenced by the higher rainfall received in the area. However, few big sized trees with large crowns still exist in areas experiencing harsh environmental conditions and sandy soils in Kitui and Mwingi. The production of fruits depends on an extensive crown of a vigorously growing tree. Identification of such trees growing in extreme marginal areas may provide future well adapted germplasm for planting.

CONCLUSIONS AND RECOMMENDATION

This study was aimed at investigating the population structure and regeneration status of *V. payos* in the drylands of Kenya. The spatial distribution pattern of mature *V. payos* trees showed an aggregation of trees both on farms and in bushes. Tree densities ranged from 1.6 trees per hectare on farmlands to 20.3 trees in the bushes. The expected mean distances between nearest neighbouring trees were higher than the observed values in all sites, confirming that *V. payos* trees were more aggregated than randomly dispersed. The mean heights of the trees varied from 5.19 to 7.05 m cross the three study districts. Tree heights in the three districts were significantly different from each other. The sampled populations were dominated by trees (55%) with a mean diameter ranging from 18.0 to 24.4 cm and mean crown diameter ranging from 4.86 to 5.8 m.

Highest densities of seedlings from all origins (seed, stumps and root sprouts) were recorded in the bushes in Kitui and Mbeere Districts with combined 101 and 78 seedlings per hectare respectively. Regeneration of seedlings was very low in Mwingi and some areas in Mbeere. Sapling densities were generally low with an overall mean of five saplings per hectare on the farmland against a mean of 16 per hectare in bushes from the three study areas. *V. payos* fruits are generally valued in some areas as a subsistence and commercial product. However, because of the high preferences of exotic fruits coupled with an increasing human population in the study areas, it was observed *V. payos* among other indigenous fruit trees continue to be cut down. Furthermore, natural regeneration was affected by livestock through browsing and trampling. Considering the adaptability and resilience of *V. payos* in these harsh environmental conditions in drylands, there is a need to promote its wide scale utilization and conservation. Superior trees in terms of tree size and the desired fruit traits should be sought to provide planting germplasm.

Conflict of Interest

The authors have not declared any conflict of interest.

REFERENCES

Campbell B (1996). Aggregation and regularity: An inclusive one tailed nearest neighbour analysis of small spatially patchy populations. Oecologia 106:206-211.

Clark PJ, Evans FC (1954). Distance to nearest neighbour as a measure of spatial relationships in populations. Ecology 35(4):445-453

Dancey CP, Reidy J (2007). Statistics without maths for Psychology. Prentice Hall. 619 pages.

Donnelly K (1978). Simulations to determine the variance and edge-effect of total nearest neighbour distance. In: Hodder, I. (ed.) Simulation methods in archaeology. Cambridge University Press, London, pp. 91-95.

Duchok R, Kent K, Khumbongmayum AD, Paul A, Khan ML (2005). Population structure and regeneration status of medicinal tree *Illicium griffithii* in relation to disturbance gradients in temperate broad-leaved forest of Arunachal Pradesh. Curr. Sci. 89(4):673-676.

Dunn JE (1991). Why can't we see the indigenous wood for the exotic Trees? Glo. Ecol. Biogeogr. Lett. 1(2):33-35.

Ekesa BN, Walingo MK, Abukutsa-Onyango MO (2009). Accessibility to and consumption of indigenous vegetables and fruits by rural households in Matungu division, Western Kenya. Afr. J. Food Agric. Dev. 9(8):1725-1738.

He FL, Legendre P, LaFrankie JV (1997). Distribution patterns of tree species in a Malaysian tropical rain forest. J. Veg. Sci. 8:105-114.

Jama BA, Mohamed AM, Mulatya J, Njui AN (2008). Comparing the Big Five": A framework for the sustainable management of indigenous fruit trees in the drylands of East and Central Africa. Ecol. Indicat. 8:170-179.

Kauter D, Lewandowski I, Claupeina W (2003). Quantity and quality of harvestable biomass from Populus short rotation coppice for solid fuel use—a review of the physiological basis and management influences. Biomass Bioenergy 24:411-427.

Khan ML, Rai JPN, Tripathi RS (1987). Population structure of some tree species in disturbed and protected sub-tropical forests of northeast India. Acta Ecol. *8*:247–255.

Kimondo JM (2010). The potential for optimization of *V. payos* as a dryland resource in Kenya. PhD thesis, University of Wales, Bangor UK.

Kindeya G (2003). Ecology and management of *Boswellia papyrifera* (Del.) Hochst. dry forests in Tigray, Northern Ethiopia. PhD thesis Georg-August-University of Göttingen, Germany.

Kleinbaum DG, Kupper LL (1978). Applied regression analysis and other multivariate methods. Duxbury Press, North Scituate, Massachusetts. 556pp.

Luoga EJ, Witkowski ETF, Balkwill K (2004). Regeneration by coppicing (resprouting) of miombo (African savanna) trees in relation to land use. For. Ecol. Manage. 189:23-35.

Maundu PM, Ngugi GW, Kabuye CHS (1999). Traditional food plants of Kenya. English Press Ltd. Nairobi, Kenya. National Museums of Kenya.

Mbora A, Jamnadass R, Lillesø JPB (2008). Growing high priority fruits and nuts in Kenya: Uses and management. Nairobi. The World Agroforestry Centre. 61 pages.

Morris RJ (2010). Anthropogenic impacts on tropical forest biodiversity: a network structure and ecosystem functioning perspective. Phil. Trans. R. Soc. B 365:3709–3718.

Neke KS, Owen-Smith N, Witkowski ETF (2006). Comparative resprouting response of savanna woody plant species following harvesting: the value of persistence. For. Ecol. Manage. 232:114-123.

Ogbazghi W, Bongers F, Rijkers T, Wessel M (2006). Population structure and morphology of frankincense tree Boswellia papyrifera along an altitude gradient in Eriteria. J. Drylands 1(1):85-94.

Oroda ASK (2011). The impact of increased human population and sedentarisation of the pastoral communities on the land cover and the resources of Mt. Marsabit forest and the surrounding lands. MSc Thesis, Kenyatta University, 135 pages.

Sparks JC, Masters RE, Payton ME (2002). Comparative evaluation of accuracy and efficiency of six forest sampling methods. Proc. Okla. Acad. Sci. 82:49-56.

Turner RM, Ochung HA, Turner JB (1998). Kenya's Changing Landscape. The University of Arizona Press. Tucson. J. Arid Environ. 40(4):341-345.

Wanyonyi RW (2012). Impact of human population on land degradation in former Lugari District, Kakamega County, Kenya. MSc Thesis, Kenyatta University. 137 pages.

The effect of plant growth regulators and their concentration *in vitro* on mass poropagation of Myrobalan 29C rootstock

Zahra Shabani[1], Ebrahim Ganji Moghadam[2], Bahram Abedi[1] and Ali Tehranifar[1]

[1]Horticultural Department, Ferdowsi University of Mashhad, Iran.
[2]Horticultural Department of Khorasan Razavi Agricultural and Natural Resource Research Center, Mashhad, Iran.

Myrobalan 29C is one of the most important rootstocks that is widely used for plum and apricot trees. This study was conducted to determine the most suitable media culture and regulators of plant growth for micropropagation of Myrobalan 29C in Khorasan Natural Resource and Agricultural Research Centre Mashhad, Iran. 10 explants treated were sterilized in 70% Ethanol for 1 min, Mercuric chloride (0.1%) for 1, 2 and 3 min and sodium hypochlorite (3 and 10%) for 10, 20 and 30 min. Results showed that 10% sodium hypochlorite (30 min) with 2% decay was the best treatment. In this experiment, proliferation and rooting were performed in three kinds of culture media: Murashige and Skoog (MS), McCown and Lioyd (WPM) and Driver and Kuniyuki (DKW). They were supplemented with plant growth regulators (benzyl amino purine (BAP) and thiadizuron (TDZ)) of 0, 1, 2, 3, 4 mg l^{-1} in all treatments of the proliferation; and with indole-3-butyric acid (IBA) and naphthalene acetic acid (NAA) of 0, 1, 2, 3 mg l^{-1} in the rooting step. Results showed that the highest number and length of shoot respectively were 5.58 and 2.50 cm in MS medium with 2 mg l^{-1} BAP concentrations. The DKW medium in 1 mg l^{-1} of NAA, the highest percent of rooting (100%) and root length were about 14.5 cm in MS medium with 2 mg l^{-1} of NAA respectively. The acclimatization of plantlets was successful in greenhouse conditions. The survival percent in substrateperlite (100% V) was about 80%.

Key words: Tissue culture, disinfection, micropropagation, acclimatization.

INTRODUCTION

Myrobalan 29C (*Prunus cerasifera* L.) has been widely used as a fruit tree rootstock for plum and apricot because of its rusticity and adaptability to several soil conditions (Plopa et al., 2012). This rootstock is suitable for high density plum orchard. In addition, Myrobalan has also been used as a parent in several rootstock and edible plum breeding programs (Arbeloa et al., 2006). It has also a good grafting compatibility where majority of the new Romanian plum varieties resulted from. When propagated by softwood cuttings, the results are very good in respect to rooting ability (75 to 95% rooted cuttings); but a lot of cuttings on the rooting beds blossom. This finally results in very weak rooted cuttings, not suitable for grafting in nursery field. That is the reason why this valuable rootstock might be micropropagated on commercial scale. The use of *in vitro* propagation technique allows for the efficiency of propagation, but for some species the propagation efficiency depends on some factors. Micropropagation is used as a useful method for propagation of clonal rootstocks (Hossini et al., 2010).

The most proliferation of Myrobalan 29C was obtained in MS medium supplemented with 1 mgl^{-1} benzyl adenine (BA) and 0.1 mgl^{-1} IBA (Ruzick, 2013; Plopa, 2012). Myrobalan 29C in MS culture media containing 0.5 mg l^{-1} benzyl adenine (BA) had the most multiplication rate (Movsiuw, 2011). Dardi et al. (2011) reported that Mahaleb rootstock in MS medium containing BAP and NAA showed the highest proliferation and rooting. Bonjak (2012) showed that BA and TDZ concentration had significantly affected the proliferation rate of Gisela5. In the micro-propagation of *Prunus avium*, the combination of 0.5 mg l^{-1} BAP and 0.05 mg l^{-1} TDZ were suitable for proliferation and a medium containing 0.3 mg l^{-1} IBA was desirable for rooting (Hossini et al., 2010; Nazeri et al., 2010). *In vitro* micropropagation of Chinese plum in 1/2 MS media showed the highest percentage rooting and acclimatization of rooted plantlets to greenhouse conditions (Ying-Ning, 2010; Mikhilov et al., 2008; Tian et al., 2007). Tatari (2013), Xiaomei (2008) and Yao et al., 2011) reported that rooted plantlet were acclimatized after 6 weeks of growth in the laboratory and 76 to 84% of rooted plantlets survived after acclimatization in the greenhouse.

Due to the importance of achieving an efficient protocol for the mass propagation of Myrobalan 29C, this study was conducted with the purpose of evaluating the most efficient and effective micropropagation protocol for Myrobalan 29C rootstock in Iran. This would be useful in future research work as a reference for the cultivation of economically important rootstock.

MATERIALS AND METHODS

Plant materials

The explants were collected from shoots of Myrobalan 29C rootstock maintained in the experimental greenhouse of Khorasan Razavi Agricultural and Natural Resources Research Centre (Mashhad, Iran), on 25 June, 2013. The shoots were transferred to the laboratory where the tips and axillary buds were dissected with a scalpel and used as a source for explants. The explants were washed with water and dishwashing liquid to remove surface contamination and then divided into parts containing one bud. Then they were sterilized with 70% ethanol for 1 min, Mercuric chloride (0.1%) for 1, 2 and 3 min, sodium hypochlorite (3 and 10%) for 10, 20 and 30 min; and then washed 3 times with sterile distilled water and cultured in a medium. Each treatment disinfection consisted of three replications and each replicate had five samples.

Proliferation

Proliferation was performed in three kinds of culture media: MS (Murashige and Skooge, 1962), WPM (Lloyd and McCown, 1980) and DKW (Driver and Kuniyuki, 1984). They were supplemented with plant growth regulators (benzyl amino purine (BAP) and thiadiazuron) (TDZ) of 0, 1, 2, 3 and 4 mg l^{-1}. In the rooting step after three subcultures (21 days between each subculture), the numbers, length and quality of the shoots were measured. This stage was carried with four replications and each replicate had five samples.

Rooting

For rooting, *in vitro* developed shoots (2 to 3 cm long) were placed on three culture media (MS, DKW and WPM) supplemented with indole-3-butyric acid (IBA) and naphthalene acetic acid (NAA) at four levels (0, 1, 2 and 3 mgl^{-1}). The explants were transferred to the culture growth chamber at a temperature of 22°C and night temperatures of 16°C with a photoperiod of 16 h light and 8 h dark with light from 2000 to 2500 Lux. This stage was carried out in four replications and each replicate had three samples. After rooted explants were determined as the best cultured media and combination of rooting growth regulators, number and root length, leaf number and stem length and quality of explants were recorded.

Acclimatization

All plants with properly developed roots were transferred into the growth substrate coco peat and perlite (50%: 50% V) and perlite (100% V). The transferred roots were carefully washed under tap water to remove all remnants of agar. The acclimatization of the transferred plants was conducted in a mist chamber inside the greenhouse. The experiment was carried out based on completely randomized design (CRD) with four replications and each replicates had five samples. Statistical analysis of the data was carried out by using JMP software and difference among treatment means was compared by using least significance difference test (LSD).

RESULTS

Surface sterilization

The best result was achieved from the sterilization of the explants in 10% sodium hypochlorite for 30 min. The results showed that among the sterilization treatments, 3% sodium hypochlorite for 10, 20 and 30 min had the highest contamination, and among other treatments, the percentage of the contamination was 2%; sodium hypochlorite (10%) for 30 min had the least explants of active sterile (Table 1).

Proliferation

The proliferation results showed that the MS media

Table 1. The effects of surface sterilization treatments on contamination of explants Myrobalan 29C.

Treatment	Active explants (%)	Contamination (%)
Ethanol, 70%, 1 min.	8.5[c*]	80.0[ab]
Mercuric chloride, 0.1%, 1 min.	4.5[c]	5.5[c]
Mercuric chloride, 0.1%, 2 min.	5.5[c]	5.5[c]
Mercuric chloride, 0.1%, 3 min.	6.5[c]	5.5[c]
Sodium hypochlorite, 3%, 10 min.	7.5[c]	100.0[a]
Sodium hypochlorite, 3%, 20 min.	8.1[c]	100.0[a]
Sodium hypochlorite, 3%, 30 min.	1.2[c]	100.0[a]
Sodium hypochlorite, 10%, 10 min.	66.6[ab]	4.5[c]
Sodium hypochlorite, 10%, 20 min.	66.6a[b]	4.5[c]
Sodium hypochlorite, 10%, 30 min.	100.0[a]	2.0[c]

* Mean values followed by the same letters within a column are not significantly different according to least significance difference test (LSD) at 5% level.

culture supplemented with 2 mg l^{-1} BAP with mean of 5.58 numbers of shoot, 2.5 cm shoot length and 11.33 numbers of leaf is the most suitable treatment (Figures 1, 2 and 3). There was no significant difference in the quality of plantlet in all three media; however, best quality was observed in the media MS. Moreover, the considerations showed that DKW with 2 mg l^{-1} TDZ has the least numbers and length of shoot (Figures 2 and 3).

Rooting

The best root formation was observed in DKW media containing 1 mg l^{-1} NAA, which produced roots readily with 100% efficiency compared to other rooting media cultures (Table 2). The mean of numbers and length of roots in this culture media were 6.75 and 14.3 cm, respectively (Table 2). The results showed that DKW culture media without growth regulators having mean of 0.3 root numbers and the MS culture media without growth regulators having mean of 1 cm root length had the least number and length of root, respectively. However, the highest length of roots, length plant, number of leaf and plantlet quality were observed in the MS media (Table 2 and Figure 4).

DISCUSSION

In the present study, sampling date, genotype and different concentrations of sodium hypochlorite affected surface sterilization. Vujovic (2012) has reported sterilization of Gisela5, apricot and plum buds with 10% sodium hypochlorite for 15 min to be more effective than other treatments. Sulusoglu and Covusoglu (2013) investigated that rootstock micropropagation (*Prunus laurocerasus*) with 5% sodium hypochlorite for 14 min disinfection of treated explants was considered. Nacheva and Gercheva (2009) obtained 85 to 100% sterile explants after sterilization with calcium hypochlorite. For surface sterilization of explants, 70% ethanol, sodium hypochlorite (10 to 50%) and mercuric chloride were used (Asadi et al., 2009; Sana et al., 2006; Jang et al., 2008). Sodium hypochlorite causes the oxidation of cells, microorganisms and affects major components of cells, such as lipids, proteins and DNA. Ethanol damages cell membranes, proteins and causes more rapid disintergration of cell metabolism (George, 2008). Therefore, one can conclude that alternating the use of these two substances that deplete cells and bacterial pathogens is better; and the higher the concentration and time of treatment, the more its effectiveness improved. The use of mercuric chloride with minimum contamination, compared to sodium hypo-chlorite has 0% active buds. Perhaps due to the toxicity and lethality of mercuric chloride, it penetrates into plant tissue and destroys the buds in active meristem (Ruzic and Vujovic, 2013).

A significant interaction between media and growth regulators regarding shoot regeneration percentage was found at $p<0.05$. This study showed that number of shoot was increased as concentration of BAP increased to certain amounts. There is a positive correlation between BAP concentration (2 mgl) and number of shoots. The increasing concentrations of BAP inhibited shoot proliferation (Sulusoglu and Cavusoglu, 2013). The highest shoot multiplication was obtained in 2 mg l^{-1} BAP added to MS medium. Erbenova (2009) and Sulusoglu (2012) reported 50% increase in multiplication rate of the dwarf rootstocks of *Prunus* in MS media culture containing 1.5 mg l^{-1} BAP. George (2008) reported that cytokinins when added to media promote cell division. Nordstrom et al. (2004) suggested that auxins may also play a direct regulatory role in the balance of cytokinins levels by suppressing both the synthesis rate and pool size of cytokinins. These results can be related to the manipulation of exogenous cytokinins concentrations that may cause an increase in endogenous auxin concentrations.

Table 2. The effects different media, IBA and NAA concentrations on rooting parameters of Myrobalan 29C.

Culture media	Growth regulators (mg l^{-1})	Rooting (%)	Root length (cm)	Number of root	Plantlet length (cm)	Number leaf	Plantlet quality
	IBA(0)	3.5^{c}	0.26^{f}	0.4^{de}	3.00^{defghi}	10.00^{abc}	1.66^{ab}
	IBA(1)	50.00^{ab}	11^{ab}	2.25^{abcde}	5.5^{bcdefg}	13.00^{a}	1.25^{ab}
	IBA(2)	72.08^{ab}	10.50^{abc}	3.25^{ab}	$3.87^{cdefghi}$	9.25^{abc}	2.00^{a}
	IBA(3)	68.75^{ab}	13.75^{a}	2.50^{abcde}	6.75^{abc}	13.25^{a}	2.00^{a}
MS	NAA(0)	50.00^{ab}	8.50^{bcd}	1.75^{abcde}	2.75^{fghi}	10.00^{abc}	2.00^{a}
	NAA(1)	91.66^{a}	10.50^{abc}	3.25^{ab}	9.75^{a}	12.75^{a}	2.00^{a}
	NAA(2)	75.00^{ab}	14.50^{a}	3.25^{ab}	9.75^{a}	12.5^{ab}	2.00^{a}
	NAA(3)	87.50^{a}	12^{ab}	3.75^{a}	8.75^{ab}	12.75^{a}	2.00^{a}
	IBA(0)	14^{c}	0.37^{f}	0.3^{e}	2.5^{fghi}	10.00^{abc}	1.00^{b}
	IBA(1)	66.25^{ab}	2.87^{ef}	3.00^{abc}	$4.75^{cdefghi}$	10.25^{abc}	1.75^{ab}
	IBA(2)	77.08^{ab}	4.50^{def}	2.25^{abcde}	6.25^{bcde}	9.75^{abc}	1.00^{b}
	IBA(3)	52.94^{ab}	3.75^{ef}	1.75^{abcde}	$4.5^{cdefghi}$	10.5^{abc}	1.00^{b}
DKW	NAA(0)	66.50^{ab}	2.12^{ef}	1.50^{abcde}	2.37^{fghi}	9.50^{abc}	1.00^{b}
	NAA(1)	100^{a}	6.00^{cde}	4.00^{a}	6.25^{abcd}	12.75^{a}	1.00^{b}
	NAA(2)	68.75^{ab}	2.25^{ef}	2.25^{abcde}	$3.87^{cdefghi}$	9.50^{abc}	1.00^{b}
	NAA(3)	81.25^{ab}	3.62^{ef}	2.75^{abcd}	5.75^{bcdef}	10.50^{abc}	1.00^{b}
	IBA(0)	62.5	1.00^{f}	0.75^{bcde}	1.87^{hi}	6.00^{cd}	2.00^{a}
	IBA(1)	25.00^{ab}	1.12^{f}	1.00^{bcde}	1.75^{hi}	6.00^{cd}	2.00^{a}
	IBA(2)	50.00^{ab}	2.00^{ef}	1.75^{abcde}	2.37^{fghi}	6.00^{cd}	2.00^{a}
	IBA(3)	43.75^{ab}	2.12^{ef}	1.75^{abcde}	2.12^{ghi}	7.25^{bcd}	1.50^{ab}
WPM	NAA(0)	25.00^{ab}	1.5^{ef}	0.50^{cde}	1.87^{hi}	5.25^{cd}	1.50^{ab}
	NAA(1)	93.75^{a}	3.20^{ef}	3.75^{a}	3.00^{efghi}	6.75^{cd}	2.00^{a}
	NAA(2)	68.75^{ab}	2.62^{ef}	2.75^{abcd}	2.5^{fghi}	6.00^{cd}	2.00^{a}
	NAA(3)	54.16^{ab}	1.25^{f}	2.00^{abcde}	1.25^{i}	3.50^{d}	1.25^{ab}

^{a}– explants strong growth, with no signs of verification, necrosis of leaf are yellowing terminal meristem, ^{b}– less than15% have the symptoms of verification, necrosis of leaf are yellowing terminal meristem, and ^{c}– explant weak, 15-30% have the symptoms of verification, necrosis of leaf are yellowing terminal meristem.

This is probably induced by an inhibition of free IAA conjugates due to the presence of exogenous cytokinins (Jaramillo et al., 2008; George (2008).

It is observed that the auxin to cytokinins ratio represents an important signal in the formation of cell phenotype and also in the onset and maintenance of the process of cell division. Since auxins are capable of imitating cell division involved in the formation of meristems, giving rise to either unorganized tissue or defined organs.

Tatari and Mosavi (2013) reported that low shoot growth with necrotic leaf could be due to the effects of high concentrations of BAP and hormonal imbalances in the explants. Considering the MS medium contrast, WPM and DKW media have no Cl^- in the culture media while chlorotic plants increased stability. The medium increases the amount of Cl^- as plants balance is disturbed at different nutrient absorption. The MS media culture contains more calcium than the two

Figure 1. Effects of media and plant growth regulator on proliferation of Myrobalan29C (a) (MS), (b) (DKW) and (c) (WPM).

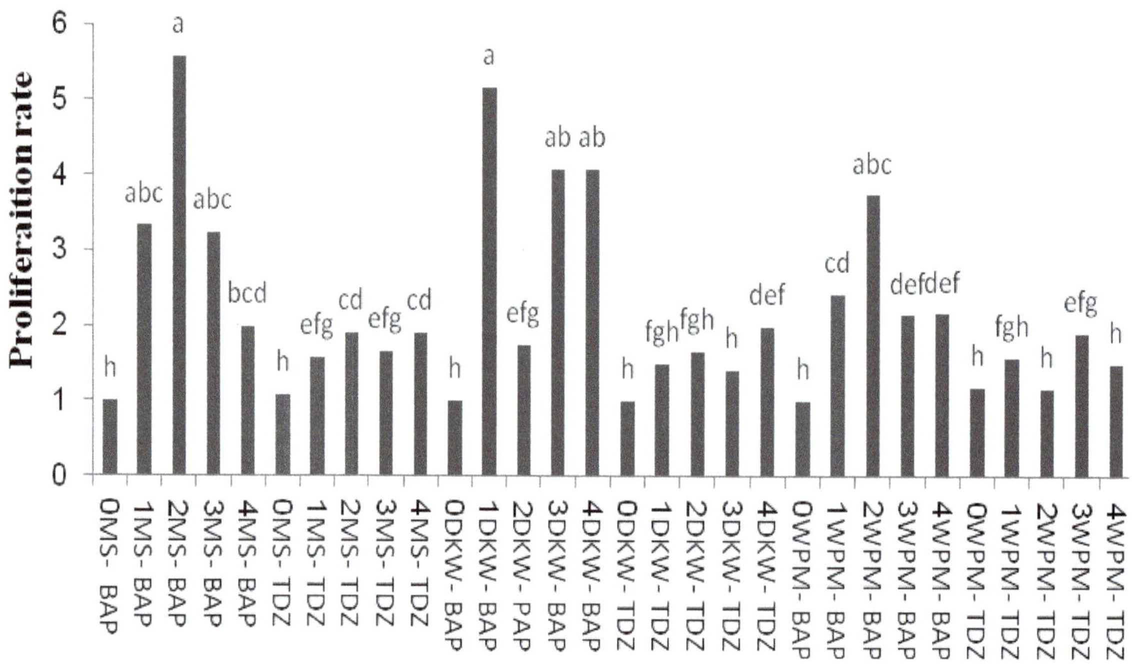

Figure 2. The effects of different media, BAP and TDZ concentrations on propagation coefficient of Myrobalan 29C.

other media which play an important role in the stability and strength of the plantlet. Nutrient concentration of N, K, Mg and macro nutrients is lower in WPM and DKW media than in MS media which is effective in reducing the proliferation
of the media culture.

The best root formation was observed in DKW media containing 1 mg l^{-1} NAA. Roots formation in tissue culture can be induced by exogenous auxins such as IBA, NAA and IAA and their interaction with endogenous auxins (Thorpe et al., 2008). Among the three auxins, NAA was found superior to IAA and IBA, though thick and callused rooted were obtained in one- step procedure. Low levels of NAA resulted in the highest rooting, whereas high concentration gave rise to more root initials, which eventually developed into callus rather than rootlets. They are only required at an early stage to emerge new formed roots. The MS culture media with reduced concentration of macro and microelements by ½ gave a maximum rooting percentage of 100% (Plopa et al., 2012). Vujovic et al. (2012) reported that the rooting rates in media containing 1 mg/L IBA or NAA (65 and 70%,

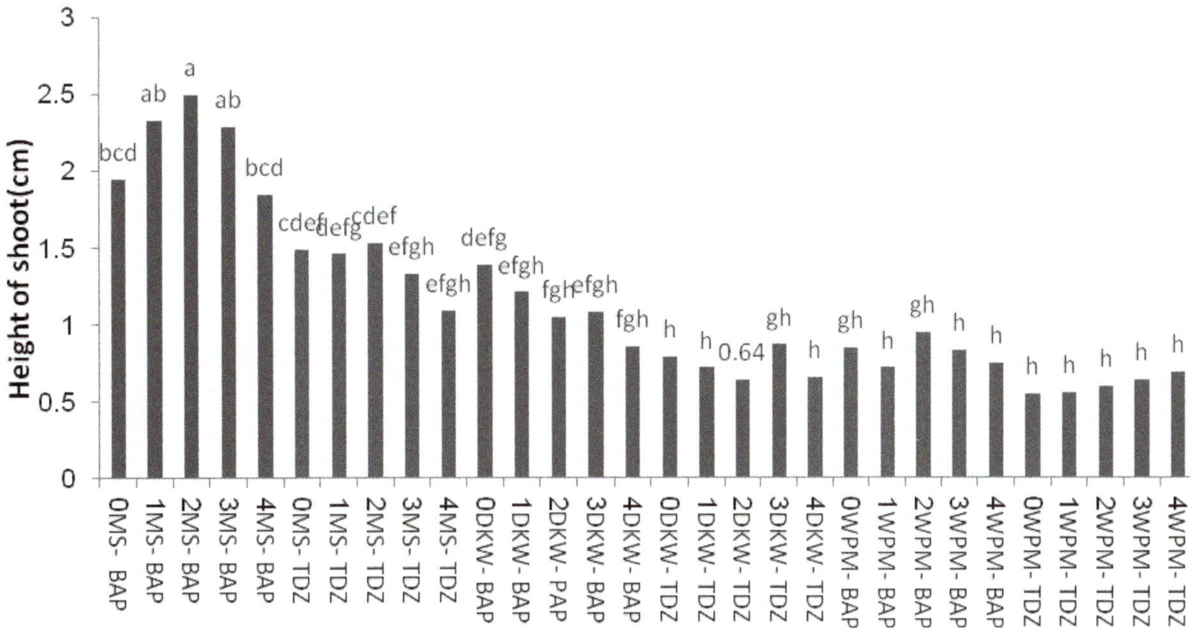

Figure 3. Effects BAP and TDZ concentrations on shoot proliferation and elongation of Myrobalan 29C.

Figure 4. Root formation of Myrobalan29c. (a) (MS), (b) (DKW) and (c) (WPM).

respectively); although number of roots and plant height was significantly higher when IBA was supplemented. Hossini et al. (2010) reported that IBA concentration increase caused shortening of the root; 80% rooting and 2.9 cm root length were obtained in culture medium with the addition of 0.5 mg l^{-1} IBA and greater percentage of rooting (88.6%) was attained when rooting medium was supplemented with 1.0 mg l^{-1} NAA. Root percentage of the regenerated shoots was 38.2% by addition of 10.74 µM NAA in the medium (Demiral and Ulger, 2008; Mohamed, 2012).

Balla and Kirilla (2006) reported that following a nice growth during the multiplication and elongation phases, difficulties arose during the rooting phase and large differences were found in the nutrient demand of the rooting phase. In spite of the high percentage of rooting, widespread shoot tip necrosis was detected and optimal levels of different macro elements, iron and sugar had to be determined in a series of experiments for the clonal rootstock cultivars.

Figure 5. Acclimatization of Myrobalan29c rootstock on *in vivo* condition.

Acclimatization was affected directly by rooting conditions. Survival was best when plantlets were transferred to pots after a short period of root emergence on rooting media. The plants with older roots and brown colour were better adapted than new plantlets with white roots. The results showed that after 2 months of acclimatization, the best substrates, perlite (100%V) was about 80% (Figure 5). Therefore, acclimatization directly affected rooting plants that have high quality; best rooting had rate induction (Mahdaviyan et al., 2010). Rooted plantlets survived Gisela 6 after acclimatization with the greenhouse was successful (80 and 79.2%) (Hossini et al., 2010; Sulusoglu and Cavusoglu, 2013).

Conclusion

The results of this research showed that Myrobalan 29C rootstock can be reproduced *in vitro*. According to this research, MS and DKW media including BAP with 2 mg l^{-1} and NAA with 1 mg l^{-1} growth regulators are most suitable for micro propagation.

Conflict of Interest

The authors have not declared any conflict of interest.

ACKNOWLEDGMENT

Authors have to thank all who helped in the laboratory experiments, in particular the personnel of the Husbandry and Biotechnology Department of Khorasan Razavi Agricultural and Natural Resource Research Centre and Ferdowsi University of Mashhad.

REFERENCES

Arbeloa A, GarcíaE, Andreu P, Marín J.A (2006). Rootstock Breeding Programme for Apricot through Interspecific Crosses of Myrobalan x Apricot: Significant Effect of Accidental Pollinations. Acta Hortic. (717): 133-136.

Asadi AA, Vedadi C, Rahimi M, Naserian B (2009). Effect of plant growth hormones on root and shoot regeneration in rose (*Morrasia*) under *in-vitro* conditions. Biosci. Res. 6:40-45.

Balla I, Kirilla Z (2006). Micropropagation of peach rootstocks and cultivars. Acta Hortic. 725:511-516.

Bonjak A, Keresa S, Jercicl, Baric M (2012). The effect of cytokine type and explant orientation on axillary shoot proliferation and in vitro rooting of Gisela 5 cherry rootstock. J. Food Agric. Environ. 10:616-620.

Dardi G, Vito G, Standardi A (2011). Invitromass poropagation of eleven *Prunus mahaleb* ecotypes. Acta Hortic. 410:98-100.

Demiral S, Ulger SA (2008). Research on Propagation of Gisela 5 Cherry Rootstock by Tissue Culture. Akdeniz Univ. Ziraat Fak. 21:117–121.

Driver JA, Kunyoxi H (1984).In Vitro propagation of paradox walnut rootstocks. Hort. Science.(19): 57-509.

Erbenova M, Paprstein F, Sedlak J (2009). *In vitro* propagation of dwarfed rootstocks for sweet cherry. Acta Horticulturae.(560): 477-480.

George EF, Hall MA, De Klerk GJ (2008). Plant propagation by tissue culture. Plant Cell. Reprod. 186-113.

Hossini AD, Moghadam EG, Anahid S (2010). Effects of media Cultures and Plant Growth Regulators in Micropropagation of Gisela6 Rootstock. Ann. Biol. Res. (1):135-141.

Jang HH, Ann SH, Kim MD, Kim CW (2008). Use of hydrogen peroxide as an effective disinfectant to Actinobacillus ureae. Process Biochem. 43:225-228.

Jaramillo EH, De Forero A, Cancino G, Moreno AM, Monsalve LE, Acero W (2008). *In vitro* regeneration of three chrysanthemum (*Dendrathema grandiflora*) varieties "via" organogenesis and somatic embryogenesis. Univ. Sci. (13): 118-127.

Mahdaviyan M, Bozari N, Abdolahi H (2010). Effect media culture and plant growth regulators on Proliferation and rooting of vegetative St.loci64 rootstock. Seed Plant J. (26):15-26.

Lloyd GB, McCrown BH (1980). Commerically feasible micropropagation of mountain by use of shoot tip culture. Proc. Int. Plant Propagators Soc. 30:421-437.

Mikhilov RV, Shulga OA, Dolgov SV (2008). Genetic transformation of somatic tissues and molecular biology approach for improving resistance to sharka disease of plum (*Prunus domestica* L.) varieties. Biotechnology 944:44.

Mohamed A (2012). Establishment of Regeneration system for Taif peach (*Prunus persica* L. Batsch) cultivar (Balady cultivar) in Taif, KSA. J. Am. Sci. 8:232-239.

Movsiuw A (2011). *In vitro* propagation of virus-free Myrobalan 29C rootstock. GRI (73):101-102.

Murashige T, Skooge F (1962). A revised medium for rapid growth and bio-assays with tobacco tissue culture. Physiol. Plant 15:473-497.

Nacheva L, Gercheva P (2009). Micropropagation of Gisela 5 (cherry dwarf rootstock): The effect of the type and the concentration of the carbohydrates in the nutrient medium. ISHS Acta Hortic. 825:261–281.

Nazeri MA, Yadolahi A, Moeeni A (2010). Investigate on the effect TDZ and BA on Proliferation Gisela rootstock. Special Twelfth Iranian Genet. Congr. 5:1-5.

Nordstrom AP, Tarkowski D, Tarkowska R, Norbaek C, Astot K, Dolzel G. Sandberg (2004). Auxin Regulation of cytokinins biosynthesis in the *Arabidopsis thaliana*: A factor of potential importance for auxin cytokinins regulated development. Proc. Natl. Acad. Sci. USA 101(21):8039-8044.

Plopa C, Dutu I, Isac V, Mazilu C, Ancu S (2012). Factors influencing *in vitro* Propagation of Myrobolan Dwarf Plum Rootstock II EUFRIN Plum and Prune Working Group Meeting on Present Constraints of Plum Growing in Europe. ISHS Acta Hortic. 968.

Ruzic DV, Vajovic T (2013).The effect of cytoknins types and their concentration on *in vitro* multiplication of sweet cherry. Hortic. Sci. 35:12-21.

Sana B, Ghosh D, Saha M, Mukherjee J (2006). Purification and characterization of a salt, solvent, detergent and bleach tolerant protease from a new gamma- Proteobacterium isolated from the marine environment of the *Sundarbans*. Process Biochem. 41:208-215.

Shekafandeh A, Qasemi M (2008). Effects of BA and TDZ on bud growth and somatic embryogenesis of immature almond (*Prunus dulcis* L.) varieties blooming late 7 Shahrody. J. Hortic. Sci. 2:148-152.

Sulusoglu M (2012). Development of embryo culture protocol for cherry laurel (*Prunus laurocerasus* L.). J. Food Agric. Environ. 10:47-352.

Sulusoglu M, Cavusoglu A (2013). Micropropagation of cherry laurel *Prunus laurocerasus* L. J. Food Agric. Environ. 11:576-579.

Tatari M, Mosavi A (2013). Evaluation *in vitro* culture of vegetative Tetra, Nemagard and GF677 rootstock. Byagricultural 3:103-115.

Tian L, Wen Y, Jayasankar S, Sibbald S (2007). Regeneration of *Prunus salicina* Lindl (Japanese plum) from hypocotyls of mature seeds. *In vitro* Cell. Dev. Biol. Plant 43:343-347.

Thorpe T, Stasolla C, Yeung EC, de Klerk (2008). Plant Growth Regulators. Third Ed, Vol 1., Springer. pp. 115-173.

Vujovic T, Ruzic, D, Cerovic, R (2012). *In vitro* shoot multiplication as influenced by repeated subculturing of shoots of contemporary fruit rootstocks. Hortic. Sci. 39(3):101- 107.

Xiaomei L, Paula M (2008). Plant regeneration from in vitro leaves of mature black cherry (*Prunus serotina*). Plant Cell Tiss. Organ Cult. 94:113–123.

Yao YX, Sun YW, Li GG, Li GH (2011). Regeneration of plant from *in vitro* culture of petioles in *Prunus domestica* Lindl (European plum). Biotechnology 10:5504.

Ying-Ning Z (2010). Micropropagation of Chinese plum (*Prunus salicina* Lindl.) Using Mature Stem Segments. Agrobotany (38):214-218.

Timber species from Afram arm of the Volta Lake in Ghana: Planing and sanding properties

Francis Wilson Owusu , Felix Boakye and Godwin Zorve

CSIR, Forestry Research Institute of Ghana (FORIG), Kumasi-Ashanti, Ghana.

Volumes of timber logs (dead trees) in the Volta Lake in Ghana, that had become a dead trap for water transportation, were being extracted for utilization to increase the timber resource base. Unfortunately, their properties were not known for efficient promotion and utilization. The influence and relative significance of machine surface planing and sanding parameters in the production of good quality timber surface finish on four underwater timber species from the Volta Lake were investigated. The preparation of test specimens, testing, evaluation and grading of the tests specimens were conducted according to ASTM-D 143, 1666-87 and DIN 4768. A cutting depth of 2 mm was made constant. Wood specimens were planed and sanded at 12% moisture content and the surface qualities obtained were visually graded. Surface quality performance increased with decreasing rake angle and feed speed. Low rake angle of 15° with 6 m/min and 9 m/min feed speeds resulted in a high planing quality. The degree of magnitude of the chipped/torn grain defects decreased with decreasing rake angle and feed speed. Grit sizes of sand paper had significant effect on the surface quality of the four species. The chipped/torn grain defects observed were eliminated from all the wood species with grit size 40. High surface sanding quality was registered for all the species with grit 150.

Key words: Feed speed, planing quality, rake angle, sanding quality, surface quality.

INTRODUCTION

The flooding of the Volta Lake in Ghana after its creation in 1965 submerged thousands of hardwood trees (Figure 1), which have been posing danger to water transportation, but at the same time could be useful as timber when harvested. As Ghana's timber resource base is dwindling, there is the need to find alternative resources to feed the timber industry, hence attention being turned to the large volumes of timber under the Volta Lake. However, like the lesser-used and lesser-known timber species, the underwater species need to be promoted to enhance their utilization, but it is not known whether their properties remain the same as the species from the natural forest. It is therefore necessary to establish their properties, which include machining, for their efficient promotion. The study was to determine the planing and sanding characteristics of four timber species harvested from the Volta Lake and recommend some potential uses. Some of the timber logs that had been

Figure 1. Some hardwood trees in the Volta Lake.
Source: Field (2007).

harvested and identified as *Manilkara multinervis* (Berekankum), *Erythrophyleum guinense* (Potrodom), *Diospyros mespiliformis* (Savannah Ebony) and *Cylicodiscus gabunensis* (Denya / Okan) with their scientific and local names respectively, were used for the study.

The results from the study will facilitate the promotion of the new timber resource, thereby making more timber logs available for processing and utilization by the timber industry and revive some of the timber companies that are at the point of collapse. This will lead to reduction of the rate of accident on the Volta Lake, generation of employment which will enhance the livelihood of the local communities around the Volta Lake, save the Ghanaian forests from total depletion in the near future and increase in the country's annual revenue through export of timber products.

One of the most important advantages of wood is its easy machinability in contrast to metal and plastic products (Sofuoğlu and Kurtoğlu, 2014). Karpat and Özel (2007) have indicated that for manufacturers to remain competitive on both local and international markets, they must increase their productivity while maintaining, if not improving, product quality. Any surface defects due to an improper machining process will also reduce the quality of the final product, resulting in an increase in the cost of the manufactured unit (Malkoçoğlu and Özdemir, 2006).

In the machining operations, production rate, cost, and product quality are three incompatible objectives (Cus and Zuperl, 2006). They continue that as the machining industry welcomes the introduction of new materials and cutting tools, it finds itself undergoing a rapid development which is giving rise to processes of highly complex and non-linear phenomena.

Ambrogio et al. (2008) and Tamizharasan et al. (2006) have stated that surface quality in machining is an essential consumer requirement because of its impact on product performance. Also the characteristics of machined surfaces that significantly influence its physical properties have been pointed out by Basheer et al. (2008). Sogutlu and Togay (2011) have also stated that surface roughness is an important factor in the quality of the internal and external decorative elements produced from wood. Chang and Lu (2006) and Oktem et al. (2006) have indicated that surface roughness plays an important role in determining the quality of a machined product.

Surface roughness has been defined by Benardos and Vosniakos (2003) as the superimposition of deviations from a nominal surface from the third to the sixth order where the orders of deviation are defined by international standards, DIN 4760 (1982). Correa et al. (2009) have also defined surface roughness as the functional behavior of a part. Roughness is a measure of the fine irregularities on a surface. The shape and dimensions of the irregularities on a surface determine surface quality of a product. According to Hiziroglu and Kosonkorn (2006) the surface roughness of a wood panel plays an important role since any surface irregularities may show through thin overlays reducing the final quality of the wood panel. Roughness is thus an indicator of process performance and must be controlled within suitable limits for particular machining operations and that the factors leading to roughness formation are complex (Pal and Chakraborty, 2005). Benardos and Vosniakos (2002) and Tan et al. (2012) have declared that statistically significant in roughness formation are the absolute values of cutting parameters such as depth of cut, feed speed, and components of cutting force.

Various techniques, according to Gurau et al. (2005); Hendarto et al. (2005) and Sandak and Tanaka (2005) that have been used to determine surface roughness include optical, pneumatic, electrical, ultrasonic, photographic and stylus methods. It has also been found by Sieminski and Skarzynska (1989) that the stylus-type profilometer was suitable within the tested methods for the measurement of the surface roughness of wood.

Surface roughness is influenced by the rays, pore sizes, knots and reaction wood (Taylor et al., 1999). Skaljic et al. (2009) have also reported that physical and mechanical properties and anatomical structure of wood affect the surface roughness. As the number of blades increases, the surface roughness values decrease, and as the feed rate increases, the surface roughness also increases (Ors and Baykan, 1999; Roger and Cool, 2008).

In planing, the different cutting angles and number of cutters had different effects on surface quality in terms of wood species (Ratnasingam and Scholz, 2007; Malkoçoğlu and Özdemir, 2006). For good machining quality, the cutters used in the machining of the material should be sharpened and machinery should be maintained and stablized (Sofuoğlu and Kurtoğlu, 2014).

In general, better machining performance is obtained with decreasing feed speed (Skaljic et al., 2009). Moreover, increasing feed speed has been reported to cause strong machining defects (Malkoçoðlu, 2007). Farrokhpayam et al. (2010) have also stated that small differences between the different levels of the feed rate affect surface quality and that the slowest feed rate of 8 m min $^{-1}$ gives the best results, with progressively poorer work as fastest feed speed is used. Fujiwara et al. (2001) have confirmed in various studies that if variables such as the feed rate are not selected correctly, then the desired surface quality will not be obtained. Sogutlu and Togay (2011) have reported that when the Oriental beech (*Fagus orientalis* Lipsky) and Scotch pine (*Pinus silvestris* L.) wood specimens were prepared by planing, the surface roughness was less in the tangential cut as compared to the radial cut. Furthermore, the surface roughness was less in planing with 4 blades as compared to planing with 2 blades. Best surface quality could be achieved where the depth of the final cut is from 1 to 2 mm and that excess stock removal of more than 2 mm will cause chipped/torn grain or pickup defects (Tjernlund, 1984).

In planing and sanding, the degree of surface roughness is dependent on the properties of the raw material and production processes (Hendarto et al., 2005; Nemli et al., 2005, 2007). A study by Nemli et al. (2007) shows that grit sizes, feed speed of panels and the feed power of the heads of the sander are the main considerations for a successful sanding operation. Also a report by Beaty (1983) and NPA (1993) indicate that sanding equipment, coated abrasive belt specifications, production scheduling and correct machine setup are the main considerations for successful sanding operation. Sanding is the most common and most influential operation for achieving surface quality during the phase of surface preparation (Skaljic et al., 2009). There is the need to recognize and understand the key factors and variables that affect selection of sanding system, sanding equipment, sanding quality and the ultimate performance of the finished product.

The study sought to investigate the effect of the process parameters that exert significant influence on the surface quality of Savanna Ebony, Berekankum, Denya/Okan and Potrodom (extracted from the Volta Lake in Ghana) after planing and sanding operations. Specifically, the effects of the rake/hook angles, spindle speeds, feed speeds and grit sizes of sand paper on wood planing and sanding qualities were investigated.

MATERIALS AND METHODS

Collection of samples

Four (4) timber species used for the study were *Manilkara multinervis* (Berekankum), *Erythrophyleum guinense* (Potrodom), *Diospyros mespiliformis* (Savannah Ebony) and *Cylicodiscus gabunensis* (Denya / Okan) in their scientific and local names respectively. These were the timber species that had been harvested from the Volta Lake by Clark Sustainable Resource Development (CSRD) at the time of the study. According to their utilization status, *M. multinervis* was a lesser-known timber species while the remaining three were lesser-used timber species (Oteng-Amoako, 2006). Using through and through sawing pattern, the logs were milled into lumber at Forestry Research Institute of Ghana (FORIG) under the Council for Scientific and Industrial Research (CSIR), Fumesua near Kumasi in the Ashanti region. The lumber pieces obtained from each timber species were stacked for air drying.

Preparation of tests specimens

ASTM D 143-94, ASTM D 1666-87 (ASTM, 2004; 2007) and DIN 4768 with some required adjustments were used to prepare the planing and sanding tests specimens. Some lumber boards of 25 mm thickness were taken from the stacked lumber, kiln dried to 10% moisture content for the preparation of 300 specimens of dimensions 25 × 100 × 900 mm. They were then conditioned to 12% equilibrium moisture content at 27°C and 65% relative humidity

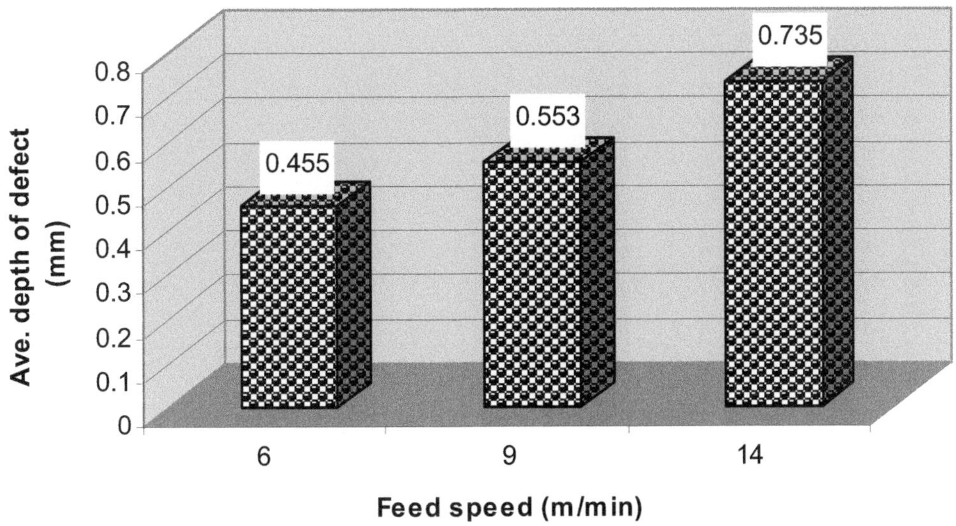

Figure 2. Average depth of chipped/torn grain defect with two cutting angles.

for ten days before the planing test.

Planing test

Planing was done using a combined surfacing and thicknessing machine of type 610 mm × 230 mm 'D.A.A' at a spindle speed of 5,200 rpm. A cutter head with a pair of planer knives were used. The selected feed speeds and rake/hook angles were 6 m/min, 9 m/min and 14m/min and 15° and 20° respectively. With six planing operations, 25 test specimens per operation were made to pass through the planer until all the operations were completed before they were graded in accordance with ASTM D 1666-87. A constant depth of cut of 2 mm per face of two for each specimen was set.

Grading

The specimens were then examined visually, with the aid of a hand lens, evaluated and classified, according to ASTM D 1666-87, on the basis of five quality grades [grade 1 = Excellent or defect-free; grade 2 = good; grade 3 = fair; grade 4 = poor and grade 5 = rejected or very poor. (Grades 2-5 were dependent on the degree of the defect observed)]. Percentage defect-free specimens was calculated for each rake angle and feed speed, as shown in Table 1. The depth of each chipped/torn grain (which is a machining defect), observed on the surfaces of the specimens, was measured and the average for each operation (a rake angle and feed speed) estimated (Table 2).

Sanding test

Sanding operation was undertaken with a vertical belt sander of model CL300. Five grit sizes of sand paper, P40, P60, P100, P120 and P150 were used. The grits were with open-coat paper-backed aluminum oxide sandpapers coated with anti-static zinc stearate. Sanding was carried out along the grain. The defective specimens (chipped/torn grain) were selected from the planed specimens and sanded with grit sizes 40 and 60 to assess the possible elimination of the defect. The specimens without the defects were sanded with grits 100, 120 and 150 for the assessment of the final surface quality for the application of finishes. The degree of scratches observed on all the specimens per species with grits 100, 120 and 150 were noted. The same grading procedure as stated above was used to grade the specimens and the percentage defect-free specimens (grade 1) were estimated for both sanding activities per species (Tables 3 and 4).

RESULTS AND DISCUSSION

Planing properties

The results on planing properties were based on clear specimens (ASTM D 143-94 and 1666-87, 1994). To identify the best planing performance, each planing condition was examined separately. Table 1 shows the percentage defect-free specimens with respect to feed speeds and or rake/hook angles for the four timber species. Analysis of variance (ANOVA) was used to test for the significant differences among the three feed speeds, two rake angles and the species. At P = 0.05, the significant differences in planing quality existed among the different feed speeds. The percentage defect-free specimens increased with decreasing feed speed. This means that the lower the feed speed the better the surface quality. This was consistent with all the timber species as feed speed changed from 6 m/min to 14 m/min (Table 1). This is also reflected in Table 2 and Figure 2 where the average depth of chipped/torn grain generated after planing decreased with decreasing feed speed. Therefore the best surface quality was generated with a feed speed of 6 m/min (72-100%) followed by 9 m/min (68-92%) and then 14 m/min (44-72%). Figure 3 shows the average percentage defect-free specimens for

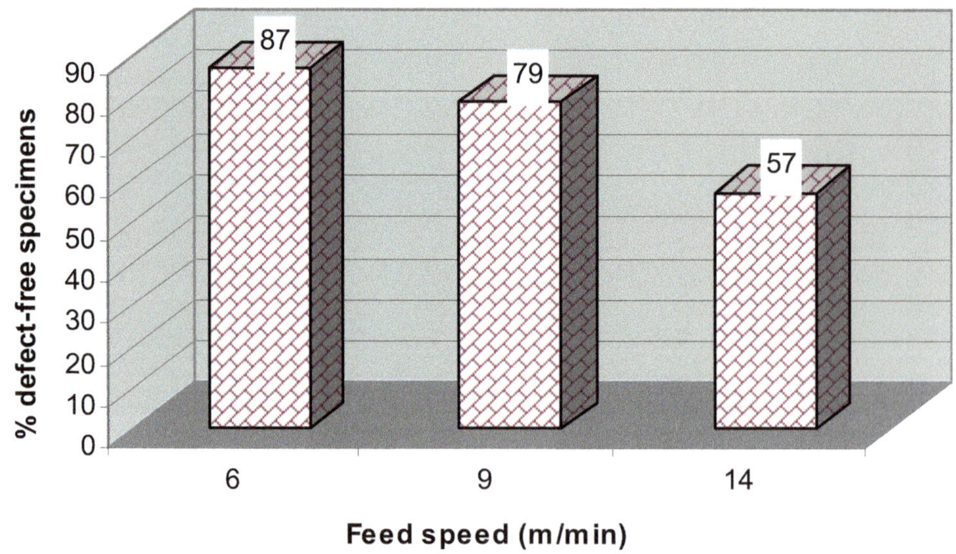

Figure 3. Average percentage defect-free specimens for two rake angles.

the two rake angles. The trend is in agreement with Ors and Baykan (1999) and Roger and Cool (2008) whose reports indicate that as the feed speed increases, the surface roughness also increases. It also confirms the statement by Fujiwara (2001) that if variables such as feed speed are not selected correctly, the desired surface quality will not be obtained. Results of the average percentage defect-free specimens indicate that feed speeds 6, 9 and 14 m/min could be rated as high, medium and low surface quality respectively

The surface quality of all the timber species was affected by the rake angle, as shown in Table 1. The results indicate that the percentage defect-free specimens increased with decreasing rake angle and that 15° rake angle recorded higher percentages than with 20°. The average percentages of defect-free specimens estimated for 15 and 20° rake angles are 81% (high surface quality) and 67% (medium surface quality) respectively. This implies that surface quality was higher with the lower rake angle of 15° than with 20°. The results showed significant differences between the rake angles at $P=0.05$. A similar trend/pattern has been recorded by Fortin (2001) that "the proportion of defect-free pieces increased as the rake angle decreased from 20 to 15°" and therefore has proposed a rake angle of 15° for planing White spruce wood from plantation forests. Gilmore and Barefoot (1974) have also proposed 15° rake angle for planing some tropical wood species from South America. From Table 2, as the rake angle decreased from 20 to 15° (for each feed speed) the depth of the chipped/torn grain defect also decreased in magnitude. For example, the average depth of the defect at 15° hook angle varied from 0.49 mm (for Ebony) to 0.56 mm (for Potrodom) while that at 20° also varied from 0.57 mm (for Ebony) to 0.65 mm (for Potrodom). The minimum and maximum average magnitudes of the depth of the defect at 15 and 20° with the three feed speeds were estimated to be 0.36-0.73 and 0.44-0.8 mm respectively. On the average 0.6 mm maximum depth was estimated for every species. Therefore the average magnitude of the defect for the three feed speeds was lower with 15° rake angle than 20° as shown in Figure 4.

Generally the planing results have shown that feed speed and rake angle are among the major factors that control wood failure during planing as have been recorded by Sogutlu and Togay (2011) and Fortin (2001). Hence feed speeds of 6, 9 and 14 m/min with a rake angle of 15° will be more appropriate for the planing of the selected timber species.

Comparing species in terms of surface quality, the average percentage defect-free specimens for all the conditions ranged between 67% (for Berekankum) and 80% (for Savanna Ebony). At 95% confidence interval, there was no significant difference between the surface quality of Denya/Okan and Potrodom which recorded an average of 74 and 75% defect-free specimens respectively. This means that the two timber species may have some common anatomical characteristics; hence planing them with the same machine setting will generate equal surface planing quality. According to Oteng-Amoako (2006), Denya and Potrodom are both coarse textured species and therefore confirms the findings of Tailor et al. (1999) that surface roughness is also influenced by rays and pore sizes. Figure 5 shows grade 1 planed specimens of Ebony, Denya, Berekankum and Potrodom. Introducing performance rate classification whereby 80-100% = high surface quality; 60-9%= medium surface quality; 40-59% = low surface quality and below 40% = poor surface quality, Savanna Ebony, on the average, recorded a high surface quality while the

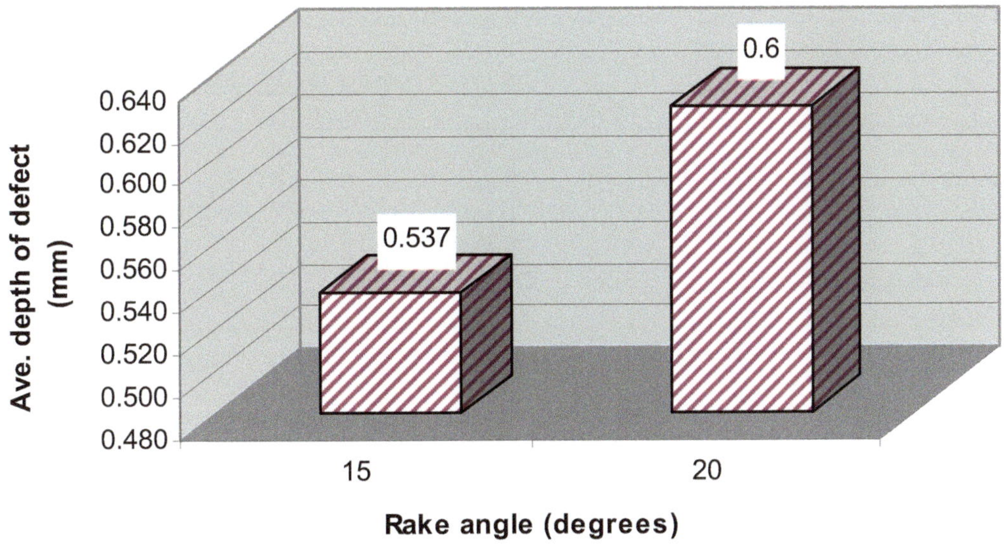

Figure 4. Average depth of chipped/torn grain defect with three feed speeds.

Figure 5. Grade 1 specimens of the four timber species.

three other species attained medium surface quality (Table 1). The average percentage surface quality for all the species with the planing tests conditions was estimated to be 74.2%, which from the machining rating introduced, could be classified as medium surface quality for every species. A report by Farmer (1972) shows that a rake angle of 10° is required for satisfactory planing of Denya (from the natural forest) when it is quarter-sawn, but the feed speed(s) and the percentage defect-free specimens were not specified. Denya (from underwater) with an average score of 75% defect-free specimens (medium surface quality rating) may be equivalent in

Table 1. Percentage defect-free specimens at two hook angles with three feed speeds

Species	Percentage defect-free specimens						
	6 m/min		9 m/min		14 m/min		Average on species basis
	15°	20°	15°	20°	15°	20°	
Berekankum *(Manilkara multinervis)*	84	72	76	68	60	44	*67.3*
Potrodom (*Erythrophyleum guinense)*	92	84	88	68	64	48	*74.0*
Savannah Ebony (*Diospyros mespiliformis)*	100	88	92	76	72	52	*80.0*
Denya / Okan (*Cylicodiscus gabunensis)*	92	84	88	72	64	52	*75.3*
Average % per species	*92*	*82*	*86*	*71*	*65*	*49*	*74.2*

Table 2. Average depth of chipped/torn grain machining defect generated on each species with the planing conditions studied.

Species	Magnitude of defect in millimeters (depth)						
	6 m/min		9 m/min		14 m/min		Average on species basis
	15°	20°	15°	20°	15°	20°	
Berekankum (*M. multinervis)*	0.42	0.50	0.50	0.62	0.71	0.78	0.6
Potrodom (*E. guinense)*	0.43	0.52	0.52	0.65	0.73	0.79	0.6
Savannah Ebony (*D. mespiliformis)*	0.36	0.44	0.47	0.55	0.65	0.72	0.5
Denya / Okan (*C. gabunensis)*	0.44	0.53	0.51	0.60	0.70	0.80	0.6
Average % per species	0.41	0.50	0.50	0.61	0.70	0.77	0.6

Table 3. Percentage defect-free specimen after sanding to remove defects with two different grit sizes.

Species	Percentage defect-free specimen	
	Grit size P60	Grit size P40
Berekankum (*M. multinervis*)	80	100
Potrodom (*E. guinense)*	80	100
Savannah Ebony (*D. mespiliformis*)	100	100
Denya / Okan (*C. gabunensis)*	80	100

Table 4. Scratching tendencies of specimen after sanding with three different grit sizes.

Species	Percentage scratching tendency		
	Grit size P100	Grit size P120	Grit size P150
Berekankum (*M. multinervis*)	56	72	80
Potrodom (*E. guinense)*	60	76	88
Savannah Ebony (*D. mespiliformis*)	48	60	80
Denya / Okan (*C. gabunensis)*	60	80	92

planing quality to that from the natural forest.

Sanding properties

Table 3 shows the percentage defect-free sanded specimens when sand paper of grit sizes 60 and 40 were used in that order. At P = 0.05, there were significant differences between grit sizes of sand papers used. With the chipped/torn grain defects that were generated after planing, grit size 60 was able to remove them from Savanna Ebony, hence scoring 100%; while Berekankum, Potrodom and Denya with 80% defect-free specimens, still had some degree of the defects on the surfaces of some of the specimens. Using sandpaper with grit size 40, the entire chipped/torn grain defects were eliminated from the wood specimens. This implies that sand paper with grit size 40 is capable of eliminating any chipped/torn grain defect from the wood members of the four timber species before further sanding could be under taken for the application of finishes. It was therefore established that the various degree of magnitude of the defect that was eliminated with grit 60 were comparatively shallower and did not affect the thickness of the specimens involved. Such shallower defect was classified by the study as chipped grain while torn grain was referred to that defect which reduced the thickness of the wood specimens above 1.5 mm. Such a reduction could affect the service life of a dimensioned wood member that is put in service (for instance, in production of furniture or roofing). According to Owusu et al. (2011), the sanding quality of Denya (from the natural forest) with grit P60 at 12% moisture content was rated as high, which is comparable to the results obtained in this study.

Grit sizes of 100, 120 and 150 were used to finally prepare the surfaces of the specimens for the application of finishes. This means that scratches formation on wood surfaces was possible at different degrees and hence has to be brought to the barest minimum. The degree of scratches generated on the wood surfaces was observed to decrease with increasing grit size and that the degree of scratches was minimal with grit size 150 than 120 and 100. Therefore higher grit size (in this case, 150) of sand paper is recommended for better finishing in order to effectively add value to tertiary wood products. This shows that grit sizes of sand paper affect surface quality and hence supports the research findings observed by Nemli et al. (2007). The scratches per grit size were observed to be prevalent on Savanna Ebony followed by Berekankum, Potrodom and Denya/Okan. This could be attributed to the texture of the species, and that scratches were obvious on fine textured species than course textured. Therefore the statement by Hendarto *et al.*, (2005); Nemli *et al.*, (2007) and Nemli *et al.*, (2005) that the degree of roughness during sanding, as machining activity, is dependent on the properties of the raw material and the production processes is confirmed. It also supports the earlier finding by Owusu et al. (2011) that "the finer the texture of the species, the more pronounced the scratching tendency". The result on the scratching tendency of Denya (from underwater) with grit P120 is equivalent to Denya (from natural forest) at 12% moisture content (Owusu et al., 2011).

Potential uses of the species

Based on the results achieved, the following potential uses are proposed for the four timber species extracted from the Volta Lake: Berekankum *(M. multinervis)*: Furniture products and could be used in replace of African Mahogany. Savannah Ebony (*D.mespiliformis*): Furniture products and walking sticks and could be a good substitute for *M. altissima*. Potrodom (*E. guinense) and* Denya / Okan (*Cylicodiscus gabunensis)*: Interior carpentry and walking sticks

Conclusions

The study has revealed that the four timber species harvested from the Volta Lake have good planing and sanding performances. The surface planing quality increased with decreasing rake angle and feed speed. The best planing condition obtained was 15° rake angle with feed speeds of 6m/min and 9 m/min. The machining defects observed were chipped and torn grain. The average range of magnitudes of the defect in terms of depth with 6, 9 and 14 m/min feed speeds for the four species at 15 and 20° rake angles were 0.36-0.73 and 0.44-0.80 mm respectively. Shallower defect was classified by the study as chipped grain while torn grain was referred to that defect which reduced the thickness of the wood specimens above 1.5 mm. Consistently, Savanna Ebony performed better in all the planing operations.

Sand paper with grit size 40 was capable of eliminating any chipped/torn grain defect from the wood specimens of the four timber species. The degree of scratches generated on the wood surfaces decreased with increasing grit size and that was minimal with grit size 150 than 120 and 100. Therefore higher grit size (in this case, 150) is recommended for final sanding to achieve better finishing in order to effectively add value to tertiary wood products. Scratches were obvious on fine textured species than course textured.

The results obtained for Denya/Okan are comparable (if not better) to that from the literature, which come from the natural forest. Since the machining performance of Potrodom (*Erythrophyleum guinense)*, Berekankum *(Manilkara multinervis)* and Savanna Ebony (*Diospyros mespiliformis*) are equivalent to Denya/Okan (*Cylicodiscus gabunensis)*, it implies that their machining properties could also be comparable with those from the natural forest.

RECOMMENDATIONS

The results indicate that the four species from the Volta Lake recorded, at least, medium machining surface quality, hence a higher potential for processing and utilization. Wood users should therefore patronize them when seen on the timber market.

Conflict of Interest

The authors have not declared any conflict of interest.

ACKNOWLEDGEMENT

The authors are grateful to Dr. J. R. Cobbinah through whose effort the work was carried out at CSIR-FORIG in collaboration with Clark Sustainable Resource Development (CSRD). They are also thankful to Dr. Oteng-Amoako and his team that identified the timber species extracted from the Volta Lake for the study, which made it easier for the analysis of the results. Sincere thanks go to the Management and Staff of the Sawmilling section, FC-WITC at Akyawkrom near Ejisu, where the logs were sawn into lumber and to Mr. Solomon Adjei for cross-cutting the logs into shorter forms. Again, the authors extend much appreciation to Messrs G.K Zorve, Felix Boakye, P.L Arthur and Richard Boamah of the wood processing workshop at CSIR-FORIG, who worked assiduously during preparation and testing of the specimens.

REFERENCES

Ambrogio,G, Filice L, Shivpuri R and Umbrello D (2008). Application of NN technique for predicting the in-depth residual stresses during hard machining of AISI 52100 steel. Int. J. Mater. Forming 1:39–45.

American Society for Testing and Materials (2007). Standard test methods for small clear specimens of timber. ASTM D 143-94. Annual Book of ASTM Standards, 2008. Section 4, Construction Vol. 04.10. West Philadelphia, PA, USA. pp. 20-52

American Society for Testing and Materials (2004). Standard test methods for conducting machining tests of wood and wood-base materials. ASTM D 1666-87. Annual Book of ASTM Standards 2008. Section 4, Construction Vol. 04.10. West Philadelphia, PA, USA. pp. 201-219.

Basheer AC, Dabade UA, Suhas SJ, Bhanuprasad VV (2008). Modeling of surface roughness in precision machining of metal matrix composites using ANN. J. Mater. Process Technol. 197:439–444

Beaty WT (1983). Preparing wood for finishing: Finishing eastern hardwoods. Forest Products Research Society, Wisconsin, USA. pp. 15–28.

Benardos PG, Vosniakos GC (2002). Prediction of surface roughness in CNC face milling using neural networks and Taguchi's design of experiments. Robot Comput. Integr. Manuf. 18:343–354

Benardos PG, Vosniakos GC (2003). Predicting surface roughness in machining: a review. Int. J. Machine Tools Manuf. 43:833–844

Chang CK, Lu HS (2006). Study on the prediction model of surface roughness for side milling operations. International Journal of Advanced Manufacturing Technology 29:867–878

Correa M, Bielza C, Pamies-Teixeira J (2009). Comparison of Bayesian networks and artificial neural networks for quality detection in a machining process. Expert Syst. Appl. 36 (3):7270–7279

Cus F, Zuperl U (2006). Approach to optimization of cutting conditions by using artificial neural networks. J. Mater. Process Technol. 173:281–290.

DIN 4760 (1982). Form deviations, concepts, classification system. Deutches Institut Fuer Normung, Ev, Berlin, Germany.

DIN 4768 (1990). Determination of values of surface roughness parameters R_a, R_z, R_{max} using electrical contact (stylus) instruments, concepts and measuring conditions. Deutsches Institut für Norming, Berlin, Germany.

Farmer RH (1972). Handbook of hardwoods, 2nd ed. Department of Environment, Building Research Establishment, Princess Risborough Laboratory. HMSO, London. pp. 3-152.

Farrokhpayam SR, Ratnasingam J, Bakar ES, Tang SH (2010). Characterizing Surface Defects of Solid Wood of Dark Red Meranti (*Shorea* sp.), Melunak (*Pentace* sp.) and Rubberwood (*Hevea brasiliensis*) in Planing Process. J. Appl. Sci. 10:915-918.

Field A (2007). Harvesting an Underwater Forest. Sloan Fellows find a triple-bottom-line business in West Africa. Stanford Graduate School of Business News. 76(1):12-14. Available at: http://public-prod-acquia.gsb.stanford.edu/news/bmag/sbsm0711/feature-harvest.html, http://www.gsb.stanford.edu/sites/default/files/documents/Stanford_NOV2007_LR.pdf

Fortin Y (2001). Wood machining properties of white spruce from plantation forests. For. Prod. J. 51(6): 4-15.

Fujiwara Y, Fujii Y, Sawada Y and Okumura S (2001). Development of a parameter to reflect the roughness of a wood surface that corresponds to tactile toughness: A novel filter to exclude local valley effects. Holz als Roh-und Werkstoff 59(5): 351-355.

Gilmore R, Barefoot AC (1974). Evaluation of some tropical woods imported into the United States from South America. Forest Prod. J. 24(2):24-28

Gurau L, Mansfield-Williams H and Irle M (2005). Processing roughness of sanded wood surfaces. Holz als Roh-und Werkstoff. 63:43-52.

Hendarto B, Shayan E, Ozarska B, Carr R (2006). Analysis of roughness of a sanded wood surface. Int. J. Adv. Manufacturing Technol. 28: 775-780

Hiziroglu S, Kosonkorn P (2006). Evaluation of surface roughness of Thai medium density fiberboard (MDF). Building and Environment 41(4): 527–533.

Karpat Y, Özel T (2007). Multi-objective optimization for turning processes using neural network modeling and dynamic neighborhood particle swarm optimization. Int. J. Adv. Manuf. Technol. 35:234–247

Malkoçoğlu A (2007). Machining properties and surface roughness of various wood species planed in different conditions. Build. Environ. 42(7):2562–2567.

Malkoçoğlu A, Özdemir T (2006). The machining properties of some hardwoods and softwoods naturally grown in Eastern Black Sea region of Turkey. J. Mater. Process. Technol. 173(3):315–320.

Nemli G, Akbulut T, Zekoviç E (2007). Effects of some sanding factors on the surface roughness of particleboard. Silva Fennica 41(2): 373–378.)

Nemli G, Ozturk I, Aydın I (2005). Some of the parameters influencing surface roughness of particleboard. Build. Environ. 40(10):1337–1340.

NPA (1993). From start to finish particleboard. National Particleboard Association. Gaithersburg, MD 20879.

Oktem H, Erzurumlu T, Erzincanli F (2006). Prediction of minimum surface roughness in end milling mold parts using neural network and genetic algorithm. Mater. Des. 27:735–744.

Ors Y, Baykan I (1999). The effect of planing and sanding on the surface roughness. Turk. J. Agric. For. 23(3): 577-582.

Oteng-Amoako AA (2006). 100 Tropical African Timber trees from Ghana: Tree description and wood identification with notes on distribution, ecology, silviculture, ethnobotany and wood uses. Graphic Packaging, Accra, Ghana. 304pp.

Owusu FW, Ayarkwa J, Frimpong-Mensah K (2011). The sanding properties of seven Ghanaian lesser-used timber species. Ghana J. For. 28(1):1-14.

Pal SK, Chakraborty D (2005). Surface roughness prediction in turning using artificial neural network. Neural Comput. Appl. 14:319–324

Ratnasingam J, Scholz F (2007). Characterizing surface defects in machine-planing of rubberwood (*Hevea brasiliensis*). Holz Roh-und Werkst65: 327–327.

Roger EH, Cool J (2008). Effects of cutting parameters on surface quality of paper birchwood machined across the grain with two planing techniques. Holz Roh-und Werkst. 66: 147-154.

Sandak J, Tanaka C (2005). Evaluation of surface smoothness using a light sectioning shadow scanner. J. Wood Sci. 51(3): 270-273.

Sieminski R, Skarzynska A (1989). Surface roughness of different species of wood after sanding. For. Prod. J. 32:98-107.

Skaljic N, Beljo-Lucic R, Cavlovic A, Obucina M (2009). Effect of Feed Speed and Wood Species on Roughness of Machined Surface. Drvna Industrija 60(4):229-234.

Sofuoğlu SD, Kurtoğlu A (2014). Some machining properties of four wood species grown in Turkey. Turk. J. Agric. For. 38:420-427.

Sogutlu C, Togay A (2011). The effect of the process parameters in the planing processes on the surface roughness of cherry and pear woods. Afr. J. Biotechnol. 10(21):4392-4399.

Tamizharasan T, Sevaraj T, Haq AN (2006). Analysis of tool wear and surface finish in hard turning. Int. J. Adv. Manuf. Technol. 28:671–679.

Tan PL, Safian S, Izman S (2012). Roughness models for sanded wood surfaces. Wood Sci. Technol. 46(1-3):129-142.

Taylor JB, Carrano AL, Lemaster RL (1999). Quantification of process parameters in a wood sanding operation. For. Prod. J. 49(5): 41-46.

Tjernlund C (1984). Guidance on machining Whitewood. A report on work carried out for the Swedish Institute for Wood Technology Research. Swedish-Finnish Timber Council, Retford, Notts, England, U.K. 45pp.

Influence of rooting media and number of nodes per stem cutting on nursery performance of vanilla (*Vanilla planifolia* Andr. *syn. Vanilla fragrans*)

Mesay Adugna[1], Derebew Belew[2] and Digafie Tilahun[3]

[1]College of Agriculture and Natural Resource, Dilla University, P. O. Box 419 Dilla, Ethiopia.
[2]College of agriculture and Veterinary Medicine, Jimma University, P. O. Box 307 Jimma, Ethiopia.
[3]Teppi National Spice Research Center, P. O. Box 34 Teppi, Ethiopia.

The effect of rooting media and number of nodes per cutting on nursery performance of vanilla fragrance were evaluated in South-western Ethiopia in 2011/ 2012. Six rooting media (forest soil, decomposed animal manure, fine sand, 1:1 mixture of forest soil: fine sand, 1:1 mixture of decomposed animal manure: fine sand and 1:1:1 mixture of forest soil: decomposed animal manure: fine sand) and four levels of node number (two, three, four, and five node cutting) were used in this experiment. Treatments were arranged in randomized complete block design with three replications. The result revealed that interaction between the two factors was significant ($P \leq 0.05$) for all parameters studied except sprouting percentage and root number. The highest shoot length, shoot girth, shoot fresh weight, shoot dry weigh, leaf number, leaf area, leaf fresh weight, leaf dry weight and root to shoot ratio were obtained from four node cuttings grown on a 1:1:1 mixture of forest soil: decomposed animal manure: fine sand rooting media, with the exception of the highest root length and rooting percentage of cuttings obtained from the rooting media containing pure sand. In addition highest root fresh weight, root volume and root dry weight were obtained from five nodal cuttings. Two nodal cutting grown on decomposed animal manure and pure fine sand media showed lower root initiation and shoot growth.

Key words: Vanilla fragrance, propagation, rooting media, node number.

INTRODUCTION

Vanilla (*Vanilla planifolia* Andr.) belongs to the family Orchidaceae, of which some 110 species have been reported so far (Talubnak and Soytong, 2010). Etymologically, vanilla is derived from the Spanish word "vainilla" meaning *little pod* (James and Ackerman, 2003). Vanilla grows best under hot humid climate from sea level to an elevation of 1500 m. Most of its production is carried out between 10°- 20° above and below the equator. The ideal growing conditions include moderate rainfall of 1500 to 3000 mm evenly distributed through 10 months of the year; temperatures of 15 to 30°C during the day and 15 to 20°C during the night and relative humidity of around 80% (Bianchessi, 2004).

Ethiopia has favorable environment for vanilla production and the crop has a huge local and international market (Zerihun et al., 2009). Currently, three accessions of vanilla introduced at different times are under adaptation trial in South-western Ethiopia. The first accession (Van.1/93) was introduced from Bali Island, Indonesia in 2001; the second accession (Van.3/04) was introduced from Uganda in 2004 and the third accession (Van.2/05) was introduced from Mauritius in 2005 (Girma et al., 2008). Currently, all three accessions are very well adapted at Teppi and their seedlings are being multiplied at Teppi National Spices Research Center (TNSRC) for further study.

Predominantly, vanilla is propagated through the vegetative means mainly from stem cuttings collected from a healthy and vigorously growing mother plant (Carlos and Balakrishnan, 1991; KAU, 2002; Purseglove, 1973; Sengupta, 2003). Different experiences are available from various locations about the number of nodes per stem cutting and ideal rooting media for propagation of vanilla. The type of rooting medium used can have a major influence on the rooting capacity of cuttings (Hartman et al., 1990). Vanilla stem cuttings are adversely affected by the availability of water and nutrients in the medium used during propagation. The length of cuttings used for vegetative propagation of vanilla is commonly determined by the availability of mother plants (Palama et al., 2010). Preliminary study in TNSRC indicated that vanilla cuttings with a minimum of a single node, cuttings with two, three and more numbers or more can be used for propagation (Girma et al., 2011). There was also an increasing trend in mean values of most planting materials growth parameters such as numbers of leaves, length of vine and root dry weight as the number of nodes per original stem cutting was increased (Girma, 2012). In addition, the use of short cuttings (30 cm long with 3 to 4 nodes) is becoming widespread due to shortage and unavailability of the preferred planting materials (Namirembe-Ssonkko et al., 2005). Therefore, the objective of this research is to determine influence of different rooting media and number of nodes per stem cutting on nursery performance of vanilla.

MATERIALS AND METHODS

Description of the study area

The experiment was conducted in South-western Ethiopia, Teppi National Spices Research Center (TNSRC) at nursery condition in 2011/2012. The site is located within latitude 7°3' N and longitude 35°0' E in an altitude of 1200 m a.s.l (metres above sea level) and receives a mean annual rainfall of 1688 mm. The mean maximum and minimum temperature of the area is 29.5 and 15.3°C, respectively. The soil type of the study area is dystric nitisols, eutric vertisols and vertic gleysols dominated by forest soil (Abayneh and Ashenafi, 2005). Minimum and maximum temperature of the experimental periods is presented in Table 1. In general, the area is characterized by high rainfall, temperature and humidity and representative hot humid low lands (EIAR, 2010).

Experimental materials

This experiment comprises two factors (rooting media composition and vanilla cuttings length). Semi hard wood cuttings of introduced vanilla accession (Van1. /93) with different node number (2, 3, 4 and 5) and six types of rooting media (fine sand, forest soil, decomposed animal manure, 1:1 mixture of forest soil and fine sand, fine sand: decomposed animal manure and 1:1:1 mixture of forest soil, decomposed animal manure and fine sand) were used for this experiment.

Preparation of rooting media and cuttings

The materials used for the preparation of the potting mixture were fine sand (2.5 mm), forest soil and decomposed animal manure. Forest soil was collected from the upper 15 cm layer of the soil under a forest canopy at the TNSRC. Decomposed animal manure was also collected from the Centre and fine sand was purchased from the local market. All rooting media were sieved through a 2.5 mm mesh and fine sand thoroughly washed to remove soil particles and other undesirable materials (Plate 2). The pH of each potting mix was determined using JENWAY digital electrode pH meter on a 1:2.5 basis of soil and water (Table 2), respectively at the soil laboratory of the Jimma University College of Agriculture and Veterinary Medicine.

Semi hard wood slant cuttings of vanilla vine with 2, 3, 4 and 5 nodes per stem cutting were taken from healthy mother plants using a pruning shear disinfected with alcohol. All leaves were removed from the vine before planting (Plate 1) and the cuttings were planted in the respective growth media followed by watering. After planting, the necessary nursery management practices such as watering and weeding were applied as per the recommendation of Girma et al. (2011) (Plate 3).

Experimental design and treatments

The study was conducted as a factorial experiment (6 media types × 4 different node numbers) in a randomized complete block design (RCBD) with three replications. Each treatment consisted of ten cuttings and a total of 720 cuttings were used in the study.

Data collection and analyses

The parameters evaluated in this experiment were shoot length (cm), shoot girth (mm), shoot fresh and dry mass (g), leaf number, total leaf area ($cm^2 \cdot plant^{-1}$), leaf fresh and dry mass (g), root number, root fresh and dry mass (g), root volume (ml), root length (cm) per rooted cutting, root to shoot ratio in terms of dry mass (g) as well as percentages of rooting and sprouting (out of ten cuttings within a plot). In all cases, roots were gently washed with tap water before measuring root growth parameters. Total leaf area was measured using leaf area meter (ADC Bio scientific Ltd Area Meter AM 200, England) in cm^2. Root volume was determined by the displacement method using a graduated cylinder half filled with water. For dry mass determination, both roots and shoots were oven dried at 80°C until the samples attained a constant mass (72 h). Except for percentage of rooted and sprouted cuttings the results are presented for discussion per plant basis.

The collected data on different growth parameters were analysed by Analysis of Variance (ANOVA). The ANOVA was done by using SAS version 9.2 Computer software (SAS Institute Inc., 2008).

Table 1. The weekly average minimum and maximum temperature of the nursery.

	Temperature	
Week	Min	Max
1	14.40	24.58
2	12.32	23.48
3	12.60	23.50
4	14.20	23.48
5	14.80	24.04
6	15.94	25.32
7	13.41	24.02
8	13.40	22.96
9	12.50	22.30
10	12.03	23. 30
11	13.00	25.50
12	12.70	24.90
13	14.50	25.30
14	15.89	24.10
15	11.60	22.85
16	13.20	24.90
17	12.32	23.60
18	11.40	23.48
19	15.20	24.80
Mean	13.44	24.06

Table 2. pH analysis of the media used in potting mix

Growing media	pH
S	6.48
F	5.09
M	7.88
F:S	6.48
M:S	7.67
F:M:S	7.57

F -Forest soil, FMS- Forest soil: decomposed animal manure: fine sand, FS- Forest soil: Fine sand, M- Decomposed animal manure:, MS -Decomposed animal manure: fine sand , S- Fine sand.

When ANOVA showed significant differences, mean separation was carried out using Least Significant Difference (LSD) test at 5% level of significance (Gomez and Gomez, 1984). All the figures and tables were generated by Excel computer program.

RESULTS AND DISCUSSION

All planting materials root and shoot growth parameters were significantly influenced by the type of rooting media and number of node per stem cuttings used ($p \leq 0.05$). However, root numbers were not significant. The highest root length (17.5 cm) (Plate 7) and percentage of rooting (99.3%) were obtained from four nodal cuttings which are grown in fine sand rooting media followed by five nodal cuttings (16.99 cm and 93.33% respectively) which were grown in 1:1:1 mixture of forest soil, decomposed animal manure and fine sand media. This could be due to the fact that low bulk density of fine sand medium invariably allows for greater root penetration leading to formation of longer roots. In accordance with this, Fagge et al. (2011) reported that influence of the medium is felt even before rooting occurs due to water retention and aeration properties and that the percentage and quality of roots in terms of number, length and weight can in many ways be directly linked to the rooting medium itself. Higher rooting percentage observed from longer cuttings could be linked to better initial carbohydrate reserves stored of longer cuttings (Ky-Dembele et al., 2011). In addition fine sand medium provides adequate aeration and drainage leading to increased porosity to promote better root initiation. This is supported by the work of Hans and Gisler (1983) where increasing air content of the rooting media improved rooting in Poinsettia and Hydrangea. While the lowest root length (6.05 cm) (Plate 8) and percentage of rooting (44.33%) were obtained from two nodal cuttings which were grown in decomposed animal manure (Table 3).

Maximum root fresh weight (1.16 g), root dry weight (0.13) and root volume (1.72 mm) was observed from five node cuttings grown on 1:1:1 mixture of forest soil: decomposed animal manure: fine sand. This may reveal that longer cuttings denote longer roots. Longer roots can have an ability of deeper penetration and therefore greater capacity in water and nutrients absorption leading to heavier root fresh weights in this media (1:1:1 mixture of forest soil: decomposed animal manure: fine sand) as compared to the decomposed animal manure medium. The result of the present investigation is in agreement with the work of Girma et al. (2012). Whereas the lowest root fresh weight (0.34 g), root dry weight (0.03) and root volume (0.78 mm) was observed from two nodal cutting grown on decomposed animal manure (Table 3). Similar trends in root fresh and dry weight were reported by Umesha et al. (2011).

The result of this experiment on the percentage of cuttings that develop new shoot (sprouting percentage) showed that there was no significant difference between interaction of factors (rooting media and node number). However, there was significant ($p<0.05$) difference between node number treatments. The early sprouting and highest sprouting percentage of vanilla cuttings was recorded from four nodal cuttings (81.94%) which are statistically at par with five nodal cuttings (78.89%), whereas the lowest sprouting percentage of vanilla was recorded from cutting with two and three nodes (68.06 and 72.5%, respectively) (Figure 1). The current result supported by Umesha et al. (2011) who observed early sprouting and better shoots growth and development from longer vanilla cuttings than single and three nodal cuttings.

Table 3. Interaction effect of different rooting media and number of nodes per cutting on the percentage of rooting (PR), root length (RL), root fresh weight (RFW), root dry weight (RDW) and root volume (RV) of *Vanilla fragrans.*

Number of nodes per stem cutting	Rooting media	Parameter				
		PR(%)	RL(cm)	RFW(g)	RDW(g)	RV(g)
2	F	46.67^{jk}	10.64^{ijk}	0.42^{ghi}	0.057^{ghijk}	0.94^{e}
	FMS	53.33^{hijk}	8.67^{lm}	0.55^{fg}	0.08^{cdef}	0.83^{fg}
	FS	50.00^{ijk}	7.63^{mn}	0.49^{fgh}	0.047^{jkl}	0.83^{fg}
	M	43.33^{k}	6.05^{n}	0.34^{i}	0.03^{l}	0.78^{g}
	MS	46.67^{jk}	12.82^{fgh}	0.54^{fgh}	0.05^{hijkl}	1.05^{cdefg}
	S	83.33^{bcd}	11.98^{hij}	0.78^{de}	0.04^{kl}	1.11^{cdef}
3	F	56.67^{ghij}	11.86^{hij}	0.42^{ghi}	0.06^{fghijk}	0.96^{defg}
	FMS	66.67^{efg}	11.47^{hij}	0.78^{de}	0.08^{cdef}	1.31^{bc}
	FS	46.67^{jk}	12.90^{efgh}	0.76^{de}	0.07^{efghij}	0.87^{fg}
	M	56.67^{ghij}	11.31^{hij}	0.61^{ef}	0.05^{ijkl}	0.80^{g}
	MS	60.00^{ghi}	9.34^{klm}	0.35^{hi}	0.05^{hijkl}	1.07^{cdefg}
	S	76.67^{de}	9.02^{klm}	1.00^{abc}	0.08^{cdef}	1.11^{cdef}
4	F	80.00^{cd}	14.81^{cde}	1.08^{ab}	0.09^{bcde}	1.46^{ab}
	FMS	93.33^{ab}	10.13^{jkl}	1.08^{ab}	0.09^{cde}	1.48^{ab}
	FS	63.33^{fgh}	14.81^{cde}	0.95^{bcd}	0.08^{defg}	1.36^{bc}
	M	56.67^{ghij}	12.24^{hi}	0.87^{cd}	0.06^{fghijk}	1.30^{bc}
	MS	73.33^{def}	14.51^{def}	0.78^{de}	0.08^{cdef}	1.27^{bcd}
	S	99.30^{a}	17.50^{a}	0.62^{ef}	0.12^{ab}	0.94^{efg}
5	F	66.67^{efg}	14.22^{defg}	0.90^{bcd}	0.06^{ghijk}	1.06^{cdefg}
	FMS	93.33^{ab}	16.99^{ab}	1.16^{a}	0.13^{a}	1.72^{a}
	FS	66.67^{efg}	15.15^{bcd}	1.08^{ab}	0.10^{bcd}	1.33^{bc}
	M	76.67^{de}	13.11^{efg}	0.90^{bcd}	0.11^{abc}	1.24^{bcde}
	MS	73.33^{def}	12.33^{ghi}	0.68^{ef}	0.07^{efghi}	1.43^{ab}
	S	90.00^{abc}	16.58^{abc}	0.91^{bcd}	0.08^{defgh}	1.46^{ab}
LSD (5%)		13.08	1.93	0.20	0.31	0.026
CV (%)		11.82	9.53	15.83	15.88	21.41

Means followed by the same letter(s) are not significantly different at p=0.05, F = forest soil , FMS = forest soil : decomposed animal manure : fin sand , FS = forest soil : fine sand , M = decomposed animal manure , MS = decomposed animal manure : fine sand , S = fine sand.

The highest leaf number (9.2), leaf fresh weight (7.16 g), leaf dry weight (2.127 g), shoot length (57.60 cm) and shoot fresh weight (33.38 g) was observed from four nodal cutting grown on 1:1:1 mixture of forest soil: decomposed animal manure: fine sand rooting media. Consequently, the highest mean values from four nodal cutting were supportive to a report by Girma et al. (2012). Whereas the lowest leaf number (5.40) was observed from two nodal cuttings which are grown on sole fine sand but statistically on par with three nodal cuttings. In addition the lowest leaf fresh weight (6.34 g), leaf dry weight (2.127 g), shoot length (25.04 cm) and shoot fresh weight (14.53 g) was observed from two nodal cutting grown on sole fine sand (Table 4). Maximum shoot girth (0.43 mm) and shoot dry weight (4.16 g) was recorded from four and five node cutting grown on 1:1:1 mixture of forest soil: decomposed animal manure: fine sand; whereas the lowest shoot girth (0.307 mm) observed from two nodal cutting grown on sole fine sand medium and the lowest shoot dry weight (2.56 g) observed from two nodal cutting grown on sole fine sand medium but statistically on par with three node cutting grown on fine sand medium (Table 4). Generally, there was an increasing trend in shoot growth and development as the node number per cutting increased.

Maximum leaf area (78.85 and 78.51 cm^2) was observed from four and five node cuttings grown on 1:1:1 mixture of forest soil: decomposed animal manure: fine

Table 4. Interaction effect of different rooting media and number of nodes per cutting on the leaf number (LN), leaf fresh weight (LFW), leaf dry weight (LDW), shoot length (SL) and shoot fresh weight (SFW) of *V. fragrans*.

Number of nodes per stem cutting	Rooting media	Parameter				
		LN	LFW(g)	LDW(g)	SL(cm)	SFW(g)
2	F	6.27^{fgh}	6.42^{ghi}	2.057^{ghijk}	28.58^{jkl}	19.98^{ijkl}
	FMS	7.48^{bcdef}	6.55^{fg}	2.08^{cdef}	41.49^{efg}	19.74^{jkl}
	FS	7.6^{abcdef}	6.78^{de}	2.04^{kl}	26.40^{k}l	16.28^{mn}
	M	6.68^{dcfgh}	6.49^{fghi}	2.053^{hijkl}	35.92^{ghij}	18.52^{lm}
	MS	5.63^{gh}	6.54^{fgh}	2.053^{hijkl}	30.79^{ijkl}	19.69^{jkl}
	S	5.40^{h}	6.34^{i}	2.03^{l}	25.04^{l}	14.53^{n}
3	F	7.80^{abcdef}	6.42^{ghi}	2.06^{fghijk}	33.96^{hij}	26.84^{cde}
	FMS	8.18^{abcde}	6.78^{de}	2.08^{cdef}	44.69^{cde}	20.27^{ijkl}
	FS	7.48^{bcdef}	6.76^{de}	2.07^{efghij}	43.86^{cde}	23.91^{fgh}
	M	7.25^{cdefg}	6.61^{ef}	2.05^{ijkl}	33.07^{hijk}	21.50^{hij}
	MS	8.45^{abcd}	6.35^{hi}	2.047^{jkl}	33.09^{hijk}	22.75^{jkl}
	S	5.48^{h}	6.00^{abc}	2.083^{cdef}	41.40^{efg}	19.98^{ijkl}
4	F	7.93^{abcdef}	7.08^{ab}	2.09^{bcde}	42.63^{def}	28.81^{bcd}
	FMS	9.20^{a}	7.16^{a}	2.127^{a}	57.60^{a}	33.38^{a}
	FS	7.45^{bcdef}	6.95^{bcd}	2.08^{defg}	52.07^{ab}	26.02^{def}
	M	8.80^{abc}	6.87^{cd}	2.063^{fghijk}	53.07^{ab}	26.84^{cde}
	MS	8.45^{abcd}	6.78^{de}	2.083^{cdef}	55.79^{ab}	29.00^{bc}
	S	8.45^{abcd}	6.62^{ef}	2.09^{cde}	40.19^{efgh}	24.94^{efg}
5	F	8.93^{ab}	6.90^{bcd}	2.057^{ghijk}	42.63^{def}	22.74^{ghi}
	FMS	9.11^{ab}	6.90^{bcd}	2.10^{bcd}	53.90^{ab}	30.66^{ab}
	FS	9.00^{ab}	7.08^{ab}	2.12^{ab}	49.90^{bc}	24.71^{efg}
	M	6.68^{fgh}	7.08^{ab}	2.107^{abc}	52.83^{ab}	18.69^{klm}
	MS	8.60^{abc}	6.68^{ef}	2.073^{efghi}	28.323^{jkl}	21.50^{hijk}
	S	6.53^{fgh}	6.91^{bcd}	2.077^{defgh}	35.92^{fghi}	26.26^{cdef}
LSD (5 %)		1.67	0.196	0.026	7.25	2.80
CV (%)		13.34	15.826	21.411	10.71	10.03

Means followed by the same letter(s) are not significantly different at P=0.05 level significance , F = forest soil , FMS = forest soil : decomposed animal manure : fin sand , FS = forest soil : fine sand , M = decomposed animal manure , MS = decomposed animal manure : fine sand , S = fine sand.

sand and 1:1 mixture of forest soil: fine sand rooting media, respectively; whereas the lowest leaf area (46.98 cm^2) was observed from two node cuttings raised on pure fine sand. However, root number showed no significant difference ($p > 0.05$) between interaction of rooting media and number of nodes per stem cutting. Highest root to shoot dry weight ratio (0.133 g) was recorded from four node cutting grown on 1:1:1 mixture of forest soil: decomposed animal manure: fine sand followed by five nodal cutting (0.103) grown on 1:1:1 mixture of forest soil: fine sand: decomposed animal manure and 1:1 mixture of forest soil: fine sand whereas the lowest root to shoot dry weight (0.04 g) ratio is recorded from two and three node cutting grown on sole fine sand rooting media (Table 5). On the other hand, an increasing pattern of root to shoot ratio was obtained from two nodes to four nodes cutting while a decreasing order was recorded from four nodes to five node cutting.

In general, longer cuttings grown on 1:1:1 mixture of forest soil: fine sand: decomposed animal manure resulted in highest root and shoot growth. However, cuttings from the medium composed of pure sand showed poor shoot growth and development and cuttings grown in pure decomposed animal manure showed lowest root growth (Plates 4, 5 and 6). The current findings in vanilla fragrance are also supported by the

Table 5. Interaction effect of different rooting media and number of nodes per cutting on the leaf area (LA), shoot girth (SG), shoot dry weight (SDW) and root to shoot ratio (RS) of *V. fragrans*.

Number of nodes per stem cutting	Rooting media	Parameter			
		LA(cm^2/plant)	SG	SDW(g)	RS
2	F	51.20^{lm}	0.4^{abcde}	2.81^{i}	0.073^{defgh}
	FMS	53.55^{kl}	0.317^{hi}	2.83^{i}	0.047^{ij}
	FS	57.19^{ijkl}	0.36^{efghi}	2.89^{ghi}	0.067^{efghi}
	M	66.42^{cdefg}	0.40^{abcde}	2.88^{hi}	0.060^{ghij}
	MS	61.11^{ghij}	0.34^{fghi}	2.81^{i}	0.067^{efghi}
	S	46.98^{m}	0.307^{i}	2.56^{j}	0.040^{j}
3	F	60.88^{ghij}	0.387^{abcdef}	2.82^{i}	0.0767^{cedfg}
	FMS	68.75^{bcdef}	0.37^{bcdefg}	2.83i	$0.10b^{c}$
	FS	62.86^{fghi}	0.40^{abcde}	3.05^{efg}	0.067^{efghi}
	M	54.91^{jkl}	0.35^{efghi}	2.81^{i}	0.067^{efghi}
	MS	65.62^{cdefg}	0.33^{ghi}	3.03^{fg}	0.053^{ghij}
	S	57.78^{ijk}	0.417^{abcd}	2.603^{j}	0.040^{j}
4	F	69.71^{bcd}	0.36^{defgh}	2.82^{i}	$0.043i^{j}$
	FMS	78.85^{a}	0.43^{a}	3.14^{ef}	0.133^{a}
	FS	63.41^{efghi}	0.40^{abcde}	4.16^{a}	0.067^{efghi}
	M	66.07^{cdefg}	0.35^{efghi}	3.15^{ef}	0.093^{bcd}
	MS	69.27^{bcde}	0.37^{cdefgh}	3.56^{b}	0.06^{ghij}
	S	63.11^{efghi}	$0.38^{abcdefg}$	3.41^{bc}	0.067^{efghi}
5	F	71.24^{ab}	0.42^{abc}	3.14^{ef}	0.05^{hij}
	FMS	73.75^{ab}	0.43^{a}	4.16^{a}	0.103^{b}
	FS	78.51^{a}	0.40^{abcde}	3.4^{c}	0.103^{b}
	M	60.28^{ghij}	0.40^{abcde}	2.82^{i}	0.090^{bcde}
	MS	64.27^{defgh}	0.37^{cdefgh}	2.88^{hi}	0.087^{bcdef}
	S	58.09^{hijk}	0.387^{abcdef}	3.27^{de}	0.06^{fghij}
LSD (5%)		6.237	0.06	0.16	0.0257
CV (%)		5.977	8.62	8.95	21.83664

Means followed by the same letter(s) are not significantly different at p=0.05 level significance, F = forest soil, FMS = forest soil: decomposed animal manure: fin sand, FS = forest soil: fine sand, M = decomposed animal manure, MS = decomposed animal manure: fine sand, S = fine sand.

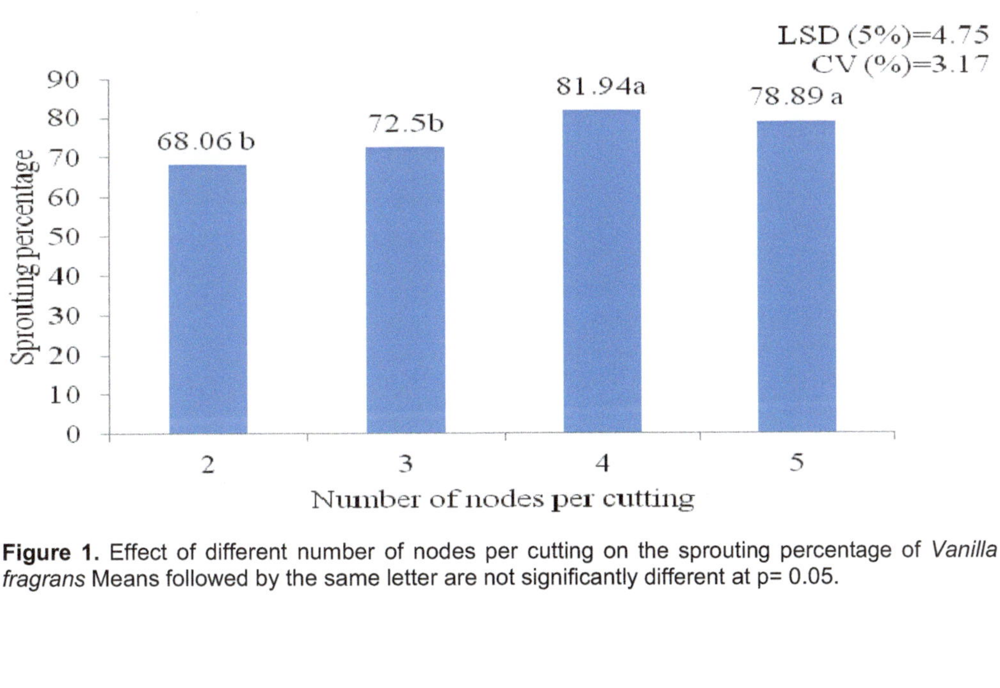

Figure 1. Effect of different number of nodes per cutting on the sprouting percentage of *Vanilla fragrans* Means followed by the same letter are not significantly different at p= 0.05.

Plate 1. Pots ready for cutting insertion.

Plate 2. Cuttings of introduced vanilla accession (Van1. /9) with different node number.

Plate 3. Overview of cuttings on the rooting bed.

Plate 4. Overview of cuttings after two month.

reports of Digafie (2006) who observed higher shoot growth response in black pepper cuttings raised on a multi-component media than those grown on single component. The results of using four node numbers for vanilla planting material preparation was also supporting to a recommendation by Girma et al. (2012). The authors stated that size and uniformity of planting materials from four node cutting were very promising. They also recommended using four node vanilla (*V. fragrans*) cutting is economical as well as environmental to raise nursery rooted planting materials in South western Ethiopia. According to Namirembe -Ssonkko et al. (2005) *Vanilla fragrans* can be easily and successfully propagated in the nursery using short stem cuttings (30cm, approximately 4 nodes) inserted in a medium with a high OM content and a good water holding capacity. Umesha *et al.* (2011) reported that root initiation is faster and vigorous in three to five node cuttings.

Plate 5. Overview of cuttings after four month.

Plate 6. Overview of cuttings at the end of nursery work.

CONCLUSION AND RECOMMENDATION

Rooting media type showed considerable effect on all root and shoot development and growth parameters (except sprouting percentage, shoot girth and root number). In most of the case leaf number (9.2), leaf fresh (7.157 g) and dry weight (2.127 g), shoot length (57.60 cm) , shoot fresh weight (33.383 g) and root to shoot ratio (0.133 g) characters evaluated, better results were obtained from four node vanilla cuttings grown on a 1:1:1 mixture of forest soil: decomposed animal manure : fine

Plate 7. Root of four node cutting grown on fine sand rooting media.

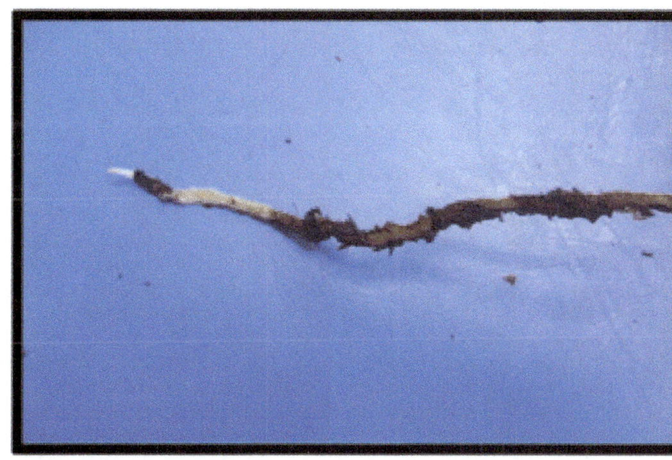

Plate 8. Root of two node cutting grown on decomposed animal manure rooting media.

sand, with the exception of the biggest root length (17.5 cm) and rooting percentage (99.3%) of cuttings obtained from the medium containing pure sand. cuttings obtained from the medium containing pure sand. 1:1:1 mixture of forest soil: decomposed animal manure: fine sand rooting media give rise to branched roots having better chance of adaptation after transplanting, but those grown from the medium composed of pure sand resulted in long, coarse and brittle roots that were not preferable. The highest value for some root parameters (root volume, root fresh and dry weight) was obtained from five node cuttings grown on a 1:1:1 mixture of forest soil: decomposed animal manure: fine sand. In addition early and highest sprouting percentage were obtained from longer (four and five node) cuttings. Higher mean values of leaf area, shoot girth and shoot dry weight were obtained from four and five node cuttings grown on a 1:1:1 mixture of forest soil: decomposed animal manure: fine sand. Therefore, the study reveals that vegetative propagation of vanilla could be successfully attained by raising four node cuttings on 1:1:1 mixture of forest soil: decomposed animal manure: fine sand rooting media.

Conflict of Interest

The authors have not declared any conflict of interest.

ACKNOWLEDGEMENTS

The authors thank Teppi National Spices Research Center for their cooperation in providing necessary assistance and experimental materials and Jimma University College of Agriculture and Veterinary Medicine for granting the research expenses. Special gratitude goes to Mrs. Kassaye Tolosa and Mr.Girma Hailemichael for their unreserved advice and encouragement during the study time

REFERENCES

Abayneh E, Ashenafi A (2005). Soils of Tepi and Haru Agricultural Research Sub centers. National Soil Research Center, Ethiopian Institute of Agricultural Research, Addis Ababa, Ethiopia. pp. 1-34.

Bianchessi P (2004). Vanilla agriculture and curing techniques, a photographic handbook for vanilla farmers. Venue vanilla Co Ltd. Vanuatu. pp. 20-32.

Carlos T, Juan, Balakrishnan S (1991). South pacific perennial spice production developments and prospects. CTA, ADB and IRETA.

Digafie T (2006). Influence of Growth Media and Physiological Conditions of The Stock Plant on Root and Sprout Development of Black Pepper (*Piper nigrum* L.). A Master of Science Thesis Submitted to the Department of Plant Sciences, School of Graduate Studies, Haramaya. P 70.

EIAR (Ethiopian Institute of Agricultural Research), 2009. One Step ahead in Spices Research. EIAR, Addis-Ababa.

Fagge AA, Manga AA (2011). Effect of Sowing Media and Gibbrellic Acid on the Growth and Seedling Establishment of *Bougainvillea glabra* , *Ixora coccinea* and *Rosa chinensis* Characters. 2: Root Characterization. Bayero J. Pure Appl. Sci. 4(2):155 – 159

Girma H, Digafie T, Edossa E, Belay YB, Weyessa G (2008). Spices research achievements, revised edition. Ethiopian Institute of Agricultural Research, Addis Ababa Ethiopia. pp. 43-45.

Girma H, Digafie T, Habtewold K, Haimanot M (2011). Effect of Node Number on the Nursery performance of vanilla (*Vanilla fragrans*) Cutting in South Western Ethiopia. Proceeding of the third biennial Ethiopian Horticultural Science Society (EHSS) Conference on Improving Quality Production Horticultural Crops for Sustainable Development, Jimma, Ethiopia. Jimma University College of Agriculture and Veterinary Medicine.

Girma H, Digafie T, Habtewold K, Haimanot M (2012). The Effect of Different Node Number cuttings on Nursery performance of vanilla (*Vanilla planifolia* syn. *fragrans*) Cuttings in South Western Ethiopia. Int. Res. J. Agric. Sci. Soil Sci. 2(9):408-412.

Gomez K, Gomez AA (1984). Statistical procedure for agricultural research, 2nd Edn. John Wiley and Sons, New York, USA. P 704.

Hans R, Gisler D (1983). Physical conditions of propagation media and their influence on the rooting of cuttings. III. The effect of air content and temperature in different propagation media on the rooting of cuttings. Plant Soil 75:1-14.

Hartman HT, Kester DE, Davies FT (1990). Plant propagation: Principles and Practices, 5th ed. Prentice-Hall Publishers, Englewood Cliff, NJ, USA. P 647.

James D, Ackerman (2003). "Vanilla". Flora South America 26(4):507-516

KAU (Kerala Agricultural University) (2002). Package of practices and recommendations: Crops. 12th Edition Kerala Agricultural University. pp. 278-290.

Ky-Dembele CK, Tigabu M, Bayala J, Savadogo P, Boussim IJ, Od´en PC (2011). Clonal Propagation of *Khaya senegalenss*. The Effects of Stem Length, Leaf Area, Auxins, Smoke Solution, and Stock plant Age. Int. J. Forest. Res. 20(11): 269-281

Namirembe-Ssonkko R, Lule A, Kyakimwa F (2005). Evaluation of Rooting Media for Commercial Nursery Propagation of *Vanilla fragrans* in Uganda. Proceedings of the Third Horticulture Workshop on Sustainable Horticultural Production in the Tropics. Maseno, Kenya, Maseno University, MSU.

Palama T, Menard P, Fock I, Choi Y, Bourdon E, Govinden-Soulange J, Bahut M, Payet B, Verpoorte R, Kodja H (2010). Shoot differentiation from protocorm callus cultures of *Vanilla planifolia* (Orchidaceae): Proteomic and Metabolic responses at early stage. BMC Plant Biol. 10:82-94.

Purseglove JW (1973). General agronomic aspects of spice crops. In: Purseglove JW (eds.), Tropical Products Institute Proceedings of the Conference on Spices, London School of Pharmacy. pp. 85-90.

SAS (Statistical Analysis System) (2008). SAS Institute. Cary, North Carolina, U.S.A.

Sengupta R (2003). Vanilla. The spice of life for Indian farmers. pp. 5-10.

Talubnak C, Soytong K (2010). Biological control of vanilla anthracnose using *Emericella nidulans*. J. Agric. Technol. 6(1):47-55.

Umesha K, Murthy G, Smitha GR (2011). Environmental conditions and type of cuttings on rooting and growth of vanilla (*Vanilla planifolia* Andrews). J. Trop. Agric. 49(1-2):121-123.

Zerihun A, Ayelign M, Alemayehu T, Wondyfraw T (2009). Efficient *in vitro* multiplication protocol for *Vanilla planifolia* using nodal explants in Ethiopia. Afr. J. Biotechnol. 8(24):6817-6821.

Effect of mulching and amount of water on the yield of tomato under drip irrigation

Baye Berihun

Adet Agricultural Research Center, P. O. Box 08, Bahir-Dar, Ethiopia. E-mail: betibaye@yahoo.com.

The main challenge confronting both rain fed and irrigated agriculture is to improve WUE and sustainable water use for agriculture. An experiment was conducted to evaluate the effect of mulch and amount of water on the yield of tomato under drip irrigation system and to assess the potential of deficit irrigation to improve the economic efficiency of tomato production at Adet Agricultural Research Center, horticultural crops trial site (Woramit) (North Western Ethiopia) from 2006 to 2007. A factorial combination of three levels of water (namely 315, 440 and 565 mm) combined with three mulch treatments [namely without mulch (WM), black plastic mulch (PM) and straw or crop residue mulch (STM)] amid three replications and two days irrigation interval was used. Amount of water significantly affected the number of fruits per plant, average weight of fruits marketable and total fruit yield/ha. Significant difference was also shown between mulch treatments on number of fruits, unmarketable, marketable and total fruit yield/ha. Based on the partial budget analysis, the highest net benefit was obtained via 440 mm water with straw mulch amid a net benefit (52,959.40 birr/ha and a marginal rate of return (MRR) 690%. Therefore, application of 440 mm/ha water in two days interval with straw mulch is found to be economically and agronomically feasible and is recommended for Woramit and its surrounding and other similar agro-ecologies under drip irrigation system.

Key words: Tomato, drip, mulching, water levels, marketable fruit yield, total fruit yield, marginal rate.

INTRODUCTION

The advent of increasing water scarcity in this century will observe less increase in irrigated land availability for food production than in the past. Novel irrigation technologies need to be tested under local environments and particularly in agricultural production systems of developing countries. While irrigation can benefit yields and enhance water use efficiency (WUE) in water limited environments, the potential for full irrigation is decreasing, with increased competition from the domestic and industrial sectors. Thus, the main challenge confronting both rain fed and irrigated agriculture is to improve WUE and sustainable water use for agriculture.

Ethiopia is facing a tremendous challenge in meeting the food needs of rapidly growing population. There are small, medium and large scale irrigation systems in Ethiopia (FAO, 1995). To this end, both irrigated and dry land cropping areas will have to be developed or improved in the future. However, these tasks will not be easy, the cost of developing large scale and medium scale level irrigation is by now sky rocketing. Therefore, efficient utilization of water resources and development of small scale irrigation schemes at family level is crucial for countries like Ethiopia, which has a huge water resource: yet their population is chronically food insecure

Micro irrigation system was found to result in 30 to 70% water savings in various orchard crops and vegetables along with 10 to 60% increases in yield as compared to conventional methods of irrigation. It is prudent to make efficient use of water and bring more area under irrigation through available water resources. This can be achieved by introducing advanced methods of irrigation and improved water management practices (Zaman et al., 2001). Drip irrigation in combination with mulch is one of the best irrigation methods, which can improve the water management practice significantly. Surface mulches have been used to improve soil water retention, reduce soil temperature and reduce wind velocity at the soil surface and arid lands (Kay, 1978; Jalota and Prihar, 1998). Surface mulches can also improve water penetration by impeding runoff and protecting the soil from raindrop-splash and reducing soil crusting (Munshower, 1994).

Tomato is the leading vegetable crop in the world. It is

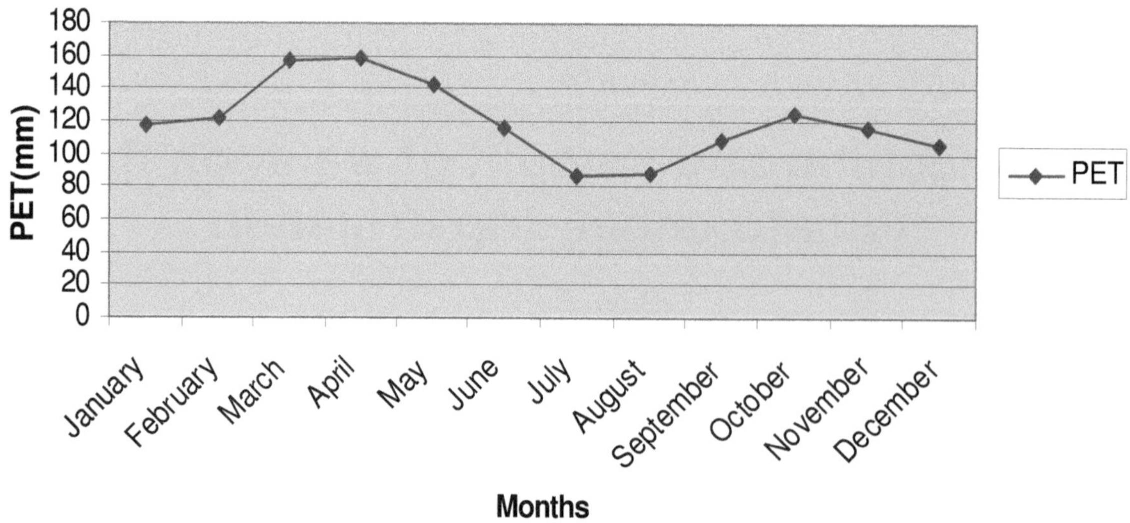

Figure 1. Monthly potential evapotranspiration at Woramit. Source: FAO Climatic Database (2005).

also one of the dominant vegetable crops in Ethiopia that is best suited for drip irrigation in combination with mulch. However, no work has been done to study the effect of drip irrigation in combination with mulch in the area. The present study was planned to evaluate the effect of mulch and amount of water on the yield of tomato under drip irrigation system and to assess the economic feasibility in relation to mulch used in tomato production.

MATERIALS AND METHODS

Experimental site

An experiment was conducted to evaluate the effect of mulch and amount of water on the yield of tomato under drip irrigation system and to assess the potential of deficit irrigation to improve the economic efficiency of tomato production at Woramit from 2006 to 2007. Woramit is located in the Northwestern part of Ethiopia, Bahirdar, at Adet Agricultural Research Center, horticultural crops trial site at an altitude of 1800 m above sea level. It is regarded as a warm temperate climatic zone where there is distinct dry months in winter. The soil is deep with red-brown color (netosol). The mean daily maximum temperature is 23.5°C in August and 29.5°C in April. The mean daily minimum temperature is 6.2°C in January and 13.3°C in May. The area receives a total annual rainfall of 800 to 1250 mm. The average potential evapo-transpiration (4.08 mm/day) and evapo-transpiration of the crop (ETc) value is (6.73 mm/day) (Figure 1).

Experimental materials and design

Factorial combination of water levels (namely 315, 440 and 565 mm) and three mulch treatments [namely without mulch, black plastic mulch (BPM) and straw mulch (SM)] were tested with two days irrigation interval under drip irrigation with three replications. A total of 9 treatment combinations were evaluated. The variety of Melka-salsa (processing type tomato variety) was used as the testing material. Seedlings were raised before a month and undamaged, reasonably uniform and clean seedlings were selected and transplanted to properly prepare plots on November 13, 2006 and December 8, 2007. Each plot consisted of two rows that are 3 m long with row spacing of 1 m. It was planted at a spacing of 30 cm between plants accommodating 20 plants per plot.

Irrigation system

Chapin bucket kits drip irrigation system was used. A 20 L bucket, which serve as a container (water source) was mounted 1 m above the ground on a stand constructed from wood (Figure 2). The drip tape was stretched on each row and installation of the system was made as per the instructions of Chapin bucket kits drip irrigation system. Measured volumes of water were filled in the bucket manually. The system was trickling water at every 30 cm space for each plant with a discharge rate of 0.03 liters/second for the dripper.

RESULTS AND DISCUSSION

Tomato plants are sensitive to water stress and they show high correlation between evapo-transpiration and crop yield. The statistical analysis in Table 1 indicates that average weight of fruits, marketable and total fruit yield were significantly (P=0.01) affected by the amount of water applied. However, the effect on plant height, average weight of fruits per plant, and unmarketable fruit yield was not significant. Even though the effect is statistically non-significant, the maximum plant height and average weight of fruits were recorded at 565 mm water level.

The statistical analysis in Table 2 shows that average weight of fruits, marketable and total fruit yield are significantly affected by the amount of irrigation water. The overall year combined effect of the amount of water as indicated in Table 3, on average weight of fruits was significant (P= 0.05) and highly significant (P=0.01) on marketable and total fruit yield. The highest marketable and total fruit yield (50.94 and 57.51 tone/ha) was

Figure 2. Chapin bucket kits drip irrigation system mounted 1 m above the ground on a stand constructed from wood.

obtained via 565 mm (WL), while the lowest marketable and total fruit yield (39.97 and 46.16 tone/ha) was recorded via 315 mm (WL) (Table 3).

The result indicates that, even with a minimal amount of water, we can get a reasonable fruit yield. According to Amhara Region Bureau of Agriculture (BOA) 2002 unpublished report, the blanket average water requirement of tomato in the low and mid altitude areas of Amhara region is up to 830 mm. This result showed that drip irrigation system reduced the water requirement of the crop by 47 to 62% as compared to furrow irrigation. As Fekadu and Teshome (1997) cited Pruitt, drip irrigation increased the yield of tomato and water use efficiency (WUE) by 19 and 20%, respectively as compared to furrow irrigation. Similarly, Stein et al. (1996) and Raina et al. (1998) reported that drip irrigation system significantly reduced the water requirement of field pea and cantaloupe as compared to furrow irrigation, because it properly managed the drip systems supply of adequate moisture to the root zone and do not wet the area between beds. This technique applies the water when and where needed, while maintaining a dry area that can absorb excess moisture during heavy rain.

Water saving using drip irrigation on crops can be as much as 80% when compared to other irrigation techniques (Bogle and Hartz, 1986). Raina et al. (1998) reported that drip irrigation besides giving a saving of 32% water resulted in 49.5% higher yield as compared to surface irrigation. The analysis of variance shows that mulch has significant influence on yield and yield components of tomato. In view of that, plant height, number of fruits per plant, unmarketable yield, marketable and total fruit yield were significantly affected by mulch in the first year of study (Table 4), while in the second year, the influence of mulch was significant on plant height, marketable and total fruit yield (Table 2). The highest marketable and total fruit yield (48.02 and 55.32 tons/ha) in the first and (65.44 and 70.85 tons/ha) second year, respectively were obtained through black plastic mulch. The second maximum marketable and total fruit yield (38.92 and 47.72) in the first and (50.02 and 59.0 tons/ha) second year were recorded via straw mulch, respectively.

The overall year combined effect of mulch on number of fruits, unmarketable, marketable and total fruit yield was significant. Consequently, among the mulch applications, the maximum marketable and total fruit yield (56.43 and 63.0 tons/ha) were obtained via black plastic mulch (BPM) followed by straw mulch. This result is disparate with the investigation of Levent et al. (2001), who reported that the highest fruit yield was obtained from wheat straw mulch followed by transparent and black polyethylene mulch, respectively.

The highest and lowest unmarketable yield (7.65 and

Table 1. Effect of amount of water, mulch and interaction of mulch and amount of water on the mean fruit yield and yield components of tomato in 2006.

Water level treatments	Plant height (PH) (cm)				Average weight of fruits (g)				Marketable fruit yield (MFY) (Ton/ha)				Total fruit yield (TFY) (Ton /ha)			
	Mulch treatments				Mulch treatments				Mulch treatments				Mulch treatments			Means
	WM	BM	SM	Means	WM	BM	SM	Means	WM	BM	SM	Means	WM	BM	SM	
315	44.13	49.80	46.47	46.80	27.38	29.84	27.93	28.38	22.52	47.45	32.22	34.07^b	27.95	55.32	39.72	40.99^b
440	47.27	52.93	47.87	49.36	30.14	28.94	29.50	29.53	40.88	49.06	41.82	43.92^a	47.24	56.04	52.56	51.95^a
565	51.73	50.13	47.07	49.64	31.55	29.36	27.71	29.54	46.83	47.54	42.73	45.70^a	56.04	54.60	51.06	53.89^a
Means	47.7b	50.96a	47.13b	-	29.69	29.38	28.38		36.74b	48.02a	38.92b		43.74b	55.32a	47.72ab	
C.V%		5.75				8.23				26.43				16.65		
LSD 5% LxM	NS	NS	NS		NS	NS	NS		NS	NS	NS		NS	NS	NS	
LSD 5% WL				NS								**				**
LSD 5% M		*				NS		NS		**				**		

M = Mulch, WL = water level, WM = without mulch, BM = black plastic mulch, SM = straw mulch, w = water level, M = mulch, NS = Non significant, * = significant at P = 0.05, ** = significant at P = 0.01

Table 2. Effect of amount of water, mulch and interaction of mulch and amount of water on the mean fruit yield and yield components of tomato in 2007.

Water level treatments	Plant height (PH) (cm)				Average weight of fruits (g)				Marketable fruit yield (MFY) (Ton/ha)				Total fruit yield (TFY) (Ton /ha)			
	Mulch treatments				Mulch treatments				Mulch treatments				Mulch treatments			
	WM	BM	SM	Means	WM	BM	SM	Means	WM	BM	SM	Means	WM	BM	SM	Mean s
315	**48.47**	57.07	54.00	53.18	**18.43**	23.92	24.02	22.12b	**28.71**	61.85	47.98	45.88b	32.70	67.08	54.21	51.33b
440	50.40	58.73	55.20	54.78	26.27	25.29	24.35	25.30ab	56.56	64.37	49.54	56.17a	62.65	70.01	56.11	62.92a
565	54.00	58.27	53.20	55.16	26.05	26.87	25.24	26.05a	45.87	70.11	52.53	56.82a	50.52	75.47	57.39	61.13a
Means	50.96b	58.02a	54.13b		23.59	25.36	24.54		43.41b	65.44a	50.02b		48.62b	70.85a	59.90b	
C.V%		8.90				13.26				24.01				15.40		
LSD5%WLxM	NS	NS	-		NS	NS	NS	-	NS	NS	NS	-	NS	NS	NS	-
LSD 5%W	-	-	NS		-	-	-	*	-	-	-	*	-	-	-	*
LSD5%M		*		-						**		-		**		

M = Mulch, WL = water level, WM = without mulch, BM = black plastic mulch, SM = straw mulch, w = water level, M = mulch, NS = Non significant, * = significant at P = 0.05, ** = significant at P = 0.01.

5.99 tone/ha) were recorded through straw mulch and without mulch treatments, respectively. In this case, the vegetative growth of the crop was vigor in straw and black plastic mulch and this high vegetative growth may favor the occurrence of insect pest and fruit decay that may lead to high unmarketable yield. However, from the total unmarketable fruit yield, 49.18% was due to fruit decay, 37.04% was due to insect damage and the rest 13.77% was due to crack. The effect of mulch in respect of weed control was non significant. Even though there is no significant difference between mulch treatments on number of weeds/m^2, mulching reduced the incidence of weed from 38 to 50% as compared to the control

Table 3. Overall year combined ANOVA of water level and mulching effect on fruit yield and yield components of tomato at Woramit in 2006 and 2007.

Water level treatments (mm)	Plant height (cm)				Average weight of a fruit (g)				Marketable fruit yield (MFY) (Ton/ha)				Total fruit yield (TFY) (Ton /ha)			
	Mulching				Days											
	WM	BM	SM	Means	WM	BM	SM	Means	WM	BM	SM	Means	WM	BM	SM	Means
315	**46.30**	53.43	50.23	49.99	22.91	26.88	25.97	25.25b	**29.24**	47.23	38.84	38.44b	34.39	53.78	45.51	44.56b
440	48.83	55.83	52.53	52.07	28.20	27.11	26.92	27.41a	49.68	56.72	52.19	52.86a	56.96	63.02	59.03	59.67a
565	52.87	54.20	50.13	52.40	28.80	28.11	26.48	27.79a	51.31	61.93	51.79	55.01a	565.63	68.14	57.87	60.86a
Means	49.33	54.50	50.63	-	26.64	27.37	26.46	-	43.41c	55.29a	47.61b	-	49.30c	61.65a	54.14b	-
C.V%		7.54				10.90				27.94				11.77		
P.L WxM	NS	NS	NS	-	NS	NS	NS	-	*	*	*	-	*	*	*	-
PL W	-	-	-	NS	-	-	-	*	-	-	-	**	-	-	-	**
PL M		NS		-		NS		-		NS		-		**		-

M = Mulch, WL = water level, WM = without mulch, BM = black plastic mulch, SM = straw mulch, w = water level, M = mulch, NS = Non significant, * = significant at P = 0.05, ** = significant at P = 0.01.

(without mulch treatment). Mulching is not only important to reduce weed incidence, but also, it improves the soil micro-environment (Dickerson, 1996), which indicated that organic mulches help to cool the soil, conserve soil moisture, reduced annual weed production and return nutrients to the soil through decomposition.

Maximum control of the soil environment including water conservation can be obtained with the use of drip irrigation under either organic or plastic mulch (Lamont, 1991). Generally, the result points out that mulch have significantly increase the growth and fruit yield of tomato. Similarly, Geber et al. (1988) and Salman et al. applications increase the soil temperature so that vegetative development and fruit yield of tomato increased. The interaction effect of the amount of water and mulch in both years was not significant in any parameter. However, the highest marketable and total fruit yield (61.93 and 68.13 tons/ha) were obtained via the interaction effect of 565 mm (WL) with black plastic mulch. The second maximum marketable and total fruit yield (56.72 and 63.02 tone/ha) respectively were recorded through the interaction of 440 mm (WL) with black plastic mulch. In straw mulch, the maximum marketable yield (47.63 ton/ha) was obtained via 440 (WL). Drip irrigation combined with mulch has a momentous influence on the water use efficiency of tomato. Water use efficiency (WUE) is agronomically, simply the efficiency in which water is used to produce an economic yield. Water use efficiency of each treatment was determined after the marketable yield was obtained. The total water utilized was calculated taking the actual application volume (315, 440 and 565 mm) of irrigation levels, while the water use efficiency was calculated by dividing marketable yield by the volume of applied water. The highest water level (565 mm) combined with black plastic mulch gave the maximum fruit yield. However, the highest water use efficiency value was recorded at the lowest water level (315 mm) with black plastic mulch, whereas the lowest WUE (9.08 kg/m^3) was obtained at 565 mm without mulch treatment, which indicated that the plastic mulch distinctly improve the water use efficiency of tomato (Table 5).

The result is in line with the findings of Seyfi et al. (2007), which showed that drip irrigation with black plastic mulch markedly decreased the amount of water applied, increased water use efficiency (WUE) and increased crop yield (cantaloupe) due to increase in number of fruits per plant, fruit weight and fruit thickness. Similarly, Stein et al. (1996) reported that the use of mulch combined with drip reduced supplemental water needs 29 to 36% as compared to non mulched plots. Tomato, grown under plastic mulch, significantly increased earliness, fruit quality and control weeds by about 30 to 90% and increased total fruit yield and marketable fruit yield by about 20 and 24%, respectively (http://www.actahort.org). This might be due to the frequent application of water resulting in more even distribution of soil moisture in crop root zone, sufficient moisture conservation, and proper temperature control owing to presence of mulch, better utilization of nutrients and negligible infestation of weeds.

Based on the biological yield data, 565 mm of water combined with black plastic mulch gave the (1992) pointed out that mulch and tunneling maxi-mum marketable fruit yield(61.93 ton/ha) (Table 6). However, in order to recommend this result for farmers, it is necessary to estimate the minimum

Table 4. Effect of mulch and/or amount of water on water use efficiency of tomato.

Water levels (mm)	Marketable yield (ton/ha)			Volume of water/ha ($m^{3/LGP}$)	Water use efficiency (kg /m^3)		
	Mulch type				Mulch type		
	WM	BM	SM		WM	BM	SM
315	**29.24**	47.23	38.84	3150	9.28	**15.00**	12.33
440	49.68	56.72	52.19	4400	11.29	**12.89**	11.86
565	51.31	61.93	51.79	5650	9.08	**10.96**	9.17

WM = without mulch, BM = black plastic mulch, SM = straw mulch.

Table 5. Partial budget, dominance and marginal rate of return (MRR) analysis at Woramit with 10% sensitivity (considering the cost of tap water as cost of irrigation water).

Water m^3/Mulch levels	Unadjusted marketable Yield/ha (tone)	Adjusted marketable Yield/ha (tone)	Gross benefit	Total variable cost (birr)	Net benefit (birr)	Marginal rate of return (MRR%)
3150/WM	29.236	26.312	35,521.20	12,530.90	22,991.20	-
4400/WM	47.232	42.509	57,387.15	17710.00	39,677.15	322.20
3150/SM	49.683	44.715	60,365.25	18,030.90	42,334.35	828.05
5650/WM	38.844	34.956	47,190.60	22,522.50	24,668.10	D
4400/SM	56.715	51.044	68,909.40	23,210.00	45699.40	3059.10
5650/SM	52.192	46.973	63,413.55	28,022.50	35,391.05	D
3150/BM	51.313	46.182	62,345.70	235,204.52	-	-
4400/BM	61.931	55.738	75,246.30	240,383.62	-	-
5650/BM	51.788	46.609	62,922.15	245,196.12	-	-

WM - without mulch, SM - Straw mulch, BPM - black plastic much. D: Stands for dominated treatment; the marketable fruit yield was adjusted by 10% adjustment coefficient. During the experimentation period, the field price of tomato, straw, black plastic and water was 1.50 birr/kg, 9birr/m^3, 18.22 birr/kg and 1.5 bir/m^3, respectively.

Table 6. Assumption of the partial budget, dominance and MRR analysis that there is no water cost with 10% for other variable costs increment.

Water m^3/Mulch levels	Unadjusted marketable Yield/ha(tons)	Adjusted marketable Yield/ha(tons)	Gross benefit	Total variable cost	Net benefit	Marginal rate of return (MRR%)
3150/WM	29.236	26.312	35,521.20	7333.40	28,187.80	-
4400/WM	47.232	42.509	57,387.15	10,450.00	46,937.15	601.60
3150/SM	49.683	44.715	60,365.25	12,833.40	47,531.85	25.00
5650/WM	38.844	34.956	47,190.60	13,200.00	33,990.60	D
4400/SM	56.715	51.044	68,909.40	15,950.00	52,959.40	690.00
5650/SM	52.192	46.973	63,413.55	18,700.00	44,713.55	D
3150/BM	51.313	46.182	62,345.70	230,007.02	-	-
4400/BM	61.931	55.738	75,246.30	233,123.62	-	-
5650/BM	51.788	46.609	62,922.15	235,873.62	-	-

rate of return acceptable to farmers in the recommendation domain. According to CIMMYT (1988), the minimum acceptable marginal rate of return (MRR) should be between 50 and 100%. So far, there is no estimated cost for irrigation water in our country. Therefore, we did the partial budget analysis in two ways: considering the cost of tap water as cost of irrigation water and assuming the cost of irrigation water as zero. Based on the partial budget analysis, considering the cost of tap water as a cost of irrigation water with 10% variable cost increment and 10% produce price reduction, the highest net benefit was obtained via 440 mm with straw mulch amid a net benefit (45,699.40 birr/ha) and a marginal rate of return (MRR) 3059.10%. The result indicates that even when the cost of drinking water (tap water) is used as the cost of irrigation water, it is still gainful to produce tomato under drip irrigation.

Considering the cost of irrigation water as zero with 10% prices increment of other variable costs, the highest net benefit was obtained via 440 mm with straw mulch amid a net benefit (52,959.40 birr/ha) and a marginal rate of return (MRR) 690.00%. The marketable fruit yield advantage of 440 mm/with SM over 315 mm/without mulch was 94%.

Based on the biological data, the highest water level (565 mm) combined with black plastic mulch gave the maximum fruit yield and the highest water use efficiency value was recorded at the lowest water level (315 mm) with black plastic mulch. Even though, we have obtained a higher fruit yield and better water use efficiency value from plastic mulch treatments, they were not economically feasible. Therefore, application of 440 mm/ha water with straw mulch amid a net benefit (52,959.40 birr/ha) and marginal rate of return (MRR) (690.00%) is found to be economically and agronomically feasible and is recommended around Bahirdar and other similar agro-ecologies under drip irrigation.

SUMMARY AND CONCLUSION

Novel irrigation technologies need to be tested under local environments and particular agricultural production systems. Thus, the main challenge confronting both rain fed and irrigated agriculture is to improve WUE and sustainable water use for agriculture. Drip irrigation increased fruit yield of tomato and improved WUE due to consumption of less water. However, integrated use of drip irrigation and straw mulch was more appropriate and profitable.

The interaction effect of the two factors had shown non-significant difference on all parameters. The highest marketable and total fruit yield (61.93 and 68.14ton/ha) were obtained via the interaction effect of 565 mm (WL) with black plastic mulch. Based on the partial budget analysis, the highest net benefit was obtained via 440 mm with straw mulch amid a net benefit of 52,959.40 birr/ha and a marginal rate of return (MRR) of 690%. The marketable fruit yield advantage of 440 mm water with straw mulch over 315 mm without mulch was also very high (94%).

In conclusion, the present study points out that 440 mm/ha of water with straw mulch are economically more profitable than the other mulch treatments around Bahirdar and similar areas. Therefore, it is the subject of future investigations, to consider water levels below 315 mm and between 315 and 440 mm combined with straw mulch under drip irrigation, especially in drought prone areas where water is very scarce to produce crops.

ACKNOWLEDGEMENTS

The author gratefully acknowledges Amhara Regional Agricultural Research Institute (ARRI) for allocating the budget that aided the successful accomplishment of this work. Also, he acknowledges Adet Agricultural Research Center, Horticulture research team, especially Mr. Mehateme-work Mamo, for their unreserved technical assistance in the entire period of the experiment, of which without their assistance and hospitality this work may not have been materialized.

REFERENCES

Bogle O, Hartz (1986). Comparison of drip and furrow irrigation forMuskmelon production. Hortic. Sci., 21: 242-244.

CIMMYT (1988). From agronomic data to farmer recommendations. An economics training manual. Completely revised edition. Mexico.DF.

Dickerson GW (1996). Mulches for Gardens and land scapes. Guide H-121, cooperative extension services, NMSU. Las.Cruces. Nm.

Fekadu Y, Teshome H (1997). Effect of drip and furrow irrigation and plant spacing on yield of tomato at Dire Dawa, Ethiopia.

FAO (1995). Production Year book. Rome, Italy.

FAO (2005). New Loc Clim local climate estimator. Environments and natural resources working paper No 20. http//www.acta.hort.org.

Geber JM, Mohd KL, Splittoesser WE (1988). Low tunnel effect on growth, yield and fruit quality of bell pepper. Hortic. Sci., 26(3-4): 191-197

Jalota SK, Prihar SS (1998). Reducing soil water evaporation with Tillage and Mulching. Iowa State University Press, Ames.IA.142P

Kay BL (1998). Mulching and chemical stabilizers for land reclamation in dry regions. pp 467-483.In.F.W.Schaller and P.sutten, editors. Reclamation of Drastically Distributed lands. Am. Soc. Agronomy Madison. WI.

Lamont Jr (1991). Drip irrigation part of a complete vegetable production package.

Munshower FF (1994). Practical hand book of disturbed land revegetation lewis. Publishers. Boca. Raton. FL.

Raina JN, Thakur BC, Bhandaria R (1998). Effect of drip irrigation and plastic mulch on yield, water use efficiency and benefit-cost ratio of pea cultivation. Indian J. Soil Sci.

Salman SK, Abou-Hadid AF, Beltagy IMJ, Beltagy AS (1992). Plastic house micro climate as affected by low tunnels and plastic mulch. Egyptian. J. Hortic., 2: 111-119.

Seyfi K, Rashidi M (2007). Effect of drip irrigation and plastic mulch on crop yield and yield components of Cantaloupe. Int. J. Agric. Biol., 9(2).

Stein L, White K, Dainello F (1996). Drip irrigation and Plastic mulch conserve water, while maintaing cantaloupe yields and quality. Texas A&M Agricultural Research and extension center at Uvalde.

Zaman WU, Arshad M, Saleem K (2001). Distribution of nitrate – nitrogen in the soil profile under different irrigation methods. International, J. Agri. Biol., 2: 208-9.

Permissions

All chapters in this book were first published in JHF, by Academic Journals; hereby published with permission under the Creative Commons Attribution License or equivalent. Every chapter published in this book has been scrutinized by our experts. Their significance has been extensively debated. The topics covered herein carry significant findings which will fuel the growth of the discipline. They may even be implemented as practical applications or may be referred to as a beginning point for another development.

The contributors of this book come from diverse backgrounds, making this book a truly international effort. This book will bring forth new frontiers with its revolutionizing research information and detailed analysis of the nascent developments around the world.

We would like to thank all the contributing authors for lending their expertise to make the book truly unique. They have played a crucial role in the development of this book. Without their invaluable contributions this book wouldn't have been possible. They have made vital efforts to compile up to date information on the varied aspects of this subject to make this book a valuable addition to the collection of many professionals and students.

This book was conceptualized with the vision of imparting up-to-date information and advanced data in this field. To ensure the same, a matchless editorial board was set up. Every individual on the board went through rigorous rounds of assessment to prove their worth. After which they invested a large part of their time researching and compiling the most relevant data for our readers.

The editorial board has been involved in producing this book since its inception. They have spent rigorous hours researching and exploring the diverse topics which have resulted in the successful publishing of this book. They have passed on their knowledge of decades through this book. To expedite this challenging task, the publisher supported the team at every step. A small team of assistant editors was also appointed to further simplify the editing procedure and attain best results for the readers.

Apart from the editorial board, the designing team has also invested a significant amount of their time in understanding the subject and creating the most relevant covers. They scrutinized every image to scout for the most suitable representation of the subject and create an appropriate cover for the book.

The publishing team has been an ardent support to the editorial, designing and production team. Their endless efforts to recruit the best for this project, has resulted in the accomplishment of this book. They are a veteran in the field of academics and their pool of knowledge is as vast as their experience in printing. Their expertise and guidance has proved useful at every step. Their uncompromising quality standards have made this book an exceptional effort. Their encouragement from time to time has been an inspiration for everyone.

The publisher and the editorial board hope that this book will prove to be a valuable piece of knowledge for researchers, students, practitioners and scholars across the globe.

List of Contributors

S. A. Enebak
School of Forestry and Wildlife Sciences, 602 Duncan Drive, Auburn University, 36849, USA

T. E. Starkey
School of Forestry and Wildlife Sciences, Southern Forest Nursery Management Cooperative, Auburn University, AL 36849, USA

M. Quicke
School of Forestry and Wildlife Sciences, Southern Forest Nursery Management Cooperative, Auburn University, AL 36849, USA

A. S. Oyekale
Department of Agricultural Economics, University of Ibadan, Ibadan, Nigeria

D. O. Ajesi
Department of Agricultural Economics, University of Ibadan, Ibadan, Nigeria

Asrar Sarwar
Research Officer, Rural Business Development Center, Lok Sanjh Foundation, Faisalabad, Pakistan

Zahoor Aslam
Project Director NABSAB, Ministry of Agriculture, Islamabad, Pakistan

Endrias Geta
Haramaya University, Ethiopia

Asfaw Kifle
Areka Agricultural Research Center, SNNPRS, Ethiopia

Ferdinand E. Klas
Department of Plant Pathology and Plant-Microbe Biology, Cornell University, New York State Agricultural Experiment Station, Geneva, NY, 14456, USA
Anton de Kom University van Suriname Faculteit der Technologische Wetenschappen, Universiteitscomplex Leysweg, Building XVII, P. O. Box 12292, Paramaribo Suriname, USA

Marc Fuchs
Department of Plant Pathology and Plant-Microbe Biology, Cornell University, New York State Agricultural Experiment Station, Geneva, NY, 14456, USA

Dennis Gonsalves
Department of Plant Pathology and Plant-Microbe Biology, Cornell University, New York State Agricultural Experiment Station, Geneva, NY, 14456, USA
USDA-ARS, Pacific Basin Agricultural Research Center, 64 Nowelo Street, Hilo, HA 96720, USA

Thomas Graham
School of Environmental Sciences, University of Guelph, Ontario, N1G 2W1, Canada

Ping Zhang
School of Environmental Sciences, University of Guelph, Ontario, N1G 2W1, Canada

Michael Dixon
School of Environmental Sciences, University of Guelph, Ontario, N1G 2W1, Canada

V Girijashankar
ITC R&D Centre, Shameerpet, Hyderabad, Andhra Pradesh, India

Satid Pinmanee
The Uplands Program, Faculty of Agriculture, Chiang Mai University, Chiang Mai, Thailand

Wolfram Spreer
The Uplands Program, Faculty of Agriculture, Chiang Mai University, Chiang Mai, Thailand
Institute of Agricultural Engineering, University of Hohenheim, Stuttgart, Germany

Klaus Spohrer
Institute of Agricultural Engineering, University of Hohenheim, Stuttgart, Germany

Somchai Ongprasert
The Uplands Program, Faculty of Agriculture, Chiang Mai University, Chiang Mai, Thailand
Department of Soil Resources and Environment, Mae Jo University, Chiang Mai, Thailand

Joachim Müller
The Uplands Program, Faculty of Agriculture, Chiang Mai University, Chiang Mai, Thailand
Institute of Agricultural Engineering, University of Hohenheim, Stuttgart, Germany

P. E Ogbonna
Department of Crop Science, University of Nigeria, Nsukka, 410001, Nigeria

L. C. Odo
Department of Crop Science, University of Nigeria, Nsukka, 410001, Nigeria

Samir Mohamed Ali Hassan Alredaisy
Department of Geography, Faculty of Education, University of Khartoum, 406 Omdurman, Sudan

Mohamed Ahmed Haj Ali Zubair
Department of Geography, Faculty of Education, University of Khartoum, 406 Omdurman, Sudan

S. O. Ochieng
Jomo Kenyatta University of Agriculture and Technology, P. O. Box 62000-00100 Nairobi, Kenya

P. W. Nderitu
Department of Seed, Crop and Horticultural Sciences, Moi University, P. O. Box 1125-30100 Eldoret, Kenya

E. A. Egbe
Department of Plant and Animal Sciences, Faculty of Science, University of Buea, P. O. Box 63, Buea, Cameroon

G. B. Chuyong
Department of Plant and Animal Sciences, Faculty of Science, University of Buea, P. O. Box 63, Buea, Cameroon

A. B. Fonge
Department of Plant and Animal Sciences, Faculty of Science, University of Buea, P. O. Box 63, Buea, Cameroon

B. L. Tata
Department of Plant and Animal Sciences, Faculty of Science, University of Buea, P. O. Box 63, Buea, Cameroon

P. T. Tabot
Department of Plant and Animal Sciences, Faculty of Science, University of Buea, P. O. Box 63, Buea, Cameroon

Tenaw Workayehu
Awassa Agricultural Research Center, P. O. Box 366, Awassa, Ethiopia

Waga Mazengia
Awassa Agricultural Research Center, P. O. Box 366, Awassa, Ethiopia

Legesse Hidoto
Awassa Agricultural Research Center, P. O. Box 366, Awassa, Ethiopia

Anandhi Selvarasu
Department of Medicinal and Aromatic Crops, Horticultural College and Research Institute, Tamil Nadu Agricultural University, Coimbatore-641 003, Tamil Nadu, India

Rajamani Kandhasamy
Department of Medicinal and Aromatic Crops, Horticultural College and Research Institute, Tamil Nadu Agricultural University, Coimbatore-641 003, Tamil Nadu, India

Gabriel M. Muturi
Kenya Forestry Research Institute, P. O. Box 20412, 00200-Nairobi, Kenya
Forest Ecology and Forest Management Group, Wageningen University, P. O. Box 47, AA Wageningen, The Netherlands

Jason G. Kariuki
Kenya Forestry Research Institute, P. O. Box 20412, 00200-Nairobi, Kenya

Lourens Poorter
Forest Ecology and Forest Management Group, Wageningen University, P. O. Box 47, AA Wageningen, The Netherlands

Godefridus M. J. Mohren
Forest Ecology and Forest Management Group, Wageningen University, P. O. Box 47, AA Wageningen, The Netherlands

M. O. Kheyar
Département du Génie Rural, Ecole Nationale Supérieure Agronomique El Harrach Alger (Algérie)

B. Mouhouche
Département du Génie Rural, Ecole Nationale Supérieure Agronomique El Harrach Alger (Algérie)

M. Mahdi
Département du Génie Rural, Ecole Nationale Supérieure Agronomique El Harrach Alger (Algérie)

G. M. Anjah
Department of Plant Biology, University of Dschang, Bp 67, Dschang, Cameroon

O. Y. Ogunsanwo
Department of Forest Resources Management, University of Ibadan, Nigeria

S. O. Jimoh
Department of Forest Resources Management, University of Ibadan, Nigeria

J. N. Forjoh
Department of Plant Biology, University of Dschang, Bp 67, Dschang, Cameroon

F. MTsombou
Department of Plant Biology, University of Dschang, Bp 67, Dschang, Cameroon

Sandeep K. Chauhan
FHEL, CONCOR, Government of India

K. S. Thakur
Department of Postharvest Technology, Dr Y.S. Parmar University of Horticulture and Forestry, Nauni-173 230, India

N. K. Jawa
FHEL, CONCOR, Government of India

K. P. Thakur
FHEL, CONCOR, Government of India

M. Moghtader
Department of Biodiversity, Institute of Science and High Technology and Environmental Sciences, Graduate University of Advanced Technology, Kerman, Iran

J. K. Githiomi
Kenya Research Institute-Forest Products Research Centre, P.O. Box 64636-00620 Nairobi, Kenya

J. B. Kungú
Department of Environmental Sciences, Kenyatta University, P.O. Box 43844-00100 Nairobi, Kenya

D. N. Mugendi
Department of Environmental Sciences, Kenyatta University, P.O. Box 43844-00100 Nairobi, Kenya

Scott A. Enebak
Southern Forest Nursery Management Cooperative, Forest Health Dynamics Laboratory, School of Forestry and Wildlife Sciences, 602 Duncan Drive, Auburn University, Alabama, 36849, USA

D. Paul Jackson
Department of Agricultural Sciences, Louisiana Tech University, P. O. Box 10198, Ruston, LA 71272, USA

Tom E. Starkey
Southern Forest Nursery Management Cooperative, Forest Health Dynamics Laboratory, School of Forestry and Wildlife Sciences, 602 Duncan Drive, Auburn University, Alabama, 36849, USA

Marie Quicke
Southern Forest Nursery Management Cooperative, Forest Health Dynamics Laboratory, School of Forestry and Wildlife Sciences, 602 Duncan Drive, Auburn University, Alabama, 36849, USA

A. R. Sardashti
Department of Chemistry, Faculty of Sciences, University of Sistan and Baluchestan, P. O. Box 98167 Zahedan, Iran

J. Valizadeh
Department of Biology, Faculty of Sciences, University of Sistan and Baluchestan, P. O. Box 98167, Zahedan, Iran

Y. Adhami
Department of Chemistry, Faculty of Sciences, University of Sistan and Baluchestan, P. O. Box 98167 Zahedan, Iran

P. Srimathi
Department of Seed Science and Technology, Seed Centre, TNAU, Coimbatore-641003, Tamil Nadu, India

N. Mariappan
Vanavarayar Institute of Agriculture, Pollachi, Coimbatore-642103, Tamil Nadu, India

L. Sundaramoorthy
Vanavarayar Institute of Agriculture, Pollachi, Coimbatore-642103, Tamil Nadu, India

M. Paramathma
Centre of Excellence for Biofuel, Forest College and Research Institute, TNAU, Mettupalayam-641103, Tamil Nadu, India

P. Lampert
Department of Horticulture and Food Engineering, University of Applied Sciences Weihenstephan-Triesdorf, Germany

K. Menrad
Department of Horticulture and Food Engineering, University of Applied Sciences Weihenstephan-Triesdorf, Germany

J. Schoeps
Department of Horticulture and Food Engineering, University of Applied Sciences Weihenstephan-Triesdorf, Germany

D. Paul Jackson
School of Forestry and Wildlife Sciences, Auburn University, Alabama 36849, USA

Scott A. Enebak
School of Forestry and Wildlife Sciences, Auburn University, Alabama 36849, USA

David B. South
School of Forestry and Wildlife Sciences, Auburn University, Alabama 36849, USA

Khair Tuwair Said Al-Busaidi
Directorate General of Agriculture and Livestock Research, Ministry of Agriculture and Fisheries, Crop Production Research Centre, Fruit Research Department, P. O. Box 343 Postal Code 615, Samail, Sultanate of Oman

M. K. Yadav
Department of Horticulture, N. M. College of Agriculture, Navsari Agricultural University, Dandi Road, Navsari-396450, India

B. R. Parmar
Department of Horticulture, N. M. College of Agriculture, Navsari Agricultural University, Dandi Road, Navsari-396450, India

George Muthike
Kenya Forestry Research Institute (KEFRI) Forest Products Research Centre, P. O. Box. 64636-00620, Kenya

Christopher Kanali
Department of Biomechanical and Environmental Engineering, Jomo Kenyatta University of Agriculture and Technology, Nairobi, Kenya

Douglas Shitanda
MultiMedia University of Kenya, Kenya

Dolly Parmar Kotwal
Government Post Graduate College, Barwani, M. P. India

M. Moghtader
Department of Biodiversity, Institute of Science and High Technology and Environmental Sciences, Graduate University of Advanced Technology, Kerman, Iran

Ochiki Sophia
Department of Crops, Horticulture and Soils Sciences, Egerton University, P. O. Box 536-20115, Egerton-Kenya

Gesimba Morwani Robert
Department of Crops, Horticulture and Soils Sciences, Egerton University, P. O. Box 536-20115, Egerton-Kenya

Wolukau Joseph Ngwela
Department of Crops, Horticulture and Soils Sciences, Egerton University, P. O. Box 536-20115, Egerton-Kenya

Zuzana Jureková
Faculty of European Studies and Regional Development, Slovak University of Agriculture in Nitra, Marianská 10, 949 01 Nitra, Slovakia

Kristína Németh-Molnár
Faculty of European Studies and Regional Development, Slovak University of Agriculture in Nitra, Marianská 10, 949 01 Nitra, Slovakia

Viera Paganová
Faculty of Horticulture and Landscape Engineering, Slovak University of Agriculture in Nitra, Tulipánová 7, 949 76 Nitra, Slovakia

Eliseo S. Gimenes
Department of Crop Science, Federal University of Santa Maria, RS, Brazil

Paula Kielse
Department of Crop Science, Federal University of Santa Maria, RS, Brazil

Kelen L. Haygert
Department of Crop Science, Federal University of Santa Maria, RS, Brazil

Frederico D. Fleig
Department of Forestry Science, Federal University of Santa Maria, RS, Brazil

Daniel E. Keathley
Department of Horticulture, Michigan State University, East Lansing, MI, USA

Dilson A. Bisognin
Department of Crop Science, Federal University of Santa Maria, RS, Brazil

James Munga Kimondo
Kenya Forestry Research Institute, P. O. Box 20412-00200, Nairobi, Kenya

Jacob Godfrey Agea
College of Agricultural and Environmental Sciences, Makerere University, P. O. Box 7062 Kampala, Uganda

Clement Akais Okia
ICRAF Uganda Country Office P. O. Box 26416 Kampala, Uganda

Andrew Woiso Dino
Department of Biological Science, Sokoine University of Agriculture, P. O. Box 3038, Morogoro, Tanzania

Refaat Atalla Ahmed Abohassan
Faculty of Meteorology, Environment and Arid Land Agriculture, King Abdulaziz University, P. O. Box 80208 Jeddah, 21589 Saudi Arabia

Jackson Mulatya
Kenya Forestry Research Institute, P. O. Box 20412-00200, Nairobi, Kenya

Zewge Teklehaimanot
School of Environment, Natural Resources and Geography, Bangor University, Bangor-Gwynedd LL57, UK

Zahra Shabani
Horticultural Department, Ferdowsi University of Mashhad, Iran

Ebrahim Ganji Moghadam
Horticultural Department of Khorasan Razavi Agricultural and Natural Resource Research Center, Mashhad, Iran

Bahram Abedi
Horticultural Department, Ferdowsi University of Mashhad, Iran

Ali Tehranifar
Horticultural Department, Ferdowsi University of Mashhad, Iran

Francis Wilson Owusu
CSIR, Forestry Research Institute of Ghana (FORIG), Kumasi-Ashanti, Ghana

Felix Boakye
CSIR, Forestry Research Institute of Ghana (FORIG), Kumasi-Ashanti, Ghana

Godwin Zorve
CSIR, Forestry Research Institute of Ghana (FORIG), Kumasi-Ashanti, Ghana

Mesay Adugna
College of Agriculture and Natural Resource, Dilla University, P. O. Box 419 Dilla, Ethiopia

Derebew Belew
College of agriculture and Veterinary Medicine, Jimma University, P. O. Box 307 Jimma, Ethiopia

Digafie Tilahun
Teppi National Spice Research Center, P. O. Box 34 Teppi, Ethiopia

Baye Berihun
Adet Agricultural Research Center, P. O. Box 08, Bahir-Dar, Ethiopia

Fourier Transform: Signal Processing

Fourier Transform: Signal Processing

Editor:

Olga Moreira, Ph.D.

ArclerPress

www.arclerpress.com

Fourier Transform: Signal Processing

Editor: Olga Moreira, Ph.D.

708 3rd Avenue, 6th Floor
New York
NY 10017
United States of America
www.arclerpress.com

ISBN: 978-1-68094-395-5

Library of Congress Control Number: 2016951048

About the Editor

Olga Moreira, Ph.D.

Olga Moreira obtained her Ph.D. in Astrophysics from the University of Liege (Belgium) in 2010, her BSc. in Physics and Applied Mathematics from the University of Porto (Portugal). Her post-graduate travels and international collaborations with the European Space Agency (ESA) and European Southern Observatory (ESO) led to great personal and professional growth as a scientist. Currently, she is working as an independent researcher, technical writer, and editor in the fields of Mathematics, Physics, Astronomy and Astrophysics.

Contents

List of Contributors

P. L. Butzer
Lehrstuhl A für Mathematik, RWTH Aachen University, 52056 Aachen, Germany

M. M. Dodson
Department of Mathematics, University of York, York YO1O 5DD, UK

P. J. S. G. Ferreira
IEETA/DETI, Universidade de Aveiro, 3810-193 Aveiro, Portugal

J. R. Higgins
I.H.P., 4 rue du Bary, 11250 Montclar, France

G. Schmeisser
Department of Mathematics, University of Erlangen-Nuremberg, 91058 Erlangen, Germany

R. L. Stens
Lehrstuhl A für Mathematik, RWTH Aachen University, 52056 Aachen, Germany

Francisco J. Mendoza-Torres
Facultad de Ciencias Físico-Matemáticas, Benemérita Universidad Autónoma de Puebla, Puebla, Mexico

Ma. Guadalupe Morales-Macías
Departamento de Matemáticas, Universidad Autónoma Metropolitana-Iztapalapa, México, D. F., Mexico

Salvador Sánchez-Perales
Instituto de Física y Matemáticas, Universidad Tecnológica de la Mixteca, Oaxaca, Mexico

Juan Alberto Escamilla-Reyna
Facultad de Ciencias Físico-Matemáticas, Benemérita Universidad Autónoma de Puebla, Puebla, Mexico

Jiufei Luo
School of Advanced Manufacture Engineering, Chongqing University of Posts and Telecommunications, Chongqing 400065, People's Republic of China

Shuaicheng Hou
School of Automation, Chongqing University, Chongqing 400044, People's Republic of China

Xinyi Li
Department of Mechanical Engineering, Chongqing University, Chongqing 400044, People's Republic of China.

Qi Ouyang
School of Automation, Chongqing University, Chongqing 400044, People's Republic of China

Yi Zhang
School of Advanced Manufacture Engineering, Chongqing University of Posts and Telecommunications, Chongqing 400065, People's Republic of China

Jesús Ramón Lerma-Aragón
Facultad de Ciencias, Universidad Autónoma de Baja California, México

Josué Álvarez-**Borrego**
CICESE, División de Física Aplicada, Departamento de Óptica, México

Tao Guan
College of Electronic Science and Engineering, National University of Defense Technology, Kaifu, Changsha 410073, Hunan, China

Dongxiang Zhou
College of Electronic Science and Engineering, National University of Defense Technology, Kaifu, Changsha 410073, Hunan, China

Chao Xu
College of Electronic Science and Engineering, National University of Defense Technology, Kaifu, Changsha 410073, Hunan, China

Yunhui Liu
Department of Mechanical and Automation Engineering, The Chinese University of Hong Kong, Shatin, Hong Kong, China.

Lebing Pan
Key Laboratory of Wireless Sensor Networks, Shanghai Institute of Microsystem and Information Technology, Chinese Academy of Sciences, 865 Changning Road, Shanghai 200050, China
University of Chinese Academy of Sciences, No.19A Yuquan Road, Beijing 100049, China

Shiliang Xiao
Key Laboratory of Wireless Sensor Networks, Shanghai Institute of Microsystem and Information Technology, Chinese Academy of Sciences, 865 Changning Road, Shanghai 200050, China
University of Chinese Academy of Sciences, No.19A Yuquan Road, Beijing 100049, China

Xiaobing Yuan
Key Laboratory of Wireless Sensor Networks, Shanghai Institute of Microsystem and Information Technology, Chinese Academy of Sciences, 865 Changning Road, Shanghai 200050, China

Baoqing Li
Key Laboratory of Wireless Sensor Networks, Shanghai Institute of Microsystem and Information Technology, Chinese Academy of Sciences, 865 Changning Road, Shanghai 200050, China

Siwar Rekik
Université de Bretagne Occidentale, Brest, France
Canadian University of Dubai, Dubai, UAE

Driss Guerchi
Canadian University of Dubai, Dubai, UAE

Sid-Ahmed Selouani
University of Moncton, Shippagan, NB, Canada

Habib Hamam
University of Moncton, Moncton, NB, Canada.

Vinod Chandran
School of Electrical Engineering and Computer Science Queensland University of Technology

Qiong Wu, Qilian Liang
Department of Electrical Engineering University of Texas at Arlington

Xiumei Li
School of Information Science and Engineering, Hangzhou Normal University, China

Guoan Bi
School of Electrical and Electronic Engineering, Nanyang Technological University, Singapore

Shenghong Li
School of Electronic Information and Electrical Engineering, Shanghai Jiao Tong University, China

Jiexiao Yu
School of Electronic Information Engineering, Tianjin University, Tianjin, China

Liang Zhang
School of Electronic Information Engineering, Tianjin University, Tianjin, China

Kaihua Liu
School of Electronic Information Engineering, Tianjin University, Tianjin, China

Deliang Liu
School of Electronic Information Engineering, Tianjin University, Tianjin, China

Steffen Weimann
Institute of Applied Physics, Abbe School of Photonics, Friedrich-Schiller-Universit t Jena, Max-Wien Platz 1, 07743 Jena, Germany

Armando Perez-Leija
Institute of Applied Physics, Abbe School of Photonics, Friedrich-Schiller-Universit t Jena, Max-Wien Platz 1, 07743 Jena, Germany

Maxime Lebugle
Institute of Applied Physics, Abbe School of Photonics, Friedrich-Schiller-Universit t Jena, Max-Wien Platz 1, 07743 Jena, Germany

Robert Keil
Institut fu¨r Experimentalphysik, Universita¨t Innsbruck, Technikerstrae 25, 6020 Innsbruck, Austria

Malte Tichy
Department of Physics and Astronomy, University of Aarhus, 8000 Aarhus, Denmark

Markus Grfe
Institute of Applied Physics, Abbe School of Photonics, Friedrich-Schiller-Universit t Jena, Max-Wien Platz 1, 07743 Jena, Germany

Rene´ Heilmann
Institute of Applied Physics, Abbe School of Photonics, Friedrich-Schiller-Universit t Jena, Max-Wien Platz 1, 07743 Jena, Germany

Stefan Nolte
Institute of Applied Physics, Abbe School of Photonics, Friedrich-Schiller-Universit t Jena, Max-Wien Platz 1, 07743 Jena, Germany

Hector Moya-Cessa
INAOE, Coordinacion de Optica, Luis Enrique Erro No. 1, Tonantzintla, Puebla 72840, Mexico

Gregor Weihs
Institut fu¨r Experimentalphysik, Universita¨t Innsbruck, Technikerstrae 25, 6020 Innsbruck, Austria

Demetrios N. Christodoulides
CREOL, The College of Optics & Photonics, University of Central Florida, Orlando, Florida 32816, USA

Alexander Szameit
Institute of Applied Physics, Abbe School of Photonics, Friedrich-Schiller-Universit t Jena, Max-Wien Platz 1, 07743 Jena, Germany

Katherine M. M. Tant
Department of Mathematics and Statistics

Anthony J. Mulholland
Department of Mathematics and Statistics

Matthias Langer
Department of Mathematics and Statistics

Anthony Gachagan
Centre for Ultrasonic Engineering, University of Strathclyde, Glasgow, UK

Wen-Qin Wang
School of Communication and Information EngineeringUniversity of Electronic Science and Technology of China

Said Assous
Weatherford, East Leake, Leicestershire LE12 6JX, UK

Boualem Boashash
Perinatal Research Centre and UQ Centre for Clinical Research, Royal Brisbane and Women's Hospital, The University of Queensland, Hertson QLD 4029, Australia
College of Engineering, Qatar University, Doha, Qatar

Ata Mahjoubfar
Department of Electrical Engineering, University of California, Los Angeles, California 90095, USA
California NanoSystems Institute, Los Angeles, California 90095, USA

Claire Lifan Chen
Department of Electrical Engineering, University of California, Los Angeles, California 90095, USA
California NanoSystems Institute, Los Angeles, California 90095, USA

Bahram Jalali
Department of Electrical Engineering, University of California, Los Angeles, California 90095, USA
California NanoSystems Institute, Los Angeles, California 90095, USA
Department of Bioengineering, University of California, Los Angeles, California 90095, USA. Correspondence and requests for materials should be addressed to A.M

Dariusz Komorowski
Faculty of Biomedical Engineering, Department of Biosensors and Biomedical Signals Processing, Silesian University of Technology, 40 Roosevelt'a street, 44-800 Zabrze, Poland

Stanislaw Pietraszek
Institute of Electronics, Division of Biomedical Electronics, Silesian University of Technology, 16 Akademicka street, 44-100 Gliwice, Poland

Muhammad Sajid
Geoscience and Petroleum Engineering Department, Universiti Teknologi PETRONAS, Tronoh, Malaysia

Deva P. Ghosh
Geoscience and Petroleum Enginseering Department, Universiti Teknologi PETRONAS, Tronoh, Malaysia

I. A. Satti
Geoscience and Petroleum Engineering Department, Universiti Teknologi PETRONAS, Tronoh, Malaysia

Preface

Contrary to the rest of the book where signal and image processing FT methods are introduced from an application-oriented perspective, the first two chapters of this book will introduce several fundamental theorems and important definitions behind the Fourier analysis and other signal processing techniques from a more mathematics-oriented perspective.

The FT decomposes functions into a frequency spectra which contain information about the original signals. In signal processing, any periodic signal (with period P) can be represented as summation of an infinite number of instances of an aperiodic function that are shifted by integer multiples of P. Alternatively, the periodic function can be represented as a complex Fourier series where the coefficients are proportional to the sampling of the Continuous Fourier Transform (CFT) at intervals of 1/P. Sampling theorems are fundamental in digital signal processing because it establishes a relationship between continuous-time signals and discrete-time signals. When Fourier coefficients are samples of the periodic function at constant time intervals, it becomes equivalent to a periodic summation of the CFT known as a discrete-time Fourier transform (DFT), and its computation is normally called of Fourier transform (FFT) algorithm.

After the mathematics-oriented chapters, the book focuses on FT and DFT spectrum analysis techniques and concepts such as frequency estimation, spectral leakage, Mellin and Scale transforms which are of great usefulness in stationary signal processing. However, FT data processing techniques alone are inadequate methods for tracking changes in signal magnitude, frequency or phase, i.e. for the analysis of non-stationary signals. Short-time Fourier transform (STFT) methods are the most widely used in analyzing the time-frequency properties of non-stationary signals. The remaining chapters focus on the data analysis techniques based on linear transformations generalizing the FT and STFT such as the Fractional Fourier transform (FrFT) and the Local Polynomial Fourier Transform (LPFT), respectively. The last chapters are focused on alternative methods such as the Time Stretch Dispersive Fourier Transform (a.k.a the warped-stretch transform) used in real-time spectroscopy as well as Wavelet, Radon, Fan-beam, Ridgelet and Curvelet transforms used seismic data analysis.

Editor
Olga Moreira, Ph.D.

INTRODUCTION

In mathematics, Fourier analysis (English pronunciation) is the study of the way general functions may be represented or approximated by sums of simpler trigonometric functions. Fourier analysis grew from the study of Fourier series, and is named after Joseph Fourier, who showed that representing a function as a sum of trigonometric functions greatly simplifies the study of heat transfer.

Today, the subject of Fourier analysis encompasses a vast spectrum of mathematics. In the sciences and engineering, the process of decomposing a function into oscillatory components is often called Fourier analysis, while the operation of rebuilding the function from these pieces is known as Fourier synthesis. For example, determining what component frequencies are present in a musical note would involve computing the Fourier transform of a sampled musical note. One could then re-synthesize the same sound by including the frequency components as revealed in the Fourier analysis. In mathematics, the term Fourier analysis often refers to the study of both operations.

The decomposition process itself is called a Fourier transformation. Its output, the Fourier transform, is often given a more specific name, which depends on the domain and other properties of the function being transformed. Moreover, the original concept of Fourier analysis has been extended over time to apply to more and more abstract and general situations, and the general field is often known as harmonic analysis. Each transform used for analysis (see list of Fourier-related transforms) has a corresponding inverse transform that can be used for synthesis.

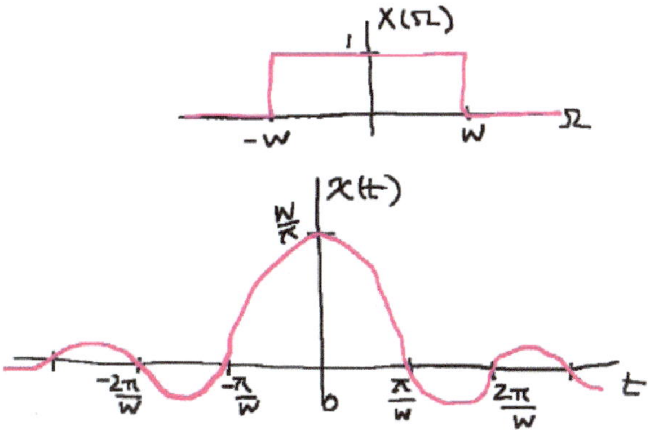

Applications

Fourier analysis has many scientific applications – in physics, partial differential equations, number theory, combinatorics, signal processing, imaging, probability theory, statistics, forensics, option pricing, cryptography, numerical analysis, acoustics, oceanography, sonar, optics, diffraction, geometry, protein structure analysis, and other areas.

This wide applicability stems from many useful properties of the transforms:

- The transforms are linear operators and, with proper normalization, are unitary as well (a property known as Parseval's theorem or, more generally, as the Plancherel theorem, and most generally via Pontryagin duality).

- The transforms are usually invertible.

- The exponential functions are eigenfunctions of differentiation, which means that this representation transforms linear differential equations with constant coefficients into ordinary algebraic ones (Evans). Therefore, the behavior of a linear time-invariant system can be analyzed at each frequency independently.

- By the convolution theorem, Fourier transforms turn the complicated convolution operation into simple multiplication, which means that they provide an efficient way to compute convolution-based operations such as polynomial multiplication and multiplying large numbers (Knuth).

- The discrete version of the Fourier transform (see below) can be evaluated quickly on computers using Fast Fourier Transform (FFT) algorithms.

In forensics, laboratory infrared spectrophotometers use Fourier transform analysis for measuring the wavelengths of light at which a material will absorb in the infrared spectrum. The FT method is used to decode the measured signals and record the wavelength data. And by using a computer, these Fourier calculations are rapidly carried out, so that in a matter of seconds, a computer-operated FT-IR instrument can produce an infrared absorption pattern comparable to that of a prism instrument.

Fourier transformation is also useful as a compact representation of a signal. For example, JPEG compression uses a variant of the Fourier transformation (discrete cosine transform) of small square pieces of a digital image. The Fourier components of each square are rounded to lower arithmetic precision, and weak components are eliminated entirely, so that the remaining components can be stored very compactly. In image reconstruction, each image square is reassembled from the preserved approximate Fourier-transformed components, which are then inverse-transformed to produce an approximation of the original image.

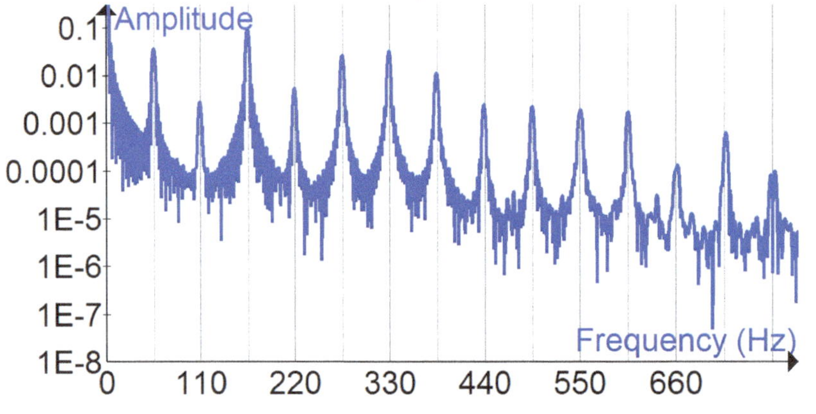

Applications in signal processing

When processing signals, such as audio, radio waves, light waves, seismic waves, and even images, Fourier analysis can isolate narrowband components of a compound waveform, concentrating them for easier detection or removal. A large family of signal processing techniques consist of Fourier-transforming a signal, manipulating the Fourier-transformed data in a simple way, and reversing the transformation.[2]

Some examples include:

- Equalization of audio recordings with a series of band pass filters;
- Digital radio reception without a super heterodyne circuit, as in a modern cell phone or radio scanner;
- Image processing to remove periodic or anisotropic artifacts such as jaggies from interlaced video, stripe artifacts from strip aerial photography, or wave patterns from radio frequency interference in a digital camera;
- Cross correlation of similar images for co-alignment;
- X-ray crystallography to reconstruct a crystal structure from its diffraction pattern;
- Fourier transform ion cyclotron resonance mass spectrometry to determine the mass of ions from the frequency of cyclotron motion in a magnetic field;
- Many other forms of spectroscopy, including infrared and nuclear magnetic resonance spectroscopies;
- Generation of sound spectrograms used to analyze sounds;
- Passive sonar used to classify targets based on machinery noise.

Variants of Fourier analysis

Fourier transform

Most often, the unqualified term Fourier transform refers to the transform of functions of a continuous real argument, and it produces a continuous function of frequency, known as a frequency distribution. One function is transformed into another, and the operation is reversible. When the domain of the input (initial) function is time (t), and the domain of the output (final) function is ordinary frequency, the transform of function s(t) at frequency *f* is given by the complex number:

$$S(f) = \int_{-\infty}^{\infty} s(t) \cdot e^{-i2\pi ft} dt.$$

Evaluating this quantity for all values of *f* produces the frequency-domain function. Then s(t) can be represented as a recombination of complex exponentials of all possible frequencies:

$$s(t) = \int_{-\infty}^{\infty} S(f) \cdot e^{i2\pi ft} df,$$

which is the inverse transform formula. The complex number, S(*f*), conveys both amplitude and phase of frequency *f*.

See Fourier transform for much more information, including:

- conventions for amplitude normalization and frequency scaling/units
- transform properties
- tabulated transforms of specific functions
- an extension/generalization for functions of multiple dimensions, such as images.

The Fourier Transform´s ability to represent time-domain data in the frequency domain and vice-versa has many applications. One of the most frequent applications is analyzing the spectral (frequency) energy contained in data that has been sampled at evenly-spaced time intervals. Other applications include fast computation of convolution (linear systems responses, digital filtering, correlation (time-delay estimation, similarity measurements) and time-frequency analysis.

The fast version of this transform, the Fast Fourier Transform (or FFT) was first developed by Cooley and Tukey[2] and later refined for even greater speed and for use with different data lengths through the "mixed-radix" algorithm. Igor computes the FFT using a fast multidimensional prime factor decomposition Cooley-Tukey algorithm.

While the Fourier Transform is mathematically complicated, Igor´s Fourier Transforms dialog makes it easy to use:

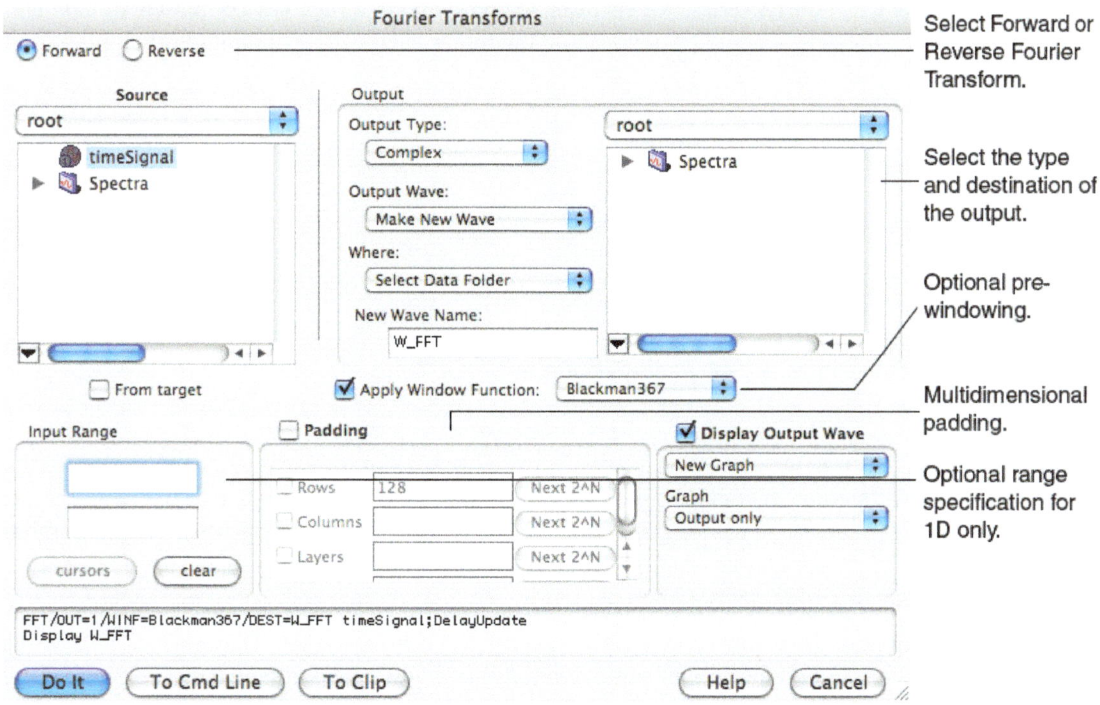

Spectral Windowing

The FFT computation presumes that the input data repeats over and over. This is important when the initial and final values of your data are not the same: the discontinuity causes aberrations in the spectrum computed by the FFT. "Windowing" smooths the ends of the data to eliminate these aberrations.

Power Spectra

"Power Spectra" answer the question "which frequencies contain the signal´s power?" The answer is in the form of a distribution of power values as a function of frequency, where "power" is considered to be the average of the signal². In the frequency domain, this is the square of FFT´s magnitude.

Power spectra can be computed for the entire signal at once (a "periodogram") or periodograms of segments of the time signal can be averaged together to form the "power spectral density".

Hilbert Transform

The Hilbert Transform computes a time-domain signal that is 90 degrees "out of phase" with the input signal. One-dimensional applications include computing the envelope of a modulated signal and the measurement of the decay rate of an exponentially decaying sinusoid often encountered in underdamped linear and non-linear systems.

Time Frequency Analysis

When you compute the Fourier spectrum (or Power Spectra) of a signal you dispose of all the phase information contained in the Fourier transform. You can find out which frequencies a signal contains but you do not know when these frequencies appear in the signal. For example, consider the signal:

$$f(t) = \begin{cases} \sin(2\pi f_1 t) & 0 \le t < t_1 \\ \sin(2\pi f_2 t) & t_1 \le t < t_2 \end{cases}.$$

Fourier series

The Fourier transform of a periodic function, $s_p(t)$, with period P, becomes a Dirac comb function, modulated by a sequence of complex coefficients:

$$S[k] = \frac{1}{P} \int_P s_P(t) \cdot e^{-i2\pi \frac{k}{P} t}\, dt \qquad \text{for all integer values of k,}$$

and where $\int_P$ is the integral over any interval of length P.

The inverse transform, known as Fourier series, is a representation of $s_p(t)$ in terms of a summation of a potentially infinite number of harmonically related sinusoids or complex exponential functions, each with an amplitude and phase specified by one of the coefficients:

$$s_P(t) = \sum_{k=-\infty}^{\infty} S[k] \cdot e^{i2\pi \frac{k}{P} t} \quad \stackrel{\mathcal{F}}{\Longleftrightarrow} \quad \sum_{k=-\infty}^{+\infty} S[k]\, \delta\left(f - \frac{k}{P}\right).$$

When $s_p(t)$, is expressed as a periodic summation of another function, s(t):

$$s_P(t) \stackrel{\text{def}}{=} \sum_{k=-\infty}^{\infty} s(t - kP),$$

the coefficients are proportional to samples of S(*f*) at discrete intervals of 1/P:

$$S[k] = \frac{1}{P} \cdot S\left(\frac{k}{P}\right)$$

A sufficient condition for recovering s(t) (and therefore S(*f*)) from just these samples is that the non-zero portion of s(t) be confined to a known interval of duration P, which is the frequency domain dual of the Nyquist–Shannon sampling theorem.

Convolution and Correlation

You can use convolution to compute the response of a linear system to an input signal. The linear system is defined by its impulse response. The convolution of the input signal and the impulse response is the output signal response. Digital filtering is accomplished by defining a linear system´s impulse response that when convolved with the signal accomplishes the desired result (low-pass or high-pass filter).

The correlation algorithm is very similar mathematically to convolution, but is used for different purposes. It is most frequently used to identify the time delay at which two signals "line up", or are "most similar".

Smoothing

Smoothing removes short-term variations, or "noise" to reveal the important underlying form of the data.

The simplest form of smoothing is the "moving average" which simply replaces each data value with the average of neighboring values. (Other terms for this kind of smoothing are "sliding average", "box smoothing", or "boxcar smoothing".)

Igor´s Smooth operation performs box smoothing, "binomial" (Gaussian) smoothing, and Savitzky-Golay (polynomial) smoothing. The different smoothing algorithms compute weighted averages that multiply neighboring values by differing weights or "coefficients" to compute the smoothed value.

Digital Filters

Digital filters are a natural tool when data is already digitized. Reasons for applying digital filtering to data include:

- Elimination of unwanted signal components ("noise")
- Enhancing of desired signal components
- Detecting the presence of certain signals
- Simulation of linear systems (compute the output signal given the input signal and the system´s "transfer function")
- Digital filters generally come in two flavors: Finite Impulse Response (FIR) and Infinite Impulse Response (IIR) filters.

Igor implements FIR digital filtering primarily through time-domain convolution using the Smooth or Smooth Custom commands. (In spite of its name, Smooth Custom convolves data with user-supplied filter coefficients to implement any kind of FIR filter, low-pass, high-pass, band-pass, etc.)

Design of the FIR filter coefficients used with Smooth Custom is most easily accomplished using the Igor Filter Design Laboratory (a separate product which also requires Igor Pro).

CHAPTER 1 SEVEN PIVOTAL THEOREMS OF FOURIER ANALYSIS, SIGNAL ANALYSIS, NUMERICAL ANALYSIS AND NUMBER THEORY

P. L. Butzer, M. M. Dodson, P. J. S. G. Ferreira, J. R. Higgins, G. Schmeisser and R. L. Stens

Keywords: Bandlimited and non-bandlimited functions, Sampling theorem, Parseval formula, Reproducing kernel formula, Poisson's summation formula, Riemann's zeta function, Euler–Maclaurin summation formula

INTRODUCTION

A provoking result of Fourier analysis has turned up in the past 15 years, namely

Generalized Parseval Decomposition Formula (GPDF)

For $f \in F^2 \cap S_w^1,\ w > 0$ and $g \in F^2$, there holds $R_w f \in L^2(\mathbb{R})$ and

$$\int_{\mathbb{R}} f(u)\overline{g}(u)\,du = \frac{1}{w}\sum_{k\in\mathbb{Z}} f\left(\frac{k}{w}\right)\overline{g}\left(\frac{k}{w}\right)$$

$$-\frac{1}{w}\sum_{k\in\mathbb{Z}} f\left(\frac{k}{w}\right)\frac{1}{\sqrt{2\pi}}\int_{|v|\geq\pi w}\widehat{\overline{g}}(v)e^{ikv/w}\,dv + \int_{\mathbb{R}}(R_w f)(u)\overline{g}(u)\,du, \tag{1.1}$$

where

$$(R_w f)(t) := \frac{1}{\sqrt{2\pi}}\sum_{k\in\mathbb{Z}}\left(1 - e^{-i2\pi kwt}\right)\int_{(2k-1)\pi w}^{(2k+1)\pi w}\widehat{f}(v)e^{ivt}\,dv \quad (t \in \mathbb{R}). \tag{1.2}$$

Observe that $\lim_{w\to\infty}(R_w f)(t)=0$ uniformly for $t\in\mathbb{R}$, since

$$\left|(R_w f)(t)\right| \leq \sqrt{\frac{2}{\pi}}\int_{|v|\geq\pi w}|\widehat{f}(u)|\,du. \tag{1.3}$$

Here F^2 and $F^2\cap S^1_w$ are suitable subspaces of $L^2(\mathbb{R})$, and $\hat{f}$ denotes the Fourier transform (for exact definitions see Sect. 2.1).

Formula (1.1), first established in [17, 18], can be said to be intermediate between the classical General Parseval/Plancherel formula (or power/energy theorem) for $L^2(\mathbb{R})$-functions, namely

---

Citation: Butzer, P.L., Dodson, M.M., Ferreira, P.J.S.G. et al., "Seven pivotal theorems of Fourier analysis, signal analysis, numerical analysis and number theory," Bull. Math. Sci. (2014) 4: 481. doi:10.1007/s13373-014-0057-3.

$$\int_{\mathbb{R}} f(u)\overline{g}(u)\,du = \int_{\mathbb{R}} \widehat{f}(v)\overline{\widehat{g}}(v)\,dv \quad (f, g \in L^2(\mathbb{R})),$$

and the special case of (1.1) for bandlimited $L^2(R)$-functions, i. e., for functions f,g∈ $\widehat{B}^2_{\pi w}$, namely,

$$\int_{\mathbb{R}} f(u)\overline{g}(u)\,du = \frac{1}{w}\sum_{k\in\mathbb{Z}} f\left(\frac{k}{w}\right)\overline{g}\left(\frac{k}{w}\right) = \int_{\mathbb{R}} \widehat{f}(v)\overline{\widehat{g}}(v)\,dv \quad (f, g \in \widehat{B}^2_{\pi w}). \tag{1.4}$$

In fact, formula (1.1) is the extension of (1.4) from $\widehat{B}^2_{\pi w}$ to a larger subclass of $L^2(R)$.

Note that if g∈$\widehat{B}^2_{\pi w}$, then the hypotheses of GPDF are satisfied and the second series on the right-hand side of (1.1) vanishes, and so (1.1) reduces to

$$\int_{\mathbb{R}} f(u)\overline{g}(u)\,du = \frac{1}{w}\sum_{k\in\mathbb{Z}} f\left(\frac{k}{w}\right)\overline{g}\left(\frac{k}{w}\right) + \int_{\mathbb{R}} (R_w f)(u)\overline{g}(u)\,du. \tag{1.5}$$

Moreover, if, in addition, f∈$\widehat{B}^2_{\pi w}$, then the integral on the right-hand side of (1.5) also vanishes since (Rwf)(u)=0, and so (1.5) reduces to the (classical) General Parseval formula (GPF) for f, g∈$\widehat{B}^2_{\pi w}$, w>0, already stated in (1.4). In this sense the latter two terms in GPDF (1.1) can be regarded as remainder terms.

In this instance the assumption f∈S^1_w is implicitly contained in the hypothesis f∈$\widehat{B}^2_{\pi w}$; see Sect. 2.1.

The *second* formula to be considered is the approximate sampling formula (ASF) treated in Weiss [83], Brown [11, 12] and Butzer-Splettstößer [22]; see also [84, pp. 64–66].

Approximate Sampling Formula (ASF)

For $f \in F^2 \cap S^1_w$, w>0, we have

$$f(t) = \sum_{k\in\mathbb{Z}} f\left(\frac{k}{w}\right)\operatorname{sinc}(wt-k) + (R_w f)(t) \qquad (t \in \mathbb{R}), \tag{1.6}$$

the series converging absolutely and uniformly on RR, where

$$\operatorname{sinc} t := \begin{cases} \dfrac{\sin \pi t}{\pi t}, & t \in \mathbb{R}\backslash\{0\}, \\ 1, & t = 0. \end{cases}$$

For the important role played by the sincsinc-function in mathematics see [80, 81].

The ASF for not necessarily bandlimited functions generalizes the classical version

$$f(t) = \sum_{k\in\mathbb{Z}} f\left(\frac{k}{w}\right)\operatorname{sinc}(wt-k) \qquad (t \in \mathbb{R}) \tag{1.7}$$

of the sampling formula for f∈$\widehat{B}^2_{\pi w}$.

In signal analysis, convergence the sampling series is usually defined via the limit of the symmetric partial sums, i. e., $\lim_{N\to\infty}\sum_{|k|\le N} f(k/w)$sinc(wt–k) [see (2.20)], reflecting the mode of convergence of complex Fourier series. In particular, it is the usual approach in sampling theory, where the ASF holds for functions in F^2. In the present case, however, the focus is on the equivalence of the validity of the ASF with other results, in particular with the PSF.

Whereas to proceed from (1.7) to (1.6) one adds the "error" term $(R_w f)(t)$, to go from (1.4) to (1.1) one has to add two "error" terms.

To make the foregoing generalization fully clear, the ASF yields the GPDF and, conversely, the GPDF for just one special function will yield the ASF; thus the two formulae are truly equivalent to another, as already shown in [18].

A *third* formula belonging to our grouping of six equivalences in Figure 1 brings us into contact with the theory of Hilbert spaces with reproducing kernel, an important branch of functional analysis and complex function theory having many areas of applications, with a long history. See, e. g., [3, 37, 46, 62].

It is well known that $\widehat{B}^2_{\pi w}$is a sub-Hilbert space of $L^2(R)$ and is, furthermore, an example of a Hilbert space with reproducing kernel. It is a special case of Saitoh's presentation (see, e. g., [41, Ch. 3], [77, pp. 9, 194]). The reproducing kernel is (t,u)↦w sinc w(t–u) and the reproducing kernel formula (RKF), expressing the reproduction of f∈$\widehat{B}^2_{\pi w}$ from itself, that is, all of its values, is

$$f(t) = w\int_{\mathbb{R}} f(u)\,\mathrm{sinc}\,w(t-u)\,du = \int_{\mathbb{R}} f\left(\frac{\kappa}{w}\right)\mathrm{sinc}(wt-\kappa)\,d\kappa \quad (t\in\mathbb{R}). \tag{1.8}$$

The second integral in (1.8) is just the first integral rearranged to give a precise integral analogue of the series in (1.7), which may be thought of as a "discrete reproducing kernel formula", expressing the reproduction of f from a discrete subset of its values.

In view of the foregoing equivalence GPDF⇔ASF, the question arises whether there can be established a generalized version of RKF, thus one for non-bandlimited $L^1(R)$-functions, which is equivalent to the former two. In fact, one basic aim of this paper is to introduce the new approximate reproducing kernel formula, namely

Approximate Reproducing Kernel Formula (ARKF)

Let $f\in F^2\cap S^1_w$ with w>0. Then $R_w f\in L^2(R)$ and

$$f(t) = w\int_{\mathbb{R}} f(u)\,\mathrm{sinc}\,w(t-u)\,du + (R_w f)(t) - w\int_{\mathbb{R}}(R_w f)(u)\,\mathrm{sinc}\,w(t-u)\,du. \tag{1.9}$$

Clearly, if $f\in\widehat{B}^2_{\pi w}$, then again $R_w f=0$, and so we obtain the (classical) reproducing kernel formula RKF, namely (1.8). Thus (1.9) represents the classical RKF plus two remainder terms.

There is a *fourth* formula which is fundamental in a variety of mathematical fields, namely the truly well-known Poisson summation formula (PSF). But for purposes of precision let us state the version we are using.

Poisson's Summation Formula (PSF)

Let $f\in L_1(R)$ such that $\widehat{f}\in S^1_w$ for some w>0; then

$$\sqrt{2\pi}\,w\sum_{k\in\mathbb{Z}} f(x+2k\pi w) = \sum_{k\in\mathbb{Z}}\widehat{f}\left(\frac{k}{w}\right)e^{ikx/w} \quad (a.e.). \tag{1.10}$$

Now to the *fifth* formula which is actually a different version of GPDF but has a structure similar to that of the Poisson summation formula, however not for one function f, but for the product of two, namely $f\overline{g}$ on the left-hand side, and the product of their Fourier transforms on the right-hand side.

Parseval Decomposition–Poisson Summation Formula (PDPS)

For $f\in F^2\cap S^1_w$, w>0,σ>0, and $g\in F^2$, there holds

$$\begin{aligned}\frac{1}{w}\sum_{k\in\mathbb{Z}} f\left(\frac{k}{w}\right)\overline{g}\left(\frac{k}{w}\right)e^{ikx/w} &= \sum_{k\in\mathbb{Z}}\int_{-\sigma}^{\sigma}\widehat{f}(v-x+2k\pi w)\,\overline{\widehat{g}}(v)\,dv\\ &+\frac{1}{w}\sum_{k\in\mathbb{Z}} f\left(\frac{k}{w}\right)\frac{1}{\sqrt{2\pi}}\int_{|v|\geq\sigma}\overline{\widehat{g}}(v)e^{ik(v+x)/w}\,dv\end{aligned} \tag{1.11}$$

almost everywhere on RR. Furthermore, if rather $g\in F^1$ instead of $g\in F^2$, then the formula holds for all x∈R.

PDPS can in fact be interpreted as an *approximate* form of PSF for a product $f\overline{g}$ in which g takes the role of an *approximately* bandlimited function. Indeed, the second series on the right-hand side of PDPS may be seen as an error term depending on the deviation of g from a bandlimited function. If g is bandlimited to [−σ,σ], i. e. $g\in\widehat{B}^2_\sigma$, then PDPS reduces to

$$\frac{1}{w}\sum_{k\in\mathbb{Z}} f\left(\frac{k}{w}\right)\overline{g}\left(\frac{k}{w}\right)e^{ikx/w} = \sum_{k\in\mathbb{Z}}\int_{-\sigma}^{\sigma}\widehat{f}(v-x+2k\pi w)\,\overline{\widehat{g}}(v)\,dv \tag{1.12}$$

holding *almost everywhere* on R when $g\in F^2$ and *everywhere* if rather $g\in F^1$.

A particular form of PDPS had already been used in [18] in one of the proofs, namely the form

$$\begin{aligned}\frac{1}{w}\sum_{k\in\mathbb{Z}} f\left(\frac{k}{w}\right)\overline{g}\left(\frac{k}{w}\right) &= \sum_{k\in\mathbb{Z}}\int_{-\pi w}^{\pi w}\widehat{f}(v+2k\pi w)\overline{\widehat{g}}(v)\,dv\\ &+\frac{1}{w}\sum_{k\in\mathbb{Z}} f\left(\frac{k}{w}\right)\frac{1}{\sqrt{2\pi}}\int_{|v|\geq\pi w}\overline{\widehat{g}}(v)e^{ikv/w}\,dv,\end{aligned} \tag{1.13}$$

provided $f\in F^2\cap S^1_w$ and $g\in F^1$. It is the case x=0, σ=πw of (1.11), noting

$$\int_{-\sigma}^{\sigma} \widehat{f}(v-x+2k\pi w)\overline{\widehat{g}}(v)\,dv = \int_{2k\pi w-x-\sigma}^{2k\pi w-x+\sigma} \widehat{f}(v)\overline{\widehat{g}}(v+x-2k\pi w)\,dv. \tag{1.14}$$

In fact, PDPS in the form (1.13) is just a different representation of GPDF; see [18] for the details.

Of interest is a version of (1.13) with g replaced by $\hat{g}$, the Fourier transform of a function $g \in L^1(R) \cap L^2(R)$, thus $\hat{g} \in F^2$. The result can be regarded as a discretization of the classical Fourier exchange formula (2.5).

Just as for GPDF, the authors have also never met PDPS in the literature; both formulae also do not seem to be particular cases of specific trace formulae known to the authors.

Next to our *sixth* formula, the famous

Functional Equation of the Riemann zeta Function (FERZ)

For the Riemann zeta function, defined by

$$\zeta(s) := \sum_{k=1}^{\infty} \frac{1}{k^s} \quad (s = \sigma + i\tau \in \mathbb{C}, \sigma > 1), \tag{1.15}$$

and for $\sigma \le 1$ by analytic continuation, there holds for all $s \in C$ the equation

$$\pi^{-\frac{s}{2}}\Gamma\left(\frac{s}{2}\right)\zeta(s) = \pi^{-\frac{1-s}{2}}\Gamma\left(\frac{1-s}{2}\right)\zeta(1-s) \quad (s \in \mathbb{C})$$

or, equivalently,

$$2(2\pi)^{-s}\cos\left(\frac{\pi s}{2}\right)\Gamma(s)\zeta(s) = \zeta(1-s). \tag{1.16}$$

In fact, we shall reprove the well-known equivalence FERZ ⇔ PSF in great detail, in particular in regard to the interchanges of the orders of summation and/or integration occurring. What is remarkable about this equivalence is that in the direction FERZ ⇒ PSF, a very particular result, namely the functional equation for $\zeta(s)$, implies a very general result, the PSF for a large class of functions. In other words, a formula of number theory, (externally) unrelated to periodicity and Fourier series, can yield a basic, general formula of analysis, one connecting Fourier series and integrals.

It is well-known that the FERZ is equivalent to the second famous functional equation, namely FEJT

$$\vartheta\left(-\frac{1}{z}\right) = (-iz)^{\frac{1}{2}}\vartheta(z), \tag{1.17}$$

satisfied by Jacobi's ϑ-function

$$\vartheta(z) = 1 + 2\sum_{k=1}^{\infty}\exp(i\pi k^2 z) = \sum_{k=-\infty}^{\infty}\exp(i\pi k^2 z). \tag{1.18}$$

The function ϑ, defined on the upper half-plane $H:=\{z \in C; Iz>0\}$, is analytic in HH and periodic with period 2.

As to the proofs of FERZ ⇒ FEJT see e. g. Hamburger [36, pp. 136–137], and for the converse Riemann [74]. In the matter see also [44].

It is well known that an application of PSF to the Gauß kernel of (2.12) yields FEJT (see [20, p. 204]).

Furthermore, Dedekind's ηη-function (1877) of elliptic modular functions,

$$\eta(z) = \exp\left(\frac{i\pi z}{12}\right)\prod_{k=1}^{\infty}\left(1-\exp(i2\pi kz)\right), \tag{1.19}$$

analytic on HH, satisfies the same functional equation (1.17), namely FEDE (cf. [2, pp. 47–50]),

$$\eta\left(-\frac{1}{z}\right) = (-iz)^{\frac{1}{2}}\eta(z).$$

Finally to our *seventh* formula, discovered by Euler in connection with the so-called Basel problem, i. e., with determining $\zeta(2)$ in modern terminology.

Euler–Maclaurin Summation Formula (EMSF)

For $n,r \in N$ and $f \in C^{(2r)}[0,n]$, we have

$$\sum_{k=0}^{n} f(k) = \int_0^n f(x)\,dx + \frac{1}{2}[f(0)+f(n)] + \sum_{k=1}^{r}\frac{B_{2k}}{(2k)!}[f^{(2k-1)}(n) - f^{(2k-1)}(0)]$$
$$+(-1)^r\sum_{k=1}^{\infty}\int_0^n \frac{e^{i2\pi kt}+e^{-i2\pi kt}}{(2\pi k)^{2r}} f^{(2r)}(t)\,dt, \tag{1.20}$$

where B_{2k} are the Bernoulli numbers.

The implications to be established in this paper are indicated by the arrows in Figure 1. For a proof of ASF ⇔⇔ EMSF the reader is referred to [16]. Thus each of the seven formulae GPDF, ASF, ARKF, PSF, PDPS, FERZ and EMSF is equivalent to each other, in the sense that each is a corollary of each of the others.

This means that three different, particular results of number theory (FERZ, FEJT, FEDE) are equivalent to six general summation formulae of Fourier analysis, signal analysis and numerical analysis, namely PSF, PDPS, GPDF, ASF, ARKF, EMSF.

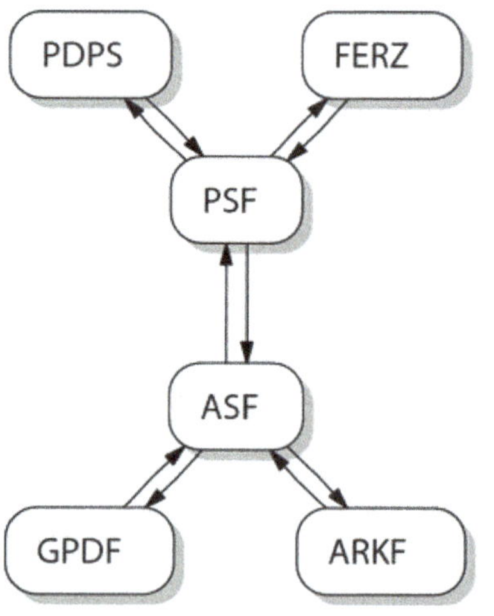

Figure 1: The implications to be proved.

Section 2 is devoted to notations and the side results needed. In Sect. 3 we prove a first grouping of equivalences, namely, GPDF⇔ASF, and then ASF⇔ ARKF, PSF⇔PDPS, and finally ASF⇔PSF. The latter three equivalences are new results of this paper. Section 4 is concerned with the equivalence of PSF and FERZ, and in Sect. 5 gives a proof of ASF independent of the other formulae of this paper.

The epilogue of Sect. 6, with the topic abstract settings, deals with the possibility of presenting this paper in a setting based on reproducing kernel (r.k.) theory, and also in terms of abstract harmonic analysis, using locally compact abelian groups.

In the present paper, dealing with the grouping (A): GPDF, ASF, ARKF, PSF, PDPS, FERZ, as well as the Euler–Maclaurin summation formula EMSF (handled in [16])—all for non-bandlimited functions—the basic result is that all seven theorems are equivalent amongst themselves. On the other hand, in our joint paper [15], treating the grouping (B), namely the classical sampling theorem CSF, the general Parseval formula GPF (i. e., the GPDF for bandlimited functions), the reproducing kernel formula GPF (i. e., GPDF for bandlimited functions) and the PSF—all for bandlimited functions—it was shown that the latter four theorems are also all equivalent amongst themselves.

However in our article [14] it was shown that the approximate sampling theorem, the ASF, is equivalent to the classical sampling theorem CSF. Thus the conclusion of the present paper, together with the papers [15, 16] and [14], is the full equivalence of the groupings (A) and (B). In other words, the seven theorems for non-bandlimited functions are all equivalent to the corresponding ones for the classical bandlimited functions.

Whereas the majority of the assertions of the theorems in question have been stated and established in earlier papers by a (part) of the present authors, the proofs of the many delicate interconnections between those of grouping (A), the chief topic of the paper, are new in their essential steps and conclusions, as referred to.

Electrical engineers use bandlimited functions as a mathematical model which is a rather severe restriction. In fact there do not exist signals which are simultaneously bandlimited and duration-limited (time-limited). Hence engineers work intuitively with approximately bandlimited functions; these are covered by our non-bandlimited theorems. Thus our paper can also be regarded as a further justification of what electrical engineers do in their real life work.

NOTATIONS AND SIDE RESULTS

Notations

The Fourier transform of $f \in L^p(\mathbb{R})$ with p=1 or p=2 is defined by

$$\widehat{f}(v) := \frac{1}{\sqrt{2\pi}} \int_{\mathbb{R}} f(u) \mathrm{e}^{-ivu}\, du \qquad (v \in \mathbb{R}),$$

the integral being understood as the limit in the $L^2(\mathbb{R})$-norm for p=2. We shall consider the following function spaces, namely

$$F^p := \{ f \in L^p(\mathbb{R}) \cap C(\mathbb{R});\ \widehat{f} \in L^1(\mathbb{R}) \} \qquad (p = 1, 2),$$

where C(R) denotes the space of all uniformly continuous and bounded functions on R, and

$$S_w^p := \{ f : \mathbb{R} \to \mathbb{C};\ \left(f(\tfrac{k}{w})\right)_{k \in \mathbb{Z}} \in \ell^p(\mathbb{Z}) \} \qquad (w > 0, 1 \le p \le$$

There holds $F^1 \subset F^2$ and $S_w^1 \subset S_w^{p_1} \subset S_w^{p_2} \subset S_w^\infty$ for $1 \le p_1 \le p_2 \le \infty$.

The Bernstein spaces $\widehat{B}_\sigma^p$ for p=1,2 and σ>0σ>0 are defined in terms of the Fourier transform via

$$\widehat{B}_\sigma^p := \{ f \in L^p(\mathbb{R});\ \widehat{f}(v) = 0\ a.e. \text{ outside } [-\sigma, \sigma] \}.$$

Since the Fourier transform of a function $f \in \widehat{B}_\sigma^p$ has support on a set of finite measure, those functions are also called "bandlimited". One has the inclusions $\widehat{B}_\sigma^1 \subset \widehat{B}_\sigma^2$ and $\widehat{B}_\sigma^p \subset F^p \cap S_w^1 \subset F^p \cap S_w^p$, p=1,2; see [64, pp. 123,126].

Results on Fourier Analysis

Let us list some well-known results from Fourier analysis; see any textbook in the matter, e. g., [20, Chapt. 5.1 and 5.2].

Proposition 1

(a) If $f \in L^p(\mathbb{R})$, p=1,2, then for each $h \in \mathbb{R}$,

$$[f(\cdot + h)]^\wedge(v) = \widehat{f}(v) e^{ihv} \tag{2.1}$$

$$[e^{-ih\cdot} f(\cdot)]^\wedge(v) = \widehat{f}(v + h) \tag{2.2}$$

for all $v \in \mathbb{R}$ in case p=1 and a. e. in case p=2.

(b) If $f \in L^1(\mathbb{R})$, $g \in L^p(\mathbb{R})$, p=1,2, then the convolution

$$(f * g)(t) := \frac{1}{\sqrt{2\pi}} \int_{\mathbb{R}} f(u) g(t - u)\, du \tag{2.3}$$

belongs to $L^p(\mathbb{R})$, and for the Fourier transform one has the convolution theorem

$$(f * g)^\wedge(v) = \widehat{f}(v)\, \widehat{g}(v) \tag{2.4}$$

for all $v \in \mathbb{R}$ in case p=1 and a. e. in case p=2.

(c) If $f, g \in L^p(\mathbb{R})$, p=1,2, then there holds the exchange formula

$$\int_{\mathbb{R}} f(v) \widehat{g}(v)\, dv = \int_{\mathbb{R}} \widehat{f}(v) g(v)\, dv. \tag{2.5}$$

d) If $f \in L^2(\mathbb{R})$, then $\widehat{f}$ also belongs to $L^2(\mathbb{R})$ and $\|f\|_{L^2(\mathbb{R})} = \|\widehat{f}\|_{L^2(\mathbb{R})}$. Furthermore, there holds the inversion formula

$$f(t) = \int_{\mathbb{R}} \widehat{f}(v) e^{ivt}\, dv = \widehat{\widehat{f}}(-t) \qquad a.e., \tag{2.6}$$

where the integral is again understood as the limit in $L^2(\mathbb{R})$-norm. If $f \in F^p$ with p=1 or p=2, then the integral in (2.6) exists as an ordinary Lebesgue integral and both equalities hold for all $t \in \mathbb{R}$.

Lemma 1

There hold the formulae

$$\frac{1}{\sqrt{2\pi}\, w} \operatorname{rect}\left(\frac{\cdot}{w}\right)^\wedge(v) = \operatorname{sinc}(wv) \qquad (v \in \mathbb{R}), \tag{2.7}$$

$$\operatorname{sinc}(w\cdot)^\wedge(v) = \frac{1}{\sqrt{2\pi}\, w} \operatorname{rect}\left(\frac{v}{w}\right) \qquad (v \in \mathbb{R}) \tag{2.8}$$

with the rectangle function

$$\operatorname{rect}(t) := \begin{cases} 1, & |t| < \pi, \\ \frac{1}{2}, & |t| = \pi, \\ 0, & |t| > \pi. \end{cases}$$

Furthermore, one has

$$\int_{\mathbb{R}} \operatorname{sinc}(u-s)\operatorname{sinc}(u-t)\,du = \operatorname{sinc}(s-t) \qquad (s, t \in \mathbb{R}), \tag{2.9}$$

and, in particular,

$$\int_{\mathbb{R}} \operatorname{sinc}(u-j)\operatorname{sinc}(u-k)\,du = \begin{cases} 1, & j = k, \\ 0, & j \neq k. \end{cases} \qquad (j, k \in \mathbb{Z}). \tag{2.10}$$

Proof

Equation (2.7) follows by a simple integration, and (2.8) follows from (2.7) by the Fourier inversion formula. (For an elementary proof of (2.8) without using the inversion formula see [10, p. 13 f].) As to (2.9) one has by Proposition 1 (a), (c), and by (2.7) and (2.8)

$$\begin{aligned}
\int_{\mathbb{R}} \operatorname{sinc}(u-s)\operatorname{sinc}(u-t)\,du &= \int_{\mathbb{R}} \left[\frac{1}{\sqrt{2\pi}} \operatorname{rect}(\cdot)e^{i\cdot s}\right]^{\wedge}(u)\operatorname{sinc}(u-t)\,du \\
&= \int_{\mathbb{R}} \left[\frac{1}{\sqrt{2\pi}} \operatorname{rect}(v)e^{ivs}\right] \operatorname{sinc}(\cdot - t)^{\wedge}(v)\,du \\
&= \int_{\mathbb{R}} \left[\frac{1}{\sqrt{2\pi}} \operatorname{rect}(v)e^{ivs}\right]\left[\frac{1}{\sqrt{2\pi}} \operatorname{rect}(v)e^{-ivt}\right] dv \\
&= \frac{1}{2\pi}\int_{\mathbb{R}} \operatorname{rect}(v)e^{-iv(t-s)}\,dv = \frac{1}{\sqrt{2\pi}}\operatorname{rect}^{\wedge}(t-s) \\
&= \operatorname{sinc}(t-s) = \operatorname{sinc}(s-t).
\end{aligned}$$

Finally, (2.10) follows from (2.9) noting that $\operatorname{sinc}(j-k)=\delta_{j,k}$.

We shall also need the following result for the approximate identities g_ρ of Gauß-Weierstraß and χ_ρ of Fejér; see [20, Sec. 3.1.3, 3.1.2, 3.2].

Proposition 2

(a) For $f \in C(R)$, we have

$$\lim_{\rho \to 0+} \frac{1}{\sqrt{2\pi}} \int_{\mathbb{R}} f(u) g_\rho(t-u)\,du = f(t) \tag{2.11}$$

uniformly for $t \in R$, where

$$g_\rho(u) := \frac{1}{\sqrt{2\rho}} \exp\left(\frac{-u^2}{4\rho}\right) \quad (u \in \mathbb{R};\ \rho > 0) \tag{2.12}$$

is the Gaussian kernel, its Fourier transform being

$$\widehat{g}_\rho(v) = e^{-\rho v^2} \quad (v \in \mathbb{R};\ \rho > 0). \tag{2.13}$$

The same result is valid for $f \in L^p(R)$, $1 \le p < \infty$; the convergence in (2.11) holding now in Lp(R)-norm as well as a. e. on R.

(b) The assertions of part (a) remain valid, if the Gaussian kernel is replaced by Fejér's kernel

$$\chi_\rho(u) := \rho\sqrt{2\pi}\operatorname{sinc}^2(\rho u) \qquad (u \in \mathbb{R}, \rho > 0), \tag{2.14}$$

having Fourier transform

$$\widehat{\chi}_\rho(v) = \left(1 - \frac{|v|}{2\pi\rho}\right)_+ \qquad (v \in \mathbb{R}, \rho > 0), \tag{2.15}$$

where

$h_+(u) := h(u)$ *if* $h(u) \ge 0$, *and* $h_+(u) = 0$ *if* $h(u) < 0$.

For $f := \operatorname{sinc}(\cdot - u)$, we obtain from ASF the following sinc summation formula:

$$\operatorname{sinc}(t-u) = \sum_{k \in \mathbb{Z}} \operatorname{sinc}(t-k)\operatorname{sinc}(u-k). \tag{2.16}$$

However, we need not invoke ASF for deriving (2.16). It is known that (2.16) is a simple consequence of the famous, classical cotangent expansion. For the latter, easy elementary proofs are known; see e. g. [43], [15, p. 448]

It follows from (2.16) that

$$\sum_{k\in\mathbb{Z}} \operatorname{sinc}^2(t-k) = 1 \qquad (t \in \mathbb{R}). \tag{2.17}$$

The following lemma will be useful; for a proof see [40, p. 207].

Lemma 2

Let f∈L^p(R), 1<p<∞, and $(fk)_{k\in N}$ be a sequence in L^p(R). If there exists a constant M such that ‖fk‖L^p(R)≤M for all k∈N and if limk→∞fk(u)=f(u) a. e. in RR, then for every g∈Lq(R), 1/p+1/q=1,

$$\lim_{k\to\infty} \int_{\mathbb{R}} f_k(u)g(u)\,du = \int_{\mathbb{R}} f(u)g(u)\,du.$$

The Sampling Series $S_w f$

Concerning the convergence of the series

$$(S_w f)(t) := \sum_{k\in\mathbb{Z}} f\left(\frac{k}{w}\right) \operatorname{sinc}(wt-k) \qquad (t \in \mathbb{R}), \tag{2.18}$$

occurring in (1.6), (1.7), we have

Lemma 3

Let f∈S^1_w for some w>0. The series $(S_w f)(t)$ converges absolutely and uniformly on R. Furthermore, $S_w f$ converges in L^p(R), 1<p<∞. In particular, $S_w f$∈L^p(R), and

$$\lim_{n\to\infty} \int_{\mathbb{R}} (S_{w,n} f)(u)\overline{g}(u)\,du = \int_{\mathbb{R}} (S_w f)(u)\overline{g}(u)\,du \tag{2.19}$$

for all g∈L^q(R), 1/p+1/q=1, where

$$(S_{w,n} f)(t) := \sum_{|k|\le n} f\left(\frac{k}{w}\right) \operatorname{sinc}(wt-k) \tag{2.20}$$

denote the partial sums of S_wf.

Proof

For simplicity take w=1. For 0<n_1<n_2 there holds

$$\begin{aligned}\left\| S_{1,n_1} f - S_{1,n_1} f \right\|_{L^p(\mathbb{R})} &\le \sum_{n_1<|k|\le n_2} |f(k)|\, \|\operatorname{sinc}(\cdot - k)\|_{L^p(\mathbb{R})} \\ &= \|\operatorname{sinc}\|_{L^p(\mathbb{R})} \sum_{n_1<|k|\le n_2} |f(k)|,\end{aligned} \tag{2.21}$$

where the latter term tends to zero for n_1,n_2→∞ since f∈S^1_1. The same estimate holds for the space C(R), giving the absolute and uniform convergence of (S_1,nf)(t) to (S_1f)(t). Further, the completeness of L^p(R) and estimate (2.21) guarantee that S_1,nf tends to some function g∈L^p(R) which must be equal to the uniform limit S1f.

Finally, by Hölder's inequality

$$\left| \int_{\mathbb{R}} (S_{w,n} f)(u)\overline{g}(u)\,du - \int_{\mathbb{R}} (S_w f)(u)\overline{g}(u)\,du \right| \le \|S_{w,n} f - S_w f\|_{L^p(\mathbb{R})} \|g\|_{L^q(\mathbb{R})},$$

which yields (2.19).

Remarks on Poisson's Summation Formula

As to the PSF, let $h(x) := \sqrt{2\pi}\,w \sum_{k\in\mathbb{Z}} |f(x+2k\pi w)|$; then

$$\begin{aligned}\int_{-\pi w}^{\pi w} h(x)\,dx &= \sqrt{2\pi}\,w \sum_{k\in\mathbb{Z}} \int_{-\pi w}^{\pi w} |f(x+2k\pi w)|\,dx \\ &= \sqrt{2\pi}\,w \sum_{k\in\mathbb{Z}} \int_{(2k-1)\pi w}^{(2k+1)\pi w} |f(x)|\,dx = \sqrt{2\pi}\,w \int_{\mathbb{R}} |f(x)|\,dx < \infty,\end{aligned} \tag{2.22}$$

showing that h is integrable over the interval [−πw,πw], and hence over any compact interval in view of the 2πw-periodicity of h.

Now one has for arbitrary a,b∈R that

$$\int_a^b \left|\sqrt{2\pi w}\sum_{k=m}^{n} f(x+2k\pi w)\right| dx \le \int_a^b h(x)\,dx < \infty \quad (m,n\in\mathbb{Z}), \tag{2.23}$$

and so it follows that the left-hand side of (1.10) converges a. e. on R, and the convergence is dominated by the locally integrable function h.

When we wish to deduce PSF from any of the other formulae, it suffices to prove it for a dense subspace of $X\subset L^1(R)$ with $\widehat{g}\in S^1_w$ for all g∈X. Indeed, let $f\in L^1(R)$ with $\widehat{f}\in S^1_w$, and let $(f_n)_{n\in N}$ be a sequence in X with $\lim_{n\to\infty}\|f_n-f\|_{L^1(\mathbb{R})}=0$. Denoting $f^*(x):=\sqrt{2\pi w}\sum_{k\in\mathbb{Z}} f(x+2k\pi w)$, $f_n^*(x):=\sqrt{2\pi w}\sum_{k\in\mathbb{Z}} f_n(x+$, then (cf. (2.23), (2.22)),

$$\int_{-\pi w}^{\pi w} |f_n^*(u)-f^*(u)|\,du \le \|f_n-f\|_{L^1(\mathbb{R})} \quad (n\in\mathbb{N}). \tag{2.24}$$

This yields that

$$\lim_{n\to\infty}\widehat{f_n^*}(k)=\widehat{f^*}(k) \quad (k\in\mathbb{Z}), \tag{2.25}$$

where $\widehat{f_n^*}(k)$ and $\widehat{f^*}(k)$ denote the trigonometric Fourier coefficients of the 2πw-periodic function f_n^* and f^*, respectively.

On the other hand, PSF applied to the fn states that

$$f_n^*(x)=\sum_{k\in\mathbb{Z}}\widehat{f_n}\left(\frac{k}{w}\right)e^{ikx/w} \quad a.e.,$$

meaning, in view of the uniform convergence of the series on the right, that $\widehat{f_n}$ (k/w), k∈Z, are the Fourier coefficients of the 2πw-periodic function on the left. This implies that

$$\widehat{f^*}(k)=\lim_{n\to\infty}\widehat{f_n^*}(k)=\lim_{n\to\infty}\widehat{f_n}\left(\frac{k}{w}\right)=\widehat{f}\left(\frac{k}{w}\right) \quad ((k\in\mathbb{Z}), \tag{2.26}$$

the last equality being valid in view of

$|\widehat{f_n}(v)-\widehat{f}(v)|\le\|f_n-f\|_{L^1(\mathbb{R})}$ for all v∈R. Hence the Fourier coefficients of f* are given by $\widehat{f^*}(k)=\widehat{f}(k/w)$, k∈Z, i. e.

$$f^*(x)=\sum_{k\in\mathbb{Z}}\widehat{f}\left(\frac{k}{w}\right)e^{ikx/w} \quad a.e.$$

The equality a. e. here holds in view of $\widehat{f}\in S^1_w$, giving the uniform convergence of the Fourier series of f*. This is PSF for all $f\in L^1(R)$ with $\widehat{f}\in S^1_w$.

EQUIVALENCE OF THE DECOMPOSITION FORMULA AND FOUR SUMMATION FORMULAE

In this section we shall deduce the basic new ARKF theorem for $L^2(R)$-function by means of the ASF. Even more so, the ASF itself is a consequence of ARKF. Thus the two are equivalent. On top, three other basic equivalences of this paper are also established in this section.

When we want to deduce one of the formulae GPDF, ASF or ARKF, it suffices to carry it out for w=1. Indeed, the general formula is obtained from that special case by replacing f by f(·/w), t by wt and noting that

$$\left[(R_1 f)\left(\frac{\cdot}{w}\right)\right](wt)=(R_w f)(t).$$

GPDF and ASF

But first to the equivalence of GPDF with ASF. The proof of the implication GPDF⇒⇒ASF is essentially that of Thm 1.3 in [18]. We present it here again for convenience.

Proof of GPDF⇒ASF for $f\in F^2\cap S^1_w$. In regard to the proof which follows, it is appropriate to rewrite (1.1) as

$$\int_{\mathbb{R}} f(u)\overline{g}(u)\,du=\frac{1}{w}\sum_{k\in\mathbb{Z}} f\left(\frac{k}{w}\right)\frac{1}{\sqrt{2\pi}}\int_{|v|\le\pi w}\widehat{\overline{g}}(v)\,e^{ikv/w}\,dv$$
$$+\int_{\mathbb{R}}(R_w f)(u)\overline{g}(u)\,du. \tag{3.1}$$

Indeed, one has in view of the Fourier inversion formula (2.6) for g

$$\frac{1}{w}\sum_{k\in\mathbb{Z}} f\left(\frac{k}{w}\right)\overline{g}\left(\frac{k}{w}\right) - \frac{1}{w}\sum_{k\in\mathbb{Z}} f\left(\frac{k}{w}\right)\frac{1}{\sqrt{2\pi}}\int_{|v|\geq\pi w}\widehat{\overline{g}}(v)e^{ikv/w}\,dv$$
$$=\frac{1}{w}\sum_{k\in\mathbb{Z}} f\left(\frac{k}{w}\right)\frac{1}{\sqrt{2\pi}}\left\{\int_{-\infty}^{\infty}\widehat{\overline{g}}(v)e^{ikv/w}\,dv - \int_{|v|\geq\pi w}\widehat{\overline{g}}(v)e^{ikv/w}\,dv\right\}.$$

A substitution of this identity into (1.1) yields (3.1).

Now we restrict the matter to w=1w=1 and apply the GPDF in the form (3.1) to $f\in F^2\cap S_1^1$ and $g=(2\pi)^{-1/2}g_\rho(t-\cdot)\in F^2$ of (2.12). Noting that $\widehat{g_\rho(t-\cdot)}(v)=e^{-\rho v^2}e^{-itv}$ by (2.1) and (2.13), we obtain

$$\frac{1}{\sqrt{2\pi}}\int_{\mathbb{R}} f(u)g_\rho(t-u)\,du=\sum_{k\in\mathbb{Z}} f(k)\frac{1}{2\pi}\int_{|v|\leq\pi} e^{-\rho v^2}e^{iv(k-t)}\,dv$$
$$+\frac{1}{\sqrt{2\pi}}\int_{\mathbb{R}}(R_1 f)(u)g_\rho(t-u)\,du.$$

In this equation, we let $\rho\to 0+$. Since f and $R_1 f$ belong to C(R), and since the infinite series is uniformly convergent in view of $f\in S_1^1$, Proposition 2 yields

$$f(t)=\sum_{k\in\mathbb{Z}} f(k)\frac{1}{2\pi}\int_{|v|\leq\pi} e^{iv(k-t)}\,dv+(R_1 f)(t)\qquad(t\in\mathbb{R}).$$

This is ASF, because the integral equals 2πsinc(t−k). The absolute and uniform convergence of the series $(S_1 f)(t)$ follows from Lemma 3.

The proof of the converse direction ASF⇒GPDF is similar to the corresponding one in [18] for non-uniform sampling.

Proof of ASF⇒GPDF for f$f\in F^2\cap S_w^1,\ g\in F^2$. Let w=1, multiply ASF by $\overline{g}$ and integrate over R. Then

$$\int_{\mathbb{R}} f(u)\overline{g}(u)\,du=\int_{\mathbb{R}}(S_1 f)(u)g(u)\,du+\int_{\mathbb{R}}(R_1 f)(u)\overline{g}(u)\,du. \tag{3.2}$$

Since $f\in L^2(R)$ by assumption, we also have $S_1 f\in L^2(R)$ in view of Lemma 3, and hence $R_1 f=f-S_w f\in L^2(R)$. Thus all integrals in (3.2) exists. For the first term on the right-hand side of (3.2) one obtains

$$\int_{\mathbb{R}}(S_1 f)(u)\overline{g}(u)\,du=\sum_{k\in\mathbb{Z}} f(k)\int_{\mathbb{R}}\overline{g}(u)\,\mathrm{sinc}(u-k)\,du, \tag{3.3}$$

the interchange of integration and summation being justified, since $(f(k))\in l^1(\mathbb{Z})$ and the integrals on the right-hand side are bounded by $\|g\|_{L^2(\mathbb{R})}\cdot\|\mathrm{sinc}\|_{L^2(\mathbb{R})}$. Now, using the convolution theorem (2.4), formula (2.8), and the Fourier inversion formula (2.6), we can rewrite the latter integral as

$$\int_{\mathbb{R}}\overline{g}(u)\,\mathrm{sinc}(u-k)\,du=\sqrt{2\pi}(\overline{g}*\mathrm{sinc})(k)=\frac{1}{\sqrt{2\pi}}\int_{\mathbb{R}}\widehat{\overline{g}}(v)\,\mathrm{rect}(v)e^{ikv}\,dv$$
$$\int\ =\frac{1}{\sqrt{2\pi}}\left\{\int_{\mathbb{R}}-\int_{|v|\geq\pi}\right\}\widehat{\overline{g}}(v)e^{ikv}\,dv=\overline{g}(k)-\frac{1}{\sqrt{2\pi}}\int_{|v|\geq\pi}\widehat{\overline{g}}(v)e^{ikv}\,dv$$

Inserting this into (3.3), and then the resulting equation into (3.2) yields the assertion (1.1) for w=1.

Thus we have established the equivalence ASF⇔GPDF.

ASF and ARKF

Now to the basic equivalence ASF⇔ARKF mentioned above.

Proof of ASF⇒ARKF for $f\in F^2\cap S_w^1,\ g\in F^2$. The fact that $R_w f=f-S_w f\in L^2$ follows again from Lemma 3. Now, let w=1 and let f be as in ARKF. We consider ASF (1.6), replace t by u, multiply both sides by sinc(t−u) and integrate with respect to u. This gives

$$\int_{\mathbb{R}} f(u)\,\mathrm{sinc}(t-u)\,du=\int_{\mathbb{R}}\sum_{k\in\mathbb{Z}} f(k)\,\mathrm{sinc}(u-k)\,\mathrm{sinc}(t-u)\,du$$
$$+\int_{\mathbb{R}}(R_1 f)(u)\,\mathrm{sinc}(t-u)\,du. \tag{3.4}$$

By Lemma 3 we can interchange integration and summation on the right-hand side and we obtain by (2.9)

$$\int_{\mathbb{R}} \sum_{k\in\mathbb{Z}} f(k)\,\mathrm{sinc}(u-k)\,\mathrm{sinc}(t-u)\,du$$

$$= \sum_{k\in\mathbb{Z}} f(k) \int_{\mathbb{R}} \mathrm{sinc}(u-k)\,\mathrm{sinc}(t-u)\,du = \sum_{k\in\mathbb{Z}} f(k)\,\mathrm{sinc}(t-k).$$

Applying ASF once more, the latter series can be replaced by f(t)−(R_1f)(t), and hence

$$\int_{\mathbb{R}} \sum_{k\in\mathbb{Z}} f(k)\,\mathrm{sinc}(u-k)\,\mathrm{sinc}(t-u)\,du = f(t) - (R_1 f)(t).$$

Inserting this into (3.4) yields the ARKF in the form (1.9).

Proof of ARKF⇒ASF for $f \in F^2 \cap S^1_1$. It suffices to show that

$$\int_{\mathbb{R}} [f(u) - (R_1 f)(u)]\,\mathrm{sinc}(t-u)\,du = \sum_{k\in\mathbb{Z}} f(k)\,\mathrm{sinc}(t-k), \tag{3.5}$$

since the integral equals f(t)−(R1f)(t). For this purpose, we shall use the sinc summation formula (2.16) on the left-hand side and interchange integration and summation. For justifying the interchange, we want to employ Lemma 2. We have to verify that the hypotheses of that result are satisfied in the present situation.

Clearly, $f - R_1 f$ and $\mathrm{sinc}(\cdot - t)$ belong to $L^2(\mathbb{R})$. Next, setting

$$\mathrm{sinc}_N(t-u) := \sum_{|k|\le N} \mathrm{sinc}(u-k)\,\mathrm{sinc}(t-k),$$

we have $\lim_{N\to\infty} \mathrm{sinc}_N(t-u) = \mathrm{sinc}(t-u)$. Furthermore, by employing (2.9) and (2.17), we find that

$$\|\,\mathrm{sinc}_N(t-\cdot)\|^2_{L^2(\mathbb{R})}$$

$$= \sum_{|k|\le N} \sum_{|l|\le N} \left\{ \int_{\mathbb{R}} \mathrm{sinc}(u-k)\,\mathrm{sinc}(u-l)\,\mathrm{d}u \right\} \mathrm{sinc}(t-k)\,\mathrm{sinc}(t-l)$$

$$= \sum_{|k|\le N} \sum_{|l|\le N} \delta_{k,l} \cdot \mathrm{sinc}(t-k)\,\mathrm{sinc}(t-l)$$

$$= \sum_{|k|\le N} \mathrm{sinc}^2(t-k) \le \sum_{k\in\mathbb{Z}} \mathrm{sinc}^2(t-k) = 1,$$

where Kronecker's delta has been used. Now Lemma 2 allows us to conclude that

$$\int_{\mathbb{R}} [f(u) - R_1 f(u)]\,\mathrm{sinc}(t-u)\,\mathrm{d}u$$

$$= \lim_{N\to\infty} \int_{\mathbb{R}} [f(u) - R_1 f(u)]\,\mathrm{sinc}_N(t-u)\,\mathrm{d}u$$

$$= \lim_{N\to\infty} \sum_{|k|\le N} \left\{ \int_{\mathbb{R}} [f(u) - R_1 f(u)]\,\mathrm{sinc}(u-k)\,\mathrm{d}u \right\} \mathrm{sinc}(t-k).$$

By ARKF for t=k, the expression in braces is equal to f(k)−R_1f(k). Since R_1f vanishes at the integers, we see that (3.5) holds. The statement regarding the convergence of the series is verified as in the proof of GPDF⇒ASF.

The equivalence assertion ASF ⇔ ARKF, just established, is another new result of this paper.

PSF and PDPS

The following concerns the final grouping of this section.

Proof of PSF⇒PDPS It suffices again to consider the case w=1w=1. First we assume that g∈F^2 only and define

$$\overline{g}_1(u) := \frac{1}{\sqrt{2\pi}} \int_{-\sigma}^{\sigma} \widehat{\overline{g}}(v) e^{iuv}\,dv \quad \text{and} \quad \phi(u) := \left(f\overline{g}_1\right)^{\widehat{}}(-u),$$

which implies that $\widehat{\phi} = f\overline{g}_1$. Then $\widehat{\phi} \in L^1(\mathbb{R})$ since it is a product of two functions from L^2(R), and $\widehat{\phi} \in S^1_1$ since f∈ S^1_1 and $\overline{g}_1$ is bounded. Furthermore, ϕ∈C(R) since ϕ(−·) is the Fourier transform of a function from L^1(R). Now one has by Proposition 1 (a), (c) that

$$\phi(-\xi) = \left(f\overline{g}_1\right)\widehat{f}(\xi) = \frac{1}{\sqrt{2\pi}}\int_{\mathbb{R}} f(v)\,\overline{g}_1(v)e^{-iv\xi}\,dv$$
$$= \frac{1}{\sqrt{2\pi}}\int_{\mathbb{R}}\left[f(\cdot)e^{-i\xi\cdot}\right]\widehat{f}(-v)\,\overline{g}_1(v)\,dv$$
$$= \frac{1}{\sqrt{2\pi}}\int_{\mathbb{R}}\widehat{f}(\xi - v)\,\widehat{\overline{g}}_1(v)\,dv. \tag{3.6}$$

Hence $\phi(-\cdot)$ is the convolution of $\widehat{f} \in L^1(\mathbb{R})$ and $\widehat{\overline{g}}_1 \in L^1(\mathbb{R})$, and so it follows by Proposition 1 b) that $\phi \in L^1(R)$. Altogether we have shown that ϕ satisfies the hypotheses of PSF for w=1, and so

$$\sqrt{2\pi}\sum_{k\in\mathbb{Z}}\phi(x + 2k\pi) = \sum_{k\in\mathbb{Z}}\widehat{\phi}(k)e^{ikx} = \sum_{k\in\mathbb{Z}} f(k)\overline{g}_1(k)e^{ikx} \quad a.e. \tag{3.7}$$

Employing (3.6) and noting that $\widehat{\overline{g}}_1(v) = \widehat{\overline{g}}_1(-v) = \widehat{\overline{g}}(-v)\,\mathrm{rect}(v\pi/\sigma)$, we find that

$$\phi(\xi) = \left(f\overline{g}_1\right)\widehat{f}(-\xi) = \frac{1}{\sqrt{2\pi}}\int_{-\sigma}^{\sigma}\widehat{f}(v - \xi)\,\widehat{\overline{g}}(v)\,dv.$$

Substituting this into the left-hand side of (3.7) yields

$$\sum_{k\in\mathbb{Z}}\int_{-\sigma}^{\sigma}\widehat{f}(v - x - 2k\pi)\,\widehat{\overline{g}}(v)\,dv = \sum_{k\in\mathbb{Z}} f(k)\overline{g}_1(k)e^{ikx} \quad a.e. \tag{3.8}$$

Using the Fourier inversion formula to obtain

$$\overline{g}_1(u) = \overline{g}(u) - \frac{1}{\sqrt{2\pi}}\int_{|v|\geq\sigma}\widehat{\overline{g}}(v)e^{iuv}\,dv$$

for eliminating $\overline{g}_1$ on the right-hand side of (3.8), we arrive at

$$\sum_{k\in\mathbb{Z}}\int_{-\sigma}^{\sigma}\widehat{f}(v - x - 2k\pi)\,\widehat{\overline{g}}(v)\,dv$$
$$= \sum_{k\in\mathbb{Z}} f(k)\,\overline{g}(k)e^{ikx} - \sum_{k\in\mathbb{Z}} f(k)\frac{1}{\sqrt{2\pi}}\int_{|v|\geq\sigma}\widehat{\overline{g}}(v)e^{ik(v+x)}\,dv \quad a.e. \tag{3.9}$$

The desired formula is obtained by replacing k by −k on the left-hand side.

When $g\in F^1$, then $\widehat{\overline{g}} \in C(\mathbb{R})$. Since also $f\in S_1^1$ and $\widehat{\overline{g}} \in L^1(\mathbb{R})$, we observe that the two series on the right-hand side of (3.9) depend continuously on xx.

As to the left-hand side, choose $n\in N$ such that $n\pi \geq \sigma$, then,

$$\left|\sum_{k\in\mathbb{Z}}\int_{-\sigma}^{\sigma}\widehat{f}(v - 2k\pi)\,\widehat{\overline{g}}(v)dv\right| \leq \|\widehat{g}\|_{L^\infty(\mathbb{R})}\sum_{k\in\mathbb{Z}}\int_{-n\pi}^{n\pi}\left|\widehat{f}(v - 2k\pi)\right|dv$$
$$= \|\widehat{g}\|_{L^\infty(\mathbb{R})}\int_{-n\pi}^{n\pi}\sum_{k\in\mathbb{Z}}\left|\widehat{f}(v - 2k\pi)\right|dv. \tag{3.10}$$

Here and in the next displayed formula the interchange of summation and integration is justified by Beppo Levi's theorem. Since the integrand in the last line of (3.10) is a 2π-periodic function, the integral equals n-times the integral from $-\pi$ to π, and hence

$$\left|\sum_{k\in\mathbb{Z}}\int_{-\sigma}^{\sigma}\widehat{f}(v - 2k\pi)\,\widehat{\overline{g}}(v)dv\right| \leq n\|\widehat{g}\|_{L^\infty(\mathbb{R})}\int_{-\pi}^{\pi}\sum_{k\in\mathbb{Z}}\left|\widehat{f}(v - 2k\pi)\right|dv$$
$$= n\|\widehat{g}\|_{L^\infty(\mathbb{R})}\sum_{k\in\mathbb{Z}}\int_{(-2k-1)\pi}^{(-2k+1)\pi}\left|\widehat{f}(v)\right|dv = n\|\widehat{g}\|_{L^\infty(\mathbb{R})}\int_{-\infty}^{\infty}\left|\widehat{f}(v)\right|dv.$$

This inequality, with $\widehat{f}(\cdot)$ replaced by $\widehat{f}(\cdot - x) - \widehat{f}(\cdot - y)$, yields

$$\left|\sum_{k\in\mathbb{Z}}\int_{-\sigma}^{\sigma}\widehat{f}(v-x-2k\pi)\overline{\widehat{g}}(v)dv-\sum_{k\in\mathbb{Z}}\int_{-\sigma}^{\sigma}\widehat{f}(v-y-2k\pi)\overline{\widehat{g}}(v)dv\right|$$
$$\leq n\|\widehat{g}\|_{L^\infty(\mathbb{R})}\int_{-\infty}^{\infty}\left|\widehat{f}(v-x)-\widehat{f}(v-y)\right|dv,$$

and since the right side of this inequality tends to zero for y→x by the continuity in the mean of $\widehat{f}\in L^1(\mathbb{R})$, we have shown that the left-hand side of (3.9) is a continuous function of x∈R. Thus, both sides being continuous, (3.9) must hold for all x∈R.

Proof of PDPS⇒PSF for f∈L¹(R) *with* $\widehat{f}\in S_1^1$. First assume f∈F¹(R) with $\widehat{f}\in S_1^1$. We apply PDPS with σ=π, f replaced by $\widehat{f}\in F^1\cap S_1^1\subset F^2\cap S_1^1$, and g(t) replaced by $\widehat{g}_\rho(t)e^{-ixt}\in F^1$ of (2.12), (2.13). Noting that g_ρ, $\widehat{g}_\rho$ are real and even, (2.6), and

$$\overline{\left[\widehat{g}_\rho(t)e^{-ixt}\right]^{\wedge}}(v)=\overline{\left[g_\rho(t-x)\right]^{\widehat{\widehat{}}}}(v)=\overline{g}_\rho(-v-x)=g_\rho(v+x),$$
$$\left[\widehat{g}_\rho(t)e^{-ixt}\right]^{\wedge}(v)=\left[\widehat{g}_\rho(v)e^{ixt}\right]^{\wedge}(v)=\left[g_\rho(t+x)\right]^{\widehat{\widehat{}}}(v)=g_\rho(x-v),$$

we obtain for each x∈R,

$$\sum_{k\in\mathbb{Z}}\widehat{f}(k)\widehat{g}_\rho(k)e^{ixk}=\sum_{k\in\mathbb{Z}}\int_{-\pi}^{\pi}\widehat{\widehat{f}}(v+2k\pi)g_\rho(v+x)\,dv$$
$$+\sum_{k\in\mathbb{Z}}\widehat{f}(k)\frac{1}{\sqrt{2\pi}}\int_{|v|\geq\pi}g_\rho(x-v)e^{ikv}\,dv. \tag{3.11}$$

Now let ρ→0+. Since $\lim_{\rho\to0+}\widehat{g}_\rho(k)=1$ by (2.13), it follows that

$$\lim_{\rho\to0+}\sum_{k\in\mathbb{Z}}\widehat{f}(k)\widehat{g}_\rho(k)e^{ixk}=\sum_{k\in\mathbb{Z}}\widehat{f}(k)e^{ixk}\quad(x\in\mathbb{R}),$$

where the interchange of summation and integration is allowed since the series is uniformly convergent with respect to ρ>0 because of $\{\widehat{f}(k)\}\in\ell^1(\mathbb{Z})$ and $0<\widehat{g}_\rho(k)\leq1$ for all k∈Z and ρ>0.

As to the right-hand side of (3.11), the first term can be rewritten as (cf. (2.6)),

$$\sum_{k\in\mathbb{Z}}\int_{-\pi}^{\pi}\widehat{\widehat{f}}(v+2k\pi)g_\rho(v+x)\,dv=\sum_{k\in\mathbb{Z}}\int_{-\pi}^{\pi}f(v-2k\pi)g_\rho(x-v)\,dv$$
$$=\int_{\mathbb{R}}\left\{\sum_{k\in\mathbb{Z}}f(v+2k\pi)\right\}\mathrm{rect}(v)g_\rho(x-v)\,dv,$$

where the interchange of summation and integration is justified in view of the dominated convergence of the series. An application of Proposition 2 shows that the latter integral tends to $\sqrt{2\pi}\sum_{k\in\mathbb{Z}}f(v+2k\pi)$ a. e. on (−π,π) for ρ→0+.

The integral in the second term on the right-hand side of (3.11) can be rewritten as

$$\frac{1}{\sqrt{2\pi}}\int_{\mathbb{R}}g_\rho(x-v)e^{ikv}\{1-\mathrm{rect}(v)\}dv,$$

which tends to zero for ρ→0+ a. e. on (−π,π) by Proposition 2. Hence we obtain

$$\lim_{\rho\to0+}\sum_{k\in\mathbb{Z}}\widehat{f}(k)\frac{1}{\sqrt{2\pi}}\int_{|v|\geq\pi}g_\rho(v-x)e^{ikv}\,dv=0\quad\left(\text{a.e. on }(-\pi,\pi)\right),$$

noting again that $\{\widehat{f}(k)\}\in\ell^1(\mathbb{Z})$.

Altogether it follows from (3.11) for ρ→0+,

$$\sum_{k\in\mathbb{Z}}\widehat{f}(k)e^{ixk}=\sqrt{2\pi}\sum_{k\in\mathbb{Z}}f(v+2k\pi)\quad\left(\text{a.e. on }(-\pi,\pi)\right). \tag{3.12}$$

Since both sides of (3.12) define a 2π-periodic function, equation (3.12) holds even a. e. on R, which is PSF for the space $\{f\in F^1;\widehat{f}\in S_1^1\}\subset L^1(\mathbb{R})$.

In order to extend PSF to the whole of L¹(R) with $\widehat{f}\in S_1^1$, we have only to note the remarks on PSF in Sec. 2.4, since {f∈F1;f^∈S11} is dense in L¹(R). For example, the convolution $f_\rho:=f*g_\rho$ of f∈L¹(R) with gρ of (2.12) belongs to F1, $\widehat{f_\rho}\in S_1^1$ for all ρ>0, and $\lim_{\rho\to\infty}f_\rho=f$ in L1(R)-norm (Proposition 2).

The equivalence PSF ⇔ GPDF was established in [18] with the PSF-version

$$\sqrt{2\pi}\sum_{k\in\mathbb{Z}}\widehat{f}(x+2k\pi w)=\sum_{k\in\mathbb{Z}}f\left(\frac{k}{w}\right)e^{ikx/w}\quad (x\in\mathbb{R}).$$

whereby the roles of f and $\hat{f}$ have been exchanged, under the additional assumption $\widehat{f}\in BV(\mathbb{R})$.

The present proof is modeled on that of [18], but the lack of the BV-assumption in our case causes some additional arguments.

PSF and ASF

As to the equivalence of PSF and ASF, it was already established in [23, 24], however with different versions of PSF.

Proof of PSF⇒ASF for $f\in F^1\cap S^1_1$. We again apply PSF (1.10) to the function $\widehat{f}\in F^1$, noting that $(\widehat{\widehat{f}}(k))=(f(-k))\in\ell^1(\mathbb{Z})$, giving

$$\sqrt{2\pi}\sum_{k\in\mathbb{Z}}\widehat{f}(v+2k\pi)=\sum_{k\in\mathbb{Z}}f(k)e^{-ikv}\quad a.e.,$$

Hence, recalling (2.7),

$$\frac{1}{\sqrt{2\pi}}\int_{-\pi}^{\pi}\sum_{k\in\mathbb{Z}}\widehat{f}(v+2k\pi)\,dv=\frac{1}{2\pi}\int_{-\pi}^{\pi}\sum_{k\in\mathbb{Z}}f(k)e^{-ikv}e^{ivt}\,dv$$
$$=\frac{1}{2\pi}\sum_{k\in\mathbb{Z}}f(k)\int_{-\pi}^{\pi}e^{-ikv}e^{ivt}\,dv=\frac{1}{2\pi}\sum_{k\in\mathbb{Z}}f(k)\int_{-\infty}^{\infty}\mathrm{rect}(v)e^{-ikv}e^{ivt}\,dv$$
$$=\sum_{k\in\mathbb{Z}}f(k)\,\mathrm{sinc}(t-k)=(S_1f)(t)\qquad(t\in\mathbb{R}),$$

the interchange of summation and integration being justified in view of the absolute and uniform convergence of the series involved.

On the other hand, we have

$$\int_{-\pi}^{\pi}\sum_{k\in\mathbb{Z}}\widehat{f}(v+2k\pi)e^{ivt}\,dv=\sum_{k\in\mathbb{Z}}\int_{(2k-1)\pi}^{(2k+1)\pi}\widehat{f}(v)e^{it(v-2k\pi)}\,dv\qquad(t\in\mathbb{R}),$$

yielding

$$(S_1f)(t)=\frac{1}{\sqrt{2\pi}}\sum_{k\in\mathbb{Z}}e^{-i2k\pi t}\int_{(2k-1)\pi}^{(2k+1)\pi}\widehat{f}(v)e^{ivt}\,dv\qquad(t\in\mathbb{R}).\tag{3.13}$$

But by the Fourier inversion integral (2.6),

$$f(t)=\frac{1}{\sqrt{2\pi}}\int_{\mathbb{R}}\widehat{f}(v)e^{ivt}\,dv=\frac{1}{\sqrt{2\pi}}\sum_{k\in\mathbb{Z}}\int_{(2k-1)\pi}^{(2k+1)\pi}\widehat{f}(v)e^{ivt}\,dv\quad(t\in\mathbb{R}).\tag{3.14}$$

Subtracting (3.13) from (3.14) yields $f(t)-(S_1f)(t)=(R_1f)(t)$, thus the approximate sampling theorem.

Proof of ASF⇒PSF Noting (3.14) it follows immediately from ASF with remainder (1.2) that for f∈F^2 and w=1

$$\sum_{k\in\mathbb{Z}}f(k)\,\mathrm{sinc}(t-k)=\frac{1}{\sqrt{2\pi}}\sum_{k\in\mathbb{Z}}e^{-i2k\pi t}\int_{(2k-1)\pi}^{(2k+1)\pi}\widehat{f}(v)e^{ivt}\,dv\quad(t\in\mathbb{R}).\tag{3.15}$$

Using (2.5) and (2.7) one can rewrite the terms on the right-hand side as a convolution product, namely

$$\frac{e^{-i2k\pi t}}{\sqrt{2\pi}}\int_{(2k-1)\pi}^{(2k+1)\pi}\widehat{f}(v)e^{itv}\,dv=\frac{e^{-i2k\pi t}}{\sqrt{2\pi}}\int_{\mathbb{R}}\widehat{f}(v)\,\mathrm{rect}(v-2k\pi)e^{itv}\,dv$$
$$=\int_{\mathbb{R}}f(u)e^{-i2k\pi u}\,\mathrm{sinc}(t-u)\,du=\sqrt{2\pi}\big[(f(\cdot)e^{-i2k\pi\cdot})*\mathrm{sinc}(\cdot)\big](t)\quad(t\in\mathbb{R}).$$

For the following, we first assume f∈$\widehat{B}^1_\sigma$ for some σ>0. Then the terms on the right of (3.15) vanish for $|k|>\sigma_0:=(\sigma/\pi+1)/2$, and (3.15) can be rewritten as

$$\sum_{k\in\mathbb{Z}} f(k)\,\mathrm{sinc}(t-k) = \sqrt{2\pi}\sum_{|k|\le\sigma_0}\left[(f(\cdot)e^{-i2k\pi\cdot}) * \mathrm{sinc}(\cdot)\right](t) \quad (t\in\mathbb{R}).$$

Now, taking the L^2-Fourier transform of this equation, noting that the series on the left converges with respect to $L^2(R)$-norm (Lemma 3), Proposition 1 and (2.8), we obtain

$$\frac{1}{\sqrt{2\pi}}\sum_{k\in\mathbb{Z}} f(k)\,\mathrm{rect}(v)e^{ikv} = \sum_{|k|\le\sigma_0}\widehat{f}(v+2k\pi)\,\mathrm{rect}(v) \quad a.e. \tag{3.16}$$

Since $f\in\widehat{B}^1_\sigma$, $\widehat{f}$ can be regarded as the L^1-Fourier transform, which is continuous. Then (3.16) holds even for every v in a neighborhood of the origin, since all functions involved are continuous there. So one may take v=0 to obtain

$$\frac{1}{\sqrt{2\pi}}\sum_{k\in\mathbb{Z}} f(k) = \sum_{|k|\le\sigma_0}\widehat{f}(2k\pi).$$

Replacing now $f(\cdot)$ by $2\pi f(x+2\pi\cdot)$, and noting Proposition 1 (a) gives PSF.

It remains to remove the restriction $f\in\widehat{B}^1_\sigma$ for some $\sigma>0$. According to Sect. 2.4 it suffices to show that $\bigcup_{\sigma>0}\widehat{B}^1_\sigma$ is dense in $L^1(R)$. To see this, one may take $f*\chi_\rho$, $\rho>0$, with $f\in L^1(R)$ and χ_ρ of (2.14). This convolution belongs to $\widehat{B}^1_{2\pi\rho}$, noting (2.15) and Proposition 1 (b), and $\lim_{\rho\to\infty} f*\chi_\rho = f$ in $L^1(R)$-norm (Proposition 2 (b).

PSF AND THE FUNCTIONAL EQUATION FOR Z(S)

First some remarks concerning the ζ-function. Since the series (1.15) is locally uniformly convergent in the half-plane $\sigma>1$, ζ(s) is analytic there. To extend it to a function which is analytic in the whole complex plane C except s=1, one can employ the representation

$$\zeta(s) = s\int_1^\infty u^{-s-1}\left(\mu(u)+\frac{1}{2}\right)du + \frac{1}{s-1}+\frac{1}{2} \tag{4.1}$$

with the measure $\mu(u):=\lfloor u\rfloor-u$, where $\lfloor u\rfloor$ denotes the largest integer less than or equal to u; see Titchmarsh [82, (2.1.4)].

Noting that $\left|\int_1^u(\mu(t)+\frac{1}{2})\,dt\right|$ is bounded by 1/8 on R, one can easily show by partial integration that the integral is locally uniformly convergent in the half-plane $\sigma>-1$. Hence (4.1) defines an extension of (1.15) to a function which is analytic for $\sigma>-1$ except for the point s=1, where there is a simple pole with residue 1. This function can be continued further to a function which is analytic in the whole complex plane except for s=1.

From (4.1) one can also deduce by partial integration the estimate

$$|\zeta(s)| \le |s| + \frac{1}{|s-1|}+\frac{1}{2} \quad (s=\sigma+i\tau\in\mathbb{C}\backslash\{1\},\ \sigma>-1). \tag{4.2}$$

In the strip $0<\sigma<1$ the representation (4.1) can be rewritten as (see Tichmarsh (2.1.5))

$$\zeta(s) = s\int_0^\infty u^{-s-1}\mu(u)du \quad (s=\sigma+i\tau,\ 0<\tau<1), \tag{4.3}$$

noting that

$$\int_0^1 u^{-s-1}\mu(u)du = \int_0^1 u^{-s}du = \frac{1}{s-1}.$$

Now we turn to the equivalence PSF ⇔ FERZ.

Proof of PSF⇒FERZ If f belongs to the Schwartz space S of rapidly decreasing functions, the PSF (1.10), in its simplest form for w=1, x=0, reads

$$\sqrt{2\pi}\sum_{k\in\mathbb{Z}} f(2k\pi) = \sum_{k\in\mathbb{Z}}\widehat{f}(k).$$

In fact, under the given assumption $f\in S$, Eq. (1.10) holds for all $x\in R$, since both series are uniformly convergent, at least on compact subsets of R, and hence define continuous functions.

If f is even, so will be $\widehat{f}$, and PSF turns into

$$\sqrt{2\pi}\left\{\sum_{k=1}^{\infty} f(2\pi k) - \int_0^{\infty} f(2\pi u)\,du\right\} = \int_0^{\infty} \widehat{f}(v)\,dv - \sum_{k=1}^{\infty} \widehat{f}(k),$$

noting that $\frac{1}{2}\widehat{f}(0) = (1/\sqrt{2\pi})\int_0^{\infty} f(x)\,dx = \sqrt{2\pi}\int_0^{\infty} f(2\pi u)\,du$ and, by the Fourier inversion formula (2.6), $\frac{1}{2}\sqrt{2\pi}f(0) = \int_0^{\infty}\widehat{f}(v)\,dv$

This equation can be conveniently be written as

$$\sqrt{2\pi}\int_0^{\infty} f(2\pi u)\,d\mu(u) = \int_0^{\infty}\widehat{f}(v)\,d\mu(v), \tag{4.4}$$

Further, as f∈S implies $f', \widehat{f}' \in L^1(\mathbb{R})$ and $\lim_{u\to\infty} f(u) = \lim_{v\to\infty}\widehat{f}(v) = 0$,, the two integrals can be converted into ordinary Lebesgue integrals by partial integration, namely,

$$\begin{aligned}\sqrt{2\pi}\int_0^{\infty} f(2\pi u)\,d\mu(u) &= \sqrt{2\pi}f(2\pi u)\mu(u)\Big|_0^{\infty} - (2\pi)^{3/2}\int_0^{\infty}\mu(u)f'(2\pi u)\,du\\ &= -(2\pi)^{3/2}\int_0^{\infty}\mu(u)f'(2\pi u)\,du,\end{aligned}$$

and similarly,

$$\int_0^{\infty}\widehat{f}(v)\,d\mu(v) = -\int_0^{\infty}\mu(v)\widehat{f}'(v)\,dv.$$

Hence we obtain the following "integrated form" of PSF,

$$(2\pi)^{3/2}\int_0^{\infty}\mu(u)f'(2\pi u)\,du = \int_0^{\infty}\mu(v)\widehat{f}'(v)\,dv. \tag{4.5}$$

For the Gauß–Weierstraß kernel g_ρ of (2.12), (2.13) with $\rho=\pi t^2$, $t>0$, this identity reads,

$$-2\pi t^{-3}\int_0^{\infty} u\,\mu(u)e^{-\pi t^{-2}u^2}\,du = -2\pi t^2\int_0^{\infty} v\,\mu(v)e^{-\pi t^2 v^2}\,dv. \tag{4.6}$$

Now we multiply both sides of (4.6) by $-t^{-\sigma}$, $\sigma>0$, integrate with respect to t from 0 to ∞, and interchange the order of integration, which is allowed by the Fubini–Tonelli theorem [40, (21.13)]. This yields for the left-hand side by a simple change of variables

$$2\pi\int_0^{\infty} u\,\mu(u)\int_0^{\infty} t^{-3-\sigma}e^{-\pi t^{-2}u^2}\,dt\,du = \frac{1}{2}\pi^{-\sigma/2}\Gamma\left(\frac{\sigma}{2}\right)\sigma\int_0^{\infty} u^{-\sigma-1}\mu(u)\,du,$$

and for the right-hand side

$$\begin{aligned}&2\pi\int_0^{\infty} v\,\mu(v)\int_0^{\infty} t^{2-\sigma}e^{-\pi t^2v^2}\,dt\,dv\\ &= \frac{1}{2}\pi^{-\frac{1-\sigma}{2}}\Gamma\left(\frac{1-\sigma}{2}\right)(\sigma-1)\int_0^{\infty} v^{\sigma-2}\mu(v)\,dv.\end{aligned}$$

Using the representation (4.3) for the ζ-function, the functional equation follows for $s=\sigma$ with $0<\sigma<1$, and analytic continuation yields the general result.

Proof of FERZ⇒PSF It is well-known that an infinite series can be transformed into an integral involving ζ(s) (as well as vice versa!) using Mellin transform methods; see e. g. [19]. Formally one proceeds as follows: To sum

$$g(x) := \sum_{k=1}^{\infty} f(kx) \quad (x \in \mathbb{R}_+),$$

one first applies the Mellin transform

$$\mathcal{M}(f)(s) \equiv \widehat{f}_M(s) := \int_0^{\infty} f(u)u^{s-1}\,du \quad (s = \sigma + i\tau \in \mathbb{C})$$

on both sides to yield

$$\widehat{g}_M(s) = \sum_{k=1}^{\infty}\frac{1}{k^s}\widehat{f}_M(s) = \widehat{f}_M(s)\zeta(s), \tag{4.7}$$

since the Mellin transform of f(kx), where k>0, is $k^{-s}\widehat{f}_M(s)$. The Mellin inversion formula

$$f(x) = \frac{1}{2\pi i}\int_{c-i\infty}^{c+i\infty} \widehat{f_M}(s)x^{-s}\,ds \quad (c > 0)$$

would then lead to

$$g(x) = \sum_{k=1}^{\infty} f(kx) = \frac{1}{2\pi i}\int_{c-i\infty}^{c+i\infty} \widehat{f_M}(s)\zeta(s)x^{-s}\,ds \quad (c > 1). \tag{4.8}$$

In our situation, to work with the sum (4.8) we need to move the integration contour from c+it to 1−c+it, −∞<t<+∞. The equation for x=2π then becomes

$$\sum_{k=1}^{\infty} f(2k\pi) = \frac{1}{2\pi i}\int_{1-c-i\infty}^{1-c+i\infty} \widehat{f_M}(s)\zeta(s)(2\pi)^{-s}\,ds + S,$$

where S accounts for the sum of the residues of the integrand due to the contour shift. The change of variable s↦1−s in the integral causes the appearance of a term ζ(1−s) in the integrand. The functional equation is used at this point, to make the term ζ(s) reappear. But, as Eq. (4.8) shows, the inverse Mellin transform of a product involving ζ(s) represents a series ∑h(k), for some function h. This function will turn out to be the Fourier transform of f, so that

$$\sum_{k=1}^{\infty} f(2k\pi) = \sum_{k=1}^{\infty} f^{\wedge}(k) + S.$$

A computation will show that the residues are $R_0 = -\frac{1}{2}f(0)$ and $R_1 = \frac{1}{2\pi}\int_0^{\infty}$ f(x)dx, and PSF will then follow.

To carry out the above sketch of the proof and to justify these steps, consider a locally integrable function f(x). Let it be O(x−a) as x→0 and O(x−b) as x→∞ for some a<b. The Mellin transform $\widehat{f_M}(s)$ then exists for any complex s=σ+iτ in the fundamental strip a<σ<b. This follows from

$$\left|\int_0^{\infty} f(u)u^{s-1}du\right| \le \int_0^1 |f(u)|u^{\sigma-1}du + \int_1^{\infty} |f(u)|u^{\sigma-1}du$$
$$\le A\int_0^1 u^{-a+\sigma-1}du + B\int_1^{\infty} u^{-b+\sigma-1}du,$$

since the first integral exists if σ>a and the second if σ<b. If, in addition, f(y) is of bounded variation in a neighborhood of y=x, the Mellin inversion formula holds, in the form

$$\frac{f(x^+) + f(x^-)}{2} = \frac{1}{2\pi i}\int_{\sigma-i\infty}^{\sigma+i\infty} \widehat{f_M}(s)x^{-s}\,ds,$$

for a<σ<b. To justify it one may appeal to the relation between the Mellin and the Fourier transform, which is revealed by the change of variables x=e−y, since, as we have seen, $f(x)x^{\sigma-1} \in L^1(0,\infty)$under the stated hypothesis.

Now, if the intersection of the fundamental strip a<σ<b with the half-plane σ>1 is nonempty, and if s belongs to that intersection, then (4.7) holds because there exists under the stated conditions

$$\sum_{k=1}^{\infty}\left|\frac{1}{k^s}\right|\int_0^{\infty} |u^{s-1}f(u)|du.$$

If g is a continuous function of bounded variation in a neighbourhood of x, the application of the Mellin inversion formula to (4.7) then leads to the identity (4.8), that is,

$$g(x) = \sum_{k=1}^{\infty} f(kx) = \frac{1}{2\pi i}\int_{\sigma-i\infty}^{\sigma+i\infty} \widehat{f_M}(s)\zeta(s)x^{-s}\,ds.$$

To continue with the proof, let f be an even function in the Schwartz space S. Then, the Mellin transform of f exists for all s=σ+it with σ>0. Below we will have to integrate $\widehat{f_M}(s)$ in a region contained in the left semiplane, so we need to investigate its analytic extension there. The necessary information can be obtained through an integration by parts of the Mellin integral,

$$\widehat{f_M}(s) = -\frac{1}{s}\int_0^{\infty} f'(x)x^s\,dx = -\frac{1}{s}\widehat{f'}_M(s+1).$$

The procedure can be repeated to give

$$\widehat{f}_M(s) = \frac{(-1)^r}{s(s+1)\cdots(s+r-1)}\widehat{[f^{(r)}]}_M(s+r) \quad (r \in \mathbb{N}). \tag{4.9}$$

This provides the desired analytic extension and shows that $\widehat{f}_M(s)$ may have simple poles at the non-positive integers. The residue at s=0s=0 is given by

$$\lim_{s\to 0} s\left(-\frac{1}{s}\widehat{f'}_M(s+1)\right) = -\widehat{f'}_M(1) = -\int_0^\infty f'(x)\,dx = f(0). \tag{4.10}$$

Although we will not need any other residues, one could easily check that the residue at $s=-n$ is $\frac{1}{n!}f^{(n)}(0)$.

We are now ready to return to the fundamental transformation (4.8) with x=2π. We take an arbitrary 1<c<2 and change the contour,

$$\frac{1}{2\pi i}\int_{c-i\infty}^{c+i\infty}\widehat{f}_M(s)\zeta(s)(2\pi)^{-s}\,ds = \frac{1}{2\pi i}\int_{1-c-i\infty}^{1-c+i\infty}\widehat{f}_M(s)\zeta(s)(2\pi)^{-s}\,ds + S,$$

where S is the sum of the residues of $\widehat{f}_M(s)\zeta(s)(2\pi)^{-s}$ at the poles in the strip 1−c<σ<c. There are two such poles: one at s=1, due to ζ(s), and another at s=0, due to $\widehat{f}_M(s)$. The residue of the integrand at the former is

$$R_1 = \frac{1}{2\pi}\widehat{f}_M(1) = \frac{1}{2\pi}\int_0^\infty f(x)\,dx$$

and at the latter

$$R_0 = f(0)\zeta(0) = -\frac{1}{2}f(0),$$

in view of (4.10). As a result, (4.8) becomes

$$\sum_{k=1}^{\infty} f(2k\pi) = \frac{1}{2\pi i}\int_{1-c-i\infty}^{1-c+i\infty}\widehat{f}_M(s)\zeta(s)(2\pi)^{-s}\,ds + \frac{1}{2\pi}\int_0^\infty f(x)\,dx - \frac{1}{2}f(0).$$

We now replace s↦1−s in the integral

$$\sum_{k=1}^{\infty} f(2k\pi) = \frac{1}{2\pi}\int_0^\infty f(x)\,dx - \frac{1}{2}f(0) + \frac{1}{2\pi i}\int_{c-i\infty}^{c+i\infty}\widehat{f}_M(1-s)\zeta(1-s)(2\pi)^{s-1}\,ds$$

and use the functional equation in the form (1.16) to deduce

$$\sum_{k=1}^{\infty} f(2k\pi) \tag{4.11}$$

$$= \frac{1}{2\pi}\int_0^\infty f(x)\,dx - \frac{1}{2}f(0) + \frac{1}{2\pi i}\int_{c-i\infty}^{c+i\infty}\widehat{f}_M(1-s)2\cos\left(\frac{\pi s}{2}\right)\Gamma(s)\zeta(s)\,ds. \tag{4.12}$$

According to the principle expressed by (4.8), the last integral represents the sum of a series,

$$\frac{1}{2\pi i}\int_{c-i\infty}^{c+i\infty}\widehat{f}_M(1-s)2\cos\left(\frac{\pi s}{2}\right)\Gamma(s)\zeta(s)\,ds = \sum_{k=1}^{\infty} h(k). \tag{4.13}$$

The question is: what is the relation between h and f? The elementary Mellin transform formula

$$\int_0^\infty \cos(ax)x^{s-1}\,dx = a^{-s}\cos\left(\frac{\pi s}{2}\right)\Gamma(s) \qquad (0 < \Re s < 1) \tag{4.14}$$

shows that

$$\begin{aligned}\int_0^\infty \widehat{f}(v)v^{s-1}\,dv &= 2\int_0^\infty\left(\frac{1}{\sqrt{2\pi}}\int_0^\infty f(t)\cos(vt)\,dt\right)v^{s-1}\,dv\\ &= \frac{2}{\sqrt{2\pi}}\int_0^\infty f(t)\left(\int_0^\infty \cos(vt)v^{s-1}\,dv\right)dt\\ &= \frac{2}{\sqrt{2\pi}}\cos\left(\frac{\pi s}{2}\right)\Gamma(s)\int_0^\infty f(t)t^{-s}dt\\ &= \frac{2}{\sqrt{2\pi}}\cos\left(\frac{\pi s}{2}\right)\Gamma(s)\widehat{f}_M(1-s).\end{aligned} \tag{4.15}$$

In order to justify the interchange of the integration order, first note that

$$\begin{aligned} I &:= \int_0^\infty f(t)\left(\int_0^\infty \cos(vt)v^{s-1}\,dv\right)dt \\ &= \int_0^\infty f(t)\lim_{R\to\infty}\left(\int_0^R \cos(vt)v^{s-1}\,dv\right)dt \\ &= \lim_{R\to\infty}\int_0^\infty f(t)\left(\int_0^R \cos(vt)v^{s-1}\,dv\right)dt. \end{aligned} \tag{4.16}$$

Here the interchange of the limit with the outer integral is justified by Lebesgue's dominated convergence theorem, since

$$\begin{aligned} \left|f(t)\int_0^R \cos(vt)v^{s-1}\,dv\right| &= \left|t^{-s}f(t)\int_0^{Rt}\cos(u)u^{s-1}\,du\right| \\ &\le c(s)t^{-\sigma}|f(t)| \in L^1(0,\infty), \end{aligned}$$

where the constant c(s)c(s) is an upper bound for the continuous and bounded function $x \mapsto \left|\int_0^x \cos(u)u^{s-1}\,du\right|$ on [0,∞). Furthermore, the order of integration in the last line of (4.16) can be interchanged by Fubini's theorem, and one obtains

$$I = \lim_{R\to\infty}\int_0^R\left(\int_0^\infty f(t)\cos(vt)dt\right)v^{s-1}\,dv,$$

which proves the desired interchange of integrals in (4.15).

Hence we have shown that

$$\int_0^\infty \widehat{f}(v)v^{s-1}\,dv = \frac{2}{\sqrt{2\pi}}\cos\left(\frac{\pi s}{2}\right)\Gamma(s)\widehat{f}_M(1-s). \tag{4.17}$$

Since we have used the elementary formula (4.14) to deduce this equation, it holds at first for 0<Rs<1. On the other hand, both sides of (4.17) define analytic functions for Rz>0, and hence (4.17) is even valid for Rz>0.

Now, equation (4.17) reveals that the factor that multiplies ζ(s) in the integrand of (4.13) is the Mellin transform of the Fourier transform $\hat{f}$ of f. Thus (4.13) becomes

$$\frac{1}{2\pi i}\int_{c-i\infty}^{c+i\infty}\widehat{f}_M(1-s)2\cos\left(\frac{\pi s}{2}\right)\Gamma(s)\zeta(s)\,ds = \frac{1}{\sqrt{2\pi}}\sum_{k=1}^\infty \widehat{f}(k).$$

Replacing it in (4.11) finally yields

$$\sum_{k=1}^\infty f(2k\pi) = \frac{1}{2\pi}\int_0^\infty f(x)\,dx - \frac{1}{2}f(0) + \frac{1}{\sqrt{2\pi}}\sum_{k=1}^\infty \widehat{f}(k).$$

Noting that $\frac{1}{2\pi}\int_0^\infty f(x)\,dx = \frac{1}{2}\frac{1}{\sqrt{2\pi}}\widehat{f}(0)$, the PSF (1.10) for even functions in the Schwarz space S and x=0 follows.

To extend this to arbitrary functions f∈S, one decomposes f in the form f(t)= $\frac{1}{2}[f(t)+f(-t)]+\frac{1}{2}[f(t)-f(-t)]$, i. e., in an even and an odd part. Since both sums in (1.10) vanish for odd functions and x=0, PSF for f∈S and x=0 follows. To obtain PSF for arbitrary x∈R, one replaces f(·) by 2πf(x+2π·) as in the proof of ASF⇒PSF. Finally, one uses the fact that SS is dense in L^1(R) to deduce the general PSF for f∈L^1(R); cf. Sect. 2.4.

Let us remark that a proof of PSF yields FERZ is to be found in Mordell [57], and the converse in Ferrar [30]. In his treatise Titchmarsh [82] presented seven different methods of proof of the functional equation; see also Kahane and Mandelbrojt [50], Bellman [8], Ivić [49], Patterson [66, Chapter 2], Karatsuba and Voronin [51], Rooney [75], Flajolet et al. [31], Brüdern [13, pp. 58–66], Newman [63]. Latter proofs are Knopp and Robins [55], Schuster [78], Murty [58] and Higgins [45].

Whereas several of the above mentioned books or papers contain a proof of PSF ⇒ FERZ, proofs of the converse FERZ ⇒ PSF, an important part of our procedure, are rarer. In fact, there are other proofs of FERZ ⇔ PSF, but ours are fully detailed and more than just polished versions of others. Especially the delicate interchanges of the orders of summation and/or integration needed in the proofs are justified in great detail.

In his exhaustive report [44] on four papers of Hamburger [33–36] of 1921–1922 in connection with Riemann's functional equation, R. Higgins gave a complete review of Hamburger's collection of five equivalent results, namely that Riemann's functional equation for the zeta function, Jacobi's transformation formula for the elliptic theta function, the partial fractions expansion for the cotangent function, Poisson's summation formula and a special Fourier series are all equivalent. He also discussed these in the light of later contributions.

Here the basic earlier work being the "Lehrbuch der Thetafunktionen" by Krazer [56], papers by Siegel [79], those of Mordell [57] and Ferrar [30], as well as more results in the light of work by Doetsch [27] and Klusch [52, 53]. Klusch proves many implications, and collects others, between well-known results in signal analysis, number theory and applied mathematics. In particular, he gives a direct proof of FERZ implies ASF, an open question at the time. These results do not seem to be at all well known.

A PROOF OF ASF

In the previous section we have shown that the formulae of Sects. 1 or 3 can all be deduced from each other. Hence it suffices to prove one of them in order to verify all. We will now prove ASF under the weaker hypothesis that f∈F^2 and with a weaker conclusion on the convergence of the series. From that result, the statement (1.6) for $f \in F^2 \cap S^1_w$ given in Sect. 1 is easily deduced.

Theorem 1

For w>0 and f∈F^2, we have

$$f(t) = \sum_{k\in\mathbb{Z}} f\left(\frac{k}{w}\right) \operatorname{sinc}(wt-k) + R_w f(t),$$

where the series converges pointwise as the symmetric sum.

Proof

It suffices to consider the case w=1. Employing the Fourier inversion theorem for f∈F^2, see [20, p. 214, Proposition 5.2.16], we may rewrite the remainder (1.2) as

$$\begin{aligned} R_1 f(t) &= \frac{1}{\sqrt{2\pi}}\left[\int_{\mathbb{R}} \widehat{f}(v)e^{ivt}\,dt - \sum_{k\in\mathbb{Z}} \int_{(2k-1)\pi}^{(2k+1)\pi} \widehat{f}(v)e^{it(v-2\pi k)}\,dv\right] \\ &= f(t) - \frac{1}{\sqrt{2\pi}}\int_{\mathbb{R}} \widehat{f}(v)g(t,v)\,dv, \end{aligned} \tag{5.1}$$

where g(t,·) is the function obtained by restricting e^{it} to the interval [−π,π) and extending it to R by 2π periodic continuation. By a simple calculation we obtain the Fourier expansion

$$g(t,v) = \sum_{k\in\mathbb{Z}} \operatorname{sinc}(t-k)e^{ikv}. \tag{5.2}$$

Since g(t,·) is of bounded variation, the Fourier series converges and (5.2) holds at each point of continuity. Moreover, the partial sums

$$g_N(t,v) := \sum_{|k|\le N} \operatorname{sinc}(t-k)e^{ikv}$$

are uniformly bounded, that is

$$|g_N(t,v)| \le C \qquad (v\in\mathbb{R}, N\in\mathbb{N})$$

for some constant C; see [85, p. 90, Thm. 3.7]. Since $\widehat{f} \in L^1(\mathbb{R})$, we see that $\widehat{f}(\cdot)g_N(t,\cdot)$ has an absolutely integrable majorant for all N∈N. Therefore, Lebesgue's dominated convergence theorem allows us to conclude that

$$\begin{aligned} \frac{1}{\sqrt{2\pi}}\int_{\mathbb{R}} \widehat{f}(v)g(t,v)\,dv &= \lim_{N\to\infty} \frac{1}{\sqrt{2\pi}}\int_{\mathbb{R}} \widehat{f}(v)g_N(t,v) \div \\ &= \lim_{N\to\infty} \sum_{|k|\le N} \left\{\frac{1}{\sqrt{2\pi}}\int_{\mathbb{R}} \widehat{f}(v)e^{ikv}\,dv\right\} \operatorname{sinc}(t-k) \\ &= \lim_{N\to\infty} \sum_{|k|\le N} f(k)\operatorname{sinc}(t-k), \end{aligned}$$

where again the Fourier inversion theorem (2.6) has been used in the last step. Now the proof is completed by substituting the last expression in (5.1).

Let us mention that in the meantime some of the authors have shown that ASF is equivalent to the classical sampling theorem (1.7); see [14]. The above proof of ASF is, on the other hand, a Fourier analytic proof, fully independent of CSF and all the theorems of this paper.

EPILOGUE: SOME POSSIBLE GENERALIZATIONS

The natural question arises as to whether the results above can be extended to higher dimensions or other settings. There are (at least) two ways : via functional analysis, based on r.k. theory, and via abstract harmonic analysis, using locally compact abelian (LCA) groups.

In the first of these generalizations, sampling itself is placed in the more general context of reproducing kernel Hilbert spaces of functions defined on an abstract set. In the second, the real line RR, which can be thought of as time domain, is replaced by an abstract LCA topological group GG, as usual assumed to be Hausdorff. Functions are now defined on a LCA group, itself a fruitful generalization of the real line.

Reproducing Kernel Theory

The purpose of this section is to give some very brief and necessarily quite simplified background to Hilbert spaces with reproducing kernel. The material in this section, proofs etc., can be found in [42, 77]. There are several studies in the literature of sampling to be found in the general area of reproducing kernel spaces, and we can ask whether results of the present paper might be given more general forms in that theory. For example, Nashed and his associates have given sampling theorems in certain Hilbert, Banach, Sobolev and translation invariant spaces, and addressed such probems as the construction of reproducing kernel Hilbert and Banach spaces with a given sampling set (see, e. g., [59]). Much of this work is quite technical and uses, e. g., iterative methods. Saitoh's theory of linear transformations of Hilbert space offers another possibility for this kind of extension (see, e. g., [77]).

Hilbert Spaces with Reproducing Kernel

Let KK be a separable Hilbert space with inner product denoted by ⟨·,·⟩K and norm denoted by ‖·‖K.

Definition 1

Let KK consist of complex valued functions defined on E, an abstract set. Then K is said to have *reproducing kernel* if there exists a function k:E×E↦C, such that k(·,t)∈K for every t∈E, and the *reproducing equation*

f(t)=⟨f,k(·,t)⟩K

holds for every f∈ K . Such a space is called a *reproducing kernel Hilbert space (RKHS)* and k(s,t) is its *reproducing kernel.*

A Summary of the Relevant Saitoh Theory

We now quote that part of Saitoh's fundamental theory of linear transformations of Hilbert space which provides a suitable setting for the present discussion ([77, Ch. 2]), but first some background is needed.

Let H be a separable Hilbert space with inner product denoted by ⟨·,·⟩H, and for each t belonging to an abstract set E, let κt be a mapping of E into H. Then k(s,t):=⟨κt,κs⟩H is defined on E×E and is called the *kernel function* of the map κt. This kernel function is a *positive matrix* [77, Ch. 2, Sect. 2] and as such it determines one and only one Hilbert space for which it is the reproducing kernel. This Hilbert space is denoted by Rk; it turns out to be the set of images of H under the transformation

$$(\mathcal{L}\varphi)(t) := \langle \varphi, \kappa_t \rangle_{\mathfrak{H}}, \tag{6.1}$$

and has reproducing kernel k(s,t)=(Lκt)(s). This situation is governed by:

Theorem 2

(Saitoh) With the notations established above, Rk is a Hilbert space which has the reproducing kernel k(·,·), and is uniquely determined by this kernel. We have, for f∈Rk,

$$\|f\|_{R_k} = \|\mathcal{L}\omega\|_{R_k} \leq \|\omega\|_{\mathfrak{H}}, \tag{6.2}$$

and there exists a unique member, ω_0 say, of the class of all ω's satisfying (6.2) such that

$$f(t) = \langle \omega_0, \kappa_t \rangle_{\mathfrak{H}}, \qquad (t \in E),$$

and

$$\|f\|_{R_k} = \|\omega_0\|_{\mathfrak{H}}.$$

The reproducing equation for Rk is

$$f(t) = \langle f, k(\cdot, t) \rangle. \tag{6.3}$$

It is often supposed that {κt}, (t∈E) is complete in H, a rather mild restriction. This means that the only possible ω in (6.2) is ω_0, because from (6.1) the null space of L is {θ}. Then L:H↦Rk is an isometry and therefore bounded. It is clearly linear, one-to-one and 'onto'. Hence by the bounded inverse theorem, L^{-1} is bounded. These properties show that L is an isometric isomorphism of HH onto Rk.

Possibilities for Extension

The r.k. theory outlined above is in some ways more general and in some ways less general than the LCA group setting. It is tied to Hilbert space methods and no extension beyond these methods is known to the authors.

However, in order to find some process of extension beyond Hilbert space there might be a way forward if a regime such as the following could be established. To describe such a regime we will adopt the following notations.

Suppose that X is a linear space, and let X^{ex} denote a linear space that extends X; that is, X is a subspace of X^{ex}. Again, suppose that P denotes a proposition associated with X; it might be, for example, the assertion that a certain formula holds for all members of X. Let P^{ex} denote an extension of P, that is, a proposition associated with X^{ex} so chosen that it reduces to P for members of X.

With these notations, the following extension procedure can be suggested for the r.k. theory. Let Hex denote a linear space which extends H in such a way as to extend the domain of L, thus, L on this larger domain induces an extension $R^{ex}k$ of R_k. Let P denote a proposition associated with R_k, and P^{ex} a proposition associated with $R^{ex}k$, asserting that PP holds for Re but only in an approximate form when associated with $R^{ex}_k \setminus R_k$.

A scheme of this kind might afford some way of generalizing results found in the main body of the paper, such as the passage from the classical to the approximate sampling theorem (1.7)/(1.6), or the reproducing equation (1.8) to the approximate reproducing equation (1.9). But the construction of such a scheme is an open problem, no doubt a rather difficult one.

Table 1: A classical-reproducing kernel theory dictionary

Classical	r.k. theory
$\mathbb{R}$	E, an abstract set
$L^2(-\pi,\pi)$	$\mathfrak{H}$, a separable Hilbert space
$(2\pi)^{-1/2}e^{-it\cdot}\chi_{[-\pi,\pi]}(\cdot)$	$\kappa_t \in \mathfrak{H}$, $(t \in E)$
reproducing kernel: $\mathrm{sinc}(s-t)$	$k(s,t) = \langle \kappa_t, \kappa_s \rangle_{\mathfrak{H}}$
$\mathcal{F}^{-1}$, inverse Fourier transform	$\mathcal{L}$
$\hat{B}^p_\sigma$	R_k
reproducing equation $f(t) = \langle f, \mathrm{sinc}(\cdot - t)\rangle$	$f(t) = \langle f, k(\cdot,t)\rangle_{R_k}$

A Dictionary

The question arises as to whether the ASF of (1.6) and ARKF of (1.9), introduced in this paper, as well as GPDF of (1.1) can be built into the more recent reproducing kernel Banach space theory (see e. g. [1, 32, 37, 47, 60, 61]), since the Hilbert space approach may not suffice. Its *non-constructed* but practical applications would be precisely the present six theorems under discussion. Perhaps an extension of the Banach space theory may also be necessary. It is seemingly a wide open field (Table 1).

Abstract Harmonic Analysis

In abstract harmonic analysis, the real line in classical Fourier analysis is replaced by an LCA group. A very brief description is now given, following Rudin's notation and terminology in [76]. Further details can also be found in [4, 25, 38, 39, 73] and briefer accounts are in [6, 7, 28]. The website [65] of the Numerical Harmonic Analysis Group (NuHAG) at the University of Vienna is a comprehensive and up-to-date source of information about all aspects of sampling theory.

Locally Compact Abelian Groups

The property of being an LCA group is often preserved under processes that are subject to natural technical restrictions. Thus closed subgroups of an LCA group are also LCA, as are quotient groups of closed subgroups and so on. LCA groups enjoy a translation invariant measure m_G, called Haar measure, which is unique up to a multiplicative constant (for the real line, the Haar measure is the familiar Lebesgue measure). As with Lebesgue measure, Haar measure gives rise to an integral $\int_G f(x)dm_G(x) = \int_G f$; the notation $f \in L^p(G)$ is used if $\int_G |f|^p < \infty, p \geq 1$. As usual the statement that a property holds for (Haar) almost all points in a subset of G means that it holds for all points in the set except for a set of (Haar) measure 0.

The Abstract Fourier Transform

The exponential function $t \mapsto e^{iut}$ is replaced by a continuous homomorphism or *character* $\gamma: G \to S1$, which takes values on the unit circle. Under pointwise multiplication, these homomorphisms form an abelian group $G\wedge := \Gamma$ which is locally compact with the compact-open topology. Greek letters such as γ and λ will be used for elements in Γ, which is called the *dual* group of GG and corresponds to the frequency domain. The dual of Γ is isomorphic to G. This duality allows the value $\gamma(x)$, $x \in G$, of a character γ to be written as an ordered pair (x,γ), where $(x,\gamma)(x',\gamma)=(x+x',\gamma)$ and similarly for γ.

The Fourier transform $\widehat{f}: \Gamma \to \mathbb{C}$ of the function $f \in L^1(G)$, defined by

$$\widehat{f}(\gamma) = \int_G f(x)(x, -\gamma)dm_G(x), \tag{6.4}$$

is continuous and vanishes at ∞. For each ψ in $L^1(\Gamma)$, the inverse Fourier transform function $\psi^\vee: G \to C$ is defined by

$$\psi^\vee(x) = \int_\Gamma \psi(\gamma)(x, \gamma)dm_\Gamma(\gamma). \tag{6.5}$$

The Haar measure $m\Gamma$ on Γ can be normalized so that the inversion formula

$$f(x) = \int_\Gamma \widehat{f}(\gamma)(x, \gamma)dm_\Gamma(\gamma) \tag{6.6}$$

holds almost always for suitable $f = \psi^\vee$ (see [76, Sect. 1.5] or [39, Thm. 31.17]).

These ideas are the basis of a beautiful and broad abstract analogue of classical Fourier analysis that retains the principal results of classical theory. Thus the abstract Fourier–Plancherel transform can be defined on $L^2(G)$ and analogues of the classical Fourier–Plancherel theorem $(\int_G |f|^2 = \int_\Gamma |f^\wedge|^2)$ [39, Sect. 31], [76] and the Parseval theorem hold (here and where appropriate ^^ is the Fourier–Plancherel transform). The abstract Poisson summation formula also holds under certain integrability conditions [76].

The general and unifying framework offered by abstract Fourier analysis includes n-dimensional Euclidean space as an important special case. Some other concrete examples of LCA groups, their duals, measures and characters are provided in Table 2; Zn is the finite additive group of residues mod n and $T^r = R^r/Z^r$ is the r-dimensional torus which we will take to be $[-1/2, 1/2]^r$, the r-fold product of the unit interval with endpoints identified. In particular, $T^1 \cong S^1$, the unit circle. This normalization differs slightly from that adopted in Table 3.

Table 2: Some LCA groups, their duals, measures and characters

Group G	measure m_G	dual Γ	measure m_Γ	character (x, γ)
$\mathbb{R}$	Lebesgue	$\mathbb{R}$	Lebesgue	$e^{iv\xi}$
$\mathbb{R}^r$	Lebesgue	$\mathbb{R}^r$	Lebesgue	$e^{i\mathbf{v}\cdot\boldsymbol{\xi}}$
$\mathbb{Z}^s$	Point measure	$\mathbb{T}^s$	induced Lebesgue	$e^{i\mathbf{k}\cdot\boldsymbol{\xi}}$
$\mathbb{Z}_n^s$	Point measure	$\mathbb{Z}_n^s$	point measure	$e^{i\mathbf{k}\cdot\boldsymbol{\nu}}$

Table 3: A classical-abstract dictionary for sampling theory

Classical	Abstract
Time domain $\mathbb{R}$	LCA group G
Frequency domain $\mathbb{R}$	dual LCA group$G^\wedge = \Gamma$
Lebesgue measure $\lvert\cdot\rvert$	Haar measure m_G
e^{ixu}	character (x, γ)
$\widehat{f}(u) := \int_{\mathbb{R}} f(x)e^{-ixu}dx$	$\widehat{f}(\gamma) := \int_G f(x)(x, -\gamma)dm_G(x)$
$2w\mathbb{Z}$ discrete subgroup of $\mathbb{R}$	Λ discrete subgroup (lattice) of Γ
$\mathbb{R}/(2w\mathbb{Z}) \cong \mathbb{S}^1$	Γ/Λ compact abelian group
$(-\pi w, \pi w]$ transversal of $\mathbb{R}/(2\pi w\mathbb{Z})$	Ω transversal of Γ/Λ
$\mathbb{Z}/w = (2w\pi\mathbb{Z})^\perp$ sampling set	$H = \Lambda^\perp = \{h \in G: (h, \lambda) = 1, \lambda \in \Lambda\}$
$\lvert(-\pi w, \pi w]\rvert = 2\pi w$	$m_\Gamma(\Omega) < \infty$
$\sum_{k\in\mathbb{Z}} g(k/2w)$	$\int_H g(h)dm_H(h) = m_H(\{0\}) \sum_{h\in H} g(h)$
$\widehat{B}^p_{\pi w}$ band-limited signals	$\widehat{B}^p(G)$
F^p includes non band-limited signals	$F^p(G)$
$\ell^p(\mathbb{Z}/w)$	$\ell^p(H)$

Abstract Sampling Theory

Sampling theory fits naturally into the abstract Fourier analysis setting, as the table above indicates, with different LCA groups giving rise to a variety of seemingly disparate sampling results [4]. Of course, there are some limitations to be expected in such a general theory. To explain it more fully, some additional definitions and notation are needed.

Let Λ be a *lattice in* Γ, i. e., a countable discrete subgroup of Γ with compact quotient Γ/Λ. Note that it is appropriate to work with the dual group Γ here, as it corresponds to the frequency domain. The lattice Λ corresponds to the sampling set in time domain.

The annihilator

$\Lambda^\perp$:=H={h∈G:(h,λ)=1 for all λ∈Λ}

of Λ is a closed subgroup which we will write H and which satisfies $H^\perp=\Lambda^{\perp\perp}=\Lambda$ [76, Lemma 2.1.3]. The annihilator HH of Λ is isomorphic to the dual of Γ/ΛΓ, i. e., H≅(Γ/Λ)^(algebraically and topologically) [76, Thm. 2.1.2] and is identified with the dual (Γ/Λ)^. The annihilator H plays the role of the sampling set in the case of the LCA group GG. When G=Γ=R with lattice Λ=2πwZ, H reduces to the sampling set Z/w in (1.7).

The Haar measure mΛmΛ on ΛΛ will be normalized so that the *Weil coset decomposition formula*

$$\int_\Gamma \varphi(\gamma)\,dm_\Gamma(\gamma) = \int_{\Gamma/\Lambda}\int_\Lambda \varphi(\gamma+\lambda)\,dm_\Lambda(\lambda)\,dm_{\Gamma/\Lambda}([\gamma]) \tag{6.7}$$

holds for suitable φ:Γ→C (see [39, Sect. 28.54 (iii)] or [73, Sect. 2.7.3]). The Haar measure mΓ/Λ of Γ/Λ is normalized so that the corresponding inversion formula (6.6) holds (with G replaced by Γ/Λ and Γ by H). The Haar measures of H and G/H are also normalized so that the Weil formula (6.7) holds for G and H and so that the corresponding inversion formula (6.6) holds (with GG replaced by G/H and Γ by Λ).

The quotient group Γ/Λ has a complete set of coset representatives, also referred to as a *transversal*, the terminology we will use. Transversals are not unique and there is always a measurable one [29] which we will choose. Note that while the quotient group is compact, the transversal is only of finite measure; indeed unbounded transversals are possible (see [5, Sect. 4]). By definition, a transversal Ω consists of just one point from each distinct coset [γ]=Λ+γ, i. e., Ω∩(Λ+γ) consists of a single point in Ω. Thus translates of Ω by non-zero elements in ΛΛare disjoint from Ω. Using (6.7), it is straightforward to verify that the Haar measure of the transversal Ω of Γ/Λ satisfies $m_\Gamma(\Omega) = m_\Lambda(\{0\})m_{\Gamma/\Lambda}(\Gamma/\Lambda) = m_\Lambda(\{0\})/m_H(\{0\}) < \infty$.

When the group G is discrete and countable, the integral over G reduces to a sum. This is particularly exploited for the subgroup Λ of Γ which is discrete and assumed countable, so that the integral

$$\int_\Lambda \varphi(\lambda)dm_\Lambda(\lambda) = m_\Lambda(\{0\}) \sum_{\lambda\in\Lambda} \varphi(\lambda),$$

where mΛ({0}) is the (discrete) measure of {0}.

Exact Sampling in Abstract Harmonic Analysis

Kluvánek [54] was the first to place sampling theory in the LCA group setting by establishing the abstract analogue of the classical sampling theorem (i. e., the exact sampling formula (1.7)). He considered LCA groups GG, each with dual Γ having a lattice Λ (so that Λ is a discrete, countable subgroup of Γ and Γ/Λ is compact). A measurable transversal Ω, with necessarily finite Haar measure $m_\Gamma(\Omega)$, is taken to represent the spectrum of f. Then Kluvánek took as an analogue of the class $\widehat{B}^2_{\pi w}$ (defined in Sect. 2.1 and called PW_w elsewhere) of band-limited functions, the class

$$\widehat{B}^2_\Omega(G) := \{f \in L^2(G) \cap C(G)\colon\ \operatorname{supp}\widehat{f} \subseteq \Omega\}. \tag{6.8}$$

Kluvánek showed that functions f in $\widehat{B}^2(G)$ have the representation $f=S_Hf$, where S_Hf is the abstract analogue of the classical sampling series S_wf (2.18) and is given by

$$(S_H f)(x) := \frac{1}{m_\Gamma(\Omega)} \sum_{h\in H} f(h)\chi_\Omega(x-k),$$

where convergence is in the L^2 sense and uniform. He further showed that $\|f\|_G^2 = \int_G |f|^2 = \sum_{h\in H} |f(h)|^2$, the abstract sampling analogue for Plancherel's theorem, equivalent to that for Parseval's theorem: $\int_G f\overline{g} = \sum_{h\in H} f(h)\overline{g}(h)$ and obviating the need for the $\ell^2(H)$ summability condition.

The discrete Fourier transform, involving essentially Z and S^1, also fits naturally into this framework. The abstract analogue would involve a discrete abelian group and a compact group respectively. A similar pattern holds for multi-dimensional signals in higher dimensions.

Approximate Sampling in Abstract Harmonic Analysis

More recently, abstract analogues of the approximate or generalized sampling formula (ASF) have been established. Faridani [28] proved a very general approximation theorem under an integrability condition. In [7, 26], an abstract approximate sampling theorem has been established under the square summability condition $f\in\ell^2(H)$ (in Sect. 2.1, $f\in S^2w=\ell^2(Z/w)$. The generalization of the classical sets F^p and S^p_w, p=1,2p (defined in Sect. 2.1), which are appropriate for extending approximate sampling theory to LCA groups, are

$$F^P(G) := \{f \in L^P(G) \cap C(G) \colon \widehat{f} \in L^1(\Gamma)\}$$

and

$$S_H^p(G) := \{f : G \to \mathbb{C} \colon f \in \ell^p(H)\} = \ell^p(H).$$

The abstract analogue of the ASF (1.6), is now given.

Theorem 3

Suppose that the dual Γ of the locally compact abelian group G has a discrete subgroup ΛΛ with Γ/Λ compact and that Ω is a measurable transversal of Γ/Λ. Then each $f\in F^2(G)\cap\ell^2(H)$ has a representation

$$f = S_H f + R_H f,$$

where $H = \Lambda^{\perp}$ *and* $|(R_H f)(x)| \le 2\int_{\Gamma\setminus\Omega} |\widehat{f}|$.

An asymptotic formula needs more information [7, Thms. 5, 6].

A norm result for f corresponding to Kluvánek's theorem is not known, although putting f=g in the generalized or approximate Parseval formula (1.1) suggests that modulo some summability conditions

$$\int_G |f|^2 = \sum_{h\in H} |f(h)|^2 - E_1 + E_2, \tag{6.9}$$

where

$$E_1 = \frac{1}{m_\Gamma(\Omega)} \sum_{h\in H} f(h) \int_{\Gamma\setminus\Omega} (\overline{f})^\wedge(\gamma)\,(h,\gamma)d\gamma,\ E_2 = \int_G (R_H f)\overline{f}.$$

This would imply an abstract Parseval type result and merits investigation. However, the proof in [18] of the formula (1.1) uses some subtle Fourier analysis similar to that employed by Brown [11, 12] and Boas [9] so that extra hypotheses would be very likely be needed.

Turning to the question of extending the equivalence results in Sect. 3, it could be conjectured possible that under suitable hypotheses, the abstract analogue of ASF (1.6) is equivalent to those of ARKF (1.9) and GPDF (1.1). However, the proof in [45] that the functional equation of the zeta function ζ is equivalent to the classical exact sampling theorem is far from direct, requiring long chains of implications and many results from complex analysis. Thus proving an LCA group result would appear to be very difficult, undoubtedly involving extra hypotheses. Moreover, even defining a suitable zeta function for LCA groups seems problematic as the natural numbers NN form a multiplicative semigroup generated by primes. Nevertheless, the classical situation suggests some interesting lines of inquiry and is a useful guide to the further understanding of abstract Fourier analysis.

Sampling in Hilbert Spaces, on Riemannian Manifolds and Graphs

Let L be a self-adjoint operator in a Hilbert space HH and P_λ,λ∈R, its spectral resolution of identity. With every vector f in H the family of projectors P_λ associates a measure on RR which is given by the formula $\langle P_\lambda f,f\rangle$, where $\langle\cdot,\cdot\rangle$ is the inner product in HH. Using this framework I. Pesenson introduced Paley–Wiener-type subspaces $PW_\omega(L)$,ω>0, , by saying that f belongs to $PW_\omega(L)$ if and only if the measure $\langle P_\lambda f,f\rangle$ has support in [−ω,ω]. It turned out that these subspaces enjoy all the basic properties of the classical Paley–Wiener spaces and in particular they are suitable for development of a rich sampling and approximation theories. On the level of abstract Hilbert spaces it was done by Pesenson in [67, 70]. Moreover, he specified this abstract setup in a number of very important situations such as: Riemannian compact and non-compact manifolds of bounded geometry [69], stratified Lie groups [68], combinatorial and quantum graphs [71, 72].

ACKNOWLEDGMENTS

The authors would like to thank Professor Janine Splettstößer, Wolfgang's oldest daughter, for her contribution to the biography. Paulo Ferreira was partially supported by FCT under PTDC/EEA-TEL/108568/2008.

REFERENCES

1. Aldroubi, A., Gröchenig, K.: Nonuniform sampling and reconstruction in shift-invariant spaces. SIAM Rev. 43(4), 585–620 (2001). doi:10.1137/S0036144501386986(electronic)
2. Apostol, T.M.: Modular Functions and Dirichlet Series in Number Theory, 2nd edn. Springer, New York (1990). doi:10.1007/978-1-4612-0999-7
3. Aronszajn, N.: Theory of reproducing kernels. Trans. Am. Math. Soc. 68, 337–404 (1950)
4. Beaty, M.G., Dodson, M.M.: Abstract harmonic analysis and the sampling theorem. In: Higgins, J.R., Stens, R.L. (eds.) Sampling Theory in Fourier and Signal Analysis, vol 2: Advanced Topics, chap. 10, pp. 233–265. Oxford University Press, Oxford (1999)
5. Beaty, M.G., Dodson, M.M.: The Whittaker-Kotel'nikov-Shannon theorem, spectral translates and Plancherel's

formula. J. Fourier Anal. Appl. 10(2), 179–199 (2004). doi:10.1007/s00041-004-8010-6

6. Beaty, M.G., Dodson, M.M., Eveson, S.P.: A converse to Kluvánek's theorem. J. Fourier Anal. Appl. 13(2), 187–196 (2007). doi:10.1007/s00041-006-6025-x

7. Beaty, M.G., Dodson, M.M., Eveson, S.P., Higgins, J.R.: On the approximate form of Kluvánek's theorem. J. Approx. Theory 160(1–2), 281–303 (2009). doi:10.1016/j.jat.2009.02.013

8. Bellman, R.: A Brief Introduction to Theta Functions. Holt, Rinehart and Winston, New York (1961)

9. Boas Jr, R.P.: Summation formulas and band-limited signals. Tôhoku Math. J. 2(24), 121–125 (1972). doi:10.2748/tmj/1178241524

10. Bochner, S.: Vorlesungen über Fouriersche Integrale. Chelsea Publishing Co., New York (1948)

11. Brown Jr, J.L.: On the error in reconstructing a non-bandlimited function by means of the bandpass sampling theorem. J. Math. Anal. Appl. 18(1), 75–84 (1967). doi:10.1016/0022-247X(67)90183-7

12. Brown Jr, J.L.: Erratum to: "On the error in reconstructing a nonbandlimited function by means of the bandpass sampling theorem". J. Math. Anal. Appl. 21, 699 (1968). doi:10.1016/0022-247X(68)90271-0

13. Brüdern, J.: Einführung in die analytische Zahlentheorie. Springer, Berlin (1995)

14. Butzer, P.L., Schmeisser, G., Stens, R.L.: The classical and approximate sampling theorems and their equivalence for entire functions of exponential type. J. Approx. Theory 179, 94–111 (2014). doi:10.1016/j.jat.2013.11.010

15. Butzer, P.L., Ferreira, P.J.S.G., Higgins, J.R., Schmeisser, G., Stens, R.L.: The sampling theorem, Poisson's summation formula, general Parseval formula, reproducing kernel formula and the Paley-Wiener theorem for bandlimited signals - their interconnections. Appl. Anal. 90(3–4), 431–461 (2011). doi:10.1080/00036811003627567

16. Butzer, P.L., Ferreira, P.J.S.G., Schmeisser, G., Stens, R.L.: The summation formulae of Euler–Maclaurin, Abel–Plana, Poisson, and their interconnections with the approximate sampling formula of signal analysis. Results Math. 59(3–4), 359–400 (2011). doi:10.1007/s00025-010-0083-8

17. Butzer, P.L., Gessinger, A.: A decomposition theorem for Parseval's equation; Connections with uniform and nonuniform sampling. In: Bilinskis, I., Cain, G., Marvasti, F. (eds.) SampTA'95, pp. 100–107. Institute of Electronics and Computer Science, Riga (1995)

18. Butzer, P.L., Gessinger, A.: The approximate sampling theorem, Poisson's sum formula, a decomposition theorem for Parseval's equation and their interconnections. Ann. Numer. Math. 4(1–4), 143–160 (1997)

19. Butzer, P.L., Jansche, S.: A direct approach to the Mellin transform. J. Fourier Anal. Appl.3(4), 325–376 (1997). doi:10.1007/BF02649101

20. Butzer, P.L., Nessel, R.J.: Fourier Analysis and Approximation. Academic Press, New York; Birkhäuser Verlag, Basel (1971)

21. Butzer, P.L., Schmeisser, G., Stens, R.L.: An introduction to sampling analysis. In: Marvasti, F. (ed.) Nonuniform Sampling: Theory and Practice, pp. 17–121. Kluwer/Plenum, New York (2001). doi:10.1007/978-1-4615-1229-5

22. Butzer, P.L., Splettstößer, W.: Approximation und Interpolation durch verallgemeinerte Abtastsummen. Forschungsberichte des Landes Nordrhein-Westfalen Nr. 2708. Westdeutscher, Opladen (1977)

23. Butzer, P.L., Stens, R.L.: The Euler–MacLaurin summation formula, the sampling theorem, and approximate integration over the real axis. Linear Algebra Appl. 52(53), 141–155 (1983). doi:10.1016/0024-3795(83)80011-1

24. Butzer, P.L., Stens, R.L.: The Poisson summation formula, Whittaker's cardinal series and approximate integration. In: Ditzian, Z., Meir, A., Riemenschneider, S.D., Sharma, A. (eds.) Second Edmonton Conference on Approximation Theory, pp. 19–36. American Mathematical Society, Providence (1983)

25. Deitmar, A.: A First Course in Harmonic Analysis, 2nd edn. Springer, New York (2005)

26. Dodson, M.M.: Approximating signals in the abstract. Appl. Anal. 90(3–4), 563–578 (2011). doi:10.1080/00036811003627575

27. Doetsch, G.: Summatorische Eigenschaften der Besselschen Funktionen und andere Funktionalrelationen, die mit der linearen Transformationsformel der Thetafunktion äquivalent sind. Compositio Math. 1, 85–97 (1935)

28. Faridani, A.: A generalized sampling theorem for locally compact abelian groups. Math. Comp. 63(207), 307–327 (1994). doi:10.2307/2153577

29. Feldman, J., Greenleaf, F.P.: Existence of Borel transversals in groups. Pac. J. Math. 25, 455–461 (1968)

30. Ferrar, W.L.: Summation formulae and their relation to Dirichlet's series. Compositio Math.1, 344–360 (1935). http://www.numdam.org/item?id=CM_1935__1__344_0

31. Flajolet, P., Gourdon, X., Dumas, P.: Mellin transforms and asymptotics: harmonic sums. Theoret. Comput. Sci. 144(1–2), 3–58 (1995). doi:10.1016/0304-3975(95)00002-E

32. García, A.G., Portal, A.: Sampling in reproducing kernel Banach spaces. Mediterr. J. Math. Published online: 28 Nov 2012. doi:10.1007/s00009-012-0234-0

33. Hamburger, H.: Über die Riemannsche Funktionalgleichung der ξξ-Funktion. Math. Z.10(3–4), 240–254 (1921). doi:10.1007/BF01211612

34. Hamburger, H.: Über die Riemannsche Funktionalgleichung der ζζ-Funktion. Math. Z.11(3–4), 224–245 (1921). doi:10.1007/BF01203626

35. Hamburger, H.: Über die Riemannsche Funktionalgleichung der ζζ-Funktion. Math. Z.13(1), 283–311 (1922). doi: 10.1007/BF01485292

36. Hamburger, H.: Über einige Beziehungen, die mit der Funktionalgleichung der Riemannschenζζ-Funktion

äquivalent sind. Math. Ann. 85(1), 129–140 (1922). doi:10.1007/BF01449611

37. Han, D., Nashed, M.Z., Sun, Q.: Sampling expansions in reproducing kernel Hilbert and Banach spaces. Numer. Funct. Anal. Optim. 30(9–10), 971–987 (2009). doi:10.1080/01630560903408606
38. Hewitt, E., Ross, K.A.: Abstract Harmonic Analysis, vol. I: Structure of Topological Groups. Integration Theory, Group Representations. Springer, Berlin (1963)
39. Hewitt, E., Ross, K.A.: Abstract Harmonic Analysis, vol. II: Structure and Analysis for Compact Groups. Analysis on Locally Compact Abelian Groups. Springer, New York (1970)
40. Hewitt, E., Stromberg, K.: Real and Abstract Analysis. A Modern Treatment of the Theory of Functions of a Real Variable. Springer, Berlin (1965)
41. Higgins, J.R.: Sampling Theory in Fourier and Signal Analysis. Clarendon Press, Oxford (1996)
42. Higgins, J.R.: A sampling principle associated with Saitoh's fundamental theory of linear transformations. In: Saitoh, S., Hayashi, N., Yamamoto, M. (eds.) Analytic Extension Formulas and Their Applications (Collected papers 2nd Internat. Congress of the Internat. Society for Analysis, its Applications and Computation (ISAAC'99), Fukuoka, Japan, Aug. 16–21, 1999, and Research Meeting on Applications of Analytic Extensions, Kyoto, Japan, Jan. 11–13, 2000). Kluwer Academic Publishers, Dordrecht, pp. 73–86 (2001)
43. Higgins, J.R.: Two basic formulae of Euler and their equivalence to Tschakalov's sampling theorem. Sampl. Theory Signal Image Process. 2(3), 259–270 (2003)
44. Higgins, J.R.: H. Hamburger's collection of five equivalent results, and some later developments, Technical Report, p. 31 (2007)
45. Higgins, J.R.: The Riemann zeta function and the sampling theorem. Sampl. Theory Signal Image Process. 8(1), 1–12 (2009)
46. Higgins, J.R.: Paley–Wiener spaces and their reproducing formulae. In: Ruzhansky, M., Wirth, J. (eds.) Progress in Analysis and Its Applications (Proceedings of 7th Internat. ISAAC Congress, London, Jul. 13–18, 2009, pp. 273–279. World Scientific Publisher, Hackensack (2010). doi:10.1142/9789814313179_0036
47. Hong, Y.M., Kim, J.M., Kwon, K.H.: Sampling theory in abstract reproducing kernel Hilbert space. Sampl. Theory Signal Image Process. 6(1), 109–121 (2007)
48. Ismail, M.E.H., Nashed, M.Z., Zayed, A.I., Ghaleb, A.F., eds. Mathematical Analysis, Wavelets, and Signal Processing, Proceedings of Internat. Conference on Mathematical Analysis and Signal Processing, Cairo, Jan. 3–9, American Mathematical Society, Providence, 1995, (1994). doi:10.1090/conm/190
49. Ivić, A.: The Riemann Zeta-Function. Wiley, New York (1985)
50. Kahane, J.P., Mandelbrojt, S.: Sur l'équation fonctionnelle de Riemann et la formule sommatoire de Poisson. Ann. Sci. École Norm. Sup. 3(75), 57–80 (1958)
51. Karatsuba, A.A., Voronin, S.M.: The Riemann Zeta-Function. Walter de Gruyter & Co., Berlin (1992). doi:10.1515/9783110886146
52. Klusch, D.: The sampling theorem, Dirichlet series and Bessel functions. Math. Nachr. 154, 129–139 (1991). doi:10.1002/mana.19911540111
53. Klusch, D.: The sampling theorem, Dirichlet series and Hankel transforms. J. Comput. Appl. Math. 44(3), 261–273 (1992). doi:10.1016/0377-0427(92)90001-E
54. Kluvánek, I.: Sampling theorem in abstract harmonic analysis. Mat.-Fyz. Časopis Sloven. Akad. Vied 15, 43–48 (1965)
55. Knopp, M., Robins, S.: Easy proofs of Riemann's functional equation for ζ(s)ζ(s) and of Lipschitz summation. Proc. Am. Math. Soc. 129(7), 1915–1922 (2001). doi:10.1090/S0002-9939-01-06033-6
56. Krazer, A.: Lehrbuch der Thetafunktionen. Chelsea Publishing Co., New York (1970)
57. Mordell, L.J.: Poisson's summation formula and the Riemann zeta function. J. Lond. Math. Soc.S1–4(4), 285 (1929). doi:10.1112/jlms/s1-4.4.285
58. Murty, M.R.: Problems in Analytic Number Theory, 2nd edn. Springer, New York (2008)
59. Nashed, M.Z.: Inverse problems, moment problems, signal processing: un menage a trois. In: Siddiqi, A.H., Singh, R.C., Manchanda, P. (eds.) Mathematics in Science and Technology (Proceedings of Satellite Conference of the Internat. Congress of Mathematicians, New Delhi, India, Aug. 14–17, 2010). World Scientific Publishers, Hackensack, pp. 2–19 (2011)
60. Nashed, M.Z., Sun, Q.: Sampling and reconstruction of signals in a reproducing kernel subspace of Lp(Rd) Lp(Rd). J. Funct. Anal. 258(7), 2422–2452 (2010). doi:10.1016/j.jfa.2009.12.012
61. Nashed, M.Z., Sun, Q., Xian, J.: Convolution sampling and reconstruction of signals in a reproducing kernel subspace. Proc. Am. Math. Soc. 141(6), 1995–2007 (2013). doi:10.1090/S0002-9939-2012-11644-2
62. Nashed, M.Z., Walter, G.G.: Reproducing kernel Hilbert spaces from sampling expansions. In: Ismail, M.E.H., Nashed, M.Z., Zayed, A.I., Ghaleb, A.F. (eds.) Mathematical Analysis, Wavelets, and Signal Processing, pp. 221–226 (1994). doi:10.1090/conm/190/02304
63. Newman, D.J.: Analytic Number Theory. Springer, New York (1998)
64. Nikol'skiĭ, S.M.: Approximation of Functions of Several Variables and Imbedding Theorems. Springer, New York (1975)
65. NuHAG (Numerical Harmonic Analysis Group). http://www.univie.ac.at/nuhag-php/home
66. Patterson, S.J.: An Introduction to the Theory of the Riemann Zeta-Function. Cambridge University Press, Cambridge (1988). doi:10.1017/CBO9780511623707
67. Pesenson, I.: Best approximations in a space of the representation of a Lie group (russian). Dokl. Akad. Nauk SSSR 302(5), 1055–1058 (1988). Trans. Soviet Math. Dokl. 38(2), 384–388 (1989)

68. Pesenson, I.: Sampling of Paley–Wiener functions on stratified groups. J. Fourier Anal. Appl. 4(3), 271–281 (1998). doi:10.1007/BF02476027

69. Pesenson, I.: A sampling theorem on homogeneous manifolds. Trans. Am. Math. Soc.352(9), 4257–4269 (2000). doi:10.1090/S0002-9947-00-02592-7

70. Pesenson, I.: Sampling of band-limited vectors. J. Fourier Anal. Appl. 7(1), 93–100 (2001). doi:10.1007/s00041-001-0007-9

71. Pesenson, I.: Polynomial splines and eigenvalue approximations on quantum graphs. J. Approx. Theory 135(2), 203–220 (2005). doi:10.1016/j.jat.2005.04.007

72. Pesenson, I.: Sampling in Paley–Wiener spaces on combinatorial graphs. Trans. Am. Math. Soc. 360(10), 5603–5627 (2008). doi:10.1090/S0002-9947-08-04511-X

73. Reiter, H.: Classical Harmonic Analysis and Locally Compact Groups. Clarendon Press, Oxford (1968)

74. Riemann, B.: Über die Anzahl der Primzahlen unter einer gegebenen Größe. Monatsber. Königl. Preuss. Akad. Wiss. Berlin Nov. 1859, 671–680. Reprinted. In: Riemann, B., Narasimhan, R. (eds.) Gesammelte mathematische Werke, wissenschaftlicher Nachlass und Nachträge, (Based on the edition by Weber, H., Dedekind, R.). Springer, Berlin (1990)

75. Rooney, P.G.: Another proof of the functional equation for the Riemann zeta function. J. Math. Anal. Appl. 185(1), 223–228 (1994). doi:10.1006/jmaa.1994.1244

76. Rudin, W.: Fourier Analysis on Groups. Interscience Publishers, New York (1962)

77. Saitoh, S.: Integral Transforms. Reproducing Kernels and Their Applications. Longman, Harlow (1997)

78. Schuster, W.: Ein kurzer Beweis der Funktionalgleichung der Riemannschen Zetafunktion (a short proof of the functional equation of the riemann zeta function). Aequationes Math.70(1–2), 191–194 (2005). doi:10.1007/s00010-004-2743-9

79. Siegel, C.: Bemerkung zu einem Satz von Hamburger über die Funktionalgleichung der Riemannschen Zetafunktion. Math. Ann. 86(3–4), 276–279 (1922). doi:10.1007/BF01457989

80. Stenger, F.: Numerical Methods Based on sinc and Analytic Functions. Springer, New York (1993)

81. Stenger, F.: Sinc convolution-a tool for circumventing some limitations of classical signal processing. In: Ismail, M.E.H., Nashed, M.Z., Zayed, A.I., Ghaleb, A.F. (eds.) Mathematical Analysis, Wavelets, and Signal Processing, pp. 227–240 (1994). doi:10.1090/conm/190/02305

82. Titchmarsh, E.C.: The Theory of the Riemann Zeta-Function, 2nd edn. Clarendon Press, New York (1986)

83. Weiss, P.: An estimate of the error arising from misapplication of the sampling theorem. Notices Am. Math. Soc. 10, 351 (1963). Abstract No. 601–54

84. Zayed, A.I.: Advances in Shannon's Sampling Theory. CRC Press, Boca Raton (1993)

85. Zygmund, A.: Trigonometric Series. vols. I, II, 2nd edn., reprinted with corrections and some additions (two volumes bound as one) edn. Cambridge University Press, London (1968)

CHAPTER 2

HENSTOCK-KURZWEIL INTEGRAL TRANSFORMS AND THE RIEMANN-LEBESGUE LEMMA

Francisco J. Mendoza-Torres, Ma. Guadalupe Morales-Macías, Salvador Sánchez-Perales, and Juan Alberto Escamilla-Reyna

INTRODUCTION

Let f be a function defined on a closed interval [a, b] in the extended real line $\overline{R}$, its Fourier transform at s ∈ R is defined as

$$\hat{f}(s) = \int_a^b e^{-ixs} f(x)dx. \tag{1}$$

The classical Riemann-Lebesgue Lemma states that

$$\lim_{|s|\to\infty} \int_a^b e^{-ixs} f(x)dx = 0, \tag{2}$$

whenever $f \in L^1([a, b])$.

We consider important to study analogous results about this lemma due to the following reasons:

- The classical Riemann-Lebesgue Lemma is an important tool used when proving several results related with convergence of Fourier Series and Fourier transform. In turn, these theorems have applications in the Harmonic Analysis which has many applications in the physics, biology, engineering and others sciences. For example, it is directly applied to the study of periodic perturbations of a class of resonant problems.
- An important problem is to consider an orthogonal basis, different to the trigonometric basis, and study the Fourier expansion of a function with respect to this basis. In this case, it is obtained an expression in this way

$$\int_a^b h(xs)f(x)dx. \tag{3}$$

- In some cases the expression (1) exists and the expression (2) is true even if the function f is not Lebesgue integrable.

Thus, a variant of the Riemann-Lebesgue Lemma is to get conditions for the functions f and h which ensure that (3) is well defined and satisfies

$$\lim_{|s|\to\infty} \int_a^b h(xs)f(x)dx = 0. \tag{4}$$

Some results of this type and related results are found in [1], [2], [3], and [4].

In the space of Henstock-Kurzweil integrable functions over R, HK(R), the Fourier transform does not always exist. In [5] was proven that $e^{-i(\cdot)s} f$ is Henstock-Kurzweil integrable under certain conditions and that, in general, does not satisfy the Riemann-Lebesgue Lemma. Subsequently, it was shown in [6], [3] and [4] that the Fourier transform exists and the equation (2) is true when $-\infty = a$, $b = \infty$ and f belongs to $BV_0(R)$, the space of bounded variation functions that vanish at infinity. A special case arises when f is in the intersection of functions of bounded variation and Henstock-Kurzweil integrable functions.

There exist Henstock-Kurzweil (HK) integrable functions f which $f \in HK(I) \setminus L^1(I)$ such that (2) is not fulfilled, when I is a bounded interval. In [7], Zygmund exhibited Henstock-Kurzweil integrable functions such that their Fourier coefficients do not tend to zero. In [8] are given necessary and sufficient conditions in order to $\int_a^b f(x)g_n(x)dx \longrightarrow$

Citation: Francisco J. Mendoza-Torres, Ma. Guadalupe Morales-Macías,Salvador Sánchez-Perales and Juan Alberto Escamilla-Reyna (2015). Henstock-Kurzweil Integral Transforms and the Riemann-Lebesgue Lemma, Fourier Transform - Signal Processing and Physical Sciences, Dr.Ing. Salih Salih (Ed.), InTech, DOI: 10.5772/59766.

$\int_a^b f(x)g(x)dx$, for all f ∈ HK([a, b]). Thus, we will prove that the Fourier transform has the asymptotic behavior:

$$\hat{f}(s) = o(s), \text{ as } |s| \to \infty,$$

where f ∈ HK(I) \ L^1 (I)

Moreover in [9], Titchmarsh proved that it is the best possible approximation for functions with improper Riemann integral.

This chapter is divided into 5 sections; we present the main results we have obtained in recent years: [3], [4], [6] and [10]. In this section we introduce basic concepts and important theorems about the Henstock-Kurzweil integral and bounded variation functions. In the second part of this study we prove some generalizations about the convergence of integrals of products in the completion of the space HK([a, b]), $\widehat{HK([a,b])}$, where [a, b] can be a bounded or unbounded interval. As a consequence, some results related to the Riemann-Lebesgue Lemma in the context of the Henstock-Kurzweil integral are proved over bounded intervals. Besides, for elements in the completion of the space of Henstock-Kurzweil integrable functions, we get a similar result to the Riemann-Lebesgue property for the Dirichlet Kernel, as well as the asymptotic behavior of the n-th partial sum of Fourier series.

In the third section, we consider a complex function g defined on certain subset of R^2. Many functions on functional analysis are integrals of the form $\Gamma(s) = \int_{-\infty}^{\infty} f(t)g(t,s)dt$. We study the function Γ when f belongs to BV_0(R) and g(t, ·) is continuous for all t. The integral we use is Henstock-Kurzweil integral. There are well known results about existence, continuity and differentiability of Γ, considering the Lebesgue theory. In the HK integral context there are results about this too, for example, Theorems 12.12 and 12.13 from [11]. But they all need the stronger condition that the function f(t) g(t,s) is bounded by a HK integrable function. We give more conditions for existence, continuity and differentiability of Γ. Finally we give some applications such as some properties about the convolution of the Fourier and Laplace transforms.

In section 4, we exhibit a family of functions in HK(R) included in BV_0(R) \ L^1 (R). At the last section we get a version of Riemann-Lebesgue Lemma for bounded variation functions that vanish at infinity. With this result we get properties for the Fourier transform of functions in BV_0(R): it is well defined, is continuous on R − {0}, and vanishes at ±∞, as classical results. Moreover, we obtain a result on pointwise inversion of the Fourier transform.

Basic Concepts and Nomenclature

We will refer to a finite or infinite interval if its Lebesgue measure is finite or infinite. Let I ⊂ R be a closed interval, finite or infinite. A partition P of I is a increasing finite collection of points $\{t_0, t_1, ..., t_n\}$ ⊂ I such that if I is a compact interval [a, b], then t_0 = a and t_n = b; if I = [a, ∞), t_0 = a; and if I = (−∞, b] then t_n = b.

Let us consider I ⊂ R as a closed interval finite. A tagged partition of I is a set of ordered pairs $\{[t_i-1, t_i], s_i\}$ n i=1 where it is assigned a point si ∈ [ti−1 , ti], which is called a tag of [ti−1 , ti]. With this concept we define the Henstock-Kurzweil integral on finite intervals in R.

Definition 1. The function f : [a, b] → R is Henstock-Kurzweil integrable if there exists H ∈ R which satisfies the following: for each ε > 0 exists a function γ_ε : [a, b] → (0, ∞) such that if P = $\{([t_i-1, t_i], s_i)\}$ n i=1 is a tagged partition such that

$$[t_{i-1}, t_i] \subset [s_i - \gamma_\varepsilon(s_i), s_i + \gamma_\varepsilon(s_i)] \quad for\ i = 1, 2, ..., n., \tag{5}$$

then

$$|\Sigma_{i=1}^n f(s_i)(t_i - t_{i-1}) - H| < \varepsilon.$$

H is the integral of f over [a, b] and it is denoted as

$$H = \int_a^b f = \int_a^b f dt$$

It is said that a tagged partition is called γ_ε−fine if satisfies (5).

This definition can be extended on infinite intervals as follows.

Definition 2. Let γ : [a, ∞] → (0, ∞) be a function, we will say that the tagged partition P = $\{([t_{i-1}, t_i], s_i)\}_{i=1}^{n+1}$ is γ−fine if:

- t_0 = a, t_{n+1} = ∞.
- [ti−1 , t_i] ⊂ [$s_i - \gamma_\varepsilon(s_i)$, $s_i + \gamma_\varepsilon(s_i)$] for i = 1, 2, ..., n.
- [t_n, ∞] ⊂ [1/γ(∞), ∞]. Put f(∞) = 0 and f(−∞) = 0.

This allows us define the integral of f over infinite intervals.

Definition 3. It is said that the function f : [a, ∞] → R is Henstock-Kurzweil integrable if it satisfies the Definition 1, but the partition P must be γε−fine according to Definition 2.

For functions defined on [−∞, a] or [−∞, +∞] the integral is defined analogously. We will denote the vector space of Henstock-Kurzweil integrable functions on I as HK(I)

The space of Henstock-Kurzweil integrable functions on the interval I = [a, b], finite or infinite interval, is a semi-normed space with the Alexiewicz semi-norm

$$||f||_A = \sup_{a \le x \le b} \left| \int_a^x f(t)dt \right|. \tag{6}$$

We denote the space of functions in HK([c, d]) for each finite interval [c, d] in I as HK_{loc}(I).

Definition 4. A function f : I → R is a bounded variation function over I (finite interval) if exists a M > 0 such that

$$\text{Var}(f, [a, t]) \le N,$$

for all t ≥ a. The total variation of f on I is equal to

$$\text{Var}(f, [a, \infty)) = \sup\{\text{Var}(f, [a, t]) : a \le t\}. \tag{7}$$

For I = (−∞, b] the considerations are analogous.

The set of bounded variation functions over [a, b] is denoted as BV([a, b]) and we will denote the space of functions f such that f ∈ BV([c, d]) for each compact interval [c, d] in R as BV_{loc}(R). We will refer to BV_0(R) as the subspace of functions f belong to BV(R) such that vanishing at ±∞.

At the Lemma 25 we prove that: HK(R) ∩ BV(R) ⊂ BV_0(R). It is not hard prove that $BV_0(R) \subsetneq L^1(R)$ and $BV_0(R) \subsetneq HK(R)$. Furthermore, there are functions in HK(R) or L 1 (R) but they are not in BV_0(R). For example, the function f(t) defined by 0 for t ∈ (−∞, 1) and 1/t for t ∈ [1, ∞) belongs to BV_0(R) but does not belong to L^1 (R), neither in HK(R).

In addition, other examples are the characteristic function of Q on a compact interval and g(t) = $t^2 \sin(\exp(t^2))$ are in HK(R) \ BV_0(R). We consider the completion of HK([a, b]) as

$$\{[\{f_k\}] : \{f_k\} \text{ is a Cauchy sequence in } HK([a,b])\},$$

where the convergence is respect Alexiewicz norm, and will be denoted by $\widehat{HK([a,b])}$. It is possible to prove that $\widehat{HK([a,b])}$ is isometrically isomorphic to the subspace of distributions each of which is the distributional derivative of a continuous function, see [12]. The indefinite integral of f = [{ fk}] ∈ $\widehat{HK([a,b])}$ is defined as

$$\int_a^x f = \lim_{k\to\infty} \int_a^x f_k.$$

Thus, $\widehat{HK([a,b])}$ is a Banach space with the Alexiewicz norm (6). The completion is also defined in [13]. Besides, basic results of the integral continue being true on the completion. More details see [12].

To facilitate reading, we recall the following results. The first one is a well-known result, and it can be found for example in [14] and [15].

Theorem 5. Let f be a real function defined on N × N. If $\lim_{n\to n0}$ f(k, n) = ψ(k) exists for each k, and $\lim_{k\to k0}$ f(k, n) = ϕ(n) converges uniformly on n, then

$$\lim_{k\to k_0}\lim_{n\to n_0} f(k,n) = \lim_{n\to n_0}\lim_{k\to k_0} f(k,n).$$

Theorem 6. [16, Theorem 33.1] Suppose X is a normed space, Y is a Banach space and that {T_n} is a sequence of bounded linear operators from X into Y. Then the conditions: i) {||T_n||} is bounded and ii) {T_n(x)} is convergent for each x ∈ Z, where Z is a dense subset on X implies that for each x ∈ X, the sequence (T_n(x)) is convergent in Y and the linear operator T : X → Y defined by T(x) = $\lim_{n\to\infty}$ T_n(x) is bounded.

Theorem 7. [17] If g is a HK integrable function on [a, b] ⊆ R and f is a bounded variation function on [a, b], then f g is HK integrable on [a, b] and

$$\left|\int_a^b fg\right| \le \inf_{t\in[a,b]}|f(t)|\left|\int_a^b g(t)dt\right| + \|g\|_{[a,b]}\mathrm{Var}(f,[a,b]).$$

Theorem 8. [11, Hake's Theorem] ϕ ∈ HK([a, ∞]) if and only if for each b, ε such that b > a b − a > ε > 0, it follows that ϕ ∈ HK([a + ε, b]) and $\lim_{\varepsilon\to 0,\, b\to\infty} \int_{a+\varepsilon}^b \varphi(t)dt$ exists. In this case, this limit is $\int_a^\infty \varphi(t)dt$.

Theorem 9. [11, Chartier-Dirichlet's Test] Let f and g be functions defined on [a, ∞). Suppose that

- g ∈ HK([a, c]) for every c ≥ a, and G defined by G(x) = $\int_a^x g$ is bounded on [a, ∞).
- f is of bounded variation on [a, ∞) and $\lim_{x\to\infty} f(x) = 0$.

Then f g ∈ HK([a, ∞)).

Moreover, by Multiplier Theorem, Hake's Theorem and Chartier-Dirichlet Test, we have the following lemma.

Lemma 10. Let f , g : [a, ∞] → R. Suppose that f ∈ BV0([a, ∞]), ϕ ∈ HK([a, b]) for every b > a, and $\Phi(t) = \int_a^t \varphi du$ is bounded on [a, ∞). Then ϕ f ∈ HK([a, b]),

$$\int_a^\infty \varphi f dt = -\int_a^\infty \Phi(t)df(t)$$

and

$$\left|\int_a^\infty \varphi f dt\right| \le \sup_{a<t} |\Phi(t)|\,\mathrm{Var}(f,[a,\infty]).$$

Similar results are valid for the cases [−∞, ∞] and [−∞, a].

Let I = [a, b] and E ⊂ I. We say that the function F : I → R is in AC_δ (E) if for each e > 0 there exist $\eta_\varepsilon > 0$ and a gauge δ_ε on E such that if {(x_i , y_i)} N $_i$ is a (δ_ε , E)−fine subpartition of E such that Σ_i^N ($y_i - x_i$) < η_ε , then Σ_1^N |F(x_i) − F(y_i)| < e. On the other hand, F belongs to the class ACG_δ (I) if there exists a sequence $\{En\}_1^\infty$ of sets in I such that $I = \cup_{n=1}^\infty E_n$ and F ∈ AC_δ (E_n) for each n ∈ N. A characterization of this type of functions is the following.

Theorem 11. A function f ∈ HK(I) if and only if there exists a function F ∈ ACGδ such that F 0 = f a.e.

Theorem 12. [18, Theorem 4] Let a, b ∈ R. If h : R × [a, b] → C is such that

- h(t, ·) belongs to ACG_δ on [a, b] for almost all t ∈ R;
- and h(·,s) is a HK integrable function on R for all s ∈ [a, b].

Then $H := \int_{-\infty}^{\infty} h(t,\cdot)dt$ belongs to ACG_δ on [a, b] and $H'(s) = \int_{-\infty}^{\infty} D_2h(t,s)dt$ for almost all s ∈ (a, b), iff,

$$\int_s^t \int_{-\infty}^{\infty} D_2h(t,s)dtds = \int_{-\infty}^{\infty} \int_s^t D_2h(t,s)dsdt$$

for all [s, t] ⊆ [a, b]. In particular,

$$H'(s_0) = \int_{-\infty}^{\infty} D_2h(t,s_0)dt$$

when $H_2 := \int_{-\infty}^{\infty} D_2h(t,\cdot)dt$ is continuous at s_0.

FOURIER COEFFICIENTS FOR FUNCTIONS IN THE HENSTOCK-KURZWEIL COMPLETION

For finite intervals, the Theorem 12.11 of [19] tells us that: In order that $\int_a^b fg_n \to \int_a^b fg$, n → ∞, whenever f ∈ HK([a, b]), it is necessary and sufficient that: i) g_n is almost everywhere of bounded variation on [a, b] for each n; ii) $\sup\{||g_n||_\infty + ||g_n||_{BV}\} < \infty$; iii) $\int_c^d g_n \to \int_c^d g$, $n \to \infty$, for each interval (c, d) ⊂ (a, b). The Theorem 3 of [8] proves that above theorem is valid for infinite intervals. In this section we show that [19, Theorem 12.11] and [8, Theorem 3] are true for functions belonging to the completion of the Henstock-Kurzweil space. First, we need to prove the next lemma. The class of step functions on [a, b] will be denoted as K([a, b]).

Lemma 13. Let [a, b] be an infinite interval. The set K([a, b]) is dense in HK([a, b]).

Proof. Let f ∈ HK([a, ∞]) and e > 0 be given. By Hake's Theorem, exists N ∈ N such that for each x ≥ N,

$$\left|\int_x^{\infty} f\right| < \frac{\epsilon}{2}. \tag{8}$$

Since K([a, N]) is dense in HK([a, N]), by Theorem 7 of [20], there exists a function h ∈ K([a, N]) such that

$$||f-h||_{A,[a,N]} = \sup_{x\in[a,N]}\left|\int_a^x (f-h)\right| < \frac{\epsilon}{2}. \tag{9}$$

Defining h0 ∈ K([a, ∞]) as

$$h_0(x) = \begin{cases} h(x) \text{ if } x \in [a,N] \\ 0 \text{ if } x \in (N,\infty]. \end{cases}$$

It follows, by (8) and (9), that

$$||f-h_0||_A \le \epsilon.$$

Similar arguments apply for intervals as [−∞, a] or [−∞, ∞].

The Convergence of Integrals of Products in the Completion

The following result appears in [3]. Here, we present a detailed proof.

Theorem 14. Let [a, b] ⊂ R. In order that

$$\int_a^b fg_n \to \int_a^b fg, \quad n \to \infty, \tag{10}$$

whenever f ∈ $\widehat{HK([a,b])}$, it is necessary and sufficient that: i) g_n is almost everywhere of bounded variation on [a, b] for each n; ii) sup{$||g_n||\infty$ + $||g_n||$BV} < ∞; iii) $\int_c^d g_n \to \int_c^d g$, n → ∞, for each interval (c, d) ⊂ (a, b).

Proof. The necessity follows from [19, Theorem 12.11]. Now we will prove the sufficiency condition. Define the linear functionals T, Tn : $\widehat{HK([a,b])}$ → R by

$$T_n(f) = \int_a^b fg_n \text{ and } T(f) = \int_a^b fg. \tag{11}$$

Supposing i) and ii), we have, by Multiplier Theorem, that the sequence {Tn} is bounded by sup{$||g_n||\infty$ + $||g_n||$BV}. Owing to Lemma 13, the space of step functions is dense in $\widehat{HK([a,b])}$, then considering the Theorem 6 it is sufficient to prove that {Tn(f)} converge to T(f), for each step function f. First, let f(x) = $\chi_{(c,d)}$ (x) be the characteristic function of (c, d) ⊂ [a, b]. Thus,

$$T_n(f) = \int_a^b \chi_{(c,d)}g_n = \int_c^d g_n,$$

by the hypothesis iii), we have that {$T_n(\chi_{(c,d)})$} converges to T($\chi_{(c,d)}$), as n → ∞. Now, let f be a step function. Being that each T_n is a linear functional, then {T_n(f)} converges to T(f), as n → ∞. Thus, the result holds.

Remark 15. On HK([a, b]). The hypothesis iii) can be replaced by: g_n converges pointwise to g, then the result follows from Corollary 3.2 of [21]. The result on the completion holds by Theorem 6. Note that the conditions i), ii) and iii) do not imply converges point wise from {g_n} to g, see example 2 of [8].

For the case of functions defined on a finite interval we get Theorem 16, and a lemma of Riemann-Lebesgue type for functions in the Henstock-Kurzweil space completion.

Theorem 16. Let [a, b] be a finite interval. If i) g_n converges to g in measure on [a, b], ii) each g_n is equal to h_n almost everywhere, a normalized bounded variation function and iii) there is M > 0 such that Var(h_n, [a, b]) ≤ M, n ≥ 1, then for all f ∈ $\widehat{HK([a,b])}$,

$$\int_a^b fg_n \to \int_a^b fg, \quad n \to \infty.$$

Proof. Let f ∈ $\widehat{HK([a,b])}$ be given, where f = [{ fk}], we want to prove that

$$\lim_{n\to\infty}\lim_{k\to\infty}\int_a^b f_kg_n = \int_a^b fg.$$

Define f(k, n) = $\int_a^b f_kg_n$. By the hypothesis i) about (g_n) we have

$$\lim_{n\to\infty}\int_a^b f_k g_n = \int_a^b f_k g.$$

Moreover

$$\lim_{k\to\infty}\lim_{n\to\infty}\int_a^b f_k g_n = \int_a^b fg,$$

by the integral definition on the completion. We will prove that $\lim_{k\to\infty} f(k,n) = \int_a^b fg_n$ converges uniformly on n. Let e > 0 be given, there exists k_0 such that || fk − f ||A ≤ e, if k ≥ k_0. Besides, if k ≥ k_0,

$$\left|\int_a^b f_k g_n - \int_a^b f g_n\right| \le ||f_k - f||_A Var(g_n,[a,b]) \le M\epsilon.$$

Therefore, by Theorem 5,

$$\lim_{n\to\infty}\lim_{k\to\infty}\int_a^b f_k g_n = \int_a^b fg.$$

The following result is a "generalization" of Riemann-Lebesgue Lemma on the completion of the space $\widehat{HK([a,b])}$, over finite intervals, it also appears in [3].

Corollary 17. If ϕ : R → R such that ϕ› exists, is bounded and ϕ(s) = o(s), as |s| → ∞, then for each f ∈ $\widehat{HK([a,b])}$ we have the next asymptotic behavior

$$\int_a^b \varphi(st)f(t)dt = o(s), \quad as \ |s| \to \infty.$$

Proof. For each s ≠ 0 define ϕ_s : R → R as $\phi_s(t)$ = ϕ(st)/s. In order to prove

$$\lim_{s\to\infty}\int_a^b \frac{\varphi(st)}{s} f(t)dt = 0,$$

it is sufficient to show that j_s fulfills the hypothesis of Theorem 16. Now, we will check item by item. i) Because of j (s) = o(s), as |s| → ∞ and the interval [a, b] is finite, it follows that j_s converges in measure to 0. ii) Owing to ϕ' is bounded then, by the Mean Value Theorem, we have that j_s ∈ BV([a, b]). iii) Var(j_s, [a, b]) is bounded uniformly by upper bound of j' and a − b.

Riemann-Lebesgue Property

This property establishes that $\int_r^\pi f(t)D_n(t)dt \to 0$, for each $f \in L^1[-\pi,\pi]$ and r ∈ (0, π], where $D_n(t) = \frac{\sin(n+1/2)t}{\sin(t/2)}$ denotes the n-th Dirichlet Kernel of order n. Now, we provide an analogous result concerning the Henstock-Kurzweil completion.

Theorem 18. For any f ∈ $\widehat{HK([-\pi,\pi])}$, and r ∈ (0, π],

$$\lim_{n\to\infty}\frac{1}{n}\int_r^\pi f(t)D_n(t)dt = 0. \quad (12)$$

Proof. Note that the function g(t) = 1/ sin(t/2) is in BV([r, π]). Moreover, by Multiplier Theorem we have f g ∈ $\widehat{HK([r,\pi])}$. Hence, by Corollary 17, we get

$$\int_r^\pi f(t)g(t)\sin(n+1/2)tdt = o(n), \quad |n|\to\infty.$$

Considering an similar argument from above proof, it follows that

$$\int_r^\pi f(t)\frac{\sin(n+1/2)t}{t/2}dt = o(n), \quad |n|\to\infty. \quad (13)$$

For n ∈ NU {0}, we define the function $\Phi_n(t) = \frac{\sin(n+1/2)t}{t/2}$ for t ≠ 0 and $\Phi_n(0)$ = 2n +1, it is called the discrete Fourier Kernel of order n. This kernel provides a very good approximation to the Dirichlet Kernel D_n for |t| < 2, but Φ_n decreases more rapidly than D_n, see [1].

Theorem 19. Let f ∈ $\widehat{HK([0,\pi])}$ and r ∈ (0, π]. Then, assuming that any of next limits exist,

$$\lim_{n\to\infty}\frac{1}{n}\int_0^r f(t)D_n(t)dt = \lim_{n\to\infty}\frac{1}{n}\int_0^r f(t)\frac{\sin(n+1/2)t}{t/2}dt.$$

Proof. Define g : [0, π] → R by

$$g(t) = \begin{cases} \frac{1}{\sin(t/2)} - \frac{1}{t/2} & \text{for } t \in (0,\pi] \\ 0 & \text{for } \quad t = 0. \end{cases}$$

Since g ∈ BV([0, π]), f g ∈ $\widehat{HK([0,\pi])}$. By Corollary 17, we have

$$\lim_{n\to\infty}\frac{1}{n}\int_0^\pi f(t)\left(\frac{1}{\sin(t/2)} - \frac{1}{t/2}\right)\sin(n+1/2)tdt = 0. \quad (14)$$

Now, by (14), we have

$$\lim_{n\to\infty}\frac{1}{n}\int_0^\pi \left(f(t)D_n(t) - f(t)\frac{\sin(n+1/2)t}{t/2}\right)dt = 0.$$

Then

$$\lim_{n\to\infty}\frac{1}{n}\left[\int_0^r \left(f(t)D_n(t) - f(t)\frac{\sin(n+1/2)t}{t/2}\right)dt + \int_r^\pi \left(f(t)D_n(t) - f(t)\frac{\sin(n+1/2)t}{t/2}\right)dt\right] = 0.$$

By Theorem 18 and (13),

$$\lim_{n\to\infty}\frac{1}{n}\int_r^\pi \left(f(t)D_n(t) - f(t)\frac{\sin(n+1/2)t}{t/2}\right)dt = 0.$$

Therefore, assuming that any of the limits exist, we have

$$\lim_{n\to\infty} \frac{1}{n} \int_0^r f(t)D_n(t)dt = \lim_{n\to\infty} \frac{1}{n} \int_0^r f(t)\frac{\sin(n+1/2)t}{t/2}dt.$$

The following result is a characterization of the asymptotic behavior of n − th partial sum of the Fourier series, it can be found in [3].

Corollary 20. Let f ∈ $\widehat{HK([-\pi,\pi])}$ be 2π− periodic. The n − th partial sum of the Fourier series at t has the following asymptotic behavior $S_n(f, t) = o(n)$, when |n| → ∞ iff

$$\int_0^\pi [f(t+u)+f(t-u)]\frac{\sin(n+1/2)u}{u}du = o(n),$$

if |n| → ∞.

Proof. Since $S_n(f,t) = \int_{-\pi}^{\pi} f(t+u)D_n(u)du$, realizing a change of variable (see section 6 of [13]), then by Theorem 19 we get the result.

HENSTOCK-KURZWEIL INTEGRAL TRANSFORM

The results in this section are based for functions in the vector space $BV_0(R)$, and they have to [10] as principal reference.

We will introduce some additional terminology in order to facilitate the following results.

If g : R × R → C is a function and s_0 ∈ R, we say that s0 fulfills hypothesis (H) relative to g if:

(H) there exist $\delta = \delta(s_0) > 0$ and M = M(s0) > 0, such that, if $|s - s_0| < \delta$ then

$$\left|\int_u^v g(t,s)dt\right| \le M,$$

for all [u, v] ⊆ R.

This condition plays a significant role in the following results. Also, the next theorems can be found in [10].

Theorem 21. Let f : R → R and g : R × R → C be functions. Assume that f ∈ $BV_0(R)$, and s_0 ∈ R fulfills Hypothesis (H) relative to g, then

$$\Gamma(s) = \int_{-\infty}^{\infty} f(t)g(t,s)dt$$

is well defined for all s in a neighborhood of s0.

Proof. Applying Theorem 9 the result holds.

Theorem 22. Let f : R → R and g : R × R → C be functions assume that

f belongs to $BV_0(R)$, g is bounded, and

1. G(T, ·) IS CONTINUOUS FOR ALL T ∈ R.

If s_0 ∈ R fulfills Hypothesis (H) relative to g, then the function Γ is continuous at s0.

Proof. By Hypothesis (H), there exist $\delta_1 > 0$ and M > 0, such that, if $|s - s_0| < \delta_1$ then

$$\left|\int_u^v g(t,s)dt\right| \le M \tag{15}$$

for all [u, v] ⊆ R. From Theorem 21, Γ(s) exists for all s ∈ $B_{\delta 1}(s_0)$.

Let an arbitrary e > 0, by Hake's theorem, there exists $K_1 > 0$ such that

$$\left|\int_{|t|\ge u} f(t)g(t,s_0)dt\right| < \frac{\epsilon}{3} \tag{16}$$

for all $u \ge K_1$. On the other hand, as

$$\lim_{t\to-\infty} \mathrm{Var}(f,(-\infty,t]) = 0 \quad\text{and}\quad \lim_{t\to\infty} \mathrm{Var}(f,[t,\infty)) = 0,$$

there is $K_2 > 0$ such that for each $t > K_2$,

$$\mathrm{Var}(f,(-\infty,-t]) + \mathrm{Var}(f,[t,\infty)) < \frac{\epsilon}{3M}.$$

Let $K = \max\{K_1, K_2\}$. From Theorem 7, it follows that for every v ≥ K and every s ∈ $B\delta_1(s_0)$,

$$\left|\int_K^v f(t)g(t,s)dt\right| \le \|g(\cdot,s)\|_{[K,v]}\left[\inf_{t\in[K,v]}|f(t)| + \mathrm{Var}(f,[K,v])\right]$$
$$\le M\left[|f(v)| + \mathrm{Var}(f,[K,\infty))\right],$$

where the second inequality is true due to (15). This implies, since limt→∞ | f(t)| = 0, that

$$\left|\int_K^\infty f(t)g(t,s)dt\right| \le M\cdot \mathrm{Var}(f,[K,\infty)).$$

Analogously we have that

$$\left|\int_{-\infty}^{-K} f(t)g(t,s)dt\right| \le M\cdot \mathrm{Var}(f,(-\infty,-K]).$$

Therefore, for each s ∈ $B\delta_1(s_0)$,

$$\left|\int_{|t|\ge K} f(t)g(t,s)dt\right| \le M\left[\mathrm{Var}(f,(-\infty,-K])f + \mathrm{Var}(f,[K,\infty))\right]$$
$$< M\frac{\epsilon}{3M} = \frac{\epsilon}{3}. \tag{17}$$

Since f is $L^1[-K, K]$, g is bounded and g(t, ·) is continuous for all t ∈ R. For example, using Theorem 12.12 of [11], it is easy to show that the function

$$\Gamma_K(s) = \int_{-K}^{K} f(t)g(t,s)dt, \quad s \in \mathbb{R},$$

is continuous at s_0. This implies that there is $\delta_2 > 0$ such that for every s ∈ $B\delta_2(s_0)$,

$$\left|\int_{-K}^{K} f(t)[g(t,s) - g(t,s_0)]dt\right| < \frac{\epsilon}{3}. \tag{18}$$

Let $\delta = \min\{\delta_1, \delta_2\}$. Then for all s ∈ $B_\delta(s_0)$,

$$|\Gamma(s) - \Gamma(s_0)| \le \left|\int_{-K}^{K} f(t)[g(t,s) - g(t,s_0)]dt\right|$$
$$+ \left|\int_{|t|\ge K} f(t)g(t,s)dt\right| + \left|\int_{|t|\ge K} f(t)g(t,s_0)dt\right|.$$

Thus, from (16), (17) and (18), |Γ(s) − Γ(s0)| < $\frac{\epsilon}{3} + \frac{\epsilon}{3} + \frac{\epsilon}{3} = \epsilon$, for all s ∈ $B_\delta(s_0)$.

Theorem 23. Let a, b ∈ R. If f : R → R and g : R × [a, b] → C are functions such that

1. f ∈ $BV_0(R)$, g is measurable, bounded and
2. for all s ∈ [a, b], s satisfies Hypothesis (H) relative to g.

Then

$$\int_a^b \int_{-\infty}^{\infty} f(t)g(t,s)dtds = \int_{-\infty}^{\infty} \int_a^b f(t)g(t,s)dsdt$$

Proof. From (2) and since [a, b] is compact, there exists M > 0 such that, for every s ∈ [a, b] and for all $[u,v] \subseteq \mathbb{R} : |\int_u^v g(t,s)dt| \leq M.$

For r > 0 and s ∈ [a, b], let $\Gamma_r(s) = \int_{-r}^{r} f(t)g(t,s)dt$. By Theorem 7, we notice that

$$\begin{aligned}|\Gamma_r(s)| &= \left|\int_{-r}^{r} f(t)g(t,s)dt\right| \\ &\leq \|g(\cdot,s)\|_{[-r,r]}\left[\inf_{t\in[-r,r]}|f(t)| + V_{[-r,r]}f\right] \\ &\leq M[|f(0)| + Vf]\end{aligned}$$

for all s ∈ [a, b].

So, for each r > 0, Γr is HK integrable on [a, b] and is bounded for a fixed constant. Moreover, by Theorem 21 and Hake's theorem

$$\lim_{r\to\infty}\Gamma_r(s) = \Gamma(s)$$

for all s ∈ [a, b].

Using the Lebesgue Dominated Convergence Theorem, we have that Γ is HK integrable on [a, b] and

$$\int_a^b \Gamma(s)ds = \lim_{r\to\infty}\int_a^b \Gamma_r(s)ds.$$

Now, because of f is Lebesgue integrable on [−r, r]; g is measurable and bounded; and by Fubini's theorem, it follows that

$$\int_a^b \int_{-r}^{r} f(t)g(t,s)dtds = \int_{-r}^{r} \int_a^b f(t)g(t,s)\, dsdt.$$

Consequently

$$\lim_{r\to\infty}\int_{-r}^{r}\int_a^b f(t)g(t,s)dsdt = \lim_{r\to\infty}\int_a^b \Gamma_r(s)ds = \int_a^b \Gamma(s)ds.$$

So by Hake's theorem,

$$\int_{-\infty}^{\infty}\int_a^b f(t)g(t,s)\, dsdt = \int_a^b \Gamma(s)ds = \int_a^b\int_{-\infty}^{\infty} f(t)g(t,s)dtds.$$

Theorem 24. Let f ∈ BV_0(R) and g : R × R → C be a function such that its partial derivative D_2g is bounded and continuous on R × R. If s_0 ∈ R is such that

there is K > 0 for which $\|g(\cdot,s_0)\|_{[u,v]} \leq$ K for all [u, v] ⊆ R, and

s_0 satisfies Hypothesis (H) relative to D_2g.

Then Γ is derivable at s_0, and

$$\Gamma'(s_0) = \int_{-\infty}^{\infty} f(t)D_2g(t,s_0)dt. \tag{19}$$

Proof. Using conditions (1) and (2) and the Mean Value theorem, there exist δ > 0 and M > 0 such that, for each s ∈ (s_0 − δ,s_0 + δ),

$$\left|\int_u^v D_2g(t,s)dt\right| < M \text{ and } \left|\int_u^v g(t,s)dt\right| < M, \tag{20}$$

for all [u, v] ⊆ R.

Let a, b be real numbers with $s_0 - \delta < a < s_0 < b < s_0 + \delta$. We use Theorem 12 to prove (19). The function f(t)g(t, ·) is differentiable on [a, b] for each t ∈ R, therefore f(t)g(t, ·) is ACG_δ on [a, b] for all t ∈ R. By (20) and Theorem 9, f(·)g(·,s) is HK-integrable on R for all s ∈ [a, b]. Then

$$\Gamma'(s_0) = \int_{-\infty}^{\infty} f(t)D_2g(t,s_0)dt$$

when, if

$$\Gamma_2 := \int_{-\infty}^{\infty} f(t)D_2g(t,\cdot)dt$$

is continuous at s_0, and

$$\int_s^t\int_{-\infty}^{\infty} f(t)D_2g(t,s)dtds = \int_{-\infty}^{\infty}\int_s^t f(t)D_2g(t,s)dsdt$$

for all [s, t] ⊆ [a, b]. The first affirmation is true by (20) and Theorem 22, and the second affirmation is true due to (20) and Theorem 23

Some Applications

An important work about the Fourier transform using the Henstock-Kurzweil integral: existence, continuity, inversion theorems etc. was published in [5]. Nevertheless, there are some omissions in that results that use the Lemma 25 (a) of [5]. Also the authors of this book chapter in [6], [3] and [4] have studied existence, continuity and Riemann-Lebesgue lemma about the Fourier transform of functions belong to HK(R) ∩ BV(R) and BV_0(R). Following the line of [6], in Theorem 26 we include some results from them as consequences of theorems above section.

Let f and g be real-valued functions on R. The convolution of f and g is the function f*g defined by

$$f * g(x) = \int_{-\infty}^{\infty} f(x-y)g(y)dy$$

for all x such that the integral exists. Several conditions can be imposed on f and g to guarantee that f ∗ g is defined on R. For example, if f is HK- integrable and g is of bounded variation.

Lemma 25. For f ∈ HK(R) ∩ BV(R), $\lim_{|x|\to\infty} f(x) = 0$.

Proof. Since f is a bounded variation function on R then the limit of f(x), as |x| → ∞, exists. Suppose $\lim_{|x|\to\infty} f(x) = \alpha \neq 0$. Take 0 < e < |α|. There exists A > 0 such that α − e < f(x), for all |x| > A. Observe that f(x) > 0 on [A, ∞), so f ∈ L([A, ∞)). Therefore the constant function α − e is Lebesgue integrable on [A, ∞), which is a contradiction.

Observe, as consequence of above Lemma, we have that the vector space HK(R) ∩ BV(R) is contained in BV0(R). So the next theorem is an immediately consequence of above section.

Theorem 26. If f ∈ HK(R) ∩ BV(R), then

1. $\hat{f}$ exists on b R.
2. $\hat{f}$ is continuous on b R \ {0}.
3. If g(t) = t f(t) and g ∈ HK(R) ∩ BV(R) , then f is differentiable on b R \ {0}, and

$$\hat{f}'(s) = -i\hat{g}(s), \quad for\ each\ s \in \mathbb{R} \setminus \{0\}.$$

4. For h ∈ L ¹ (R) ∩ BV(R), $\widehat{f * h}(s) = \hat{f}(s)\hat{h}(s)$ for all s ∈ R.

Proof. ***We observe that***

$$\left|\int_u^v e^{-its}dt\right| \leq \frac{2}{|s|}, \tag{21}$$

for all [u, v] ⊆ R. Thus, each $s_0 \neq 0$ satisfies Hypothesis (H) relative to e −its .

- Theorem 21 implies that $\hat{f}$ (s_0) exists for all $s_0 \neq 0$ and, since f ∈ HK(R), fb(0) exists. Therefore $\hat{f}$ exists on R.
- By Theorem 22, $\hat{f}$ is continuous at s_0, for all $s_0 \neq 0$.
- It follows by Theorem 12 in similar way to the proof of Theorem 24.
- Let k(x, y) = f(y − x)e $^{-iys}$, where s is a fixed real number. We get, for each y ∈ R and all [u, v] ⊆ R,

$$\left|\int_u^v k(x,y)dx\right| = \left|\int_u^v f(y-x)dx\right| = \left|\int_{y-u}^{y-v} f(z)dz\right| \leq \|f\|_A.$$

So, every real number y satisfies Hypothesis (H) relative to k. Now, observe that h ∈ BV0(R) and k is measurable and bounded. Thus, by Theorem 23,

$$\int_{-a}^{a}\int_{-\infty}^{\infty} h(x)k(x,y)dxdy = \int_{-\infty}^{\infty}\int_{-a}^{a} h(x)k(x,y)dydx, \tag{22}$$

for all a > 0.

On the other hand,

$$\left|h(x)\int_{-a}^{a} f(y-x)e^{-iys}dy\right| \leq |h(x)|\left|\int_{-a-x}^{a-x} f(z)e^{-izs}dz\right| \leq |h(x)|\|f(\cdot)e^{-i(\cdot)s}\|_A.$$

Since h ∈ L(R), using Dominated Convergence theorem, it follows that

$$\hat{f}(s)\hat{h}(s) = \int_{-\infty}^{\infty} h(x)\int_{-\infty}^{\infty} f(y-x)e^{-iys}dydx = \lim_{a\to\infty}\int_{-\infty}^{\infty} h(x)\int_{-a}^{a} f(y-x)e^{-iys}dydx.$$

Moreover, from (22), we have

$$\hat{f}(s)\hat{h}(s) = \lim_{a\to\infty}\int_{-a}^{a}\int_{-\infty}^{\infty} h(x)f(y-x)e^{-iys}dxdy = \lim_{a\to\infty}\int_{-a}^{a}(f*h)(y)e^{-iys}dy.$$

We conclude, by Hake's theorem, that

$$\widehat{f*h}(s) = \hat{f}(s)\hat{h}(s).$$

Recall that the Laplace transform, at z ∈ C, of a function f : [0, ∞) → R is defined as

$$L(f)(z) = \int_0^{\infty} f(t)e^{-zt}dt.$$

Theorem 27. If f ∈ HK([0, ∞)) ∩ BV([0, ∞)), then

L(f)(z) exists for all z ∈ C.

If F(x, y) = L(f)(x + iy), then F(·, y) is continuous on R for all $y \neq 0$, and F(x, ·) is continuous on R for all $x \neq 0$..

A SET OF FUNCTIONS IN HK(R) ∩ BV_0(R) \ L¹ (R)

Taking into account Lemma 25, the set HK(R) ∩ BV(R) is included in BV0(R) and does not have inclusion relations with L 1 (R). Since the step functions belong to HK(R) ∩ BV(R), then by Lemma 13, we have that HK(R) ∩ BV(R) is dense in HK(R). In this section we exhibit a set of functions in HK(R) ∩ BV0(R) \ L 1 (R).

Proposition 28. Let b > a > 0. Suppose that f : [a, ∞) → R is not identically zero, is continuous and periodic with period b − a. Let F(x) = $\int_a^x f(t)dt$ be bounded on [a, ∞). Moreover, assume that j : [a, ∞) → R is a nonnegative and monotone decreasing function which satisfies the next conditions:

limt→∞ j (t) = 0,

j ∈/ HK([a, ∞)).

Then the product j f ∈ HK([a, ∞)) \ L ¹ ([a, ∞)).

Proof. We take to ∈ (a, b), δ_o > 0 and γ > 0 such that

$$\gamma \leq |f(t)| \quad \text{for each} \quad t \in [t_o - \delta_o, t_o + \delta_o] \subset (a,b).$$

Periodicity of f gives

$$\gamma \leq |f(t)|$$

for each t ∈ $\bigcup_{k=0}^{\infty}$ [to − δo + k(b − a), to + δo + k(b − a)]. Therefore,

$$\begin{aligned}\int_a^{b+n(b-a)} \varphi(t)|f(t)|dt &\geq \gamma \sum_{k=0}^{n} \int_{t_o-\delta_o+k(b-a)}^{t_o+\delta_o+k(b-a)} \varphi(t)dt \\ &\geq \gamma \sum_{k=0}^{n} \int_{t_o-\delta_o+k(b-a)}^{t_o+\delta_o+k(b-a)} \varphi(t_o+\delta_o+k(b-a))dt \\ &= \gamma(2\delta_o) \sum_{k=0}^{n} \varphi(t_o+\delta_o+k(b-a)).\end{aligned} \quad (23)$$

Also,

$$\begin{aligned}\int_a^{b+n(b-a)} \varphi(t)dt &\leq \sum_{k=0}^{n} \int_{a+k(b-a)}^{b+k(b-a)} \varphi(t)dt \\ &\leq \sum_{k=0}^{n} \varphi(a+k(b-a)) \int_{a+k(b-a)}^{b+k(b-a)} dt \\ &\leq (b-a)\varphi(a) \\ &\quad +(b-a)\sum_{k=1}^{n} \varphi(t_o+\delta_o+(k-1)(b-a)).\end{aligned} \quad (24)$$

Because of j ∉ HK([a, ∞]), we get $\lim_{n\to\infty} \int_a^{b+n(b-a)} \varphi(t)dt = \infty$. Thus, equations (23) and (24) imply j f ∉ L^1 ([a, ∞)]. On the other hand, by Chartier-Dirichlet's Test of [11], the function j f belongs to HK([a, ∞)).

Corollary 29. Let α, β be positive numbers such that α + β > 1 with β ≤ 1. Suppose a > 0 and f : [a, ∞) → R obeys the hypotheses of Proposition 28. Then, the function $f_{\alpha,\beta} : [a^{1/\alpha}, \infty) \to \mathbb{R}$ defined by

$$f_{\alpha,\beta}(t) = \frac{f(t^\alpha)}{t^\beta} \quad (25)$$

is in HK([a $^{1/\alpha}$, ∞)) \ L^1 ([a $^{1/\alpha}$, ∞)).

Proof. The change of variable u = t $^\alpha$ gives,

$$\int_{a^{\frac{1}{\alpha}}}^{\infty} \frac{f(t^\alpha)}{t^\beta} dt = \int_a^{\infty} \frac{f(u)}{u^{\frac{\beta-1}{\alpha}+1}} du. \quad (26)$$

The hypotheses for α, β imply that the function $\varphi(u) = u^{-[\frac{(\beta-1)}{\alpha}+1]}$ satisfies the conditions of Proposition 28. Then, j f ∈ HK([a, ∞) \ L^1 ([a, ∞)), satisfying the statement of the corollary.

Proposition 30. Let β > α > 0 be fixed with β + α > 1. Suppose f : [a, ∞) → R is a bounded and continuous function, with bounded derivative. Then the function $f_{\alpha,\beta}$: [a $^{1/\alpha}$, ∞) → R, defined by $f_{\alpha,\beta}(t)$ = f(t α)/t β , belongs to the space BV([a $^{1/\alpha}$, ∞)).

Proof. Let M1 and M2 be bounds for f and f', respectively. We have,

$$f'_{\alpha,\beta}(t) = \frac{\alpha f'(t^\alpha)}{t^{\beta-\alpha+1}} - \frac{\beta f(t^\alpha)}{t^{\beta+1}},$$

which gives

$$\left|f'_{\alpha,\beta}(t)\right| \leq \frac{\alpha M_2}{t^{\beta-\alpha+1}} + \frac{\beta M_1}{t^{\beta+1}}.$$

Now, take x > a $^{1/\alpha}$. Since β − α > 0, then

$$\frac{\alpha M_2}{t^{\beta-\alpha+1}} + \frac{\beta M_1}{t^{\beta+1}} \in L^1([a^{\frac{1}{\alpha}}, x)).$$

A straightforward application of the Theorem 7.7 of [11] implies $f'_{\alpha,\beta} \in L^1([a^{1/\alpha}, x))$. Moreover

$$\begin{aligned}\int_{a^{\frac{1}{\alpha}}}^{x} |f'_{\alpha,\beta}(t)|dt &\leq \alpha M_2 \int_{a^{\frac{1}{\alpha}}}^{x} t^{-\beta+\alpha-1}dt \\ &\quad +\beta M_1 \int_{a^{\frac{1}{\alpha}}}^{x} t^{-\beta-1}dt \\ &= -\frac{\alpha M_2}{\beta-\alpha}\left(\frac{1}{x^{\beta-\alpha}} - \frac{1}{a^{\frac{\beta-\alpha}{\alpha}}}\right) \\ &\quad -M_1\left(\frac{1}{x^\beta} - \frac{1}{a^{\frac{\beta}{\alpha}}}\right) \\ &\leq \frac{\alpha M_2}{\beta-\alpha}\frac{1}{a^{\frac{\beta-\alpha}{\alpha}}} + \frac{M_1}{a^{\frac{\beta}{\alpha}}}.\end{aligned}$$

These estimates together with the Theorem 7.5 of [11] imply

$$V(f_{\alpha,\beta};\ [a^{\frac{1}{\alpha}}, x)) \leq \frac{\alpha M_2}{\beta-\alpha}\frac{1}{a^{\frac{\beta-\alpha}{\alpha}}} + \frac{M_1}{a^{\frac{\beta}{\alpha}}}.$$

(27)

If x tends to ∞, one gets $f_{\alpha,\beta} \in BV([a^{1/\alpha}, \infty))$.

Corollary 29 and Proposition 30 provide us Henstock-Kurzweil integrable functions defined on unbounded intervals which are not Lebesgue integrable.

Corollary 31. Let a, α, β be such that: 0 < a, 0 < α < β ≤ 1 and 1 < β + α. Suppose that f : [a, ∞) → R satisfies both the hypotheses of Corollary 29 and Proposition 30. Then, the function fα,β belongs to HK([a $^{1/\alpha}$, ∞)) ∩ BV([a $^{1/\alpha}$, ∞)) \ L 1 ([a $^{1/\alpha}$, ∞)). Taking into account the above functions we have the following corollary.

Corollary 32. Let a, α, β be such that: $0 < a$, $0 < \alpha < \beta \leq 1$ and $1 < \beta + \alpha$, and let h in BV($[-a^{1/\alpha}, a^{1/\alpha}]$). Suppose that $f : [a, \infty) \to R$ satisfies both the hypotheses of Corollary 29 and Proposition 30. Then $f : R \to R$ defined by

$$g(t) = \begin{cases} h(t) & if\ t \in (-a^{1/\alpha}, a^{1/\alpha}), \\ \dfrac{f(|t|^\alpha)}{|t|^\beta} & if\ f \in (-\infty, -a^{1/\alpha}] \cup [a^{1/\alpha}, \infty) \end{cases}$$

is in HK(R) ∩ BV(R) \ L(R).

Example 33. Let us consider the trigonometric functions sin(t) and cos(t). Then the following family of functions satisfies the hypotheses of Theorem 34.

$$\sin_\beta^\alpha : \mathbb{R} \to \mathbb{R}; \quad \sin_\beta^\alpha(t) = \chi_{1,\alpha}(t)\frac{\sin(t^\alpha)}{t^\beta},$$

$$\cos_\beta^\alpha : \mathbb{R} \to \mathbb{R}; \quad \cos_\beta^\alpha(t) = \chi_{2,\alpha}(t)\frac{\cos(t^\alpha)}{t^\beta}.$$

Here χ_1,α and χ_2,α are the characteristic functions of the intervals [π 1/α , ∞) and [(π/2) 1/α , ∞), respectively. The numbers α, β are taken as in Corollary 31.

From the above example belongs to HK(R) ∩ BV(R) \ L(R). By the Multiplier theorem it follows that HK(R) ∩ BV(R) ⊂ L^2 (R), so the above function is in BV0(R) ∩ L^2 (R) \ L(R). Therefore, there exist functions in L^2 (R) \ L(R) such that their Fourier transforms exist as in (1), as an integral in HK sense.

THE RIEMANN-LEBESGUE LEMMA AND THE DIRICHLET-JORDAN THEOREM FOR BV_0 FUNCTIONS

The Riemann-Lebesgue lemma is a fundamental result of the Harmonic Analysis. An novel aspect is the validity of this lemma for functions which are not Lebesgue integrable, since this fact could help to expand the space of functions where the inversion of the Fourier transform is possible. In this section we prove a generalization of the Riemann-Lebesgue Lemma for functions of bounded variation which vanish at infinity. As consequence, it is obtained a proof of the Dirichlet-Jordan theorem for this kind of functions. This theorem provides a pointwise inversion of the Fourier transform. We observe that the implications 1 and 2 of Theorem 26 are particularizations of the next result.

We observe that the implications 1 and 2 of Theorem 26 are particularizations of the next result.

Theorem 34 (Generalization of Riemann-Lebesgue Lemma). Let j ∈ HKloc(R) be a function such that Φ(t) = $\int_0^t \varphi(x)dx$ is bounded function on R. If f ∈ BV_0(R), then the function H(w) = $\int_{-\infty}^{\infty} f(t)\varphi(wt)dt$ is defined on R \ {0} , it is continuous and

$$\lim_{|w|\to\infty} H(w) = 0.$$

Proof. Given w ∈ R, we define $\square_w(t)$ = j (wt). Because of j ∈ HKloc(R), then j and $\square_w$ are in HK([0, b]), for b > 0. By Jordan decomposition, there exist functions f1 and f2 which are nondecreasing functions belonging to BV_0(R) such that $f = f_1 - f_2$. Hence, by Chartier-Dirichlet's Test, f □w ∈ HK([0, ∞]). By applying the Multiplier Theorem and supposing w ≠ 0, it follows

$$\begin{aligned}\int_0^\infty f(t)\varphi(wt)dt &= -\int_0^\infty \frac{\Phi(wt)}{w}df(t) \\ &= -\int_0^\infty \frac{\Phi(wt)}{w}df_1(t) \\ &\quad + \int_0^\infty \frac{\Phi(wt)}{w}df_2(t),\end{aligned} \tag{28}$$

where d f_i (t) is the Lebesgue-Sieltjes measure generated by f_i, i = 1, 2.

Let β a positive number and let M the upper bound of |Φ|. For w ∈ [β, ∞) we have that

$$\left|\frac{\Phi(wt)}{w}\right| \leq \frac{M}{\beta}. \tag{29}$$

Since Φ(wt)/w is continuous over [β, ∞) and the measures d f_i (t) are finite, then by the Dominated Convergence Theorem applied to right side integrals in (28), it follows that

$$\lim_{w\to w_0} H(w) = H(w_0),$$

for each w_0 ∈ [β, ∞). Since β is arbitrary, we obtain the continuity of H on (0, ∞).

Moreover, by (28), we have for w ∈ (0, ∞) that

$$\left|\int_0^\infty f(t)\varphi(wt)dt\right| \leq \frac{M}{|w|}Var(f;[0,\infty]).$$

Thus, we conclude that

$$\lim_{|w|\to\infty} \int_0^\infty f(t)\varphi(wt)dt = 0.$$

To complete the proof, we use similar arguments for the interval (−∞, 0].

The above theorem confirms that H ∈ C0(R \ {0}), for each f ∈ BV0(R). As corollary we have the Riemann-Lebesgue Lemma.

Corollary 35. If f ∈ BV_0(R), then $\hat{f} \in C_0(R \setminus \{0\})$.

We know that if g, h ∈ BV([a, ∞]) then gh ∈ BV([a, ∞]). Employing this fact and Theorem 34 we get the following corollary.

Corollary 36. Suppose that δ, α > 0 and f ∈ BV(R), then

$$\lim_{M\to\infty} \int_\delta^\infty \frac{f(t)}{t^\alpha}e^{-iMt}dt = 0.$$

The Sine Integral is defined as

$$Si(x) = \frac{2}{\pi}\int_0^x \frac{\sin t}{t}dt,$$

which has the properties:

Si(0) = 0, $\lim_{x\to\infty}$ Si(x) = 1 and

Si(x) ≤ Si(π) for all x ∈ [0, ∞].

We use the Sine Integral function in the proof of the following lemma.

Lemma 37. Let δ > 0. If f ∈ BV_0(R), then

$$\lim_{\varepsilon\to 0}\int_{\delta}^{\infty} f(t)\frac{\sin\varepsilon t}{t}dt = 0.$$

Proof. By Lemma 10 we have

$$\left|\int_{\delta}^{\infty}\frac{\sin\varepsilon t}{t}f(t)dt\right| \le \left|\int_{\delta}^{\infty}\left(\int_{\delta\varepsilon}^{t\varepsilon}\frac{\sin u}{u}du\right)df(t)\right|. \tag{30}$$

Since for each t ∈ [a, ∞): $\lim_{\varepsilon\to 0}\int_{\delta\varepsilon}^{t\varepsilon}\frac{\sin u}{u}du = 0$ and $\left|\int_{\delta\varepsilon}^{t\varepsilon}\frac{\sin u}{u}du\right| \le \pi Si(\pi)$ for all ε > 0. Then, we obtain the result applying the Lebesgue Dominated Convergence theorem to the integral on the right in (30).

Lemma 38. Suppose that 0 < α < β or α < β < 0. If f ∈ BV_0(R), then for all s ∈ [α, β] we have

$$\lim_{\substack{a\to-\infty\\ b\to\infty}}\int_{\alpha}^{\beta} e^{ixs}\int_{a}^{b} f(t)e^{-ist}dtds = \int_{\alpha}^{\beta} e^{ixs}\int_{-\infty}^{\infty} f(t)e^{-ist}dtds. \tag{31}$$

Proof. We will do the proof for 0 < α < β. Let $\widehat{f}_{0b}(s) = \int_0^b f(t)e^{-ist}dt$ and $\widehat{f}_0(s) = \int_0^\infty f(t)e^{-ist}dt$, which are continuous on R r {0}. Therefore the integrals in (31) exist. We know that there is R > 0 such that | f(t)| ≤ R for all t ∈ R, and that for any b > 0 : V(f ; [0, b]) ≤ V(f ; [0, ∞)). For each s ∈ [α, β] the Multiplier theorem implies

$$\left|\widehat{f}_{0b}(s)\right| \le \frac{2}{\alpha}\{R + V(f;[0,\infty))\} = N.$$

This inequality implies that for any b > 0 and all $s\in[\alpha,\beta] : \left|e^{ixs}\widehat{f}_{0b}(s)\right| \le N$, for each x ∈ R. Applying the theorem of Hake we have: $\lim_{b\to\infty}\widehat{f}_{0b}(s) = \widehat{f}_0(s)$. Then, by the Lebesgue Dominated Convergence theorem,

$$\lim_{b\to\infty}\int_{\alpha}^{\beta} e^{ixs}\widehat{f}_{0b}(s)ds = \int_{\alpha}^{\beta} e^{ixs}\widehat{f}_0(s)ds.$$

To conclude the proof, we follow a similar process over the interval [a, 0] leading a to minus infinity.

To obtain the Dirichlet-Jordan theorem we state the following lemma [22, Theorem 11.8].

Lemma 39. Let δ > 0. If g is of bounded variation on [0, δ], then

$$\lim_{M\to\infty}\frac{2}{\pi}\int_0^{\delta} g(t)\frac{\sin Mt}{t}dt = g(0+)$$

Theorem 40 (Dirichlet-Jordan Theorem). If f is a function in BV_0(R), then, for each x ∈ R,

$$\lim_{\substack{M\to\infty\\ \varepsilon\to 0}}\frac{1}{2\pi}\int_{\varepsilon<|s|<M} e^{ixs}\widehat{f}(s)ds = \frac{1}{2}\{f(x+0)+f(x-0)\}. \tag{32}$$

Proof. Let g(x, t) = f(x − t) + f(x + t) and suppose that δ > 0. By Fubini's theorem for the Lebesgue integral [22, Theorem 15.7] at [−M, −ε] × [a, b] and [ε, M] × [a, b] and Lema 38, we have

$$\begin{aligned}\int_{\varepsilon<|s|<M} e^{ixs}\int_{-\infty}^{\infty} f(t)e^{-ist}dtds &= \lim_{\substack{a\to-\infty\\ b\to\infty}}\left(\int_{-M}^{-\varepsilon}+\int_{\varepsilon}^{M}\right)e^{ixs}\int_a^b f(t)e^{-ist}dtds\\ &= \lim_{\substack{a\to-\infty\\ b\to\infty}}\int_a^b f(t)\left(\int_{-M}^{-\varepsilon}+\int_{\varepsilon}^{M}\right)e^{is(x-t)}dsdt\\ &= \int_{-\infty}^{\infty} f(t)\left(\int_{-M}^{-\varepsilon}+\int_{\varepsilon}^{M}\right)e^{is(x-t)}dsdt\\ &= 2\int_0^{\infty}\frac{g(x,t)}{t}(\sin M\,t-\sin\varepsilon t)dt\\ &= 2\int_{\delta}^{\infty}\frac{g(x,t)}{t}(\sin M\,t-\sin\varepsilon t)dt\\ &\quad +2\int_0^{\delta}\frac{g(x,t)}{t}(\sin M\,t-\sin\varepsilon t)dt.\end{aligned}$$

In [δ, ∞], by Corollary 36 and Lemma 37, we get

$$\lim_{M\to\infty,\varepsilon\to 0}\int_{\delta}^{\infty}\frac{g(x,t)}{t}(\sin M\,t-\sin\varepsilon t)dt = 0. \tag{33}$$

In [0, δ], applying the Lebesgue Dominate Convergence theorem,

$$\lim_{\varepsilon\to 0}\int_0^{\delta}\frac{g(x,t)}{t}\sin\varepsilon t dt = 0. \tag{34}$$

Now, by Lemma 39,

$$\lim_{M\to\infty}\int_0^{\delta} g(x,t)\frac{\sin Mt}{t}dt = g(x,0+) = \frac{\pi}{2}\left[f(x-0)+f(x+0)\right].$$

We conclude the proof combining (33), (34) and the above expression.

We observe that the classical theorem of Dirichlet-Jordan on L(R) is a particular case of Theorem 40. Taking into account that HK(R) ∩ BV(R) ⊂ BV_0(R), then from Theorem 34 and Theorem 40 we get that if f ∈ HK(R) ∩ BV(R), then its Fourier transform $\widehat{f}$ (s) exists in each s ∈ R; $\widehat{f}$ ∈ C_0(R\ {0}), and the expression (32) holds for each x ∈ R.

Corollary 41. There exist functions in L^2 (R) \ L(R) such that their Fourier transforms exist as in (1) and, for each x ∈ R, the expression (32) is true.

CONCLUSIONS

We present theorems about convergence of integrals of products in the completion of HK(I), which those we have a version of Riemann-Lebesgue Lemma (over compact intervals) and analogous results at Riemann-Lebesgue property, a characterization of behavior of n-th partial sum of the Fourier series. Moreover, we have gotten basic properties (existence as integral, continuity, asymptotic behavior) about Fourier transform using Henstock-Kurzweil Integral, for this was necessary to get a generalization of Riemann-Lebesgue Lemma over BV_0(R), in particular those characteristics are valid over HK(R) ∩ BV(R). This intersection does not have relation inclusion with Lebesgue integrable functions space, we give a set of functions such that it belongs to HK(R) ∩

BV(R) \ L(R). Finally we have a generalization of Dirichlet-Jordan over $BV_0(R)$.

ACKNOWLEDGMENTS

The authors acknowledge the support provided by the VIEP-BUAP, Puebla, Mexico.

REFERENCES

1. Bachman G, Narici L, Beckenstein E. Fourier and Wavelet Analysis. New York: Springer-Verlag; 1991.
2. Kahane Ch. Generalizations of the Riemann-Lebesgue and Cantor-Lebesgue lemmas. Czech. Math. J. 1980;30(1):108-117. Available from: http://tinyurl.com/myrs3wp
3. Mendoza-Torres FJ, Morales-Macías MG, Escamilla-Reyna JA, Arredondo-Ruiz JH. Several aspects around the Riemann-Lebesgue Lemma. J. Adv. Res. Pure Math. 2012 Nov; 5(3), 33-46.
4. Mendoza-Torres FJ, On pointwise inversion of the Fourier transform of BV0 functions, Ann. Funct. Anal. 2010 Dic; 1 2010(2); 112-120. Available from: http://tinyurl.com/kgtq3vd
5. Talvila E. Henstock-Kurzweil Fourier transforms. Illinois J. Math. 2002; 46: 1207-1226. Available from: http://tinyurl.com/kbf2llg
6. Mendoza-Torres FJ, Escamilla Reyna JA, Sánchez-Perales S. Some results about the Henstock-Kurzweil Fourier transform. Math. Bohem. 2009; 134(4): 379–386. Available from: http://tinyurl.com/pwtsekm
7. Zygmund A. Trigonometric Series: Vols. 1 and 2. London-New York: Cambridge University Press; 1968.
8. Mohanty P, Talvila E. A product convergence theorem for Henstock-Kurzweil integral. Real Anal. Exch. 2003-2004;29(1): 199-204. Available from: http://tinyurl.com/nb5c7a6
9. E.C. Titchmarsh. The order of magnitude of the coefficients in a generalized Fourier series. Proc. Lond. Math. Soc. (2) 1923-1924; 22: xxv-xxvi, doi:10.1112/plms/s2-22.1.1-s
10. Sánchéz-Perales S. et al. Henstock-Kurzweil integral transforms. Int. J. Math. Math. Sci. 2012(2012), Article ID 209462, 11 pages. Available from: http://www.hindawi.com/journals/ijmms/2012/209462/ref/
11. Bartle R. A Modern Theory of Integration. Graduate Studies in Mathematics, Vol. 32. Providence: American Mathematical Society; 2001.
12. Bongiorno B., Panchapagesan TV. On the Alexiewicz topology of the Denjoy space. Real Anal. Exch. 1995-1996;21(2): 604–614. Avialable from: http://tinyurl.com/m2rbdr4
13. Talvila E. The Distributional Denjoy integral. Real Anal. Exch. 2007;33(1): 51-84. Available from: http://tinyurl.com/m4zjn9b
14. Kadelburg Z, Marjanovi´c M. Interchanging two limits. Enseign. Math. 2005(8); 15-29. Availabre from: http://tinyurl.com/kyjy6p2
15. Rudin W. Principles of Mathematical Analysis. United States of America: McGraw Hill; 1964
16. Heuser H. Functional Analysis. New York: John Wiley and Sons; 1982.
17. Riesz M, Livingston AE. A Short Proof of a Classical Theorem in the Theory of Fourier Integrals. Amer. Math. Monthly 62(6) Jun. - Jul. 1955, 434-437. Available from: http://www.jstor.org/stable/2307003?seq=1
18. Talvila E. Necessary and sufficient conditions for differentiating under the integral sign. Amer. Math. Monthly 2001(108): 544-548. Available from: http://arxiv.org/pdf/math/0101012.pdf
19. Peng-Yee L. Lanzhou Lectures on Henstock Integration. Series in Real Analysis Vol.2. Singapore: World Scientific; 1989.
20. Swartz Ch. Introduction to Gauge Integrals. Singapore: World Scientific; 2001.
21. Talvila E. Limits and Henstock integrals of products. Real Anal. Exch. 1999/00;25(2): 907-918. Available from: http://tinyurl.com/ke4gfkh
22. Apostol T. Mathematical Analysis. Massachusetts: Addison-Wesley Publishing Company Reading; 1974.

CHAPTER 3

GENERALIZATION OF INTERPOLATION DFT ALGORITHMS AND FREQUENCY ESTIMATORS WITH HIGH IMAGE COMPONENT INTERFERENCE REJECTION

Jiufei Luo, Shuaicheng Hou, Xinyi Li, Qi Ouyang and Yi Zhang

Keywords: Frequency estimation, Discrete Fourier transform, Spectral leakage, Image frequency interference

INTRODUCTION

Spectral analysis based on the discrete Fourier transform (DFT) and implemented by the fast Fourier transform (FFT) has been widely used in many fields for several decades. However, there are two drawbacks in the classic version: the spectral leakage effect (SLE) caused by the lack of periodicity and the picket fence effect (PFE) due to the frequency sampling [1]. These effects will lead to significant errors in spectral analysis such as parameter estimation [2, 3]. In order to obtain accurate estimates of signal parameters, a lot of solutions were proposed [4–14]. The interpolation discrete Fourier transform (IpDFT) algorithm is one of the most popular algorithms.

Early in 1970s, interpolation algorithms based on the moduli of two FFT spectral bins were presented by Rife et al. [4]. Various improved algorithms have been put forward in the following decades, such as the weighted interpolated DFT (WIDFT) proposed by Agrez [9] and the multi-point interpolated DFT approach (WMIpDFT) by Belega and Dallet [12]. Specifically, simple analytical solutions can be obtained when the maximum sidelobe decay windows (MSDW, also known as Rife–Vincent class I windows) are adopted [11]. However, the algorithms mentioned above are all established on a very important assumption that the leakage coming from the image component plays a minor role and could be ignored [11–13]. In fact, if signals contain a small number of cycles, the negative frequency component usually will exercise great influences on the estimators [15]. Although the multi-point IpDFT methods can reduce the sensitivity to some extent, frequency estimation errors due to the spectral interference from the image component still remain significant [15], especially for the signals with a few cycles. Furthermore, it should be emphasized that the reduction of systematic errors by the weighted interpolated DFT or multi-point IpDFT methods is at the expense of worse noise properties [9, 12]. In engineering practice, noise properties usually are more important than systematic errors. The accuracy could be improved with longer records as well, but the cost is increased response time [16]. As a result, we often have to make a tradeoff between the estimation accuracy and the overall system responsiveness. Consequently, it is of great significance to propose a new and simple algorithm by which accurate parameter estimates can be achieved with short intervals, especially for those fields where real-time response is required [16–18].

Recently, Belega et al. proposed an improved three-point IpDFT which exhibits a high rejection capability with respect to the interference from negative frequency [15]. In this paper, we proposed a novel frequency estimator by which the leakage coming from image component can be further reduced compared with the algorithm proposed by Belega et al. More importantly, it also keeps good noise properties due to a two-point-based mechanism. The remaining parts are organized as follows. In Section 2, we present a concise

Citation: Jiufei Luo, Shuaicheng Hou, Xinyi Li, Qi Ouyang, Yi Zhang, "Generalization of interpolation DFT algorithms and frequency estimators with high image component interference rejection" EURASIP J. Adv. Signal Process. (2016) 2016: 30. doi:10.1186/s13634-016-0330-6.

summary of the traditional interpolation algorithms. Symbols and basic equations used throughout the paper are defined in this section as well. In Section 3, the analytical expressions of the interference from the negative frequency are derived. With the properties of the derived expressions, we generalize the interpolation DFT algorithms. In Section 4, the frequency estimators with high interference rejection of image component are proposed. In Section 5, some computer simulations are carried out and the performance of the proposed algorithms are compared with some other state-of-the-art IpDFT methods. Finally, main conclusions are drawn in Section 6.

THEORETICAL BACKGROUND

In order to explain the basis of the frequency-estimating procedure, first, let $x_{raw}(n)$ be the samples of a discrete cosine wave in the form

$$x_{\text{raw}}(n) = A_0 \cos(2\pi f_0 n\Delta T + \varphi_0) \qquad n = 0, 1, 2, \ldots \tag{1}$$

where A_0, f_0, and φ_0 are the amplitude, frequency, and phase, respectively. ΔT denotes the sampling interval and n is the index of the samples. Sampling rate $f_s = 1/\Delta T$ is supposed to fulfill the Nyquist sampling theorem so that aliasing of spectrum does not occur. When N samples are acquired, f_0 is normalized by the frequency resolution $\Delta f = f_s/N$ and is expressed as

$$\lambda_0 = \frac{f_0}{\Delta f} = \frac{f_0}{f_s/N}, \tag{2}$$

where λ_0 is the normalized frequency expressed in bins [12]. Usually, samples are weighted by a window function $w_N(n)$ before DFT

$$x(n) = x_{\text{raw}}(n) w_N(n) \tag{3}$$

The DFT of N weighted samples at the spectral line k is given by

$$X(k) = \frac{A_0}{2} e^{j\varphi_0} W_N(k - \lambda_0) + \frac{A_0}{2} e^{-j\varphi_0} W_N(k + \lambda_0), \tag{4}$$

where $W_N(\cdot)$ denotes the DTFT of the window $w_N(n)$. If the bin number with the largest magnitude is ll, then the largest magnitude is given by

$$|X(l)| = \left| \frac{A_0}{2} e^{j\varphi_0} W_N(l - \lambda_0) + \frac{A_0}{2} e^{-j\varphi_0} W_N(l + \lambda_0) \right| \tag{5}$$

The second largest magnitude is given by

$$|X(l \pm 1)| = \left| \frac{A_0}{2} e^{j\varphi_0} W_N(l - \lambda_0 \pm 1) + \frac{A_0}{2} e^{-j\varphi_0} W_N(l + \lambda_0 \pm 1) \right|, \tag{6}$$

where it takes the negative if $1 - \lambda_0 > 0$ and the positive if $1 - \lambda_0 < 0$. The second terms on the right in (5) and (6) represent the image component in the spectrum. The interference is very small compared with the first term as long as l is far from zero frequency and Nyquist frequency. In this situation, $W_N(l + \lambda_0)$ can be ignored and (5) is reduced to

$$|X(l)| = \frac{A_0}{2} |W_N(l - \lambda_0)| \tag{7}$$

Similarly, (6) can be rewritten as

$$|X(l \pm 1)| \cong \frac{A_0}{2} |W_N(l - \lambda_0 \pm 1)| \tag{8}$$

For the two-point (2p) interpolation algorithm [6, 8, 11], we introduce α defined as the ratio of the two largest magnitudes

$$\alpha = \frac{|X(l \pm 1)|}{|X(l)|} \cong \frac{W_N(l - \lambda_0 \pm 1)}{W_N(l - \lambda_0)} \tag{9}$$

It can be clearly seen from Eq. (9) that the ratio α only depends on the normalized frequency λ_0 if the data window is already known. In particular, λ_0 can be determined by simple and explicit forms when the maximum sidelobe decay windows are used [11],

$$\lambda_0 = l \pm \frac{H\alpha - H + 1}{\alpha + 1} \tag{10}$$

In (10), H denotes the number of terms in the maximum sidelobe decay window. For other windows, λ_0 can be obtained by the polynomial approximation method [13].

Similarly, with proper combination of three or more spectral lines, a ratio α can be obtained which only depends on the selected window and λ_0 [9, 12]. Once the data window is chosen, λ_0 can be solely determined by α,

$$\lambda_0 = h(\alpha) \tag{11}$$

If the maximum sidelobe decay windows are selected, α can be obtained from (10). Once α is determined, the normalized frequency can be computed and the frequency can be worked out in Eq. (2).

As shown above, the second terms on the right in Eqs. (5) and (6) representing the interferences from the image component in the spectrum have been neglected. The approximation will be reasonable if $\lambda_0 > 5$ and $\lambda_0 < N/2 - 5$. However, if λ_0 were out of the specified ranges, significant errors would be generated [15]. Therefore, it is of great importance and necessary to deduce a simple algorithm that is applicable even when λ_0 is in the extreme ranges.

GENERALIZATION OF INTERPOLATION DFT ALGORITHMS

In Section 2, it was already known that

$$X(l) = \frac{A_0}{2} e^{j\varphi_0} W_N(-\delta) + \frac{A_0}{2} e^{-j\varphi_0} W_N(2l + \delta), \tag{12}$$

where $W_N(\cdot)$ is the DTFT of the window $w_N(n)$ and

$$\delta = \lambda_0 - l \tag{13}$$

It should be pointed out that due to the actual processing requirements of causality, $w_N(n)$ is a time-shifting window, where $n = 0, 1 \cdots N - 1$. It can be obtained by

$$w_N(n) = w(n - N/2) \tag{14}$$

w(n) is assumed to be a DFT-even window [2], which is symmetric with respect to the origin. According to the time-shifting property of DFT, we have

$$W_N(k) = W(k) e^{-jk\pi}, \tag{15}$$

where W(k) is the DTFT of the window w(n) and the complex exponential factor corresponds to the time shift. W(k) is also symmetric and real because w(n) is symmetric and real. Substituting Eq. (15) into Eq. (12), X(l) can be rewritten as

$$X(l) = \frac{A_0}{2} e^{j(\delta\pi+\varphi_0)} W(-\delta) + \frac{A_0}{2} e^{-j(2l\pi+\delta\pi+\varphi_0)} W(2l+\delta) \tag{16a}$$

Because the period of the complex sinusoidal is 2π, Eq. (16a) can be further expressed as

$$X(l) = \frac{A_0}{2} e^{j(\delta\pi+\varphi_0)} W(-\delta) + \frac{A_0}{2} e^{-j(\delta\pi+\varphi_0)} W(2l+\delta) \tag{16b}$$

The real and imaginary parts are expressed as

$$X_R(l) = \frac{A_0}{2} \cos(\varphi_0 + \delta\pi)[W(-\delta) + W(2l+\delta)], \tag{17a}$$

and

$$X_I(l) = \frac{A_0}{2} \sin(\varphi_0 + \delta\pi)[W(-\delta) - W(2l+\delta)], \tag{17b}$$

respectively. Accordingly, we have

$$X_R(l \pm 1) = -\frac{A_0}{2} \cos(\varphi_0 + \delta\pi)[W(-\delta \pm 1) + W(2l+\delta \pm 1)], \tag{18a}$$

and

$$X_I(l + 1) = -\frac{A_0}{2} \sin(\varphi_0 + \delta\pi)[W(-\delta \pm 1) - W(2l+\delta \pm 1)] \tag{18b}$$

Now, we introduce two variables α_R, α_I defined as

$$\alpha_R = \left| \frac{X_R(l \pm 1)}{X_R(l)} \right| = \frac{W(-\delta \pm 1) + W(2l+\delta \pm 1)}{W(-\delta) + W(2l+\delta)}, \tag{19a}$$

and

$$\alpha_I = \left| \frac{X_I(l \pm 1)}{X_I(l)} \right| = \frac{W(-\delta \pm 1) - W(2l+\delta \pm 1)}{W(-\delta) - W(2l+\delta)} \tag{19b}$$

It is assumed, similar to that in Section 2, that λ_0 is far enough from the origin so that the leakage terms $W(2l+\delta)$ and $W(2l+\delta\pm 1)$ are very small compared to $W(-\delta)$ and $W(-\delta\pm 1)$ and could be ignored. With this assumption, we can get the following relationship

$$\alpha_R \cong \alpha_I \cong \alpha \cong \frac{W(-\delta \pm 1)}{W(-\delta)} \tag{20}$$

The relationship indicates that the ratio of the real or imaginary part can also be used for estimation if the interference from the image part can be neglected. Note that a very important phenomenon is implied from Eqs. (19a) and (19b) that the ratio is phase independent, i.e., the phase does not affect the result of frequency estimation. In other words, the frequency estimate only depends on the frequency itself.

We proceed to analyze the traditional algorithm without making approximation. The moduli of X(l) and X(l± 1) can be obtained by the square roots of J(l) and J(l+ 1), respectively, according to the results in Eqs. (17a), (17b), (18a), and (18b).

$$J(l) = \frac{A_0^2}{4} \left[W^2(-\delta) + W^2(2l+\delta) \right] + \frac{A_0^2}{2} \cos(2\varphi_0 + 2\delta\pi) W(-\delta) W(2l+\delta) \tag{21a}$$

$$J(l \pm 1) = \frac{A_0^2}{4} \left[W^2(-\delta \pm 1) + W^2(2l+\delta \pm 1) \right] + \frac{A_0^2}{2} \cos(2\varphi_0 + 2\delta\pi) W(-\delta \pm 1) W(2l+\delta \pm 1) \tag{21b}$$

Without making approximation, Eq. (10) becomes

$$\bar{\alpha} = \frac{|X(l \pm 1)|}{|X(l)|} = \frac{\sqrt{J(l \pm 1)}}{\sqrt{J(l)}} \tag{22}$$

Obviously, $\bar{\alpha}$ is phase dependent. Also, for Hanning window, (namely two-term MSD window) it can be proved that

$$\lambda_R \geq \bar{\lambda} \geq \lambda_I, \tag{23}$$

where λ_R, λ_I, and $\bar{\lambda}$ are corresponding estimation values by α_R, α_I and $\bar{\alpha}$. If $\delta \neq 0$, the first equality sign is only for $\varphi_0 + \delta\pi = 0$ or π and the second equality sign is only for $\varphi_0 + \delta\pi = \pm\pi/2$. Systematic error will be introduced no matter which of the three ratios is used in formula (11). The difference is that α_R is only affected by the real part of negative frequency and α_I is only affected by the imaginary part of the negative frequency component, while $\bar{\alpha}$ is affected by both the real and imaginary parts.

Inspired by modulus-based ratio, we can extend the ratio to a more general notation. With a proper combination of the real and imaginary parts of X(l) and X(l± 1), we can get various kinds of ratios. For example, $\hat{\alpha}$ can be defined as

$$\hat{\alpha} = \frac{|X_R(l \pm 1)| + |X_I(l \pm 1)|}{|X_R(l)| + |X_I(l)|} \tag{24}$$

Similar to $\bar{\alpha}$, we have

$$\hat{\alpha} \cong \frac{W(-\delta \pm 1)}{W_N(-\delta)}, \tag{25}$$

and for Hanning window, we can also obtain

$$\lambda_R \geq \hat{\lambda} \geq \lambda_I, \tag{26}$$

where $\hat{\lambda}$ is estimation value by $\hat{\alpha}$. It is indicated that a new estimator can be obtained if α in formula (11) is replaced by $\hat{\alpha}$. The proof of Eq. (26) is similar to that of Eq. (23). Frequency estimation errors of the four estimators when $1 < \lambda_0 < 2$ (samples are weighted by Hanning window) are shown in Figs. 1, 2, 3, and 4. In Figs. 1 and 2, it can be seen that for a certain frequency, the frequency error is constant in spite of the variable phase. It is confirmed that the frequency estimators, based on the α_R, α_I separately, are independent of the signal phase, while the estimators based on $\bar{\alpha}$, $\hat{\alpha}$ are the functions of both the frequency and the phase. Comparing the estimated errors in Figs. 1, 2, 3, and 4, we can also see that the simulation results coincide well with the theoretical analysis in Eqs. (23) and (26). The difference between the two estimators in Figs. 3 and 4 is that the developing trend of errors in Figure 3 is smoother than that in Figure 4. We now have to emphasize again that maximum absolute errors are obtained when the phase and the offset satisfy the relationship $\varphi_0+\delta\pi=0$ or $\varphi_0+\delta\pi=\pi/2$. For a certain offset, the maximum error is equal to the error shown in Figs. 1 and 2. In addition, we can also construct other types of ratio to create new estimators by combining the real and imaginary parts of X(l) and X(l± 1) properly.

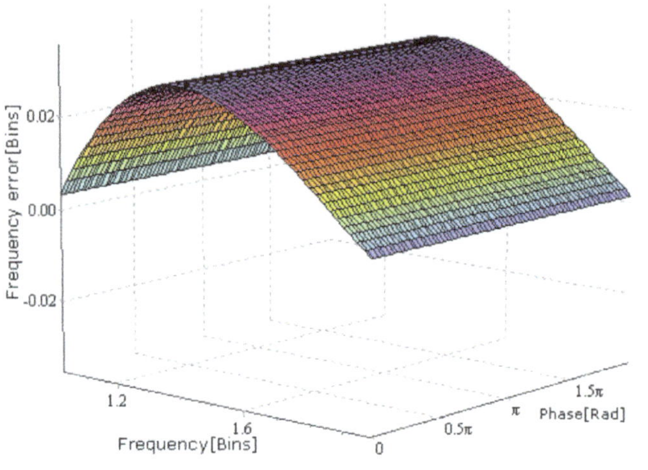

Figure 1: Frequency estimation errors when Hanning window is used ($\alpha=\alpha_R$).

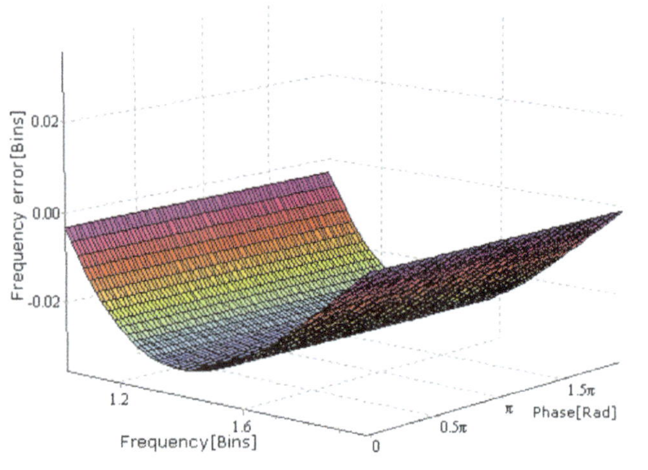

Figure 2: Frequency estimation errors when Hanning window is used ($\alpha=\alpha_I$).

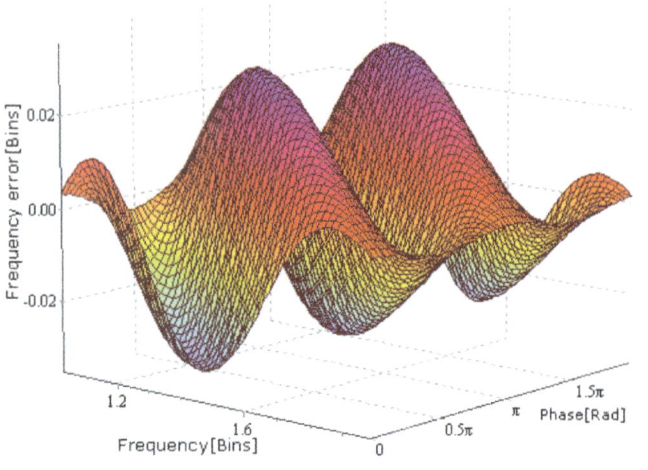

Figure 3: Frequency estimation errors when Hanning window is used ($\alpha=\bar{\alpha}$).

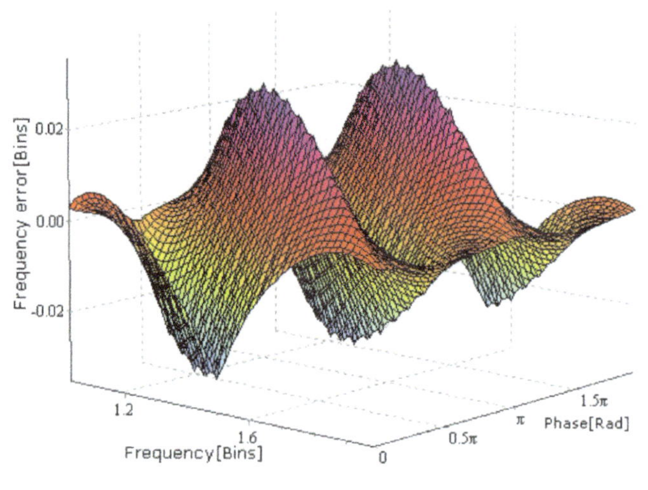

Figure 4: Frequency estimation errors when Hanning window is used ($\alpha=\hat{\alpha}$).

ALGORITHMS WITH HIGH IMAGE COMPONENT INTERFERENCE REJECTION

As we can see in Figs. 1, 2, 3, and 4, the maximum error due to negative frequency can reach as high as 0.04 for all the estimators. Simulation results show that it can be even up to nearly 0.1 for the rectangle window and up to 0.2 for the three-term maximum decay window. As a result, it is of great significance to reduce the interference resulting from the negative frequency. A new interpolated algorithm which has strong resistance against interference from the negative frequency component is proposed in this section.

Simple Algorithms with High Image Component Interference Rejection

Recalling Eqs. (19a) and (19b) and observing Figs. 1 and 2, we can infer that there is nearly a complementary relationship between the two estimators. A proper combination of the two estimators can probably result in an intrinsic rejection of negative frequency leakage. Hence, we introduce α_1 defined as the arithmetic mean value of α_R and α_I,

$$\alpha_1 = (\alpha_R + \alpha_I)/2 = \frac{W(-\delta \pm 1) - W(2l+\delta \pm 1)W(2l+\delta)/W(-\delta)}{W(-\delta) - W^2(2l+\delta)/W(-\delta)} \tag{27a}$$

Now, the interference terms are $W(2l+\delta\pm 1)W(2l+\delta)/W(-\delta)$ and $W^2(2l+\delta)/W(-\delta)$, much smaller than the original terms $W(2l+\delta\pm 1)$ and $W(2l+\delta)$. Most errors are eliminated by a simple arithmetic average operation. Similarly, we can also introduce α_2, α_3, and α_4 defined as the geometric mean value, the harmonic mean value, and the quadratic mean of α_R and α_I, respectively, as indicated in

$$\alpha_2 = \sqrt{\alpha_R \alpha_I} = \sqrt{\frac{W^2(-\delta \pm 1) - W^2(2l+\delta \pm 1)}{W^2(-\delta) - W^2(2l+\delta)}}, \tag{27b}$$

$$\alpha_3 = \frac{2}{1/\alpha_R + 1/\alpha_I} = \frac{\alpha_2^2}{\alpha_1}, \tag{27c}$$

and

$$\alpha_4 = \sqrt{(\alpha_R^2 + \alpha_I^2)/2} = \sqrt{2\alpha_1^2 - \alpha_2^2} \quad (27d)$$

Similar to α_1, the interference terms of α_2 become much smaller as well. α_3 and α_4 can be written as simple functions of α_1 and α_2, respectively. The values of the four means are approximately equal. To demonstrate the ability of negative frequency leakage rejection, the frequency errors of the four estimators are displayed when $1<\lambda_0<2$. Note that the four estimators are also phase independent because no phase information is involved in the ratios shown in Eqs. (27a)–(27d). Consequently, the phase was just set to zero. Figures 5 and 6 show the estimation results as a function of the frequency. It can be seen that the frequency errors sharply decrease compared with the modulus-based algorithms. In particular, the remaining error for the one based on the harmonic mean value is less than 10^{-3}, which is small enough for the engineering practice.

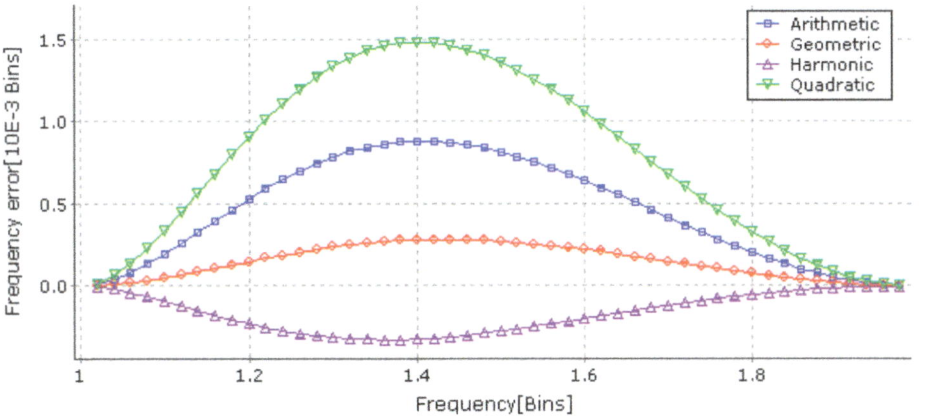

Figure 5: Frequency estimation errors based on different means of α_R and α_I when Hanning window is used.

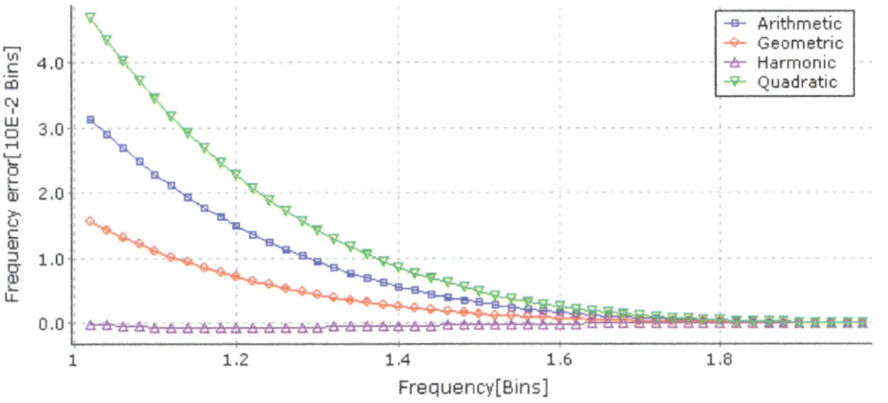

Figure 6: Frequency estimation errors based on different means of α_R and α_I when three-term MSD window is used.

Further Improved Interpolation Algorithms with Slide DFT

However, there is a serious defect in the above algorithms in which the weighed ratio is used. The algorithms may become quite vulnerable if $\cos(\varphi_0+\delta\pi)\approx 0$ or $\sin(\varphi_0+\delta\pi)\approx 0$. Under such two circumstances, the imaginary parts or the real parts would be so small that even a small disturbance would lead to a dramatic change in α_R or α_I, resulting of significant errors in the final frequency estimates. Theoretical cosine waves corrupted by low-level random noise were generated to confirm the defect, and the vulnerability of two frequency estimators based on $\alpha_R$and α_I is shown in Figs. 7 and 8, respectively. It is clearly shown that radical changes appear when $\cos(\varphi_0+\delta\pi)\approx 0$ for α_R and $\sin(\varphi_0+\delta\pi)\approx 0$ for α_I. Sharp peaks along these lines are observed in the surface of the frequency errors. They look like knife-edge arêtes so we call them "Luo–arêtes." In contrast, for the frequencies that are not in the vicinity of these lines, errors are very small and remain stable.

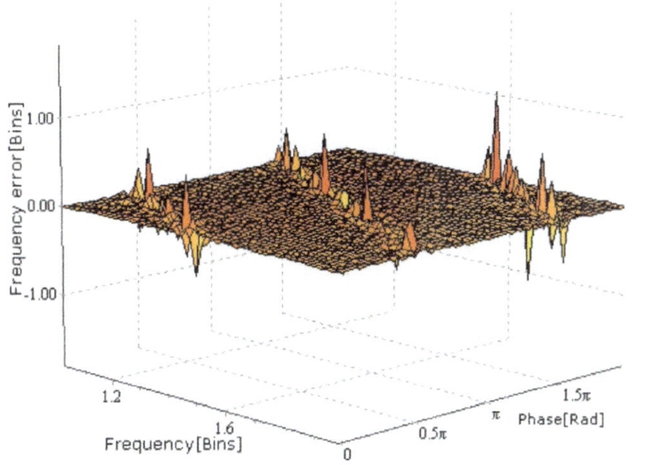

Figure 7: The phenomenon of "Luo–arêtes" in the α_R-based estimator (Hanning window).

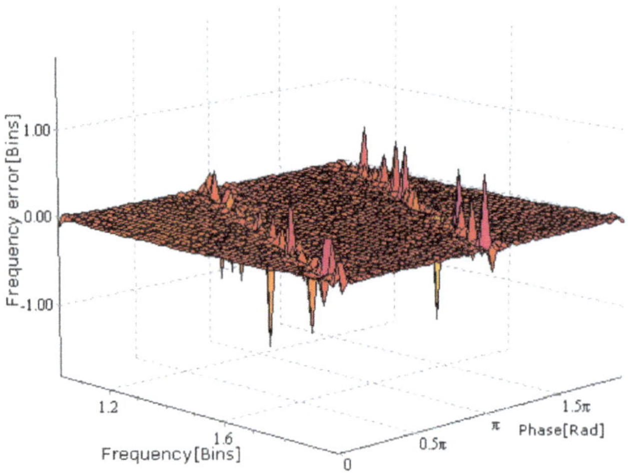

Figure 8: The phenomenon of "Luo–arêtes" in the α_I-based estimator (Hanning window).

To avoid the "Luo–arêtes," a further improved algorithm has been proposed. The fact that $\alpha_R$and α_I (including various kinds of mean values of the two) are phase independent and the change of phase has no influences on frequency estimates help us get a robust ratio by time-shifting technique. Now, consider a discrete sequence with N samples $x(0), x(1)\cdots x(N-1)$ and assume that its observed phase is φ_1. After delaying for L samples, we get a time-shifted sequence $x(L), x(L+1)\cdots x(N+L-1)$ and its observed phase φ_2 can be expressed as

$$\phi_2 = \phi_1 + 2\pi f_0 L\Delta T = \phi_1 + 2\pi f_0 L/f_s = \phi_1 + 2\pi\lambda_0 L/N \quad (28)$$

In the above equation, ΔT is the sampling interval, f_0 is the theoretical frequency and λ_0 is the normalized frequency scaled by frequency resolution. As λ_0 is unknown, we can use its largest bin number l, instead.

$$L \approx N\frac{\phi_2 - \phi_1}{2\pi l} \quad (29)$$

The actual observed phase of the time delay sequence is $\varphi'_2 = \varphi_1 + 2\pi lL/N$ and the phase error is

$$\phi_2' - \phi_2 = \Delta\phi_2 = 2\pi\delta L/N \quad (30)$$

Generally, we have $l > 1$ and $\varphi_2 - \varphi_1 < \pi/2$ so that $L/N < 1/4$. Considering $\delta \in [-0.5, 0.5)$, the absolute phase error is less than $\pi/4$. For α_R, if φ_2 was set to 0 or π, φ'_2 was limited in the range of $(-\pi/4, \pi/4)$ or $(3\pi/4, 5\pi/4)$. For α_I, if φ_2 was set to $\pi/2$ or $-\pi/2$, φ'_2 was limited in the range of $(\pi/4, 3\pi/4)$ or $(-3\pi/4, -\pi/4)$. It is indicated that we can adjust the observed phase by the time-shifting technique so that we can get a robust ratio to avoid the "Luo–arêtes." In addition, we can obtain the spectral lines l and $l \pm 1$ of the time-delayed sequence by means of the sliding discrete Fourier transform (SDFT) [19, 20]. It will be more efficient compared with another separate FFT or DFT.

COMPARISON WITH OTHER STATE-OF-THE-ART METHODS

In this section, some computer simulations were conducted to verify the effectiveness and the accuracy of the proposed algorithms. For conciseness, only the results of the algorithm based on the harmonic mean of α_R and α_I were shown. All simulation results are returned by the algorithm proposed in Section 4.2. In addition, the results of the traditional IpDFT algorithm (2pIpDFT) [5, 8, 11], the classical three-point IpDFT algorithm (3pIpDFT) [9, 12], Quinn's two-point-based estimator (Quinn1, only applicable for Hanning window), and Quinn's three-point-based optimal estimator (Quinn2, only applicable for Hanning window) [21–23], the estimator proposed by Macleod in 1998 (Macleod, only applicable for Hanning window) [24], the three-point complex spectrum-based estimator (Jacobsen, only applicable for Hanning window) [25] referred by Jacobsen and Kootsookos in 2007, and the improved three-point IpDFT algorithm with high image frequency interference rejection capability recently proposed by Belega (Belega14) [15], were also displayed for comparison. In all the considered IpDFT-based algorithms, both the two-term and three-term MSD windows were adopted. For simplicity but without loss of generality, parameters used in all the simulation experiments were as follows, the amplitude of the cosine wave $A = 1$, the number of samples $N = 512$, and the sampling rate $f_s = 512$. The results shown in this section were scaled by frequency resolution and expressed in bins.

Theoretical Cosine Wave without Noise

To demonstrate the excellent rejection capability against the interference from negative frequency, theoretical signals contain a small number of cycles. The normalized frequency λ_0 is varied in the range (1, 11) with a step of 1/8. For each frequency, the phase θ is varied in the range $[-\pi, \pi)$ with a step of $\pi/72$. The maximum absolute frequency errors $|\delta|_{max}$ are shown in Figs. 9 and 10 as a function of λ_0 for the two-term and three-term MSD windows, respectively. It is clearly revealed that the errors due to negative frequency interference are remarkably reduced. In general, both the proposed method and the improved three-point IpDFT method outperform other estimators throughout the entire range of considered λ_0. When the Hanning window is adopted, Jacobsen's estimator has the worst performance. The traditional IpDFT algorithm, Quinn's two-point-based estimator, and Quinn's three-point-based optimal estimator have similar trend. The classical three-point IpDFT algorithm and Macleod's estimator provide better results than the above three estimators. If $\lambda_0 < 4.5$ (especially $\lambda_0 < 1.5$) and $\delta > 0$, the new method provides better performance than the improved three-point IpDFT and the opposite holds if $\delta < 0$. When $\lambda_0 > 4.5$, the improved three-point IpDFT shows a small advantage than the proposed algorithm. When the three-term MSD window is adopted, the new estimator has overwhelming advantage over the rest. It should be pointed out that regardless of the adopted window type and the value of λ_0, the maximum frequency error of the new algorithm never goes above 10^{-3}, even though only one or two cycles are obtained.

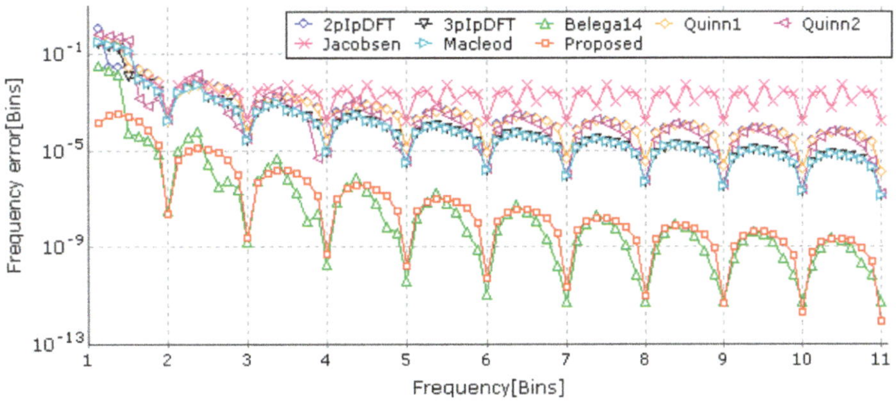

Figure 9: Maximum estimation error as a function of λ_0 for different estimators without noise (Hanning window).

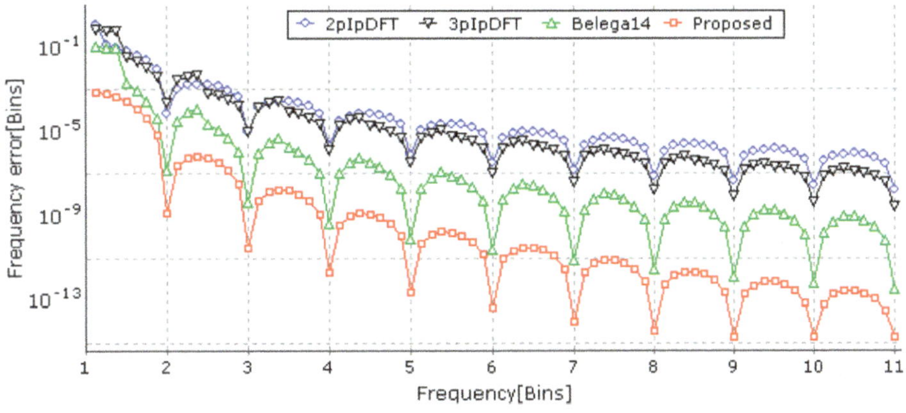

Figure 10: Maximum estimation error as a function of λ_0 for different estimators without noise (three-term MSD window).

Theoretical Cosine Wave Corrupted by Additive Noise

In this subsection, we considered the ideal cosine wave contaminated with the additive Gaussian noise. Similar to [15], we investigated the RMSE of estimates returned by the considered estimators as a function of λ_0 for certain SNRs. For each frequency, 50,000 instances were generated with a random phase. For conciseness, four algorithms are considered, including the traditional IpDFT algorithm (2pIpDFT), the classical three-point IpDFT algorithm (3pIpDFT), the improved three-point IpDFT algorithm with high image frequency interference rejection capability (Belega14), and the proposed algorithm. Results with SNR=30 dB and 50 dB for Hanning window and three-term MSD window were shown in Figs. 11, 12, 13, and 14, respectively. The results for Hanning window agree well with those in [15]. As shown in Figs. 11and 12, the estimated RMSE of the proposed method is always in a low level with a small fluctuation and essentially the novel method has no competitor for $\lambda_0<3$. When λ_0 becomes larger, the estimated RMSE of four estimators tend to be in a similar level and it is interesting to find that the two-point-based algorithms show a better performance at the worst incoherent sampling condition ($\delta\approx0.5$), while the three-point-based algorithms provide better results whenλ_0 is close to integer values ($\delta\approx0$).

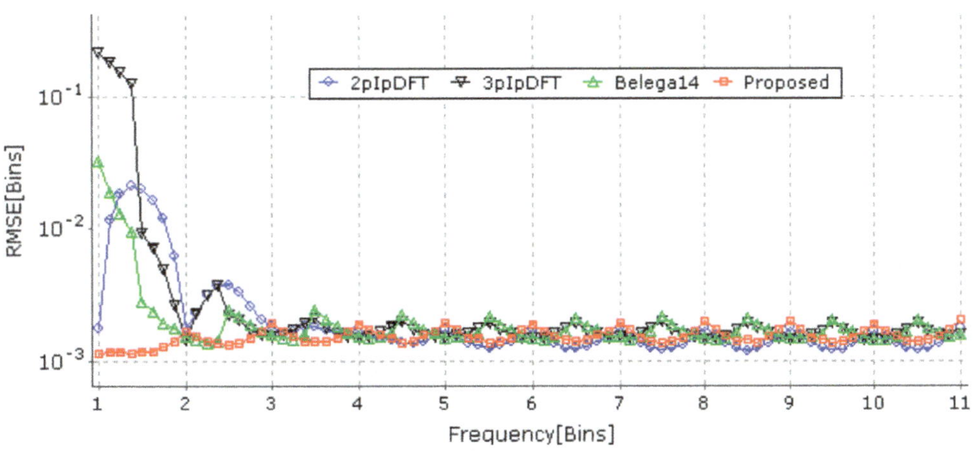

Figure 11: RMSE for different algorithms (SNR=30 dB, Hanning window).

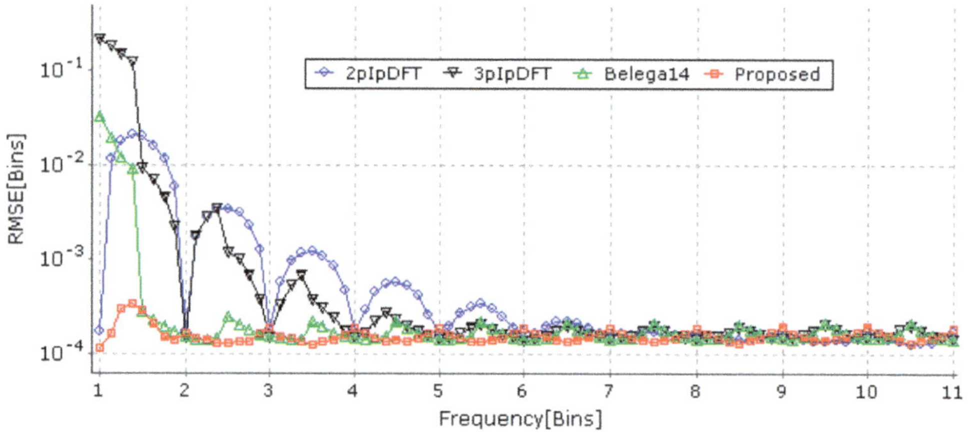

Figure 12: RMSE for different algorithms (SNR=50 dB, Hanning window).

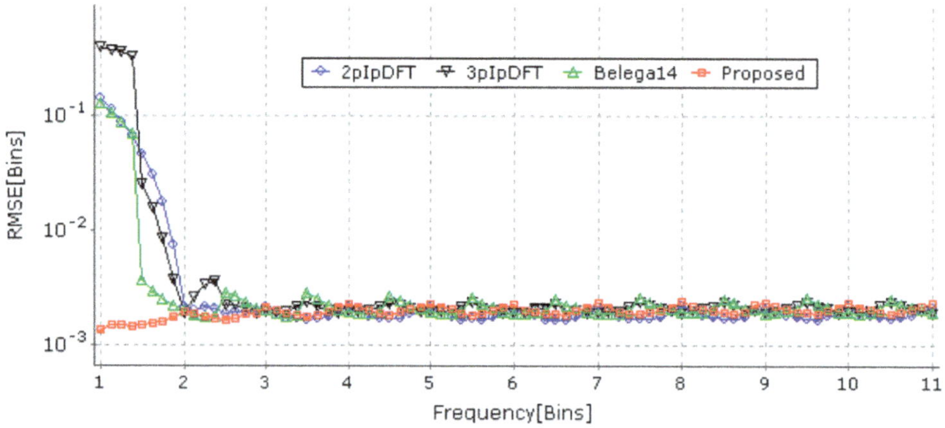

Figure 13: RMSE for different algorithms (SNR=30 dB, three-term MSD window).

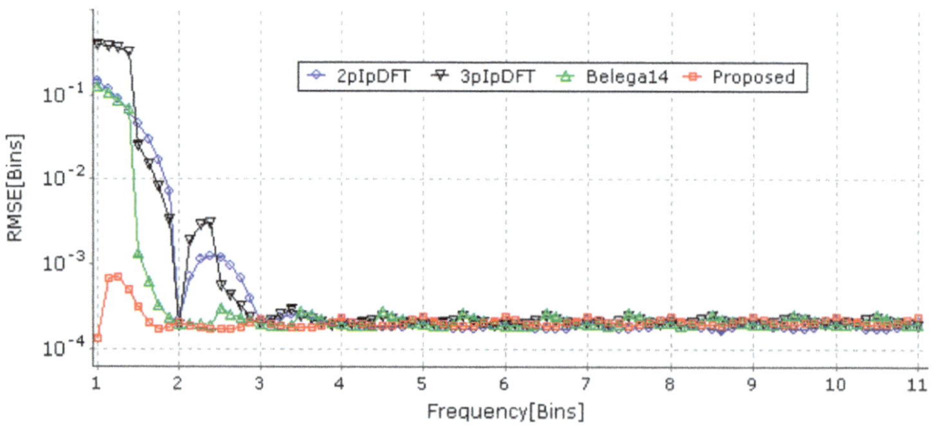

Figure 14: RMSE for different algorithms (SNR=50 dB, three-term MSD window).

When the three-term MSD window is adopted, the overall trend is similar to that of Hanning window. For $\lambda_0<2$, the performance of the traditional 2pIpDFT and 3pIpDFT is worse than Hanning window due to the wider mainlobe of the three-term MSD window. For $\lambda_0>2$, they have better performances because of the faster sidelobe decay rate. For the same SNR, the RMSE values of the two algorithms employing three-term MSD window decrease faster than that employing Hanning window before reaching the ultimate stable level. Meanwhile, the ultimate stable level is higher than that employing Hanning window because of the worse noise properties of the three-term MSD window. The performance of the proposed algorithm maintains its superiority to the other three algorithms for $\lambda_0<3$. To sum up, the traditional two-point- or three-point-based algorithms are very good choices because of their simplicity when random noise has a significant influence on the uncertainty of estimates. The proposed algorithm is strongly recommended when the image frequency interference is the main error source especially when a small number of cycles are contained in samples.

CONCLUSIONS

Frequency estimation by the IpDFT method is studied in this paper. We have quantitatively analyzed the influences of the interference from the image component and generalized the interpolated DFT algorithms. Based on the analysis, novel frequency estimators are proposed, which have strong rejection against the high image component interference. Accuracy of the novel algorithm has been confirmed by simulations. Comparative studies reveal that the proposed algorithm has a better performance than the traditional algorithms when the spectral interference from negative frequency component is significant, especially for the cycles less than one and a half. This proposed algorithm is simple to understand, easy to implement, and very suitable for real-time analysis.

ACKNOWLEDGEMENTS

The authors want to thank the anonymous reviewers for their helpful comments, which significantly improved the quality of the paper. This research was supported by the Science Foundation for Young Scientists (Grant No. E010A2015063) and Startup Fund for Doctors (Grant No. E010A2015037) of Chongqing University of Posts and Telecommunications, Chongqing Science and Technology Commission (Grant No. cstc2015jcyjB0241) and also was partly supported by the National Natural Science Foundation of China (Grant No. 51374264).

REFERENCES

1. KF Chen, JT Jiang, S Crowsen, Against the long-range spectral leakage of the cosine window family. Comput Phys Commun 180, 904–911 (2009)
2. FJ Harris, On the use of windows for harmonic analysis with the discrete Fourier transform. Proc IEEE 66, 51–83 (1978)
3. J Luo, M Xie, Phase difference methods based on asymmetric windows. Mech Syst Signal Process 54, 52–67 (2015)
4. DC Rife, G Vincent, Use of the discrete Fourier transform in the measurement of frequencies and levels of tones. Bell Syst Tech J 49, 197–228 (1970)
5. VK Jain, WL Collins, DC Davis, High-accuracy analog measurements via interpolated FFT. Instrumentation and Measurement, IEEE Transactions on 28, 113–122 (1979)
6. T Grandke, Interpolation algorithms for discrete Fourier transforms of weighted signals. Instrumentation and Measurement, IEEE Transactions on 32, 350–355 (1983)
7. M Xie, K Ding, Corrections for frequency, amplitude and phase in a fast Fourier transform of a harmonic signal. Mech Syst Signal Process 10, 211–221 (1996)
8. J Luo, Z Xie, M Xie, Frequency estimation of the weighted real tones or resolved multiple tones by iterative interpolation DFT algorithm. Digital signal processing 41, 118–129 (2015)MathSciNet
9. D Agrez, Weighted multipoint interpolated DFT to improve amplitude estimation of multifrequency signal. Instrumentation and Measurement, IEEE Transactions on 51, 287–292 (2002)
10. J Luo, Z Xie, M Xie, Interpolated DFT algorithms with zero padding for classic windows. Mech Syst Signal Process 70–71, 118–129 (2016)
11. D Belega, D Dallet, Multifrequency signal analysis by Interpolated DFT method with maximum sidelobe decay windows. Measurement 42, 420–426 (2009)
12. D Belega, D Dallet, D Petri, Accuracy of sine wave frequency estimation by multipoint interpolated DFT approach. Instrumentation and Measurement, IEEE Transactions on 59, 2808–2815 (2010)
13. J-R Liao, C-M Chen, Phase correction of discrete Fourier transform coefficients to reduce frequency estimation bias of single tone complex sinusoid. Signal Process 94, 108–117 (2014)
14. J-R Liao, S Lo, Analytical solutions for frequency estimators by interpolation of DFT coefficients. Signal Process 100, 93–100 (2014)
15. D Belega, D Petri, D Dallet, Frequency estimation of a sinusoidal signal via a three-point interpolated DFT method with high image component interference rejection capability. Digital Signal Processing 24, 162–169 16.
16. D Belega, D Petri, Accuracy analysis of the multicycle synchrophasor estimator provided by the interpolated DFT algorithm. IEEE Trans Instrum Meas 62, 942–953 (2013)
17. P Castello, M Lixia, C Muscas, PA Pegoraro, Impact of the model on the accuracy of synchrophasor measurement. Instrumentation and Measurement, IEEE Transactions on61, 2179–2188 (2012)
18. Y Tu, H Zhang, Method for CMF signal processing based on the recursive DTFT algorithm with negative frequency contribution. Instrumentation and Measurement, IEEE Transactions on 57, 2647–2654 (2008)
19. E Jacobsen, R Lyons, The sliding DFT. Signal Processing Magazine, IEEE 20, 74–80 (2003)
20. K Duda, Accurate, guaranteed stable, sliding discrete Fourier transform [DSP tips & tricks]. Signal Processing Magazine, IEEE 27, 124–127 (2010)
21. B.G. Quinn, Frequency estimation using tapered data, in Proceedings of IEEE International Conference on Acoustics, Speech and Signal Processing, ICASSP 2006, (IEEE, Piscataway, 2006), pp. 73–76.
22. B.G. Quinn, in Handbook of Statistics: Time Series Analysis: Methods and Applications, T.S. Rao, S.S. Rao, C.R. Rao (Eds.). The estimation of frequency, (Elsevier Press, North Holland,2012), pp. 585–621.
23. B.G. Quinn, E.J. Hannan, The Estimation and Tracking of Frequency, (Cambridge University Press, New York, 2001), pp. 180–206.
24. MD Macleod, Fast nearly ML estimation of the parameters of real or complex single tones or resolved multiple tones. IEEE Trans Signal Process 46, 141–148 (1998)
25. E Jacobsen, P Kootsookos, Fast, accurate frequency estimators [DSP tips & tricks]. IEEE Signal Process Mag 24, 123–125 (2007)

CHAPTER 4 VECTORIAL SIGNATURES FOR PATTERN RECOGNITION

Jesús Ramón Lerma-Aragón and Josué Álvarez-Borrego

INTRODUCTION

Due to the variety of shapes and sizes that present both living organisms and static objects, the necessity to look for automated systems of identification, both in industry and scientific research has arisen. During the 1960s, the scientific community in the field of optics has used the Fourier transform and other types of mathematical transformations for pattern recognition, taking advantage of their different properties; for example, invariance to position, rotation and scale.

Many invariant descriptors use the Fourier transform to extract invariant features. The Fourier transform has been a powerful tool for pattern recognition. One important property of the Fourier transform is that a shift in the time domain causes no change in the magnitude spectrum. This can be used to extract invariant features in pattern recognition.

In the last few years, they have been used as a tool in the digital processing of pattern recognition. Recently several books have been published (Pratt, 2007, Gonzalez et al., 2008 and 2009, Cheriet et al., 2007, and Obinata, 2007), that show us a general view, the progress and development of pattern recognition, as well as different tools for image pre-processing, extraction, selection and creation of features, classification methods, using different types of transforms, etc. However, some studies have focused on solving the problem of invariance to rotation and scale (Cohen, 1993, Casasent, 1976a, Pech et al, 2001 and Pech et al, 2003, Solorza and Álvarez-Borrego, 2010, Solorza and Álvarez-Borrego, 2011).

For practical implementation of the theory it should be considered, according to Casasent and Psaltis (1976a, 1976b, 1976c), a system that is invariant to scale through the manipulation of the Fourier transform directly changing the input function. We must also consider the practical realization of the Mellin transformation given by a logarithmic mapping of the input stage followed by a Fourier transform.

Schwartz (1994) found that there is strong evidence in many physiological and psychophysical visual systems (including human) use of such logarithmic mapping between the retina and visual cortex. The motion of changing a scaling to a shift by a logarithmic transformation of the coordinates occurs in many areas. In fact, the log-polar coordinate system described above seems to have a biological analogue. Since biological systems can shift objects to the center of the field of view by movement of the eyes, it seems logical that a mapping that facilitates scale and rotation invariant recognition would be most useful. This is not to say that biological systems also compute Fourier transforms of these representations.

Other methods can take advantage of representation in which transformations are reduced to shifts.

The different techniques used today have difficulties, the main one being that the calculations involve a high computational complexity. Moreover, in this case, the complexity of the operation performed is directly proportional to the resolution of the images used. The computational cost associated inevitably increase the time required for comparison. Emerging applications of computer vision and pattern recognition require the development of new algorithms. The

Citation: Jesus Ramon Lerma-Aragon and Josue Alvarez-Borrego (2012). Vectorial Signatures for Pattern Recognition, , Dr Salih Salih (Ed.), InTech, DOI: 10.5772/36964.

vectorial signatures techniques have an important role to play in this context, given their simplicity and low computational requirements.

In this chapter we will focus on examining the capabilities of vectorial signatures, as a method of recognizing objects, and we will consider inside this algorithm invariance to position, rotation and scale. It will be applied here to the recognition of alphabetical letters in Arial typeface, copepod species and some butterflies species (color images). However, this algorithm can be applied to any kind of object. The system was developed by taking advantage of the properties of the Fourier, Mellin and Scale transforms.

MATHEMATICAL DESCRIPTION

In this section we present some basic concepts, such as the definition of Fourier, Mellin and Scale transforms, tools that are of great usefulness in audio and image processing, noise reduction signals (such as white noise) frequency analysis of any discrete signal, materials analysis and statistical synthesis by inverse Fourier transform, and so on.

Fourier Transform

The Fourier transform F(u) of a continuous function f(x) of a single variable is defined

$$F(u)=\int_{-\infty}^{\infty} f(x)e^{-j2\pi ux}dx \tag{1}$$

Where $j=\sqrt{-1}$. Consequently, the inverse Fourier transform (IFT) of F(u) results in the original function, described by

$$f(x)=\int_{-\infty}^{\infty} F(u)e^{j2\pi ux}du \tag{2}$$

The Fourier-Mellin transform

Mellin transform is especially useful for the scale invariant (Bracewell, 1978). This has been applied to the spatially varying image restoration and in the analysis of networks that vary with time, among others.

The two dimensional Mellin transform M(ju,jv) of a function f(x,y) along the imaginary axis is defined by Casasent and Psaltis (1976a)

$$M(ju,jv)=\int_0^{\infty}\int_0^{\infty} f(x,y)x^{-ju-1}y^{-jv-1}dxdy \tag{3}$$

The Scale Transform

Cohen (1993) introduced the "scale transform." This transform is said to be scale invariant, thus meaning that the signals differing just by a scale transformation (compression or expansion with energy preservation) have the same transform magnitude distribution. Cohen showed that the scale transform is a restriction of the Mellin transform on the vertical line p = −jc + 1/2, with c ∈ R.

We choose the scale transform, which is a restriction of the Fourier-Mellin transform (Casasent, 1976c). The main property of the scale transform is the scale invariant property.

If we call c the scale variable, then in two dimensions the scale transform and its inverse is given by Cristobal & Cohen (1996).

$$D(c_x,c_y)=\frac{1}{\sqrt{2\pi}}\int_0^{\infty}\int_0^{\infty} f(x,y)e^{\left(-jc_x\ln x-jc_y\ln y\right)}\frac{dxdy}{\sqrt{xy}}, \tag{4}$$

$$f(x,y)=\frac{1}{\sqrt{2\pi}}\int_{-\infty}^{\infty}\int_{-\infty}^{\infty} D(c_x,c_y)e^{\left(jc_x\ln x+jc_y\ln y\right)}\frac{dc_xdc_y}{\sqrt{xy}}. \tag{5}$$

IMPLEMENTATION

In this section we will use two-dimensional scale transformations now written in polar coordinates $D(c_r,c_\theta)$, because this transform is invariant to changes in size and rotation

$$D(c_r,c_\theta)=\frac{1}{\sqrt{2\pi}}\int_0^{\infty}\int_0^{2\pi} f(r,\theta)r^{\left(-jc_r-\frac{1}{2}\right)}e^{(-jc_\theta\theta)}drd\theta, \tag{6}$$

and taking the logarithm of the radial coordinate, that is λ= lnr, we get

$$D(c_\lambda,c_\theta)=\frac{1}{\sqrt{2\pi}}\int_0^{\infty}\int_0^{2\pi} f(\lambda,\theta)r^{\left(-jc_r-\frac{1}{2}\right)}e^{\frac{\lambda}{2}}e^{(\lambda c_\theta+\theta c_\theta)}d\lambda d\theta. \tag{7}$$

Applying the transformation described above, we built a new algorithm. Fig. 1 shows the block diagram of the methodology used. The target (I i) to be recognized is denoted by the function f(x,y) (step 1) and the modulus of its Fourier transform (F) is obtained (step 2). A parabolic filter to the modulus of F is applied (step 3). In this way, low frequencies are attenuated and high frequencies are enhanced in proportion to $(w_x)^2$, $(w_y)^2$ (Pech-Pacheco et al., 2003), where w_x and w_y are the angular frequencies. The scale factor $\sqrt{r}$, (step 4) is applied to the results of step 3 (r is the radial spatial frequency). This process is what differentiates the scale transform from the Mellin transform. After these steps, we mapped the Cartesian coordinates to polar-logarithmic coordinates (step 5), to obtain invariance to rotation. In this step we introduced a bilinear interpolation of the first data of coordinate conversion. This is done to avoid the aliasing due to the log-polar sampling. A subimage denoted by M(λ,θ) (Lerma-Aragón & Álvarez-Borrego 2008) is chosen from step 5, which contains more information of the target and a bilinear interpolation is done again in order to resize the subimage data to its original size (step 6). From the subimage, we obtain two 1-D vectors by projecting M(λ,θ) onto the x and y axes (steps 7 and 8). In other words, we compute the marginals, according to the equations

$$M(\lambda_m)=\sum_{k=1}^{n} M(\lambda_m,\theta_k), \tag{8}$$

$$M(\theta_n)=\sum_{k=1}^{m} M(\lambda_k,\theta_n). \tag{9}$$

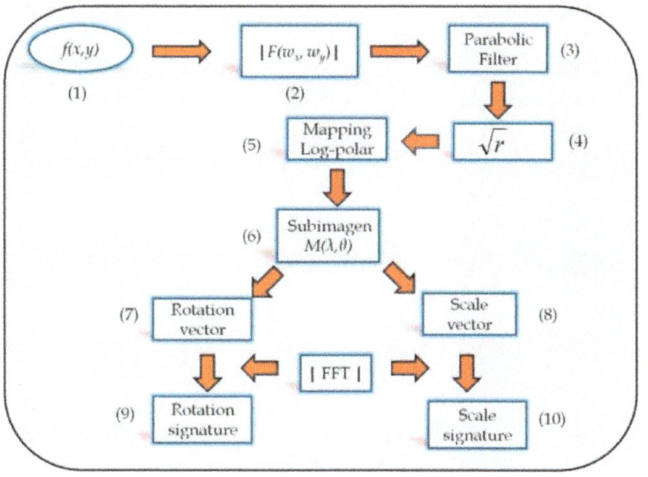

Figure 1. Block diagram of used procedure.

A modulus of the Fourier transform is calculated in order to obtain the rotation and scale signatures (equations 10 and 11), which will be the two unidimensional vectors for the target (steps 9 and 10).

$$V_1(w_\lambda) = \left|F\left[M(\lambda_m)\right]\right|, \tag{10}$$

$$V_2(w_\theta) = \left|F\left[M(\theta_n)\right]\right|, \tag{11}$$

To evaluate this algorithm, the target image (I^i) and an image problem (I^j) (I^j is denoted by the function g(x,y) either rotated or scaled) are selected, to which we apply the procedure described above. To determine the similarity between the images I^i and I^j in the database, using the Euclidean distance (E_d) is calculated among their average firms using the following equation

$$E_d = \sqrt{\sum\left\{\left[V_{1f}(w_\lambda) - V_{1g}(w_\lambda)\right]^2 + \left[V_{2f}(w_\theta) - V_{2g}(w_\theta)\right]^2\right\}}. \tag{12}$$

Thus, the signatures of all the j images of the database can be compared with any image i to be recognized.

Application of the Algorithm

To apply the algorithm based on the theory described above, is considered primarily a target image, in this particular case, we consider a simple image (rectangle) of size 256 X 256 pixels and black background, as shown in Fig. 2(a), which is calculated the magnitude of the Fourier transform, which can be seen in Fig. 2(b). We used Matlab colormap command to change to a color figure and of this way clearly visualize each step of the procedures. When these values are scaled linearly for display, the brightest pixels will dominate the display, at the expense of lower (and just as important) values of the spectrum. If, instead of displaying the values in this manner, we apply logarithmic transformations (Gonzalez, 2009) to the spectrum values, then the range of values of the result is a more manageable number. Fig. 3 shows the result of scaling this new range linearly and displaying the spectrum. The wealth of detail visible in this image compared to a straight display of the spectrum is evident from these pictures.

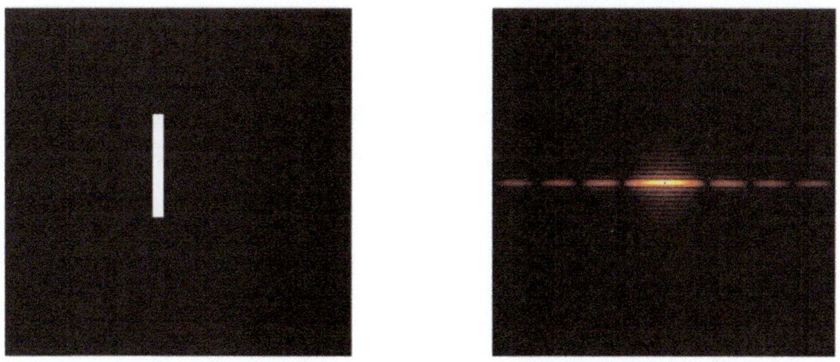

(a) (b)

Figure 2. (a) Simple image of a rectangle. (b) The corresponding centered spectrum.

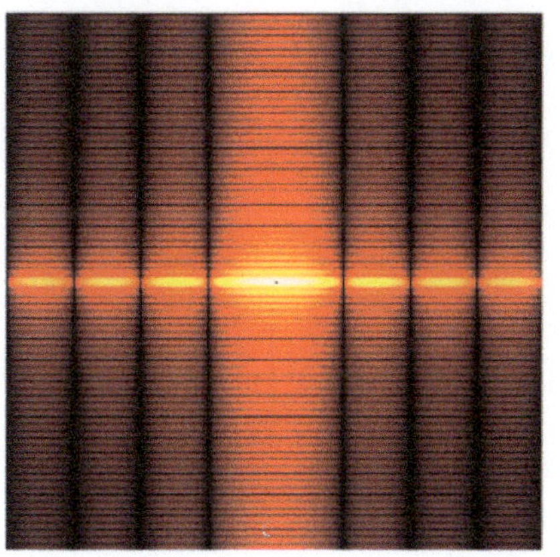

Figure 3. Result showing increased detail after a log transformation.

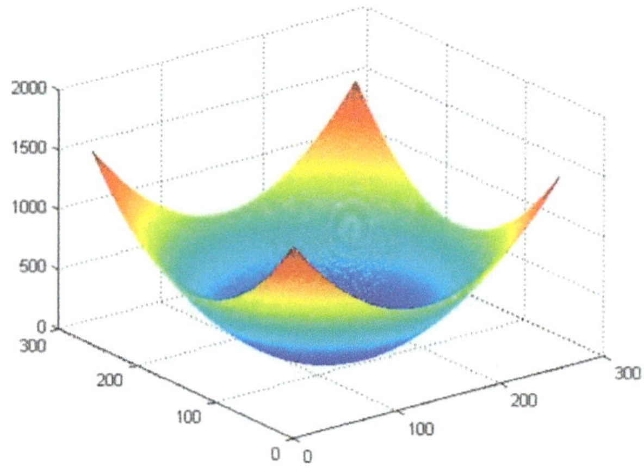

Figure 4. Parabolic filter.

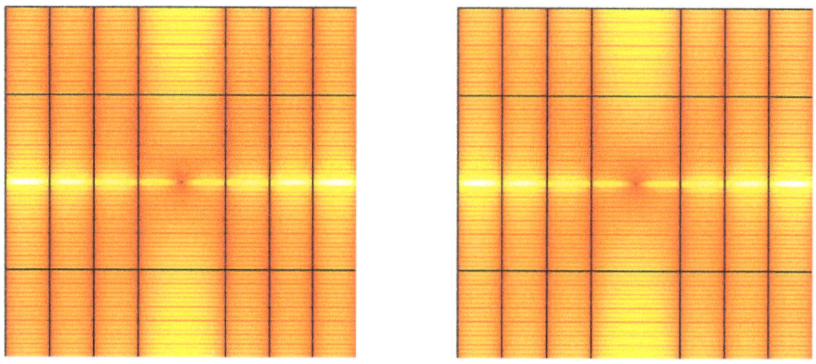

(a) (b)

Figure 5. (a) Modulus with parabolic-effect. (b) Image with scale factor.

A parabolic filter (Fig. 4) to the modulus of F is applied, in this way, low frequencies are attenuated and high frequencies are enhanced as shown in Fig. 5(a). The scale factor $\sqrt{r}$, is applied to the previous results (Fig. 5b). A subimage is chosen from Fig. 6(a) corresponding to the zone that contains the major quantity of information. In the Fig. 6(b) results are observed.

From the subimage M(λ,θ) seen in Fig. 6(b), we compute the marginal functions, using the following equations

$$g(x) = \sum_{\theta} M(\lambda,\theta), \quad (13)$$

$$h(x) = \sum_{\lambda} M(\lambda,\theta)., \quad (14)$$

In the Fig. 7(a) the subimage (Fig. 6b) appears, in three dimensions, and in 7(b) and 7(c) the marginal functions can be observed, with which scale and rotation vectors are obtained.

Finally, Fig. 8 shows the modulus of the Fourier transform of the vectors showed in the Fig. 7(b) and 7(c) respectively, which defines the vectorial signatures of the image, and they are used for comparison with respect to another image.

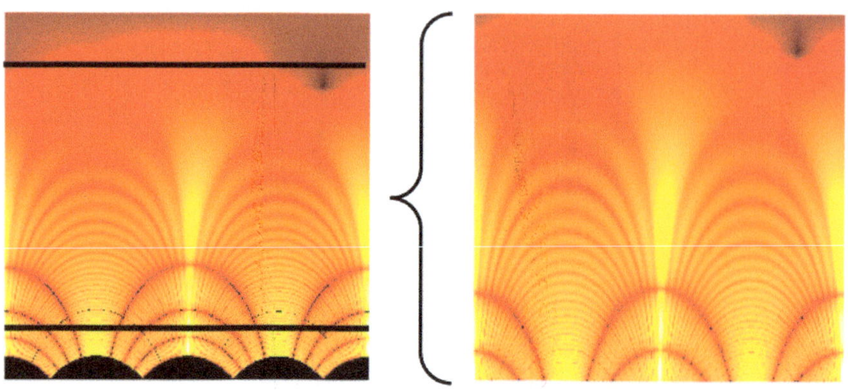

(a) (b)

Figure 6. (a) Image 5(b) in polar-logarithmic coordinates. (b) Selected subimage.

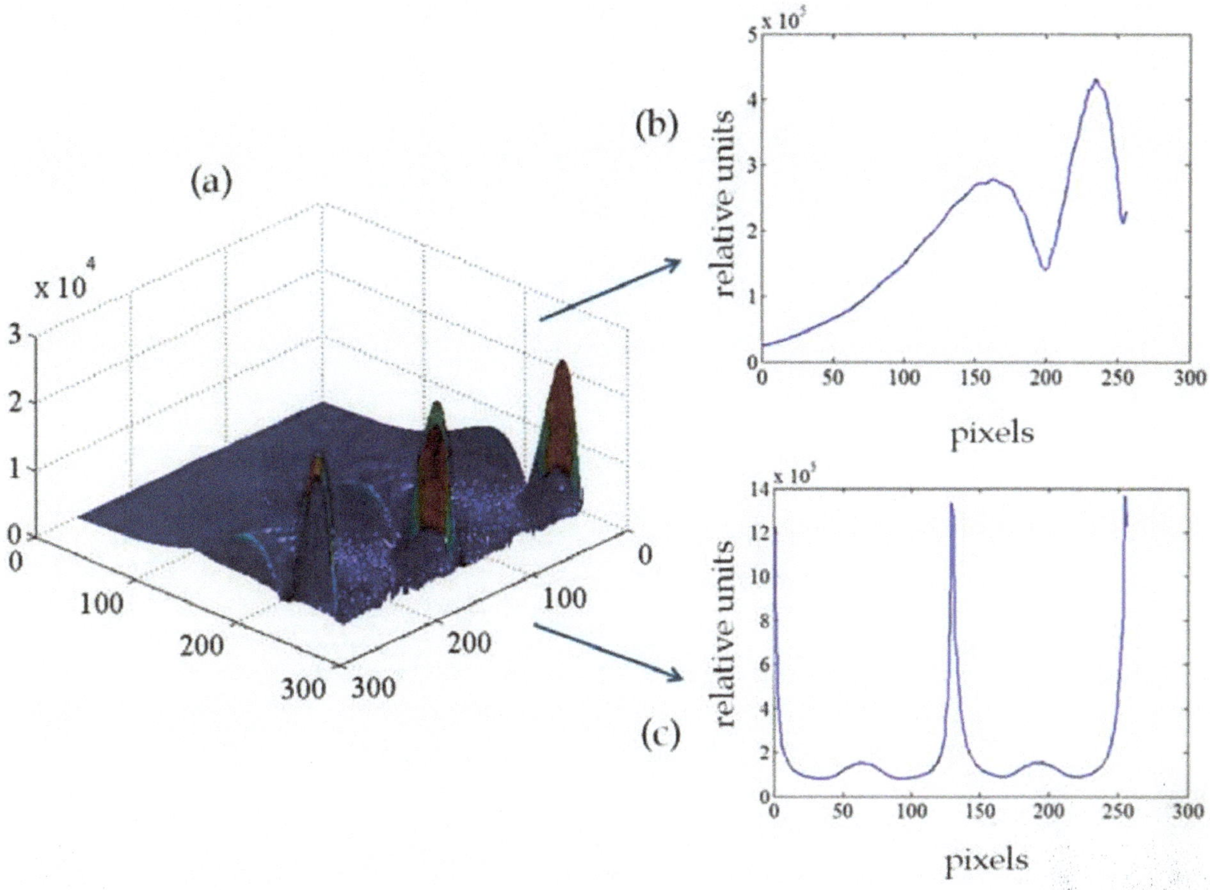

Figure 7. (a) Tridimensional image of Fig. 6(b). (b) Vector of the summation on the rotation axis. (c) Vector of the summation on the scale axis.

COMPUTER SIMULATIONS IN GRAYSCALE IMAGES

To evaluate the performance of the vectorial signatures, images of different types were used as input images.

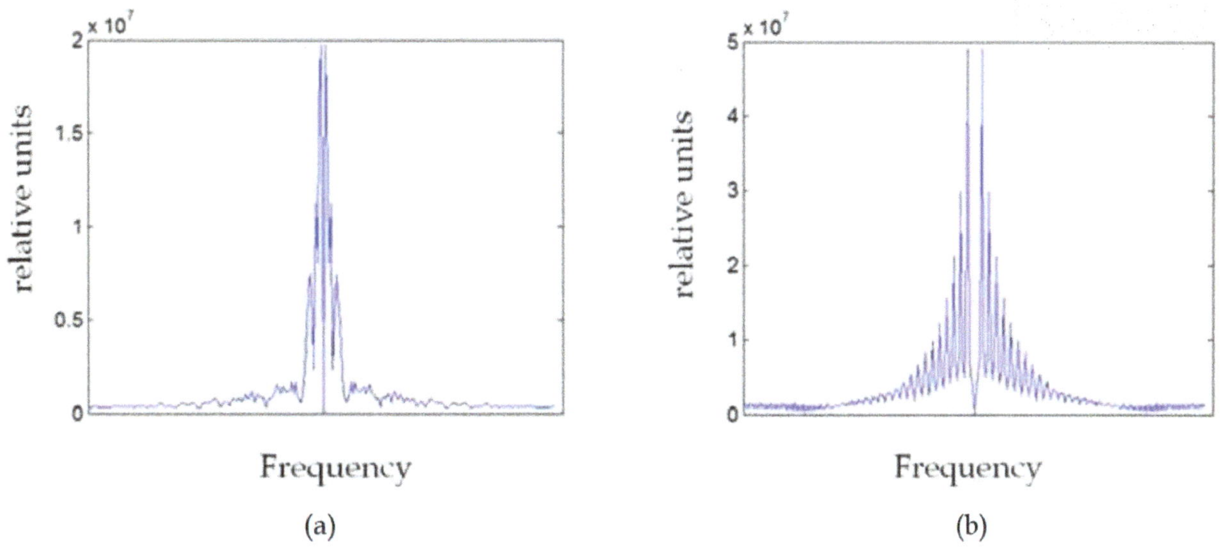

Figure 8. (a) Scale signature vector. (b) Rotation signature vector.

Simulation Using Letters

The first case analyzed corresponds to images of the different letters of the alphabet. Each letter is an image of 256 X 256 pixels of black background with a centered white Arial letter size 72. The methodology described above was applied to these letters. The behavior of the Euclidean distance (E_d) for the target E versus itself with respect to rotation from 0 to 359° with variations of one degree was studied. Because of the letter distortions, when is rotated, due to the square pixels, the curve representing the E_d is cyclic and with symmetry of 180° (Fig. 9).

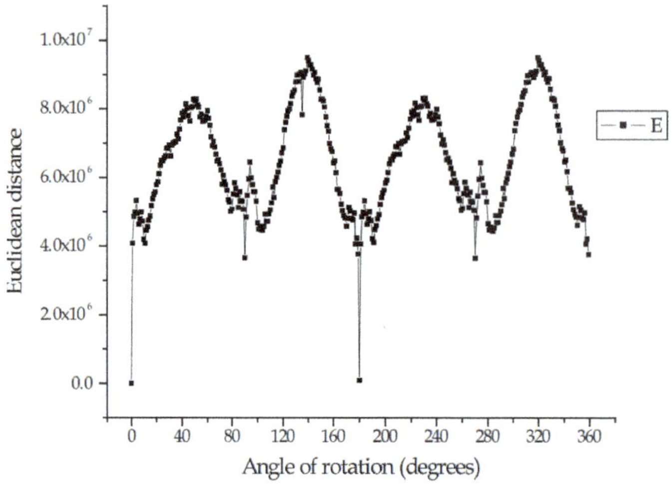

Figure 9. Behavior of the Euclidean distance where the target and the input scene is the letter E.

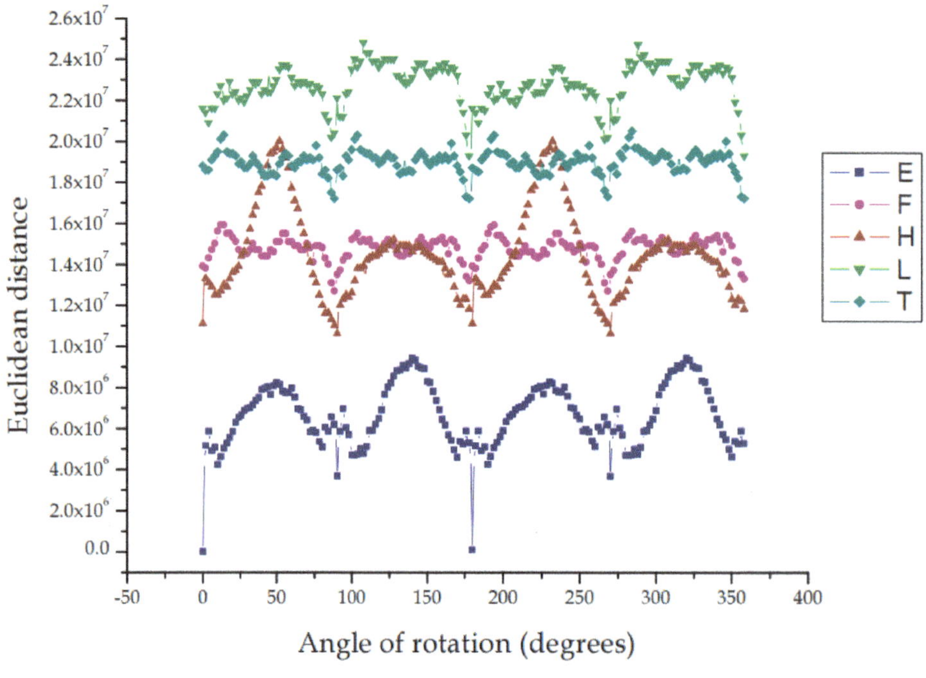

Figure 10. Euclidean distance of the letter E compared with other letters.

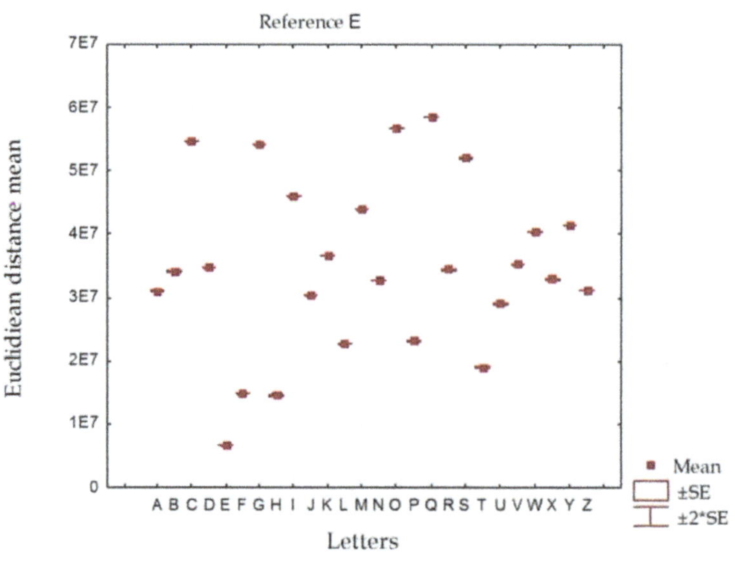

Figure 11. Statistical distance behavior.

To evaluate the distinguish ability between the 26 letters of the alphabet; each one was rotated 360 degrees, with a variation of one degree. Simulation was performed to determine the difference between them, using the 9360 images (26x360). In Figure 10 shows the behavior of the letters that have greater similarity with the reference (E); the result shows a very clear separation between the values of the Euclidean distance of the letter E compared with the rest of the letters, which allows being able to identify it. When an image is similar to another one, the E_d has a minimum value. Statistic was performed and the mean value ± 2SE (two standard error) was calculated. This algorithm has at least a 95.4% level of confidence for this case (Fig. 11 shows the comparison with all the letters).

To analyze the behavior to changes in scale, changes were tested by 90% to 110% with respect to the target image, with increments of 0.5%. Fig. 12. shows the results obtained by reference to the letter E and compared again with all the letters of the alphabet.

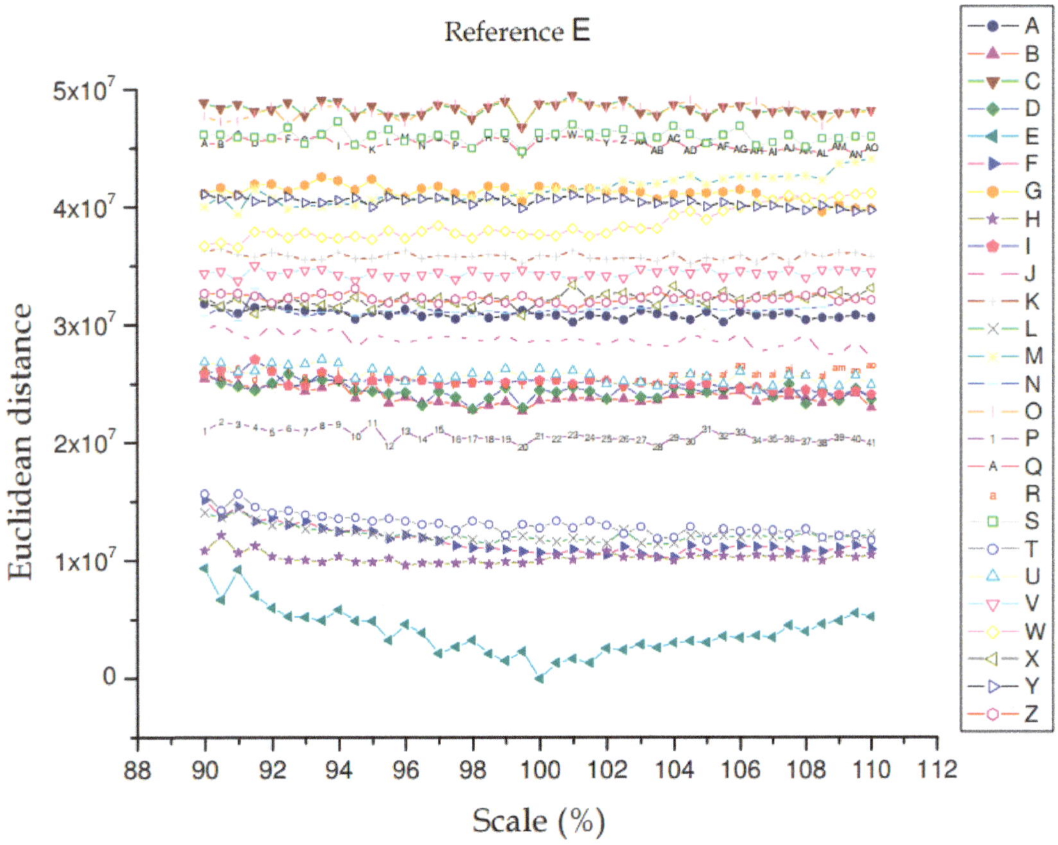

Figure 12. Changes of distance versus scale.

Simulation Using Copepod Species

Copepods are the dominant group in marine zooplankton; they constitute at least 70% of the planktonic fauna. This group of organisms has a great diversity. There are about 11,500 species of copepods described by Humes (1994) and the number is steadily increasing with the description of new species, such as those from anchialine caves (Fosshagen, 1991), hydrothermal vents (Humes, 1991), and those previously reported under the name of another species (Bradford, 1976; Soh & Suh, 2000).

Copepods are of prime importance in marine ecosystems. The majority of copepods feed on phytoplankton, forming a direct link between primary production and commercially important fish, such as sardine, herring, and pilchard. Copepods are also the main food source for a great variety of invertebrates. Studies on copepod abundance and species composition are particularly relevant, because most larvae of commercial fish feed on copepods. Hence, changes in the abundance of these plankters from year to year, may determine interannual population fluctuations of the commercially exploited fish stocks in a particular region.

For this study, adult stages of different copepod species were separated from several plankton samples. The specimens were observed with an optical microscope and their images were digitally captured using a charge coupled device camera (CCD). We used 14 images from seven different species of copepods each with male and female samples (Fig. 13). Since the background noise in all the images is mostly repetitive, the images where cleaned using some methods described en Guerrero & Álvarez-Borrego (2009). Fig. 14 shows the original and its pre-processed image for a female specimen *Rhincalanus nasutus*.

Figure 13. Species of copepod used in the study: (a) *Calanus pacificus* female; (b) *C. pacificus* male; (c) *Rhincalanus nasutus* female; (d) *R. nasutus* male; (e) *Centropages furcatus* female; (f) *C. furcatus* male; (g) *Pleuromamma gracilis* female; (h) *P. gracilis* male; (i) *Temora discaudata* female; (j) *T. discaudata* male; (k) *Acartia tonsa* female; (l) *A. tonsa* male; (m) *Centropages hamatus* female; (n) *C. hamatus* male.

The image of one of the calanoides *Calanus pacificus* female Fig. 13(a) was used as a target to discriminate between copepod species and sex from the rest of the organisms, the image size is 256 X 256 pixels of black background with a centered copepod. The image was rotated 360° in increments of 1° to recognize a target in an input scene; signatures were compared using the Euclidean distance (E_d) between the target and the input image and, because there is no variation in the size of copepod adults, we did not analyze the changes of scale.

In order to see if this methodology has a good performance for recognizing the target image when compared with other images that don't correspond to the copepod chosen as reference, numerical calculations were performed for the entire set of 14 images showed in Fig. 13.

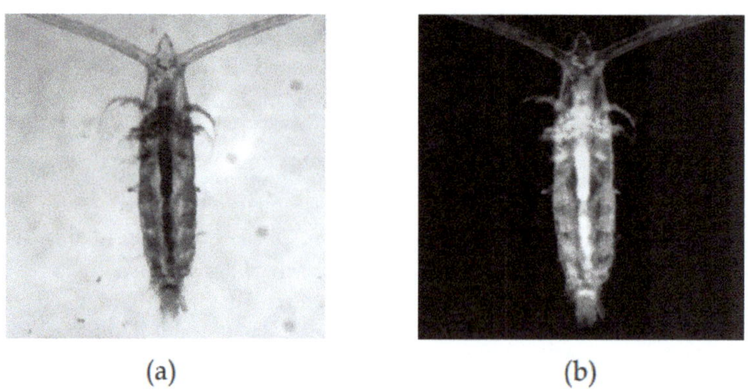

Figure 14. (a) Original image. (b) Cleaned images.

Fig. 15. shows the performance of E_d for the fourteen pictures rotated 360 degrees. The Fig. 13(a) was used as a target, corresponding to a female *Calanus pacificus*. We can see a very good separation of the Ed for the target with respect to the calculation of the E_d for the nontarget images. As can be seen in Fig. 15, the curve of the copepod (Fig. 13a) represents an E_d with minimum values, which distinguish it from the rest of the copepods. Statistic was realized and the mean value ±2 SE was calculated. We can see this algorithm has at least a 95.4% level of confidence for this case (Fig. 16).

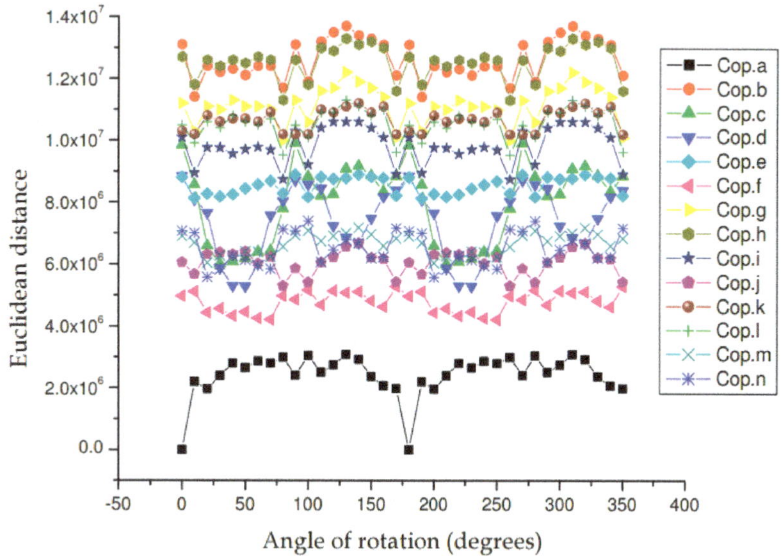

Figure 15. Change the angle of rotation against E_d.

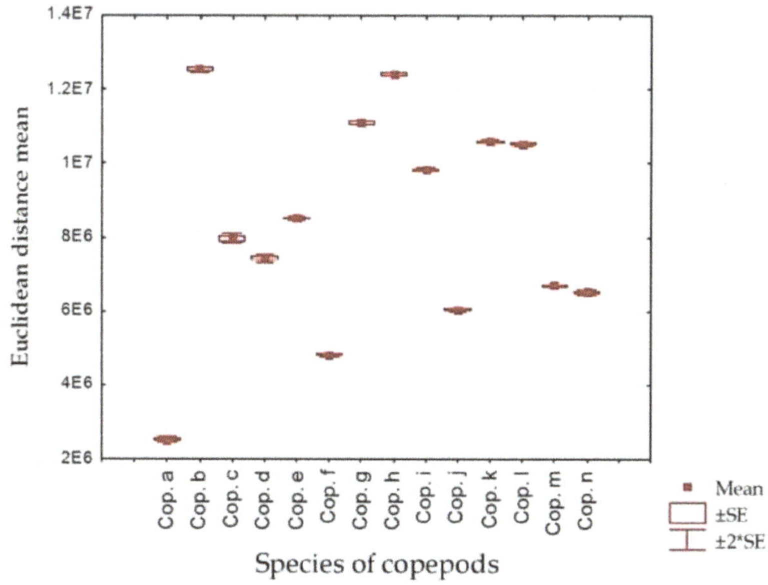

Figure 16. Statistical behavior of the E_d, for a *Calanus pacificus* female specimen.

Average Signature

Like a second set of test images 30 females and 30 males from each of the 7 different species of copepods were used, thus forming a database of 420 images. To have an identification system that contains information from several copepods, vector composed signatures were designed, which is derived from adding the signatures of 10 different copepods and to get its average, which is now the reference vectorial signature (Fig. 17).

In Fig. 18 is shown the application of this method, we can see an excellent separation of the Euclidean distance mean, which corresponds to *Centropages furcatus* female (Group e), used like target, with respect to the calculation of the Ed from the rest of the groups. Statistic was realized and the mean value ± 2SE was calculated. We can identify the species and sex of the target with the confidence level was about 95.4%.

Comparison with Other Algorithms

To evaluate the performance of the algorithm, we compare it with respect to that published by Álvarez-Borrego and Castro-Longoria, considering that both are similar in application to pattern recognition and methodology based on the properties of the Fourier transform, in the second method is pattern matching using the correlation.

Comparing the results of the two algorithms applied to the same sets of images, we observed that in the second is not possible to distinguish the selected image as reference, with respect to the others. Regarding the computational cost, the processing time of our procedure is approximately 20% lower.

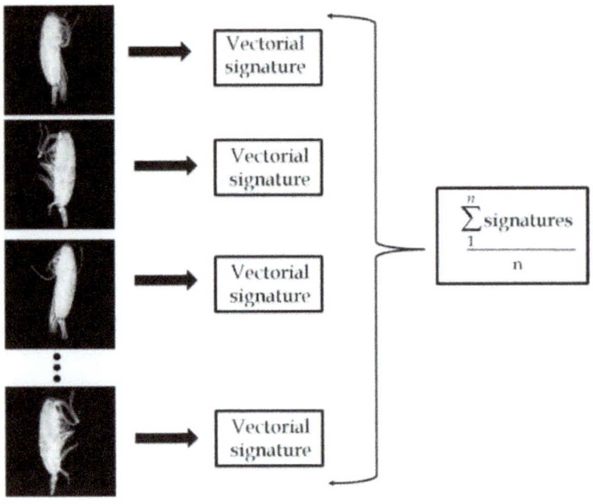

Figure 17. Average vectorial signature.

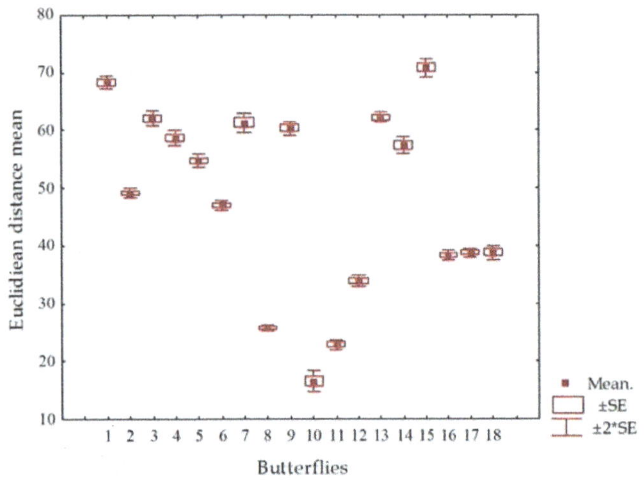

Figure 18. Statistical behavior of the Euclidean distance, in the case of vectorial signatures composite of 10 images of *Centropages furcatus* female (Group e).

COMPUTER SIMULATIONS IN COLOR IMAGES

One of the most important characteristics, which account for visual pattern recognition is the color, and has been used in a wide range of applications in different areas of knowledge, such as evaluation and identification of textiles, flowers, microscopic images, surface corrosion, species recognition, and so on. In our daily life, our vision and actions are influenced by a variety of shapes and colors. The introduction of color increases the amount of information in pattern recognition, so that discrimination can be improved considerably. Color is a perceived phenomenon and not a physical dimension like length or temperature. A suitable form of representation must be found for storing, displaying, and processing color images. This representation must be well suited to the mathematical demands of a color image processing algorithm, to the technical conditions of a camera, printer, or monitor, and to human color perception as well. These various demands cannot be met equally well simultaneously. For this reason, different representations are used in color image processing according to the goal. For these use different models that are known as color space (Gonzalez & Woods, 2008, Kosch & Abidi, 2008, Westland & Ripamonti, 2004) were used. One of the most used color space is RGB, this model is based on the additive mixture of the three primary colors: red (R), green (G) and blue (B). When dealing with color images, there are several concepts that are inherently quite different. For example, if we treat the RGB signals at a pixel as a three-dimensional vector, a color image becomes a vector field, while a monochromatic image is a scalar field.

Methodology

Fig. 19. shows the block diagram of the methodology applied to RGB color spaces: in the first step, is select the color image to be used like a target (I i) which is denoted by fi(x, y), where i = R,G,B, then each image is decomposed into its respective RGB channels [$f_R(x, y)$, $f_G(x, y)$, $f_B(x, y)$]. To have an identification system that contains information from the components $f_i(x,y)$, a vector-composed signatures were obtained, which is now the average vectorial signature $V_{1f}(w_\lambda)$ and $V_{2f}(w_\theta)$, for the test image (I j) represented by gj(x, y) is performed the same process where were obtained $V_{1g}(w_\lambda)$ and $V_{2g}(w_\theta)$. To determine the similarity between the two images I^i and I^j , the value of the Ed is calculated.

Simulation for Butterflies

To test this algorithm a set of images shown in Fig. 20 was used. With different types of butterflies, each image has a size of 256 x 256 pixels. In addition it is estimated that the number of species of butterflies in the world varies between 15.000 and 20.000, which according to the North American Butterfly Association (NABA), about 2000 are located in México.

In Fig. 21 we can see the statistical behavior of the Euclidean distance for the case of rotation of the image 360 degrees in one degree increments, using as reference the butterfly (10), which corresponds to a specimen *Archaeoprepona amphimachus amphiktion*, which at first glance is very similar to others butterflies, for example with butterfly (11) belongs to same genus but different species. As expected, the minimum value of the distance corresponds to the target. We can see that this algorithm has a very good performance in to recognize the target when is compared with the others butterflies. Numerical simulations were performed considering as a reference each of the 18 butterflies and in all cases could be distinguished from the rest.

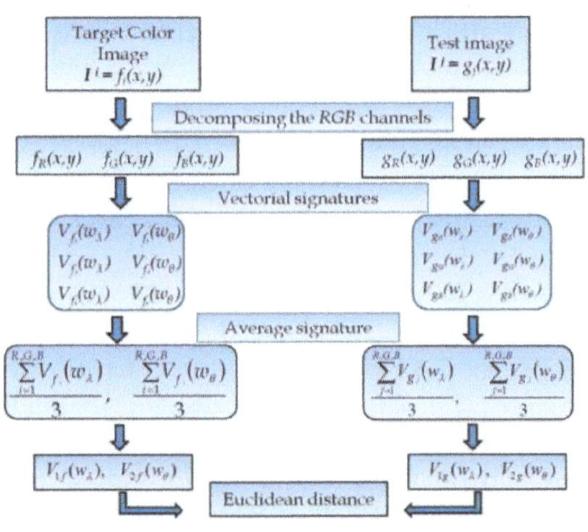

Figure 19. Block diagram of used procedure.

Figure 20. Butterflies used. (1) Actinote guatemalena guerrerensis. (2) Actinote stratonice oaxaca. (3) Agraulis vanillae incarnata. (4) Agrias amydon oaxacata. (5) Anae aidea. (6) Ancyluris inca mora. (7) Anetia thirza. (8) Ansyluris jurgennseni. (9) Arawacus sito. (10) Archaeoprepona amphimachus amphiktion. (11) Archaeoprepona demophon centralis. (12) Baeotus baeotus. (13) Basilarchia archippus. (14) Biblis hyperia aganissa. (15) Bolboneura Sylphis sylphis. (16) Caligo eurylochus sulanos. (17) Caligo memnon. (18) Caligo oileus Scamander.

To analyze the invariance to scale changes were made from 70% to 130%, with variations of one percent each of the test images. The analysis of the statistical behavior of the distance is shown in Fig. 22. for the case where the butterfly (3) is used like target. The results indicate that there is separation of the reference image with the rest of the butterflies, which can be identified without difficulty, regardless of changes of scale.

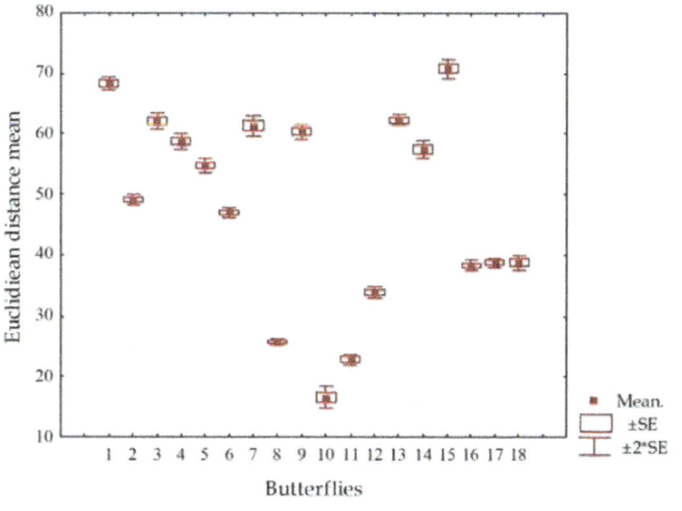

Figure 21. Statistical behavior of the Ed, for an *Archaeoprepona amphimachus amphiktion* specimen.

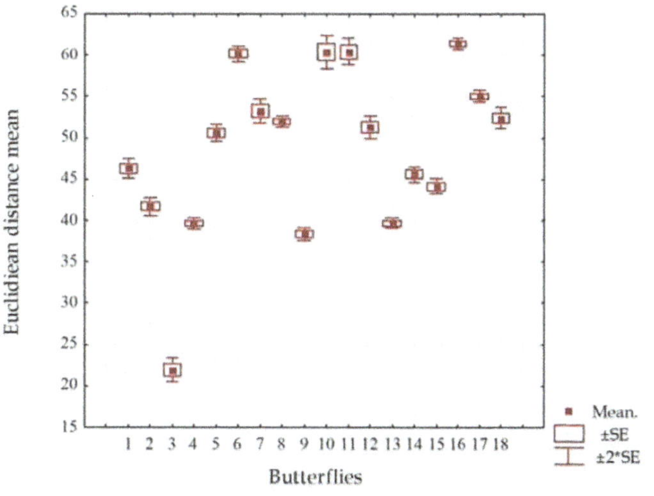

Figure 22. Statistical behavior of the E_d, for a *Agraulis vanillae incarnata* specimen.

CONCLUSION

In this chapter it was developed a system for pattern recognition with the utilization of vectorial signatures, based on the well-known relation between scale and Fourier transform, and has been performed to be practical and accurate. To test the built model, simulations were made and the results were analyzed. Conducting various numerical tests with letters of the alphabet, with variations in the scale and rotation, the results obtained from a comparison of the Euclidean distances show that it is possible for identification and discrimination with a high confidence level (at least 95.4%). The system was applied to the recognition of copepod species and sex. With vectorial signatures it is possible to identify, with a confidence level of 95.4%, the species and sex of various species of copepods. A modification of the algorithm in color image recognition, using average vector signature from the information contained in the RGB channels was done. For this case images of real butterflies were used, finding that the system can identify the species you choose as a reference, when compared with images of different butterflies, for any angle of rotation that contains, as well as changes in the scale of 70% to 130%.

This work contributes to increase the potential recognition systems for use in applications requiring assessment and interpretation of data, applications that were traditionally performed by human vision or specialists trained technicians and are still far from automation.

ACKNOWLEDGMENT

This document is based on work partially supported by UABC and CONACYT under Project grants 102007 and 169174.

REFERENCES

1. Álvarez-Borrego J. & Castro-Longoria E., (2003). Discrimination between Acartia (Copepoda: Calanoida) species using their diffraction pattern in a position, rotation invariant digital correlation, Journal of Plankton Research, Vol. 25, No. 2, (February, 2003), pp 229-233, ISSN: 0142-7873.
2. Bracewell, R. N. (1999). The Fourier transform and its applications, (3th ed), McGraw-Hill, ISBN-10: 0073039381, New York.
3. Bradford, J. M. (1976). Partial revision of the Acartia subgenus Acartiura (Copepoda: Calanoida: Acartidae), New Zealand Journal of Marine and Freshwater Research., Vol. 10, No. 1, (March, 1976), pp. 159-202, ISSN: 0028-8330.
4. Casasent, D. & Psaltis, D. (1976a). Scale invariant optical correlation using Mellin transforms. Optics Communication, Vol. 17, No. 1, (April, 1976), pp. 59-63, ISSN: 0030-4018.
5. Casasent, D. &Psaltis, D. (1976b). Scale invariant optical transforms. Optical Engineering, Vol. 15, (May-June, 1976), pp. 258-261, ISSN: 0091-3286.
6. Casasent, D. & Psaltis, D. (1976c). Position, rotation, and scale invariant optical correlation. Applied Optics Vol. 15, No. 7, (July, 1976), pp. 1795-1799, ISSN: 0003-6935.
7. Cheriet, M., Kharma N., Liu C., Suen, C. (2007). Character recognition systems. A guide for students and practioners, John Wiley & Sons, Inc., ISBN: 978-0-471-41570-1, Hoboken, New Jersey.
8. Cohen, L. The Scale Representation, IEEE Transactions on signal processing, Vol. 41, No. 12, (December 1993), pp. 3275-3292. ISSN: 1053-587X
9. Cohen, L.(1993). The scale representation, IEEE Transactions on Signal Processing, Vol. 41, No. 12, (December 1993), pp.3275-3291, ISSN: 1053-587X.
10. Cristobal, G. & Cohen, L. (1996). Scale in image, SPIE Proceedings. Vol. 2846, pp. 251-261, Denver, CO, USA, August 1996.
11. Fosshagen, A. & Iliffe T, (1991), A new genus of calanoid copepod from an anchialine caves in Belize, Bull. Plankton Soc. Japan, Spec.Vol, pp. 339-346, ISSN: 0387-8961.
12. Gonzalez, R. & Woods, R. (2008). Digital Image Processing, (3rd Edition), Pearson Prentice Hall, ISBN: 978-0-13-168728-8, Upper Saddle River. New Jersey.
13. Gonzalez, R. C. & Woods, R.E. Eddins, S. L. (2009). Digital Image Processing Using MATLAB®, (2th ed), Gatesmark Publishing, ISBN: 978-0-9820854-0-0, USA.
14. Guerrero, R.E. & Álvarez-Borrego, J. (2009). Nonlinear composite filter performance, Optical Engineering. Vol. 48, No. 6, pp. 067201 1-11, ISSN: 0091-3286.
15. Humes, A. G. (1991). Zoogeography of copepods at hydrothermal vents in the eastern Pacific Ocean, Bull. Plankton Soc. Japan, pp. 383-389, ISSN: 0387-8961.
16. Humes, A. G. (1994). How many copepods?, Hydrobiologia, Vol. 292-293, No. 1, pp. 1-7, ISSN: 0018-8158.
17. Koschan, A. & Abidi, M. (2008). Digital Color Image Processing (First Edition), John Wiley & Sons, Inc., ISBN 9780470147085, Hoboken, New Jersey.
18. Lerma-Aragón, J. & Álvarez-Borrego, J. (2009). Vectorial signatures for invariant recognition of position, rotation and scale pattern recognition, Journal of Modern Optics, Vol. 56, No 14, pp. 1598-1606, ISSN 0950-0340.
19. Obinata, G. & Morris, G. (2007). Vision systems, segmentation and pattern recognition, I-Tech Education and Publishing, ISBN 978-3-902613-01-1, Vienna, Austria.
20. Pech, J., Cristobal, G., Álvarez-Borrrego, J., and Cohen, L., (2001), Automatic system for phytoplanktonic algae identification, Limnetica. Vol. 20, No. 1, pp. 143-158, ISSN 0213-8409
21. Pratt, W. K. (2007). Digital image processing, (4th ed), John Wiley & Sons, Inc., ISBN: 978-0- 471-76777-0, Hoboken, New Jersey.
22. Schwartz, E. (1994). Topographic mapping in primate visual cortex: anatomical and computation approaches, Visual Science and Engineering. Models and Applications, Ed. Marcel Dekker, New York. ISBN: 0824791851.
23. Soh, H. Y. & Suh, H.L. (2000). A new species of Acartia (Copepoda, Calanoida) from the Yellow Sea, Journal of

Plankton Research, Vol. 22, (February 2000), pp. 321-337, ISSN: 0142-7873.

24. Solorza, S. & Álvarez-Borrego, J., (2010). Digital system of invariant correlation to position and rotation, Optics Communications, Vol. 283, No. 19 (October 2010), pp. 3613–3630, ISSN: 0030-4018

25. Solorza, S. & Álvarez-Borrego, J., (2011). Digital system of invariant correlation to position and scale using adaptive ring masks and unidimensional signatures, Proceedings of SPIE, 22nd Congress of the International Commission for Optics: Light for the Development of the World, ISBN: 9780819485854, Vol. 8011, Puebla, México, August 2011.

26. Westland, S. & Ripamonti, C. (2004). Computational Colour Science using MATLAB, John Wiley & Sons Ltd, ISBN 0-470-84562-7, West Su.

CHAPTER 5

FOURIER TRANSFORM-BASED COLOR SPACE FOR OPTICAL MICROSCOPIC IMAGE PROCESSING

Tao Guan, Dongxiang Zhou, Chao Xu and Yunhui Liu

Keywords: RGB color space, HSV color space, Fourier transform, K-means

BACKGROUND

Color image processing is one of the hot topics in image processing and computer vision. To effectively represent colors, various color spaces have been presented in the literature, such as the RGB (R, G, and B) color space, the HSV (hue, saturation, and value) color space, the CIELAB color space, and the YCbCr color space. A detailed survey of these color spaces can be found in [1]. In recent years, a number of new methods have also been proposed for describing colors more effectively. Chengjun Liu and Jian Yang [2] introduced the independent component analysis (ICA) into the analysis of color features and proposed an ICA-based color space. Margarita Bratkova et al. developed an oRGB color space [3] based on the opponent color theory. Ingmar Lissner and Philipp Urban derived a perception-based color space [4] that has high perceptual uniformity. For color image processing algorithms, it is of vital importance to select the optimal color space [1],[2],[5]-[7]. However, most existing color spaces, such as the RGB color space and the HSV color space, are application dependent. Once the image varies, the used color space may decrease its performance. In microscopic image processing, such as the cervical cell image segmentation, most algorithms are still based on classical color spaces, which cannot best describe the color features of microscopic images [8],[9],[7],[10]-[13]. In the existing studies of how to effectively represent colors, we found that few methods analyze colors from the frequency domain. Therefore, we investigate colors in microscopic images based on the frequency analysis of the R, G, and B components of pixels' colors in this paper.

Color images are generally stored in the format of the RGB color space. Through linear or nonlinear transforms, the RGB color space can be transformed to other classical color spaces. However, these classical color spaces mostly treat the R, G, and B components as three independent elements. In this paper, on contrary to classical color spaces, we treat the R, G, and B components as a sequence of a spectral signal, called a RGB signal. Then, based on the methodology of frequency spectrum analysis in the signal processing area, we combine the characteristics of 1-D discrete Fourier transform [14] with the color features of microscopic images and propose a RGB Fourier transform-based color space, namely an IF_1F_2 color space, to describe colors of microscopic images. In this color space, similar colors have uniform data distributions whereas different colors have obviously different data distributions.

METHODS

Color can be represented by a combination of tristimuli R (red), G (green), and B (blue) [1]. Actually, the R, G, and B components are obtained from three separate color filters, namely red filter S_R, green filter S_G, and blue filter S_B, according to the equations below:

Citation: Guan et al, Fourier transform-based color space for optical microscopic image processing, doi:10.1186/s40638-014-0016-1.

$$R = \int P(\lambda)S_R(\lambda)d\lambda$$
$$G = \int P(\lambda)S_G(\lambda)d\lambda$$
$$B = \int P(\lambda)S_B(\lambda)d\lambda \quad (1)$$

where P(λ) is the light radiance with the wavelength λ.

From Equation 1, we may find that the R, G, and B components are not independent but actually correlated with each other. Unlike many existing classical color spaces that treat the R, G, and B as independent elements, we treat them as a sequence of a sampled spectral signal that has three samples with the value of R, G, and B in order. In the area of signal processing, we know that similar signals have a similar Fourier frequency spectrum [14]. Consequently, if we treat the R, G, and B as signal samples, similar colors in the image may have a similar signal frequency spectrum, whereas different colors have a different signal frequency spectrum. Then, the next question is how to describe the color features (the intensity and the chromatic information) based on the Fourier frequency spectrum? We know that in the RGB color space, the gray level (without chromatic information) from black to white is represented by R=G=B={0~255} [1]. To some extent, the chromatic information is caused by the difference between R, G, and B. When the R, G, and B components are arranged into a sequence and become a signal, the high frequency component of the signal's Fourier transform can describe the difference between R, G, and B. Inspired by this idea, we introduce the 1-D discrete Fourier transform into the color feature analysis of microscopic images as follows.

Assume that a pixel's RGB value is (r, g, b), then (r, g, b) is arranged into a sequence s, where s(0)=r, s(1)=g, s(2)=b. We assume s to be a sampled signal, which has three samples with the value r, g, and b, respectively. We name this signal a RGB signal. According to the definition of 1-D discrete Fourier transform (DFT) [14], the DFT of s is:

$$X(k) = \mathrm{DFT}\{s\} = \sum_{n=0}^{2} s(n)\cdot \exp\left(-j\frac{2\pi}{3}nk\right),\ k = 0, 1, 2 \quad (2)$$

where X is called the RGB Fourier transform of the pixel's color.

Then, the proposed color space is constituted of three components:

$$I = \mu\cdot \mathrm{abs}(X(0))$$
$$F_1 = \mathrm{real}(X(2))$$
$$F_2 = \mathrm{imag}(X(2)) \quad (3)$$

where X(0) is the low frequency (zero frequency) component of X and X(2) is the high frequency component of X. μ is a scaling parameter used to scale I into a proper numerical range for specific image processing applications, abs(X(0)) is the complex magnitude of X(0), real(X(2)) is the real part of X(2), and imag(X(2)) is the imaginary part of X(2). Since the RGB signal only has three samples, we use 3-point DFT to calculate its discrete Fourier spectrum. For a 3-point DFT, X(1) and X(2) both belong to the high frequency components. The magnitudes ofX(1) and X(2) are equivalent, while their phases are different. Therefore, we can either choose X(1) or X(2) as the chromatic component of the proposed color space. If two colors are slightly different, the major difference between them may be caused by how the RGB values of each color vary. In the methodology of frequency spectrum analysis, the high frequency component of a signal's Fourier spectrum represents the differences between the signal's samples, and these differences, to some extent, describe how the signal's samples vary. Therefore, we use the high frequency component of the RGB signal to describe the color features of microscopic images in this paper.

Then, we construct a vector **F**=[F_1,F_2], and then I and F are the intensity component and the chromatic component of the proposed color space, respectively. This new color space is called an IF_1F_2 color space. We use a color cervical cell image as an example to analyze the characteristics of different colors in the IF_1F_2 color space. In Figure 1, the second column shows the RGB signals. Each RGB signal corresponds to a sampled pixel's color (as the arrow shows in Figure 1) in the cell image. The data distribution map in the middle part of Figure 1 shows the data distribution of vector **F** corresponding to each RGB signal in the second column. We can find that, for similar colors, the data distribution of vector **F** is concentrated, whereas for different colors, the data distribution of vector **F** is dispersed. In addition, we also provide the data distribution map of the chromatic components (the H and S components) in the HSV color space concerning the same pixels in the left cell image, as shown in the right part of Figure 1. To make fair comparisons, we rescaled the H and S values to the same range as that of the values in vector **F**. We can see that the data distributions of H and S are not as concentrated as that of vector **F** for the red points in Figure 1.

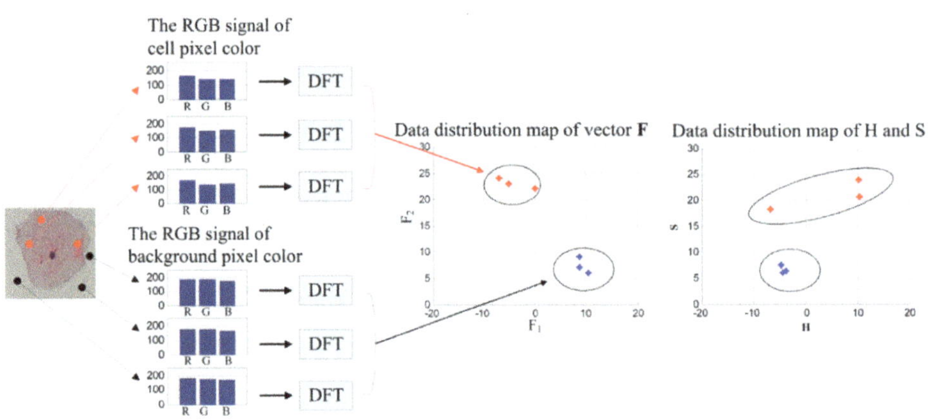

Figure 1: RGB signals of different colors. Their corresponding data distribution map of vector F as well as the H and S components in the HSV color space is also shown.

In addition, we provide the color vector field of a cervical cell image to further demonstrate the effectiveness of the IF_1F_2 color space, as shown in Figure 2. In the color vector field of the image, each point corresponds to one image pixel and the vector on that point corresponds to the color vector **F** of that pixel. From Figure 2, we may observe that the vectors within the cell region are similar, whereas the vectors in the cell region are obviously different from those in the background region.

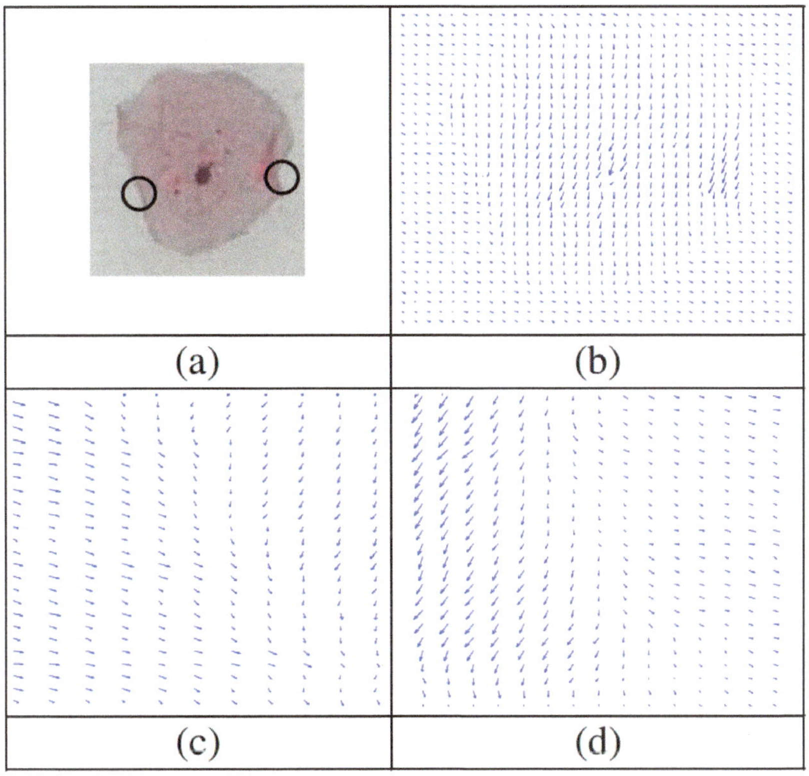

Figure 2: Color vector field of a cervical cell image. (a) Original color cell image. (b) Color vector field of (a). (c, d) Color vector fields in high resolution corresponding to the left and right circles in **(a)**, respectively.

RESULTS AND DISCUSSION

Our experiments focused on validating the superiority of the proposed IF_1F_2 color space over the classical RGB and HSV color spaces for describing the color features of microscopic images. Because K-means algorithm is a widely used clustering algorithm, which can merge data points with high similarity into the same class and split data points with low similarity into different classes, we adopted K-means clustering algorithm to make rough segmentation of the cell images and test the discriminating power of different color spaces.

Two optical microscopic image datasets including the Herlev dataset [15] containing 100 normal cervical cell images (Dataset 1) and the single-cell cervical smear image dataset of our lab containing 100 images (Dataset 2) have been tested. The cell images of our lab were acquired by an automatic microscopic image acquisition system with a CCD camera (DA-HENG, Beijing, China), an optical microscope (Olympus CX31, Olympus Corporation, Shinjuku-ku, Japan), and a 40× magnification lens.

In each cell image, there are typically two kinds of pixels: the pixels in the cell region (with approximately red or blue color) and the pixels in the background (with approximately white color). The aim of the rough segmentation is to split the pixels into two classes: the cell (foreground) and the background. The colors of the pixels within the same region (the cell or the background) are similar, whereas the major difference between the cell and the background pixels is caused by the chromatic components of pixels' colors. Therefore, we chose the chromatic components of the HSV and IF_1F_2 color spaces as the input feature vector to the K-means algorithm. For the RGB color space, because there are no chromatic components of the RGB space, we chose all the R, G, and B components. Since the color difference between the nuclei and the cytoplasm is mainly caused by the intensity contrast, whereas their chromatic components are similar, they are supposed to be within the same class.

We employed the same metric, ZSI (Zijdenbos similarity index) [16], as used in [8] to test the accuracy of the rough segmentation results in different color spaces. The ZSI is defined as

$$\text{ZSI} = 2\frac{\#\{A_1 \cap A_2\}}{\#\{A_1\} + \#\{A_2\}} \tag{4}$$

where A_1 is the set of ground truth foreground pixels, A_2 is the set of segmented foreground pixels, and #{} is the number of pixels in the set. This index is similar to the Dice [17] similarity coefficient in the literature. The ground truth cell regions are manually segmented by an expert cytopathologist. The ZSI results of the IF_1F_2 color space and the other two color spaces are provided in Table 1 and Figure 3.

Table 1: The ZSI mean μ_{ZSI} and the deviation σ_{ZSI} of the clustering results in different color spaces

	$\mu_{ZSI} \pm \sigma_{ZSI}$		
	RGB	**HSV**	$\mathbf{IF_1F_2}$
Dataset 1	0.8090±0.0880	0.8237±0.0838	0.8801±0.0543
Dataset 2	0.8121±0.0856	0.8103±0.0809	0.8910±0.0508

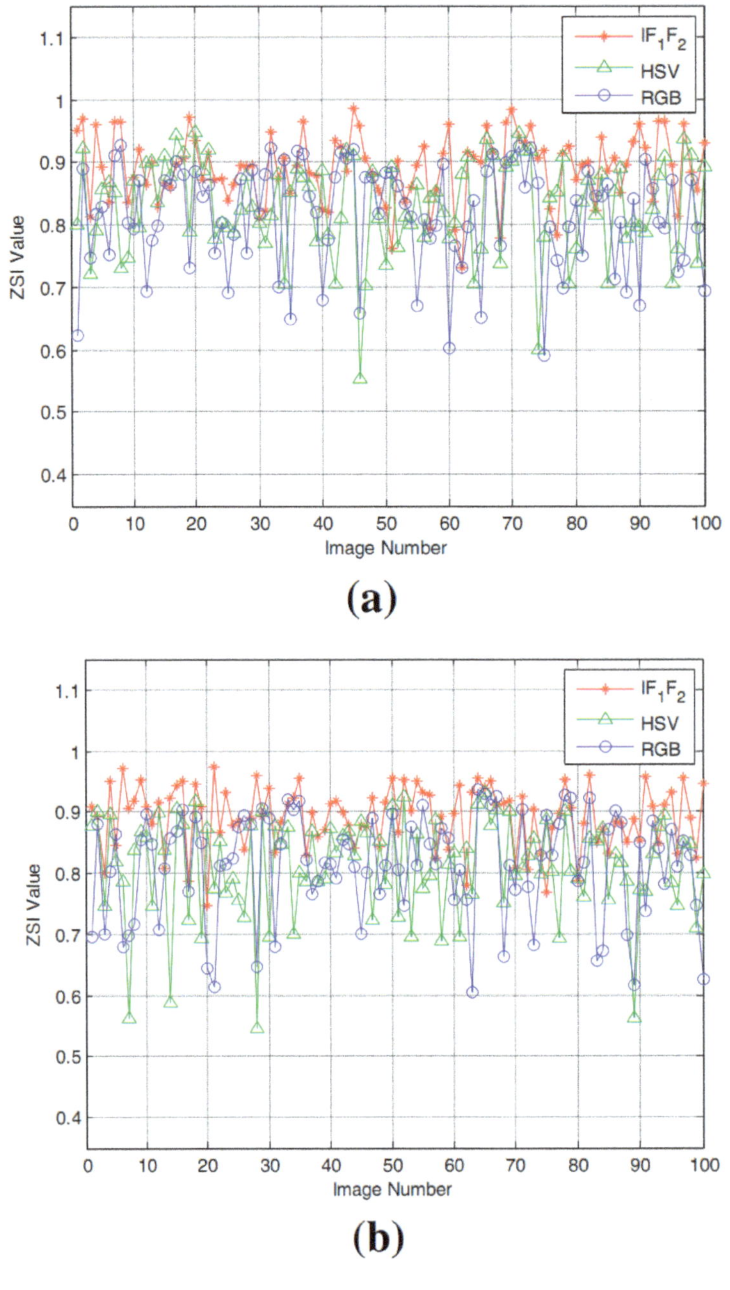

Figure 3: The statistical ZSI results. (a) Dataset 1. (b) Dataset 2.

Figure 4 shows several rough segmentation results. The first three cell images are from Dataset 1, and the other images are from Dataset 2. We can find that the foreground regions (black) of the clustering results in the IF_1F_2 color space are more approximated to the real cell regions than the RGB and HSV color spaces. Many pixels are classified into the wrong class in the RGB and HSV color spaces. In Table 1, we can find that the ZSI for our method has a mean larger than 0.89 and standard deviation smaller than 0.06, which indicates that the accuracy of the rough segmentation results in the IF_1F_2 color space is higher than that in the other two color spaces. This demonstrates that the discriminating power of the IF_1F_2 color space is higher than that of the other two classical color spaces. Because the high frequency component of the RGB signal's Fourier spectrum is, to some extent, the reflection of the difference between the R, G, and B, it may be more suitable for describing colors of the microscopic images than classical color spaces. From Table 1 and Figure 4, it can be found that the performance of the proposed IF_1F_2 color space is better than that of the other two color spaces.

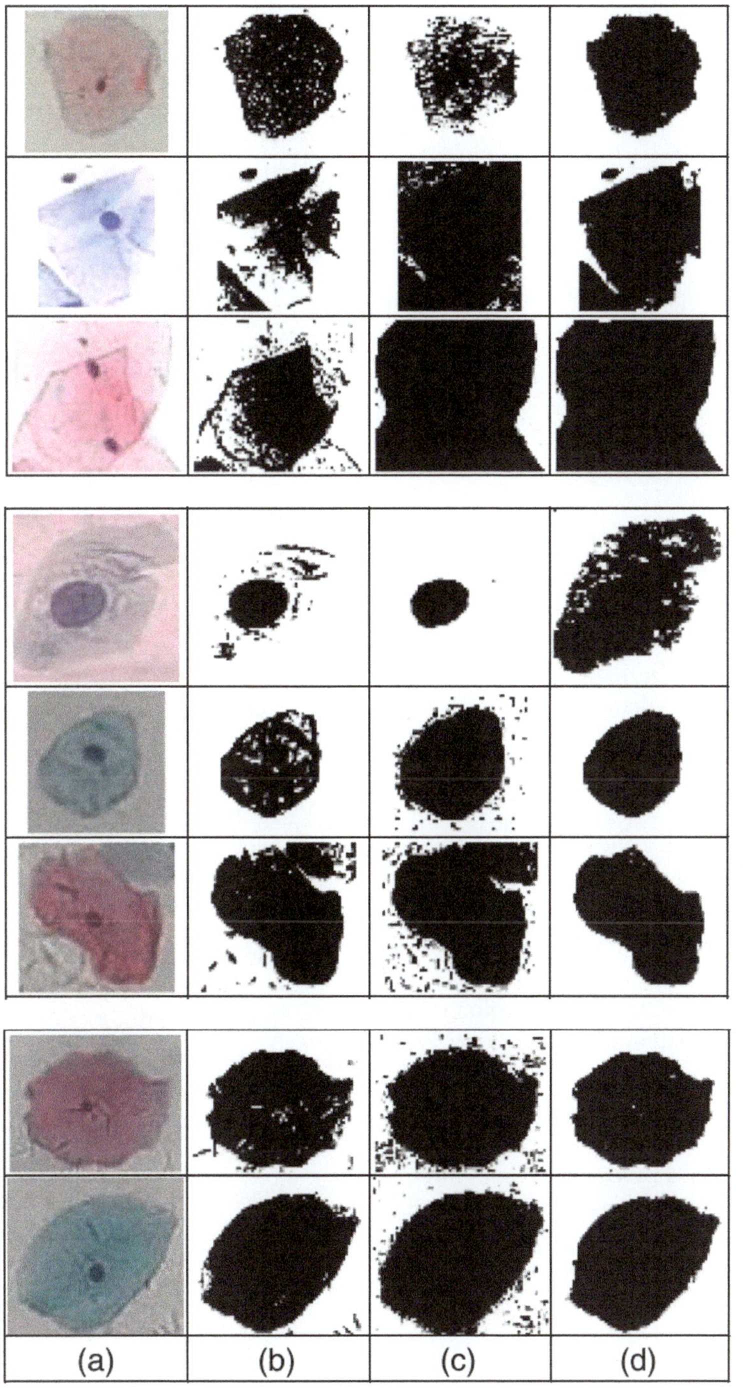

Figure 4: K-means clustering results. (a) Original microscopic images. (b) Results of the RGB space. (c) Results of the HSV space. **(d)** Results of the proposed color space.

CONCLUSIONS

In conclusion, we proposed a novel color space for describing colors of the microscopic images in this paper. On contrary to classical color spaces, we treat the R, G, and B components as a sampled spectral signal. Based on the Fourier spectrum analysis of the signal, we propose a RGB Fourier transform-based color space, namely IF_1F_2 color space. K-means clustering results of two microscopic image datasets show that the performance of the proposed IF_1F_2 color space is better than that of the classical RGB and HSV color spaces. The proposed IF_1F_2 color space may be also useful for other kinds of color images and the relative automatic medical systems.

In this paper, we treat the R, G, and B components of a pixel's color as a sampled signal. Since this signal only has three samples, we use 3-point DFT to calculate its discrete Fourier transform. For a 3-point DFT, the Fourier transform is just expended to the second order. However, we think that higher orders of the Fourier transform may be useful for specific applications, and we will investigate this topic in our future work.

AUTHORS' CONTRIBUTIONS

TG carried out the study of the proposed color space, participated in the programming of the experiments, and drafted the manuscript. DZ carried out the study of using color vector field for analyzing the effectiveness of the proposed color space. CX performed the statistical analysis in the experiments. YL conceived of the study and helped to draft the manuscript. All authors read and approved the final manuscript.

REFERENCES

1. Cheng H, Jing X, Sun Y, Wang J: Color image segmentation: advances and prospects.Pattern Recogn 2001, 34(12):2259–2281. 10.1016/S0031-3203(00)00149-7
2. Liu C, Yang J: ICA color space for pattern recognition. IEEE Trans Neural Netw 2009, 20(2):248–257. 10.1109/TNN.2008.2005495
3. Bratkova M, Boulos S, Shirley P: oRGB: a practical opponent color space for computer graphics. IEEE Comput Graph Appl 2009, 29(1):42–55. 10.1109/MCG.2009.13
4. Lissner I, Urban P: Toward a unified color space for perception-based image processing.IEEE Trans Image Process 2012, 21(3):1153–1168. 10.1109/TIP.2011.2163522
5. Asmare MH, Asirvadam VS, Iznita L: Color space selection for color image enhancement applications. Paper presented at the 2009 international conference on signal acquisition and processing, Kuala Lumpur, Malaysia, 3–5 Apr 2009. 2009.
6. Jiao C, Gao M, Shi Y: Color image segmentation in a novel dynamic color space. Paper presented at the proceedings of the 7th world congress on intelligent control and automation, Chongqing, China, 25–27 June 2008. 2008.
7. Zhang C, Wang P: A new method of color image segmentation based on intensity and Hue clustering. Paper presented at the proceedings of the 15th international conference on pattern recognition, Barcelona, Spain, 3–7 Sept 2000. 2000.
8. Li K, Lu Z, Liu W, Yin J: Cytoplasm and nucleus segmentation in cervical smear images using radiating GVF snake. Pattern Recogn 2012, 45(4):1255–1264. 10.1016/j.patcog.2011.09.018
9. Plissiti ME, Nikou C, Charchanti A: Automated detection of cell nuclei in Pap smear images using morphological reconstruction and clustering. IEEE Trans Inf Technol Biomed 2011, 15(2):233–241. 10.1109/TITB.2010.2087030
10. Genctav A, Aksoy S, Onder S: Unsupervised segmentation and classification of cervical cell images. Pattern Recogn 2012, 45(12):4151–4168. 10.1016/j.patcog.2012.05.006
11. Plissiti ME, Nikou C, Charchanti A: Accurate localization of cell nuclei in Pap smear images using gradient vector flow deformable models. Paper presented at the proceedings of 3rd international conference on bio-inspired signals and systems (BIOSIGNALS 2010), Valencia, Spain, 20–23 Jan 2010. 2010.
12. Plissiti ME, Nikou C, Charchanti A: Watershed-based segmentation of cell nuclei boundaries in Pap smear images. Paper presented at the proceedings of the IEEE/EMBS Region 8 international conference on information technology applications in biomedicine, Corfu, Greece, 3–5 Nov 2010. 2010.
13. Plissiti ME, Nikou C, Charchanti A: Combining shape, texture and intensity features for cell nuclei extraction in Pap smear images. Pattern Recogn Lett 2011, 32(6):838–853. 10.1016/j.patrec.2011.01.008
14. Lathi BP: Signal processing linear systems. Oxford University Press, New York; 1998.
15. Jantzen J, Dounias G: Analysis of Pap-smear image data. Paper presented at the proceedings of the nature-inspired smart information systems 2nd annual symposium, Tenerife, Spain, 29 Nov–1 Dec 2006. 2006.
16. Zijdenbos A, Dawant B, Margolin R, Palmer A: Morphometric analysis of white matter lesions in MR images: method and validation. IEEE Trans Med Imaging 1994, 13(4):716–724. 10.1109/42.363096
17. Dice LR: Measures of the amount of ecologic association between species. Ecology 1945, 26(3):297–302. 10.2307/1932409

CHAPTER 6

JOINT MULTIBAND SIGNAL DETECTION AND CYCLIC SPECTRUM ESTIMATION FROM COMPRESSIVE SAMPLES

Lebing Pan, Shiliang Xiao, Xiaobing Yuan and Baoqing Li

Keywords: Multiband signal detection, Parameter extraction, Cyclic spectrum estimation, Compressive sampling

INTRODUCTION

Signal detection and parameter estimation of wideband radio signals are increasingly important aspects of communication systems such as electronic countermeasures[1], carrier tracking[2], and spectrum sensing[3]. However, conventional detection techniques using standard analog-to-digital converters (ADCs) have been primarily designed to capture narrowband signals[4]. Wideband sensing based on Nyquist sampling is similarly unsuitable in such applications, often yielding either an insupportable sampling burden or an unacceptable sampling complexity. Additionally, because of this, wideband receivers require a high sampling rate, which typically results in high power consumption. In contrast, analog front-ends have often been used to capture sparse signals in scenarios such as cognitive radio networks and radar detection systems. Consequently, compressive sampling (also known as compressive sensing[5]), conducted using low rate ADCs, has been extensively employed to detect wideband signals. With the underlying assumption of signal sparsity, this technology is capable of complete signal recovery from a small number of linear measurements[6].

A number of methods for radio signal detection using compressive sampling have been presented in the literature. Several studies have focused on use of blind multiband signal reconstruction[7] and power spectrum sensing[8–11]. These works aim to detect active sub-bands distributed over a wide frequency spectrum, which is a crucial step in dynamic spectrum access[12, 13]. Other literature has proposed methods for cyclic spectrum estimation using sub-Nyquist sampling. Also known as cyclic feature detection, these methods pertain not only to spectrum sensing but also to symbol rate estimation, carrier frequency estimation, and modulation identification[14–17]. In[18], the spectrum correlation function (SCF) (also known as cyclic spectrum.) of each narrowband signal was analyzed individually. This scheme recovered signals in the time domain using compressive sampling and then computed the SCF at the Nyquist rate. The linear relationship between the SCF and the time-varying covariance function was first derived in[19] and[20]. This operation makes the sparse approximation algorithms generated by compressive sensing available for cyclic feature detection. The scheme in[21] used the symmetry property of the cyclic autocorrelation function for blind spectrum sensing. On another side, the time smoothed algorithm known as the fast Fourier transform (FFT) accumulation method (FAM) was presented in[22] for effective cyclic spectrum estimation based on the Nyquist sampling. A practical implementation of this algorithm for spectrum sensing, presented in[23], showcased its high efficiency and low complexity. In this paper, a two-step compressive sampling FAM (CS-FAM) scheme that employs multiple measurement vectors (MMVs) is presented. Multiband detection is accomplished in the first step by means of a modified version of the periodogram method using sub-Nyquist samples, while cyclic spectrum estimation is performed in the second step.

Citation: Pan et al.: Joint multiband signal detection and cyclic spectrum estimation from compressive samples. EURASIP Journal on Wireless Communications and Networking 2014 2014:218. doi:10.1186/1687-1499-2014-218.

The main characteristics of CS-FAM are as follows:

- based on the FAM framework, this scheme has low complexity, which is important because cyclic spectrum analysis can lead to high complexity due to the use of two-dimensional spectrum correlation computations, especially from compressive samples. For example, the methods in[20] required matrices proportional to N^4. However, in implementations such as high-resolution cyclic spectrum estimation by the order of $N \times N$, the memory requirements are quite large, making cyclic spectrum analysis especially difficult.
- Conventional cyclic spectrum estimation methods require front-end components capable of high rate sampling. In contrast, in this work, the detection problem is addressed by using a low rate sampling device to capture sub-Nyquist samples. Another advantage of CS-FAM is that performing multiuser detection in the first step eliminates the effect of signal superposition in the multiuser cyclic spectrum, whereas conventional methods require an additional signal separation process.
- Due to the use of power spectrum estimation, the multiband signal detection process is robust, compared to other methods based on signal recovery from compressive sampling, which are sensitive typically to noise. For example, the power spectrum blind sampling (PSBS) method used in[9] and[10] demonstrated poor estimation performance in inactive bands, which also led to high complexity in matrix operations.
- The process of CS-FAM is flexible in regards to practical application. Multiuser detection is achieved in only one step and, based on the results of the first step, parameter estimation is accomplished in the second step by searching for the peaks in the cyclic spectrum rather than by setting a threshold.

The rest of the paper is organized as follows: Section 2 provides the system model along with the FAM algorithm and compressive sampling framework. CS- FAM is then developed within the framework of FAM. Signal detection is addressed in Section 3. A simple signal recovery algorithm, simultaneous orthogonal matching pursuit (SOMP) is employed for multiband detection. A vector indicating the power distribution of the analog input is then computed from the iteration of SOMP. Following this, active sub-bands are identified using a constant false-alarm detector, and multiuser detection is achieved using a bandwidth constraint. If the signal is reconstructed, the second step of CS-FAM, cyclic feature detection, can be performed. Simulations are presented in Section 4 to demonstrate the detection performance of this scheme, followed by a summary in Section 5.

SYSTEM MODEL

Modulated signals have a built-in periodicity, which is characterized as cyclostationary[24]. The auto-correlation function $R_x(t,\tau)=E\{x(t)*x(t+\tau)\}$ (* is the complex conjugate operator), which is periodic in the time domain with a period T_0, is given by

$$R_x(t+T_0,\tau)=R_x(t,\tau). \quad (1)$$

The cyclic auto-correlation function (CAF) is represented in terms of Fourier coefficients as

$$R_x^\alpha(\tau)=\frac{1}{T_0}\int_{-T_0/2}^{T_0/2} R_x(t,\tau)e^{-j2\pi\alpha t}dt, \quad (2)$$

where $\alpha \triangleq \{k/T_0\}_{k\in Z}$ represents the cyclic frequencies. The SCF is defined by

$$S_x^\alpha(f)=\int_{-\infty}^{\infty} R_x^\alpha(\tau)e^{-i2\pi f\tau}d\tau. \quad (3)$$

Wide-sense stationary signals can be detected from stationary interference because general stationary signals do not possess cyclic stationary characteristics. The advantage of cyclostationary feature detection lies in its ability to separate the signal of interest (SOI) from noise and interference in the spectral correlation plane, making it well suited for modulation recognition and parameter estimation. The framework of CS-FAM is illustrated in Figure 1. Because the detection method in this paper is based on a process of cyclic spectrum estimation using compressive sampling, we will introduce the FAM algorithm and compressive sampling framework first.

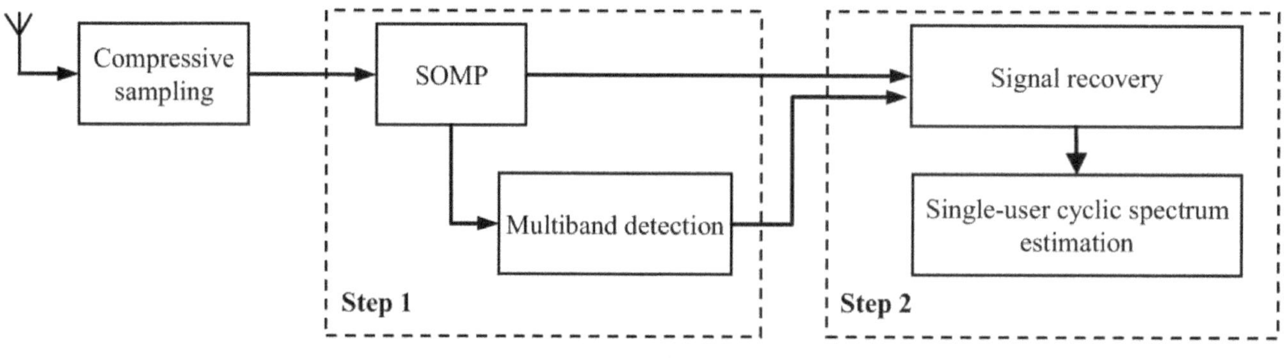

Figure 1: Flexible signal detection by CS-FAM.

The FAM Algorithm

Compared to other methods used to estimate SCF, time smoothed algorithm is computationally efficient for general cyclic spectral analysis [22]. Based on the time smoothed cyclic periodogram, the parameter estimation is calculated with a data tapering window of the length T sliding over the data for a time span of Δt and $\Delta t = N'T_s$ (N' is the length of x[n]), where T_s is the sample interval. The SCF of the digital signal x[n] is defined as $S_x^\alpha(f)=\sum_{b=-\infty}^{\infty} R_x^\alpha(b)e^{-j2\pi fb}$ with the discrete Fourier transform (DFT). The cyclic periodogram[25] is defined as

$$S_x^\alpha(n,f) = X_T(n,f_1)X_T^*(n,f_2), \tag{4}$$

where $f=(f_1+f_2)/2$ and $\alpha=f_2-f_1$. $X_T(n, f)$ denotes a complex demodulator, which is expressed as

$$X_T(n,f) = \sum_{r=-N'/2}^{N'/2-1} w(r)x(n-r)e^{-i2\pi f(n-r)/N'}, \tag{5}$$

where w(n) is a data tapering window, which is well accomplished by using a Hamming window. The limited cyclic spectrum with the discrete-time span of N' is defined in terms of the time smoothed cyclic periodogram as

$$S_x^\alpha(n,f)_{N'} = \sum_{r=n-N'/2}^{n+N'/2-1} X_T(r,f_1)X_T^*(r,f_2)g(n-r). \tag{6}$$

where g(n) is a data tapering window of length N'. An implementation based on time smoothed cyclic periodogram is illustrated in Figure 2. As shown in[25] and[26], if the time windows w(n) and g(n) are properly normalized, the time smoothed cyclic periodogram converges to the cyclic spectrum in the limit as follows:

$$\lim_{\Delta f\to 0}\lim_{N'\to\infty} S_x^\alpha(n,f)_{N'} = S_x^\alpha(f). \tag{7}$$

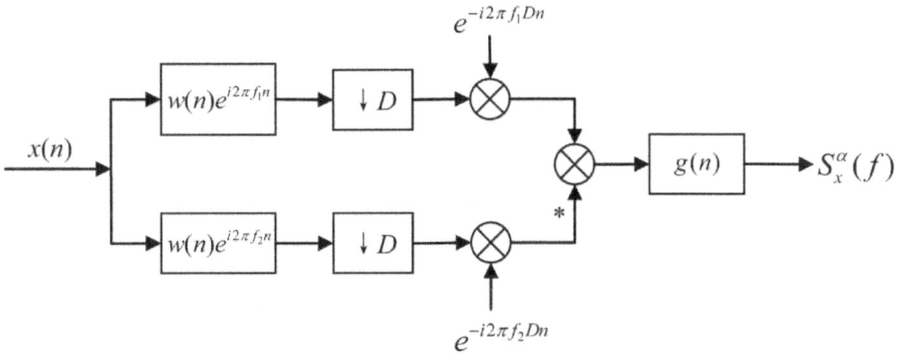

Figure 2: Implementation of time smoothed cyclic periodogram.

In order to obtain a substantial reduction in random effects using a time-smoothed approach, the time-spectrum resolution product must significantly exceed unity

$$\Delta t\Delta f >> 1. \tag{8}$$

Thus, substantial observation time is needed to obtain a reliable SCF estimate, where a smaller spectral resolution is required to resolve the individual features of the SCF. The output filter g(n) is not as crucial as the input filter w(n). Therefore, for simplicity, g(n) is taken to be a rectangular window, as is commonly done in FAM algorithms. The computational efficiency of (6) can be improved by decimating the outputs of the filter w(n). With a decimating factor D, (6) is modified to

$$S_x^\alpha(Dn,f)_{N'} = \sum_{r=-L/2}^{L/2-1} X_T(rD,f_1)X_T^*(rD,f_2), \tag{9}$$

L=N'/D. The filter w(r) ensures that no aliasing takes place after D-decimation[27–29]. Since the outputs of channelization are oversampled by a factor of N, the sampling rate can be reduced to f_s/D, $D\leq N$ before aliasing occurs[22, 28]. The effects of these parameters on the FAM algorithm are analyzed in[22], which discusses resolution, reliability, and computation reduction. In this paper, we have selected D=N to satisfy the compressive sampling framework expressed by (12). However, the filter is always not ideal. Then, the leakage will influence the estimation performance. While the selection D=N leads to a slight increase in cycle leakage, it significantly reduces complexity. Considering the frequency shifting of the production sequence from α to α+qΔα, where q is an integer, (9) becomes

$$S_x^{\alpha+q\Delta\alpha}(n,f_i)_{N'} = \sum_{r=-L/2}^{L/2-1} X_T(r,f_l)X_T^*(r,f_k)e^{-i2\pi rq/L}. \tag{10}$$

Here, f_l and f_k are the frequency values in correlating two spectral components. The FAM estimate computed in (10) corresponds to $\alpha=(f_l-f_k)$ and $f_j=(f_l+f_k)/2$. The length of w(n) is Nand the frequency resolution is $\Delta f=f_s/N$, where f_s is the sampling frequency. The other dimension α describes the cyclic frequency and the resolution $\Delta\alpha=f_s/N'$. The FAM is a computationally efficient algorithm by adopting N-point FFT operation to (5) after decimating and L-point FFT to (10), respectively.

Figure 3a presents a block diagram of the FAM algorithm. The first step includes channelization using data tapering and the N-point FFT operation. Data channelization divides the input x[n] into L segments where the size of each segment is N. The complex demodulator of (5) after decimating is computed by the N-point FFT operation in the first step of Figure 3, which is derived as

$$\begin{aligned}\mathbf{X}[f]_{N\times L} &= \mathbf{FWX}_0[n]\\ &= \mathbf{FW}\left(\mathbf{F}^{-1}\mathbf{F}\right)\mathbf{X}_0[n]\\ &= \bar{\mathbf{W}}\mathbf{X}_0[f],\end{aligned} \tag{11}$$

where n and f denotes the data is expressed in the time domain and in the frequency domain, respectively. **F** is the N-points of the DFT matrix and $\mathbf{F}_{(i,j)}=[e^{-j2\pi ij/N}]$. $\mathbf{X}_0[f]=\mathbf{FX}_0[n]$. $\bar{\mathbf{W}}=\mathbf{FWF}^{-1}$ and $\mathbf{F}^{-1}$ is the N-points of the IDFT matrix. **W** is a diagonal matrix whose diagonal elements are determined from the N-point Hamming window. $\mathbf{X}_0[n]$ is the data matrix of size N × L, which is generated from x[n] using channelization. x[n] is the digital signal from the Nyquist sampling.

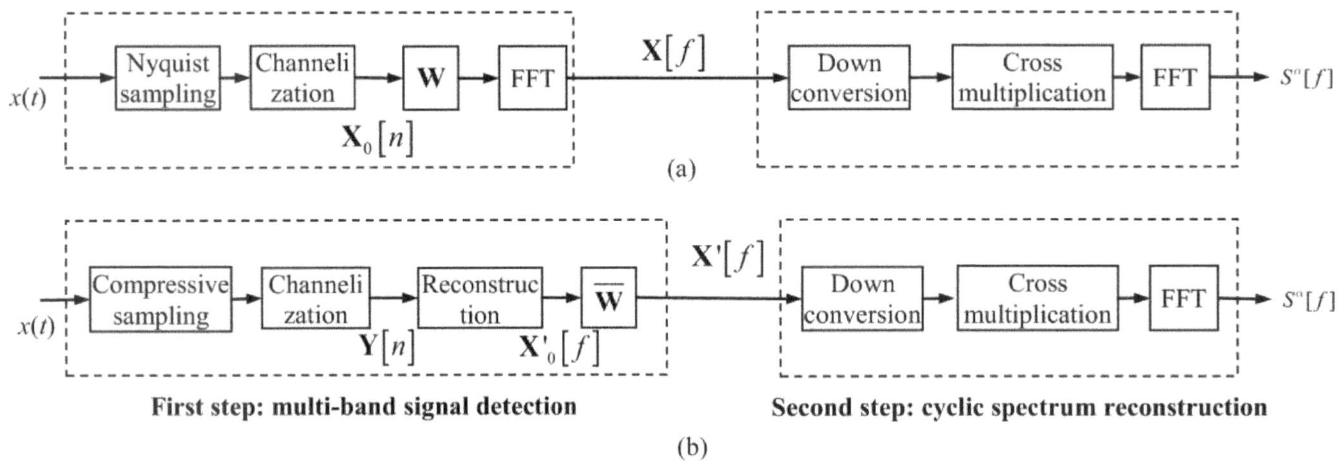

Figure 3: The scheme of CS-FAM. (a) FAM base on Nyquist sampling. (b) CS-FAM base on compressive sampling.

There are three operations in the second step of FAM as shown in Figures 2 and3. The complex demodulators are downshifted in frequency to baseband by down conversion. The SCD function is then estimated by multiplying its complex conjugate. The second FFT is a smoothing operation, which is executed by L-point FFT. More details on this process are given in[22,24–26]. In this paper, the proposed method is based on the framework of FAM from sub-Nyquist samples as presented in Figure 3b. The process of obtaining $\mathbf{X}_0[f]$ from compressive samples in the first step of CS-FAM is discussed in following section, while the merits of CS-FAM's second step are presented in Section 3.2.

Compressive Sampling

According to compressive sampling theory, a small amount of signal data obtained through inner-product operations can be used to reconstruct the original signal with overwhelming accuracy[5, 6]. The holy grail of compressed sampling is an acquisition device that exploits signal structure in order to reduce the sampling rate. An example of this is the analog-to-information converter (AIC)[30], which is used in place of the ADC in the front-end of a sampling device. As illustrated in Figure 3b, using an AIC, the output from compressive sampling is expressed as

$$\mathbf{Y}[n] = \mathbf{A}\mathbf{X}_0[n], \tag{12}$$

where **A** is a measurement matrix of the size $M \times N$ ($M<N$), the elements of which are randomly selected. A measurement matrix obeying the restricted isometry property (RIP) in[6] can be used for CS-FAM (e.g., Gaussian random matrix or ± 1 distribution matrix obeys the RIP with high probability). In this paper, the elements are selected from {-1, 1}, which facilitates hardware implementation by using integrate-and-dump processes[31]. The average sampling rate is equal to the Nyquist rate multiplied by M/N. **F** is invertible. Therefore, we can have

$$\begin{aligned}\mathbf{Y}[n] &= \mathbf{A}\mathbf{F}^{-1}\mathbf{F}\mathbf{X}_0[n] \\ &= \mathbf{\Phi}\mathbf{X}_0[f],\end{aligned} \tag{13}$$

where $\mathbf{\Phi} = \mathbf{A}\mathbf{F}^{-1}$. The output of the first step in CS-FAM is derived by $\mathbf{X}'[f] = \bar{\mathbf{W}}\mathbf{X}_0'[f]$ from (11) as illustrated in Figure 3b. $\mathbf{X}_0'[f]$ is the sparse approximate matrix of $\mathbf{X}_0[f]$, which is recovered from compressive samples **Y**[n] and only contains the signal in active bands. $\mathbf{X}_0'[f]$ is computed in Section 3.2.

JOINT SIGNAL PROCESSING FROM COMPRESSIVE SAMPLES

In this article, multiband detection and parameter extraction are performed in the frequency domain. The process of signal reconstruction using the MMV model will be introduced first. Due to the sparsity of wideband signals, only a few rows of $\mathbf{X}_0[f]$ contain large values, denoted as signals with additive noise. In the other rows, where only noise is present, the values are very small. In this study, we let K be the number of rows where the signal is present in $\mathbf{X}_0[f]$, and the value of K is small because there are only a few active sub-bands in the whole frequency range. This has given rise to the problem of sparse signal recovery using the MMV model[31] to analyze joint row sparsity, as shown in Figure 4 and which will be addressed below. However, it has been shown that signal recovery rate is greatly increased by using the MMV rather than the single measurement vector (SMV) model. The results in[32–34] reveal the advantages of the MMV model, including the key assumption that the support (i.e., the indexes of the signal entries) is identical in every column of $\mathbf{X}_0[f]$. The literature also endorses increasing L for better recovery performance.

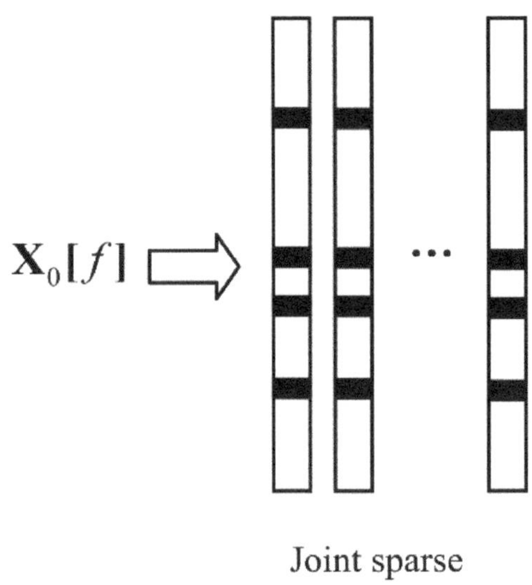

Figure 4: Joint row sparse based on MMV model.

Many effective algorithms have been proposed for solving the MMV problem, such as convex relaxation methods using mixed norms, greedy search mechanisms, and sparse Bayesian learning. Compared to other methods, the algorithms based on greedy search mechanisms have the advantageous characteristic of low complexity. Therefore, we selected the simple algorithm SOMP [35], which has proven that under certain conditions, the OMP algorithm can find the sparsest representations from the MMV model with computational efficiency. The SOMP algorithm is modified in this paper as follows:

SOMP algorithm

Input: $\mathbf{Y}$, $\mathbf{\Phi}=\mathbf{AF}^{-1}$, iteration times l.

Initialization: residual $\mathbf{R}_0=\mathbf{Y}$, $\mathbf{S}_0=\varnothing$, $i=0$, $\mathbf{d}=0_{1\times N}$.

While $(i<l)$

1: $i := i+1$. i-th iteration.

2: Compute column index k, $k=\underset{k}{\operatorname{argmax}}\left\|\mathbf{\Phi}_k^T\mathbf{R}_{i-1}\right\|$. Find the best matching.

3: $d[k]=\left\|\mathbf{\Phi}_k^T\mathbf{R}_{i-1}\right\|^2$. Update $\mathbf{d}$.

4: $\mathbf{S}_i=[\mathbf{S}_{i-1},\mathbf{\Phi}_k^T]$. Update $\mathbf{S}$ with new canonical basis vector.

5: $\mathbf{X}_i=\underset{\mathbf{X}_i}{\operatorname{argmin}}\left\|\mathbf{Y}-\mathbf{X}_i\mathbf{S}_i\right\|$. Compute approximation matrix $\mathbf{X}_i$.

6: $\mathbf{R}_i=\mathbf{Y}-\mathbf{X}_i\mathbf{S}_i$. Compute the new residual.

end

Output: $\mathbf{S}$, $\mathbf{X}$ and $\mathbf{d}$

where $(.)^T$ denotes the matrix (or vector) transpose and $\mathbf{\Phi}_k$ is the k-th row of $\mathbf{\Phi}$. $\mathbf{X}$ is used to obtain the sparse approximate matrix $\mathbf{X}_0'[f]$ in Section 3.2. The iteration times l should be larger than K. If K is unknown, l should be chosen such that it is sufficiently large. To distinguish between signal and noise measurements, in this paper, the threshold λ_k is computed by (23). Therefore, in practice, we can give the iteration a termination condition of $d[k]<\lambda_k$. When the best matching column in $\mathbf{\Phi}$ is found in step 2 of SOMP, $\left\|\mathbf{\Phi}_k^T\mathbf{R}_{i-1}\right\|$ indicates the value of $\|\mathbf{X}_k\|$, where $\|\mathbf{X}_k\|$ is the k-th row in $\mathbf{X}$. Therefore, it is possible to use $\mathbf{d}$ to identify the active sub-bands using a technique that is similar to the periodogram method, as shown in Figure 5b.

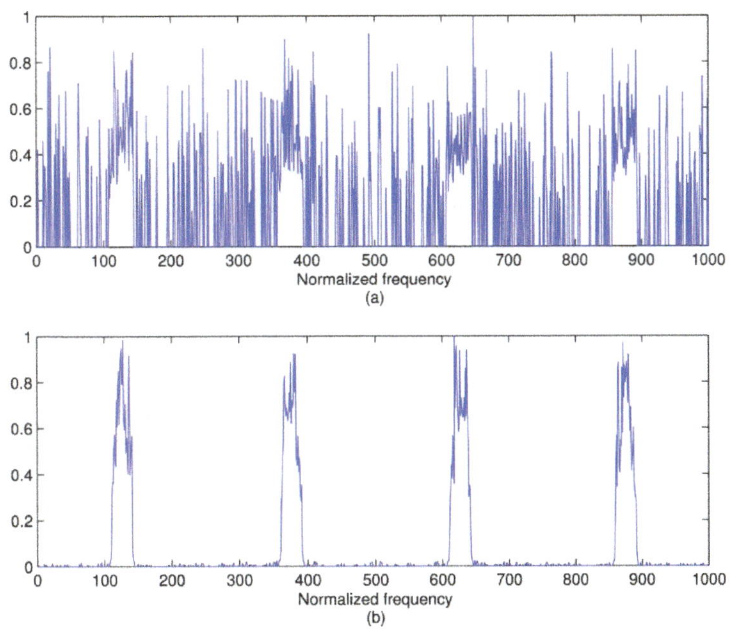

Figure 5: The decision statistic by SOMP algorithm. (a) The results computed by (14). (b)The results computed by the step.3 of SOMP.

Multiband Signal Detection from Compressive Samples

If M is large enough ($M>2K$ is an empirical condition for SOMP), $\mathbf{X}$ is accurately recovered by SOMP with high signal-to-noise ratio (SNR). Constant false-alarm rate energy detection is employed for multiband detection using the periodogram method[36]. The decision statistic for energy detection in the k-th row of $\mathbf{X}$ is

$$\tilde{d}[k]=\frac{1}{L}\sum_{l=0}^{L-1}\left|X_{(k,l)}\right|^2 \qquad (k=1,2,\ldots,N), \tag{14}$$

where $X_{(k,l)}$ is the (k, l)-th element in $\mathbf{X}$. In the following, the detection model is represented by

$$\begin{array}{lll} \mathrm{H}_0: & \tilde{d}_0=n_0 & \text{signal absent} \\ \mathrm{H}_1: & \tilde{d}_1=s+n_0 & \text{signal present,} \end{array} \tag{15}$$

where s is signal sample, and n_0 is circular complex zero-mean Gaussian independent and identically distributed (i.i.d.) noise with a variance of σ^2. With this model, $\tilde{d}_0$ is the sum of the squares of L independent identically distributed zero-mean Gaussian random variables. Therefore, the model produces a random variable using a chi-squared distribution that has L degrees of freedom. To achieve predefined sensing performance, the sample frame number L is always large. Therefore, we can use the central limit theorem to approximate $\tilde{d}_0$ as the Gaussian function

$$\tilde{d}_0 \sim \mathcal{CN}\left(\sigma^2, \frac{2\sigma^4}{L}\right). \tag{16}$$

A binary hypothesis test is then formulated using a decision rule, resulting in

$$\begin{aligned} \mathrm{H}_0 &: \quad \tilde{d}[k] < \lambda \\ \mathrm{H}_1 &: \quad \tilde{d}[k] \geq \lambda. \end{aligned} \tag{17}$$

The threshold λ is chosen according to a target false-alarm probability P_{FA} where $P_{FA} = P_r\{\tilde{d} > \lambda | \mathrm{H}_0\}$ and is given by

$$\lambda = \left(1 + \sqrt{\frac{2}{L}}\right)\sigma^2 Q^{-1}(P_{FA}). \tag{18}$$

where Q(.) stands for the tail probability of the Gaussian distribution

$$Q(x) = \int_x^{\infty} \frac{1}{2\pi} \exp\left(-\frac{u^2}{2}\right) du. \tag{19}$$

However, it must be noted that the recovery performance of compressive sensing is sensitive to the number of sampling branches and to the SNR[37]. When M <2K or the noise level is large, **X** cannot be accurately recovered by SOMP and errors may be introduced into $\tilde{d}$, as illustrated in Figure 5a. Performance is even worse if the SNR<0 dB. Therefore, in our scheme, we employ the vector **d** from SOMP as the input to ensure testing accuracy. The approximate value of the residual $\mathbf{R}_i$ is obtained using steps 4 to 6 of SOMP.

$$\mathbf{R}_i = \mathbf{BX}, \tag{20}$$

B is a submatrix of **Φ** by removing the columns same with **S** (**S** is derived from SOMP algorithm). Then,

$$d[k] = \left\|\mathbf{\Phi}_k^T \mathbf{BX}\right\|^2, \tag{21}$$

where $\mathbf{\Phi}_k$ is the k-th column of **Φ**. The correlation between two different columns of the sampling matrix is weak, i.e., $\mathbf{\Phi}_k^T \mathbf{\Phi}_k \gg \mathbf{\Phi}_k^T \mathbf{\Phi}_{k'}$, (k'≠k), which satisfies the basic RIP condition for compressive sampling[6]. The element in **A** is selected from {1, -1}, so the sign of the elements in **Φ** is usually random and $\mathbf{\Phi}_k^T \mathbf{\Phi}_{k'}$ is close to zero. We then have

$$\begin{aligned} d[k] &\approx \sum_{l=0}^{L-1} \left\|\mathbf{\Phi}_k^T \mathbf{\Phi}_k X_{(k,l)}\right\|^2 \\ &= \rho_k \sum_{l=0}^{L-1} \left\|X_{(k,l)}\right\|^2, \end{aligned} \tag{22}$$

where $X_{(k,l)}$ is the (k, l) -th element of **X**, ρk= $\left\|\mathbf{\Phi}_k^T \mathbf{\Phi}_k\right\|^2$. The threshold of decision statistic d[k] is

$$\lambda_k = \left(1 + \sqrt{\frac{2}{L}}\right)\rho_k \sigma^2 Q^{-1}(P_{FA}). \tag{23}$$

Taking this into account, the binary hypothesis test (16) is rewritten as

$$\begin{aligned} \mathrm{H}_0 &: \quad d[k] < \lambda_k \\ \mathrm{H}_1 &: \quad d[k] \geq \lambda_k. \end{aligned} \tag{24}$$

In a multiuser network, a single user signal usually occupies successive frequency bands in high-resolution spectrums. Therefore, **d** can be divided into N_g groups where the length of i-th group is b_i, the number of users is N_g, and the i-th user occupies bandwidth B_i. To eliminate false spurs in the decision results with the threshold (23), a multiband signal bandwidth constraint is given by

$$\begin{aligned} B_i &= \frac{b_i}{N} f_s \geq \frac{b_{mw}}{N} f_s \\ \Delta B &\geq \frac{b_{mi}}{N} f_s, \end{aligned} \tag{25}$$

where b_{mw} is minimum frequency bin and b_{mi} is minimum interval frequency bin. $\frac{b_{mw}}{N} f_s$ is minimum bandwidth of a user and $\frac{b_{mi}}{N} f_s$ is minimum bandwidth interval between two adjacent users.

Cyclic Spectrum Estimation

Another objective of the SOMP algorithm is to reconstruct $\mathbf{X}_0$[f], which involves signal recovery from compressive samples in the frequency domain. However, using CS-FAM for signal detection differs from using the SOMP algorithm. SOMP directly assigns $\mathbf{X}_0$[f] = **X**, where **X** is the output of SOMP. In contrast, according to the first step of CS-FAM, the row index of $\mathbf{X}_0$[f] indicating only the presence of signal is known. The sparse approximate matrix $\mathbf{X}_0$' [f] is computed by

$$\mathbf{X}_0'[f] = \operatorname{argmin}_{\mathbf{X}} \left\|\mathbf{Y} - \mathbf{X}\mathbf{\Phi}'\right\| \quad \text{s.t.} \quad \mathbf{X}_k = \mathbf{0}_{1\times L}, \; k \in \mathbf{V}. \tag{26}$$

$\mathbf{X}_k$ is the k-th row of **X**, the elements of which are set to zero, as it indicates only the presence of noise. **Φ**‹ is the submatrix of **Φ** with removing the columns whose index k supports k∈**V**. The element of set **V** are defined by (24) as the row index of k, s.t. d[k]<λ_k. Equation (26) is then solved using the least square (LS) method assuming M>K. It is obvious that if the iteration of SOMP algorithm is ended by the termination condition of d[k]<λ_k, we can have $\mathbf{X}_0$' [f]=**X**, where **X** is the output of SOMP. If the approximate matrix $\mathbf{X}_0$' [f] is obtained, cyclic spectrum estimation is processed in the second step. Cyclic feature detection in CS-FAM aims to extract parameters rather than to merely detect whether a signal is present or not.

The complexity of the second step is greatly reduced because only K rows are nonzero following (26). Compared to FAM, which is used to estimate the whole span of the spectrum, CS-FAM computes solely the occupied frequency regions of each user. The operation complexity of cyclic spectrum estimation in the second step is illustrated in Table 1. Due to $\sum_{i=1}^{N_g} b_i = K \ll N$, this complexity of the second step is substantially reduced.

Table 1: Operation complexity of the second step

	FAM	CS-FAM
Down conversion	$2LN$	$2KL$
Cross multiplication	LN^2	$L\sum_{i=1}^{N_g} b_i^2$
FFT	$N^2\frac{L}{2}\log(L)$	$\frac{L}{2}\log(L)\sum_{i=1}^{N_g} b_i^2$

Different modulation formats result in different type of spectral peaks. The pairs of the peaks exhibited by for some common digital modulations[38] are summarized in Table 2, where f_c is carrier frequency and $1/T_b$ is symbol rate. If the occupied frequency areas have been identified in the first step, the peaks can be detected in the second by searching within the support range rather than by setting a threshold with the allowable false-alarm rate. For example, if a user with band of (f_c-f_a, f_c+f_a) is detected from the first step of CS-FAM, the support range of its cyclic spectrum[24] is given by

$$S_z^{\alpha}(f) = \begin{cases} 0 & ||f|-|\alpha/2|| \le f_c - f_a \quad or \\ & ||f|+|\alpha/2|| \le f_c + f_a \\ S_x^{\alpha}(f) & \text{otherwise} \end{cases} \qquad (27)$$

Table 2: Peak pairs of digital modulation

Modulation	Peak pairs (f, α)
BPSK	$(f_c, \frac{1}{T}), (0, 2f_c), (0, 2f_c \pm \frac{1}{T})$
MSK	$(f_c, \frac{1}{T}), (0, 2f_c \pm \frac{1}{T})$
AM	$(0, 2f_c)$
QAM	$(f_c, \frac{1}{T})$

Once the peaks are obtained, modulation classification is performed. Some literature on the subject of automatic modulation classification based on cyclic spectrum analysis is summarized in[39]. The symbol rate and carrier frequency are obtained after the modulation format is identified. Because CS-FAM separates multi-signals in the first step, the multiuser cyclic spectrum reconstruction problem of conventional methods becomes simply multiple operations of a single user problem, avoiding signal separation from the whole cyclic spectrum. Performing multiuser detection in the first step of CS-FAM also avoids cyclic spectrum superposition of different users, which may lead to false peaks.

DETECTION PERFORMANCE AND SIMULATION RESULTS

In this section, we demonstrate the performance capacity of the proposed method with performance test results. In Section 4.1, we examine multiband signal detection performance from compressive samples, assuming that the received signals contain only zero-mean Gaussian i.i.d. noise. Section 4.2 is dedicated to examining cyclic spectrum estimation performance.

Multiband Signal Detection Performance

The process of multiband detection using CS-FAM is illustrated in Figure 6. The analog input content was provided by three binary phase shift keying (BPSK) modulated signals, N=1,000,M=400, and L=50, shaped by a raised-cosine filter. Figure 6a shows the spectrum of the received signals, which were transmitted through the additive Gaussian white noise (AWGN) channel. The SNR was 0 dB. The normalized frequency in Figure 6b, c, d is presented with the resolution of f_s/N, where f_s is the sampling frequency. The result of **d**, computed by the SOMP algorithm, is presented in Figure 6b. We then applied the constant false-alarm detector with $P_{FA}=0.05$. The signal detection results are shown in Figure 6c, where it can be seen that the false-alarm detector identified a number of false spurs. Figure 6d presents the final detection results following elimination of false spurs using a bandwidth constraint of $b_{mw}=b_{mi}=2$. All of the active sub-bands were detected and all three users identified.

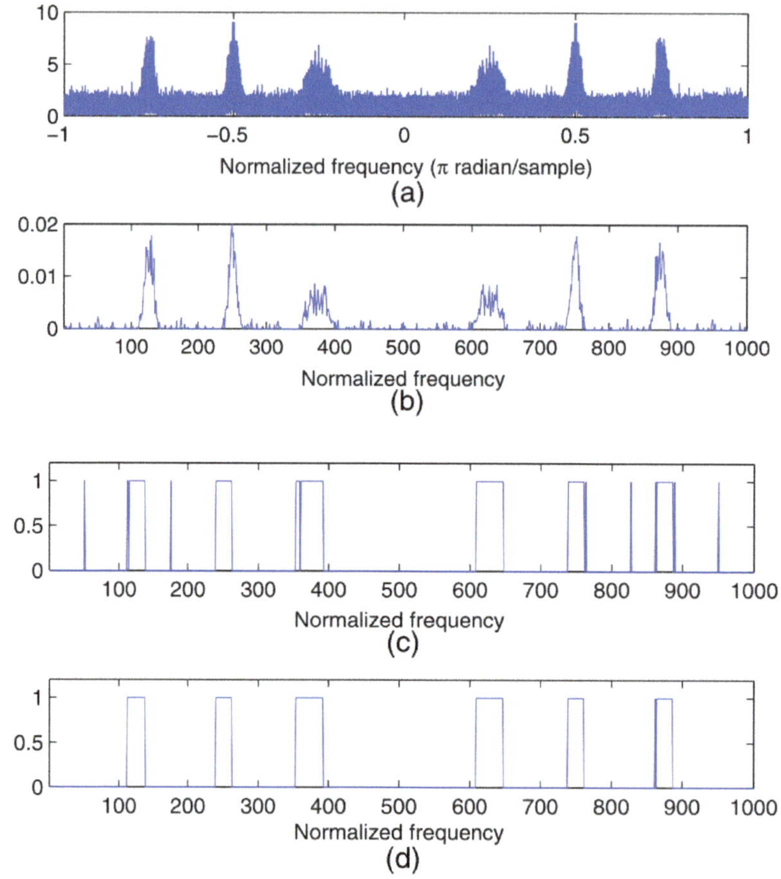

Figure 6: The process of spectrum sensing in the first step of CS-FAM. (a) The spectrum of original multiband signal, SNR=0 dB. **(b)** The results of **d** computed by SOMP algorithm. **(c)** The decision results with the constant false-alarm detector. **(d)** Eliminate the false spurs by bandwidth constraint.

An evaluation of the performance of the CS-FAM test is shown in Figure 7. The reference line of Theory in Figure 7a was set by $P'_{FA}=P_{FA}$, so it presents as a diagonal line. Because using the bandwidth constraint eliminated some of the false spurs, the probability of false-alarm occurrence was lower in the test than in the theoretical value computed by (23). In addition, the rate of false-alarm occurrence using CS-FAM was identical under various noise levels.

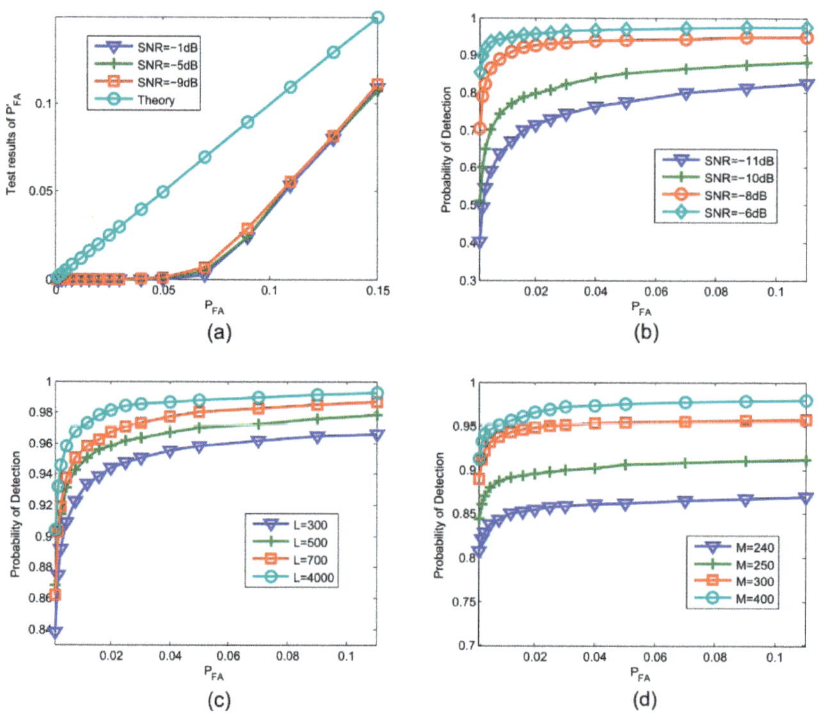

Figure 7: The spectrum sensing performance of CS-FAM. (a) The test results of probability of false-alarm for various SNR. **(b)** The detection performance for various SNR. **(c)** The detection performance for various sensing time. **(d)** The detection performance for various number of sampling branches.

Detection performance of the test is presented in Figure 7c, d. The multiband signal contained two BPSK modulated signals. The main lobe of each signal occupied the spectrum from $-0.032\pi$to 0.032π in the baseband. The threshold was set by (23) according to the target false-alarm rate. Figure 7b shows the detection performance under different noise levels, illustrating that the detector demonstrates better performance in higher SNR scenarios. The quality of detection is improved by increasing sensing time, as shown in Figure 7c. However, the results also show that this effect is limited when L is very large. Detection performance with various compression ratios is shown in Figure 7d, which indicates that the ratios are improved by increasing the number of sampling branches.

As mentioned before, the CS-FAM multiband detection method is based on power spectrum estimation. Therefore, a comparison between CS-FAM and PSBS[9, 10] is presented in Figure 8. As illustrated in Figure 8a, the probability of false alarms in the test was higher when using PSBS than when using CS-FAM, meaning that the PSBS method has poor estimation performance in the inactive bands where there is no signal. This shortcoming is clearly shown in Figure 8b. Compared to the PSBS method, CS-FAM identifies a wide disparity between signals and noise, as shown in Figure 8c, which is advantageous for signal detection using a threshold. Another drawback of PSBS is that the size of the matrix used in this method is proportional to $3M^2 \times 3N$, so the values of M and N chosen for PSBS were much smaller than those selected for CS-FAM, as shown in Figure 8. When using PSBS, a large M value leads to high complexity, which further demonstrates that the PSBS method is not suitable for high spectral resolution sensing.

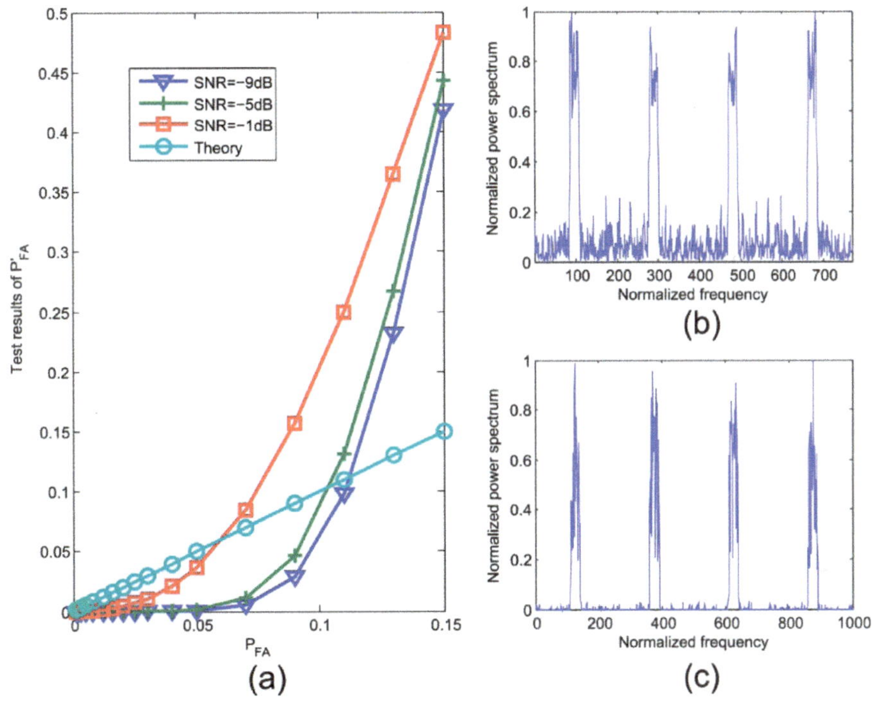

Figure 8: Spectrum sensing based on power spectrum estimation. (a) The test results of probability of false alarm for various SNR by the PSBS method. **(b)** Power spectrum estimation by PSBS, M = 100, N = 250. **(c)** Power spectrum estimation by CS-FAM, M = 400, N = 1,000.

Cyclic Spectrum Estimation Performance

Cyclic spectrum estimation is performed in the second step of CS-FAM, assuming that the active sub-bands have been detected in the first step. Here, we discuss the application of cyclic spectrum estimation from compressive samples. In the following tests, BPSK signals that exhibited spectral peaks at several (f, α) pairs in the cyclic spectrum were selected for analysis. The results showed that three distinct peaks are present at $(f_c, 1/T_b)$, $(0, 2f_c)$, and $(0, 2f_c \pm 1/T_b)$. From these spectral peaks, the carrier frequency, symbol rate, and modulation format can be obtained.

The results of cyclic spectrum estimation using compressive samples are presented in Figure 8. In this test, the analog input contains two BPSK signals and the power of the weaker signal was one quarter that of the larger one. The carrier frequencies of the weaker and stronger signals were $f_1 = 3$ kHz and $f_2 = 7.1$ kHz, respectively, where N = 660 and M = 160. Contour graphs of the estimation results are given in Figure 9. As Figure 9a shows, the cyclic spectrum of the two BPSK signals exhibits special peaks when using the FAM method with the Nyquist sampling. A similar estimation result is obtained using CS-FAM (from compressive samples without multiuser detection in the first step) is shown in Figure 9b. The number of the areas with a large magnitude was 16, including eight false areas yielded by the superposition of the two signals. Figure 9c, d presents the cyclic spectrum estimation results of each signal using CS-FAM with multiuser detection. The signal was correctly identified by the CS-FAM method, as shown in Figure 9c, d, whereas the FAM method failed to identify the peaks of the weaker signal, as shown in Figure 9a. Using FAM, a complex separating operation was required to identify the weaker signal correctly. In contrast, using the CS-FAM scheme, the smaller signal is separated during the first step and is not masked by the larger one.

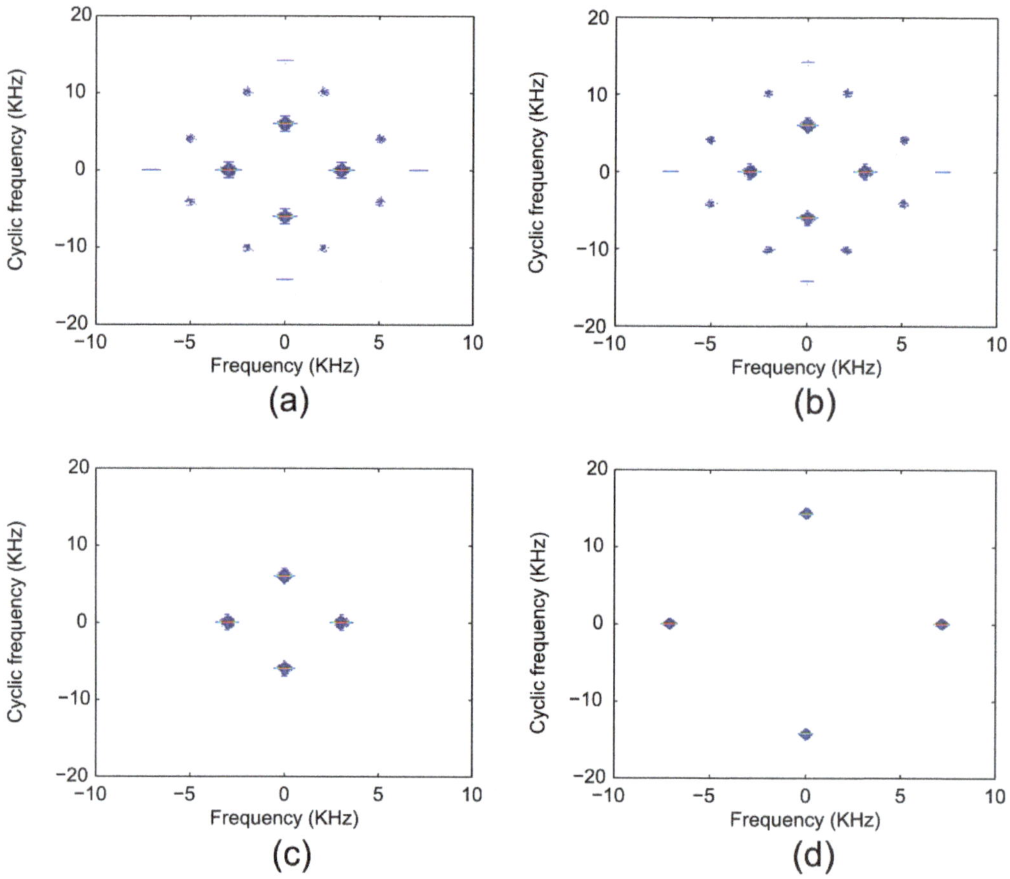

Figure 9: Cyclic feature detection by CS-FAM. (a) Cyclic spectrum estimation based on FAM from the Nyquist samples. (b) The whole cyclic spectrum estimation by CS-FAM without multiuser detection. **(c-d)** Two different BPSK signals are separated by CS-FAM with multiuser detection.

In the next experiment, we examined the cyclic feature detection performance of CS-FAM using different compressive ratios. Three BPSK signals with different carrier frequencies were selected as analog inputs. A simple binary hypothesis test statistic was employed by

$$\begin{array}{lll} H_0 : & U' \neq U & \text{format is not identified} \\ H_1 : & U' = U & \text{format is identified,} \end{array} \qquad (28)$$

where U is the reference set of a modulated signal from Table 2, i.e., $U=\{(f_c, 1/T_b), (0, 2f_c), (0, 2f_c \pm 1/T_b)\}$ if the input is a BPSK signal. U' is the set composed of the detected peaks. This matching decision method is stricter than the approaches used in[39] because the goal is to demonstrate cyclic spectrum estimation performance rather than signal classification. The resulting U' set contained four detected peaks for the BPSK signals. Figure 10 shows the symbol rates of the three BPSK signals, which were $5f_s/N$, N=500, and L=48. To reconstruct the cyclic spectrum with an order similar to $N \times N$, the methods in[20] require matrices proportional to $(500)^4$, which is unacceptable for common devices. It is clear that, when using CS-FAM, estimation quality improves with compressive ratio and that it converges towards the Nyquist rate obtained using FAM.

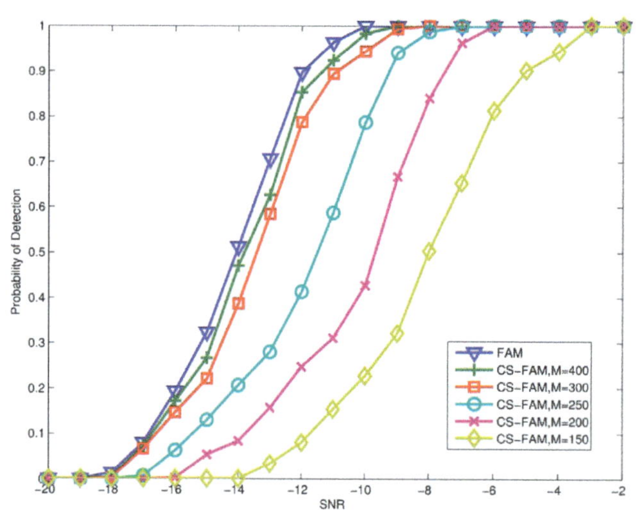

Figure 10: The BPSK identification performance of CS-FAM.

In Figure 11, a BPSK signal is selected with different bandwidth, indicated by K. SNR=0 dB and N=500. The process of cyclic spectrum estimation requires more sampling branches than multiband detection does. Therefore, if the second step is operated effectively, the assumption of the active bands been detected is stated for (26). In addition, the results indicate that both of the steps require more sampling branches to maintain high performance when the wideband signal is less sparse.

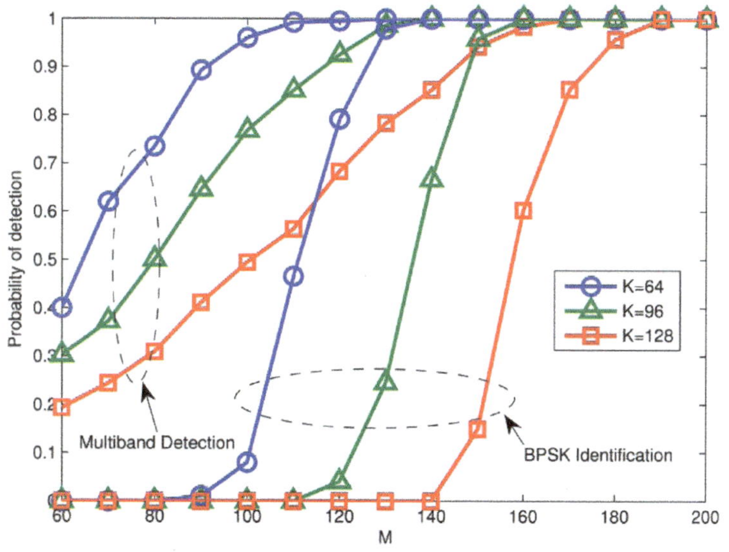

Figure 11: The performance of CS-FAM for various K and M.

CONCLUSIONS

This paper has introduced a joint scheme for low complexity cyclic spectrum estimation and robust multiband signal detection method based on compressive sampling. The process is divided into two steps. First, active sub-bands are identified from MMVs using joint sparse recovery and multiuser detection is performed using a bandwidth constraint. The proposed multiband detection method demonstrates striking robustness with sampling rate reduction and noise uncertainty. In the second step, the cyclic spectrums of each user are individually reconstructed. We also examine signal identification performance under different sampling rates and noise levels. The flexible nature of this scheme, accomplishing multiband and multiuser detection in its first step and parameter extractions in its second, allows for implementation in many practical applications.

ACKNOWLEDGEMENTS

This work was supported by the Shanghai Provincial Natural Science Foundation of China under Grant No. 14ZR1447200, and the National Defense Foundation of China under Grant No. 9140C18010211Zxxx01 and No.9140C1801021xxx18. The authors also thank the anonymous reviewers and the editor for their valuable comments.

REFERENCES

1. Neng-Jing L, Yi-Ting Z: A survey of radar ECM and ECCM. IEEE Trans. Aero Electron Syst. 1995, 31(3):1110-1120. 10.1109/7.395232
2. Lopez-Salcedo JA, Peral-Rosado JAD, Seco-Granados G: Survey on robust carrier tracking techniques. Commun. Surv. Tutor. IEEE 2014, 16(2):670-688.
3. Axell E, Leus G, Larsson EG, Poor HV: Spectrum sensing for cognitive radio: state-of-the-art and recent advances. IEEE Signal Process. Mag. 2012, 29(3):101-116.
4. Sun H, Nallanathan A, Wang C-X, Chen Y: Wideband spectrum sensing for cognitive radio networks: a survey. IEEE Wirel. Comm. Mag. 2013, 20(2):74-81.
5. Baraniuk RG: Compressive sensing [lecture notes]. IEEE Signal Process. Mag. 2007, 24(4):118-121.
6. Foucart S, Rauhut H: A ematical Introduction to Compressive Sensing. Springer, Boston; 2013.
7. Mishali M, Eldar YC: Blind multiband signal reconstruction: compressed sensing for analog signals. IEEE Trans. Signal Process. 2009, 57(3):993-1009.
8. Mehanna O, Sidiropoulos ND: Frugal sensing: wideband power spectrum sensing from few bits. IEEE Trans. Signal Process. 2013, 61(10):2693-2703.
9. Leus G, Ariananda DD: Power spectrum blind sampling. IEEE Signal Process Lett. 2011, 18(8):443-446.
10. Ariananda DD, Leus G: Compressive wideband power spectrum estimation. IEEE Trans. Signal Process. 2012, 60(9):4775-4789.
11. Miar Y, D'Amours C, Aboulnasr T: A novel reduced power compressive sensing technique for wideband cognitive radio. EURASIP J. Wirel. Commun. Netw. 2012, 2012(1):281. 10.1186/1687-1499-2012-281
12. Unnikrishnan J, Veeravalli VV: Algorithms for dynamic spectrum access with learning for cognitive radio. IEEE Trans. Signal Process. 2010, 58(2):750-760.
13. Bourdoux A, Cabric D, Ferrari G, Horlin F, Tian Z: Dynamic spectrum access: from the concept to the implementation. EURASIP J. Wirel. Commun. Netw. 2010, 2010(1):654395. 10.1155/2010/654395
14. Noguet D, Biard L, Laugeois M: Cyclostationarity detectors for cognitive radio: architectural

tradeoffs. EURASIP J. Wirel. Commun. Netw. 2010, 2010(1):526429. 10.1155/2010/526429

15. Dobre OA, Abdi A, Bar-Ness Y, Su W: Survey of automatic modulation classification techniques: classical approaches and new trends. Commun. IET 2007, 1(2):137-156. 10.1049/iet-com:20050176

16. Sun Z, Liu R, Wang W: Joint time-frequency domain cyclostationarity-based approach to blind estimation of OFDM transmission parameters. EURASIP J. Wirel. Commun. Netw.2013, 2013(1):117. 10.1186/1687-1499-2013-117

17. Bolcskei H: Blind estimation of symbol timing and carrier frequency offset in wireless OFDM systems. IEEE Trans. Commun. 2001, 49(6):988-999. 10.1109/26.930629

18. Cohen D, Rebeiz E, Jain V, Eldar YC, Cabric D: Cyclostationary feature detection from sub-nyquist samples. In 4th IEEE International Workshop (2011). IEEE, New York, USA; 2011:333-336.

19. Rebeiz E, Jain V, Cabric D: Cyclostationary-based low complexity wideband spectrum sensing using compressive sampling. In Communications (ICC). IEEE International Conference, IEEE, Ottawa, Canada; 2012:619-1623.

20. Tian Z, Tafesse Y, Sadler BM: Cyclic feature detection with sub-Nyquist sampling for wideband spectrum sensing. IEEE J Sel Top Signal Process 2012, 6(1):58-69.

21. Safatly L, Aziz B, Nafkha A, Louet Y, Nasser Y, El-Hajj A, Kabalan K, Safatly L, Aziz B, Nafkha A, Louet Y, Nasser Y, El-Hajj A, Kabalan K: Blind spectrum sensing using symmetry property of cyclic autocorrelation function: from theory to practice. EURASIP J. Wirel. Commun. Netw. 2014, 2014(1):26. 10.1186/1687-1499-2014-26

22. Roberts RS, Brown WA, Loomis HH: Computationally efficient algorithms for cyclic spectral analysis. IEEE Signal Process. Mag. 1991, 8(2):38-49.

23. MacKenzie AB, Reed JH, Athanas P, Bostian CW, Buehrer RM, DaSilva LA, Ellingson SW, Hou YT, Hsiao M, Park J-M, Patterson C, Raman S, da Silva C: Cognitive radio and networking research at Virginia Tech. Proc. IEEE 2009, 97(4):660-688.

24. Gardner WA: Exploitation of spectral redundancy in cyclostationary signals. IEEE Signal Process. Mag. 1991, 8(2):14-36.

25. Gardner WA: Statistical Spectral Analysis: A Nonprobabilistic Theory. Prentice-Hall, Englewood Cliffs, NJ; 1987.

26. Roberts RS, Brown WA Jr, Loomis HH: Cyclostationarity in Communications and Signal Processing. IEEE Press, Boston; 1994:455-479.

27. Sathe VP, Vaidyanathan PP: Effects of multirate systems on the statistical properties of random signals. IEEE Trans. Signal Process. 1993, 41(1):131-146. 10.1109/TSP.1993.193133

28. Izzo L, Napolitano A: Multirate processing of time series exhibiting higher order cyclostationarity. IEEE Trans. Signal Process. 1998, 46(2):429-439. 10.1109/78.655427

29. Sony A, Vaidyanathan PP: Bifrequency and bispectrum maps: a new look at multirate systems with stochastic inputs. IEEE Trans. Signal Process. 2000, 48(3):723-736. 10.1109/78.824668

30. Kirolos S, Laska J, Wakin M, Duarte M, Baron D, Ragheb T, Massoud Y, Baraniuk R: Analog-to-information conversion via random demodulation. In Design, Applications, Integration and Software, 2006 IEEE. CAS Workshop, Dallas; 2006:71-74.

31. Mishali M, Eldar YC: Wideband spectrum sensing at sub-Nyquist rates [applications corner]. IEEE Signal Process. Mag. 2011, 28(4):102-135.

32. Cotter SF, Rao BD, Engan K, Kreutz-Delgado K: Sparse solutions to linear inverse problems with multiple measurement vectors. IEEE Trans. Signal Process. 2005, 53(7):2477-2488.

33. Eldar YC, Mishali M: Robust recovery of signals from a structured union of subspaces.IEEE Trans. Inform. Theor. 2009, 55(11):5302-5316.

34. Chen J, Huo X: Theoretical results on sparse representations of multiple-measurement vectors. IEEE Trans. Signal Process. 2006., 54(12):

35. Tropp JA, Gilbert AC, Strauss MJ: Algorithms for simultaneous sparse approximation. Part I: greedy pursuit. Signal Process. 2006, 86(3):572-588. 10.1016/j.sigpro.2005.05.030

36. Urkowitz H: Energy detection of unknown deterministic signals. Proc. IEEE 1967, 55(4):523-531.

37. Oymak S, Khajehnejad MA, Hassibi B: Recovery threshold for optimal weight l1 minimization. In Information Theory Proceedings (ISIT), 2012 IEEE International Symposium. IEEE, Cambridge, USA; 2032-2036.

38. Gardner WA, Brown W, Chen C-K: Spectral correlation of modulated signals: part II-digital modulation. IEEE Trans. Comm. 1987, 35(6):595-601. 10.1109/TCOM.1987.1096816

39. Ramkumar B: Automatic modulation classification for cognitive radios using cyclic feature detection. IEEE Circ. Syst. Mag. 2009, 9(2):27-45.

CHAPTER 7

SPEECH STEGANOGRAPHY USING WAVELET AND FOURIER TRANSFORMS

Siwar Rekik, Driss Guerchi, Sid-Ahmed Selouani and Habib Hamam

Keywords: Audio steganography, Discrete wavelet transform, Fast Fourier transform, Data hiding, Speech steganography

INTRODUCTION

One of the concerns in the field of secure communication is the concept of information security. Today's reality is still showing that communication between two parties over long distances has always been subject to interception. Providing secure communication has driven researchers to develop several cryptography schemes. Cryptography methods achieve security in order to make the information unintelligible to guarantee exclusive access for authenticated recipients. Cryptography consists of making the signal look garbled to unauthorized people. Thus, cryptography indicates the existence of a cryptographic communication in progress, which makes eavesdroppers suspect the existence of valuable data. They are thus incited to intercept the transmitted message and to attempt to decipher the secret information. This may be seen as weakness in cryptography schemes. In contrast to cryptography, steganography allows secret communication by camouflaging the secret signal in another signal (named the cover signal), to avoid suspicion. This quality motivated the researchers to work on this burning field to develop schemes ensuring better resistance to hostile attackers.

The word steganography is derived from two Greek words: Stego (means cover) and graphy (means writing). The two combined words constitute steganography, which means covert writing, it is the art of hiding written communications. Several steganography techniques were used to send message secretly during wars through the territories of enemies. The use of steganography dates back to ancient time where it was used by romans and ancient Egyptians [1]. One technique according to Greek historian Herodotus was to shave the head of a slave, tattoo the message on the slave's scalp, and send him after his hair grew back. Another technique was to write the secret message underneath the wax of a writing tablet. A third one is to use invisible ink to write secret messages within covert letters [2].

Many techniques have been developed for hiding secret signals into other cover signals. Sridevi et al. [3] presented a method for audio steganography. It consists of substituting the least significant bit (LSB) of each sample of the cover speech signal with the secret data. While this method is easy to implement and can be used to hide larger secret messages, it cannot protect the hidden message from small modifications that can happen as a result of format conversion or compression. Hiding data in LSBs of audio samples in the time domain is one of the simplest algorithms enabling a very high data rate of inserted information. However, several steganalysis algorithms have been developed to challenge the robustness of this method. Bender et al. [4] have presented a technique for data hiding based on phase coding. This method consists of substituting the phase of the first part of an audio segment by a reference phase that represents the data. In order to conserve the relative phase between segments, an adjustment must be made in the phase of the succeeding segment. The series of steps of phase coding is as follows: (i) The original audio signal is decomposed into smaller segments such that their length is equal to the size of

Citation: Rekik et al.: Speech steganography using wavelet and Fourier transforms. EURASIP Journal on Audio, Speech, and Music Processing 2012 2012:20.

the message to be encoded; (ii) A discrete Fourier transform (DFT) is then applied on each segment leading to a phase matrix; (iii) Compute the differences between the phase of each pair of the consecutive segments; (iv) Identify the phase shifts between the consecutive segments. Although, the absolute phases of the segments may change, the relative phase differences between the consecutive segments must remain unchanged; (v) Use the new phase of the first segment and the set of original phase differences to create a new phase matrix; (vi) Regenerate the audio signal with an inverse DFT and then connect the audio segments together. This step is based on the original magnitude matrix and the newly created phase matrix. The receiver determines the length of the secret message, then applies a DFT and extract the hidden message from the cover signal. A distinctive characteristic of phase coding is the low data transmission rate due to the fact that the secret data are encoded only in the first segment of the audio signal. Controversially, any enhancement in the length of the segment may result in shifting the phase relations among the frequency elements of the segment, leading therefore to an easier detection of the existence of a secret message. Thus, the phase coding algorithm is more efficient when hiding small amount of data. Kirovski and Malvar [5] have proposed a new steganographic scheme, called Spread Spectrum (SS) coding method. This method randomly spreads the bits of the secret data message across the frequency spectrum of the audio signal. However, in contrast to LSB coding, the SS coding scheme spreads the secret message using a code independent from the concrete cover signal. The SS coding technique may outperform the LSB coding and phase coding techniques by offering a good quality for medium data transmission rates while ensuring a high level of robustness against steganalysis. However, similarly to the LSB coding technique, the SS method may introduce noise to the audio file. This is presenting a weakness since it facilitates detection by steganalysis systems.

Huang and Yeo [6] have presented an information hiding method based on echo hiding. An echo is introduced into the discrete audio signal in order to embed secret information. Similar to the SS coding method, echo hiding is used to provide a better data transmission rate and higher robustness comparing to the noise-inducing techniques. To accomplish successfully the hiding process, three fundamental parameters need to be changed from the original signal: decay rate, offset (time delay), and amplitude. These three parameters are easily defined since they are located below the human audible threshold limit which is different from the echo. Also, the offset is altered to characterize the binary message to be hidden. The first and the second offsets represent a one (binary) and a zero (binary), respectively. Shirali and Shahreza [7] present an approach for hiding information in a speech signal. This method consists of detecting the silence intervals of a speech and the corresponding length of these intervals (number of samples) and changing them with the secret information. Hiding data in silent interval of the audio samples is one of the simplest algorithms enabling a very high data rate of inserted information. However, this method is already well known and several steganalysis algorithms have been developed to defeat the robustness of this method.

Speech steganography takes advantage of the recent advancements in speech compression and data hiding. Speech is a low-pass signal; its intelligibility is retained when preserving at least the first three formants of the magnitude spectrum. In this article, we will take advantage of these speech characteristics to propose an efficient speech-in-speech hiding method. Our speech steganography system consists of embedding the secret speech parameters in the high-frequency regions of the magnitude spectrum of the cover speech. Our aim is to ensure that the stego signal obtained from combining the original phase spectrum and the modified magnitude spectrum shows similar subjective quality to the cover signal. Theoretically, the resultant stego speech is expected to be perceptually indistinguishable from the cover speech since the pertinent low-frequency components will remain intact.

Potential applications of our speech hiding scheme are reduction of speech storage and transmission overhead in electronic voice mail applications and audio streaming, speech translation, data communication secrecy, and many other web-based applications.

Objectives

Our objective is to develop a high performance speech steganography system. The design of such system consists principally of the optimization of the following attributes:

The hiding capacity, defined by the amount of the secret information (speech, text, or image) to be hidden in the cover speech signal.

The impact of the hiding process on the cover speech quality. We hope to produce a stego signal that is perceptually indistinguishable from the cover signal.

The complexity of the steganography system. Our aim is to render the steganalysis (the attempt to discover the existence of the secret message from the stego signal) by the opponent more complex.

The accuracy with which the hidden message can be recovered at the receiver. Efficient techniques are to be developed to minimize the impact of the compression on the stego signal.

We choose a speech signal as secret information to be hidden in the cover speech. Since our objective in discrete wavelet transform-fast Fourier transform (DWT-FFT)-based hiding approach is secrecy, we propose to hide the secret information within the high-frequency of the wavelet components.

The rest of the article is organized as follows: in the following section, we introduce our DWT-FFT-based approach dedicated for the steganography task. Section "Secret speech parameterization" will describe the secret speech analysis including the linear predictive coding (LPC) analysis and the line spectral frequencies (LSF) extraction procedure. In Section "Speech hiding algorithm", we proceed with the description of the used speech hiding algorithm. The general step to retrieve the secret speech signal is also included in this section. Then a description of the speech signals database used for our simulations, the parameters of our experiments, the evaluation and discussion of the results of our proposed DWT-FFT hiding approach are presented in Section "Evaluation". Finally, we conclude and suggest directions for further research in Section "Conclusions".

DWT-FFT-BASED APPROACH

Speech DWT

The wavelet transform can be considered as transforming the signal from the time domain to the wavelet domain. This new domain contains more complicated basis functions called wavelets, mother wavelets, or analyzing wavelets [8]. The fundamental idea behind wavelets is to analyze according to scale. Any signal can then be represented by translated and scaled versions of the mother wavelet. Wavelet analysis is capable of enlightening aspects of data that other signal analysis techniques are unable to perform, aspects like trends, and discontinuities in higher derivatives, breakdown points, and self-similarity.

The basic idea of DWT for one-dimensional signals is shortly described. The wavelet analysis allows the split of a signal into two parts, usually the high- and the low-frequency parts. This process is called decomposition. The edge components of the signal are largely limited to the high-frequency part. The signal is passed through a series of high-pass filters to analyze the high frequencies, and it is passed through a series of low-pass filters to analyze the low frequencies. Filters of different cutoff frequencies are used to analyze the signal at different resolutions [9,10].

The DWT involves choosing scales and positions based on powers of two, the so-called dyadic scales and positions. The mother wavelet is rescaled by powers of two and transformed by integers. Specifically, a function $f(t) \in L^2(R)$ (defines space of square integrable functions) can be represented as:

$$f(t) = \sum_{j=1}^{L} \sum_{k=-\infty}^{\infty} d(j,k)\psi(2^{-j}t-k) + \sum_{k=-\infty}^{\infty} a(L,k)\phi(2^{-L}t-k) \quad (1)$$

The function $\psi(t)$ is known as the mother wavelet, while $\phi(t)$ is known as the scaling function. The set of function $\left\{\sqrt{2^{-L}}\phi(2^{-L}t-k), \sqrt{2^{-j}}\psi(2^{-j}t-k) \middle| j \leq L, j,k,L \in Z\right\}$, where Z is the set of integers in an orthonormal basis for $L^2(R)$. The numbers $a(L,k)$ are known as the approximation coefficients at scale L, while $d(j,k)$ are identified as the detail coefficients at scale j. The approximation and detail coefficients can be expressed consecutively as:

$$a(L,k) = \frac{1}{\sqrt{2^L}} \int_{-\infty}^{\infty} f(t)\phi(2^{-L}t-k)dt \quad (2)$$

$$d(j,k) = \frac{1}{\sqrt{2^j}} \int_{-\infty}^{\infty} f(t)\psi(2^{-j}t-k)dt \quad (3)$$

For a better understanding of the above coefficients, let's consider a projection $f_l(t)$ of the function $f(t)$ that provides the best approximation (in the sense of minimum error energy) to $f(t)$ at a scale l. This projection can be constructed from the coefficients $a(L,k)$, using the equation:

$$f_l(t) = \sum_{k=-\infty}^{\infty} a(l,k)\phi(2^{-l}t-k) \quad (4)$$

As the scale l decreases, the approximation becomes finer, converging to $f(t)$ as $l \rightarrow 0$. The difference between the approximation at scale $l+1$ and that at l, $f_{l+1}(t)-f_l(t)$, is totally defined by the coefficients $d(j,k)$ using the equation of decomposition and can mathematically be expressed as follows:

$$f_{l+1}(t) - f_l(t) = \sum_{k=-\infty}^{\infty} d(l,k)\psi(2^{-l}t-k) \quad (5)$$

These given relations, $a(L,k)$ and $\{d(j,k)|j \leq L\}$, are useful for building the approximation at any scale. Hence, the wavelet transform breaks the signal up into a coarse approximation $f_L(t)$ (given $a(L,k)$) and a number of layers of detail coefficients $\{f_{j+1}-f_j(t)|j<L\}$ (given by $\{d(j,k)|j \leq L\}$). As each layer of detail is added, the approximation at the next higher scale is achieved. The original signal can be reconstructed using the Inverse DWT (IDWT), following the above procedures in the reverse order. The synthesis starts with the approximation and detail coefficients cA_j and cD_j, and then reconstructs cA_{j-1} by up sampling and filtering with the reconstruction filters [11, 12].

Speech Fourier Transform

Since speech is processed on a time-frame basis, the speech spectrum is evaluated using the DFT. The DFT of a signal $s(n)$ defined for $0 \leq n \leq M-1$ is given by

$$S(k) = \sum_{n=0}^{M-1} s(n)e^{-j2\pi kn/M}, 0 \leq k \leq M-1 \quad (6)$$

In general, $S(k)$ is a complex function of the variable k and can be expressed in polar coordinates as:

$$S(k) = |S(k)|e^{j\phi}(k) \quad (7)$$

The sequence $S(k)$ has the same number of elements as $s(n)$. However, the last $M/2$ elements of the DFT are conjugates of the first $M/2$ elements, in inverse order. Consequently, the magnitude spectrum $|S(k)|$ could be defined uniquely by the first $M/2$ frequency components since it satisfies the following symmetry:

$$|S(k)| = |S(M-k)| \quad (8)$$

This equation represents one of the DFT properties that must be maintained when hiding a message in the magnitudes. This feature is used in the fast Fourier transform (FFT) algorithm to reduce the DFT computational complexity [13]. For simplicity, we will adopt in the subsequent sections the following notations:

S=fft(s) (9)

and

s=ifft(S) (10)

where ifft, the inverse FFT, calculates the inverse DFT.

Speech Spectrum Characteristics

Speech is a baseband signal with most of the pertinent intelligibility-preserving frequency components confined to a bandwidth of 4 and 7 kHz for narrowband and wideband speech, respectively [14]. The distribution of the first three speech formants represents the primary cues to the English vowels. Most of the vowel energy is condensed below 1 kHz and decays at about −6 db/oct with frequency [15]. Figure 1 shows the wideband speech spectrum for both a liquid frame and an unvoiced fricative frame. In all vowels and most of the voiced consonants, the magnitude spectrum shows very week components at high frequencies. Even though few unvoiced fricative consonants, such as /s/, present large magnitudes at high frequencies, the intelligibility of the speech signal is negligibly affected if we do not model accurately these frequency components [14]. On the other hand, even for wideband unvoiced fricative consonants, frequencies above 7 kHz do not contribute considerably to the speech spectrum content. These two facts have motivated us to embed a separate signal in the low amplitude high-frequencies of the cover signal.

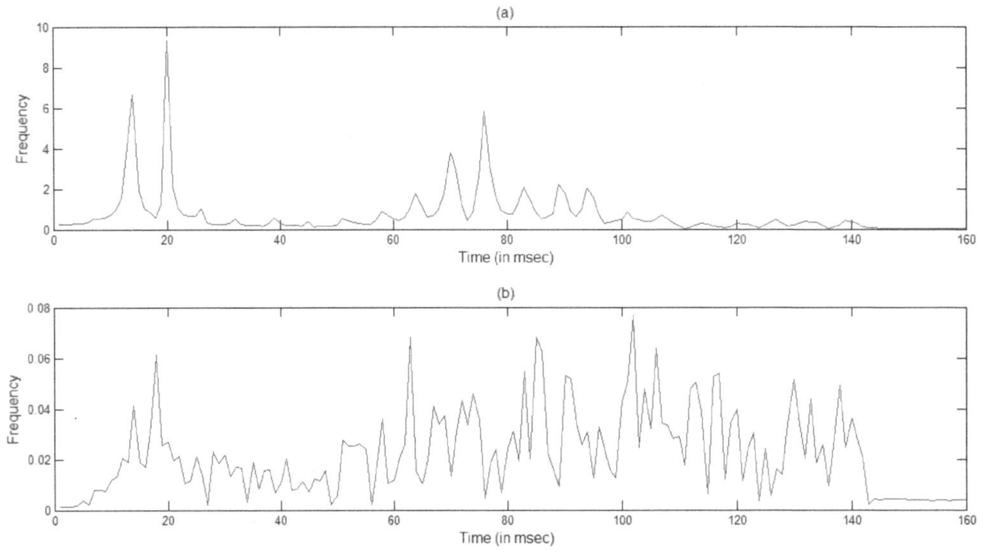

Figure 1: Magnitude spectrum of (a) a voiced frame, (b) unvoiced frame.

Secret Speech Parameterization

Many factors require the parameterization of the secret speech message before the hiding process. Among these factors, we cite the restricted number of the hiding locations in the narrowband cover speech. Speech parameterization termed as *speech analysis* is generally used in different research areas, such as automatic speech recognition and speech coding. In speech coding, the original signal is subject to a speech analysis algorithm to extract the pertinent speech parameters. In order to recreate a copy of the original signal, an inverse algorithm known as speech synthesis is used. Most of the speech analysis schemes are based on the human speech production model [15]. In this speech production model, a sequential excitation of two filters is used to produce a speech signal, a linear prediction (LP) filter is used to model the vocal tract, produces a short-term correlation present in all types of speech and a pitch filter to represent the periodicity created to the vibration of the vocal cords in voiced segments. A basic diagram of the speech production model is shown in Figure 2. The LPC is based on this diagram. The LPC schemes are usually used in the field of speech coding. For example in transmission, the speech frames are represented with a restricted number of parameters. These parameters in the receiver side are used to reconstruct a synthetic-quality speech signal. The speech analysis algorithm is based on two phases: an LP analysis to obtain p LP coefficients, $a_i (i = 1, \ldots, p)$ and a pitch analysis to extract the pitch gain g and the pitch delay d. The LP filter and the pitch filter are constructed using the LP parameters and the pitch, respectively. In the LPC model, for the unvoiced speech signal, an LP filter is used since there is no periodicity in this class of speech. The pitch filter is used for the voiced frames. Details about the speech analysis procedure are given in [16]. The LP coefficients (LPC) must be transformed to a more improved representation before any processing, since the LPC are very susceptible to errors and their direct quantization might generate an unbalanced LP filter. One of the most used representations is the LSF [17]. In this study, we adopted this representation, in the hiding process p magnitude locations are replaced by p LSF coefficients of the secret speech.

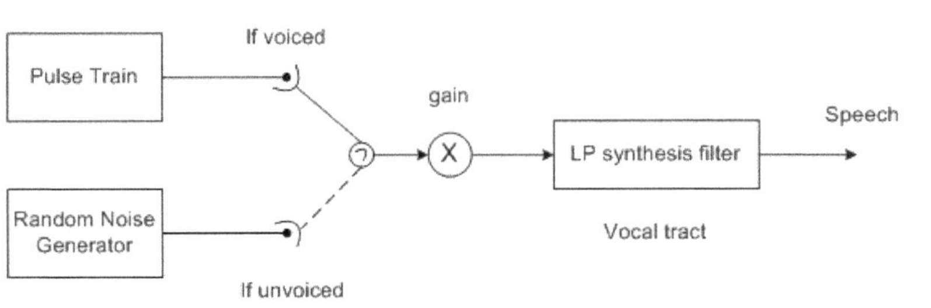

Figure 2: LP model of speech production.

Secret Speech Analysis

To perform the secret speech analysis, we will use the LP speech production model. In this model, the speech signal is subject to an LP analysis followed by pitch analysis.

LP Analysis

The LP analysis is performed every L-ms (for $M=L \times Fs$ samples), for a sampling frequency of Fs kHz, to extract p LP coefficients. These coefficients represent the vocal-track poles (or formants). To smooth the inter-frame variation of the spectral parameters, the analysis window contains more samples than the analysis frame. In addition to the current speech frame, the analysis window contains 5 ms from past speech and 5 ms from future speech. In the LP analysis, we adopt a tapered rectangular window with three parts [18]. The first part is the first half of a hamming window, the second part is a rectangular window, and the third part is the second half of a Hamming window. This window produces a narrower main lobe than the asymmetric window used in G.729 and G722.2 codec standards.

$$w(n) = \begin{cases} 54 - .46\cos\left(\frac{2\pi n}{M-1}\right), & n = 0,\ldots,\frac{M}{2}-1 \\ 1, & n = \frac{M}{2},\ldots,\frac{3M}{2} - \\ .54 - .46\cos\left(\frac{2\pi\left(n-\frac{M}{2}\right)}{M-1}\right), & n = \frac{3M}{2},\ldots,2M \end{cases} \quad (11)$$

The existence of a short-term correlation in speech signals motivates us to adopt the LP analysis. This correlation is helpful to predict a speech sample $s_2(n)$ at time n from its previous p samples $s_2(n-i)$. For each speech frame, a 10-order predictor ($p=10$) is employed on the windowed speech, $s_2(n)$, to estimate the spectral envelope. The predicted signal $\hat{s}(n)$ is given by

$$\hat{s}(n) = \sum_{i=1}^{p} a_i s(n-i) \quad (12)$$

The LP coefficients $a_i(i=1,\ldots,p)$ are predicted from the minimization (by autocorrelation method) of the error between the windowed sample $s_2(n)$ and the predicted sample $\hat{s}_2(n)$. Since the pitch and excitation analysis phases are completed in a closed-loop manner, the LP synthesis filter is required in order to reduce the error between the original speech and the synthesized speech candidates. The LP synthesis filter in the Z-domain, $H(z)$, is connected to the LPC vector by

$$H(z) = \frac{1}{\sum_{i=1}^{p} a_i z^{-1}} \quad (13)$$

The filter $H(z)$ is represented in the time domain by the impulse response function $h(n)$.

Pitch Analysis

Due to the vocal cords vibration, the voice speech segments show some long-term correlation. The vibration frequency, named pitch, is reflected in the quasi-periodicity behavior of the time domain speech waveform. An autocorrelation scheme is used to calculate the pitch lag (the inverse of the pitch frequency). Since the LP analysis frame may contain more than one pitch period, the pitch analysis is performed on sub-frame basis to extract one pitch gain and one pitch delay. One pitch gain and one pitch lag are used to represent consequently the periodicity in each speech frame [19]. In the pitch analysis algorithm, an open-loop analysis is first applied to each speech frame to estimate the pitch period. Open-loop pitch estimation is based on the weighted speech signal $s_w(n)$ which is obtained by filtering the input speech signal through the perceptual weighting filter, s_w is given by:

$$W(z) = \frac{A(Z/y_1)}{A(Z/y_2)} = \frac{1 + \sum_{i=1}^{10} y_1^i a_i z^{-1}}{1 + \sum_{i=1}^{10} y_2^i a_i z^{-1}} \quad (14)$$

That is, in a frame of size L, the weighted speech is given by:

$$S_w(n) = s(n) + \sum_{i=1}^{10} a_i y_1^i s(n-i) - \sum_{i=1}^{10} a_i y_2^i s_w(n-i), n = 0,\ldots,L-1 \quad (15)$$

Residual Excitation

The signal e(n) after removing the long-term and short term redundancies has a noise-like shape with a flat spectrum. Figure 3 shows the residual signal after removing the long and short correlations. This signal could be modulated by a random signal. Since the random signal has no correlation, this residual will be generated at the receiver side using a random signal generator. By this, we reduce the amount of information to be hidden in the cover signal. As mentioned below, the speech analysis algorithm is based on two phases: an LP analysis to obtain p LP coefficients, $a_i(i=1,\ldots,p)$ and a pitch analysis phase to extract the pitch gain g and the pitch delay d. Table 1 shows the used parameters of the LP-model for narrowband speech.

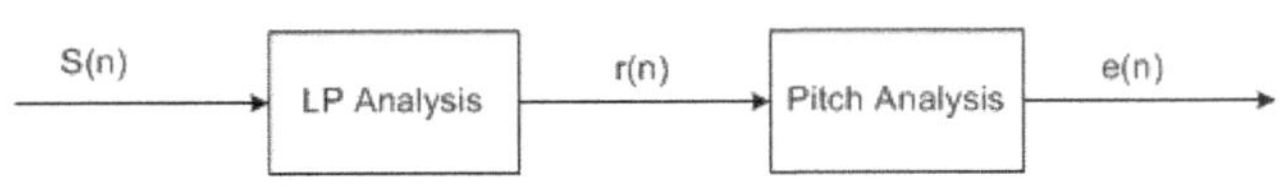

Figure 3: Residual signal after removing the long and short correlations.

Table 1: The LP model parameters.

Model parameters	Symbol	Number of parameters per frame
Pitch lag	d	1
Pitch gain	g	1
LP coefficients	$a_1, a_2, \ldots, a_p$	p
Voice/unvoiced decision	V/UV	1
Total		$p+3$

LP-model Parameters Adjustment

The spectral amplitudes must always be positive due to the absolute value applied to the speech spectrum. Direct embedding of the LP coefficients C in the magnitude spectrum will drastically destroy the cover signal since the LP parameters could have negative values. To accommodate this problem, we propose to convert the LP coefficients C to one of their frequency representations, such as LSF. As shown in the following equation, the LSF parameters w_i are ordered and are all positive.

$$0 \leq w1 \leq w_2 \leq \cdots \leq w_p \leq \pi \tag{16}$$

Since the pitch delay varies from 20 to 147 samples, direct embedding of the pitch delay in the cover speech spectrum will affect the high-frequencies small-amplitudes cover spectrum components. Hence, the need to normalize the pitch delay is by 147, the maximum pitch delay, before the hiding process. The normalized pitch delay will have a value ranging from 0 to 1. For this reason, the best location to hide these parameters is the last cover speech spectrum location since the amplitude of this last component is very small.

LSF Cues

Itakura [20] has proposed the LSF to represent the LPC. They have been demonstrated to acquire different advantageous proprieties like bounded range, sequential ordering, and ability of constancy verification [21]. Moreover, the LSFs coefficients facilitate the integration of human observation system proprieties in the frequency domain representation. According to the ITU-T Recommendation G.723.1, the extraction of the LSFs parameters is recommended in case of need to convert the LPC parameters to LSFs [22]. In LPC, the mean squared error between the original and the predicted speech is minimized over a short time interval to produce distinctive set of LP coefficients. The transfer function of the LPC filter is given by

$$H(z) = \frac{G}{1+\sum_{k=1}^{P} a_k z^{-k}} \tag{17}$$

where P the prediction order, G is the gain, and a_k is the LPC filter coefficients. The poles of this transfer function contain the poles of the vocal tract as well as those of the voice source. Solving for roots of the denominator of the transfer function gives both the formant frequencies and the poles corresponding to the voice source. Two transfer functions $Q_{p+1}(z)$ and $P_{p+1}(z)$, respectively, called difference and sum polynomials can be resulting from $H(z)$. The difference polynomial is given by:

$$Q_{p+1} = A_p(z) - z^{-(p+1)})A_p\left(z^{(-1)}\right) \tag{18}$$

and the sum polynomial is given by:

$$P_{p+1} = A_p(z) + z^{-(p+1)}A_p(z^{-1}) \tag{19}$$

where $A_p(z)$ is the denominator of $H(z)$. The polynomials contain trivial zeros for even values of$p$ at $z=-1$ and at $z=1$. These roots can be removed in order to obtain the following quantities:

$$\hat{Q}(z) = \frac{Q_{p+1}(z)}{(1+z)} = \beta_0 z^p + \beta_1 z^{p-1} + \cdots + \beta_p, \tag{20}$$

and

$$\hat{P}(z) = \frac{P_{p+1}(z)}{(1+z)} = \alpha_0 z^p + \alpha_1 z^{p-1} + \cdots + \alpha_p. \tag{21}$$

The LSFs are the roots of $\hat{Q}(z)$ and $\hat{P}(z)$ and alternate with each other on the unit circle. Note that $Q_{p+1}(z)$ is an antisymmetric polynomial and $P_{p+1}(z)$ is a symmetric polynomial. The polynomials $\hat{Q}(z)$ and P$\hat{P}(z)$ derived from $Q_{p+1}(z)$ and $P_{p+1}(z)$ are symmetrical. Therefore, for even values of p we can derive the following property:

$$\alpha_i = \alpha(p-i), 0 \leq i \leq \frac{p}{2} \tag{22}$$

Consequently (20) and (21) can be written as follows:

$$\hat{Q}(z) = z^{p/2}\Big[\beta_0\left(z^{p/2} + z^{-p/2}\right) + \beta_1\left(z^{p/2-1} + z^{-(p/2-1)}\right) + \cdots + \beta_{p/2}\Big], \tag{23}$$

and

$$\hat{P}(z) = z^{p/2}\Big[\alpha_0\left(z^{p/2} + z^{-p/2}\right) + \alpha_1\left(z^{p/2-1} + z^{-(p/2-1)}\right) + \cdots + \alpha_{p/2}\Big] \tag{24}$$

By putting $z=e^{jw}$ and then $z+z^{-1}=2\cos(w)$, we obtain the equations to be solved in order to find the LSFs according to the real root scheme ITU-T Recommendation G.723.1:

$$\hat{Q}(e^{jw}) = 2e^{jpw/2}\left[\beta_0 \cos\left(\frac{p}{2}w\right) + \beta_1 \cos\left(\frac{p-2}{2}w\right) + \cdots + \frac{1}{2}\beta_{p/2}\right] \quad (25)$$

and

$$\hat{P}(e^{jw}) = 2e^{jpw/2}\left[\alpha_0 \cos\left(\frac{p}{2}w\right) + \alpha_1 \cos\left(\frac{p-2}{2}w\right) + \cdots + \frac{1}{2}\alpha_{p/2}\right] \quad (26)$$

Input speech is segmented to different frames. Additionally, each frame is subdivided into four sub-frames. On these sub-frames, the LPC analysis is performed. The conversion of the p LPC coefficients into their p corresponding LSFs is performed in the last sub-frame. For the three of the sub-frames, the LSFs are obtained by executing linear interpolation between the LSFs of the current and the previous frame.

To achieve this purpose, the unit circle is then divided into 512 equal intervals, each of length$\pi/256$. The roots (LSFs) of $Q(z)$ and $P(z)$ polynomials are searched along the unit circle from 0 to π. A linear interpolation is performed on intervals where a sign change is observed in order to find the zeros of the polynomials. According to [20], if a sign change appears between intervals $l$and $l-1$, a first-order interpolation is executed as follows:

$$\hat{l} = l - 1 + \frac{|P(z)_{l-1}|}{|P(z)_{l-1}| + P(z)_l} \quad (27)$$

where l$\hat{l}$ is the interpolated solution index, $|P(z)_l|$ is the absolute magnitude of the result of sum polynomial evaluation at interval l (similarly for $l-1$). Since the LSFs are interlacing in the region from 0 to π, only one zero is evaluated on $P(z)$ at each step. The search for the next solution is performed by evaluating the different polynomial $Q(z)$, starting from the current solution [23, 24]. Therefore, two main reasons motivated our choice to consider the LSFs representation. The first reason is related to the fact that LP coefficients are very sensitive to errors. The direct quantization of these coefficients might produce an unstable LP filter. The second reason is related to the fact that LSFs are widely used in conventional coding schemes. This avoids the incorporation of new parameters that may require significant and costly modifications to current devices and codecs.

Speech Hiding Algorithm

We propose a new method for speech signal steganography, the secret speech signal is embedded into the coefficients in the wavelet domain. The DWT decomposes the cover speech signal into low- and high-frequency components. For speech signals, the low-frequency component is the most significant part for speech perception. On the other hand, the high-frequency component impacts flavor or nuance (noise) to the signals. Let's consider the human voice. If we remove the high-frequency components, the voice sounds different, but we can still tell what's being said. However, if we remove sufficient amount of the low-frequency components, we hear gibberish and we cannot understand what's being said. For this reason, we decide to hide information in the high-frequency in the wavelet domain. Furthermore, in wavelet analysis, we can divide the speech signal in approximations and details. The approximations are the high-scale, low-frequency components of the signal. The details are the low-scale, high-frequency components. As shown in Figure4 after passing through two complementary filters, two signals emerge from the original signal.

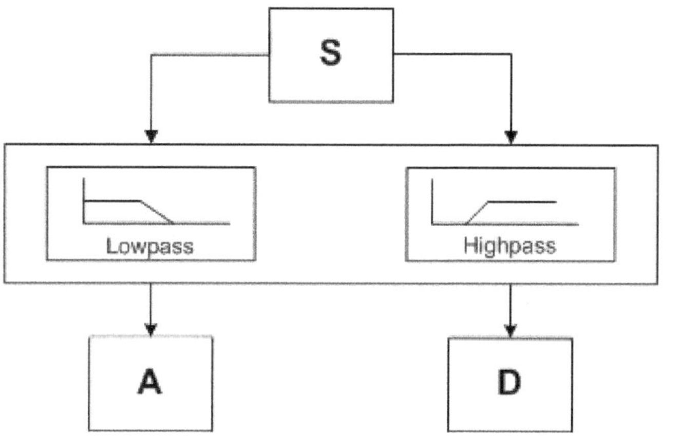

Figure 4: Level 1 decomposition of DWT coefficients.

A variety of wavelets can be used depending on the expected results. Each family of wavelets (such as Haar or Daubechies family) are wavelet subclasses distinguished by the number of filter coefficients and the level of iteration. In steganography, whatever the used algorithm for hiding data, we need to reconstruct the speech signals after embedding the message in the original signal. After that, performance measure can be used to compare the original speech signal and the stego-speech. In our method, after using the DWT to decompose the speech signals for hiding a message speech signals, we use the IDWT to reconstruct the signal. The speech-in-speech hiding algorithm is illustrated in Figure5. Both of secret and cover speech must be pre-processed in order to facilitate the hiding process. The cover speech is partitioned into L-ms frames. The DFT of each time-frame s_1(m) defined for $0 \leq m \leq m-1$ is computed using the DWT-FFT method. The obtained speech spectrum is decomposed into magnitude and phase spectra. Each L-ms of the secret message s_2(m) is embedded in the low-amplitude high-frequency region of the magnitude spectrum of the cover signal.

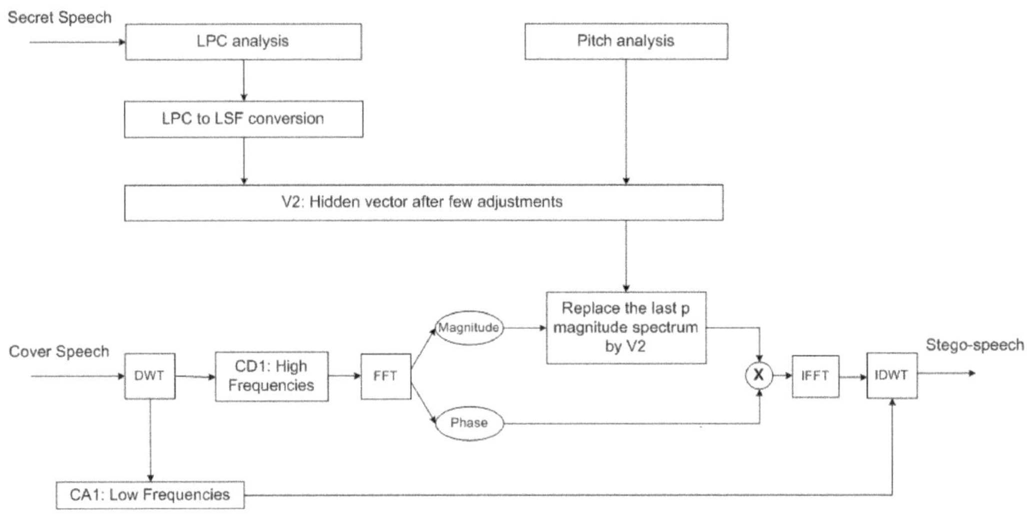

Figure 5: Block diagram showing the general steps to embed a secret speech signal $s_2$inside another cover speech signal$s_1$to produce a stego-speech signals_3.

Secret Speech Hiding

In order to hide the secret speech, the DWT is applied to the speech cover speech frame to separate the high- and the low-frequency regions. Then the FFT is applied to the high-frequency wavelets part producing a spectrum $S_1(k)$ $(k=0,\dots,M-1)$. The spectrum is decomposed into magnitude spectrum $|S_1(k)|$ and phase spectrum □$_1(k)$.

The magnitude spectrum is symmetric. The hiding process consists of representing the L last elements of the first half of $|S_1(k)|$ by the LP parameters V_2 of the secret speech$s_2(m)$.

The resulting magnitude spectrum, denoted by $|S_3(k)|$, is defined by the following expressions:

$$|S_3(k)| = \begin{cases} |S_1(k)|, & k=0,\dots,\frac{M}{2}-p-3 \\ V_2\left(k-\frac{M}{2}-p-2\right), & k=\frac{M}{2}-p-2,\dots,\frac{M}{2}-1 \\ V_2\left(\frac{M}{2}+p+1-k\right), & k=\frac{M}{2},\dots,\frac{M}{2}+p+1 \\ |S_1(k)|, & k=\frac{M}{2}+p+2,\cdots,M-1 \end{cases} \tag{28}$$

The third right-hand term in the above equation is included to preserve the DFT symmetry. These modifications lead to a new speech signal s_3. Its spectrum is a simple combination of the magnitude spectrum $|S_3(k)|$ and the cover phase spectrum $\phi_1(k)$,

$$S_3(k) = |S_3(k)|e^{j\varphi 1(k)}k = 0,\dots,M-1 \tag{29}$$

The time-frame composite (stego) signal $s_3(m), m=0,\dots,M-1$, is obtained by the IDWT,

$$s_3=\text{IDWT}(S_3) \tag{30}$$

The stego signal $s_3(m)$ is a composite signal since it contains the L-ms cover speech $s_1(m)$ and the L-ms secret signal $s_2(m)$.

Energy Normalization

In order to improve the speech quality, we preserved the speech energy by normalizing all the hidden parameters by the total energy of the original spectrum magnitudes. However, the energy preservation requires the hiding of the energy as side information. At the receiver, this energy will be used to rescale the hidden information to its original values. The scaling coefficient a is given by

$$a = \sqrt{\frac{E_c}{1+E_{\text{LSF}\ secret}}} \tag{31}$$

where E_c is the energy of the cover speech spectrum and E_{LSF} is the energy of the LSF vector.

Secret Speech Reconstruction

The secret speech is reconstructed from the stego speech by subsequent the hiding algorithm in overturn order. Figure6 illustrates the pursued steps to extract the hidden information and reconstruct the secret speech message. The first step consists of performing the DWT. Transforming by FFT the high frequencies obtained with the DWT to its corresponding spectrum. The magnitude spectrum is then acquired from the speech spectrum. The secret speech parameters are extracted from the same locations they were embedded in the spectral magnitude of the stego speech signal. The LSF vector is converted back to a P-order LPC vector $(a_1,\dots,a_p)$ to build the LP synthesis filter $H(z)$.

$$H(z) = \frac{1}{1-\sum_{i=1}^{10} a_i z^{-i}} \tag{32}$$

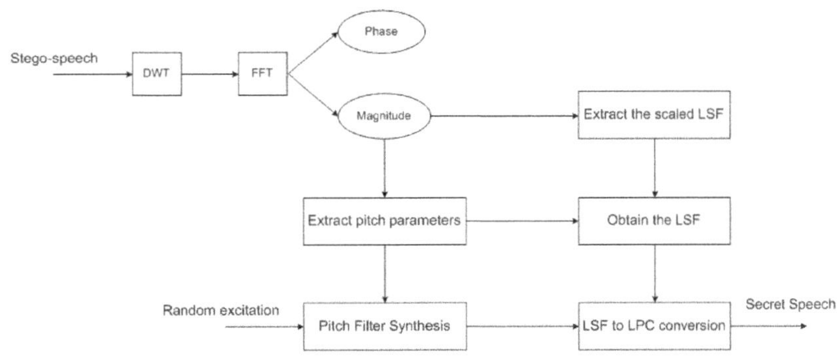

Figure 6: Block diagram showing the general steps to retrieve the secret speech signal$S_2$from the stego-speech S_3.

A random excitation signal $e(n)$ is applied to the series of the pitch and LP synthesis filters. The signal $\hat{s}(n)$, at the output of the LP synthesis filter, is a reproduction of the original secret message $s(n)$. Since the LPC-model parameter values that are extracted from the stego speech have approximately the same exact values as the embedded parameters, the reconstructed secret speech signal is not affected by the hiding process. The minor degradations noticed in this signal, when compared with the original secret signal, are resulting from the LPC model and the LSF conversion.

EVALUATION

Experimental Setup

To evaluate the performance of the proposed hiding technique, we conducted several simulations using NOIZEUS database [25–27]. This corpus contains 30 sentences from the IEEE sentence database, recorded in a sound-proof booth using Tucker Davis Technologies recording equipment. The sentences are produced by three male and female speakers. The 30 sentences: 15 male and 15 female speakers include all phonemes in the American English language. The sentences were originally sampled at 25 kHz and down-sampled to 8 kHz. The length of the speech file varies between 0.02 and 0.03 ms. In the comparative evaluation, we conducted four sets of tests. In the first set of simulations, we embedded each of the 15 male speech files in each of the 15 female speech files. In the second set of tests, we hide each of the 15 female speech files in each of the 15 male speech files. In the third set of tests, we embedded each of the 15 male speech signals in the remaining 14 male speech files. In the last sets of tests, we hide each of the 15 female speech segments in the remaining same gender speech files. Each set is iterated for five different wavelet families (Haar, Daubechies, Symlets, Coiflets, and BiorSplines). In total, we conducted 4,210 computer simulations ((15*15*2+14*14*2)*5).

In order to evaluate the impact of the DWT-FFT technique, we conducted two different comparative experiments using DWT-FFT method and then using FFT only.

Evaluation Outcomes

One of the performance measures of any steganographic system is the comparison between the cover and the stego signals. In this study, we used subjective and objective performance measures. In the subjective measures, we conducted several informal listening comparative tests. In these simulations, we played in a random order the cover speech $s_1(m)$ and the stego signal $s_3(m)$ to several listeners. Each listener had to identify the better quality speech file among the cover and the stego signals. The majority of listeners could not distinguish between the two speech files. As an objective measure, we used the segmental signal-to-noise ratio (SegSNR) and the perceptual evaluation of speech quality (PESQ). PESQ measurement provides an objective and automated method for speech quality assessment. The SegSNR is defined by

$$\mathrm{SegSNR(dB)} = 10\log_{10}\left(\frac{\sum_{m=0}^{159}[s_1(m)]^2}{\sum_{m=0}^{159}[s_1(m)-s_3(m)]^2}\right) \tag{33}$$

where s_1 and s_3 are the cover and the stego speech files, respectively. In this study, we segmented the speech files into frames of 20 ms (L=20) (or 160 samples (M=160)). In Table 2, we present the average SegSNR values for each of the four different sets of tests using DWT-FFT algorithm. In Table 3, we present the average SegSNR of the same set of tests using the FFT only. The quality of the stego signal produced by the FFT is better than the one produced by the DWT-FFT. However, the DWT-FFT increases the robustness of the hiding algorithm against steganalysis techniques. We used some of the existing wavelets to compare the impact of the different wavelet on the speech quality. The decomposition of all used wavelets is done with one level. Table 4 shows the result of different wavelets for the four different sets of tests. As can be noticed, different wavelets have almost similar results; therefore, this method is not depending on a particular type of wavelet. The SegSNR value did not differ a lot for different wavelets. The SegSNR is just an indicative performance measure. The PESQ is a more reliable method to assess the performance of our hiding technique. The PESQ measurement provides an objective and automated technique for speech quality evaluation. The degradation of the speech sample can be predicted using the PESQ algorithm with subjective opinion score. In general, the PESQ returns a score from 0.5 to 4.5, with higher scores signifying better quality [28, 29]. The PESQ method is used in our experiments to evaluate the stego speech. The reference signal refers to an original (cover) signal and the degraded signal refers to the stego signal with the hidden secret message. In Table 5, we present the average PESQ values for male and female speakers obtained by the two hiding techniques (using DWT-FFT and FFT only).

Figure 7 shows variations of PESQ for 20 speech signals of the 2 hiding approach. The hiding method achieves 3.68 and 4.14 PESQ average for DWT-FFT and FFT algorithms, respectively. Figure 8 shows the magnitude spectrum of the cover signal and the corresponding of stego speech after hiding the LPC parameters of the secret signal. The PESQ analysis shows that the stego and cover speech provide similar subjective quality. This result is supported by the resemblance between the cover and stego speech spectrograms in Figure 9. The objective and subjective performance measures show that the proposed hiding technique attracts no suspicion about the existence of a hidden message in the stego speech, while being able to recover an intelligible copy of the original secret message at the receiver side. The informal listening test to the original and the reassembled secret speech message advocate the result of the other objective performance measurement. The reconstructed secret speech $\hat{s}(n)$ (from both DWT-FFT and FFT hiding approaches) still completely comprehensible, even some perceptual distortions are simply noticeable. What concerns us is the speech intelligibility since the objective is to convey the secret message to the intended receiver. Table 6 shows the impact of the hiding algorithms on the secret speech in terms of the SegSNR.

Table 2: SNR of the DWT-FFT-based hiding approach

Cover signals	Secret signals	SegSNR (dB)
Female	Male	31.86
Male	Female	32.70
Male	Male	34.45
Female	Female	31.13
Average		32.54

Table 3: SNR of FFT-based hiding approach

Cover signals	Secret signals	SegSNR (dB)
Female	Male	51.46
Male	Female	52.62
Male	Male	54.37
Female	Female	51.09
Average		52.39

Table 4: Different wavelets results of DWT-FFT-based steganography systems

Wavelet name		Haar	Daubechies (db1)	Symlets (sym1)	Coiflets (coif1)	BiorSplines (bior1.1)
Cover signals	Secret signals	SegSNR (dB)	SegSNR (dB)	SegSNR (dB)	SegSNR (dB)	SegSNR (dB)
Female	Male	31.53	31.86	31.48	31.41	31.39
Male	Female	31.98	32.70	31.86	31.96	31.91
Male	Male	34.12	34.35	34.08	34.08	34.04
Female Female		30.79	31.13	30.68	30.76	30.71
Average		32.11	32.51	32.03	32.05	32.01

Table 5: PESQ of DWT-FFT and FFT-based hiding approach

Speaker	PESQ	
	DWT-FFT	FFT
Female	3.58	4.12
Male	3.78	4.16
Average	3.68	4.14

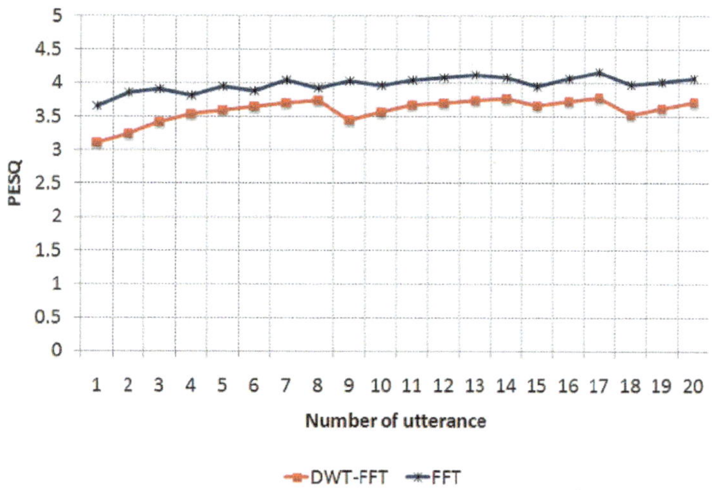

Figure 7: Comparison of the PESQ variations between DWT-FFT and FFT-based steganography systems: PESQ scores of cover speech $s_1(m)$ and the stego signal $s_3(m)$ utterances using DWT-FFT and FFT hiding approach.

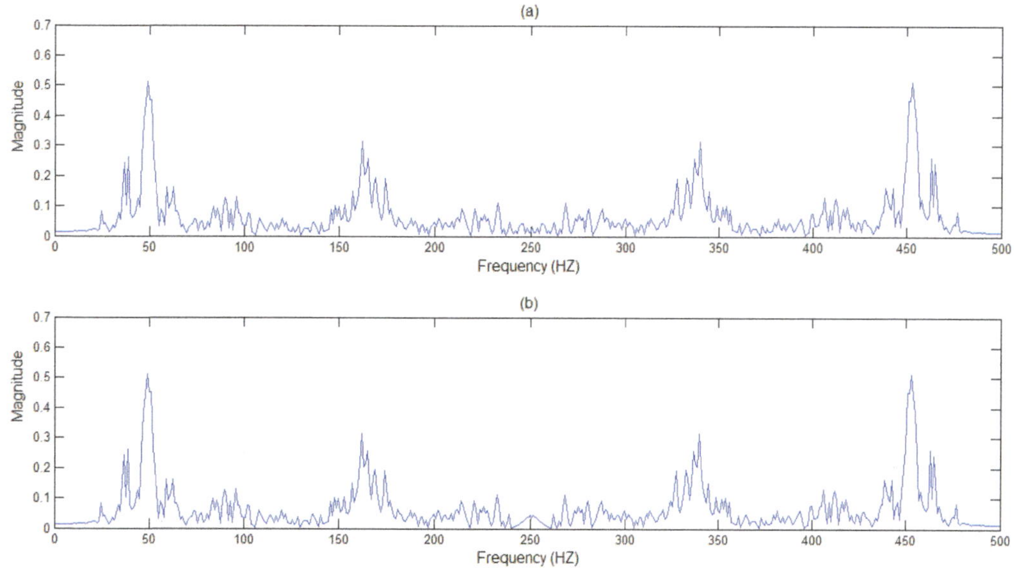

Figure 8: Magnitude spectrum of (a) the cover speech$s_1(m)$, (b) the stego speech.

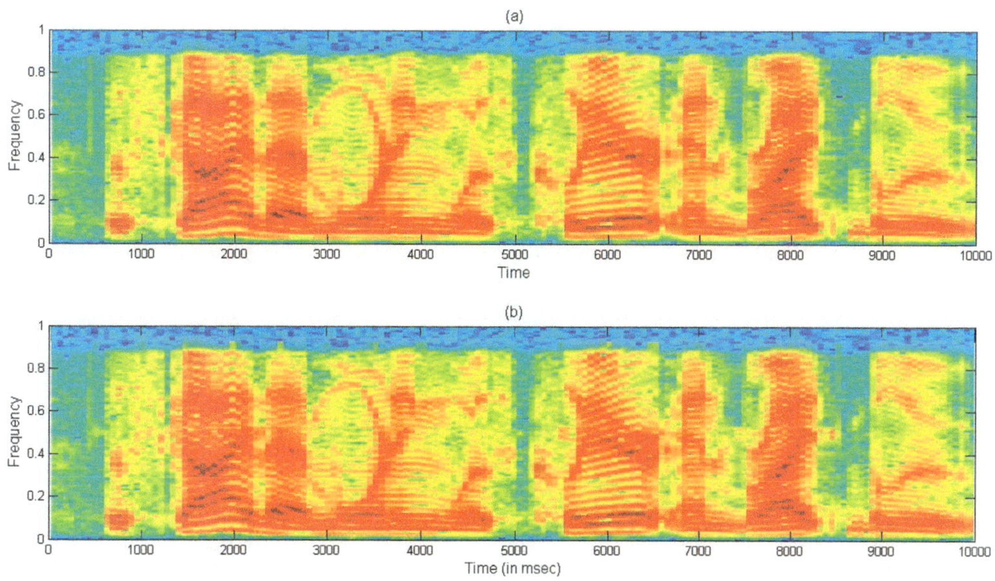

Figure 9: Spectrogram of (a) the cover speech, (b) the corresponding stego speech after hiding a male secret message within the cover speech.

Table 6: Impact of the hiding process on the secret speech in terms of SegSNR

Speaker	SegSNR(dB)	
	DWT-FFT	FFT
Female	21.76	24.64
Male	23.89	26.28
Average	22.83	25.46

CONCLUSIONS

In this article, we presented a new steganography system for secrecy applications. The proposed hiding method produces stego speech files that are indistinguishable from their equivalent cover speech files. Moreover, the complexity of our hiding technique is so high any eavesdropper cannot extract the hidden information even after suspecting the existence of a secret message. Since our aim is to render the steganalysis (the attempt to extract the secret message from the stego signal) by the opponent more complex. Our method exploits first the high frequencies using a DWT, then exploits the low-pass spectral properties of the speech magnitude spectrum to hide another speech signal in the low-amplitude high-frequencies region of the cover speech signal. Experimental simulations on both female and male speakers showed that our approach is capable of producing a stego speech that is indistinguishable from the cover speech. The receiver is still able to recover an intelligible copy of the secret speech message. In the future work, we will endeavor to extend our approach to applications involving Voice-over IP speech secrecy, which involves compressing the stego speech before transmission. This opens up the issue of preserving the secret speech after decoding the compressed stego speech.

Competing Interests

The authors declare that they have no competing interests.

REFERENCES

1. Kahn D: *The History of Steganography. Lecture Notes in Computer Science*. 1174 edition. Springer, New York; 1996:11023.
2. Johnson NF, Jajodia S: Exploring steganography: seeing the unseen. *IEEE Comput.* 1998, 31(2):26-34.
3. Sridevi R, Damodaram A, Narasimham SVL: Efficient method of audio steganography by modified LSB algorithm and strong encryption key with enhanced security. *J. Theor. Appl. Inf. Technol.* 2009, 5(6):768-771.
4. Bender W, Gruhl D, Morimoto N: Techniques for data hiding. *IBM Syst. J.* 1996, 35(3):313-336.
5. Kirovski D, Malvar H: Spread-spectrum watermarking of audio signals. *IEEE Trans. Signal Process.* 2003, 51(4):1020-1033. 10.1109/TSP.2003.809384
6. Huang D, Yeo T: *Robust and Inaudible Multi-Echo Audio Watermarking, in Proceedings of the Third IEEE Pacific-Rim Conference on Multimedia, Advances in Multimedia Information Processing Taipei*. China; 2002:615-622.
7. Shirali-Shahreza S, Shirali-Shahreza M: *Steganography in Silence Intervals of Speech, Proceedings of the Fourth IEEE International Conference on Intelligent Information Hiding and Multimedia Signal Processing (IIH-MSP 2008)*. Harbin, China; August 15–17, 2008:605-607.
8. Misiti M, Misiti Y, Oppenheim G, Poggi JM: *Matlab Wavelet Toolbox (Version 4.0): Tutorial and Reference Guide The Mathworks*. Natick, USA; janv 2007.
9. Lin B, Nguyen B, Olsen ET Signal Processing Methods for Audio, Images and Telecommunications. In *Orthogonal Wavelets and Signal Processing*. Edited by: Clarkson PM, Stark H. Academic, London; 1995:1-70. ed. by
10. Mallat S: *A Wavelet Tour of Signal Processing*. Academic, San Diego, CA; 1998.
11. Nievergelt Y: *Wavelets Made Easy*. Birkhäuser, Boston; 1999.
12. Ooi J, Viswanathan V: *Applications of Wavelets to Speech Processing. Modern Methods of Speech Processing*. Edited by: Ramachandran RP, Mammone R. Kluwer Academic Publishers, Boston; 1995:449-464.
13. Elliott DF, Rao KR: *Fast Transforms: Algorithms*. Analyses, Applications (Academic, New York; 1982.
14. Andreas S, Ed PT, Venkatraman A: *Audio Signal Processing and Coding*. Wiley-Interscience Publication, USA; 2006. ISBN 978-0-471-79147-8, TK5102.92.S73
15. Strange W, Edman TR, Jenkins JJ: Acoustic and phonological factors in vowel identification. *J. Exp. Psychol. Hum. Percept. Perform.* 1979, 5(4):643-656.
16. CY Espy-Wilson: Acoustic measures for linguistic features distinguishing the semivowels in American English. *J. Acoust. Soc. Am* 1992, 92: 736-757. 10.1121/1.403998
17. Childers DG, Hahn M, Larar JN: Silent and voiced/unvoiced/mixed excitation (four-way) classification of speech. *IEEE Trans. ASSP* 1989, 37(11):1771-1774. 10.1109/29.46561
18. O'Shaughnessy D: *Speech Communications: Human and Machine*. 2nd edition. Wiley-IEEE Press, New York, NY; 1999.
19. Makhoul J: Linear prediction: a tutorial review. *Proc. IEEE* 1975, 63(5):561-580.
20. Itakura F: Line spectrum representation of linear predictive coefficients of speech signals.*J. Acoust. Soc. Am* 1975, 57(1):S35. 10.1121/1.380398

21. Oppenheim AV, Schafer WR, Buck AJ: *Discrete-Time Signal Processing*. Prentice Hall, Upper Saddle River, NJ; 1999:468-471. ISBN 0-13-754920-2

22. Hess W: *Pitch Determination of Speech Signals*. Springer, Berlin; 1983.

23. Soong F, Juang B: *Line spectrum pair (LSP) and speech data compression. Proceedings of IEEE International Conference on Acoustics, Speech and Signal Processing (ICASSP '84)*. San Diego, Calif, USA 9; March 1984:37-40.

24. 2ITU-T: *Recommendation G. 723.1. Dual rate speech coder for multimedia communications transmitting at 5.3 and 6.3 kbit/s*. 1996.

25. 2Hu Y, Loizou P: Subjective evaluation and comparison of speech enhancement algorithms.*Speech Commun* 2007, 49: 588-601. 10.1016/j.specom.2006.12.006

26. Hu Y, Loizou P: Evaluation of objective quality measures for speech enhancement. *IEEE Trans. Speech Audio Process*. 2008, 16(1):229-238.

27. Ma J, Hu Y, Loizou P: Objective measures for predicting speech intelligibility in noisy conditions based on new band-importance functions. *J. Acoust. Soc. Am*. 2009, 125(5):3387-3405. 10.1121/1.3097493

28. ITU: *Perceptual Evaluation of Speech Quality (PESQ), and Objective Method for End-to-End Speech Quality Assessment of Narrowband Telephone Networks and Speech Codecs, ITU-T Recommendation 862*. 2000.

29. ITU-T Recommendation: *Methods for Subjective Determination of Speech Quality International Telecommunication Union*. Geneva; 2003:800.

CHAPTER 8

TIME-VARYING BISPECTRAL ANALYSIS OF VISUALLY EVOKED MULTI-CHANNEL EEG

Vinod Chandran

Keywords: EEG, HOS, Human vision, Bicoherence, Time-varying

INTRODUCTION

Bispectral analysis of EEG data has been the subject of a number of studies. Some have used single channel data only. Others have used multi-channel EEG ensembles but few have investigated multi-channel EEG using higher order spectral analysis in a time-varying manner. This article revisits the theoretical foundations to justify such analysis and provides new results from the application of time-varying bispectral analysis to evoked EEG responses.

BACKGROUND

Research on bispectral analysis of EEG signals dates back to the 1970s, not long after higher order spectral analysis emerged as a branch of study in the 1960s. Dumermuth et al. [1] demonstrated that there exists significant phase locking between alpha and beta components in intracranial EEG. Barnett et al. [2] used bispectral analysis to examine waking and sleeping states and found significant quadratic phase coupling only in the EEG of wakeful subjects with high alpha activity. These early studies used steady state potentials. Bullock et al. [3] used bicoherence analysis of intracranial and subdural EEG in a time-varying framework in an attempt to classify the onset of epileptic seizures. They analyzed EEG from sleep, wakefulness and seizure states. Their results were not conclusive on the effectiveness of the bicoherence descriptor. They found the bicoherence to fluctuate abruptly within a few seconds. The fluctuations were not consistent across subjects during the seizure period although statistically significantly higher levels of bicoherence were observed. Muthuswamy et al. [4] modeled paroxysmal burst EEG as a non-linear time-invariant process and showed that the bicoherence in the delta-theta band of EEG bursts is significantly higher than baseline waveforms in animal subjects recovering from a brain trauma. It has been shown that the bispectrum of the EEG correlates with changes in consciousness level and the bispectral index (BIS) [5, 6] derived from EEG bispectral parameters was developed as a clinical tool to monitor depth of anaesthesia during surgery. Tang and Norcia [7] used the bispectrum to study steady state visually evoked potentials. They called their method the coherent bispectrum. They used oscillatory visual stimulii and reported the presence of intermodulation frequencies and evidence of nonlinear interactions. Shen et al. [8, 9] investigated time-varying bispectral analysis of non-stationary EEG data considering piece-wise third order stationary segments and non-Gaussian autoregressive modelling. Minfen et al. [10, 11] used higher-order spectral analysis of EEG for classification of brain functional states.

In this study, time-varying bispectral analysis is applied to transient EEG responses evoked by a stimulus or related to a sensory event. A classical Fourier approach is adopted and ergodicity is only assumed over short time intervals around windows that are at the same offset with respect to a stimulus-locked time reference. Many previous EEG studies such as [12] have used stimulus-locked time references and

Citation: Vinod Chandran, Time-varying bispectral analysis of visually evoked multi-channel EEG. EURASIP Journal on Advances in Signal Processing 2012 2012:140.

stimulus-locked time averaging but most of them use grand averaging in time and some use spectral analysis. They have not investigated time-varying bispectral analysis in the manner described in this work. Bicoherence changes are tracked in this study with millisecond resolution, better tracking resolution than in earlier studies such as [1]. Auto-bicoherence is mapped simultaneously for multiple channels to obtain a spatio-temporal view of the EEG response at selected locations in the bifrequency plane, providing enhanced processing and visualization capabilities compared to any previously reported work. Such analysis will be useful in understanding the neuronal activity involved in visual and auditory perception, motor planning and movements. It can provide new features for diagnosing neurological conditions and sensory impairment.

METHODS

In this section, some equations defining higher order spectra [13–16], are revisited to provide a context and justification for the adoption of a classical short-time Fourier approach to time-varying bispectral analysis.

Random Processes

Consider a real-valued random process, x(t), that varies with time, as a signal model for any channel of EEG. An ensemble of many realizations of the random process can be used to define statistical averages or expected values that are deterministic quantities. At any given time instant t⊠, x(t⊠) is a random variable. For a first-order stationary random process, the probability density function p[x(t⊠)] is independent of the time t⊠. Descriptions of the random process that depend only on the statistics of one random variable such as the mean value are examples of first order statistics. The mean or first order moment of the process is

$$E[x(t')] = \int x(t')p[x(t')]\,dx = m_1^x \tag{1}$$

A random process is not fully characterized by its first order statistics alone. The joint probability density function p[x(t⊠),x(t⊠⊠)] provides a second order description of the random process. For a second-order stationary process, this probability density is independent of the absolute value of the time instants and depends only on the time offset τ = t⊠⊠–t⊠. The autocorrelation or second order moment of the random process is

$$m_2^{xx}(\tau) = E[x(t')x(t'') = E[x(t')x(t' + \tau)] \tag{2}$$

Here E stands for the expectation operation over an ensemble of realizations of the process. Very often, only a single realization is available. If it is sufficiently long and the process is stationary to the second order, an estimate of the autocorrelation can be computed by averaging over time rather than over the ensemble. It is given by

$$\tilde{m}_2^{xx}(\tau) = E_t[x(t')x(t'')] = \int x(t')x(t' + \tau)dt' \tag{3}$$

For an ergodic process ensemble statistics (E) are equal to time statistics (E_t). Ergodicity also implies that time statistics do not change with time and an ergodic process is necessarily stationary. If these properties hold true up to n-th order statistics, the process is said to be n-th order ergodic. Ergodicity is not guaranteed for all processes. At best it is an assumption that holds fairly well in practice to allow reliable estimates of statistical parameters that characterize the process. If the process is ergodic, a single long realization may be divided into several shorter ones for statistical expectation computation. This division into blocks of time creates an ensemble of shorter realizations of the random process. As a trade-off, the range of possible time offset values (τ) is reduced. Assume that an estimate of the autocorrelation, $m_2^{xx}(\tau)$ has thus been obtained. In practice, this autocorrelation will usually tend towards zero for large offsets and the block size can be suitably chosen to be large enough for the autocorrelation to have decayed to nearly zero. If that is the case, the autocorrelation function will be absolutely integrable and its Fourier transform will exist. The Fourier transform of the autocorrelation function is the power spectral density referred to as the power spectrum of the process, $S_2^{xx}(f)$, where f represents frequency in cycles per second or Hertz when the independent variable time of the random process and the offset τ are measured in seconds. For deeper understanding of stationarity and ergodicity in random processes and stochastic calculus the reader is referred to [17]. The power spectrum reveals the harmonic structure or frequency components in the random process. If the autocorrelation function is an impulse or Dirac delta function δ(τ) (implying that values of the process separated even infinitesmally in time are uncorrelated) the power spectrum $S_2^{xx}(f) = 1$ is constant and the process is referred to as white. In this case, the power spectrum does not reveal much about the process. Noise processes like that produced by thermal fluctuations of electrons in a material tend to be broadband in their spectrum and nearly white. There are no selectively resonant structures that would concentrate power over particular frequencies or frequency bands. If the process is also a result of superposition of many uncorrelated processes, such as the potential difference produced across a resistor owing to the thermal fluctuations of millions of electrons, the Central limit theorem dictates that the probability density function will tend towards the Gaussian distribution. This is also true of joint probability density functions of all orders. The process is then referred to as a Gaussian process. A Gaussian probability density function in D dimensions is given by

$$p(X) = \frac{1}{(2\pi)^{D/2}[\det(C_X)]^{D/2}} \times \exp\left[-\frac{1}{2}(X - \mu_X)C_X{}^{-1}(X - \mu_X)^T\right] \tag{4}$$

where μ_X is the mean vector and C_X is the covariance matrix of the D-dimensional vector X. det represents the determinant of a matrix. For D-th order statistics of a random process, these dimensions come from D time instants at which the process is considered, each a scalar random variable. Yet, only first order (mean value) and second order (pair-wise in the covariance matrix) suffice for a complete description of the Gaussian probability density. The white Gaussian random process model is often employed as a simplified model for noise. A Gaussian process is completely described by first and second order statistics because a Gaussian probability density function of any order can be represented using a mean vector and a covariance matrix alone. Deviations from Gaussian distributions can be detected or measured by using higher than second order moments. For example, skewness is computed from third moments and kurtosis from fourth order moments. Skewness is a measure of the asymmetry of the probability density and kurtosis is an indicator of the deviation of the tails of the distribution from those of a Gaussian. If a deviation from Gaussian is detected, it may be owing to particular frequency components in the process.

A Fourier representation for higher order moments was proposed by Shiryaev [13]. The third order spectrum or bispectrum was first applied to study nonlinear interactions in ocean waves [18]. The estimation of such higher order spectra was placed on a firm mathematical and statistical foundation by Brillinger and Rosenblatt [14, 15, 19] in the same decade. The fast Fourier transform (FFT) algorithm and advances in computers facilitated the numerical estimation of higher order spectra and gave impetus to higher order spectral analysis research and its application in the latter half of the twentieth century. More review articles can be found in [16, 20, 21].

The Bispectrum

The third order moment or triple correlation of a random process x(t), assumed stationary, is defined as

$$m_3^{xxx}(\tau_1,\tau_2) = E[x(t)x(t+\tau_1)x(t+\tau_2)] \quad (5)$$

For many processes this function of two offset variables will decay with increasing delay or offset. If this function satisfies the Dirichlet conditions for existence of a Fourier spectrum, such as being absolutely integrable, its Fourier transform may be defined and is known as the bispectrum of the process,

$$B(f_1,f_2) = S_3^{xxx}(f_1,f_2) = \int_{-\infty}^{\infty}\int_{-\infty}^{\infty} m_3^{xxx}(\tau_1,\tau_2) \times e^{-j2\pi(f_1\tau_1+f_2\tau_2)}d\tau_1 d\tau_2 \quad (6)$$

The subscript 3 refers to a third order parameter. The superscripts xxx are a reminder that the spectrum is an auto-spectrum computed from the same random process rather than a cross-spectrum computed from different random processes. The bispectrum of a one dimensional random process is a function of two frequencies as shown above. The pair of frequencies is referred to as a bifrequency. The bispectrum of a Gaussian random process is zero because the third moment is zero. Moments of higher order are not all zero for a Gaussian process. But cumulants [16, 21] of order three and greater are zero and it is conventional to define higher order spectra in terms of cumulants of a random process rather than moments. This work is restricted to the third order and third order moment and cumulant spectra are identical, both referring to the same bispectrum. This definition of the bispectrum may be extended to include moment functions from harmonic random processes that are not absolutely integrable but have finite power, or finite energy within a period, similar to such extension for the power spectrum. The spectrum will then theoretically comprise of impulses or Dirac delta functions rather than being continuous.

Estimation of the bispectrum

It has been shown that the bispectrum can be estimated in the Fourier domain [19]. Strictly speaking, a stationary random process does not have a Fourier transform because existence conditions are not satisfied. Any single realization of the process will have finite power but infinite energy. A Fourier transform is therefore defined using the Fourier–Stieltjes integral and Cramer spectral representations as

$$x(t) = \int_{-\infty}^{\infty} e^{j2\pi ft}dX(f) \quad (7)$$

where

$$E[dX(f)] = 0 \quad (8)$$

It can be shown [16] that for real-valued x(t)

$$m_2^{xx}(\tau) = \begin{cases} \int_{-\infty}^{\infty} e^{j2\pi(f\tau)}E[dX(f)dX^*(f)] & f_1 = f_2 = f \\ 0 & f_1 \neq f_2 \end{cases} \quad (9)$$

where ∗ represents the complex conjugate operator. The power spectrum of a random process can thus be expressed as an expected value in the frequency domain.

$$S_2^{xx}(f)df = E[dX(f)dX^*(f)] \quad (10)$$

The power over a small interval in the frequency f is the expected value of a product of Fourier–Stieltjes spectral representations that are complex conjugates. The power spectrum is real-valued and does not have any phase information. The zero value for $f_1 \neq f_2$ can be subjected to some scrutiny. If it were not the case, the second moment, m_2^{xx} would turn out to be a function of t which is a contradiction to the assumption of a stationary random process, x(t). However, it holds because the infinitesimal bandwidth processes, $dX(f_1)$ and $dX(f_2)$ are complex-valued and in a phasor representation such as Rice's representation [17] they are not "phase-locked" when $f_1 \neq f_2$. The phase of the product will be uniformly distributed in the interval [0,2Π) and the expected value of the product will be zero in this case. When $f_1 = f_2 = f$, the product becomes real-valued and can have a non-zero expected value. A different but formal proof is given in [22] where it is shown that dX(f) is an orthogonal process. The proof starts by making a single realization of the random process periodic outside a finite interval. In the same manner it can be shown [16] that the bispectrum is a triple product of such representations.

$$m_3^{xxx}(\tau_1,\tau_2) = \begin{cases} \int_{-\infty}^{\infty}\int_{-\infty}^{\infty}\int_{-\infty}^{\infty} e^{j2\pi(f_1\tau_1+f_2\tau_2)}E[dX(f_1)dX(f_2) \\ \quad \times dX^*(f_1+f_2)] & f_1+f_2=f_3 \\ 0 & f_1+f_2 \neq f_3 \end{cases} \quad (11)$$

and by comparing with Equation (6) and the inverse Fourier relationship,

$$S_3^{xxx}(f_1,f_2)df_1df_2 = E[dX(f_1)dX(f_2)dX^*(f_1+f_2)] \quad (12)$$

The reason for a non-zero expected value when the resonance condition, $f_1 + f_2 = f_3$ is satisfied is similar to that for the power spectrum as explained above except that the triple product here can become "phase-locked" and does not have to be real-valued. The term "phase-locked" is used with inverted commas because the phase is not necessarily constant in every realization. The bispectrum is a complex-valued function of two frequency variables and retains relative phase information between Fourier components.

A difficulty in the use of the frequency domain for estimation of the power spectrum or the bispectrum of a random process lies in the fact that differentiability of the spectral representation cannot be guaranteed. However, estimates of higher order statistical or spectral parameters have to be computed numerically for practical purposes.

Sampling of Random Processes

Sampling of stationary random processes is discussed in [17] where Rice's representation for stationary random processes in terms of in-phase and quadrature components is utilized in a theoretically rigorous treatment. A discussion is also found in [23]. For the sake of simplicity a descriptive justification is provided here. Assume that the random process has a finite bandwidth or has been filtered using a linear phase filter that does not disturb the higher order spectral parameters of interest and is thus made band limited. The process can now be sampled at the Nyquist rate of twice the bandwidth without any loss of information. It is now a discrete-time random process whose power spectrum and higher order spectra of interest are still the same within the relevant band of frequencies. Criteria to prevent aliasing in higher order frequency domains are not exactly identical as that for the power spectrum because of the multiple frequency dimensions, and the interested reader is referred to [24–26]. Here, attention is focussed only in the regions in higher order frequency space where the conditions imposed by the Nyquist criterion hold and suffice. Sampling is assumed to be without any aliasing in the power spectrum or bispectrum in the range of frequencies of interest. Sampling also imposes a limit on the resolution with which statistical parameters can be estimated in the time domain because the offset variables cannot be smaller than the sampling interval. Assume further that higher order statistical parameters of interest, such as the third moment, can be estimated to the desired accuracy using expected values, in practice, averaging over finite length epochs. Theoretically, these parameters are defined in the limit of the epoch length approaching infinity. In practice, a long enough epoch length will ensure that the estimates converge to a desired degree for processes whose moments converge asymptotically. Harmonic random processes have ensemble mean and autocorrelation functions that are periodic in time and are a class of cyclostationary random processes [17]. For cyclostationary processes higher order spectral parameters do not asymptotically converge as the time goes to infinity. Correlations for such processes are cyclic and do not decay to zero. Theoretically rigorous treatment of the calculus of various classes of random processes can be found in [17]. If a cyclostationary process were divided into sub-processes using blocks that are multiples of a period, the random-phase assumption for indidvidual harmonic components will not hold. This can be addressed by phase randomization procedures, for example by using blocks with random shifts. A phase-randomized [17] version of the cyclostationary process is a stationary process with the same moments as that computed for the cyclostationary process over one period. There are pitfalls in the phase randomization approach pointed out in [27] and ergodicity can sometimes be destroyed. For estimation of the bispectrum, it is assumed here that an ensemble of realizations is available from a number of trials and phase randomization is not necessary.

Let there be N epochs. Let each epoch be processed using blocks of M samples, appropriately large such that estimates in time are averaged over a large number of periods and the inaccuracy from averaging over a non-integral number of periods is small. A finite length block of samples satisfies the conditions for existence of a discrete-time Fourier transform (DTFT)

$$X(f) = \sum_{k=0}^{M-1} x[kT]\, e^{-j2\pi kfT} = \sum_{k=-\infty}^{\infty} x[kT]\, \Pi[k/M]\, e^{-j2\pi kfT} \quad (13)$$

where the rectangular window in time is given by

$$\Pi[k/M] = \begin{cases} 1 & 0 \le k < M \\ 0 & k \ge M \end{cases} \quad (14)$$

and T is the sampling interval. The DTFT can be computed at any desired frequency and uniquely between 0 and 1/T. With higher order spectral analysis, it is not merely a question of reducing spectral leakage from windowing but understanding the effect that leakage will have on estimates of relevant parameters. Spectral leakage and windowing in the estimation of the bispectrum are discussed in [28, 29]. In general, spectral leakage from statistically independent or random phase components will have a similar effect on higher order spectra as the addition of white random noise. It will lower the fraction of power that is phase-coupled. Tapered windows will introduce a modulation effect and spurious phase-locked low frequency components unless the mean value is removed from each block before application of the window and computation of the discrete Fourier transform (DFT). Assume that each realization of the discrete-time random process repeats itself outside the M samples. It is then converted into a hypothetical cyclostationary process which has nearly the same higher order moments of a given order provided M is sufficiently large. This assumption is different from the harmonizability of deterministic moment functions [21]. The FFT algorithm to compute the DFT can now be employed to get samples of the spectrum as the expected value.

$$X(p) = X(f)\Big|_{k=\frac{p}{T}} = \sum_{k=0}^{M-1} x[p]\, e^{-j2\pi kp} \quad (15)$$

Bispectrum Estimate in the Fourier Domain

The bispectrum of the process can then be obtained in the Fourier domain as the expected value of a triple product of Fourier coefficients using the DTFT or the DFT. It can be shown [16] that

$$E[X(f_1)X(f_2)X^*(f_1+f_2)] = MS_3^{xxx}(f_1,f_2) \quad (16)$$

Periodicity outside the M-length interval and third order stationarity are utilized in the simplification. The bispectrum can be therefore estimated as an average of a triple product of Fourier coefficients over an ensemble of realizations. It can be seen from Equation (12) that the bispectral density is actually a triple product divided by the term $df_1 df_2 = (\frac{1}{MT})^2$ but since this is a constant for a given sampling interval and block length it may be ignored when bispectral values are compared and M and T remain constant. It can also be removed when the bispectrum is normalized with power spectrum values at the frequency components involved as it will appear both in the numerator and the denominator.

For the rest of the article, $B(f_1,f_2)$ is used instead of $S_3^{xxx}(f_1,f_2)$ to refer the the auto-bispectrum of x(t) and P(f) is used instead of $S_2^{xx}(f)$ to refer to the auto power spectrum.

The bispectrum of a real-valued process satisfies a number of symmetry properties in the bi-frequency domain. Assume that time is expressed in sampling intervals and frequencies are normalized by the sampling frequency. The principal domain or non-redundant region [16, 30] of computation of the bispectrum for such a process is given by the triangular region (shown in Figure 1) in bi-frequency defined by

$$f_1 \geq 0; f_1 = f_2; f_1 + f_2 \leq 1 \quad (17)$$

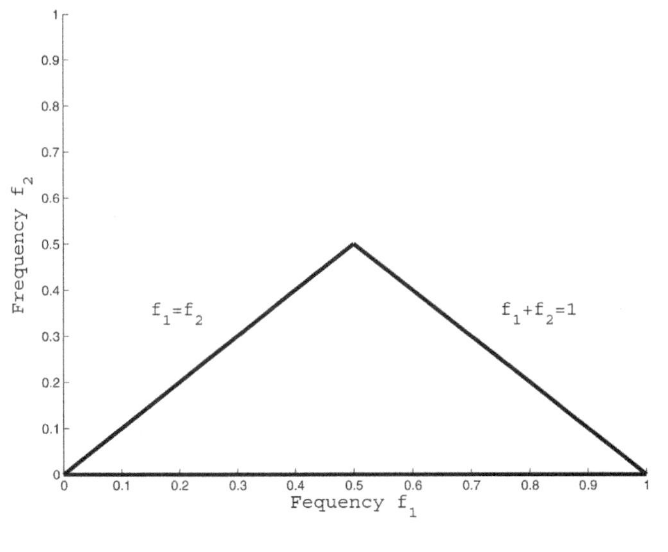

Figure 1: The principal domain of the auto-bispectrum of a sampled signal. Frequencies are normalized by one-half of the sampling frequency.

Bispectrum Magnitude and Bicoherence

Expressing the bispectrum estimate in polar form

$$B(f_1,f_2) = E[\,|X(f_1)X(f_2)X^*(f_1+f_2)|e^{j\phi(f_1)+\phi(f_2)-\phi(f_1+f_2)}] \quad (18)$$

If the phases, $\phi(f_1),\phi(f_2),\phi(f_1+f_2)$ are independent and uniform random in the interval $[0,2\Pi)$ the bispectrum will be zero. If there is perfect phase coupling between the Fourier components and $\phi(f_1)+\phi(f_2)-\phi(f_1+f_2)$ is zero for every realization, the bispectrum will be non-zero. The bispectrum has been used to study non-linear wave coupling and in this context it has been normalized to assume values between 0 and 1 similar to coherence in second order statistics. The squared magnitude of the normalized bispectrum [31] is referred to as bicoherence by

$$b^2(f_1,f_2) = \frac{|B(f_1,f_2)|^2}{P(f_1)P(f_2)P(f_1+f_2)} \quad (19)$$

Although this bicoherence spectrum measures the degree of phase coupling at the particular bi-frequency, it is not guaranteed to be between 0 and 1 when numerically estimated from a finite number of realizations. An alternative normalization, that satisfies this property better is given in [32] by

$$b^2(f_1,f_2) = \frac{|B(f_1,f_2)|^2}{E[\,|X(f_1)X(f_2)|^2]\,E[\,|X^*(f_1+f_2)|^2]} \quad (20)$$

The bicoherence is a measure of the fraction of power at the component $f_1 + f_2$ that is owing to phase-coupled Fourier components at f_1 and f_2 as opposed to arising from random-phase components, or from random additive broadband noise that has non-zero power spectral density at these frequencies. If there is a random phase component at any of the frequencies f_1,f_2 or $f_1 + f_2$ the bicoherence will reduce because at least one of the denominator terms in Equation (20) will increase. Because the bicoherence is a ratio it is sensitive to small values of the denominator that are close to zero. When the denominator is zero, the ratio should actually be 0/0 but can be a large value subject to precision in the numerical computations. This will rarely occur with broad band random processes but can occur with harmonic random processes that have line spectra. One means of avoiding this is to add white Gaussian random noise of small amplitude to the input prior to processing. This will have the undesirable effect of lowering all bicoherence values from their true value. If the power of the noise is small compared to the power of sinusoidal components at the frequencies of interest its negative bias on the true value of bicoherence will be small and negligible.

Estimation of Bicoherence

The procedure to estimate bicoherence using the Equation (20) is as follows.

1. Collect N epochs in an ensemble. Let the epochs be processed as M sample blocks. The sampling frequency, f_s, should be above twice the highest frequency of interest. An anti-aliasing filter may be applied provided it does not disturb phase relationships in the range of frequencies of interest. The frequency resolution is set by the length of each block, M, and given by $\Delta f = \frac{1}{MT}$ where $T = \frac{1}{f_s}$ is the sampling interval.
2. Estimate the signal power and add low amplitude white Gaussian noise keeping the signal to noise ratio maintained high.
3. For each block, remove the mean value.
4. Obtain the Fourier coefficients using the FFT.
5. Form the products of Fourier coefficients required, $X(f_1)X(f_2)X^*(f_1+f_2)$ and $X(f_1)X(f_2)$, and retain $X^*(f_1+f_2)$. Do this for every bi-frequency in the non-redundant region of computation.
6. Average the intermediate products above over all N realizations.

7. Divide as shown in the Equation (20) to obtain an estimate of the bicoherence.

Statistical reliability of Bicoherence Estimates

In general, the estimation of higher order spectra becomes progressively unreliable as the order increases. Increasingly larger numbers of epochs are required to achieve estimates with similar variance. An asymptotic theory of these estimates is discussed in [19]. The bias and variance of any higher order spectrum will actually depend on the true value and exact expressions are not known for arbitrary true values [33].

The estimate of the bicoherence from a finite number of realizations has a bias and variance and it must be determined whether the value is statistically significantly different from zero. For a finite value of N, it has been shown that the bicoherence of a Gaussian process is Chi-squared distributed [16]. 95% of the bicoherence values are expected to be below $\frac{3}{N}$. Statistics of the higher order coherence are discussed in [33, 34].

The development of higher order spectral analysis above assumes that the process is ergodic. This is not satisfied by the EEG signal. Statistics may be stationary over a short window in time but ergodicity may still not be satisfied. A framework for using short-time higher order spectral analysis is developed next.

Non-Stationary Random Processes and Time-Varying Spectra

Extensions of the theory of sampling to non-stationary random processes have been made by Gardner [27, 35], Garcia et al. [36] and others. Higher order spectra have been analyzed in time-varying frameworks using Wigner-Ville distributions and polynomial phase modelling in [37–42]. Essentially, Wigner-Ville distributions transform the second order correlation with respect to the time offset variable τ but leave the time variable $t^{\boxtimes}$in Equation (2) to take the non-stationarity of the signal into account. A symmetric form $x(t^{\boxtimes}-\tau/2)x(t^{\boxtimes} + \tau/2)$ is used in the definition of the autocorrelation. Higher order forms have been defined with multiple time offset variables. In order for ergodicity to hold, it is necessary that the time $t^{\boxtimes}$ refer to the same time instant in every realization of the non-stationary process. This is only possible if the process is time-locked to the reference $t^{\boxtimes}= 0$ such that there is correspondence of this time in every realization of the process with regard to the non-stationarity. This is not true of arbitrary non-stationary random processes but it can be imposed on data acquisition as done in this work. Phase is meaningless in the Wigner-Ville approach unless the cross Wigner-Ville distribution is used to access the phase (see [43]). Better resolution in the time-frequency plane is possible with the Wigner-Ville approach than with classical short-time Fourier analysis.

Wavelet transforms have also been used in defining transformed representations for higher order cumulants [44]. It may be noted here that in general the wavelet transform is quite well suited to capture transient information in the time-varying scalogram but not all mother wavelets lend themselves to representations that involve phase or utilize correlations. A real-valued mother wavelet is usually defined to be symmetric and does not have phase. Morlet wavelets (discussed in [45]) are complex-valued and are in fact complex exponentials at a central frequency modulated by a Gaussian window in time. Retaining both the real and imaginary parts of the transform, a phase can be defined as a function of the central frequency and time. It is in this respect similar to a short-time Fourier transform (STFT) with a Gaussian window. However, the approach is not well suited for phase coupling investigations because expectation by averaging over short time windows is not the same as ensemble averaging and triads of frequencies that satisfy resonance conditions for sum or difference interactions are not as easily identified in scale space as in a linear frequency space. A wavelet transform based bicoherence is developed in [45]. This work reported that lack of ensemble averaging resulted in spurious coherence. Filtering and application of thresholds based on global maxima are used to remove unwanted noise and extract evolutionary characteristics from the bicoherence.

Another approach to time-varying bispectral analysis of non-stationary EEG signals is presented in [9]. A non-stationary signal is modelled using a non-Gaussian autoregressive model with time-varying parameters under the assumption of piece-wise stationarity. This approach is suitable for anlaysis of different states but not particularly for tracking evolutionary characteristics with time.

A classical STFT approach with a sliding window is presented here. This methodology is preferable here over other methods because it provides a direct phase coupling interpretation of the results and permits relatively easy means of establishing a confidence interval to indicate whether the bicoherence observed is significantly different from zero or not. Further, the methodology developed here examines EEG frequency components in the delta, alpha and beta bands considering narrow band frequencies of interest such as 4, 8 or 20Hz. The STFT is an appropriate tool for such analysis unlike broadband analysis.

Short time Fourier analysis

The Fourier transform defined as an infinite integral for continuous signals or an infinite summation (DTFT) for discrete-time signals is not a useful tool for the analysis of non-stationary signals. Spectral representations of non-stationary signals change with time. A short-time Fourier spectrum may be computed using a window in time centred at a given time instant. As a function of time, this spectrum provides information about the signal in a time-varying manner. The squared modulus of the STFT is known as the spectrogram. The main disadvantage of the STFT is that the frequency representation is uniformly sampled. The frequency bins of the same size regardless of whether high frequencies or low frequencies are being analysed. A logarithmic spacing of frequency, with the frequency step size proportional to the frequency, would permit better analysis of most real-world signals. The time-frequency bins of the STFT are rectangular. The main advantage is that it can be computed fast using the FFT algorithm. In order to keep the processing simple and to facilitate the establishment of values that are significantly different from zero, DFT using a sliding window is adopted as the method of analysis for non-stationary random processes here. It is assumed that the statistics of the process are stationary up to the third order within a window and change slowly as the window is translated. The process is also assumed to be ergodic provided a correspondence between windows is established across multiple epochs. The correspondence across realizations can be achieved by using a reference time that is constant for all epochs with regard to the physical processes and state changes that are taking place. For evoked EEG this reference is the time of application of the stimulus. Segments of each epoch that are equally offset in time from the reference can be assumed to have similar statistics and constitute an ensemble of epochs.

Time-Varying Bicoherence

Assume that N epochs of a nonstationary random process are obtained along with a separate signal that provides a stimulus-lock time reference. Assume that the epochs are sampled and the sampling interval is T. The sample at the reference time is indexed 0. The epochs extend from $-M_1$ to $+ M_2$ samples with respect to the reference. The objective is to examine the bicoherence of the process as a function of time around the reference. This can provide useful information about the process prior to, during and after the stimulus indicating the degree of phase-coupling between harmonic components in the process. A sliding window of W = Msamples is used and the window is centred at t_w. If the window is short enough compared to the transient changes in the underlying process, an assumption of stationarity within the window will hold. The ensemble of epochs for any given t_w can be assumed ergodic if the process can be assumed to behave similarly in each epoch at a given offset t_w from the stimulus-locked reference. A time-varying version of the bispectrum can then be defined using Fourier transform coefficients with the STFT centred at time t_w,

$$B(f_1,f_2,t_w) = E[X(f_1,t_w)X(f_2,t_w)X^*(f_1+f_2 \tag{21}$$

and a time-varying bicoherence can be defined using

$$b^2(f_1,f_2,t_w) = \frac{|B(f_1,f_2,t_w)|^2}{E[\,|X(f_1,t_w)X(f_2,t_w)|^2]\,E[\,|X^*(f_1+f_2,t_w)|^2]} \tag{22}$$

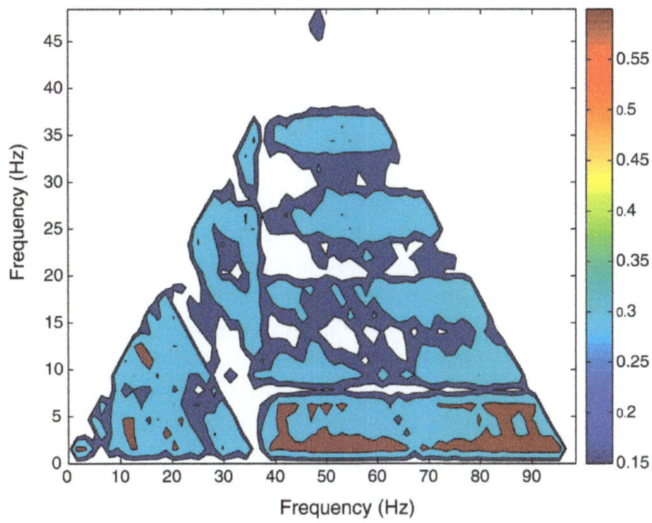

Figure 2: Bicoherence of visual evoked EEG. Bicoherence at channel O2 plotted over the principal domain in bifrequency space computed using a 128 point window around a point in time 10ms after the presentation of a flash visual stimulus to the right eye. The sampling frequency is 200Hz. Only Bicoherence values above the 95% significance level of 0.15 are coloured and the colour bar is shown on the right. It can be noted that bicoherence is significantly high over many frequency combinations and the values appearing as brown are particularly high around 0.6. This picture can change with time.

Estimation of Time-Varying Bicoherence

The procedure to estimate time-varying bicoherence is as follows.

1. Vary the position of the centre of the window from $-M'_1+W/_2$ to $M'_2-W/_2$skipping S samples between windows. For each position, consider the ensemble of windowed segments as a separate random process.
2. Estimate the bicoherence for this process.

Bicoherence can be visualized in the bi-frequency plane as an image. Time varying bicoherence, therefore, will need to be visualized as a montage of images or as an animated set of frames. A contour map with contours starting from the 95% significance level value is useful. Such a map is shown in Figure 2. For a multi-channel signal such as an EEG montage, topographic maps that show the parameter of interest simultaneously at all channels is often employed. They provide useful information about relative values and the spatial distribution. Such a map is shown in Figure 3.

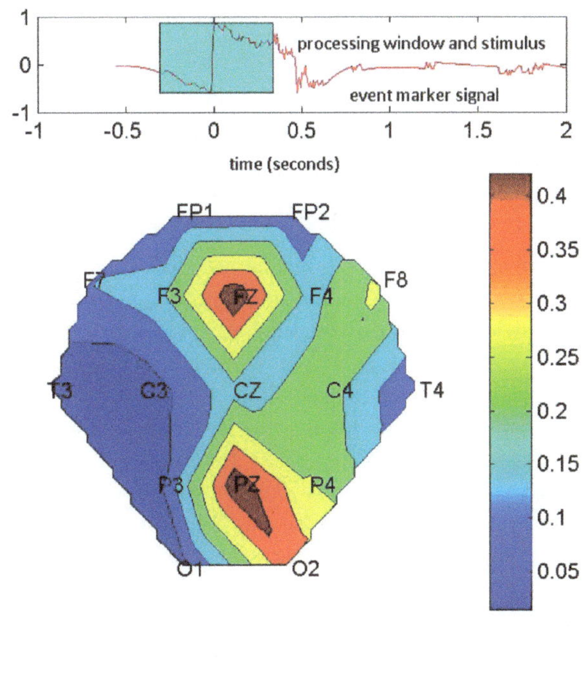

Figure 3: Bicoherence mutliple channel map. Bicoherence at (8, 8Hz) at each of 17 EEG electrodes as a spatial distribution on the scalp. A colour bar is shown on the right and values above 0.15 are statistically significantly above zero bicoherence. Bicoherence is computed using a 128 point window which extends in time over the blue region as shown on the 1D plot on top and the window is centred 10ms after the presentation of a flash visual stimulus to the right eye. The time of presentation of the stimulus is indicated by the transition in the 1D plot and this is used as the zero time reference in each epoch processed.

Only one value of bicoherence at each channel can be visualized in this manner at a time from one plot. Visualization of the time evolution requires the display of a sequence of two-dimensional plots.

Application of Time-varying HOS to evoked EEG Response

Higher order spectra, as spectral representations of cumulants or moments, are strictly valid only for ergodic random processes. These requirements are not satisfied by evoked EEG potentials. However, if a sufficiently small observation window in time is employed, the process can be assumed to be stationary within the window. The time of application of the stimulus serves as a reference. Windows at a given offset from this reference can be assumed to yield consistent averages over an ensemble if the speed of response does not vary. For spontaneous responses such as reaction to an auditory or visual stimulus this can be expected. For learned responses such as memory recall it may not hold true.

Let $x_c, W_{,t_W}(t)$ be an evoked EEG response signal from channel c over a window W centred on time t_W at time t relative to the stimulus. W is short enough for the process to be considered stationary. An array of EEG channels is processed simultaneously. Bicoherence values are obtained for each channel as a function of time t_W. For the sake of simplicity of notation, bicoherence will be referred to as $B(f_1,f_2)$ and with the channel names mentioned on the plots or tables rather than included as a subscript. Time will implicity refer to t_W and the subscript W is omitted as the window does not change except for sliding across in time.

Assume that each epoch of this process extends periodically outside the window with the same statistics up to third order. Short time higher order spectra are the computed using ensemble averaging. If N epochs are used in the ensemble average, bicoherence values for white Gaussian noise are Chi-squared distributed and the 95% confidence level is $\frac{3}{N}$[33, 34].

Response to a given stimulus is recorded several times in different epochs. In practice, the assumption of ergodicity over a window will only hold if the response is at the same speed. This can be expected for most involuntary responses such as those to a visual stimulus or an auditory stimulus from the same individual or a set of individuals who are similar in this regard. For reponses such as memory-recall, the time delays can vary considerably and the stimulus-locked ergodicity assumption may not hold even for the same individual. For a voluntary movement such as a hand squeeze, the assumption will hold provided each subject complies with the request to perform the movement as instructed. The time of application of the stimulus is recorded synchronously with the EEG channels using an auxiliary channel. The length of the window, MT, should be long enough to get the desired frequency resolution $\Delta f = \frac{1}{MT}$ and yet short enough for stationarity up to order 3 for bispectral analysis to hold true. This can be a challenge. Exactly how quickly, phase synchronization and desynchronization occurs between frequency components in the response of neural populations to stimulii is not fully known. Whether such phenomena last tens of cycles or hundreds of cycles of the component frequencies and whether they are consistent and correlated with other known markers of such responses is not yet clearly understood. Experimental analysis with real EEG data is a means to provide answers to such questions.

A phase coupling interpretation of bicoherence will strictly hold only for rhythmic components of the EEG such as alpha and beta waves. Transient or paroxysmal events such as bursts will result in high bicoherence if they occur at nearly the same offset from the stimulus onset and retain their waveform characteristics across an ensemble. In this case, they will always fall within some processing window(s) and they will be phase synchronized to the envelope signal that defines the burst.

Simulations

In order to demonstrate the usefulness of the time-varying bicoherence analysis with phase synchronized waves and bursts in the context of EEG signal processing, three simulations are performed. In each case, 20 realizations of a signal are generated with sampling frequency 200Hz and length 128 samples extending from t=0 to t=0.635s. Each realization is a rectangular windowed block centred at time t_w=0.315s of a random signal in the time domain. Three types of input signal are simulated: (a) phase-coupled and random phase sinusoids, (b) alpha burst with jitter and (c) beta burst with jitter.

Phase-Coupled and Random Phase Sinusoids

The input signal comprised of sinusoids at 8, 16, 20, and 40Hz. The sinusoid at 16Hz had a phase which was the sum of the phases of two sinusoids at 8Hz in each realization. The sinusoids at 8Hz had a random individual phase in each realization. A similar triad of two components at 20 and 40Hz, however, was random phase and had individual phases that were random in each realization and there was no relationship between their three phases. The amplitude of each sinusoid was the same and constant, equal to 1. The time domain waveform for a single realization of this signal is shown in Figure 4a. Gaussian random noise was added to each realization with 10dB SNR. Twenty realizations were generated. A power spectrum estimate for the signal using an averaged periodogram is shown in Figure 4b. As expected there are peaks in the power spectrum at 8, 16, 20, and 40Hz, and the peaks at 8 and 20 Hz are approximately twice the other because there are two random phase components at each of these frequencies. The bicoherence for this signal is shown in Figure 4c. Note the high value of bicoherence at (8Hz, 8Hz) because of the phase-coupled triad of components (8, 8, and 16Hz). The bicoherence at (20, 20) is however low because the triad of Fourier components (20, 20, and 40Hz) is not phase-coupled. How do we know whether a value is low or high? If 20 realizations are averaged in the estimate, the 95% signifance level for bicoherence (see Table 1) is 0.15. This implies that if the input were white Gaussian noise, for which theoretically the bicoherence is 0, the distribution of bicoherence [34] is such that around 95% of the values will be less than 0.15. Therefore, a value above 0.15 can be considered non-zero with 95% confidence. Why is the bicoherence not equal to 1 at a perfectly phase-coupled triad? The bicoherence is lowered by the presence of random phase components at each of the frequencies involved arising from additive random noise. Why would a triad of Fourier components in a real world signal be phase-coupled? When an input signal passes through a non-linear system, any quadratic non-linearity will generate sum and difference frequencies of input sinusoidal components which will then be phase coupled. Synchronized firing of pulses can also generate phase relationships.

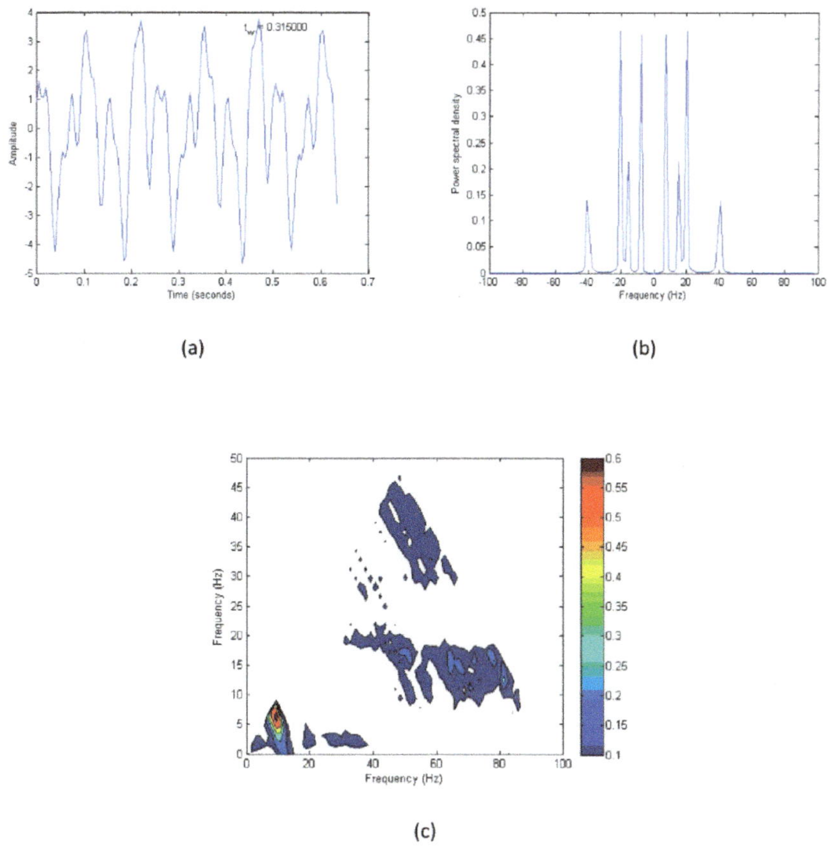

Figure 4: Phase-coupled and random phase sinusoids (a) time domain signal. This corresponds to a window of length 128 samples centred around 0.315s, **(b)** power spectrum, **(c)** bicoherence. Note that the triad (8, 8, 16Hz) is phase-coupled and shows up with high bicoherence around (8, 8) while the triad (20, 20, 40Hz) is random phase and the bicoherence at (20, 20) is not significantly high. The power spectrum is unable to reveal this difference.

Table 1: Significance levels of bicoherence

Number of epochs	b^2_{95}
10	0.3
20	0.15
100	0.03

The expected value of bicoherence at any bifrequency (f_1,f_2) is 0 if the harmonic components are random-phase. This also holds for a Gaussian random process. However, when a finite number, N, of epochs is used in estimation, bicoherence values will be non-zero. 95% of the values will lie below b^2_{95}.

Alpha Burst with Jitter

The input signal comprised of a short burst of an alpha frequency component at 8Hz. The envelope of the burst was the convolution of a short, unit-amplitude rectangular pulse of four samples width with itself four times. The width of the burst was thus 13 samples or 65ms and its waveform was a coarse approximation to a Gaussian. The width of the burst is only long enough to about one-half of a cycle of the alpha component and therefore the signal looks similar to a spike. Jitter was introduced to this burst and its position varied uniformly randomly by four samples or 20ms on either side from realization to realization. The alpha component was synchronized to the envelope of the burst such that it maintained its phase relative to the start of the burst. The amplitude of the alpha component was constant, equal to 1. Again, 20 realizations were generated and Gaussian noise was added to each realization keeping SNR at −15dB. A typical time domain waveform of this input signal is shown in Figure 5a. The power spectrum is shown in Figure 5b. Note that there is no sharp peak at 8Hz and the power is spread in the frequency domain as expected from the confinement in time owing to the burst. The power spectrum, however, would be the same regardless of any synchronization between the burst envelope and the alpha frequency sinusoid. The bicoherence, shown in Figure 5c shows significant high values around (8, 8). If there was no phase relationship between the envelope and the sinusoid, some spikes would be positive and others negative of varying amplitudes, depending on which part of a cycle of the alpha wave coincides with the burst envelope, and the bicoherence would tend to zero as the Fourier phases are random relative to each other. But that is not the case here. The spread of bicoherence values around (8,8) is similar to what would arise from leakage when a window is applied to each realization except that here the window is much shorter. This simulation example shows that information from transient events can be captured in the time-varying bicoherence. Application of time-domain windows of short duration as may occur with wavelet transforms at different scales can have a significant impact on the bispectrum. Although a wavelet transform based approach might appear to be suited to detection of such transient waveforms, it must be pointed out that in the presence of jitter, ensemble averaging of time-domain signal realizations can destroy any signature of the event. Without such averaging, a wavelet transform based feature can capture the event in each realization separately at appropriate scales. However, when the signal to noise ratio falls very low, as is the case with EEG signals, it will not be possible to obtain a signature of the event from any single realization and ensemble averaging will not help because of the jitter. Bicoherence, however, as computed here, will be significantly high provided enough realizations are averaged in the estimate and it is tolerant of any jitter as long as the event is within the processing window in all the realizations. This is owing to the translation invariance property of the bispectrum; linear frequency dependent phase shifts arising from translation cancel out in the triple product of Fourier coefficients in Equation 16 or 18.

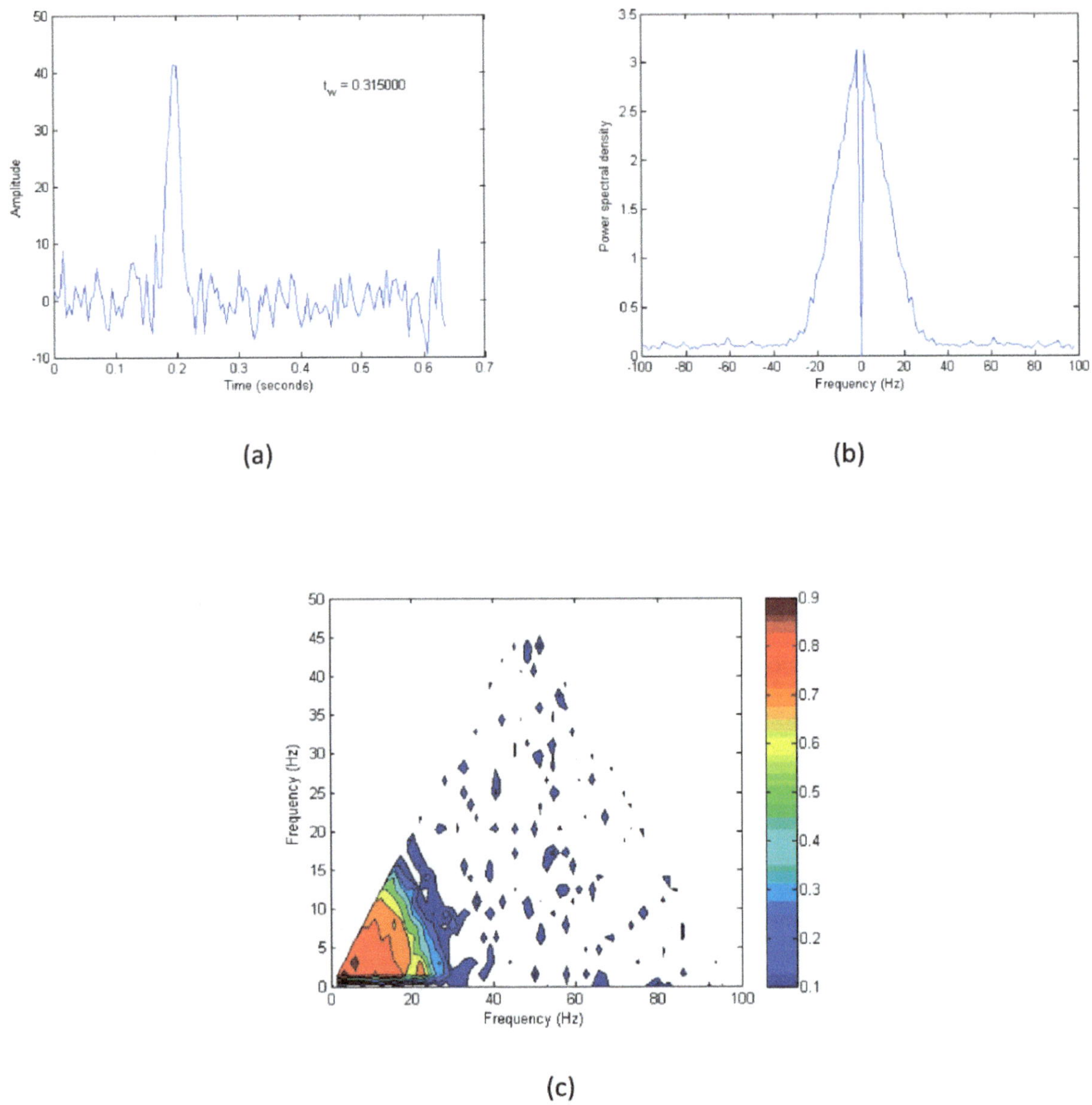

Figure 5: Alpha burst with jitter.**(a)** time domain signal, **(b)** power spectrum, **(c)** bicoherence.

Beta Burst with Jitter

The input signal here comprised of a short burst of beta frequency at 20Hz. The rest of the details are the same as above for the alpha frequency burst except that the width of the burst can hold more than one cycle and therefore the typical time domain waveform shown in Figure6a is no longer just a single spike. The power spectrum shown in Figure 6b is again spread out in the frequency domain but has its peak around 20Hz. The bicoherence shown in Figure 6c shows significantly high bicoherence at (20, 20) and in the region around it, especially towards lower frequencies. This simulation example shows that bicoherence can detect transient events that show phase synchronization characteristics and also reveal differences between them.

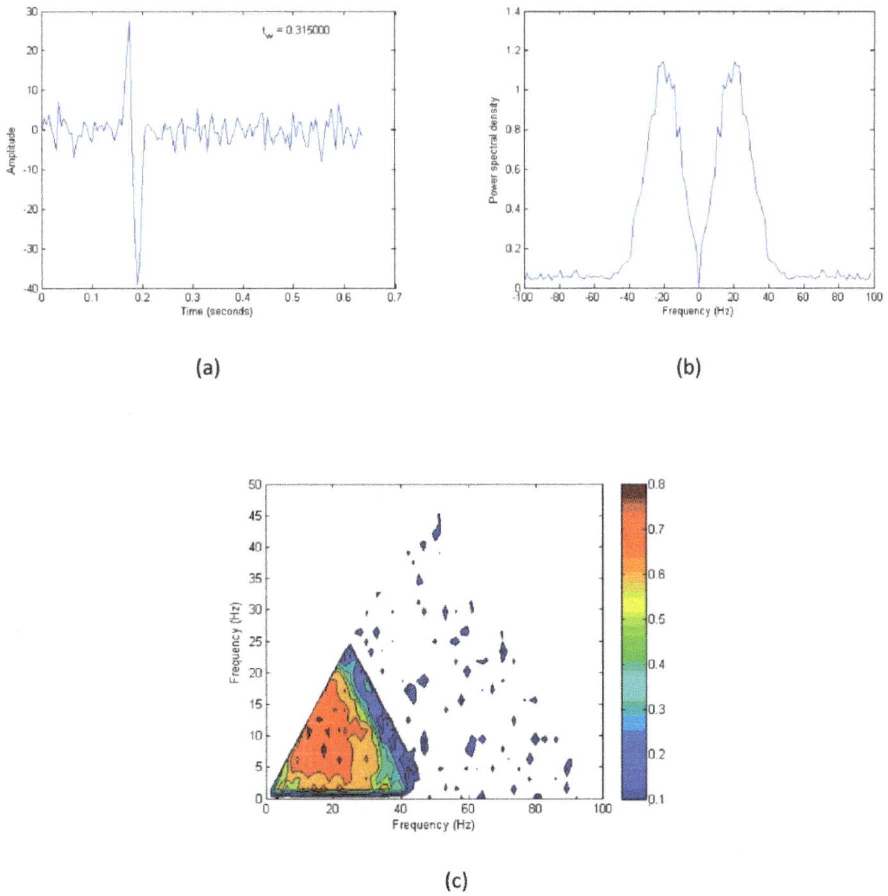

Figure 6: Beta burst with jitter.**(a)** time domain signal, **(b)** power spectrum, **(c)** bicoherence.

In practice, when time-varying bicoherence is computed, the processing window will slide across into the event, around it and then away from it. The event is not tracked with high resolution in bifrequency space with the processing parameters used above, but that is not always necessary. The methodology developed here is intended to use information from one or more such events—wave or transient—that occur int the EEG signal, and allow channels to be grouped based on the similarity of their activity in response to a stimulus.

It is difficult to validate the methodology with real EEG data because there does not exist a signal model for evoked response that would predict significant bicoherence or quadratic coupling at particular bifrequencies which can then be compared to experimental findings. Confidence in the usefulness of the methodology can be improved if the spatial and temporal patterns of bicoherence can be correlated with simultaneously acquired data from a complementary methodology like functional magnetic resonance imaging. Such data are not available for this work.

In this study, tests are reported that examine whether the bicoherence evolution patterns can be used to establish relationships between different regions of the brain. EEG electrodes from the array are segmented into groups based on their bicoherence evolution in the evoked response. Signals from two electrodes that have the same bicoherence as a function of time following the stimulus will have the same fraction of power that is quadratically phase-coupled. Whether this suggests involvement in the particular response or joint inactivity is not clearly established by this procedure alone. The segment maps of electrode groups thus produced is analyzed together with knowledge about different lobes and cortical regions involved in the particular response. For the visual response, the primary visual cortex is in the occipital region and other cortical regions and information pathways have been identified by neuroscience research [46]. This is a degree of validation of the methodology. Such segmentation can be done using various bifrequencies.

Application to Real Data

EEG data were collected using 19 EEG channels. 17 of these channels are shown in Figure 3. The standard 10–20 system of EEG electrode placement (explained in [47]) was used. Auxiliary electrodes were placed above and below one eye and to the right and left of the eyes to capture eye movements in the horizontal and vertical directions. These signals were used in rejecting epochs where eye blinking occurred because the interest is in visual processing and the EEG signal is desired to be free of motor activity related potentials. An LED light flash was used as the stimulus. A stimulus-locked signal was obtained synchronously with the EEG. This signal provides a sharp transition at the time of application of the stimulus. It is used to segment a continuously recorded EEG ensemble into epochs.

Left hemisphere EEG electrodes were referenced to an electrode behind the left ear and right electrodes similary to a reference behind the right ear. These reference electrodes are not expected to vary significantly as a result of the stimulus.

Medial electrodes were referenced to ground. A 50Hz power supply frequency rejection filter is provided with the system. Ag–AgCl electrodes with conductive brine Gel were used with the KT88-2400 Contec 24 channel EEG system for data acquisition. Electrode contact impedances were checked to be low enough using the LED indicator on the system.

Data were captured with each subject blindfolded in the left eye for one session and then in the right eye in the following session. Samples were continuously collected at 200Hz and segmented to yield 512 sample epochs. The epoching was done offline. Epochs were screened and selected free of artifacts such as eye blinks. The initial screening process was software driven and it was followed by manual examination of the output to finalize the selection process. All epochs processed were ensured to be free of noise from eye blinking. Each epoch represented 2.56s starting 0.56s prior to the application of the stimulus. Data were collected from five subjects. A variable number of usable epochs were obtained from each session following the screening process. More than 20 usable epochs from each session were available from three out of five subjects, labelled as E01, E02 and E05. These are processed in the work reported here.

Processing

Processing decisions must be made with regard to the type of window, the length of the window and the time duration over which the window centre will be moved. Although windows other than rectangular, such as Hanning, Hamming or Blackman, are routinely employed in spectral analysis and provide better suppression of leakage, they are not preferred in this context. These windows are essentially multiplying the data samples in time and will result in a convolution of the true spectrum with the spectrum of the window. This introduces additional complexity in analysis and interpretation and is less desirable than some spectral leakage in this context. A rectangular window is therefore adopted. Each epoch was segmented to contain data for 2s following a visual stimulus. This is enough time for the response to a flash visual stimulus to be registered. Visual evoked potential markers identified from grand averages lie within about one second from the stimulus. Further, data are to be clean from eye blinks and larger time intervals make this more difficult. Since data are sampled at 200Hz, this is about 400 samples. If time-varying bispectra are to be computed over an interval of at least one second following the response, the upper limit on the window length becomes one-half this interval or 200 samples. A lower limit is set by the frequency resolution desired. For a frequency resolution better than 2Hz, the window length must be greater than 0.5s or 100 samples. A window length of 128 samples makes it convenient to use an FFT algorithm. If the window length is increased, the frequency resolution will be better but if the underlying phenomenon is transient and short-lived, the trade-off in time resolution will be a problem. If the window is made shorter, the frequency resolution is too poor for EEG frequencies of interest such as the alpha at around 8Hz. A rectangular sliding window of 128 samples was therefore used. This yields a frequency resolution of 1.5625Hz in the output of each short-time FFT. In order to move this window with its centre starting from the zero time reference derived from the stimulus, each epoch must contain at least 64 samples before the reference. There can be a small offset between the time reference and the actual stimulus event. In order to account for this and to facilitate some processing before the onset of the stimulus additional samples prior to the zero time reference were included in each epoch. This number was selected to be 112 such that each epoch was 512 samples, a multiple of 2. This resulted in epochs between −0.56 and 2s with respect to the time reference. For the results presented here, the time duration prior to the onset of the stimulus is not important. The centre of the rectangular window is moved from −0.07 to 1.2s, covering the region of time over which the response is likely to manifest itself significantly and allowing for any small negative offset. It may be noted that even when the window is centred at -0.07 a number of data samples come from post-stimulus measurements. Twenty epochs are processed here providing ensemble averaging. Bicoherence above 0.15 implies that there is 95% confidence that the true value is not zero as shown in Table 1.

A large number of plots can be generated using the methodology and bicoherence can be visualized as a function of a number of different variables—over the bifrequency plane, with time and with channels. In this study, the alpha wave and its harmonics are of particular interest. The alpha frequency can vary over a few Hertz with age and from individual to individual. It is typically around 8Hz for an adult. A number of adjacent frequencies were examined in the tests but 8, 4, and 20Hz and their bifrequency combinations are selected for figures in this work. Only auto-bicoherence is reported in this study. A frequency of 8Hz actually implies the frequency closest to it when the 128 point DFT is computed. This is the 5-th frequency bin at 7.8125Hz. Similarly 9Hz refers to the closest DFT bin at 9.375Hz. For the sake of clarity, frequencies are kept as integers in the discussion here, with the effect of finite resolution implicit.

Bicoherence is tracked with time for all EEG channels. It is computed with two sample skips or every 10ms. Bicoherence patterns are compared across pairs of electrodes. Time-varying bicoherence plots are shown in Figure 7. A mean of absolute difference of bicoherence (BMAD) is calculated for each pair of patterns. Each bicoherence lies between 0 and 1. The BMAD value also lies between 0 and 1. If two bicoherence patterns are exactly identical it will be zero.

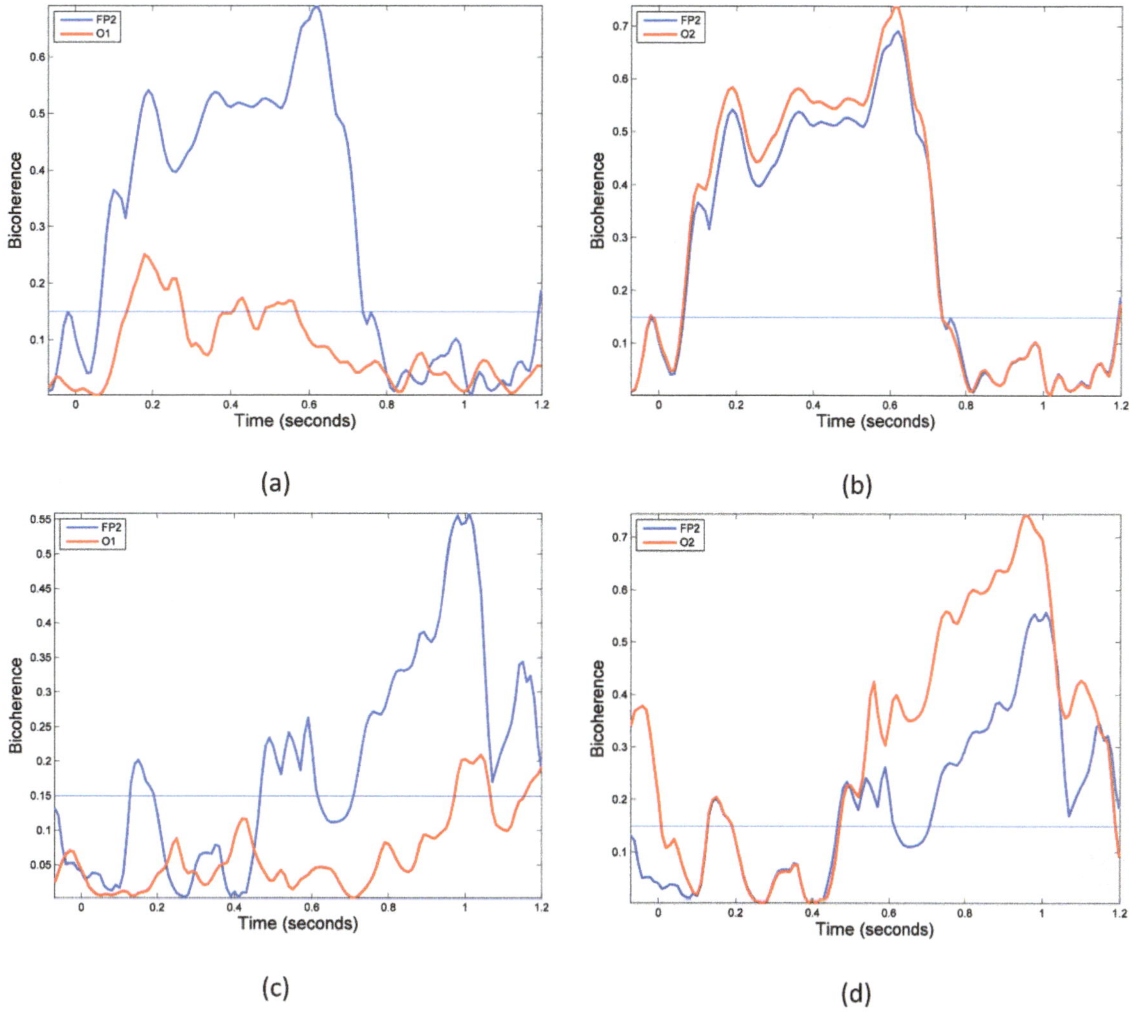

Figure 7: Bicoherence evolutions at (4, 4). Bicoherence evolutions at (4, 4) as functions of time for subject E01: **(a)** over Fp2 and O1 for a flash stimulus viewed through the left eye,**(b)** over Fp2 and O2 and the left eye, **(c)** over Fp2 and O1 and the right eye, **(d)** over Fp2 and O2 and the right eye. There is high correlation in **(b)** and the BMAD values will be small. In **(d)**, the correlation immediately following the stimulus with a delay of about 100ms is nearly perfect for the next 400ms.

Electrodes are separated into segments based on using the following procedure:

1. Group all electrodes, x = 1…K, into one group labelled i = 1.
2. Take the first electrode, say x, at level i and examine BMAD(x,y) for all y > x.
3. Apply a threshold in the range, ($BMAD_{max}$,$BMAD_{min}$), to each where the range of threshold is between the maximum and minimum values.
4. If BMAD(x,y) > Th, then change the label of y to i + 1.
5. Replace i with i + 1 and go to step 2 until all electrodes are processed.
6. If the number of segments is greater than 5 terminate else lower the threshold and proceed to step 1.
7. Repeat the above procedure until the lowest possible threshold is reached.

The above segmentation results in greater than 5 segments in most but not all cases. It imposes the constraint that segment separation be achieved using the same threshold for separating every group from its parent. Sometimes there is no single threshold that will necessarily produce five or more segments. The attempt is to divide the electrodes into a small number of regions that might have a correspondence with known cortical regions and it will be difficult to interpret what is happening if there are too many segments.

Electrode groups are plotted using different colours based only on the order in which they get separated. All electrodes in a given segment have the same colour. No value is attached to the colour and it is not a measure. An indication of the BMAD between two electrodes within a segment is provided by drawing a line between the two and making the thickness of the line logarithmically proportional to the closeness of their BMAD value to zero. This is done from an arbitrarily chosen (depending on the first electrode to separate into that segment) electrode for that segment to all other electrodes in the segment. In these plots, electrodes are represented by three concentric circles of a given colour corresponding to the segment. Such maps are shown in Figure 8.

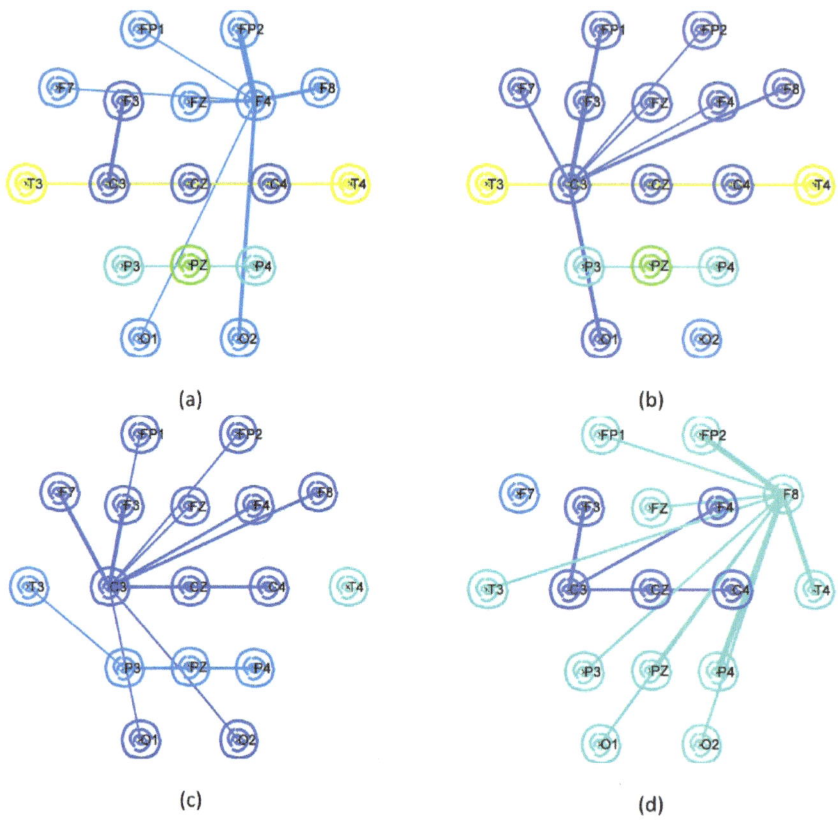

Figure 8: Segmentation maps (4, 4). Segmentation of electrodes based on bicoherence at (4, 4):**(a)** for E01 and the left eye, **(b)** for E01 and the right eye, **(c)** for E02 and the left eye,**(d)** for E02 and the right eye. In each region, electrodes are linked to a representative electrode using lines whose thickness is indicative of the inverse of the BMAD value on a logarithmic scale. The thicker the line, the closer the two electrodes are in terms of their bicoherence fluctuation.

RESULTS

Time-varying bicoherence obtained on the data is analyzed at different frequencies and over different channels and as a function of time.

Bicoherence Distribution over Frequencies

Significant bicoherence was observed over large regions in the bifrequency plane and this pattern changed with time. A typical plot of bicoherence over the entire principal domain in bifrequency space is shown in Figure 2. This plot corresponds to the EEG channel O2, computed with a window around time t = 0.10s. This plot is for subject E01. Significantly high bicoherence (above 0.15) extends over the several EEG frequency bands. Particularly high values above 0.5 are observed in the delta band (around 2–4Hz), alpha band (around 8–12Hz), beta band (around 20Hz) and extending to much higher harmonics.

It is quite possible that actual phase-coupling occurs at discrete frequencies and their harmonics and what is observed over larger frequency bands is owing to spectral leakage arising from the finite length DFT. If an attempt were being made to hypothesize or validate a signal model this would be of consequence. If it is desired to use the bicoherence values without any assumption of a parametric signal model for the purpose of analyzing spatial-temporal functioning of the brain it does not matter whether there is leakage as long as the processing remains consistent across any comparisons. The latter is the case here. Specific bifrequencies are selected and the bicoherence fluctuation with time is computed for all channels.

Bicoherence Distribution over Channels

Figure 3 shows the distribution of bicoherence at (8,8) for E01 computed with a window around time t = 0.10 seconds. These are the same conditions as those in Figure 2 except that only one bifrequency is selected. Bicoherence is plotted here for 17 channels and is interpolated in between to yield a contour-filled colour plot. It can be observed that some electrodes such as Fz, Pz and O2 show very high bicoherence. These plots can be hard to interpret because high bicoherence values can emerge and fade away if they were viewed as a sequence with time, and there is no knowledge of the involvement of different regions of the brain at different times following a stimulus to validate this. There are also a number of such plots possible, one for each bifrequency, and there is no signal model or precise understanding of the role of quadratic non-linearity in the firing of neural populations to guide any selection based on bifrequency. The problem can be simplified to some extent by making pair-wise channel comparisons, examining how bicoherence evolves with time for two electrodes. It can then be determined whether there is strong correlation or similarity between the bicoherence variations with time at the two channels.

Bicoherence Evolution with Time

Figure 7a through Figure 7d plot bicoherence at (8,8) for subject E01, at two electrodes each, as a function of time. Figure 7a shows the bicoherence at Fp2 and O1 for a flash stimulus to the left eye. Fp2 is in the frontal region and close to the eyes whereas O2 (and O1) is close to the occipital region, known for visual processing activity in the human brain. Figure 7b shows bicoherence at electrodes Fp2 and O2 following a left stimulus. Figure 7c,d are similar bicoherence plots for a flash stimulus to the right eye.

It can be observed that the two bicoherence variations in Figure 7b (left eye) are very highly correlated. The mean absolute difference of bicoherence (BMAD) between Fp2 and O2 in this plot is very small. In Figure 7d (right eye) there is very high correlation between Fp2 and O2 only during the 100 to 500ms duration following the stimulus. After that the two functions separate although the trends are still quite positively correlated as judged visually. The BMAD taken over the entire 1.27s will be higher than that for Figure 7b.

Figure 7a,c show that O1 is not highly correlated with Fp2 for either left or right eye responses and the BMAD values will be much higher. Based on these plots, Fp2 will be grouped with O2 for the left eye response even with a low threshold on BMAD. They will be grouped in the same segment for right eye response only if the threshold of separation is large.

In order to tell left from right, markers may need to be computed over significant intervals in time such as the 100 to 500ms duration. In this study, attention is focussed on holistic differences over the entire time response from −0.07 to 1.2s. These are used to segment the electrode array into groups. Segment maps are examined for comparison with existing knowledge about visual signal processing in the human brain.

Segment Maps

Figures 8, 9 and 10 show segment maps that divide the electrode array into groups according to the temporal evolution of bicoherence using the algorithm described above. The objective is to divide the electrodes into groups and examine (a) whether they correspond to lobes—frontal, central, parietal and occipital, and (b) the nature of the connection between frontal regions close to the sensors (eyes) and the occipital region known to be involved in visual processing.

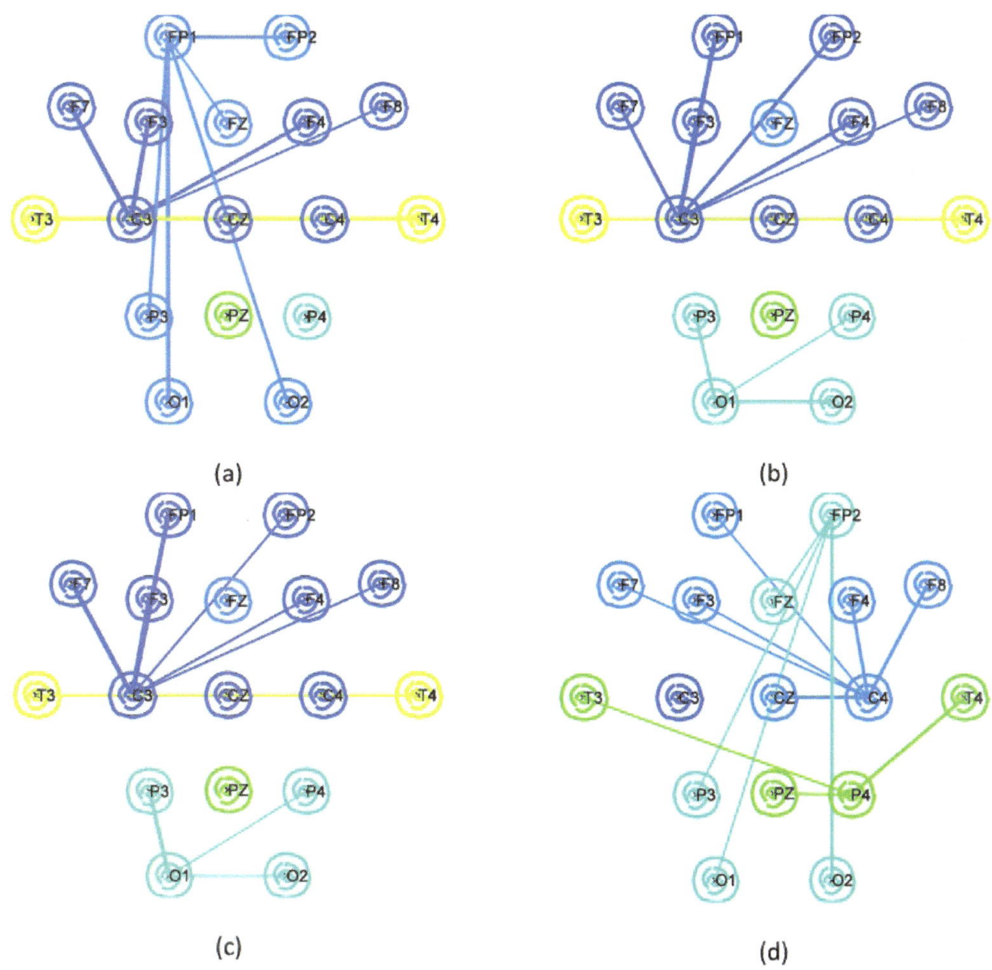

Figure 9: Segmentation maps (8, 8). Segmentation of electrodes based on bicoherence at (8, 8): **(a)** for E01 and the left eye, **(b)** for E01 and the right eye, **(c)** for E05 and the left eye,**(d)** for E05 and the right eye.

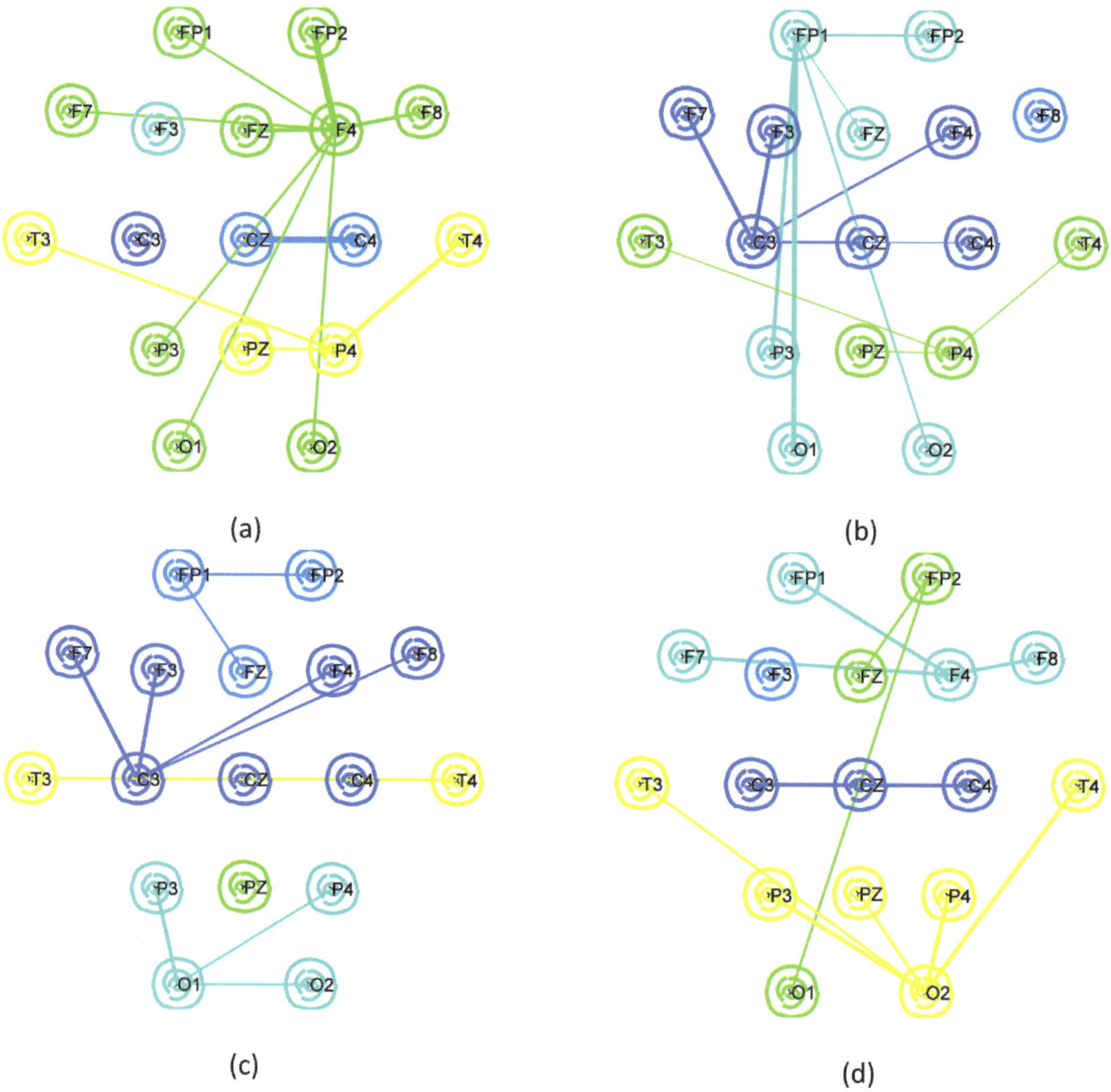

Figure 10: Segmentation maps (20, 20). Segmentation of electrodes based on bicoherence at (20, 20): **(a)** for E01 and the left eye, **(b)** for E01 and the right eye, **(c)** for E05 and the left eye, **(d)** for E05 and the right eye.

Figure 8a through Figure 8d use bicoherence at (4,4) and show maps for subjects E01 and E02 viewing a flash stimulus through the left eye and the right eye. All the figures indicate a relationship between the frontal and occipital regions as being grouped in the same segment. Figure 8c groups parietal lobes and a temporal lobe in the same segment. Figure 8d groups parietal and temporal lobes also with the frontal and occipital but leaves the central electrodes separate. This requires further investigation.

Figure 9a through Figure 9d show a set of segment maps for subjects E01 and E05 viewing a flash stimulus through the left and right eyes, using bicoherence at (8,8). Figure 9a,d show frontal and occipital electrodes in the same segment but Figure 9b,c do not. It is likely that because the frequencies 8 and 16Hz, are also involved in motor activity, the latter is influencing bicoherence evolution phenomena. In Figure 9b,c, the BMAD values between Fp1, Fp2 and O1, O2 are relatively small although they appear in different segments. In Figure 9b, BMAD(Fp2,O2) is 0.060, BMAD(Fp1,O1) is 0.028 and BMAD(O1,O2) is 0.099. In Figure 9c BMAD(Fp2,O2) is 0.110, BMAD(Fp1,O1) is 0.064, BMAD(O1,O2) is 0.127.

Figure 10a through Figure 10d show similar segment maps for subects E01 and E05 using bicoherence at the much higher frequencies of (20,20). In Figure 10a,b,d frontal and occipital electrodes are connected. FZ is also included in the same segment. In Figure 10c, they form two separate segments. This is partly owing to the segmentation algorithm. Electrodes separate from parent groups based on one threshold distance from a representative electrode and the procedure is repeated until a suitable threshold is found to create at least five segments if this is possible. The maps are only a rough guide. More information is obtained by closer examination of bicoherence as a function of time at electrodes. This is done for the map of Figure 10c in Figure 11a through Figure 11d.

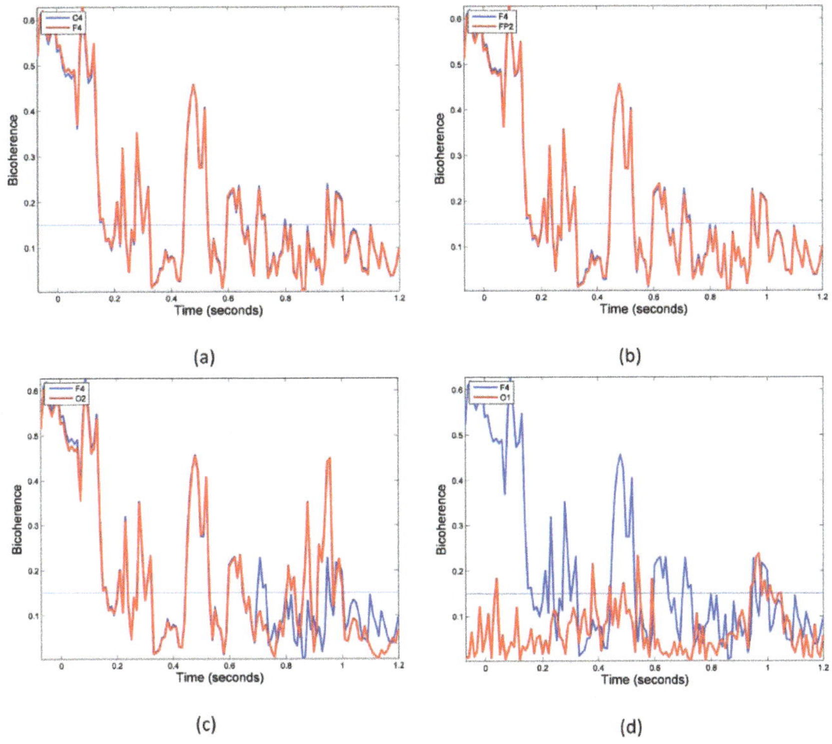

Figure 11: Bicoherence evolutions at (20, 20). Bicoherence evolutions at (20, 20) as functions of time for subject E05 for a flash visual stimulus viewed through the left eye: **(a)** over F4 and C4, **(b)** over F4 and Fp2, **(c)** over F4 and O2 and **(d)** over F4 and O1. There is near perfect correlation in **(a)** and **(b)**—almost eerie for signals that are fluctuating so much. There is near perfect correlation in **(c)** for about 700ms following the stimulus. The correlation is weak in **(d)**, vastly different from the others. It serves to provide confidence that the near perfect correlations are not artefacts of data collection or processing. O1 and O2 are adjacent channels; F4 is not near either of them.

At the beta band frequency of 20Hz and its second harmonic, frontal electrodes have bicoherence evolution highly correlated with central electrodes as shown in Figure 11a for F4 and C4. BMAD(F4,C4) is 0.003. Although they appear in different segments, F4 also is highly correlated with Fp2 as shown in Figure 11b and Fp2 is only separated because the segmentation algorithm attempts to find at least five segments if possible with a single threshold. BMAD(F4,Fp2) is 0.003. Figure 11c,d shows a comparison between bicoherence evolutions at F4 and O2 and F4 and O1, respectively. Some phenomena do not last the entire time duration used in the computation of the BMAD. F4 is highly correlated with O2 for 700 milliseconds following the stimulus as shown in Figure 11c. BMAD(F4,O2) is 0.043. The BMAD between them is primarily dependent on the duration from 0.7 to 1.2s. Bicoherence at O1 on the other hand is rarely above the 95% statistically siginificant level of 0.15 and is not highly correlated with F4. BMAD(F4,O1) is 0.178.

Relationship to Existing Knowledge

Dynamics of visual recognition in the human brain are discussed in [46] which gives a recent review of the area. A great deal is known about the visual cortex from studies on humans and primates. The occipital region is mapped in this study by electrodes O1 and O2 and the visual cortex by O2. As explained in [46] (refer Figure 1 in [46]), the cortex is further divided into regions labelled as V1, V2, V3 and V4. More is known about V1, the primary visual cortex, than any other region. Information from the right and left eye is combined in V1. There are two main pathways from V1 leading to the frontal cortex and on to the hippocampus and surrounding structures. There are the dorsal pathway and the ventral pathway. The dorsal pathway includes the medial temporal area (MT) and the medial superior temporal area (MST) that are known to be involved in processing spatial information—object motion, position and depth perception from stereo vision. The ventral pathway includes the inferior temporal cortex (ITC) which is involved in discriminating colours and shapes. The EEG electrode array used in this study does not have the spatial resolution to differentiate all these regions—nor is it within the domain of the simple flash visual stimulus experimentation. This methodology is not intended to produce the same level of detail in understanding as neuro-anatomical and neuro-physiological studies. However, it does have the advantage of working on scalp EEG and does not involve the use of intra-cranial electrodes or invasive surgical procedures on animals. Nevertheless, it is encouraging that segmentation of electrodes using this methodology establishes connections between the frontal and occipital regions and also indicates involvement of other areas in the parietal lobe.

The experiments here are intended to be preliminary validation of the use of temporal evolution of bicoherence in better understanding the functioning of the human brain using scalp EEG. The methodology does have the advantage that different EEG frequency bands can be analyzed using the same set of experimentally recorded data.

CONCLUSIONS

In this work, a methodology for time-varying bispectral analysis of a multi-channel signal during a stimulus locked transient response is developed. Theoretical foundations of higher order spectral analysis are revisited in this context to examine its validity. It is applied to scalp EEG data. Time-varying bispectral analysis provides information additional to what can be obtained from grand averages (first order) or power spectral (second order) techniques. It is observed that some EEG channels exhibit time-varying bicoherence in response to a stimulus with remarkable correlation over time intervals and even similarity in magnitude lasting hundreds of milliseconds. A new methodology based on the mean absolute difference of bicoherence over such intervals is developed and applied to segment channels into sets that are more similar in their response to the stimulus than others. It is shown that channel segmentation using time-varying bicoherence of the EEG response evoked by visual stimulii is consistent with knowledge about involvement of the occipital region and frontal regions in visual processing. Future work possible includes validation of the methodology using other sensory stimulii such as auditory, the use of higher spatial resolution EEG, analysis of test data from subjects with known visual system impairments such as partial blindness, prosopagnosia (inability to recognize faces), achromatopsia (inability to perceive colours) and comparison of the markers obtained via this methodology with results obtained using functional magnetic resonance imaging. If the methodology can be developed for practical diagnostic and monitoring applications, it is a much less invasive and less expensive option than many others because it is based on scalp EEG recordings.

Acknowledgements

The author is grateful to Mr. M. Gnananthan and Mr. W. Maier at the Queensland University of Technology (QUT) for collecting visually evoked EEG data. This research was not funded by any external research grant. Computing support and publication costs were provided by QUT. The author retains copyright and rights to all commercial uses of intellectual property in accordance with employment regulations.

Competing Interests

The author declares that he has no competing interests.

REFERENCES

1. Dumermuth G, Huber P, Kleiner B, Gasser T: Numerical analysis of electroencephalographic data. IEEE Trans. Audio Electroacoust 1970, 18(4):404-411. 10.1109/TAU.1970.1162144
2. Barnett T, Johnson LC, Naitoh P, Hicks N, Nute C: Bispectral analysis of elecroencephalogram signals during waking and sleeping. Science 1971, 172: 401-402. 10.1126/science.172.3981.401
3. Bullock T, Achimowicz J, Duckrow R, Spencer S, Iragai-Madoz V: Bicoherence of intracranial eeg in sleep, wakefulness and seizure. Electroencephalogr, Clin. Neurophysiol1997, 103: 661-678. 10.1016/S0013-4694(97)00087-4
4. Muthuswamy J, Sherman D, Thakor N: Higher-order spectral analysis of burst patterns in eeg. IEEE Trans. Biomed. Eng 1999, 46(1):92-99. 10.1109/10.736762
5. Pomfrett C, Pearson A: Eeg monitoring using bispectral analysis,. In IEEE Colloquium on New Measurements and Techniques in Intensive Care (Digest No. 1996/179). Dec 1996. London, UK; pp. 5/1-5/3.
6. Greenwald S, Smith C, Sigl J, Cai H, Devlin P: The eeg bispectral indextm (bistm): development and utility,. In [Engineering in Medicine and Biology, 1999. 21st Annual Conf. and the 1999 Annual Fall Meeting of the Biomedical Engineering Soc.] BMES/EMBS Conference, 1999. Proceedings of the First Joint. vol. 1, Atlanta, USA; 1999:-p. 443.
7. Tang Y, Norcia A: Coherent bispectral analysis of the steady-state vep,. In IEEE 17th Annual Conf. Eng. in Medicine and Biology Society. vol 2, Montreal, Canada; 1995:pp. 901-902.
8. Shen M, Sun L: The analysis and classification of phonocardiogram based on higher-order spectra,. In Higher-Order Statistics, 1997. Proceedings of the IEEE Signal Processing Workshop. July 1997, Banff, Canada; pp. 29-33.
9. Shen M, Liu Y, Chan F, Beadle P: Novel approach for time-varying bispectral analysis of non-stationary eeg signals,. In 27th Annual International Conference of the Engineering in Medicine and Biology Society, 2005. IEEE-EMBS 2005. Jan 2005, Shanghai, China; pp. 829-832.
10. Minfen S, Lisha S, Congtao X, Guoping Z: Application of higher-order statistics for the analysis of electroencephalogram in different brain functional states,. In 6th International Conference on Neural Information Processing, 1999. Proceedings. ICONIP '99. 2, Perth, Australia; 1999:pp. 622-626.
11. Minfen S, Lisha S, Beadle P: Parametric bispectral estimation of eeg signals in different functional states of brain,. In Advances in Medical Signal and Information Processing, 2000. First International Conference on (IEE Conf. Publ. No. 476). Briston, UK; 2000:pp. 66-72.
12. Tallon-Baudry C, Bertrand O, Delpuech C, Pernier J: Stimulus specificity of phase-locked and non phase-locked 40 hz visual responses in humans. Neuroscience 1996, 16(13):4240-4249.
13. Shiryaev A: Some problems in the spectral theory of higher order moments. Theor. Prob ,Appl. 1960, 5: 265-284. 10.1137/1105026
14. Brillinger D: An introduction to polyspectra. Ann. Math. Stat. 1965, 36: 1351-1374. 10.1214/aoms/1177699896
15. Brillinger DR, Rosenblatt M: Computation and Interpretation of k-th Order Spectra. InSpectral Analysis of Time Series. Edited by: Harris B. Wiley, New York; 1967. pp. 189–232
16. Nikias SS, Petropulu A: Higher Order Spectral Analysis. Prentice Hall, Englewood Cliffs; 1993.
17. Gardner WA: Introduction to Random Processes with Applications to Signals and Systems. MacMillan, New York; 1986.
18. Hasselman K, Munk W, MacDonald G: Bispectra of Ocean Waves. In Time Series Analysis. Edited by: Rosenblatt M. Wiley, New York; 1963. pp. 125–130
19. Brillinger DR, Rosenblatt M: Asymptotic Theory of Estimates of k-th Order Spectra. InSpectral Analysis of Time Series. Edited by: Harris B. Wiley, New York; 1967.

20. Nikias C, Raghuveer M: Bispectrum estimation: a digital signal processing framework.Proc. IEEE 1987, 75(7):869-891.
21. Boashash B, Powers E, Zoubir A: Higher Order Statistical Signal Processing. Longman Australia, Melbourne; 1995.
22. Priestley MB: Spectral Analysis and Time Series. Academic Press, San Diego; 1981.
23. Franks LE: Signal Theory. Dowden and Culver, Stroudsburg; 1981.
24. Hinich M, Messer H: On the principal domain of the discrete bispectrum of a stationary signal. IEEE Trans. Signal Process 1995, 43(9):2130-2134. 10.1109/78.414775
25. Vixie K, Wolinsky M, Sigeti D: The bispectral aliasing test: a clarification and some key examples. Proceedings of the Fifth International Symposium on Signal Processing and Its Applications, 1999. ISSPA '99, vol. 1, 1999, pp. 255–258
26. Schreier P: A note on aliasing in higher order spectra. Communications Theory Workshop, 2005. Proceedings. 6th Australian Feb 2005, pp. 184–188
27. Gardner W: Common pitfalls in the application of stationary process theory to time-sampled and modulated signals. IEEE Trans. Commun. 1987, 35(5):529-534.
28. Chandran V, Elgar S: Mean and variance of estimates of the bispectrum of a harmonic random process-an analysis including leakage effects. IEEE Trans. Signal Process. 1991, 39(12):2640-2651. 10.1109/78.107414
29. Huber P, Kleiner B, Gasser T, Dumermuth G: Statistical methods for investigating phase relations in stationary stochastic processes. IEEE Trans. Audio Electroacoust. 1971, 19(1):78-86. 10.1109/TAU.1971.1162163
30. Chandran V, Elgar S: A general procedure for the derivation of principal domains of higher-order spectra. IEEE Trans. Signal Process. 1994, 42(1):229-233. 10.1109/78.258147
31. Haubrich R: Earth noises, 5 to 500 millicycles per second 1. J. Geophys. Res. 1965, 70: 1415-1427. 10.1029/JZ070i006p01415
32. Kim Y, Powers E: Digital bispectral analysis and its applications to nonlinear wave interactions. IEEE Trans. Plasma Sci 1979, PS-7(2):120-131.
33. Chandran V, Elgar S, Vanhoff B: Statistics of tricoherence. IEEE Trans. Signal Process.1994, 42: 3430-3440. 10.1109/78.340777
34. Elgar S, Guza R: Statistics of bicoherence. IEEE Trans. Acoust. Speech Signal Process.1988, 36: 1667-1668. 10.1109/29.7555
35. Gardner W: A sampling theorem for nonstationary random processes (corresp.). IEEE Trans. Inf. Theory 1972, 18(6):808-809. 10.1109/TIT.1972.1054917
36. Garcia F, Lourtie I, Buescu J: L2(r) nonstationary processes and the sampling theorem.IEEE Signal Process. Lett. 2001, 8(4):117-119.
37. Frazer G, Boashash B: Detection of underwater transient acoustic signals using time-frequency distributions and higher-order spectra. 1991 Conference Record of the Twenty-Fifth Asilomar Conference on Signals, Systems and Computers, 1991, vol. 2, Nov 1991 pp. 1103-1107.
38. Boashash B, Ristich B: Polynomial wigner-ville distributions and time-varying higher-order spectra. Time-Frequency and Time-Scale Analysis, 1992, Proceedings of the IEEE-SP International Symposium, Oct 1992 pp. 31-34.
39. Boashash B, Frazer G: Time-varying higher-order spectra, generalised wigner-ville distribution and the analysis of underwater acoustic data. IEEE International Conference on Acoustics, Speech, and Signal Processing, 1992. ICASSP-92, vol. 5, Mar 1992 193-196.
40. Boashash B, Ristic B: Time-varying higher-order cumulant spectra: application to the analysis of composite fm signals multiplicative and additive noise. IEEE International Conference on Acoustics, Speech, and Signal Processing, 1994. ICASSP-94, vol. iv, Apr 1994 pp. IV/325-IV/328.
41. Boashash B, O'Shea P: Polynomial wigner-ville distributions and their relationship to time-varying higher order spectra. IEEE Trans. Signal Process. 1994, 42(1):216-220. 10.1109/78.258143
42. Ristic B, Roberts G, Boashash B: Higher-order scale spectra and higher-order time-scale distributions. International Conference on Acoustics, Speech, and Signal Processing, 1995. ICASSP-95, vol. 3, May 1995 pp. 1577–1580
43. Boles P, Boashash B: The cross wigner-ville distribution—a two dimensional analysis method for the processing of vibroseis seismic signals,. Intl. Conf. Acoustics, Speech and Signal Processing (ICASSP), April 1988 pp. 904–90
44. El-Jaroudi A, Akgul T, Simaan M: Application of higher order spectra to multi-scale deconvolution of sensor array signals. IEEE International Conference on Acoustics, Speech, and Signal Processing, 1994. ICASSP-94, vol. iv, Apr 1994 pp. IV/413–IV/416
45. Gurley K, Kijewski T, Kareem A: First- and higher-order correlation detection using wavelet transform. J. Eng. Mech. 2003, 129: 188-201. 10.1061/(ASCE)0733-9399(2003)129:2(188)
46. Blumberg J, Kreiman G: How cortical neurons help us see: visual recognition in the human brain. J. Clin. Invest. 2010, 120(9):3054-3063. 10.1172/JCI42161
47. Niedermeyer E, Lopez da Silva F: Electroencephalography: Basic Principles, Clinical Applications and Related Fields. 2004.

CHAPTER 9

NONSTATIONARY SIGNAL IN RADAR SIGNAL PROCESSING

Qiong Wu and Qilian Liang

Keywords: Coprime sampling, Non-stationary signal, Short-time, Fourier transform, Radar signal processing

INTRODUCTION

Both of the designs of radar system and sensor network could be attribute to obtaining sufficient samples to generate the correlation function so that a good ambiguity scale or spectrum estimation could be obtained[1]. The design of radar system needs to take advantage of the ambiguity function (AF) between received signal and transmitted signal to determine the resolution of the radar, side lobe behavior, and ambiguities in both time and Doppler domains. AF is calculated via the convolution of transmitted signal with received signal, which contains the copy of transmitted signal, noise, and Doppler shift caused by the movement of the target. Furthermore, considering cost of deployment in broad range, many applications of sensor network require to distribute the sensor elements sparsely. The power spectral density (PSD) acquired by these sensors could describe the power incidents for the given direction and area, and PSD is the Fourier transform of autocorrelation function of received signal or correlation function among the signal received in different sensors in the array. Hence, both scenarios could benefit from sparse sensing a rapidly changing signal sequence with optimal performance in terms of retaining the resolution or detecting ability compared with dense sampling.

The degree of freedom (DoF) of sampling defined the minimum number of sample points, which could specify certain properties of the sequence as a whole[2]. Before the research of coprime samplers, the available sensors were considered as a signal array and increasing DoF could be achieved by performing an augmentation algorithm on the covariances obtained via minimum redundancy arrays (MRA)[3], which consisted uniformly linear arrays with maximum possible aperture. Bedrosian[4] extended the linear array to non-uniformly distribution such that their pairwise differences could generate full coverage for certain span, the article also enumerated the array size M from 3 to 11 to achieve full coverage as much as $M(M-1)/2$. The algorithm proposed in[5] could find near-optimal integer sensor locations that maximized the number of distinct nonnegative integers, but it also restated the fact that location of elements in an MRA could only be approximated rather than specified in closed form. Besides, there were other ways to generate extra freedoms, including higher order statistics based methods, Khatri-Rao product based methods, and nested array[6]. Besides, the article[7] developed the application of nested array beyond focusing on the DoF, finding nested array could improve the spectrum efficiency.

Coprime sampling first had been used for identifying sinusoids in noise[8] along with other methods proposed for synthetic aperture radar locating and imaging of moving targets[9]. Further research explored the properties and applications of coprime sampling and array in both time and frequency domains. The article[10] used coprime samplers to increase the dimensions of DFT filter banks after sensor arrays as well as to estimate the power spectrum density of received signal. In the article[11], the multidimensional coprime sensing extended the previous implementations to acquire densely sampled domain. The article[12] proposed spatial smoothing algorithm together with coprime sampling to estimate frequencies of sinusoids buried in noise and directions-of-arrival of impinging signals on a sensor array.

Citation: Qiong Wu, Qilian Liang, Nonstationary signal in radar signal processing, DOI: 10.1186/1687-1499-2013-58.

Note that the article presenting coprime sampling[10] strictly confines discussion within the underlying assumption of wide-sense stationary signal so that the expectation of autocorrelation could approach the real value via multi-times averaging. This increased delay is used to compensate the variation introduced by sub-Nyquist sampling. On the other hand, however, in the real-world application, just as the description in the first paragraph, the working scenarios of many applications involves non-stationary signal. The sampled points could not simply ascribe to independent and identical distribution either. Consequently, the autocorrelation coefficients might change dramatically during a short period. In this article, we deal with this inconsistency and discuss the coprime sampling for non-stationary signal to obtain its second order statistic properties. In general, the classic point of view for processing non-stationary signal regards it as piece-wise stationary signal, but as these two theories combining together there are many research problems such as stability of estimation, coverage of second order derivatives, and so on. In the following content, we will discuss these problems and our tentative solutions in detail.

The rest of this article is organized as follows, we first quickly revisit the basic concepts and properties about coprime sampling in Section Theory and properties for coprime sampling. In Section STFT for coprime sampling non-stationary signal, we propose and simulate the algorithm of two-steps coprime sampling especially used for the non-stationary signal. In Section Implementation in radar signal processing, we extend the implementation scenario to radar signal processing and discuss several critical trade-offs in designing the radar signal processing system with coprime sampling. Finally, we conclude the research discussion in Section Conclusions and future research.

THEORY AND PROPERTIES FOR COPRIME SAMPLING

The algorithm of coprime sampling was introduced in[10]. The input signal is $S(T)$. Original sample rate is T_s, and the down sampling rate for two sample streams are M and N whose greatest common divisor is one. Then, except the beginning point, the two generated sample streams do not have any overlap in origin signal sequence.

$$x_1[n_1] = S(MT_s)$$
$$x_2[n_2] = S(NT_s) \tag{1}$$

Definition 1

The difference co-array $x_k[n_1,n_2]$ is generated by two sample sequences $x_1[n_1]$ and $x_2[n_2]$ coprime sampled from input signal. Its index k satisfies

$$k = Nn_1 - Mn_2$$

$$-n_1 \in \left[0, \left\lfloor \frac{L}{M} \right\rfloor\right], n_2 \in \left[0, \left\lfloor \frac{L}{N} \right\rfloor\right] \tag{2}$$

The markers $\lfloor Z \rfloor$stand for the largest integer less than certain values Z, and L stands for the total length of the signal segment. The coprimality of M and N can be used to show that the range of distinct value in $x_k[n_1,n_2]$ is the product of the coprime factors[10]. That is

$$-MN + 1 \leq Nn_1 - Mn_2 \leq MN - 1 \tag{3}$$

First of all, the physical meaning of this difference co-array is that via this difference co-array between the two coprime sampled steams the correlation of the original sequence could be calculated at all lags. Note that it does not confine the rate of down sampling, which might result the sample rate way below the Nyquist-sampling restriction. That is, the sampling might be arbitrarily sparse. On the other hand, however, there are two major drawbacks relevant with large values of coprime pairs: the latency in the time domain and the resolution range in the frequency domain. We will discuss them in detail in the following Section.

Besides, the minor differences in value ranges of coprime sampled signal streams generate different coverage of difference co-array and result in different coverage of autocorrelation coefficients.

Property 1

With n_1 and n_2 restricted to the range $0 \leq n_1 \leq N-1$ and $0 \leq n_2 \leq M-1$, index of the resulting difference co-array $k=Mn_1-Nn_2$ will have MN distinct values in the range $-(M-1)N \leq k \leq (N-1)M$, which also indicates that there are absent values in the given range of k.

Property 2

If the ranges of n_1 and n_2 are $0 \leq n_1 \leq N-1$ and $-M+1 \leq n_2 \leq M-1$, the resulting index of difference co-array will achieve full coverage for $0 \leq k \leq MN-1$.

The detailed demonstration of two properties above could be found in[10]. Furthermore, in this article, we implement coprime sampling beyond the limit of $MN-1$, which leads to the following property.

Property 3

Given sample points in the range $(-L,L)$, the largest coprime pair that it could have is M and Nsubject to $MN<L$, such that n_1 and n_2 restricted to the range $0 \leq n_1 \leq \lfloor L/M \rfloor$ and $-\lfloor L/N \rfloor \leq n_2 \leq \lfloor L/N \rfloor$, the resulting index of difference co-array $k=Mn_1-Nn_2$ will achieve full coverage in the range $0 \leq k \leq L-1$.

Proof

Following from the Euclid's Theorem[13], we could conclude that with any integer k in the range $[0,L-1]$, there are always integers n_1' and n_2' such that $k=Mn_1'-Nn_2'$.

Adding lMN to both terms in the right hand side of the formula with proper selection of variable$l$, we could let $n_1=n_1' + lN$ such that $n_1 \in [0, \lfloor L/M \rfloor]$. Then we have

$$k = M(n_1' + lN) - N(n_2' + lM) \tag{4}$$

$$N(n_2' + lM) = M(n_1' + lN) - K \tag{5}$$

Since we have already known that $k \in [0,L-1]$ and $Mn_1 \in [0,L]$, the range of $N(n_2' + lM)$ should be $[-L,L]$. Let $n_2=n_2' + lM$, we could have $n_2 \in [-\lfloor L/N \rfloor \lfloor L/n \rfloor]$ which concludes the proof.

Moreover, in the range $-MN+1 \leq k \leq 0$, there are still absent values. But based on the symmetry property of autocorrelation, these results could be used for averaging the expectation of the symmetric positive counterpart.

STFT FOR COPRIME SAMPLING NON-STATIONARY SIGNAL

Short time Fourier transform with Coprime Sampling

The presumption to generate autocorrelation from the coprime sampled sequence based on the previous chapter is that the second-order expectations of the sequence remain unchanged over time, which is essentially the wide-sense stationary (WSS) signal. In the application of radar signal processing, however, this criteria cannot hold anymore. In this Section, we will discuss how to combine coprime sampling with short time Fourier transform (STFT-CS) to process non-stationary signal, and demonstrate this algorithm is useful to preserve both the original quality of the signal and at the same time dramatic decrease the sample rate.

The choice of short time Fourier transform (STFT) is because this method is widely used in analyzing the time-frequency properties of non-stationary signals. In an STFT, the signal is segmented by a window function and performed Fourier transform within the window. The width of the window is a trade-off between temporal resolution and frequency resolution–better time resolution is achieved by narrow window while wider window could achieve better frequency resolution. In addition, in the scenario of coprime sampling, based on the Property 3, the window size also dictates the upper bound of the values of coprime pairs. Consequently, it determines the trade-off between stability of the estimation and the computational complexity of STFT-CS.

First of all, there is one definition to simplify the description of algorithm. Because the number of available autocorrelation estimation is changing along with the choice of coprime pairs, we define the procedure of finding the average as a single operator.

Definition 2

$E(R_{xy}(k))$ stands for mathematical expectation of autocorrelation $R(k)$ for a given k using all available estimations. The value of k is determined by two independent index variables of the input sequence x and y.

The algorithm involves several important independent variables listed in Table1.

Table 1: Variables used in STFT-CS

Operators	Description
M and N	the coprime pair used in STFT-CS
L	window size for slicing signal, in which the signal could be consider as aperiodic and stationary
P	the window size of STFT
$w_c[n]$	the window function used in STFT
Q	the processed length of autocorrelation for each slicing window

Based on STFT, within every slicing window we consider the sequence

$$x[n]=\begin{cases} s[n] & 0\le n\le L-1,\\ 0 & \text{otherwise,}\end{cases} \tag{6}$$

The estimate of autocorrelation is

$$\hat{\varphi}_{xx}[m]=E(c_{xx}[m]), \tag{7}$$

where $c_{xx}[-m]=c_{xx}[m]$,

$$c_{xx}[m]=\begin{cases} \sum_{n=0}^{L-|m|-1} x[n]x[n+|m|] & m\le L-1,\\ 0 & \text{otherwise,}\end{cases} \tag{8}$$

The implementation of this estimate could be implemented via using fast N-point DFT algorithm three times.

$$X[k]=\sum_{n=0}^{N-1} x[n]\,e^{-j(2\pi/N)kn} \tag{9}$$

$$|X[k]|^2=X[k]X^*[k] \tag{10}$$

$$c_{xx}[m]=\frac{1}{N}\sum_{k=0}^{N-1}|X[k]|^2 e^{j(2\pi/N)km} \tag{11}$$

Finally, we could calculate the PSD of input signal via estimate of autocorrelationφ^xx[m]$\hat{\varphi}_{xx}[m]$.

$$s[m]=\begin{cases} \hat{\varphi}_{xx}[m]\,w_c[m] & 0\le m\le Q-1\\ 0 & Q\le m\le P-Q\\ \hat{\varphi}_{xx}[P-m]\,w_c[P-m] & 1\le P-m\le Q-1\end{cases} \tag{12}$$

The resulted PSD for given sliced signal is

$$S[k]=\sum_{m=0}^{Q-1} s[m]\,e^{-j(2\pi/Q)km} \tag{13}$$

Along with the moving of slicing window, we can acquire the spectrogram of input signal via STFT-CS.

We implement the algorithm with linear frequency modulation (LFM) to test it validity. The sample rate of the signal is 8000 Hz, sweeping frequency from 0 Hz to 4000 Hz in ten seconds, which can be observed from the top row of Figure1. The configurations of important variables corresponding to the Table1 include: the length of slicing window is 256 sample points (sp), the length of STFT is 512 sp, the processed length of autocorrelation is 255 sp, the window function is Hamming window with window size equal to the size of Fourier transform.

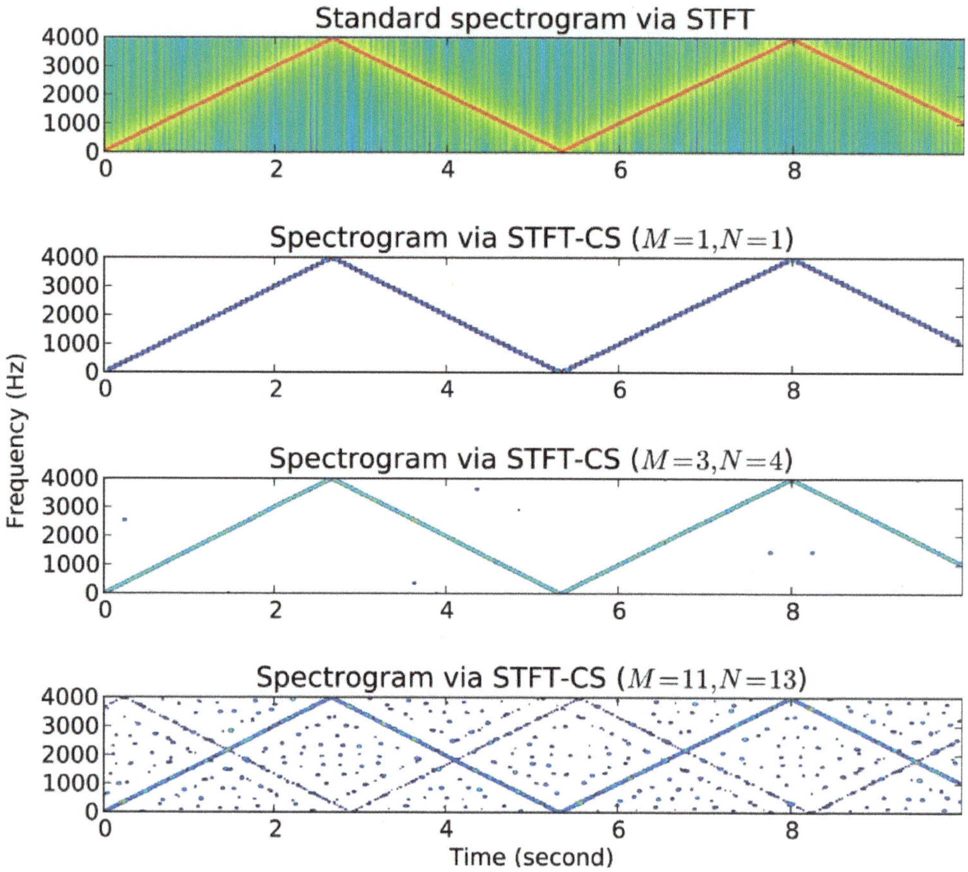

Figure 1: Comparison of different coprime pairs. The vertical axes of all the four rows in this figure are frequency in Hz, and all horizontal axes indicate time in second. The comparison among these rows show the validity of the algorithm STFT-CS and its properties.

As shown in the Figure1, the first row is the standard algorithm to calculate STFT generating spectrogram, and the other rows are using the algorithm STFT-CS mentioned above. We can see that both standard STFT and STFT-CS could accurately trace the change of frequency.

Besides, based on the comparison of the lower three sub-figures in Figure1 using SFTF-CS, we can see that as the increase of coprime pairs, there are more and more traces of aliasing frequency appearing in the spectrogram. This is because as the algorithm select less sample points to estimate the autocorrelation, there will be more variation.

On one hand, the decreasing of sample points is desirable for signal processing. For example, the fourth row in the Figure1 only utilize about 17 percent of the sample points to achieve the same instantaneous PSD estimation with minor quality degeneration. On the other hand, however, the variation become more obvious if we continue increasing the values of coprime pair. This is the motivation for us to develop the 2-step STFT-CS presented in the following section.

2-steps STFT Coprime Sampling

As the spectrogram described above, large values of coprime pair could generate lots of noise. An intuitive method to identify fundamental frequency buried under noise is to calculate its autocorrelation. Then, it becomes an interesting procedure of iterative autocorrelation, that is, estimating the autocorrelation via using convolution three times.

In time domain, we calculate the autocorrelation function based on (7)

$$\hat{\varphi}'_{xx}[m] = \hat{\varphi}_{1xx}[m] * \hat{\varphi}^*_{2xx}[m] = \sum_{n=0}^{L-|m|-1} \hat{\varphi}_{1xx}[n]\,\hat{\varphi}^*_{2xx}[n+|m|] \quad (14)$$

where the $\varphi_{1xx}[m]$ and $\varphi_{2xx}[m]$ are two autocorrelation estimations which could be either same or different values of coprime pairs. The counterpart in frequency domain is straightforward. It is the product of PSD generated by two coprime pairs.

$$S'[k] = DFT(\hat{\varphi}_{1xx}[m] * \hat{\varphi}^*_{2xx}[m]) = S_1[k]\,S_2[k] \quad (15)$$

In Figure2, we show the result of 2-steps STFT-CS together comparing with three results of 1-step STFT-CS with different configurations. The first row lists STFT-CS without coprime sampling as benchmark. The second and third rows are consistent with what we found in the previous section. When the coprime pair increases to 17 and 19, we can hardly distinguish the real trace of spectrogram from the noise aliasing. The fourth row is the result of 2-steps STFT-CS using $M_1=17$, $N_1=19$, and $M_2=11$, $N_2=13$. The resulting sequence has roughly the same degree of down-sampling rate (about 27 percent of the original sample points) as the experiment in second row. But we can observe that via the 2-step autocorrelation the false positive PSD estimates are obviously decreased.

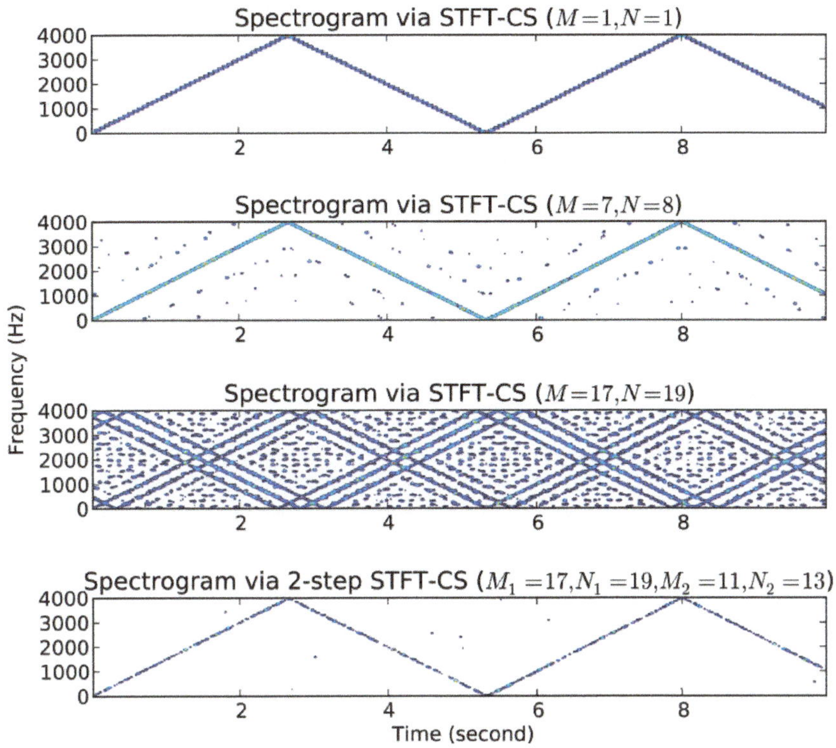

Figure 2: Comparison of STFT-CS and 2-step STFT-CS. This figure has the same units in both axes as Figure1. It indicates the validity and superiority of 2-step STFT-CS comparing with STFT-CS with the same rate of down sampling.

Variation Analysis for Estimating Autocorrelation

In the article[10], the coprime sampling is the method dealing with the sub-Nyquist sampling frequency. Though it does provide promising potential of dramatically decreasing the sampling rate via coprime pair, the estimation is inherently suffering the problem of taking much longer latency. While in the non-stationary scenario, this situation would raise the major problem generating pronounced estimation variation for the reason that only a small piece of samples could be considered as stationary and processed once with autocorrelation estimation in STFT-CS. There is not enough latency permitted for averaging.

In other words, the statistical stability is sacrificed negatively proportional to the degree of coprime sampling. As the choice of coprime pair increases, the density of differential array generated would decrease correspondingly, though the coprime sampling might still calculate the full coverage of all lag by satisfying the Property 4. Then the correlation estimates at that lag could be deteriorated offsetting from the real values.

The article[14] examined the error of estimating autocorrelation and the article[15] linked the variation with sampling rate and refined it in the form of mean-square error. Besides, this article also advocated that for short data records, whose sample points were less than 50 or the product of bandwidth and sampling period is less than 25, the preferred sampling rate was the twice of Nyquist rate. Otherwise, there would be obvious increases in the variance of the estimation.

Comparing this claim with the scenario of experiments in this article, the sampling periods would fall into the category of short-data records while the sampling rate should be regarded as sub-Nyquist rate which is much lower than the desired rate in this criteria. Hence, the estimation will definitely suffer from significant variance.

The method of statistical differential could be used for estimating the covariances of autocorrelation coefficients [16]. For convenience of analysis, we could treat the LFM as piecewise stationary signal and define it as

$$x[n] = \sum_{s=0}^{L} h_s \varepsilon_{n-s}, \tag{16}$$

where the series $\sum_{n=0}^{\infty} h_n$ are absolutely convergent, and ϵ_n is a WSS process with zero means and variance δ^2, that is

$$E(\varepsilon_n) = 0, \quad E(\varepsilon_n^2) = \delta^2 \tag{17}$$

$$E(\varepsilon_n \varepsilon_m) = 0 \quad for\, m \neq n \tag{18}$$

Then, the real value of autocorrelation is

$$R_k = \mathrm{cov}(x_n, x_n + k) = \delta^2 \sum_{s=0}^{L} h_s h_{s+k} \tag{19}$$

and the estimation of autocorrelation is

$$\hat{R}_{k,L} = E(x_{n_1 M} x_{n_2 N}) \tag{20}$$

standing for averaging all of the available values of $x_{n_1 M} x_{n_2 N}$ to calculate the autocorrelation k within the range L.

Assume $h_t = 0$, we could calculate the covariance based on (16)

$$\mathrm{cov}\left(x_n x_{m+k}, x_p x_{q+k}\right) = \kappa_4 \sum_{r=-\infty}^{+\infty} h_{n-r} h_{m+k-r} h_{p-r} h_{q+k-r} + R_{n-p} Rm - q + R_{n-q-k} R_{m-p+k}, \tag{21}$$

where $\kappa_4 = E(\epsilon^4) - 3\delta^4$.

Therefore, we could have[17]

$$\lim_{L\to\infty} L\,\text{cov}(\hat{R}_{k,L};\hat{R}_{l,L}) = \frac{\kappa_4}{\delta^4}R_kR_l + \sum_{q=-\infty}^{+\infty}(R_qR_{q+k-l}+R_{q+k}R_{q-l}) = \nu_{kl} \quad (22)$$

and the particular case is the variance of autocorrelation

$$\lim_{L\to\infty} L\,\text{var}(\hat{R}_{k,L}) = \frac{\kappa_4}{\delta^4}R_k^2 + \sum_{q=-\infty}^{+\infty}(R_q^2+R_{q+k}R_{q-k}) = \nu_{kk} \quad (23)$$

Another estimator for the autocorrelation is

$$C_{k.L} = \frac{1}{L-k}\sum_{l=1}^{L-k} x_t x_{t+k} \quad (24)$$

which confines estimate only based on the available sample points.

Similarly to (22), (23), we have

$$\lim_{L\to\infty} L\,\text{cov}(C_{k,L};C_{l,L}) = \nu_{kl} \quad (25)$$

$$\lim_{L\to\infty} L\,\text{var}(C_{k,L}) = \nu_{kk} \quad (26)$$

Compared with (22), we could have

$$\text{var}(C_{k,L}-\hat{R}_{k,L}) = O\left(\frac{1}{L^2}\right) \quad (27)$$

Based on (23), (26) with Schwarz Inequality, we could have two measures for the variation of autocorrelation estimation with the length of available sample points.

$$\text{cov}(C_{k,L};C_{l,L}) - \text{cov}(\hat{R}_{k,L};\hat{R}_{l,L}) = O\left(\frac{1}{L^{3/2}}\right) \quad (28)$$

$$E(\hat{R}_{k,L}-R_k) = O\left(\sqrt{\frac{\nu_{kk}}{L}}\right) \quad (29)$$

From (28), (29) we can see why the estimate variation is increase as the decrease of sample points. This is an inherent problem confining the choices of coprime pairs in processing non-stationary signal using coprime sampling.

IMPLEMENTATION IN RADAR SIGNAL PROCESSING

The working principle of matched filter in radar signal processing is to output the cross-correlation of target-plus-noise signal and transmitted signal [18]. So, it is possible to implement the matched filter as a correlation process. When the signal-to-noise (SNR) ratio is large, the output of the matched filter can usually be approximated be the autocorrelation function of the transmitted signal. Hence, we could use much less sampling points via coprime sampling to estimation the output of matched filter.

In this section, we still consider the typical LFM waveform, which is consistent with the previous section and also used as a basic waveform in radar transmission because it could independently control pulse energy through its duration and range resolution through its bandwidth[19]. Thus, if the transmitted signal could be processed to have long duration and narrowly concentrated autocorrelation, both good range resolution and good energy can be obtained simultaneously.

Considering a modified waveform $x'(t)$ by modulating $x(t)$ with a LFM complex chirp and compute its complex ambiguity function

$$x'(t) = x(t)e^{j\pi\beta t^2/\tau} \quad (30)$$

The instantaneous frequency of this waveform is the derivative of the phase function

$$F_i(t) = \frac{1}{2\pi}\frac{d\theta(t)}{dt} = \frac{\beta}{\tau}t \quad (31)$$

in which the $\beta\tau$ is called time-bandwidth product of the LFM pulse. The time-delay measurement error is proportional to τ and the frequency measurement error is proportional to $1/\tau$.

In many radar application, the moving target generate Doppler shift in its echo signal, which makes the output of the matched filter should be considered as the cross correlation between the Doppler-shifted received signal and the transmitted signal. In this case, we use ambiguity function (AF) to generate the behavior of a waveform paired with its matched filter. Based on the analysis of AF, we could easily examining resolution, side lobe behavior, and ambiguities in both time and Doppler domains.

Assume the Doppler frequency is F_D, then the input waveform with a Doppler-shifted response is $x(t)e^{j2\pi F_D t}$. Also assume that the filter is designed to peak at $T_M=0$, which means that the time axis at the filter output is relative to the expected peak output time for the range of a target. Assuming M and N are the coprime pair and T_s is the sampling rate. Then the AF could be defined as

$$\begin{aligned}\hat{A}'(k,F_D) &= \int_{-\infty}^{\infty} x'(MT_s)x'^*(NT_s)e^{j2\pi MT_sF_D}ds \\ &= \int_{-\infty}^{\infty} x(MT_s)x^*(NT_s)e^{j\pi\beta(M^2-N^2)T_s^2/\tau+j2\pi MT_sF_D}ds \\ &= e^{-j\pi\beta k^2/\tau}\int_{-\infty}^{\infty} x(MT_s)x^*(NT_s)e^{j2\pi(F_D+\beta k/\tau)MT_s}ds \\ &= e^{-j\pi\beta k^2/\tau}\hat{A}(k,F_D+\frac{\beta}{\tau}k)\end{aligned} \quad (32)$$

where k is the difference between two sample points, and $\hat{A}(k,F_D)$ is the original complex ambiguity function for the simple pulse signal

$$\hat{A}(k,F_D) = \frac{e^{j2\pi F_D k/2}}{\tau j2\pi F_D}\left(e^{j2\pi F_D\frac{\tau-k}{2}} - e^{-j2\pi F_D\frac{\tau-k}{2}}\right) \quad (33)$$

And its amplitude is

$$A(k,F_D) = \left|\hat{A}(k,F_D)\right| = \left|\frac{\sin[\pi F_D(\tau-|k|)]}{\tau\pi F_D}\right| \quad (34)$$

Then we can have the amplitude for the AF of the LFM waveform

$$A'(k,F_D) = |\hat{A}'(k,F_D)| = \left|\frac{\sin[\pi(F_D+\beta k/\tau)(\tau-|k|)]}{\tau\pi(F_D+\beta k/\tau)}\right| \quad (35)$$

The zero-Doppler cut of the LFM ambiguity function, which is just the matched filter output when there is no Doppler mismatch, is

$$A'(k,0) = \left| \frac{\sin\left[\pi\beta k(1-|k|/\tau)\right]}{\pi\beta k} \right| \qquad -\tau \leq k \leq \tau \tag{36}$$

and the zero-delay response is

$$A'(0,F_D) = \left| \frac{\sin(\pi F_D \tau)}{\pi F_D \tau} \right| \tag{37}$$

In the experiment, we use coprime sampling on both transmitted signal in matched filter and received signal. Because the length of the chirp is predefined and need to fully analyze, based on Property 2, we could only have the difference co-array of index with missing values. But since the missing values will be more often for the autocorrelation with larger values, and we have already assumed $T_M=0$ making the AF located relative to the time axis, there is not obvious effect of the missing values for the image generated by coprime sampled AF. The following simulation also confirms this claim.

From Figure3, we can see that when we use small values of coprime pair in the upper right plotting, the resulting AF has inconspicuous degradation comparing with the upper left one, which is derived directly from formula. But as the values of coprime pair increase, there will be duplicated aliasing parts getting closer to the correct estimation. When we choose $M=9$ and $N=7$, the aliasing parts could still be easily eliminated, but when the pair becomes $M=10$ and $N=11$, or even bigger, the resulting AF is unable to use because all of the estimations overlap with each other.

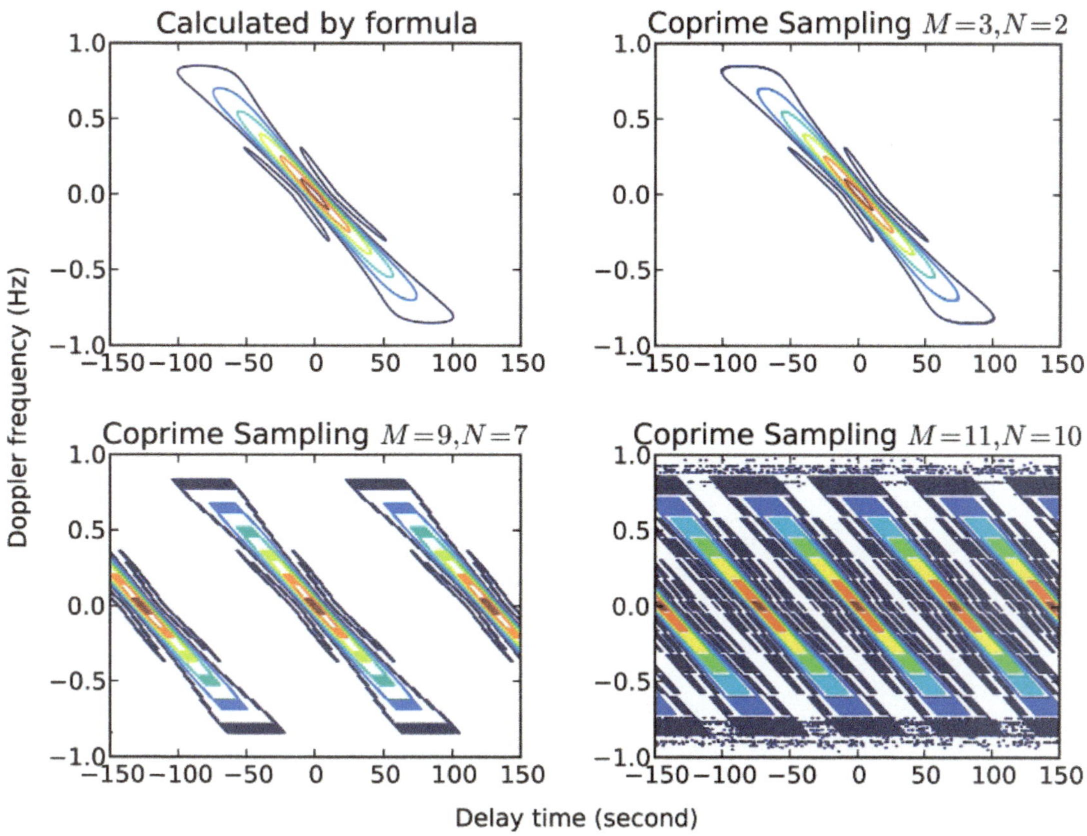

Figure 3: Comparison of different choices of coprime sampling for AF. The vertical axes of all the four rows in this figure are Doppler frequency in Hz, and all horizontal axes indicate time delay in second. The comparison among these rows show the validity of using coprime sampling with AF and its properties.

Then, based on Figures4 and5, we can observe different effects of the coprime sampling to the estimate of Doppler shift and time delay. Both of them are generated simultaneously with Figure3. In Figure4, because the coprime sampling is implemented in the time domain, the variation becomes more and more obvious as the increase of coprime factors. We have thoroughly discuss the reason of this phenomenon in the previous section. In Figure5, since we keep the iteration along the Doppler axis the same, there is no variation existing. As the values of coprime pair increase, however, the distance between Doppler shift becomes smaller and smaller. Hence, we can conclude that as the increase of values of coprime pair, it will have deleterious effects including amplifying variation along time axis and decrease the scope of Doppler shift frequency.

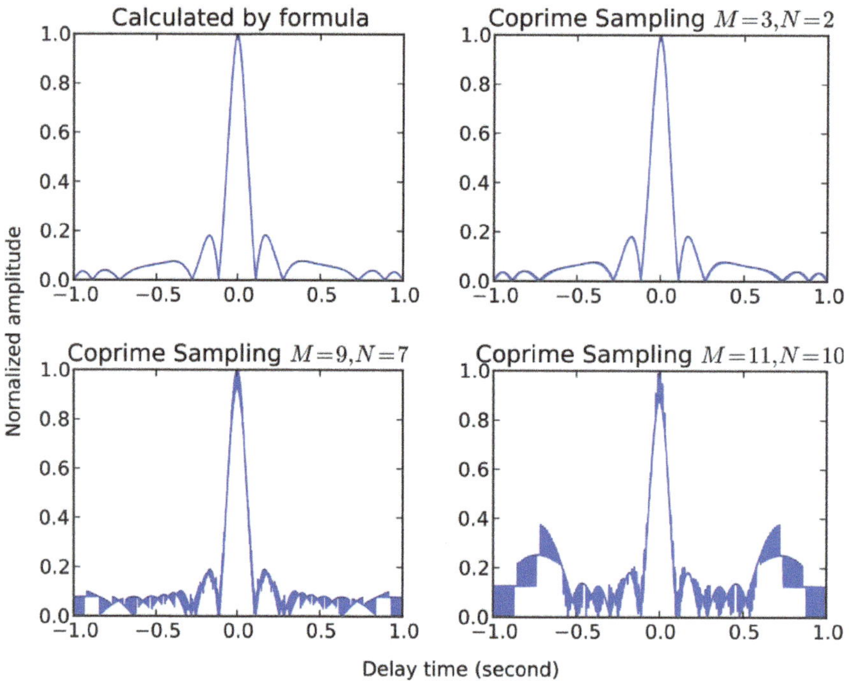

Figure 4: Zero Doppler shift ambiguity function. The vertical axes of all the four rows in this figure are normalized amplitude, and all horizontal axes indicate delay time in second. The comparison among these rows show the variation becomes more and more obvious as the increase of coprime factors.

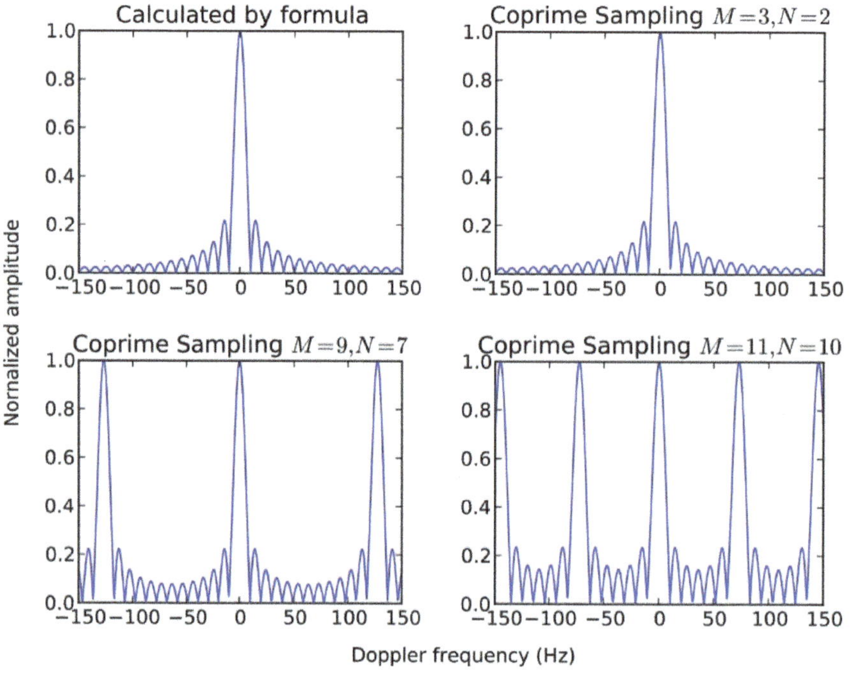

Figure 5: Zero-delay ambiguity function. The vertical axes of all the four rows in this figure are normalized amplitude, and all horizontal axes indicate Doppler frequency in Hz. The comparison among these rows show that as the increase of values of coprime pair, the scope of Doppler shift frequency decreases obviously.

To further quantify the effect of coprime sampling, we enumerate all coprime combinations under 17. The reason that we choose the threshold as 17 is because if the values of pair above this threshold severe overlaping of aliasing parts make the output useless. Besides, as shown in the following experiments, we find most of the results could be consistently arranged according to the products of coprime pairs. That is, four out of five important properties of coprime sampling AF are relevant with the product of coprime pairs rather than the value of either factor.

The distance between main lobes in Doppler axis determines the scope of Doppler frequency. From Figure6, we can see that this distance is decreasing monotonically from out-of-scope to about 33 Hz along with the increase of the product of coprime pair. Considering the width of main lobe provided in Figure7, for the case of 33 Hz distance, the second lobes of two AF estimations would overlap together. Note that for product less than 50, there will be no duplicated main lobe in the scope. For the worst case, the largest side lobes of each duplicate have overlapped together.

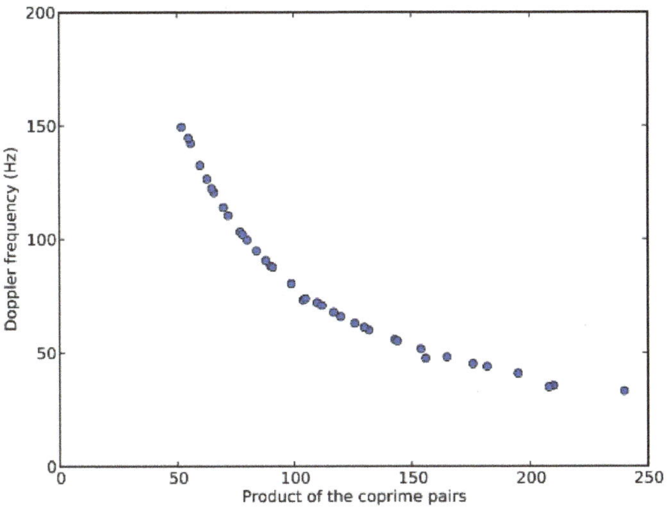

Figure 6: Distance between nearest main lobes in Doppler axis. The vertical axis of this figure is Doppler frequency in Hz, and horizontal axis indicates the product of coprime pair. It shows that the distance between main lobes of Doppler estimation is decreasing monotonically from out-of-scope to about 33 Hz along with the increasing of the product of coprime pair.

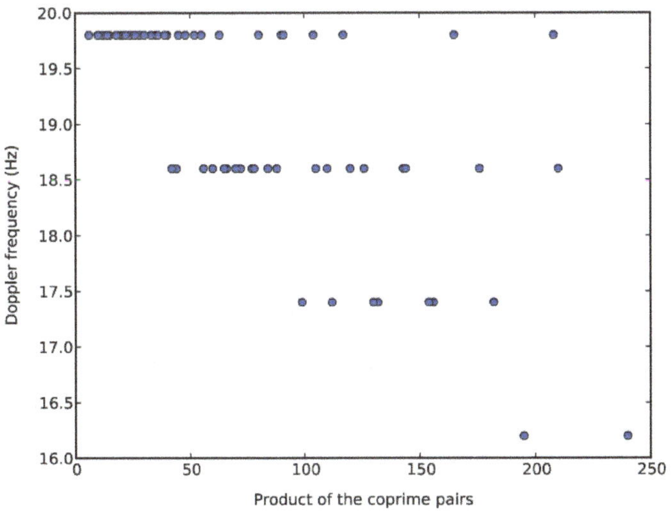

Figure 7: The width of major main lobe in Doppler axis. The vertical axis of this figure is Doppler frequency in Hz, and horizontal axis indicates the product of coprime pair. It shows that the width of main lobe of Doppler estimation is decreasing discretely along with the increasing of the product of coprime pair.

The width of the main lobe in Doppler axis determines the Doppler resolution. In the Figure7, its range is from 19.8 to 16.2 Hz. The width has only three discrete possible values and does not directly relevant with the product of coprime pair, though the general trend of width is getting smaller with larger products. This finding is instructive to find such coprime pair with narrow main lobe width but also less variation in time domain and longer distance among main lobes in Doppler axis.

Despite the largest side lobes in the Doppler axis become larger along with increasing coprime pairs, as shown in Figure8, this is still not the major challenge comparing with the main lobes approaching to each other shown in Figure6. Note that there is one abnormal value generated by M=14 and N=11. But it is more like a cutting-off main lobe located in the edge of scope rather than a real side lobe.

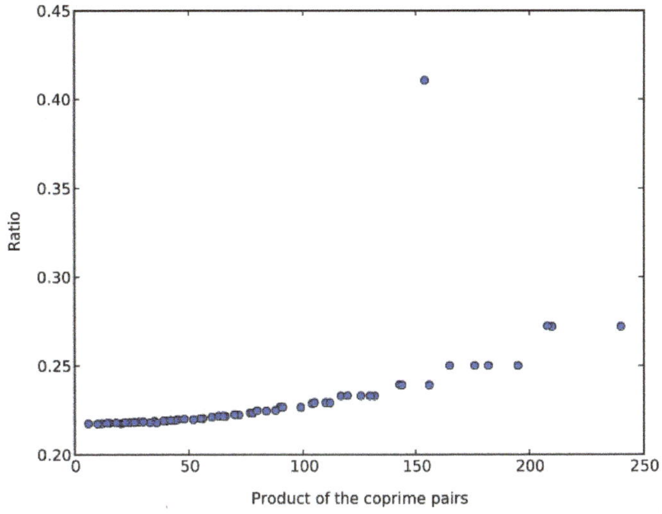

Figure 8: Ratio of 2nd side lobes to main lobes in Doppler axis. The vertical axis of this figure is the proportion of the largest side lobe to main lobe in Doppler axis, and horizontal axis indicates the product of coprime pair. It shows that the increase of side lobes is not obvious in Doppler axis along with the increasing of the product of coprime pair.

In Figure9, the radiated shape shows no obvious relationship between the trend of main lobe and the choice of coprime pairs.

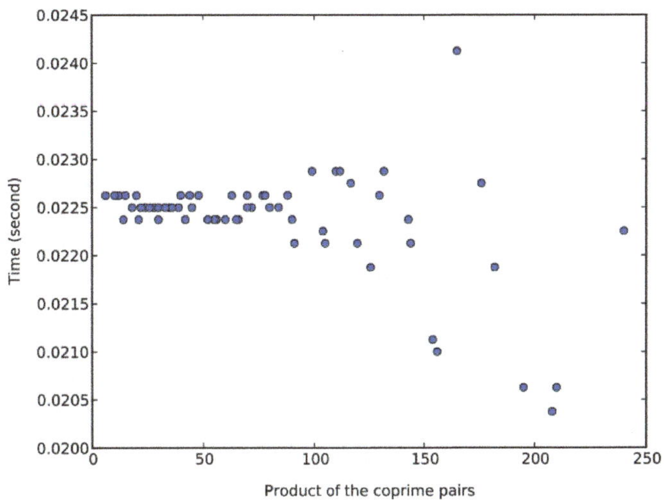

Figure 9: The width of the main lobe in time domain. The vertical axis of this figure is delay time in second, and horizontal axis indicates the product of coprime pair. It shows that the width of main lobe of estimation about delay time does not follow the increasing of the product of coprime pair.

Comparing Figure10 with Figure8, we can see the main problem in time domain is caused by the variation, which in turn make the largest side lobes comparable to the main lobe. Note that there is a turning point in the production of 88 for the ratio changing from stable around 18 percent to increasing with the production.

CONCLUSIONS AND FUTURE RESEARCH

In this article, we develop the algorithm STFT-CS to deal with non-stationary signal. The decreasing of processed data is favorable for sparse sampling as well as decreasing the computation complexity, but the cost is increasing estimate variation. To alleviate the side-effects, we introduce two-steps STFT-CS. The simulation indicates it is effective to eliminate aliasing estimations.

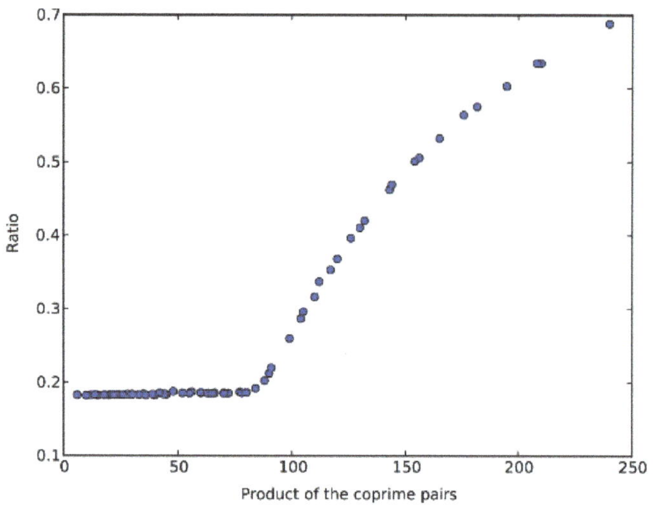

Figure 10: Ratio of 2nd side lobes to main lobes in time domain. The vertical axis of this figure is the proportion of the largest side lobe to main lobe in delay time axis, and horizontal axis indicates the product of coprime pair. It shows that the width of side lobes retain in certain level and then increase with the increasing of the product of coprime pair.

Besides, we also implement the coprime sampling with the matched filter of radar signal processing, and quantify the effect of coprime sampling in such process. Based on our analysis, one could integrate the coprime sampling in radar system to detect targets, and choose the suitable configuration based on specific circumstance and needs.

The future research directions include further optimizing the algorithm and using it with real-world radar data. Besides, coprime sampling and coprime sensor array do have many interesting features which might be useful for other applications, such as wireless communication or image/audio signal processing. Moreover, just as using STFY-CS converting time domain signal to more meaningful PSD representation, coprime sampling could be regarded as preprocessing for contaminant data to restore the fundamental information.

ACKNOWLEDGEMENTS

This study was supported by Office of Naval Research under Grants N00014-13-1-0043, N00014-11-1-0071, N00014-11-1-0865, and U.S. National Science Foundation under Grants CNS-1247848, CNS-1116749, CNS-0964713.

Competing Interests

The authors declare that they have no competing interests.

REFERENCES

1. Lang S, Duckworth G, McClellan J: Array design for MEM and MLM array processing. In*IEEE International Conference on Acoustics, Speech, and Signal Processing, ICASSP'81*. Atlanta, Georgia; 1981:145-148.
2. Hastie T, Tibshirani R, Friedman J: *The Elements of Statistical Learning: Data Mining, Inference, and Prediction*. Springer, Berlin: Springer Science+Business Media Deutschland GmbH, Heidelberger Platz 3, 14197; 2009.
3. Moffet A: Minimum-redundancy linear arrays. *IEEE Trans. Antennas Propag* 1968, 16(2):172-175. 10.1109/TAP.1968.1139138
4. Bedrosian S: Nonuniform linear arrays: Graph-theoretic approach to minimum redundancy. *Proc. IEEE* 1986, 74(7):1040-1043.
5. Pearson D, Pillai S, Lee Y: An algorithm for near-optimal placement of sensor elements.*IEEE Trans. Inf. Theory* 1990, 36(6):1280-1284. 10.1109/18.59928
6. Pal P, Vaidyanathan P: Nested arrays: a novel approach to array processing with enhanced degrees of freedom. *IEEE Trans. Signal Process* 2010, 58(8):4167-4181.
7. Chen J, Liang Q, Wang J, Choi HA, Wang X, Zheng R: Spectrum efficiency of nested sparse sampling. In *Wireless Algorithms, Systems, and Applications vol. 7405 of Lecture Notes in Computer Science*. Edited by: Jing T, Xing K. Berlin Heidelberg: Springer; 2012:574-583. 10.1007/978-3-642-31869-6_50
8. Xia XG: On estimation of multiple frequencies in undersampled complex valued waveforms. *IEEE Trans. Signal Process* 1999, 47(12):3417-3419. 10.1109/78.806088
9. Li G, Xu J, Peng YN, Xia XG: Location and imaging of moving targets using nonuniform linear antenna array SAR. *IEEE Trans. Aerosp. Electron. Syst* 2007, 43(3):1214-1220.
10. Vaidyanathan P, Pal P: Sparse sensing with co-prime samplers and arrays. *IEEE Trans. Signal Process* 2011, 59(2):573-586.
11. Vaidyanathan P, Pal P: Theory of sparse coprime sensing in multiple dimensions. *IEEE Trans. Signal Process* 2011, 59(8):3592-3608.
12. Pal P, Vaidyanathan P: Coprime sampling and the music algorithm. In *2011 IEEE Digital Signal Processing Workshop and IEEE Signal Processing Education Workshop (DSP/SPE)*. Sedona, Arizona; 2011:289-294.
13. Nagell T: *Introduction to Number Theory*. Providence, RI: American Mathematical Society; 2001.
14. Marriott F, Pope J: Bias in the estimation of autocorrelations. *Biometrika* 1954, 41: 390-402.
15. Kay S: The effect of sampling rate on autocorrelation estimation. *IEEE Trans. Acoustics Speech Signal Process* 1981, 29(4):859-867. 10.1109/TASSP.1981.1163634
16. Lomnicki Z, Zaremba S: On the estimation of autocorrelation in time series. *Annals Math. Stat* 1957, 28: 140-158. 10.1214/aoms/1177707042
17. Bartlett M: On the theoretical specification and sampling properties of autocorrelated time-series. *Suppl. J. Royal Stat. Soc* 1946, 8: 27-41. 10.2307/2983611
18. Skolnik M: *Introduction to Radar Systems*. P.O. Box 182605, Columbus, OH 43218: McGraw-Hill Higher Education's Mathematics for Science/Engineering and Advanced Mathematics, McGraw-Hill Education; 2002.
19. Richards MA: *Fundamentals of Radar Signal Processing*. P.O. Box 182604, Columbus, OH 43272: The McGraw-Hill Companies; 2005.

CHAPTER 10

UNCERTAINTY PRINCIPLE OF THE LOCAL POLYNOMIAL FOURIER TRANSFORM

Xiumei Li, Guoan Bi and Shenghong Li

Keywords: Uncertainty principleshort-time, Fourier transformlocal polynomial, Fourier transform, Wigner-Ville distribution

INTRODUCTION

The Fourier transform is an important tool to obtain signal representation in the frequency domain. However, it is not suitable to deal with time-varying signals that contain frequencies changing with time. To describe the frequency characteristics of such signals, the short-time Fourier transform (STFT), defined as

$$\mathrm{STFT}(t,\omega) = \int_{-\infty}^{+\infty} s(\tau)h(\tau - t)\exp\left\{-j\omega\tau\right\}d\tau, \quad (1)$$

is widely used.[a] The basic assumption of the STFT is that the frequencies in each signal segment are not changed with time so that the frequency variations with time are approximately described by the Fourier transforms of the successive signal segments. Because this assumption is not generally true, however, the resolution of the STFT in the time-frequency domain is often limited.

To overcome this drawback, other time-frequency transforms such as fractional Fourier transform [1, 2] and local polynomial Fourier transform (LPFT) [3] were proposed. As a generalization of the STFT, the kernel of the LPFT uses extra parameters to approximate the phase of the signal into a polynomial form. With these parameters, or the polynomial coefficients, the LPFT can describe the time-varying frequencies with a better accuracy, and therefore the resolution of signal representation in the time-frequency domain can be significantly improved compared with that achieved by the STFT. Because the LPFT is a linear transform, it is free from the cross terms which may exist in some other time-frequency representations. Due to its desirable advantages, the LPFT has found applications in various areas. For example, it has been used to improve radar images of fast maneuvering targets [4,5]. It has been applied for non-stationary interference suppression in noise radar systems [6], as well as in spread spectrum communications [7], to achieve improved performance compared with that obtained by the STFT. In [8], the LPFT-based beamformer is derived for source localization and tracking in nonstationary environment, which can resolve closely spaced sources when their velocities are sufficiently different. Moreover, the LPFT can be used as an instantaneous frequency estimator, and the corresponding asymptotic covariance matrix and bias of the estimates are studied in [3]. A comprehensive review on the LPFT and its applications can be referred to [9].

In the classical time-frequency domain, the product of time duration and bandwidth cannot be arbitrarily small and is generally described in the form of the uncertainty principle [10]. The uncertainty principle plays an important role in signal analysis [10], and studies on the uncertainty principle of various transforms have been reported in the literature [1, 11–16]. For example, it was reported that the STFT is limited by the uncertainty principle [17]. The local uncertainty principle was demonstrated in [12] by defining local quantities as conditional standard deviations, while the standard formulation of the uncertainty principle is in terms of global standard deviation. It shows that the local uncertainty product of the spectrogram

Citation: Li, X., Bi, G. & Li, S. EURASIP J., Uncertainty Principle of the Local Polynomial Fourier Transform, doi:10.1186/1687-6180-2012-120.

indeed has a lower bound due to the windowing approach but it is not a property of the signal or a fundamental limit. For a large class of bilinear time-frequency distributions, the local uncertainty product is always less than or equal to the global uncertainty product and can be arbitrarily small [12].

It has been observed that the resolution of the LPFT in the time-frequency domain is also influenced by the window length which controls the tradeoff of bias and variance [3, 18]. Therefore, it is believed that there must exist some form of the uncertainty principle based on the LPFT. The uncertainty principle of the second order LPFT was only recently discussed in [19]. It was shown that the uncertainty product of the second order LPFT for the second order PPS is time-independent when Gaussian window is used to segment the signal. However, a comprehensive study on the uncertainty principle for an arbitrary order LPFT has not been reported in the literature.

Similar to the steps used to derive the uncertainty principle of the STFT [17], this article firstly reports the mathematical derivations of the uncertainty principles of the second, third, fourth order and thereafter M th order LPFTs. It is found that the uncertainty product of an arbitrary order LPFT is related to the parameters of the signal and the window function, as well as the errors of estimating the polynomial coefficients. When the polynomial coefficients of the transform kernel are accurately estimated, the uncertainty product becomes a constant if a Gaussian window is used to segment the input signal. Based on the derived signal duration and bandwidth, we can obtain the uncertainty product. Then we will discuss the effects, in terms of the resolution in the time-frequency domain and the minimization of the required computational complexity, of window width, order mismatch, i.e., using the second order LPFT to process higher-order PPS, and the length of overlap between signal segments. Comparisons are also made on the resolutions of signal representations in the time-frequency domain achieved by using the STFT, the Wigner-Ville distribution (WVD) and the second order LPFT.

The rest of the article is organized as follows. With the similar terminologies used in deriving the uncertainty principle of the STFT in Appendix [17], the uncertainty principles of the LPFTs of various orders are given in Section 2. Section 3 discusses the effects of various issues on obtaining desirable signal representations in the time-frequency domain. Signal concentrations of different time-frequency representations are also compared. Finally conclusions are given in Section 4.

UNCERTAINTY PRINCIPLES OF THE LPFTS

In this section, the definition of the LPFT is given and some important issues that affect the computational complexity are addressed. Then we will give the derivation of the uncertainty principles of different order LPFTs and achieve the general expression of the uncertainty principle for the Mth order LPFT.

The LPFT

The LPFT is a generalization of the STFT, and the M th-order LPFT is defined as [3, 9]:

$$\begin{aligned} \mathrm{LPFT}(t,\omega) &= \int s(t+\tau)h(\tau)\exp\left\{-j\omega\tau - j\sum_{m=2}^{M}\frac{\omega_{m-1}\tau^m}{m!}\right\}d\tau, \\ &= \int s(\tau)\exp\left\{j\omega t - j\sum_{m=2}^{M}\frac{\omega_{m-1}(\tau-t)^m}{m!}\right\}h(\tau-t)\exp\{-j\omega\tau\}\,d\tau, \end{aligned} \tag{2}$$

where M is the order of the polynomial function, $\omega_2 = \frac{d^2\Omega(t)}{dt^2}$, $\omega_2 = \frac{d^2\Omega(t)}{dt^2}$, .., and $\omega_{M-1} = \frac{d^{M-1}\Omega(t)}{dt^{M-1}}$ are the polynomial coefficients, and Ω(t) is the instantaneous frequency of the signal.

The LPFT is particularly suited to process the polynomial phase signals (PPSs) with a Gaussian amplitude defined by

$$s(t) = \left(\frac{\alpha}{\pi}\right)^{1/4}\exp\left\{-\frac{\alpha t^2}{2}\right\}\exp\left\{j\sum_{m=1}^{P}\frac{a_{m-1}t^m}{m!}\right\}, \tag{3}$$

where P is the order of the PPSs. To compute the LPFT in (2), the polynomial coefficients $\omega_1, \omega_2, .., \omega_{M-1}$ are first estimated and then used to form the polynomial phase. The LPFT is obtained by using an STFT procedure whose input is obtained by modulating s(τ) with $\exp\left\{j\omega t - j\sum_{m=2}^{M}\frac{\omega_{m-1}(\tau-t)^m}{m!}\right\}$.

A few important issues, that have direct impacts on the resolution of time-frequency representation and the required computational complexity, have to be addressed. It is always desired to accurately estimate ω_1, ω_2,.., and ω_{M-1} so that a high resolution of the signal representation in the time-frequency domain can be obtained. The polynomial time frequency transform (PTFT) [20, 21], which is considered as a maximum likelihood estimator, can be used to estimate the polynomial coefficients used in (2). The PTFT is a transform which converts a 1-D input signal into a multi-dimensional output from which the polynomial coefficients are estimated. The maximum number of polynomial coefficients that can be estimated is the same as the order of the PTFT. Figure 1 shows a 2D PTFT of two chirp components. From the location coordinates of the two peaks, the values of ω and ω_1 for the two chirp components can be approximately achieved.

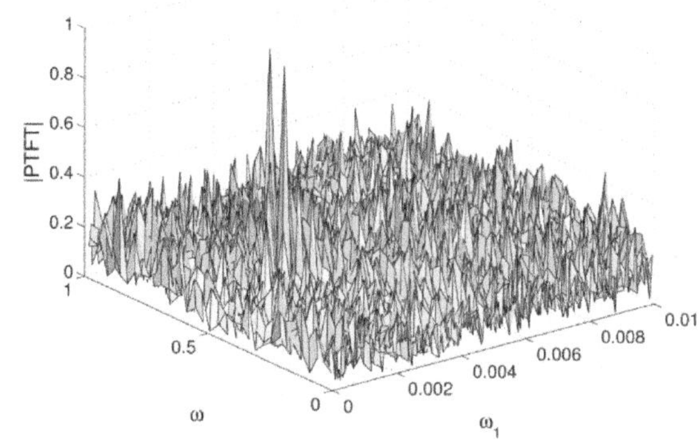

Figure 1: 2D PTFT of two chirp components. From the location coordinates of the two peaks, the values of ω and ω_1 for the two chirp components can be approximately achieved.

When M, the order of the LPFT, is large, one practical problem is that the corresponding PTFT requires a heavy computational load even using the fast algorithms reported in [22, 23]. Based on the radix-2 fast algorithm for the PTFT [22] and the fast Fourier transform (FFT) algorithm [24–26] for the STFT computation, the number of complex multiplications needed by the LPFT is approximately

$$C_{\mathrm{LPFT}}(N) \simeq \left[C^{r}_{\mathrm{PTFT}}(q) + \frac{q}{2}\log_2 q + 0.75q\right]\frac{N}{q-l}, \tag{4}$$

where N is the length of the input signal, q is the length of the window (or signal segment), andl, $0 \le l \le q - 1$, is the length of overlap between signal segments. In (4), $C^r_{PTFT}(q)$ is the number of complex multiplications needed by the r th order PTFT of length q, the second term in the square brackets is the number of complex multiplications for an FFT of length q, the third term in the square brackets is the number of complex multiplications for windowing and modulation operations, and the factor outside the square brackets is approximately the number of the segments used for the LPFT. It was reported that for the r th order PTFT of length N, where r = 2 or 3, the required number of complex multiplications are in the order of N^r [22]. Because the PTFT requires a heavy computational complexity, especially when N is large, it is important to find various methods to minimize the computational complexity for the parameter estimation by using the PTFT.

Similar to the concept of the STFT which uses Fourier transform to deal with signal segments that contain time-varying frequencies, we can use a smaller number of the polynomial coefficients in the LPFT to approximate the frequency contents of the signal segments. If the second order LPFT is used to process higher-order PPSs, as shown in [27, 28] for example, the number of polynomial coefficients to be estimated is reduced, which directly leads to the reduction of the PTFT order, or equivalently the reduction of the required computational complexity. We will discuss the side effects of this order mismatch, i.e., the order of the LPFT is smaller than the order of the PPSs, on the resolution of the signal representation in the time-frequency domain. Other practical issues, which also exist in the STFT, including the effects of window width and the length of overlap between adjacent signal segments will be discussed. The computational complexity can be significantly reduced if the overlap length can be minimized without obviously degrading the resolution of the time-frequency representation.

Uncertainty Products of the LPFTs

This subsection first presents the uncertainty principles for the lower order LPFTs, i.e., M = 2, 3 and 4. Then, a general expression of the uncertainty principle for the M th order LPFT is deduced.

It is assumed that the input signals are PPSs with the same orders of the LPFTs, i.e., M in (2) is equal to P in (3). The corresponding normalized local signal is

$$\eta_t(\tau) = \frac{s(\tau)h(\tau - t)\exp\left\{j\omega t - j\sum_{m=2}^{M}\frac{\omega_{m-1}(\tau-t)^m}{m!}\right\}}{\sqrt{\int |s(\tau)h(\tau-t)|^2 d\tau}}, \tag{5}$$

where

$$h(t) = \left(\frac{a}{\pi}\right)^{1/4}\exp\left\{-\frac{at^2}{2}\right\}, \tag{6}$$

is a Gaussian window used to obtain the signal segments and a > 0 is the parameter to control the window width.

With the above defined input signal and the window function, the normalization factor, i.e., the integral in the denominator of (5), becomes

$$\int |s(\tau)h(\tau-t)|^2 d\tau = \left(\frac{a\alpha}{\pi(a+\alpha)}\right)^{1/2}\exp\left\{-\frac{a\alpha t^2}{a+\alpha}\right\}. \tag{7}$$

Therefore, the normalized local signal is expressed as

$$\eta_t(\tau) = \left(\frac{\alpha+a}{\pi}\right)^{1/4} \cdot \exp\left\{-\frac{a^2t^2}{2(\alpha+a)}\right\}\exp\left\{-\frac{(\alpha+a)\tau^2}{2} + a\tau t + j\sum_{m=1}^{P}\frac{a_{m-1}t^m}{m!}\right\} \cdot \exp\left\{j\omega t - j\sum_{m=2}^{M}\frac{\omega_{m-1}(\tau-t)^m}{m!}\right\}. \tag{8}$$

For an arbitrary order PPS, it is easy to derive

$$|\eta_t(\tau)|^2 = \left(\frac{\alpha+a}{\pi}\right)^{1/2}\exp\left\{-(\alpha+a)\left(\tau - \frac{at}{\alpha+a}\right)^2\right\}, \tag{9}$$

which is independent of the order of the PPS. With (32) in Appendix, the mean time, $\langle\tau\rangle_t$, of the normalized local signal becomes

$$\langle\tau\rangle_t = \frac{at}{a+\alpha}.$$

Similarly, we also obtain

$$\langle\tau^2\rangle_t = \frac{1}{2(a+\alpha)} + \frac{a^2t^2}{(a+\alpha)^2}. \tag{10}$$

Therefore the time duration of the normalized local signal, which is defined in (33) in Appendix, is expressed as

$$T_t^2 = \frac{1}{2(a+\alpha)}. \tag{11}$$

Since both $\langle\tau\rangle_t$ and T_t^2 are related to $|\eta_t(\tau)|$ instead of $\eta_t(\tau)$, they are independent of the order of the input signal. It means that the signal duration in the time domain is determined by the signal parameter α and window parameter a only.

Let us now compute the mean frequency $\langle\omega\rangle_t$ and the bandwidth B_t^2 for the LPFTs of various orders. For the second order LPFT, the local signal is

$$s_t(\tau) = s(\tau)\, h(\tau - t) \exp\left\{ j\omega t - j\frac{\omega_1 (\tau - t)^2}{2} \right\}, \tag{12}$$

where s(t), the input of the second order LPFT, is the second order PPS or the chirp signal, obtained by setting P = 2 in (3), and ω_1 is estimated by using the second order PTFT. In this case, the Fourier transform of the local signal, $s_t(\tau)$, becomes the second order LPFT of s(t).

The normalized local signal for P = 2 in (8) at the time t becomes

$$\begin{aligned} \eta_t(\tau) &= \frac{s(\tau)\, h(\tau - t) \exp\left\{ j\omega t - j\frac{\omega_1(\tau - t)^2}{2} \right\}}{\sqrt{\int |s(\tau)\, h(\tau - t)|^2 d\tau}} \\ &= \left(\frac{\alpha + a}{\pi}\right)^{1/4} \exp\left\{ -\frac{(\alpha + a)\tau^2}{2} + a\tau t + j a_0 \tau + \frac{j a_1 \tau^2}{2} - \frac{a^2 t^2}{2(\alpha + a)} \right\} \\ &\cdot \exp\left\{ j\omega t - \frac{j\omega_1(\tau - t)^2}{2} \right\}. \end{aligned}$$

With the definition in (35), we have

$$\langle\omega\rangle_t = a_0 + \omega_1 t + \frac{at(a_1 - \omega_1)}{a + \alpha} = a_0 + \omega_1 t + (a_1 - \omega_1)\langle\tau\rangle_t, \tag{13}$$

in which the first term is related to the signal parameter a_0, the second term is related to ω_1, and the last term is a product of the deviation of ω_1 from a_1 and the mean time, $\langle\tau\rangle_t$.

Similarly, by using (37) in Appendix, we have

$$\langle\omega^2\rangle_t = \frac{a + \alpha}{2} + \langle\omega\rangle_t^2 + \frac{(a_1 - \omega_1)^2}{2(a + \alpha)}, \tag{14}$$

which shows that the frequency variance is determined by the parameters of the signal and the window in the first term, the square of mean frequency in the second term, and the scaled square of the estimation error of the polynomial parameter ω_1 in the last term. With the above derivation results, the bandwidth in the domain of the second order LPFT becomes

$$B_t^2 = \langle\omega^2\rangle_t - \langle\omega\rangle_t^2 = \frac{a + \alpha}{2} + \frac{(a_1 - \omega_1)^2}{2(a + \alpha)}. \tag{15}$$

Therefore

$$B_t^2 T_t^2 = \frac{1}{4} + \frac{(a_1 - \omega_1)^2}{4(a + \alpha)^2}, \tag{16}$$

which is related to the parameters of the signal and the window, and the errors of estimating the polynomial parameters ω_1.

Let us next consider the uncertainty principle of the third order LPFT. The local signal is

$$s_t(\tau) = s(\tau)\, h(\tau - t) \exp\left\{ j\omega t - j\frac{\omega_1}{2}(\tau - t)^2 - j\frac{\omega_2}{6}(\tau - t)^3 \right\},$$

where s(t), the input of the third order LPFT, is the third order PPS obtained by setting P = 3 in (3).

In this case, the local normalized signal becomes

$$\begin{aligned} \eta_t(\tau) &= \left(\frac{\alpha + a}{\pi}\right)^{1/4} \\ &\cdot \exp\left\{ -\frac{(\alpha + a)\tau^2}{2} + a\tau t + j a_0 \tau + \frac{j a_1 \tau^2}{2} + \frac{j a_2 \tau^3}{6} - \frac{a^2 t^2}{2(\alpha + a)} \right\} \\ &\cdot \exp\left\{ j\omega t - \frac{j\omega_1(\tau - t)^2}{2} - \frac{j\omega_2(\tau - t)^3}{6} \right\}. \end{aligned}$$

With the same procedure, the mean frequency, defined in (35) in Appendix, is expressed as

$$\langle\omega\rangle_t = a_0 + \omega_1 t - \frac{\omega_2 t^2}{2} + (a_1 - \omega_1 + \omega_2 t)\langle\tau\rangle_t + \frac{1}{2}(a_2 - \omega_2)\langle\tau^2\rangle_t, \tag{17}$$

and $\langle\omega^2\rangle_t$, defined in (37) in Appendix, becomes

$$\langle\omega^2\rangle_t = \frac{a+\alpha}{2} + \langle\omega\rangle_t^2 + \frac{1}{2(a+\alpha)}[(a_1 - \omega_1 + \omega_2 t) + (a_2 - \omega_2)\langle\tau\rangle_t]^2 + \frac{(a_2-\omega_2)^2}{8(a+\alpha)^2} \tag{18}$$

Thus, the bandwidth becomes

$$B_t^2 = \langle\omega^2\rangle_t - \langle\omega\rangle_t^2 = \frac{a+\alpha}{2} + \frac{[(a_1 - \omega_1 + \omega_2 t) + (a_2 - \omega_2)\langle\tau\rangle_t]^2}{2(a+\alpha)} + \frac{(a_2-\omega_2)^2}{8(a+\alpha)^2}. \tag{19}$$

Therefore

$$B_t^2 T_t^2 = \frac{1}{4} + \frac{[(a_1 - \omega_1 + \omega_2 t) + (a_2 - \omega_2)\langle\tau\rangle_t]^2}{4(a+\alpha)^2} + \frac{(a_2-\omega_2)^2}{16(a+\alpha)^3}. \tag{20}$$

It is observed that the uncertainty product, $B_t^2 T_t^2$, is affected by the errors of estimating the polynomial parameters ω_1 and ω_2 in addition to the parameters of the signal and the window function.

By setting P = 4 in (3) and using the fourth order LPFT, we can similarly obtain the corresponding mean frequency $\langle\omega\rangle_t$ and the bandwidth B_t^2, which are listed in Table 1respectively. From the expressions in the table, we can observe the rules of $\langle\omega\rangle_t$ and B_t^2 as M, the order of the LPFT, is increased. The mean frequency and the bandwidth for the M th-order LPFT are

Table 1: Expressions of $\langle\omega\rangle_t$ and B_t^2 for LPFTs of order 2, 3 and 4

M	
	$\langle\omega\rangle_t$
2	$a_0 + \omega_1 t + (a_1 - \omega_1)\langle\tau\rangle_t$
3	$a_0 + \omega_1 t - \frac{\omega_2 t^2}{2} + (a_1 - \omega_1 + \omega_2 t)\langle\tau\rangle_t + \frac{1}{2}(a_2 - \omega_2)\langle\tau^2\rangle_t$
4	$a_0 + \omega_1 t - \frac{\omega_2}{2}t^2 + \frac{\omega_3}{6}t^3 + (a_1 - \omega_1 + \omega_2 t - \frac{\omega_3}{2}t^2\langle\tau\rangle_t + \frac{1}{2}(a_2 - \omega_2 + \omega_3 t)\langle\tau^2\rangle_t + \frac{1}{6}(a_3 - \omega_3)\langle\tau^3\rangle_t$
	B_t^2
2	$\frac{a+\alpha}{2} + \frac{1}{2(a+\alpha)}(a_1 + \omega_1)^2$
3	$\frac{a+\alpha}{2} + \frac{1}{2(a+\alpha)}[(a_1 - \omega_1 + \omega_2 t) + (a_2 - \omega_2)\langle\tau\rangle_t]^2 + \frac{1}{8(a+\alpha)^2}(a_2 - \omega_2)^2$
4	$\frac{a+\alpha}{2} + \frac{1}{2(a+\alpha)}[\left(a_1 - \omega_1 + \omega_2 t - \frac{\omega_3}{2}t^2\right) + (a_2 - \omega_2 + \omega_3 t)\langle\tau\rangle_t + \frac{1}{2}(a_3 - \omega_3)\langle\tau^2\rangle_t]^2 + \frac{1}{8(a+\alpha)^2}(a_2 - \omega_2 + \omega_3 t) + (a_3 - \omega_3)\langle\tau\rangle_t]^2 + \frac{1}{48(a+\alpha)^3}(a_3 - \omega_3)^2$

The lengthy mathematical derivation of the above two equations is omitted due to the space limit. It is observed that both $\langle\omega\rangle_t$ and B_t^2 are related to the moments $\langle\tau^n\rangle_t$. The expressions of the moments $\langle\tau^n\rangle_t$, for n = 1, 2, .., 6, which can be calculated from (34), are listed in Table 2.

Table 2: Expressions of $\langle\tau^n\rangle_t$ for n = 1, 2, 3 4, 5 and 6

n	$\langle\tau^n\rangle$
1	$\frac{at}{a+\alpha}$
2	$\frac{1}{2(a+\alpha)}+\left(\frac{at}{a+\alpha}\right)^2$
3	$\frac{3at}{2(a+\alpha)^2}+\left(\frac{at}{a+\alpha}\right)^3$
4	$\frac{3}{4(a+\alpha)^3}+\frac{3(at)^2}{(a+\alpha)^3}+\left(\frac{at}{a+\alpha}\right)^4$
5	$\frac{15at}{4(a+\alpha)^3}+\frac{5(at)^3}{(a+\alpha)^4}+\left(\frac{at}{a+\alpha}\right)^5$
6	$\frac{15}{8(a+\alpha)^3}+\frac{5(at)^2}{4(a+\alpha)^4}+\frac{15(at)^4}{2(a+\alpha)^5}+\left(\frac{at}{a+\alpha}\right)^6$

In general the uncertainty product of an arbitrary order LPFT is related to the parameters of the signal and the window function, as well as the errors of estimating the polynomial coefficients. When the polynomial coefficients are estimated correctly for the M th order LPFT, that is, $\omega_1=\frac{d\Omega(t)}{dt},\ldots,\omega_{M-1}=\frac{d^{M-1}\Omega(t)}{dt^{M-1}}$, the mean frequency becomes

$$\langle\omega\rangle_t = a_0 + a_1 t + \frac{a_2}{2}t^2 + \cdots + \frac{a_{M-1}}{(M-1)!}t^{M-1} = \sum_{m=1}^{M}\frac{a_{m-1}t^{m-1}}{(m-1)!} \qquad (21)$$

which is exactly the instantaneous frequency of the M th order PPS.

Without giving further details, the bandwidth in the domain of the M th order LPFT is derived as $B_t^2=\frac{a+\alpha}{2}$ when the polynomial parameters are estimated correctly, which leads to the uncertainty product to be

$$B_t^2T_t^2=\frac{1}{4}. \qquad (22)$$

Based on the above derived results, we have the following conclusions. When using the Gaussian window to segment the signal, the signal duration of the PPS is only related to the signal parameter α, the window parameter a and is independent of the order. For M > 1, the bandwidth B_t^2 is related to the parameters of the signal and the window function, as well as the errors of estimating the polynomial parameters ω_1, ω_2 ..., and ω_{M-1}. When the polynomial coefficients are estimated accurately, the uncertainty product of the M th order LPFT, with the same order PPS as the input, is independent of time or frequency, and becomes a constant. When there are errors of estimating the polynomial coefficients, the bandwidth together with the uncertainty product, are polynomial functions of these errors, which directly affect the signal representation in the time-frequency domain.

It should be noted that in this article the uncertainty principle of the LPFT is derived based on the PPSs. For other kind of time-varying signals such as sinusoidal FM signals or parabolic FM signals, we can use the window function to segment the signals to ensure that each segment within the window can be assumed to be the PPS. As long as the signal segments can be modeled as PPSs, the above presented results are still valid. In this way, the derived uncertainty principle of the LPFT can also be extended to general time-varying signals.

DISCUSSIONS

The main objective on studying various issues of the LPFT is to achieve the desired signal resolution in the time-frequency domain with minimum computational costs. This section considers a few important issues, in terms of the uncertainty principles of the LPFTs, that have direct impacts on signal resolution and the required computational complexity.

Window Width Effects

The uncertainty relations derived in the previous section show that there exists a trade-off between the resolutions in the time and frequency domains. With the assumption that the polynomial coefficients used in the LPFT are accurately estimated, the signal duration T_t^2 is inversely proportional to the sum of window width a and the signal parameter α, while the signal bandwidth B_t^2 is directly proportional to this sum, regardless of the order of the PPS and the LPFT. With different widths of the Gaussian window, Figure 2 shows the signal concentrations in the time-frequency domain achieved by using the second order LPFT to process the PPS which contains chirp components.

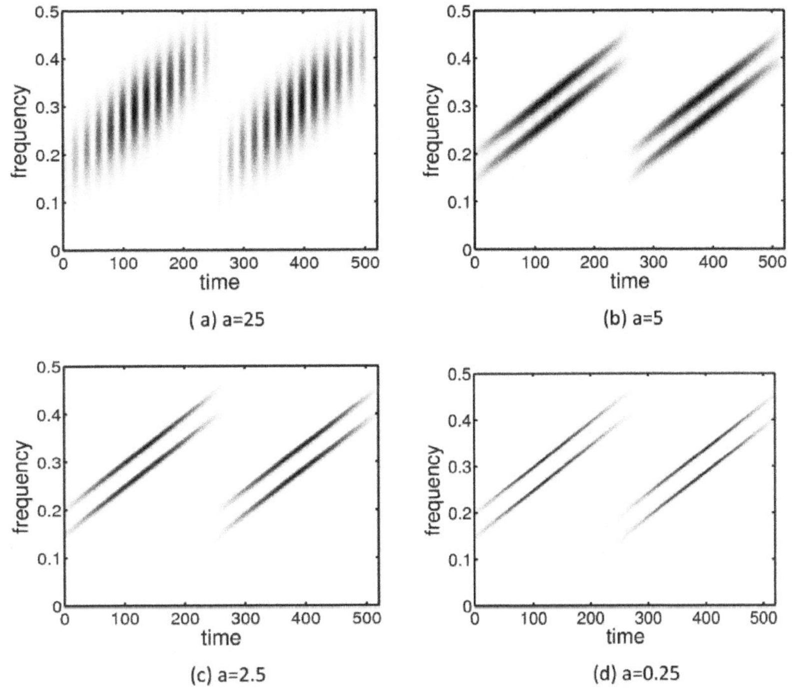

Figure 2: The second order LPFTs of a second order PPS. The window length is 1/4 of the signal length.

It is seen that, as a decreases or the window width increases, the chirp components become more concentrated in the frequency domain, or equivalently, the resolution of the signal representation in the frequency direction is increased. In contrast, the resolution in the time direction decreases. This observation is consistent with the signal duration T_t^2 and the bandwidth B_t^2 derived in the last Section. For example, decreasing the window parameter a leads to the increase of the signal duration T_t^2, or equivalently, the increase of signal spread in time. At the same time, the bandwidth B_t^2 decreases as a decreases so that the signal spread in frequency is reduced, that is, the resolution of the signal representation in the frequency direction is increased. Because it is impossible to increase the resolution in both domains at the same time, a compromise has to be made to balance the requirements of the signal representation in the time-frequency domain. When a = 2.5, for example, Figure 2c provides acceptable resolutions for both time and frequency domains, that is, the resolution in the time domain and the frequency domain is high enough to help us clearly separating the closely located chirp components.

Order Mismatch Effects

As discussed in the Section 1, it is desired to use lower order LPFTs to deal with higher-order PPSs to significantly reduce the required computational complexity. However, this mismatch between the orders of the LPFT and the PPSs will affect the performance of the signal representation in the time-frequency domain. It is necessary to find out how the signal representation in the time-frequency domain is affected and if possible, how the mismatch effects are to be minimized. Because it is difficult to draw a general conclusion for all possible cases, let us consider the use of the second order LPFT to process the third order PPSs.

By setting $\omega_2 = 0$ in (17) and (19), the mean frequency and the bandwidth become

$$\langle\omega\rangle_t = a_0 + \omega_1 t + (a_1 - \omega_1)\langle\tau\rangle_t + \frac{1}{2}a_2\langle\tau^2\rangle_t$$
$$= a_0 + \omega_1 t + \frac{at(a_1 - \omega_1)}{a+\alpha} + \frac{a_2\left(2a^2t^2 + a + \alpha\right)}{4(a+\alpha)^2},$$

$$B_t^2 = \frac{a+\alpha}{2} + \frac{1}{2(a+\alpha)}\left[(a_1 - \omega_1) + \frac{a_2at}{a+\alpha}\right]^2 + \frac{a_2^2}{8(a+\alpha)^2}.$$

When the parameter ω_1 is estimated correctly, i.e., $\omega_1 = a_1 + a_2 t$, we have

$$\langle\omega\rangle_t = a_0 + a_1 t + \frac{a_2 t^2}{2} + \frac{a^2 a_2 t^2}{2(a+\alpha)^2} + \frac{a_2}{4(a+\alpha)}, \quad (23)$$

$$B_t^2 = \frac{a+\alpha}{2} + \frac{(a_2 at)^2}{2(a+\alpha)^3} + \frac{a_2^2}{8(a+\alpha)^2}. \quad (24)$$

Therefore, the uncertainty product becomes

$$B_t^2 T_t^2 = \frac{1}{4} + \frac{1}{4(a+\alpha)^2}\left(\frac{a_2 at}{a+\alpha}\right)^2 + \frac{a_2^2}{16(a+\alpha)^3}. \quad (25)$$

In (23) or (24), the last term is the effects of the order mismatch, which are directly proportional to the polynomial coefficient, a_2. Therefore, the order mismatch may have a small effect when a_2 is small enough, which is generally true in many practical applications. Another way to reduce the effect of the order mismatch is to minimize the signal duration T_t^2 by decreasing the window width or increasing the window parameter a. For example, when a → ∞, $\langle\omega\rangle_t$ in (23) is approaching to $a_0 + a_1 t + \frac{a_2}{2}t^2$, which is the instantaneous frequency of the third order PPS. Meanwhile B_t^2 in (24) is approaching to $\frac{a+\alpha}{2}$. It means that, when the third order PPS is processed with the second order LPFT, $\langle\omega\rangle_t$ and B_t^2 of the second order LPFT approach to those of the third order LPFT as the window width is decreased. The uncertainty product in (25) also approaches to the minimum.

When the signal parameter α approaches to zero, the third order PPS with a Gaussian amplitude becomes a third order PPS with a constant amplitude. Under this situation, the mean frequency and bandwidth become

$$\langle\omega\rangle_t = a_0 + a_1 t + \frac{a_2 t^2}{2} + \frac{a_2}{4a},$$
$$B_t^2 = \frac{a}{2} + \frac{a_2^2}{8a^2}, \quad (26)$$

and the uncertainty product is

$$B_t^2 T_t^2 = \frac{1}{4} + \frac{a_2^2}{16a^3} \quad \text{as} \quad \alpha \to 0, \quad (27)$$

which increases with the increase of the polynomial coefficient a_2 and with the decrease of the window parameter a. Similarly, as a goes to infinity, $\langle\omega\rangle_t$ approaches to the instantaneous frequency, and $B_t^2 T_t^2$ approaches to $\frac{1}{4}$ which is the minimum of the uncertainty product.

From the above discussion, it is seen that increasing the window parameter a always helps to achieve a better estimation of the instantaneous frequency $\langle\omega\rangle_t$. It should be noted that, however, as a goes to infinity, that is as the window becomes narrower in the time domain, the signal bandwidth B_t^2, or the spread in frequency in (24) also approaches to infinity, which decreases the resolution in frequency. Therefore, the window width for practical applications should be properly selected to achieve a compromise between the resolutions in both time and frequency. An adaptive procedure for window length selection has been provided in [28]. Similarly, an automated procedure for window width selection is possible and will be discussed in our future study.

Effects of Overlap Lengths

It was shown in [28] that when the second order LPFT is used to process the chirp signals, the length of overlap between the adjacent signal segments can be significantly reduced without obviously degrading the signal resolution in the time-frequency domain, which allows us to further minimize the required computational complexity. For example, the required number of complex multiplications in (4) is reduced as l increases. For the third order PPS,

$$s(t) = \left(\frac{\alpha}{\pi}\right)^{1/4} e^{-\alpha t^2/2}\left[e^{j(0.4t+1.15\times10^{-3}t^2+1.48\times10^{-6}t^3)} + e^{j(0.45t+1.15\times10^{-3}t^2+1.48\times10^{-6}t^3)}\right] \quad (28)$$

with $\alpha = 10^{-4}$, Figure 3a shows the signal representation obtained by the third order LPFT without any overlap. Compared to the performance with the maximum overlap, i.e., l = q- 1, as shown in Figure 3b, no obvious degradation in signal resolution is observed compared to Figure3a. Because the total computational complexity is directly proportional to the number of signal segments being processed, the LPFT computation without overlap between signal segments makes significant savings on the computational complexity compared to that using the overlap between signal segments.

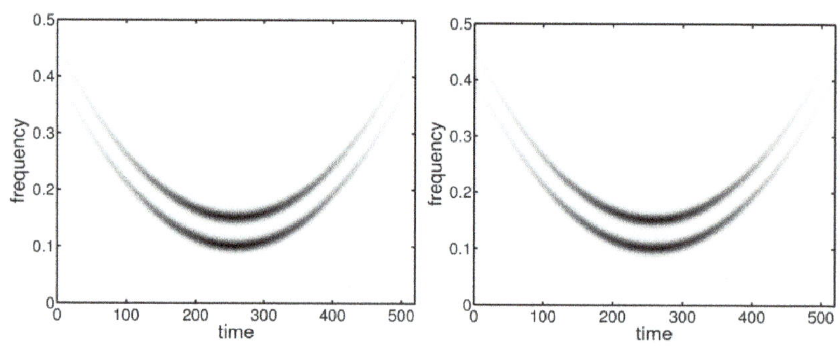

(a) (b)

Figure 3: The third order LPFTs of a third order PPS with a = 2.5 (the window length is 1/4 of the signal length). (a) The LPFT without overlap between adjacent segments, and (**b**) The LPFT with the maximum overlap.

Let us next consider the effects of using lower order LPFT with different overlap lengths. Figure4a shows the representation of the third order PPS obtained by using the second order LPFT without any overlap. Obvious unsmoothness is observed at the junctions of two adjacent segments where the values of the estimated chirp rates have sudden changes. This order mismatch effect can be minimized by increasing the length of overlap between the adjacent segments. As shown in Figure 4b, the second order LPFT with the maximum overlap achieves a representation that is comparable to that obtained by using the third order LPFT in Figure 3b.

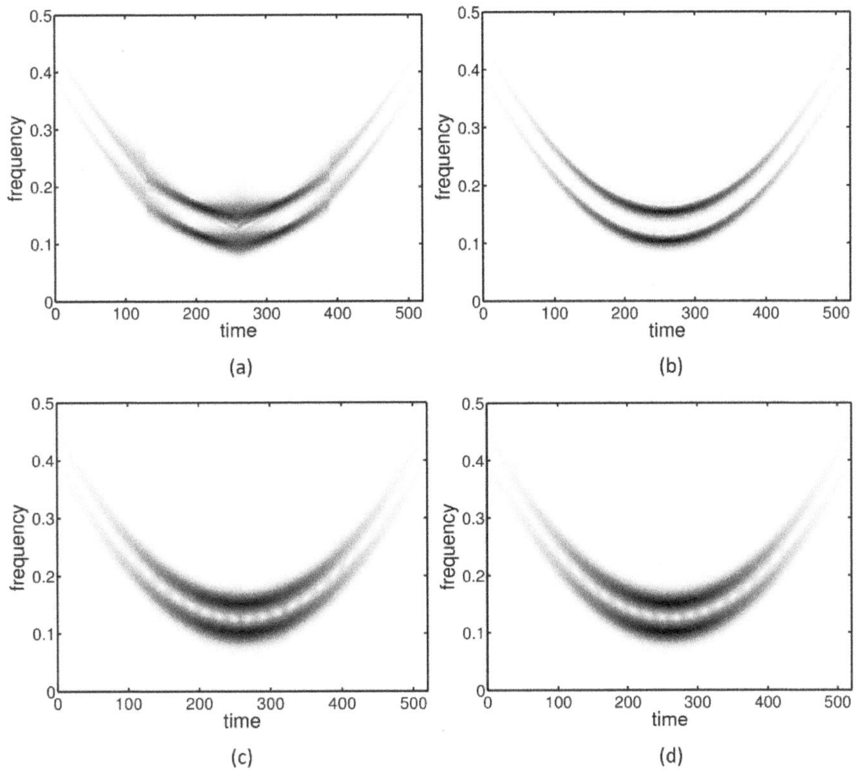

Figure 4: The LPFTs of the third order PPS with a= 2.5. (**a**) The second order LPFT without overlap, (**b**) the second order LPFT with the maximum overlap, (**c**) the second order LPFT without overlap and with a shorter window, and (**d**) the third order LPFT without overlap. The window length for (**a**) and (**b**) is 1/4 of signal length and that for (**c**) and (**d**) is 1/8 of signal length.

To process higher-order PPSs with the second order LPFT, another method for improvement on signal representation is to reduce the window length to ensure that each segment can be approximately assumed as a chirp component. Compared with Figure 4a, c shows the improvement on the smoothness of signal representation, which is achieved by reducing the window length from 1/4 to 1/8 of the signal length. It should be noted that, compared with the third order LPFT with the small window length in Figure 4d, the second order LPFT with the same window length in Figure 4c can achieve comparable result and with less computational complexity. It is worth mentioning that for any order LPFTs, the frequency resolution will become degraded as the window length is reduced, as shown in Figure 4c, d. For other higher-order PPSs, the second order LPFT can also be used, as long as the window length is small enough to ensure that each segment within the window can be approximately assumed as the chirp signal.

Resolution comparisons

Let us now consider the performances achieved by the STFT and the second order LPFT of a chirp signal obtained from (3) with P = 2. Using the STFT, the corresponding mean frequency and bandwidth are [29]

$$\langle\omega\rangle_t = a_0 + \frac{\alpha a_1 t}{a+\alpha}, \quad B_t^2 = \frac{a+\alpha}{2} + \frac{a_1^2}{2(a+\alpha)}. \tag{29}$$

The mean time $\langle\tau\rangle_t$ and the duration T_t^2 of the STFT are the same as those of the LPFT. Therefore the uncertainty product of the STFT is

$$B_t^2T_t^2 = \frac{1}{4} + \frac{a_1^2}{4(a+\alpha)^2}. \tag{30}$$

When $a_1 = 0$, the input signal becomes one whose frequencies do not change with time and the STFT achieves the lowest bound of the uncertainty product $B_t^2T_t^2 = \frac{1}{4}$. Therefore, the STFT can provide good resolution only for signals with constant frequencies. For the chirp signals, however, B_t^2 in (29) has a non-zero second term. Therefore, the uncertainty product $B_t^2T_t^2$ of the STFT of time-varying signals must be larger than $\frac{1}{4}$. For the same input signal, if the polynomial coefficient, ω_1, of the LPFT is correctly estimated, we achieve the lowest bound of the uncertainty product $B_t^2T_t^2 = \frac{1}{4}$, as seen in (22). Therefore the second order LPFT achieves a more concentrated distribution in the time-frequency domain than the STFT.

Another important issue on the LPFT is about the estimation errors on ω_1. Although the PTFT is the maximum likelihood estimator, it is still possible to have some estimation errors, especially when the signals are corrupted by heavy noises. Because the second term of (15) is proportional to $(a_1 - \omega_1)^2$, the uncertain product for the second order LPFT of chirp signals is $B_t^2T_t^2 > \frac{1}{4}$ if ω_1 is not correctly estimated. Comparing (15) and (29) reveals that as long as $|a_1 - \omega_1| < |a_1|$, i.e., the estimation error of ω_1 is smaller than $|a_1|$, the LPFT of the chirp signal achieves a smaller bandwidth than the STFT. Therefore, the second order LPFT can still achieve a more concentrated distribution in the time-frequency domain than the STFT. Furthermore, the resolution performance of the STFT cannot be further improved because a_1 in (29) or (30) is the signal parameter. In comparison, improvement on the resolution performance of the LPFT can be achieved by minimizing the error of parameter estimation. It should be noted that if the estimation error of ω_1 is larger than $|a_1|$, the resolution of the LPFT will become worse than that of the STFT although such a case rarely happens when the PTFT is used to estimate the polynomial coefficients.

Besides the STFT and LPFT, which are linear time-frequency transforms, another bilinear time-frequency transform, known as the WVD, has been often used in many applications [29]. The WVD of the chirp signal is

$$\mathrm{WVD}(t,\omega) = \frac{1}{\pi}\exp\left\{-\alpha t^2 - \frac{(\omega - a_1 t - a_0)^2}{\alpha}\right\}.$$

When is small, the WVD is concentrated on $\omega = a_0 + a_1 t$. The associated mean frequency and bandwidth are [29]

$$\langle\omega\rangle_t = a_0 + a_1 t \quad \text{and} \quad B_t^2 = \frac{\alpha}{2}.$$

It means that the WVD can provide a better frequency resolution for chirp signals as far as α is small enough. However, the WVD suffers from the cross terms for multiple-component signals. It is difficult to remove the cross terms without sacrificing the resolution of the signal representation, particularly for non-linear multiple-component chirp signals [29].

In order to clearly show the frequency resolution of the signal, Figure 5 compares signal concentration obtained from the STFT, LPFT and WVD at a particular time instant. Resolution usually means whether we can separate two closely located components in the time and/or frequency domains, while concentration describes how the transforms concentrate along the signal›s instantaneous frequency. Resolution and concentration are closely related and high concentration generally contributes to high resolution. For signals with constant frequency components, the STFT can provide the good frequency resolution, as shown in Figure 5. For chirp signals, with the same window length and window parameter a, however, the LPFT achieves a better frequency resolution as seen in Figure 5b, c. Although the WVD also has a better frequency resolution, the cross terms between the two chirp components, as shown in Figure 5d, can be easily mistaken as a valid frequency component. Therefore, the second order LPFT is the best choice for dealing with chirp signals since it is free from the cross terms and also achieves a better resolution than the STFT.

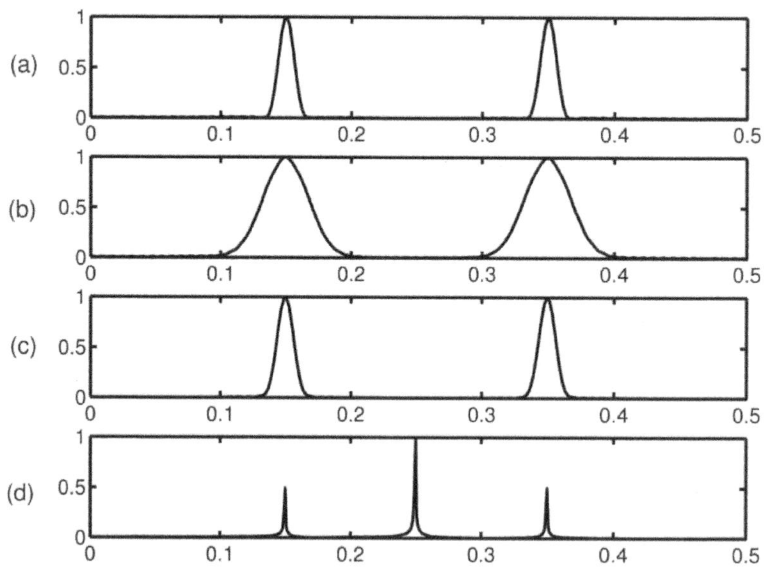

Figure 5: The signal concentration comparisons of different time-frequency rep-resentations for a two-component signal. (a) The STFT of a signal with constant frequencies; (**b**)-(**d**) The concentrations of chirp signals using the STFT, the second order LPFT, and the WVD, respectively. The horizontal axis is the normalized frequency, and the vertical axis is the normalized amplitude.

CONCLUSION

In this article, the LPFT is shown to be limited by the uncertainty principle, and the un-certainty relations of various order LPFTs are derived to show the trade-off between the resolutions of signal representation in the time and frequency domains and the required computational complexity. The uncertainty product of an arbitrary order is determined by the signal parameters, the window function and the errors of estimating the polynomial coefficients. When Gaussian window function is employed to segment the signals, the uncertainty products of the LPFT are time independent under the condition that the polynomial coefficients are accurately estimated. The effects of the window width, the estimation errors, and the order mismatch by using the second order LPFT to process higher-order PPSs are discussed. Resolution comparisons with the STFT and WVD are also provided to show the merits of the LPFT, and the LPFT is found to be a better tool to dealing with time-varying signals.

APPENDIX

Review on Uncertainty Principles of the STFT

This appendix provides an example of deriving the uncertainty principle for the STFT. The definitions of some terminologies in this appendix are also used in the main sections of this article. Let us consider the P th-order PPS defined in (3) as the input signal. By multiplying s(t) with a window function h(t), the local signal is defined as $s_t(\tau) = s(\tau)h(t - \tau)$. It is easily seen that the Fourier transform of the local signal $s_t(\tau)$ is exactly the STFT of the original signal s(t). The normalized local signal at time t is

$$\eta_t(\tau) = \frac{s(\tau)h(\tau-t)}{\sqrt{\int |s(\tau)h(\tau-t)|^2 d\tau}}, \tag{31}$$

which ensures that for any t, $\int |\eta_t(\tau)|^2 d\tau = 1$. Let us review some terminologies based on the normalized local signal $\eta_t(\tau)$ [29] which is a function of τ within a short duration, and its Fourier transform is defined as

$F_t(\omega)\ \int \eta_t(\tau)\exp\{-j\omega\tau\}d\tau,$

to reveal the spectral information around the time instant t. For this normalized local signal, all the relevant quantities, such as the mean time duration, mean frequency and bandwidth, are time dependent [29].

The mean time of the normalized local signal is defined as

$$\langle\tau\rangle_t = \int \tau|\eta_t(\tau)|^2 d\tau = \frac{\int \tau|s(\tau)h(\tau-t)|^2 d\tau}{\int |s(\tau)h(\tau-t)|^2 d\tau}, \tag{32}$$

and its time duration is defined as

$$T_t^2 \int (\tau - \langle\tau\rangle_t)^2 |\eta_t(\tau)|^2 d\tau = \langle\tau^2\rangle_t - \langle\tau\rangle_t^2, \tag{33}$$

where $\langle\tau^2\rangle_t$ is the second order moment of the normalized local signal, defined as

$$\langle\tau^2\rangle_t = \int \tau^2 |\eta_t(\tau)|^2 d\tau.$$

Similarly, the n th order moment of the normalized local signal is defined as

$$\langle\tau^n\rangle_t = \int \tau^n |\eta_t(\tau)|^2 d\tau. \tag{34}$$

The mean frequency of the normalized local signal is defined as

$$\langle\omega\rangle_t = \int \omega|F_t(\omega)|^2 d\omega = \int \eta_t^*(\tau)\frac{1}{j}\frac{d}{d\tau}\eta_t(\tau)d\tau, \tag{35}$$

and its bandwidth is

$$B_t^2 = \int (\omega - \langle\omega\rangle_t)^2 |F_t(\omega)|^2 d\omega = \langle\omega^2\rangle_t - \langle\omega\rangle_t^2, \tag{36}$$

where

$$\langle\omega^2\rangle_t = \int \omega^2 |F_t(\omega)|^2 d\omega = -\int \eta_t^*(\tau)\frac{d^2\eta_t(\tau)}{d\tau^2}d\tau. \tag{37}$$

Here the duration T_t^2 and bandwidth B_t^2 are good measures of the broadness of the local signal in the time and frequency domains, respectively. For example, T_t^2 (or) of a signal means the width of the signal concentrated around the mean time $\langle\tau\rangle_t$ (or mean frequency $\langle\omega\rangle_t$). The uncertainty product in the STFT domain is generally a function of time, the signal s(t) and the window h(t), and has a lower bound [29]

$$B_t^2 T_t^2 \geq \frac{1}{4}.$$

It should be noted that the focus of this article is on the uncertainty product obtained by multiplying the duration and bandwidth of the local signal. It is important to understand that this uncertainty product places limits on the processing techniques of the windowed transforms. Other kinds of uncertainty products, such as the global uncertainty products, can be referred to [17].

Endnote

[a]For simplicity in the rest of the article, the integral without limits implies that the integra-tion is from -∞ to ∞.

ACKNOWLEDGEMENTS

This research work was supported by the National Natural Science sFoundation of China (No. 61102164), and the Scientific Research Fund of Hangzhou Normal University (No. 2011QDL021).

Competing Interests

The authors declare that they have no competing interests.

REFERENCES

1. Ozaktas H, Zalevsky Z, Kutay M: The Fractional Fourier Transform With Applications in Optics and Signal Processing. Wiley, Chichester; 2001.
2. Zhang F, Chen YQ, Bi G: Adaptive harmonic fractional Fourier transform. IEEE Signal Process Lett 1999, 6(11):281-283.
3. Katkovnik V: A new form of Fourier transform for time-frequency estimation. Signal Process 1995, 47(2):187-200. 10.1016/0165-1684(95)00107-7
4. Djurovic I, Thayaparan T, Stankovic L: Adaptive local polynomial Fourier transform in ISAR. EURASIP J Appl Signal Process 2006, 2006: 1-15.
5. Li X, Bi G, Ju Y: Quantitative SNR analysis for ISAR imaging using local polynomial Fourier transform. IEEE Trans Aerosp Electron Syst 2009, 45(3):1241-1248.
6. Dakovic M, Thayaparan T, Djukanovic S, Stankovic L: Time-frequency-based non-stationary interference suppression for noise radar systems. IET Radar Sonar Navig 2008, 2(4):306-314. 10.1049/iet-rsn:20070137
7. Stankovic L, Djukanovic S: Order adaptive local polynomial FT based interference re-jection in spread spectrum communication systems. IEEE Trans Instrum Meas2005,54(6):2156-2162.10.1109/TIM.2005.858567
8. Katkovnic V, Gershman A: A local polynomial approximation based beamforming for source localization and tracking in nonstationary environments. IEEE Signal Process Lett2000, 7(1):3-5.
9. Li X, Bi G, Stankovic S, Zoubir A: Local polynomial Fourier transform: a review on recent development and applications. Signal Process 2011, 91: 1370-1393. 10.1016/j.sigpro.2010.09.003
10. Gabor D: Theory of communication. J Inst Electr Eng 1946, 93(26):429-457.
11. Korn P: Some uncertainty principles for time-frequency transforms of the Cohen class.IEEE Trans Signal Process 2005, 53(2):523-527.
12. Loughlin P, Cohen L: The uncertainty principle: global, local, or both? IEEE Trans Signal Process 52(5):1218-1227.
13. Shinde S, Gadre V: An uncertainty principle for real signals in the fractional Fourier transform domain. IEEE

Trans Signal Process 2001, 49(11):2545-2548. 10.1109/78.960402

14. Mustard D: Uncertainty principle invariant under fractional Fourier transform. J Austr Math Soc Ser B 1991, 33: 180-191. 10.1017/S0334270000006986

15. Sharma K, Joshi S: Uncertainty principle for real signals in the linear canonical transform domain. IEEE Trans Signal Process 2008, 56(7):2677-2683.

16. Xu G, Wang X, Xu X: The logarithmic, Heisenberg's and short-time uncertainty principles associated with fractional Fourier transform. Signal Process 2009, 89(3):339-343. 10.1016/j.sigpro.2008.09.002

17. Cohen L: The uncertainty principles of the short-time Fourier transform. Proc SPIE 1995, 2563: 80-90.

18. Katkovnikm V: Discrete-time local polynomial approximation of the instantaneous frequency. IEEE Trans Signal Process 1998, 46(10):2626-2637. 10.1109/78.720366

19. Li X, Bi G: Uncertainty principle of the second-order LPFT. In IEEE International Symposium on Circuits and Systems (ISCAS'2009). Taipei, Taiwan; 2009:325-328.

20. Ikram M, Abed-Meraim K, Hua Y: Fast quadratic phase transform for estimating the parameters of multicomponent chirp signals. Dig Signal Process 1997, 7(2):127-135. 10.1006/dspr.1997.0286

21. Xia X: Discrete chirp-Fourier transform and its applications to chirp rate estimation. IEEE Trans Signal Process 2000, 48(11):3122-3133. 10.1109/78.875469

22. Bi G, Wei Y: Radix-2 DIF fast algorithms for polynomial time frequency transforms. IEEE Trans Aerosp Electron Syst 2006, 42(4):1540-1546.

23. Ju Y, Bi G: Generalized fast algorithms for the polynomial time-frequency transform. IEEE Trans Signal Process 2007, 55(10):4907-4915.

24. Bi G, Chen Y: Fast DFT algorithms for length N = q * 2 m . IEEE Trans Circ Syst II-Analogu Dig Signal Process 1998, 45(6):685-690. 10.1109/82.686687

25. Bi G, Chen Y, Zeng Y: Fast algorithms for generalized discrete Hartley transform of composite sequence lengths. IEEE Trans Circ Syst II-Analogu Dig Signal Process 2000, 47(9):893-901. 10.1109/82.868457

26. Bi G, Chen Y: Fast generalized DFT and DHT algorithms. Signal Process 1998, 65(3):383-390. 10.1016/S0165-1684(97)00234-X

27. Djurovic I: Robust adaptive local polynomial Fourier transform. IEEE Signal Process Lett11(2):201-204.

28. Wei Y, Bi G: Efficient analysis of time-varying muliticomponent signals with modified LPTFT. EURASIP J Appl Signal Process 2005, 2005(1):1261-1268.

CHAPTER 11

SEPARATION AND LOCALIZATION OF MULTIPLE DISTRIBUTED WIDEBAND CHIRPS USING THE FRACTIONAL FOURIER TRANSFORM

Jiexiao Yu, Liang Zhang, Kaihua Liu and Deliang Liu

Keywords: Direction of arrival, Fractional Fourier transform, Distributed wideband chirp, Generalized array manifold

INTRODUCTION

The direction of arrival (DOA) estimation for wideband chirp signal has been greatly achieved under the point source assumption [1–3]. However, the signal is always affected by multipath and scattering propagation in the practical application, which will cause the angular spread, such as the local scattering source in the mobile multipath environment, the distributed target reflection wave in the low-elevation radar target tracing system, the surface and the bottom reflection signal in the sonar detection of the shallow sea, tropospheric or ionospheric propagation of radio waves, the part of detection target in the passive radar and sonar system [4–6]. In this situation, the received signal in an array can be considered as a superposition of scattered signals originating from the different direction and the estimation performance of DOA will be degraded significantly if using the traditional estimators based on the point source model [7, 8].

For obtaining the exact nominal DOA and angular spread of a spatially distributed source, the problem of distributed source model has been widely studied since the early 1990s, and a large number of methods are proposed for the parameter estimation of distributed source [9–17]. However, most of the models and the estimation algorithms can be only exploited in the case of narrowband source, because the location vector in the time domain is time varying when the incident source is wideband. To date, the estimation method for wideband source with local scattering is still in scarcity. M. Ghogho et al. proposed the distributed wideband source model [18]. Liu et al. [19] proposed a kind of maximum-likelihood (ML) estimates for finite-bandwidth distributed sources by the perturbation method. Foroozan and Asif [20] introduced time reversal-based range and DOA estimators to exploit spatial/multipath diversity existing in strong multipath environments, which has a high complexity in the multi-source scenarios. Mecklenbrauker et al. [21] extended the Bayesian approach to a distributed wideband source for the application to seismic recordings, the kind that should be sparse.

The wideband chirp signal, whose frequency is linearly increasing with time, is widely used in many applications, such as communication, radar, sonar, and biomedicine, and also be used as signal model for a good deal of natural phenomena. However, for the distributed wideband chirps, we have not found much work referring to the problem of DOA estimation.

In our previous work [22], the properties of the fractional Fourier transform (FrFT), which is a generalized form of the ordinary Fourier transform (FT), is exploited to address the problem mentioned above because it provides a compact representation for chirp signals as it is based on the decomposition of the signal on the orthonormal basis set of the chirp functions [23, 24]. The models of wideband chirp sources in both time domain and energy-concentrated domain are studied. And an estimation of the spatial parameters of

Citation: Yu, J., Zhang, L., Liu, K. et al. Separation and localization of multiple distributed wideband chirps using the fractional Fourier transform, doi:10.1186/s13638-015-0497-9.

multiple wideband chirps with local scattering is provided.

In order to obtain better estimation precision, we will further study the problem in another fractional Fourier domain, the Dechirping domain, in this paper. The representation of location vector in the Dechirping domain is derived based on a number of good properties of chirp signal in two special fractional Fourier domains, which is only related to the angular parameter. Then, the standard MUSIC algorithm combined with the source separation technique is extended to estimate the incident angles of multiple sources in the Dechirping domain. Numerical results illustrate that the proposed algorithm could separate multiple chirp signals successfully, which is far more than the number of sensors, and estimate the DOA of each chirp accurately.

FRACTIONAL FOURIER TRANSFORM

Notation and Definition

The fractional Fourier transform as a linear integral transform with kernel $K_\alpha(u,t)$ [25, 26]:

$$X_\alpha(u) = F^\alpha[x(t)] = \int_{-\infty}^{\infty} K_\alpha(u,t)x(t)dt \tag{1}$$

where $F^\alpha[\cdot]$ denotes the FrFT operator and kernel $K_\alpha(u,t)$ is

$$K_\alpha(u,t) = \begin{cases} \sqrt{1-j\cot\alpha}\, e^{j\pi(t^2\cot\alpha - 2tu\csc\alpha + u^2\cot\alpha)}, & \alpha \neq n\pi \\ \delta(t-u), & \alpha = 2n\pi \\ \delta(t+u), & \alpha = (2n\pm 1)\pi \end{cases} \tag{2}$$

where the domain u is generally known as the fractional Fourier domain which makes the rotation angle α with the time domain.

Rather than defining the fraction of the transform as the rotation angle, α, in the interval $[-\pi,\pi]$ radians, a new variable, p, is defined as the order of the transform and is valid in the interval $[-2, 2]$. The zeroth order transform is simply the function itself, whereas the first order transform is its Fourier transform.

Time-Frequency Rotating Properties

According to the time-frequency rotating properties of the FrFT, the Wigner-Ville distribution (WVD) of the FrFT of a signal can be interpreted as the coordinate rotating form of the WVD of this signal [27]. The energy spectrum of a finite chirp signal shows a fin-shape line in the fractional Fourier domain, as shown in Fig. 1 and two special fractional Fourier domains are particularly noteworthy: the Dechirping domain, whose axis u coincides with the line, and its perpendicular domain, energy-concentrated domain, in which the energy distribution of chirp signal shows an obvious peak.

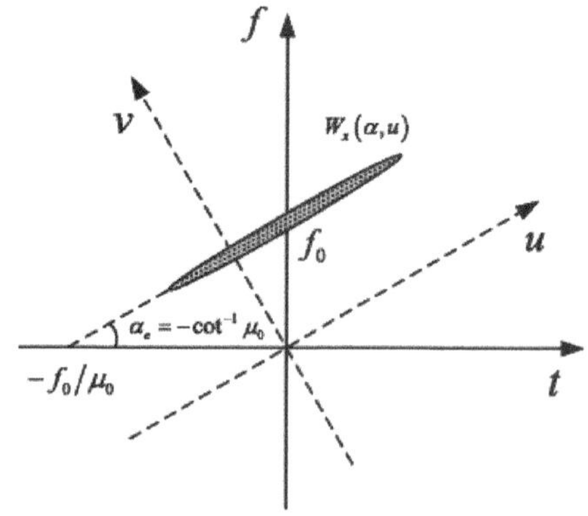

Fig. 1

The energy spectrum of a chirp signal with a finite length.

We assume that a chirp signal is modeled as

$$y(t) = \beta e^{j\pi(2f_0 t + \mu_0 t^2)} \tag{3}$$

where $\beta = a_0 e^{j\varphi_0}$ is a constant, a_0 symbolizes the amplitude of the chirp signal, φ_0 is the initial phase, f_0 is the initial frequency, and μ_0 is the chirp rate.

Then, the rotation angles of the chirp signal in the Dechirping domain and the energy-concentrated domain can be written as

$$\alpha_d = tan^{-1}\mu_0 \tag{4}$$

$$\alpha_e = -\cot^{-1}\mu_0 \tag{5}$$

The relationship between these two rotation angles can be represented by

$$\alpha_d = \alpha_e + \pi/2 \tag{6}$$

$$Y_{\alpha_d}(u) = F[Y_{\alpha_e}(u)] \tag{7}$$

where $F^\alpha[\cdot]$ denotes the FT operator.

It can be seen that the FrFT of a chirp signal in the Dechirping domain equals to the FT of the FrFT of this signal in the energy-concentrated domain. As a result, when the chirp rate μ_0 is unknown which is satisfied in most instances, the rotation angle a_d can be acquired using (6) after the determination of the rotation angle a_e in the energy-concentrated domain by searching the coordination of spectral peak in two-dimensional plane (α,u).

Time Delay Representation in Dechirping Domain

According to (3), we can obtain the $y(t-\tau)$ as follows [25]:

$$\begin{aligned} y(t-\tau) &= \beta e^{j\pi\left[2f_0(t-\tau)+\mu_0(t-\tau)^2\right]} \\ &= \beta e^{j\pi(-2f_0\tau+\mu_0\tau^2)} e^{j2\pi(f_0-\mu_0\tau)t} e^{j\pi\mu_0 t^2} \end{aligned} \tag{8}$$

where τ is the time delay.

Compared with (3), it can be seen that the chirp rate of the signal is invariable after the time delay performs on this signal. Therefore, the rotation angle of the time-delayed signal is the same as that of the original signal in the Dechirping domain.

In Dechirping domain, the FrFT of a chirp signal given by (3) can be written as:

$$Y^{\alpha_d}(u) = F^{\alpha_d}[y(t)] = Ce^{j2\pi f_0 \cos\alpha_d u} \tag{9}$$

where

$$C = \beta \cos\alpha_d \sqrt{1 + j\tan\alpha_d} e^{\left(-j\pi f_0^2 \sin\alpha_d \cos\alpha_d\right)} \tag{10}$$

According to the time shifting property of the FrFT, the FrFT of $y(t-\tau)$ in the same Dechirping domain is:

$$\begin{aligned} F^{\alpha_d}[y(t-\tau)] &= e^{(j2\pi\tau^2 \sin\alpha_d \cos\alpha_d)} e^{(-j2\pi u\tau \sin\alpha_d)} Y^{\alpha_d}(u-\tau\cos\alpha_d) \\ &= Ce^{j2\pi(f_0\cos\alpha_d - \tau\sin\alpha_d)u} e^{j\pi[\tau^2 \sin\alpha_d \cos\alpha_d - 2f_0 \cos^2\alpha_d \tau]} \end{aligned} \tag{11}$$

where τ is the time delay.

In the digital calculation process in FrFT [28], the chirp rate will be significantly reduced because of the dimensional normalization. As a result, the numerical value of the rotation angleα_d is far from $\pi/2$. If $\tau << f_0$, which is satisfied in most practical applications, we can obtain that

$$|\tau \sin\alpha_d| << |f_0 \cos\alpha_d| \tag{12}$$

Thus, (11) can be approximated to

$$F^{\alpha}[y(t-\tau)] \approx Ce^{j2\pi f_0 \cos\alpha u} e^{j\pi[\tau^2 \sin\alpha \cos\alpha - 2f_0 \cos^2\alpha\tau]} \tag{13}$$

Comparing (8) and (12), we can obtain:

$$F^{\alpha}[y(t-\tau)] = A(\tau)F^{\alpha}[y(t)] \tag{14}$$

where

$$A(\tau) = e^{j\pi[\tau^2 \sin\alpha_d \cos\alpha_d - 2f_0 \cos^2\alpha_d \tau]} \tag{15}$$

In (15), $A(\tau)$is only related with the time delay τ , and the influence of the independent variable u has been eliminated. This useful conclusion will play an important role in deducing data model in Dechirping domain in the next section.

DATA MODEL

Data model in Time Domain

A scenario with Q uncorrelated sources transmitted wideband chirp signals is considered. Due to multipath propagation, each source can be seen as a superposition of N_q scattered point-source components. The uniform linear array (ULA) with P sensors is taken as an example for derivation, and the similar conclusions are easy to be generalized to the arrays with other types. The complex envelope of the output vector in the array can be modeled as

$$\boldsymbol{x}(t) = \sum_{q=1}^{Q}\sum_{i=1}^{N_q} \beta_{qi} \mathbf{a}\left(\theta_q + \tilde{\theta}_{qi}, t\right) s_q(t-v_{qi}) + \boldsymbol{n}(t) \tag{16}$$

where

β_{qi}: the complex amplitude of the ith scattered source from the qth source;

$\mathrm{a}^{\left(\theta_i + \tilde{\theta}_{qi}, t\right)}$: the time-varying location vector of the array, θ_q is the nominal DOA of the qth source and $\theta_i + \tilde{\theta}_{qi}$ is the DOA of the ith scattered components from the qth source;

$s_q(t)$: the qth transmitted signal;

v_{qi}: the time delay associated with the ith scattered source from the qth source;

$\boldsymbol{n}(t)$: the additive zero-mean noise vector, which is assumed to be spatially white and independent of the transmitted signals.

When the received signals are wideband and with local scattering, there is no one-to-one relationship between the DOA parameter and the location vector $\mathrm{a}^{\left(\theta_i + \tilde{\theta}_{qi}, t\right)}$ which is also dependent on the time and deviation angle variables. Therefore, traditional parameter estimation methods cannot work well in the case of distributed wideband chirp. The time-frequency properties of chirp signal in the two fractional Fourier domain mentioned above can help us eliminate the influence of time and deviation angle variable on the location vector. The detailed derivation will be given in the next section.

APPROXIMATING THE SPATIAL SIGNATURE

Because the majority of scattered signals distribute in the vicinity of the transmitted source, the time delay v_{qi} is relatively small so that (12) is satisfied and the quadratic term of v_{qi} can be ignored. Hence, the ith scattered component of the qth source received on the conference sensor in the Dechirping domain is given by

$$\begin{aligned} F^{\alpha_d}\left[s_q(t-v_{qi})\right] &\approx e^{j\pi\left[(v_{qi})^2 \sin\alpha_d \cos\alpha_d - 2f_q \cos^2\alpha_d v_{qi}\right]} F^{\alpha_d}\left[s_q(t)\right] \\ &\approx e^{j\pi\left(-2f_q \cos^2\alpha_d v_{qi}\right)} F^{\alpha_d}\left[s_q(t)\right] \end{aligned} \tag{17}$$

Then, the FrFT with a rotation angle α_d performs on the output vector in (16), and the result can be given by

$$\mathbf{X}(u) = \sum_{q=1}^{Q}\sum_{i=1}^{N_q} \beta_{qi} e^{-j2\pi f_q \cos^2\alpha_d v_{qi}} \mathbf{A}_q\left(\theta_q + \tilde{\theta}_{qi}\right) S_q^{\alpha_d}(u) + \mathbf{N}(u) \tag{18}$$

where $\mathbf{A}_q\left(\theta_q + \tilde{\theta}_{qi}\right)$ is the location vector of the qth source, $S_q^{\alpha_d}(u)$ is the FrFT of $sq\ (t)$ and $\mathrm{N}(u)$ is the FrFT of $n(t)$ in the Dechirping domain.

Because FrFT of multiple sources or that of one single source's multi-scattering components does not result in cross-term interference, the conclusion in [31] could be exploited here directly as follows:

$$\mathbf{A}_q\left(\theta_q + \tilde{\theta}_{qi}\right) = \left[1, A_2\left(\theta_q + \tilde{\theta}_{qi}\right), \cdots, A_P\left(\theta_q + \tilde{\theta}_{qi}\right)\right]^T \tag{19}$$

where T is the transpose and

$$A_p\left(\theta_q + \tilde{\theta}_{qi}\right) = e^{j\pi\left[(\tau_p)^2 \sin\alpha_d \cos\alpha_d - 2f_0 \cos^2\alpha_d \tau_p\right]} \tag{20}$$

where $\tau_p = (p-1)\, d\sin\left(\theta_q + \tilde{\theta}_{qi}\right)/c$ represents the time delay on the pth sensor, d is the inter-sensor distance of ULA, and c is the transmission speed.

In order to estimate θ_q, [29] proposed an approximate model called the generalized array manifold (GAM) and a corresponding algorithm using a Vandermonde structure. Due to the deviation angle $\tilde{\theta}_{qi}$ of each scattered signal is relatively small, $\mathbf{A}_q\left(\theta_q + \tilde{\theta}_{qi}\right)$ in (20) can be approximated through the first-order Taylor series expansion:

$$\mathbf{A}_q\left(\theta_q + \tilde{\theta}_{qi}\right) \approx \mathbf{A}_q\left(\theta_q\right) + \tilde{\theta}_{qi} d\left(\theta_q\right) \quad (21)$$

where

$$\mathbf{d}(\theta) = \frac{\partial \mathbf{A}_q(\theta)}{\partial \theta} \quad (22)$$

Then, we define a variable v_q, that is

$$v_q = \sum_{i=1}^{N_q} c_{qi}\mathbf{A}_q\left(\theta_q + \tilde{\theta}_{qi}\right) = \sum_{i=1}^{N_q} c_{qi}\left(\mathbf{A}_q(\theta_q) + \tilde{\theta}_{qi}\mathbf{d}(\theta_q)\right)$$
$$= \left(\sum_{i=1}^{N_q} c_{qi}\right)\mathbf{A}_q(\theta_q) + \left(\sum_{i=1}^{N_q} c_{qi}\tilde{\theta}_{qi}\right)\mathbf{d}(\theta_q) = \mathbf{A}_q(\theta_q) + \phi_q\mathbf{d}(\theta_q) \quad (23)$$

where $c_{qi} = \beta_{qi} e^{-j2\pi f_q \cos^2\alpha_d v_{qi}}$, $\phi_q = \sum_{i=1}^{N_q} c_i\tilde{\theta}_{qi}$ and we assume that $\sum_{i=1}^{N_q} c_{qi} = 1.$

Hence, the compact matrix notation V can be given by

$$V = \mathbf{A}\left(\underline{\theta}, \underline{\phi}\right) = \mathbf{A}(\underline{\theta}) + \mathbf{D}\left(\underline{\theta}\right)\mathbf{\Phi}(\underline{\phi}) \quad (24)$$

where

$$\begin{aligned} \mathbf{A}(\underline{\theta}) &= \left[\mathbf{A}_1(\theta_1), \ldots, \mathbf{A}_Q(\theta_Q)\right] \\ \mathbf{D}(\underline{\theta}) &= \left[\mathbf{d}(\theta_1), \ldots, \mathbf{d}(\theta_Q)\right] \\ \mathbf{\Phi}\left(\underline{\phi}\right) &= \mathrm{diag}\{\phi_1, \ldots, \phi_Q\} \end{aligned} \quad (25)$$

$$\underline{\theta} = \left[\theta_1, \ldots, \theta_Q\right]^{\mathrm{T}} \underline{\phi} = \left[\phi_1, \ldots, \phi_Q\right]^{\mathrm{T}} \quad (26)$$

Finally, the model (18) can be approximated as

$$X(u) = \mathbf{V}\mathbf{S}(u) + \mathbf{N}(u) \quad (27)$$

where

$\mathbf{S}(u) = \left[S_1^{\alpha_d}(u) \ \cdots \ S_Q^{\alpha_d}(u)\right]^T$ and $\mathbf{N}(u) = [N_1^{\alpha_d}(u) \ \cdots \ N_Q^{\alpha_d}(u)]^T$.

According to (24), we know that the spatial parameters $\underline{\theta}$ and $\underline{\phi}$ are needed to estimate in the approximated model.

PARAMETER ESTIMATION ALGORITHM

Single Source Estimation Algorithm

Firstly, we consider one chirp source (Q=1 in (16)). Many DOA estimation algorithms for the narrowband point source model can also be applied here because the location matrix in the Dechirping domain has a one-to-one relationship with the spatial parameters. In this paper, the classical MUSIC algorithm in the time domain is extended to estimate spatial parameter of wideband chirp source with local scattering. The spatial estimation spectrum of standard MUSIC algorithm is

$$V_{\mathrm{MUSIC}}(\theta) = \frac{\mathbf{a}^*(\theta)\hat{\mathbf{E}}_n\hat{\mathbf{E}}_n^*\mathbf{a}(\theta)}{\mathbf{a}^*(\theta)\mathbf{a}(\theta)} \quad (28)$$

In the Dechirping domain, the calibrated location vector is exploited to replace the location vector in (28), that is

$$\mathbf{A}(\theta) + \phi\mathbf{d}(\theta) = \bar{\mathbf{A}}(\theta)\bar{\phi} \quad (29)$$

where $\bar{\mathbf{A}}(\theta) = [\mathbf{A}(\theta) \quad \phi\mathbf{d}(\theta)]$ and $\bar{\phi}(\theta) = [1 \quad \phi]^{\mathrm{T}}$. Hence, the estimation spectrum for the proposed model should be revised to

$$\bar{V}_{\mathrm{MUSIC}}(\theta, \phi) = \frac{\phi^*\bar{\mathbf{A}}^*(\theta)\hat{\mathbf{E}}_n\hat{\mathbf{E}}_n^*\bar{\mathbf{A}}(\theta)\bar{\phi}}{\mathbf{A}^*(\theta)\mathbf{A}(\theta)} \quad (30)$$

The parameters θ and ϕ can be determined by the spectral peak position.

Separation of Different Chirp Sources and the Source Number Determination

As (5) shows, chirp signals with different chirp rates have different rotation angles in their energy-concentrated domain in which chirp signal shows an obvious spectrum peak. The energy-concentrated domain of each signal can be determined by the two dimension search in the time-frequency plane (α, u). In each energy-concentrated domain, only one chirp signal can acquire the best energy-concentrated property while the energy distributions of the other signals and noise are dispersing, therefore, this chirp signal can be separated from others through a band-pass filter. Then, the parameter estimation algorithm proposed above can be applied to filtered single source after that is transformed into the corresponding Dechirping domain. The detailed separation process could be found in [31]. It is noted that the separation process cannot only resist the interference of noise and other signals, but also solve the problem of source number restriction in many traditional DOA estimators. For example, the MUSIC algorithm cannot be applied when the source number exceeds the number of the sensors in the array. However, the proposed algorithm avoids this problem through the separation processing. The parameters of each source can be estimated respectively.

SIMULATION STUDY AND RESULTS

To demonstrate the performance of the proposed algorithm, an ULA with eight sensors is employed to estimate the spatial parameters of the proposed model. In the simulation, two uncorrelated wideband chirp sources are considered as signal emitters. The initial parameters of the first wideband chirp source are f_1 = 100 Hz and μ_1 = 10 Hz/s. And that of the second one are f_2=200 Hz and μ_2=−10 Hz/s. Besides, the central angle of the first source is taken as θ_1=30° with the extension width Δ_1=2° from a uniform angular distribution, and the central angle of the second source is taken as θ_2=−45° with the extension width Δ_2=2° from a Gaussian angular distribution of width, respectively. The signal-to-noise ratio (SNR) is set at 10 dB, and the number of snapshots is 500.

The two-dimensional searching results of two sources on the reference sensor are shown in Fig. 2. According to the coordinates of two distinct spectrum peaks, the rotation angles of their energy-concentrated domain can be determined, and the source number can be easily obtained by the peak number.

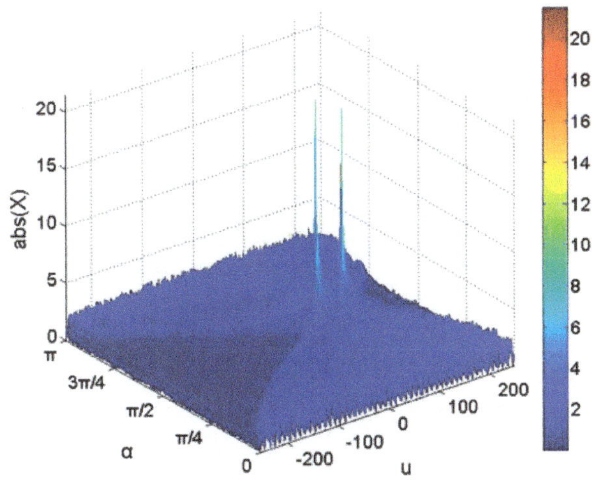

Figure 2: Two-dimensional spectrum of two wideband chirp sources in the plane (α, u).

In Fig. 3a, b, the energy spectrums of two sources on the first sensor in the corresponding energy-concentrated domain show an obvious energy concentration respectively. In this energy-concentrated domain, the chirp signal forms a sinc function and the majority of the energy spectrum focuses on its support, and the single peak will appear on the other sensors as well. Therefore, the desired chirp signal on each sensor can be separated conveniently using the band-pass filtering, and the filtered results are illustrated in Fig. 3c, d.

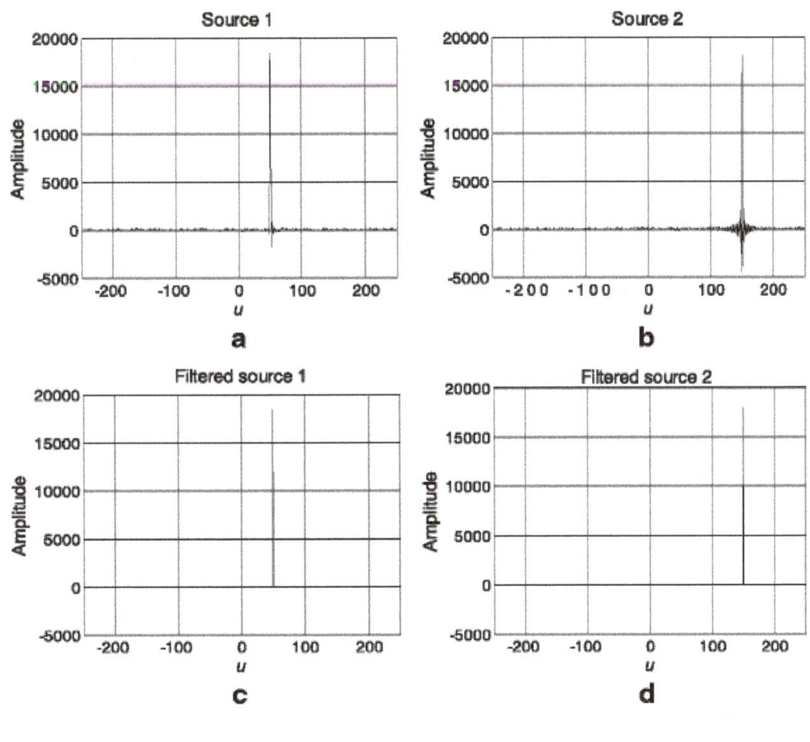

Figure 3: The energy spectrums of the chirp signal and noise on the first sensor in the energy-concentrated domain. **a** Source 1. **b** Source 2. **c** Filtered source 1. **d** Filtered source 2.

The FrFT-MUSIC algorithm proposed in [30], which is applied to estimate the DOA of the wideband chirp source based on the point source model, and the method in [22] is chosen as the comparing algorithm to demonstrate the performance of the proposed algorithm for the wideband chirp source with local scattering. For each figure, 100 Monte Carlo simulations were run to estimate the root mean-square error (RMSE) of the estimates. We focus on the estimation of the DOA, while the other parameters are considered mainly as nuisance parameters.

Figure 4 shows the RMSE for the DOA estimates versus the different SNRs using three algorithms. For both sources, the RMSE of the proposed algorithm is less than that for the other two algorithms. Figure 5 clearly demonstrates that the proposed estimator has excellent performance versus the different extension widths. With the increasing extension width, the proposed algorithm has better location accuracy than the others.

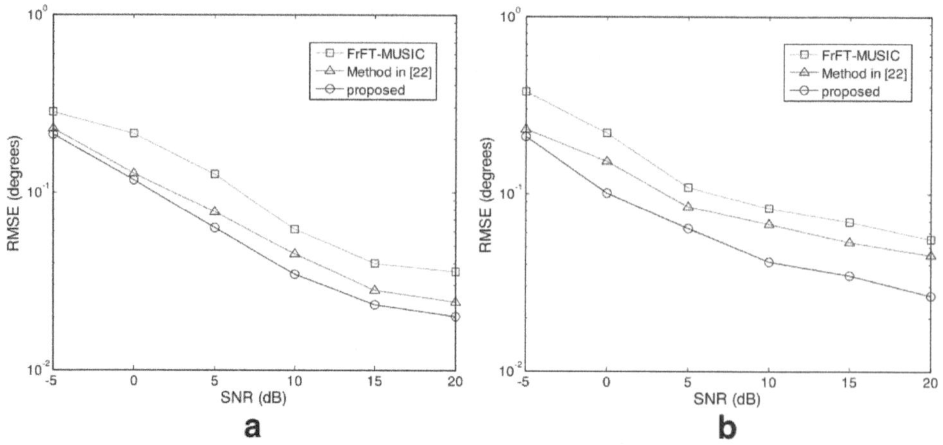

Figure 4: The estimation RMSE for the central angle versus the SNR. **a** Source 1. **b** Source 2.

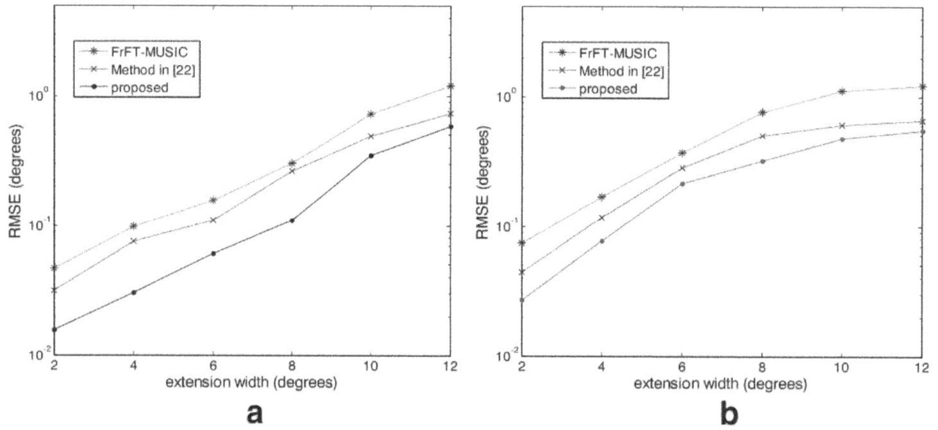

Figure 5: RMSE for the central angle versus the extension width. **a** Source 1. **b** Source 2.

Finally, the resolution performance of the proposed algorithm versus the different source number is demonstrated. In this experiment, the chirp rate interval $\Delta\mu$ of each source is chosen as 1 Hz/s. The numerical results can be seen in Fig. 6. The proposed algorithm can remain the good resolution performance even though the source number is far more than the sensor number.

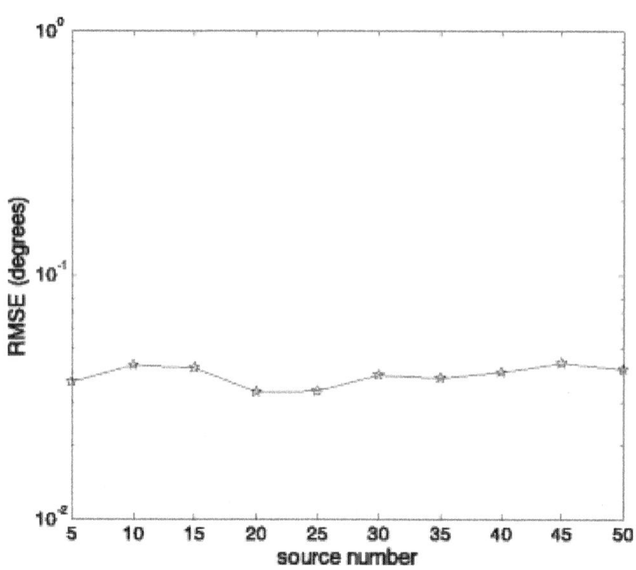

Figure 6: RMSE for the central angle versus the source number.

CONCLUSIONS

In this paper, parametric localization of multiple wideband chirp sources with local scattering have been considered. The models in both the time and fractional Fourier domain were proposed. In the Dechirping domain, a novel DOA estimator combined the properties of the chirp signal and the Taylor series expansion was addressed. The DOA of each source and the source number can be determined using the proposed method. Besides, the source number is allowed to exceed the sensor number in the array. The simulation study demonstrated that the proposed algorithm provided a superior spatial resolution performance for wideband chirp sources with local scattering than the previous FrFT-MUSIC algorithm.

ACKNOWLEDGEMENTS

This work is funded by the National Natural Science Fund of China (No. 61501322).

Competing Interests

The authors declare that they have no competing interests.

REFERENCES

1. AB Gershman, MG Amin, Wideband direction-of-arrival estimation of multiple chirp signals using spatial time-frequency distributions. Signal Process IEEE Trans. 7(6), 152–155 (2000)

2. F Sellone, Robust auto-focusing wideband DOA estimation. Signal Process. **86**(1), 17–37 (2006)
3. PH Leong, T Abhayapala, TA Lamahewa, Multiple target localization using wideband echo chirp signals. Signal Process IEEE Trans **61**(16), 4077–4089 (2013)
4. KI Pedersen, PE Mogensen, BH Fleury, Power azimuth spectrum in outdoor environments. Electron. Lett. **33**(18), 1583–1584 (1997)
5. CF Patrick, Angular propagation descriptions relevant for base station adaptive antenna operations. Wireless Pers. Commun. **11**(1), 3–29 (1999)
6. M Tapio, *Direction and Spread Estimation of Spatially Distributed Signals via the Power Azimuth Spectrum, (IEEE Acoustics, Speech, and Signal Processing (ICASSP)* (Orlando, FL, USA, 2002), pp. 3005–3008
7. D Astely, B Ottersten, The effect of local scattering on direction of arrival estimation with MUSIC. Signal Process IEEE Trans. **47**(12), 3220–3234 (1999)
8. J Tabrikian, H Messer, *Robust Localization of Scattered Sources* (IEEE Statistical Signal and Array Processing, Pcono Manor, PA, 2000), pp. 453–457
9. G Fuks, J Goldberg, H Mesor, Bearing estimation in a Ricean channel-Part 1: inherent accuracy limitations. IEEE Trans. Signal Processing **49**(5), 925–937 (2001)
10. S Valaee, B Champagne, P Kabal, Parametric localization of distributed source. Signal Process IEEE Trans. **43**(9), 2144–2153 (1995)
11. R Raic, J Goldberg, H Messer, Bearing estimation for a distributed source: modeling inherent accuracy imitation and algorithms. IEEE Trans. Signal Processing **48**(2), 429–441 (2000)
12. 1M Bengtsson, N Ottersten, Low complexity estimation for distributed sources. IEEE Trans. Signal Process. **48**(8), 2185–2194 (2000)
13. S Shahbazpanahi, S Valaee, AB Gershman, A covariance fitting approach to parametric localization of multiple incoherently distributed sources. IEEE Trans. Signal Process **52**(3), 592–600 (2003)
14. BT Sieskul, S Jitapunkul, An asymptotic maximum likelihood for joint estimation nominal angle and angular spreads of multiple spatially distributed sources. IEEE Trans. Veh. Technol. **59**(3), 1534–1538 (2006)
15. J Lee, J Joung, JD Kim, A method for the direction-of-arrival estimation of incoherently distributed sources. IEEE Trans. Veh. Technol. **57**(5), 2885–2893 (2008)
16. A Zoubir, Y Wang, Performance analysis of the generalized beamforming estimators in the case of coherently distributed sources. Signal Process **88**, 428–435 (2008)
17. K Han, A Nehorai, Nested array processing for distributed sources. IEEE Signal Process. Lett. **21**(9), 1111–1114 (2014)
18. M Ghogho, TS Durrani, Broadband direction of arrival estimation in presence of angular spread. Electronics Letters **37**(15), 986–987 (2001)
19. SJ Liu, Q Wan, J Yang, YN Peng, Asymptotic performance analysis of bearing estimate for spatially distributed source with finite bandwidth. IEEE Electron. Lett. **38**(24), 1600–1601 (2002)
20. F Foroozan, A Asif, Time reversal based active array source localization. IEEE Trans. Signal Process. **59**(6), 2655–2668 (2011)
21. CF Mecklenbrauker, P Gerstoft, H Yao, *Bayesian Sparse Wideband Source Reconstruction of Japanese 2011 Earthquake* (IEEE Computational Advances in Multi-Sensor Adaptive Processing (CAMSAP), San Juan, 2011), pp. 273–276
22. L Zhang, J Yu, K Liu, D Liu, *A Novel DOA Estimation Algorithm for Wideband LFM Source with Local Scattering* (International Conference on Communications, Signal Processing, and Systems, Hohhot, China, 2014), pp. 491–499
23. DMJ Cowell, S Freear, Separation of overlapping linear frequency modulated (LFM) signals using the fractional Fourier transform. IEEE Trans. Ultrason., Ferroelect., Freq. Contr **57**(10), 2324–2333 (2010)
24. R Chen, Y Wang, Universal FRFT-based algorithm for parameter estimation of chirp signals. Journal of Systems Engineering and Electronics **18**(4), 495–501 (2012)
25. HM Ozaktas, MA Kuta, D Mendlovic, Introduction to the fractional Fourier transform and its applications. Adv. Imag. Electron. Phys. **10**(6), 239–291 (1999)
26. V Ashok Narayanan, KMM Prabhu, The fractional Fourier transform: theory, implementation and error analysis. Microprocessors and Microsystems **27**, 511–521 (2003)
27. LB Almeida, The fractional Fourier transform and time-frequency representations. IEEE Trans. Signal Process **42**(11), 3084–3091 (1994)
28. HM Ozaktas, O Arikan, MA Kutay, G Bozdagi, Digital computation of the fractional Fourier transform. IEEE Trans. Signal Process. **44**(9), 2141–2150 (1996)
29. D Asztely, B Ottersten, AL Swindlehurst, Generalised array manifold model for wireless communication channels with local scattering. Proc. Inst. Electr. Eng.—Radar, Sonar Navigat **145**(1), 51–57 (1998)
30. R Tao, Novel method for the direction of arrival estimation of wideband linear frequency modulated sources based on fractional Fourier transform. Transaction of Beijing Institute of Technology. **25**(10), 895–899 (2005)
31. JX Yu, L Zhang, KH Liu, Coherently distributed wideband LFM source localization. IEEE Signal Processing Letters **22**(4), 504–508 (2015)

CHAPTER 12

QUANTUM AND CLASSICAL DISCRETE FRACTIONAL FOURIER TRANSFORMS

Steffen Weimann, Armando Perez-Leija, Maxime Lebugle, Robert Keil, Malte Tichy, Markus Gräfe, Rene´ Heilmann, Stefan Nolte, Hector Moya-Cessa, Gregor Weihs, Demetrios N. Christodoulides and Alexander Szameit

INTRODUCTION

Two hundred years ago, Joseph Fourier introduced a major concept in mathematics, the so-called Fourier transform (FT). It was not until 1965, when Cooley and Tukey developed the 'fast Fourier transform' algorithm, that Fourier analysis became a standard tool in contemporary sciences1. Two crucial requirements in this algorithm are the discretization and truncation of the domain, where the signals to be transformed are defined. These requirements are always satisfiable, since observable quantities in physics must be well behaved and finite in extension and magnitude.

In 1980, Namias made another significant leap with the introduction of the fractional Fourier transform (FrFT), which contains the FT as a special case2. Several investigations quickly followed, leading to a more general theory of joint time-frequency signal representations3 and fractional Fourier optics4. The vast scope of the FrFT has been demonstrated in areas such as wave propagation, signal processing and differential equations3· 5· 6· 7. So far, the FT of fractional order was realized only by single-lens systems8· 9, although other theoretical suggestions, including multi-lens systems10 or graded index fibres exist11. The aim to discretize this generalized FT led to the introduction of the discrete fractional Fourier transform (DFrFT) operating on a finite grid in a way similar to that of a discrete FT12. Along those lines, several versions of the DFrFT have been introduced5, however, without any experimental realization, so far. In this work we focus on the optical implementation of the so-called Fourier–Kravchuk transform12 that can be equally applied to the classical and quantum states. Throughout our paper, we simply refer to this transform as DFrFT whose application reaches from the demonstration of the Fourier suppression law13, N00N-state generation14· 15 and qubit storage16 to the realization of perfect discrete lenses for non-uniform input distributions.

In this work, we report on the realization of DFrFTs of one-dimensional optical signals based on an integrated lattice of evanescently coupled waveguides. In these photonic arrangements, the inter-channel couplings are designed in such a way that the system readily performs the DFrFT of any incoming signal. The signal evolution is governed by the Schrödinger equation and the associated Hamilton operator is known as the J_x-operator in the quantum theory of angular momentum or likewise as the Heisenberg XY model from the quantum theory of ferromagnetism. We foresee that the inherent versatility of this approach will make other realizations of the DFrFT, the FT and the fast Fourier transform recognizably simple and thus may open the door to many interesting applications in integrated quantum computation17.

RESULTS

Theoretical Approach

Similarly to its continuous counterpart, the DFrFT can be interpreted physically as a continuous rotation of the associated wave functions through an angle Z in phase space (see Fig. 1a)18. The idea is thus to construct finite circuits that are capable of imprinting such rotations to any light field.

Citation: Weimann, S. et al. Quantum and classical discrete fractional Fourier transforms. Nat. Commun. 7:11027 doi: 10.1038/ncomms11027 (2016).

In quantum mechanics, three-dimensional spatial rotations of complex state vectors are generated via operations of the angular momentum operators J_k (k=x, y, z) on the Hilbert space of the associated system19. In particular, the rotation imprinted by the J_x-operator turns out to be an elaborated definition of the DFrFT (see Methods section for discussion). These concepts can be readily translated to the optical domain by mapping the matrix elements of the J_x-operator over the inter-channel couplings of engineered waveguide arrays (Fig. 1b–e)20. The coupling matrix of such waveguide arrays is thus given by $(J_x)_{m,n}=\kappa_0\left(\sqrt{(j-m)(j+m+1)}\delta_{m+1,n}+\sqrt{(j+m)(j-m+1)}\delta_{m-1,n}\right)/2$ (ref. 19). Here, κ_0 is a scaling factor introduced for experimental reasons. The indices m and n range from −j to j in unit steps. Meanwhile, j represents an arbitrary positive integer or half-integer that determines the total number of waveguides via N=2j+1 (Fig. 1b).

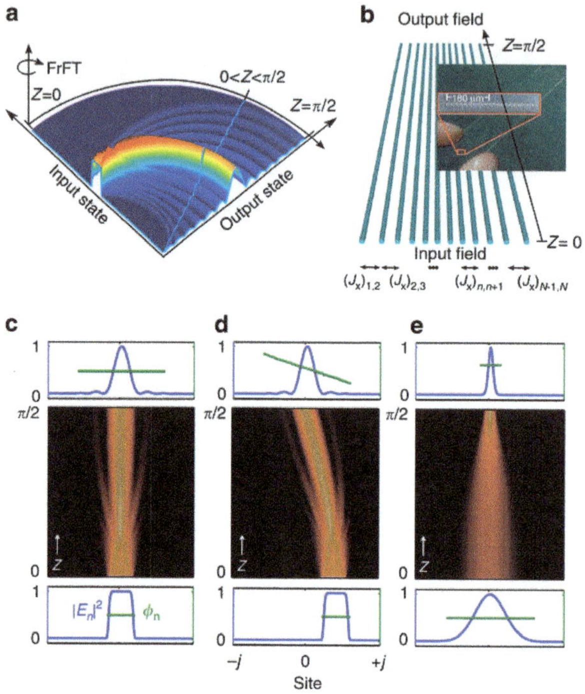

Figure 1: Discretization of the FrFT. (**a**) Pictorial view of actual fractional Fourier transforms exemplified as continuous rotations in phase space. (**b**) Schematic representation of a pre-engineered J_x-array. (**c–e**) Top views of continuous 'rotations' of a rectangular (**c**), displaced rectangular (**d**) and Gaussian (**e**) optical wave functions in a J_x-array with N=151. The bottom and top plots show the intensities of the ingoing and outgoing wave packets, respectively. The green lines describe the magnitude of phase distributions of the optical fields, that is, the phase jumps of π due to a change of sign of the signal's amplitude are not shown.

Coupled mode theory states that the evolution of light in the J_x-waveguide array is governed by the following set of equations20

$$i\frac{d}{dZ}E_m(Z)=\frac{1}{\kappa_0}\sum_{n=-j}^{j}(J_x)_{m,n}E_n(Z). \tag{1}$$

Here, $E_n(Z)$ denotes the complex electric field amplitude at site n. In the quantum optics regime, single photons traversing such devices are governed by a set of Heisenberg equations that are isomorphic to equation (1). The only difference is that in the quantum case $E_n(Z)$ must be replaced by the photon creation operator $a_n^{\dagger}(Z)$. In a spintronic context, the evolution parameter Z is associated with time, whereas in the framework of integrated quantum optics, Z represents the propagation distance, see Fig. 1b–e. A spectral decomposition of the J_x-matrix yields the eigenvectors $u_n^{(m)}=2^n\sqrt{\frac{(j+n)!(j-n)!}{(j+m)!(j-m)!}}P_{j+n}^{(m-n,-m-n)}(0)$14,19 14, 19, which in combination with the eigenvalues, $\beta_m=-j,\ldots,j$, render the closed-form point-spread function

$$G_{p,q}(Z)=i^{p-q}\sqrt{\frac{(j+p)!(j-p)!}{(j+q)!(j-q)!}}\left[\sin\left(\frac{Z}{2}\right)\right]^{q-p}\left[\cos\left(\frac{Z}{2}\right)\right]^{-q-p}P_{j+p}^{(q-p,-q-p)}(\cos(Z)). \tag{2}$$

Note that q and p represent the excited and observed sites, respectively, and $P_n^{(A,B)}(x)$ are the Jacobi polynomials of order n (see Methods section for discussion). Using equation (2), we can compute the response of the system to any input signal, which in turn gives the DFrFT12. Accordingly, DFrFT of any particular order arises at one specific propagation distance Z lying between 0 and $\pi/2$. In the limit N→∞, the eigenvectors of J_x, $u^{(m)}_{x\sqrt{2/N}}$, become the continuous Hermite–Gauss polynomials $H_m(x)$, which are known to be the eigenfunctions of the fractional Fourier operator12. As a result, in the continuous limit, the DFrFT described by equation (2) converges to the continuous FrFT5, 12; and the standard FT is recovered at Z=π/2 (Fig. 1c–e). Note that in general, the DFrFT obtained in our devices and the usual DFT become equal only in the continuous limit N→∞ and at Z=π/2.

Experiments with Classical Light

To experimentally demonstrate the functionality of the suggested waveguide system, we use N=21 waveguides to perform FTs of simple wave packets. We first consider a Gaussian wave packet with a full-width at half-maximum (FWHM) covering the five central sites (Fig. 2a). The input signal is prepared by focusing a Gaussian beam from a HeNe laser onto the front facet of the sample. By exploiting the fluorescence from colour centres within the waveguides21, we monitor the full intensity evolution from the input to the output plane. The fluorescence image, Fig. 2a, shows a gradual transition from an initially narrow Gaussian distribution at the input to a broader one at the Fourier plane (left and right panels Fig. 2a), demonstrating that narrow signals in space correspond to broad signals in Fourier space. For intermediate propagation distances ($Z \in [0, \pi/2]$) we extract other orders of the DFrFT, simultaneously. For comparison, we plot the continuous FrFT produced by the corresponding continuous Gaussian profile (red curves Fig. 2a). The agreement between the computed FrFT and the experimental DFrFT proves that for the considered Gaussian input signal, N=21 is sufficient to achieve the continuous limit. We now shift the input Gaussian beam by six channels towards the edge. Since the separations between adjacent waveguides at the edges are bigger than the separations between adjacent waveguides in the centre, the discretization grid is not perfectly homogeneous. Strictly speaking, the discretized shifted Gaussian just at the input plane covers slightly less than five waveguides FWHM. We observe that the well-approximated off-centre Gaussian travels to the centre at $Z=\pi/2$ (Fig. 2b), hereby showing the famous shift theorem. In additional experiments, extended signals, for example, a shifted top-hat function, are found to be well transformed according to equation (2) as well. However, we find that for this type of excitation N>21 would be required to discuss the continuous limit.

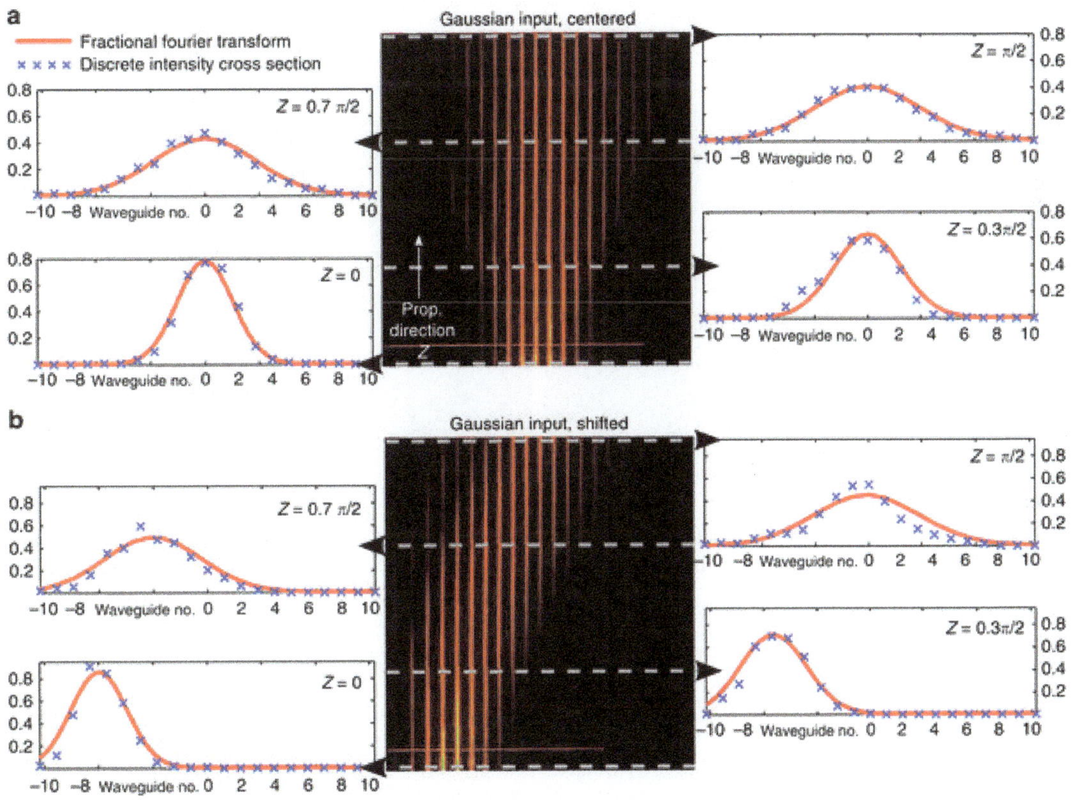

Figure 2: DFrFT of classical light. (**a**) Transformation of a Gaussian input into a Gaussian profile of larger width along the evolution in the J_x-array. The FT is obtained at $Z=\pi/2$. The experimental data (blue crosses) is compared with the numeric FrFT (red curves). (**b**) A shifted input Gaussian profile evolves towards the centre of the array and acquires the same width as in **a**.

An unequivocal criterion, for the functionality of devices that perform the DFrFT, equation (2), can be formulated by evaluating $G_{p,q}(\frac{\pi}{2})$. At this particular distance, point-like excitations will give rise to signal magnitudes that perfectly resemble the magnitudes of one of the eigenstates of the transform. More specifically, for transforms such as equation (2), one finds that an excitation of theqth site excites the qth system eigenstate up to local phases $(G_{p,q}(\frac{\pi}{2})=i^{p-q}u_p^{(q)})$ (see the Methods section for explanations). The experimental demonstration of this intriguing effect is shown in the subpanels of Fig. 3a–d along with the theoretical predictions. It can be argued that for any point-like excitation, the continuous limit cannot be met experimentally (see the Methods for discussion). Instead, equation (2) creates a non-uniform amplitude distribution with a phase difference of $\pi/2$ between adjacent sites. Nevertheless, in the continuous limit, $G_{p,q}(\frac{\pi}{2})$ tends to the usual FT kernel12. At this point, it is worth emphasizing the formal equivalence to the quantum HeisenbergXY model in condensed matter physics22, 23. In this respect, our observations demonstrate the capability of the here-presented systems to store quantum information in XY Hamiltonians by converting specific inputs into eigenstates of the system16. To our knowledge, this rather rare property has never been thoroughly investigated before.

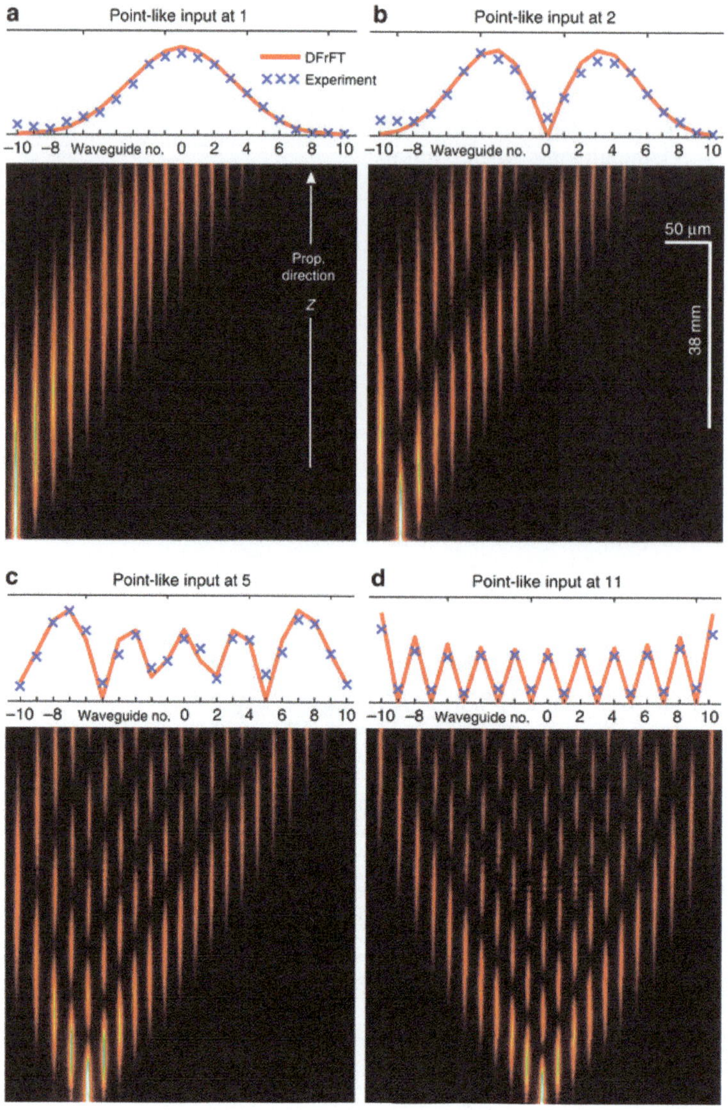

Figure 3: Experimental visualization of the discrete Hermite–Gauss polynomials. (**a–d**) Evolution of single-site inputs into the magnitudes of the respective eigensolutions, as predicted theoretically (methods). The experimental data (blue crosses) is compared with the analytic DFrFT (red curves).

Quantum Experiments

To demonstrate the applicability of our approach in the quantum domain, we now analyse intensity correlations of separable and path-entangled photon pairs propagating through these Fourier transformers. To do so, we fabricated J_x-arrays involving N=8 channels. The importance of exploring FTs of such states has been highlighted in several investigations, demonstrating interesting effects such as suppression of states and portraying biphoton spatial correlations24·25·26.

In this discrete quantum optical context, pure separable two-photon states are readily produced by coupling pairs of indistinguishable photons into two distinct lattice sites (m, n), this state is mathematically described by $|\Psi(0)\rangle = a_m^\dagger a_n^\dagger |0\rangle$. Conversely, path-entangled two-photon states are created by simultaneously launching both photons at either site m or n with exactly the same probability, that is, $|\Psi(0)\rangle = [(a_m^\dagger)^2 + (a_n^\dagger)^2]|0\rangle/2$. Furthermore, the probability of observing one of the photons at site k and its twin at site l is given by the intensity correlation matrix $\Gamma_{k,l}(Z) = \langle a_k^\dagger a_l^\dagger a_l a_k \rangle$ (ref. 27). An intriguing and unique property of the J_x-systems is that at $Z=\pi/2$ the correlation matrices are given in terms of the eigenstates, as noticed above. Hence, for the separable case, $|\Psi(0)\rangle = a_m^\dagger a_n^\dagger |0\rangle$, the correlation matrices are given by $\Gamma_{k,l} = |u_k^{(m)} u_l^{(n)} + u_k^{(n)} u_l^{(m)}|^2$, whereas for the path-entangled state, $|\Psi(0)\rangle = [(a_m^\dagger)^2 + (a_n^\dagger)^2]|0\rangle/2$, we have $\Gamma_{k,l} = |(-i)^{2m} u_k^{(m)} u_l^{(m)} + (-i)^{2n} u_k^{(n)} u_l^{(n)}|^2$. Of particular interest is the separable case, where the photons are symmetrically coupled into the outermost waveguides, $|\Psi(0)\rangle = a_j^\dagger a_{-j}^\dagger |0\rangle$. In this scenario, only the correlation matrix elements for which (k+l)=odd are nonzero, and are given by $\Gamma_{k,l} = 4^{k+l+1}(j+k)!(j-k)!(j+l)!(j-l)!(P_{j+k}^{(j-k,-j-k)}(0)P_{j+l}^{(-j-l,j-l)}(0))/[(N-1)!]^2$. These effects are demonstrated for the initial state $|\Psi(0)\rangle = a_{-\frac{7}{2}}^\dagger a_{\frac{7}{2}}^\dagger |0\rangle$ in Fig. 4a, where concentration and absence of probability in the correlation matrix clearly show that some states are completely suppressed—a hallmark of any Fourier unitary process13. An estimation of the statistical significance of the data set, along with a short discussion on incoherence effects can be found in the Supplementary Note 2 involving Supplementary Figs 2 and 3.

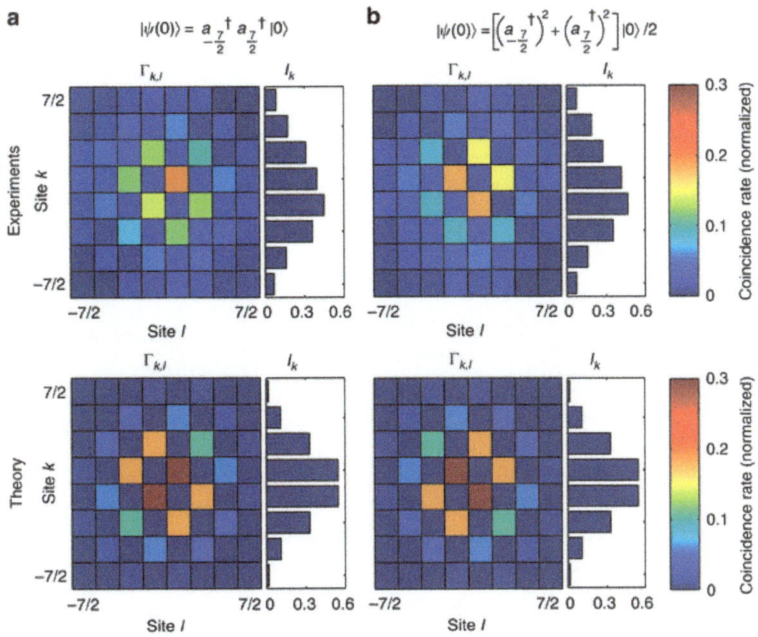

Figure 4: DFrFT of quantum light. Correlation maps $\Gamma_{k,l}$ of a two-photon state either prepared (**a**) in a product state or (**b**) in a path-entangled state after propagating through a J_x-array. The photon density I_k at the output is shown on the right side of each map. The evaluation of the s.d. of the coincidence rates is presented in the Supplementary Note 2.

As a second case, we consider a fully symmetric path-entangled two-photon state of the form $|\Psi(0)\rangle = [(a_j^{\dagger})^2 + (a_{-j}^{\dagger})^2]|0\rangle/2$. Physically, both photons are entering together into the array at either site j or −j with equal probability28, 29, 30. The correlations are determined by $\Gamma_{k,l} = |u_k^{(j)} u_l^{(j)} + u_k^{(-j)} u_l^{(-j)}|^2$, from which we infer that the probability of measuring photon coincidences at coordinates (k, l) vanishes at sites where the sum (k+l) is odd. In contrast, at coordinates where (k+l) is even, the correlation function collapses to the expression $\Gamma_{k,l} = 4^{k+l+1}(j+k)!(j-k)!(j+l)!(j-l)!(P_{j+k}^{(j-k,-j-k)}(0)P_{j+l}^{(-j-l,j-l)}(0))^2/[(N-1)!]^2$. This indicates that in this path-entangled case the correlation map appears rotated by 90° with respect to the matrix obtained with separable two-photon states. We performed an experiment to demonstrate these predictions using states of the type $|\Psi(0)\rangle = [(a^{\dagger}_{-\frac{7}{2}})^2 + (a^{\dagger}_{\frac{7}{2}})^2]|0\rangle/2$, which were prepared using a 50:50 directional coupler acting as a beam splitter29. The whole experiment is achieved using a single chip containing both the state preparation stage followed by a J_x-system, yielding high interferometric control over the field dynamics (Fig. 5). The experimental measurements are presented in Fig. 4b. Similarly, suppression of states occurs as a result of destructive quantum interference. As predicted, a closer look into the correlation pattern reveals that indeed the correlation map appears rotated by 90° with respect to the matrix obtained with separable two-photon states.

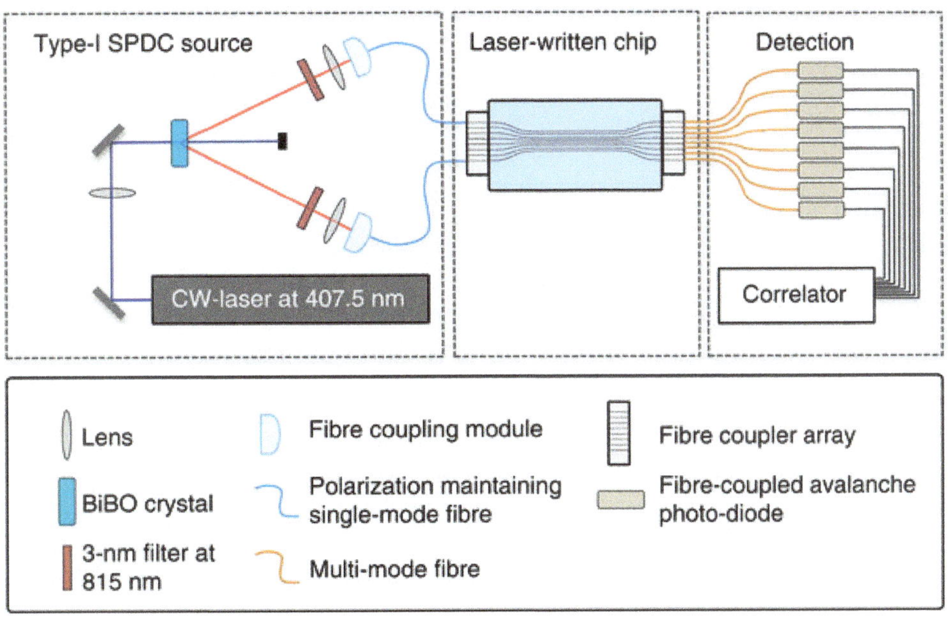

Figure 5: Set-up used to carry out spatial correlation measurements of photonic quantum states. The set-up consists of three parts—the state preparation, the execution of the DFrFT and the correlation measurement.

DISCUSSION

We emphasize that our quantum measurements feature interference fringes akin to the ones observed in quantum Young's two-slit experiments of biphoton wave functions in free space as demonstrated in ref. 24. In such free-space experiments, however, far-field observations were carried out using lenses and the two slits were emulated by optical fibres24. Along those lines, we have created a fully integrated quantum interferometer to observe fundamental quantum mechanical features25. This additionally suggests an effective way to generate quantum states containing only even (odd) non-vanishing inter-particle distance probabilities for the separable input state (symmetric path-entangled state). In addition, the eigenfunctions associated with the Hamiltonian system explored in our work are specific Jacobi polynomials, which are well known as the optimal basis for quantum phase-retrieval algorithms31 and these eigenstates can be retrieved by limited phase operations. Knowing a quantum wavefunction and its FT, a phase-retrieval algorithm for signals that are a superposition of a finite number of Hermite–Gauss polynomials has been introduced32. This phase-retrieval algorithm might be implemented employing our system, since its discrete character automatically possesses a finite number of polynomials that are closely connected to the Hermite–Gauss polynomials. Another potential application is the realization of the Radon–Wigner transform given by the squared modulus of the FrFT6, 33. The Radon–Wigner function is a basic tool for the reconstruction of Wigner quasi-probability distributions in quantum optics34, 35. Also, FrFTs appear naturally in optics as free-space propagation between two spherical reference planes in general4. Like in our system, the order of the FrFT is proportional to the propagation distance. On that basis, complex spatial filtering involving several fractional Fourier planes was suggested36. In this description, the optical signal is discretized and thus described by a vector of field amplitudes at certain sites. In our approach this is inherently realized.

In conclusion, we have successfully demonstrated a universal discrete optical device capable of performing classical and quantum DFrFTs. Our studies might find applications in developing a more general quantum suppression law13 and perhaps in the development of new quantum algorithms.

METHODS

The Fractional Fourier Transform in quantum harmonic oscillations

In this section, we briefly describe the relation between the continuous FrFT operator and the Hamiltonian of the quantum harmonic oscillator2.

The FrFT operator $\hat{F}_Z$ is defined by the following eigenvalue equation involving the Hermite–Gauss polynomials of order n and the eigenvalues $\lambda_n=\exp(inZ)$

$$\hat{F}_Z\{\exp(-x^2/2)H_n(x)\} = \exp(inZ)\exp(-x^2/2)H_n(x), \quad (3)$$

where $Z\in\mathbb{R}$. Concurrently, one can interpret equation (3) as quantum time evolution from timet=0 to t=Z. We write $\hat{F}_Z$ as $\exp(iZ\hat{A})$.

$$\exp(iZ\hat{A})\{\exp(-x^2/2)H_n(x)\} = \exp(inZ)\exp(-x^2/2)H_n(x). \quad (4)$$

To show that $\hat{A}$ is the Hamilton operator of the harmonic oscillator, we differentiate both sides of equation (4) with respect to Z and evaluate the result at Z=0. We obtain

$$\hat{A}\exp(-x^2/2)H_n(x) = n\exp(-x^2/2)H_n(x). \quad (5)$$

To find the spatial representation of $\hat{A}$, consider the differential equation

$$\left(-\frac{1}{2}\frac{d^2}{dx^2}+x\frac{d}{dx}\right)H_n(x) = nH_n(x), \quad (6)$$

for the Hermite polynomials $H_n(x)$ of order n. Using the identities $1=\exp(x^2/2)\exp(-x^2/2)$ and $\exp(\xi x^2/2)\left(\frac{d^n}{dx^n}\right)\exp(-\xi x^2/2)=\left(\frac{d}{dx}-\xi x\right)^n$, one can show that equation (6) can be written as

$$\left(-\frac{1}{2}\frac{d^2}{dx^2}+\frac{1}{2}x^2-\frac{1}{2}\right)\exp(-x^2/2)H_n(x) = n\exp(-x^2/2)H_n(x), \quad (7)$$

where we have used the commutator $[x,\frac{d}{dx}]=-1$. Comparing equation (5) and equation (7), one can see that $\hat{A}$ becomes

$$\hat{A} = \left(-\frac{1}{2}\frac{d^2}{dx^2}+\frac{1}{2}x^2-\frac{1}{2}\right). \quad (8)$$

In summary, the FrFT operator $\hat{F}_Z$ can be written as

$$\hat{F}_z = \exp\left(\frac{iz}{2}\left(-\frac{d^2}{dx^2}+x^2-1\right)\right). \quad (9)$$

Because of this one-to-one correspondence between the dynamics of the quantum harmonic oscillator and the fractional Fourier operator, the implementation of such transform is immediate using harmonic oscillator systems37.

J_x-photonic Lattices as Discrete Harmonic Oscillators

Our aim in this section is to show that in the continuous limit N→∞, the eigenvalue equation for J_x-arrays becomes the eigenvalue equation of the quantum harmonic oscillator. Consider the matrix representation of the J_x-operator again. For convenience we take, in this section only, $\kappa_0=1$.

$$\begin{aligned}(J_x)_{m,n} &= \tfrac{1}{2}\left(\sqrt{(j-m)(j+m+1)}\delta_{n,m+1}+\sqrt{(j+m)(j-m+1)}\delta_{n,m-1}\right)\\ &= \tfrac{1}{2}\left(\sqrt{j(j+1)-m(m+1)}\delta_{n,m+1}+\sqrt{j(j+1)-m(m-1)}\delta_{n,m-1}\right).\end{aligned} \quad (10)$$

The indices m and n range from −j to j in unit steps and j is an arbitrary positive integer or half-integer. The dimension of the J_x-matrix is N=2j+1. We now introduce the variable $\gamma=j(j+1)=(N^2-1)/4$, which implies that $(J_x)_{m,n}$ can be written as

$$(J_x)_{m,n}=\frac{\sqrt{\gamma}}{2}\left(\sqrt{1-\frac{1}{\gamma}m(m+1)}\delta_{n,m+1}+\sqrt{1-\frac{1}{\gamma}m(m-1)}\delta_{n,m-1}\right). \quad (11)$$

Let us consider the eigenvalue equation for this matrix

$$\frac{\sqrt{\gamma}}{2}\left(\sqrt{1-\frac{1}{\gamma}m(m+1)}\psi_{m+1}+\sqrt{1-\frac{1}{\gamma}m(m-1)}\psi_{m-1}\right)=\beta_m\psi_m. \quad (12)$$

Considering the region $m \ll j$, since $\gamma \propto N^2$, in the limit N $\to\infty$, the terms $m(m \pm 1)/\gamma \ll 1$. Hence, in the domain far from the edge of the array a Taylor expansion yields

$$\sqrt{\gamma}\sqrt{1-\frac{m}{\gamma}(m\pm 1)} \approx \sqrt{\gamma}\left(1-\frac{m}{2\gamma}(m\pm 1)-\frac{m^2}{8\gamma^2}(m\pm 1)^2\right). \quad (13)$$

By defining $m=x\gamma^{1/4}$ (or $x=m/\gamma^{1/4}$), we obtain

$$\sqrt{\gamma}\sqrt{1-\frac{m}{\gamma}(m\pm 1)} \approx \gamma^{1/2}-\frac{x^2}{2}\mp\frac{x}{2\gamma^{1/4}}-\frac{x^4}{8\gamma^{1/2}} \quad (14)$$

Plugging this expression into equation (12)

$$\left(\gamma^{1/2}-\frac{x^2}{2}-\frac{x}{2\gamma^{1/4}}-\frac{x^4}{8\gamma^{1/2}}\right)\psi_{m+1}+\left(\gamma^{1/2}-\frac{x^2}{2}+\frac{x}{2\gamma^{1/4}}-\frac{x^4}{8\gamma^{1/2}}\right)\psi_{m-1} \approx 2\beta_m\psi_m \quad (15)$$

We redefine the functions

$\psi_m=\psi(x)=\psi(\frac{m}{\gamma^{1/4}})$ and $\psi_{m+1}=\psi(x+\frac{1}{\gamma^{1/4}})=\psi(\frac{m}{\gamma^{1/4}}+\frac{1}{\gamma^{1/4}})$ such that we can introduce the Taylor series

$$\psi_{m\pm 1}=\psi\left(x\pm\frac{1}{\gamma^{1/4}}\right)=\psi(x)\pm\frac{1}{\gamma^{1/4}}\psi'(x)+\frac{1}{2\gamma^{1/2}}\psi''(x), \quad (16)$$

where, again, we have kept only terms up to second order in $1/\gamma^{1/4}$. Substituting equation (16) into equation (15), and using the limit

$$\lim_{N\to\infty}\sqrt[4]{\left(\frac{4}{N^2}\right)\left(\frac{1}{1-\frac{1}{N^2}}\right)}=0.$$

We obtain the time-independent Schrödinger equation for the harmonic oscillator

$$\left(-\frac{1}{2}\frac{d^2}{dx^2}+\frac{1}{2}x^2\right)\psi(x)=(\sqrt{\gamma}-\beta)\psi(x). \quad (17)$$

Therefore, in the continuous limit N$\to\infty$, the difference equation describing J_x-photonic lattices becomes the time-independent Schrödinger equation for the quantum harmonic oscillator. Note, however, that due to the importance of the condition $m \ll j$ in this derivation, this statement is only valid when dealing with signals that are square integrable in the continuous limit. Thus, the operator equation (11) can be used to define the discrete version of the quantum harmonic oscillator and thus the DFrFT.

The Point-spread Function for J_x-Photonic Lattices

In this section, it is shown that at $Z=\pi/2$, the green function of J_x-systems becomes proportional to the amplitude of one of the eigenstates. The evolution of light in J_x-arrays is governed by the set of N coupled differential equations (equation (1)).

The normalized propagation coordinate Z is given by $Z=\kappa_0 z$, where z is the actual propagation distance and κ_0 is an arbitrary scale factor. The quantity $E_n(Z)$ denotes the mode field amplitude at site n. A spectral decomposition of the J_x-matrix yields the eigensolutions

$$u_n^{(m)}=(2)^n\sqrt{\frac{(j+n)!(j-n)!}{(j+m)!(j-m)!}}P_{j+n}^{(m-n,-m-n)}(0). \quad (18)$$

$P_n^{(A,B)}(x)$ are the Jacobi polynomials of order n. And the corresponding eigenvalues are integers or half-integers, $\beta_m=-j, \ldots, j$, depending on the parity of N14, 19. Using the eigenvectors and eigenvalues we obtain the point-spread function

$$G_{p,q}(Z)=\sum_{r=-j}^{j}u_q^{(r)}u_p^{(r)}\exp(irZ). \quad (19)$$

$G_{p,q}(Z)$ represents the amplitude at site p after an excitation of site q. Using equation (18) and the properties of the Jacobi polynomials one can show that equation (19) reduces to the closed-form expression

$$G_{p,q}(Z)=(-i)^{q-p}\sqrt{\frac{(j+p)!(j-p)!}{(j+q)!(j-q)!}}\left[\sin\left(\frac{Z}{2}\right)\right]^{q-p}\left[\cos\left(\frac{Z}{2}\right)\right]^{-q-p}P_{j+p}^{(q-p,-q-p)}(\cos(Z)). \quad (20)$$

Evaluation of equation (20) at $Z=\pi/2$ yields

$$G_{p,q}\left(\frac{\pi}{2}\right)=(-i)^{q-p}(2)^p\sqrt{\frac{(j+p)!(j-p)!}{(j+q)!(j-q)!}}P_{j+p}^{(q-p,-q-p)}(0)=(-i)^{q-p}u_p^{(q)}. \quad (21)$$

Equation (21) shows that at $Z=\pi/2$ the point-spread function becomes proportional to the amplitude of the corresponding eigenstates depending on the excited site. In other words, there is a one-to-one correspondence between the excited site number and the eigenstates of the system: excitation of the qth site excites the qth eigenstate up to well-defined local phases.

Devices Fabrication and Specifications

Our devices are fabricated in bulk-fused silica samples (Corning 7980, ArF grade) using the femtosecond laser direct-write approach21. The transparent material is modified within the focal region due to nonlinear absorption resulting in a local increase of the refractive index. Effectively, the waveguides possess only the fundamental mode. The coupling between neighbouring waveguides depends on their separation within the glass chip. Regarding the theoretical description of the waveguide array, the validity of equation (1) is only given if the fundamental modes of neighbouring waveguides have a negligible overlap. For a given N and a maximum length of the glass wafers of Z/κ_0, this can only be ensured for a certain range of κ_0. Outside this range of κ_0, coupled mode theory will break down and errors are introduced to the implementation of the DFrFT. In our experiments these errors are kept small but impossibly perfectly zero.

For the fabrication of the devices used to transform classical light, we employed an Yb-doped fibre laser (Amplitude Systèmes) operating at a wavelength of 532 nm, a repetition rate of 200 kHz and a pulse length of 300 fs. Waveguides were written with 300 nJ pulses focused by a 20x objective. The sample was moved at a velocity of 200 mm min^{-1} by high-precision positioning stages (ALS 130, Aerotech Inc.) with a positioning error of ±0.1 µm. From this random positioning error, the realized inter-channel couplings inherit a relative error of 2%. The mode field diameters of the guided mode were 4 µm × 7 µm at 632 nm. In the classical experiments, the Fourier plane lies at Z/κ_0=7.48 cm, that is, κ_0=0.21 cm^{-1}. The desired nearest-neighbor couplings $(J_x)_{n\pm 1,n}$ determine the separations of the waveguides n and n±1. The largest separation of 22.8 µm between adjacent waveguides occurs at the edges. In the centre, the separation is 17.6 µm.

For the samples illuminated with single-photon states of light, we used a RegA 9,000 seeded by a Mira Ti:Sa femtosecond laser oscillator. The amplifier produced 150 fs pulses centred at 800 nm at a repetition rate of 100 kHz, with energy of 450 nJ. The structures were permanently inscribed with a 20x objective while moving the sample at a constant speed of 60 mm min^{-1}, using the positioning system described above. The mode field diameters of the guided mode were 18 µm × 20 µm at 815 nm. All structures were designed with fan-in and fan-out sections arranged in a three-dimensional geometry, and located prior and after the J_x-lattice, as illustrated in Fig. 5. This effectively suppresses any unwanted crosstalk between the guides and permits easy coupling to fibre arrays with a standard spacing of 127 µm. In the presented device used to transform quantum states, we have κ_0=0.6 cm^{-1}, that is, the Fourier plane is located 2.62 cm after the beginning of the J_x-array.

Experiment on the Characterization of two-Photon Correlations

A BiB_3O_6 nonlinear crystal was pumped with a 70 mW continuous wave pump laser emitting at 407.5 nm, which provided pairs of indistinguishable photons due to type-I spontaneous parametric down-conversion, see Fig. 5. Photon pairs with a central wavelength of 815 nm were filtered by 3 nm (FWHM) interference filters. They were further coupled to the chip via fibre arrays through polarization maintaining fibres, and subsequently fed into single-photon detectors (avalanche photodiodes). The two-photon correlation function was determined by analysing the twofold coincidences recorded between all output channels with the help of an electronic correlator card (Becker & Hickl: DPC230). The spatial correlation results presented in Fig. 4 were extracted from a data set with total integration time of 5 min. The coincidences were then analysed with a time window set at 5 ns and are corrected for detector efficiencies. To assess the statistical consistency of the results in Fig. 4, we discuss in the Supplementary Note 2 the data set presented in terms of correlation event numbers before normalization. For both measurements, the detector clicks data set initially consisted of ~1.5 × 10^6 events in total, which after post selection was reduced to a set of ~5 × 10^4 coincidence events in total, for both input states. Accidental coincidences due to simultaneous detection of two photons not coming from the same pair are estimated to occur with a negligible rate of <2 × 10^{-6}s^{-1}. Non-deterministic number-resolved photon detection was achieved using fibre beam splitters. This set-up thus allows the determination of all 36 two-photon coincidence events occurring in photonic lattices consisting of eight channels. Furthermore, at the wavelength of interest, propagation losses and birefringence are estimated to be 0.3 dB cm^{-1} and in the order of 10^{-7}, respectively. Additional discrepancies in between the measured correlation matrices and the theoretical ones may appear due to imperfect excitations, asymmetric output coupling losses (or detector efficiencies) or limited indistinguishability of the photons.

ACKNOWLEDGEMENTS

We gratefully acknowledge financial support by the German Ministry of Education and Research (Center for Innovation Competence programme, grant no. 03Z1HN31) and the Deutsche Forschungsgemeinschaft (grant no. NO462/6-1 and SZ276/7-1). M.L. thanks the Initial Training Network PICQUE (grant no. 608062) within the Seventh Framework Programme for Research of the European Commission for funding.

Supplementary Information

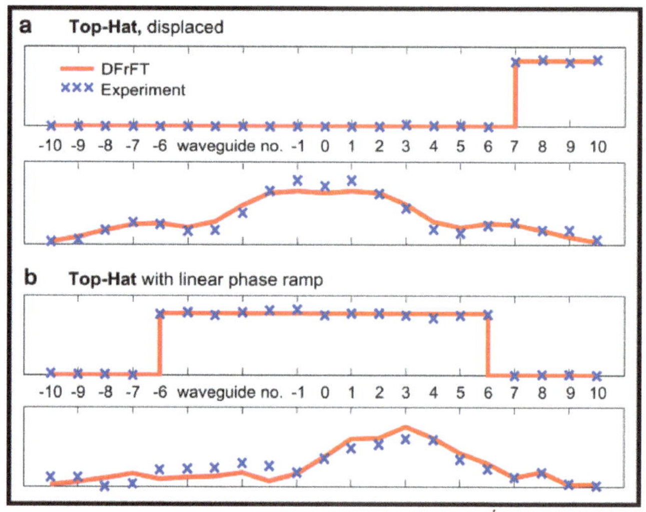

Supplementary Figure 1: Experimental observation of DFrFTs of top-hat functions (a) launched at the edge and (b) launched at the center (with a linear phase ramp) of a J_x -photonic lattice. Upper panels in a and b show the respective input intensity distributions while the lower panels show the output intensity distributions.

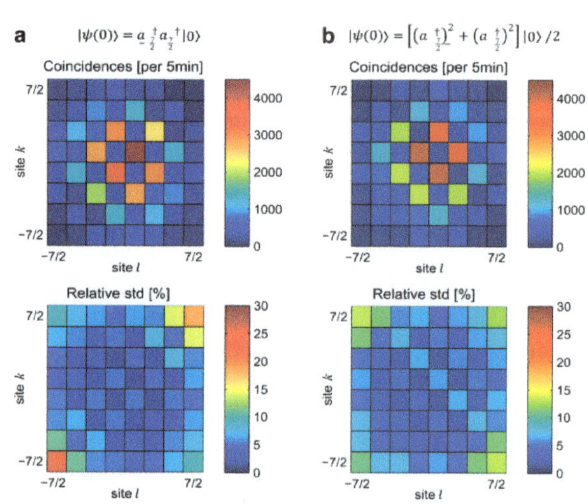

Supplementary Figure 2: Statistical consistency of the obtained quantum measurements (a) for the product state and (b) for the entangled state considered in the study. Top matrices show the same data set than Fig. 4 in the main text without normalizing as correlation function, that is, expressed in correlation event numbers (see colorbar). Bottom matrices show an estimator of the relative standard deviation (std.) of the measurement, expressed in percent (see colorbar).

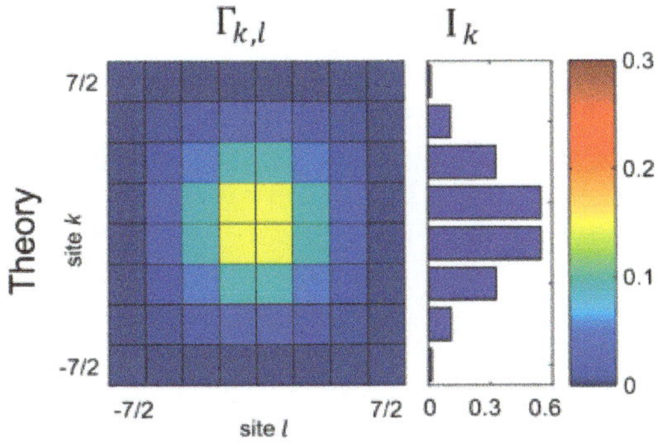

Supplementary Figure 3: Correlations of distinguishable photons

Theoretical correlation maps $\Gamma_{k,l}$ for a pair of distinguishable photons after propagating through a J_x-lattice. The photon density I_k at the output is shown on the right side of the map.

REFERENCES

1. Cooley, J. & Tukey, J. W. An algorithm for the machine calculation of complex Fourier series. Math. Comput. 19, 297–301 (1965).
2. Namias, V. The fractional order Fourier transform and its application to quantum mechanics. J. Inst. Maths. Applics 25, 241–265 (1980).
3. Almeida, L. B. The fractional Fourier transform and time-frequency representations. IEEE Trans. Signal Process. 42, 3084–3091 (1994).
4. Ozaktas, H. M. & Mendlovic, D. Fractional Fourier Optics. J. Opt. Soc. Am. A 12, 743–751 (1995).
5. Ozaktas, H. M., Zalevsky, Z. & Kutay, M. A. The Fractional Fourier Transform with Applications in Optics and Signal Processing (Wiley, 2001).
6. Lohmann, A. W. & Soffer, B. H. Relationship between the Radon–Wigner and fractional Fourier transforms. J. Opt. Soc. Am. A 11, 1798–1801 (1994).
7. Man'ko, M. A. Fractional Fourier transform in information processing, tomography of optical signal, and green function of harmonic oscillator. J. of Russ. Laser Research 20, 226–228 (1999).
8. Dorsch, R. G., Lohmann, A. W., Bitran, Y., Mendlovic, D. & Ozaktas, H. M. Chirp filtering in the fractional Fourier domain. Appl. Opt. 33, 7599–7602 (1994).
9. Lohmann, A. W. & Mendlovic, D. Fractional Fourier transform: photonic] implementation. Appl. Opt. 33, 7661–7664 (1994).
10. Lohmann, A. W. Image rotation, Wigner rotation, and the Fractional Fourier transform. J. Opt. Soc. Am. A 10, 2181–2186 (1993).
11. Mendlovic, D. & Ozaktas, H. Fractional Fourier transforms and their optical implementation: I. J. Opt. Soc. Am. A 10, 1875–1881 (1993).
12. Atakishiyev, N. M. & Wolf, K. B. Fractional Fourier-Kravchuk transform. J. Opt. Soc. Am. A 14, 1467–1477 (1997).
13. Tichy, M. C., Mayer, K., Buchleitner, A. & Molmer, K. Stringent and efficient assessment of Boson Sampling Devices. Phys. Rev Lett. 113, 020502–020506 (2014).
14. Perez-Leija, A., Keil, R. & Moya-Cessa, H. Perfect transfer of path-entangled photons in Jx-photonic lattices, A. Szameit, and D. N. Christodoulides. Phys. Rev. A 87, 022303 (2013).
15. Humphreys, P. C., Barbieri, M., Datta, A. & Walmsley, I. A. Quantum enhanced multiple phase estimation. Phys. Rev. Lett. 111, 070403 (2013).
16. Perez-Leija, A. et al. Eigenstate-assisted longitudinal quantum state transfer and qubit-storage in photonic and spin lattices, 45-th annual meeting of the APS Division of Atomic. Molecular and optical Physics 59, J4.010 (2014).
17. Aspuru-Guzik, A., Dutoi, A. D., Love, P. J. & Head-Gordon, M. Simulated Quantum computation of molecular energies. Science 309, 1704–1707 (2005).
18. Wolf, K. B. Integral transforms in science and engineering Vol. 11 (Mathematical Concepts and Methods for Science and Engineering, 1979).
19. Narducci, L. M. & Orzag, M. Eigenvalues and Eigenvectors of angular momentum operator Jx without theory of rotations. Am. J. of Phys 40, 1811–1814 (1972).
20. Christodoulides, D. N., Lederer, F. & Silberberg, Y. Discretizing light behavior in linear and nonlinear waveguide lattices. Nature 424, 817–823 (2003).
21. Szameit, A. & Nolte, S. Discrete optics in femtosecond-laser-written photonic structures. J. Phys. B 43, 163001 (2010).
22. Lieb, E., Schultz, T. & Matts, D. Two soluble models of an antiferromagnetic chain. Ann. Phys. 16, 407–466 (1961).
23. Cormick, C., Bermudez, A., Huelga, S. F. & Plenio, M. Preparation of the ground state of a spin chain by dissipation in a structured environment. New J. Phys. 15, 073027 (2013).
24. Bobrov, I. B., Kalashnikov, D. A. & Krivitsky, L. A. Imaging of spatial correlations of two-photon states. Phys. Rev. A 89, 043814 (2014).
25. Peeters, W. H., Renema, J. J. & van Exter, M. P. Engineering of two-photon spatial quantum correlations behind a double slit. Phys. Rev. A 79, 043817 (2009).
26. Poem, E., Gilead, Y., Lahini, Y. & Silberberg, Y. Fourier processing of quantum light. Phys. Rev. A 86, 023836 (2012).
27. Bromberg, Y., Lahini, Y., Morandotti, R. & Silberberg, Y. Quantum and classical correlations in waveguide lattices. Phys. Rev. Lett. 102, 253904 (2009).
28. Marshall, G. D. et al. Laser written waveguide photonic quantum circuits. Opt. Exp. 17, 12546–12554 (2009).
29. Lebugle, M. et al. Experimental observation of N00N state Bloch oscillations. Nat. Commun. 6, 8273 (2015).
30. Hong, C. K., Ou, Z. Y. & Mandel, L. Measurement of subpicosecond time intervals between two photons by interference. Phys. Rev. Lett. 59, 2044 (1987).

31. Wiseman, H. M. & Milburn, G. J. Quantum Measurement and control (Cambridge University Press, 2010).

32. Orlowski, A. & Paul, H. Phase retrieval in quantum mechanics. Phys. Rev. A 50, R921–R924 (1994).

33. Wood, J. & Barry, D. T. Radon transformation of time-frequency distributions for analysis of multicomponent signals. IEEE Transac. Signal Process. 42, 3166–3177 (1994).

34. Vogel, K. & Risken, H. Determination of quasiprobability distributions in terms of probability distributions for the rotated quadrature phase. Phys. Rev. A 40, 2847 (1989).

35. Raymer, M. G., Beck, M. & McAlister, D. F. Quantum Optics VI. (eds Harvey, J. D. & Wall, D. F.) (Springer-Verlag, 1994).

36. Ozaktas, H. M. & Mendlovic, D. Fractional Fourier transforms and their optical implementations. J. Opt. Soc. Am. A 10, 2522–2531 (1993).

37. Marhic, M. E. Roots of the identity operator and optics. J. Opt. Soc. Am. A 12, 1448 (1995).

CHAPTER 13

FRACTIONAL FOURIER TRANSFORM ANALYSIS

Katherine M. M. Tant, Anthony J. Mulholland, Matthias Langer and Anthony Gachagan

INTRODUCTION

Non-destructive testing (NDT) is an umbrella term for a wide and varied group of analysis techniques used to evaluate and characterize materials non-invasively [1]. Of particular interest to this paper are ultrasonic phased array systems, which have become increasingly popular as tools for flaw detection and characterization within the NDT industry. They provide improved resolution and coverage by transmitting and receiving ultrasound signals over multiple elements, which, when fired in predefined sequences, can provide increased control of beam directivity [2]. The existence and location of flaws are then deduced via images generated by post-processing the data captured by these arrays [3–5].

The examination of welds is of particular interest to the NDT community, given their role in safety critical structures in nuclear power plants, aeroengines, pipelines, etc. They are subject to cyclic loads and, as with any type of bond, constitute the weak point of the structure. Austenitic steel welds are notoriously difficult to inspect [1,6]. Owing to thermal effects as the weld is forming, a spatially heterogeneous structure is formed by local fluctuations in the crystal orientation. This complex internal geometry is highly scattering and results in low signal-to-noise ratio (SNR) levels, which can subsequently lead to the obscuration of defects. To combat this problem, two approaches are suggested and combined in this paper: the use of chirp excitations and the post-processing of the collected data in the time–frequency domain.

(a) Chirp excitations

Coded excitations are an effective way of delivering large amounts of energy using relatively low acoustic pressure amplitudes. Inspiration for coded signal design can be drawn from bioacoustics; bats and dolphins use frequency-modulated sweeps to navigate and hunt [**7,8**]. The use of coded excitations in signal processing has been shown to improve SNR and lessen trade-offs between sample penetration and image resolution [**9,10**]. Chirps contain multiple frequencies which vary in time, typically in a linear or exponential manner. Their broad frequency content increases the likelihood of reaching the resonant frequency of a defect, in turn causing stronger vibrations and consequently improving the probability of detection. For the purposes of this work, a Gaussian-modulated linear chirp in time (t) of the form

$$q(t) = \exp[-2\pi i f_1(t + mt^2)] \exp\left[-\frac{(t-t_1)^2}{\sigma^2}\right] \tag{1.1}$$

has been used, where m is the gradient of the chirp (the rate at which it sweeps through a prescribed range of frequencies), f_1 is the initial frequency, t_1 is the centre of the Gaussian envelope and σ is its standard deviation. By varying f_1, m and σ, the bandwidth of the chirp can be altered. Setting m=0 Hz, the chirp reverts back to a time harmonic signal with frequency f_1. For a fair comparison of the chirp with a continuous gated waveform (the typical signal emitted by the transducer), it is imperative both are optimized for the same transducer and hence use its full bandwidth. **Figure 1** shows one such matched pair. In the time domain, the gated continuous sine

Citation: Tant KMM, Mulholland AJ, Langer M, Gachagan A. 2015 Fractional Fourier Transform Analysis. Soc. A 471: 20140958. http://dx.doi.org/10.1098/rspa.2014.0958.

wave (in black) spans a far shorter time interval than that of the Gaussian-modulated linear chirp (both signals have the same peak amplitude). Studying the plots of their respective Fourier transforms in **figure 1***b*, it is apparent that they have similar −6 dB bandwidths (approx. 50%) but observe that the chirp contains a far greater amount of energy.

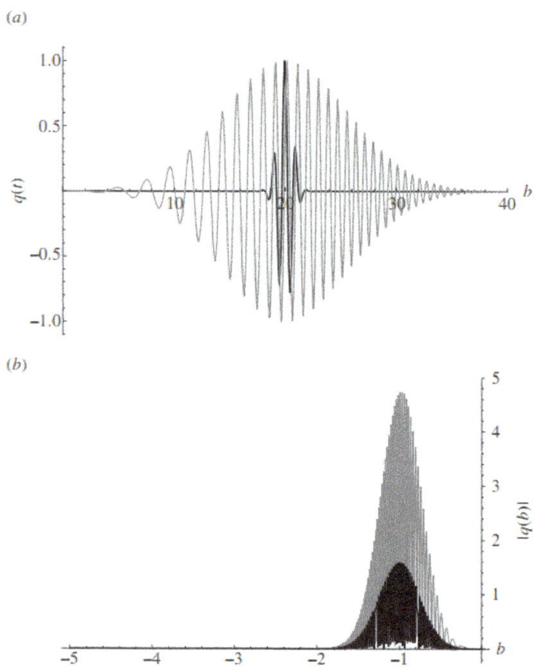

Figure 1: (*a*) A gated continuous sine wave with frequency f=1 MHz modulated by a Gaussian envelope with parameters σ=1 µs, t_1= 20 µs (in black) and a Gaussian-modulated linear chirp with parameters σ=8 µs, t_1=20 µs, m=0.22 MHz and f_1= 0.1 MHz (in grey). Their respective Fourier transforms (power spectra) are shown in (*b*).

(b) The Fractional Fourier Transform

An alternative approach to differentiating between flaw scattering and noise can be taken via the analysis of the collected data in the time–frequency domain. In an ideal case, a received ultrasound signal would exhibit signs of scattering at the time interval pertaining to the location of the flaw, facilitating detection and subsequent characterization. However, in practice, scattering by the microstructure of the host media can dominate the signal. To improve identification of defects it is suggested that the scattered signals are analysed for their frequency content; the frequency spectrum of the wave scattered by a flaw should be different from that scattered by a heterogeneity. However, time information must be retained in order to locate the flaw. This can be achieved via (i) the time windowed Fourier transform [**11**], where the discrete Fourier transform is applied to short time intervals allowing the frequency content at that specific time to be analysed independently of the rest of the signal, or (ii) the fractional Fourier transform (FrFT) [**12–14**], which enables continuous movement between the time and frequency domains, allowing the simultaneous retention of both frequency and time domain information. In this work, the FrFT is employed to analyse the benefits of chirp excitation over gated continuous wave excitation.

As a generalization of the ordinary Fourier transform, the FrFT is more flexible in its applications and hence of potential interest to any area in which the Fourier transform is frequently implemented. Its main advantage is that it allows continuous movement between the time and frequency domains, retaining information from each, thus presenting an alternative to using the time windowed Fourier transform. There exist several conventions for defining the FrFT, each of which gives a slightly different physical interpretation. For the purposes of this work, the FrFT of order a is given as the linear integral transform [**12**]

$$\mathcal{F}_a(u) \equiv \int_{-\infty}^{\infty} K_a(u,u')f(u')\,\mathrm{d}u', \tag{1.2}$$

where

$$K_a(u,u') \equiv \sqrt{1-\mathrm{i}\cot\alpha}\,\exp[\mathrm{i}\pi(\cot\alpha u^2 - 2\csc\alpha uu' + \cot\alpha u'^2)] \tag{1.3}$$

and α=$a\varpi$/2. When a is an integer, it denotes the number of repeated applications of the ordinary Fourier transform. Hence, setting a=1 (and, consequently, α=π/2), equation (1.2]) simplifies to

$$\mathcal{F}_1(u) \equiv \int_{-\infty}^{\infty} f(u')\,\mathrm{e}^{-\mathrm{i}2\pi uu'}\,\mathrm{d}u', \tag{1.4}$$

the ordinary Fourier transform.

SOLVING THE INHOMOGENEOUS WAVE EQUATION IN TIME–FREQUENCY SPACE

To build a mathematical framework to allow for analysis of the scattering of an ultrasonic chirp by a flaw, the wave equation with a time-dependent forcing function, $q(t)$, will be solved in the time–frequency space. Note that $q(t)$ is spatially independent and hence acts like a body force, affecting the whole flaw domain simultaneously. To justify this, it is assumed that the length of the flaw is smaller than the wavelength. This is in keeping with the low-frequency assumption made in the Born approximation [**15**] which is used later. A second assumption inherent to the Born approximation is that the transmission and reception of waves takes place at a distance from the flaw which is much larger than the wavelength: the far-field assumption. To begin, consider the non-homogeneous wave equation

$$\frac{\partial^2}{\partial t^2}f(\mathbf{x},t) - c^2\nabla^2 f(\mathbf{x},t) = q(t), \tag{2.1}$$

where x∈R^3, t∈R and c is the wave speed. We demand that the solution is bounded in time and space and satisfies the Sommerfeld radiation condition

$$\frac{\partial f}{\partial r} - \mathrm{i}kf = \mathcal{O}(r^{-1}) \quad \text{for } r = |\mathbf{x}| \to \infty, \tag{2.2}$$

guaranteeing that the waves are outgoing and decay sufficiently fast so there exist no sources at infinity. It is also assumed that an initial pressure amplitude of h_0 is present at the ultrasonic array. This simplified case will of course cover the detection of an object in a fluid host but is also relevant to certain restricted classes within elastodynamics such as the propagation of horizontal shear waves in an isotropic solid. As mentioned above, the applications of interest will involve heterogeneous materials (with spatially dependent wave speeds) but this is a reasonable model with which to start this investigation and the methodology will extend naturally to these more general cases.

By **[12]**, the FrFT, taken with respect to time, of a derivative of a function is given by

$$\mathcal{F}_a\left(\left[(2\pi\mathrm{i})^{-1}\frac{\mathrm{d}}{\mathrm{d}u}\right]^n f(u)\right) = \left[u\sin\alpha + \cos\alpha(2\pi\mathrm{i})^{-1}\frac{\mathrm{d}}{\mathrm{d}u}\right]^n f_a(u). \tag{2.3}$$

Hence, taking the FrFT of every term in equation (2.1) gives

$$(2\pi\mathrm{i})^2\sin^2\alpha u^2 f_a(\mathbf{x},u) + 4\pi\mathrm{i}\sin\alpha\cos\alpha u\frac{\partial}{\partial u}f_a(\mathbf{x},u) + \cos^2\alpha\frac{\partial^2}{\partial u^2}f_a(\mathbf{x},u) = c^2\nabla^2 f_a(\mathbf{x},u) + q_a(u), \tag{2.4}$$

the non-homogeneous wave equation in time–frequency space.

(a) The homogeneous solution

To solve equation (2.4), we start by finding the solution to the homogeneous differential equation via separation of variables. The solution is written in product form

$$f_a^h(\mathbf{x},u) = h(\mathbf{x})g_a(u) \tag{2.5}$$

and substituted into the homogeneous wave equation to give

$$(2\pi\mathrm{i})^2\sin^2\alpha u^2 + 4\pi\mathrm{i}\sin\alpha\cos\alpha u\frac{g_a'(u)}{g_a(u)} + \cos^2\alpha\frac{g_a''(u)}{g_a(u)} = c^2\frac{\nabla^2 h(\mathbf{x})}{h(\mathbf{x})} = -b^2, \tag{2.6}$$

for some b∈Rb∈R. This can then be separated into two equations

$$\left(-4\pi^2\sin^2\alpha u^2 + \mathrm{i}4\pi\sin\alpha\cos\alpha u\frac{\mathrm{d}}{\mathrm{d}u} + \cos^2\alpha\frac{\mathrm{d}^2}{\mathrm{d}u^2}\right)g_a(u) = -b^2 g_a(u) \tag{2.7}$$

and

$$c^2\nabla^2 h(\mathbf{x}) = -b^2 h(\mathbf{x}), \tag{2.8}$$

from which the temporal and spatial components of the homogeneous solution can be derived. To solve equation (2.7), a chirp-like ansatz of the form

$$g_a(u) = \exp[\gamma u^2 + \beta u] \tag{2.9}$$

is chosen. Substituting this into equation (2.7) gives

$$[-(4\pi^2\sin^2\alpha\, u^2 - b^2) + 4\pi\mathrm{i}\sin\alpha\cos\alpha u(2\gamma u + \beta) + \cos^2\alpha(2\gamma + (2\gamma u + \beta)^2)]\,\mathrm{e}^{\gamma u^2 + \beta u} = 0, \tag{2.10}$$

that is,

$$[u^2(4\gamma^2\cos^2\alpha + 8\pi\mathrm{i}\gamma\sin\alpha\cos\alpha - 4\pi^2\sin^2\alpha) + u(4\gamma\beta\cos^2\alpha + 4\pi\mathrm{i}\beta\sin\alpha\cos\alpha) + (2\cos^2\alpha\gamma + \beta^2\cos^2\alpha + b^2)] = 0. \tag{2.11}$$

Equating coefficients of powers of u, one can calculate γ and β as

$$\gamma = -\pi\mathrm{i}\tan\alpha \quad \text{and} \quad \beta = \pm\sqrt{-b^2\sec^2\alpha + 2\pi\mathrm{i}\tan\alpha}, \tag{2.12}$$

where the square root is taken so as the real part is positive. Hence

$$g_a(u) = d_1\exp\left[u(-\pi\mathrm{i}\tan\alpha u - \sqrt{-b^2\sec^2\alpha + 2\pi\mathrm{i}\tan\alpha})\right] + d_2\exp\left[u(-\pi\mathrm{i}\tan\alpha u + \sqrt{-b^2\sec^2\alpha + 2\pi\mathrm{i}\tan\alpha})\right]. \tag{2.13}$$

However, to ensure that the solution is bounded in u, d_2 is set equal to zero.

Turning our attention now to the spatially dependent component of the homogeneous solution, equation (2.8) can be recognized as the homogeneous Helmholtz equation

$$\nabla^2 h(\mathbf{x}) + \hat{k}^2 h(\mathbf{x}) = 0, \tag{2.14}$$

where $\hat{k}$ =b/c (c is the plane wave speed and b is analogous to the circular frequency). An explicit approximation of the scattered wave at a specified b, $h(\mathbf{x}, b)$, is derived via the Born approximation **[15]**, giving

$$h(\mathbf{y},b) = h_0\frac{\mathrm{e}^{\mathrm{i}\hat{k}r_s}}{r_s}\frac{a_1a_2a_3[\gamma_\lambda - \gamma_\rho(\mathbf{e}_i\cdot\mathbf{e}_s)]}{|\mathbf{e}_i - \mathbf{e}_s|^2 r_e^2}\left[\frac{\sin(\hat{k}|\mathbf{e}_i - \mathbf{e}_s|r_e) - \hat{k}|\mathbf{e}_i - \mathbf{e}_s|r_e\cos(\hat{k}|\mathbf{e}_i - \mathbf{e}_s|r_e)}{\hat{k}|\mathbf{e}_i - \mathbf{e}_s|r_e}\right], \tag{2.15}$$

where a_1, a_2 and a_3 represent the flaw dimensions (the flaw is modelled as an ellipsoid),$\gamma\rho=1-\rho_0/\rho_1$ (ρ_0 and ρ_1 are the material densities of the host and flaw materials, respectively),$\gamma_\lambda=1-\lambda_0/\lambda_1$ (λ_0 and λ_1 are the bulk moduli of the host and flaw materials, respectively), r_s is the distance of the flaw from the array, h_0 is the initial pressure amplitude, $\mathbf{e}_i$ and $\mathbf{e}_s$ are unit vectors in the incident and scattered wave directions, respectively, and r_e is the effective radius of the flaw, given by

$$r_e = \sqrt{a_1^2(\mathbf{e}_q \cdot \mathbf{u}_1)^2 + a_2^2(\mathbf{e}_q \cdot \mathbf{u}_2)^2 + a_3^2(\mathbf{e}_q \cdot \mathbf{u}_3)^2}, \tag{2.16}$$

where

$$\mathbf{e}_q = \frac{\mathbf{e}_i - \mathbf{e}_s}{|\mathbf{e}_i - \mathbf{e}_s|} \tag{2.17}$$

and the unit vectors $\mathbf{u}_1$, $\mathbf{u}_2$ and $\mathbf{u}_3$ lie along the axes of the flaw. Thus, a solution to the homogeneous equation can be written as

$$\begin{aligned} f_a^h(\mathbf{x},u,b) &= d_1 \exp\left[u(-\mathrm{i}\pi \tan\alpha u - \sqrt{-b^2 \sec^2\alpha + \mathrm{i}2\pi \tan\alpha})\right] \\ &\times h_0 \frac{\mathrm{e}^{\mathrm{i}\hat{k}r_s}}{r_s} \frac{a_1 a_2 a_3[\gamma_\lambda - \gamma_\rho(\mathbf{e}_i \cdot \mathbf{e}_s)]}{|\mathbf{e}_i - \mathbf{e}_s|^2 r_e^2} \\ &\times \left[\frac{\sin(\hat{k}|\mathbf{e}_i - \mathbf{e}_s|r_e) - \hat{k}|\mathbf{e}_i - \mathbf{e}_s|r_e \cos(\hat{k}|\mathbf{e}_i - \mathbf{e}_s|r_e)}{\hat{k}|\mathbf{e}_i - \mathbf{e}_s|r_e}\right]. \end{aligned} \tag{2.18}$$

Because any superposition of solutions of a linear, homogeneous PDE is again a solution, the dependence on b can be removed by integrating over b

$$\begin{aligned} f_a^h(\mathbf{x},u) &= \int_{-\infty}^{\infty} d_1 \exp\left[u(-\mathrm{i}\pi \tan\alpha u - \sqrt{-b^2 \sec^2\alpha + \mathrm{i}2\pi \tan\alpha})\right] \\ &\times h_0 \frac{\mathrm{e}^{\mathrm{i}\hat{k}r_s}}{r_s} \frac{a_1 a_2 a_3[\gamma_\lambda - \gamma_\rho(\mathbf{e}_i \cdot \mathbf{e}_s)]}{|\mathbf{e}_i - \mathbf{e}_s|^2 r_e^2} \\ &\times \left[\frac{\sin(\hat{k}|\mathbf{e}_i - \mathbf{e}_s|r_e) - \hat{k}|\mathbf{e}_i - \mathbf{e}_s|r_e \cos(\hat{k}|\mathbf{e}_i - \mathbf{e}_s|r_e)}{\hat{k}|\mathbf{e}_i - \mathbf{e}_s|r_e}\right] \mathrm{d}b. \end{aligned} \tag{2.19}$$

However, to ensure that this integral is finite, d_1 must be chosen as some function of b with which to bound the solution. For this work, the function

$$d_1(b) = A \exp\left[\frac{-(b-b_1)^2}{\sigma'^2}\right] \tag{2.20}$$

has been chosen, where A, b_1 and σ' are chosen to mimic the Gaussian envelope observed in the plot of the power spectrum of the forcing function (**figure 1***b*). Thus, the solution is effectively summed over the range of frequencies received by the transducer array.

(b) The inhomogeneous solution

To find a time-dependent inhomogeneous solution $f_a^p(u)$ to equation (2.1), an ansatz of the form

$$f_a^p(u) = \int_{-\infty}^{\infty} v_a(b) s_a(u,b)\, \mathrm{d}b \tag{2.21}$$

is chosen, where

$$s_a(u,b) = \sqrt{1 + \mathrm{i}\tan\alpha}\, \exp[-\mathrm{i}\pi(u^2 \tan\alpha + 2ub \sec\alpha + b^2 \tan\alpha)]. \tag{2.22}$$

This form is chosen since, using the identities $\alpha = -\cot(\alpha + \pi/2)$ and $\sec\alpha = \csc(\alpha + \pi/2)$, one can rewrite $s_a(u,b)$ as

$$\begin{aligned} s_a(u,b) &= \sqrt{1 - \cot\left(\alpha + \frac{\pi}{2}\right)} \\ &\quad \times \exp\left[\pi \mathrm{i}\left(u^2 \cot\left(\alpha + \frac{\pi}{2}\right) - 2ub \csc\left(\alpha + \frac{\pi}{2}\right) + b^2 \cot\left(\alpha + \frac{\pi}{2}\right)\right)\right] \\ &= K_{a+1}(u,b), \end{aligned} \tag{2.23}$$

the kernel of the FrFT of order a+1. The derivatives of $sa(u,b)$ are given by

$$\begin{aligned} \frac{\partial}{\partial u} s_a(u,b) &= \sqrt{1 + \mathrm{i}\tan\alpha}(-2\pi \mathrm{i}u \tan\alpha - 2\pi \mathrm{i}b \sec\alpha) \\ &\quad \times \exp[-\pi \mathrm{i}(u^2 \tan\alpha + 2ub \sec\alpha + b^2 \tan\alpha)] \end{aligned} \tag{2.24}$$

and

$$\frac{\partial^2}{\partial u^2} s_a(u,b) = \sqrt{1+\mathrm{i}\tan\alpha}[(-2\pi\mathrm{i}u\tan\alpha - 2\pi\mathrm{i}b\sec\alpha)^2 - 2\pi\mathrm{i}\tan\alpha] \\ \times \exp[-\pi\mathrm{i}(u^2\tan\alpha + 2ub\sec\alpha + b^2\tan\alpha)]. \tag{2.25}$$

Substituting these into equation (2.4) shows that the left-hand side can be written as

$$\begin{aligned}&(2\pi\mathrm{i})^2\sin^2\alpha u^2 s_a(u,b) + \mathrm{i}4\pi\sin\alpha\cos\alpha u\frac{\partial}{\partial u}s_a(u,b) + \cos^2\alpha\frac{\partial^2}{\partial u^2}s_a(u,b)\\ &= \sqrt{1+\mathrm{i}\tan\alpha}\exp[-\pi\mathrm{i}(u^2\tan\alpha + 2ub\sec\alpha + b^2\tan\alpha)]\\ &\quad\times[-4\pi^2\sin^2\alpha u^2 + 8\pi^2u^2\sin^2\alpha + 8\pi^2ub\sin\alpha - 2\pi\mathrm{i}\sin\alpha\cos\alpha\\ &\quad - 4\pi^2u^2\sin^2\alpha - 8\pi^2ub\sin\alpha - 4\pi^2b^2]\\ &= \sqrt{1+\mathrm{i}\tan\alpha}\exp[-\pi\mathrm{i}(u^2\tan\alpha + 2ub\sec\alpha + b^2\tan\alpha)](-4b^2\pi^2 - \pi\mathrm{i}\sin 2\alpha)\\ &= s_a(u,b)(-4b^2\pi^2 - \pi\mathrm{i}\sin 2\alpha).\end{aligned} \tag{2.26}$$

Substituting equation (2.21) into equation (2.4) then gives

$$\int_{-\infty}^{\infty}(-4b^2\pi^2 - \pi\mathrm{i}\sin 2\alpha)v_a(b)s_a(u,b)\,\mathrm{d}b = q_a(u). \tag{2.27}$$

Note that, when u lies along the a=0 axis, it is analogous to the original time domain signal. In the work below, $q_a(u)$ will be written as $q(t)$ and $q_1(u)$ as $q(b)$, to allow for easier physical interpretation. By letting $D(b) = (-4b^2\pi^2 - \mathrm{i}\pi\sin 2\alpha)v_a(b)$ (and remembering $s_a(u,b)=Ka_{+1}(u,b)$) equation (2.27) can be written as

$$\mathcal{F}_{a+1}(D(b)) = \mathcal{F}_a(q(t)), \tag{2.28}$$

which holds for all orders a. Hence, by the index additivity rules it is shown that

$$\mathcal{F}_1(D(b)) = q(t). \tag{2.29}$$

By taking the FrFT of order a=−1 (analogous with the inverse Fourier transform) of both sides, it follows that

$$D(b) = \int_{-\infty}^{\infty} q(t)\,\mathrm{e}^{\mathrm{i}2\pi tb}\,\mathrm{d}t, \tag{2.30}$$

and the unknown function $va(b)$ of the integral equation (2.27) can thus be found.

(c) The inverse Fourier Transform of a Gaussian-modulated linear chirp

An analytical expression for the inverse Fourier transform of a Gaussian-modulated linear chirp must now be derived. To evaluate equation (2.30), the chirp given by equation (1.1) must be rewritten in the form

$$q(t) = E_0\exp[-(p+\mathrm{i}r)t^2 - \mathrm{i}w_0t], \tag{2.31}$$

where $E_0 = \exp(-t_1^2/\sigma^2)$, $p = 1/\sigma^2$, $r = 2\pi f_1 m$ and $w_0 = 2\pi f_1 + 2t_1\mathrm{i}/\sigma^2$. Applying the inverse Fourier transform gives

$$D(b) = \int_{-\infty}^{\infty} E_0\exp[-((p+\mathrm{i}r)t^2 - \mathrm{i}(2\pi b - w_0)t)]\,\mathrm{d}t. \tag{2.32}$$

The standard integral for a general quadratic exponent is calculated using

$$\int_{-\infty}^{\infty}\exp[-ax^2 + bx + c]\,\mathrm{d}x = \sqrt{\frac{\pi}{a}}\exp\left[\frac{b^2}{4a} + c\right], \tag{2.33}$$

assuming that the real part of a is positive. It follows that

$$D(b) = \frac{\sqrt{\pi}}{\sqrt{p+\mathrm{i}r}}E_0\exp\left(\frac{-(2\pi b - w_0)^2}{4(p+\mathrm{i}r)}\right), \tag{2.34}$$

and thus

$$v_a(b) = \frac{\sqrt{\pi}}{\sqrt{p+\mathrm{i}r}}\frac{E_0}{(-4b^2\pi^2 - \mathrm{i}\pi\sin 2\alpha)}\exp\left(\frac{-(2\pi b - w_0)^2}{4(p+\mathrm{i}r)}\right). \tag{2.35}$$

Finally, by equations (2.22) and (2.35), the inhomogeneous solution of the wave equation in time–frequency space is given by

$$f_a^p(u) = \frac{\sqrt{\pi}}{\sqrt{p+\mathrm{i}r}} \int_{-\infty}^{\infty} \frac{E_0}{(-4b^2\pi^2 - \mathrm{i}\pi \sin 2\alpha)} \exp\left(\frac{-(2\pi b - w_0)^2}{4(p+\mathrm{i}r)}\right) \times \sqrt{1+\mathrm{i}\tan\alpha}\, \exp(-\mathrm{i}\pi(u^2\tan\alpha + 2ub\sec\alpha + b^2\tan\alpha))\, \mathrm{d}b. \tag{2.36}$$

(d) The General Solution

Using equations (2.19) and (2.36), the general expression for the scattering from a flaw owing to the excitation by a linear chirp forcing function is given by

$$\begin{aligned} f_a(\mathbf{x},u) = \int_{-\infty}^{\infty} & d_1(b) \exp\left[u(-\mathrm{i}\pi \tan\alpha u - \sqrt{-b^2\sec^2\alpha + \mathrm{i}2\pi\tan\alpha})\right] \\ & \times h_0 \frac{\mathrm{e}^{\mathrm{i}\hat{k}r_s}}{r_s} \frac{a_1a_2a_3[\gamma_\lambda - \gamma_\rho(\mathbf{e}_i\cdot\mathbf{e}_s)]}{|\mathbf{e}_i - \mathbf{e}_s|^2 r_e^2} \\ & \times \left[\frac{\sin(\hat{k}|\mathbf{e}_i - \mathbf{e}_s|r_e) - \hat{k}|\mathbf{e}_i - \mathbf{e}_s|r_e\cos(\hat{k}|\mathbf{e}_i - \mathbf{e}_s|r_e)}{\hat{k}|\mathbf{e}_i - \mathbf{e}_s|r_e}\right] \\ & + \frac{\sqrt{\pi}}{\sqrt{p+\mathrm{i}r}} \frac{E_0}{(-4b^2\pi^2 - \mathrm{i}\pi\sin 2\alpha)} \exp\left(\frac{-(2\pi b - w_0)^2}{4(p+\mathrm{i}r)}\right) \\ & \times \sqrt{1+\mathrm{i}\tan\alpha}\, \exp(-\mathrm{i}\pi(u^2\tan\alpha + 2ub\sec\alpha + b^2\tan\alpha))\, \mathrm{d}b. \end{aligned} \tag{2.37}$$

Note that the spatial component of the solution, which gives rise to the scattering profile of the defect and encapsulates the geometry of the flaw, is amplified by the chirp excitation and this suggests that an increased SNR will result.

COMPARISON OF A GATED CONTINUOUS WAVE EXCITATION WITH A CHIRP EXCITATION

A general solution to the inhomogeneous wave equation in the time–frequency domain with a Gaussian-modulated linear chirp forcing function has been derived. It can be observed that the additive inhomogeneous solution does not carry flaw shape information but does involve the chirp excitation parameters. Hence, to contrast the excitation by a chirp with that of a gated continuous wave, attention can be restricted to the inhomogeneous solution only.

To compute the inhomogeneous solution numerically, the infinite integral over b is approximated by the integral over the bandwidth of the transducer. This is justified as it can be assumed that the integral is zero outside this interval (the received signal will not contain any frequencies outside of the bandwidth). This approach also resolves any issues around the singularity in the integrand when b=0 and α=0. The inhomogeneous solution is plotted as a time-order plot in **figure 2** for the cases where the forcing function is set as (i) a gated continuous wave and (ii) a Gaussian-modulated linear chirp. Each point within the plot is the additional amplitude at that point in time–frequency space obtained by using the particular excitation. It can be seen that, since the linear chirp contains more energy, it provides a marked increase in the scattering amplitude over the entire time–frequency space. In the case of the gated continuous wave in **figure 2***a*, some orders of the FrFT (for example a=0.55) offer no additive amplification at any point along u. To further demonstrate the benefits of excitation by the chirp, the corresponding frequency domain scattering matrices for a crack-like flaw (generated by the spatial component of the homogeneous solution over a range of transmit/receive directions, **e***i* and **e***s*, which mimic inspection by a linear array) have been summed over the frequency range of the hypothetical transducer. The amplitudes along the a=1 axes in **figure 2***a*,*b* are used to amplify the scattering matrices at the corresponding frequencies for the continuous gated wave excitation and the linear chirp excitation, respectively. The results are shown in **figure 3**. Geometrically, the scattering matrices are identical; however, it can be observed that the scattering profile of the flaw is amplified in the case of chirp insonification (**figure 3***b*). It is clear that, if noise were present in the signal, the higher amplitudes exhibited in the case of chirp excitation would provide an increased SNR.

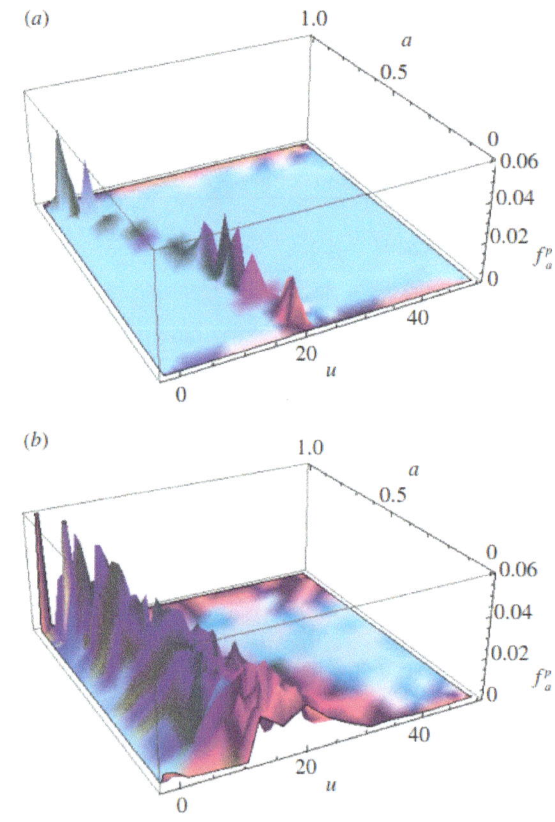

Figure 2. Three-dimensional surface plots of the inhomogeneous solution given by equation (2.36) with chirp parameters (*a*) f_1=1 MHz, m=0 MHz, t_1=20 µs and σ=1 µs (gated continuous wave) and (*b*) f_1=0.1 MHz, m=0.22 MHz, t_1=20 µs and σ=8 µs (Gaussian-modulated linear chirp). (Online version in colour.)

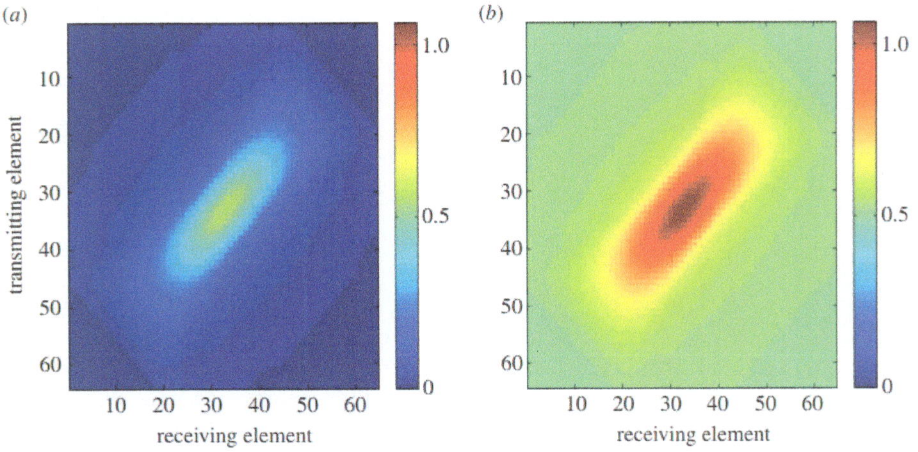

Figure 3. Sum of the scattering matrices as generated by the Born approximation arising from (*a*) gated continuous wave excitation and (*b*) linear chirp insonification. (Online version in colour.)

(a) Choosing the optimal order *a*

It is shown in [**16**] that the FrFT of a Gaussian function has the form of a Gaussian for all orders *a*. The standard deviation of these Gaussian functions, σ_a, varies in time–order space, with the narrowing of the function being synonymous with an increase in maximum amplitude. Hence, it can be concluded that the optimal value of *a* at which to employ the FrFT is the value at which the minimum σ_a occurs. As the chirp rate *m* increases, this maximum peak (which occurs at minσaminσa) moves further away from the frequency domain. This is explained schematically in**figure 4**. The long-dashed horizontal line in **figure 4***a* represents a continuous wave (where $m=0$) and hence results in a single value on the frequency axis. As the gradient increases (see the dotted and solid lines), the breadth of the frequency spectrum increases. The curves in **figure 4***b* demonstrate how the width of the Gaussian changes in the fractional Fourier domain (of course, the Gaussian is infinite but here the width is approximated by $6\sigma_a$ as 99.73% of the signal lies within this interval). The fractional order which exhibits the widest Gaussian is orthogonal to the order with the narrowest distribution (i.e. there is a difference of 1 between the orders at which these extremes occur). It then follows that, as the bandwidth of the chirp decreases, the order at which the narrowest Gaussian form arises approaches the frequency domain. Now, plotting the rate of frequency change (with respect to time) of the linear chirp results in the plot as seen in **figure 5**. The angle made with the frequency axis can thus be calculated as $\alpha = \tan^{-1}(\frac{1}{2}f_1 m)$. This provides the optimum angle at which to take the FrFT [**17**] and translates to order $a = 2\tan^{-1}(\frac{1}{2}f_1 m)/\pi$. To assess the formula's success in predicting this order in regards to the inhomogeneous solution derived above, $f_a^p(u)$ (as defined in equation (2.36)) is plotted over orders $-2 \le a \le 2$ in **figure 6**. As the Gaussian function is not centred at zero, the plot appears skewed; however, the narrowing phenomena can still be observed. Owing to the low gradient of the chirp excitation, the location of the narrowest distribution approaches the frequency domain as predicted. The simple algebraic formula derived for the optimal order of the FrFT of a Gaussian-windowed linear chirp exhibits an error in application to the inhomogeneous solution and does not incorporate the maximum amplitude (which is circled in black). However, the error is small (within 0.1 of the order at which the maximum does occur) and the formula could potentially guide the implementation of the discrete FrFT [**18**], effectively reducing the neighbourhood (and, subsequently, the computational expense) over which the FrFT is taken. It is hoped that the general solution as derived above (equation (2.37)) will act as a basis for further work on improving the extraction of the optimal order *a* at which to implement the FrFT for signals which have encountered a defect and been subsequently scattered, thus eventually reducing the time–order space to one dimension for numerical implementations.

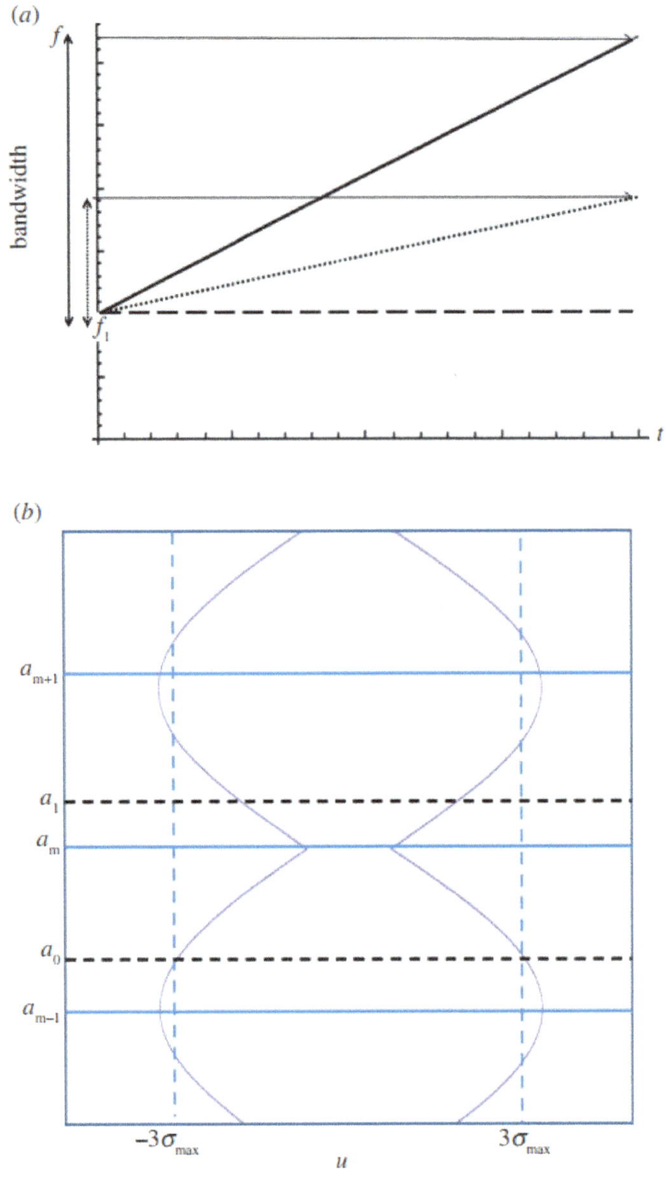

Figure 4. (*a*) Schematic demonstrating the increase in bandwidth with the increase of chirp rate *m*. The long-dashed line is representative of a single frequency pulse and its bandwidth is a delta function at that frequency. The dotted and solid lines represent linear chirps with increasing gradients which result in increased bandwidths. (*b*) Schematic demonstrating the change in σa in the fractional Fourier domain. (Online version in colour.)

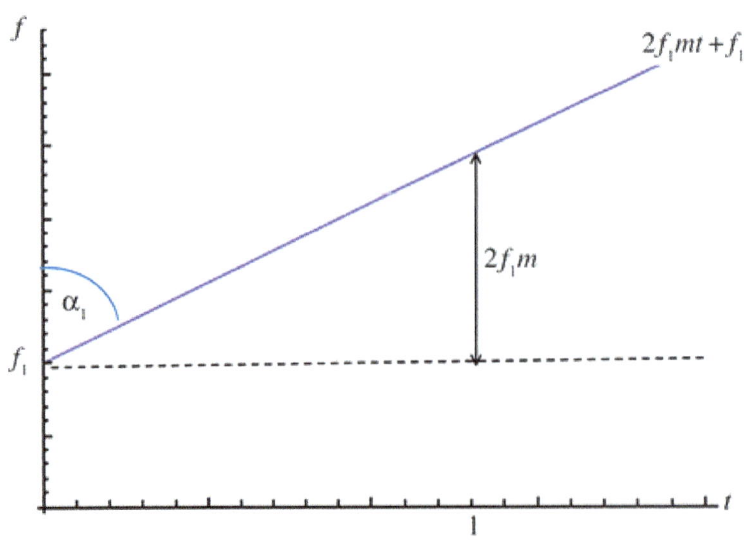

Figure 5. The geometrical interpretation of the optimal α at which to take the FrFT, dependent on *m* and f_1. (Online version in colour.)

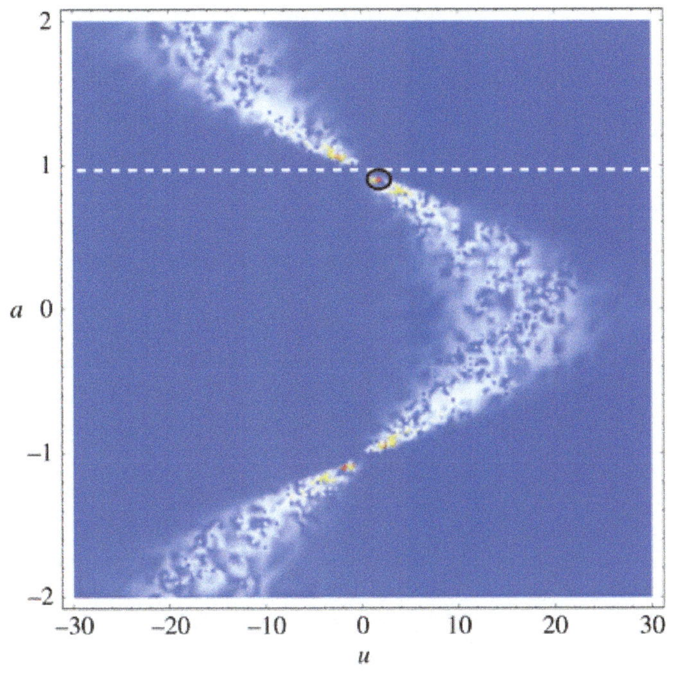

Figure 6. The inhomogeneous solution as given by equation (2.36), arising from linear chirp excitation with parameters m=0.22 MHz, f_1= 0.1 MHz, σ=8 µs and t_1=20 µs. The dashed line marks the order a at which the formula derived in [**17**] predicts the optimal value of a should occur. The predicted optimal a does not correspond to the order at which the maximum peak occurs (which is circled in black). (Online version in colour.)

CONCLUSION

A wealth of information regarding the detection, imaging and sizing of flaws is contained within the scattering matrices as constructed by data arising from ultrasonic phased array inspections. Of course, the successful extraction of this information relies on the received data having a reasonable SNR. This paper examined the use of chirp excitation as a means of improving the SNR by increasing the amplitude of the recovered signal. A general solution to the inhomogeneous wave equation, and the subsequent scattering by a flaw, in the time–frequency domain, with a Gaussian-modulated linear chirp forcing function has been derived. This was achieved by taking the FrFT of the inhomogeneous wave equation and finding the homogeneous solution via separation of variables and the Born approximation. The inhomogeneous solution was obtained by choosing an ansatz that, once substituted into the inhomogeneous wave equation, resulted in a linear integral equation which could be solved by formulating a Fourier transform pair for a Gaussian-modulated linear chirp. Since the excitation parameters are exclusively contained in the additive term provided by the inhomogeneous solution, the comparison between gated continuous wave excitation and chirp excitation can be drawn by focusing solely on that term. It was plotted in a time–order plot for the cases where the forcing function was set as (i) a gated continuous wave and (ii) a Gaussian-modulated linear chirp and it was shown that, since the linear chirp contained more energy, there was a marked increase in the scattering amplitude. This was reinforced by plotting and comparing the corresponding scattering matrices for a chosen peak in the time–order plot, which further demonstrated the increased amplification provided by the chirp. Thus it is anticipated that an improved SNR will result when applied to experimental data.

Recent studies on the use of the FrFT with regard to chirp signals [**16,17**] suggest a potential methodology for studying the wave scattering problem considered in this paper. Indeed, the findings from that work were assessed for their applicability in this paper and, whilst the discrepancies were clear, they did provide a reasonable estimate of the optimal order at which to implement the FrFT. It is envisaged that the analytical formulation of the general solution in §2d will allow for improved extraction of the optimal a for the more complicated case of a chirp which has been scattered by a defect. The benefit in doing so would be the effective reduction of the time–frequency space to one dimension, thus enabling reduced computational cost in NDT applications.

Author contributions

K.M.M.T., A.J.M. and M.L. conceived the mathematical models and interpreted the results. A.G. was involved in the conceptual design of the study. All authors gave final approval for publication.

Funding statement

This work was funded through the UK Research Centre in NDE Targeted Programme by the Engineering and Physical Sciences Research Council (grant no. EP/I019731/1), AMEC, National Nuclear Laboratory, Rolls Royce, Shell and Weidlinger.

REFERENCES

1. Charlesworth JP, Temple JAG. 2001 Engineering applications of ultrasonic time-of-flightdiffraction. Baldock, UK: Research Studies Press.
2. Duxbury D, Russell J, Lowe M. 2013 The effect of variation in phased array element performance for non-destructive evaluation (NDE). Ultrasonics 53, 1065–1078. (doi:10.1016/j.ultras.2012.07.008)
3. Holmes C, Drinkwater BW, Wilcox PD. 2005 Post processing of the full matrix of ultrasonic transmit receive array data for non destructive evaluation. NDT E Int. 38, 701–711. (doi:10.1016/j.ndteint.2005.04.002)
4. Zhang J, Drinkwater BW, Wilcox PD. 2013 Comparison of ultrasonic array imaging algorithms for nondestructive evaluation. IEEE TUFFC 60, 1732–1745. (doi:10.1109/TUFFC.2013.2754)
5. Drinkwater BW, Wilcox PD. 2006 Ultrasonic arrays for non-destructive evaluation: a review. NDT E Int. 39, 525–541. (doi:10.1016/j.ndteint.2006.03.006)
6. Chassignole B, Villard D, Dubuget M, Baboux JC, El Guerjouma R. 2000 Characterization of austenitic stainless steel welds for ultrasonic NDT. Rev. Prog. Quant. Nondestruct. Eval. 19, 1325–1332. (doi:10.1063/1.1307835)
7. Griffin DR, Webster FA, Michael CR. 1960 The echolocation of flying insects by bats. Anim. Behav. 8, 141–154. (doi:10.1016/0003-3472(60)90022-1)
8. Au WWL, Moore PWB, Pawloski DA. 1988 Detection of complex echoes in noise by an echolocating dolphin. J. Acoust. Soc. Am. 83, 662–668. (doi:10.1121/1.396161)

9. O'Donnell M. 1992 Coded excitation system for improving the penetration of real-time phased-array imaging systems. IEEE TUFFC 39, 341–351. (doi:10.1109/58.143168)

10. Chiao RY, Hao X. 2005 Coded excitation for diagnostic ultrasound: a system developer's perspective. IEEE TUFFC 52, 160–170. (doi:10.1109/TUFFC.2005.1406543)

11. Kaiser G. 2010 A friendly guide to wavelets. Boston, MA: Birkhäuser.

12. Ozaktas HM, Zalevsky Z, Kutay MA. 2001 The fractional Fourier transform with applications in optics and signal processing. Chichester, UK: John Wiley & Sons Ltd.

13. Namias V. 1980 The fractional order Fourier transform and its applications to quantum mechanics. IMA J. Appl. Math. 25, 241–264. (doi:10.1093/imamat/25.3.241)

14. McBride AC, Kerr FH. 1987 On Namias's fractional Fourier transform. IMA J. Appl. Math. 39, 159–175. (doi:10.1093/imamat/39.2.159)

15. Schmerr LW. 1998 Fundamentals of ultrasonic nondestructive evaluation: a modelling approach. New York, NY: Plenum Press.

16. Capus C, Brown K. 2003 Fractional Fourier transform of the Gaussian and fractional domain signal support. IEE Proc. Vis. Image Signal Process. 150, 99–106. (doi:10.1049/ip-vis:20030313)

17. Capus C, Rzhano Y, Linnett L. 2000 The analysis of multiple linear chirp signals. In Proc. IEE Seminar on Time-Scale and Time-Frequency Analysis and Applications, London, UK, 29 February 2000. Stevenage, UK: The Institution of Engineering and Technology.

18. Saxena R, Singh K. 2013 Fractional Fourier transform: a novel tool for signal processing. J. Indian Inst. Sci. 85, 11–26.

CHAPTER 14

SIMPLIFIED FRACTIONAL FOURIER TRANSFORM

Wen-Qin Wang

Keywords: Fractional Fourier transform (FrFT), Simplified FrFT, Along-track interfer-ometry (ATI), Displaced phase center antenna (DPCA), Ground moving targets detection (GMTD), Synthetic aperture radar (SAR), Three-antenna SAR

INTRODUCTION

Ground moving target indication (GMTI) is of great interest for surveillance and reconnaissance [1–4], but it is not an easy job because separating the moving targets' returns from stationary clutter is a technical challenge [5]. Moving target indication is twofold [6]: one is the detection of moving targets within severe ground clutter, and the other is the estimation of their parameters such as velocity and location. As such, radar clutter has received much recognition in recent years. Several clutter suppression approaches have been proposed [7], but they often require high pulsed repeated frequency (PRF), which is not desirable to avoid excessive data rate and PRF ambiguity problem.

It is well known that the moving target with a slant range velocity will generate a differential phase shift. This phase may be detected by interferometric combination of the signals from a two-channel along-track interferometry (ATI) synthetic aperture radar (SAR) system. The ATI SAR was initially proposed for detecting ground moving targets [8–10], which uses two antennas to detect targets by providing essentially two identical views of the illuminated scene but at slightly different time. Several interferometry SAR (InSAR)-based moving targets detection algorithms have been proposed previously [11–14]. However, the stationary clutter unavoidably corrupts the interferometric phase of the targets depending on its signal-to-clutter environment. Consequently, the imaged moving targets will be displaced in azimuth according to its radial velocity.

There have been several studies on the clutter effects on the intended signals [15, 16]. But there remains still many unresolved problems, e.g., how to reliably estimate the target's true interferometric phase from the clutter. Moreover, in a nonhomogeneous terrain, the degree of physical overlap of the target with a bright stationary point clutter may also influence the estimation accuracy. In order to accurately estimate the target's true velocity, clutter contamination on the signal must be minimized. Precise knowledge of the interferogram's phase and amplitude statistics is very important for distinguishing the moving targets from the clutter. A straightforward approach to clutter cancelation is the displaced phase center antenna (DPCA) technique [17]. For one two-antenna DPCA system, the additional freedom provided by the second antenna can be used to cancel the clutter; however, it can no longer be used to estimate the moving targets' position information.

Moreover, estimating the moving targets' Doppler parameters is often required, but the Wigner-Vill distribution-based algorithms will generate cross-terms [18], particularly in the presence of multiple moving targets. In this case, the fractional Fourier transform (FrFT) is a powerful tool. But the conventional FrFT is redundant for moving targets detection [19]. This article presents a simplified FrFT (SFrFT) and three-antenna SAR combined GMTI approach. After canceling the stationary clutter using three-antenna ATI SAR, two-channel signals through which moving targets can be detected are formed. Next, one SFrFT-based algorithm is presented to estimate the Doppler parameters of the moving targets. Finally, the moving targets are located through two-channel

Citation: Wen-Qin Wang, Simplified Fractional Fourier Transform. (2011) 2011: 117, doi:10.1186/1687-6180-2011-117.

interferometric processing algorithm. The remaining sections are organized as follows. Section II introduces the SFrFT and its mathematical properties. Section III describes the system scheme of DPCA-based three-antenna ATI SAR for GMTI applications. Next, the SFrFT-based detection algorithm is detailed in Section IV, followed by decorrelation discussion in Section V. Finally, Section VI concludes the whole paper.

SIMPLIFIED FRACTIONAL FOURIER TRANSFORM (FRFT)

The FrFT is a generalization of regular Fourier transform in that the Fourier transform transforms a signal from time-domain to frequency-domain, the FrFT transforms it into a fractional Fourier domain, which is a hybridized time-frequency domain. The transform kernel of the conventional FrFT is defined as [20]

$$K_\alpha(t,\mu) = \begin{cases} \sqrt{\frac{1-j\cot\alpha}{2\pi}}e^{j\frac{t^2+\mu^2}{2}\cot\alpha - j\mu t\csc\alpha}, \alpha \neq n\pi, \\ \delta(t-\mu), \alpha = 2n\pi, \\ \delta(t+\mu), \alpha = (2n+1)\pi. \end{cases} \quad (1)$$

where n denotes an integer, and α indicates the rotation angle in FrFT domain. This operation can be considered as a generalized form of Fourier transform that corresponds to a rotation over an arbitrary angle $\alpha = a\varpi/2$ with a∈R ∈ ℜ, as shown in Figure 1. The forward and inverse FrFT of $x(t)$ are defined, respectively, by

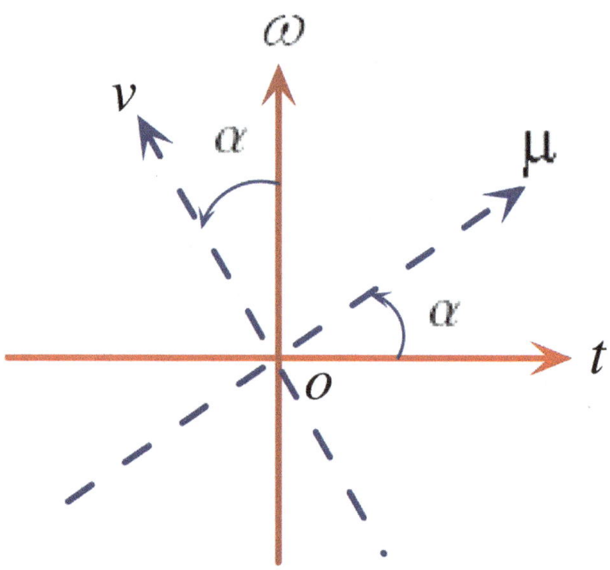

Figure 1: Time-frequency plane and a set of coordinates (μ,v) rotated by an angle$a$relative to the original coordinates (t, ω).

$$\chi_\alpha(\mu) = \int_{-\infty}^{\infty} x(t)K_\alpha(t,\mu)\mathrm{d}t, \quad (2)$$

$$x(t) = \int_{-\infty}^{\infty} \chi_\alpha(\mu)K_{-\alpha}(t,\mu)\mathrm{d}u. \quad (3)$$

Given that **F** is the Fourier transform operator and F_r^α is the fractional Fourier transform operator, then the FrFT possesses the following important properties.

1. Zero rotation: $F_r^0 = I$ rotation: $F_r^{2\pi} = I$.
2. Consistency with Fourier transform: $F_r^{\pi/2} = F$
3. Additivity of rotations: $F_r^\beta \cdot F_r^\alpha = F_r^{\beta+\alpha}$.
4. Linearity:

$$F_r^\alpha[c_1 f(t) + c_2 g(t)] = c_1 F_r^\alpha(f(t)) + c_2 F_r^\alpha(g(t)).$$

Additional properties can be found in the Ref. [20]. The domains $0 < \alpha < \pi/2$ are called as the fractional Fourier domains. The FrFT of a function $x(t)$, with an angle α, can be computed as the following steps.

Step 1. A product by a chirp:

$$g(t) = x(t)e^{-j\frac{1}{2}t^2\tan\left(\frac{\alpha}{2}\right)} \quad (4)$$

Step 2. A Fourier transform (with its argument scaled by csc(α)) or a convolution:

$$g^*(t) = g(t) \circledast e^{-j\frac{1}{2}t^2\csc(\alpha)} = \int_{-\infty}^{\infty} x(t)e^{-j\frac{1}{2}(\mu-t)^2\csc(\alpha)}\mathrm{d}t. \quad (5)$$

where ⊛ denotes a convolution operator.

Step 3. Another product by a chirp:

$$f(t) = e^{-j\frac{1}{2}\mu^2\tan\left(\frac{\alpha}{2}\right)}g^*(t). \quad (6)$$

Step 4. A product by a complex amplitude factor:

$$F_r^\alpha[x(t)] = \sqrt{\left(\frac{1-j\cot(\alpha)}{2\pi}\right)}f(t). \quad (7)$$

As the Steps 3 and 4 are redundant for signal detection, we name the FrFT without the Steps 3 and 4 as simplified FrFT (SFrFT). Note this SFrFT is different the simplified FrFT proposed in [21]. The SFrFT is also a linear transform and continuous in the angle α. This provides us a powerful tool for detecting SAR moving targets, particularly when there are multiple moving targets.

THREE-ANTENNA DPCA-BASED SAR OPERATION SCHEME

Background

Consider an ATI SAR (see Figure 2) consists of two antennas moving along the azimuth direction X and suppose that the two antennas are separated along the azimuth direction. For simplicity and without loss of generality, we assume that there are two interfering point targets, one moving and one stationary. Suppose they are perfectly compressed in the two SAR images, the signals reflected from the moving target and received by the two antennas can be expressed, respectively, by [15]

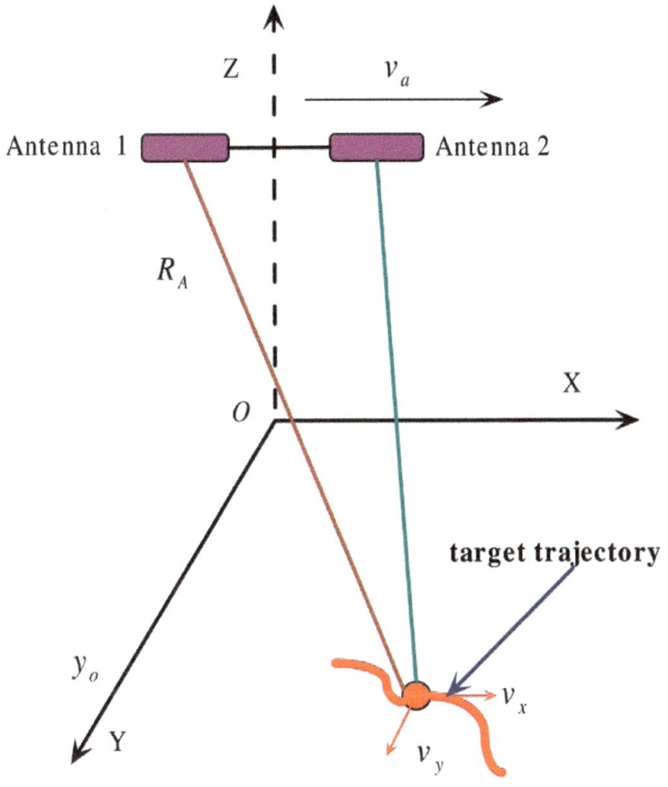

Figure 2: Along-track interferometry SAR geometry.

$$s_{t1} = A_{t1}\delta(x - X_{t1})\delta(y - Y_{t1}) \exp\left(j4\pi\frac{R_{t1}}{\lambda}\right), \quad (8)$$

$$s_{t2} = A_{t2}\delta(x - X_{t2})\delta(y - Y_{t2}) \exp\left(j4\pi\frac{R_{t2}}{\lambda}\right), \quad (9)$$

where A_{ti} is the combined gain with azimuth-compression gain and backscatter coefficient, (X_{ti}, Y_{ti}) is the target position in the imaged images, λ is wavelength, and R_{ti} is the instantaneous slant range modified by convolving with the azimuth reference function. Note that in an actual system the moving point target cannot be well focused with the stationary-terrain matched filter. Similarly, the signals reflected from the stationary target and received by the two antennas are

$$s_{c1} = A_{c1}\delta(x - X_{c1})\delta(y - Y_{c1}) \exp\left(j4\pi\frac{R_{c1}}{\lambda}\right), \quad (10)$$

$$s_{c2} = A_{c2}\delta(x - X_{c2})\delta(y - Y_{c2}) \exp\left(j4\pi\frac{R_{c2}}{\lambda}\right). \quad (11)$$

Suppose the two point targets are overlaped with each other, i.e., $X_{ti} = X_{ci}$ and $Y_{ti} = Y_{ci}$, after registering the SAR image interferometric processing yields

$$\begin{aligned} s_{ATI} &= (s_{t1} + s_{c1}) \cdot (s_{t2} + s_{c2})^* \\ &= A_{t1}A_{t2} \exp\left(j4\pi\left(\frac{R_{t1} - R_{t2}}{\lambda}\right)\right) + A_{c1}A_{c2} \exp\left(j4\pi\left(\frac{R_{c1} - R_{c2}}{\lambda}\right)\right) \\ &\quad + A_{t1}A_{c2} \exp\left(j4\pi\left(\frac{R_{t1} - R_{c2}}{\lambda}\right)\right) + A_{c1}A_{t2} \exp\left(j4\pi\left(\frac{R_{c1} - R_{t2}}{\lambda}\right)\right), \end{aligned} \quad (12)$$

where * denotes a conjugate operator. The first term is the moving target's interferogram that we are wanted. The second term is the stationary target's interferogram. Its phase should be equal to zero because a stationary scene does not change with time, i.e., $R_{c1} = R_{c2}$. The remaining two terms are cross-terms, which come from the clutter contamination at the SAR image formation stage. As the phase angle is 2π periodic, the two cross-terms may have different phase values; hence, the effects of cross-terms on the total along-track interferometric phase are not easily predictable.

As ATI SAR output is signal power, slowly moving targets will not attenuated along with the stationary clutter when we utilize magnitude and phase information for target extraction. In the case of low signal-to-clutter ratio (SCR), the ATI SAR will lose its ability to detect slowly moving targets and to correctly estimate their velocities because the system noise (additive thermal noise and multiplicative radar phase noise) scatters the stationary clutter signal around the real axis in the complex plane. If the clutter contribution is not negligible when compared to the signal power, the estimation of the target radial velocity from the contaminated interferometric phase may lead to erroneous results. When the SCR is small, this effect will become more serious for slowly moving targets and the moving targets will be indistinguishable from the clutter. Moreover, in this case, the targets› impulse responses are not normal delta functions, particularly for the moving targets which are poorly focused because of the unmatched azimuth-compression filter. This leads to a point target›s response overlaps with several neighboring resolution cells. This also means a varying SCR across the target›s response, which in turn affects its interferometric phase. Therefore, the ATI SAR is a clutter limited moving targets detector and applying some efficient clutter suppression or cancelation techniques is necessary.

Three-Antenna ATI SAR Scheme

To improve the performance of slowly moving targets detection, it is necessary to apply some clutter suppression or cancelation suppression techniques. The DPCA is just proposed for this aim, which synthesizes a static antenna system allowing cancelation of static returns on a pulse-to-pulse basis. However, for the two-antenna DPCA system, after clutter cancelation it can not be used to locate the moving targets. As such, this article uses one three-antenna DPCA SAR scheme, as shown in Figure 3. An antenna of length L is used as a single aperture in transmission and is split into three receiving sub-apertures in reception. In this case, the receive phase centers are displaced in the along-track direction by $L/3$.

It is then effective to define the 'two-way' phase centers as the mid-points between the phase center of the whole transmit antenna and the phase center of the receive sub-apertures. So, all goes as if the radar samples were collected by the transmit and receive antennas with phase centers co-located in the two-way phase centers. To reach aim, it is assumed that the pulse repeated frequency (PRF) is matched to the platform velocity and the distance between the three receive sub-aperture phase centers. Under this condition, the two-way phase center of the trailing sub-aperture occupies the same location of the two-way phase center of the leading sub-aperture one pulse repetition interval (PRI) later.

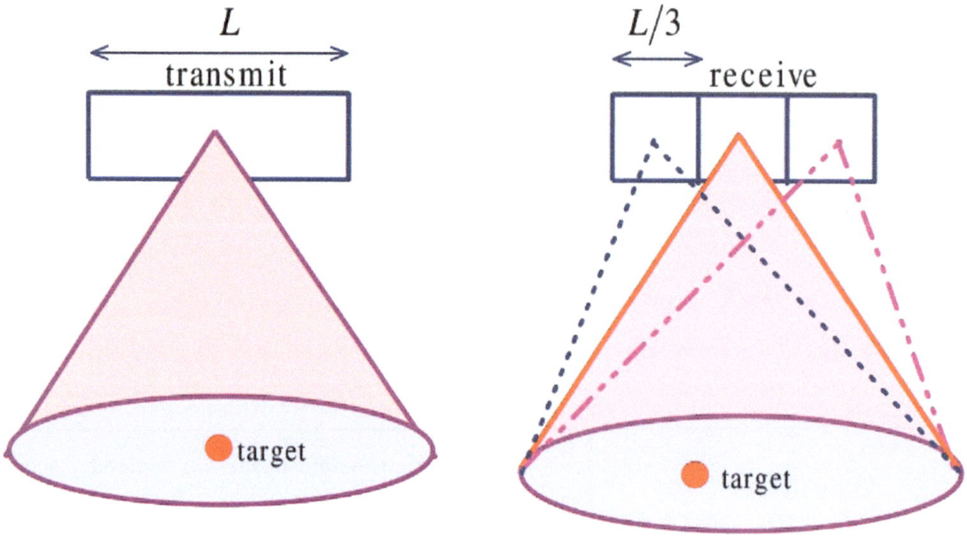

Figure 3: Three-antenna DPCA-based moving target detection scheme.

When the DPCA condition is matched, the clutter cancelation can then be performed by subtracting the samples of the radar returns received by two-way phase centers in the same spatial position, which are temporally displaced. The radar returns corresponding to stationary objects like the clutter from natural scenes are canceled, while the returns backscattered by moving targets have a different phase in the two acquisitions and remain uncanceled. Therefore, all static clutter scatterers are canceled, leaving only moving targets and a much simplified target detection problem (which is detailed in the next section). If the DPCA condition is not matched, the collected azimuth samples will be spaced nonuniformly. This problem can be solved using the reconstruction filtering algorithm detailed in [22].

Signal Models

We make the assumptions of far-field, flat earth, free space, and single polarization for our model. Although DPCA SAR can be realized with airborne and spaceborne platforms, we restrict ourselves to airborne platform only. Figure 4 shows the geometry of the three-antenna DPCA-based moving target detection (MTD) system. The platform altitude and velocity are h and v_a, respectively.

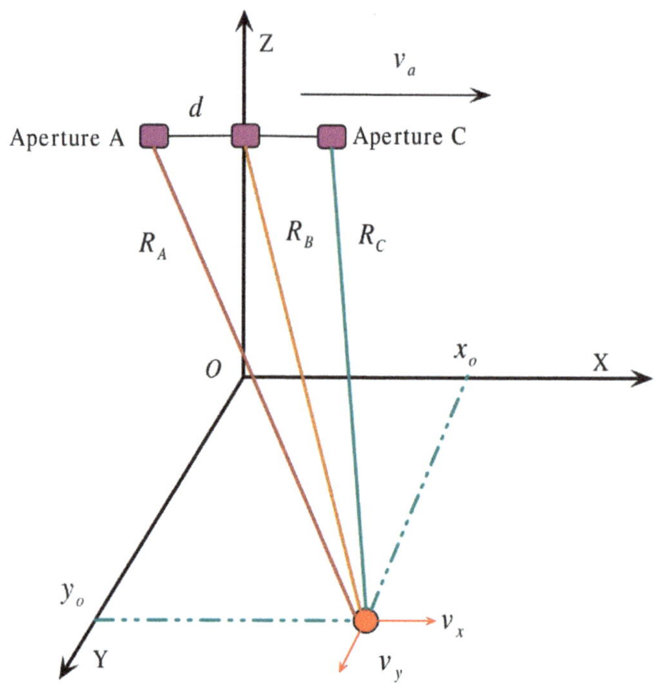

Figure 4: Geometry of a three-antenna DPCA-based MTD system.

The range history from the central aperture to a specific moving target located in (x_o, y_o) with velocity (v_x, v_y) can be represented by

$$R_c(t_a) = \sqrt{(x_o + (v_a - v_x)t_a)^2 + (y_o + v_y t_a)^2 + h^2} \approx R_o + v_y t_a + \frac{(x_o + (v_a - v_x)t_a)^2 + (v_y t_a)^2}{2R_o}, \tag{13}$$

$$R_o = \sqrt{y_o^2 + h^2}. \tag{14}$$

In an alike manner, the range histories of the left aperture and the right aperture can be represented, respectively, by

$$R_l(t_a) \approx R_o + v_y t_a + \frac{(x_o + d + (v_a - v_x)t_a)^2 + (v_y t_a)^2}{2R_o}, \tag{15}$$

$$R_r(t_a) \approx R_o + v_y t_a + \frac{(x_o - d + (v_a - v_x)t_a)^2 + (v_y t_a)^2}{2R_o}, \tag{16}$$

with $d = L/3$. Suppose the transmitted radar signal is

$$s(t_r) = rect\left[\frac{t}{T_p}\right] \exp\left[j2\pi\left(f_c t_r + \frac{1}{2}k_r t_r^2\right)\right], \tag{17}$$

where T_p is the pulse duration, f_c is the carrier frequency, k_r is the chirp rate, and t_r is the range time. After range compressing the three-antenna SAR data, we have [23]

$$S_l(t_r, t_a) = T_p \exp[-j2\pi f_c \xi_l - j\pi k_r (t_r - \xi_l)^2] \cdot \frac{\sin(k_r \pi (t_r - \xi_l)T_p)}{k_r \pi (t_r - \xi_l)T_p}, \tag{18a}$$

$$S_c(t_r, t_a) = T_p \exp[-j2\pi f_c \xi_c - j\pi k_r (t_r - \xi_c)^2] \cdot \frac{\sin(k_r \pi (t_r - \xi_c)T_p)}{k_r \pi (t_r - \xi_c)T_p}, \tag{18b}$$

$$S_r(t_r, t_a) = T_p \exp[-j2\pi f_c \xi_r - j\pi k_r (t_r - \xi_r)^2] \cdot \frac{\sin(k_r \pi (t_r - \xi_r)T_p)}{k_r \pi (t_r - \xi_r)T_p}, \tag{18c}$$

Where $\xi_l = (R_l(t_a) + R_c(t_a))/c_o$, $\xi_c = 2R_c(t_a)/c_o$ and $\xi_r = (R_c(t_a) + R_r(t_a))/c_o$, with c_o is the speed of light. To cancel the clutter, we perform

$$S_{cl}(t_r, t_a) = G_1 \cdot S_c(t_r, t_a) - S_l(t_r, t_a + t_d), \tag{19a}$$

$$S_{rc}(t_r, t_a) = S_r(t_r, t_a) - G_1 \cdot S_c(t_r, t_a + t_d), \tag{19b}$$

where $t_d = d/v_a$ is the relative azimuth delay between two apertures, and G_1 is used to compensate the corresponding phase shift

$$G_1 = \exp\left(-j\frac{2\pi d^2}{4R_o\lambda}\right). \tag{20}$$

where λ is the carrier wavelength. As there is

$$A_o = T_p \frac{\sin(k_r\pi(t_r - \xi_l)T_p)}{k_r\pi(t_r - \xi_l)T_p} \approx T_p \frac{\sin(k_r\pi(t_r - \xi_c)T_p)}{k_r\pi(t_r - \xi_c)T_p} \approx T_p \frac{\sin(k_r\pi(t_r - \xi_r)T_p)}{k_r\pi(t_r - \xi_r)T_p}. \tag{21}$$

Equation (19a) can be expanded into [24]

$$S_{cl}(t_r, t_a) = A_o \exp\left\{-j\frac{4\pi}{\lambda}(R_o + v_y t_a)\right\} \cdot \exp\left\{\frac{x_o^2 + \frac{d^2}{4}((v_a - v_x)t_a)^2 - 2x_o(v_a - v_x)t_a + (v_y t_a)^2}{2R_o}\right\} \cdot \left[1 - \exp\left(-j\frac{2\pi}{\lambda}2v_y t_d\right)\right] \tag{22}$$

From Eq. (22) we can notice that, if $v_y = 0$, there is $|S_{cl}(t_r, t_a)| = 0$; hence, the clutter has been successfully canceled by this method. The remaining problem is to detect the moving targets.

From Eq. (22) we can derived that its Doppler frequency center and Doppler chirp rate are represented, respectively, by

$$f_{dc} = -\frac{2}{\lambda}\left[v_y - \frac{(v_a - v_x)x_o}{R_o}\right], \tag{23}$$

$$k_d = -\frac{2}{\lambda R_o}[(v_a - v_x)^2 + v_y^2]. \tag{24}$$

Then, equation (22) can be rewritten as

$$S_{cl}(t_r, t_a) = A'_o \exp\left[2j\pi\left(f_{dc}t_a + \frac{1}{2}k_d t_a^2\right) + \phi_1\right], \tag{25}$$

where

$$\phi_1 = -\frac{4\pi}{\lambda}\left(R_o + \frac{x_o^2 + \frac{d^2}{4}}{2R_o}\right), \tag{26}$$

$$A'_o = A_o\left[1 - \exp\left(-j\frac{4\pi}{\lambda}v_y t_d\right)\right]. \tag{27}$$

Thus, once the Doppler parameters described in the Eq. (25) are estimated, the target velocity (v_x, v_y) can then be determined from the Eqs. (23) and (24).

To utilize the ATI technique, after compensating the phase terms caused by the relative aperture displace between $S_{cl}(t_r, t_a)$ and $S_{cr}(t_r, t_a)$ by multiplying

$$G_2 = \exp\left[j\frac{2\pi}{\lambda}\left(\frac{v_a t_a d}{R_o}\right)\right]. \tag{28}$$

From Eq. (19b) we have

$$S_{rc}(t_r, t_a) = A'_o \exp\left[2j\pi\left(f_{dc}t_a + \frac{1}{2}k_d t_a^2\right) + \phi_2\right], \tag{29}$$

with

$$\phi_2 = -\frac{4\pi}{\lambda}\left(R_o + \frac{x_o^2 - x_o d + \frac{d^2}{2}}{2R_o}\right). \tag{30}$$

Then $S_{cl}(t_r, t_a)$ and $S_{rc}(t_r, t_a)$ can be interfered with each other by conjugate multiplication

Scl(tr,ta)S∗rc(tr,ta)=||||A′o||||2exp(−2π(xod−d24)λRo)=||||A′o||||2exp(Φ),

$$S_{cl}(t_r, t_a)S_{rc}^*(t_r, t_a) = |A'_o|^2 \exp\left(-\frac{2\pi\left(x_o d - \frac{d^2}{4}\right)}{\lambda R_o}\right) = |A'_o|^2 \exp(\Phi) \tag{31}$$

where * is the complex conjugate operator, and Φ is the interferometric phase. The azimuth position of the moving target can be determined by

$$R_o = \frac{c_o\hat{t}_r}{2} \tag{32}$$

As the velocity of the moving target is relative small, from Eq. (13) R_o can be approximated by

$$R_o = \frac{c_o t_r}{2} \tag{33}$$

where

$$\frac{\sin(k_r\pi(\hat{t}_r - \xi_l)T_p)}{k_r\pi(\hat{t}_r - \xi_l)T_p} = MAX\left\{\frac{\sin(k_r\pi(t_r - \xi_l)T_p)}{k_r\pi(t_r - \xi_l)T_p}\right\} \tag{34}$$

As there is $-W_a/2 \geq x_o \leq W_a/2$, $W_a \gg d$ with W_a is the synthetic aperture length in azimuth, the interferometric phase is limited by

$$\Phi| \ll \pi\left|\frac{\pm W_a d - \frac{d^2}{2}}{R_{o\lambda}}\right| \approx |\pi\frac{W_a}{R_o\frac{\lambda}{d}}| \leq \pi. \tag{35}$$

The interferometric phase Φ ∈ [-π, π] is unambiguous; hence, the unambiguous x_o can be obtained in this way. Once R_o and x_o are determined, the y_o can then be derived from the Eq. (14) because the h is known from the inboard motion sensors.

SFrFT-BASED MOVING TARGET DETECTION

To estimate the Doppler parameters, applying the SFrFT to the Eq. (20), we get

$$\chi_\alpha(\mu) = A'_o \exp(j\phi_1).$$

$$\int_{-T_p/2}^{T_p/2} \exp\left[j2\pi f_{dc}t_a + j\pi k_d t_a^2 + j\frac{t_a^2}{2}\cot(\alpha) - j\mu t_a\csc(\alpha) + j\phi_1\right]dt_a \tag{36}$$

It arrives its maximum at [19]

$$\{\hat{\alpha}_0, \hat{\mu}_0\} = \arg\max_{\alpha,\mu}|X_p(\mu)|^2 \tag{37}$$

$$\begin{cases} \hat{k}_d = -cot(\hat{\alpha}_0), \\ \hat{\alpha}_0 = \frac{|X_{\hat{p}}(\hat{\mu}_0)|}{\Delta\tau}, \\ \hat{f}_{dc} = \hat{\mu}_0\csc(\hat{\alpha}_0). \end{cases} \tag{38}$$

This condition forms the basis for estimating the moving targets' parameters. In the SFrFT domain with a proper α, the spectra of any strong moving target will concentrate to a narrow impulse, and that of the clutter will be spread. If we can construct a narrow band-stop filter in the SFrFT domain whose center frequency around at the center of the narrowband spectrum of a strong moving target, then the signal component of this moving target can be extracted from the initial signal. With this method, the strong moving targets can be extracted iteratively, thereafter the weak moving targets may be detectable. This method can be regarded as an extension of the CLEAN algorithm [25] to the SFrFT.

Therefore, after canceling the stationary clutter, the identification of moving targets can be implemented with SFrFT in the following steps:

Step 1. Apply one SFrFT to the data in which the clutter has been canceled with different α, and from the maximal peak get the numerical estimation of $(\hat{\mu}, \hat{\alpha})$.

Step 2. Apply $F_r^{\hat{\alpha}}$ to the same data, we have

$$X_{\hat{\alpha}}(\mu) = \chi_{\hat{\alpha}}(\mu). \tag{39}$$

Step 3. After identifying the first moving target, we then construct a narrow band-stop filter $M(\mu)$ to notch the narrow band-stop spectrum of this moving target

$$X'_{\hat{\alpha}}(\mu) = X_{\hat{\alpha}}(\mu)M(\mu) \tag{40}$$

Step 4. The filtered signals are then rotated back to time-domain by an inverse SFrFT.

Step 5. Repeat the operations from Step 1 to Step 4 until all the desired moving targets are identified.

Once the Doppler parameters of each target are obtained, substituting them into Eq. (23), we can get the v_x

$$v_y = \frac{(v_a - v_x)x_o}{R_o} - \frac{\lambda f_{dc}}{2} \approx \frac{v_a x_o}{R_o} - \frac{\lambda f_{dc}}{2}. \quad (41)$$

In this step, since v_a, x_o, R_o, λ and f_{dc} are all known variables, the v_y can be determined successfully. In a like manner, substitute Eq. (41) into Eq. (24), we can get the v_x

$$v_x = v_a - \sqrt{-\frac{\lambda k_d R_o}{2} - v_y^2}. \quad (42)$$

Note that, here $k_d \leq 0$ is assumed. Now, the parameters (v_x, v_y) and (x_o, y_o) are all determined successfully. Next, the moving targets can then be focused with one uniform image formation algorithm, such as Range-Doppler (RD) [26] and Chirp Scaling algorithms [27]. The corresponding processing steps are given in Figure 5.

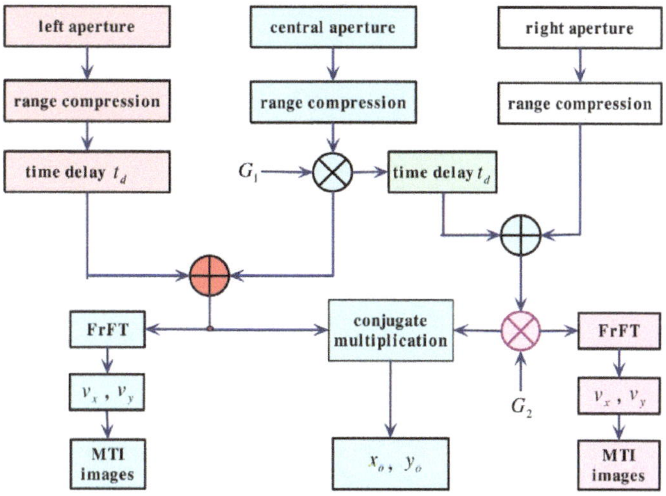

Figure 5: The flow chart of the SFrFT and DPCA combined parameters estimation algorithm.

Simulation Results

To evaluate the performance of the described processing algorithm, three-antenna DPCA stripmap SAR data from three point targets, one moving target and two stationary targets, are simulated using the parameters listed in Table 1. Figure 6 shows the processing results using the general range-Doppler imaging algorithm. It can be noticed that the imaged moving target B is overlapped with the stationary target A. The moving target cannot be identified from this figure, because we cannot discern which is the moving one and which is the stationary one. Figure 7shows the processing results after clutter cancelation by the DPCA operation. The clutter and static returns have been canceled successfully, leaving only moving targets and a much simplified target detection problem. However, the moving target is not focused due to the improper Doppler parameters used in the range-Doppler imaging algorithm. Moreover, the imaged target position is also drifted. To focus the moving target, the accurate Doppler parameters are required.

Table 1: Simulation parameters

Parameters	Values	Units
carrier frequency	1.25	GHz
pulse repeated frequency	360	Hz
flying altitude	7000	m
flying velocity	180	m/s
pulse duration	5	*fJ,S*
range resolution	5	m
antenna length of each aperture	1	m
position of the target A	(x = 50,y = 12000)	m
position of the target B	(x = 58,y = 12000)	m
position of the target C	(x = 50,y = 12250)	m

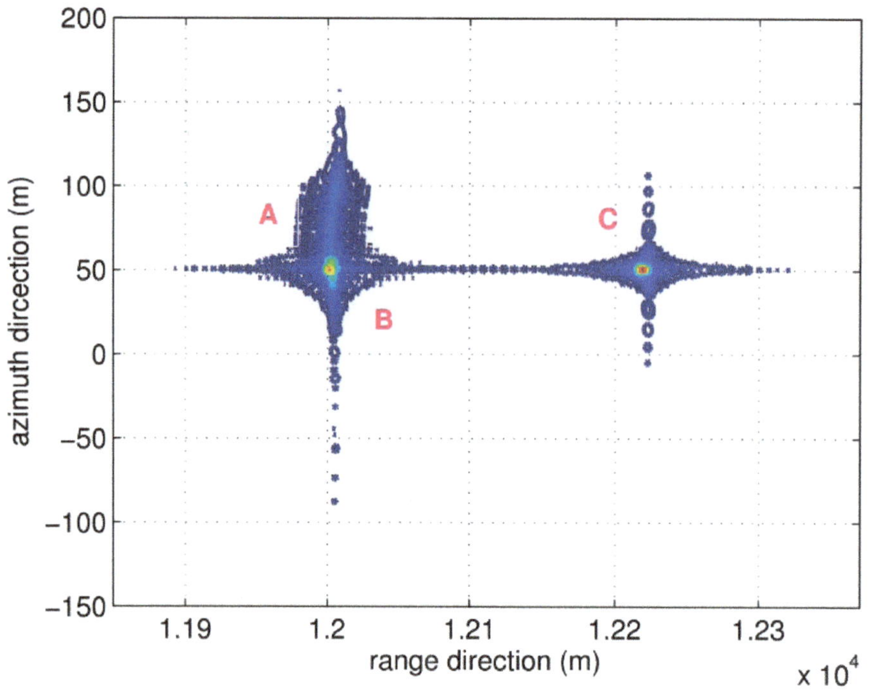

Figure 6: Processing results before clutter cancelation by DPCA operation.

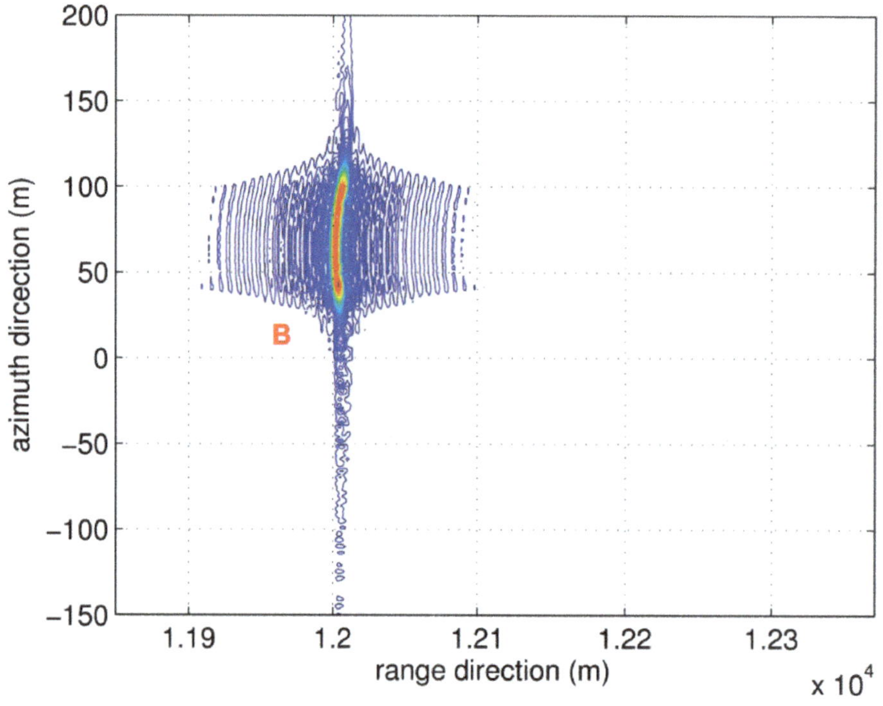

Figure 7: Processing results after clutter cancelation by DPCA operation.

Many algorithms considering linearly frequency modulated (LFM) signal detection, such as Wigner-Ville distribution (WVD) and Radon-Wigner transform, have been proposed. These algorithms are developed primarily for detecting single LFM signal in noise. However, they may generate cross-terms, particularly in the presence of multiple moving targets. Since cross-terms tend to oscillate more rapidly in time-frequency plane than signal auto-components, two-dimensional smoothing suppresses cross-term artifacts at the expense of decreased localization. From Figure 8, we can notice that the results contain both auto-terms and cross-terms, which makes possible false detection (one more target is detected). In contrast, since FrFT is a linear transform, there will be no cross-terms. We can easily locate the peak in the result of FrFT shown in Figure 9. From the peak we get ($\mu = 0.2567$, $\alpha = 0.0092$). Accordingly, the estimated Doppler parameters of the moving target are ($k_d = -17.34$, $f_{dc} = 4.45$). Using these estimated parameters, Figure 10 gives the corresponding imaging results. It is shown that the moving targets can be successfully focused in this way.

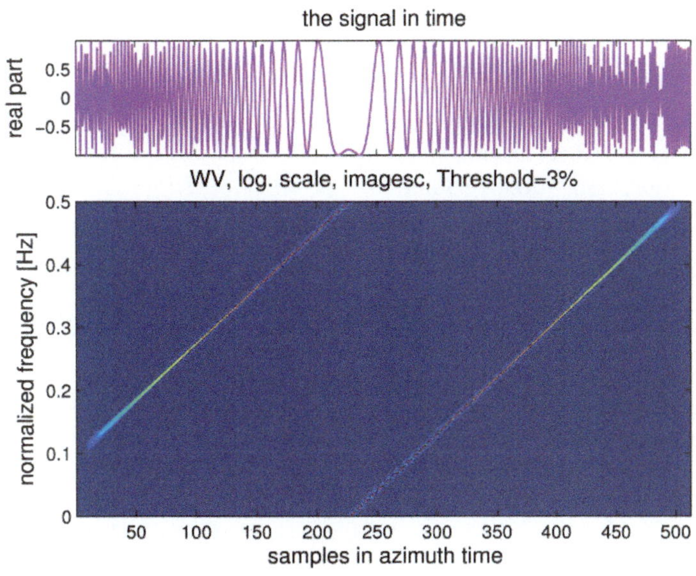

Figure 8: WVD distribution of the return of the single moving target.

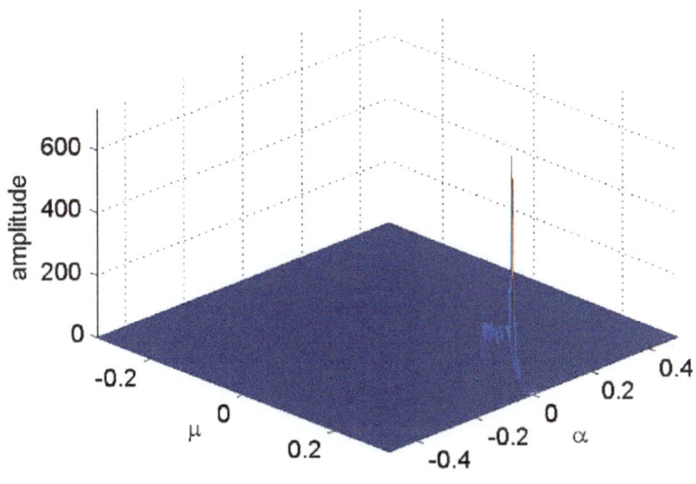

Figure 9: FrFT domain of the return of the single moving target.

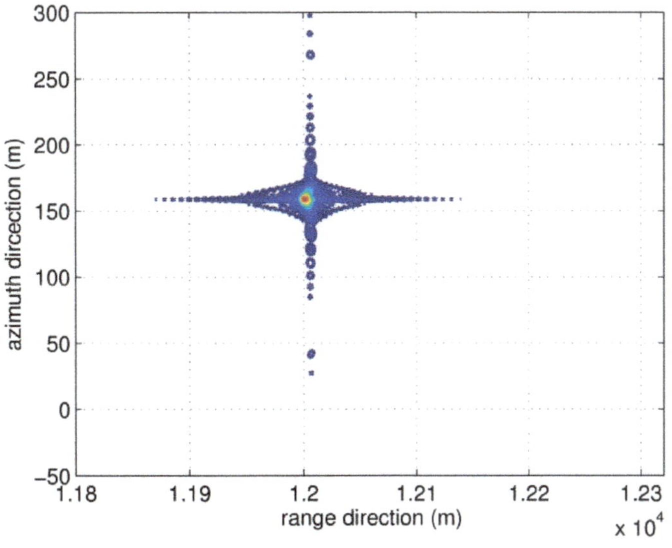

Figure 10: The focused image of the single moving target.

DISCUSSIONS

In this article, the clutter cancelation is performed between two DPCA antennas. Hence, the clutter cancelation performance mainly depends on the correlation characteristics of the signals from fore and aft antennas. But phase center offset and antenna deformation may cause decorrelation. So, decorrelation analysis is necessary.

Suppose the clutter signals from the two DPCA antennas are represented by

$$s_1=c_1+n_1,\ s_2=c_2+n_2, \quad (43)$$

where c_i denotes clutter signals from the i th antenna, and n_i denotes additive noise in the i th antenna. The covariance matrix between the two DPCA antennas can then be represented by

$$R = E\{\begin{bmatrix} s_1{}^* \cdot s_1, & s_1{}^* \cdot s_2 \\ s_2{}^* \cdot s_1, & s_2{}^* \cdot s_2 \end{bmatrix}\} = \begin{bmatrix} E\{c_1{}^* \cdot c_1 + n_1{}^* \cdot n_1\}, & E\{c_1{}^* \cdot c_2\} \\ E\{c_2{}^* \cdot c_1\}, & E\{c_2{}^* \cdot c_2 + n_2{}^* \cdot n_2\} \end{bmatrix}. \quad (44)$$

The multiplicative random phase noise that decorrelates the antenna 2 from the antenna 1 can be modeled as

$$c_2=c_1\exp(j\varphi), \quad (45)$$

where c_1 is deterministic and φ is normally distributed. Suppose φ with a probability density function of $f(\varphi) = N(0, \sigma_\varphi^2)$ we can get

$$E\{c_1{}^* \cdot c_1\} = \xi_c^2, \qquad E\{n_1{}^* \cdot n_1\} = \xi_n^2,$$
$$E\{c_1{}^* \cdot c_2\} = E\{c_2{}^* \cdot c_1\} = \xi_c^2 \exp\left(-\sigma_\varphi^2/2\right). \quad (46)$$

Then, Equation (44) can be further simplified as

$$R = \begin{bmatrix} \xi_c^2 + \xi_n^2, & \xi_c^2 \exp(-\sigma_\varphi^2/2) \\ \xi_c^2 \exp(-\sigma_\varphi^2/2), & \xi_c^2 + \xi_n^2 \end{bmatrix} = \begin{bmatrix} \xi^2, & \xi^2 \cdot \rho \\ \xi^2 \cdot \rho, & \xi^2 \end{bmatrix}, \quad (47)$$

where ρ is the correlation coefficient, which can be determined by [28]

$$\rho = \frac{\xi_c^2/\xi_n^2}{1 + \xi_c^2/\xi_n^2} \exp\left(-\frac{\sigma_\varphi^2}{2}\right). \quad (48)$$

Figure 11 shows several example correlation coefficients. We can notice that the decorrelation characteristics depend on both additive noise and multiplicative phase noise, but the DPCA clutter cancelation performance depends only on multiplicative phase noise. Thus, DPCA SAR detection performance is noise limited. In contrast, ATI SAR detection performance is clutter limited. In this article, the advantage of DPCA SAR and that of ATI SAR are combined; hence, the moving target detection performance can be improved.

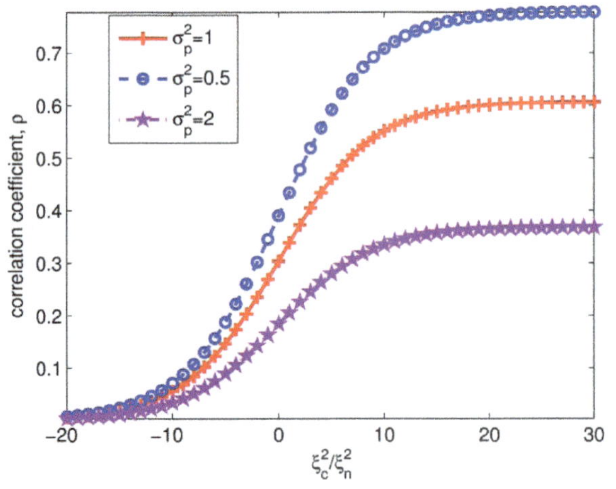

Figure 11: Example correlation coefficients as a function of clutter-to-noise ratio.

CONCLUSION

In this article, one SFrFT and DPCA combined approach is proposed for GMTI applications. This approach realizes target location and velocity estimation with three antennas. After canceling by the three antennas, two-channel signals through which moving targets can be detected are formed. Next, the Doppler parameters of the moving targets are estimated with the SFrFT algorithm. Finally, the moving targets are focused with one uniform image formation algorithm. In this way, both target location and target velocity are acquired, and high-resolution moving target SAR images are obtained. Simulation results show its validity. While compared to conventional approaches, this approach is more effective and robust. In particular, it is not dependent on a target's across-track velocity component or its Doppler shift, which is difficult to determine due to insufficient freedom degrees. This approach depends only on target's Doppler rate, and this is shown to be measurable with a high degree of robustness. In contrast, the conventional approaches like ATI SAR depend not only on a target's Doppler rate but also on its across-track velocity component. Moreover, the selection of matched filter length directly affects the measured ATI phase. These additional unknowns make the SAR ATI a less desirable method for estimating target parameters than the SFrFT and DPCA combined approach, which also allows the estimation of target's true azimuth position directly from its measured position in the final SAR images, particularly when there are multiple moving targets. Therefore, the SFrFT and DPCA combined method is elegant and effective in moving target identification.

ACKNOWLEDGEMENTS

This work was supported in part by the Specialized Fund for the Doctoral Program of Higher Education for New Teachers under Contract number *200806141101*, and the open funds of the Key Laboratory of Ocean Circulation and Waves, Chinese Academy of Sciences under contract number *KLOCAW1004*.

Competing Interests

The author declares that they have no competing interests.

REFERENCES

1. Li G, Xia XG, Peng YN: Doppler keystone transform: an approach suitable for parallel implementation of SAR moving target imaging. *IEEE Geo-science and Remote Sensing Letters* 2008,5(4):573-577.C
2. Wu D, Zhu DY, Zhu ZD: Knowledge-aided multichannel adaptive SAR/GMTI processing: algorithm and experimental results. *EURASIP Journal on Advances in Signal Processing* 2010, 12. Article ID 164187
3. Bergin JS, Techau PM: Multiresolution signal processing techniques for ground moving target detection using airborne radar. *EURASIP Journal on Advances in Signal Processing* 2006, 16. Article ID 47534
4. Page D, Owirka G: Knowledge-aided STAP processing for ground moving target indication radar using multilook data. *EURASIP Journal on Advances in Signal Processing* 2006, 16. Article ID 74838
5. Wang WQ: An approach for multiple moving targets detection and velocity estimation. In *Proceedings of IEEE Radar Conference*. New York; 2006:749-753.
6. Ender JHG, Gierull CH, Maori DC: Improved space-based moving target indication via alternate transmission and receiver switching. *IEEE Transactions on Geoscience and Remote Sensing* 2008,46(12):3960-3974.C
7. Li XM, Feng DZ, Liu HW, Xing MD, Luo D: Two-dimensional pulse-to-pulse canceler of ground clutter in airborne radar. *IET Radar, Sonar and Navigation*2009,3(2):133-143. 10.1049/iet-rsn:20080108C
8. Budillon A, Pascazio V, Schirinzi G: Estimation of radial velocity of moving targets by along-track interferometric SAR systems. *IEEE Geoscience and Remote Sensing Letters* 2008,5(3):349-353.C
9. Dong Z, Cai B, Liang DN: Detection of ground moving targets for two-channel spaceborne SAR-ATI. *EURASIP Journal on Advances in Signal Processing* 2010, 9. Article ID 230785
10. Yang L, Wang T, Bao Z: Ground moving target indication using an In-SAR system with a hybrid baseline. *IEEE Geoscience and Remote Sensing Letters* 2008,5(3):373-377.C
11. Guarbieri AM, Tebaldini S: On the exploitation of target statistics for SAR interferometry applications. *IEEE Transactions on Geoscience and Remote Sensing*2008,46(11):3436-3443.C
12. Romeiser R, Runge H, Suchandt S, Sprenger J, Weilbeer H, Sohrmann A, Stammer D:Current measurements in rivers by spaceborne along-track In-SAR. *IEEE Transactions on Geoscience and Remote Sensing* 2007,45(12):4019-4031.C
13. Romeiser R, Runge H: Theoretical evaluation of several possible along-track InSAR modes of TerraSAR-X for ocean current measurements. *IEEE Transactions on Geoscience and Remote Sensing* 2007,45(1):21-35.C
14. Chiu S, Dragosevic MV: Moving target indication via RADARSAT-2 multichannel synthetic aperture. *EURASIP Journal on Advances in Signal Processing* 2010, 19. Article ID 740130
15. Chiu S: Clutter effects on ground moving target velocity estimation with SAR along-track interferometry. In *Proceedings of International IEEE Geo-science and Remote Sensing Symposium*. Toulouse, France; 2003:1314-1319.
16. Durak A, Gierull CH: Clutter effects on the interferometric phase of ground moving targets. 2005.
17. Lombardo P, Colone F, Pastina D: Monitoring and surveillance potentialities obtained by splitting the antenna of the COSMO-SkyMed SAR into multiple sub-apertures. *IEE Proceedings on Radar, Sonar and Navigation* 2006,153(2):104-116. 10.1049/ip-rsn:20045122C
18. Shui PL, Shang HY, Zhao YB: Instantaneous frequency estimation based on directionally smoothed pseudo-Wigner-Vill distribution bank. *IET Radar, Sonar and Navigation* 2007,1(4):317-325. 10.1049/iet-rsn:20060123C
19. Wang WQ: Approach of multiple moving targets detection for microwave surveillance sensors. *International Journal of Information Acquisition* 2007,4(1):57-68. 10.1142/S0219878907001162C
20. Almeida LB: The fractional Fourier transform and time-frequency representations. *IEEE Transactions on Signal Processing* 1992,42(10):3084-3091.
21. Pei SC, Ding JJ: Simplified fractional Fourier transform. *J Opt Soc Amer*2000,17(13):2355-2367.
22. Gebert N, Krieger G, Moreira A: SAR signal reconstruction from nonuni-form displaced phase centre sampling in the presence of perturbations. In*Proceedings of IEEE Geoscience and Remote Sensing Symposium*. Seoul, Kroea; 2005:1034-1037.
23. Wang WQ, Ding CB, Liang XD: Time and phase synchronization via direct-path signal for bistatic synthetic aperture radar systems. *IET Radar Sonar and Navigation* 2008,2(1):1-11. 10.1049/iet-rsn:20060097C
24. Zheng MJ, Yang RL, Zhou JC: A new method of moving targets detection and imaging based on multi-phase center antenna. *Modern Radar In Chinese*2003,25(12):55-57.
25. Tsao J, Steinber BD: Reduction of sidelobe and speckle artifacts in microwave imaging. *IEEE Transactions on Antennas and Propagation* 1988,36(4):543-556. 10.1109/8.1144C
26. Neo YL, Wong FH, Cumming IG: Processing of azimuth-invariant bistatic SAR data using the range Doppler algorithm. *IEEE Transactions on Geo-science and Remote Sensing* 2008,46(1):14-21.C
27. Wong FH, Yeo TS: New applications of nonlinear chirp scaling in SAR data processing. *IEEE Transactions on Geoscience and Remote Sensing* 2001,39(5):946-953. 10.1109/36.921412C
28. Chiu S, Livingstone C: A comparison of displaced phase centre antenna and along-track interferometry techniques for RADARSAT-2 ground moving indication. *Cana J Remote Sens* 2005,33(1):27-51.

CHAPTER 15

MODIFIED S-TRANSFORM FOR TIME-FREQUENCY SYNCHRONY ANALYSIS

Said Assous and Boualem Boashash

Keywords: Time-frequency coherence, cross time-frequency, modified S-transform, phase synchrony, array processing, EEG signal, time-frequency signal analysis, Quadratic Time-Frequency Distributions, instantaneous frequency

INTRODUCTION

Time-Frequency Methods

Non-stationary signals have statistical properties that vary with time and hence the traditional time averaged amplitude spectrum obtained using Fourier transform is inadequate to track changes in signal magnitude, frequency or phase. In analyzing non-stationary and multi-component signals, time-frequency-based techniques were shown to outperform classical techniques based on either time or frequency domains [1] (Chapter 1). The basic idea of time-frequency analysis is to understand and describe situations where the frequency content of a signal is changing in time. Although time-frequency analysis had its origin almost 50 years ago, significant advances have occurred in the past 20 years or so. In particular, the time-frequency representation has received considerable attention as a powerful high resolution and precision tool for analyzing a variety of bio-signals and systems such as speech, ECG, EEG, PCG, EMG, as well as signals arising from other fields [2]. A time-frequency distribution (TFD) is used to analyze and process non-stationary signals in the joint time-frequency domain. Several TFDs exist in the literature [1]. Most of them are based on the Wigner-Ville distribution (WVD) [3], as all the other TFDs can be expressed as a smoothed version of the WVD. A popular candidate of this class is the spectrogram, which is the square modulus of the short time Fourier transform (STFT). The spectrogram is the WVD smoothed in time and frequency by the ambiguity function of the window used in the STFT [4]; and all the quadratic TFDs (QTFD) are 2D-smoothed versions of the WVD (and therefore of the spectrogram which is the square modulus of the STFT). The spectrogram has been widely used as an initial investigative tool [1] as it has the property of being a cross-terms free TFD; but it also suffers from the undesirable trade-off between time concentration and frequency concentration. To address the problem of cross-terms suppression, while keeping a high time-frequency resolution, other TFDs have been proposed. Among these, one can cite the smoothed pseudo WVD (SPWVD) [1], the Cone shaped kernel TFD (ZAMD) [5], Gaussian TFD [6] and the Modified B-distribution (MBD) [1] (Chapter 3) just to name a few. These methods form a class commonly referred to as QTFDs or Bilinear TFDs, as they represent and distribute the energy of a signal in the time-frequency domain. A different method, the continuous wavelet transform (CWT) has a wide area of application and can be seen as an extension of the spectrogram in a wide sense, except for representing signals in time-scale space instead of time-frequency space. As it is linear and based on amplitude decomposition, it is also a useful technique for filtering in the time-scale space. A more recently developed method called S-transform (ST) is conceptually a hybrid of the STFT and CWT. The ST uses a variable analyzing window length but preserves the phase information by using a Fourier kernel in the signal decomposition [7].

The above indicates that various methods of performing a time-frequency analysis exist. No method is superior to another, since they all have benefits for specific types of signals and applications to which they are adapted, depending on signal characteristics such as spectral contents.

Citation: Assous, S. & Boashash, B, Modified S-Transform for Time-Frequency Synchrony Analysis (2012) 2012: 49, doi:10.1186/1687-6180-2012-49.

The novel contributions in this article are mostly based on

- The design of the modified ST (a signal-dependent version of the standard ST with an improved time-frequency resolution).
- The design and application of the cross-modified ST to investigate the phase synchrony between signals in the time-frequency plane and source location.

Phase Synchrony and Source Location

Phase synchrony analysis is a useful measure of linear dependence between two stochastic signals. This approach is based on the concept of phase synchronization of chaotic oscillators studied by Rosenblum et al. [8]. The phase synchrony (coefficient) takes on values between 0, for two signals at different frequencies, and 1, for signals that exhibit a constant difference in instantaneous phase (representing the situation where a signal and its time-shifted version are observed). So, phase synchrony refers to the interdependence between the instantaneous phases of two signals; the instantaneous phases may be strongly synchronized even when the amplitudes of the two signals are statistically independent [8]. Estimating this measure requires pre-filtering at a frequency of interest. The interesting properties of the phase synchrony coefficient are its independence of the signal amplitudes and that no assumptions about the nature of the signals are made. Another useful measure of linear dependence between two stochastic signals is the coherence function, which quantifies linear correlations in frequency domain based measure between two wide sense stationary signals [9, 10]. The magnitude squared coherence function "measures the degree to which one process can be represented as the output of a linear filter operating on the other process" [10] and varies from 0--for two statistically independent processes--to 1, when one process is the result of linear filtering performed on the other. The phase coherence is usually interpreted as a phase lead of one signal over the other; it finds many useful applications but the results based on coherence depend on several factors like stationarity of the signal, segment length, number of segments, etc. [11]. It is necessary to emphasize that although phase coherence and phase synchrony are quite similar and often are mixed up; they are two principally different measures. Phase coherence can be interpreted as phase shifts and amplitude changes over frequency between two correlated sequences, while phase synchrony indicates whether the phase shift is close to a constant over the specified time interval. This interpretation justifies narrow-band filtering in the case of phase synchrony. The concept of a phase shift (either lead or lag) between two signals is only applicable when both signals are at the same frequency. One way to accomplish phase synchrony is via the Hilbert transform (HT); which consists of transforming the original signal, using an auxiliary function, into a complex-valued signal called an analytical form, from which the instantaneous phase is easily obtained [12]. This method relies on the assumption that the signal is composed of a narrow band of frequencies. Hence, it requires the bandpass filtering of the signal around a frequency of interest and then applies the HT to obtain the instantaneous phase. Another approach to estimate the phase synchrony is by performing a time-varying complex energy spectrum using either the CWT with a complex Morlet wavelet [13] or the short-time Fourier transform (STFT) [14]. Being Gaussian and well localised in time and frequency domains, the Morlet wavelet has an optimal time and frequency resolution [13]. Basically, both CWT and STFT outperform the HT-based methods with the prior giving higher resolution phase synchrony estimates over time and frequency, especially at the low frequency range [8], as they take into account the non-stationarity of the signal. So, in contrast to HT, the wavelet-based function has compact support in such way that it is defined only for those frequencies close to the frequency of interest, so it is equivalent to band-pass filtering at this frequency, which makes the pre-filtering unnecessary. However, the phase information resulting from the wavelet transform is currently not fully understood, and is largely dependent on the specific wavelet, as it is essentially a time-scale method and not a time-frequency method; it has the useful property of being linear in time for sinusoidal signals, and can thus be applied for phase synchrony. Although the wavelet and STFT-based phase synchrony approaches consider the nonstationarity issue, they suffer from a number of drawbacks. In the case of the wavelet transform, where we have a progressive time-frequency resolution the estimate won't be the same for the whole bandwidth of the signal, which means a non uniform time-frequency tiling on the analyzed signal and thus results in biased energy representations and corresponding phase estimates. In the case of STFT, there is a trade-off between time and frequency resolution due to the window function. For these reasons, there is a need for a higher time-frequency resolution phase distributions that can track dynamic changes in phase synchrony over the whole time-frequency plane. As cited above, and given a measurements made from an array of electrodes or sensors, phase synchronization methods are used to detect the phase difference between two signals from two different electrodes or sensors, this can be applied for source location. First, appropriate neighborhood of each channel or electrode is constructed. Then phase synchronization is measured between the channel and each neighbor. Only those channels have been identified as closest to the potential sources where strong phase synchronization is detected will be considered for source location.

The article is organized as follow: The Section 2 introduces the ST definition, its key properties and provides a brief comparison between ST, CWT, and QTFD will be also reviewed. The Section 3 presents the concept and formulation of generalized ST and shows the improved resolution using the modified ST on synthetic examples. The Section 4 discusses the Cross-MST for time-frequency coherence analysis on synthetic examples. The Section 5 demonstrates the application of the cross-MST for phase synchrony detection to simulated EEG seizure and real newborn EEG data.

ADVANTAGES AND LIMITATIONS OF THE ST

The advent of time-frequency analysis techniques using QTFDs, STFT, and wavelet transforms made the analysis of non-stationary signals more precise. The low resolution of the STFT and the absence of phase information in the CWT led to the development of the ST, which has the property of retaining the absolute phase information, while preserving a good time-frequency resolution for all frequencies. Even though the ST has better time-frequency resolution compared to STFT, the resolution is far from perfect and needs improvement. The CWT uses a basis function which dilates and contracts with frequency; it does not retain the absolute phase information; further, the visual analysis of the time-scale plots that are produced by the CWT is intricate. The ST combines the good features of STFT and CWT and can be viewed either as a variable sliding window STFT or as a phase-corrected CWT [7], as detailed below.

Definition and Key Relationships

Given a time series x(t), the local spectrum at time t = τ can be determined by multiplying x(t) with a Gaussian window located at t = τ and taking the Fourier transform of the product. This defines the ST as follows: [7]

$$S(t,f,\sigma) = \int_{-\infty}^{+\infty} x(\tau)g(t-\tau,\sigma)e^{-j2\pi f\tau}d\tau \qquad (1)$$

The original ST uses g(t - τ, σ) as the scaled Gaussian window whose midpoint is τ = t. At any time t and frequency f, the ST can be seen as a set of localised Fourier coefficients, obtained by considering only the portion of the primary function lying within a few cycles on either side of τ= t. The scaled contraction of g(t - τ, σ) causes the relevant range of τ to become more localised around t as f increases.

$$g(t-\tau,\sigma) = \frac{1}{\sqrt{2\pi}\sigma}e^{\frac{-(t-\tau)^2}{2\sigma^2}} \qquad (2)$$

If we select $\sigma = \frac{1}{|f|}$, Equation (2) becomes

$$g(t-\tau,1/f) = \frac{|f|}{\sqrt{2\pi}}e^{\frac{-(t-\tau)^2f^2}{2}} \qquad (3)$$

and its Fourier transform with respect to τ (which is gaussian as well) becomes

$$G(\nu,f) = e^{\frac{-2\pi^2\nu^2}{f^2}} \qquad (4)$$

By choosing σ=1|f| $\sigma = \frac{1}{|f|}$ makes g(t - τ, σ) with two control parameters rather than three and add some constraints to the frequency. Hence, the ST can be defined as an adaptative STFT or as a CWT with a specific mother wavelet-like term multiplied by a phase factor.

$$S(t,f) = e^{-j2\pi ft}WT(t,d) \qquad (5)$$

where the scale factor d is inversely proportional to the frequency f and

$$WT(t,d) = \int_{-\infty}^{+\infty} x(\tau)g(t-\tau,d)d\tau \qquad (6)$$

is the CWT of a function x(t) with a mother wavelet-like term g(t - τ, *f*). Using Equations (3) and (5), this term can be defined as

$$g(t-\tau,f) = \frac{|f|}{\sqrt{2\pi}}e^{-\frac{-f^2(t-\tau)^2}{2}}e^{-j2\pi f(t-\tau)} \qquad (7)$$

The ST separates the mother wavelet-like into two parts, the slowly varying envelope (the Gaussian function) which localises in time the amplitude modulated (AM) component, and the oscillatory exponential kernel $e^{-j2\pi ft}$which selects the frequency being localised, the frequency modulated (FM) component. Hence, this separate AM-FM decomposition is similar in concept to the empirical mode decomposition (EMD), which is a local and fully data-driven technique aimed at decomposing non-stationary multicomponent signals in «intrinsic» AM-FM contributions [15]. The ST analyzing window is actually not a mother wavelet in a strict sense. It is the time localizing Gaussian parameter that is translated while the oscillatory exponential kernel remains stationary. By not translating the oscillatory exponential kernel in Equation (7), the ST localises the real and the imaginary components of the spectrum independently, therefore localizing the phase spectrum as well as the amplitude spectrum. This is referred to as absolutely referenced phase information [7]. Unlike the CWT, the ST produces a time-frequency representation instead of a time-scale representation.

The ST and Signal Spectrum

Integrating the ST over time results in the Fourier transform. This direct relation to the Fourier transform and the linearity property of the ST make both the inversion to time domain and the filtering process in the time-frequency domain possible. The ST is thus a representation of local spectrum; The relationship between the ST and the signal spectrum X(f) can be expressed as follows:

$$\int_{-\infty}^{+\infty} S(t,f)dt = X(f) \qquad (8)$$

where X(f) is the Fourier transform of x(t). Hence, x(t) can be retrieved from ST by:

$$x(t) = \int_{-\infty}^{+\infty}\left\{\int_{-\infty}^{+\infty} S(t,f)dt\right\}e^{j2\pi ft}df \qquad (9)$$

This relationship shows that the concept of the ST is different from both the wavelet transform and QTFDs. The concept of recovering back the signal from the ST time-frequency plane is due to the fact that the phase is referenced at the origin, which means that the phase information given by the ST refers to the argument of the cosinusoid at zero time as it is the case for Fourier transform. In contrast, The phase spectrum of the signal can be extracted from its wavelet transform CWT, as a convolution of the signal with a complex wavelet, where the phase of the CWT is relative to the center (in time) of the analyzing wavelet (the mother wavelet). Thus as the wavelet translates, the reference point of the phase translates giving a locally referenced phase which is distinguishable from the phase properties of the ST and the signal can't be recovered from the time-scale plane. On the other hand, the fact that QTFDs are bilinear and real valued energy distributions describing the energy of the signal over time and frequency, simultaneously, they cannot be used for estimating the phase of an individual signal and the phase synchrony between two signals. Moreover, the inverse process also is not straightforward, as it requires some initial conditions.

The relation between the ST and the signal spectrum X(f) is given in [7] and can be written as

$$S(t,f) = \int_{-\infty}^{+\infty} X(\nu + f) e^{-\frac{2\pi^2 \nu^2}{f^2}} e^{j2\pi\nu t} d\nu, \quad f \neq 0 \tag{10}$$

The equivalent discrete version of Equation (10) can be used to compute the discrete ST by taking advantage of the computational efficiency of the fast Fourier transform (FFT) and the convolution theorem.

Comparison of ST, CWT, and QTFDs

The similarities and differences between the ST, CWT, and QTFDs are reviewed below.

(1) ST and Morlet wavelet: Although the differences between the ST and Morlet wavelet are suitable, the only difference between ST and Morlet wavelet decomposition is that the ST time-frequency function is scaled by the carrier frequency f. Additionally, the amplitude modulation term f smoothes the ST spectrum, but may cause computational artifacts in the Morlet spectrum at low frequencies. Furthermore, the phase correction term in ST enable the ST to preserve absolute referenced phase information of a signal. Also, ST approach has advantages over the Morlet wavelet approach in terms of easy interpretation and fast computation.

(2) Sampling frequency: The discrete Fourier transform (DFT) has a very well defined sampling of the frequencies, in order to be both complete and orthonormal. The discrete ST has the identical sampling of the frequency space and retains the sampling of the signal in the time domain, Similar to a STFT this is redundant. On the other hand, CWT normally employs an octave scaling for frequencies, which results in an oversam-pled representation at the low frequencies and an under sampled representation at the higher frequencies. QTFDs have a sampling that is similar to the DFT.

(3) Invertibility and recoverable signal parameters: As the ST output is complex, we can read the amplitude A(t) = abs(S(t, f)), the frequency f, and the phase information $\phi(t) = \text{atan}\left(\frac{\text{Im }(S(t,f))}{\text{Re }(S(t,f))}\right)$ for each time step from the ST. This allows us to extract the corresponding signal by reproducing it in the time domain as follows: signal (t) = A(t)cos($2\pi f$(t)t+ ϕ(t)). The signal parameters were recoverable due to the combination of absolutely referenced phase information and frequency invariant amplitude of the ST, and such direct extraction cannot be done with CWT-based method. Moreover, the direct measurement of phase information makes the ST a potential candidate to estimate the phase synchrony between two signals.

For QTFDs, a signal x(t) can be recovered exactly, apart from a complex scaling factor [1, p. 61].

(4) ST phase: The ST retains the absolute phase information, where as the phase information is lost in the CWT. The absolutely referenced (at time t = 0) phase of the ST leads to a generalization of the instantaneous frequency to broadband signals and can be used as a local peak-finding algorithm [16, 17], for example. In contrast, in the wavelet approach; the phase is relative to the center (in time) of the analyzing wavelet, and as the wavelet translates, the reference point of the phase translates, hence the relative phase becomes meaningless. With the ST, the sinusoidal component of the basis function remains stationary, while the Gaussian envelope translates in time. Thus, the reference point for the phase remains stationary and the phase has the same meaning as in the Fourier domain [18]. For QTFDs, the phase information is linked to the time delay or instantaneous frequency [12]. The phase information can be extracted using the cross-WVD or related methods [19].

(5) ST amplitude: The time domain localizing window (the Gaussian function) in Equation (7) is normalized by the factor $|f|2\pi^{\frac{|f|}{2\pi}}$ to the unit area. In contrast to the CWT, this makes the amplitude response of the ST invariant to the frequency, which means that for a sinusoid with an amplitude A (x(t) = A cos($2\pi ft$)), the ST returns an amplitude A regardless of the frequency fin a similar concept as the amplitude of the Fourier transform. On the other hand, the amplitude of the CWT is large for the lower frequencies and diminishes at the higher frequency components. This lower amplitude estimation at the high frequencies is mainly due to the normalization of the CWT [18]. For QTFDs, it is the signal energy that is distributed in t and f, not the signal amplitude. Hence, the interpretation is different. The spectrogram is the square modulus of the STFT; and as the QTFDS are a 2D convolution of the spectrogram, the spectrumX(f) can also be recovered subject to the initial complex constant X(f = 0).

(6) Cross-ST analysis: As the ST is complex and its phase characteristic is referenced at the origin, it can be employed in a cross spectrum analysis in a local manner to estimate the phase synchrony between two signals. Consider two signals measured by two receivers (transducers) separated by a known distance. Let a sinusoidal wave propagate through the medium of view of both receivers. What we will get is the same signal in both sensors with an additive noise (assuming a controlled environment conditions), but also with a time shift between the two signals. Since the ST is a linear operation on the signal and localizes spectral components in time, the cross correlation of specific events on two spatially separated STs will give the phase difference, hence the time delay between the two signals can be measured [18]. This concept is also found in cross-TFDs using quadratic methods, where the signal and its complex conjugate are used to calculate the cross-spectrum [19].

(7) Co-ST and quadrature-ST: As the local phase information can be extracted from the ST, we can use the cross-ST function to analyze the in-phase and the out-of-phase components in time-frequency space. This is a very useful characteristic for cross-spectral and phase synchrony. This property is exploited later for an array of EEG signals to estimate the phase synchrony between signals recorded from different electrodes and assess the source location of the brain activities. The cross-ST phase can be used to analyze this synchrony or asynchrony, thus providing a significant discriminating

feature for potential abnormal brain activity. As with classical co-spectrum analysis, the real part of the cross-ST function gives the in-phase components of the local spectra. The imaginary part of the cross-ST function gives the in-quadrature components. This property is discussed in detail in Section 3.

The above paragraphs indicate that, unlike the QTFDs, the ST is linear with progressive resolution (see Section 3.3.2 for more details). Unlike the CWT and STFT, the ST has a variable analysis window and an absolute phase reference. These useful properties of the ST make it a natural candidate to study the phase synchrony between two different spatially recorded signals.

Limitations of the ST

The ST was defined with two unnecessary restrictions on the window function detailed below.

- Firstly, only a Gaussian window g(t, σ) is considered.
- Secondly, frequency dependence of the analyzing window of the ST has been through horizontal and vertical dilations of the Gaussian window.
- Also, that window has no parameters to allow its width in time or frequency to be adjusted, as shown in Equation (2).

The resolution, on the ST, of the onset times of events can be improved by using a narrower window, for example, using a better controlled parameters of the Gaussian window. However, when a window is narrowed in the time domain, it inevitably widens in the frequency domain, with consequent loss of resolution in the frequency direction on the ST compromising the identification of the whole event. One way of addressing this problem is to use extra parameters controlling the scale and the shape of the analyzing window rather than just only making the frequency inversely proportional to the standard deviation σ.

EXTENSION OF THE ST FOR RESOLUTION IMPROVEMENT

Concept and Definition

To remedy the limitations discussed in the previous section, a variant of the original ST, introduced by Mansinha et al. [20], replaces f with $\frac{f}{\gamma}$ in Equation (7). By doing this, one standard deviation of the Gaussian window contains γ wavelengths of the Fourier sinusoid at all frequencies. This allows a better control of the time and frequency resolution of the ST on the time-frequency plane. More recently, McFadden et al. [21] presented a generalized ST which includes windows which are asymmetrical. They applied an asymmetrical window to analyze gearbox vibration signals. Later on, Pinnegar and Mansinha [22] have extended the generalization idea in [21] of the ST to include windows which have complicated scaling properties, including frequency-dependent shape and applied this approach to a broadband earthquake seismogram contaminated by noise. This generalized form retains the property of invertibility of the original S-transform [7] and its relation to the Fourier spectrum. The generalized ST is parameterized by a width control adjustment coefficient γ; it is given by

$$S(t,f,\gamma) = \int_{-\infty}^{+\infty} x(\tau)g(t-\tau,f,\gamma)e^{-j2\pi f\tau}d\tau \tag{11}$$

where g(t - τ, f, γ) is the generalized window function of the ST and γ denotes the set of parameters that determine the shape and the properties of the window function, and the parameter t controls the position of the generalized window on the time axis. The Gaussian window is given by

$$g(t-\tau,f,\gamma) = \frac{|f|}{\gamma\sqrt{2\pi}}e^{\frac{-f^2(t-\tau)^2}{2\gamma^2}} \tag{12}$$

Note that, as in Equation (10), the generalized ST can also be obtained from the signal spectrum as shown below.

$$S(t,f,\gamma) = \int_{-\infty}^{+\infty} X(\nu+f)G(\nu,f,\gamma)e^{j2\pi\nu t}d\nu \tag{13}$$

Key Properties

The generalized window g(t - τ, f, γ) satisfies the following normalization criteria,

$$\int_{-\infty}^{+\infty} g(t-\tau,f,\gamma)dt = 1 \tag{14}$$

This ensures that the time averaging of the ST spectrum S(t, f, γ) yields the spectrum X(f), as shown in (15) below

$$\int_{-\infty}^{+\infty} S(t,f,\gamma)dt = \int_{-\infty}^{+\infty}\int_{-\infty}^{+\infty} x(\tau)g(t-\tau,f,\gamma)e^{-j2\pi f\tau}d\tau dt \tag{15}$$

$$= \int_{-\infty}^{+\infty} x(\tau)e^{-j2\pi f\tau}\int_{-\infty}^{+\infty} g(t-\tau,f,\gamma)dtd\tau \tag{16}$$

$$= \int_{-\infty}^{+\infty} x(\tau)e^{-j2\pi f\tau}d\tau = X(f) \tag{17}$$

where X(f) is the Fourier transform of the signal x(t).

Modified ST

Formulation of the Modified ST

Based on this generalized form, we can examine the Gaussian window at a different scaling rule [23, 24]. This can be achieved by getting the scaling parameter γ defined above to vary linearly with the frequency to get better progressive control of the window width as follows

$$\gamma(f)=mf+k \tag{18}$$

where m is the slope and k is the intercept for a linear change in frequency. Hence, the resolution in time and in frequency will be tuned depending on both the parameters m and k. These parameters will be able to control the Gaussian window length and variance to be appropriate for localizing the low and high frequencies. Figure 1 shows the Gaussian windows with a different value of γ obtained by varying the two parameters m and k. Using this new formulation, the ST becomes

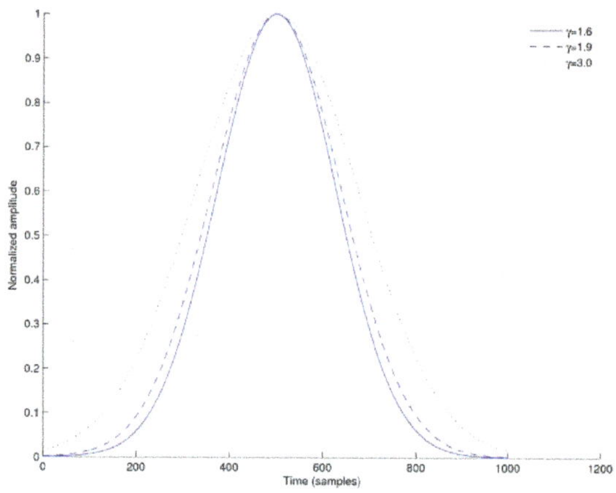

Figure 1: The Gaussian window varying with the parameter γ for frequency f= 10Hz.

$$MST(t,f,m,k) = \int_{-\infty}^{+\infty} x(\tau)g(t-\tau,f,m,k)e^{-j2\pi f\tau}d\tau \tag{19}$$

where g(t - τ, f, m, k) denotes the window function of the modified ST (MST), given as follows

$$g(t-\tau,f,m,k) = \frac{|f|}{(mf+k)\sqrt{2\pi}}e^{\frac{-f^2(t-\tau)^2}{2(mf+k)^2}} \tag{20}$$

Using Equations (19) and (20), the modified ST is given by:

$$MST(t,f,m,k) = \int_{-\infty}^{+\infty} x(\tau)\frac{|f|}{(mf+k)\sqrt{2\pi}}e^{\frac{-f^2(t-\tau)^2}{2(mf+k)^2}}e^{-j2\pi f\tau}d\tau \tag{21}$$

The MST also satisfies the normalization condition for ST window defined in Equation (14) and hence is invertible.

$$\int_{-\infty}^{+\infty} \frac{|f|}{(mf+k)\sqrt{2\pi}}e^{\frac{-f^2(t-\tau)^2}{2(mf+k)^2}}d\tau = 1 \tag{22}$$

Improved Resolution of the MST

In a similar concept to [20], The parameter $\frac{f}{\gamma}$ represents the number of cycles (periods) of a frequency that can be contained within one standard deviation σ of the Gaussian window given by Equation (2). Hence, we have a progressive improved resolution in this case. When too small, the Gaussian window retains very few cycles of the sinusoid and the frequency resolution degrades at low frequencies. In contrast, if it is too large, the window retains more cycles within it and in consequence, the time resolution degrades at high frequencies. This trade-off between time and frequency resolution is governed by the «Heisenberg» uncertainty principle in a similar way as in the FT, ST or the CWT.

The value of m and k need to be selected properly to get better resolution depending on the properties of the signal under consideration. Selected values of k equals to $\frac{1}{N}$ and m equals to four times the variance of the signal were found empirically to be appropriate; where N is the length of the signal in samples. The method is therefore data dependent or adaptive. As an example for illustration, let's consider a three components signal composed of: a low frequency (100Hz), medium frequency (200Hz) and a high frequency burst (400Hz) as shown in Figure2a. Figure 2b,c represent the ST and the MST, respectively, of the signal in Figure 2a.

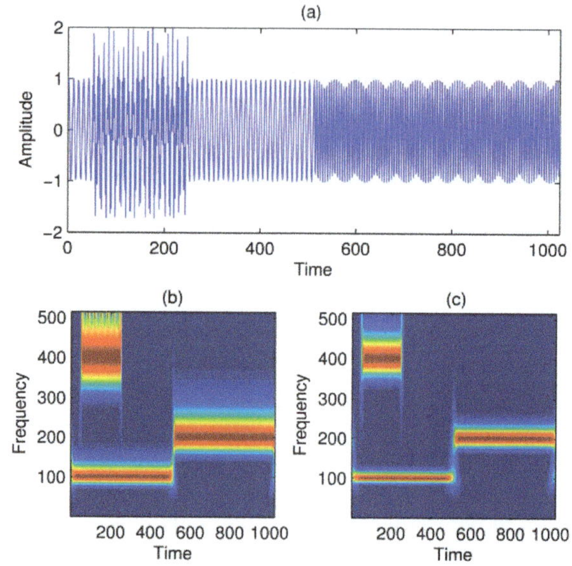

Figure 2: ST and MST comparaison. (a) Time plot of a three-component signal. **(b)** The ST of the signal in **(a)**. **(c)** The MST of the signal in **(a)**.

Figure 2b,c show clearly the difference in time-frequency resolution when the ST and the MST are applied. Using the MST leads to an improvement of almost twofold in resolution over the ST.

Simulated Results, Discussion and Interpretation

In the previous example, the values of m and k were chosen depending on the length and the variance of the signal to get a better resolution (see Appendix 1 for more details). This indicates that as is the case with power spectral estimation, the approach is not restricted to a Gaussian window and any relevant window or apodising function may be employed to improve resolution. The method is not restricted to a Gaussian window and can be generalized in a similar concept to the CWT for which different windows can be used. Hence, the ST can be seen as a special case of the CWT using a Morlet-type wavelet, with a phase and amplitude correction. This generalized form of ST is more flexible, as it uses two added parameters for varying the frequency in a linear way, resulting in adjustable windows for improved time or frequency resolution. Similar ideas led to the development of transforms based on groups generated by translations, modulations, and dilation of a mother wavelet

[25]. All of these approaches give a more versatile choice of transform suitable to particular cases and applications. However, the particular characteristic of absolute phase, make the MST more attractive than wavelet when phase need to be estimated (e.g., in coherence analysis and cross-spectral analysis). It can simultaneously estimate the local amplitude spectrum and the local phase spectrum, whereas a wavelet approach is only capable of probing the local amplitude/power spectrum. The MST fully estimates the amplitude of the signal, in contrast to the CWT which attenuates high frequencies. The MST can be further improved by dealing with issues such as fast algorithms, redundancy of representations, Hilbert space properties, resolution of the identity, among others [7].

CROSS-MST AND PHASE SYNCHRONY

Motivation and Illustration

As mentioned above, the MST phase preservation characteristic is an advantage that allows a cross spectrum analysis in a local manner. Let's consider two signals measured by two sensors separated by a known distance. What these two sensors see is the same signal, but with a time shift in the arrivals as well as noise. Since the MST localizes spectral components in time, the cross correlation of specific events on two spatially separated MSTs gives the phase difference and hence the phase synchrony can be estimated. The amplitude of the cross-MST indicates coincident or partly overlapped signals. The phase of the cross-MST at local maxima indicates the phase difference between them. The cross-MST of two signal x(t) and y(t) is defined as follows [18]

$$\mathrm{crossMST}(t,f) = \mathrm{MST}_x(t,f).\mathrm{MST}_y(t,f)^* \tag{23}$$

where ()* denotes the complex conjugate.

The phase of the cross-MST is given by

$$\arg(\mathrm{crossMST}) = \phi_x(t,f) - \phi_y(t,f) \tag{24}$$

Three key properties of the MST that make the cross-MST useful are detailed below.

(1) The MST translation property in time and frequency: As mentioned in [18], the ST (or MST) time translation property is similar to what Fourier transform states that is: if

$$x(t) \Leftrightarrow \mathrm{MST}(t,f) \tag{25}$$

As MST preserve the same translation property as Fourier transform, if we translate x(t) by an amount r we obtain

$$x(t-r) \Leftrightarrow \mathrm{MST}(t-r,f)e^{-j2\pi fr} \tag{26}$$

The above equation shows that the phase difference between two signals is equivalent to the cross-power spectrum. By Taking the inverse MST transform of the representation in the frequency domain, the displacement between two signals can be easily obtained (phase correlation technique).

(2) The MST phase shift: According to [18], if we consider a sinusoidal signal x(t)

$$x(t) = e^{j2\pi f_0 t} \tag{27}$$

with a spectrum

$$X(\nu) = \frac{1}{2}\delta(\nu - f_0] \tag{28}$$

Using Equation (13), The MST of x(t), is given by

$$\mathrm{MST}_{x(t)}(t,f) = \int \delta(\nu + f - f_0) e^{\frac{-2\pi^2\nu^2}{f^2}} e^{j2\pi\nu t} d\nu \tag{29}$$

which results in:

$$\mathrm{MST}_{x(t)}(t,f) = \frac{1}{2} e^{\frac{-(f-f_0)^2}{\left(\frac{f}{2\pi}\right)^2}} e^{j2\pi(f-f_0)t} \tag{30}$$

(3) Phase in the cross-MST: if we introduce a constant phase shift into x(t), we get

$$y(t) = x(t)e^{j\phi} = e^{j2\pi f_0 t + j\phi} \tag{31}$$

with a resulting spectrum

$$Y(\nu) = \frac{1}{2}\delta(\nu - f_0)e^{j\phi} \tag{32}$$

this will also introduce a phase shift in the modified ST

$$\mathrm{MST}_{y(t)}(t,f) = \frac{1}{2} e^{\frac{-(mf+k-f_0)^2}{\left(\frac{mf+k}{2\pi}\right)}} e^{j2\pi(mf+k-f_0)t} e^{j\phi} \tag{33}$$

The three above properties indicate that we can perform a cross spectral analysis between x(t) and y(t) by multiplying the MST of x(t) with the complex conjugate of the modified ST of y(t) as follows

$$\mathrm{crossMST}(t,f) = \frac{1}{2} e^{\frac{-(mf+k-f_0)^2}{\left(\frac{mf+k}{2\pi}\right)^2}} e^{j2\pi(mf+k-f_0)t} \; \frac{1}{2} e^{\frac{(mf+k-f_0)^2}{\left(\frac{mf+k}{2\pi}\right)^2}} e^{-j2\pi(mf+k-f_0)t} e^{-j\phi} \tag{34}$$

Which simplifies to:

$$\mathrm{crossMST}(t,f) = \frac{1}{4} e^{-j\phi} \tag{35}$$

hence, the phase of cross-MST is clearly given by

$$\mathrm{Phase}\{\mathrm{MST}_x \mathrm{MST}_y^*\} = -\phi \tag{36}$$

The above derivation shows that we can use the cross spectral property to detect a time lag between the two signal as a function of t and f. A similar attempt was made in [26, 27] by using the cross-WVD for representing non-stationary processes.

Illustration on Simulated Signals

Let's consider the signal x(t) given in Figure 2 with 1,000 samples length and sampled at 1 kHz; it includes three lower (100Hz), medium (200Hz), and high frequency (400Hz) components; consider also another signal y(t) with similar three components except that the higher component is translated and partly overlaps in time with the higher component of x(t) (see Figure 2a). The higher components are out of phase (the x(t) component is a cosine function, the y(t) component is a sine function). The lower and medium components of the two signals are in phase. Figure 3 shows the two signals x(t) and y(t) and their related MSTs. Figure 4 shows the cross-MST of the signals x(t) and y(t) corresponding to the two MSTs in Figure 3c,d. The figure indicates that, for the higher component, only the overlapped part in time is detected, as expected.

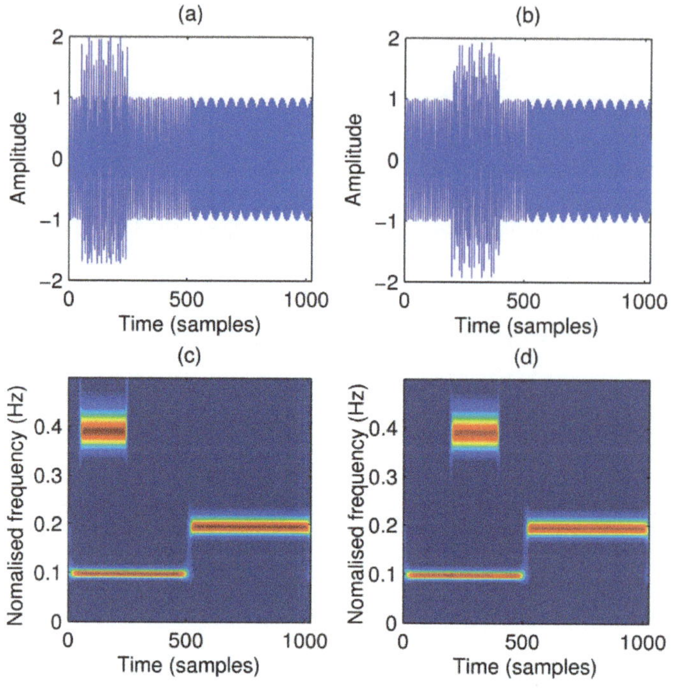

Figure 3: **MST for the two signals*x(t)* and*y(t)*. (a)** The signal x(t) with three-components (100, 200, and 400 Hz). **(b)** The signal y(t) with three-components (100, 200, and 400 Hz). **(c)** The MST of the signal x(t) given in **(a)**. **(d)** The MST of the signal y(t) given in**(b)**.

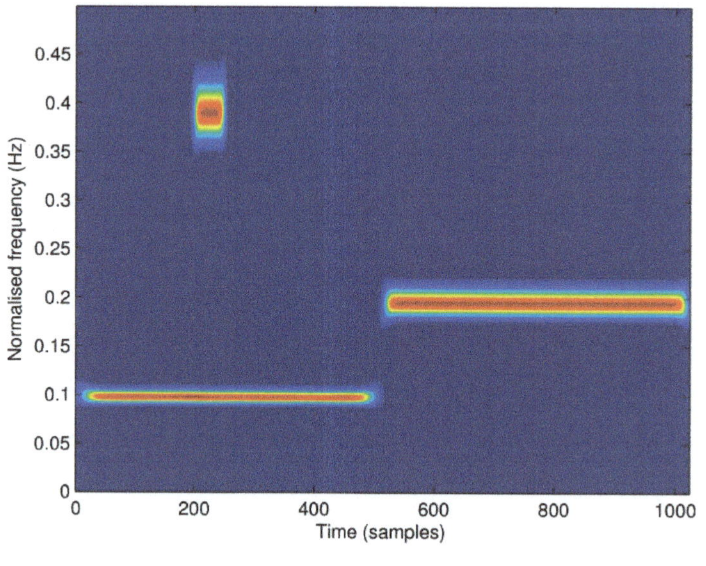

Figure 4: Cross-MST of the two signals x(t) and y(t) MSTs given in Figure 3c,d, respectively.

In a similar way to co-spectrum analysis, the real part of the cross-MST gives the in-phase components of the local spectra and the imaginary part of the cross-MST gives the in-quadrature components. Figures 5a,b show the Co-MST and the Quadrature-MST amplitudes corresponding to the two signals x(t) and y(t), respectively. In the co-MST (Figure 5a), only the features that are in phase are present--those being the low and the medium frequency components. The higher frequency (which is out of phase) does not appear. In the quadrature-MST plot, only the out-of-phase components appear, which is the higher frequency component. As the example above was only for illustration, where the used signals have straight and linear IF components and there is no noise; these results show that it is possible to study the coherence with sufficient high resolution using the cross-MST, whether for amplitude or phase coupling between two different spatially recorded components or signals.

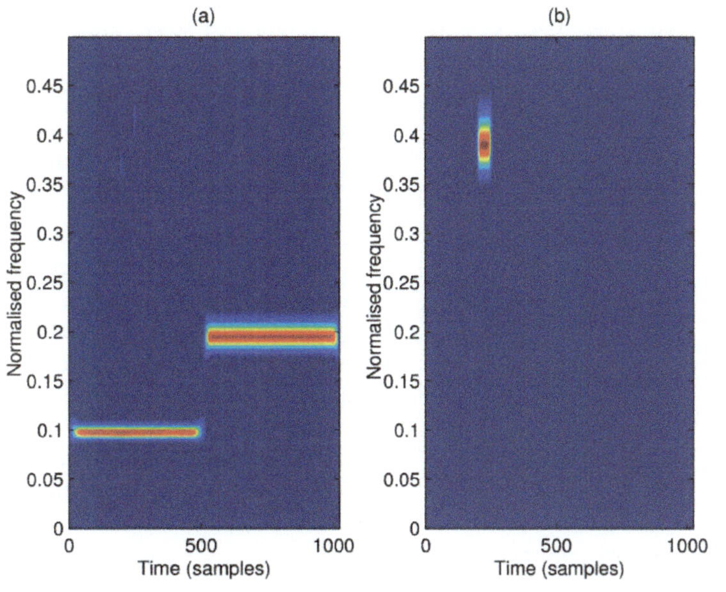

Figure 5: **co-MST and quadrature-MST**. **(a)** The extracted in-phase components of x(t) andy(t). **(b)** The extracted out-of-phase components of x(t) and y(t).

APPLICATION TO EEG SEIZURE PREDICTION

Phase Synchrony Information

The nonlinear interdependencies between fluctuations of brain physiological activities EEG-based recordings were intensively studied in pairs, and the synchrony flow differences were compared [28]. It was shown in [29] that a neurological disorder in the brain can be detected by the random occurrence of its clinical manifestations, i.e., the seizure. There is also evidence that the nonseizure (inter-ictal) to seizure (ictal) transition is not an abrupt phenomenon [29]. Hence, to provide a valuable insight into such mechanism, identification of early changes in EEG signals [28] can be used for prediction, prevention, and control of upcoming seizures. Different approaches have been employed for seizure prediction in the past. Some limited techniques based on visual inspection of the EEG signals or on linear methods have failed to detect specific and sustained changes preceding seizures [28]. Other methods reported some changes based on the spectral contents [29], complexity [30], or spatio-temporal patterns of spikes [31]. Other recent approaches applied to newborn EEG-based seizure detection have been proposed such as time-frequency signal processing [32, 33], and nonlinear signal processing [34, 35].

Neuroscientists claim that when a focal seizure is generated, synchronized brain activity is initially observed only in a small area of the brain; and, from this focus, the activity spreads spatially out to other brain areas in the temporal lobe over time [29]. Involving the seizure focus spread into its surrounding, a hyper-synchronous state happens and neighboring areas lose their synchronization with the other cortical regions around them [36]. In consequence, the seizure focus becomes isolated from the rest of the brain dynamics, making the considered population of neurons inactive. Methods to track these changes using coherence and synchronization are being explored using EEG, for the purpose of predicting an impending seizure [36]. Phase synchronization was also shown to be a sensitive indicator of coupling between signals and as an important factor in the genesis of epileptic phenomenon [36]. Many studies suggest an underlying correlation between neuronal synchronization and seizure development and onset [36, 37]. Thus, the concept of phase synchrony helped to measure the synchrony evolution while the amplitude of the signals remained uncorrelated. In contrast with cross-correlation that measures linear relationships only, a phase coupling approach is able to show the presence of non-linear coupling [38]. Moreover, phase synchrony can be used for non-periodic and for chaotic signals such as EEG [8, 39]. Some authors used HT to measure phase synchrony between two signals [37, 38, 40]. However, the use of the HT assumes the signals have a narrow frequency band [1, p. 14], and it is not straightforward to extend the same analysis for broadband data. This assumption is usually seen to be ignored in the context of biomedical signals like the EEG [41]. As EEG signals are sometimes broadband (1-100 Hz) [41], the HT may not be able to correctly estimate the instantaneous phase of broadband signals. This raises the concern that the broadband phase synchronization analysis may lead to a misinterpretation of the results [41]. Therefore, when the signal is broadband it is necessary to pre-filter it in the frequency band of interest before applying the HT, in order to get an better estimate of the phase [42]. Some authors used wavelet-based approach to analyze phase synchrony between pairs of EEG signals [13, 43–45]. These time-varying measures of phase synchrony using wavelet or HTs are similar in their results. Both give a sharper phase synchrony estimates over time and frequency, especially at the low frequency range [46]. Although the wavelet and Hilbert based phase synchrony estimates address the issue of non-stationarity of EEG signals, they suffer from limited time-scale resolution due to the limited number of available scales in the case of wavelets and the narrowband assumption in the case of HT. Moreover, the role of phase in wavelet analysis is not as well understood as it is for the Fourier transform, especially for orthonormal wavelet representations. Current complex wavelets such as Daubechies, dual tree, and Shannon wavelets do not have a direct relationship to the signal spectrum [18]. For these reasons, this article proposes to use the MST and the cross-MST to analyze the phase synchrony with the required resolution and with meaning, as the phase in the ST is meaningful and has the same concept as in the Fourier transform approach.

Coherence Analysis using Cross-MST for Simulated EEG Seizure

Due to its non-stationarity characteristic, the newborn EEG seizure was modelled in the time-frequency domain [47–49]. using time-frequency characteristics previously identified in the newborn EEG seizure. Three models of newborn EEG seizure simulation were previously proposed. One model is based on some physiological parameters of the brain and utilises a stationary sawtooth waveform [50]. This approach is extended in [47] to incorporate a single linear frequency modulation (LFM) signal. Another piecewise LFM model was defined for seizure based on non-stationary inputs to a nonlinear model [51]. We propose here a method of newborn EEG seizure analysis using the piecewise LFM pattern outlined in [47–50] to simulate the nonstationary behavior of the newborn EEG seizures.

The simulated signal is composed of several piecewise LFM components with different LFM slope parameters. Figure 6a represents three LFM piecewise signal with added Gaussian noise simulating the EEG seizure and the corresponding MST analysis. Figure 6b represents another three LFM piecewise signal with its MST, similar to the one in Figure 5a except that the last component is out of phase compared to the two other components which are in phase in the two signals (More details are given in Appendix 2).

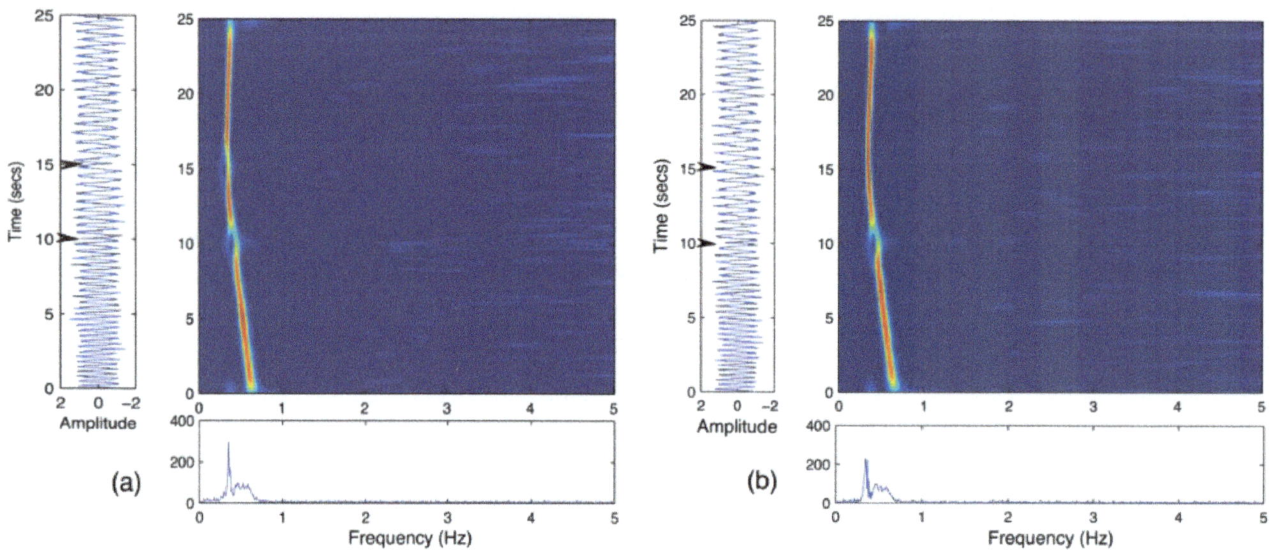

Figure 6: **Simulated EEG seizure and its MST analysis**. The third component in **(a)** is a cosine function, the third component in **(b)** is a sine function, the meeting points are shown by arrows in the signal time plot around 10 and 15s.

Key parts of the commented Matlab® code written to implement this method are given in Appendix 2.

Figure 7a represents the real part of the cross-MST of the two MSTs in Figure 6. We can see clearly that only the in-phase components are detected. However, in Figure 7b, only the out-of-phase component is detected (the third component). Key parts of the commented Matlab® code written to implement this method are given in Appendix 1. Cross-MST of the Simulated EEG seizure showing the in-phase components (a) and the out-of-phase component (b).

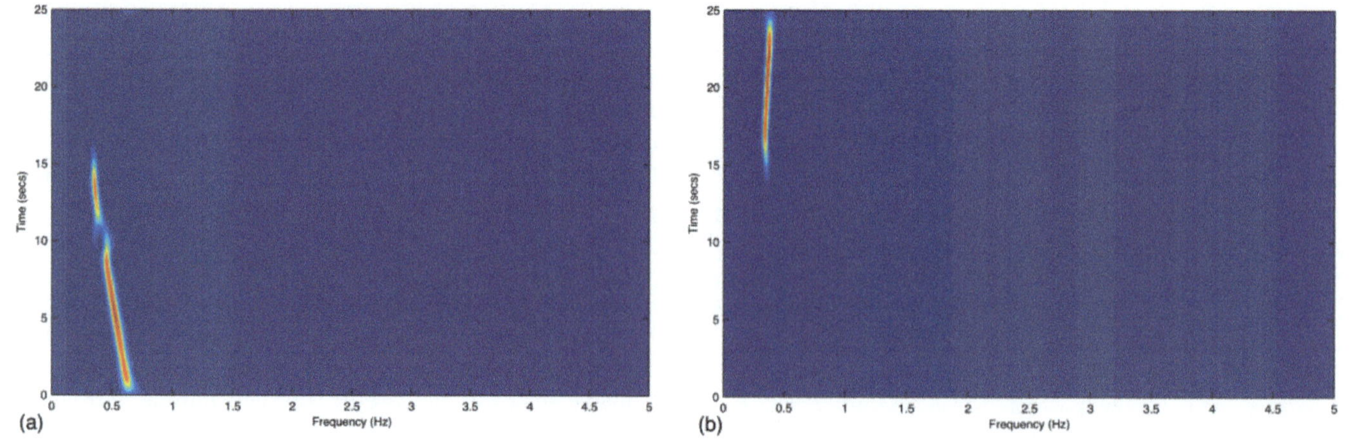

Figure 7: **Cross-MST EEG**. Cross-MST of the simulated EEG seizure showing the in-phase components **(a)** and the out-of-phase component **(b)**.

Coherence Analysis using cross-MST for new-born EEG data

Background and Analysis

The data was acquired using a 20-channel Medelec Profile system (Medelec, Oxford Instruments, Old Woking, UK) at 256Hz sampling rate and marked by a pediatric neurologist from the Royal Children's Hospital, Brisbane, Australia. The experiment was conducted as part of a long term collaboration between Prof. Boashash, Director of the SPRC and Prof. Paul Colditz, Director of the Perinatal Research Centre at the University of Queensland. Measurements were made from a number of sick babies and this work was previously reported in [52]. Five monopolar channels (O1, O2, P3, P4, Cz) out of the 14 recorded signals according to the 10-20 standard [53] modified for newborn subjects were selected from newborn EEG dataset. The data were analysed to investigate the coherence and phase coupling interhemispheric and intrahemispheric interactions during an EEG seizure period. Figure 8ashows the 10-20 standard map of the 20-channel EEG recorder.

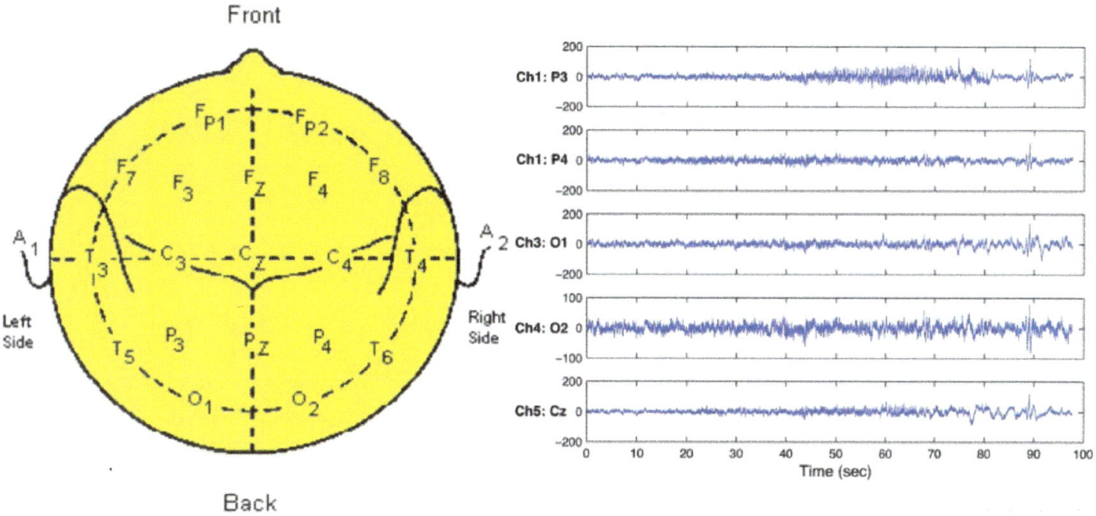

Figure 8: **EEG 5 channels.** **(a)** The 10-20 standard map for 20-channel EEG recorder (taken from: http://faculty.washington.edu/chudler/1020.html), **(b)** the recorded signals from the five channels.

To study the phase coupling between the five channels; we apply the MST to the five signals recorded from the five channels corresponding to P3, P4, O1, O2, and Cz, respectively. The location of the electrodes for the five channels is shown in Figure 8a. Then, we apply the cross-MST function to analyze the in-phase and out-of-phase components in time-frequency space between the five channels. As mentioned previously, the real part of the cross-MST function gives the in-phase components of the local spectra; and the imaginary part of the cross-MST function gives the in-quadrature components. Figure 8 show the in-phase components for the connectivity between channels. From the all 25 possible connectivities shown, Figure 9 indicates that a strong phase synchrony exist in the last few seconds of the recording. This is shown by a peak in the real cross-MST images around 90s. This is in agreement with the recorded signals shown in the time domain (see Figure 8b) as at the corresponding time (90 s) a peak with high amplitude is appearing in the signals. This may explain that at this time (around 90s) the pulses coming from different group of neurone are in phase and when they are added together, the sum results in high peak amplitude. This interpretation will not be possible if we just look at the signals amplitude from Figure 8b. The time signals recorded amplitude does not give a phase information to justify the synchrony; however, the cross-MST shows clearly this phase synchrony by a high peak amplitude in the real part present in the connectivity images (Figure9).

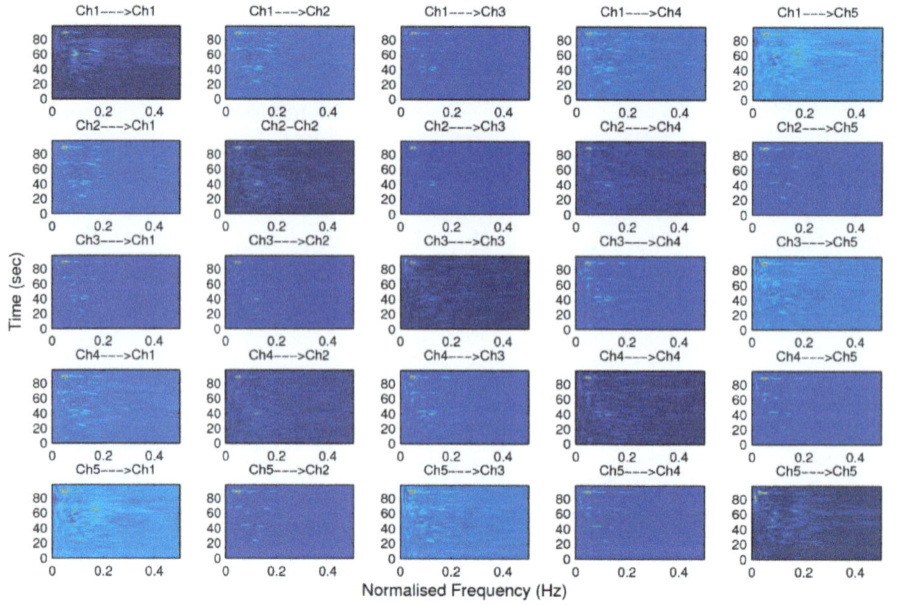

Figure 9: The phase synchrony between the five channels using the real part of the cross-MST.

Figure 10 represents the imaginary part of the cross-STs which reveal the out-phase behavior between the five channels. Note that there is no activity in the diagonal images of Figure 10. This is because the diagonal represents the self-coherence between the five channels, hence there is no out-of-phase in this case; which is consistent as a signal can't be out-of-phase with itself. From the off-diagonal images, we can see that most of the channels exhibit an out of phase behavior mostly in the period 40-80 s corresponding probably to the important duration of the appearance of the seizure. Some out-of-phase components are also present before this 40-80 s period in some channels, which can be explained by the fact that the seizure is not appearing in the same time and with the same way in the spatially different parts of the brain from where the EEG signals are recorded. The end of seizure can be explained by the disappearance of the out-of-phase components and the appearance of the out-of-phase components.

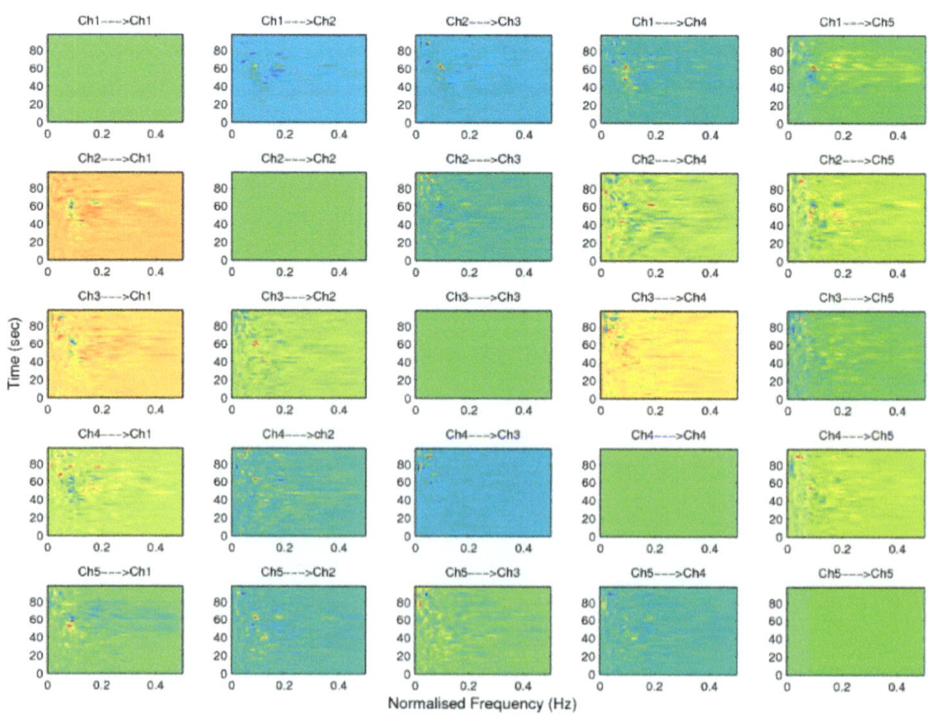

Figure 10: The out-off-phase (asynchrony) between the five channels using the imaginary part of the cross-MST.

Results, Discussion and Interpretation

The results on the real newborn EEG data show that the cross-ST can be used as a effective tool to study the phase coupling between channels. These results suggest that during the seizure period, all the components lose synchrony and are out of phase. This can be seen like a disorder in the neuronal activity during the seizure interval. This observed irregularity in the time-frequency patterns during the seizure period (40-80 s) between two hemispheres in the posterior area suggest that different regions and not only one specific region may be causing the seizure. In contrast, the components (neural activity) became synchronized again after the seizure activity vanishes. This EEG seizure application demonstrates that the time-frequency coherence is an appropriate tool to study the phase coupling between different signals recorded from different spatially separated electrodes. This study is performed by the cross-MST which has a referenced phase at the origin, a property that allows the study of phase coupling and time-frequency coherence between the neuronal activities. Although the cross-MST represent a good candidate tool to analyze such coupling behavior, this methodology can be extended to other cross-time-frequency methods with high time-frequency resolution such as the QTFDs. In particular, the time-frequency coherence using the cross-WVD has a useful property for cross spectral analysis, and can be interpreted as a time and frequency dependent correlation coefficients [26].

CONCLUSIONS AND PERSPECTIVES

This article applies the concept of time-frequency coherence using the cross-MST which is a time-varying cross-spectral analysis method obtained by extending the S-transform. The article demonstrates the ability of this time-frequency coherence using the cross-MST to study the functional neuronal phase synchrony between channels. The performance of the phase synchrony estimation using the cross-MST are evaluated both by simulated EEG seizure using a piecewise LFM signal to extract the in and out-of-phase components, and through the analysis of seizure EEG abnormalities in the newborn, where the dynamics of brain changes rapidly and the neurones lose their synchrony during seizure. Both the simulated and newborn results show cross-MST phase and synchrony estimation to be more robust to noise. The resolution obtained by the MST shows almost twofold improvement over the standard ST approach. Future work will concentrate on a comparison study between the proposed method and other existing time-frequency techniques such as Wigner-Ville and QTFDs based methods, for both simulated and real EEG data to develop more efficient time-varying phase and synchrony estimation for nonstationary signals, and such results will appear elsewhere. The proposed method can be extended to other physiological problems where the phase coupling is relevant, such in cardio-respiratory relationship, for example, where the coupling can be affected by cardio-vascular diseases.

ACKNOWLEDGEMENTS

The Authors wish to thank Prof Paul Colditz, Director of the Perinatal Research Centre, and his team for providing the real newborn EEG data presented in this paper. The second author acknowledges funding by QNRF, Qatar Foundation under grant NPRP 09 465 2 174. The first author also acknowledges Weatherford for giving permission to publish this work.

APPENDIX

1 Computing the MST

```
% Compute the MST
[M,N]=size(sig);            % get the length of the signal
N2=fix(N/2); j = 1;
if N2*2==N;  j = 0;end
f=[0:N2 -N2+1-j:-1]/N;      %frequency
MST=zeros(N2+1,N);          %allocate memory for positive frequencies of MST.
SIG=fft(sig,N);             %compute the signal spectrum
g=(1/N)*f+4*var(sig);       %parameter gamma
 for i = 2:N2
   SIGs=circshift(SIG,[0,-(i-1)]);     %circshift the spectrum SIG
   W=(g(i)/f(i))*2*pi*f;               % Scale Gaussian
   G=exp((-W.^2)/2);                   %W in Fourier domain
   MST(i,:)=ifft(SIGs.*G);             %Compute the complex values of MST
 end
 imagesc(abs(MST)');
   set(gca,'Ydir','Normal'); %default Y axis in imagesc is inverted.
```

2 Generation of piece-wise LFM

```
%Piecewise LFM chirp generation for EEG Seizure simulation
fs = 50;    %Sampling frequency
t1=(-5:1/50:5); t2=(-2.5:1/50:2.5); t3=(-5:1/50:5); % time for the first,
                                                     % the second and the
                                                     % third segment
t=[t1 t2 t3];                  % total time for the Piecewise LFM.
f00 = 2.626; f01 = 1.83; f10 = 1.751;   % start frequencies
a1=-0.7; a2=-0.6; a3 = 0.2;             % chirp rates or slope
ph1 = 2*pi*f00*t1+0.5*a1*t1.*t1;        % quadratic phase of piecewise LFM1
ph2 = 2*pi*f01*t2+0.5*a2*t2.*t2;        % quadratic phase of piecewise LFM2
ph3 = 2*pi*f10*t3+0.5*a3*t3.*t3;        % quadratic phase of piecewise LFM3
lfm1=exp(1i*ph1);lfm2=exp(1i*ph2);lfm3=exp(1i*ph3); %generate piecewise LFMs
lfm=[real(lfm1) real(lfm2) imag(lfm3)];    % put all in one signal LFM.
sig=lfm+0.2*randn(length(lfm),1)';   % add gaussian noise
```

3 Computing the cross-MST

```
%The cross-MST, co-MST and quadrature-MST
%Computing the Cross-MST of two signal x(t) and y(t)
```

```
%having as modified ST: MST1 and MST2, respectively.
% 1.Compute MST1 and MST2 using the related code in appendix A.
% 2.Compute the cross-MST between MST1 and MST2
cross MST = MST1.*conj(MST2); %Get the cross-MST.
% 3. Get the out-of-phase components
imagesc(imag(cross_MST)); set(gca,'Ydir','Normal')
% 4. Get the in-phase components
imagesc(real(cross_MST)); set(gca,'Ydir','Normal')
```

Competing Interests

The authors declare that they have no competing interests.

REFERENCES

1. Boashash B: Time Frequency Signal Analysis and Processing: A Comprehensive Reference. Elsevier, Oxford; 2003.
2. Imberger J, Boashash B: Application of the Wigner-Ville distribution to temperature gradient microstructure: a new technique to study small-scale variations. Phys Oceanograp1986, 16(12):1997-2012. 10.1175/1520-0485(1986)016<1997:AOTWDT>2.0.CO;2
3. Boashash B: Note on the use of the Wigner-Ville distribution for time-frequency signal analysis. IEEE Trans Acoust Speech Signal Process 1988, 36(9):1518-1521. 10.1109/29.90380
4. Lu W-k, Zhang Q: Deconvolutive short-time fourier transform spectrogram. IEEE Signal Process Lett 2009, 16(7):576-579.
5. Zhao Y, Atlas LE, Marks RJ: The use of cone-shape kernels for generalized time-frequency representations of nonstationary signals. IEEE Trans Acoust Speech Signal Process 1990, 38(7):1084-1091. 10.1109/29.57537
6. Choi H, Williams WJ: Improved time-frequency representation of multicomponent signals using exponential kernels. IEEE Trans Acoust Speech Signal Process 1989, 37(6):862-871. 10.1109/ASSP.1989.28057
7. Stockwell RG, L Mansinha L, Lowe RP: Localisation of the complex spectrum: the Stransform. IEEE Trans Signal Process 1996, 44(4):998-1001. 10.1109/78.492555
8. Rosenblum M, Pikovsky A, Kurths J: Phase synchronization of chaotic oscillators. Phys Rev Lett 1996, 76(11):1804-1807. 10.1103/PhysRevLett.76.1804
9. Gardner WA: Introduction to Random Processes: with Applications to Signals and Systems. McGraw-Hill Inc., US; 1990.
10. Koopmans LH: The Spectral Analysis of Time Series; in Probability and ematical Statistics. Volume 22. Academic Press, London, UK; 1995.
11. Bendat JS, Piersol AG: Random Data Analysis and Measurement Procedures. Wiley, New York; 2000.
12. Boashash B: Estimating and interpreting the instantaneous frequency of a signal. I. Fundamentals. IEEE, Proceedings 1992, 80(4):520-538. 10.1109/5.135376
13. Li D, Jung R: Quantifying coevolution of nonstationary biomedical signals using time-varying phase spectra. Ann Biomed Eng 2000, 28(9):1101-1115.
14. Gochenig K: Foundations of Time-Frequency Analysis. Birkhauser, Boston; 2001.
15. Huang NE, Shen Z, Long SR, ML Wu, Shih HH, Zheng Q, Yen NC, Tung C, Liu HH: The empirical mode decomposition and Hilbert spectrum for nonlinear and non-stationary time series analysis. Proc R Soc Lond A 1998, 454: 903-1005. 10.1098/rspa.1998.0193
16. Lin W, Xiaofeng M: An adaptative generalized S -transform for instantaneous frequency estimation. Signal Process 2011, 91(8):1876-1886. 10.1016/j.sigpro.2011.02.010
17. Sejdic E, Stancovic L, Dakovic M, Jiang J: Instantaneous frequency estimation using the S -transform. IEEE Trans Signal Process Lett 2008, 15: 309-312.
18. Stockwel RG: Why use the S-transform. Volume 52. Edited by: A Wong. Fields Institute Communications, Providence; 2007:279-309. Pseudo-Differential Operators: PDEs and Time-Frequency Analysis
19. Boashash B, O'Shea PJ: Use of the cross Wigner-Ville distribution for estimation of instantaneous frequency. IEEE Trans Signal Process 1993, 41(3):1439-1445. 10.1109/78.205752
20. Mansinha L, Stockwell RG, Lowe RP: Pattern analysis with two-dimensional spectral localisation: applications of two dimensional S -transforms. Phys A 1997, 239: 286-295. 10.1016/S0378-4371(96)00487-6
21. McFadden PD, Cook JG, Forster LM: Decomposition of gear vibration signals by the generalized S transform. Mechan Syst Signal Process 1999, 13: 691-707. 10.1006/mssp.1999.1233
22. Pinnegar CR, Mansinha L: The S -transform with windows of arbitrary and varying shape.Geophysics 2003, 68: 381-385. 10.1190/1.1543223
23. Kondaveeti G, Reddy MJB, Mohanta DK: Power quality analysis on EHV transmission line using modified S -transform. In 9th International Conference on Environment and Electrical Engineering (EEEIC2010). Prague, Czech Republic, IEEE-PES; 2010:337-340.
24. Nithin GV, Sitanshu SS, Mansinha L, Tiampo KF, Panda G: Time localised band filtering using modified S -transform. In Proceedings of the International Conference on Signal Processing Systems (ICSPS09). Singapore; 2009:42-46.
25. Czaja W, Kutyniok G, Speegle D: The geometry of sets of parameters of wave packet frames.Appl Comput Harmon Anal 2006, 20: 108-112. 10.1016/j.acha.2005.04.002
26. 2White LB, Boashash B: Cross spectral analysis of nonstationary processes. IEEE Trans Inf Theory 1990, 36(4):830-835. 10.1109/18.53742
27. Boles P, Boashash B: The cross Wigner-Ville distribution--a two dimensional analysis method for the processing of vibroseis seismic signals. In Proc International Conference on Acoustics, Speech, and Signal Processing (ICASSP), IEEE-ICASSP. NY, USA; 1988:904-907.

28. Chavez M, Quyen MLV, Navarro V, Baulac M, Martinerie J: Spatio-temporal dynamics prior to neocortical seizures: amplitude versus phase couplings. IEEE Trans Biomed Eng2003, 50(5):571-583. 10.1109/TBME.2003.810696

29. Alessandro MD, Esteller R, Vachtsevanos G, Hinson A, Echauz J, Litt B: Epileptic seizure prediction using hybrid feature selection over multiple intracranial EEG electrode contacts: a report of four patients. IEEE Trans Biomed Eng 2003, 50(5):603-615. 10.1109/TBME.2003.810706

30. Andrzejak RG, Mormann F, Widman G, Kreuz T, Elger CE, Lehnertz K: Improved spatial characterization of the epileptic brain by focusing on nonlinearity. Epilepsy Res 2006, 69(2):30-44.

31. Gupta D, James CJ, Gray WP: Phase synchronization with ica for epileptic seizure onset detection in the long term EEG. In Proc 4th IET Int Conf on Intelligent, Advanced in Medicine, Signal and Information Processing (MEDSIP), IET 2008. Seattle, USA; 2008:1-4.

32. Hassanpour H, Mesbah M, Boashash B: Timefrequency based newborn EEG seizure detection using low and high frequency signatures. Physiol Meas 2004, 25: 935-944. 10.1088/0967-3334/25/4/012

33. Rankine M, Mesbah LV, Boashash B: Automatic newborn EEG seizure spike and event detection using adaptive windowoptimization. In Proc ISSPA 2005, IEEE 2005. Sydney, Australia; 2005:187-190.

34. Roessgen M, Zoubir AM, Boashash B: Seizure detection ofnewborn eeg using a model-based approach. IEEE Trans Biomed Eng 1998, 45(6):673-685. 10.1109/10.678601

35. Altenburg J, Vermeulen R, Strijers R, Fetter W, Stam C: Seizure detection in the neonatal EEG with synchronization likelihood Clin Neurophysiol. 2003, 114: 50-55.

36. Quyan MLV: Anticipating epileptic seizures: from ematics to clinical applications. C R Biol 2005, 328(2):187-198. 10.1016/j.crvi.2004.10.014

37. Netoff TI, Schiff SJ: Decreased neuronal synchronization during experimental seizures. J Neurosci 2002, 22(16):7297-7307.

38. Rosenblum M, Pikovsky A, Kurths J, Schafer C, Tass PA: Phase synchronization: from theory to data analysis. In Neuro-Informatics and Neural Modeling. Edited by: F Moss, S Gielen. Elsevier, North-Holland; 2001:279-321.

39. Mormann F, Lehnertz K, David P, Elger CE: Mean phase coherence as a measure for phase synchronization and its application to the EEG of epilepsy patients. Phys D Nonlinear Phen2000, 144(3-4):358-369. 10.1016/S0167-2789(00)00087-7

40. De-Clercq W, Lemmerling P, Van-Paesschen W, Van-Huffel S: Characterization of in-terictal and ictal scalp EEG signals with the Hilbert transform. In Proc 25th Annu International Conf of the IEEE-EMBS. Cancun, Mexico; 2003:2459-2462.

41. Gupta D, James CJ: Narrowband vs. broadband phase synchronization analysis applied to independent components of ictal and interictal EEG. In Proceedings of the 29th Annual International Conference of the IEEE-EMBS. Edited by: Y Smith. Lyon, France; 2007:3864-3867.

42. Angelini L, Tommaso LMD, Guido M, PC Hu, Ivanov K, Marinazzo D, Nardulli G, Nitti L, Pellicoro M, Pierro C, Stramaglia S: Steady-state visual evoked potentials and phase synchronization in migraine patients. Phys Rev Lett 2004, 93(3):38103-38106.

43. Adeli H, Zhou Z, Dadmehr N: Analysis of EEG records in an epiliptic patient using wavelet transform. Neurosci Methods 2003, 123: 69-87. 10.1016/S0165-0270(02)00340-0

44. Bob P, Susta M, Glaslova K: EEG phase synchronization in patients with paranoid schizophrenia. Neurosci Lett 2008, 447: 73-77. 10.1016/j.neulet.2008.09.055

45. Lachaux JP, Lutz A, Rudrauf D, Cosmelli D, Le-Van-Quyen M, Martinerie J, Varela F: Estimating the time course of coherence between single-trial brain signals: an introduction to wavelet coherence. Neurophysiol Clin 2002, 32(1-3):1-18.

46. Quyan MLV, Foucher J, Lachaux E, Rodriguez JP, Lutz A, Martinerie J, Varela FJ: Comaprison of Hilbert transform and wavelet methods for the analysis of neural synchrony. Neurosci Methods 2001, 111(2):83-98. 10.1016/S0165-0270(01)00372-7

47. Boashash B, Mesbah M: Time-frequency methodology for newborn electroencephalo-graphic seizure detection. In Applications in Time-frequency Signal Processing. Edited by: A Papandreou-Suppappola. CRC Press LLC, Boca Raton; 2002:339-369.

48. Boashash B, Mesbah M: Time-frequency approach for newborn EEG seizure detection.IEEE Eng Med Biol Mag 2001, 20(5):54-64.

49. Boashash B, Mesbah M, Colditz P: Newborn EEG seizure pattern characterisation using time-frequency analysis. In Proc International Conference on Acoustics, Speech, and Signal Processing (ICASSP), IEEE-ICASSP 2001. Salt Lake City, USA; 2001:1041-1044.

50. Roessgen M, Zoubir A, Boashash B: Seizure detection of newborn EEG using a model based approach. IEEE Trans Biomed Eng 1998, 45(6):673-685. 10.1109/10.678601

51. Stevenson NJ, Mesbah M, Boylan GB, Colditz PB, Boashash B: A nonlinear model of newborn EEG with nonstationary inputs. Ann Biomed Eng 2010, 38(9):3010-3021. 10.1007/s10439-010-0041-3

52. Celka P, Boashash B, Colditz P: Pre-processing and time-frequency analysis of newborn EEG seizures. IEEE Eng Med Biol Mag 2001, 20(5):30-38. 10.1109/51.956817

53. Niedermeyer E, Lopes-da Silva F: Electroencephalography: Basic Principles, Clinical Applications and Related Fields. Lippincott williams & Wilkins, Philadelphia; 2004.

CHAPTER 16 DESIGN OF WARPED STRETCH TRANSFORM

Ata Mahjoubfar, Claire Lifan Chen and Bahram Jalali

INTRODUCTION

Time stretch dispersive Fourier transform[1,2,3] addresses the analog-to-digital converter (ADC) bottleneck in real-time acquisition of ultrafast signals. It leads to fast real-time spectral measurements of wideband signals by mapping the signal into a waveform that is slow enough to be digitized in real-time. Combined with temporal or spatial encoding, time stretch dispersive Fourier transform has been used to create instruments that capture extremely fast optical phenomena at high throughput. By doing so, it has led to the discovery of optical rogue waves[4], the creation of a new imaging modality known as the time stretch camera[5], which has enabled detection of cancer cells in blood with record sensitivity[6,7,8], a portfolio of other fast real-time instruments such as an ultrafast vibrometer[9,10], and world record performance in analog-to-digital conversion[11,12]. The key feature that enables fast real-time measurements is not the Fourier transform, but rather the time stretch. For example, direct frequency-to-time mapping can be replaced by phase retrieval[13] or coherent detection after the dispersion[14] followed by back propagation.

Using warped group delay dispersion as a photonic hardware accelerator[15], an optical signal's intensity envelope can be engineered to match the specifications of the data acquisition back-end[16,17,18]. One can slow down an ultra-fast burst of data, and at the same time, achieve data compression by exploiting sparsity in the original data[19]. Also called anamorphic stretch transform[16,17], the warped stretch transform performs a nonuniform frequency-to-time mapping followed by a uniform sampler. The combined effect of the transform is that the signal's Fourier spectrum is sampled at a nonuniform rate and resolution. By designing the group delay profile according to the sparsity in the spectrum of the input signal, more samples are allocated to the information-rich portions of the spectrum and fewer to the information-sparse regions where they would be redundant. The only prior information needed is the sparsity of the signal's spectral features, i.e. information about the ensemble of the signal spectrum. No instantaneous feature detection is required as long as the signal's spectral sparsity is within the design range. As a primary application, the utility of this method has been recently demonstrated in real-time optical image compression[19].

In conventional time stretch dispersive Fourier transform, a linear group delay profile is used to impose a non-zero constant group delay dispersion over the full bandwidth of the optical signal as shown in Fig. 1 (orange color). This profile as the kernel of the transform generates a linearly increasing frequency-dependent temporal shift across the bandwidth, which maps the optical spectrum into a temporal waveform detectable with a single-pixel photodetector. In other words, temporal dispersion stretches the optical signal in time into its Fourier transform. If the optical signal is a train of ultrafast pulses such as the output of a mode-locked laser, the spectrum of each individual pulse is mapped into a temporal waveform filling the gaps between pulses. To analyze the signal in digital domain, an analog-to-digital converter (ADC) samples the output of the photodetector at a constant rate, which can be interpreted as a uniform Fourier domain sampling of the signal

Citation: Mahjoubfar A. et al, Design of Warped Stretch Transform. Sci. Rep. 5, 17148; doi: 10.1038/srep17148 (2015).

spectrum. It is important to note that the value of each sample of the temporal waveform corresponds to the integral of the optical spectrum over a spectral resolvable window. For a linear group delay profile, the width of this spectral window is fixed, and it does not change over the bandwidth (see orange stripes in Fig. 1).

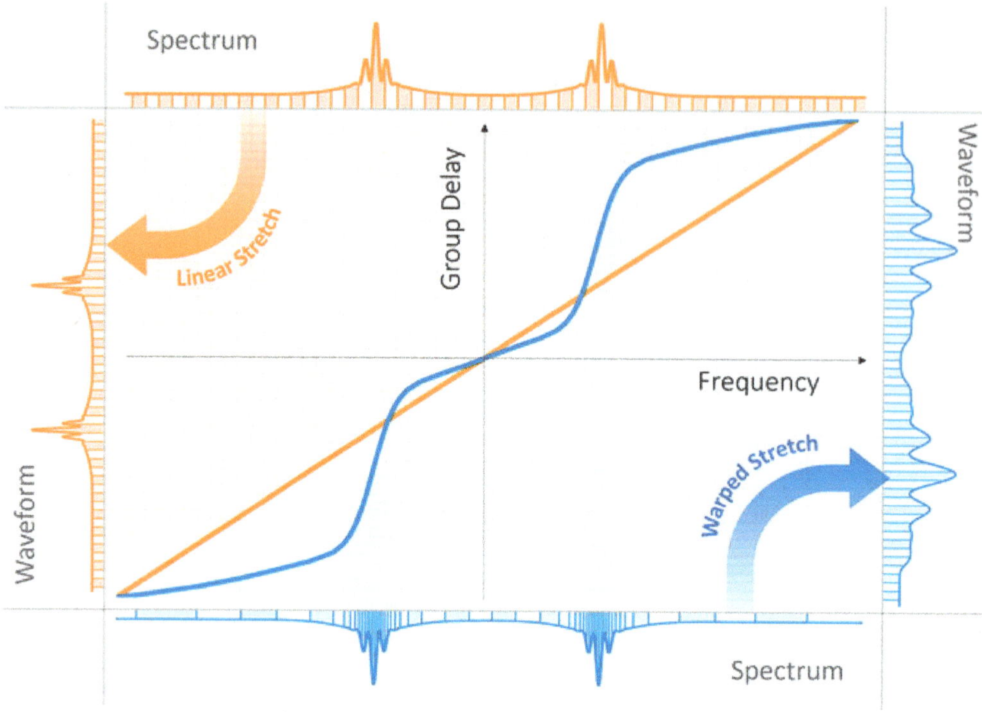

Figure 1: Linear and warped time stretch dispersive Fourier transforms. In linear time stretch (orange plots), a linear group delay profile with significant group delay over the signal bandwidth is used as the kernel to delay various spectral components differently, but with a constant group delay dispersion (slope of the group delay profile). If the input signal is a pulse train, pulse spectra are linearly mapped to the silent times in between the pulses. If a digitizer with constant sampling rate is used to capture the stretched pulse waveforms, the spectral sampling resolution is uniform across the bandwidth (see orange bars). In contrast, a nonlinear group delay profile (blue plots) with varying group delay dispersion over the bandwidth (different slopes) can stretch the signal spectrum nonlinearly, in which parts of the spectrum are stretched more than the others. This can be used, even with a constant rate sampler, to increase the spectral resolution at regions of the bandwidth where higher resolution is required and to reduce the resolution where the spectral features are sparse (see blue bars). If the sparsity of the spectral features, that is, regions of spectrum with fast varying magnitude or higher frequency of spectrum, is known, the nonlinear group delay profile can be designed accordingly to stretch those regions more than the linear case. Conversely, regions with low bandwidth for the frequency of spectrum can be stretched less than the linear profile. In this way, the spectrum is warp stretched into a waveform with the same temporal duration as the linear profile, but smaller bandwidth. In other words, for signals with spectral sparsity, the envelope time bandwidth product can be reduced.

A warped time stretch dispersive Fourier transform is achieved by a nonlinear group delay profile as the kernel of the transform, which imposes a frequency-dependent group delay dispersion onto the spectral components. An example of a nonlinear group delay profile is shown in blue plots of Fig. 1. Here, the optical spectrum is mapped nonuniformly to time, stretching parts of the spectrum more than the other. In the example shown, the dispersion (slope of the group delay profile) in the central and peripheral parts of the spectrum is smaller than the linear profile, leading to a stretch which is less than the linear profile. However, the profile has also two regions of high dispersion, which map the spectrum into a longer and more-detailed temporal waveform compared to the linear case. It is a remarkable fact that uniform sampling of the warped time stretch output signal corresponds to nonuniform sampling of the spectrum with a frequency-dependent spectral resolvable window (see blue stripes in Fig. 1).

RESULTS

The warped time stretch dispersive Fourier transformation can be contemplated as a spectrotemporal operation, where its effectiveness in capturing spectral details is dictated by the sparsity and the redundancy of the input signal spectrum. Namely, sparsity in the spectrum is the attribute that influences and guides the design of the group delay dispersion profile. Note that the traditional notion of sparsity, i.e. sparsity in time or spectrum, is not pertinent here. Instead, sparsity here refers to the absence of the spectral features, i.e. abrupt variations of spectrum magnitude. The spectrum only needs to be feature-sparse; it does not need to be narrowband or contain limited number of spectral components.

We describe the spectrotemporal operation of the warped time stretch with a set of examples. Without loss of generality, we initially assume that any chirp in the input signal is negligible compared to the applied group delay dispersion chirp; for example, the total temporal duration of each pulse is much shorter than the overall group delay of the dispersion profile over the pulse bandwidth. In Fig. 2a, the envelope of an optical field as the input signal of the stretch transform is shown. The spectrum of this envelope (Fig. 2b) has fast variations in the central region (feature-dense) and is relatively smooth in the wings (feature-sparse). The spectrum of the envelope can also be viewed as the spectrum of the input optical field downshifted to baseband. To examine the local properties of the spectrum, we use the short-term Fourier transform of the spectrum and form the spectrogram of the spectrum. This is equivalent to viewing the spectrum as a temporal waveform and plotting its short-time Fourier transform (Fig. 2c). As a result, the horizontal axis is the input frequency and the vertical axis is the local frequency of the variations in spectrum magnitude. We call this the frequency of spectrum, which corresponds to the period of variations in the spectrum magnitude or the temporal distance of frequency components of the signal. Hence, the local frequency bandwidth is broad in the central region and narrow in the wings (Fig. 2c). A linear group delay profile (Fig. 2d) performs conventional time stretch dispersive Fourier transform, in which the spectrum is uniformly mapped into a temporal waveform (Fig. 2e). The short-time Fourier transform of this temporal waveform (Fig. 2f) resembles that of the spectrum (Fig. 2c). To reduce the required acquisition bandwidth or time duration (memory), the spectrotemporal distribution of the signal can be reshaped by a nonlinear group delay profile as a filter, whose characteristics conform to the local frequency patterns. In the regions where the spectrum magnitude has fast variations, the filter should present a high group delay dispersion (slope) resulting in larger stretching in time than the slow varying regions of the spectrum magnitude. For the signal spectrum shown in Fig. 2b, a desired group delay profile is shown in Fig. 2g. The frequency-to-time mapping and temporal stretching are warped in such a manner that the sparse wings of the spectrum are squeezed relative to the dense central region (Fig. 2h). The sparse wings are squeezed in so that they occupy a shorter time duration after the frequency to time mapping. As a result, the reshaped spectrotemporal distribution (Fig. 2i) has the same bandwidth as the linear case, but a compressed time duration (compare dot-dashed boxes of Figs. 2f-2i). We will show that the blue dot-dashed box in Fig. 2i, which depicts the acquisition time-bandwidth limit, translates into a frequency-dependent effective bandwidth on the frequency of spectrum as shown with the blue dot-dash contour in Fig. 2c, and the spectrotemporal feature sparsity of the signal can be used to reduce the envelope acquisition time-bandwidth product in time stretch dispersive Fourier transform.

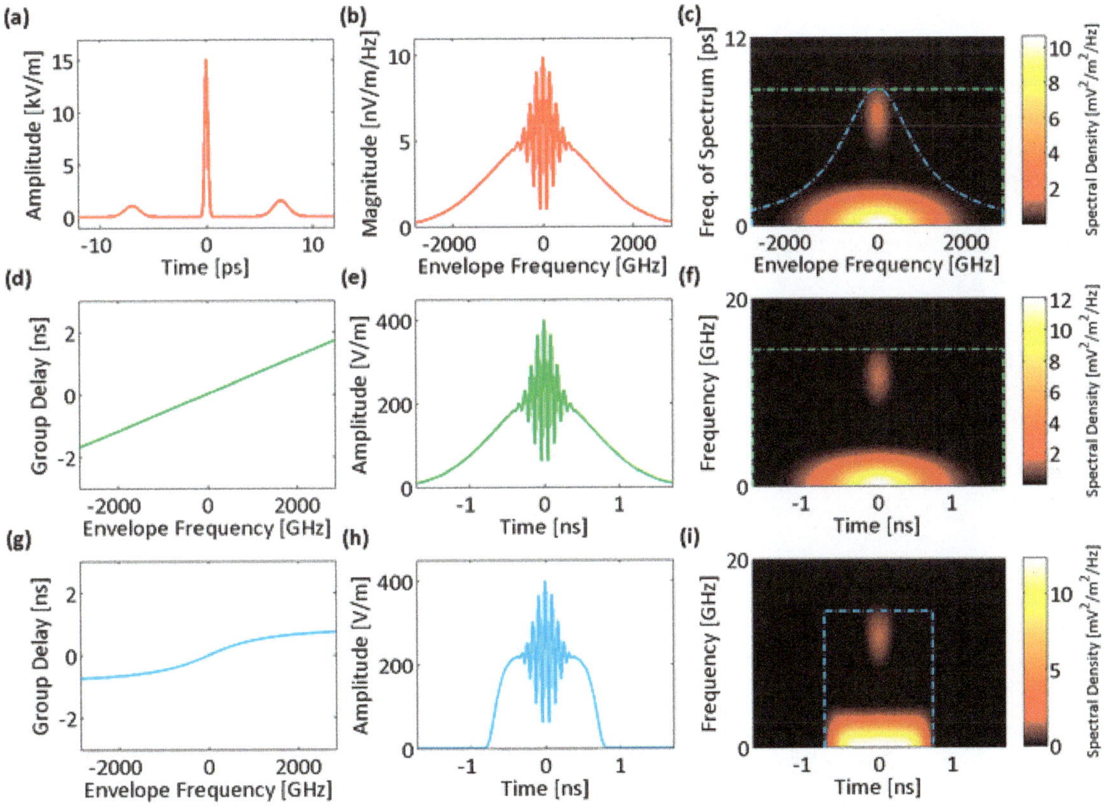

Figure 2: Group delay design based on spectrotemporal sparsity at the spectrum peripheries. (**a**) Envelope of the electric field of an input optical signal. (**b**) The spectrum magnitude of the input envelope. (**c**) Spectrogram of the spectrum magnitude formed by short-term Fourier transform. Here, the window of the short-term Fourier transform is slid over the envelope frequencies for the input envelope spectrum magnitude, and the Fourier transform of the signal in the window gives the local frequencies of spectrum. At the center of the spectrum magnitude, there are fast oscillations (see Fig. 2b), which result in high frequencies of spectrum. (**d**) If a temporally dispersive element with a linear group delay profile over the optical bandwidth is used to stretch the input optical field, (**e**) the spectrum maps uniformly to temporal envelope of the electric field. (**f**) Spectrogram of the envelope waveform amplitude resembles that of the spectrum magnitude (shown in Fig. 2c). The electronic acquisition time and bandwidth are marked with the green dot-dash box. Also, equivalent limitations on the envelope optical frequency and the frequency of spectrum are marked with a similar green dot-dash box on Fig. 2c. (**g**) If a nonlinear group delay profile with lower dispersion (slope of the group delay profile) at the sides of the bandwidth is used to stretch the optical pulse, (**h**) the spectrum is nonlinearly mapped to the electric field envelope in time with the sides of the spectrum stretched less than the central part. This is desirable as the sides of the spectrum magnitude do not have fast oscillations, and to capture them with a limited acquisition bandwidth, there is no need to stretch them as much as the central part. Essentially, warped stretch avoids overstretching of the spectrum peripheries unlike the linear profile (see Fig. 2e). (**i**) The spectrogram of the electric field envelope amplitude after the nonuniform dispersion shows that the power in the spectrum peripheries is squeezed toward the center, and a shorter temporal window is required to capture the waveform with the same acquisition bandwidth (see the spectrotemporal electronic acquisition window marked by the blue dot-dash box).

Upon uniform temporal sampling, the nonuniform mapping of warped stretch causes the information dense portion of the spectrum to effectively receive higher sampling resolution than the information sparse regions, leading to nonuniform spectral sampling. The local spectral sampling rate is basically designed to match the signal's spectrum sparsity. We note that this nonuniform sampling is performed not by a hard-to-reach variable rate sampler, but with a uniform sampler preceded by warped spectrotemporal reshaping. This approach offers similar functionality as compressive sensing20·21 albeit it achieves it via an entirely different approach, namely that of warped time stretch dispersive Fourier transform. If the input signal is significantly chirped, the group delay profile should correspond to the difference between the desired profile for the transform limited version of the input signal (inverse Fourier transform of the input signal spectrum modulus) and the chirp of the input signal.

Spectral Resolution

The resolution of the nonuniform spectral sampling is determined by the sampling resolution of the temporal waveform and the spectral resolution of dispersive Fourier transform. The temporal resolution of the samples is itself limited by the photodetector electrical bandwidth and the Nyquist bandwidth of the analog-to-digital converter. If we assume a resistor-capacitor circuit model for the photodetector output and use its 10% to 90% rise time as the temporal resolution22, the photodetector spectral resolution limit is

$$\delta\omega_{PD}(\omega) = \frac{0.35}{B_{PD} \cdot \left|\frac{d\tau_g(\omega)}{d\omega}\right|} \quad (1)$$

where, $\tau_g(\omega)$ is group delay profile and B_{PD} is the electrical bandwidth of the photodetector. Also, the resolution limit in Fourier domain set by the Nyquist bandwidth of the analog-to-digital converter23, B_{ADC}, is

$$\delta\omega_{ADC}(\omega) = \frac{0.5}{B_{ADC} \cdot \left|\frac{d\tau_g(\omega)}{d\omega}\right|} \quad (2)$$

The spectral resolution of the dispersive Fourier transform imposed by ambiguity in frequency-to-time mapping is

$$\delta\omega_{DFT}(\omega) = \sqrt{\frac{4\pi}{\left|\frac{d\tau_g(\omega)}{d\omega}\right|}} \quad (3)$$

which is derived by stationary phase approximation24. Finally, the overall spectral sampling resolution is limited by the largest of these three at each frequency. Therefore,

$$\delta\omega_{Total}(\omega) = max\{\delta\omega_{PD}(\omega), \delta\omega_{ADC}(\omega), \delta\omega_{DFT}(\omega)\} \quad (4)$$

Clearly, the resolution of the nonuniform spectral sampling is frequency-dependent for a nonlinear group delay profile. Fig. 3a,c show the spectral resolutions and their limiting components for both the linear and the nonlinear group delay profiles of Fig. 2, respectively. In low group delay dispersions, temporal resolution limits, $\delta\omega_{PD}$ or $\delta\omega_{ADC}$, are mainly the limiting factors, whereas in high group delay dispersions, the resolution of the dispersive Fourier transform dominantly limits the spectral resolution. The spectral sampling resolution can also be translated into an effective bandwidth, B_ω, for the frequency of spectrum. This effective bandwidth, calculated as

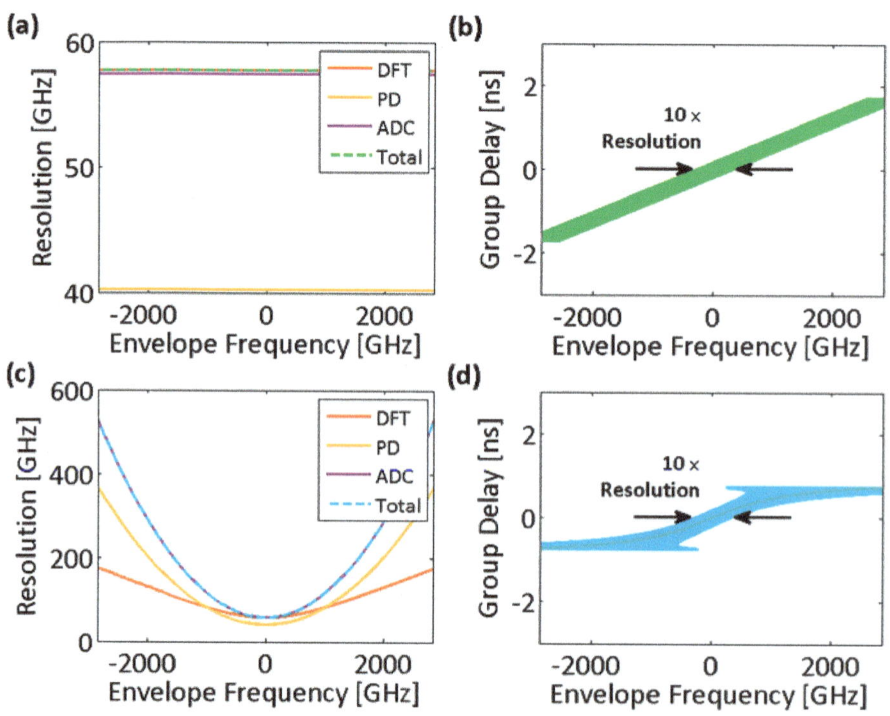

Figure 3: Spectral sampling resolution of time stretch dispersive Fourier transform tuned for the spectrum center. Resolution of the spectral sampling using time stretch dispersive Fourier transform is the maximum of resolution limits imposed by the temporal resolution of the photodetector, the bandwidth of the analog-to-digital converter, and the ambiguity in the frequency-to-time mapping of the dispersive Fourier transform. (**a**) Spectral resolution limits for the linear group delay profile and the acquisition system of Fig. 2d–f (photodetector and analog-

to-digital converter Nyquist bandwidths are 14.5 GHz). The overall spectral sampling resolution of the linear time stretch is independent of the envelope optical frequency and limited by the ambiguity in the frequency-to-time mapping of the dispersive Fourier transform. (**b**) The spectral sampling resolution of the linear group delay profile magnified ten times (for visual clarity) and overlapped on the profile. One tenth of the overlay width at each group delay corresponds to the set of the optical frequencies that are captured at the same delay and are indistinguishable in the temporal waveform. (**c**) Spectral resolution limits for the nonlinear group delay profile and the acquisition system of Fig. 2g–i, unlike the linear stretch, depend of the envelope optical frequency (photodetector and analog-to-digital converter Nyquist bandwidths are 14.5 GHz). The overall spectral sampling resolution is limited by the ambiguity in the frequency-to-time mapping of the dispersive Fourier transform at the center of the spectrum and by the Nyquist bandwidth of the analog-to-digital converter at the peripheries of the spectrum. (**d**) Magnified spectral sampling resolution of the warped time stretch overlapped on its group delay profile clearly shows the ambiguity grows at the spectrum peripheries. DFT: dispersive Fourier transform; PD: photodetector; ADC: analog-to-digital converter; Total: overall spectral sampling resolution.

$$B_\omega(\omega) = \frac{0.5}{\delta\omega_{Total}(\omega)} \tag{5}$$

is shown with dot-dashed lines in Fig. 2c. The effective bandwidth illustrates whether the characteristics of the group delay profile match the spectrotemporal sparsity of the signal, and it proficiently guides the profile design. Fig. 3b,d show the group delay profiles of Fig. 2 overlaid with their spectral resolutions. The widths of the curves at each frequency correspond to ten times the spectral resolution. The magnification factor, ten, is used to make the subtle changes in spectral resolution more noticeable next to the group delay profile, which prominently determines it.

As another example, we consider a test signal with opposite sparsity compared to that shown in Fig. 2. Figure 4 shows a signal that is sparse in the central region of the spectrum and feature-dense in the wings. The group delay profile that matches this waveform has higher dispersion, i.e., temporal stretch factor, in the wings. Here, the required acquisition bandwidth is lower for the warped group delay case compared to the linear case. Figure 5 shows the nonuniform spectral resolution corresponding to this group delay design.

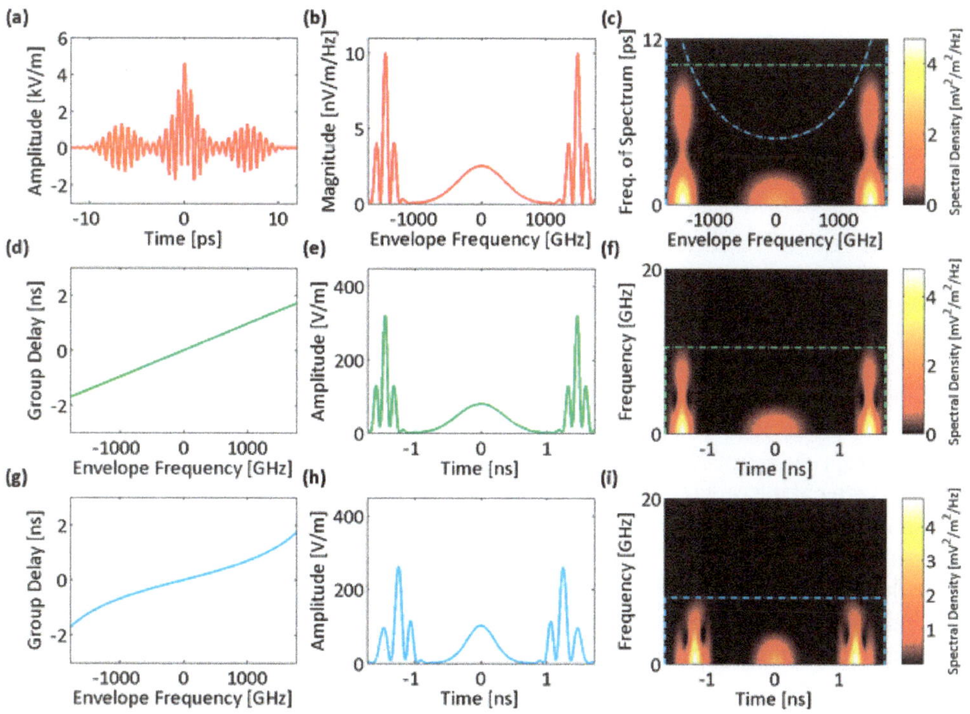

Figure 4: Group delay design based on spectrotemporal sparsity at the spectrum center. (**a**) Envelope of the electric field of an input optical signal. (**b**) The spectrum magnitude of the input envelope. (**c**) Spectrogram of the spectrum magnitude formed by short-term Fourier transform. At the peripheries of the spectrum magnitude, there are fast oscillations (see Fig. 4b), which result in high frequencies of spectrum. (**d**) If a temporally dispersive element with a linear group delay profile over the optical bandwidth is used to stretch the input optical field, (**e**) the spectrum maps uniformly to temporal envelope of the electric field. (**f**) Spectrogram of the envelope waveform amplitude resembles that of the spectrum magnitude (shown in Fig. 4c). The electronic acquisition time and bandwidth are marked with the green dot-dash box. Also, equivalent limitations on the envelope optical frequency and the frequency of spectrum are marked with a similar green dot-dash box on Fig. 4c. (**g**) If a nonlinear group delay profile with lower dispersion (slope of the group delay profile) at the center of the bandwidth is used to stretch the optical pulse, (**h**) the spectrum is nonuniformly mapped to the electric field envelope in time with the center of the spectrum stretched less than the peripheral parts. This is desirable as the center of the spectrum magnitude does not have fast oscillations, and to capture it with a limited acquisition time, there is no need to stretch it as much as the peripheral parts. Essentially, the warped profile avoids overstretching of the spectrum center and understretching of the spectrum peripheries, unlike linear profile (see Fig. 4e). (**i**) The spectrogram of the electric field envelope amplitude after the nonuniform dispersion shows that the power at high frequencies is squeezed toward the lower frequencies, and a smaller acquisition bandwidth is required to capture the waveform with the same temporal sampling duration (see the spectrotemporal electronic acquisition window marked by the blue dot-dash box). This time-bandwidth limit translates into a frequency-dependent effective bandwidth for the frequency of spectrum as shown with the blue dot-dash region in Fig. 4c.

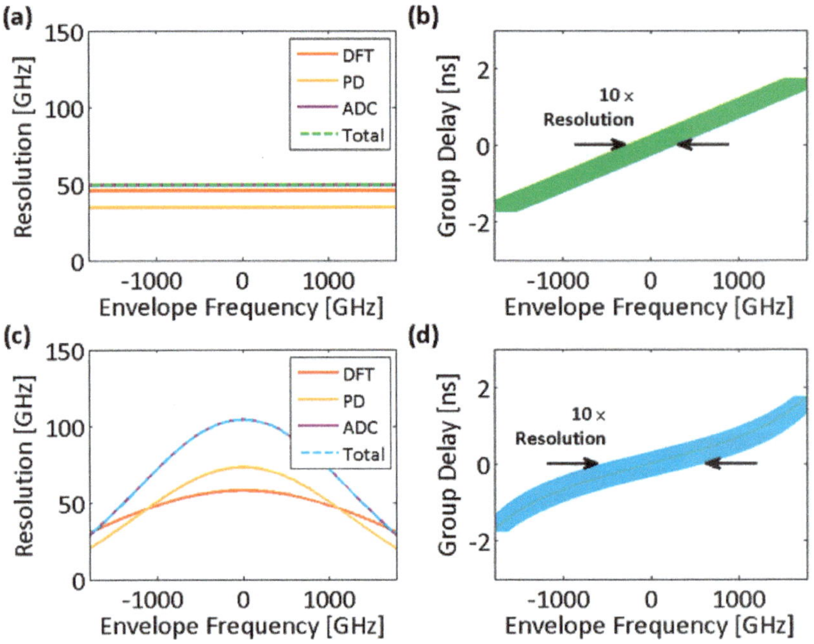

Figure 5: Spectral sampling resolution of time stretch dispersive Fourier transform tuned for the spectrum peripheries.

(**a**) Spectral resolution limits for the linear group delay profile and the acquisition system ofFig. 4d–f (photodetector and analog-to-digital converter Nyquist bandwidths are 10.5 GHz). The overall spectral sampling resolution of the linear time stretch is independent of the envelope optical frequency and limited by the Nyquist bandwidth of the analog-to-digital converter. (**b**) The spectral sampling resolution of the linear group delay profile magnified ten times (for visual clarity) and overlapped on the profile. (**c**) Spectral resolution limits for the nonlinear group delay profile and the acquisition system of Fig. 4g–i (photodetector and analog-to-digital converter Nyquist bandwidths are 8 GHz). The overall spectral sampling resolution is limited by the ambiguity in the frequency-to-time mapping of the dispersive Fourier transform at the peripheries of the spectrum and by the Nyquist bandwidth of the analog-to-digital converter at the center of the spectrum. (**d**) Magnified spectral sampling resolution of the warped time stretch overlapped on its group delay profile clearly shows the ambiguity grows at the spectrum center. DFT: dispersive Fourier transform; PD: photodetector; ADC: analog-to-digital converter; Total: overall spectral sampling resolution.

Group Delay Profile Design

The concept of effective bandwidth can be used to design an ideal group delay profile that maximally exploits the spectrotemporal sparsity of a signal. Given an acceptable signal-to-noise ratio (a tolerable spectrotemporal power loss level e.g. the noise floor of the spectrogram), the important features of the spectrogram can be contoured. At each envelope frequency, the maximum frequency of spectrum on the contour line corresponds to the desired effective bandwidth, B_ω, for the ideal group delay profile. It follows from Eq. (1, 2, 3), and Eq. (5) that the dispersion of the group delay profile is

$$\left|\frac{d\tau_g(\omega)}{d\omega}\right| = max\left\{\frac{0.7B_\omega}{B_{PD}}, \frac{B_\omega}{B_{ADC}}, 16\pi B_\omega^2\right\} \quad (6)$$

Since the ideal desired group delay profile is monotonic, it is easily derived as

$$\tau_g(\omega) = \int\left|\frac{d\tau_g(\omega)}{d\omega}\right| d\omega \quad (7)$$

The spectrogram that is used here to design the ideal group delay profile, does not display the effect of the input signal chirp because it is formed from the spectrum magnitude. If the input chirp is not negligible, it must be subtracted from the Eq. 7 to get the total group delay profile, which performs the desired frequency-to-time mapping. Figure 6a shows a chirped input signal with a spectrum magnitude (Fig. 6b) same as that of Fig. 4b. The spectrogram is contoured at −30 dB of the peak power density (blue dot-dash line in Fig. 6c), which specifies the required effective bandwidth of the frequency of spectrum for an ideal group delay profile. We compare using a linear design for the group delay profile (Fig. 6d) with designing the ideal group delay profile according the effective bandwidth by Eq. (7) (Fig. 6g). In either case, the chirp of the input signal is subtracted from the group delay design to cancel the effect of the input chirp. Thus, the total group delay profiles preserve the desired forms of frequency-to-time mappings (Fig. 6e,h). Clearly, the power in the sparse regions of the spectrogram is concentrated by the nonlinear profile (Fig. 6i) compared to that by the linear profile (Fig. 6f), which corresponds to a reduction in the required acquisition time and bandwidth. Figure 7 shows the spectral resolution of the linear and warped stretch transforms with the profiles in Fig. 6. The linear profile has a uniform resolution across the bandwidth, whereas the resolution of the nonlinear profile is frequency-dependent and designed to match the sparsity of the spectrotemporal distribution.

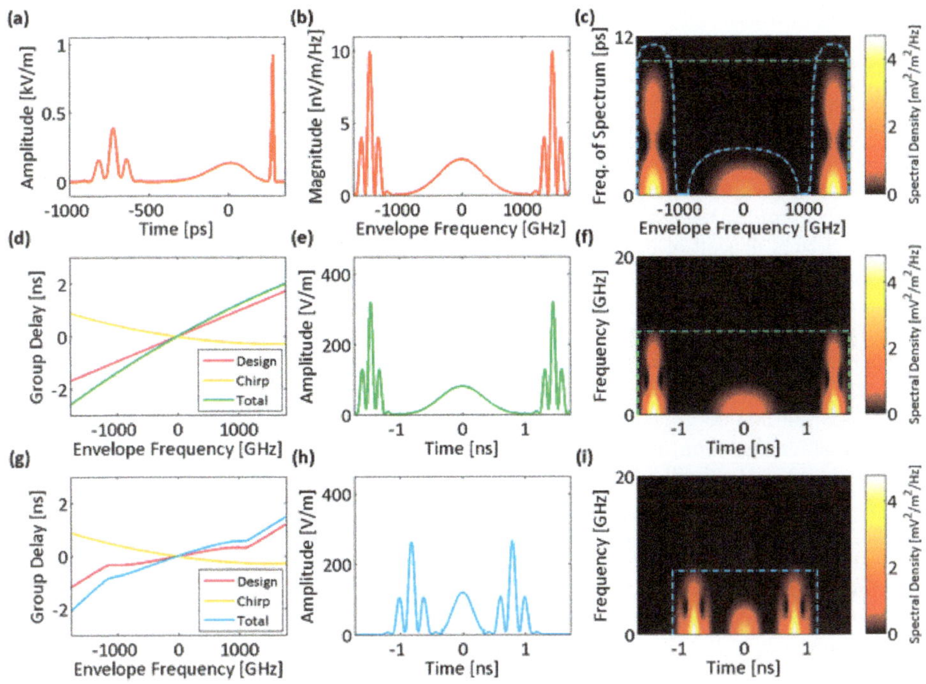

Figure 6: Design of an ideal group delay profile based on spectrotemporal sparsity and input signal chirp. (**a**) Envelope of the electric field of an input optical signal. (**b**) The spectrum magnitude of the input envelope. (**c**) Spectrogram of the spectrum magnitude formed by short-term Fourier transform. The region of the spectrogram with power density 30 dB less than the peak power density is contoured with a blue dot-dash line. This contour is considered as the necessary effective bandwidth to design an ideal group delay profile. (**d**) To perform uniform frequency-to-time mapping, a linear group delay design minus the input signal chirp should be used to stretch the input optical field. (**e**) The spectrum maps uniformly to temporal envelope of the electric field by the total group delay profile. (**f**) Spectrogram of the envelope waveform amplitude resembles that of the spectrum magnitude (shown in Fig. 6c). The electronic acquisition time and bandwidth are marked with a green dot-dash box. Also, equivalent limitations on the envelope optical frequency and the frequency of spectrum are marked with a similar green dot-dash box on Fig. 6c. (**g**) If a nonlinear group delay profile is designed based on the blue dot-dash contour in Fig. 6c and the input signal chirp, (**h**) the spectrum is nonlinearly mapped to the electric field envelope in time in such a way that the acquisition time is minimized for the set acquisition bandwidth (8 GHz) and the enforced spectrotemporal accuracy level (−30 dB). Essentially, the warped profile ideally avoids overstretching of the spectrum center and understretching of the spectrum peripheries, unlike the linear profile (see Fig. 6e). (**i**) The spectrogram of the electric field envelope amplitude after the nonuniform dispersion shows that the power in the spectrum peripheries is squeezed toward the center, and a shorter temporal window and acquisition bandwidth is required to capture the waveform compared to the linear case (acquisition window is marked by the blue dot-dash box). This time-bandwidth limit translates into the frequency-dependent effective bandwidth as shown with the blue dot-dash box in Fig. 6c, used for the design of the ideal group delay profile.

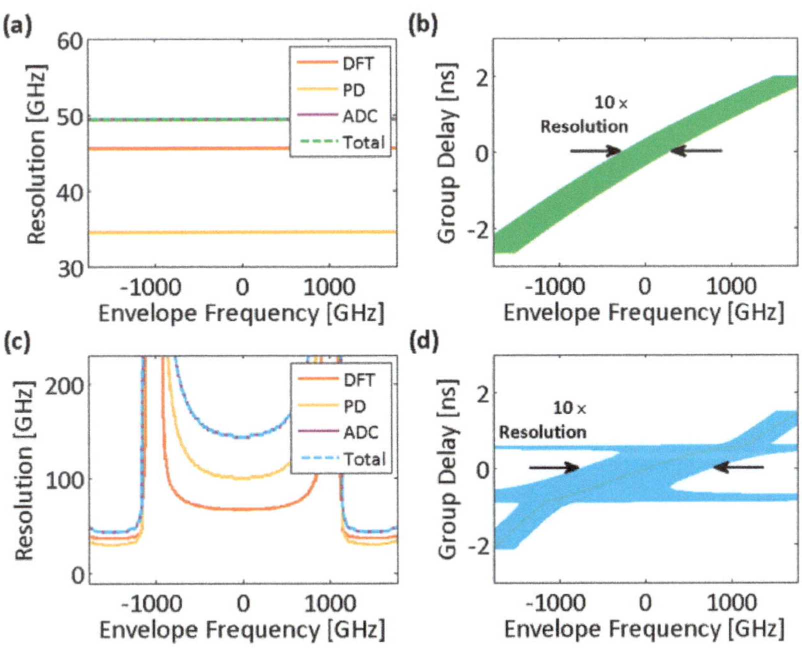

Figure 7: Spectral sampling resolution of time stretch dispersive Fourier transform designed for ideal exploitation of the spectrotemporal sparsity. (**a**) Spectral resolution limits for the chirp compensated linear group delay profile and the acquisition system of Fig. 6d–f (photodetector and analog-to-digital converter Nyquist bandwidths are 10.5 GHz). The overall spectral sampling resolution of the linear time stretch

is independent of the envelope optical frequency and limited by the Nyquist bandwidth of the analog-to-digital converter. (**b**) The spectral sampling resolution of the linear group delay profile minus chirp magnified ten times (for visual clarity) and overlapped on the profile. One tenth of the overlay width at each group delay corresponds to the set of the optical frequencies that are captured at the same delay and are indistinguishable in the temporal waveform. (**c**) Spectral resolution limits for the nonlinear group delay profile and the acquisition system of Fig. 6g–i, unlike the linear stretch, depend of the envelope optical frequency (photodetector and analog-to-digital converter Nyquist bandwidths are 8 GHz). The overall spectral sampling resolution is limited by the Nyquist bandwidth of the analog-to-digital converter. (**d**) Magnified spectral sampling resolution of the warped time stretch overlapped on its group delay profile clearly shows the ambiguity grows at the regions of the spectrum that do not contain significant information. DFT: dispersive Fourier transform; PD: photodetector; ADC: analog-to-digital converter; Total: overall spectral sampling resolution.

To further show the applicability of our design method, we consider a signal (Fig. 8a) with asymmetric spectrum about the carrier frequency (Fig. 8b). This corresponds to a signal with complex temporal envelope (In Fig. 8a, we are showing the absolute value of the complex envelope). Using our design algorithm, the spectrogram is contoured at −30 dB of the peak power density (blue dot-dash line in Fig. 8c), specifying the required effective bandwidth of the frequency of spectrum for an ideal group delay profile. If we use a chirp-compensated linear group delay profile (Fig. 8d), the spectrum to time mapping would be uniform. Of course, this results in an asymmetric temporal waveform (Fig. 8e), which resembles the input signal spectrum (Fig. 8b). The vertical flip is due to higher frequencies receiving larger group delays, lagging more behind. The spectrogram of the temporal envelope shows the same type of flip (Fig. 8f). If we use the nonlinear group delay profile designed with our technique (Fig. 8g) to stretch the signal nonuniformly (Fig. 8h), compared to the linear case (Fig. 8f), a shorter time duration and a smaller acquisition bandwidth are sufficient (Fig. 8i). Clearly, even in the case of a signal with asymmetric spectrum, the design algorithm leads to a nonlinear group delay profile, which efficiently reduces the time-bandwidth product of the envelope by warped stretch transform. For the group delay profiles shown in Fig. 8, the spectral resolutions of the linear and warped stretch transforms are depicted in Fig. 9. The linear profile has a fixed resolution across the bandwidth (Fig. 9a,b), but the resolution of the nonlinear profile is frequency dependent (Fig. 9c,d). For example, the algorithm has designed the nonlinear profile in such a way that it allocates better resolution to the parts of the spectrum that contain fast variations (feature-dense regions).

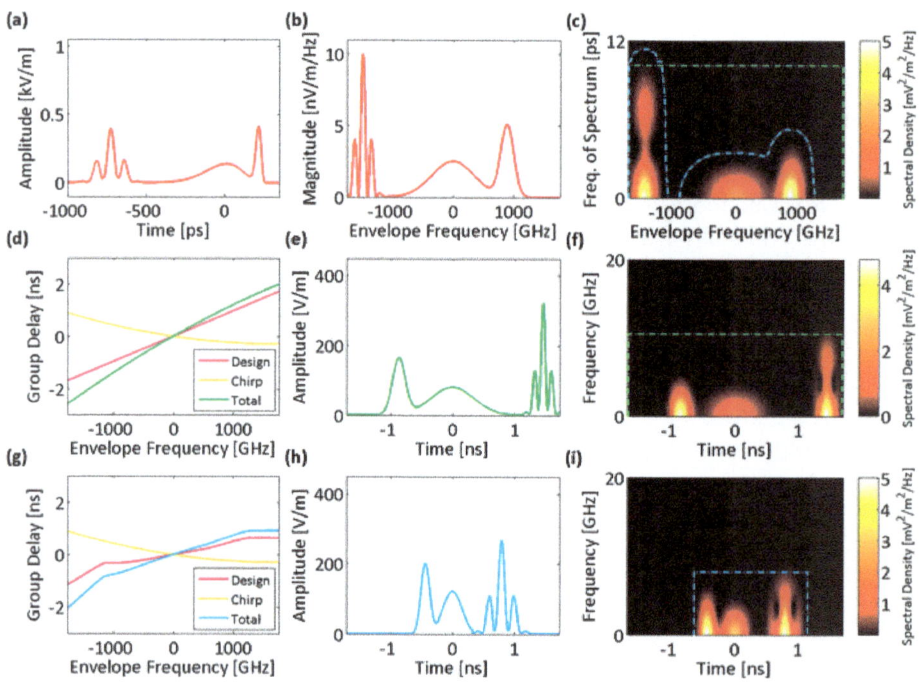

Figure 8: Design of an ideal group delay profile for a signal with asymmetric spectrum about the carrier frequency. (**a**) Envelope of the electric field of an input optical signal. (**b**) The spectrum magnitude of the input envelope is asymmetric. (**c**) Spectrogram of the spectrum magnitude formed by short-term Fourier transform. The region of the spectrogram with power density 30 dB less than the peak power is contoured with a blue dot-dash line. This contour is considered as the necessary effective bandwidth to design an ideal group delay profile. Note that this contour also becomes asymmetric about the carrier frequency. (**d**) To perform uniform frequency-to-time mapping, a linear group delay design minus the input signal chirp should be used to stretch the input optical field. (**e**) The spectrum maps uniformly to temporal envelope of the electric field by the chirp-corrected linear group delay profile. The frequencies higher than the center frequency experience relatively positive group delays meaning they lag behind the center frequency. (**f**) Spectrogram of the envelope waveform amplitude resembles that of the spectrum magnitude (shown in Fig. 8c). The electronic acquisition time and bandwidth are marked with a green dot-dash box. Also, equivalent limitations on the envelope optical frequency and the frequency of spectrum are marked with a similar green dot-dash box on Fig. 8c. (**g**) If a nonlinear group delay profile is designed based on the blue dot-dash contour in Fig. 8c and the input signal chirp, (**h**) the spectrum is nonuniformly mapped to the electric field envelope in time in such a way that the acquisition time is minimized for the set acquisition bandwidth (8 GHz) and the enforced spectrotemporal accuracy level (−30 dB). (**i**) The spectrogram of the electric field envelope amplitude after the nonuniform dispersion shows that a shorter temporal window and a smaller acquisition bandwidth are sufficient to capture the waveform compared to the linear case (the blue dot-dash box). This time-bandwidth acquisition window translates into the frequency-dependent effective bandwidth as shown with the blue dot-dash contour in Fig. 8c, which was used to design the nonlinear group delay profile.

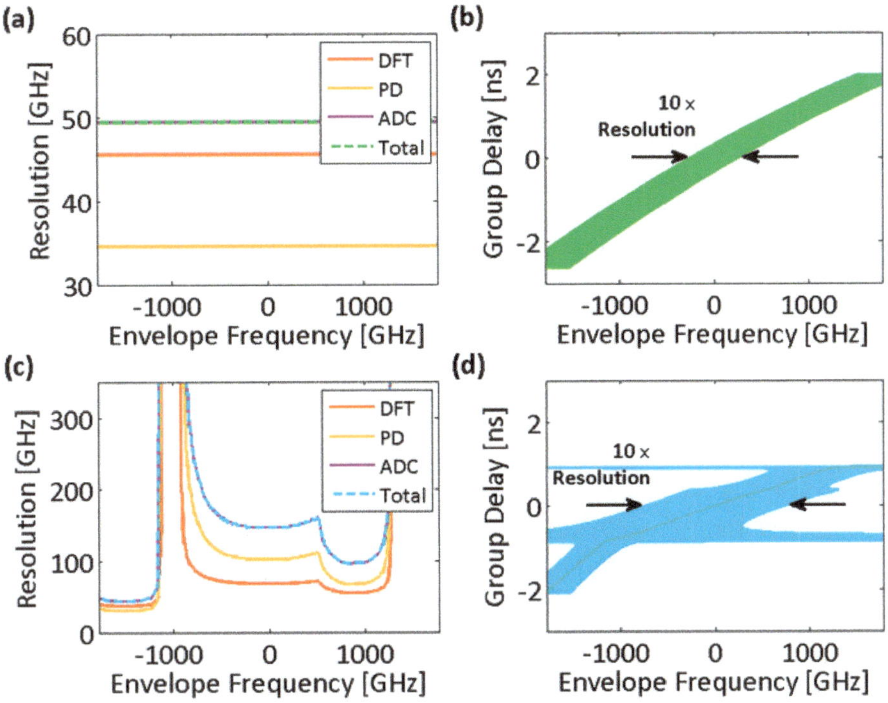

Figure 9: Spectral sampling resolution of time stretch dispersive Fourier transform designed according to the spectrotemporal sparsity for a signal with asymmetric spectral features. (**a**) Spectral resolution limits for the chirp compensated linear group delay profile and the acquisition system of Fig. 8d–f (photodetector and analog-to-digital converter Nyquist bandwidths are 10.5 GHz). The overall spectral sampling resolution of the linear time stretch is independent of the envelope optical frequency and limited by the Nyquist bandwidth of the analog-to-digital converter. (**b**) The spectral sampling resolution of the linear group delay profile minus chirp magnified ten times (for visual clarity) and overlapped on the profile. (**c**) Spectral resolution limits corresponding to the nonlinear group delay profile and the acquisition system of Fig. 8g–i, which unlike the linear stretch, depend of the envelope optical frequency (photodetector and analog-to-digital converter Nyquist bandwidths are 8 GHz). In this case, the overall spectral sampling resolution is limited by the Nyquist bandwidth of the analog-to-digital converter. (**d**) Magnified spectral sampling resolution of the warped time stretch overlapped on its group delay profile clearly shows the ambiguity grows at the regions of the spectrum that do not contain sharp spectral features. DFT: dispersive Fourier transform; PD: photodetector; ADC: analog-to-digital converter; Total: overall spectral sampling resolution.

DISCUSSION

The nonuniform sparse Fourier domain sampling described above may be used for data compression. This works when some frequencies carry more information than others. Such frequencies are coded with fine resolution preserving features of spectrum at these frequencies. On the other hand, less important frequencies are coded with a coarser resolution. Naturally, some of the finer details of less important frequencies will be lost in the coding.

The information of interest is usually encoded into the magnitude of the spectrum, therefore a simple unwarping of the time-to-spectrum map is sufficient for reconstruction. For a more general case where the information is contained in both the amplitude and phase, reconstruction requires either coherent detection or recovery of phase from amplitude measurements. The input signal is then recovered by simulation of back propagation through the dispersive profile (filter). Generally known as a phase retrieval method, there are numerous digital algorithms available for recovering the complex amplitude from intensity-only measurements[25,26].

The reconstruction accuracy and lossy nature of this compression have been analyzed previously[18]. The system reshapes the spectrotemporal structure of the signal such that nearly all the signal energy is within the bandwidth of the photodetector and the real-time digitizer of the acquisition system. Because of the limited resolution of dispersive Fourier transform by ambiguity in frequency-to-time mapping, the limited bandwidths of the photodetector and the digitizer, and the limited resolution of the digitizer, as measured by its effective number of bits (ENOB), the reconstruction will never be ideal, and therefore, this is a lossy compression method. In general, for any time-limited pulse, the spectrum is not bandlimited, and the signal reconstruction will suffer from the loss of out-of-band spectral components in the acquisition system. If the temporal width of the input signal is small enough, so that the spectral resolution is sufficient for capturing details of the input signal spectrum (i.e. in the far field), the bandwidth limitations imposed by the acquisition system can be considered as a frequency-dependent effective bandwidth on the input signal frequency of spectrum as shown in Figs. 2c, 4c, 6c and 8c. The effective bandwidth interpretation facilitates the design of group delay profile for a set of target signals with known spectral characteristics and determines the minimum amount of loss in the compression process.

As a consequence of dispersion, temporal features are transformed and slowed down in time. The amount by which a particular temporal feature is stretched is proportional to the bandwidth of the feature and the overall dispersion over the bandwidth. Fast temporal features have larger bandwidth, and as a result, they are transformed and stretched more than slow temporal features. This feature selective stretch has been referred to as a type of self adaptivity through which the output adapts to the input even when the transfer function of the system is static27.

The group delay dispersion profile is designed according to the spectral sparsity of the input signal as described, i.e. smart stretching. Beyond that, the group delay is static; in other words, it does not need to be dynamically varied according to the instantaneous behavior of the signal. This is why the time stretch dispersive Fourier transform as well as its warped counterpart were called self-adaptive27. However, if the spectral characteristics of the signal slowly varies over time, a feedback mechanism can be used to adapt the group delay profile, and subsequently the effective bandwidth to the sparsity requirements. The sophisticated group delay profiles designed by our algorithm can be readily implemented by chirped fiber Bragg grating (CFBG) technology16,19. Another implementation option is to use chromo-modal dispersion (CMD) device28,15, which uses the large modal dispersion of multimode waveguides in conjunction with the angular dispersion of diffraction gratings to create huge chromatic dispersion. For either technology, the implemented group delay profile will have some deviations from the design. A numerical study of the tolerance to profile nonidealities is performed previously17.

Time stretch dispersive Fourier transform maps the spectrum of the pulses in a burst-mode signal to the silent intervals in between them. In order to use the time stretch transform for the acquisition of the spectrotemporal evolution of a continuous-time signal, the signal needs to be segmented into multiple pulse trains in a process, which is called virtual time gating29. The pulse trains are independently dispersed by linear or warped time stretch systems in parallel, and the acquired signals are digitally stitched together to reveal the spectral features of the continuous-time signal as it varies with time. If the temporal duration of each time gate window is very small, that is, the sliced segments of the continuous-time signal are very short, the resolution of the linear or warped dispersive Fourier transform can be limited by the bandwidth of the gating window. Also, the temporal durations of the time gate windows can be different.

Linear time stretch dispersive Fourier transform acquires the spectrum of each pulse with uniform spectral resolution up to frequencies far beyond the electrical acquisition bandwidth of the analog-to-digital converter and photodetector (Fig. 10a). Warped time stretch dispersive Fourier transform has the same properties as its linear counterpart, but its spectral resolution is not uniform across the bandwidth (Fig. 10b). This nonuniformity can be designed to match the spectrotemporal sparsity of the signal and therefore, increase the spectral resolution at desired frequencies under the same envelope time-bandwidth product. Both linear and warped time stretch dispersive Fourier transforms can be used in conjunction with virtual time gating technique for acquisition of the continuous-time signals (Fig. 10c,d). In virtually time gated warped stretch transform, gates can have dissimilar group delay profiles corresponding to different distributions of the nonuniform spectral resolution and be suitable for various types of spectral sparsity (Fig. 10d).

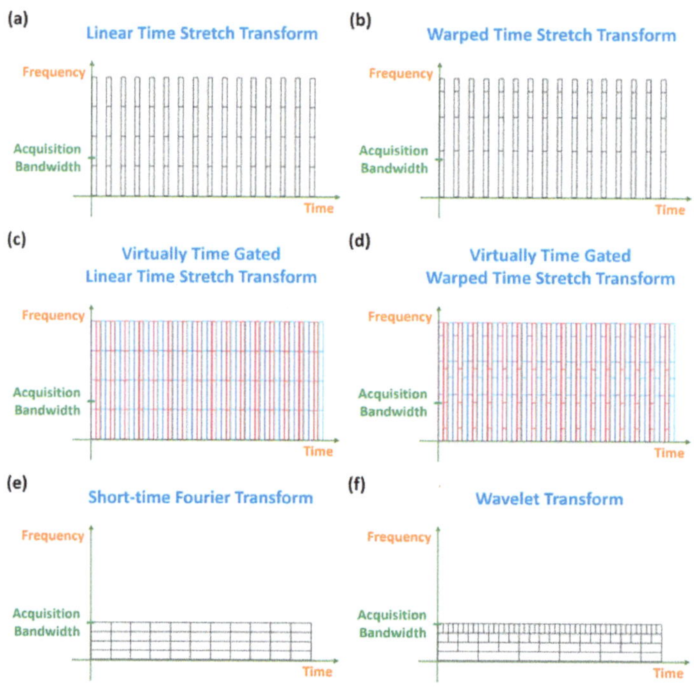

Figure 10: Spectrotemporal resolution of time stretch dispersive Fourier transform vs short-time Fourier transform and wavelet transform. For burst mode signals, time stretch dispersive Fourier transform takes advantage of the empty intervals between pulses to map their spectra into temporal waveforms and captures the signals with a bandwidth far beyond the acquisition bandwidth of the electronic back-end e.g. analog-to-digital converter and photodetector. (**a**) In linear time stretch, the spectral resolution is uniform, and the temporal resolution is same as the input pulse width. (**b**) In warped time stretch, the spectral resolution is nonuniform as discussed earlier, and the temporal resolution is again same as the input pulse width. (**c**) To use time stretch dispersive Fourier transform for analysis of continuous-time signals, the signal should be segmented into multiple burst mode signals. This process is called virtual time gating and can be used for both linear and warped stretch transforms. The temporal durations of the gates should not be very short to limit the spectral sampling resolution, but can be different. In virtually time gated time stretch transform, all of the gates (shown with different colors) have the same uniform spectral resolution if identical dispersions and back-end electronics are used. (**d**) In addition, for virtually time gated warped time stretch transform, the gates can have different distributions

of nonuniform spectral resolutions. (**e**) Short-time Fourier transform can also be used to digitally generate the spectrotemporal distribution of an already acquired signal, but its bandwidth is limited to that of the electronic acquisition system. (**f**) The bandwidth of digitally-implemented wavelet transform is also restricted to the electronic acquisition bandwidth, but its temporal resolution can be nonuniform.

Spectrotemporal characteristics of a signal can also be analyzed digitally by capturing the signal using an analog-to-digital converter and performing short-time Fourier transform or wavelet transform on the samples. However, in these cases, the maximum frequency that can be measured is limited by the electronic acquisition bandwidth (Fig. 10e,f). The wavelet transform can also generate nonuniform temporal resolution for the spectrotemporal distribution of the signal while keeping the spectral resolution fixed (Fig. 10f).

CONCLUSION

Time stretch dispersive Fourier transform is an indispensable tool for acquisition and analysis of the wideband signals at frequencies far beyond the acquisition bandwidth of the electronic back-end i.e. digitizer and photodetector. The more general form of it, warped time stretch, offers nonlinear mapping of spectrum to time, which leads to a nonuniform sampling of the spectrum. We analyzed the spectral resolution of the warped time stretch dispersive Fourier transform and defined an effective bandwidth for the transform, which guides the design of a proper group delay profile based on the spectral sparsity of the signal. Finally, linear and warped time stretch transforms are compared to other methods such as wavelet transform for spectrotemporal analysis of continuous-time signals.

METHODS

Simulation Model

We used a discrete-time complex-envelope simulation model for the analysis of warped stretch transform. The carrier frequency is assumed to be 200 THz, resembling an optical wavelength of 1.5 μm. The signal spectra are downshifted from carrier frequency to baseband, easing the required temporal resolution of the simulations. The temporal and spectral resolutions at baseband simulations are set at 0.1 ps and 150.15 MHz, respectively. The output complex envelope spectrum, $\widetilde{E}_{out}(\omega)$, is calculated as the product of the input complex envelope spectrum, $\widetilde{E}_{in}(\omega)$, and the impulse response of the downshifted dispersion profile:

$$\widetilde{E}_{out}(\omega) = H(\omega) \cdot \widetilde{E}_{in}(\omega) \tag{8}$$

Here, ω is the modulation frequency, and the impulse response, $H(\omega)$, is a frequency-dependent phase shift filter;

$$H(\omega) = exp\{j\varphi(\omega)\} \tag{9}$$

where the phase shift, $\varphi(\omega)$, corresponds to the integral of the group delay profile; i.e.,

$$\varphi(\omega) = \int \tau_g(\omega)\, d\omega \tag{10}$$

Furthermore, linear interpolation is employed to generalize designed group delay to arbitrary frequencies. If any, the chirp of the signal is derived by a moving-window short-time Fourier transform.

Spectrograms

Spectrograms depend on the estimation method of power spectral density, e.g. the window size of the short-term Fourier transform. If the width of the short-time Fourier transform window is reduced, the time resolution of the spectrogram improves, but its frequency resolution degrades. For the frequency-of-spectrum spectrogram, these changes in window size alter the effective bandwidth contour and lead to variations in the design of the group delay profile. In other words, the width of the spectral window should be small enough to capture localized fluctuations of the spectral sparsity, but not too narrow to overestimate the effective bandwidth. Alternatives to spectrogram, for example instantaneous frequency estimated from the analytic form of the signal calculated by the Hilbert transformation, can also be used to determine the required effective bandwidth and design the group delay profile.

ACKNOWLEDGEMENTS

We would like to thank our colleague, Jacky Chan, at University of California, Los Angeles for helpful discussions and a review of this manuscript. This work was supported by Office of Naval Research Multidisciplinary University Research Initiatives (ONR MURI) program on Optical Computing.

REFERENCES

1. Goda, K. & Jalali, B. Dispersive fourier transformation for fast continuous single-shot measurements. Nat. Photonics 7, 102–112 (2013).
2. Coppinger, F., Bhushan, A. & Jalali, B. Time magnification of electrical signals using chirped optical pulses. Electron. Lett. 34, 399–400 (1998).
3. Kelkar, P., Coppinger, F., Bhushan, A. & Jalali, B. Time-domain optical sensing. Electron. Lett. 35, 1661–1662 (1999).
4. Solli, D., Ropers, C., Koonath, P. & Jalali, B. Optical rogue waves. Nature 450, 1054–1057 (2007).
5. Goda, K., Tsia, K. & Jalali, B. Serial time-encoded amplified imaging for real-time observation of fast dynamic phenomena. Nature 458, 1145–1149 (2009).
6. Goda, K. et al. High-throughput single-microparticle imaging flow analyzer. Proc. Natl. Acad. Sci. 109, 11630–11635 (2012).
7. Mahjoubfar, A., Chen, C., Niazi, K. R., Rabizadeh, S. & Jalali, B. Label-free high-throughput cell screening in flow. Biomed. Opt. Express 4, 1618–1625 (2013).
8. Chen, C. et al. Hyper-dimensional analysis for label-free high-throughput imaging flow cytometry. In Conference on Lasers and Electro-Optics (CLEO): Applications & Technology, AW3L-2 (Optical Society of America (OSA), San Jose, California, USA, 2014).
9. Mahjoubfar, A. et al. High-speed nanometer-resolved imaging vibrometer and velocimeter. Appl. Phys. Lett. 98, 101107 (2011).
10. Mahjoubfar, A. et al. 3d ultrafast laser scanner. In Frontiers in Ultrafast Optics: Biomedical, Scientific, and

Industrial Applications XIII, 86110N–86110N (SPIE, San Francisco, California, USA, 2013).

11. Gupta, S. & Jalali, B. Time-warp correction and calibration in photonic time-stretch analog-to-digital converter. Opt. Lett. 33, 2674–2676 (2008).
12. Ng, W., Rockwood, T., Sefler, G. & Valley, G. Demonstration of a large stretch-ratio (m= 41) photonic analog-to-digital converter with 8 enob for an input signal bandwidth of 10 ghz. IEEE Photon. Technol. Lett 24, 1185–1187 (2012).
13. Solli, D., Gupta, S. & Jalali, B. Optical phase recovery in the dispersive fourier transform. Appl. Phys. Lett. 95, 231108 (2009).
14. DeVore, P. T., Buckley, B. W., Asghari, M. H., Solli, D. R. & Jalali, B. Coherent time-stretch transform for near-field spectroscopy. IEEE Photonics J. 6, 3300107 (2014).
15. Jalali, B. & Mahjoubfar, A. Tailoring wideband signals with a photonic hardware accelerator. Proc. IEEE 103, 1071–1086 (2015).
16. Asghari, M. H. & Jalali, B. Experimental demonstration of optical real-time data compression. Appl. Phys. Lett. 104, 111101 (2014).
17. Jalali, B., Chan, J. & Asghari, M. H. Time-bandwidth engineering. Optica 1, 23–31 (2014).
18. Chan, J., Mahjoubfar, A., Asghari, M. & Jalali, B. Reconstruction in time-bandwidth compression systems. Appl. Phys. Lett. 105, 221105 (2014).
19. Chen, C. L., Mahjoubfar, A. & Jalali, B. Optical data compression in time stretch imaging. PLOS ONE 10, e0125106 (2015).
20. Bosworth, B. T. & Foster, M. A. High-speed ultrawideband photonically enabled compressed sensing of sparse radio frequency signals. Opt. Lett. 38, 4892–4895 (2013).
21. Valley, G. C., Sefler, G. A. & Shaw, T. J. Compressive sensing of sparse radio frequency signals using optical mixing. Opt. Lett. 37, 4675–4677 (2012).
22. Liu, J.-M. Photonic devices (Cambridge University Press, 2005).
23. Oppenheim, A. V. & Schafer, R. W. Discrete-time signal processing (Prentice Hall, 2009).
24. Goda, K., Solli, D. R., Tsia, K. K. & Jalali, B. Theory of amplified dispersive fourier transformation. Phys. Rev. A 80, 043821 (2009).
25. Walmsley, I. A. & Dorrer, C. Characterization of ultrashort electromagnetic pulses. Adv. Opt. Photonics 1, 308–437 (2009).
26. Jaganathan, K., Oymak, S. & Hassibi, B. Recovery of sparse 1-d signals from the magnitudes of their fourier transform. In International Symposium on Information Theory Proceedings (ISIT), 1473-1477 (IEEE, Cambridge, Massachusetts, USA, 2012).
27. Jalali, B. & Asghari, M. H. The anamorphic stretch transform: Putting the squeeze on “big data”. Opt. and Photon. News 25, 24–31 (2014).
28. Diebold, E. D. et al. Giant tunable optical dispersion using chromo-modal excitation of a multimode waveguide. Opt. Express 19, 23809–23817 (2011).
29. Han, Y. & Jalali, B. Continuous-time time-stretched analog-to-digital converter array implemented using virtual time gating. IEEE Trans. Circuits Syst. I-Regul. Pap. 52, 1502–1507 (2005).

CHAPTER 17

THE USE OF CONTINUOUS WAVELET TRANSFORM BASED ON THE FAST FOURIER TRANSFORM IN THE ANALYSIS OF MULTI-CHANNEL ELECTROGASTROGRAPHY RECORDINGS

Dariusz Komorowski and Stanislaw Pietraszek

Keywords: Electrogastrography, Continuous wavelet transform, Dominant frequencyIndex of normogastria

INTRODUCTION

Electrogastrography is a research method designed for noninvasive assessment of gastric slow wave propagation [1–4]. One or multichannel EGG signals are obtained from the disposable electrodes, appropriately arranged on the surface of the abdomen of the patient's stomach [5–7]. It is assumed that the frequency range of EGG signal is from 0.015 to 0.15 Hz and the amplitude of it is about 100–400 μV [6, 8]. The typical EGG examination takes about 2 h and consists of three parts: the first one - preprandial, usually no longer than 30 min, it is a stage before a meal (person under investigation should be fasted). The second part takes about 5 to 15 min, including time when the person accepts a standardized meal, and the third part - postprandial, about 60–120 min, when the meal is digested. The standard of a meal depends on the examining center. Most frequently three types of meals are used: non-caloric meal e.g., 400 ml water, liquid meal e.g., 250 ml of yogurt and the caloric meal, e.g., pancake with jam prepared according to a well-defined recipe [9, 10].

The initial analysis of EGG signals involves calculating dominant frequency and dominant power of slow waves [2, 4, 6, 11–13]. In the case of EGG examination the frequency is typically calculated in cycles per minute (cpm), as a medical standard [5]. According to the DF values, the EGG rhythm is classified to: bradygastria (*0.5–2.0 cpm*), normal rhythm (*2.0–4.0 cpm*) or tachygastria (*4.0–9.0 cpm*) [4, 5]. Due to very high level of disturbances and interferences in EGG signals while receiving a meal, the DF values are calculated only for preprandial and postprandial parts. On the basis of the rhythm classification, the normogastria index is calculated [6]. This index is expressed as the amount of DF values in the range of normal rhythm to the total amount of the DF values [4, 5].

The parameters DF and MP are usually calculated by means of the spectral analysis. The spectral analysis of EGG is done for short segments (1 to 5 min length) of the signals. The values of DF and MP are calculated for each segment. The length of the segments depends on the limitations of the used method and is a compromise between accuracy and resolution of calculated frequency and its time location in the analyzed signal. Segments of 3 to 5 min length are used to calculate spectrum of the EGG signal, using the short-time Fourier transform (STFT), and for the nonparametric methods (for example AR, ARMA modeling), the minimum length of the segment is about 60 s [4, 6, 11, 12]. In the case of EGG signal analysis, the process of calculation the spectrum of consecutive or overlapped fragments is often referred as a running spectrum analysis (RSA) [8, 14]. Figure 1 shows a 1-min segment of the signal before and after application of Tukey window (left), its power spectrum (middle) with marked both DF and MP values, and an example of RSA analysis of one channel EGG signal (right).

Citation: Komorowski D. & Pietraszek S., The Use of Continuous Wavelet Transform Based on the Fast Fourier Transform in the Analysis of Multi-channel Electrogastrography Recordings 40: 10, doi:10.1007/s10916-015-0358-4.

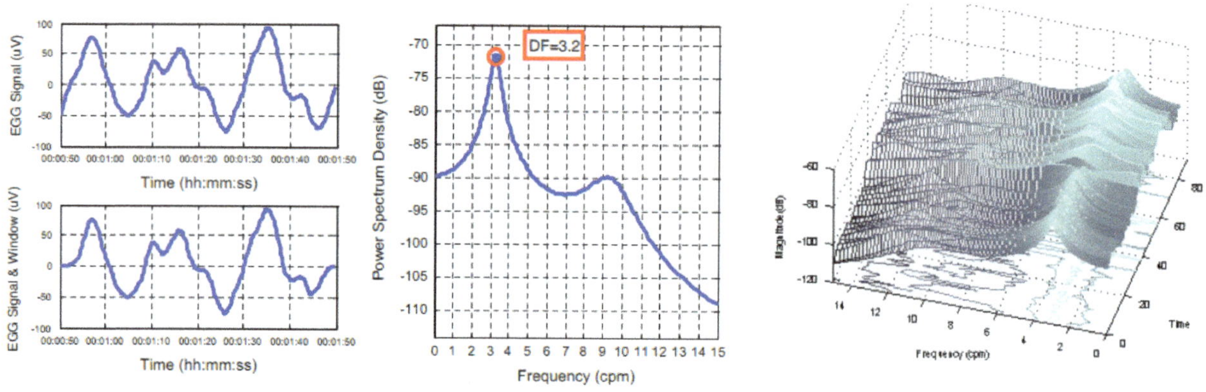

Figure 1: An example of 60s segment of EGG signal (*left*), its spectrum (*middle*) with marked DF and MP and an example of RSA analysis of one-channel EGG signal (*right*).

The EGG signals analysis based on CWT is widely presented in the literature [15–19]. In this study, the CWT algorithm based on FFT was applied to determine the dominant frequency of the slow wave in the EGG signal. In the literature, this algorithm is referred to as CWTFT [20–22]. An application of CWT allows to reduce limitations of classical methods of spectrum analysis (e.g., FFT, STFT) and facilitates to determine instantaneous frequencies and its location in the time domain [23–25].

The continuous wavelet transform is a powerful tool for analyzing nonstationary time series signals in the time-frequency domain and substantially differs from the STFT method that allows clear localization on the time axis of the frequency components, existing in the analyzed signals. Such an analysis is possible using of variable width window which corresponds to the scale of observation (analysis). Wavelet analysis allows using long time windows when we need more precise low frequency information, and shorter when we need high frequency information. It should be noted that the wavelet analysis does not include the area of time-frequency (as in the case of STFT), but the area: the time-scale. The time-scale area can be converted to the appropriate area of pseudo-frequency - time, where the pseudo-frequency is the characteristic central frequency of the wavelet. Another important distinction from the STFT is that the CWT is not limited to sinusoidal analyzing function and do not requires, that the signal meets fairly strict criteria, which are required in the classical Fourier analysis [22].

METHOD

The CWT reflects the correlation between the analyzed continuous-time signal $x(t)$ and a function referred to as wavelets and is defined by the following formula

$$Cw(a,b) = \int_{-\infty}^{+\infty} x(t)\psi_{a,b}{}^{*}dt = \frac{1}{\sqrt{a}}\int_{-\infty}^{+\infty} x(t)\psi^{*}\left(\frac{t-b}{a}\right)dt, \quad (1)$$

where: $Cw(a, b)$ is the function of the parameters a and b.

The a parameter is the dilation of wavelet (scale) and b defines a translation of the wavelet and indicates the time localization, $\psi^*(t)$ is the complex conjugate of the analyzing mother wavelet $\psi(t)$ [22, 26]. The coefficient $\frac{1}{\sqrt{a}}$ is an energy normalized factor (the energy of the wavelet must be the same for different a value of the scale). Moreover, to be classified as a basic permissible wavelet, a wavelet function must satisfy the following mathematical criteria [27–30]:

The wavelet must have finite energy

$$E = \int_{-\infty}^{+\infty} |\psi(t)|^2 dt < \infty \quad (2)$$

The following condition must hold true

$$C_{\psi} = \int_{-\infty}^{+\infty} \frac{|\psi^{\wedge}(\omega)|^2}{\omega} d\omega < \infty, \quad (3)$$

where

$$\hat{\psi}(\omega) = \int_{-\infty}^{+\infty} \psi(t)e^{-i\omega t}dt \quad (4)$$

is the Fourier transform of the $\psi(t)$ function and $\omega=2\pi f$ is the circular frequency. This condition is defined as a condition of admissibility and can be interpreted as a requirement that $|\psi(\omega)|^2$ decay endeavored (head) to zero faster than $1\omega1\omega$. This condition means that the wavelet has no zero frequency component, that is

$$\hat{\psi}(0) = 0 \quad (5)$$

Otherwise, the wavelet must have the zero mean value. In the literature the parameter C_{ψ} is called the admissibility constant. The value of C_{ψ} depends on the chosen wavelet [27–29].

In the case of complex wavelets, Fourier transform must have both a real component and a value of zero for negative frequencies [22].

As a result of CWT transform we obtain the two-dimensional function $E(a,b)=|Cw(a,b)|^2$called scalogram, which presents energy distribution of signals for used scales a and the time position b (locations). In practice, all functions that differ from $|Cw(a,b)|^2$ only by the constant multiplicative factor are also called scalograms [22]. The CWT can be considered as a transform that converts the signal from the time domain to the scale-time domain. The scale as mentioned above can be converted to a frequency value (pseudo-frequency), the value of which depends on the center frequency of the applied wavelets and the scale value a

$$f_a = \frac{f_c}{a}, \tag{6}$$

where: f_a is the frequency associated with the wavelet at the specific a scale, while f_c is the characteristic frequency of mother wavelet at scale $a = 1$, and time position $b = 0$. There is a very important distinction to be made here: "The characteristic frequency f_c of the wavelet used in the wavelet transform is representative of the whole frequency makeup of the wavelet. The wavelet does not contain a single frequency, and the signal is not decomposed according to numerous single (sinusoidal) frequencies; this is not Fourier analysis!" [22].

It is known, if the wavelet transform coefficients are given, it is possible to reconstruct the original signal by the inverse wavelet transform described by the following equation [20]

$$x(t) = \frac{1}{K_\psi}\int_{a=0}^{+\infty}\left(\int_{b=-\infty}^{+\infty} Cw(a,b)\psi_{a,b}(t)\frac{db}{a^2}\right)da = \frac{1}{K_\psi}\int_0^{+\infty} D(a,t)da, \tag{7}$$

where K_ψ is a constant factor that depends on the applied wavelet function and the details function $D(a, t)$ is given by the following equation

$$D(a,t) = \int_{b=-\infty}^{+\infty}\frac{1}{a^2}Cw(a,b)\psi_{a,b}(t)db \tag{8}$$

In practical CWT applications usually discrete values of the a scale in the range of continuous values are used. As the result the wavelet coefficients are obtained (called wavelet series). The choice of appropriate wavelet function is crucial to obtain good results during signal analysis. The classic CWT transform is time consuming and it requires considerable computing power to apply it to the analysis of long signals.

In recent years, the new efficient algorithms have been developed for significant acceleration of CWT calculation. One of them uses well-known FFT algorithms to calculate the CWT [20, 22,23].

If we define

$$\psi_a(t) = \frac{1}{\sqrt{a}}\psi\left(\frac{t}{a}\right) \tag{9}$$

and

$$\psi_{ab}(t) = \frac{1}{\sqrt{a}}\psi\left(\frac{t-b}{a}\right) \tag{10}$$

the definition of CWT transform (1) can be rewritten in the following form

$$Cw(a,b) = \int_{-\infty}^{+\infty} x(t)\psi_{ab}^*(t)dt = \int_{-\infty}^{+\infty} x(t)\psi_a^*(b-t)dt, \tag{11}$$

which clearly indicates that the CWT can be treated as a convolution of the signal and wavelets. Consequently CWT can be expressed as an inverse Fourier transform

$$Cw(a,b) = \frac{1}{2\pi}\int_{-\infty}^{+\infty}\hat{x}(\omega)\hat{\psi}^*_{a,b}(\omega)d\omega, \tag{12}$$

where:

$$\hat{\psi}^*_{a,b}(\omega) = \sqrt{a}\hat{\psi}^*(a\omega)e^{i\omega b} \tag{13}$$

denote the Fourier transform of the analyzed wavelet at scale a and location b and

$$\hat{x}(\omega) = \int_{-\infty}^{+\infty} x(t)e^{-i\omega t}dt \tag{14}$$

is the Fourier transform of the analyzed signal $x(t)$.

In the case of discrete signals (which are typical for the signals processing problems), assuming that the input signal $x(n)$ includes N samples, the discrete versions of the convolution can be represented as

$$W_a(b) = \sum_{n=0}^{N-1} x(n)\psi_a^*(b-n) \tag{15}$$

We can easily notice that in order to obtain the CWT we have to calculate the convolution of the signal and wavelets for each value of the location b and repeat the calculations for each value of the scale a. In the case of two periodic sequences (signals) we can use the property described by Eq. (15) and apply fast algorithms for determining the discrete Fourier transform (DFT) to calculate the circular convolution [31]

$$\sum_{n=0}^{N-1} x(n)\psi_a^*(b-n) = \frac{1}{N}\sum_{k=0}^{N-1}\hat{x}(k)\hat{\psi}_a^*(k)e^{i\frac{2\pi}{N}kb}, \quad b = 0,1,2 \ldots N-1 \tag{16}$$

where the discrete Fourier transform of $x(n)$ signal is given by formula

$$\hat{x}(k) = \sum_{n=0}^{N-1} x(n)e^{-i\frac{2\pi}{N}nk}, \quad k = 0,1,2\ldots N-1 \tag{17}$$

$\hat{\psi}_a$ is the discrete Fourier transform of the wavelet ψ_a

$$\hat{\psi}_a(k) = \sum_{n=0}^{N-1}\psi_a(n)e^{-i\frac{2\pi}{N}nk}, \quad k = 0,1,2\ldots N-1 \tag{18}$$

where k is an index of frequency.

Assuming that the signal is sampled at a frequency $f=f_s$, the sampling period is $\Delta t = 1/f_s$ and in order to obtain the unit energy for each scale a, the wavelet function is normalized by the following formula [23]

$$\hat{\psi}_a(a\omega_k) = \sqrt{\frac{2\pi a}{\Delta t}}\hat{\psi}(a\omega_k), \tag{19}$$

where:

$$\omega_k = \frac{2\pi k}{N\Delta t} \tag{20}$$

The CWT can be expressed as the products of the inverse Fourier transform

$$W_a(b) = \frac{1}{N}\sqrt{\frac{2\pi a}{\Delta t}}\sum_{k=0}^{N-1}\hat{x}\left(\frac{2\pi}{N\Delta t}k\right)\hat{\psi}^*\left(a\frac{2\pi}{N\Delta t}k\right)e^{i\frac{2\pi}{N}kb} \tag{21}$$

The above described method was used for the EGG signals analysis. The calculations were made for various wavelets e.g.,: Morlet, Derivative of Gausian and Paul. During initial tests, authors have examined all available in the applied version of the Matlab (R2013b) types of wavelets for CWTFT algorithm: m-th order derivative of a Gaussian wavelet, analytic Morlet wavelet, non-analytic Morlet wavelet, non-analytic Morlet wavelet with zero mean, Mexican hat wavelet. The non-analytic Morlet wavelet was chosen because the obtained scalograms provided similar information consistent with the commonly known properties of EGG signals as the occurrence of the characteristic frequencies (e.g., 3 cpm) and typical changes in parameters after administration of the meal [3].

In our work all the presented results were obtained for the non-analytic Morlet wavelet with zero mean, specified in the Fourier transform domain by the following formula

$$\hat{\psi}(a\omega) = \frac{1}{\sqrt[4]{\pi}}\left(e^{-\frac{(a\omega-\omega_0)^2}{2}} - e^{-\frac{\omega_0^2}{2}}\right), \tag{22}$$

where ω_0 is the non-dimensional frequency parameter and its value was set to 6 and 15. If $\omega_0=6$ the admissibility condition is satisfied [23, 32].

The dependence of the pseudo-frequency scale on the chosen wavelet is given by

$$f_a = \frac{1}{a\Delta t\lambda}, \tag{23}$$

where λ is the Fourier wavelength (frequency Fourier factor) for the Morlet wavelet and it is given by the following formula [23]

$$\lambda = \frac{4\pi a}{\omega_0 + \sqrt{2+\omega_0{}^2}} \tag{24}$$

During calculations the range of the $a_0=\omega_0\Delta t$ scale (1.5 and 3.75) to 50 in steps of 0.15 were used, which correspond to the range of the pseudo-frequency values 0.6453–0.0040 Hz for $\omega_0=6$ and 0.638–0.0039 Hz for $\omega_0=15$, respectively. The examples of the applied Morlet wavelet are shown in the Fig. 2.

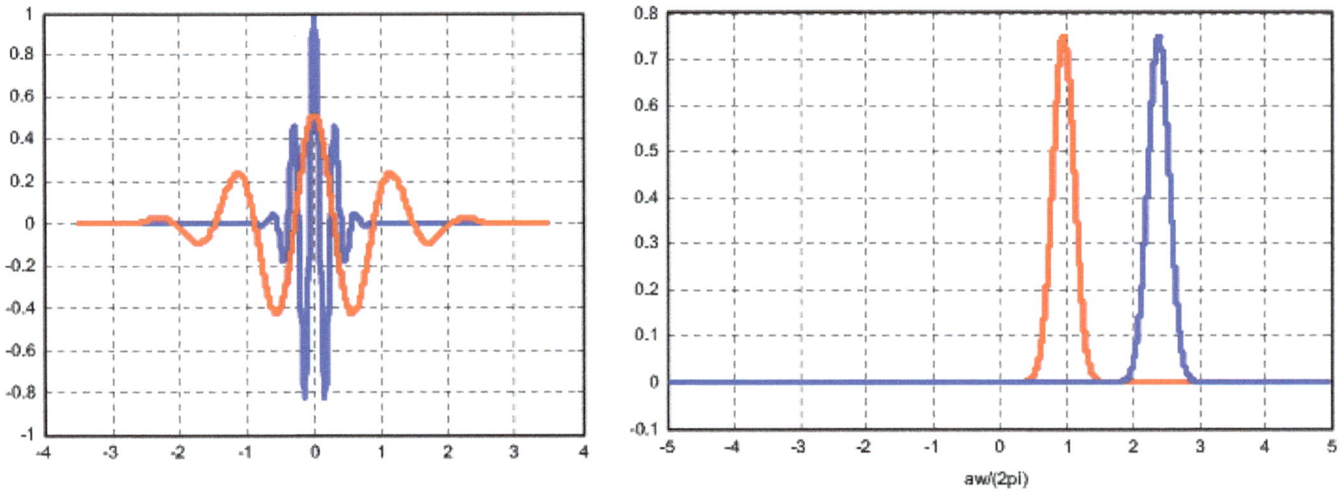

Figure 2: Morlet wavelet (*blue* $a=1$, *red* $a=3.75$) in the time domain (*left*) and in the frequency domain(*red* $\omega_0=6$, *blue* $\omega_0=15$) (*right*).

The Fig. 3. depicts the deference between STFT and CWTFT transforms for an example of a chirp signal (from *0.005 to 0.025Hz*, sampled at $f_s=4Hz$).

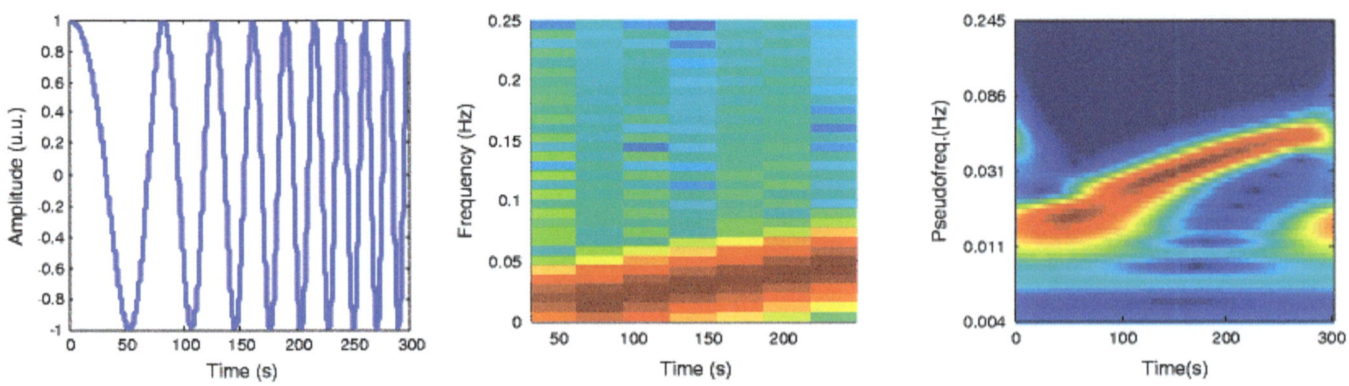

Figure 3: Segment of a chirp signal (*left*) associated with STFT transform (*medium*) and CWTFT (Morlet wavelet, $\omega_0=6$) transform (*right*).

APPLICATION TO EGG

As the aim of this work is to show the application of new algorithms for EGG signal analysis, the paper presents the results of research carried out for only four subjects (women), volunteered to participate in the study. Their average age was 25.75 years (range: 24–31) and average BMI 19.83 (range: 18.6–21.1). Every volunteer gave a written consent to participate in the study. The research project was approved by the Bioethics Committee of the Silesian Medical University.

The duration of EGG study was in the range of 120 to 170 min and it consisted of three parts: preprandial (30–40 min), meal (5–10 min) and postprandial (90–120 min). Before the test, all participants were in a fasting state.

The calculations of the CWT coefficients (absolute values of elements W_{ab} further denoted $|W_{ab}|$) were performed for four-channels EGG signals, sampled at the frequency 4 Hz and the resolution of 12 bits, using the 4-channel prototype biomedical amplifier with the input range± 1 *mV* and gain *5000.*

In order to assess the accuracy of the obtained results, the values of normogastria index (NI) were calculated for each EGG channel using the following procedure. The signal was divided into 60 s length segments (with overlap 50 %) and the $|W_{ab}|$ matrices were calculated. The $|W_{ab}|$ matrices were reshaped in order to remove the components which corresponded to the pseudo-frequency greater than 9 cpm (0.15Hz), because the analysis of EGG signals above this frequency is not carried out [2, 4, 5]. Then, for each segment of the modified $|W_{ab}|$ matrices the maximum value which corresponds to the pseudo-maximum power *pMP(l)* was found and its corresponding value of the dominant pseudo-frequency *pDF(l)*, where *l* is the number of 1-min segment of the EGG signal. Finally *NI* index as the ratio of the number of *pDF* values in the range of 2–4 cpm to all values was calculated, for all channels of recorded EGG signal [2, 5, 6,10]. Then, evaluated normogastria indexes were compared with those calculated by means of the classical method [4, 6].

RESULTS

The described method was applied to the analysis of over 2 h, four-channel EGG recordings, using a light non caloric liquid meal (400 ml cold water). Figure 4 shows the result of the CWT analysis using the non-analytic Morlet wavelet for $\omega_0 = 6$. The scale values were converted to the corresponding pseudo-frequency values. This figure clearly shows that the maximum value of the energy in the signal occurs about 3 cpm - the typical frequency of the slow wave in the EGG signal [4, 8]. Especially in the first part of the examination (preprandial) we can see clearly that the frequency is almost constant or oscillates near the 3 cpm. Figure 5 shows the frequency values corresponding to the maximum energy of scalogram. This process (curve) can be treated as a continuous frequency of the slow waves in the EGG signal. Figures 6 and 7 illustrate respectively the *pDF* and *pMP* for the sequent of 60 s segments of EGG signal. These values were also obtained based on the analysis of the CWTFT coefficients. Figures 6 and 7 were presented to compare the obtained results against the results of the classical EGG signals analysis (Fig. 12). The Figs. 8, 9, 10, and 11 show results of analysis using the Morlet wavelet with $\omega_0 = 15$.

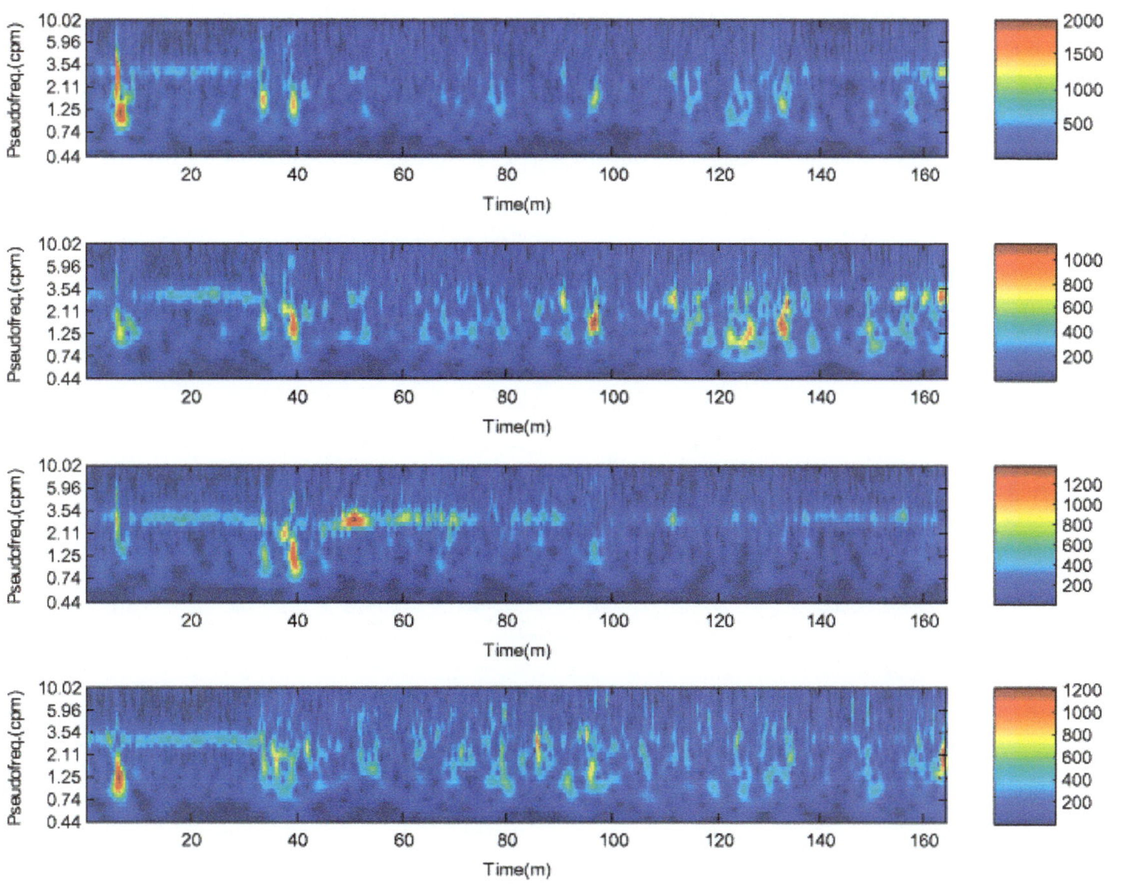

Figure 4: Time-frequency analysis with CWTFT (non-analytic Morlet wavelet, $\omega_0 = 6$) for four-channels EGG signal.

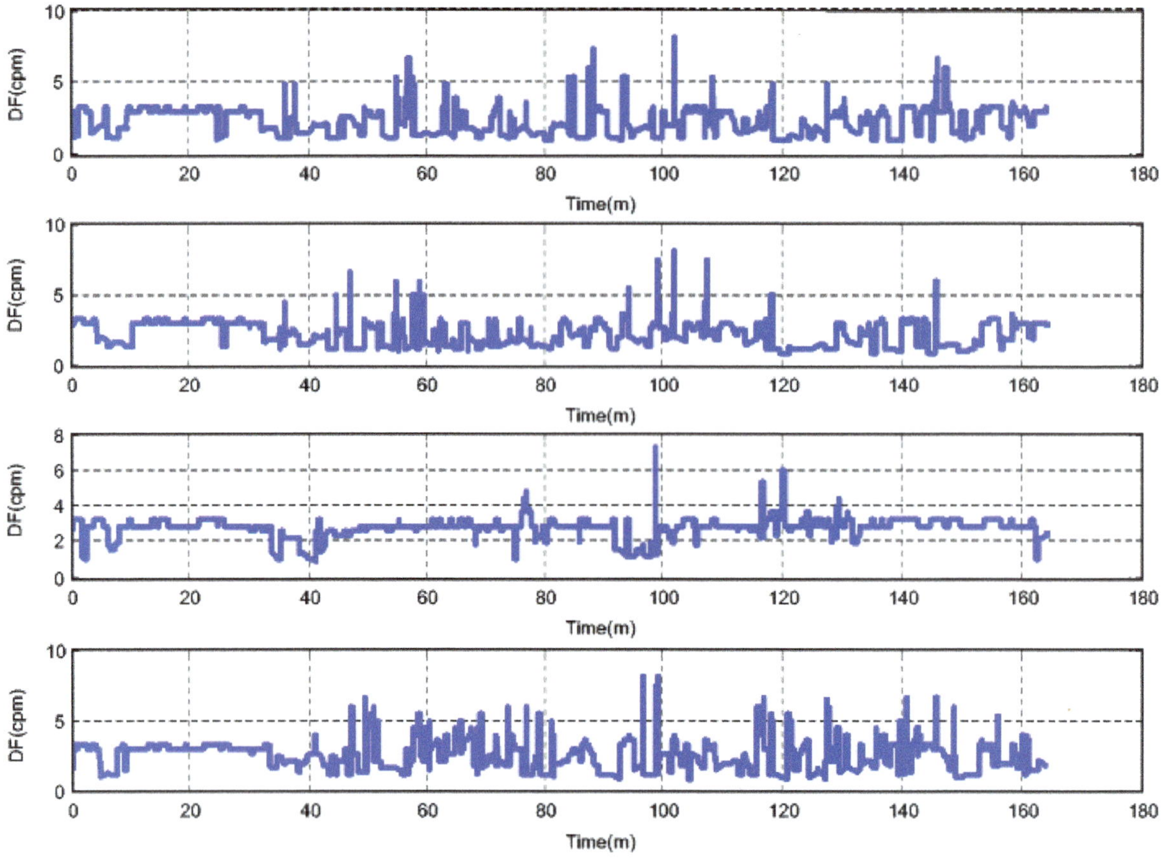

Figure 5: Instantaneous dominant frequency obtained by means of the CWTFT (non-analytic Morlet wavelet, $\omega_0=6$) analysis for four-channels EGG signal.

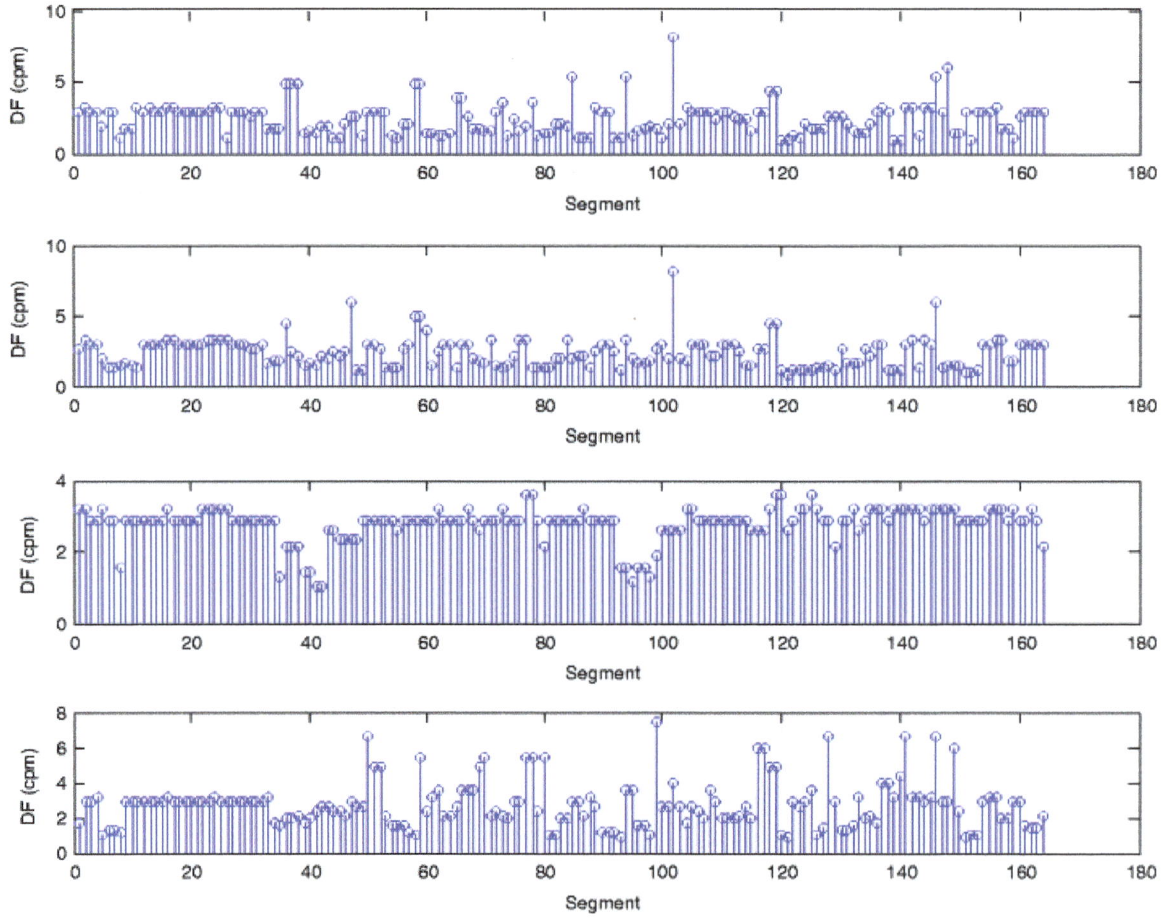

Figure 6: Dominant frequency (pDF) obtained by means of the CWTFT (non-analytic Morlet wavelet, $\omega_0=6$) analysis for four-channels EGG signal (for 60s segments).

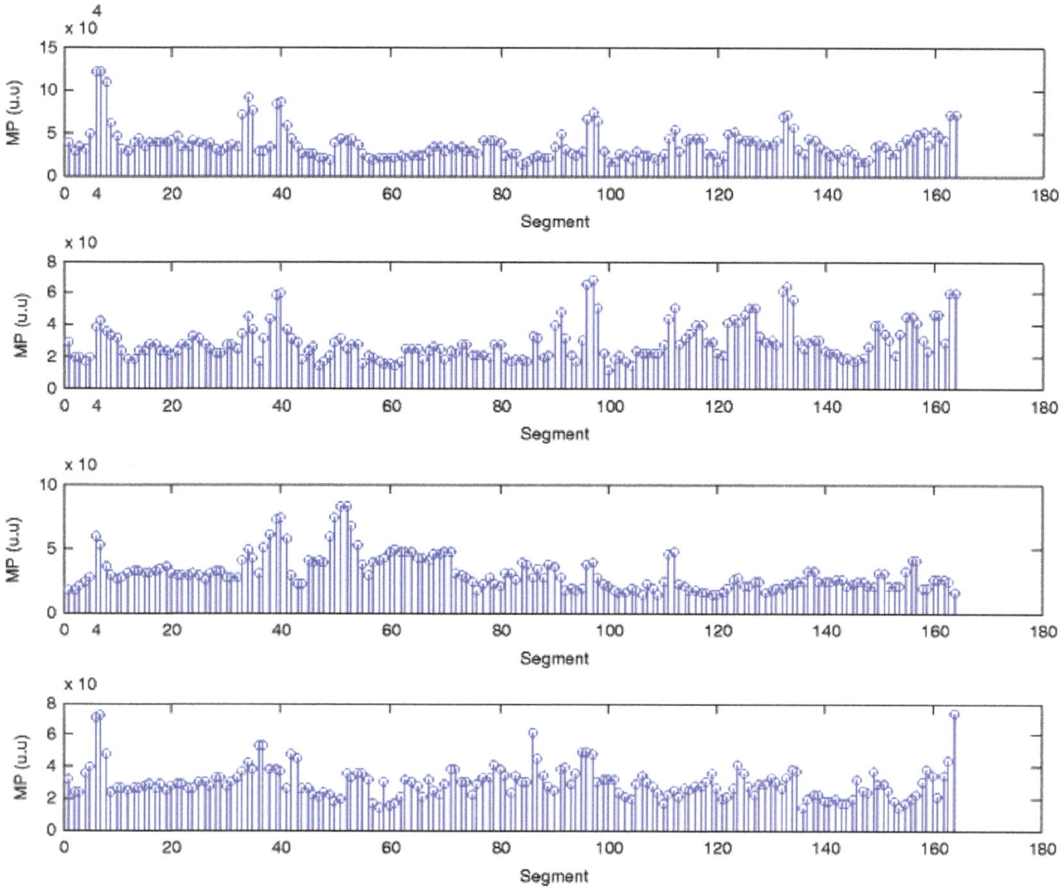

Figure 7: Maximum Power (pMF) obtained by means of the CWTFT (non-analytic Morlet wavelet, $\omega_0=6$) analysis for four-channels EGG signal (for 60s segments).

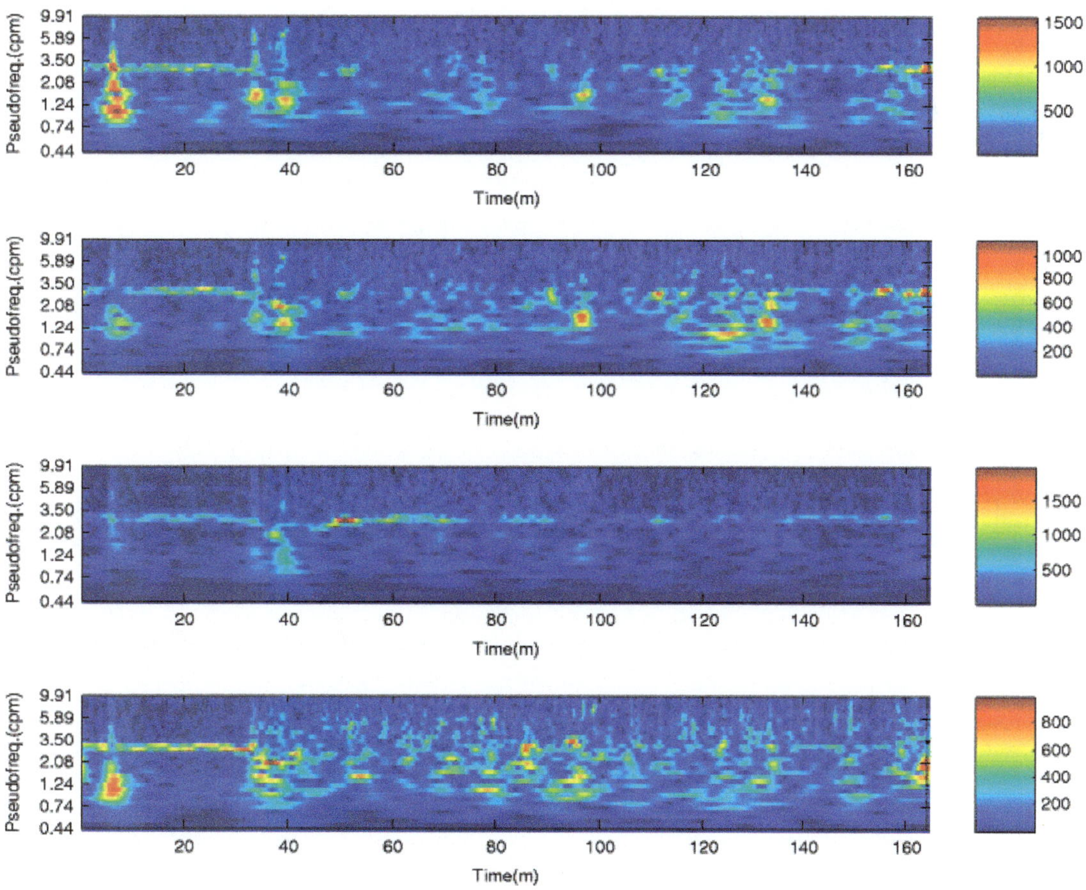

Figure 8: Time-frequency analysis with the CWTFT (non-analytic Morlet wavelet, $\omega_0=15$) for four-channels EGG signal.

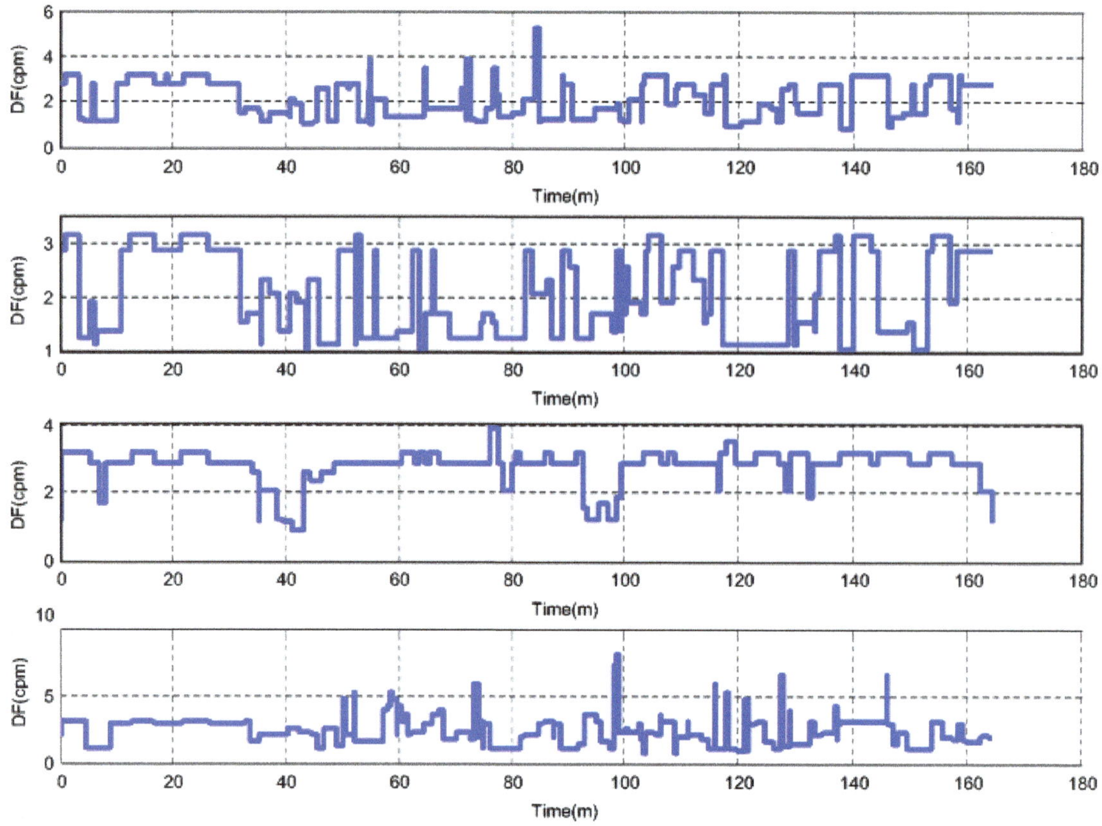

Figure 9: Instantaneous dominant frequency obtained by means of the CWTFT (non-analytic Morlet wavelet, $\omega_0=15$) analysis for four-channels EGG signal.

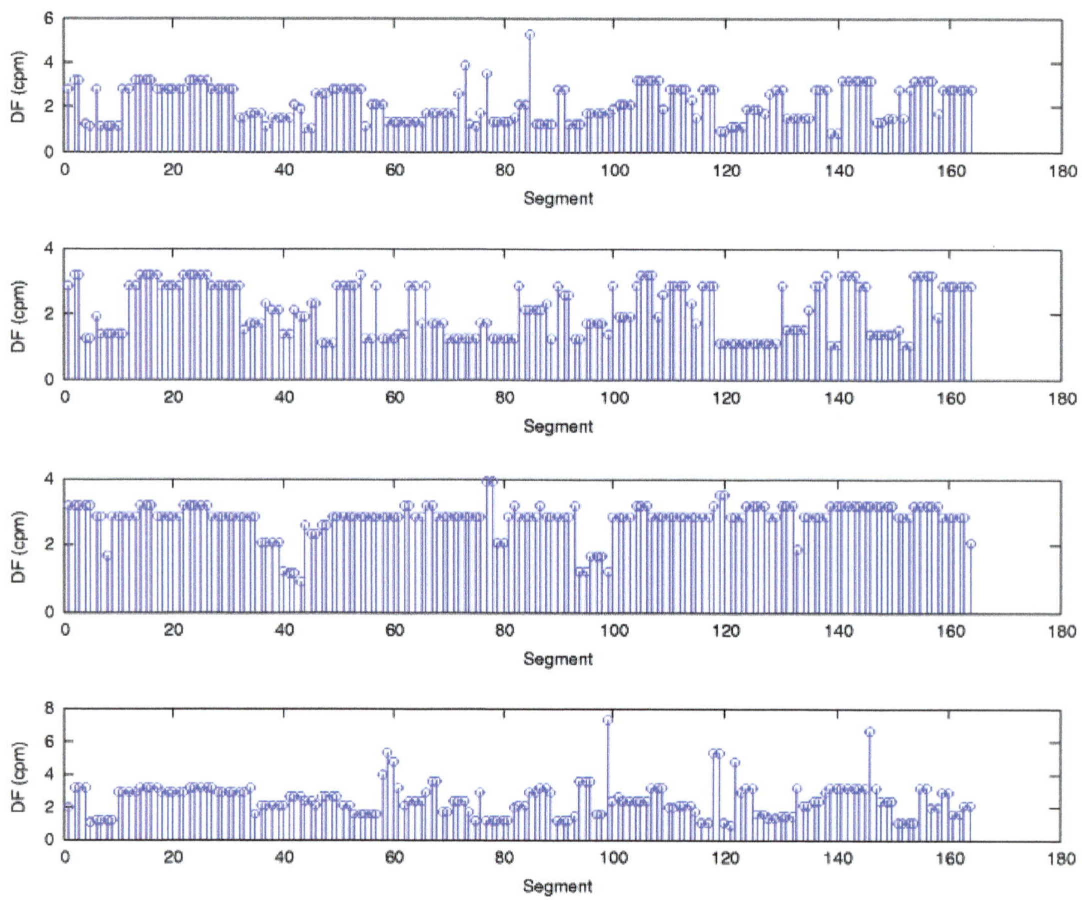

Figure 10: Dominant frequency (pDF) obtained by means of the CWTFT (non-analytic Morlet wavelet, $\omega_0=15$), analysis for four-channels EGG signal (for 60s segments).

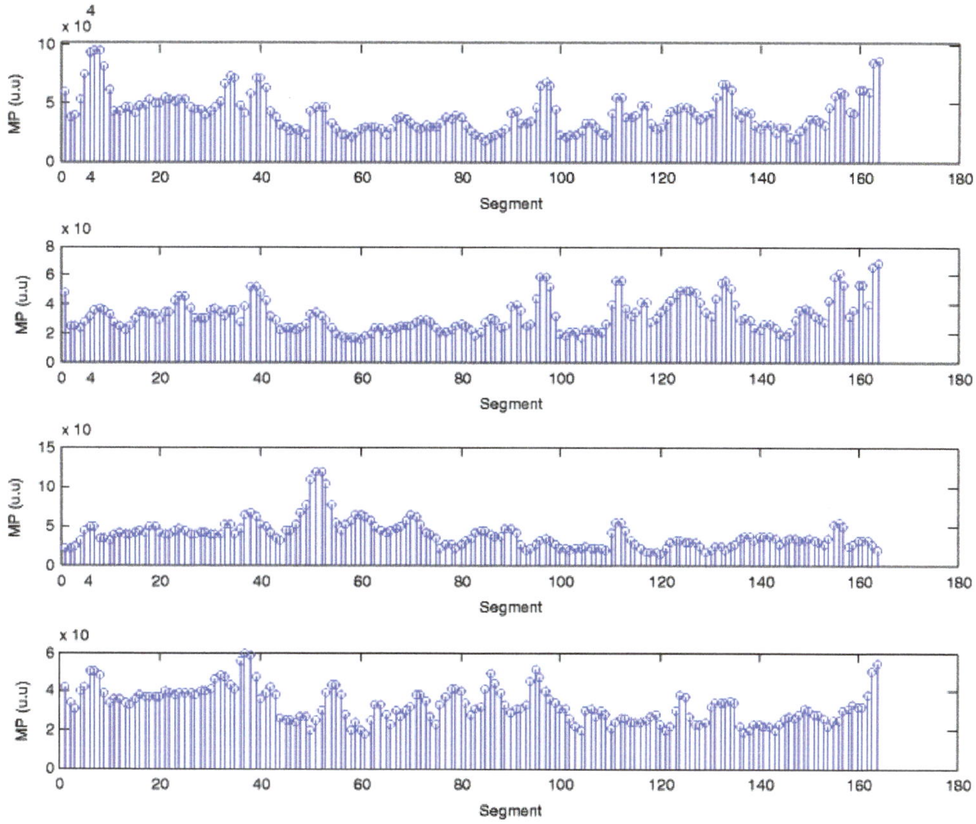

Figure 11: Maximum power (pMP) obtained by means of the CWTFT (non-analytic Morlet wavelet, $\omega_0 = 15$) analysis for four-channels EGG signal (for 60s segments).

The Fig. 12 shows the values of the DF and the MP for the same signal derived from the classical EGG analysis for the same EGG signals.

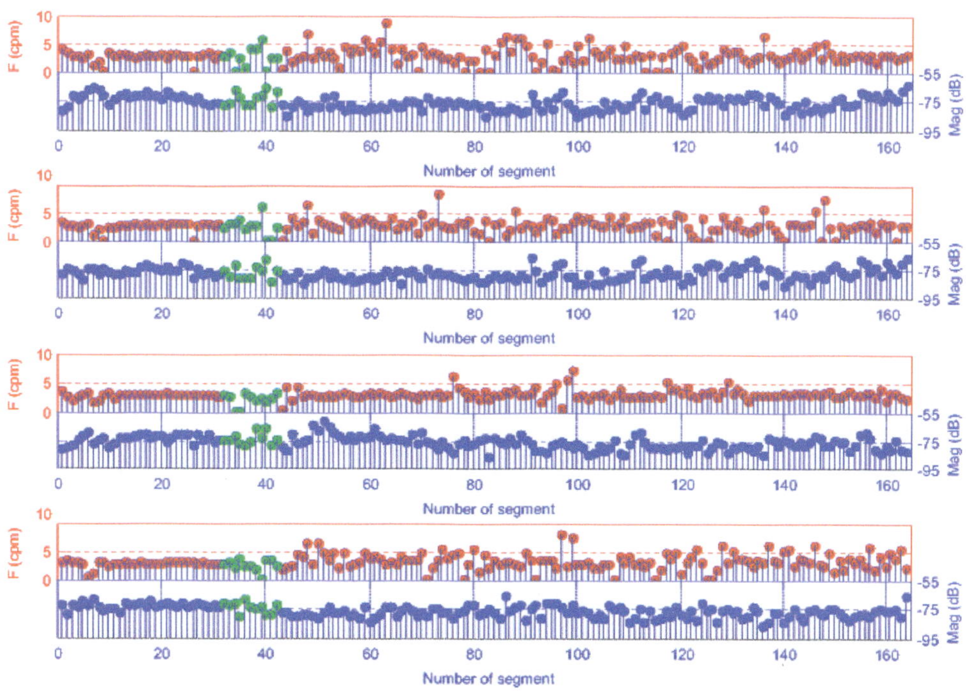

Figure 12: Dominant frequency (top) and MP (bottom) obtained by means of the classical method of EGG signal analysis (4-channels EGG signals, 60s segments) (the green color indicates the meal period).

Normogastria indexes $Norm_I$ and $Norm_{60}$ (corresponding to NI index) were calculated by the proposed method while $Norm_c$ was obtained by the means of classical method. These values of normogastria indexes for four EGG recordings are presented in Table 1 The normogastria indexes obtained on the basis of CWTFT were calculated by two methods: $Norm_I$ was calculated on the basis of all the instantaneous frequency (corresponding to the maximum energy) and$Norm_{60}$ was based on the average value for the 60 s segments.

Table 1: The comparison of normogastria indexes

Signal		CWT(FT) (Morlet, ω_0=6)		CWT(FT) (Morlet, ω_0=15)		Classical method
		$Norm_I$	$Norm_{60}$	$Norm_I$	$Norm_{60}$	$Norm_c$
NS01A	ch1	0.549	0.594	0.589	0.616	0.752
	ch2	0.441	0.474	0.472	0.519	0.677
	ch3	0.224	0.286	0.190	0.210	0.534
	ch4	0.570	0.662	0.665	0.729	0.789
NS02A	ch1	0.819	0.854	0.908	0.915	0.939
	ch2	0.691	0.805	0.800	0.829	0.756
	ch3	0.791	0.866	0.803	0.817	0.829
	ch4	0.956	0.963	0.986	0.988	0.951
NS03A	ch1	0.740	0.754	0.779	0.821	0.873
	ch2	0.796	0.844	0.834	0.866	0.881
	ch3	0.845	0.858	0.883	0.873	0.896
	ch4	0.820	0.851	0.878	0.888	0.858
NS04A	ch1	0.257	0.307	0.289	0.331	0.551
	ch2	0.375	0.472	0.300	0.315	0.567
	ch3	0.908	0.937	0.971	0.976	0.921
	ch4	0.975	0.976	1.000	1.000	0.866

Figures 13, 14, and 15 show the results of the analysis for another EGG signal, which can clearly depict an immediate increase of power in the signal EGG after the meal and its gradual reduction during the postprandial phase. This is the typical phenomenon which can be often observed during the EGG examination [13].

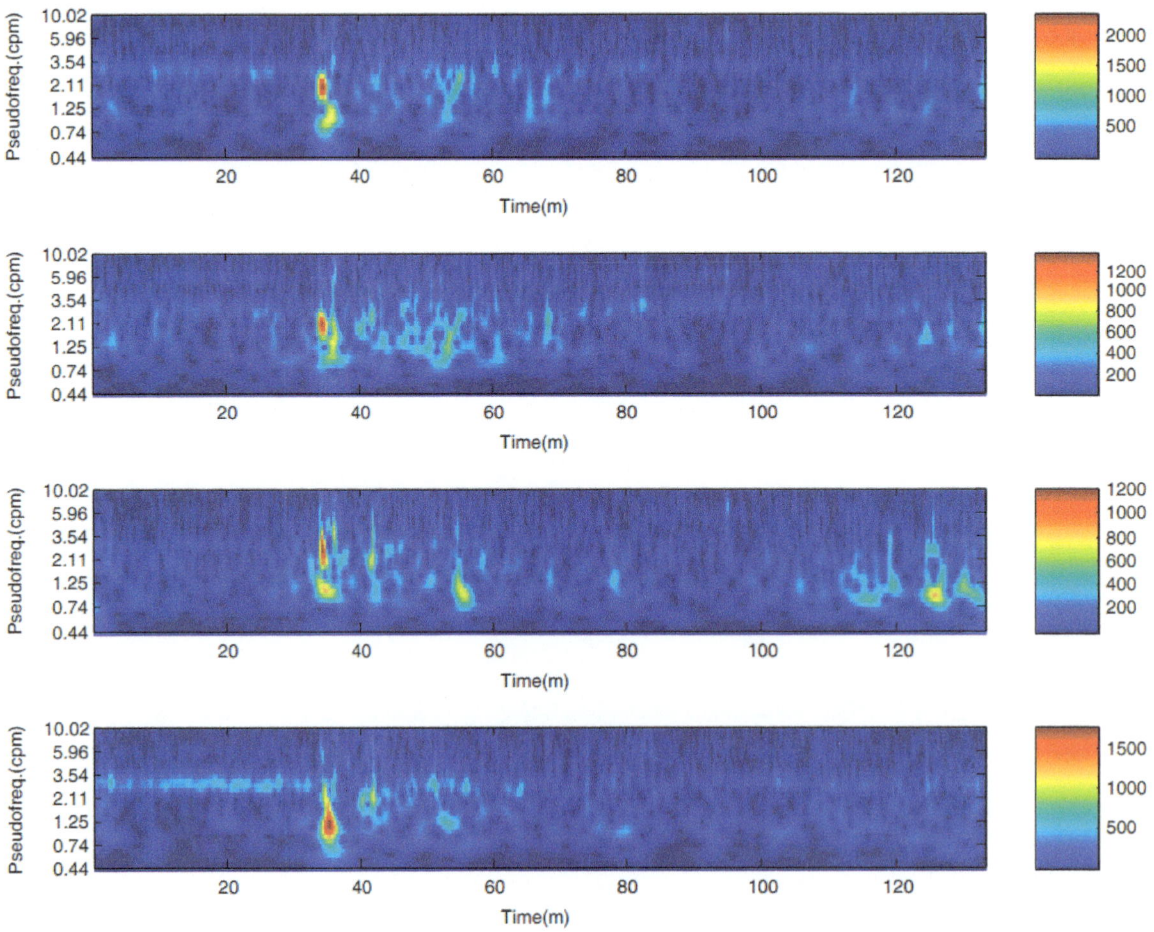

Figure 13: Time-frequency analysis with the CWTFT (non-analytic Morlet wavelet, ω_0=6) for four-channels EGG signal.

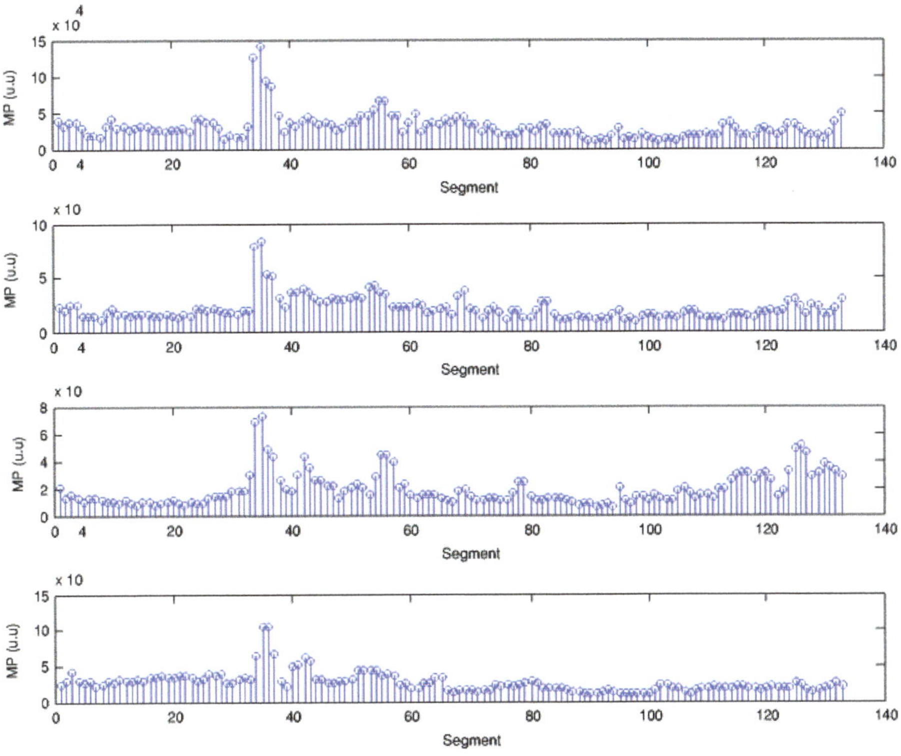

Figure 14: Maximum power obtained by means of the CWTFT (non-analytic Morlet wavelet, $\omega_0=6$) analysis for four-channels EGG signal (for 60s segments).

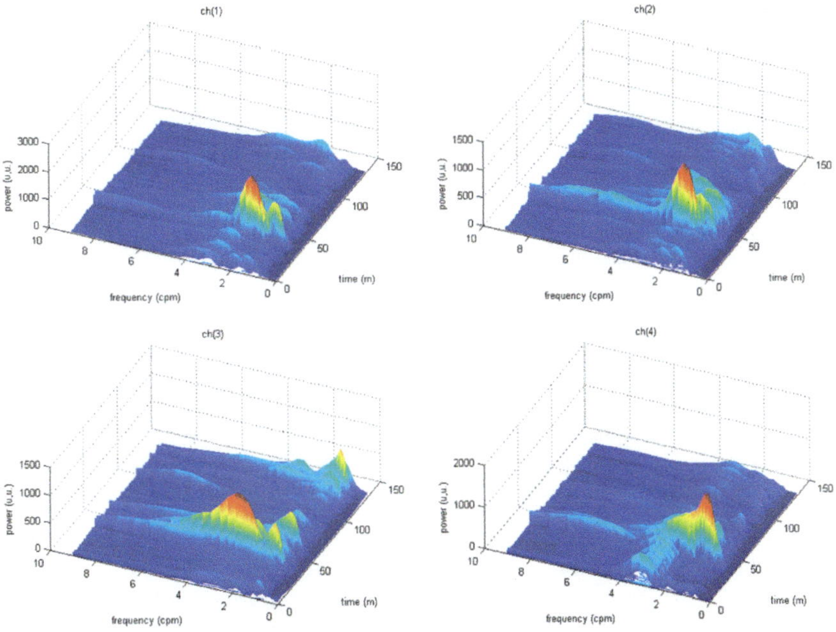

Figure 15: Pseudo-RSA (obtained by means of the CWTFT) of four-channels EGG signal.

CONCLUSION

In this paper, the method of multi-channels EGG signals analysis by means of the CWTFT was described. Thanks to FFT applying in the CWT algorithm computation, the results for relatively long records of EGG can be obtained in a fairly short time (much faster than using the classical methods based on RSA analysis, e.g., using AR or ARMA models).

In this study, for the first time, the authors show the possibility of a parametric analysis of multi-channel EGG signals, using continuous wavelet transform. The normogastria indexes obtained by means of the proposed method, have similar values to the coefficients calculated by means of the classical analysis. However, it was noticed that the largest differences occurred in the signals for which the value of normogastria index was below 0.6. This phenomenon requires further investigation and necessary medical verification.

We can notice that results obtained by the analysis of the CWT are dependent on the mother wavelet function, which significantly affects the quality of signal analysis and results [16]. According to our tests, the Morlet wavelet gives the best results of the EGG signal analysis moreover preliminary results show the robustness of the method and its large potential in the future analysis of the EGG signals. Additionally, the presented method allows to determine the instantaneous values of the dominant frequency and maximum energy which was not possible with the classical EGG signal analysis.

The possibility of continuous observation of the dominant frequency and the dominant power (and other coefficients calculated on this basis) gives opportunities for a wider application of the proposed method in the medical diagnosis of digestive systems. From medical point of view, the described method must be clinically verified, which requires a sufficiently long time, adequate resources and a commitment of medical environments.

REFERENCES

1. Alvarez, W. C., The electrogastrogram and what it shows. *JAMA* 78:1116–1119, 1922.
2. Yin, J., and Chen, J. D. Z., Electrogastrography: Methodology, validation and applications.*J. Neurogastroenterol. Motil* 19:5–17, 2013. doi:10.5056/jnm.2013.19.1.5.Central
3. Riezzo, G., Russo, F., and Indrio, F., *Electrogastrography in adults and children: The strength, pitfalls, and clinical significance of the cutaneous recording of the gastric electrical activity*. BioMed Research International, vol. 2013, Article ID 282757, 14 pages, 2013. doi:10.1155/2013/282757.
4. Chen, J., and McCallum, R. W., Electrogastrography: Measurement, analysis and prospective applications. *Med. Biol. Eng. Comput.* 29(4):339–350, 1991.
5. Parkman, H. P., Hasler, W. L., Barnett, J. L., and Eaker, E. Y., Electrogastrography: A document prepared by the gastric section of The American Motility Society Clinical GI Motility Testing Task Force. *Neurogastroenterol. Motil.* 15:89–102, 2003.
6. Medtronic A/S. Polygram NetTM Reference Manual. Skovlunde (Denmark), 2002.
7. Matsuura, Y., and Takada, H., Form and its nonlinear analysis for the use of electrogastrogram as a gastrointestinal motility test. *Forma* 26(1):39–50, 2011.
8. Koch, K. L., and Stern, R. M., *Handbook of electrogastrography*. Oxford University Press, New York, 2004.
9. Tomczyk, A., and Jonderko, K., Multichannel electrogastrography as a non-invasive tool for evaluation of the gastric myoelectrical activity a study on reproducibility of electrogastrographic parameters before and after a meal stimulation. *Ann. Acad. Med. Siles.* 61:5, 2007.
10. Jonderko, K., et al., How reproducible is cutaneous electrogastrography? An in-depth evidence-based study. *Neurogastroenterol. Motil.* 17(6):800–809, 2005.
11. Komorowski, D., and Pietraszek, S., Preprocessing for spectral analysis of electrogastrogram. world congress on medical physics and biomedical engineering 2009 (WC2009), Munich, Germany, September 7 – 12, 2009, Issue on CD.
12. Pietraszek, S., and Komorowski, D., Heart rate analysis in the EGG examination. Advances in Intelligent and Soft Computing 69, Information Technologies in Medicine Volume 2, Springer-Verlag Berlin Haidelberg 2010. ISBN 978-3-642-13104-2, e-ISBN 978-3-642-13105-9, ISNN 1867-5662, p. 283–292.
13. Hamilton, J. W., Bellahsene, B. E., Reichelderfer, M., Webster, J. G., and Bass, P., Human electrogastrograms: Comparison of surface and mucosal recordings. *Dig. Dis. Sci.*31(1):33–39, 1986.
14. 1Schee, E. J., and Grashuis, J. L., Running spectrum analysis as an aid in the representation and interpretation of electrogastrographic signals. *Med. Biol. Eng. Comput.* 25(1):57–62, 1987.
15. Tokmakçi, M., Analysis of the electrogastrogram using discrete wavelet transform and statistical methods to detect gastric dysrhythmia. *J. Med. Syst.* 31(4):295–302, 2007.
16. de Sobral Cintra, R. J., Tchervensky, I. V., Dimitrov, V. S., and Mintchev, M. P., Optimal wavelets for electrogastrography. *Conf. Pro.c IEEE Eng. Med. Biol. Soc.* 1:329–332, 2004.
17. Liang, H., and Lin, Z., Multiresolution signal decomposition and its applications to electrogastric signals. Recent research developments in biomedical engineering, Vol. 1, 2002, pp. 15–31.
18. Liang, H., and Lin, Z., Stimulus artifact cancellation in the serosal recordings of gastric myoelectrical activity using wavelet transform. *IEEE Trans. Biomed. Eng.* 49(7), 2002.
19. Qiao, W., Sun, H. H., Chey, W. Y., and Lee, K. Y., Continuous wavelet analysis as an aid in the representation and interpretation of electrogastrographic signals. *Ann. Biomed. Eng.*26(6):1072–1081, 1998.
20. Montejo, L. A., and Suarez, L. E., An improved CWT-based algorithm for the generation of spectrum-compatible records. *Int. J. Adv. Struct. Eng.* 5:26, 2013.
21. Liang-Chuan, L., A new method of wavelet transform based on FFT for signal processing. Intelligent Systems (GCIS), 2010 Second WRI Global Congress on, vol.3, no., pp. 203,206, 2010. doi: 10.1109/GCIS.2010.22.
22. Addison, P., Walker, J., and Guido, R., Time-frequency analysis of biosignals. *IEEE Eng. Med. Biol. Mag.* 28(5):14–29, 2009. doi:10.1109/MEMB.2009.934244.
23. Torrence, C., and Compo, G. P., A practical guide to wavelet analysis. *Bull. Am. Meteorol. Soc.* 79:61–78, 1998.
24. Rioul, O., and Vetterli, M., Wavelets and signal processing. IEEE Signal Processing Magazine, 1991, pages14–38.
25. Allen, J. B., Short term spectral analysis, synthesis, and modification by discrete Fourier transform. *IEEE Trans. Acoust. Speech Signal Process.* 25(3):235–238, 1977.
26. Najmi, A. H., and Sadowsky, J., The continuous wavelet transform and variable resolution time- frequency analysis. *J. Hopkins APL Tech. Dig.* 18(1):134–140, 1997.

27. Steinbuch, M., and van de Molengraft, M. J. G., Eindhoven University of Technology, Control Systems Technology Group Eindhoven, Wavelet Theory and Applications, a literature study. R.J.E. Merry, DCT, 2005.53.

28. Bialasiewicz, J. T., Muljadi, E., Nix, R. G., and Drouilhet, S., Renewable energy power system modular simulator RPMSim user's guide. NREL/TP-500-29721, 2001, 171 pages.

29. Bialasiewicz, J. T., Falki i aproksymacje (Wavelets and Approximations). Wydawnictwa Naukowo-Techniczne, Warsaw, Poland (in Polish), 2000, 288 pages.

30. Mallat, S.G., A theory for multiresolution signal decomposition: The wavelet representation. *IEEE Trans. Pattern Anal. Mach. Intell.* 11(7), 1989.

31. Oppenheim, A. V., Schafer, R. W., and Buck, J. A., Discrete-time signal processing. Upper Saddle River, N.J.: Prentice Hall. ISBN 0137549202, 1999.

32. 3Farge, M., Wavelet transforms and their applications to turbulence. *Annu. Rev. Fluid Mech.* 24:395–457, 1992.

CHAPTER 18

COMPARATIVE STUDY OF NEW SIGNAL PROCESSING TO IMPROVE S/N RATIO OF SEISMIC DATA

Muhammad Sajid, Deva P. Ghosh and I. A. Satti

Keywords: Seismic processing Signal-to-noise ratio, Signal processing, Seismic resolution, Noise attenuation

INTRODUCTION

Fourier analysis is a well-known tool to decompose a signal into its orthogonal components of Sine and Cosine waves (Eq. 1). It transforms the signal from the time domain to the frequency domain. The amplitude and phase spectrum obtained through the Fourier analysis is the solution of a stationary signal. But, the seismic wavelet is non-stationary and changes its shape and frequency contents as it propagates in subsurface. Short Time Fourier Transform (STFT) is the windowed version of Fourier transform (Eq. 2) (Donoho 1995; Jacobsen and Lyons 2003,2004). It tries to accommodate the deficiency of the Fourier transform, but the selection of the type and size of the window has its own limitations. A large window size gives more frequency resolution but less time resolution as the product of time-resolution and frequency-resolution is constant. It is not possible to have both high time resolution and frequency resolution at the same instant. The selection of the size of the window is a trade between either high frequency resolution or time resolution. This phenomenon is known as the Gabor–Heisenberg Uncertainty Principle.

$$F(f) = \int_{-\infty}^{\infty} x(t)\mathrm{e}^{-j2\pi ft}\,\mathrm{d}t \tag{1}$$

$$F(\tau, f : h) = \int_{-\infty}^{\infty} x(t)h^*(t-\tau)\mathrm{e}^{-j2\pi ft}\,\mathrm{d}t \tag{2}$$

Where $x(t)$ is the time domain signal, f the frequency, t the time, $h^*(t)$, the decompostion window, and τ is the translation of window along time axis.

The Wavelet transform provides one of the solutions using the finite duration wavelets with a variable length which defines its central frequency. Both of these techniques (i.e., Short Time Fourier Transform and Wavelet transform) come in the category of spectral decomposition and work on one trace at a time. The rest of the other techniques discussed in the paper are at least two dimensional. Radon and Fan-beam transforms provide the opportunity of the angle-dependent removal of noise from seismic data. The Radon transform computes projections of a two-dimensional image matrix on a straight or curved line. This projection on the line is calculated for 0°–180°. Whereas in Fan-beam, this projection can be on a straight or curved line. The projection for each direction is calculated by taking the line integral of the straight lines emerging from a single vertex. This vertex must reside outside of a two-dimensional image matrix. The last two techniques are Ridgelet and Curvelet transform. The Ridgelet transform is obtained by taking the Radon transform of a 2D Fourier transform seismic section. It extends the capability of the Radon transform for the removal of noise from the seismic data through the reconstruction of the filtered coefficients of Ridgelet. Curvelet coefficients are calculated through the application of the Ridgelet transform on each scaled (2D Wavelet transform) version of the seismic data.

Citation: Sajid, M., Ghosh, D.P. & Satti, I.A. J Petrol Explor Prod Technol, Comparative study of new signal processing to improve S/N ratio of seismic data 4: 87, doi:10.1007/s13202-013-0092-8.

METHODOLOGY

Signal processing techniques are tested on both synthetic and real seismic data. Testing includes two levels. In the first level, an algorithm has to reconstruct all the seismic features (i.e., linear, inclined and curved features without any distortion) without preconditioning the input data. After the first-level filtering, the qualified algorithm is tested for its improvement of the signal-to-noise ratio using coefficient filtering.

These algorithms have been tested on synthetic and real seismic data. To make the procedure simple, the synthetic seismic data were created using a convolution technique on a thin-bed reflectivity model. This reflectivity model contained both horizontal and inclined layers. Random noise with a signal-to-noise ratio of 3 was added to the reflectivity model before the convolution operation. Figure 1 shows the synthetic seismic data used in the algorithm testing. Whereas, the curved feature was tested on real seismic data as shown in Fig. 2.

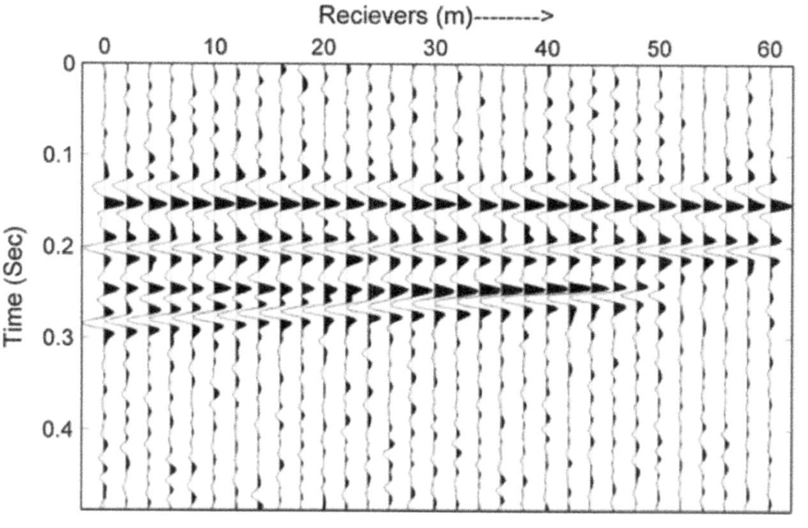

Figure 1: Synthetic seismic data with a signal-to-noise ratio of 3.

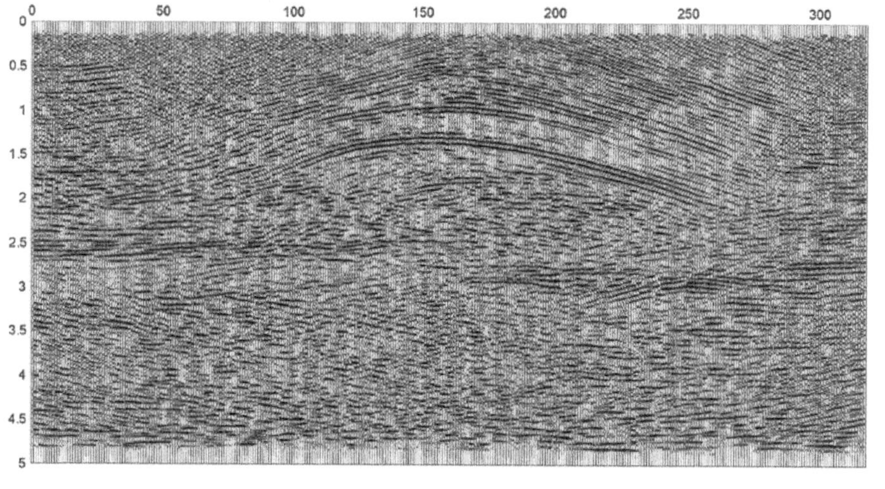

Figure 2: Real seismic section contained curved features and random noise.

Wavelet Transform

The Wavelet transform can be calculated using Eq. (3). In this analysis, the Dabachi wavelet 6 was used which satisfied the admissibility condition (Eq. 4). Input 2D synthetic seismic data were decomposed into 4 levels. The first level decomposed the seismic data into detailed (high frequency) and approximation (low frequency) seismic sections. In the second level, the approximation was decomposed into its details and approximation; it was similarly performed for the 3rd and 4th level as shown in Fig. 3a (Mallat 1989; Torrence and Compo 1998).

$$X_w(a,b) = \frac{1}{\sqrt{b}} \int_{-\infty}^{\infty} x(t)\psi^*\left(\frac{t-a}{b}\right) \mathrm{d}t, \begin{cases} a\epsilon(-\infty,\infty) \\ b\epsilon(0,\infty) \end{cases} \tag{3}$$

where, $x(t)$, is the time domain signal, a, the location parameter of wavelet translation, b, the wavelet scaling parameter, and $\psi^*(t)$ is the complex conjugate of the mother wavelet$\psi(t)$

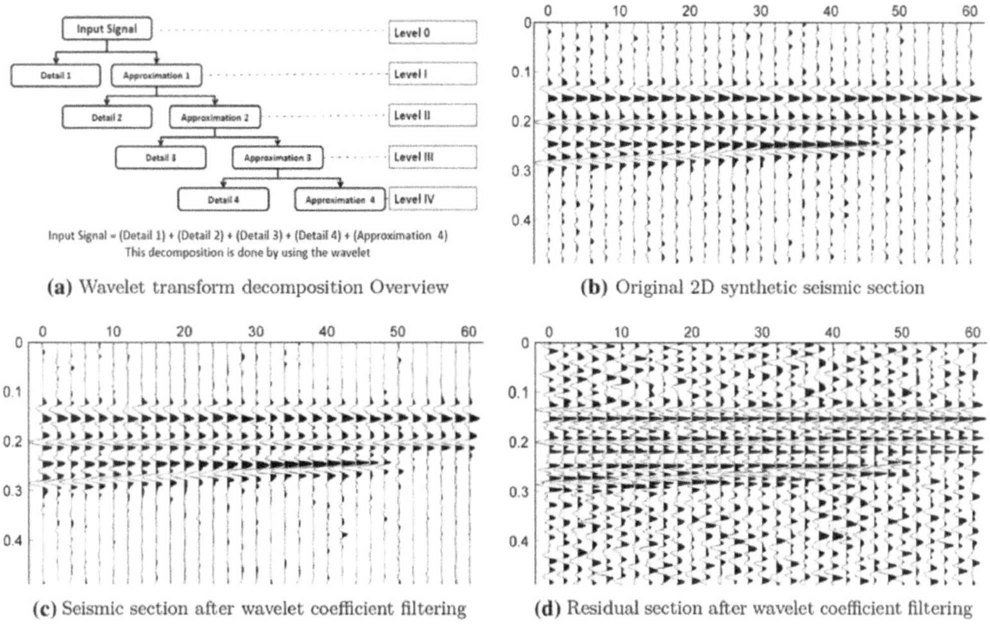

(a) Wavelet transform decomposition Overview

(b) Original 2D synthetic seismic section

(c) Seismic section after wavelet coefficient filtering

(d) Residual section after wavelet coefficient filtering

Figure 3: Application of the Wavelet transform on a 2D Synthetic seismic.

Stretching of the mother wavelet gave a long duration with a low central frequency; whereas, compression led to a short analysis wavelet with a high central frequency.

$$C_\psi = \int_0^\infty \frac{|\Psi(f)|^2}{|f|} \, df < \infty \tag{4}$$

$\Psi(f)$ is the Fourier transform of the wavelet $\psi(t)$. Since C_ψ has to be finite, the integral defining C_ψ should be integrable at $f = 0$. This implies that $\Psi(0)=0$. This means that the average value of the wavelet should be zero i.e. $\int_{\infty}^{-\infty} \psi(t)\,dt = 0$

This decomposition provided the opportunity to filter the noise at a desired level and the seismic data were able to be reconstructed back from its filtered coefficients. The simple idea of the threshold of the Wavelet coefficients was used in this analysis. Figure 3b–d show the input synthetic seismic data, denoised seismic data after the application of DWT and the residual data, respectively. It can be observed that the seismic data reconstructed from the filtered coefficients of the Wavelet transform had a better signal-to-noise ratio but the residual contained energy of the signal, which can cause seismic resolution problems. Figure 4 shows the real seismic data reconstructed after the threshold filtering of the Wavelet transformed coefficients. In comparison with the original seismic section shown in Fig. 1, the signal-to-noise ratio was improved after the Wavelet transform coefficient filtering, and three features, i.e. horizontal, inclined (Fig. 3) and curved (Fig. 4) were recovered.

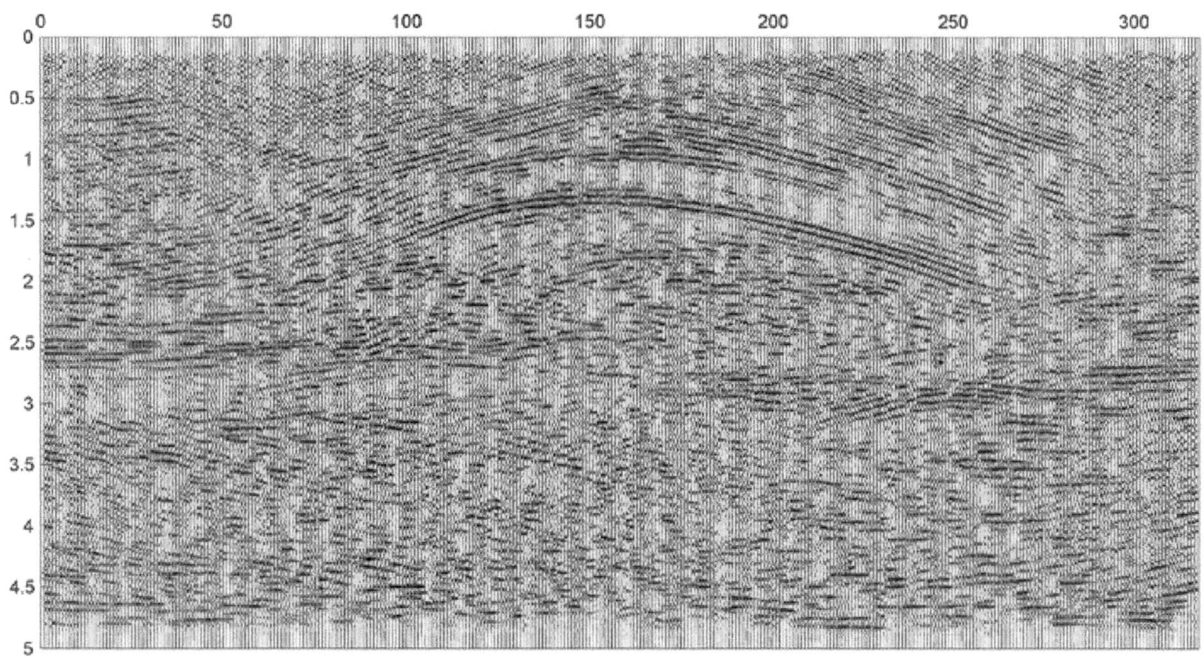

Figure 4: Real seismic data after the threshold filtering of the Wavelet transform coefficients.

Radon Transform

The Radon transform computed the projections of an image matrix along the specified directions. A projection of the two dimensional function *f*(*x*, *y*) was a set of line integrals as shown in Fig. 5. Radon transform computed the line integral of parallel beams for different angle ranges from 0° to 180° with respect to the center of the image (Eq. 5) (Lim and Jae 1990; Bracewell and Ronald 1995). The result of the 2D image was a 2D matrix where each column represented the angle of the beam, the rows represented the distance from the center of the image.

$$Rf(\theta, t) = \int f(x_1, x_2)\delta(x_1 cos\theta + x_2 sin\theta - t)\, dx_1\, dx_2 \quad (5)$$

where, δ, is the Dirac distribution, θ, the angle (0°−180°), and t is the translation.

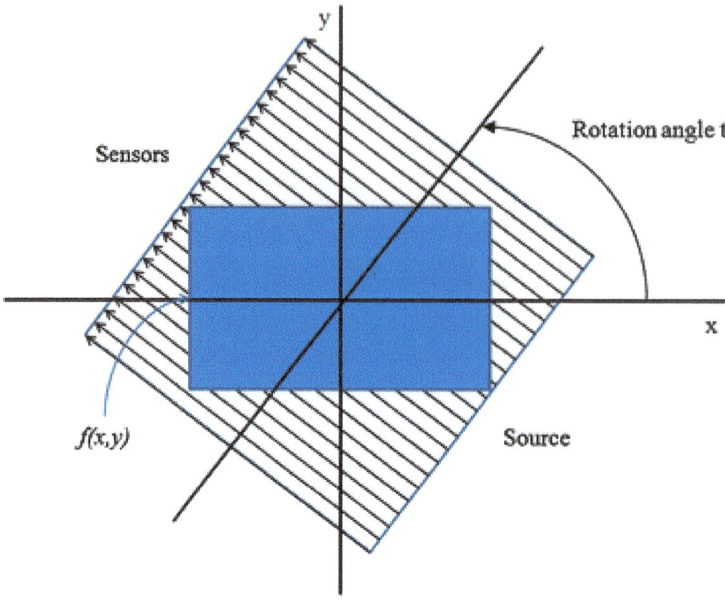

Figure 5 Radon transform projected the 2D image matrix on a straight line through the line integration of the parallel straight line for each angle of rotation.

The Radon transform provided the opportunity of the angle-dependent removal of the noise from the seismic data. Figure 6a shows the seismic section with noise. The Radon transform was calculated with the angle range from 0° to 180° with an interval of 1°. In the Radon domain, the angle ranges from 0° to 20° and 155° to 180° were filtered. Figure 6b shows the filtered section and Fig. 6c shows the seismic section with an improved S/N ratio.

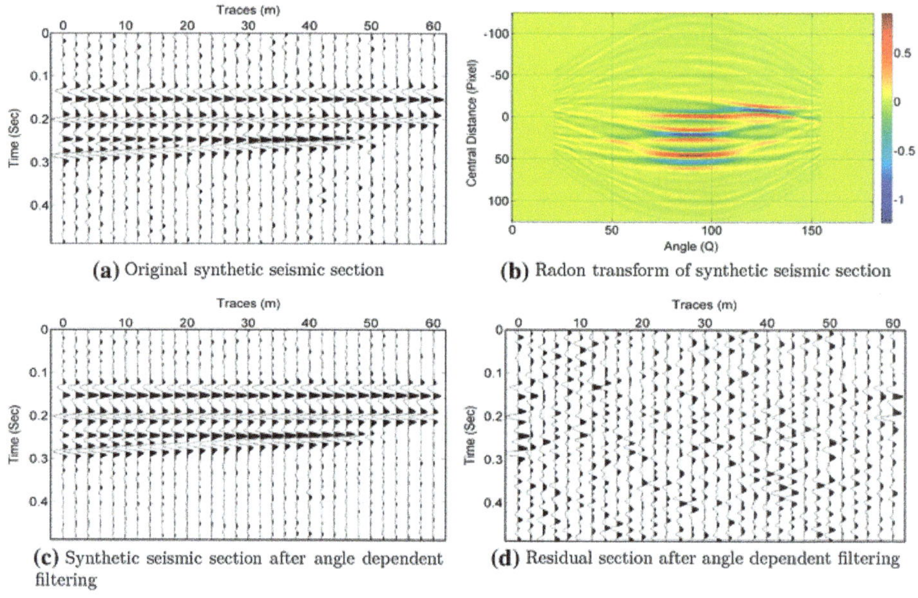

Figure 6: Application of the Radon transform on the synthetic seismic section.

As the Radon transform is a two dimensional operation and uses line integrals at each angle of rotation, the reconstruction of the curved feature becomes challenging and produces noise. Figure 7 shows the real seismic data reconstructed back from the Radon domain without any filtering. In comparison with the original seismic data, the signal-to-noise ratio was decreased and the curved features were adversely affected by the transformation process.

Fan-Beam Transform

Fan-beam used a special X-ray procedure. In its coefficient calculation, the angle of the projection ranged from 0° to 360° (Kak and Slaney 1988), from the center of the image (2D seismic) rather than 0° to 180° for the Radon transform discussed above. This projection can be on a straight or curved line (Matlab 2013). So, Fan-beam provided more flexibility of the angle-dependent filtering than the Radon transform. In this analysis, the Fan-beam coefficients were calculated by projecting on a straight line. The seismic data were reconstructed back after the filtering of the angle range of 0°–25°, 155°–205°, 335°–360° as shown in Fig. 8. This angle-dependent filtering improved the signal-to-noise ratio of the linear and inclined features of the seismic data; whereas, testing on the real seismic section (Fig. 9) shows that the curved feature of the seismic data was adversely affected by the transformation process. This led to a decrease in the signal-to-noise ratio and false seismic features.

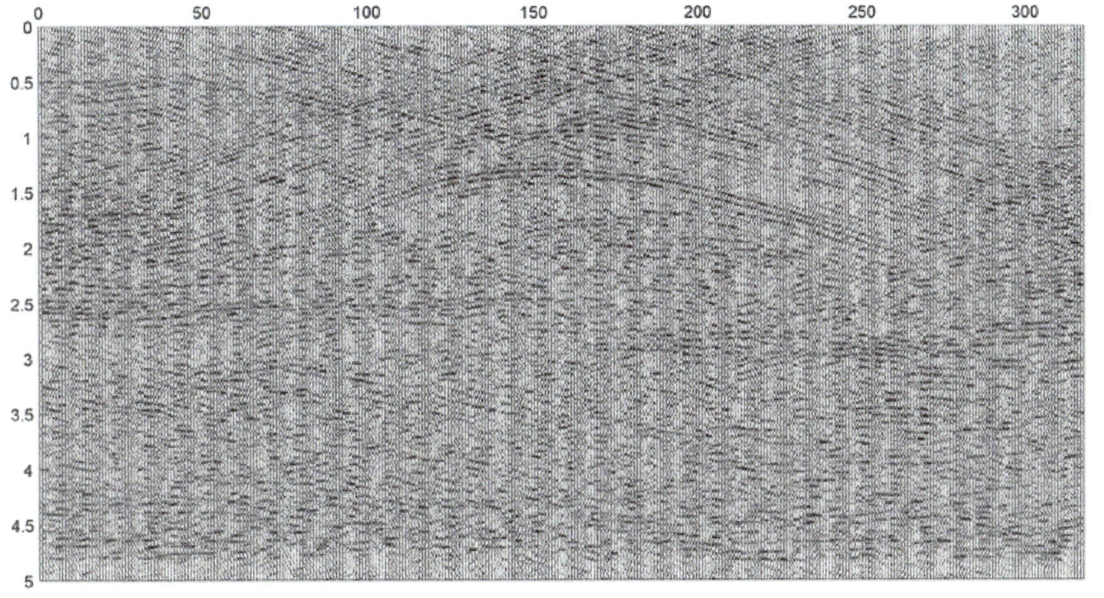

Figure 7: Curved features of seismic data were not properly reconstructed back from the Radon domain and the signal-to-noise ratio was decreased.

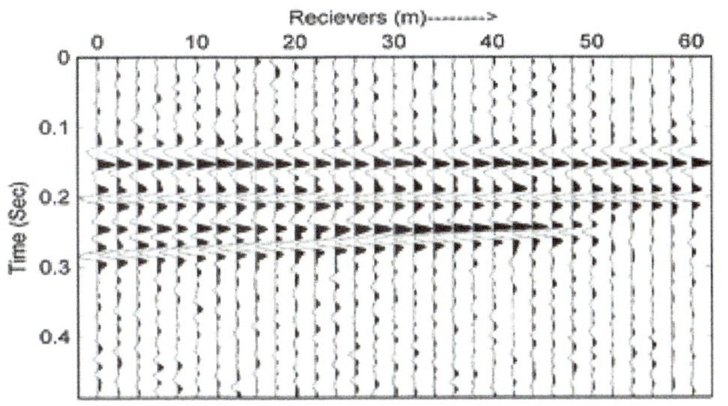

(a) Synthetic seismic section with noise

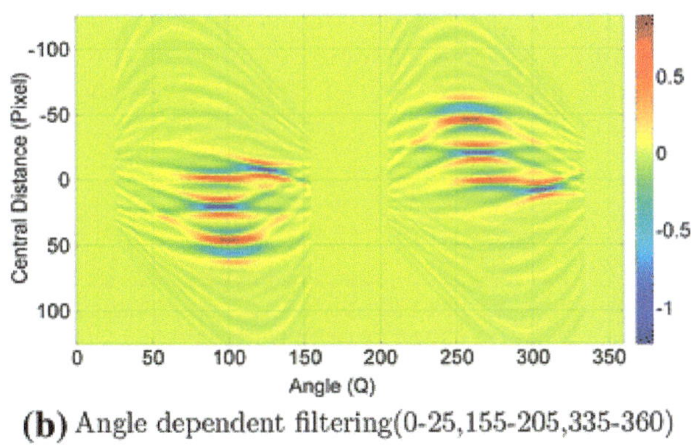

(b) Angle dependent filtering(0-25,155-205,335-360)

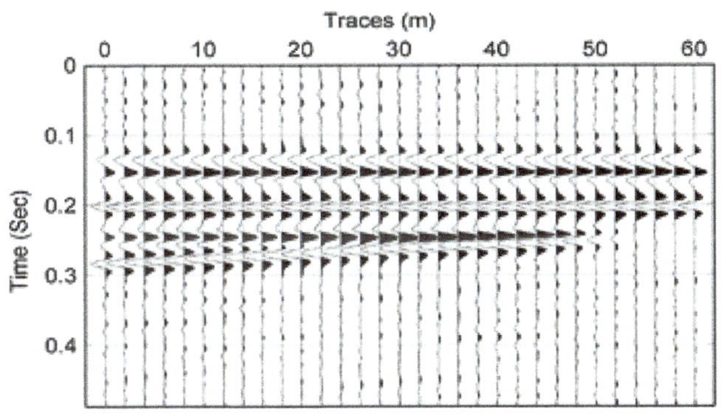

(c) Synthetic seismic section after Fanbeam transform, shows decrease in noise

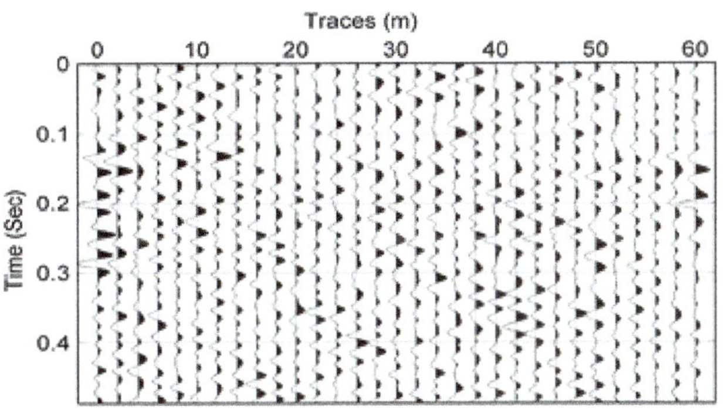

(d) Residual section after angle dependent filtering

Figure 8: Application of Fan-beam transform on the real seismic section.

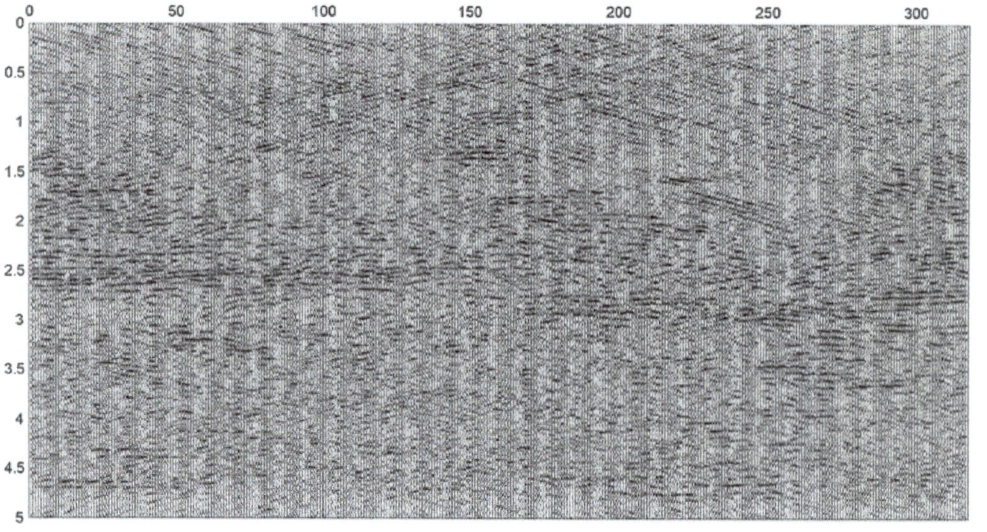

Figure 9: Curved features of the seismic data were badly effected by the Fan-beam transform.

Ridgelet Transform

The Ridgelet transform is good for identification of line discontinuities in seismic data and it can handle edge effects better than the Wavelet transform

$$Rf(\theta, t) = \int f(x_1, x_2)\delta(x_1 cos\theta + x_2 sin\theta - t)\, dx_1 dx_2$$

$$R_f(a, b, \theta) = \int Rf(\theta, t).a^{-\frac{1}{2}}\psi\left(\frac{t-b}{a}\right) dt \tag{6}$$

where, δ, is the Dirac distribution, t, the translation, θ, the angle (0°−180°),$Rf(\theta, t)$, the Radon transform, $R_f(a, b, \theta)$, the Ridgelet transform, ψ, the wavelet, a, the wavelet scale factor, b, the wavelet translation factor.

Ridgelet coefficients can be obtained by taking the Wavelet transform of the Radon transformed seismic data (Eq. 6) (Do and Vetterli 2003; Jean-Luc et al. 2002). Figure 10a shows the input seismic section along with the wavelet used for its decomposition. Figure 10b–d show the reconstructed seismic section from the filtered Ridgelet coefficients, residual section and reconstructed seismic section after the application of the wiener filter, respectively. The analysis shows that the Ridgelet transform had an adverse effect on the inclined features as observed in Fig. 10c (where the residual section contained the energy of inclined features). The wiener filter (which is a low pass-filter) was applied to filtered seismic section to remove the effect of jaggedness. It used the pixel-wise adaptive wiener method based on the statistics (mean & standard deviation) estimated from a location's neighborhood. In this study, the neighborhood was defined by the window. The Ridgelet threshold coefficient filtering produced noise in the seismic section which led to a decrease in the signal-to-noise ratio as shown in Fig. 11.

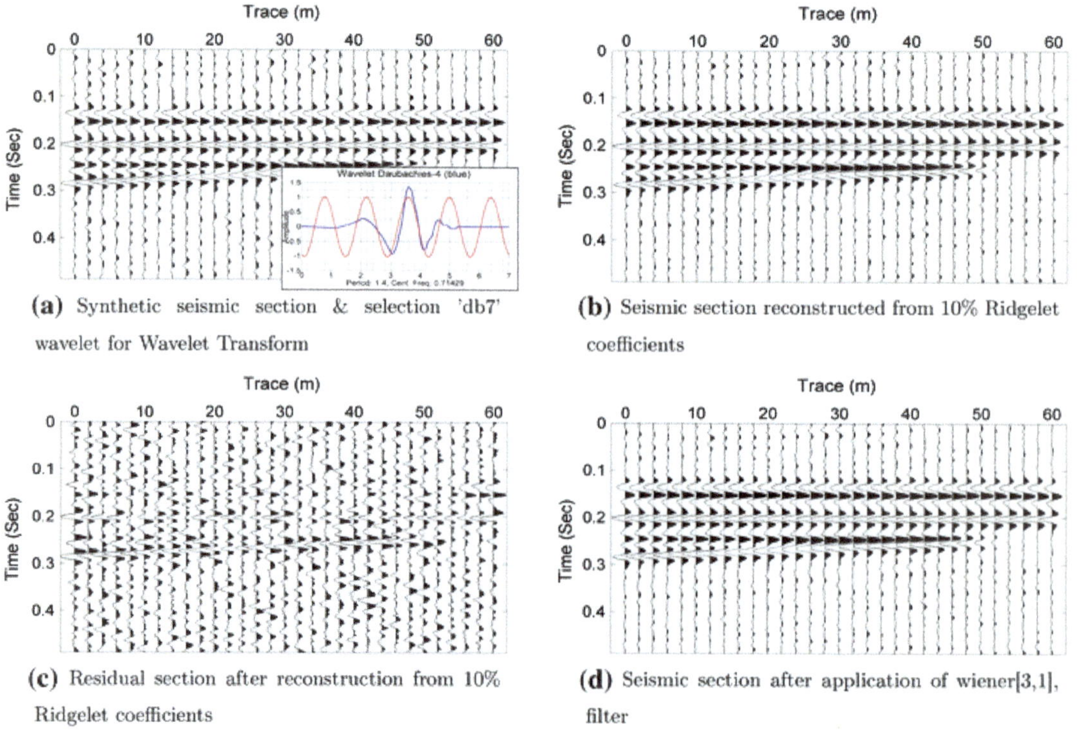

(a) Synthetic seismic section & selection 'db7' wavelet for Wavelet Transform

(b) Seismic section reconstructed from 10% Ridgelet coefficients

(c) Residual section after reconstruction from 10% Ridgelet coefficients

(d) Seismic section after application of wiener[3,1], filter

Figure 10: Application of the Ridgelet transform on the synthetic seismic data.

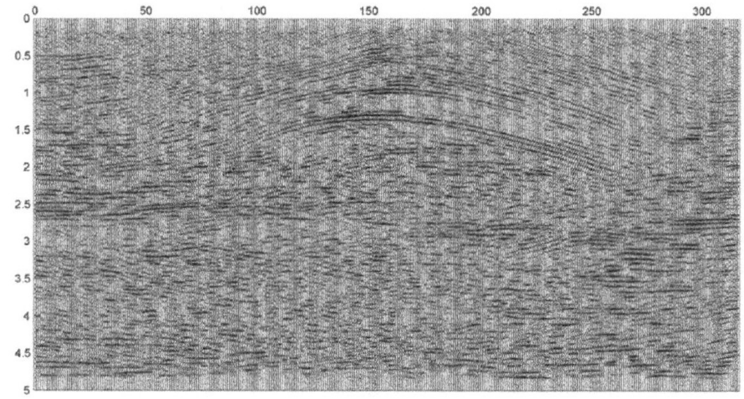

Figure 11: Ridgelet transform reconstructed the curved feature of the seismic section but the threshold coefficient filtering decreased the signal-to-noise ratio.

Curvelet Transform

The Curvelet transform is a powerful tool to improve the signal-to-noise ratio of seismic data. The Curvelet transform used the 2D Wavelet transform before the implementation of the Ridgelet transform on each scaled version of the seismic section (Do and Vetterli 2003). Figure 12 shows the application of the Wrapping Curvelet transform on the noisy data. Figure 12a shows the synthetic seismic data with no noise; whereas 12b shows the same synthetic seismic data with the addition of normally distributed noise with an S/N ratio of 3. The Curvelet transform of the synthetic seismic data was calculated and the synthetic data were reconstructed with the 0.008 % Curvelet coefficients shown in Fig. 12c. It can be observed that random noise was considerably removed from the synthetic seismic section. The Curvelet threshold coefficient filtering effectively removed the random noise from the seismic section without much effect to the curved features of the seismic section, as shown in Fig. 13.

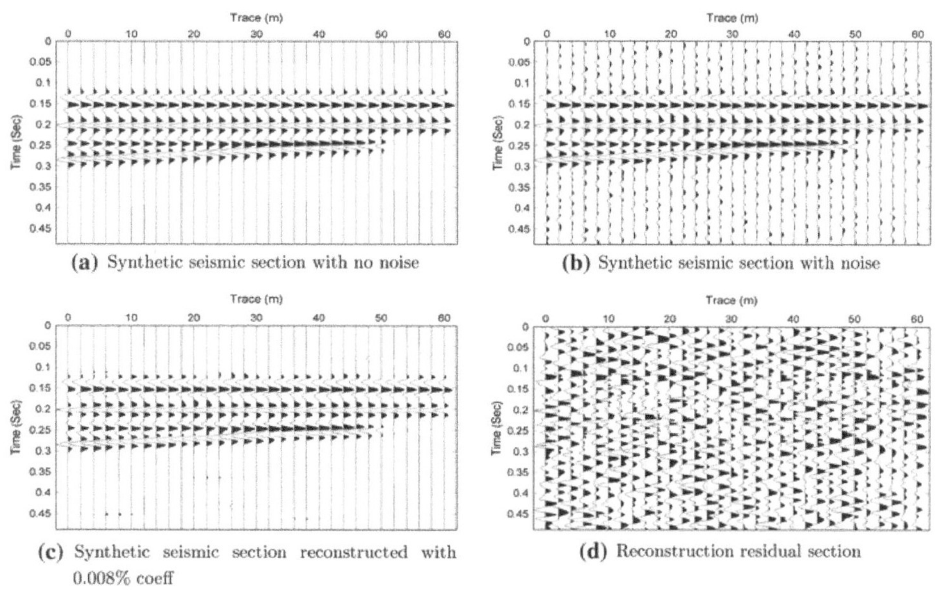

(a) Synthetic seismic section with no noise

(b) Synthetic seismic section with noise

(c) Synthetic seismic section reconstructed with 0.008% coeff

(d) Reconstruction residual section

Figure 12: Application of the wrapping Curvelet transform on the synthetic seismic data.

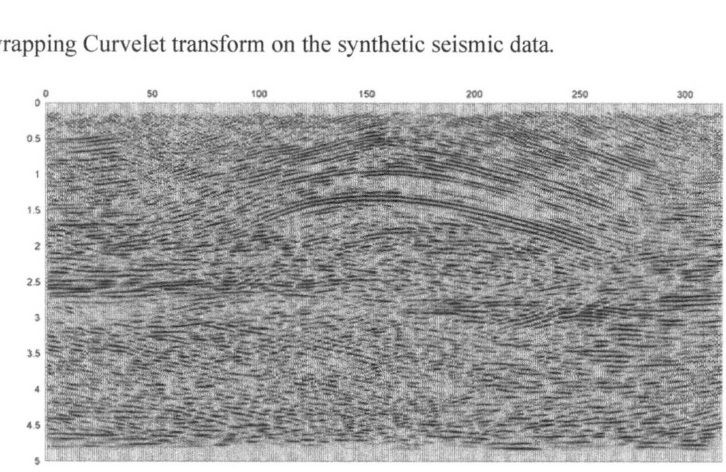

Figure 13: Reconstructed seismic section from the filtered Curvelet coefficients effectively reduced the random noise.

RESULTS AND DISCUSSION

A comparative study between new signal processing techniques has been performed to improve the signal-to-noise ratio of seismic data. The Fourier transform is a powerful tool to transform the signal from the time domain to the frequency domain, but it has a lack of time localization of the seismic event in the Fourier domain. So, in non-stationary seismic signals, the Fourier transform is not the solution. The Short-Time-Fourier-Transform, STFT, has extended the capability of the Fourier transform by adding the time dimension. This has helped to filter the desire frequency at a known time; but, STFT decomposition depends upon the size and shape of the spectral decomposing window. One of the solutions to this uncertainty principle is a Wavelet transform. But the Wavelet transform cannot handle edge effects when the seismic data are reconstructed back from its filtered coefficients because the edges in the seismic data are repeated scale after scale. So, filtering though the Wavelet transform can cause resolution problems in seismic data. The Radon transform and Fan-beam transform provide the opportunity of angle- dependent filtering using a technique of projection of a beam on line. These techniques have been found to be good at improving the signal-to-noise ratio, but have adverse effects on curved features of seismic data. Secondly, both of the techniques use interpolation for the reconstruction from their coefficients which causes resolution problems in seismic data. The Ridgelet transform, which uses the Radon transform in calculation of Ridgelet coefficients, solves the problem of edges in the seismic data but produces adverse effects on the inclined features and curved features when seismic data are reconstructed back from their filter coefficients. The Curvelet transform (uses the 2D Wavelet transform before the application of the Ridgelet transform on each scaled version of the seismic data) is more robust to handle inclined and curved features. It handles the horizontal, inclined and curved features of seismic data more effectively when a seismic section is reconstructed back from its filter coefficients. But this technique requires a long processing time and memory. So, a faster and more memory efficient algorithm is needed to be developed so that it can be implemented on a large seismic volume.

ACKNOWLEDGMENTS

We are thankful to Universiti Teknologi PETRONAS (UTP) and PETRONAS for support in this research. I would will to thank University COMSATS for providing me the opportunity to carry out this research in UTP. In this reserach we used Matlab, and some tool boxes like CREWES, Curvelab and frit toolbox.

REFERENCES

1. Bracewell RN (1995) Two-dimensional imaging. Prentice Hall, Englewood Cliffs
2. Do MN, Vetterli M (2003) The finite ridgelet transform for image representation. IEEE Trans Image Process 12(1):16–28
3. Donoho DL (1995) De-noising by soft-thresholding. IEEE Trans Inf Theory 41(3): 613–627
4. Jacobsen E, Lyons R (2003) The sliding DFT. IEEE Signal Process Mag 20(2):74–80
5. Jacobsen E, Lyons R (2004) An update to the sliding DFT. IEEE Signal Process Mag 21(1):110–111
6. Jean-Luc S, Candes EJ, Donoho DL (2002) The curvelet transform for image denoising. IEEE Trans Image Process 11(6):670–684
7. Kak AC, Slaney M (1988) Principles of computerized tomographic imaging. IEEE Press, New York
8. Lim JS (1990) Two-dimensional signal and image processing. Prentice Hall, Englewood Cliffs
9. Mallat SG (1989) A theory for multiresolution signal decomposition: the wavelet representation. IEEE Trans Pattern Anal Mach Intell 11(7):674–693
10. Matlab (2013) Help documentation [Online].http://www.mathworks.com/help/images/ref/fanbeam.html?searchHighlight=fanbeam
11. Torrence C, Compo GC (1998) A practical guide to wavelet analysis. Bull Am Meteorol Soc 79:61-78

INDEX

Q

R

S

T

U

W